Dmitrii S. Silvestrov
American-Type Options

De Gruyter Studies in Mathematics

Volume 56

Dmitrii S. Silvestrov

American-Type Options

Stochastic Approximation Methods
Volume 1

DE GRUYTER

Mathematics Subject Classification 2010
Primary: 91G20, 91G60, 60J05, 60G40, 60J22; Secondary: 60J10, 60G15, 60G50, 65C40, 62L15

Author
Prof. Dr. Dmitrii S. Silvestrov
Stockholm University
Department of Mathematics
SE-106 91 Stockholm
Sweden
silvestrov@math.su.se

ISBN 978-3-11-032967-4
e-ISBN 978-3-11-032982-7
Set-ISBN 978-3-11-032983-4
ISSN 0179-0986

Library of Congress Cataloging-in-Publication Data
A CIP catalog record for this book has been applied for at the Library of Congress.

Bibliographic information published by the Deutsche Nationalbibliothek
The Deutsche Nationalbibliothek lists this publication in the Deutsche Nationalbibliografie; detailed bibliographic data are available in the Internet at http://dnb.dnb.de.

Typesetting: le-tex publishing services GmbH, Leipzig
Printing and binding: Hubert & Co. GmbH & Co. KG, Göttingen
♾Printed on acid-free paper
Printed in Germany

www.degruyter.com

Preface

American-type options are one of the very important financial instruments and at the same time one of the most interesting and popular objects for research in financial mathematics.

The main mathematical problems connected with such options relate to finding of the optimal expected option rewards, in particular, fair prices of options, as well as finding the optimal strategies for buyers of options that are optimal stopping times for execution of options.

In this way, the theory of American-type options is connected with optimal stopping problems for stochastic processes, which play an important role in the theory of stochastic processes and its applications in sequential analysis, control theory and many other applied areas.

As is well known, analytical solutions for American-type options are available only in some special cases and, even in such cases, the corresponding formulas are not easily computable. These difficulties dramatically increase in the case of multivariate log-price processes and nonstandard pay-off functions.

Approximation methods are a reasonable alternative that can be used in cases, where analytical solutions are not available. The main classes of approximation methods are: stochastic approximation methods based on approximation of the corresponding stochastic log-price processes by simpler processes, for which optimal expected rewards can be effectively computed; integrodifferential approximation methods based on approximation of integrodifferential equations that can be derived for optimal expected rewards by their difference analogs; and Monte Carlo methods based on simulation of the corresponding log-price processes.

Stochastic approximation methods have important advantages in comparison with other two methods. They let one usually impose weaker smoothness conditions on transition probabilities and pay-off functions in comparison with integrodifferential approximation methods and they are also computationally more effective in comparison with Monte Carlo-based methods.

This book is devoted to systematical presentation of stochastic approximation methods for models of American-type options with general pay-off functions for discrete time Markov log-price processes.

The principal novelty of our studies is that we systematically consider discrete time multivariate modulated Markov log-price processes and general pay-off functions, which can depend not only on price but also an additional stochastic index component with a general phase space. We also impose minimal conditions of smoothness on transition probabilities and pay-off functions as well as minimal moment compactness conditions on log-price processes and restrictions on the rate of growth for pay-off function in price arguments.

The book contains ten chapters.

In Chapter 1, we introduce models of multivariate modulated Markov log-price processes and consider main examples of such processes, including log-price processes represented by multivariate modulated random walks and various autoregressive and autoregressive moving average-type log-price processes as well as autoregressive and autoregressive and moving average stochastic volatility models.

In Chapter 2, we define American-type options and present basic examples such as call and put options, exchange of assets options, digital options, and knockout options, and their portfolios. We also introduce basic objects connected with American-type options, namely reward functions, optimal expected rewards, optimal stopping times and present related basic results, namely, backward recurrence relations for reward functions and results concerning hitting structure of optimal stopping times for American-type options.

In Chapter 3, we investigate the structure of backward recurrence relations for log-price processes represented by atomic Markov chains, which transition probabilities are concentrated on finite sets. Such log-price processes play the role of approximating processes in the corresponding stochastic approximation reward algorithms. The so-called recombining conditions, which guarantee that the number of time–space nodes for trajectories of such atomic Markov chains has not more than polynomial rate of growth as a function of number of time steps, play here the key role. We investigate in details these conditions, for univariate and multivariate binomial and trinomial models both homogeneous and inhomogeneous in time and space, as well as for general discrete random walks and atomic Markov chains.

In Chapter 4, we give upper bounds for reward functions and moments of random rewards for American-type options for multivariate modulated Markov log-price processes with bounded and unbounded characteristics and with general pay-off functions, which have not more than polynomial rate of growth in price arguments. We also specify these results for log-price processes represented by modulated random walks and multivariate Markov Gaussian processes with bounded and unbounded drift and volatility coefficients.

In Chapters 5 and 6, we present our main convergence results for rewards of American-type options with general perturbed pay-off functions for perturbed multivariate modulated Markov log-price processes, respectively, with bounded and unbounded characteristics. It is important that we impose minimal conditions of smoothness on the limiting transition probabilities and pay-off functions. For the basic case, where the transition probabilities have densities with respect to some pivotal Lebesgue-type measure, it is usually required that the sets of weak discontinuity for the limiting transition probabilities, as functions of initial points, and the sets of discontinuity for pay-off functions are zero sets with respect to the above pivotal measure. In fact, such assumptions make it possible for the transition probabilities and pay-off functions to be very irregular.

In Chapter 7, we preset general results about so-called space-skeleton approximations for rewards of American-type options with general pay-off functions for multivariate modulated Markov log-price processes. In this model, the space discretization of underlying Markov log-price processes is used. The approximating atomic Markov chains have finite numbers of nodes located on some regular finite greeds for each moment of time. The corresponding backward recurrence relations let one effectively compute the approximating reward functions and optimal expected rewards. Their convergence to the corresponding limiting quantities is derived using the general convergence theorems given in Chapters 5 and 6.

In Chapter 8, we present results about convergence of option rewards and space-skeleton reward approximations for multivariate modulated Markov Gaussian log-price processes with bounded and unbounded drift and volatility coefficients. We also give results of almost sure and weak consistency of estimates for rewards of American-type options for Markov Gaussian log-price processes with estimated coefficients.

In Chapters 9 and 10, we present results about fitting of coefficients and convergence for reward functions of American-type options for binomial and trinomial tree approximations for univariate and multivariate Markov Gaussian log-price processes homogeneous and inhomogeneous in time and space.

The bibliography, which contains more than 500 references, is also preceded by brief remarks.

It should be noted that models of log-price processes with discrete time have their own very important value. However, they can also be imbedded in models of log-price processes with continuous time. In such framework, American-type options for discrete time model should be considered as Bermudian type options for the corresponding continuous time models. The discrete time models are also used as a main tool in approximation reward algorithms for continuous time models.

This book is the first volume of a comprehensive two volumes monograph. Let us shortly present the content of the second volume. It will include: results about space-skeleton approximations for rewards of American-type options for various autoregressive and autoregressive moving average-type log-price processes and autoregressive and autoregressive moving average stochastic volatility log-price processes; Monte Carlo-based approximations for rewards of American-type options; results about multithreshold structure of stopping domains for optimal stopping times; time skeleton reward approximations connecting discrete and continuous time models; convergence and approximation results for option rewards for continuous time multivariate modulated Markov log-price processes, log-price processes represented by modulated Lévy processes, and multivariate modulated diffusion log-price processes; applications to European type options, American-type options connected with reselling of European options, and knockout options; as well as results of experimental studies.

The presentation of material in this book is organized in a way that will hopefully be appreciated by readers. Each chapter has a preamble in which the main results are outlined and the content of the chapter is thereafter presented in sections. Each section is broken up into titled subsections.

I would also like to comment on the notation system used in the book. Throughout the text I make use of several basic classes of conditions. Conditions that belong to a specific class are denoted by the same letter. For example, the letter **B** is used for conditions restricting the rate of growth of pay-off functions in price arguments, the letter **C** for moment compactness conditions imposed on log-price processes, and so forth. Conditions belonging to a specific class have subscripts numbering conditions in the class. A list of all conditions is given in the Index.

Chapters and sections have, respectively, a singular and a double numeration. Formulas also have a double numeration. For example, label (1.2) refers to formula 2 in Chapter 1. Subsections, theorems, lemmas, definitions and remarks have a triple numeration. For example, Theorem 1.2.3 means Theorem 3 in Section 1.2.

I hope that the publication of this new book and the comprehensive bibliography of works related to problems of stochastic approximation methods, optimal stopping, and convergence for American-type options will be a useful contribution to the continuing intensive studies in this actual area of financial mathematics. In addition to its employment for research and reference purposes, the book can be used in special courses related to stochastic approximation methods, optimal stopping problems and option pricing and as a complementary reading in general courses on stochastic processes. In this respect, it may be useful for specialists as well as doctoral and advanced undergraduate students.

I am much indebted to Dr. Evelina Silvestrova for her continuous encouragement and support of various aspects of my work on this book.

I would like to thank my collaborators for their fruitful contribution at different stages of my research studies in the area; Professor Alexander Kukush, Professor Raimondo Manca, Dr. Henrik Jönson, Dr. Robin Lundgren, Dr. Anatoliy Malyarenko, Dr. Evelina Silvestrova, and Dr. Fredrik Srenberg for the fruitful cooperation.

I would also like to thank all my colleagues at the Department of Mathematics, Stockholm University, Stockholm, Sweden, and the School of Education, Culture and Communication, Mälardalen University, Västerås, Sweden, for creating the inspiring research environment and friendly atmosphere which so stimulated my work.

I also would like to thank the Riksbankens Jubileumsfond for the very important and stimulating support of my research in the area and work on this book.

Stockholm, March 2013 *Dmitrii Silvestrov*

Contents

1 Multivariate modulated Markov log-price processes (LPP)

In this chapter, we introduce models of multivariate modulated Markov log-price and price processes.

In Section 1.1, we introduce models of multivariate modulated Markov log-price and price processes. In particular, we discuss and comment the sense of introducing a stochastic modulating index component and consider different variants of modulation. We also discuss variants of definition for multivariate modulated Markov log-price and price processes in a stochastic dynamic form.

In Section 1.2, we present different models of multivariate modulated Markov log-price and price processes represented, respectively, by usual and exponential random walks. We show in which way a multivariate modulated Markov log-price and price processes inhomogeneous in time can be interpreted as multivariate modulated Markov random walks inhomogeneous in time and space.

Then, we consider different particular variants of such models. These are models of log-price and price processes represented, respectively, by usual or exponential multivariate random walks with independent jumps, multivariate modulated Markov random walks with inhomogeneous in time and space scale-location parameters, and multivariate modulated random walks with Markov and semi-Markov modulation.

In Sections 1.3 and 1.4, we present various models of modulated autoregressive and autoregressive moving average-type log-price and price processes and autoregressive and autoregressive and moving average stochastic volatility-type log-price and price processes. These are nonmodulated and modulated autoregressive (AR), autoregressive moving average (ARMA), Cox, Ingersoll, Ross (CIR), centered autoregressive conditional heteroskedastic (ARCH), autoregressive/centered autoregressive conditional heteroskedastic (AR/ARCH), and generalized autoregressive conditional heteroskedastic (GARCH) models. We show in which way the above autoregressive models of log-price and pice processes can be imbedded in the class of multivariate modulated Markov log-price and price processes.

1.1 Markov LPP

In this section, we introduce models of discrete time multivariate modulated Markov--type log-price and price processes that are studied in this book.

1.1.1 Discrete time multivariate modulated log-price and price processes

Let $\mathbb{R}_k$ be a k-dimensional Euclidean space and $\mathcal{B}_k$ is the corresponding Borel σ-algebra of subsets of $\mathbb{R}_k$.

Also let $\mathbb{X}$ be a measurable space with a σ-algebra of measurable subsets $\mathcal{B}_{\mathbb{X}}$.

A typical example is, where $\mathbb{X}$ is a complete, separable, metric space (Polish space) with a metric $d_{\mathbb{X}}(x', x'')$ and $\mathcal{B}_{\mathbb{X}}$ be the corresponding Borel σ-algebra of subsets of $\mathbb{X}$ (the minimal σ-algebra containing all balls $R_d(x) = \{x' \in \mathbb{X} : d_{\mathbb{X}}(x, x') \leq d\}$ in the space $\mathbb{X}$).

Let us consider the space $\mathbb{Z} = \mathbb{R}_k \times \mathbb{X}$ with the σ-algebra of measurable subsets $\mathcal{B}_{\mathbb{Z}} = \mathcal{B}_k \times \mathcal{B}_{\mathbb{X}}$, which is the minimal σ-algebra containing all the sets $B \times C$, where $B \in \mathcal{B}_k$, $C \in \mathcal{B}_{\mathbb{X}}$.

In the case, where $\mathbb{X}$ is a Polish metric space with a metric $d_{\mathbb{X}}(x', x'')$, the space $\mathbb{Z} = \mathbb{R}_k \times \mathbb{X}$ is also a Polish metric space with the natural metric $d_{\mathbb{Z}}(\vec{z}\,', \vec{z}\,'') = \sqrt{|\vec{y}\,' - \vec{y}\,''|^2 + d_{\mathbb{X}}(x', x'')^2}$, $\vec{z}\,' = (\vec{y}\,', x')$, $\vec{z}\,'' = (\vec{y}\,'', x'') \in \mathbb{Z}$. Let $\mathcal{B}_{\mathbb{Z}}$ be the corresponding Borel σ-algebra of subsets of $\mathbb{Z}$.

We also use the space $\mathbb{R}_k^+ = \{\vec{s} = (s_1, \ldots, s_k) : s_1, \ldots, s_k > 0\}$ that is the k-dimensional product of the interval $\mathbb{R}_1^+ = (0, \infty)$, and the space $\mathbb{V} = \mathbb{R}_k^+ \times \mathbb{X}$,. The space $\mathbb{V}$ is also a Polish metric space with the metric $d_{\mathbb{V}}(\vec{v}\,', \vec{v}\,'') = \sqrt{|\vec{s}\,' - \vec{s}\,''|^2 + d_{\mathbb{X}}(x', x'')^2}$, $\vec{v}\,' = (\vec{s}\,', x')$, $\vec{v}\,'' = (\vec{s}\,'', x'') \in \mathbb{V}$. The corresponding Borel σ-algebras for the spaces $\mathbb{R}^+$ and $\mathbb{V}$ are denoted, respectively, as $\mathcal{B}_k^+$ and $\mathcal{B}_{\mathbb{V}}$.

Let us consider a random sequence (discrete time stochastic process) $\vec{Y}_n = (Y_{n,1}, \ldots, Y_{n,k})$, $n = 0, 1, \ldots$ with the phase space $\mathbb{R}_k$, a random sequence X_n, $n = 0, 1, 2, \ldots$ with the phase space $\mathbb{X}$, and the random sequence $\vec{Z}_n = (\vec{Y}_n, X_n)$, $n = 0, 1, 2, \ldots$ with the phase space $\mathbb{Z}$.

We always assume that the random sequences $\vec{Y}_n$ and X_n and, thus, Z_n as well are defined on the same probability space $\langle \Omega, \mathcal{F}, \mathsf{P} \rangle$.

We also consider the random sequence $\vec{S}_n = (S_{n,1}, \ldots, S_{n,k})$, $n = 0, 1, \ldots$ connected with the random sequence $\vec{Y}_n = (Y_{n,1}, \ldots, Y_{n,k})$, $n = 0, 1, \ldots$, by the following relations:

$$\vec{S}_n = e^{\vec{Y}_n}, \quad \vec{Y}_n = \ln \vec{S}_n\,, \quad n = 0, 1, \ldots\,. \tag{1.1}$$

Here and henceforth, the notations, $e^{\vec{y}} = (e^{y_1}, \ldots, e^{y_k})$, $\vec{y} = (y_1, \ldots, y_k) \in \mathbb{R}_k$, and $\ln \vec{s} = (\ln s_1, \ldots, \ln s_k)$, $\vec{s} = (s_1, \ldots, s_k) \in \mathbb{R}_k^+$ are used.

Let us also consider the random sequence $\vec{V}_n = (\vec{S}_n, X_n)$, $n = 0, 1 \ldots$.

By the definition, the processes $\vec{Y}_n$, X_n, $\vec{Z}_n$, $\vec{S}_n$, and $\vec{V}_n$ have, respectively, the phase spaces, $\mathbb{R}_k$, $\mathbb{X}$, $\mathbb{Z}$, $\mathbb{R}_k^+$, and $\mathbb{V}$.

We interpret $\vec{Y}_n$ as a multivariate log-price process with discrete time, X_n as a stochastic index modulating the log-price process $\vec{Y}_n$, and $\vec{Z}_n$ as a multivariate modulated log-price process (the extended log-price process supplemented by the additional index component X_n).

$\vec{S}_n$ is interpreted as a multivariate price process and $\vec{V}_n$ as a multivariate modulated price process (the extended price process supplemented by the additional index component X_n).

We do prefer to use log-price processes as initial objects instead of price processes. This is because of log-price processes usually have an additive structure of increments, while price processes usually have a multiplicative structure of increments. The additive structure of increments is, as a rule, simpler and more convenient for analysis than the multiplicative one.

The character of increments mentioned above also explains using of exponential transformation which connects log-price processes with the corresponding price processes.

1.1.2 Information filtrations generated by log-price and price processes

Let $\mathcal{F}'_n = \sigma[\vec{Y}_0, \dots, \vec{Y}_n]$, $n = 0, 1, \dots$, be a natural filtration generated by the log-price process $\vec{Y}_n$ that is the family of σ-algebras of random events generated by the random variables $\vec{Y}_0, \dots, \vec{Y}_n$, for $n = 0, 1, \dots$. Also let $\mathcal{F}_n = \sigma[\vec{Z}_0, \dots, \vec{Z}_n]$, $n = 0, 1, \dots$, be a natural filtration generated by the process $\vec{Z}_n$.

It is useful to note that $\mathcal{F}'_n$ and $\mathcal{F}_n$ coincide with the natural filtrations generated, respectively, by the processes $\vec{S}_n$ and $\vec{V}_n$.

By the definition, $\mathcal{F}'_n \subseteq \mathcal{F}_n$, $n = 0, 1, \dots$, i.e. filtration $\mathcal{F}_n$ is an extension of filtration $\mathcal{F}'_n$.

Thus, the component X_n represents an additional market information that become available at the moment n additionally to the value of the log-price process $\vec{Y}_n$ or the corresponding price process $\vec{S}_n$.

This information can be supplied by some additional log-price or price process X_n, which values are observable but not included in the vector log-price process $\vec{Y}_n$ or the price process $\vec{S}_n$.

Another variant is where X_n represents stochastic dynamics of some parameters for the log-price process $\vec{Y}_n$, for example, its stochastic volatility.

The third variant is where X_n is, indeed, a market index, for example, a global price index "controlling" market prices, or a process representing some market regime index, for example, indicating growing, declining, or stable market situation.

1.1.3 Markov log-price and price processes

In what follows, we always assume that $\vec{Z}_n$, $n = 0, 1, \dots$, is an inhomogeneous in time Markov chain with a phase space $\mathbb{Z}$, initial distribution $P_0(A) = \mathsf{P}\{\vec{Z}_0 \in A\}$ defined for $A \in \mathcal{B}_{\mathbb{Z}}$ and the one-step transition probabilities defined for $\vec{z} \in \mathbb{Z}, A \in \mathcal{B}_{\mathbb{Z}}$,

$n = 1, 2, \dots,$

$$P_n(\vec{z}, A) = \mathsf{P}\left\{\vec{Z}_n \in A/\vec{Z}_{n-1} = \vec{z}\right\} . \tag{1.2}$$

Obviously, $\vec{V}_n$ is also an inhomogeneous Markov chain with the phase space $\mathbb{V}$.

The initial distribution $\dot{P}_0(B)$, and the transition probabilities $\dot{P}_n(\vec{v}, B)$ of the Markov chain $\vec{V}_n$ are connected with the initial distribution and the transition probabilities of the Markov chain $\vec{Z}_n$ by the following relations, for $\vec{v} = (\vec{s}, x) \in \mathbb{V}, B \in \mathcal{B}_{\mathbb{V}}$, $n = 1, 2, \dots,$

$$\dot{P}_0(B) = P(A_B)\,, \quad \dot{P}_n(\vec{v}, B) = P_n(\vec{z}_{\vec{v}}, A_B)\,, \tag{1.3}$$

where $\vec{z}_{\vec{v}} = (\ln \vec{s}, x) \in \mathbb{Z}$, $A_B = \{\vec{z}_{\vec{v}} : \vec{v} \in B\} \in \mathcal{B}_{\mathbb{Z}}$.

It is worth noting that the log-price process $\vec{Y}_n$ and the corresponding price process $\vec{S}_n$ themselves may not be Markov processes. The component X_n represents information which addition to information supplied by the log-price process $\vec{Y}_n$ or the price process $\vec{S}_n$ makes the corresponding extended processes $\vec{Z}_n = (\vec{S}_n, X_n)$ and $\vec{V}_n = (\vec{S}_n, X_n)$ Markov processes.

An important case is where the Markov chains $\vec{Z}_n$ and $\vec{V}_n$ are homogeneous in time. This is the case, where the transition probabilities $P_n(\vec{z}, A) = P(\vec{z}, A)$ and $\dot{P}_n(\vec{v}, A) = \dot{P}(\vec{v}, A)$ do not depend on time $n = 0, 1, \dots$.

1.1.4 Dynamic representations for Markov log-price and price processes

The Markov process $\vec{Z}_n = (\vec{Y}_n, X_n)$ can be given in a stochastic dynamic form, i.e. be defined by the following stochastic transition dynamic relation:

$$\vec{Z}_n = A_n(\vec{Z}_{n-1}, U_n)\,, \quad n = 1, 2, \dots\,, \tag{1.4}$$

where (a) $\vec{Z}_0$ is a random variable taking values in the space $\mathbb{Z}$, (b) U_n, $n = 1, 2, \dots$, is a sequence of "noise" independent random variables taking values in some measurable space $\mathbb{U}$ with a σ-algebra of measurable subsets $\mathcal{B}_{\mathbb{U}}$, (c) the random variable $\vec{Z}_0$ and the sequence of random variables U_n, $n = 1, 2, \dots$, are independent, and (d) $A_n(\vec{z}, u)$, $n = 1, 2, \dots$, are the measurable functions acting from the space $\mathbb{Z} \times \mathbb{U}$ to the space $\mathbb{Z}$.

Usually, the initial state can also be represented in the form $\vec{Z}_0 = A_0(U_0)$, where (e) U_0 is a random variable taking values in the space $\mathbb{U}$, (f) the random variable U_0 and the sequence of random variables U_n, $n = 1, 2, \dots$, are independent, and (g) $A_0(u)$, $n = 1, 2, \dots$, is a measurable function acting from the space $\mathbb{U}$ to the space $\mathbb{Z}$.

If the Markov chain $\vec{Z}_n$ has a Polish phase space $\mathbb{Z}$ and is given in the standard form by its initial distribution and transition probabilities, it is always possible to construct on some probability space a Markov chain $\vec{Z}'_n$, which is stochastically equivalent to the Markov chain $\vec{Z}_n$ in the distributional sense and is given in the above dynamic form.

As is shown, for example, in Gikhman and Skorokhod (1977), in this case, it is possible to construct the measurable functions $A_n(\vec{z}, u)$, $n = 1, 2, \dots$, acting from the space $\mathbb{Z} \times [0, 1]$ to the space $\mathbb{Z}$ and a function $A_0(u)$ acting from the interval $[0, 1]$ to the space $\mathbb{Z}$ such that, for any sequence of the i.i.d. random variables U_n, $n = 0, 1, \dots$, uniformly distributed in the interval $[0, 1]$, the random sequence $\vec{Z}'_n = A_n(Z'_{n-1}, U_n)$, $n = 1, 2, \dots$ $\vec{Z}'_0 = A_0(U_0)$ is a Markov chain with the space space $\mathbb{Z}$, the initial distribution $P_0(A)$, and the transition probabilities $P_n(\vec{z}, A)$.

The above remarks, let us, in most of the cases, define a Markov log-price process $\vec{Z}_n$ in the dynamic form without any loss of generality.

The transition dynamic functions $A_n(\vec{z}, u)$ can be represented in the form $A_n(\vec{z}, u) = (B_n(\vec{y}, x, u), C_n(\vec{y}, x, u)), \vec{z} = (\vec{y}, x) \in \mathbb{Z}, u \in \mathbb{U}, n = 1, 2, \dots$, where $B_n(\vec{y}, x, u)$ and $C_n(\vec{y}, x, u)$ are the components of functions $A_n(\vec{z}, u)$ acting from the space space $\mathbb{R}_k \times \mathbb{X} \times \mathbb{U}$, respectively, to the space $\mathbb{R}_k$ and $\mathbb{X}$.

Analogously, the function $A_0(u)$ can be represented in the form $A_0(u) = (B_0(u), C_0(u))$, $u \in \mathbb{U}$, where $B_0(u)$ and $C_0(u)$ are the components of the functions $A_0(u)$ acting from the space space $\mathbb{U}$, respectively, to the space $\mathbb{R}_k$ and $\mathbb{X}$.

The stochastic transition dynamic relation (1.4) can be rewritten in the following form:

$$\begin{cases} \vec{Y}_n = B_n(\vec{Y}_{n-1}, X_{n-1}, U_n), \\ X_n = C_n(\vec{Y}_{n-1}, X_{n-1}, U_n), \\ \quad n = 1, 2, \dots, \end{cases} \tag{1.5}$$

where (a) $\vec{Z}_0 = (\vec{Y}_0, X_0)$ is a random variable taking values in the space $\mathbb{Z}$, (b) U_n, $n = 1, 2, \dots$, is a sequence of "noise" independent random variables taking values in some measurable space $\mathbb{U}$ with a σ-algebra of measurable subsets $\mathcal{B}_{\mathbb{U}}$, (c) the random variable $\vec{Z}_0$ and the sequence of random variables U_n, $n = 1, 2, \dots$, are independent, (d) $B_n(\vec{y}, x, u)$, $n = 1, 2, \dots$, are the measurable functions acting from the space $\mathbb{R}_k \times \mathbb{X} \times \mathbb{U}$ to the space $\mathbb{R}_k$, and (e) $C_n(\vec{y}, x, u)$, $n = 1, 2, \dots$, are the measurable functions acting from the space $\mathbb{R}_k \times \mathbb{X} \times \mathbb{U}$ to the space $\mathbb{X}$.

The stochastic transition dynamic relation (1.5) can also be rewritten in terms of the extended price process $\vec{V}_n$,

$$\begin{cases} \vec{S}_n = e^{B_n(\ln \vec{S}_{n-1}, X_{n-1}, U_n)}, \\ X_n = C_n(\ln \vec{S}_{n-1}, X_{n-1}, U_n) \\ \quad n = 1, 2, \dots . \end{cases} \tag{1.6}$$

Note that the transition probabilities $P_n(\vec{z}, A)$ of the Markov chain $\vec{Z}_n$ are connected with the dynamic transition functions $A_n(\vec{z}, u)$ by the following relation, for $\vec{z} \in \mathbb{Z}, A \in \mathcal{B}_{\mathbb{Z}}, n = 1, 2, \dots$:

$$P_n(\vec{z}, A) = \mathsf{P}\{\vec{Z}_n \in A / \vec{Z}_{n-1} = \vec{z}\} = \mathsf{P}\{A_n(\vec{z}, U_n) \in A\}. \tag{1.7}$$

1.1.5 Modulated Markov log-price and price processes

The index component X_n can also be interpreted as a modulator of the corresponding log-price process $\vec{Y}_n$ or the price process $\vec{S}_n$.

To see this, let us first consider the case, where the phase space $\mathbb{X} = \{1, 2, \dots, m\}$ is a finite set. In this case, as usual, $\mathcal{B}_{\mathbb{X}}$ is the σ-field of all subsets of $\mathbb{X}$.

Let us introduce conditional marginal transition probabilities for the index X_n,

$$\begin{aligned} P_{n,\vec{y}}(x, x') &= \mathsf{P}\left\{X_n = x' / \vec{Y}_{n-1} = \vec{y}, X_{n-1} = x\right\} \\ &= P_n((\vec{y}, x), \mathbb{R}_k \times \{x'\}) \end{aligned} \tag{1.8}$$

and conditional marginal transition probabilities for the log-price process $\vec{Y}_n$

$$\begin{aligned} P_{n,xx'}(\vec{y}, B) &= \mathsf{P}\left\{\vec{Y}_n \in B / \vec{Y}_{n-1} = \vec{y}, X_{n-1} = x, X_n = x'\right\} \\ &= P_n((\vec{y}, x), B \times \{x'\}) / P_{n,\vec{y}}(x, x') . \end{aligned} \tag{1.9}$$

Relation (1.9) defines the conditional probabilities $P_{n,xx'}(\vec{y}, B)$ in the cases, where the probabilities $P_{n,\vec{y}}(x, x') > 0$. Otherwise, an arbitrary probability measure (as a function of B) can play the role of this conditional transition probability.

The transition probabilities $P_n(z, A)$, $z = (\vec{y}, x) \in \mathbb{Z}$, $A \in \mathcal{B}_{\mathbb{Z}}$ are uniquely determined by their values on cylindric sets $A = B \times C$, where $B \in \mathcal{B}_k$, $C \in \mathcal{B}_{\mathbb{X}}$. For such sets, these probabilities can be represented via the corresponding marginal transition probabilities introduced above,

$$P_n((\vec{y}, x), B \times C) = \sum_{x' \in C} P_{n,xx'}(\vec{y}, B) P_{n,\vec{y}}(x, x') . \tag{1.10}$$

In this representation let us consider the index X_n as a modulator of the log-price component $\vec{Y}_n$. In fact, relation (1.10) shows that some kind of mutual cross modulation for processes $\vec{Y}_n$ and X_n takes place.

There is, however, the important case where the index X_n modulates the log-price process $\vec{Y}_n$ but not vice versa. This is the case, where the marginal transition probabilities for the index do not depend on values of the log-price process, i.e.

$$P_{n,\vec{y}}(x, x') = P_n(x, x') . \tag{1.11}$$

In what follows, we always assume that the corresponding relation holds for all admissible values of arguments if this is not specified in this relation.

It is easy to show that, in this case, the index X_n is a Markov chain with the phase space $\mathbb{X}$ and the one-step transition probabilities $P_n(x, x')$.

If the model is defined in the dynamic form given by relation (1.5) then the following relation replaces (1.11):

$$C_n((y, x), U_n) = C_n(x, U_n) . \tag{1.12}$$

Let us now consider the case, where $\mathbb{X}$ is a general space with the σ-algebra of measurable subsets $\mathcal{B}_{\mathbb{X}}$.

Let us introduce conditional marginal transition probabilities for the stochastic index X_n

$$\begin{aligned} P_{n,\vec{y}}(x, C) &= \mathsf{P}\left\{X_n \in C / \vec{Y}_{n-1} = \vec{y}, X_{n-1} = x\right\} \\ &= P_n\left((\vec{y}, x), \mathbb{R}_k \times C\right) . \end{aligned} \tag{1.13}$$

The transition probabilities $P_n(\vec{z}, A), \vec{z} = (\vec{y}, x) \in \mathbb{Z}, A \in \mathcal{B}_{\mathbb{Z}}$ are uniquely determined by their values on cylindric sets $A = B \times C$, where $B \in \mathcal{B}_k, C \in \mathcal{B}_{\mathbb{X}}$.

By the definition, $P_n((\vec{y}, x), B \times C) \leq P_n((\vec{y}, x), \mathbb{R}_k \times C), C \in \mathcal{B}_{\mathbb{X}}$, for every $(\vec{y}, x) \in \mathbb{Z}, B \in \mathcal{B}_k$. Thus, by the Radon–Nikodym theorem, $P_n((\vec{y}, x), B \times C)$, can be represented in the following form:

$$P_n((\vec{y}, x), B \times C) = \int_C P_{n,x,x'}(\vec{y}, B) P_{n,\vec{y}}(x, dx') , \tag{1.14}$$

where $P_{n,x,x'}(\vec{y}, B)$ is the Radon–Nikodym derivative of measure $P_n((\vec{y}, x), B \times C), C \in \mathcal{B}_{\mathbb{X}}$ with respect to measure $P_n((\vec{y}, x), \mathbb{R}_k \times C), C \in \mathcal{B}_{\mathbb{X}}$.

Under some minor conditions, for example, if $\mathbb{X}$ is a Polish space and $\mathcal{B}_{\mathbb{X}}$ is the corresponding Borel σ-algebra of subsets of $\mathbb{X}$, there exists the regular variant of the derivative $P_{n,x,x'}(\vec{y}, B)$, which is a measurable function in argument $(x, x', \vec{y}) \in \mathbb{X} \times \mathbb{X} \times \mathbb{R}_k$, for every $B \in \mathcal{B}_k$ and a probability measure in $B \in \mathcal{B}_k$ for every $(x, x', \vec{y}) \in \mathbb{X} \times \mathbb{X} \times \mathbb{R}_k$. This derivative is a regular conditional distribution

$$P_{n,xx'}(\vec{y}, B) = \mathsf{P}\left\{\vec{Y}_n \in B / \vec{Y}_{n-1} = \vec{y}, X_{n-1} = x, X_n = x'\right\} . \tag{1.15}$$

One can consider the index X_n as a modulator of the log-price component $\vec{Y}_n$. In fact, relation (1.15) shows that processes $\vec{Y}_n$ and X_n, in some sense, cross-modulate each other.

There is, however, the important case, where the index X_n modulates the log-price process $\vec{Y}_n$ but not vice versa. This is the case, where the marginal transition probabilities for the index X_n do not depend on the values of the log-price process, i.e.

$$P_{n,\vec{y}}(x, C) = P_n(x, C) . \tag{1.16}$$

It is possible to show that, in this case, the index X_n is a Markov chain with the phase space $\mathbb{X}$ and the one-step transition probabilities $P_n(x, C)$.

The stochastic transition dynamic relation (1.5) takes, in this case, the following simpler form:

$$\begin{cases} \vec{Y}_n = B_n(\vec{Y}_{n-1}, X_{n-1}, X_n, U_n') , \\ X_n = C_n(X_{n-1}, U_n'') , \\ n = 1, 2, \dots , \end{cases} \tag{1.17}$$

where (a) $\vec{Z}_0 = (\vec{Y}_0, X_0)$ is a random variable taking values in the space $\mathbb{Z}$, (b) U_n', U_n'', $n = 1, 2, \dots$ is a family of independent random variables taking values in some measurable space $\mathbb{U}$ with a σ-algebra of measurable subsets $\mathcal{B}_{\mathbb{U}}$, (c) the random variable $\vec{Z}_0$ and the family of random variables U'_n, U''_n, $n = 1, 2, \ldots$, are independent, (d) $B_n(\vec{y}, x, x', u)$, $n = 1, 2, \ldots$, are the measurable functions acting from the space $\mathbb{R}_k \times \mathbb{X} \times \mathbb{X} \times \mathbb{U}$ to $\mathbb{R}_k$, and (e) $C_n(x, u)$, $n = 1, 2, \ldots$, are the measurable functions acting from the space $\mathbb{X} \times \mathbb{U}$ to the space $\mathbb{X}$.

Relation (1.17) takes the following exponential form:

$$\begin{cases} \vec{S}_n = e^{B_n(\ln \vec{S}_{n-1}, X_{n-1}, X_n, U'_n)}, \\ X_n = C_n(X_{n-1}, U''_n), \\ n = 1, 2, \ldots, \end{cases} \tag{1.18}$$

1.2 LPP represented by random walks

In this section, we consider a number of typical examples of discrete time modulated Markov log-price and price processes represented by multivariate modulated random walks.

1.2.1 Log-price and price processes represented by Markov random walks inhomogeneous in space and time

A modulated Markov chain $\vec{Z}_n = (\vec{Y}_n, X_n)$ can always be represented in the form of the modulated Markov random walk, given by the following stochastic transition dynamic relation,

$$\vec{Z}_n = (\vec{Y}_n, X_n) = \begin{cases} (\vec{Y}_0, X_0) & \text{if } n = 0, \\ (\vec{Y}_{n-1} + \vec{W}_n, X_n) & \text{if } n = 1, 2, \ldots, \end{cases} \tag{1.19}$$

where

$$\vec{W}_n = \vec{Y}_n - \vec{Y}_{n-1}, \quad n = 1, 2, \ldots \tag{1.20}$$

In this case, it is convenient to operate with the transition jump probabilities $\tilde{P}_n(\vec{y}, x, A) = \mathsf{P}\{(\vec{W}_n, X_n) \in A / \vec{Y}_{n-1} = \vec{y}, X_{n-1} = x\}$, which are connected with the transition probabilities of the Markov chain $\vec{Z}_n$ by the following relations, for $\vec{z} = (\vec{y}, x) \in \mathbb{Z}, A \in \mathcal{B}_{\mathbb{Z}}, n = 0, 1, \ldots$:

$$\begin{aligned} \tilde{P}_n(\vec{y}, x, A) &= \mathsf{P}\Big\{(\vec{W}_n, X_n) \in A / \vec{Y}_{n-1} = \vec{y}, X_{n-1} = x\Big\} \\ &= \mathsf{P}\Big\{(\vec{Y}_n - \vec{y}, X_n) \in A / \vec{Y}_{n-1} = \vec{y}, X_{n-1} = x\Big\} = P_n(\vec{z}, A_{[\vec{y}]}), \end{aligned} \tag{1.21}$$

where $A_{[\vec{y}]} = \{\vec{z}\,' = (\vec{y}\,', x) : (\vec{y}' - \vec{y}, x) \in A\}$.

A dynamic analog of this representation can be obtained by simple transformation of the stochastic transition dynamic relation (1.5). Indeed, by introducing a new dynamic transition function $\tilde{B}_n(\vec{y}, x, u) = B_n(\vec{y}, x, u) - \vec{y}$, one can rewrite relation (1.5) in the following form:

$$\begin{cases} \vec{Y}_n = \vec{Y}_{n-1} + \tilde{B}_n(\vec{Y}_{n-1}, X_{n-1}, U_n)\,, \\ X_n = C_n(\vec{Y}_{n-1}, X_{n-1}, U_n)\,, \\ n = 1, 2, \dots\,, \end{cases} \tag{1.22}$$

where (a) $\vec{Z}_0 = (\vec{Y}_0, X_0)$ is a random variable taking values in the space $\mathbb{Z} = \mathbb{R}_k \times \mathbb{X}$, (b) U_n, $n = 1, 2, \dots$, is a sequence of independent random variables taking value in some measurable space $\mathbb{U}$ with a σ-algebra of measurable subsets $\mathcal{B}_{\mathbb{U}}$, (c) the random variable $\vec{Z}_0$ and the random sequence U_n, $n = 1, 2, \dots$, are independent, (d) $\tilde{B}_n(\vec{y}, x, u)$, $n = 1, 2, \dots$, are the vector functions in which components are the measurable functions acting from the space $\mathbb{R}_k \times \mathbb{X} \times \mathbb{U}$ to $\mathbb{R}_k$, (e) $C_n(\vec{y}, x, u)$, $n = 1, 2, \dots$, are the measurable functions acting from the space $\mathbb{R}_k \times \mathbb{X} \times \mathbb{U}$ to the space $\mathbb{X}$.

As far as the modulated price process $\vec{V}_n$ is concerned, it will be given in this case by the following multiplicative analog of stochastic transition dynamic relation: (1.22):

$$\begin{cases} \vec{S}_n = \vec{S}_{n-1} \cdot e^{\tilde{B}_n(\ln \vec{S}_{n-1}, X_{n-1}, U_n)}\,, \\ X_n = C_n(\ln \vec{S}_{n-1}, X_{n-1}, U_n)\,, \\ n = 1, 2, \dots\,. \end{cases} \tag{1.23}$$

Here and henceforth, the vector product $\vec{c} = \vec{a} \cdot \vec{b}$ should be understood as the vector with the components $c_i = a_i b_i$, $i = 1, \dots, k$.

In the following subsections, we consider typical examples of log-price and price processes represented, respectively, by random walks and exponential random walks.

1.2.2 Log-price and price processes represented by standard random walks

The simplest example is where a multivariate Markov log-price $\vec{Y}_n$ is a standard random walk

$$\vec{Y}_n = \vec{Y}_{n-1} + \vec{W}_n\,, \quad n = 1, 2, \dots\,, \tag{1.24}$$

where (a) $\vec{Y}_0 = (Y_{0,1}, \dots, Y_{0,k})$ is a random vector taking values in the space $\mathbb{R}_k$, (b) $\vec{W}_n = (W_{n,1}, \dots, W_{n,k})$, $n = 1, 2, \dots$, is a sequence of independent random vectors taking values in $\mathbb{R}_k$, and (c) the random vector $\vec{Y}_0$ and the random sequence $\vec{W}_n$, $n = 1, 2, \dots$, are independent.

In this case, the model without index component is considered. In fact, one can always define and include in the model a "virtual" index component X_n, which takes a constant value $x_0 \in \mathbb{X}$, for every $n = 0, 1, \dots$.

Relation (1.24) defines the process $\vec{Y}_n$ in the dynamic form. Also note that in this case, the role of "noise" random variable U_n is played by the random vector $\vec{W}_n$, for $n = 1, 2, \dots$. Thus, the space $\mathbb{U} = \mathbb{R}_k$.

As far as the price process $\vec{S}_n$ is concerned, it will be given in this case by the following multiplicative analog of relation (1.24):

$$\vec{S}_n = \vec{S}_{n-1} \cdot e^{\vec{W}_n}, \quad n = 1, 2, \dots . \tag{1.25}$$

The process $\vec{S}_n$ is called a standard exponential random walk.

It is useful to note that the random walks $\vec{Y}_n$ and $\vec{S}_n$ are homogeneous in time if the random vectors $\vec{W}_n$, $n = 1, 2, \dots$, are identically distributed.

There are several the most frequently used models.

The first and the simplest one is the binomial model, where components $W_{n,1}, \dots, W_{n,k}$ of random vectors $\vec{W}_n$, $n = 1, 2, \dots$, are binary random variables, i.e. random variables taking just two different values. Usually, one of these values is positive while another is negative. In this case, $\vec{Y}_n$ and $\vec{S}_n$ are, respectively, a multivariate binomial random walk and a multivariate exponential binomial random walk.

Slightly more complex is the trinomial model, where the components $W_{n,1}, \dots, W_{n,k}$ of the random vectors $\vec{W}_n$, $n = 1, 2, \dots$, are trinary random variables, i.e. random variables taking just three different values. Usually one of these values is positive, one is zero and one is negative. In this case, $\vec{Y}_n$ and $\vec{S}_n$ are, respectively, a multivariate trinomial random walk and a multivariate exponential trinomial random walk.

Also, variants of the model with more general random vectors $\vec{W}_n$, $n = 1, 2, \dots$, with a discrete sets of possible values are considered.

Discrete time models are often appear as a result of some time discretization procedures applied to the continuous time models of log-price processes with independent increments. In such cases, the following three variants of log-price processes represented by random walks appear in a natural way.

For example, application of the time discretization procedure mentioned above to a continuous time log-price process represented by a multivariate Brownian motion, results with the model of discrete time log-price processes $\vec{Y}_n$, $n = 1, 2, \dots$, represented by a standard random walk with jumps $\vec{W}_n$, $n = 1, 2, \dots$, which have multivariate normal distributions. In this case, $\vec{Y}_n$ and $\vec{S}_n$ are discrete time analogs of, respectively, a multivariate Brownian motion and a multivariate exponential Brownian motion.

In the case, where the corresponding continuous time model is represented by a so-called jump-diffusion log-price process, application of the time discretization procedure mentioned above results with the model of a discrete time log-price processes $\vec{Y}_n$, $n = 1, 2, \dots$, represented by a standard random walk with jumps $\vec{W}_n = \vec{W}'_n + \vec{W}'_n$, $n = 1, 2, \dots$. Here, the additive components of the jumps, $\vec{W}'_n$ and $\vec{W}'_n$, are independent, the random vectors $\vec{W}'_n$ have a multivariate normal distributions while the random vectors $\vec{W}''_n$ have vector compound Poisson distributions (i.e., $\vec{W}''_n$ can be represented in the form of random sum, where summands are i.i.d. random vectors independent of a random index, which counts number of summands and has a Poisson distribution).

Finally, if the corresponding continuous time model is represented by the Lévy process, application of the time discretization procedure mentioned above results with the model of a discrete time log-price processes $\vec{Y}_n$, $n = 1, 2, \dots$, with jumps $\vec{W}_n$, $n = 1, 2, \dots$, possessing infinitely divisible distributions.

1.2.3 Log-price and price processes represented by inhomogeneous in time and space random walks

This model is also convenient to introduce using the stochastic transition dynamic relation

$$\vec{Y}_n = \vec{Y}_{n-1} + \tilde{B}_n(\vec{Y}_{n-1}, U_n), \quad n = 0, 1, \dots, \tag{1.26}$$

where (a) $\vec{Y}_0$ is a random vector taking values in the space $\mathbb{R}_k$, (b) U_n, $n = 1, 2, \dots$, is a sequence of independent random variables taking value in some measurable space $\mathbb{U}$ with a σ-algebra of measurable subsets $\mathcal{B}_{\mathbb{U}}$, (c) the random vector $\vec{Y}_0$ and the random sequence U_n, $n = 1, 2, \dots$, are independent, (d) $\tilde{B}_n(\vec{y}, u)$ are the measurable functions acting from the space $\mathbb{R}_k \times \mathbb{U}$ to $\mathbb{R}_k$.

As given in the previous subsection, the model without index component is considered. In fact, one can always to define a “virtual” index component X_n which takes a constant value x_0, for every $n = 0, 1, \dots$.

As far as the price process $\vec{S}_n$ is concerned, it is given in this case by the following formula:

$$\vec{S}_n = \vec{S}_{n-1} \cdot e^{\tilde{B}_n(\ln \vec{S}_{n-1}, U_n)}, \quad n = 1, 2, \dots. \tag{1.27}$$

One of the important variants is the model, where the log-price process is a random walk with drift and volatility functional coefficients inhomogeneous in space and time,

$$\vec{Y}_n = \vec{Y}_{n-1} + \vec{\mu}_n(\vec{Y}_{n-1}) + \Sigma_n(\vec{Y}_{n-1})\vec{W}_n, \quad n = 1, 2, \dots, \tag{1.28}$$

where (a) $\vec{Y}_0 = (Y_{0,1}, \dots, Y_{0,k})$ is a k-dimensional random vector, (b) $\vec{W}_n = (W_{n,1}, \dots, W_{n,k})$, $n = 1, 2, \dots$, is a sequence of k-dimensional independent random vectors, (c) the random vector $\vec{Y}_0$ and the random sequence $\vec{W}_n$, $n = 1, 2, \dots$, are independent, (d) $\vec{\mu}_n(\vec{y}) = (\mu_{n,1}(\vec{y}), \dots, \mu_{n,k}(\vec{y}))$, $n = 1, 2, \dots$, are the vector functions in which components are the measurable functions acting from $\mathbb{R}_k$ to $\mathbb{R}_1$, and (e) $\Sigma_n(\vec{y}) = \|\sigma_{n,ij}(\vec{y})\|$, $n = 1, 2, \dots$, are the $k \times k$ matrix functions with elements $\sigma_{n,ij}(\vec{y})$ which are the measurable functions acting from $\mathbb{R}_k$ to $\mathbb{R}_1$.

It is useful to note that here and henceforth we interpret vectors $\vec{b} = (b_1, \dots, b_k)$ as column vectors and use an ordinary definition for a product of a $k \times k$ matrix $A = \|a_{i,j}\|$ by a vector $\vec{b}$ as a column vector $A\vec{b} = \vec{c} = (c_1, \dots, c_k)$, where $c_i = \sum_{j=1}^{k} a_{i,j} b_j$, $i = 1, \dots, k$.

As far as the price process $\vec{S}_n$ is concerned, it will be given in this case by the following multiplicative analog of relation (1.28):

$$\vec{S}_n = \vec{S}_{n-1} \cdot e^{\vec{\mu}_n(\ln \vec{S}_{n-1}) + \Sigma_n(\ln \vec{S}_{n-1})\vec{W}_n}, \quad n = 1, 2, \dots . \tag{1.29}$$

The log-price process $\vec{Y}_n$ and the price process $\vec{S}_n$ described above have no index component. These processes are discrete time Markov chains.

These Markov chains are homogeneous in time if the random vectors $\vec{W}_n$, $n = 1, 2, \dots$, are identically distributed and functions $\vec{\mu}_n(\vec{y}) = \vec{\mu}(\vec{y})$ and $\Sigma_n(\vec{y}) = \Sigma(\vec{y})$ do not dependent on n.

Relation (1.28) can be rewritten in the form of stochastic difference equation

$$\vec{Y}_n - \vec{Y}_{n-1} = \vec{\mu}_n(\vec{Y}_{n-1}) + \Sigma_n(\vec{Y}_{n-1})\vec{W}_n, \quad n = 1, 2, \dots . \tag{1.30}$$

Equation (1.30) can be considered as a discrete time analog of a stochastic differential equation.

If it is additionally assumed that random vectors $\vec{W}_n$, $n = 1, 2, \dots$, have normal distributions, the process $\vec{Y}_n$, $n = 0, 1, \dots$, is the discrete time analog of a diffusion process, while $\vec{S}_n$, $n = 0, 1, \dots$, is the discrete time analog of an exponential diffusion process.

The following two examples of the above model appear in many applications.

In the one-dimensional case, this is a discrete time analog of an Ornstein–Uhlenbeck-type process given by the following stochastic difference equation:

$$Y_n - Y_{n-1} = \lambda''_n(Y_{n-1} - \lambda'_n) + \sigma_n W_n, \quad n = 1, 2, \dots , \tag{1.31}$$

where (a) Y_0 is the real-valued random variable, (b) W_n, $n = 1, 2, \dots$, is a sequence of real-valued independent random variables, (c) the random variable Y_0 and the random sequence W_n, $n = 1, 2, \dots$, are independent, and (d) $\lambda'_n, \lambda''_n, \sigma_n$, $n = 1, 2, \dots$, are the real-valued constants.

It is useful to note that the process Y_n possesses so-called mean-reverse property if the constants $\lambda''_n \leq 0$, $n = 1, 2, \dots$.

In the multivariate case, a discrete time Ornstein–Uhlenbeck-type process is given by the following vector stochastic difference equation:

$$\vec{Y}_n - \vec{Y}_{n-1} = \Lambda_n(\vec{Y}_{n-1} - \vec{\lambda}_n) + \Sigma_n \vec{W}_n, \quad n = 1, 2, \dots , \tag{1.32}$$

where (a) $\vec{Y}_0 = (Y_{0,1}, \dots, Y_{0,k})$ is a k-dimensional random vector with real-valued components, (b) $\vec{W}_n = (W_{n,1}, \dots, W_{n,k})$, $n = 1, 2, \dots$, is a sequence of independent random vectors taking values in the space $\mathbb{R}_k$, (c) the random vector $\vec{Y}_0$ and the random sequence $\vec{W}_n$, $n = 1, 2, \dots$, are independent, (d) $\vec{\lambda}_n = (\lambda_{n,1}, \dots, \lambda_{n,k})$, $n = 1, 2, \dots$, are the k-dimensional nonrandom vectors with real-valued components, and (e) $\Lambda_n = \|\lambda_{n,ij}\|$, $\Sigma_n = \|\sigma_{n,ij}\|$, $n = 1, 2, \dots$, are the $k \times k$ matrices with real-valued elements.

The Ornstein–Uhlenbeck-type process $\vec{Y}_n$ is a homogeneous in time Markov chain, if the random vectors $\vec{W}_n$, $n = 1, 2, \dots$, are identically distributed and the parameters $\vec{\lambda}_n = \vec{\lambda}$, $\Lambda_n = \Lambda$, $\Sigma_n = \Sigma$, $n = 1, 2, \dots$, do not depend on n.

1.2.4 Log-price and price processes represented by multivariate modulated random walks

This model can be introduced by the following stochastic transition dynamic relation: which is a particular case of relation (1.22):

$$\begin{cases} \vec{Y}_n = \vec{Y}_{n-1} + \tilde{B}_n(X_{n-1}, U_n)\,, \\ X_n = C_n(X_{n-1}, U_n)\,, \\ \quad n = 1, 2, \ldots\,, \end{cases} \tag{1.33}$$

where (a) $(\vec{Y}_0, X_0) = ((Y_1, \ldots, Y_k), X_0)$ is a random vector taking values in space $\mathbb{Z}$, (b) U_n, $n = 1, 2, \ldots$, is a sequence of independent random variables taking value in some measurable space $\mathbb{U}$ with a σ-algebra of measurable subsets $\mathcal{B}_{\mathbb{U}}$, (c) the random vector $(\vec{Y}_0, X_0)$ and the random sequence U_n, $n = 1, 2, \ldots$, are independent, (d) $\tilde{B}_n(x, u) = (\tilde{B}_{n,1}(x, u), \ldots, \tilde{B}_{n,k}(x, u))$, $n = 1, 2, \ldots$, are the measurable functions acting from $\mathbb{X} \times \mathbb{U}$ to $\mathbb{R}_k$, and (e) $C_n(x, u)$, $n = 1, 2, \ldots$, are the measurable functions acting from $\mathbb{X} \times \mathbb{U}$ to $\mathbb{X}$.

As far as the price process $\vec{V}_n = (\vec{S}_n, X_n)$ is concerned, it will be given in this case by the following multiplicative analog of formula (1.33):

$$\begin{cases} \vec{S}_n = \vec{S}_{n-1} \cdot e^{\tilde{B}_n(X_{n-1}, U_n)}\,, \\ X_n = C_n(X_{n-1}, U_n)\,, \\ \quad n = 1, 2, \ldots\,. \end{cases} \tag{1.34}$$

Note that, in this case, the transition probabilities $\tilde{P}_n(\vec{y}, x, A)$ take the following form, for $\vec{z} = (\vec{y}, x) \in \mathbb{Z}$, $A \in \mathcal{B}_{\mathbb{Z}}$, $n = 1, 2, \ldots$,

$$\begin{aligned} \tilde{P}_n(\vec{y}, x, A) &= \mathsf{P}\left\{(\vec{W}_n, X_n) \in A / \vec{Y}_{n-1} = \vec{y}, X_{n-1} = x\right\} \\ &= \mathsf{P}_{\vec{z},n}\left\{(\tilde{B}_n(X_{n-1}, U_n), C_n(X_{n-1}, U_n)) \in A\right\} \\ &= \mathsf{P}\left\{(\tilde{B}_n(x, U_n), C_n(x, U_n)) \in A\right\} = \tilde{P}_n(x, A). \end{aligned} \tag{1.35}$$

In this case, the transition dynamic functions $\tilde{B}_n(x, u)$ and $C_n(x, u)$ as well as the transition probabilities $\tilde{P}_n(\vec{y}, x, A) = \tilde{P}_n(x, A)$ do not depend on the log-price argument $\vec{y}$.

Due to this, the index component X_n itself is a Markov chain modulating the log-price component $\vec{Y}_n$.

The transition probabilities of the Markov chain X_n take the following form, for $x \in \mathbb{X}$, $C \in \mathcal{B}_{\mathbb{X}}$, $n = 1, 2, \ldots$:

$$\begin{aligned} P_n(x, C) &= \mathsf{P}\{X_n \in C / X_{n-1} = x\} \\ &= \mathsf{P}\{C_n(x, U_n) \in C\} = \tilde{P}_n(x, \mathbb{R}_k \times C)\,. \end{aligned} \tag{1.36}$$

The random sequences $\vec{Z}_n = (\vec{Y}_n, X_n)$ and $\vec{V}_n = (\vec{S}_n, X_n)$ are Markov chains.

These Markov chains are homogeneous in time if the random variables U_n, $n = 1, 2, \dots$, are identically distributed and functions $\tilde{B}_n(x, u) = \tilde{B}(x, u)$ and $C_n(x, u) = C(x, u)$ do not depend on n.

In general, random variables $\tilde{B}_n(x, U_n)$ and $C_n(x, U_n)$ are dependent.

However, there is a case, where these random variables are independent. Let us assume that the space $\mathbb{U} = \mathbb{U}' \times \mathbb{U}''$ is the product of two spaces and that the σ-algebras of measurable subsets $\mathcal{B}_{\mathbb{U}}$ is the minimal σ-algebra that contains all sets $A \times B$, $A \in \mathcal{B}_{\mathbb{U}'}$, $B \in \mathcal{B}_{\mathbb{U}''}$, where $\mathcal{B}_{\mathbb{U}'}$ and $\mathcal{B}_{\mathbb{U}''}$, are σ-algebras of measurable subsets, respectively, for spaces $\mathbb{U}'$ and $\mathbb{U}''$.

In this case, the random variables $U_n = (U'_n, U''_n)$, $n = 1, 2, \dots$, can be represented in the form of random vectors with components U'_n and U''_n, which are random variables taking values, respectively, in the spaces $\mathbb{U}'$ and $\mathbb{U}''$.

Let us also assume that the transition dynamic functions $\tilde{B}_n(x, u) = \tilde{B}'_n(x, u')$ and $C_n(x, u) = C'_n(x, u'')$ for $u = (u', u'') \in \mathbb{U}$, $u' \in \mathbb{U}'$, $u'' \in \mathbb{U}''$, i.e. these functions depend only of the corresponding components of the vector argument $u = (u', u'')$.

In this case, the stochastic transition dynamic relation (1.33) takes the following form:

$$\begin{cases} \vec{Y}_n = \vec{Y}_{n-1} + \tilde{B}'_n(X_{n-1}, U'_n)\,, \\ X_n = C'_n(X_{n-1}, U''_n)\,, \\ n = 1, 2, \dots\,. \end{cases} \tag{1.37}$$

If the random variables U'_n and U''_n are independent for every $n = 1, 2, \dots$, then random variables $\tilde{B}'_n(x, U'_n)$ and $C'_n(x, U''_n)$ will be also independent random variables for every $n = 0, 1, \dots$, and, therefore, the transition probabilities $\tilde{P}_n(x, B \times C)$ take the following form, for $x \in \mathbb{X}$, $B \in \mathcal{B}_{\mathbb{R}_k}$, $C \in \mathcal{B}_{\mathbb{X}}$, $n = 1, 2, \dots$,

$$\tilde{P}_n(x, B \times C) = \mathsf{P}\{\tilde{B}'_n(x, U'_n) \in B\}\mathsf{P}\{C'_n(x, U''_n) \in C\}\,. \tag{1.38}$$

An important variant is the model, where the log-price process $\vec{Y}_n$ is a random walk with modulated scale-location parameters.

In this model, the modulated log-price process $\vec{Y}_n$ is given by the following stochastic transition dynamic relation:

$$\begin{cases} \vec{Y}_n = \vec{Y}_{n-1} + \vec{\mu}_n(X_{n-1}) + \Sigma_n(X_{n-1})\vec{W}_n\,, \\ X_n = C_n(X_{n-1}, W_n)\,, \\ n = 1, 2, \dots\,, \end{cases} \tag{1.39}$$

where (a) $\vec{Y}_0 = (Y_{0,1}, \dots, Y_{0,k})$ is a random vector taking values in space $\mathbb{R}_k$, (b) $\vec{W}_n = (W_{n,1}, \dots, W_{n,k})$, $n = 1, 2, \dots$, is a sequence independent random vectors taking values in space $\mathbb{R}_k$, (c) X_0 is a random variable taking values in space $\mathbb{X}$, (d) W_n, $n = 1, 2, \dots$, is a sequence of independent random variables taking value in some measurable space $\mathbb{W}$ with a σ-algebra of measurable subsets $\mathcal{B}_{\mathbb{W}}$, (e) the random vector $(\vec{Y}_0, X_0)$ and the random sequences $(\vec{W}_n, W_n)$, $n = 1, 2, \dots$, are independent,

(f) $\vec{\mu}_n(x) = (\mu_{n,1}(x), \dots, \mu_{n,k}(x))$, $n = 1, 2, \dots$, are the vector functions which components are the measurable functions acting from the space $\mathbb{X}$ to $\mathbb{R}_1$, (g) $\Sigma_n(x) = \|\sigma_{n,ij}(x)\|$, $n = 1, 2, \dots$, are $k \times k$ matrix functions with elements $\sigma_{n,ij}(x)$ which are the measurable functions acting from the space $\mathbb{X}$ to $\mathbb{R}_1$, and (h) $C_n(x, w)$, $n = 1, 2, \dots$, are the measurable functions acting from the space $\mathbb{X} \times \mathbb{W}$ to the space $\mathbb{X}$.

In this case, the role of random vectors $U_n = (U'_n, U''_n)$ is played by the random vectors $(\vec{W}_n, W_n)$ taking values in the space $U = \mathbb{R}_k \times \mathbb{W}$.

As far as the price process $\vec{S}_n$ is concerned, it will be given in this case by the following multiplicative analog of the first formula in (1.39):

$$\vec{S}_n = \vec{S}_{n-1} \cdot e^{\vec{\mu}_n(X_{n-1}) + \Sigma_n(X_{n-1})\vec{W}_n}, \quad n = 1, 2, \dots . \tag{1.40}$$

The random sequences $\vec{Z}_n = (\vec{Y}_n, X_n)$ and $\vec{V}_n = (\vec{S}_n, X_n)$ are Markov chains.

These Markov chains are homogeneous in time if the random vectors $(\vec{W}_n, W_n)$, $n = 1, 2, \dots$, are identically distributed and the transition functions $\vec{\mu}_n(x) = \vec{\mu}(x)$, $\Sigma_n(x) = \Sigma(x)$ and $C_n(x, w) = C(x, w)$ do not depend on n.

1.2.5 Log-price and price processes represented by random walks modulated by Markov chains

For simplicity, let us restrict consideration by the case when the phase space $\mathbb{X} = \{1, 2, \dots, \}$ is a discrete set, and the log-price process $\vec{Y}_n$ is given by the following stochastic transition dynamic relation:

$$\vec{Y}_n = \vec{Y}_{n-1} + \vec{W}_n(X_{n-1}), \quad n = 1, 2, \dots , \tag{1.41}$$

where (a) $\vec{Y}_0 = (Y_{0,1}, \dots, Y_{0,k})$ is a k-dimensional random vector, (b) $\vec{W}_n(x) = (W_{n,1}(x), \dots, W_{n,k}(x))$, $n = 1, 2, \dots, x \in \mathbb{X}$ is a family of k-dimensional independent random vectors, (c) X_n, $n = 0, 1, \dots$, is a Markov chain with the phase space $\mathbb{X}$, an initial distribution $P(C) = \mathsf{P}\{X_0 \in C\}$ and the one-step transition probabilities $P_n(x, y) = \mathsf{P}\{X_n = y / X_{n-1} = x\}$, $x, y \in X$, $n = 1, 2, \dots$, and (d) the random vector $\vec{Y}_0$, the family of random variables $\vec{W}_n(x)$, $n = 1, 2, \dots, x \in \mathbb{X}$, and the Markov chain X_n, $n = 0, 1, \dots$, are independent.

As far as the price process $\vec{S}_n$ is concerned, it will be given in this case by the following multiplicative analog of relation (1.41),

$$\vec{S}_n = \vec{S}_{n-1} \cdot e^{\vec{W}_n(X_{n-1})}, \quad n = 1, 2, \dots . \tag{1.42}$$

1.2.6 Log-price and price processes represented by random walks modulated by semi-Markov indices

Let us consider another example, where the log-price process $\vec{Y}_n$ is a random walk modulated by a semi-Markov stochastic index.

Let us first introduce a discrete time semi-Markov process X_n, $n = 0, 1, \ldots$. We consider a two-dimensional Markov chain (J_n, T_n), $n = 0, 1, \ldots$, with a phase space $\mathbb{X} \times \mathbb{N}$, where $\mathbb{N} = \{0, 1, \ldots\}$, an initial distribution $Q(C) = \mathsf{P}\{J_0 \in C\} = \mathsf{P}\{J_0 \in C, T_0 = 0\}$ and the one-step transition probabilities $Q_n(x, C, r) = \mathsf{P}\{J_n \in C, T_n \leq r / J_{n-1} = x, T_{n-1} = l\}$, $n = 1, 2, \ldots$, which do not depend on the current state of the second component. Such Markov chain can be referred to as a discrete time Markov renewal process.

Then a discrete time semi-Markov process X_n can be defined in the following way:

$$X_n = J_{N(n)}\,, \quad n = 0, 1, \ldots\,, \tag{1.43}$$

where

$$N_n = \max(r : T_1 + \cdots + T_r \leq n)\,, \quad n = 0, 1, \ldots\,. \tag{1.44}$$

The random variables J_n are sequential states of the semi-Markov process X_n at moments of jumps, while the random variables T_n are the corresponding interjump times.

An usual assumption $Q_n(x, \mathbb{X}, 0) = 0$, $x \in \mathbb{X}$, $n = 1, 2, \ldots$, is also made in order to exclude instant jumps. In this case, $N_n \leq n$, $n = 0, 1, \ldots$, with probability 1.

Note that we introduced above an inhomogeneous in time Markov renewal process (J_n, T_n). Consequently, the corresponding semi-Markov process X_n defined above is also inhomogeneous in time.

The semi-Markov process X_n is homogeneous in time if the Markov renewal process (J_n, T_n) is homogeneous in time Markov chain. This is where its transition probabilities $Q_n(x, C, r) = Q(x, C, r)$, $x \in \mathbb{X}$, $C \in \mathcal{B}_{\mathbb{X}}$, $n, r = 1, 2, \ldots$, do not depend on n.

As is well known, the semi-Markov process X_n is not in general case a Markov chain. However, one can add additional components to the random process X_n which will make the corresponding process Markov.

Let us define the process that counts time from the moment of last jump before moment n and n,

$$M_n = n - T_{N_n}\,, \quad n = 0, 1, \ldots\,. \tag{1.45}$$

In this case, the discrete time process $\bar{X}_n = (X_n, M_n, N_n)$, $n = 0, 1, \ldots$, is a homogeneous Markov chain with the phase space $\bar{\mathbb{X}} = \mathbb{X} \times \mathbb{N} \times \mathbb{N}$, the initial distribution $P(C) = \mathsf{P}\{X_0 \in C\} = \mathsf{P}\{X_0 \in C, M_0 = 0, N_0 = 0\}$ and the one-step transition probabilities

$$\begin{aligned} & P(x, m, r, C, m', r') \\ & \quad = \mathsf{P}\left\{X_{n+1} \in C, M_{n+1} = m', N_{n+1} = r' / X_n = x, M_n = m, N_n = r\right\} \\ & \quad = \begin{cases} I(x \in C)\, \frac{1 - Q_{r+1}(x, \mathbb{X}, m)}{1 - Q_{r+1}(x, \mathbb{X}, m-1)}\,, & \text{if } m' = m + 1, r' = r\,, \\ \frac{Q_{r+1}(x, C, m) - Q_{r+1}(x, C, m-1)}{1 - Q_{r+1}(x, \mathbb{X}, m-1)}\,, & \text{if } m' = 0, r' = r + 1\,, \\ 0\,, & \text{otherwise}\,. \end{cases} \end{aligned} \tag{1.46}$$

If the semi-Markov chain X_n is homogeneous in time, the component N_n may be omitted. In this case, the discrete time process $\bar{X}_n = (X_n, M_n)$, $n = 0, 1, \dots$, is a homogeneous Markov chain with the phase space $\bar{\mathbb{X}} = \mathbb{X} \times \mathbb{N}$, the initial distribution $Q(C) = \mathsf{P}\{X_0 \in C\} = \mathsf{P}\{X_0 \in C, M_0 = 0\}$ and the one-step transition probabilities

$$
\begin{aligned}
P(x, m, C, m')& \\
&= \mathsf{P}\left\{X_{n+1} \in C, M_{n+1} = m' / X_n = x, M_n = m\right\} \\
&= \begin{cases} I(x \in C)\, \frac{1-Q(x,\mathbb{X},m)}{1-Q(x,\mathbb{X},m-1)}, & \text{if } m' = m+1, \\ \frac{Q(x,C,m)-Q(x,C,m-1)}{1-Q(x,\mathbb{X},m-1)}, & \text{if } m' = 0, \\ 0, & \text{otherwise}. \end{cases}
\end{aligned}
\tag{1.47}
$$

A random walk modulated by a semi-Markov stochastic index can be introduced in the way analogous to those used in stochastic transition relation (1.39) defining a random walk modulated by a Markov stochastic index,

$$
\vec{Y}_n = \vec{Y}_{n-1} + \vec{\mu}_n(X_{n-1}) + \Sigma_n(X_{n-1})\vec{W}_n, \quad n = 1, 2, \dots, \tag{1.48}
$$

where (a) $\vec{Y}_0$ is a k-dimensional random vector, (b) $\vec{W}_n = (W_{n,1}, \dots, W_{n,k})$, $n = 1, 2, \dots$, is a sequence of k-dimensional independent random vectors with real-valued components, (c) X_n, $n = 0, 1, \dots$, is a discrete time semi-Markov process defined in (1.43) and (1.44), (d) the semi-Markov process X_n, $n = 0, 1, \dots$, the random vector $\vec{Y}_0$, and the random sequence $\vec{W}_n$, $n = 1, 2, \dots$, are independent, (e) $\vec{\mu}_n(x) = (\mu_{n,1}(x), \dots, \mu_{n,k}(x))$, $n = 1, 2, \dots$, are the vector functions which components are the measurable functions acting from the space $\mathbb{X}$ to $\mathbb{R}_1$, and (f) $\Sigma_n(x) = \|\sigma_{n,ij}(x)\|$, $n = 1, 2, \dots$, are the measurable matrix functions with elements $\sigma_{n,ij}(x)$ which are the measurable functions acting from the space $\mathbb{X}$ to $\mathbb{R}_1$.

As far as the price process $\vec{S}_n$ is concerned, it will be given in this case by the following multiplicative analog of formula (1.48):

$$
\vec{S}_n = \vec{S}_{n-1} \cdot e^{\vec{\mu}_n(X_{n-1}) + \Sigma_n(X_{n-1})\vec{W}_n}, \quad n = 1, 2, \dots. \tag{1.49}
$$

Note that the process $\vec{Z}_n = (\vec{Y}_n, X_n)$ is not a Markov chain. In order to have this property, one should consider the process $\bar{X}_n$ as the stochastic index modulating the corresponding log-price or price process. The processes $\vec{Z}_n = (\vec{Y}_n, \bar{X}_n)$ and $\vec{V}_n = (\vec{S}_n, \bar{X}_n)$ are Markov chains.

These Markov chains are homogeneous in time if the random vectors $\vec{W}_n$, $n = 1, 2, \dots$, are identically distributed and the transition functions $\vec{\mu}_n(x) = \vec{\mu}(x)$ and $\Sigma_n(x) = \Sigma(x)$ do not depend on n. Note that, the Markov chain $\vec{X}_n$ is homogeneous in time due to involvement of additional counting components.

1.3 Autoregressive LPP

In this section, we consider a number of typical examples of discrete time modulated Markov log-price and price processes of the autoregressive type.

1.3.1 Autoregressive-type log-price and price processes

Let us consider a model known as an autoregressive $AR(p)$ model. For simplicity, we consider a univariate log-price process defined by the following stochastic transition dynamic relation:

$$Y_n = a_0 + a_1 Y_{n-1} + \cdots + a_p Y_{n-p} + \sigma W_n\,, \quad n = 1, 2, \ldots\,, \tag{1.50}$$

where (a) $\vec{Y}_0 = (Y_0, \ldots, Y_{-p+1})$ is a p-dimensional random vector with real-valued components, (b) $W_1, W_2, \ldots,$ is a sequence of real-valued i.i.d. random variables, (c) the random vector $\vec{Y}_0$ and the random sequence $W_1, W_2, \ldots,$ are independent, (d) p is a positive integer number, and (e) $a_0, a_1, \ldots, a_p, \sigma$ are the real-valued constants.

The corresponding price process, which can be referred to as an exponential autoregressive process, is defined by the following relation:

$$S_n = e^{Y_n}\,, \quad n = 0, 1, \ldots\,. \tag{1.51}$$

The standard case is where the random variables $W_1, W_2, \ldots,$ have the standard normal distribution with the parameters 0 and 1.

It is useful to note that the above autoregressive model can be defined in the equivalent form of stochastic difference equation

$$\begin{aligned} Y_n - Y_{n-1} = a_0 &+ (a_1 - 1) Y_{n-1} + \cdots \\ &+ a_p Y_{n-p} + \sigma W_n\,, \quad n = 1, 2, \ldots\,. \end{aligned} \tag{1.52}$$

Also note that the above stochastic difference equation can be represented in the equivalent alternative form

$$\begin{aligned} Y_n - Y_{n-1} = a'_0 &+ a'_1 (Y_{n-1} - Y_{n-2}) + \cdots \\ &+ a'_{p-1}(Y_{n-p+1} - Y_{n-p}) + \sigma' W_n\,, \quad n = 1, 2, \ldots\,. \end{aligned} \tag{1.53}$$

Indeed, the stochastic difference equation (1.53) can be rewritten in the form of equation (1.52) with the coefficients $a_0 = a'_0$, $a_1 = a'_1 + 1$, $a_2 = a'_2 - a'_1, \ldots, a_{p-1} = a'_{p-1} - a'_{p-2}$, $a_p = -a'_{p-1}$, $\sigma = \sigma'$.

The log-price process Y_n is not a Markov chain. However, this process can be considered as the first component of some p-dimensional Markov chain.

Let us introduce the p-dimensional vector process

$$\vec{Y}_n = (Y_{n,1}, \dots, Y_{n,p}) = (Y_n, \dots, Y_{n-p+1}), \quad n = 1, 2, \dots, \tag{1.54}$$

and the sequence of p-dimensional i.i.d. random vectors

$$\vec{W}_n = (W_{n,1}, \dots, W_{n,p}) = (W_n, 0, \dots, 0), \quad n = 1, 2, \dots. \tag{1.55}$$

Using the above formula and relations (1.50), (1.54), and (1.55), we get the following vector stochastic transition dynamic relation (written as a system of stochastic transition dynamic relations):

$$\begin{cases} Y_{n,1} = a_0 + a_1 Y_{n-1,1} + \cdots + a_p Y_{n-1,p} + \sigma W_{n,1}, \\ Y_{n,2} = Y_{n-1,1} + W_{n,2}, \\ \dots \quad \dots \\ Y_{n,p} = Y_{n-1,p-1} + W_{n,p}, \\ \quad n = 1, 2, \dots. \end{cases} \tag{1.56}$$

This dynamic relation can be rewritten in the equivalent form of vector autoregressive stochastic difference equation (written as a system of stochastic difference equations),

$$\begin{cases} Y_{n,1} - Y_{n-1,1} = a_0 + (a_1 - 1) Y_{n-1,1} + \cdots + a_p Y_{n-1,p} + \sigma W_{n,1}, \\ Y_{n,2} - Y_{n-1,2} = Y_{n-1,1} - Y_{n-1,2} + W_{n,2}, \\ \quad \dots \quad \dots \\ Y_{n,p} - Y_{n-1,p} = Y_{n-1,p-1} - Y_{n-1,p} + W_{n,p}, \\ \qquad n = 1, 2, \dots. \end{cases} \tag{1.57}$$

The vector stochastic difference equation (1.57) is a particular case of the vector stochastic difference equation (1.32).

Thus, the process $\vec{Y}_n$, $n = 0, 1, \dots$, is a discrete time homogeneous multivariate Ornstein–Uhlenbeck-type process, with the initial random value $\vec{Y}_0$ and vector and matrix parameters $\vec{\lambda}$ and Λ, Σ, determined by relations (1.50) and (1.54)–(1.57).

1.3.2 Nonlinear autoregressive-type log-price and price processes

Let us consider a model that can be interpreted as a nonlinear autoregressive-type process. For simplicity, we consider a univariate log-price process defined by the following stochastic transition dynamic relation:

$$Y_n = A_n(Y_{n-1}, \dots, Y_{n-p}, W_n), \quad n = 1, 2, \dots, \tag{1.58}$$

where (a) $\vec{Y}_0 = (Y_0, \dots, Y_{-p+1})$ is a p-dimensional random vector with real-valued components, (b) $W_1, W_2, \dots,$ is a sequence of real-valued independent random variables, (c) the random vector $\vec{Y}_0$ and the random sequence $W_1, W_2, \dots,$ are independent, (d) p is a positive integer number, and (e) $A_n(y_1, \dots, y_p, w)$, $n = 1, 2, \dots,$ are the measurable functions acting from the space $\mathbb{R}_{p+1}$ to $\mathbb{R}_1$.

The corresponding price process, which can be referred to as a nonlinear exponential autoregressive-type process, is defined by the following relation:

$$S_n = e^{Y_n}, \quad n = 0, 1, \dots . \tag{1.59}$$

The log-process Y_n is not a Markov chain. Let us show in which way this model can be imbedded in a model of multivariate Markov log-price process.

As given in Subsection 1.3.1, let us introduce the p-dimensional vector process

$$\vec{Y}_n = (Y_{n,1}, \dots, Y_{n,p}) = (Y_n, \dots, Y_{n-p+1}), \quad n = 1, 2, \dots, \tag{1.60}$$

and the sequence of p-dimensional independent random vectors

$$\vec{W}_n = (W_{n,1}, \dots, W_{n,p}) = (W_n, 0, \dots, 0), \quad n = 1, 2, \dots . \tag{1.61}$$

Using relations (1.58), (1.60), and (1.61), we get the following vector stochastic transition dynamic relation (written as a system of stochastic transition dynamic relations):

$$\begin{cases} Y_{n,1} = A_n(Y_{n-1,1}, \dots, Y_{n-1,p}, W_{n,1}), \\ Y_{n,2} = Y_{n-1,1} + W_{n,2}, \\ \dots \quad \dots \\ Y_{n,p} = Y_{n-1,p-1} + W_{n,p}, \\ \quad n = 1, 2, \dots . \end{cases} \tag{1.62}$$

The vector transition dynamic relation (1.62) is a particular case of the vector transition dynamic relation (1.4).

Thus, the vector process $\vec{Y}_n$, $n = 0, 1, \dots,$ is an inhomogeneous multivariate Markov chain with the initial random value $\vec{Y}_0$ and the corresponding vector transition dynamic functions determined by relations (1.58) and (1.62).

A particular case of the above model is an inhomogeneous in time autoregressive-type model, where the log-price process Y_n is given by the following transition dynamic relation:

$$Y_n = a_{n,0} + a_{n,1}Y_{n-1} + \cdots + a_{n,p}Y_{n-p} + \sigma_n W_n, \quad n = 1, 2, \dots, \tag{1.63}$$

where (a) $\vec{Y}_0 = (Y_0, \dots, Y_{-p+1})$ is a p-dimensional random vector with real-valued components, (b) $W_1, W_2, \dots,$ is a sequence of independent real-valued random variables, (c) the random vector $\vec{Y}_0$ and the random sequence $W_1, W_2, \dots,$ are independent, (d) p is a positive integer number, and (e) $a_{n,0}, a_{n,1}, \dots, a_{n,p}, \sigma_n$, $n = 1, 2, \dots,$ are the real-valued constants.

1.3.3 Modulated autoregressive log-price and price processes

Let us consider a process that can be interpreted as a modulated nonlinear autoregressive-type process. For simplicity, we consider a log-price process $Z_n = (Y_n, X_n)$ with a univariate log-price component Y_n.

Here and henceforth $\mathbb{X}^{(r)} = \mathbb{X} \times \cdots \times \mathbb{X}$ is r-times product of the space $\mathbb{X}$, where $\mathbb{X}$ is a measurable space with a σ-algebra of measurable subsets $\mathcal{B}_{\mathbb{X}}$.

The log-price process Z_n is defined by the following stochastic transition dynamic relation:

$$\begin{cases} Y_n = A_n\left(Y_{n-1}, \dots, Y_{n-p}, X_{n-1}, \dots, X_{n-r}, W_n\right), \\ X_n = C_n\left(X_{n-1}, \dots, X_{n-r}, U_n\right), \\ n = 1, 2, \dots, \end{cases} \tag{1.64}$$

where (a) $\vec{Y}_0 = (Y_0, \dots, Y_{-p+1})$ is a p-dimensional random vector with real-valued components, (b) $\vec{X}_0 = (X_0, \dots, X_{-r+1})$ is a r-dimensional random vector with components taking values in a measurable space $\mathbb{X}^{(r)}$, (c) $(W_1, U_1), (W_2, U_2) \dots,$ is a sequence of independent random vectors taking values in the space $\mathbb{R}_1 \times \mathbb{U}$, where $\mathbb{U}$ is a measurable space with a σ-algebra of measurable subsets $\mathcal{B}_{\mathbb{U}}$, (d) the random vector $\vec{Z}_0 = (\vec{Y}_0, \vec{X}_0)$ and the random sequence $(W_1, U_1), (W_2, U_2) \dots,$ are independent, (e) p and r are the positive integer numbers, (f) $A_n(y_1, \dots, y_p, x_1, \dots, x_r, w)$, $n = 1, 2, \dots,$ are the measurable functions acting from the space $\mathbb{R}_p \times \mathbb{X}^{(r)} \times \mathbb{R}_1$ to $\mathbb{R}_1$, and (g) $C_n(x_1, \dots, x_r, u)$, $n = 1, 2, \dots,$ are the measurable functions acting from the space $\mathbb{X}^{(r)} \times \mathbb{U}$ to the space $\mathbb{X}$.

The corresponding modulated price process, which can be referred to as a modulated nonlinear exponential autoregressive type process, is defined by the following relation:

$$V_n = (S_n, X_n) = (e^{Y_n}, X_n), \quad n = 0, 1, \dots. \tag{1.65}$$

The log-process Y_n is not a Markov chain. Let us show in which way this model can be imbedded in a model of the multivariate Markov log-price process.

As given in Subsection 1.3.2, let us introduce the p-dimensional vector process, $\vec{Y}_n = (Y_{n,1}, \dots, Y_{n,p}) = (Y_n, \dots, Y_{n-p+1})$, $n = 1, 2, \dots,$ and the sequence of p-dimensional independent random vectors $\vec{W}_n = (W_{n,1}, \dots, W_{n,p}) = (W_n, 0, \dots, 0)$, $n = 1, 2, \dots,$ respectively, according to relations (1.60) and (1.61).

Also, let us introduce the r-dimensional vector process

$$\vec{X}_n = (X_{n,1}, \dots, X_{n,r}) = (X_n, \dots, X_{n-r+1}), \quad n = 1, 2, \dots. \tag{1.66}$$

Using relations (1.60), (1.61), (1.64), and (1.66), we get the following vector stochastic transition dynamic relation (written as a system of stochastic transition dynamic relations):

$$\begin{cases} Y_{n,1} = A_n(Y_{n-1,1}, \dots, Y_{n-1,p}, X_{n-1,1}, \dots, X_{n-1,r}, W_{n,1})\,, \\ Y_{n,2} = Y_{n-1,1} + W_{n,2}\,, \\ \dots \quad \dots \\ Y_{n,p} = Y_{n-1,p-1} + W_{n,p}\,, \\ X_{n,1} = C_n(X_{n-1,1}, \dots, X_{n-1,r}, U_n)\,, \\ X_{n,2} = X_{n-1,1}\,, \\ \dots \quad \dots \\ X_{n,r} = X_{n-1,r-1}\,, \\ \quad n = 1, 2, \dots\,. \end{cases} \tag{1.67}$$

The vector stochastic transition dynamic relation (1.67) is a particular case of the vector transition dynamic relation (1.4).

Thus, the vector process $\vec{Z}_n = (\vec{Y}_n, \vec{X}_n)$, $n = 0, 1, \dots$, is inhomogeneous multivariate Markov chain with the initial random value $\vec{Z}_0$ and the corresponding vector transition dynamic functions determined by relations (1.64) and (1.67).

Note that in this case, the role of the index component is played by the process $\vec{X}_n$, which has the phase space $\mathbb{X}^{(r)}$.

A particular case of the above model is an inhomogeneous in time autoregressive-type model, where the modulated log-price process $Z_n = (Y_n, X_n)$ is given by the following stochastic transition dynamic relation:

$$\begin{cases} Y_n = a_{n,0}(X_{n-1}, \dots, X_{n-r}) + a_{n,1}(X_{n-1}, \dots, X_{n-r})Y_{n-1} \\ \qquad + \dots + a_{n,p}(X_{n-1}, \dots, X_{n-r})Y_{n-p} \\ \qquad + \sigma_n(X_{n-1}, \dots, X_{n-r})W_n\,, \\ X_n = C_n(X_{n-1}, \dots, X_{n-r}, U_n)\,, \\ \quad n = 1, 2, \dots\,, \end{cases} \tag{1.68}$$

where (a) $\vec{Y}_0 = (Y_0, \dots, Y_{-p+1})$ is a p-dimensional random vector with real-valued components (b) $\vec{X}_0 = (X_0, \dots, X_{-r+1})$ is a r-dimensional random vector with components taking values in the space $\mathbb{X}^{(r)}$, (c) $(W_1, U_1), (W_2, U_2) \dots$, is a sequence of independent random vectors taking values in the space $\mathbb{R}_1 \times \mathbb{U}$, (d) the random vector $\vec{Z}_0 = (\vec{Y}_0, \vec{X}_0)$ and the random sequence $(W_1, U_1), (W_2, U_2) \dots$, are independent, (e) p and r are positive integer numbers, (f) $a_{n,i}(x_1, \dots, x_r)$, $\sigma_n(x_1, \dots, x_r)$, $i = 1, \dots, p$, $n = 1, 2, \dots$, are the measurable functions acting from the space $\mathbb{X}^{(r)}$ to $\mathbb{R}_1$, and (g) $C_n(x_1, \dots, x_r, u)$, $n = 1, 2, \dots$, are the measurable functions acting from the space $\mathbb{X}^{(r)} \times \mathbb{U}$ to the space $\mathbb{X}$.

1.3.4 Autoregressive moving average-type log-price and price processes

Let us consider a model known as an autoregressive moving average model, ARMA(p, q). For simplicity, we consider a univariate log-price process defined by the following stochastic transition dynamic relation:

$$\begin{aligned} Y_n = {} & a_0 + a_1 Y_{n-1} + \cdots + a_p Y_{n-p} \\ & + b_1 W_{n-1} + \cdots + b_q W_{n-q} + \sigma W_n \,, \quad n = 1, 2, \ldots \,, \end{aligned} \tag{1.69}$$

where (a) $\vec{Y}_0 = (Y_0, \ldots, Y_{-p+1}, W_0, \ldots, W_{-q+1})$ is a $(p+q)$-dimensional random vector with real-valued components, (b) $W_1, W_2, \ldots,$ is a sequence of real-valued i.i.d. random variables, (c) the random vector $\vec{Y}_0$ and the random sequence $W_1, W_2, \ldots,$ are independent, (d) p and q are the positive integer numbers, and (e) $a_0, a_1, \ldots, a_p, b_1, \ldots, b_q, \sigma$ are the real-valued constants.

The standard case is where the random variables $W_1, W_2, \ldots,$ have the standard normal distribution with the parameters 0 and 1.

The corresponding price process, which can be referred to as an exponential autoregressive moving average-type process, is defined by the following relation:

$$S_n = e^{Y_n} \,, \quad n = 0, 1, \ldots \,. \tag{1.70}$$

The log-price process Y_n is not a Markov chain. However, this process can be represented as the first component of some $(p+q)$-dimensional Markov chain.

Let us introduce the $(p+q)$-dimensional vector process

$$\begin{aligned} \vec{Y}_n &= (Y_{n,1}, \ldots, Y_{n,p}, Y_{n,p+1}, \ldots, Y_{n,p+q}) \\ &= (Y_n, \ldots, Y_{n-p+1}, W_n, \ldots, W_{n-q+1}) \,, \\ n &= 1, 2, \ldots \,, \end{aligned} \tag{1.71}$$

and the sequence of $(p+q)$-dimensional i.i.d. random vectors

$$\begin{aligned} \vec{W}_n &= (W_{n,1}, W_{n,2}, \ldots, W_{n,p}, W_{n,p+1}, W_{n,p+2}, \ldots, W_{n,p+q}) \\ &= (W_n, 0, \ldots 0, W_n, 0, \ldots, 0) \,, \\ n &= 1, 2, \ldots \,. \end{aligned} \tag{1.72}$$

Using this formula and relations (1.69), (1.71), and (1.72), we get the following vector stochastic transition dynamic relation (written as a system of stochastic transition dynamic relations):

$$
\begin{cases}
Y_{n,1} = a_0 + a_1 Y_{n-1,1} + \cdots + a_p Y_{n-1,p} \\
\qquad + b_1 Y_{n-1,p+1} + \cdots + b_q Y_{n-1,p+q} + \sigma W_{n,1}, \\
Y_{n,2} = Y_{n-1,1} + W_{n,2}, \\
\ldots \quad \ldots \\
Y_{n,p} = Y_{n-1,p-1} + W_{n,p}, \\
Y_{n,p+1} = W_{n,p+1}, \\
Y_{n,p+2} = Y_{n-1,p-1} + W_{n,p+2}, \\
\ldots \quad \ldots \\
Y_{n,p+q} = Y_{n-1,p+q-1} + W_{n,p+q}, \\
n = 1, 2, \ldots .
\end{cases}
\tag{1.73}
$$

This dynamic relation can be rewritten in the equivalent form of vector stochastic difference equation (written as a system of stochastic difference equations),

$$
\begin{cases}
Y_{n,1} - Y_{n-1,1} = a_0 + (a_1 - 1) Y_{n-1,1} + \cdots + a_p Y_{n-1,p} \\
\qquad + b_1 Y_{n-1,p+1} + \cdots + b_q Y_{n-1,p+q} + \sigma W_{n,1}, \\
Y_{n,2} - Y_{n-1,2} = Y_{n-1,1} - Y_{n-1,2} + W_{n,2}, \\
\ldots \quad \ldots \\
Y_{n,p} - Y_{n-1,p} = Y_{n-1,p-1} - Y_{n-1,p} + W_{n,p}, \\
Y_{n,p+1} - Y_{n-1,p+1} = -Y_{n-1,p+1} + W_{n,p+1}, \\
Y_{n,p+2} - Y_{n-1,p+2} = Y_{n-1,p-1} - Y_{n-1,p+2} + W_{n,p+2}, \\
\ldots \quad \ldots \\
Y_{n,p+q} - Y_{n-1,p+q} = Y_{n-1,p+q-1} - Y_{n-1,p+q} + W_{n,p+q}, \\
n = 1, 2, \ldots .
\end{cases}
\tag{1.74}
$$

The vector stochastic difference equation (1.74) is a particular case of the vector stochastic difference equation (1.32).

Thus, the process $\vec{Y}_n$, $n = 0, 1, \ldots$, is also a discrete time homogeneous multivariate Ornstein–Uhlenbeck-type process, with the initial random value $\vec{Y}_0$ and, respectively, vector and matrix parameters $\vec{\Lambda}$ and Λ, Σ determined by relations (1.69) and (1.74).

1.3.5 Nonlinear autoregressive moving average-type log-price and price processes

Let us consider a model that can be interpreted as a nonlinear autoregressive moving average-type process. For simplicity, we consider a univariate log-price process defined by the following stochastic transition dynamic relation:

$$Y_n = A_n(Y_{n-1}, \dots, Y_{n-p}, W_{n-1}, \dots, W_{n-q}, W_n), \quad n = 1, 2, \dots, \tag{1.75}$$

where (a) $\vec{Y}_0 = (Y_0, \dots, Y_{-p+1}, W_0, \dots, W_{-q+1})$ is a $(p+q)$-dimensional random vector with real-valued components, (b) $W_1, W_2, \dots,$ is a sequence of real-valued independent random variables, (c) the random vector $\vec{Y}_0$ and the random sequence $W_1, W_2, \dots,$ are independent, (d) p and r are positive integer numbers, and (e) $A_n(y_1, \dots, y_p, w_1, \dots, w_q, w)$, $n = 1, 2, \dots,$ are the measurable functions acting from the space $\mathbb{R}_{p+q+1}$ to $\mathbb{R}_1$.

The corresponding price process, which can be referred to as a nonlinear exponential autoregressive moving average-type process, is defined by the following relation:

$$S_n = e^{Y_n}, \quad n = 0, 1, \dots. \tag{1.76}$$

The log-price process Y_n is not a Markov chain. However, this process can be considered as the first component of some $(p+q)$-dimensional Markov chain.

As given in Subsection 1.3.4, let us introduce the $(p+q)$-dimensional vector process

$$\begin{aligned} \vec{Y}_n &= (Y_{n,1}, \dots, Y_{n,p}, Y_{n,p+1}, \dots, Y_{n,p+q}) \\ &= (Y_n, \dots, Y_{n-p+1}, W_n, \dots, W_{n-q+1}), \\ n &= 1, 2, \dots, \end{aligned} \tag{1.77}$$

and the sequence of $(p+q)$-dimensional independent random vectors

$$\begin{aligned} \vec{W}_n &= (W_{n,1}, W_{n,2}, \dots, W_{n,p}, W_{n,p+1}, W_{n,p+2}, \dots, W_{n,p+q}) \\ &= (W_n, 0, \dots 0, W_n, 0, \dots, 0), \\ n &= 1, 2, \dots. \end{aligned} \tag{1.78}$$

Using relations (1.75), (1.77), and (1.78), we get the following vector stochastic transition dynamic relation (written as a system of stochastic transition dynamic relations), which is a particular case of the vector transition dynamic relation (1.4):

$$\begin{cases} Y_{n,1} = A_n(Y_{n-1,1}, \dots, Y_{n-1,p+q}, W_{n,1}) \\ Y_{n,2} = Y_{n-1,1} + W_{n,2}, \\ \dots \quad \dots \\ Y_{n,p} = Y_{n-1,p-1} + W_{n,p}, \\ Y_{n,p+1} = W_{n,p+1}, \\ Y_{n,p+2} = Y_{n-1,p-1} + W_{n,p+2}, \\ \dots \quad \dots \\ Y_{n,p+q} = Y_{n-1,p+q-1} + W_{n,p+q}, \\ n = 1, 2, \dots. \end{cases} \tag{1.79}$$

Thus, the process $\vec{Y}_n$, $n = 0, 1, \dots$, is an inhomogeneous in time multivariate Markov chain with the initial random value $\vec{Y}_0$ and the corresponding vector dynamic transition functions determined by relations (1.75) and (1.79).

A particular case of the above model is an inhomogeneous in time autoregressive-type model, where the log-price process Y_n is given by the following transition dynamic relation:

$$\begin{aligned} Y_n &= a_{n,0} + a_{n,1}Y_{n-1} + \cdots + a_{n,p}Y_{n-p} \\ &\quad + b_{n,1}W_{n-1} + \cdots + b_{n,q}W_{n-q} + \sigma_n W_n\,, \quad n = 1, 2, \dots\,, \end{aligned} \tag{1.80}$$

where (a) $\vec{Y}_0 = (Y_0, \dots, Y_{-p+1}, W_0, \dots, W_{-q+1})$ is a $(p+q)$-dimensional random vector with real-valued components, (b) $W_1, W_2, \dots$, is a sequence of real-valued independent random variables, (c) the random vector $\vec{Y}_0$ and the random sequence $W_1, W_2, \dots$, are independent, (d) p and q are the positive integer numbers, and (e) $a_{n,0}, a_{n,1}, \dots, a_{n,p}, b_{n,1}, \dots, b_{n,q}, \sigma_n$, $n = 1, 2, \dots$, are the real-valued constants.

1.3.6 Modulated mixed autoregressive moving average-type log-price and price processes

Let us consider a model that can be interpreted as a modulated nonlinear autoregressive moving average-type process. For simplicity, we consider a log-price process $Z_n = (Y_n, X_n)$ with a univariate log-price component Y_n. The log-price process Z_n is defined by the following stochastic transition dynamic relation:

$$\begin{cases} Y_n = A_n(Y_{n-1}, \dots, Y_{n-p}, W_{n-1}, \dots, W_{n-q}, X_{n-1}, \dots, X_{n-r}, W_n)\,, \\ X_n = C_n(X_{n-1}, \dots, X_{n-r}, U_n)\,, \\ \quad n = 1, 2, \dots\,, \end{cases} \tag{1.81}$$

where (a) $\vec{Y}_0 = (Y_0, \dots, Y_{-p+1}, W_0, \dots, W_{-q+1})$ is the $(p+q)$-dimensional random vector with real-valued components, (b) $\vec{X}_0 = (X_0, \dots, X_{-r+1})$ is a r-dimensional random vector with components taking values in space $\mathbb{X}^{(r)}$, (c) $(W_1, U_1), (W_2, U_2)$, $\dots$, is a sequence of independent random vectors taking values in the space $\mathbb{R}_1 \times \mathbb{U}$, where $\mathbb{U}$ is a measurable space with a σ-algebra of measurable subsets $\mathcal{B}_{\mathbb{U}}$, (d) the random vector $\vec{Z}_0 = (\vec{Y}_0, \vec{X}_0)$ and the random sequence $(W_1, U_1), (W_2, U_2), \dots$, are independent, (e) p, q, and r are the positive integer numbers, (f) $A_n(y_1, \dots, y_p, w_1, \dots, w_q, x_1, \dots, x_r, w)$, $n = 1, 2, \dots$, are the measurable functions acting from the space $\mathbb{R}_{p+q} \times \mathbb{X}^{(r)} \times \mathbb{R}_1$ to $\mathbb{R}_1$, and (g) $C_n(x_1, \dots, x_r, u)$, $n = 1, 2, \dots$, are the measurable functions acting from the space $\mathbb{X}^{(r)} \times \mathbb{U}$ to the space $\mathbb{X}$.

The corresponding price process, which can be referred to as a modulated nonlinear exponential autoregressive moving average-type process, is defined by the following relation:

$$V_n = (S_n, X_n) = (e^{Y_n}, X_n)\,, \quad n = 0, 1, \dots\,. \tag{1.82}$$

The log-price process Y_n is not a Markov chain. However, this process can be represented as the first component of some $(p+q)$-dimensional Markov chain.

As given in Subsection 1.3.5, let us introduce the $(p+q)$-dimensional vector process $\vec{Y}_n = (Y_{n,1}, \dots, Y_{n,p}, Y_{n,p+1}, \dots, Y_{n,p+q}) = (Y_n, \dots, Y_{n-p+1}, W_n, \dots, W_{n-q+1})$, $n = 1, 2, \dots$, and the sequence of $(p+q)$-dimensional i.i.d. random vectors, $\vec{W}_n = (W_{n,1}, W_{n,2}, \dots, W_{n,p}, W_{n,p+1}, W_{n,p+2}, \dots, W_{n,p+q}) = (W_n, 0, \dots 0, W_n, 0, \dots, 0)$, $n = 1, \dots$, according, respectively, relations (1.77) and (1.78), and, also, the r-dimensional vector process,

$$\vec{X}_n = (X_{n,1}, \dots, X_{n,r}) = (X_n, \dots, X_{n-r+1}), \quad n = 1, 2, \dots. \tag{1.83}$$

Using relations (1.77), (1.78), (1.81), and (1.83), we get the following vector stochastic transition dynamic relation (written as a system of stochastic transition dynamic relations):

$$\begin{cases} Y_{n,1} = A_n(Y_{n-1,1}, \dots, Y_{n-1,p+q}, X_{n-1,1}, \dots, X_{n-1,r}, W_{n,1}) \\ Y_{n,2} = Y_{n-1,1} + W_{n,2}, \\ \dots \quad \dots \\ Y_{n,p} = Y_{n-1,p-1} + W_{n,p}, \\ Y_{n,p+1} = W_{n,p+1}, \\ Y_{n,p+2} = Y_{n-1,p-1} + W_{n,p+2}, \\ \dots \quad \dots \\ Y_{n,p+q} = Y_{n-1,p+q-1} + W_{n,p+q}, \\ X_{n,1} = C_n(X_{n-1,1}, \dots, X_{n-1,r}, U_n), \\ X_{n,2} = X_{n-1,1}, \\ \dots \quad \dots \\ X_{n,r} = X_{n-1,r-1}, \\ \quad n = 1, 2, \dots. \end{cases} \tag{1.84}$$

The vector transition dynamic relation (1.84) is a particular case of the vector transition dynamic relation (1.4).

Thus, the vector process $\vec{Z}_n = (\vec{Y}_n, \vec{X}_n)$, $n = 0, 1, \dots$, is an inhomogeneous in time multivariate Markov chain with the initial random value $\vec{Z}_0$ and the corresponding vector dynamic transition functions determined by relations (1.81) and (1.84).

A particular case of the above model is an inhomogeneous in time autoregressive moving average-type model, where the modulated log-price process $Z_n = (Y_n, X_n)$ is given by the following stochastic transition dynamic relation:

$$\begin{cases} Y_n = a_{n,0}(X_{n-1}, \dots, X_{n-r}) + a_{n,1}(X_{n-1}, \dots, X_{n-r})Y_{n-1} \\ \qquad + \cdots + a_{n,p}(X_{n-1}, \dots, X_{n-r})Y_{n-p} \\ \qquad + b_{n,1}(X_{n-1}, \dots, X_{n-r})W_{n-1} \\ \qquad + \cdots + b_{n,q}(X_{n-1}, \dots, X_{n-r})W_{n-q} \\ \qquad + \sigma_n(X_{n-1}, \dots, X_{n-r})W_n\,, \\ X_n = C_n(X_{n-1}, \dots, X_{n-r}, U_n)\,, \\ \quad n = 1, 2, \dots\,, \end{cases} \tag{1.85}$$

where (a) $\vec{Y}_0 = (Y_0, \dots, Y_{-p+1})$ is a p-dimensional random vector with real-valued components, (b) $\vec{X}_0 = (X_0, \dots, X_{-r+1})$ is a r-dimensional random vector with components taking values in the space $\mathbb{X}^{(r)}$, (c) $(W_1, U_1), (W_2, U_2) \dots$, is a sequence of independent random vectors taking values in the space $\mathbb{R}_1 \times \mathbb{U}$, (d) the random vector $\vec{Z}_0 = (\vec{Y}_0, \vec{X}_0)$ and the random sequence $(W_1, U_1), (W_2, U_2), \dots$, are independent, (e) p, q, and r are positive integer numbers, (f) $a_{n,i}(x_1, \dots, x_r)$, $b_{n,j}(x_1, \dots, x_r)$, $\sigma_n(x_1, \dots, x_r)$, $i = 1, \dots, p$, $j = 1, \dots, q$, $n = 1, 2, \dots$, are the measurable functions acting from the space $\mathbb{X}^{(r)}$ to $\mathbb{R}_1$, and (g) $C_n(x_1, \dots, x_r, u)$, $n = 1, 2, \dots$, are the measurable functions acting from the space $\mathbb{X}^{(r)} \times \mathbb{U}$ to the space $\mathbb{X}$.

1.4 Autoregressive stochastic volatility LPP

In this section, we consider a number of typical examples of discrete time modulated Markov log-price and price processes of autoregressive stochastic volatility type.

1.4.1 Centered autoregressive conditional heteroskedastic-type log-price and price processes

Let us consider a model known as a centered autoregressive conditional heteroskedastic ARCH(p) type model. For simplicity, we consider a univariate log-price process, which is defined by the following stochastic difference equation:

$$Y_n - Y_{n-1} = g_\kappa(\sigma_n)W_n\,, \quad n = 1, 2, \dots\,, \tag{1.86}$$

where

$$\begin{aligned} \sigma_n = (&d_0 + d_1(Y_{n-1} - e_1 Y_{n-2})^2 \\ &+ \cdots + d_{p-1}(Y_{n-p+1} - e_{p-1}Y_{n-p})^2)^{\frac{1}{2}}\,, \quad n = 1, 2, \dots\,, \end{aligned} \tag{1.87}$$

and (a) $\vec{Y}_0 = (Y_0, \dots, Y_{-p+1})$ is a p-dimensional random vector with real-valued components, (b) $W_1, W_2, \dots,$ is a sequence of real-valued i.i.d. random variables, (c) the random vector $\vec{Y}_0$ and the random sequence $W_1, W_2, \dots,$ are independent, (d) p is a positive integer number, (e) $d_0, d_1, \dots, d_{p-1}$ are nonnegative constants, (f) $e_1, \dots, e_{p-1}$ are the constants taking values in the interval $[0, 1]$, and (g) $g_\kappa(\cdot)$ is a function from the class $\mathcal{G}_\kappa$, for some $\kappa \geq 0$.

Here and henceforth, $\mathcal{G}_\kappa = \{g_\kappa(\cdot)\}$ is, for every $\kappa \geq 0$, the class of measurable functions acting from $[0, \infty)$ to $[0, \infty)$ such that $\sup_{y \geq 0} \frac{g_\kappa(y)}{1+y^\kappa} < \infty$.

The case with the parameter $\kappa = 1$ corresponds to $ARCH(p)$-type models. In particular, the standard model corresponds to the case where function $g_\kappa(y) \equiv y$ and constants $e_i = 1, i = 1, \dots, p - 1$.

The case with the parameter $\kappa = \frac{1}{2}$ corresponds to $CIR(p)$ (Cox, Ingersoll, Ross) autoregressive-type models. In particular, the standard model corresponds to the case where function $g_\kappa(y) \equiv \sqrt{y}$, parameter $p = 2$ and the constants $d_0 = 0$, $d_1 = 1$, $e_1 = 0$.

The standard assumption is also that the random variables $W_1, W_2, \dots,$ have the standard normal distribution with parameters 0 and 1.

The volatility transformation $g_\kappa(\cdot)$ penetrating relation (1.86) plays a role of smoothing transformation. It prevents the volatility of log-price process Y_n from taking too large values that can cause in some cases appearance of infinite values for expected log-rewards.

In particular, such problems can appear for the standard $ARCH(p)$ models with Gaussian noise variables W_n, but they do not appear for standard $CIR(p)$ models with Gaussian noise variables W_n. The corresponding examples are given in the second volume of the book.

The corresponding centered autoregressive conditional heteroskedastic price process is defined by the following relation:

$$S_n = e^{Y_n}, \quad n = 0, 1, \dots . \tag{1.88}$$

The log-price process Y_n is not a Markov chain. However, this process can be represented as the first component of some p-dimensional Markov chain.

Let us introduce the p-dimensional vector process

$$\vec{Y}_n = (Y_{n,1}, \dots, Y_{n,p}) = (Y_n, \dots, Y_{n-p+1}), \quad n = 1, 2, \dots , \tag{1.89}$$

and the sequence of p-dimensional i.i.d. random vectors

$$\vec{W}_n = (W_{n,1}, \dots, W_{n,p}) = (W_n, 0, \dots, 0), \quad n = 1, 2, \dots . \tag{1.90}$$

Using relations (1.86), (1.87), (1.89), and (1.90), we get the following vector stochastic transition dynamic relation (written as a system of stochastic transition dynamic relations):

$$\begin{cases} Y_{n,1} = Y_{n-1,1} + g_\kappa(\sigma(Y_{n-1,1}, \dots, Y_{n-1,p}))W_{n,1}\,, \\ Y_{n,2} = Y_{n-1,1} + W_{n,2}\,, \\ \dots \quad\quad \dots \\ Y_{n,p} = Y_{n-1,p-1} + W_{n,p}\,, \\ \quad n = 1, 2, \dots\,, \end{cases} \tag{1.91}$$

where the functions $\sigma(y_1, \dots, y_p)$ defined for $y_1, \dots, y_p \in \mathbb{R}_1$ by the following formula:

$$\sigma(y_1, \dots, y_p) = (d_0 + d_1(y_1 - e_1 y_2)^2, \dots, d_{p-1}(y_{p-1} - e_{p-1}y_p)^2)^{\frac{1}{2}}\,. \tag{1.92}$$

The vector transition dynamic relation (1.91) is a particular case of the vector transition dynamic relation (1.4).

Thus, the vector process $\vec{Y}_n$, $n = 0, 1, \dots$, is a homogeneous in time multivariate Markov chain with the initial random value $\vec{Y}_0$ and the corresponding vector transition dynamic functions determined by relations (1.91) and (1.92).

1.4.2 Autoregressive conditional heteroskedastic-type log-price and price processes

Let us consider a model known as the autoregressive conditional heteroskedastic $AR(p')/ARCH(p)$ type model. For simplicity, we consider an univariate log-price process, which is defined by the following stochastic autoregressive difference equation:

$$\begin{aligned} Y_n - Y_{n-1} = a_0 + a_1(Y_{n-1} - f_1 Y_{n-2}) + \cdots \\ + a_{p'-1}(Y_{n-p'+1} - f_{p-1}Y_{n-p'}) + g_\kappa(\sigma_n)W_n\,, \quad n = 1, 2, \dots\,, \end{aligned} \tag{1.93}$$

where

$$\begin{aligned} \sigma_n = (d_0 + d_1(Y_{n-1} - e_1 Y_{n-2})^2 \\ + \cdots + d_{p-1}(Y_{n-p+1} - e_{p-1}Y_{n-p})^2)^{\frac{1}{2}}\,, \quad n = 1, 2, \dots\,, \end{aligned} \tag{1.94}$$

and (a) $\vec{Y}_0 = (Y_0, \dots, Y_{-\max(p',p)+1})$ is a $\max(p, p')$-dimensional random vector with real-valued components, (b) $W_1, W_2, \dots$, is a sequence of real-valued i.i.d. random variables, (c) the random vector $\vec{Y}_0$ and the random sequence $W_1, W_2, \dots$, are independent, (d) p' and p are the positive integer numbers, (e) $a_0, a_1, \dots, a_{p'-1}$ are the real-valued constants, (f) $d_0, d_1, \dots, d_{p-1}$ are nonnegative constants, (g) $e_1, \dots, e_{p-1}$, $f_1, \dots, f_{p-1}$ are constants taking values in the interval $[0, 1]$, and (h) $g_\kappa(\cdot)$ is a function from the class $\mathcal{G}_\kappa$, for some $\kappa \geq 0$.

Respectively, the corresponding autoregressive conditional heteroskedastic-type price process is defined by the following relation:

$$S_n = e^{Y_n}\,, \quad n = 0, 1, \dots\,. \tag{1.95}$$

Note first of all that the consideration can be restricted by the case $p' = p$. Indeed, this can be achieved by choosing coefficients $d_{p'}, \ldots, d_{p-1} = 0$ if $p' < p$ or coefficients $a_p, \ldots, a_{p'-1} = 0$ if $p < p'$.

The log-price process Y_n is not a Markov chain. However, this process can be represented as the first component of some p-dimensional Markov chain.

As given in Subsection 1.4.1, let us introduce the p-dimensional vector process

$$\vec{Y}_n = (Y_{n,1}, \ldots, Y_{n,p}) = (Y_n, \ldots, Y_{n-p+1}), \quad n = 1, 2, \ldots, \tag{1.96}$$

and the sequence of p-dimensional i.i.d. random vectors

$$\vec{W}_n = (W_{n,1}, \ldots, W_{n,p}) = (W_n, 0, \ldots, 0), \quad n = 1, 2, \ldots. \tag{1.97}$$

Using relations (1.93), (1.94), (1.96), and (1.97), we get the following vector stochastic transition dynamic relation (written as a system of stochastic transition dynamic relations)

$$\begin{cases} Y_{n,1} = Y_{n-1,1} + a_0 + a_1(Y_{n-1,1} - f_1 Y_{n-1,2}) \\ \qquad\quad + \cdots + a_{p-1}(Y_{n-1,p-1} - f_{p-1} Y_{n-1,p}) \\ \qquad\quad + g_\kappa(\sigma(Y_{n-1,1}, \ldots, Y_{n-1,p})) W_{n,1}, \\ Y_{n,2} = Y_{n-1,1} + W_{n,2}, \\ \ldots \qquad \ldots \\ Y_{n,p} = Y_{n-1,p-1} + W_{n,p}, \\ \quad n = 1, 2, \ldots, \end{cases} \tag{1.98}$$

where

$$\sigma_n(y_1, \ldots, y_p) = (d_0 + d_1(y_1 - e_1 y_2)^2 + \cdots + d_{p-1}(y_{p-1} - e_{p-1} y_p)^2)^{\frac{1}{2}}, \quad n = 1, 2, \ldots. \tag{1.99}$$

The vector stochastic transition dynamic relation (1.98) is a particular case of the vector transition dynamic relation (1.4).

Thus, the vector process $\vec{Y}_n$, $n = 0, 1, \ldots$, is a homogeneous in time multivariate Markov chain with the initial random value $\vec{Y}_0$ and the corresponding vector dynamic transition functions determined by relations (1.98) and (1.99).

1.4.3 Nonlinear autoregressive conditional heteroskedastic-type log-price and price processes

For simplicity, we consider a univariate log-price process, which is defined by the following stochastic transition dynamic relation:

$$Y_n = A'_n(Y_{n-1}, \ldots, Y_{n-p}) + A''_n(Y_{n-1}, \ldots, Y_{n-p}) W_n, \quad n = 1, 2, \ldots, \tag{1.100}$$

where (a) $\vec{Y}_0 = (Y_0, \dots, Y_{-p+1})$ is a p-dimensional random vector with real-valued components, (b) $W_1, W_2, \dots,$ is a sequence of real-valued independent random variables, (c) the random vector $\vec{Y}_0$ and the random sequence $W_1, W_2, \dots,$ are independent, (d) p is a positive integer number, and (e) $A'_n(y_1, \dots, y_p)$, $A''_n(y_1, \dots, y_p)$, $n = 1, 2, \dots,$ are the measurable functions acting from the space $\mathbb{R}_p$ to $\mathbb{R}_1$.

This model is, in fact, the particular case of the model introduced in the Subsection 1.3.2, with the transition dynamic functions $A_n(y_1, \dots, y_p, w) = A'_n(y_1, \dots, y_p) + A''_n(y_1, \dots, y_p)w$, $n = 1, 2, \dots$.

The particular variant of this model is an inhomogeneous in time autoregressive conditional heteroskedastic-type model given by the following stochastic difference equation:

$$
\begin{aligned}
Y_n - Y_{n-1} = a_{n,0} + a_{n,1}(Y_{n-1} - f_{n,1}Y_{n-2}) + \cdots \\
+ a_{n,p-1}(Y_{n-p+1} - f_{n,p-1}Y_{n-p}) + g_\kappa(\sigma_n)W_n\,, \quad n = 1, 2, \dots\,,
\end{aligned}
\tag{1.101}
$$

where

$$
\begin{aligned}
\sigma_n = (d_{n,0} + d_{n,1}(Y_{n-1} - e_{n,1}Y_{n-2})^2 \\
+ \cdots + d_{n,p-1}(Y_{n-p+1} - e_{n,p-1}Y_{n-p})^2)^{\frac{1}{2}}\,, \quad n = 1, 2, \dots\,,
\end{aligned}
\tag{1.102}
$$

and (a) $\vec{Y}_0 = (Y_0, \dots, Y_{-p+1})$ is a p-dimensional random vector with real-valued components, (b) $W_1, W_2, \dots,$ is a sequence of real-valued independent random variables, (c) the random vector $\vec{Y}_0$ and the random sequence $W_1, W_2, \dots,$ are independent, (d) p is a positive integer number, (e) $a_{n,0}, a_{n,1}, \dots, a_{n,p-1}$, $n = 1, 2, \dots,$ are the real-valued constants, (f) $d_{n,0}, d_{n,1}, \dots, d_{n,p-1}$, $n = 1, 2, \dots,$ are nonnegative constants, (g) $e_{n,1}, \dots, e_{n,p-1}, f_{n,1}, \dots, f_{n,p-1}$, $n = 1, 2, \dots,$ are constants taking values in the interval $[0, 1]$, and (h) $g_\kappa(\cdot)$ is a function from the class $\mathcal{G}_\kappa$ for some $\kappa \geq 0$.

1.4.4 Modulated nonlinear autoregressive conditional heteroskedastic-type log-price and price processes

Let us consider a process that can be interpreted as a modulated nonlinear autoregressive conditional heteroskedastic-type process. For simplicity, we consider a log-price process $Z_n = (Y_n, X_n)$ with a univariate log-price component Y_n. The process Z_n is defined by the following stochastic transition dynamic relation:

$$
\begin{cases}
Y_n = A'_n(Y_{n-1}, \dots, Y_{n-p}, X_{n-1}, \dots, X_{n-r}) \\
\qquad + A''_n(Y_{n-1}, \dots, Y_{n-p}, X_{n-1}, \dots, X_{n-r})W_n\,, \\
X_n = C_n(X_{n-1}, \dots, X_{n-r}, U_n)\,, \\
n = 1, 2, \dots\,,
\end{cases}
\tag{1.103}
$$

where (a) $\vec{Y}_0 = (Y_0, \dots, Y_{-p+1})$ is a p-dimensional random vector with real-valued components, (b) $\vec{X}_0 = (X_0, \dots, X_{-r+1})$ is a r-dimensional random vector with components taking values in the space $\mathbb{X}^{(r)}$, (c) $(W_1, U_1), (W_2, U_2), \dots$, is a sequence of independent random vectors taking values in the space $\mathbb{R}_1 \times \mathbb{U}$, where $\mathbb{U}$ is a measurable space with a σ-algebra of measurable subsets $\mathcal{B}_{\mathbb{U}}$, (d) the random vector $\vec{Z}_0 = (\vec{Y}_0, \vec{X}_0)$ and the random sequence $(W_1, U_1), (W_2, U_2), \dots$, are independent, (e) p and r are the positive integer numbers, (f) $A'_n(y_1, \dots, y_p, x_1, \dots, x_r), A''_n(y_1, \dots, y_p, x_1, \dots, x_r)$, $n = 1, 2, \dots$, are the measurable functions acting from the space $\mathbb{R}_p \times \mathbb{X}^{(r)}$ to $\mathbb{R}_1$, and (g) $C_n(x_1, \dots, x_r, u)$, $n = 1, 2, \dots$, are the measurable functions acting from the space $\mathbb{X}^{(r)} \times \mathbb{U}$ to $\mathbb{X}$.

The corresponding modulated price process, which can be referred to as a modulated nonlinear exponential autoregressive-type process, is defined by the following relation:

$$V_n = (S_n, X_n) = (e^{Y_n}, X_n), \quad n = 0, 1, \dots . \tag{1.104}$$

This model is, in fact, the particular case of the model introduced in Subsection 1.3.3, with the transition dynamic functions $A_n(y_1, \dots, y_p, x_1, \dots, x_r, w) = A'_n(y_1, \dots, y_p, x_1, \dots, x_r) + A''_n(y_1, \dots, y_p, x_1, \dots, x_r)w$ and $C_n(x_1, \dots, x_r, u)$, $n = 1, 2, \dots$.

A particular case of the above model is an inhomogeneous in time autoregressive-type model, where the modulated log-price process $Z_n = (Y_n, X_n)$ is given by the following vector stochastic difference equation:

$$\begin{cases} Y_n - Y_{n-1} = a_{n,0}(X_{n-1}, \dots, X_{n-r}) \\ \qquad + a_{n,1}(X_{n-1}, \dots, X_{n-r})(Y_{n-1} - f_{n,1}(X_1, \dots, X_r)Y_{n-2}) \\ \qquad + \cdots + a_{n,p-1}(X_{n-1}, \dots, X_{n-r})(Y_{n-p+1} \\ \qquad - f_{n,p-1}(X_1, \dots, X_r)Y_{n-p}) \\ \qquad + g_\kappa(\sigma_n)W_n , \\ X_n = C_n(X_{n-1}, \dots, X_{n-r}, U_n) , \\ n = 1, 2, \dots , \end{cases} \tag{1.105}$$

where

$$\begin{aligned} \sigma_n = {} & (d_{n,0}(X_{n-1}, \dots, X_{n-r}) \\ & + d_{n,1}(X_{n-1}, \dots, X_{n-r})(Y_{n-1} - e_{n,1}(X_1, \dots, X_r)Y_{n-2})^2 \\ & + \cdots + d_{n,p-1}(X_{n-1}, \dots, X_{n-r})(Y_{n,p+1} \\ & - e_{n,p-1}(X_1, \dots, X_r)Y_{n-p})^2)^{\frac{1}{2}} , \\ n = {} & 1, 2, \dots , \end{aligned} \tag{1.106}$$

and (a) $\vec{Y}_0 = (Y_0, \dots, Y_{-p+1})$ is a p-dimensional random vector with real-valued components, (b) $\vec{X}_0 = (X_0, \dots, X_{-r+1})$ is a r-dimensional random vector takin values in the space $\mathbb{X}^{(r)}$, (c) $(W_1, U_1), (W_2, U_2) \dots$, is a sequence of independent random vectors taking values in the space $\mathbb{R}_1 \times \mathbb{U}$, (d) the random vector $\vec{Z}_0 = (\vec{Y}_0, \vec{X}_0)$ and the random sequence $(W_1, U_1), (W_2, U_2) \dots$, are independent, (e) p and r are the positive integer numbers, (f) $a_{n,i}(x_1, \dots, x_r)$, $i = 0, \dots, p-1$, $n = 1, 2, \dots$, are the measurable functions acting from the space $\mathbb{X}^{(r)}$ to $\mathbb{R}_1$, (g) $d_{n,i}(x_1, \dots, x_r)$, $i = 0, \dots, p-1$, $n = 1, 2, \dots$, are the measurable functions acting from the space $\mathbb{X}^{(r)}$ to the interval $[0, \infty)$, (h) $e_{n,i}(x_1, \dots, x_r), f_{n,i}(x_1, \dots, x_r)$, $i = 0, \dots, p-1$, $n = 1, 2, \dots$, are the measurable functions acting from the space $\mathbb{X}^{(r)}$ to the interval $[0, 1]$, (i) $C_n(x_1, \dots, x_r, u)$, $n = 1, 2, \dots$, are the measurable functions acting from the space $\mathbb{X}^{(r)} \times \mathbb{U}$ to the space $\mathbb{X}$, and (j) $g_\kappa(\cdot)$ is a function from the class $\mathcal{G}_\kappa$ for some $\kappa \geq 0$.

1.4.5 Generalized autoregressive conditional heteroskedastic-type log-price and price processes

Let us consider a model that can be referred to as a autoregressive conditional heteroskedastic model GARCH(p, q) type model. For simplicity, we consider a univariate log-price process Y_n defined by the stochastic difference equation

$$Y_n - Y_{n-1} = g_\kappa(\sigma_n) W_n\,, \quad n = 1, 2, \dots\,, \tag{1.107}$$

where

$$\begin{aligned}\sigma_n = \big(&d_0 + d_1(Y_{n-1} - e_1 Y_{n-2})^2 + \cdots + d_{p-1}(Y_{n-p+1} - e_{p-1} Y_{n-p})^2 \\ &+ b_1 \sigma_{n-1}^2 + \cdots + b_q \sigma_{n-q}^2\big)^{\frac{1}{2}}\,, \quad n = 1, 2, \dots\,,\end{aligned} \tag{1.108}$$

and (a) $\vec{Y}_0 = (Y_0, \dots, Y_{-(p+q)+1})$ is a $(p+q)$-dimensional random vector with real-valued components, (b) $W_1, W_2, \dots$, is a sequence of real-valued i.i.d. random variables, (c) the random vector $\vec{Y}_0$ and the random sequence $W_1, W_2, \dots$, are independent, (d) p and q are the positive integer numbers, (e) $d_0, d_1, \dots, d_{p-1}$ and $b_1, \dots, b_q$ are nonnegative constants, (f) $e_1, \dots, e_{p-1}$ are constants taking values in the interval $[0, 1]$, and (g) $g_\kappa(\cdot)$ is a function from the class $\mathcal{G}_\kappa$, for some $\kappa \geq 0$.

The corresponding price process is defined by the following relation:

$$S_n = e^{Y_n}\,, \quad n = 0, 1, \dots\,. \tag{1.109}$$

The standard case is where random variables $W_0, W_1, \dots$, have the standard normal distribution with parameters 0 and 1.

Let us introduce the $(p+q)$-dimensional vector process

$$\begin{aligned}\vec{Y}_n &= (Y_{n,1}, \dots, Y_{n,p}, Y_{n,p+1}, \dots, Y_{n,p+q}) \\ &= (Y_n, \dots, Y_{n-p+1}, \sigma_n, \dots, \sigma_{n-q+1})\,, \quad n = 1, 2, \dots\,,\end{aligned} \tag{1.110}$$

and the sequence of $(p+q)$-dimensional i.i.d. random vectors,

$$\begin{aligned} \vec{W}_n &= (W_{n,1}, \dots, W_{n,p+q}) \\ &= (W_n, 0, \dots, 0), \quad n = 1, 2, \dots . \end{aligned} \tag{1.111}$$

Using relations (1.107), (1.108) (1.110), and (1.111), we get the following vector stochastic transition dynamic relation (written as a system of stochastic transition dynamic relations):

$$\begin{cases} Y_{n,1} = Y_{n-1,1} + g_\kappa(\sigma(Y_{n-1,1}, \dots, Y_{n-1,p+q}))W_{n,1}, \\ Y_{n,2} = Y_{n-1,1} + W_{n,2}, \\ \dots \quad \dots \\ Y_{n,p+q} = Y_{n-1,p+g-1} + W_{n,p+q}, \\ \quad n = 1, 2, \dots, \end{cases} \tag{1.112}$$

where the functions $\sigma(y_1, \dots, y_{p+q})$ are defined for $y_1, \dots, y_p \in \mathbb{R}_1, y_{p+1}, \dots, y_{p+q} \in [0, \infty)$ by the following formula:

$$\begin{aligned} \sigma(y_1, \dots, y_{p+q}) = \big(&d_0 + d_1(y_1 - e_1 y_2)^2 + \cdots + d_{p-1}(y_{p-1} - e_{p-1} y_p)^2 \\ &+ b_1 y_{p+1}^2 + \cdots + b_q y_{p+q}^2\big)^{\frac{1}{2}} . \end{aligned} \tag{1.113}$$

The vector transition dynamic relation (1.112) is a particular case of the vector transition dynamic relation (1.4).

Thus, the vector process $\vec{Y}_n$, $n = 0, 1, \dots$, is a homogeneous multivariate Markov chain with the initial random value $\vec{Y}_0$ and the corresponding vector dynamic transition functions determined by relations (1.107), (1.108), and (1.112).

1.4.6 Nonlinear generalized autoregressive conditional heteroskedastic-type log-price and price processes

Let us consider a model that can be referred to as a nonlinear autoregressive conditional heteroskedastic-type model. For simplicity, we consider a univariate log-price process Y_n defined by the following stochastic transition dynamic relation:

$$Y_n = \sigma_n W_n, \quad n = 1, 2, \dots, \tag{1.114}$$

where

$$\sigma_n = A_n(Y_{n-1}, \dots, Y_{n-p}, \sigma_{n-1}, \dots, \sigma_{n-q}), \quad n = 1, 2, \dots, \tag{1.115}$$

and (a) $\vec{Y}_0 = (Y_0, \dots, Y_{-(p+q)+1})$ is a $(p+q)$-dimensional random vector with real-valued components, (b) $W_1, W_2, \dots$, is a sequence of real-valued i.i.d. random variables, (c) the random vector $\vec{Y}_0$ and the random sequence $W_1, W_2, \dots$, are independent, (d) p

and q are the positive integer numbers, and (e) $A_n(y_1, \dots, y_{p+q})$, $n = 1, 2, \dots$, are the measurable functions acting from the space $\mathbb{R}_{p+q}$ to $\mathbb{R}_1$.

As given in Subsection 1.4.5, let us introduce the $(p+q)$-dimensional vector process

$$\begin{aligned} \vec{Y}_n &= (Y_{n,1}, \dots, Y_{n,p}, Y_{n,p+1}, \dots, Y_{n,p+q}) \\ &= (Y_n, \dots, Y_{n-p+1}, \sigma_n, \dots, \sigma_{n-q+1}), \quad n = 1, 2, \dots, \end{aligned} \tag{1.116}$$

and the sequence of $(p+q)$-dimensional independent random vectors

$$\vec{W}_n = (W_{n,1}, \dots, W_{n,p+q}) = (W_n, 0, \dots, 0), \quad n = 1, 2, \dots. \tag{1.117}$$

Using relations (1.114)–(1.117), we get the following vector stochastic transition dynamic relation (written as a system of stochastic transition dynamic relations):

$$\begin{cases} Y_{n,1} = A_n(Y_{n-1,1}, \dots, Y_{n-1,p+q}) W_{n,1}, \\ Y_{n,2} = Y_{n-1,1} + W_{n,2}, \\ \quad \dots \qquad \dots \\ Y_{n,p+q} = Y_{n-1,p+g-1} + W_{n,p+q}, \\ \quad n = 1, 2, \dots. \end{cases} \tag{1.118}$$

The vector stochastic transition dynamic relation (1.118) is a particular case of the vector transition dynamic relation (1.4).

Thus, the vector process $\vec{Y}_n$, $n = 0, 1, \dots$, is a homogeneous multivariate Markov chain with the initial random value $\vec{Y}_0$ and the corresponding vector dynamic transition functions determined by relations (1.114) and (1.118).

The particular variant of the above model is an inhomogeneous in time autoregressive conditional heteroskedastic model GARCH(p, q)-type log-price process given by the following stochastic difference equation:

$$Y_n - Y_{n-1} = g_\kappa(\sigma_n) W_n, \quad n = 1, 2, \dots, \tag{1.119}$$

where

$$\begin{aligned} \sigma_n = \big(& d_{n,0} + d_{n,1}(Y_{n-1} - e_{n,1}Y_{n-2})^2 + \cdots \\ &+ d_{n,p-1}(Y_{n-p+1} - e_{n,p-1}Y_{n-p})^2 \\ &+ b_{n,1}\sigma^2_{n-1} + \cdots + b_{n,q}\sigma^2_{n-q}\big)^{\frac{1}{2}}, \quad n = 1, 2, \dots, \end{aligned} \tag{1.120}$$

and (a) $\vec{Y}_0 = (Y_0, \dots, Y_{-(p+q)+1})$ is a $(p+q)$-dimensional random vector with real-valued components, (b) $W_1, W_2, \dots$, is a sequence of real-valued i.i.d. random variables, (c) the random vector $\vec{Y}_0$ and the random sequence $W_1, W_2, \dots$, are independent, (d) p and q are the positive integer numbers, (e) $d_{n,0}, \dots, d_{n,p-1}$, $n = 1, 2, \dots$, and $b_{n,1}, \dots, b_{n,q}$, $n = 1, 2, \dots$, are nonnegative constants, (f) $e_{n,1}, \dots, e_{n,p-1}$, $n = 1, 2, \dots$, are constants taking values in the interval $[0, 1]$, and (g) $g_\kappa(\cdot)$ is a function from the class $\mathcal{G}_\kappa$, for some $\kappa \geq 0$.

1.4.7 Modulated nonlinear generalized autoregressive conditional heteroskedastic-type log-price and price processes

Let us consider a model that can be referred to as a modulated nonlinear autoregressive conditional heteroskedastic-type model. For simplicity, we consider a log-price process $Z_n = (Y_n, X_n)$ with a univariate log-price component Y_n. The process Z_n is defined by the following stochastic transition dynamic relation:

$$\begin{cases} Y_n = \sigma_n W_n \\ X_n = C_n(X_{n-1}, \dots, X_{n-r}, U_n), \\ \quad n = 1, 2, \dots, \end{cases} \tag{1.121}$$

where

$$\begin{aligned} \sigma_n = A_n(&Y_{n-1}, \dots, Y_{n-p}, \sigma_{n-1}, \dots, \sigma_{n-q}, \\ &X_{n-1}, \dots, X_{n-r}), \quad n = 1, 2, \dots, \end{aligned} \tag{1.122}$$

and (a) $\vec{Y}_0 = (Y_0, \dots, Y_{-(p+q)+1})$ is a $(p+q)$-dimensional random vector with real-valued components, (b) $\vec{X}_0 = (X_0, \dots, X_{-r+1})$ is a r-dimensional random vector taking values in the space $\mathbb{X}^{(r)}$, (c) $(W_1, U_1), (W_2, U_2) \dots$, is a sequence of independent random vectors taking values in the space $\mathbb{R}_1 \times \mathbb{U}$, where $\mathbb{U}$ is a measurable space with a σ-algebra of measurable subsets $\mathcal{B}_{\mathbb{U}}$, (d) the random vector $\vec{Z}_0 = (\vec{Y}_0, \vec{X}_0)$ and the random sequence $(W_1, U_1), (W_2, U_2) \dots$, are independent, (e) p, q and r are the positive integer numbers, (f) $A_n(y_1, \dots, y_{p+q}, x_1, \dots, x_r)$, $n = 1, 2, \dots$, are the measurable functions acting from the space $\mathbb{R}_{p+q} \times \mathbb{X}^{(r)}$ to $\mathbb{R}_1$, and (g) $C_n(x_1, \dots, x_r, u)$, $n = 1, 2, \dots$, are the measurable functions acting from the space $\mathbb{X}^{(r)} \times \mathbb{U}$ to the space $\mathbb{X}$.

The corresponding modulated price process, which can be referred to as a modulated nonlinear exponential generalized autoregressive-type process, is defined by the following relation:

$$V_n = (S_n, X_n) = (e^{Y_n}, X_n), \quad n = 0, 1, \dots. \tag{1.123}$$

As given in Subsection 1.4.6, let us introduce the $(p+q)$-dimensional vector process, $\vec{Y}_n = (Y_{n,1}, \dots, Y_{n,p}, Y_{n,p+1}, \dots, Y_{n,p+q}) = (Y_n, \dots, Y_{n-p+1}, \sigma_n, \dots, \sigma_{n-q+1})$, $n = 1, 2, \dots$, and the sequence of $(p+q)$-dimensional independent random vectors $\vec{W}_n = (W_{n,1}, \dots, W_{n,p+q}) = (W_n, 0, \dots, 0)$, $n = 1, 2, \dots$, defined, respectively, by relations (1.116) and (1.117), as well as the r-dimensional vector process,

$$\vec{X}_n = (X_{n,1}, \dots, X_{n,r}) = (X_n, \dots, X_{n-r+1}), \quad n = 1, 2, \dots. \tag{1.124}$$

Using relations (1.116), (1.117), and (1.121)–(1.124), we get the following vector stochastic transition dynamic relation (written as a system of stochastic transition dynamic relations):

$$\begin{cases} Y_{n,1} = A_n(Y_{n-1,1}, \dots, Y_{n-1,p+q}, X_{n-1,1}, \dots, Y_{n-1,r}) W_{n,1}, \\ Y_{n,2} = Y_{n-1,1} + W_{n,2}, \\ \dots \quad \dots \\ Y_{n,p+q} = Y_{n-1,p+g-1} + W_{n,p+q}, \\ X_{n,1} = C_n(X_{n-1,1}, \dots, X_{n-1,r}, U_n), \\ X_{n,2} = X_{n-1,1}, \\ \dots \quad \dots \\ X_{n,r} = X_{n-1,r-1}, \\ \quad n = 1, 2, \dots . \end{cases} \tag{1.125}$$

The vector transition dynamic relation (1.125) is a particular case of the vector transition dynamic relation (1.4).

Thus, the vector process $\vec{Z}_n = (\vec{Y}_n, \vec{X}_n)$, $n = 0, 1, \dots$, is a homogeneous multivariate Markov chain with the initial random value $\vec{Z}_0$ and the corresponding vector dynamic transition functions determined by relations (1.121) and (1.125).

The particular variant of the above model is a inhomogeneous in time modulated autoregressive conditional heteroskedastic $GARCH(p, q, r)$-type log-price process given by the following autoregressive stochastic difference equation:

$$Y_n - Y_{n-1} = g_\kappa(\sigma_n) W_n, \quad n = 1, 2, \dots, \tag{1.126}$$

where

$$\begin{aligned} \sigma_n = \Big(& d_{n,0}(X_{n-1}, \dots, X_{n-r}) \\ & + d_{n,1}(X_{n-1}, \dots, X_{n-r})(Y_{n-1} - e_{n,1}(X_{n-1}, \dots, X_{n-r}) Y_{n-2})^2 \\ & + \dots + d_{n,p-1}(X_{n-1}, \dots, X_{n-r})(Y_{n-p+1} \\ & \quad - e_{n,p-1}(X_{n-1}, \dots, X_{n-r}) Y_{n-p})^2 + b_{n,1}(X_{n-1}, \dots, X_{n-r}) \sigma^2_{n-1} \\ & + \dots + b_{n,q}(X_{n-1}, \dots, X_{n-r}) \sigma^2_{n-q} \Big)^{\frac{1}{2}}, \quad n = 1, 2, \dots, \end{aligned} \tag{1.127}$$

and (a) $\vec{Y}_0 = (Y_0, \dots, Y_{-(p+q)+1})$ is a $(p + q)$-dimensional random vector with real-valued components, (b) $\vec{X}_0 = (X_0, \dots, X_{-r+1})$ is a r-dimensional random vector taking values in the space $\mathbb{X}^{(r)}$, (c) $(W_1, U_1), (W_2, U_2) \dots$, is a sequence of independent random vectors taking values in the space $\mathbb{R}_1 \times \mathbb{U}$, (d) the random vector $\vec{Z}_0 = (\vec{Y}_0, \vec{X}_0)$ and the random sequence $(W_1, U_1), (W_2, U_2) \dots$, are independent, (e) p, q and r are the positive integer numbers, (f) $d_{n,i}(x_1, \dots, x_r)$, $i = 0, \dots, p - 1$, $n = 1, 2, \dots$, and $b_{n,j}(x_1, \dots, x_r)$, $j = 1, \dots, q$, $n = 1, 2, \dots$, are the measurable functions acting from the space $\mathbb{X}^{(r)}$ to the interval $[0, \infty)$, (g) $e_{n,i}(x_1, \dots, x_r)$, $i = 0, \dots, p-1$, $n = 1, 2, \dots$, are the measurable functions acting from the space $\mathbb{X}^{(r)}$ to the interval $[0, 1]$, and (h) $C_n(x_1, \dots, x_r, u)$, $n = 1, 2, \dots$, are the measurable functions acting from the space $\mathbb{X}^{(r)} \times \mathbb{U}$ to the space $\mathbb{X}$; (i) $g_\kappa(\cdot)$ is a function from the class $\mathcal{G}_\kappa$ for some $\kappa \geq 0$.

1.4.8 Stochastic volatility models of log-price and price processes

Let us consider a model known as a stochastic volatility $SV(p)$-type model. For simplicity, we consider a univariate log-price process Y_n defined by the following stochastic difference equation:

$$Y_n - Y_{n-1} = g_{\kappa,\nu}(\sigma_n)W'_n\,, \quad n = 1, 2, \dots\,, \tag{1.128}$$

where

$$\ln\sigma_n = a_0 + a_1\ln\sigma_{n-1} + \cdots + a_p\ln\sigma_{n-p} + cW''_n\,, \quad n = 1, 2, \dots\,, \tag{1.129}$$

and (a) $\ln\vec{\sigma}_0 = (\ln\sigma_0, \dots, \ln\sigma_{-p+1})$ is a p-dimensional random vector with real-valued components, (b) $(W'_1, W''_1), (W'_2, W''_2), \dots,$ is a sequence of i.i.d. random vectors with real-valued components, (c) the random vector $\ln\vec{\sigma}_0$ and the random sequence $(W'_1, W''_1), (W'_2, W''_2)\dots,$ are independent, (d) p is a positive integer number, and (e) $a_0, a_1, \dots, a_p, c$ are the real-valued constants, (f) $g_{\kappa,\nu}(\cdot)$ is a function from the class $\mathcal{G}_{\kappa,\nu}$, for some $\kappa, \nu \geq 0$.

Here and henceforth, $\mathcal{G}_{\kappa,\nu} = \{g_{\kappa,\nu}(\cdot)\}$ is, for every $\kappa, \nu \geq 0$, the class of functions acting from $\mathbb{R}_1$ to $[0, \infty)$ such that $\sup_{y\in\mathbb{R}_1}\frac{g_{\kappa,\nu}(e^y)}{1+|y|^\kappa e^{\nu y}} < \infty$.

The standard assumption is also that random vectors $(W'_1, W''_1), (W'_2, W''_2), \dots,$ have normal distributions.

As for autoregressive conditional heteroskedastic-type models considered in previous subsections, the volatility transformation $g_{\kappa,\nu}(\cdot)$ penetrating relation (1.86) plays a role of smoothing transformation. It prevents the volatility of log-price process Y_n from taking too large values that can cause in some cases appearance of infinite values for expected log-rewards.

In particular, such problems can appear even for the standard variant of the above $SV(p)$, where the parameters $\kappa = 0$, $\nu = 1$ and the function $g_{\kappa,\nu}(y) = y$. The corresponding examples will be given in the second volume of the book.

The corresponding price process is defined by the following relation:

$$S_n = e^{Y_n}\,, \quad n = 0, 1, \dots\,. \tag{1.130}$$

Let us show in which way this model can be imbedded in a model of multivariate Markov modulated log-price process.

Let us define the $(p+1)$-dimensional discrete time process

$$\begin{aligned}\vec{Y}_n &= (Y_{n,1}, \dots, Y_{n,p+1}) \\ &= (Y_n, \ln\sigma_n, \ln\sigma_{n-1}, \dots, \ln\sigma_{n-p+1})\,, \quad n = 1, 2, \dots\,,\end{aligned} \tag{1.131}$$

and the $(p+1)$-dimensional sequence of i.i.d. random vectors,

$$\vec{W}_n = (W_{n,1}, \dots, W_{n,p+1}) = (W'_n, W''_n, 0, \dots, 0)\,, \quad n = 1, 2, \dots\,. \tag{1.132}$$

Using relations (1.128), (1.129), (1.131), and (1.132), we get the following vector stochastic transition dynamic relation (written as a system of stochastic transition dynamic relations),

$$\begin{cases} Y_{n,1} = Y_{n-1,1} + g_{\kappa,\nu}(e^{a_0+a_1Y_{n-1,2}+\cdots+a_pY_{n-1,p+1}+cW''_n})W'_n, \\ Y_{n,2} = a_0 + a_1Y_{n-1,2} + \cdots + a_pY_{n-1,p+1} + cW''_n, \\ Y_{n,3} = Y_{n-1,2}, \\ \dots \quad \dots \\ Y_{n,p+1} = Y_{n-1,p}, \\ n = 1, 2, \dots \end{cases} \tag{1.133}$$

The vector stochastic transition dynamic relation (1.133) is a particular case of the vector transition dynamic relation (1.4).

Thus, the vector process $\vec{Y}_n$, $n = 0, 1, \dots$, is a homogeneous multivariate Markov chain with the initial random value $\vec{Y}_0$ and the corresponding vector transition dynamic functions determined by relations (1.128), (1.129), and (1.133).

1.4.9 Nonlinear autoregressive stochastic volatility models of log-price and price processes

Let us consider a model that can be referred to as a nonlinear autoregressive stochastic volatility-type model. For simplicity, we consider a univariate log-price process Y_n defined by the following stochastic transition dynamic relation:

$$Y_n = \sigma_n W'_n, \quad n = 1, 2, \dots, \tag{1.134}$$

where

$$\ln \sigma_n = A_n(\ln \sigma_{n-1}, \dots, \ln \sigma_{n-p}, W''_n), \quad n = 1, 2, \dots, \tag{1.135}$$

and (a) $\ln \vec{\sigma}_0 = (\ln \sigma_0, \dots, \ln \sigma_{-p+1})$ is a p-dimensional random vector with real-valued components, (b) $(W'_1, W''_1), (W'_2, W''_2), \dots$, is a sequence of independent random vectors with real-valued components, (c) the random vector $\ln \vec{\sigma}_0$ and the random sequence $(W'_1, W''_1), (W'_2, W''_2) \dots$, are independent, (d) p is a positive integer number, and (e) $A_n(y_1, \dots, y_p, w)$, $n = 1, 2, \dots$, are the measurable functions acting from the space $\mathbb{R}_p \times \mathbb{R}_1$ to $\mathbb{R}_1$.

The standard assumption is also that random vectors $(W'_1, W''_1), (W'_2, W''_2), \dots$, have normal distributions.

The corresponding price process is defined by the following relation:

$$S_n = e^{Y_n}, \quad n = 0, 1, \dots. \tag{1.136}$$

Let us show in which way this model can be imbedded in the model of multivariate Markov log-price process.

Let us define the $(p+1)$-dimensional discrete time process,

$$\begin{aligned}\vec{Y}_n &= (Y_{n,1}, \ldots, Y_{n,p+1}) \\ &= (Y_n, \ln \sigma_n, \ln \sigma_{n-1}, \ldots, \ln \sigma_{n-p+1}), \quad n = 1, 2, \ldots, \end{aligned} \tag{1.137}$$

and the $(p+1)$-dimensional sequence of independent random vectors

$$\begin{aligned}\vec{W}_n &= (W_{n,1}, \ldots, W_{n,p+1}) \\ &= (W'_n, W''_n, 0, \ldots, 0), \quad n = 1, 2, \ldots . \end{aligned} \tag{1.138}$$

Using relations (1.134), (1.135), (1.137), and (1.138), we get the following vector stochastic transition dynamic relation (written as a system of stochastic transition dynamic relations):

$$\begin{cases} Y_{n,1} = e^{A_n(Y_{n-1,2}, \ldots, Y_{n-1,p+1}, W''_n)} W'_n, \\ Y_{n,2} = A_n(Y_{n-1,2}, \ldots, Y_{n-1,p+1}, W''_n), \\ Y_{n,3} = Y_{n-1,2}, \\ \cdots\cdots \\ Y_{n,p+1} = Y_{n-1,p}, \\ \quad n = 1, 2, \ldots \end{cases} \tag{1.139}$$

The vector stochastic transition dynamic relation (1.139) is a particular case of the vector stochastic transition dynamic relation (1.4).

Thus, the vector process $\vec{Y}_n$, $n = 0, 1, \ldots$, is a homogeneous multivariate Markov chain with the initial random value $\vec{Y}_0$ and the corresponding vector dynamic transition functions determined by relations (1.134), (1.135), and (1.139).

The particular variant of the above model is an inhomogeneous in time stochastic volatility-type log-price process given by the following stochastic difference equation:

$$Y_n - Y_{n-1} = g_{\kappa,\nu}(\sigma_n) W'_n, \quad n = 1, 2, \ldots, \tag{1.140}$$

where

$$\begin{aligned}\ln \sigma_n &= a_{n,0} + a_{n,1} \ln \sigma_{n-1} \\ &\quad + \cdots + a_{n,p} \ln \sigma_{n-p} + c_n W''_n, \quad n = 1, 2, \ldots, \end{aligned} \tag{1.141}$$

and (a) $\ln \vec{\sigma}_0 = (\ln \sigma_0, \ldots, \ln \sigma_{-p+1})$ is a p-dimensional random vector with real-valued components, (b) $(W'_1, W''_1), (W'_2, W''_2) \ldots$, is a sequence of independent random vectors with real-valued components, (c) the random vector $\ln \vec{\sigma}_0$ and the random sequence $(W'_1, W''_1), (W'_2, W''_2) \ldots$, are independent, (d) p is a positive integer number, and (e) $a_{n,0}, a_{n,1}, \ldots, a_{n,p}, c_n$, $n = 1, 2, \ldots$, are the real-valued constants, (f) $g_{\kappa,\nu}(\cdot)$ is a function from the class $\mathcal{G}_{\kappa,\nu}$, for some $\kappa, \nu \geq 0$.

1.4.10 Modulated nonlinear autoregressive stochastic volatility log-price and price processes

Let us consider a model that can be referred to as a modulated nonlinear autoregressive stochastic volatility-type model. For simplicity, we consider a process $Z_n = (Y_n, X_n)$ with a univariate log-price component Y_n. The process Z_n is defined by the following stochastic transition dynamic relation:

$$\begin{cases} Y_n = \sigma_n W'_n\,, \\ X_n = C_n(X_{n-1}, \dots, X_{n-r}, U_n)\,, \\ \quad n = 1, 2, \dots\,, \end{cases} \tag{1.142}$$

where

$$\ln \sigma_n = A_n(\ln \sigma_{n-1}, \dots, \ln \sigma_{n-p}, \quad X_{n-1}, \dots, X_{n-r}, W''_n)\,, \quad n = 1, 2, \dots\,, \tag{1.143}$$

and (a) $\ln \vec{\sigma}_0 = (\ln \sigma_0, \dots, \ln \sigma_{-p+1})$ is a p-dimensional random vector with real-valued components, (b) $\vec{X}_0 = (X_0, \dots, X_{-r+1})$ is a r-dimensional random vector taking values in the space $\mathbb{X}^{(r)}$, (c) $(W'_1, W''_1, U_1), (W'_2, W''_2, U_2), \dots,$ is a sequence of independent random vectors taking values in the space $\mathbb{R}_1 \times \mathbb{R}_1 \times \mathbb{U}$, where $\mathbb{U}$ is a measurable space with a σ-algebra of measurable subsets $\mathcal{B}_{\mathbb{U}}$, (d) the random vector $(\ln \vec{\sigma}_0, \vec{X}_0)$ and the random sequence $(W'_1, W''_1, U_1), (W'_2, W''_2, U_2), \dots,$ are independent, (e) p and r are the positive integer numbers, (f) $A_n(y_1, \dots, y_p, x_1, \dots, x_r, w)$, $n = 1, 2, \dots,$ are the measurable functions acting from the space $\mathbb{R}_p \times \mathbb{X}^{(r)} \times \mathbb{R}_1$ to $\mathbb{R}_1$, and (g) $C_n(x_1, \dots, x_r, u)$, $n = 1, 2, \dots,$ are the measurable functions acting from the space $\mathbb{X}^{(r)} \times \mathbb{U}$ to the space $\mathbb{X}$.

The corresponding modulated price process is defined by the following relation:

$$(S_n, X_n) = (e^{Y_n}, X_n)\,, \quad n = 0, 1, \dots\,. \tag{1.144}$$

Let us show in which way this model can be imbedded in a model of modulated multivariate Markov log-price process.

As given in Subsection 1.4.9, let us define the $(p+1)$-dimensional discrete time process $\vec{Y}_n = (Y_{n,1}, \dots, Y_{n,p+1}) = (Y_n, \ln \sigma_n, \ln \sigma_{n-1}, \dots, \ln \sigma_{n-p+1})$, $n = 0, 1, \dots,$ and the $(p+1)$-dimensional sequence of independent random vectors $\vec{W}_n = (W_{n,1}, \dots, W_{n,p+1}) = (W'_n, W''_n, 0, \dots, 0)$, $n = 0, 1, \dots,$ using, respectively, relations (1.137) and (1.138), as well as define the r-dimensional vector process,

$$\vec{X}_n = (X_{n,1}, \dots, X_{n,r}) = (X_n, \dots, X_{n-r+1})\,, \quad n = 0, 1, \dots\,. \tag{1.145}$$

Using relations (1.137), (1.138), (1.142), and (1.145), we get the following vector stochastic transition dynamic relation (written as a system of stochastic transition dynamic relations),

$$\begin{cases} Y_{n,1} = e^{A_n(Y_{n-1,2},\dots,Y_{n-1,p+1},X_{n-1,1},\dots,X_{n-1,r},W''_n)} W'_n \,, \\ Y_{n,2} = A_n(Y_{n-1,2},\dots,Y_{n-1,p+1},X_{n-1,1},\dots,X_{n-1,r},W''_n)\,, \\ Y_{n,3} = Y_{n-1,2}\,, \\ \dots \quad \dots \\ Y_{n,p+1} = Y_{n-1,p}\,, \\ X_{n,1} = C_n(X_{n-1,1}\dots,X_{n-1,r},U_n)\,, \\ X_{n,2} = X_{n-1,1}\,, \\ \dots \quad \dots \\ X_{n,r} = X_{n-1,r-1}\,, \\ n = 1,2,\dots\,. \end{cases} \tag{1.146}$$

The vector stochastic transition dynamic relation (1.146) is a particular case of the vector stochastic transition dynamic relation (1.4).

Thus, the vector process $\vec{Z}_n = (\vec{Y}_n, \vec{X}_n)$, $n = 0, 1, \dots$, is a multivariate modulated Markov chain with the initial random value $\vec{Z}_0$ and the corresponding vector transition dynamic functions determined by relations (1.142), (1.143), and (1.146).

The particular variant of the above model is an inhomogeneous in time autoregressive stochastic volatility-type log-price process given by the following stochastic difference equation:

$$\begin{cases} Y_n - Y_{n-1} = g_{\kappa,\nu}(\sigma_n)\cdot W'_n\,, \\ X_n = C_n(X_{n-1},\dots,X_{n-r},U_n)\,, \\ n = 1,2,\dots\,, \end{cases} \tag{1.147}$$

where

$$\begin{aligned} \ln\sigma_n &= a_{n,0}(X_{n-1},\dots,X_{n-r}) \\ &\quad + a_{n,1}(X_{n-1},\dots,X_{n-r})\ln\sigma_{n-1} + \cdots \\ &\quad + a_{n,p}(X_{n-1},\dots,X_{n-r})\ln\sigma_{n-p} \\ &\quad + c_n(X_{n-1},\dots,X_{n-r})W''_n\,, \quad n = 1,1,\dots\,, \end{aligned} \tag{1.148}$$

and (a) $\ln\vec{\sigma}_0 = (\ln\sigma_0,\dots,\ln\sigma_{-p+1})$ is a p-dimensional random vector with real-valued components, (b) $\vec{X}_0 = (X_0,\dots,X_{-r+1})$ is a r-dimensional random vector taking values in the space $\mathbb{X}^{(r)}$, (c) $(W'_1, W''_1, U_1), (W'_2, W''_2, U_2), \dots$, is a sequence of independent random vectors taking values in the space $\mathbb{R}_1\times\mathbb{R}_1\times\mathbb{U}$, (d) the random vector $(\ln\vec{\sigma}_0, \vec{X}_0)$ and the random sequence $(W'_1, W''_1, U_1), (W'_2, W''_2, U_2), \dots$, are independent, (e) p and r are the positive integer numbers, (f) $a_{n,i}(x_1,\dots,x_r), I = 0,\dots,p$, $n = 1,2,\dots$, and $c_n(x_1,\dots,x_r)$, $n = 1,2,\dots$, are the measurable functions acting from the space $\mathbb{X}^{(r)}$ to $\mathbb{R}_1$, and (g) $C_n(x_1,\dots,x_r,u)$, $n = 1,2,\dots$, are the measurable functions acting from the space $\mathbb{X}^{(r)}\times\mathbb{U}$ to the space $\mathbb{X}$.

2 American-type options

In this chapter, we define American-type options and introduce basic objects connected with American-type options for multivariate modulated Markov price and log-price processes.

In Section 2.1, we introduce American-type options for multivariate modulated Markov price processes with general pay-off functions depending on both price and index components.

We also formulate basic optimization problems for American-type options connected with finding optimal expected rewards and optimal stopping times representing optimal stopping strategies for buyers of American options.

In Section 2.2, we present basic examples of American-type options including call and put-type options, exchange of assets options, digital options as well as option--type contracts represented by portfolios of options.

In Section 2.3, we investigate the properties of reward and log-reward functions and optimal expected rewards and give backward recurrence relations for reward and log-reward functions. Such recurrence relations play an important role in stochastic approximation algorithms for the option rewards of American-type options.

In Section 2.4, we define optimal stopping times for American-type options and investigate the hitting structure of optimal stopping times. We also give proofs of the basic theorems about backward recurrence relations for option rewards and hitting structure of optimal stopping times.

In Section 2.5, we introduce American-type knockout options and show how these options can be imbedded in the model of usual American-type options by extending the underlying price process by an additional modulating index component.

The main results are given in Theorems 2.3.1–2.3.4, which present backward recurrence relations for option rewards, and Theorems 2.4.1–2.4.4, which describe hitting structure of optimal stopping times for American-type options.

These results are variants of classical results by Chow, Robbins, and Siegmund (1971) and Shiryaev (1976) on optimal stopping for Markov processes.

2.1 American-type options

In this section, we define American-type option contracts for discrete time multivariate modulated Markov price and log-price processes.

2.1.1 American-type options for discrete time price processes

Let us consider a discrete time multivariate modulated price process $\vec{V}_n = (\vec{S}_n, X_n)$, $n = 0, 1, \dots$, and the corresponding modulated log-price process $\vec{Z}_n = (\vec{Y}_n, X_n)$, $n = 0, 1, \dots$, defined in Section 1.1.

Recall that the price process $\vec{S}_n$ and the log-price process $\vec{Y}_n$ are connected by the relation $\vec{S}_n = e^{\vec{Y}_n}$, $n = 0, 1, \dots$.

We always assume that the price process $\vec{V}_n$ is a Markov chain with a phase space $\mathbb{V}$, an initial distribution $\dot{P}_0(B) = \mathsf{P}\{\vec{V}_0 \in B\}$ and transition probabilities $\dot{P}_n(\vec{v}, B)$. In this case, the log-price process $\vec{Z}_n$ is also a Markov chain with a phase space $\mathbb{Z}$, an initial distribution $P(A)$ and transition probabilities $P_n(\vec{z}, A)$. The corresponding relations between initial and transition probabilities for the processes $\vec{V}_n$ and $\vec{Z}_n$ are given in Subsection 1.1.3.

Let also $\mathcal{F}_n = \sigma[\vec{V}_1, \dots, \vec{V}_n]$, $n = 0, 1, \dots$, be a natural filtration generated by the process $\vec{V}_n$. It coincides with the natural filtration $\mathcal{F}_n = \sigma[\vec{Z}_1, \dots, \vec{Z}_n]$, $n = 0, 1, \dots$, generated by the process $\vec{Z}_n$.

We also consider a pay-off function $g(n, \vec{s}, x)$ which is assumed to be a real-valued function defined for $(n, \vec{s}, x) \in \mathbb{N} \times \mathbb{R}_k^+ \times \mathbb{X}$, where $\mathbb{N} = \{0, 1, \dots\}$. We also assume that $g(n, \vec{s}, x)$ is a measurable function in argument $(\vec{s}, x) \in \mathbb{V}$ for every $n \in \mathbb{N}$.

The pay-off function can also be expressed in the equivalent form as $g(n, e^{\vec{y}}, x)$, and considered as a function defined for $(n, \vec{y}, x) \in \mathbb{N} \times \mathbb{R}_k \times \mathbb{X}$.

The American-type option contract is an agreement between two parties, a seller and a buyer. The contract has a price C payed by the buyer to the seller. The option contract guarantees to the buyer the possibility of executing the option contract in any moment $0 \le n \le N$ and of getting in this case the pay-off $g(n, \vec{S}_n, X_n) = g(n, e^{\vec{Y}_n}, X_n)$. The parameter $N \in \mathbb{N}$ is called a maturity.

2.1.2 Buyer aims and timing for analysis of an option contract

We analyze the contract from the position of the buyer. In this case, the American option contract can be analyzed either in the situation, when the contract is not yet bought by the buyer, or in the situation, when the contract was already bought by the buyer.

In the first case, one of the important questions is what should be the fair price of the contract?

In the second case, the goal of the buyer is to execute the contract in an optimal way according to some reasonable criteria of optimality.

One of such natural criterion is to execute the contract in the way that would maximize the expected pay-off (reward).

2.1.3 Stopping buyer's strategies

It is natural to assume that the decision of the buyer to execute or do not execute the contract at some moment $0 \leq n \leq N$ should be based only on the information about the price or log-price process and the modulating index up to moment n.

This corresponds to the assumption that the "stopping" strategy of the buyer is defined by a random stopping time τ that is a Markov moment for the price process $\vec{V}_n = (\vec{S}_n, X_n)$ or, that is equivalent, for the log-price process $\vec{Z}_n = (\vec{Y}_n, X_n)$.

This means that τ is an integer-valued random variable taking values $0, 1, \dots$, such that the random event $\{\tau = n\} \in \mathcal{F}_n$ for every $n = 0, 1, \dots$.

According the definition of the option contract, the Markov moment τ defining the stopping strategy of the buyer should also satisfy the inequality $0 \leq \tau \leq N$.

Let $\mathcal{M}_{\max,N}$ be the class of all Markov moments for the price process $\vec{V}_n$ such that $0 \leq \tau \leq N$.

The processes $\vec{V}_n$ and $\vec{Z}_n$ generate the same natural filtration $\mathcal{F}_n$, $n = 0, 1, \dots$, and, therefore, the same family $\mathcal{M}_{\max,N}$ of all Markov moments $0 \leq \tau \leq N$.

If the buyer decides to execute the option contract at a stopping moment $\tau \in \mathcal{M}_{\max,N}$, the corresponding pay-off (reward) will be a random variable $g(\tau, \vec{S}_\tau, X_\tau) = g(\tau, e^{\vec{Y}_\tau}, X_\tau)$.

The following quantity represents the expected reward corresponding to the stopping time τ:

$$\phi_\tau = \mathsf{E}g(\tau, \vec{S}_\tau, X_\tau) = \mathsf{E}g(\tau, e^{\vec{Y}_\tau}, X_\tau) . \tag{2.1}$$

2.1.4 Optimization problems connected with expected rewards

Let $\mathcal{M}_N \subseteq \mathcal{M}_{\max,N}$ be some class of Markov moments τ from the class $\mathcal{M}_{\max,N}$.

Objects of our interest are the functionals

$$\Phi(\mathcal{M}_N) = \sup_{\tau \in \mathcal{M}_N} \phi_\tau . \tag{2.2}$$

The first main optimization problem connected with the American option is to find or estimate the optimal expected reward $\Phi(\mathcal{M}_{\max,N})$.

But we shall also study functionals $\Phi(\mathcal{M}_N)$ for different classes $\mathcal{M}_N$. In particular, the class of hitting times is the subject of our special interest.

In order to simplify notations we denote

$$\Phi = \Phi(\mathcal{M}_{\max,N}) . \tag{2.3}$$

We assume that the following condition holds:

A$_1$: $\mathsf{E}|g(n, \vec{S}_n, X_n)| = \mathsf{E}|g(n, e^{\vec{Y}_n}, X_n)| < \infty , \quad n = 0, 1, \dots, N .$

In Chapter 4, we give different conditions, expressed in terms of one-step transition probabilities for price or log-price processes and pay-off functions, which are sufficient for the condition $\mathbf{A}_1$ to hold.

The following inequality holds for any random moment $0 \leq \tau \leq N$:

$$|g(\tau, \vec{S}_\tau, X_\tau)| \leq \max_{0 \leq n \leq N} |g(n, \vec{S}_n, X_n)| \leq \sum_{n=0}^{N} |g(n, \vec{S}_n, X_n)| . \tag{2.4}$$

The condition $\mathbf{A}_1$ guarantees that for any class of Markov moments $\mathcal{M}_N \subseteq \mathcal{M}_{\max,N}$,

$$\begin{aligned} |\Phi(\mathcal{M}_N)| &\leq \sup_{\tau \in \mathcal{M}_N} \mathsf{E}|g(\tau, \vec{S}_\tau, X_\tau)| \\ &\leq \mathsf{E} \max_{0 \leq n \leq N} |g(n, \vec{S}_n, X_n)| \leq \sum_{n=0}^{N} \mathsf{E}|g(n, \vec{S}_n, X_n)| < \infty . \end{aligned} \tag{2.5}$$

Thus, under the condition $\mathbf{A}_1$, the functionals ϕ_τ and $\Phi(\mathcal{M}_N)$, in particular $\Phi = \Phi(\mathcal{M}_{\max,N})$, are well defined.

Another important problem is to answer the question if there exists – and, if this the case, to find – the optimal stopping time that is the Markov moment $\tau^* \in \mathcal{M}_{\max,N}$ such that

$$\Phi = \sup_{\tau \in \mathcal{M}_{\max,N}} \phi_\tau = \phi_{\tau^*} . \tag{2.6}$$

2.2 Pay-off functions

In this section, we formulate general conditions, which we impose on pay-off functions of American-type options, and present a number of typical examples of pay-off functions.

2.2.1 General conditions imposed on pay-off functions

In this book, we study American-type options with the pay-off functions $g(n, \vec{s}, x)$ that have not more than polynomial rate of growth in the price argument $\vec{s} = (s_1, \dots, s_k) \in \mathbb{R}_k^+$ and are bounded in the time and index arguments, respectively, $0 \leq n \leq N$ and $x \in \mathbb{X}$.

This means that the following condition is assumed to hold for some vector parameter $\bar{\gamma} = (\gamma_1, \dots, \gamma_k)$ with nonnegative components:

$\dot{\mathbf{B}}_1[\bar{\gamma}]$: $\sup_{0 \leq n \leq N, (\vec{s}, x) \in \mathbb{V}} \frac{|g(n, \vec{s}, x)|}{1 + \sum_{i=1}^{k} K_{2,i}(s_i \vee s_i^{-1})^{\gamma_i}} < K_1$, for some and $0 \leq K_1, K_{2,1}, \dots, K_{2,k} < \infty$.

This condition can also be reformulated in the equivalent form, as the assumption that the pay-off function, expressed in the form $g(n, e^{\vec{y}}, x)$, has not more than exponential

rate of growth in argument $\vec{y} = (y_1, \ldots, y_k) \in \mathbb{R}_k$ and is bounded in arguments $0 \leq n \leq N$ and $x \in \mathbb{X}$:
$\mathbf{B}_1[\bar{\gamma}]$: $\sup_{0 \leq n \leq N, (\vec{y}, x) \in \mathbb{Z}} \frac{|g(n, e^{\vec{y}}, x)|}{1 + \sum_{i=1}^{k} K_{2,i} e^{\gamma_i |y_i|}} < K_1$, for some $0 \leq K_1, K_{2,1}, \ldots, K_{2,k} < \infty$.

2.2.2 Standard pay-off functions for call and put-type options

In what follows, the notation $[s]_+ = sI(s \geq 0)$ is used.

Let us describe first standard call and put contracts for univariate modulated price processes and then their different generalizations. We give below formulas for pay-off functions in more traditional terms of arguments $n, \vec{s}, x$. One can always to rewrite the corresponding formulas in terms of arguments $n, \vec{y}, x$ using the following transition relation:

$$g(n, \vec{s}, x) = g(n, e^{\vec{y}}, x), \quad \vec{s} = e^{\vec{y}}. \tag{2.7}$$

The corresponding measurability properties of pay-off functions penetrating the examples below are assumed by default.

Let us consider options for a modulated price process $V_n = (S_n, X_n)$ with a univariate price component S_n.

Pay-off functions that correspond to a standard call option and a standard put option are defined, for $n = 0, 1, \ldots, N$, $s \in \mathbb{R}_1^+$, $x \in \mathbb{X}$, by the following formulas:

$$g(n, s, x) = e^{-rn}[s - K]_+, \quad g(n, s, x) = e^{-rn}[K - s]_+, \tag{2.8}$$

where (a) $K \geq 0$ is a strike price for the corresponding option contact, (b) $r \geq 0$ is a free interest rate, and (c) $N \in \mathbb{N}$ is a maturity.

The first generalization is to consider inhomogeneous in time variants of such options. Pay-off functions that correspond to an inhomogeneous in time call option and an inhomogeneous in time put option are defined for $n = 0, 1, \ldots, N$, $s \in \mathbb{R}_1^+$, $x \in \mathbb{X}$ by the following formulas:

$$g(n, s, x) = e^{-R_n}[s - K_n]_+, \quad g(n, s, x) = e^{-R_n}[K_n - s]_+, \tag{2.9}$$

where (a) $K_n \geq 0$, $n = 0, 1, \ldots, N$ are strike prices, (b) $R_n = r_0 + \cdots + r_n \geq 0$, $n = 0, 1, \ldots, N$ are accumulated free interest rates and $r_n \geq 0$, $n = 0, 1, \ldots, N$ are instant free interest rates, and (c) N is a maturity.

The second generalization is to consider inhomogeneous in time variants of such options with strike prices depending on the index argument. Pay-off functions that correspond to an inhomogeneous in time modulated call option and an inhomogeneous in time modulated put option are defined for $n = 0, 1, \ldots, N$, $s \in \mathbb{R}_1^+$, $x \in \mathbb{X}$ by the following formulas:

$$g(n, s, x) = e^{-R_n}[s - K_n(x)]_+, \quad g(n, s, x) = e^{-R_n}[K_n(x) - s]_+, \tag{2.10}$$

where (a) $0 \leq K_n(x) \leq K < \infty, \quad x \in \mathbb{X}, n = 0, 1, \dots, N$ are strike prices, (b) $R_n = r_0 + \cdots + r_n \geq 0$, $n = 0, 1, \dots, N$ are accumulated free interest rates and $r_n \geq 0$, $n = 0, 1, \dots, N$ are instant free interest rates, and (c) N is a maturity.

It should be noted that we consider option models with pay-off functions depending only on instant values of price and index arguments. For such models, it is not possible to make the instant free interest rate to be dependent on the index argument, Indeed, the accumulated free interest rate would, in this case, depend on the whole sequence of index argument values at moments $n = 0, 1, \dots, N$.

It is also possible to combine option models described above by considering mixtures (portfolios) of the above option contracts. In this case, a pay-off function for an option portfolio is defined for $n = 0, 1, \dots, N$, $s \in \mathbb{R}_1^+$, $x \in \mathbb{X}$ by the following formula:

$$g(n, s, x) = e^{-R_n} \sum_{i=1}^{m} (a'_i(n, x)[s - K'_i(n, x)]_+ \\ + a''_i(n, x)[K''_i(n, x) - s]_+) , \tag{2.11}$$

where (a) $0 \leq K'_i(n, x), K''_i(n, x) \leq K < \infty, \quad x \in \mathbb{X}, n = 1, \dots, N, i = 1, \dots, m$ are strike prices, (b) $R_n = r_0 + \cdots + r_n$, $n = 0, 1, \dots, N$ are accumulated free interest rates and $r_n \geq 0$, $n = 0, 1, \dots, N$ are instant free interest rates, (c) $-A \leq a'_i(n, x)$, $a''_i(n, x) \leq A < \infty$, $n = 1, \dots, N$, $x \in \mathbb{X}$, $i = 1, \dots, m$ are waiting coefficients for the corresponding option subcontracts, and (d) N is a maturity.

Let us consider options for a multivariate modulated price process $\vec{V}_n = (\vec{S}_n, X_n)$.

Let us first consider a model, where an option contract is based on a linear mixture of price processes

$$S_{n,\vec{c}} = (\vec{S}_n, \vec{c}) , \quad n = 0, 1, \dots . \tag{2.12}$$

Here, the notation $(\vec{c}, \vec{s}) = c_1 s_1 + \cdots + c_k s_k$ is used for the scalar product of vectors $\vec{c} = (c_1, \dots, c_k)$ and $\vec{s} = (s_1, \dots, s_k)$.

In this case, pay-off functions that correspond to a call option and a put option are defined for $n = 0, 1, \dots, N$, $\vec{s} \in \mathbb{R}_k^+$, $x \in \mathbb{X}$ by the following formulas:

$$g(n, \vec{s}, x) = e^{-rn}[(\vec{c}, \vec{s}) - K]_+ , \quad g(n, \vec{s}, x) = e^{-rn}[K - (\vec{c}, \vec{s})]_+ , \tag{2.13}$$

where (a) $K \geq 0$ is a strike price, (b) $r \geq 0$ is a free interest rate, (c) $c_i \geq 0$, $i = 1, \dots, k$ are waiting coefficients for the corresponding components of the vector price argument, and (d) N is a maturity.

The second example is connected with the model, where an option contract is based on the maximum of price processes

$$S_n^+ = \max(S_{n,1}, \dots, S_{n,k}) , \quad n = 0, 1, \dots . \tag{2.14}$$

In this case, pay-off functions that correspond to a call option and a put option are defined for $n = 0, 1, \dots, N$, $\vec{s} \in \mathbb{R}_k^+$, $x \in \mathbb{X}$ by the following formula:

$$g(n, \vec{s}, x) = \begin{cases} e^{-rn}[\max(s_1, \dots, s_k) - K]_+, & \text{for a call option}, \\ e^{-rn}[K - \max(s_1, \dots, s_k)]_+, & \text{for a put option}, \end{cases} \tag{2.15}$$

where (a) $K \geq 0$ is a strike price, (b) $r \geq 0$ is a free interest rate, and (c) N is a maturity.

The third example is connected with a model, where an option contract is defined as a mixture (portfolio) of standard option contracts for the components of a multivariate price process. In this case, a pay-off function is defined for $n = 0, 1, \dots, N$, $\vec{s} = (s_1, \dots, s_k) \in \mathbb{R}_k^+$, $x \in \mathbb{X}$ by the following formula:

$$g(n, \vec{s}, x) = \sum_{l=1}^{k} e^{-r_l n} \sum_{i=1}^{m} (a'_{l,i}[s_l - K'_{l,i}]_+ + a''_{l,i}[K''_{l,i} - s_l]_+), \tag{2.16}$$

where (a) $K'_{l,i}, K''_{l,i} \geq 0$, $l = 1, \dots, k$, $i = 1, \dots, m$ are strike prices,, (b) $r_l \geq 0$, $l = 1, \dots, k$ are free interest rates, (c) $a'_{l,i}, a''_{l,i} \in \mathbb{R}_1$, $l = 1, \dots, i = 1, \dots, m$ are the waiting coefficients for the corresponding option subcontracts, and (d) N is a maturity.

Generalizations described above namely option contracts inhomogeneous in time and strike and waiting coefficients depending on index argument, as well as mixtures of option contracts can also be considered for the above model.

In all the above examples, pay-off functions are piecewise linear functions in the price argument $\vec{s} = (s_1, \dots, s_k)$.

In the case of the call option, the pay-off function is nonnegative and nondecreasing in s_i, $i = 1, \dots, K$. Also, the condition $\dot{\mathbf{B}}_1[\bar{1}]$ holds.

In the case of the put option the pay-off function is nonnegative and nonincreasing in s_i, $i = 1, \dots, k$. Also, the condition $\dot{\mathbf{B}}_1[\bar{0}]$ holds.

In the case of a mixture of the call and put options the nonnegativity and monotonicity properties may not hold. Also, if the mixture includes only put options, then the condition $\dot{\mathbf{B}}_1[\bar{0}]$ holds. If the mixture includes at least one call option, then the condition $\dot{\mathbf{B}}_1[\bar{1}]$ holds.

2.2.3 Nonlinear pay-off functions for call and put-type options

Let us consider options for a modulated price process $V_n = (S_n, X_n)$ with a univariate price component S_n.

In such a model, pay-off functions that correspond to a call option and a put option are defined for $n = 0, 1, \dots, N$, $s \in \mathbb{R}_1^+$, $x \in \mathbb{X}$ by the following formulas:

$$g(n, s, x) = e^{-rn}[f_n(s) - K]_+, \quad g(n, s, x) = e^{-rn}[K - f_n(s)]_+, \tag{2.17}$$

where (a) $K \geq 0$, is a strike price, (b) $f_n(s) \geq 0$, $s \in \mathbb{R}_1^+$, $n = 0, 1, \dots, N$ are some transformation functions such that $f_n(s) \leq K'_1 + K'_2 s^\gamma$, $n = 0, 1, \dots, N$, where $0 \leq K'_1, K'_2 < \infty$, $0 \leq \gamma < \infty$, (c) $r \geq 0$ is a free interest rate, and (d) N is a maturity.

The typical examples of the transformation functions are $f_n(s) = a_n + b_n s + c_n s^2$, where $a_n, b_n, \geq 0, c_n > 0$, or $f_n(s) = a_n s^\gamma$, where $a_n > 0, \gamma > 0$.

It is also worth to note that models with nonlinear functions $f_n(s)$ can also appear as a result of transformation of a price process or a log-price process.

Let $h_n(s)$ be, for every $n = 0, 1, \ldots$, a strictly monotonic function, which maps one-to-one the interval $\mathbb{R}_1^+$ to some open interval (s_n^-, s_n^+), where $0 \leq s_n^- < s_n^+ \leq \infty$. In this case, there exists the inverse function $f_n(s) = h_n^{-1}(s)$ such that $f_n(h_n(s)) = s$, $s \in \mathbb{R}_1^+$.

In some cases, the price process S_n, $n = 0, 1, \ldots$, can be simplified if one transforms it in the process $S_n^{(h)} = h_n(S_n)$, $n = 0, 1, \ldots$. The random pay-off process $g(n, S_n, X_n)$, $n = 0, 1, \ldots$, is transformed in this case as $g(n, S_n, X_n) = g(n, f_n(h_n(S_n)), X_n) = g^{(f)}(n, S_n^{(h)}, X_n)$, $n = 0, 1, \ldots$, where $g^{(f)}(n, s, x) = g(n, f_n(s), x)$ is a transformed pay-off function.

Generalizations described above, namely, portfolios of option contracts with inhomogeneous in time and strike and waiting coefficients depending on the index argument can also be considered for the above model.

In this case, a pay-off function is defined for $n = 0, 1, \ldots, N$, $\vec{s} \in \mathbb{R}_k^+$, $x \in \mathbb{X}$ by the following formula:

$$\begin{aligned} g(n, s, x) = e^{-R_n} \sum_{i=1}^{m} (a_i'(n, x)[f_i'(n, s, x) - K_i'(n, x)]_+ \\ + a_i''(n, x)[K_i''(n, x) - f_i''(n, s, x)]_+), \end{aligned} \tag{2.18}$$

where (a) $0 \leq K_i'(n, x), K_i''(n, x) \leq K < \infty$, $x \in \mathbb{X}$, $n = 1, \ldots, N$, $i = 1, \ldots, m$ are strike prices, (b) $R_n = r_0 + \cdots + r_n \geq 0$, $n = 0, 1, \ldots, N$ are accumulated free interest rates and $r_n \geq 0$, $n = 0, 1, \ldots, N$ are instant free interest rates, (c) $f_i'(n, s, x)$, $f_i''(n, s, x) \geq 0$, $n = 0, 1, \ldots, N$, $s \in \mathbb{R}_1^+, x \in \mathbb{X}$, $i = 1, \ldots, m$ are the measurable transformation functions such that $f_i(n, s, x) \leq K_1' + K_2' s^\gamma$, $n = 0, 1, \ldots, N$, $s \in \mathbb{R}_1^+, x \in \mathbb{X}$, $i = 1, \ldots, m$, where $0 \leq K_1', K_2' < \infty, 0 \leq \gamma < \infty$, (d) $-A \leq a_i'(n, x), a_i''(n, x) \leq A < \infty$, $x \in \mathbb{X}$, $n = 1, \ldots, N$, $i = 1, \ldots, m$ are waiting coefficients for the corresponding option subcontracts, and (e) N is a maturity.

Analogous examples of options with nonlinear pay-off functions and portfolios of such options for multivariate modulated price processes can also be given.

2.2.4 Pay-off functions for exchange of assets option-type contracts

Another type of option-type contracts is connected with the model of exchange of assets.

Let us first consider the simplest case, where $\vec{V}_n = (\vec{S}_n, X_n)$, where $\vec{S}_n = (S_{n,1}, S_{n,2})$, is a two-dimensional price process without an index component.

In this case, a pay-off function is defined for $\vec{s} = (s_1, s_2) \in \mathbb{R}_2^+, x \in \mathbb{X}, n = 0, 1, \ldots, N$ by the following formula:

$$g(n, \vec{s}) = e^{-rn}(s_1 - s_2)\,, \tag{2.19}$$

where (a) $r \geq 0$ is a free interest rate and (b) N is a maturity.

A natural generalization is connected with the model of exchange of k assets, where a modulated price process $\vec{V}_n = (\vec{S}_n, X_n)$ has a k-dimensional price component $\vec{S}_n = (S_{n,1}, \ldots, S_{n,k})$.

In the simplest case, a pay-off function does not depend on the index argument and is defined for $\vec{s} = (s_1, \ldots, s_k) \in \mathbb{R}_k^+, x \in \mathbb{X}, n = 0, 1, \ldots, N$ by the following formula:

$$g(n, \vec{s}, x) = e^{-rn} \sum_{i=1}^{k} a_i s_i\,, \tag{2.20}$$

where (a) $r \geq 0$ is a free interest rate, (b) $a_i \in \mathbb{R}_1, i = 1, \ldots, k$ are exchange waiting coefficients, and (c) N is a maturity.

The further generalization is connected with the model of exchange of k assets, where the waiting coefficients depend on the index argument. In this case, a pay-off function is defined for $n = 0, 1, \ldots, N, \vec{s} = (s_1, \ldots, s_k) \in \mathbb{R}_k^+, x \in \mathbb{X}$, by the following formula:

$$g(n, \vec{s}, x) = e^{-rn} \sum_{i=1}^{k} a_i(x) s_i\,, \tag{2.21}$$

where (a) $r \geq 0$ is a free interest rate,, (b) $-A \leq a_i(x) \leq A < \infty, x \in \mathbb{X}, i = 1, \ldots, k$ are exchange waiting coefficients, and (c) N is a maturity.

Also the generalization connected with inhomogeneous in time variant of such contracts is possible. In this case, a pay-off function is defined for $n = 0, 1, \ldots, N$, $\vec{s} = (s_1, \ldots, s_k) \in \mathbb{R}_k^+, x \in \mathbb{X}$, by the following formula:

$$g(n, \vec{s}, x) = e^{-R_n} \sum_{i=1}^{k} a_i(n, x) s_i\,, \tag{2.22}$$

where (a) $R_n = r_0 + \cdots + r_n, n = 0, 1, \ldots, N$ are accumulated free interest rates and $r_n \geq 0, n = 0, 1, \ldots, N$ are instant free interest rates, (b) $-A \leq a_i(n, x) \leq A < \infty, x \in \mathbb{X}, i = 1, \ldots, k, n = 0, 1, \ldots, N$ are exchange waiting coefficients, and (c) N is a maturity.

In exchange asset contracts, pay-off functions are again linear functions in argument $\vec{s}$. The nonnegativity and monotonicity properties may not hold.

In the above models, the condition $\dot{\mathbf{B}}_1[\bar{1}]$ obviously holds for pay-off functions.

As above, generalizations described above namely exchange of assets option contracts inhomogeneous in time, options with parameters depending on the index argument, and portfolios of such options for multivariate modulated price processes can also be considered.

2.2.5 Pay-off functions for digital-type options

The further examples are connected with the more exotic pay-off functions $g(n, \vec{s}, x)$. For example, so-called digital-type contracts are based on stepwise pay-off functions.

Let us restrict consideration by the simplest case, a price process $V_n = (S_n, X_n)$ has a univariate price component S_n.

In this case, a pay-off function is defined for $n = 0, 1, \dots, N$, $s \in \mathbb{R}_1^+$, $x \in \mathbb{X}$, by the following formula:

$$g(n, s, x) = e^{-rn} d(s, x)\,, \qquad (2.23)$$

where (a) $r \geq 0$ is a free interest rate, (b) $d(s, x) = \sum_{i=1}^{m} d_i(x) I(e_{i-1}(x) < s \leq e_i(x))$ is a nondiscounted stepwise (in the price argument s) pay-off function, where $-D \leq d_i(x) \leq D < \infty$, $i = 1, \dots, m$, $0 = e_0(x) < e_1(x) < \cdots < e_m(x) = \infty$, and (c) N is a maturity.

In the above model, the condition $\dot{\mathbf{B}}_1[\bar{0}]$ holds, since the pay-off functions for digital option contracts are always bounded.

As above, generalizations described above namely digital option contracts inhomogeneous in time, options with parameters of stepwise pay-off functions depending on the index argument, and portfolios of digital options for multivariate modulated price processes can also be considered for the above model.

Taking into account the great variety of different option-type contracts, we shall try to get results for general pay-off functions $g(n, s, x)$ satisfying minimal continuity conditions as well as conditions on the rate of growth for these functions in price arguments.

2.3 Reward and log-reward functions

In this section, we introduce so-called log-reward functions for American-type options for discrete time multivariate modulated Markov price and log-price processes.

2.3.1 Conditional expectations for random functionals defined on trajectories of Markov price processes

Let $\vec{V}_n = (\vec{S}_n, X_n)$ be a discrete time multivariate modulated Markov price process with a phase space $\mathbb{V}$, initial distribution $\dot{P}_0(B)$ and the transition probabilities $\dot{P}_n(\vec{v}, B)$.

Let us introduce σ-algebras $\mathcal{F}_{n,m} = \sigma[\vec{V}_n, \dots, \vec{V}_m]$, $0 \leq n \leq m < \infty$.

Let $0 \leq n \leq N$ and $\dot{W}_n$ be a real-valued random variable adapted to the σ-algebra $\mathcal{F}_{n,N}$, i.e. random variable such that events $\{\dot{W}_n \in C\} \in \mathcal{F}_{n,N}$, for any $C \in \mathcal{B}_1$.

As is known, the random variable $\dot{W}_n$ can be represented in the form $\dot{W}_n = \dot{f}_n(\vec{V}_n, \dots, \vec{V}_N)$, where $\dot{f}_n(\vec{v}_n, \dots, \vec{v}_N)$ is a measurable real-valued function acting from the space $\mathbb{V}^{(N-n+1)} = \mathbb{V} \times \cdots \times \mathbb{V}$ to $\mathbb{R}_1$ (here $\mathbb{V}^{(r)} = \mathbb{V} \times \cdots \times \mathbb{V}$ is r-times product of the space $\mathbb{V}$). Thus, $\dot{W}_n$ can be considered as a random functional defined on trajectories of the price process V_l.

In this case, there exists the conditional distribution $\dot{\mathsf{P}}_{\vec{v},n}\{\dot{W}_n \in C\}$ of the random variable $\dot{W}_n$ under the condition $\vec{V}_n = \vec{v}$, for $\vec{v} \in \mathbb{V}$. It is defined for $C \in \mathcal{B}_1$ by the following relation:

$$\begin{aligned}\dot{\mathsf{P}}_{\vec{v},n}\{\dot{W}_n \in C\} &= \mathsf{P}\{\dot{W}_n \in C/\vec{V}_n = \vec{v}\}\\ &= \int_{\mathbb{V}} \dot{P}_{n+1}(\vec{v}, d\vec{v}_{n+1}) \int_{\mathbb{V}} \dot{P}_{n+2}(\vec{v}_{n+1}, d\vec{v}_{n+2})\\ &\quad \cdots \int_{\mathbb{V}} I(\dot{f}_n(\vec{v}, \vec{v}_{n+1}, \ldots, \vec{v}_N) \in C)\dot{P}_N(\vec{v}_{N-1}, d\vec{v}_N)\,. \end{aligned} \tag{2.24}$$

Here and henceforth, we use the notation $\dot{\mathsf{P}}_{\vec{v},n}$ for conditional probabilities under the condition $\vec{V}_n = \vec{v}$.

This conditional distribution is a probability measure, as a function of $C \in \mathcal{B}_1$, for every $\vec{v} \in \mathbb{V}$, and a measurable function of argument $\vec{v} \in \mathbb{V}$, for every $C \in \mathcal{B}_1$.

Let us consider the set of integrability for the random variable W_n,

$$\dot{I}_{\dot{W}_n} = \left\{\vec{v} \in \mathbb{V} : \dot{\mathsf{E}}_{\vec{v},n}|\dot{W}_n| = \int_{\mathbb{R}_1} |w|\dot{\mathsf{P}}_{\vec{v},n}\{\dot{W}_n \in dw\} < \infty\right\}. \tag{2.25}$$

Here and henceforth we use the notation $\dot{\mathsf{E}}_{\vec{v},n}$ for conditional expectations under the condition $\vec{V}_n = \vec{v}$.

The set $\dot{I}_{\dot{W}_n} \in \mathcal{B}_{\mathbb{V}}$. In particular, the set $\dot{I}_{\dot{W}_n}$ can be empty or coincide with $\mathbb{V}$.

The conditional expectation of the random variable $\dot{W}_n$ under the condition $\vec{V}_n = \vec{v}$ is defined for $\vec{v} \in \dot{I}_{\dot{W}_n}$,

$$\begin{aligned}\dot{\mathsf{E}}_{\vec{v},n}\dot{W}_n &= \int_{\mathbb{R}_1} w\dot{\mathsf{P}}_{\vec{v},n}\{\dot{W}_n \in d\vec{v}\} = \dot{\mathsf{E}}_{\vec{v},n}\dot{f}_n(\vec{V}_n, \ldots, \vec{V}_N)\\ &= \int_{\mathbb{V}} \dot{P}_{n+1}(\vec{v}, d\vec{v}_{n+1}) \int_{\mathbb{V}} \dot{P}_{n+2}(\vec{v}_{n+1}, d\vec{v}_{n+2})\\ &\quad \cdots \int_{\mathbb{V}} \dot{f}_n(\vec{v}, \vec{v}_{n+1}, \ldots, \vec{v}_N)\dot{P}_N(\vec{v}_{N-1}, d\vec{v}_N)\,. \end{aligned} \tag{2.26}$$

In fact, the assumption $\vec{v} \in \dot{I}_{\dot{W}_n}$ enables one to first obtain the analog of relation (2.26) for the random variable $|\dot{W}_n| = |\dot{f}_n(\vec{V}_n, \ldots, \vec{V}_N)|$. This makes it possible to obtain relation (2.26) and to get the finite value for $\dot{\mathsf{E}}_{\vec{v},n}\dot{W}_n$ given by the expression on the right-hand side of this relation.

Let us assume that $\mathsf{E}|\dot{W}_n| < \infty$.

This condition does not imply that the set $\dot{I}_{\dot{W}_n} = \mathbb{V}$. But this assumption does imply that $\mathsf{P}\{\vec{V}_n \in \dot{I}_{\dot{W}_n}\} = 1$. Otherwise, the expectation $\mathsf{E}|\dot{W}_n|$ should be equal to ∞.

Note also that the assumption $\dot{I}_{\dot{W}_n} = \mathbb{V}$ does not imply that $\mathsf{E}|\dot{W}_n| < \infty$.

Under the assumption $\mathsf{E}|\dot{W}_n| < \infty$, the following relation takes place:

$$\mathsf{E}\dot{W}_n = \int_{\mathbb{V}} \dot{\mathsf{E}}_{\vec{v},n}\dot{W}_n \, \mathsf{P}\{\vec{V}_n \in d\vec{v}\}$$
$$= \int_{\mathbb{V}} \dot{P}_0(d\vec{v}_0) \int_{\mathbb{V}} \dot{\mathsf{E}}_{\vec{v},n}\dot{W}_n \dot{P}_{\vec{v}_0,0}\{\vec{V}_n \in d\vec{v}\} \, . \tag{2.27}$$

As above, the assumption $\mathsf{E}|\dot{W}_n| < \infty$ enables one first to obtain the analog of relation (2.27) for the random variable $|\dot{W}_n| = |\dot{f}_n(\vec{V}_n, \dots, \vec{V}_N)|$. This makes it possible to get relation (2.27) and to get the finite value for $\dot{\mathsf{E}}\dot{W}_n$ given by the expression on the right-hand side of this relation.

Note also that one should additionally define the function $\dot{\mathsf{E}}_{\vec{v},n}\dot{W}_n$ for $\vec{v} \in \overline{\dot{I}}_{\dot{W}_n}$, in order to be able to get relation (2.27). Since $\dot{\mathsf{P}}\{\vec{V}_n \in \overline{\dot{I}}_{\dot{W}_n}\} = 0$, this can be done in an arbitrary "measurable" way because of the Lebesgue integral over the set $\overline{\dot{I}}_{\dot{W}_n}$ will be any way equal to 0. The standard way is to define $\dot{\mathsf{E}}_{\vec{v},n}\dot{W}_n = 0$, for $\vec{v} \in \overline{\dot{I}}_{\dot{W}_n}$.

2.3.2 Reward functions for discrete time multivariate modulated Markov price processes

Let us define reward functions of an American option for a multivariate modulated Markov price process $\vec{V}_n = (\vec{S}_n, X_n)$.

Let us introduce, for $0 \leq n \leq N$, the class $\mathcal{M}_{\max,n,N}$ of all Markov moments τ for the Markov process $\vec{V}_n$ such that (a) $n \leq \tau \leq N$ and (b) event $\{\tau = m\} \in \mathcal{F}_{n,m} = \sigma[\vec{V}_n, \dots, \vec{V}_m], n \leq m \leq N$.

Obviously, the class $\mathcal{M}_{\max,0,N} = \mathcal{M}_{\max,N}$.

Let us assume the following condition:

A₂: $\dot{\mathsf{E}}_{\vec{v},n}|g(r, \vec{S}_r, X_r)| < \infty, \ \vec{v} \in \mathbb{V}, \ 0 \leq n \leq r \leq N$.

In Chapter 4, we give a number of conditions, expressed in terms of one-step transition probabilities of price or log-price processes and pay-off functions, which are sufficient for the condition **A₂** to hold.

The following inequality obviously holds, for every Markov moment $\tau_n \in \mathcal{M}_{\max,n,N}$, $n = 0, 1, \dots, N$:

$$|g(\tau_n, \vec{S}_{\tau_n}, X_{\tau_n})| \leq \max_{n \leq r \leq N} |g(r, \vec{S}_r, X_r)| \leq \sum_{r=n}^{N} |g(r, \vec{S}_r, X_r)| \, . \tag{2.28}$$

and, thus, by the condition **A₂**, the following inequality holds, for $\vec{v} \in \mathbb{V}$, $n = 0, 1, \dots, N$:

$$\dot{\mathsf{E}}_{\vec{v},n}|g(\tau_n, \vec{S}_{\tau_n}, X_{\tau_n})| \leq \sum_{r=n}^{N} \dot{\mathsf{E}}_{\vec{v},n}|g(r, \vec{S}_r, X_r)| < \infty \, . \tag{2.29}$$

Due to (2.29), the following finite measurable functions can be defined for $n = 0, 1, \dots, N$:

$$\dot{\phi}_{n,\tau_n}(\vec{v}) = \dot{\phi}_{n,\tau_n}(\vec{s}, x) = \dot{\mathsf{E}}_{\vec{v},n} g(\tau_n, \vec{S}_{\tau_n}, X_{\tau_n}), \quad \vec{v} \in \mathbb{V}. \tag{2.30}$$

Note that $g(\tau_n, \vec{S}_{\tau_n}, X_{\tau_n})$ is a random variable adapted to the σ-algebra $\mathcal{F}_{n,N}$. Thus, it can be represented in the form $g(\tau_n, \vec{S}_{\tau_n}, X_{\tau_n}) = \dot{f}_n(\vec{V}_n, \dots, \vec{V}_N)$, where $\dot{f}_n(\vec{v}_n, \dots, \vec{v}_N)$ is a measurable real-valued function defined on the space $\mathbb{V}^{(N-n+1)}$. This function has, in this case, a specific form. Since τ_n is a Markov moment taking values $n, \dots, N$, there exists the sequence of sets $B_{n,r} \in \mathcal{B}_{\mathbb{V}^{(r-n+1)}}$, $r = n, \dots N$ such that the event $\{\tau_n = r\} = \{(\vec{V}_n, \dots, \vec{V}_r) \in B_{n,r}\}$, for every $r = n, \dots, N$, and $\cup_{r=n}^N \{\tau_n = r\} = \cup_{r=n}^N \{(\vec{V}_n, \dots, \vec{V}_r) \in B_{n,r}\} = \Omega$ (here $\langle \Omega, \mathcal{F}, \mathsf{P} \rangle$ is a probability space where the process $\vec{V}_l$ is defined). Therefore, $g(\tau_n, \vec{S}_{\tau_n}, X_{\tau_n}) = \sum_{r=n}^N g(r, \vec{S}_r, X_r) I((\vec{V}_n, \dots, \vec{V}_r) \in B_{n,r})$, and, thus, $\dot{f}_n(\vec{v}_n, \dots, \vec{v}_N) = \sum_{r=n}^N g(r, \vec{s}_r, x_r) I((\vec{v}_n, \dots, \vec{v}_r) \in B_{n,r})$, where $\vec{v}_r = (\vec{s}_r, x_r)$, $r = n, \dots, N$.

The above remarks and relation (2.26) imply that the functions $\dot{\phi}_{n,\tau_n}(\vec{v})$, $n = 0, \dots, N$ are indeed measurable and that the following relation takes place, for $\vec{v} \in \mathbb{V}$, $n = 0, \dots, N$:

$$\begin{aligned} \dot{\phi}_{n,\tau_n}(\vec{v}) = \dot{\phi}_{n,\tau_n}(\vec{s}, x) &= \dot{\mathsf{E}}_{\vec{v},n} g(\tau_n, \vec{S}_{\tau_n}, X_{\tau_n}) \\ &= \sum_{r=n}^N \dot{\mathsf{E}}_{\vec{v},n} g(r, \vec{S}_r, X_r) I((\vec{V}_n, \dots, \vec{V}_r) \in B_{n,r}). \end{aligned} \tag{2.31}$$

Let us also define, for every $n = 0, 1, \dots, N$, the reward function for the American option

$$\dot{\phi}_n(\vec{v}) = \dot{\phi}_n(\vec{s}, x) = \sup_{\tau_n \in \mathcal{M}_{\max,n,N}} \dot{\phi}_{n,\tau_n}(n, \vec{v}), \quad \vec{v} = (\vec{s}, x) \in \mathbb{V}. \tag{2.32}$$

The reward functions $\dot{\phi}_n(\vec{v})$, $n = 0, \dots, N$ are finite functions. Indeed, inequalities (2.29) also imply that for every $\vec{v} \in \mathbb{V}$, $n = 0, 1, \dots, N$,

$$\begin{aligned} |\dot{\phi}_n(\vec{v})| &\le \sup_{\tau_n \in \mathcal{M}_{\max,n,N}} \dot{\mathsf{E}}_{\vec{v},n} |g(\tau_n, \vec{S}_{\tau_n}, X_{\tau_n})| \\ &\le \sum_{r=n}^N \dot{\mathsf{E}}_{\vec{v},n} |g(r, \vec{S}_r, X_r)| < \infty. \end{aligned} \tag{2.33}$$

The following theorem plays an important role in what follows.

Theorem 2.3.1. *Let the condition* $\mathbf{A}_2$ *hold. Then, the reward functions* $\dot{\phi}_n(\vec{v}) = \dot{\phi}_n(\vec{s}, x)$, $n = 0, 1, \dots, N$ *are the measurable functions of* $\vec{v} = (\vec{s}, x) \in \mathbb{V}$ *and the following backward recurrence relations hold, for every* $\vec{v}_n = (\vec{s}_n, x_n) \in \mathbb{V}$, $n = 0, 1, \dots, N$:

$$\begin{cases} \dot{\phi}_N(\vec{v}) = g(N, \vec{s}, x), \quad \vec{v} = (\vec{s}, x) \in \mathbb{V}, \\ \dot{\phi}_r(\vec{v}) = \max(g(r, \vec{s}, x), \dot{\mathsf{E}}_{\vec{v},r} \dot{\phi}_{r+1}(\vec{V}_{r+1})), \quad \vec{v} = (\vec{s}, x) \in \mathbb{V}, \\ \qquad r = N-1, \dots, n+1, \\ \dot{\phi}_n(\vec{v}_n) = \max(g(n, \vec{s}_n, x_n), \dot{\mathsf{E}}_{\vec{v}_n,n} \dot{\phi}_{n+1}(\vec{V}_{n+1})). \end{cases} \tag{2.34}$$

Theorem 2.3.1 gives a general backward reward algorithm for finding the reward functions for discrete time price processes. The algorithm is based on the recurrence relations (2.34).

This algorithm includes two rather complicated operations that are maximization and integration. This make the algorithm not too effective for general Markov price processes.

However, for some simple Markov price processes such as Bernoulli and other types of discrete random walks, the algorithm can be effectively used for finding reward functions. We present the corresponding results in Section 1.3.

Moreover, some more general discrete time Markov price processes can be approximated by simpler price processes of random walk types. The corresponding convergence results for options rewards are presented in Sections 1.5 and 1.6. This makes it possible to use the corresponding backward algorithms for finding approximations for reward functions for various discrete time Markov price processes.

Let the conditions $\mathbf{A}_1$ and $\mathbf{A}_2$ hold.

Recall the optimal expected reward $\Phi = \Phi(\mathcal{M}_{\max,N})$ introduced in Subsection 1.2.1,

$$\Phi = \sup_{\tau_0 \in \mathcal{M}_{\max,N}} \mathsf{E} g(\tau_0, \vec{S}_{\tau_0}, X_{\tau_0}) . \tag{2.35}$$

Relations (2.27) and (2.35) imply that the following relation takes place:

$$\Phi = \sup_{\tau_0 \in \mathcal{M}_{\max,N}} \int_{\mathbb{V}} \dot{\mathsf{E}}_{\vec{v},n} g(\tau_0, \vec{S}_{\tau_0}, X_{\tau_0}) \dot{P}_0(d\vec{v}) . \tag{2.36}$$

The following theorem shows that one can exchange operation of maximization and integration in formula (2.36) when computing the optimal expected reward Φ. After this exchange, computing of the expectation and then the supremum under the sign of integral yields, according to relation (2.32), the reward function $\dot{\phi}_0(\vec{v})$. Thus, the optimal expected reward Φ can be computed by integration of this reward function by the initial distribution $\dot{P}_0(B)$.

Theorem 2.3.2. *Let the conditions $\mathbf{A}_1$ and $\mathbf{A}_2$ hold. Then, the optimal expected reward Φ and the reward function $\phi_0(\vec{v})$ are connected by the following formula:*

$$\Phi = \mathsf{E}\dot{\phi}_0(\vec{V}_0) = \int_{\mathbb{V}} \dot{\phi}_0(\vec{v}) \dot{P}_0(d\vec{v}) . \tag{2.37}$$

The proofs of Theorems 2.3.1 and 2.3.2 are given in Section 2.4.

2.3.3 Conditional expectations for random functional defined on trajectories of discrete time log-price processes

Let $\vec{Z}_n = (\vec{Y}_n, X_n)$ be a discrete time multivariate modulated Markov log-price process with a phase space $\mathbb{Z}$, an initial distribution $P_0(B)$ and transition probabilities $P_n(\vec{z}, B)$.

The relations given in Subsection 2.3.1 can be rewritten in terms of the log-price process $\vec{Z}_n = (\vec{Y}_n, X_n)$ connected with the price process $\vec{V}_n = (\vec{S}_n, X_n)$ by the relation $\vec{S}_n = e^{\vec{Y}_n}$, $n = 0, 1, \ldots$.

Recall that the initial distribution and the transition probabilities for the log-price $\vec{Z}_n$ and the price process $\vec{V}_n$ are connected by the following relations, for $\vec{z} = (\vec{y}, x) \in \mathbb{Z}$, $A \in \mathcal{B}_{\mathbb{Z}}$, $n = 1, 2, \ldots$:

$$P_0(A) = \dot{P}_0(B_A)\,, \quad P_n(\vec{z}, A) = \dot{P}_n(\vec{v}_{\vec{z}}, B_A)\,, \tag{2.38}$$

where $\vec{v}_{\vec{z}} = (e^{\vec{y}}, x) \in \mathbb{V}$, $B_A = \{\vec{v}_{\vec{z}} : \vec{z} \in A\} \in \mathcal{B}_{\mathbb{V}}$.

The log-price process $\vec{Z}_n$ and the price process $\vec{V}_n$ obviously generate the same σ-algebras $\mathcal{F}_{n,m} = \sigma[\vec{Z}_n, \ldots, \vec{Z}_m] = \sigma[\vec{V}_n, \ldots, \vec{V}_m]$, $0 \le n \le m \le N$.

Let us again assume that $0 \le n \le N$ and W_n be a real-valued random variable adapted to the σ-algebra $\mathcal{F}_{n,N}$.

In this case, the random variable W_n can be represented in the form $W_n = f_n(\vec{Z}_n, \ldots, \vec{Z}_N)$, where $f_n(\vec{z}_n, \ldots, \vec{z}_N)$ is a measurable real-valued function acting from the space $\mathbb{Z}^{(N-n+1)} = \mathbb{Z} \times \cdots \times \mathbb{Z}$ to $\mathbb{R}_1$ (here $\mathbb{Z}^{(r)} = \mathbb{Z} \times \cdots \times \mathbb{Z}$ is the r-times product of the space $\mathbb{Z}$). Thus, W_n can be considered as a random functional defined on trajectories of the log-price process Z_l.

In this case, there exists the conditional distribution $\mathsf{P}_{\vec{z},n}\{W_n \in C\}$ of the random variable W_n under the condition $\vec{Z}_n = \vec{z}$, for $\vec{z} \in \mathbb{Z}$. It is defined for $C \in \mathcal{B}_1$ by the following relation:

$$\begin{aligned}\mathsf{P}_{\vec{z},n}\{W_n \in C\} &= \mathsf{P}\{W_n \in C / \vec{Z}_n = \vec{z}\} \\ &= \int\limits_{\mathbb{Z}} P_{n+1}(\vec{z}, d\vec{z}_{n+1}) \int\limits_{\mathbb{Z}} P_{n+2}(\vec{z}_{n+1}, d\vec{z}_{n+2}) \\ &\quad \cdots \int\limits_{\mathbb{Z}} I(f_n(\vec{z}, \vec{z}_{n+1}, \ldots, \vec{z}_N) \in C) P_N(\vec{z}_{N-1}, d\vec{z}_N)\,. \end{aligned} \tag{2.39}$$

Here and henceforth we use the notation $\mathsf{P}_{\vec{z},n}$ for conditional probabilities under the condition $\vec{Z}_n = \vec{z}$.

This conditional distribution is a probability measure, as a function of $C \in \mathcal{B}_1$, for every $\vec{z} \in \mathbb{Z}$, and a measurable function of argument $\vec{z} \in \mathbb{Z}$, for every $C \in \mathcal{B}_1$.

The random functionals $W_n = f_n(\vec{Z}_n, \ldots, \vec{Z}_N)$ and $\dot{W}_n = \dot{f}_n(\vec{V}_n, \ldots, \vec{V}_N)$ introduced in Subsection 2.3.1 coincide, i.e. $W_n \equiv \dot{W}_n$ if $f_n(\vec{z}_n, \ldots, \vec{z}_N) \equiv \dot{f}_n(\vec{v}_n, \ldots, \vec{v}_N)$, where $\vec{v}_r = (\vec{s}_r, x_r) = (e^{\vec{y}_r}, x_r)$, $\vec{z}_r = (\vec{y}_r, x_r) \in \mathbb{Z}$, $r = n, \ldots, N$.

In this case, relations (2.24), (2.38), and (2.39) imply that the following relation holds, for $\vec{z} \in \mathbb{Z}$, $C \in \mathcal{B}_1$, $n = 0, 1, \dots, N$:

$$\mathsf{P}_{\vec{z},n}\{W_n \in C\} = \dot{\mathsf{P}}_{\vec{v}_{\vec{z}},n}\{\dot{W}_n \in C\}\,. \tag{2.40}$$

On the left-hand side of relation (2.40), $W_n = f_n(\vec{Z}_n, \dots, \vec{Z}_N)$ is interpreted as a random functional defined on trajectories of the log-price process Z_l. On the right--hand side of relation (2.40), in fact, the same functional $\dot{W}_n = \dot{f}_n(\vec{V}_n, \dots, \vec{V}_N)$ is interpreted as a random functional defined on trajectories of the price process V_l.

Let us consider the set of integrability for the random variable W_n

$$I_{W_n} = \left\{ \vec{z} \in \mathbb{Z} : \mathsf{E}_{\vec{z},n}|W_n| = \int\limits_{\mathbb{R}_1} |w| \mathsf{P}_{\vec{z},n}\{W_n \in dw\} < \infty \right\}. \tag{2.41}$$

Here and henceforth we use the notation $\mathsf{E}_{\vec{z},n}$ for conditional expectations under the condition $\vec{Z}_n = \vec{z}$.

Set $I_{W_n} \in \mathcal{B}_{\mathbb{Z}}$. In particular, the set I_{W_n} can be empty or coincide with $\mathbb{Z}$.

Relations (2.38) and (2.40) imply that the following relations hold, for $0 \leq n \leq N$:

$$\dot{I}_{\dot{W}_n} = \{\vec{v}_{\vec{z}} : \vec{z} \in I_{W_n}\}\,, \quad I_{W_n} = \{\vec{z}_{\vec{v}} : \vec{v} \in \dot{I}_{\dot{W}_n}\}\,. \tag{2.42}$$

The conditional expectation of the random variable W_n under the condition $\vec{Z}_n = \vec{v}$ is defined for $\vec{z} \in I_{W_n}$,

$$\begin{aligned} \mathsf{E}_{\vec{z},n} W_n &= \int\limits_{\mathbb{R}_1} w \mathsf{P}_{\vec{z},n}\{W_n \in d\vec{w}\} = \mathsf{E}_{\vec{z},n} f_n(\vec{Z}_n, \dots, \vec{Z}_N) \\ &= \int\limits_{\mathbb{Z}} P_{n+1}(\vec{z}, d\vec{z}_{n+1}) \int\limits_{\mathbb{Z}} P_{n+2}(\vec{z}_{n+1}, d\vec{z}_{n+2}) \\ &\quad \cdots \int\limits_{\mathbb{Z}} f_n(\vec{z}, \vec{z}_{n+1}, \dots, \vec{z}_N) P_N(\vec{z}_{N-1}, d\vec{z}_N)\,. \end{aligned} \tag{2.43}$$

In fact, the assumption $\vec{z} \in I_{W_n}$ enables one first to get the analog of relation (2.43) for the random variable $|W_n| = |f_n(\vec{Z}_n, \dots, \vec{Z}_N)|$. This makes it possible to write relation (2.43) and to get the finite value for $\mathsf{E}_{\vec{z},n} W_n$ given by the expression on the right-hand side of this relation.

Relations (2.38) and (2.40) also imply that the following relation holds, for $\vec{z} \in I_{W_n}$, $n = 0, 1, \dots, N$:

$$\mathsf{E}_{\vec{z},n} W_n = \dot{\mathsf{E}}_{\vec{v}_{\vec{z}},n} \dot{W}_n\,. \tag{2.44}$$

As in relation (2.40), $W_n = f_n(\vec{Z}_n, \dots, \vec{Z}_N)$ is interpreted as a random functional defined on trajectories of the log-price process Z_l, on the left-hand side of relation (2.44), while, in fact, the same functional $\dot{W}_n = \dot{f}_n(\vec{V}_n, \dots, \vec{V}_N)$ is interpreted as a random functional defined on on trajectories of the price process V_l, on the right-hand side of relation (2.44).

Let us assume that $\mathsf{E}|W_n| < \infty$.

This condition does not imply that the set $I_{W_n} = \mathbb{Z}$. But this assumption does imply that $\mathsf{P}\{\vec{Z}_n \in I_{W_n}\} = 1$. Otherwise, the expectation $\mathsf{E}|W_n|$ should be equal to ∞. Note also that the assumption $I_{W_n} = \mathbb{Z}$ does not imply that $\mathsf{E}|W_n| < \infty$.

Under the assumption $\mathsf{E}|W_n| < \infty$, the following relation takes place:

$$\begin{aligned} \mathsf{E}W_n &= \int_{\mathbb{Z}} \mathsf{E}_{\vec{z},n} W_n \, \mathsf{P}\{\vec{Z}_n \in d\vec{z}\} \\ &= \int_{\mathbb{Z}} P_0(d\vec{z}_0) \int_{\mathbb{Z}} \mathsf{E}_{\vec{z},n} W_n P_{\vec{z}_0,0}\{\vec{Z}_n \in d\vec{z}\} . \end{aligned} \tag{2.45}$$

As above, the assumption $\mathsf{E}|W_n| < \infty$ enables one first to write the analog of relation (2.45) for the random variable $|W_n| = |f_n(\vec{Z}_n, \dots, \vec{Z}_N)|$. This makes it possible to obtain relation (2.45) and to get the finite value for $\mathsf{E}W_n$ given by the expression on the right-hand side of this relation.

Note also that one should additionally define the function $\mathsf{E}_{\vec{z},n}W_n$ for $\vec{z} \in \overline{I}_{W_n}$, in order to be able to obtain relation (2.45). Since $\mathsf{P}\{\vec{Z}_n \in \overline{I}_{W_n}\} = 0$, this can be done in an arbitrary "measurable" way because of the Lebesgue integral over the set $\overline{I}_{W_n}$ will be any way equal to 0. The standard way is to define $\mathsf{E}_{\vec{z},n}W_n = 0$, for $\vec{z} \in \overline{I}_{W_n}$.

Note that, in this case, relation (2.44) holds for any $\vec{z} \in \mathbb{Z}$, $n = 0, 1, \dots, N$, since we defined $\mathsf{E}_{\vec{z},n}W_n = \dot{\mathsf{E}}_{\vec{v}_{\vec{z}},n}\dot{W}_n = 0$, for $\vec{z} \in \overline{I}_{W_n}$.

Also, relations (2.27), (2.38), (2.44), and (2.45) imply, by the theorem about exchange of variables in Lebesgue integral, that the following relation holds:

$$\begin{aligned} \mathsf{E}W_n &= \int_{\mathbb{Z}} \mathsf{E}_{\vec{z},n} W_n \mathsf{P}\{\vec{Z}_n \in d\vec{z}\} \\ &= \int_{\mathbb{Z}} P_0(d\vec{z}_0) \int_{\mathbb{Z}} \mathsf{E}_{\vec{z},n} W_n P_{\vec{z}_0,0}\{\vec{Z}_n \in d\vec{z}\} \\ &= \mathsf{E}\dot{W}_n = \int_{\mathbb{V}} \dot{\mathsf{E}}_{\vec{v},n} \dot{W}_n \mathsf{P}\{\vec{V}_n \in d\vec{v}\} \\ &= \int_{\mathbb{V}} \dot{P}_0(d\vec{v}_0) \int_{\mathbb{V}} \dot{\mathsf{E}}_{\vec{v},n} \dot{W}_n \dot{P}_{\vec{v}_0,0}\{\vec{V}_n \in d\vec{v}\} . \end{aligned} \tag{2.46}$$

2.3.4 Log-reward functions for discrete time log-price processes

The above definitions and results concerned log-reward functions can also be reformulated in terms of log–log-reward functions based on the log-price processes $\vec{Z}_n = (\vec{Y}_n, X_n)$. This can be done using the relation $\vec{S}_n = e^{\vec{Y}_n}$, $n = 0, 1, \dots$, that connect the price process $\vec{S}_n$ and the log-price process $\vec{Y}_n$.

Let us recall the class $\mathcal{M}_{\max,n,N}$ of all Markov moments τ for the Markov process $\vec{Z}_n$ such that (a) $n \le \tau \le N$ and (b) event $\{\tau = m\} \in \mathcal{F}_{n,m}$, $n \le m \le N$.

As above, we assume that the condition $\mathbf{A_2}$ holds. This condition can also be expressed in terms of log-price processes:
$\mathbf{A_2}$: $\mathsf{E}_{\vec{z},n}|g(r, e^{\vec{Y}_r}, X_r)| < \infty\,, \quad \vec{z} \in \mathbb{Z}\,, \quad 0 \le n \le r \le N\,.$

The condition $\mathbf{A_2}$ implies that the following inequality analogous to (2.29) holds, for $\vec{z} \in \mathbb{Z}$, $n = 0, 1, \dots, N$:

$$\mathsf{E}_{\vec{z},n}|g(\tau_n, e^{\vec{Y}_{\tau_n}}, X_{\tau_n})| \le \sum_{r=n}^{N} \mathsf{E}_{\vec{z},n}|g(r, e^{\vec{Y}_r}, X_r)| < \infty\,. \tag{2.47}$$

Due to (2.47), the following finite measurable functions can be defined for $n = 0, 1, \dots, N$:

$$\phi_{n,\tau_n}(\vec{z}) = \phi_{n,\tau_n}(\vec{z}, x) = \mathsf{E}_{\vec{z},n} g(\tau_n, e^{\vec{Y}_{\tau_n}}, X_{\tau_n})\,, \quad \vec{z} \in \mathbb{Z}\,. \tag{2.48}$$

As was pointed out in Subsection 2.3.2, $g(\tau_n, e^{\vec{Y}_{n\tau_n}}, X_{\tau_n})$ is a variable adapted to the σ-algebra $\mathcal{F}_{n,N}$. Thus, it is can be represented in the form $g(\tau_n, e^{\vec{Y}_{\tau_n}}, X_{\tau_n}) = f_n(\vec{Z}_n, \dots, \vec{Z}_N)$, where $f_n(\vec{z}_n, \dots, \vec{z}_N)$ is a measurable real-valued function defined on the space $\mathbb{Z}^{(N-n+1)}$. This function has, in this case, a specific form. Since τ_n is a Markov moment taking values $n, \dots, N$, there exists the sequence of sets $A_{n,r} \in \mathcal{B}_{\mathbb{Z}^{(r-n+1)}}$, $r = n, \dots N$ such that event $\{\tau_n = r\} = \{(\vec{Z}_n, \dots, \vec{Z}_r) \in A_{n,r}\}$, for every $r = n, \dots, N$, and $\cup_{r=n}^{N}\{\tau_n = r\} = \cup_{r=n}^{N}\{(\vec{Z}_n, \dots, \vec{Z}_r) \in A_{n,r}\} = \Omega$ (here $\langle \Omega, \mathcal{F}, \mathsf{P}\rangle$ is a probability space where the process $\vec{Z}_l$ is defined). Therefore, $g(\tau_n, e^{\vec{Y}_{\tau_n}}, X_{\tau_n}) = \sum_{r=n}^{N} g(r, e^{\vec{Y}_r}, X_r) I((\vec{Z}_n, \dots, \vec{Z}_r) \in A_{n,r})$, and, thus, $f_n(\vec{z}_n, \dots, \vec{z}_N) = \sum_{r=n}^{N} g(r, e^{\vec{y}_r}, x_r) I((\vec{z}_n, \dots, \vec{z}_r) \in A_{n,r})$, where $\vec{z}_r = (\vec{y}_r, x_r)$, $r = n, \dots, N$.

The above remarks and relation (2.43) imply that the functions $\phi_{n,\tau_n}(\vec{z})$, $n = 0, \dots, N$ are indeed measurable and that the following relation takes place, for $\vec{z} \in \mathbb{Z}$, $n = 0, \dots, N$:

$$\begin{aligned}\phi_{n,\tau_n}(\vec{z}) &= \phi_{n,\tau_n}(\vec{y}, x) = \mathsf{E}_{\vec{z},n} g(\tau_n, e^{\vec{Y}_{\tau_n}}, X_{\tau_n}) \\ &= \sum_{r=n}^{N} \mathsf{E}_{\vec{z},n} g(r, e^{\vec{Y}_r}, X_r) I((\vec{Z}_n, \dots, \vec{Z}_r) \in A_{n,r})\,.\end{aligned} \tag{2.49}$$

Relations (2.44), (2.31), and (2.49) imply that the following relation holds, for $\vec{z} = (\vec{y}, x) \in \mathbb{Z}$, $n = 0, 1, \dots, N$:

$$\phi_{n,\tau_n}(\vec{z}) = \phi_{n,\tau_n}(\vec{y}, x) = \dot{\phi}_{n,\tau_n}(\vec{v}_{\vec{z}}) = \dot{\phi}_{n,\tau_n}(e^{\vec{y}}, x)\,. \tag{2.50}$$

Let us also define, for every $n = 0, 1, \dots, N$, the log-reward function for American option,

$$\begin{aligned}\phi_n(\vec{z}) &= \phi_n(\vec{y}, x) \\ &= \sup_{\tau_n \in \mathcal{M}_{\max,n,N}} \phi_{n,\tau_n}(n, \vec{z})\,, \quad \vec{z} = (\vec{y}, x) \in \mathbb{Z}\,.\end{aligned} \tag{2.51}$$

The log-reward functions $\phi_n(\vec{z})$, $n = 0, 1, \dots, N$ are finite functions. Indeed, inequalities (2.29) also imply that for every $\vec{z} \in \mathbb{Z}$, $n = 0, 1, \dots, N$,

$$|\phi_n(\vec{z})| \leq \sup_{\tau_n \in \mathcal{M}_{\max,n,N}} \mathsf{E}_{\vec{z},n}|g(\tau_n, e^{\vec{Y}_{\tau_n}}, X_{\tau_n})|$$
$$\leq \sum_{r=n}^{N} \mathsf{E}_{\vec{z},n}|g(r, \vec{S}_r, X_r)| < \infty \,. \tag{2.52}$$

Lemma 2.3.1. *Let the condition* $\mathbf{A}_2$ *holds. Then the following relation holds for* $\vec{z} = (\vec{y}, x) \in \mathbb{Z}$, $n = 0, 1, \dots, N$:

$$\phi_n(\vec{z}) = \phi_n(\vec{y}, x) = \dot{\phi}_n(\vec{v}_{\vec{z}}) = \dot{\phi}_n(e^{\vec{y}}, x) \,. \tag{2.53}$$

Proof. Relations (2.32), (2.50), and (2.51) imply that the following relation holds, for $\vec{z} = (\vec{y}, x) \in \mathbb{Z}$, $n = 0, 1, \dots, N$:

$$\phi_n(\vec{y}, x) = \sup_{\tau_n \in \mathcal{M}_{\max,n,N}} \phi_{n,\tau_n}(n, \vec{y}, x)$$
$$= \sup_{\tau_n \in \mathcal{M}_{\max,n,N}} \dot{\phi}_{n,\tau_n}(n, e^{\vec{y}}, x) = \dot{\phi}_n(e^{\vec{y}}, x) \,. \tag{2.54}$$

Relation (2.54) implies relation (2.53) to hold. □

Theorems 2.3.1, expressed in terms of log-reward functions, takes the following form.

Theorem 2.3.3. *Let the condition* $\mathbf{A}_2$ *holds. Then, the log-reward functions* $\phi_n(\vec{z}) = \phi_n(\vec{y}, x)$, $n = 0, 1, \dots, N$ *are the measurable functions of* $\vec{z} = (\vec{y}, x) \in \mathbb{Z}$, *and the following backward recurrence relations hold, for every* $\vec{z}_n = (\vec{y}_n, x_n) \in \mathbb{Z}$, $n = 0, 1, \dots, N$:

$$\begin{cases} \phi_N(\vec{z}) = g(N, e^{\vec{y}}, x) \,, \quad \vec{z} = (\vec{y}, x) \in \mathbb{Z} \,, \\ \phi_r(\vec{z}) = \max(g(r, e^{\vec{y}}, x), \mathsf{E}_{\vec{z},r}\phi_{r+1}(\vec{Z}_{r+1})) \,, \quad \vec{z} = (\vec{y}, x) \in \mathbb{Z} \,, \\ \qquad r = N-1, \dots, n+1 \,, \\ \phi_n(\vec{z}_n) = \max(g(n, e^{\vec{y}_n}, x_n), \mathsf{E}_{\vec{z}_n,n}\phi_{n+1}(\vec{Z}_{n+1})) \,. \end{cases} \tag{2.55}$$

The following lemma shows that the optimal expected reward Φ can be computed via the reward functions for the price process $\vec{V}_n$ or via the log-reward functions for the log-price process $\vec{Z}_n$.

Lemma 2.3.2. *Let the conditions* $\mathbf{A}_1$ *and* $\mathbf{A}_2$ *hold. Then the following relation takes place:*

$$\Phi = \sup_{\tau_0 \in \mathcal{M}_{\max,N}} \mathsf{E}g(\tau_0, e^{\vec{Y}_{\tau_0}}, X_{\tau_0}) = \sup_{\tau_0 \in \mathcal{M}_{\max,N}} \mathsf{E}g(\tau_0, \vec{S}_{\tau_0}, X_{\tau_0}) \,. \tag{2.56}$$

Proof. Relation (2.56) follows directly from the relation $\vec{S}_n = e^{\vec{Y}_n}$, $n = 1, 2, \dots,$ that links the price and the log-price processes and the fact that these processes generate the same class of Markov moments $\mathcal{M}_{\max,N}$. □

Relations (2.46) and (2.56) imply that the following relation holds:

$$\Phi = \sup_{\tau_0 \in \mathcal{M}_{\max,N}} \int_{\mathbb{Z}} \mathsf{E}_{\vec{z},n} g(\tau_0, e^{\vec{Y}_{\tau_0}}, X_{\tau_0}) P_0(d\vec{z}) . \qquad (2.57)$$

One can exchange operation of maximization and integration in formula (2.57) when computing the optimal expected reward Φ. After this exchange, computing the expectation and the supremum under the sign of integral will yield according to relation (2.51) the log-reward function $\phi_0(\vec{z})$. Thus, the optimal expected reward can be computed by integration of the corresponding log-reward function by the initial distribution $P_0(A)$.

Theorems 2.3.2, expressed in terms of log-reward functions, takes the following form.

Theorem 2.3.4. *Let the conditions* $\mathbf{A}_1$ *and* $\mathbf{A}_2$ *hold. Then, the optimal expected reward* Φ *and the log-reward function* $\phi_0(\vec{z})$ *are connected by the following formula:*

$$\Phi = \mathsf{E}\phi_0(\vec{Z}_0) = \int_{\mathbb{Z}} \mathsf{E}_{\vec{z},0}\phi_0(\vec{z}) P_0(d\vec{z}) . \qquad (2.58)$$

The proofs of Theorems 2.3.3 and 2.3.4 are given in Subsection 2.4.4.

2.4 Optimal stopping times

In this section, we describe the structure of optimal stopping times for American-type options for discrete time multivariate modulated Markov price and log-price processes.

2.4.1 Optimal stopping times for Markov price-processes

Let $\dot{D} = \langle \dot{D}_n, n = 0, 1, \dots, N \rangle$ be a sequence of measurable subsets of the space $\mathbb{V}$. We refer to $\dot{D}$ as a time–space domain.

Let us define for the multivariate, modulated Markov process $\mathbb{V}_r$ the hitting time to the time–space domain $\dot{D}$ in the discrete time interval $[n, N]$,

$$\dot{\tau}_{n,\dot{D}} = \min(r \geq n : \vec{V}_r \in \dot{D}_r) \wedge N . \qquad (2.59)$$

By $\dot{\mathcal{D}}$ we denote the family of all time–space domains $\dot{D}$ and by $\dot{\mathcal{M}}_{n,N}$ the class of all hitting times $\dot{\tau}_{n,\dot{D}}$ such that $\dot{D} \in \dot{\mathcal{D}}$.

Obviously, the classes $\dot{\mathcal{M}}_{n,N}$ and $\mathcal{M}_{\max,n,N}$ are connected by the relation $\dot{\mathcal{M}}_{n,N} \subset \mathcal{M}_{\max,n,N}$.

Let us assume that the condition $\mathbf{A}_2$ holds. Relation (2.33) implies that, in this case, the reward function $\dot{\phi}_n(\vec{s}, x)$ defined by relation (2.32) is a finite function, i.e. $|\dot{\phi}_n(\vec{s}, x)| < \infty$, for any $\vec{v} = (\vec{s}, x) \in \mathbb{V}$.

A Markov moment $\dot{\tau}_n^*$ from the class $\mathcal{M}_{\max,n,N}$ is called an optimal stopping time if it satisfies the following relation:

$$\dot{\phi}_n(\vec{s},x) = \dot{\phi}_{n,\dot{\tau}_n^*}(\vec{s},x) = \mathsf{E}_{\vec{v},n} g(\dot{\tau}_n^*, \vec{S}_{\dot{\tau}_n^*}, X_{\dot{\tau}_n^*}), \quad \vec{v} = (\vec{s},x) \in \mathbb{V}. \tag{2.60}$$

Relation (2.60) requires optimality of stopping time $\dot{\tau}_n^*$ simultaneously for all possible initial states $\vec{v} = (\vec{s},x) \in \mathbb{V}$, at moment n. The question arises: does such moment exist? The answer is affirmative. Moreover, the optimal stopping time is the hitting moment to the special time–space domain determined by the reward functions.

Since $\tau_n \equiv n$ is a Markov moment from the class $\mathcal{M}_{\max,n,N}$ the reward functions $\dot{\phi}_n(\vec{s},x)$ satisfy the following inequalities, for every $n = 0,1,\dots,N$:

$$g(n,\vec{s},x) \leq \dot{\phi}_n(\vec{s},x), \quad \vec{v} = (\vec{s},x) \in \mathbb{V}. \tag{2.61}$$

Let us denote by $\dot{D}^* = \langle \dot{D}_n^*, n = 0,1,\dots,N\rangle$ the time–space domain with the sets $\dot{D}_n^*$ defined in the following way:

$$\dot{D}_n^* = \{\vec{v} = (\vec{s},x) \in \mathbb{V} : g(n,\vec{s},x) = \dot{\phi}_n(\vec{s},x)\}, \quad n = 0,1,\dots,N. \tag{2.62}$$

The following theorem shows that the time–space domain $\dot{D}^*$ can be referred to as the optimal stopping time–space domain since it defines the hitting times $\dot{\tau}_n^* = \dot{\tau}_{n,\dot{D}^*}$, $n = 0,1,\dots,N$ such that $\dot{\tau}_n^*$ is the optimal stopping time in the class $\mathcal{M}_{\max,n,N}$, for every $n = 0,1,\dots,N$.

Theorem 2.4.1. *Let the condition* $\mathbf{A}_2$ *holds. Then, for every* $n = 0,\dots,N$, *the hitting time* $\dot{\tau}_n^* = \dot{\tau}_{n,\dot{D}^*} = \min(r \geq n : \vec{V}_r \in \dot{D}_r^*) \wedge N$ *is the optimal stopping time in the class* $\mathcal{M}_{\max,n,N}$, *for every* $n = 0,1,\dots,N$, *i.e. the following optimality relation holds for every* $\vec{v} = (\vec{s},x) \in \mathbb{V}$, $n = 0,1,\dots,N$:

$$\dot{\phi}_n(\vec{s},x) = \dot{\phi}_{n,\dot{\tau}_n^*}(\vec{s},x) = \dot{\mathsf{E}}_{\vec{v},n} g(\dot{\tau}_n^*, \vec{S}_{\dot{\tau}_n^*}, X_{\dot{\tau}_n^*}). \tag{2.63}$$

Note that Theorem 2.4.1 states that the hitting time $\dot{\tau}_{n,\dot{D}^*}$ is the optimal stopping time. However, this theorem does not guaranty that this optimal stopping time is unique.

It is worth to note that relation (2.63) implies that the reward functions $\dot{\phi}_n(\vec{s},x)$, $n = 0,1,\dots,N$ are the measurable functions in argument $(\vec{s},x)$. This does not follow directly from relation (2.32), which defines the reward functions.

Theorem 2.4.1 shows that the hitting time $\tau_0^* = \dot{\tau}_{0,\dot{D}^*}$ is an optimal stopping time simultaneously for all possible initial states $\vec{v} = (\vec{s},x) \in \mathbb{V}$, at moment 0. Thus, it is natural to expect that this hitting time should also be an optimal stopping time for the case where the initial distribution $\dot{P}_0(A)$ is not concentrated in a point.

Theorem 2.4.2. *Let the conditions* $\mathbf{A}_1$ *and* $\mathbf{A}_2$ *hold. Then, the hitting time* $\dot{\tau}_0^* = \dot{\tau}_{0,\dot{D}^*} = \min(r \geq 0 : \vec{V}_r \in \dot{D}_r^*) \wedge N$ *is the optimal stopping time which satisfies the following optimality relation:*

$$\Phi(\mathcal{M}_{\max,N}) = \mathsf{E}\dot{\phi}_0(\vec{S}_0, X_0) = \mathsf{E}g(\dot{\tau}_0^*, \vec{S}_{\dot{\tau}_0^*}, X_{\dot{\tau}_0^*}). \tag{2.64}$$

The proofs of Theorems 2.4.1 and 2.4.2 are given in Subsection 2.4.4.

2.4.2 Optimal stopping times for Markov log-price processes

The results presented in Theorems 2.4.1 and 2.4.2 can be reformulated in terms of log-price processes.

Let $D = \langle D_n, n = 0, 1, \dots, N\rangle$ be a sequence of measurable subsets of the space $\mathbb{Z}$. We refer to D as a time–space domain.

Let us define for the multivariate modulated Markov price process $\vec{Z}_n$ the hitting time to the time–space domain D in the discrete time interval $[n, N]$,

$$\tau_{n,D} = \min(r \geq n : \vec{Z}_r \in D_r) \wedge N\,. \tag{2.65}$$

By $\mathcal{D}$ we denote the family of all time–space domains D and by $\mathcal{M}_{n,N}$ the class of all hitting times $\tau_{n,D}$ such that $D \in \mathcal{D}$.

There exists the one-to-one mapping between the families of time–space domains $\dot{\mathcal{D}}$ and $\mathcal{D}$ given by the following relation connecting the corresponding time–space domains $\dot{D} = \langle \dot{D}_n, n = 0, 1, \dots\rangle$ and $D = \langle D_n, n = 0, 1, \dots\rangle$:

$$\dot{D}_n = \{(\vec{s}, x) = (e^{\vec{y}}, x) : (\vec{y}, x) \in D_n\}\,, \quad n = 0, 1, \dots, N\,. \tag{2.66}$$

This one-to-one mapping generates, due to the relation $\vec{S}_n = e^{\vec{Y}_n}$, connecting the price component of the modulated process $\vec{V}_n = (\vec{S}_n, X_n)$ and the modulated log-price process $\vec{Z}_n = (\vec{Y}_n, X_n)$, the one-to-one mapping between the classes of hitting time $\dot{\mathcal{M}}_{n,N}$ and $\mathcal{M}_{n,N}$. This mapping is given by the following relation, which connects the hitting times $\dot{\tau}_{n,\dot{D}}$ and $\tau_{n,D}$ for time space domains $\dot{D}$ and D connected by relation (2.66),

$$\dot{\tau}_{n,\dot{D}} = \tau_{n,D}\,. \tag{2.67}$$

Recall that the processes $\vec{V}_n$ and $\vec{Z}_n$ generate the same class of Markov moments $\mathcal{M}_{\max,n,N}$.

Let us assume that the condition $\mathbf{A_2}$ holds. Relation (2.52) implies that, in this case, the log-reward function $\phi_n(\vec{y}, x)$ defined by relation (2.51) is a finite function, i.e. $|\phi_n(\vec{y}, x)| < \infty$, for any $\vec{z} = (\vec{y}, x) \in \mathbb{Z}$.

A Markov moment τ_n^* from the class $\mathcal{M}_{\max,n,N}$ is called an optimal stopping time if it satisfies the following relation:

$$\phi_n(\vec{y}, x) = \phi_{n,\tau_n^*}(\vec{y}, x) = \mathsf{E}_{\vec{z},n} g(\tau_n^*, e^{\vec{Y}_{\tau_n^*}}, X_{\tau_n^*})\,, \quad \vec{z} = (\vec{y}, x) \in \mathbb{Z}\,. \tag{2.68}$$

Lemma 2.4.1. *Let the condition* $\mathbf{A_2}$ *holds. In this case, the optimal stopping time* τ_n^* *in the sense of relation* (2.60) *will also be an optimal in the sense of relation* (2.68) *and vice versa, for every* $n = 0, 1, \dots, N$.

Proof. Recall the one-to-one mapping between spaces $\mathbb{Z}$ and $\mathbb{V}$ given by the function $\vec{v}_{\vec{z}} = (\vec{s}_{\vec{z}}, x_{\vec{z}}) = (e^{\vec{y}}, x), \vec{z} = (\vec{y}, x) \in \mathbb{Z}$ and its inverse one-to-one mapping given by the function $\vec{z}_{\vec{v}} = (\vec{y}_{\vec{v}}, x_{\vec{v}}) = (\ln \vec{s}, x), \vec{v} = (\vec{s}, x) \in \mathbb{V}$.

Relations (2.50) and (2.53) imply the following relation, which holds for $\vec{z} \in \mathbb{Z}$, $n = 0, 1, \ldots, N$:

$$\phi_n(\vec{z}) = \mathsf{E}_{\vec{z},n} g(\tau_n^*, e^{\vec{Y}_{\tau_n^*}}, X_{\tau_n^*}) = \dot{\mathsf{E}}_{\vec{v}_{\vec{z}},n} g(\tau_n^*, \vec{S}_{\tau_n^*}, X_{\tau_n^*}) = \dot{\phi}_n(\vec{v}_{\vec{z}}) . \tag{2.69}$$

Since $\vec{v}_{\vec{z}}$ runs over the whole space $\mathbb{V}$ when $\vec{z}$ runs over the whole space $\mathbb{Z}$, relation (2.69) is equivalent to the following relation, which holds for $\vec{v} \in \mathbb{V}$, $n = 0, 1, \ldots, N$:

$$\dot{\phi}_n(\vec{v}) = \dot{\mathsf{E}}_{\vec{v},n} g(\tau_n^*, \vec{S}_{\tau_n^*}, X_{\tau_n^*}) . \tag{2.70}$$

Thus the optimality relation (2.60) holds for the Markov moment τ_n^*.

Analogously, relations (2.50) and (2.53) imply the following relation, which holds for $\vec{v} \in \mathbb{V}$, $n = 0, 1, \ldots, N$:

$$\dot{\phi}_n(\vec{v}) = \dot{\mathsf{E}}_{\vec{v},n} g(\dot{\tau}_n^*, \vec{S}_{\dot{\tau}_n^*}, X_{\dot{\tau}_n^*}) = \mathsf{E}_{\vec{z}_{\vec{v}},n} g(\dot{\tau}_n^*, e^{\vec{Y}_{\dot{\tau}_n^*}}, X_{\dot{\tau}_n^*}) = \phi_n(\vec{z}_{\vec{v}}) . \tag{2.71}$$

Since $\vec{z}_{\vec{v}}$ runs over the whole space $\mathbb{Z}$ when $\vec{v}$ runs over the whole space $\mathbb{V}$, relation (2.71) is equivalent to the following relation, which holds for $\vec{z} \in \mathbb{Z}$, $n = 0, 1, \ldots, N$:

$$\phi_n(\vec{z}) = \mathsf{E}_{\vec{z},n} g(\dot{\tau}_n^*, e^{\vec{Y}_{\dot{\tau}_n^*}}, X_{\dot{\tau}_n^*}) . \tag{2.72}$$

Thus, the optimality relation (2.68) holds for the Markov moment $\dot{\tau}_n^*$, for every $n = 0, 1, \ldots, N$. □

Since $\tau_n \equiv n$ is a Markov moment from the class $\mathcal{M}_{\max,n,N}$, the log-reward functions $\phi_n(\vec{s}, x)$ satisfy the following inequalities, for every $n = 0, 1, \ldots, N$:

$$g(n, e^{\vec{y}}, x) \leq \phi_n(\vec{y}, x) , \quad \vec{z} = (\vec{y}, x) \in \mathbb{Z} . \tag{2.73}$$

Let us denote by $D^* = \langle D_n^*, n = 0, 1, \ldots, N \rangle$ the time–space domain with the sets D_n^* defined in the following way:

$$D_n^* = \{\vec{z} = (\vec{y}, x) \in \mathbb{Z} : g(n, e^{\vec{y}}, x) = \phi_n(\vec{y}, x)\} , \quad n = 0, 1, \ldots, N . \tag{2.74}$$

It follows from relations (2.53) that the time–space domains $\dot{D}^*$ and D^* are connected by relation (2.66), i.e.

$$\dot{D}_n^* = \{(\vec{s}, x) = (e^{\vec{y}}, x) : (\vec{y}, x) \in D_n^*\} , \quad n = 0, 1, \ldots, N . \tag{2.75}$$

This implies that the hitting times $\dot{\tau}_{n,\dot{D}^*}$ and τ_{n,D^*} coincide for every $n = 0, 1, \ldots, N$, i.e.

$$\dot{\tau}_{n,\dot{D}^*} = \tau_{n,D^*} . \tag{2.76}$$

The following theorem shows that the time–space domain D^* defines the hitting times $\tau_n^* = \tau_{n,D^*}$, $n = 0, 1, \ldots, N$ such that τ_n^* is the optimal stopping time in the class $\mathcal{M}_{\max,n,N}$, for every $n = 0, 1, \ldots, N$.

Theorem 2.4.3. *Let the condition* $\mathbf{A}_2$ *holds. Then, for every* $n = 0, \dots, N$, *the hitting time* $\tau_n^* = \tau_{n,H^*} = \min(r \geq n : \vec{Z}_r \in D_r^*) \wedge N$ *coincides with the hitting time* $\dot{\tau}_{n,\dot{D}^*} = \min(r \geq n : \vec{V}_r \in \dot{D}_r^*) \wedge N$, *and it is the optimal stopping time in the class* $\mathcal{M}_{\max,n,N}$, *i.e. the following optimality relations hold for every* $\vec{z} = (\vec{y}, x) \in \mathbb{Z}$, $n = 0, 1, \dots, N$:

$$\phi_n(\vec{y}, x) = \phi_{n,\tau_n^*}(\vec{y}, x) = \mathsf{E}_{\vec{z},n} g(\tau_n^*, e^{\vec{Y}_{\tau_n^*}}, X_{\tau_n^*}) . \tag{2.77}$$

It is worth to note that relation (2.77) implies that the log-reward functions $\phi_n(\vec{y}, x)$, $n = 0, 1, \dots, N$ are the measurable functions in argument $(\vec{y}, x)$. This does not follow directly from relation (2.51), which defines the log-reward functions. Note also that the measurability of log-reward functions also follows from the measurability of reward functions, pointed out in the remarks following after Theorem 2.4.1, and relation (2.53) connecting reward and log-reward functions.

The following theorem, which is an analog of Theorem 2.4.2, also takes place.

Theorem 2.4.4. *Let the conditions* $\mathbf{A}_1$ *and* $\mathbf{A}_2$ *hold. Then, the hitting time* $\tau_0^* = \tau_{0,H^*} = \min(r \geq 0 : \vec{Z}_r \in D_r^*) \wedge N$ *satisfies the following optimality relation:*

$$\Phi = \mathsf{E}\phi_0(\vec{Y}_0, X_0) = \mathsf{E}g(\tau_0^*, e^{\vec{Y}_{\tau_0^*}}, X_{\tau_0^*}) . \tag{2.78}$$

2.4.3 A hitting structure of Markov stopping times

In this subsection, we give proofs of Theorem 2.3.1–2.3.4 and 2.4.1–2.4.4, which describe the structure of log-reward functions and optimal stopping times for American-type options for multivariate Markov modulated price and log-price processes.

Let $\vec{V}_n = (\vec{S}_n, X_n)$, $n = 0, 1, \dots$, be a Markov modulated price process with a phase space $\mathbb{V} = \mathbb{R}_k^+ \times \mathbb{X}$, an initial distribution $\dot{P}_0(B)$, and transition probabilities $\dot{P}_n(v, A)$.

Let $\dot{\tau}_n$ be a Markov stopping time from the class $\mathcal{M}_{\max,n,N}$, i.e. a random variable which take integer values $n \leq r \leq N$ such that the event $\{\dot{\tau}_n = r\} \in \mathcal{F}_{n,r} = \sigma[\vec{V}_n, \dots, \vec{V}_r]$ for every $n \leq r \leq N$.

Let $\mathbb{V}_l = \mathbb{V} \times \cdots \times \mathbb{V}$ be l-times product of the space $\mathbb{V}$ and $\mathcal{B}_{\mathbb{V}_l}$ be the minimal σ-algebra of subsets of space $\mathbb{V}_l$, which contains all rectangle sets $B_1 \times \cdots \times B_l$, $B_i \in \mathcal{B}_{\mathbb{V}}$, $i = 1, \dots, l$.

It follows from the above definition of the Markov stopping time $\dot{\tau}_n$ that there exist sets $B_{n,r} \in \mathcal{B}_{\mathbb{V}_{r-n+1}}$ such that the event $\{\dot{\tau}_n = r\} = \{(\vec{V}_n, \dots, \vec{V}_r) \in B_{n,r}\}$, for every $n \leq r \leq N$.

Thus let one represent the Markov stopping time $\dot{\tau}_n$ in the following hitting form:

$$\dot{\tau}_n = \min(r \geq n : (\vec{V}_n, \dots, \vec{V}_r) \in B_{n,r}) \wedge N . \tag{2.79}$$

This representation let one interpret any Markov stopping time $\dot{\tau}_n$ as some kind of generalized hitting time.

Let us connect with the Markov stopping time $\dot{\tau}_n \in \mathcal{M}_{\max,n,N}$ the shifted Markov stopping times $\dot{\tau}_n(\vec{v}), \vec{v} \in \mathbb{V}, 0 \le n < N$ defined by the following relation:

$$\dot{\tau}_n(\vec{v}) = \min(r \ge n+1 : (\vec{v}, \vec{V}_{n+1}, \dots, \vec{V}_r) \in B_{n,r}) \wedge N. \tag{2.80}$$

By the definition, the Markov stopping times $\dot{\tau}_n(v) \in \mathcal{M}_{\max,n+1,N}$, $v \in \mathbb{V}$, $0 \le n < N$.

Let $\dot{D} = \langle \dot{D}_n, n = 0, 1, \dots, N\rangle$ be a sequence of measurable subsets of the space $\mathbb{V}$.

Let us now consider the hitting time to the time–space domain $\dot{D}$ in the discrete time interval $[n, N]$,

$$\dot{\tau}_{n,\dot{D}} = \min(r \ge n : \vec{V}_r \in \dot{D}_r) \wedge N. \tag{2.81}$$

It is obvious that $\dot{\tau}_{n,\dot{D}}$ is a Markov stopping time from the class $\mathcal{M}_{\max,n,N}$.

In this case, the event $\{\dot{\tau}_{n,\dot{D}} = r\} = \{(\vec{V}_n, \dots, \vec{V}_r) \in B_{n,r}\}$, where $B_{n,r} = \mathbb{V}_{r-n} \times \dot{D}_r\}$ for $0 \le n \le r \le N$, and, therefore, the corresponding Markov stopping times $\dot{\tau}_{n,\dot{D}}(\vec{v}), \vec{v} \in \mathbb{V}, 0 \le n < N$ take the following form:

$$\dot{\tau}_{n,\dot{D}}(\vec{v}) = \min(r \ge n+1 : \vec{V}_r \in \dot{D}_r) \wedge N = \dot{\tau}_{n+1,\dot{D}}. \tag{2.82}$$

Thus, the Markov stopping times $\dot{\tau}_{n,\dot{D}}(\vec{v}) = \dot{\tau}_{n+1,\dot{D}}, \vec{v} \in \mathbb{V}, 0 \le n < N$ themselves are ordinary hitting times.

2.4.4 Proof of Theorems 2.3.1, 2.3.3, 2.4.1, and 2.4.3

Recall the time–space domains $\dot{D}^* = \langle \dot{D}^*_n, n = 0, 1, \dots, N\rangle$, where

$$\dot{D}^*_n = \{\vec{v} = (\vec{s}, x) : g(n, \vec{s}, x) = \dot{\phi}_n(\vec{v})\}, 0 \le n \le N. \tag{2.83}$$

The hitting times $\dot{\tau}^*_n = \dot{\tau}_{n,\dot{D}^*} = \min(r \ge n : \vec{V}_r \in \dot{D}^*_r) \wedge N, 0 \le n \le N$ possess the property given by relation (2.82).

Let us assume that the condition $\mathbf{A_2}$ holds.

This condition and relation (2.29) let one define the finite measurable functions, for $n = 0, 1, \dots, N$,

$$\dot{\phi}_n(\vec{v}) = \dot{\mathsf{E}}_{\vec{v},n} g(\dot{\tau}^*_n, \vec{S}_{\dot{\tau}^*_n}, X_{\dot{\tau}^*_n}), \ \vec{v} \in \mathbb{V}. \tag{2.84}$$

Note that we are going to prove that the functions defined in relation (2.84) coincide with the reward functions defined in relation (2.32) and, that is why, we use for them the same notations.

Lemma 2.4.2. *Let the condition* $\mathbf{A_2}$ *holds. Then, the functions* $\dot{\phi}_n(\vec{v}), \vec{v} \in \mathbb{V}, n = 0, 1, \dots, n$ *satisfy the following backward recurrence relations:*

$$\begin{cases} \dot{\phi}_N(\vec{v}) = g(N, \vec{s}, x), \quad \vec{v} = (\vec{s}, x) \in \mathbb{V}, \\ \dot{\phi}_m(\vec{v}) = \max\left(g(m, \vec{s}, x), \int_{\mathbb{V}} \dot{P}_{m+1}(\vec{v}, d\vec{v}\,') \dot{\phi}_{m+1}(\vec{v}\,')\right), \quad \vec{v} = (\vec{s}, x) \in \mathbb{V}, \\ \quad m = N-1, \dots, 0. \end{cases} \tag{2.85}$$

Proof. By the definition, $\dot{\tau}_N^* = N$ and, therefore, $\dot{\phi}_N(\vec{v}) = g(N, \vec{s}, x), \vec{v} = (\vec{s}, x) \in \mathbb{V}$. Thus, the first series of equalities in (2.85), for functions $\dot{\phi}_N(\vec{v})$, holds.

Let us now prove the inductive proposition that the assumption of holding equalities (2.85) for $\vec{v} = (\vec{s}, x) \in \mathbb{V}$, $m = n + 1$, for some $0 \leq n < N$ implies that these equalities also hold for $\vec{v} = (\vec{s}, x) \in \mathbb{V}$ and $m = n$.

Indeed, using the above induction assumption, the Markov property of the process V_n and relations (2.82) and (2.83), we get, for $\vec{v} \in \mathbb{V}$,

$$
\begin{aligned}
\dot{\mathsf{E}}_{\vec{v},n} g(\dot{\tau}_n^*, \vec{S}_{\dot{\tau}_n^*}, X_{\dot{\tau}_n^*}) &= I(\vec{v} \in D_n^*) g(n, \vec{s}, x) \\
&\quad + I(\vec{v} \notin D_n^*) \int_{\mathbb{V}} \dot{P}_{n+1}(\vec{v}, d\vec{v}\,') \dot{\mathsf{E}}_{\vec{v}\,',n+1} g(\dot{\tau}_{n+1}^*, \vec{S}_{\dot{\tau}_{n+1}^*}, X_{\dot{\tau}_{n+1}^*}) \\
&= I(\vec{v} \in D_n^*) g(n, \vec{s}, x) + I(\vec{v} \notin D_n^*) \int_{\mathbb{V}} \dot{P}_{n+1}(\vec{v}, d\vec{v}\,') \dot{\phi}_{n+1}(\vec{v}\,') \\
&= \max\left(g(n, \vec{s}, x), \int_{\mathbb{V}} \dot{P}_{n+1}(\vec{v}, d\vec{v}\,') \dot{\phi}_{n+1}(\vec{v}\,') \right) = \dot{\phi}_n(\vec{v}) . \qquad (2.86)
\end{aligned}
$$

By induction, equalities (2.84) hold for any $\vec{v} \in \mathbb{V}, 0 \leq m \leq N$.

In fact, first one should realize the above induction proof, in particular to get relation analogous to (2.86) for finite (due to the condition $\mathbf{A_2}$ and relation (2.29)), and measurable functions $\dot{\phi}_n^+(\vec{v}) = \dot{\mathsf{E}}_{\vec{v},n} |g(\dot{\tau}_n^*, \vec{S}_{\dot{\tau}_n^*}, X_{\dot{\tau}_n^*})|, \vec{v} \in \mathbb{V}$, $n = 0, 1, \ldots, N$. Then, one can repeat the induction proof, in particular to get relation (2.86) for the functions $\dot{\phi}_n(\vec{v}) = \dot{\mathsf{E}}_{\vec{v},n} g(\dot{\tau}_n^*, \vec{S}_{\dot{\tau}_n^*}, X_{\dot{\tau}_n^*}), \vec{v} \in \mathbb{V}$, $n = 0, 1, \ldots, N$. □

Now, we should prove that the functions $\dot{\phi}_n(\vec{v})$, $n = 0, 1, \ldots, N$ are, indeed, reward functions, i.e. they satisfy the following optimality relation, for $\vec{v} \in \mathbb{V}$, $n = 0, 1, \ldots, N$:

$$
\dot{\phi}_n(\vec{v}) = \sup_{\dot{\tau}_n \in \mathcal{M}_{\max,n,N}} \dot{\mathsf{E}}_{\vec{v},n} g(\dot{\tau}_n, \vec{S}_{\dot{\tau}_n}, X_{\dot{\tau}_n}) . \qquad (2.87)
$$

Since $\dot{\tau}_n^* \in \mathcal{M}_{\max,n,N}$, $n = 0, 1, \ldots, N$ and $\dot{\phi}_n(\vec{v}) = \dot{\mathsf{E}}_{\vec{v},n} g(\dot{\tau}_n^*, \vec{S}_{\dot{\tau}_n^*}, X_{\dot{\tau}*_n})$, for $\vec{v} \in \mathbb{V}$, $n = 0, 1, \ldots, N$, relation (2.87) follows from the following lemma.

Lemma 2.4.3. *Let the condition* $\mathbf{A_2}$ *holds. Then, the following inequality holds, for* $\vec{v} = (\vec{s}, x) \in \mathbb{V}, 0 \leq n \leq N$,

$$
\sup_{\dot{\tau}_n \in \mathcal{M}_{\max,n,N}} \dot{\mathsf{E}}_{\vec{v},n} g(\dot{\tau}_n, \vec{S}_{\dot{\tau}_n}, X_{\dot{\tau}_n}) \leq \dot{\phi}_n(\vec{v}) . \qquad (2.88)
$$

Proof. Let us take an arbitrary sequence of Markov stopping times $\dot{\tau}_n \in \mathcal{M}_{\max,n,N}$, $n = 0, 1, \ldots, N$ and use for these Markov stopping times, their representation in the form of generalized hitting time given by relation (2.79) and let also $\dot{\tau}_n(\vec{v}), \vec{v} \in \mathbb{V}$, $n = 0, 1, \ldots, N - 1$ be the corresponding shifted Markov stopping times defined in relation (2.80).

By the definition, $\dot{\tau}_N = N$ and $\dot{\phi}_N(\vec{v}) = g(N, \vec{s}, x), \vec{v} = (\vec{s}, x) \in \mathbb{V}$. Thus, relation (2.88) holds and takes, in this case, the form of equality.

Let us now prove the inductive proposition that the assumption of holding the inequalities (2.88) for $\vec{v} = (\vec{s}, x) \in \mathbb{V}$, $m = n + 1$ for some $0 \leq n < N$ implies that these inequalities also hold for $\vec{v} = (\vec{s}, x) \in \mathbb{V}$ and $m = n$.

Indeed, using the above induction assumption and the Markov property of the process V_n, we get, for any Markov stopping time $\dot{\tau}_n \in \mathcal{M}_{\max,n,N}$ and $\vec{v} \in \mathbb{V}$,

$$\begin{aligned}
\dot{\mathsf{E}}_{\vec{v},n} g(\dot{\tau}_n, \vec{S}_{\dot{\tau}_n}, X_{\dot{\tau}_n}) &= I(\vec{v} \in B_{n,n}) g(n, \vec{s}, x) \\
&\quad + I(\vec{v} \notin B_{n,n}) \int_{\mathbb{V}} \dot{P}_{n+1}(\vec{v}, d\vec{v}\,') \dot{\mathsf{E}}_{\vec{v}\,',n+1} g(\dot{\tau}_n(\vec{v}), \vec{S}_{\dot{\tau}_n(\vec{v})}, X_{\dot{\tau}_n(\vec{v})}) \\
&\leq I(\vec{v} \in B_{n,n}) g(n, \vec{s}, x) + I(\vec{v} \notin B_{n,n}) \int_{\mathbb{V}} \dot{P}_{n+1}(\vec{v}, d\vec{v}\,') \dot{\phi}_{n+1}(\vec{v}\,') \\
&\leq \max\left(g(n, \vec{s}, x), \int_{\mathbb{V}} \dot{P}_{n+1}(\vec{v}, d\vec{v}\,') \dot{\phi}_{n+1}(\vec{v}\,') \right) = \dot{\phi}_n(\vec{v}) \,. \qquad (2.89)
\end{aligned}$$

Since an arbitrary choice of the Markov stopping time $\dot{\tau}_n \in \mathcal{M}_{\max,n,N}$ and $\vec{v} \in \mathbb{V}$ in the above relation, inequality (2.88) holds for $n = m$.

By induction, inequalities (2.88) hold for any $\vec{v} \in \mathbb{V}, 0 \leq m \leq N$.

In fact, first one should realize the above induction proof, in particular to get relation analogous to (2.89) for finite (due to the condition $\mathbf{A_2}$ and relation (2.29)) and measurable functions $\dot{\mathsf{E}}_{\vec{v},n}|g(\dot{\tau}_n, \vec{S}_{\dot{\tau}_n}, X_{\dot{\tau}_n})|, \vec{v} \in \mathbb{V}, n = 0, 1, \dots, N$. Then, one can repeat the induction proof, in particular to get relation (2.89) for functions $\dot{\mathsf{E}}_{\vec{v},n} g(\dot{\tau}_n, \vec{S}_{\dot{\tau}_n}, X_{\dot{\tau}_n}), \vec{v} \in \mathbb{V}, n = 0, 1, \dots, N$. □

So, we proved that the finite measurable functions $\dot{\phi}_n(\vec{v}), \vec{v} \in \mathbb{V}, n = 0, 1, \dots, N$ defined by relation (2.84) satisfy the optimality relation (2.87) and, therefore, they coincide with the reward functions defined in relation (2.32).

Also, we proved that the hitting times $\dot{\tau}_n^* = \dot{\tau}_{n,\dot{D}^*} = \min(r \geq n : \vec{V}_r \in \dot{D}_r^*) \wedge N$ is the optimal stopping time in the class $\mathcal{M}_{\max,n,N}$, for every $n = 0, 1, \dots$.

Finally, we proved that the functions $\dot{\phi}_n(\vec{v}), \vec{v} \in \mathbb{V}, n = 0, 1, \dots, N$ satisfy the backward recurrence relations (2.85), which are equivalent to the backward recurrence relations given in Theorem 2.3.1.

Both Theorems 2.3.1 and 2.4.1 are proved. □

Theorems 2.3.3 and 2.4.3 just reformulate Theorems 2.3.1 and 2.4.1 in terms of log-price processes.

The proof analogous to those given for Theorems 2.3.1 and 2.4.1 in terms of the price process $\vec{V}_n$ can be rewritten in terms of the log-price process $\vec{Z}_n$.

However, Theorems 2.3.3 and 2.4.3 can also be considered as corollaries of Theorems 2.3.1 and 2.4.1. The proofs can be obtained by application to relations given in Theorems 2.3.1 and 2.4.1, namely, relation (2.50), which connects log-reward functions $\phi_n(\vec{z})$ and reward functions $\dot{\phi}_n(\vec{v})$; relation (2.38), which connects transition probabilities $P_n(\vec{z}, A)$ for the log-price process $\vec{Z}_n$ and transition probabilities $\dot{P}_n(\vec{v}, B)$ for the price process $\vec{V}_n$; relation (2.44), which connects the expectations of random functionals $W \equiv \dot{W}$ defined on trajectories, respectively, of log-price and price processes; relation (2.75), which connects the optimal stopping time–space domains D^* and $\dot{D}^*$; and relation (2.76), which connects the corresponding stopping times τ_{n,D^*} and $\dot{\tau}_{n,\dot{D}^*}$.

The measurability of the log-reward functions $\phi_n(\vec{z})$ follows, due to relation (2.50), from the measurability of the log-reward functions $\dot{\phi}_n(\vec{v})$.

The corresponding substitution, based on the above relations (2.50), (2.38), and (2.44), transforms the backward recurrence relations (2.34), given for reward functions $\dot{\phi}_n(\vec{v})$ in Theorem 2.3.1, in the backward recurrence relations (2.55), given for log-reward functions $\phi_n(\vec{z})$ in Theorem 2.3.3.

The substitution, based on the above relations (2.50), (2.44), (2.75), and (2.76), transforms optimality relation (2.63) given for stopping times $\dot{\tau}_{n,\dot{D}^*}$ in Theorem 2.4.1 in the analogous optimality relation (2.77) given for stopping times $\tau_{n,\dot{D}^*}$ in Theorem 2.4.3.

The above remarks complete the proofs of Theorems 2.3.3 and 2.4.3. □

2.4.5 Proof of Theorems 2.3.2, 2.3.4, 2.4.2, and 2.4.4

We assume that both the conditions $\mathbf{A_1}$ and $\mathbf{A_2}$ hold. These conditions imply that the optimal expected reward $\dot{\Phi} < \infty$.

It follows from the Markov property of the price process $\vec{V}_n$ and relation (2.45) that for any Markov stopping time $\dot{\tau}_0 \in \mathcal{M}_{\max,0,N}$,

$$\dot{\mathsf{E}}g(\dot{\tau}_0, \vec{S}_{\dot{\tau}_0}, X_{\dot{\tau}_0}) = \int_{\mathbb{V}} \dot{\mathsf{E}}_{\vec{v},0}g(\dot{\tau}_0, \vec{S}_{\dot{\tau}_0}, X_{\dot{\tau}_0})\dot{P}_0(d\vec{v})\,. \tag{2.90}$$

Relations (2.90) and (2.32) defining the reward functions $\dot{\phi}_0(\vec{v})$ imply that the following inequality holds:

$$\begin{aligned}\Phi &= \sup_{\tau_0 \in \mathcal{M}_{\max,0,N}} \dot{\mathsf{E}}g(\dot{\tau}_0, \vec{S}_{\dot{\tau}_0}, X_{\dot{\tau}_0}) \\ &\le \int_{\mathbb{V}} \sup_{\dot{\tau}_0 \in \mathcal{M}_{\max,0,N}} \dot{\mathsf{E}}_{\vec{v},0}g(\dot{\tau}_0, \vec{S}_{\dot{\tau}_0}, X_{\dot{\tau}_0})\dot{P}_0(d\vec{v}) = \int_{\mathbb{V}} \dot{\phi}_0(\vec{v})\dot{P}_0(d\vec{v})\,. \end{aligned} \tag{2.91}$$

Also, relations (2.45), (2.64) given in Theorem 2.4.2, and (2.90) imply that the following equality to holds:

$$\begin{aligned}\Phi &= \sup_{\dot{\tau}_0 \in \mathcal{M}_{\max,0,N}} \dot{\mathsf{E}}g(\dot{\tau}_0, \vec{S}_{\dot{\tau}_0}, X_{\dot{\tau}_0}) \ge \dot{\mathsf{E}}g(\dot{\tau}_0^*, \vec{S}_{\dot{\tau}_0^*}, X_{\dot{\tau}_0^*}) \\ &= \int_{\mathbb{V}} \dot{\mathsf{E}}_{\vec{v},0}g(\dot{\tau}_0^*, \vec{S}_{\dot{\tau}_0^*}, X_{\dot{\tau}_0^*})P_0(d\vec{v}) = \int_{\mathbb{V}} \dot{\phi}_0(\vec{v})\dot{P}_0(d\vec{v})\,. \end{aligned} \tag{2.92}$$

Relations (2.37) and (2.64) given in Theorems 2.3.2 and 2.4.2 follow in an obvious way from relations (2.91) and (2.92).

Theorems 2.3.4 and 2.4.4 just reformulate Theorems 2.3.2 and 2.4.2 in terms of log-price processes and do not require separate proofs. □

2.5 American-type knockout options

In this section, we extend the class of American-type options presented in this chapter. We show that so-called American-type knockout options can be imbedded into the model of ordinary American-type options by extending the corresponding price or log-price processes by a special modulating index component. The American-type knockout options include as a particular case various American-type barrier options.

2.5.1 American-type knockout options

Let $\vec{V}_n = (\vec{S}_n, X_n)$ be a Markov modulated price process with a phase space $\mathbb{V} = \mathbb{R}_k^+ \times \mathbb{X}$, an initial distribution $\dot{P}_0(A) = \dot{\mathsf{P}}\{\vec{V}_0 \in A\}$, and the transition probabilities $\dot{P}_n((\vec{s}, x), A) = \dot{\mathsf{P}}\{\vec{V}_n \in A / \vec{V}_{n-1} = (\vec{s}, x)\}$.

Let also $g(n, \vec{s}, x)$ be a pay-off function, which is assumed to be a real-valued measurable function defined for $(n, \vec{s}, x) \in \mathbb{N} \times \mathbb{R}_k^+ \times \mathbb{X}$.

Let also $\dot{H} = \langle \dot{H}_0, \dots, \dot{H}_N \rangle$ be a so-called time–space knockout domain, and $\dot{\tau}_{n,\dot{H}} = \min(r \geq n : \vec{V}_r \in \dot{H}_r) \wedge N, 0 \leq n \leq N$ be knockout stopping times.

We would like to define an American-type knockout option for the price process V_n with the pay-off function $g(n, \vec{s}, x)$ and the time–space knock-out domain $\dot{H}$.

Let us assume that the conditions $\mathbf{A_1}$ and $\mathbf{A_2}$ hold. These conditions guarantee that the reward functions and optimal expected rewards defined below are finite.

If such an option is executed at the Markov stopping time $\dot{\tau}_n \in \mathcal{M}_{\max,n,N}$ then the corresponding random reward is $g(\dot{\tau}_n, \vec{S}_{\dot{\tau}_n}, X_{\dot{\tau}_n}) I(\dot{\tau}_n < \dot{\tau}_{n,\dot{H}})$.

The corresponding knockout reward function is defined by the following relation, for $\vec{v} \in \mathbb{V}, 0 \leq n \leq N$:

$$\dot{\phi}_n(\vec{v}) = \sup_{\dot{\tau}_n \in \mathcal{M}_{\max,n,N}} \dot{\mathsf{E}}_{\vec{v},n} g(\dot{\tau}_n, \vec{S}_{\dot{\tau}_n}, X_{\dot{\tau}_n}) I(\dot{\tau}_n < \dot{\tau}_{n,\dot{H}}), \tag{2.93}$$

and the optimal knockout expected reward by the relation

$$\dot{\Phi} = \sup_{\dot{\tau}_0 \in \mathcal{M}_{\max,0,N}} \dot{\mathsf{E}} g(\dot{\tau}_0, \vec{S}_{\dot{\tau}_0}, X_{\dot{\tau}_0}) I(\dot{\tau}_0 < \dot{\tau}_{0,\dot{H}}). \tag{2.94}$$

The above American-type knockout option can also be redefined in terms of the Markov log-price process $\vec{Z}_n = (\vec{Y}_n, X_n) = (\ln \vec{S}_n, X_n)$, which is also a Markov modulated price process with a phase space $\mathbb{Z} = \mathbb{R}_k \times \mathbb{X}$, an initial distribution $P_0(B) = \mathsf{P}\{\vec{Z}_0 \in B\}$, and transition probabilities $P_n((\vec{y}, x), B) = \mathsf{P}\{\vec{Z}_n \in B / \vec{Z}_{n-1} = (\vec{y}, x)\}$.

In this case, one should consider a time–space knockout domain $H = \langle H_0, \dots, H_N \rangle$, where $H_n \in \mathcal{B}_{\mathbb{Z}}$, $n = 0, \dots, N$ and define knockout stopping times in terms of the log-price process $\vec{Z}_n$ as $\tau_{n,H} = \min(n \geq 0 : \vec{Z}_n \in H_n) \wedge N, 0 \leq n \leq N$.

The following relation define a natural one-to-one correspondence between knockout domains $\dot{H}$ and H, for $n = 0, 1, \dots, N$,

$$\dot{H}_n = \{\vec{v} = (\vec{s}, x) : \vec{s} = e^{\vec{y}}, \vec{z} = (\vec{y}, x) \in H_n\}, \tag{2.95}$$

which, obviously, implies the following relation for knockout times, for $n = 0, 1, \dots, N$:

$$\dot{\tau}_{n,\dot{H}} = \tau_{n,H}. \tag{2.96}$$

The corresponding knockout log-reward function is defined by the following relation, for $\vec{z} \in \mathbb{Z}, 0 \leq n \leq N$:

$$\phi_n(\vec{z}) = \sup_{\tau_n \in \mathcal{M}_{\max,n,N}} \mathsf{E}_{\vec{z},n} g(\tau_n, e^{\vec{Y}_{\tau_n}}, X_{\tau_n}) I(\tau_n < \tau_{n,H}). \tag{2.97}$$

and the optimal knockout expected log-reward by the relation,

$$\Phi = \sup_{\tau_0 \in \mathcal{M}_{\max,0,N}} \mathsf{E} g(\tau_0, e^{\vec{Y}_{\tau_0}}, X_{\tau_0}) I(\tau_0 < \tau_{0,H}). \tag{2.98}$$

In this case, where relation (2.95) holds, the knockout options defined via the price process or the log-price process define, in fact, the same option contract. Correspondingly, the reward functions $\dot{\phi}_n(\vec{v})$ and the log-reward functions $\phi_n(\vec{z})$ are connected by the following relation, for $\vec{z} = (\vec{y}, x) \in \mathbb{Z}, \vec{v}_{\vec{z}} = (e^{\vec{y}}, x)$, $n = 0, 1, \dots, N$:

$$\phi_n(\vec{z}) = \dot{\phi}_n(\vec{v}_{\vec{z}}), \tag{2.99}$$

while the optimal knockout expected rewards $\dot{\Phi}$ and Φ are connected by the following relation,

$$\Phi = \dot{\Phi}. \tag{2.100}$$

2.5.2 Imbedding into the model of ordinary American-type options

In the model of the American-type knockout options, the random reward $g(\tau_n, \vec{S}_{\tau_n}, X_{\tau_n}) I(\tau_n < \dot{\tau}_{n,\dot{H}})$ depends on the whole trajectory $\vec{V}_r, n \leq r \leq \tau_n \wedge N$ and the random log-reward $g(\tau_n, e^{\vec{Y}_{\tau_n}}, X_{\tau_n}) I(\tau_n < \dot{\tau}_{n,\dot{H}})$ depends on the whole trajectory $\vec{Z}_r, n \leq r \leq \tau_n \wedge N$.

However, this model can be imbedded in the model of the ordinary American-type option using an appropriate extension of the price process $\vec{V}_n$ or the log-price process $\vec{Z}_n$. Let us describe the corresponding imbedding transformation in terms of the log-price process $\vec{Z}_n$.

Let us introduce the additional index component

$$X'_n = \prod_{r=0}^{n} I(\vec{Z}_r \notin H_r), \quad n = 0, 1, \dots. \tag{2.101}$$

It is obvious that the process $\vec{Z}'_n = (\vec{Z}_n, X_n, X'_n)$, $n = 0, 1, \dots$, is also a Markov chain, with the extended phase space $\mathbb{Z}' = \mathbb{R}_k \times \mathbb{X} \times \{0, 1\}$.

In this case, the role of the modulated index process is played by the process $\vec{X}_n = (X_n, X'_n)$, with the phase space $X \times \{0, 1\}$.

Let us define the domains $G_n, 0 \leq n \leq N$ by the following relation:

$$G_n = \{(\vec{y}, x, e) : I((\vec{y}, x) \in \overline{H}_n, e = 1) + I((\vec{y}, x) \in H_n, e = 0)\}. \tag{2.102}$$

The definition of the additional index component X'_n implies that the following relation holds:

$$\mathsf{P}\{\vec{Z}'_n \in G_n,\ 0 \leq n \leq N\} = 1. \tag{2.103}$$

The initial distribution for the Markov chain $\vec{Z}'_n$ is defined by the following relation, for $A \in \mathcal{B}_{\mathbb{Z}},\ e = 0, 1$:

$$P'_0(A \times \{e\}) = \begin{cases} P_0(A \cap \overline{H}_0) & \text{if } e = 1, \\ P_0(A \cap H_0) & \text{if } e = 0. \end{cases} \tag{2.104}$$

The transition probabilities for Markov chain $\vec{Z}'_n$ are defined by the following relation, for $(\vec{y}, x, e) \in \mathbb{Z}'$, $A \in \mathcal{B}_{\mathbb{Z}}$, $e' = 0, 1$, $n = 1, 2, \dots$,

$$\begin{aligned} &P'_n((\vec{y}, x, e), A \times \{e'\}) \\ &= \begin{cases} P_n((\vec{y}, x), A \cap \overline{H}_n) & \text{if } (\vec{y}, x) \in \overline{H}_{n-1},\ e = 1,\ e' = 1, \\ P_n((\vec{y}, x), A \cap H_n) & \text{if } (\vec{y}, x) \in \overline{H}_{n-1},\ e = 1,\ e' = 0, \\ 0 & \text{if } (\vec{y}, x) \in \overline{H}_{n-1},\ e = 0,\ e' = 1, \\ P_n((\vec{y}, x), A) & \text{if } (\vec{y}, x) \in \overline{H}_{n-1},\ e = 0,\ e' = 0, \\ 0 & \text{if } (\vec{y}, x) \in H_{n-1},\ e = 1,\ e' = 1, \\ P_n((\vec{y}, x), A) & \text{if } (\vec{y}, x) \in H_{n-1},\ e = 1,\ e' = 0, \\ 0 & \text{if } (\vec{y}, x) \in H_{n-1},\ e = 0,\ e' = 1, \\ P_n((\vec{y}, x), A) & \text{if } (\vec{y}, x) \in H_{n-1},\ e = 0,\ e' = 0. \end{cases} \end{aligned} \tag{2.105}$$

Let us also introduce the transformed pay-off function defined by the following relation, for $(\vec{y}, x, e) \in \mathbb{Z}'$, $0 \leq n \leq N$:

$$g(n, e^{\vec{y}}, x, e) = g(n, e^{\vec{y}}, x) I((\vec{y}, x, e) \in G_n). \tag{2.106}$$

It is readily seen that the class of Markov stopping times $\mathcal{M}_{\max,n,N}$ coincides for the processes $\vec{Z}_n$ and $\vec{Z}'_n$, since the corresponding σ-algebras $\mathcal{F}_{n,m} = \sigma[\vec{Z}_n, \dots, \vec{Z}_m] = \sigma[\vec{Z}'_n, \dots, \vec{Z}'_m]$ for any $0 \leq n \leq m \leq N$.

Also, the following relation holds for the corresponding random rewards, for any $\tau_n \in \mathcal{M}_{\max,n,N}, 0 \le n \le N$,

$$g(\tau_n, e^{\vec{Y}_{\tau_n}}, X_{\tau_n}) I(\tau_n < \tau_{n,H}) = g(\tau_n, e^{\vec{Y}_{\tau_n}}, X_{\tau_n}, X'_{\tau_n}) . \tag{2.107}$$

Let us now introduce the log-reward functions for the ordinary American-type options for the price process $\vec{Z}'_n$ with the pay-off function $g(n, e^{\vec{y}}, x, e)$, defined for $\vec{z}\,' = (\vec{y}, x, e) \in \mathbb{Z}', 0 \le n \le N$,

$$\phi'_n(\vec{s}, x, e) = \sup_{\tau_n \in \mathcal{M}_{\max,n,N}} \mathsf{E}_{\vec{z}\,',n} g(\tau_n, e^{\vec{Y}_{\tau_n}}, X_{\tau_n}, X'_{\tau_n}) , \tag{2.108}$$

and the optimal expected reward by the relation

$$\Phi' = \sup_{\tau_0 \in \mathcal{M}_{\max,0,N}} \mathsf{E} g(\tau_0, e^{\vec{Y}_{\tau_0}}, X_{\tau_0}, X'_{\tau_0}) . \tag{2.109}$$

Relation (2.107) implies that the log-reward functions for the American-type knockout option for the log-price process $\vec{Z}_n$ (with the pay-off function $g(n, e^{\vec{y}}, x)$ and the stopping knockout time–space domain H) are connected with the log-reward functions for the ordinary American-type option for the log-price process $\vec{Z}'_n$ (with the pay-off function $g(n, e^{\vec{y}}, x, e)$) by the following relation, for $\vec{z}\,' = (\vec{y}, x, e) \in \mathbb{Z}', 0 \le n \le N$:

$$\phi'_n(\vec{y}, x, e) = \begin{cases} \phi_n(\vec{y}, x) & \text{if } (\vec{y}, x) \in \overline{H}_n,\ e = 1 , \\ 0 & \text{if } (\vec{y}, x) \in H_n,\ e = 0 , \\ 0 & \text{if } (\vec{y}, x, e) \in \overline{G}_n . \end{cases} \tag{2.110}$$

while the corresponding optimal expected rewards are connected by the following relation,

$$\Phi' = \Phi . \tag{2.111}$$

3 Backward recurrence reward algorithms

In this chapter, we investigate the structure of backward recurrence relations for log-price processes represented by atomic Markov chains, whose transition probabilities are concentrated on finite sets.

In Sections 3.1 and 3.2, we consider binomial and trinomial models, univariate and multivariate, homogeneous and inhomogeneous in time and space as well as models with Markov modulation. We study the structure of binomial and trinomial trees of nodes generated by binomial and trinomial random walks. A special attention is paid to the so-called recombining condition. This condition guarantees that the number of time–space nodes generated by trajectories of such random walks has not more than polynomial rate of growth as a function of the number of time steps. We also present recurrence systems of linear equations for log-reward functions of American-type options. Also, we present recurrent algorithms for the effective computing of binomial and trinomial probabilities penetrating the above-mentioned recurrence system of linear equations.

In Section 3.3, we consider a general random walk models with distributions of jumps concentrated on finite sets. We describe trees of nodes generated by trajectories of such random walks, formulate the corresponding variants of recombining conditions and give recurrence systems of linear equations for log-reward functions of American-type options.

In Section 3.4, we consider a general model of multivariate modulated atomic Markov chains, including space-skeleton Markov chains and space-skeleton Markov random walks. Recombining conditions are given for these models as well as recurrence system of linear equations for log-reward functions of American-type options.

The characteristic property of the models satisfying recombining conditions is that the corresponding algorithms for finding of log-reward functions have backward recurrence structure with not more than polynomial rate of growth of number of unknown values of log-reward functions as the number of the corresponding time steps. This makes these systems computationally effective.

The corresponding recurrence systems of linear equations for the log-reward functions are given for all above-listed models of atomic Markov chains in Lemmas 3.1.1–3.4.4. These lemmas are particular variants of Theorem 2.3.3 and, therefore, do not require special proofs.

3.1 Binomial tree reward algorithms

In this section, we describe backward tree algorithms for finding log-reward functions and optimal expected rewards for the binomial tree models of log-price processes. We consider classical homogeneous models, model inhomogeneous in time and space as well as models with Markov modulation. Binomial models can be effectively used for approximation of more general log-price processes represented by Gaussian random walks homogeneous and inhomogeneous in time and space.

3.1.1 A classical homogeneous binomial tree model

This is the simplest model, where a log-price Y_n is a univariate random walk, i.e.

$$Y_n = Y_{n-1} + W_n\,, \quad n = 1, 2, \dots . \tag{3.1}$$

where (a) Y_0 is a real-valued random variable with the distribution $P_0(A)$, (b) W_n, $n = 1, 2, \dots$, is a sequence of i.i.d. binary random variables, which take two values δ_+ and δ_- with probabilities, respectively, p_+ and $p_- = 1 - p_+$, (c) the random variable Y_0 and the random sequence W_n, $n = 1, 2, \dots$, are independent.

Usual assumptions also are (d) $\delta_- < \delta_+$, and (e) $0 \le p_+ \le 1$.

In this case, the index variable X_n is not included in the model. The pay-off function $g(n, e^y)$ and the corresponding log-reward functions $\phi_n(y)$ do not depend on the index argument.

Let us assume that the initial state of the log-price process $Y_0 = y_0$ with probability 1, i.e. the initial distribution $P_0(A) = I(y_0 \in A)$ is concentrated at the point $y_0 \in \mathbb{R}_1$.

The tree structure of trajectories is well visible from the recurrent algorithm, which describes sequential transitions of the log-price process Y_n. If $Y_{n-1} = y$ for some $n = 1, 2, \dots$, then, either $Y_n = y + \delta_+$ or $Y_n = y + \delta_-$.

The initial point y_0 determines the unique tree $T_{(y_0,0)}$, with the root in the node $(y_0, 0)$ in the space–time domain $\mathbb{R}_1 \times \{0, 1, \dots\}$.

According to the algorithm described above the tree $T_{(y_0,0)}$ has one initial node $(y_0, 0)$. It includes two additional nodes after the first step, then four additional nodes after the second step, etc. It seems that the tree may include 2^n additional nodes after the nth step.

If this is the case, i.e. the tree $T_{(y_0,0)}$ possesses an exponential rate of growth for the number of nodes as a function of the number of steps n, any algorithm based on such a tree would not be effective from the computational point of view.

However, this is not the case. The tree $T_{(y_0,0)}$ includes only $n + 1$ additional nodes after n steps. In fact, the tree $T_{(y_0,0)}$ satisfies the so-called recombining condition. This condition means that the random walk Y_n, which is in a position $Y_{n-1} = y$ for some $n = 1, 2, \dots$, and makes from this position first jump δ_+ and then jump δ_- should end up at the same point as if it would make the first jump δ_- and then the second jump δ_+. This should hold for any possible position $y \in \mathbb{R}_1$ and $n = 1, 2, \dots$.

Indeed, if $Y_{n-1} = y$ for some $n = 1, 2, \dots$, then $Y_{n+1} = y + \delta_+ + \delta_-$ if $W_n = \delta_+$, $W_{n+1} = \delta_-$ or $Y_{n+1} = y + \delta_- + \delta_+$ if $W_n = \delta_-$, $W_{n+1} = \delta_+$. In both cases Y_{n+1} takes the same value.

The tree $T_{(y_0,0)}$ obviously includes one initial node $(y_0, 0)$ and two additional nodes $(y_0 + \delta_+, 1)$ and $(y_0 + \delta_-, 1)$ after the first step. Due to the recombining property, the tree includes three additional nodes $(y_0 + 2\delta_+, 2)$, $(y_0 + \delta_+ + \delta_-, 2)$ and $(y_0 + 2\delta_-, 2)$ after the second step, etc.

The tree $T_{(y_0,0)}$ includes $n+1$ additional nodes after the nth step. The total number of nodes in the tree after n steps is

$$L_n = 1 + 2 + \cdots + (n+1) = \frac{(n+2)(n+1)}{2}. \tag{3.2}$$

This means that the tree possesses a quadratic rate of growth for the number of nodes in the tree after n steps. This makes algorithms based on such a tree effective from the computational point of view.

The nodes appearing in the tree after the nth step are determined by the number l of jumps δ_+ in n steps but does not depend of moments when these jumps occurred. The number l also determines the number $n - l$ of jumps δ_- in n steps. The states that can be reached by the random walk at moments $n = 0, 1, \ldots,$ are given by the following formula:

$$y_{n,l} = y_0 + \delta_+ l + \delta_-(n-l), \quad l = 0, 1, \ldots, n, \; n = 0, 1, \ldots. \tag{3.3}$$

It is worth to note that nodes appearing in the tree after n steps can alternatively be indexed by the number $r = 2l - n$ that is the difference between the number of jumps δ_+ and δ_- in n steps. We, however, do prefer to use the former indexation by the number l of jumps δ_+ in n steps. As a matter of fact, in this case of a binomial tree model, index r either takes only even values if n is an even number or only odd values if n is an odd number. This does not improve computational aspects for the corresponding backward algorithms, and is not convenient for writing the corresponding formulas.

It should be mentioned that the procedure for the construction of the tree $T_{(y_0,0)}$ described above does not take into account the actual values of probabilities for the appearance nodes in the tree generated by trajectories of the random walk Y_n.

The probability that a node $y_{n,l}$ will be reached after n steps is given by the formula

$$p_{n,l} = C_n^l p_+^l p_-^{(n-l)}, \quad l = 0, 1, \ldots, n, \; n = 0, 1, \ldots, \tag{3.4}$$

where 0^0 should be counted as 1.

Here and henceforth, $C_n^l = \frac{n!}{l!(n-l)!}$.

Note also that numbers $y_{n,l}$ satisfy the inequalities,

$$y_{n,0} < y_{n,1} < \cdots < y_{n,n}, \quad n = 0, 1, \ldots. \tag{3.5}$$

It is useful to note that all these points are located on the grid of points $G[y_0, \delta_+, \delta_-] = \{y_0 + \delta_+ l_+ + \delta_- l_-, \; l_\pm = 0, 1, \ldots\}$. Moreover, in the standard symmetric case, where $\pm\delta_\pm = \delta > 0$, this grid reduces to the simple grid of points, $G[y_0, \delta] = \{y_0 + \delta l, \; l = 0, \pm 1, \ldots\}$.

According to formula (3.3), the tree $T_{(y_0,0)}$ is the following set of nodes:

$$T_{(y_0,0)} = \langle (y_{n,l}, n), l = 0, 1, \dots, n, n = 0, 1, \dots \rangle . \quad (3.6)$$

The tree $T_{(y_0,0)}$ includes a particular point (y, n) if and only if $y_{n,l} = y_0 + \delta_+ l + \delta_-(n - l) = y$ for some $l = 0, 1, \dots, n$.

The condition $\mathbf{A_2}$ holds since, in this case, the following inequality take place for any $y \in \mathbb{R}_1$, $n = 0, 1, \dots, N$:

$$\mathsf{E}_{y,n} \max_{n \le r \le N} \left| g(r, e^{Y_r}) \right| \le \max_{n \le r \le N} \max_{m \le l \le m+r-n} \left| g(r, e^{y_{r,l}}) \right| < \infty , \quad (3.7)$$

where the points $y_{r,l}$ are given by formulas (3.3), with y_0 and $0 \le m \le n$ chosen such that $y_{n,m} = y$.

Theorem 2.3.3 takes in this case the following form.

Lemma 3.1.1. *Let a log-price process Y_n be represented by the binomial random walk defined by the stochastic transition dynamic relation* (3.1). *Then the log-reward functions $\phi_n(y_{n,l})$, for points $y_{n,l} = y_0 + \delta_+ l + \delta_-(n - l)$, $l = 0, 1, \dots, n$, $n = 0, 1, \dots, N$ given by formulas* (3.3) *are the unique solution for the following finite recurrence system of linear equations:*

$$\begin{cases} \phi_N(y_{N,l}) = g(N, e^{y_{N,l}}) , \quad l = 0, 1, \dots, N , \\ \phi_n(y_{n,l}) = \max \left(g(n, e^{y_{n,l}}), \phi_{n+1}(y_{n+1,l}) p_- \right. \\ \qquad \left. + \phi_{n+1}(y_{n+1,l+1}) p_+ \right) , \quad l = 0, 1, \dots, n , \\ \qquad n = N - 1, \dots, 0 . \end{cases} \quad (3.8)$$

This system lets one find the values of the log-reward function $\phi_n(y)$ at points $y_{n,l}$, $l = 0, 1, \dots, n$ given by formulas (3.3), recursively, for $n = N, \dots, 0$.

The particular value $\phi_n(y)$ can be calculated using this system if the initial value y_0 is chosen so that $y_{n,l} = y_0 + \delta_+ l + \delta_-(n - l) = y$ for some $l = 0, 1, \dots, n$.

It is useful to mention that the states $y_{n,l} = y_0 + \delta_+ l + \delta_-(n - l)$ are, in fact, functions of the initial state y_0. Let us use, for the moment, the notation $y_{n,l} = y_{n,l}(y_0)$ to show this explicitly.

Let us now consider the trees $T_{(y_0 + \delta_+ l_+ + \delta_- l_-, 0)}$, $0 \le l_\pm \le r_\pm$, where $0 \le r_\pm \le N$. These trees have joint nodes since the following relation hold for $0 \le l'_\pm \le r'_\pm$, $0 \le l''_\pm \le r''_\pm$, $0 \le l', l'' \le n \le N$:

$$y_{n,l'}(y_0 + \delta_+ l'_+ + \delta_- l'_-) = y_{n,l''}(y_0 + \delta_+ l''_+ + \delta_- l''_-) \text{ if } l'_\pm \pm l' = l''_\pm \pm l'' . \quad (3.9)$$

Thus, one can aggregate the backward recurrence relations (3.8) for initial nodes $(y_0 + \delta_+ l_+ + \delta_- l_-, 0)$, $0 \le l_\pm \le r_\pm$ in one system, and, in this way, reduce computations for simultaneous funding values of reward functions $\phi_n(y)$ at points $y_{n,l}(y_0 + \delta_+ l_+ + \delta_- l_-)$, $0 \le l_\pm \le r_\pm$, $0 \le l \le n \le N$.

Also it is useful to note that every node $(y_{n,m}, n), 0 \le m \le n < \infty$, in the tree $T_{(y_0,0)}$ generates its own subtree which is the following sets of nodes:

$$T_{(y_{n,m},n)} = \langle (y_{n,m} + \delta_+ l + \delta_-(r - n - l), r), \quad l = 0, \dots, r - n, r = n, n + 1, \dots \rangle . \tag{3.10}$$

Moreover, it is easily seen that the backward recurrence relations in the system (3.8) written down for points in the above-reduced tree $T_{(y_{n,m},n)}$ create the reduced system of the backward recurrence relations that let one to compute values of the log-reward function $\phi_r(y)$ at points $(y_{r,l}, r), l = m, \dots, m + r - n, r = n, \dots, N$.

As far as the optimal expected reward $\Phi(\mathcal{M}_{\max,N})$ is concerned, it is given, according to Theorem 2.3.4 by the integral, $\Phi(\mathcal{M}_{\max,N}) = \int_{\mathbb{R}_1} \phi_0(y) P_0(dy)$. In the case, where $P_0(A)$ is a discrete distribution with a finite number of atoms, i.e. $P_0(A) = \sum_{i=1}^{j} p_i I(y_i \in A)$ the above integral takes the form of the sum

$$\Phi(\mathcal{M}_{\max,N}) = \sum_{i=1}^{j} p_i \phi_0(y_i) . \tag{3.11}$$

3.1.2 A binomial tree model inhomogeneous in time and space

In this model, the log-price process is given by the following stochastic transition dynamic relation:

$$\begin{cases} Y_n = Y_{n-1} + W_n \\ W_n = \delta_n^+(Y_{n-1}) I(U_n \le p_n(Y_{n-1})) \\ \qquad + \delta_n^-(Y_{n-1}) I(U_n > p_n(Y_{n-1})) , \quad n = 1, 2, \dots , \end{cases} \tag{3.12}$$

where (a) Y_0 is a real-valued random variable with a distribution $P_0(A)$, (b) $U_n, n = 1, 2, \dots$, is a sequence of i.i.d. random variables uniformly distributed in the interval $[0, 1]$, (c) the random variable Y_0 and the random sequence $U_n, n = 1, 2, \dots$, are independent, (d) $\delta_n^+(y), \delta_n^-(y), n = 1, 2, \dots$, are measurable functions acting from $\mathbb{R}_1$ to $\mathbb{R}_1$, and (e) $p_n(y), n = 1, 2, \dots$, are the measurable functions acting from $\mathbb{R}_1$ to $[0, 1]$.

A usual assumption also is (e) $\delta_n^-(y) < \delta_n^+(y), y \in \mathbb{R}_1, n = 1, 2, \dots$.

In the above model, the index variable X_n is not included in the model. Correspondingly, a pay-off function $g(n, e^y)$ and the corresponding log-reward functions $\phi_n(y)$ do not depend on the index argument.

Let us assume that the initial state of the log-price process $Y_0 = y_0$ with probability 1, i.e. the initial distribution $P_0(A) = I(y_0 \in A)$, is concentrated at the point $y_0 \in \mathbb{R}_1$.

The tree structure of trajectories is analogous to those described above for the simplest binomial tree model. It is well visible from the following recurrent algorithm that describes sequential transitions of the log-price process Y_n. If $Y_{n-1} = y$ for some $n = 0, 1, \dots$, then, either $Y_n = y + \delta_n^+(y)$ with probability $p_n(y)$ or $Y_n = y + \delta_n^-(y)$ with probability $1 - p_n(y)$.

The initial point y_0 determines the unique tree $T_{(y_0,0)}$ with the root in the node $(y_0, 0)$ in the space–time domain $\mathbb{R}_1 \times \{0, 1, \dots\}$.

According to the algorithm described above, the tree $T_{(y_0,0)}$ may include two additional nodes after the first step, four nodes after the second steps, etc. The tree may include 2^n additional nodes after the nth step.

If this is the case, i.e. the tree $T_{(y_0,0)}$ possesses an exponential rate of growth for the number of nodes as a function of the number of steps n, any algorithm based on such a tree would not be effective from the computational point of view.

A recombining condition should hold in order to prevent such a case. As above, this condition means that the random walk Y_n, which is in a position y at a moment $n-1$, and makes first jump $\delta_n^+(y)$ and then the second jump $\delta_{n+1}^-(y + \delta_n^+(y))$ should end up at the same point as if it would make the first jump $\delta_n^-(y)$ and then the second jump $\delta_{n+1}^+(y + \delta_n^-(y))$. In this case, the recombining condition does not hold automatically. This should hold for any possible position of the random walk.

If $Y_{n-1} = y$ for some $n = 1, \dots,$, then $Y_{n+1} = y + \delta_n^+(y) + \delta_{n+1}^-(y + \delta_n^+(y))$ if $W_n = \delta_n^+(y)$, $W_{n+1} = \delta_{n+1}^-(y + \delta_n^+(y))$, while $Y_{n+1} = y + \delta_n^-(y) + \delta_{n+1}^+(y + \delta_n^-(y))$ if $W_n = \delta_n^-(y)$, $W_{n+1} = \delta_{n+1}^+(y + \delta_n^-(y))$. In order for the recombining condition to hold, Y_{n+1} should take the same value in both cases, for any $y \in \mathbb{R}_1$, $n = 1, \dots$.

Thus, the recombining condition requires the following relation to hold for any $y \in \mathbb{R}_1$, $n = 1, \dots$:

$$\delta_n^+(y) + \delta_{n+1}^-(y + \delta_n^+(y)) = \delta_n^-(y) + \delta_{n+1}^+(y + \delta_n^-(y)) . \tag{3.13}$$

This condition obviously holds for the model with values of jumps homogeneous in time and space, i.e. where there exists $\delta_- < \delta_+$ such that $\delta_n^\pm(y) = \delta_\pm$, $y \in \mathbb{R}_1$, $n = 1, 2, \dots$.

In the case, of the model with values of jumps inhomogeneous in time but homogeneous in space, i.e. where $\delta_n^\pm(y) = \delta_n^\pm, y \in \mathbb{R}_1$, $n = 1, \dots$, the recombining condition takes the form of the following relation:

$$\delta_n^+ + \delta_{n+1}^- = \delta_n^- + \delta_{n+1}^+ , \quad n = 1, 2, \dots . \tag{3.14}$$

This relation means that the difference $\delta_n^+ - \delta_n^-$ does not depend on n, i.e. there exists $\delta > 0$ such that

$$\delta_n^+ - \delta_n^- = 2\delta , \quad n = 1, 2, \dots . \tag{3.15}$$

Let us denote

$$\delta_n^\circ = \frac{\delta_n^+ + \delta_n^-}{2} , \quad n = 1, 2, \dots . \tag{3.16}$$

Then the recombining condition can be rewritten in the following form:

$$\delta_n^+ = \delta + \delta_n^\circ , \quad \delta_n^- = -\delta + \delta_n^\circ , \quad n = 1, 2, \dots . \tag{3.17}$$

As an example of the model with values of jumps inhomogeneous in time and space, let us consider a model, where values of jumps are linear functions of y, i.e. $\delta_n^{\pm}(y) = \delta_n^{\pm} y, y \in \mathbb{R}_1, n = 1, 2, \ldots$. In this case, the recombining condition takes the form of the following relation:

$$\delta_n^+ + \delta_{n+1}^- + \delta_{n+1}^- \delta_n^+ = \delta_n^- + \delta_{n+1}^+ + \delta_{n+1}^+ \delta_n^- , \quad n = 1, 2, \ldots . \tag{3.18}$$

This condition holds, for example, if coefficients $\delta_n^{\pm} = \delta^{\pm}, n = 1, 2, \ldots$, do not depend on n.

The tree $T_{(y_0,0)}$ obviously contains one initial node $(y_0, 0)$ and two nodes $(y_0 + \delta_n^+(y_0), 1)$ and $(y_0 + \delta_n^-(y_0), 1)$ after the first step. If the recombining condition holds, the tree contains three nodes $(y_0 + \delta_1^+(y_0) + \delta_2^+(y_0 + \delta_1^+(y_0)), 2)$, $(y_0 + \delta_1^+(y_0) + \delta_2^-(y_0 + \delta_1^+(y_0)), 2) = (y_0 + \delta_1^-(y_0) + \delta_2^+(y_0 + \delta_1^-(y_0)), 2)$, and $(y_0 + \delta_1^-(y_0) + \delta_2^-(y_0 + \delta_1^-(y_0)), 2)$, after the second step, etc.

If the recombining condition holds, the node with l jumps $\delta_\cdot^+(\cdot)$ in n steps does not depend of the order of such jumps in n steps. This makes it possible to index these nodes by parameter l and to use the same notation $(y_{n,l}, n), l = 0, 1, \ldots, n$ for nodes appearing in the tree after n steps.

The states $y_{n,l_n}, l_n = 0, 1, \ldots, n$, that can be reached by the log-price process Y_n with the initial value $Y_0 = y_0$ at moments $n = 0, 1, \ldots$, are given by the following recurrence relation, for $0 \leq l_n \leq n, n = 0, 1, \ldots$:

$$y_{n+1,l_{n+1}} = y_{n,l_n} + \begin{cases} \delta_{n+1}^+(y_{n,l}) & \text{if } l_{n+1} = l_n + 1 , \\ \delta_{n+1}^-(y_{n,l}) & \text{if } l_{n+1} = l_n - 1 , \end{cases} \tag{3.19}$$

where $y_{0,0} = y_0$.

Note that numbers y_{n,l_n} satisfy the inequalities

$$y_{n,0} < y_{n,1} < \cdots < y_{n,n} , \quad n = 0, 1, \ldots . \tag{3.20}$$

According to the above remarks, the tree $T_{(y_0,0)}$ is the following set of nodes:

$$T_{(y_0,0)} = \langle (y_{n,l}, n), l = 0, 1, \ldots, n, n = 0, 1, \ldots \rangle . \tag{3.21}$$

If the recombining condition holds, the number of additional nodes appearing in the tree $T_{(y_0,0)}$ after the nth step is $n + 1$. The total number of nodes in the tree $T_{(y_0,0)}$ after n steps is again $L_n = 1 + 2 \cdots + (n + 1) = \frac{(n+2)(n+1)}{2}$. This means that the tree has a quadratic rate of growth for the number of nodes in the tree as a function of the number of steps n. This makes algorithms based on such a tree effective from the computational point of view.

The tree $T_{(y_0,0)}$ includes a particular point (y, n) if and only if $y_{n,l} = y$ for some $l = 0, 1, \ldots, n$.

The condition $\mathbf{A_2}$ holds since, in this case, the following inequality take place for any $y \in \mathbb{R}_1, n = 0, 1, \ldots, N$:

$$\mathsf{E}_{y,n} \max_{n \leq r \leq N} \left| g(r, e^{Y_r}) \right| \leq \max_{n \leq r \leq N} \max_{m \leq l \leq m+r-n} \left| g(r, e^{y_{r,l}}) \right| < \infty , \tag{3.22}$$

where the points $y_{r,l}$ are given by formulas (3.19), with y_0 and $0 \le m \le n$ chosen such that $y_{n,m} = y$.

Theorem 2.3.3 takes in this case the following form.

Lemma 3.1.2. *Let the log-price process Y_n be represented by the inhomogeneous in space and time binomial random walk defined by the stochastic transition dynamic relation* (3.12). *Then the log-reward functions $\phi_n(y_{n,l})$, for points $y_{n,l}$, $l = 0, 1, \dots, n$, $n = 0, 1, \dots, N$ given by formulas* (3.19), *are the unique solution for the following finite recurrence system of linear equations:*

$$\begin{cases} \phi_N(y_{N,l}) = g(N, e^{y_{N,l}}), \quad l = 0, 1, \dots, N, \\ \phi_n(y_{n,l}) = \max\left(g(n, e^{y_{n,l}}), \phi_{n+1}(y_{n+1,l})p_n^-(y_{n,l}) \right. \\ \qquad \left. + \phi_{n+1}(y_{n+1,l+1})p_n^+(y_{n,l})\right), \quad l = 0, 1, \dots, n, \\ \qquad n = N-1, \dots, 0. \end{cases} \tag{3.23}$$

This system lets one find values of the log-reward function $\phi_n(y)$ at points $y_{n,l}$, $l = 0, 1, \dots, n$ given by formulas (3.19), recursively, for $n = N, \dots, 0$.

The particular value $\phi_n(y)$ can be calculated by using this system if the initial value y_0 is chosen so that $y_{n,l} = y$ for some $l = 0, 1, \dots, n$.

The extension and reduction of the system of backward recurrence relations (3.23) can be realized in the same way as it is described in Subsection 3.1.1, for the system of backward recurrence relations (3.8).

Also, the remarks concerned finding of optimal expected reward, analogous to those made in Subsection 3.1.1 can be made.

3.1.3 A multivariate binomial tree model

This is the model where the multivariate Markov log-price $\vec{Y}_n$ is a k-dimensional random walk,

$$\vec{Y}_n = \vec{Y}_{n-1} + \vec{W}_n, \quad n = 1, 2, \dots. \tag{3.24}$$

where (a) $\vec{Y}_0 = (Y_{0,1}, \dots, Y_{0,k})$ is a random vector, which takes values in $\mathbb{R}_k$ and has a distribution $P_0(A)$, (b) $\vec{W}_n = (W_{n,1}, \dots, W_{n,k})$, $n = 1, 2, \dots$, is a sequence of k-dimensional independent random vectors taking values $\bar{\delta}_{\bar{\imath}} = (\delta_{1,\imath_1}, \dots, \delta_{k,\imath_k})$ with probabilities $p_{\bar{\imath}}$ for $\bar{\imath} \in \mathcal{I}_k$, where $\mathcal{I}_k = \{\bar{\imath} = (\imath_1, \dots, \imath_k), \imath_1, \dots, \imath_k = \pm\}$, and (c) the random vector $\vec{Y}_0$ and the random sequence $\vec{W}_n$, $n = 1, 2, \dots$, are independent.

Also usual assumptions are (d) $\delta_{j,-} < \delta_{j,+}$, $j = 1, \dots, k$, and (e) $p_{\bar{\imath}} \ge 0$, $\bar{\imath} \in \mathcal{I}_k$, $\sum_{\bar{\imath} \in \mathcal{I}} p_{\bar{\imath}} = 1$.

In this case, the index component X_n is not included in the model. Correspondingly, a pay-off function $g(n, e^{\vec{y}})$ and the corresponding log-reward functions $\phi_n(\vec{y})$ do not depend on the index argument.

Let us assume that the initial state of the log-price process $\vec{Y}_0 = \vec{y}_0$ with probability 1, i.e. the initial distribution $P_0(A) = I(\vec{y}_0 \in A)$, is concentrated at the point $\vec{y}_0 \in \mathbb{R}_k$.

The multivariate tree structure of trajectories is analogous to those described above for the simplest binomial tree model in the one-dimensional case. It is well visible from the following recurrent algorithm that describes the sequential transitions of the log-price process $\vec{Y}_n$. If $\vec{Y}_n = \vec{y}$ for some $n = 0, 1, \dots$, then $\vec{Y}_{n+1} = \vec{y} + \bar{\delta}_{\bar{\imath}}$ with probability $p_{\bar{\imath}}$, where $\bar{\imath} \in \mathcal{I}_k$. Note that the set $\mathcal{I}_k$ contains 2^k elements.

The initial point $\vec{y}_0 = (y_{0,1}, \dots, y_{0,k})$ determines the unique tree $T_{(\vec{y}_0,0)}$ with the root in the node $(\vec{y}_0, 0)$ in the space–time domain $\mathbb{R}_k \times \{0, 1, \dots\}$.

According to the algorithm described above the tree $T_{(\vec{y}_0,0)}$ includes one initial node and may include 2^k additional nodes after the first step, 2^{2k} nodes after the second step, etc. The tree may include 2^{nk} additional nodes after the nth step.

If this is the case, i.e. the tree $T_{(\vec{y}_0,0)}$ possesses an exponential rate of growth for the number of nodes as a function of the number of steps n, any algorithm based on such a tree would not be effective from the computational point of view.

However, this is not the case. In fact, the tree includes only $(n+1)^k$ additional nodes after the nth step. This is because the tree $T_{(\vec{y}_0,0)}$ satisfies the recombining condition. This property is inherited by the multivariate tree $T_{(\vec{y}_0,0)}$ generated by the vector random walk $\vec{Y}_n$ from its one-dimensional components $T_{(y_{0,j},0)}$ generated by one-dimensional random walks $Y_{n,j}$, for every $j = 1, \dots, k$.

Let us denote $\bar{\imath}_n = (\imath_{n,1}, \dots, \imath_{n,k})$, $n = 1, 2, \dots$.

The tree $T_{(\vec{y}_0,0)}$ obviously includes one initial node $(\vec{y}_0, 0)$ and 2^k additional nodes $(\vec{y}_0 + \bar{\delta}_{\bar{\imath}_1}, 1), \bar{\imath}_1 \in \mathcal{I}_k$ after the first step. Due to the recombining property, the tree includes 3^k additional nodes $(y_0 + \bar{\delta}_{\bar{\imath}_1} + \bar{\delta}_{\bar{\imath}_2}, 2), \bar{\imath}_1, \bar{\imath}_2 \in \mathcal{I}_k$ after the second step. Indeed, the jth component of the vector $\bar{\delta}_{\bar{\imath}_1} + \bar{\delta}_{\bar{\imath}_2} = (\delta_{1,\imath_{1,1}} + \delta_{2,\imath_{2,1}}, \dots, \delta_{k,\imath_{1,k}} + \delta_{k,\imath_{2,k}})$ can take three different values $2\delta_{j,+}$, $\delta_{j,+} + \delta_{j,-}$, and $2\delta_{j,-}$. Subsequently, the tree includes 4^k additional nodes after the third step, etc.

The tree $T_{(\vec{y}_0,0)}$ includes $(n+1)^k$ additional nodes after the nth step. The total number of nodes in the tree after n steps is $L_n = 1 + 2^k + \dots + (n+1)^k$. Obviously,

$$\frac{(n+1)^k}{k+1} = \int_0^{n+1} x^k dx \le L_n \le \int_1^{n+2} x^k dx = \frac{(n+2)^k - 1}{k+1}. \tag{3.25}$$

Therefore, the tree possesses a polynomial of the order $k+1$ rate of growth for the number of nodes in the tree after n steps as a function of n. This makes algorithms based on such a tree effective from the computational point of view.

The states appearing after n steps are determined by the vector $\bar{l} = (l_1, \dots, l_k)$, where l_j is the number of jumps $\delta_{j,+}$ in n steps, for the component $Y_{n,j}$ of the random walk $\vec{Y}_n$. But these states do not depend of moments when these jumps occur. States that can be reached by this random walk at moments $n = 0, 1, \dots, N$ are given by the following relation:

$$\vec{y}_{n,\bar{l}} = (y_{n,1,l_1}, \dots, y_{n,k,l_k}), \quad \bar{l} = (l_1, \dots, l_k) \in \mathbb{L}_n, \quad n = 0, 1, \dots, \tag{3.26}$$

where the set $\mathbb{L}_n = \{\bar{l} : 0 \le l_1, \dots, l_k \le n\}$ and components $y_{n,j,l_j} = y_{0,j} + \delta_{j,+} l_j + \delta_{j,-}(n - l_j)$, $j = 1, \dots, k$.

Note also that numbers y_{n,j,l_j} satisfy the inequalities

$$y_{n,j,0} < y_{n,j,1} < \cdots < y_{n,j,n}, \quad j = 1, \dots, k, \quad n = 0, 1, \dots. \tag{3.27}$$

According to the above remarks, the tree $T_{(y_0,0)}$ is the following set of nodes:

$$T_{(\vec{y}_0,0)} = \langle (\vec{y}_{n,\bar{l}}, n), \bar{l} = (l_1, \dots, l_k) \in \mathbb{L}_n, n = 0, 1, \dots \rangle. \tag{3.28}$$

The tree $T_{(\vec{y}_0,0)}$ includes a particular point $(\vec{y}, n)$ if $\vec{y}_{n,\bar{l}} = y$ for some $\bar{l} \in \mathbb{L}_n$.

The condition $\mathbf{A_2}$ holds since, in this case, the following inequality takes place for any $y \in \mathbb{R}_1$, $n = 0, 1, \dots, N$:

$$\mathsf{E}_{\vec{y},n} \max_{n \le r \le N} \left| g(r, e^{\vec{Y}_r}) \right| \le \max_{n \le r \le N} \max_{m_j \le l_j \le m_j + r - n, \quad j=1,\dots,k} \left| g(r, e^{\vec{y}_{r,\bar{l}}}) \right| < \infty, \tag{3.29}$$

where the points $\vec{y}_{r,\bar{l}}$ are given by formulas (3.26), with $\vec{y}_0$ and $\vec{m} = (m_1, \dots, m_k)$, $0 \le m_j \le n$, $j = 1, \dots, k$ chosen such that $\vec{y}_{n,\vec{m}} = \vec{y}$.

Theorem 2.3.3 takes the following form.

Lemma 3.1.3. *Let the log-price process Y_n be represented by the multivariate binomial random walk defined by the stochastic transition dynamic relation* (3.24). *Then the log-reward functions $\phi_n(\vec{y}_{n,l})$, for points $\vec{y}_{n,\bar{l}}$, $\bar{l} = (l_1, \dots, l_k) \in \mathbb{L}_n$, $n = 0, 1, \dots, N$ given by formulas* (3.26), *are the unique solution for the following finite recurrence system of linear equations:*

$$\begin{cases} \phi_N(\vec{y}_{N,\bar{l}}) = g(N, e^{\vec{y}_{N,\bar{l}}}), \quad \bar{l} \in \mathbb{L}_N, \\ \phi_n(\vec{y}_{n,\bar{l}}) = \max\left(g(n, e^{\vec{y}_{n,\bar{l}}}), \sum_{\bar{\imath} \in \mathcal{I}_k} \phi_{n+1}(\vec{y}_{n+1,\bar{l}+\bar{e}_{\bar{\imath}}}) p_{\bar{\imath}} \right), \quad \bar{l} \in \mathbb{L}_n, \\ \qquad n = N-1, \dots, 0. \end{cases} \tag{3.30}$$

where vectors $\bar{e}_{\bar{\imath}} = (e_{\imath_1}, \dots, e_{\imath_k})$, $e_{\imath_j} = 1$ if $\imath_j = +$ or $e_{\imath_j} = 0$ if $\imath_j = -$, for $j = 1, \dots, k$.

This system of equations (3.30) let one find the values of the log-reward function $\phi_n(\vec{y})$ at points $\vec{y}_{n,\bar{l}}$, $\bar{l} \in \mathbb{L}_n$ given by formulas (3.26), recursively, for $n = N, \dots, 0$.

The particular value $\phi_n(\vec{y})$ can be found using this system if the initial value $\vec{y}_0$ is chosen so that $\vec{y}_{n,\bar{l}} = \vec{y}$ for some $\bar{l} \in \mathbb{L}_n$.

The extension and reduction of the system of backward recurrence relations (3.30) can be realized in the same way as it is described in Subsection 3.1.1 for the system of backward recurrence relations (3.8).

Also, the remarks concerning finding of optimal expected reward analogous to those made in Subsection 3.1.1 can be made.

3.1.4 A binomial tree model modulated by a Markov chain

We consider a univariate, homogeneous in time and space log-price processes represented by the binomial random walk modulated by a homogeneous in time Markov chain. We restrict consideration by the model where values of jumps do not depend while the corresponding jump probabilities do depend on states of the modulating Markov chain. We also assume that the modulating Markov chain has a finite phase space $\mathbb{X} = \{1, \dots, m\}$.

In this model, the log-price process Y_n is given by the following stochastic transition dynamic relation:

$$Y_n = Y_{n-1} + W_n(X_{n-1}), \quad n = 1, 2, \dots, \tag{3.31}$$

where (a) Y_0 is a real-valued random variable with a distribution $P_0(A)$, (b) $W_n(x), x \in \mathbb{X}, n = 1, 2, \dots,$ is a family of independent random variables taking values δ_+ and δ_- with probabilities $p_+(x)$ and $p_-(x) = 1 - p_+(x)$ which are independent of $n = 0, 1, \dots,$ but are dependent on $x \in \mathbb{X}$, (c) $X_n, n = 0, 1, \dots,$ is a homogeneous Markov chain with the phase space $\mathbb{X}$, an initial distribution $P(x) = \mathsf{P}\{X_0 = x\}, x \in \mathbb{X}$ and one-step transition probabilities $P(x', x'') = \mathsf{P}\{X_n = x''/X_{n-1} = x'\}$, $x', x'' \in \mathbb{X}, n = 1, 2, \dots$, (d) the random variable $\vec{Y}_0$, the family of random sequence $\vec{W}_n(x), x \in \mathbb{X}, n = 1, 2, \dots,$ and the Markov chain $X_n, n = 0, 1, \dots,$ are independent.

Also usual assumptions are (e) $\delta_- < \delta_+$ and (f) $0 \leq p_+(x) \leq 1, x \in \mathbb{X}$.

The Markov modulating index X_n is included in the model. Correspondingly, a pay-off function $g(n, e^y, x)$ and the corresponding log-reward functions $\phi_n(y, x)$ do depend on the index argument.

Let us assume that the initial state of the modulated log-price process $Z_0 = (Y_0, X_0) = z_0 = (y_0, x_0)$ with probability 1, i.e. the initial distribution $P_0(A) = I(y_0 \in A, x = x_0)$ is concentrated at the point $z_0 \in \mathbb{Z} = \mathbb{R}_1 \times \mathbb{X}$.

The tree structure is well visible from the following recurrent algorithm that describe the sequential transitions of the modulated log-price process $Z_n = (Y_n, X_n)$. If $Z_{n-1} = (y, x')$ for some $n = 0, 1, \dots, N-1$, then $Z_n = (y + \delta_+, x'')$ or $Z_{n+1} = (y + \delta_-, x'')$, where $x'' \in \mathbb{X}$.

The initial point $z_0 = (y_0, x_0)$ determines the unique tree $T_{((y_0,x_0),0)}$ with the root in the node $((y_0, x_0), 0)$ in the space–time domain $\mathbb{Z} \times \{0, 1, \dots\}$.

According to the algorithm described above the tree $T_{((y_0,x_0),0)}$ may include $2m$ nodes after the first step, $(2m)^2$ additional nodes after the second steps, etc. The tree may include $(2m)^n$ additional nodes after the nth step.

If this is the case, i.e. the tree $T_{((y_0,x_0),0)}$ possesses an exponential rate of growth for the number of nodes as a function of the number of steps n, any algorithm based on such a tree would not be effective from the computational point of view.

But it is not the case, since the recombining condition holds. This condition means that the modulated random walk $Z_n = (Y_n, X_n)$, which is in a state (y, x') at a moment $n-1$, and has the first transition in a state $(y + \delta_+, x'')$ and then the second transition to a state $(y + \delta_+ + \delta_-, x''')$ ends up at the same point as if it has first a transition in a state $(y + \delta_-, x'')$ and then a transition in a state $(y + \delta_- + \delta_+, x''')$. This holds for any $y \in \mathbb{R}_1$ and $x', x'', x''' \in \mathbb{X}$.

The tree $T_{((y_0,x_0),0)}$ includes one initial node $((y_0, x_0), 0)$ and $2m$ additional nodes $((y_0 + \delta_+, x_1), 1), ((y_0 + \delta_-, x_1), 1), x_1 \in \mathbb{X}$ after the first step. Due to the recombining property, the tree includes $3m$ additional nodes $((y_0 + 2\delta_+, x_2), 2), ((y_0 + \delta_+ + \delta_-, x_2), 2), ((y_0 + 2\delta_-, x_2), 2), x_2 \in \mathbb{X}$ after the second step, etc.

The states that can appear in the tree after the nth step depend on the number l of jumps δ_+ in n steps for the random walk Y_n but does not depend of moments when these jumps occurred. They are given by the following formulas:

$$y_{n,l} = y_0 + \delta_+ l + \delta_-(n - l), \quad l = 0, 1, \dots, n, \quad n = 0, 1, \dots . \tag{3.32}$$

Note also that these numbers satisfy the inequalities

$$y_{n,0} < y_{n,1} < \cdots < y_{n,n}, \quad n = 0, 1, \dots . \tag{3.33}$$

The nodes appearing in the tree $T_{((y_0,x_0),0)}$ after the n steps are given by the following formula:

$$z_{n,l} = (y_{n,l}, x_n), \quad l = 0, 1, \dots, n, \quad n = 0, 1, \dots , \tag{3.34}$$

where $y_{n,l} = y_0 + \delta_+ l + \delta_-(n - l) \in \mathbb{R}_1$, $x_n \in \mathbb{X}$.

It is useful to note that all points $z_{n,l}$ are located on the grid of points $G_m[y_0, \delta_+, \delta_-] = \{(y_0 + \delta_+ l_+ + \delta_- l_-, x),\ l_\pm = 0, 1, \dots, x \in \mathbb{X}\}$. Moreover, in the standard symmetric case, where $\pm\delta_\pm = \delta > 0$, this grid reduces to the simple grid of points, $G_m[y_0, \delta] = \{(y_0 + \delta l, x), l = 0, \pm 1, \dots, x \in \mathbb{X}\}$.

The tree $T_{((y_0,x_0),0)}$ includes $(n+1)m$ additional nodes after the nth step. The total number of nodes in the tree after n steps is $L_n = 1 + 2m + \cdots + (n + 1)m = 1 - m + \frac{(n+2)(n+1)m}{2}$. This means that the tree possesses a quadratic rate of growth for the number of nodes in the tree after n steps as a function of n. This makes algorithms based on such a tree effective from the computational point of view.

According to formulas (3.34), the tree $T_{((y_0,x_0),0)}$ is the following set of nodes:

$$T_{((y_0,x_0),0)} = \langle ((y_{n,l}, x_n), n), l = 0, 1, \dots, n, x_n \in \mathbb{X}, n = 0, 1, \dots \rangle . \tag{3.35}$$

The tree $T_{((y_0,x_0),0)}$ includes a particular node $((y, x), n)$ if $y_{n,l} = y_0 + \delta_+ l + \delta_-(n - l) = y$, $x_n = x$ for some $l = 0, 1, \dots, n$.

The condition $\mathbf{A_2}$ holds since, in this case, the following inequality takes place for any $z = (y, x) \in \mathbb{Z}$, $n = 0, 1, \dots, N$:

$$\begin{aligned} &\mathsf{E}_{z,n} \max_{n \le r \le N} \left| g(r, e^{Y_r}, X_r) \right| \\ &\quad \le \max_{n \le r \le N} \max_{m \le l \le m+r-n,\, x_r \in \mathbb{X}} \left| g(r, e^{y_{r,l}}, x_r) \right| < \infty , \end{aligned} \tag{3.36}$$

where the points $y_{r,l}$ are given by formulas (3.32), with y_0 and $0 \leq m \leq n$ chosen such that $y_{n,m} = y$.

Theorem 2.3.3 takes in this case the following form.

Lemma 3.1.4. *Let the log-price process Y_n be represented by the binomialMmodulated random walk defined by the stochastic transition dynamic relation* (3.31)*. Then the log-reward functions $\phi_n(y_{n,l})$, for points $z_{n,l} = (y_{n,l}, x_n)$, $l = 0, 1, \dots, n$, $n = 0, 1, \dots, N$ given by formulas* (3.34)*, are the unique solution for the following finite recurrence system of linear equations:*

$$\begin{cases} \phi_N(y_{N,l}, x_N) = g(N, e^{y_{N,l}}, x_N)\,, \\ \qquad l = 0, 1, \dots, N\,, \quad x_N \in \mathbb{X}\,, \\ \phi_n(y_{n,l}, x_n) = \max\Bigg(g(n, e^{y_{n,l}}, x_n)\,, \\ \qquad \sum\limits_{x_{n+1} \in \mathbb{X}} \Big(\phi_{n+1}(y_{n+1,l}, x_{n+1}) p_-(x_n) P(x_n, x_{n+1}) \\ \qquad + \phi_{n+1}(y_{n+1,l+1}, x_{n+1}) p_+(x_n) P(x_n, x_{n+1})\Big)\Bigg)\,, \\ \qquad l = 0, 1, \dots, n,\ x_n \in \mathbb{X}\,, \\ \qquad n = N-1, \dots, 0\,. \end{cases} \tag{3.37}$$

This system lets one find the values of the log-reward function $\phi_n(y, x)$ at points $(y_{n,l}, x_n)$, $l = 0, 1, \dots, n$ given by formulas (3.34), recursively, for $n = N, \dots, 0$.

The particular value $\phi_n(y, x)$ can be found using this system if the initial point y_0 is chosen so that $y_{n,l} = y_0 + \delta_+ l + \delta_-(n-l) = y$ and $x_n = x$, for some $l = 0, 1, \dots, n$.

3.2 Trinomial tree reward algorithms

In this section, we describe backward tree algorithms for finding log-reward functions and optimal expected rewards for the trinomial models of log-price processes. We consider classical homogeneous models, model inhomogeneous in time and space as well as models with Markov modulation. Trinomial models models can be effectively used to approximate more general log-price processes such as Gaussian random walks in such cases where binomial models do not work well, for example, for multivariate models inhomogeneous in time and space.

3.2.1 A classical homogeneous trinomial tree model

This is the simplest model where the log-price Y_n is a univariate random walk given by the following stochastic transition dynamic relation:

$$Y_n = Y_{n-1} + W_n\,, \quad n = 1, 2, \dots . \tag{3.38}$$

where (a) Y_0 is a real-valued random variable with the distribution $P_0(A)$, (b) W_n, $n = 1, 2, \dots$, is a sequence of i.i.d. random variables which can take three values δ_+, $\delta_\circ$ and δ_- with probabilities, respectively, $p_+, p_\circ$, and p_- and (c) the random variable Y_0 and the random sequence W_n, $n = 1, 2, \dots$, are independent.

Also, usual assumptions are (d) $\delta_- < \delta_\circ < \delta_+$, and (e) $p_+, p_\circ, p_- \geq 0$, $p_+ + p_\circ + p_- = 1$.

It is obvious that the trinomial model is a generalization of the binomial model. The trinomial model reduces to the binomial model if probability $p_\circ = 0$.

The main advantage of the trinomial model is in the presence of additional parameters that may make fitting of trinomial models to parameters of inhomogeneous in time and space models of Gaussian random walks more effective than for the binomial model.

The index X_n is not included in the model. Correspondingly, a pay-off function $g(n, e^y)$ and the corresponding log-reward functions $\phi_n(y)$ do not depend on the index argument.

Let us assume that the initial state of the log-price process $Y_0 = y_0$ with probability 1, i.e. the initial distribution $P_0(A) = I(y_0 \in A)$ is concentrated at the point $y_0 \in \mathbb{R}_1$.

The tree structure of trajectories is well visible from the following recurrent algorithm that describes sequential transitions of the log-price process Y_n. If $Y_n = y$, for some $n = 0, 1, \dots$, then, either $Y_{n+1} = y + \delta_+$ with probability p_+ or $Y_{n+1} = y + \delta_\circ$ with probability $p_\circ$, or $Y_{n+1} = y + \delta_-$ with probability p_-.

The initial point y_0 determines the unique tree $T_{(y_0,0)}$ with the root in the node $(y_0, 0)$ in the space–time domain $\mathbb{R}_1 \times \{0, 1, \dots\}$.

According to the algorithm described above the tree $T_{(y_0,0)}$ includes one initial node, may include three additional nodes after the first step, nine additional nodes after the second step, etc. It seems that the tree may include 3^n additional nodes after the nth step.

If this is the case, the tree $T_{(y_0,0)}$ possesses an exponential rate of growth for the number of nodes in the tree as a function of the number of steps, any algorithm based on such a tree would not be effective from the computational point of view.

However, this is not the case if some natural recombining condition holds.

Note that the pairwise recombining condition analogous to those formulated for binomial random walk obviously holds for any pair of possible jumps δ_+ and $\delta_\circ$, or δ_+ and δ_-, or $\delta_\circ$ and δ_-. The random walk Y_n, which makes the first jump δ_+ and then the second jump $\delta_\circ$, or makes the first a jump $\delta_\circ$ and then the second jump δ_+, will end at the same point. The similar recombining condition holds for another two pairs of jumps pointed above. These pairwise recombining conditions hold for any possible position of the random walk.

But, in the trinomial model, an additional recombining condition may also hold. This condition means that the random walk Y_n which makes the first jump δ_+ and then the second jump δ_-, or makes the first jump δ_- and then the second jump δ_+, or makes both jumps $\delta_\circ$, should end up at the same point. This should hold for any possible position of the random walk. This leads to the following recombining condition:

$$\delta_\circ = \frac{\delta_- + \delta_+}{2} . \tag{3.39}$$

Indeed, if $Y_{n-1} = y$ for some $n = 1, 2, \dots$, then $Y_{n+1} = y + \delta_+ + \delta_-$ if $W_n = \delta_+, W_{n+1} = \delta_-$ or $Y_{n+1} = y + \delta_- + \delta_+$ if $W_n = \delta_-, W_{n+1} = \delta_+$ or $Y_{n+1} = y + 2\delta_\circ = y + \delta_- + \delta_+$ if $W_n = \delta_\circ, W_{n+1} = \delta_\circ$. In all cases, Y_{n+1} takes the same value.

The tree $T_{(y_0,0)}$ obviously includes one initial node $(y_0, 0)$ and three additional nodes $(y_0 + \delta_+, 1)$, $(y_0 + \delta_\circ, 1)$, and $(y_0 + \delta_-, 1)$ after the first step. Due to the recombining property, the tree includes five additional nodes $(y_0 + 2\delta_+, 2)$, $(y_0 + \delta_+ + \delta_\circ, 2)$, $(y_0 + 2\delta_\circ, 2)$, $(y_0 + \delta_\circ + \delta_-, 2)$, and $(y_0 + 2\delta_-, 2)$ after the second step, etc.

If the recombining condition (3.39) holds, then the tree includes $2n + 1$ additional nodes after the nth step. The total number of nodes in the tree after n steps is

$$L_n = 1 + 3 + \cdots + (2n + 1) = (n + 1)^2 . \tag{3.40}$$

This means that the tree possesses a quadratic rate of growth for the number of nodes in the tree after n steps as a function of n. This makes algorithms based on such a tree effective from the computational point of view.

The nodes appearing in the tree after n steps are determined by the numbers l_+ and l_- of jumps δ_+ and δ_- in n steps, which also determine the number $l_\circ = n - l_+ - l_-$ of jumps $\delta_\circ$ in n steps. What is important is that these nodes do not depend of moments when these jumps occurred. The states that can be reached by the trinomial random walk at moments $n = 0, 1, \dots, N$ are given by the following formula:

$$y_{n,l_+,l_-} = y_0 + \delta_+ l_+ + \delta_- l_- + \delta_\circ (n - l_+ - l_-) , \tag{3.41}$$

where $l_+, l_- \geq 0$, $l_+ + l_- \leq n$, $n = 0, 1, \dots, N$.

A disadvantage of the above formula is that it does not count one-to-one possible states of the random walk. Different pairs of parameters l_+ and l_- can, in fact, determine the same state. Let us denote

$$\delta = \delta_+ - \delta_\circ = \delta_\circ - \delta_- = \frac{\delta_+ - \delta_-}{2} . \tag{3.42}$$

Now, we can rewrite formulas for y_{n,l_+,l_-} in the following form, $y_{n,l_+,l_-} = y_0 + (\delta_+ - \delta_\circ) l_+ + (\delta_- - \delta_\circ) l_- + \delta_\circ n = y_0 + \delta(l_+ - l_-) + \delta_\circ n$. This form shows that states that can be reached by the trinomial random walk at moments $n = 0, 1, \dots, N$

can be indexed by the one parameter $l = l_+ - l_-$, which is the difference between the numbers of jump δ_+ and δ_- in n steps. This parameter takes values $-n, \dots, 0, \dots, n$ and counts one-to-one states that can be reached by the trinomial random walk at moments $n = 0, 1, \dots, N$.

According to the above remarks, the states that can be reached by the trinomial random walk at moments $n = 0, 1, \dots,$ are given by the following formula:

$$y_{n,l} = y_0 + \delta l + \delta_\circ n\,, \quad l = 0, \pm 1, \dots, \pm n\,, \quad n = 0, 1, \dots\,. \tag{3.43}$$

It is useful to note that, under the recombining condition (3.39), the random jumps W_n can be represented in the form $W_n = W'_n + \delta_\circ$, $n = 1, 2, \dots,$ where W'_n, $n = 1, 2, \dots,$ are i.i.d. random variables taking values δ, $-\delta$ and 0 with probabilities, respectively, p_+, p_- and $p_\circ = 1 - p_+ - p_-$. Correspondingly, the basic relation (3.38) defining the trinomial random walk Y_n can be rewritten in the following form that is consistent with formulas (3.43):

$$Y_n = Y_{n-1} + W'_n + \delta_\circ = y_0 + W'_1 + \cdots + W'_n + \delta_\circ n\,, \quad n = 0, 1, \dots\,. \tag{3.44}$$

It should be mentioned that the procedure for the construction of the tree $T_{(y_0,0)}$ described above does not take into account values of probabilities for appearance of nodes in the tree generated by trajectories of the random walk Y_n.

The probability that a node $y_{n,l}$ will be reached after n steps is given by the following formula, for $l = 0, \pm 1, \dots, \pm n$, $n = 0, 1, \dots$:

$$p_{n,l} = \sum_{l_+, l_- \geq 0, l_- + l_+ \leq n, l_+ - l_- = l} \frac{n!}{l_+!, l_-!(n - l_+ - l_-)!} p_+^{l_+} p_-^{l_-} p_\circ^{n - l_+ - l_-}\,, \tag{3.45}$$

where 0^0 should be counted as 1.

Note also that the numbers $y_{n,l}$ satisfy the inequalities,

$$y_{n,-n} < \cdots < y_{n,0} < \cdots < y_{n,n}\,, \quad n = 0, 1, \dots\,. \tag{3.46}$$

It is useful to note that all these points are located on the grid of points $G[y_0, \delta, \delta_\circ] = \{y_0 + \delta l + \delta_\circ n,\ l = 0, \pm 1, \dots,\ n = 0, 1, \dots\}$. Moreover, in the standard symmetric case, where $\delta_\circ = 0$, this grid reduces to the simple grid of points, $G[y_0, \delta] = \{y_0 + \delta l,\ l = 0, \pm 1, \dots\}$.

According to formula (3.43), the tree $T_{(y_0,0)}$ is the following set of nodes:

$$T_{(y_0,0)} = \langle (y_{n,l}, n), l = 0, \pm 1, \dots, \pm n, n = 0, 1, \dots \rangle\,. \tag{3.47}$$

The tree $T_{(y_0,0)}$ includes a particular point (y, n), if $y_{n,l} = y_0 + \delta l + \delta_\circ n = y$ for some $l = 0, \pm 1, \dots, \pm n$.

The condition $\mathbf{A_2}$ holds since, in this case, the following inequality takes place, for $y \in \mathbb{R}_1$, $n = 0, 1, \dots, N$:

$$\mathsf{E}_{y,n} \max_{n \leq r \leq N} \left| g(r, e^{Y_r}) \right| \leq \max_{n \leq r \leq N} \max_{m-r+n \leq l \leq m+r-n} \left| g(r, e^{y_{r,l}}) \right| < \infty\,, \tag{3.48}$$

where the points $y_{r,l}$ are given by formulas (3.43), with y_0 and $-n \le m \le n$ chosen such that $y_{n,m} = y$.

Theorem 2.3.3 takes in this case the following form.

Lemma 3.2.1. *Let the log-price process Y_n be represented by the trinomial random walk defined by the stochastic transition dynamic relation* (3.38)*. Then the log-reward functions $\phi_n(y_{n,l})$, for points $y_{n,l} = y_0 + \delta l + \delta_\circ n$, $l = 0, \pm1, \dots, \pm n$, $n = 0, 1, \dots N$ given by formulas* (3.43)*, are the unique solution for the following finite recurrence system of linear equations:*

$$\begin{cases} \phi_N(y_{N,l}) = g(N, e^{y_{N,l}}), \quad l = 0, \pm1, \dots, \pm N, \\ \phi_n(y_{n,l}) = \max\big(g(n, e^{y_{n,l}}), \phi_{n+1}(y_{n+1,l-1})p_- + \phi_{n+1}(y_{n+1,l})p_\circ \\ \qquad + \phi_{n+1}(y_{n+1,l+1})p_+\big), \quad l = 0, \pm1, \dots, \pm n, \\ \qquad n = N-1, \dots, 0, \end{cases} \tag{3.49}$$

This system gives the values of the log-reward function $\phi_n(y)$ at points $y_{n,l}$, $l = 0, \pm1, \dots, \pm n$, $n = 0, 1, \dots, N$ given by formulas (3.43).

The particular value $\phi_n(y)$ can be calculated using this system if the initial value y_0 is chosen so that $y_{n,l} = y_0 + \delta_+ l + \delta_-(n - l) = y$ for some $l = 0, \pm1, \dots, \pm n$.

Remarks concerning simultaneous solution of the system (3.49) for the set of initial states $y_0 + \delta l$, $r_- \le l \le r_+$, subtrees with an initial node $y_{n,l}$ and the corresponding subsystems of equations for log-reward functions generated by such subtrees as well as formulas for finding optimal expected rewards for a given initial distribution, analogous to those made in Subsection 3.1.1 for the binomial tree model, can be translated in an obvious way to the case of the trinomial model.

3.2.2 A multivariate modulated trinomial tree model inhomogeneous in time and space

In this subsection, we consider the model of a discrete time k-dimensional log-price process represented by a inhomogeneous trinomial random walk modulated by a Markov chain with a finite phase space $\mathbb{X} = \{1, \dots, m\}$.

We would like to define the model in a dynamic form.

In order to do this, we first describe, in which way one can construct, for a k-dimensional trinary random vector $\vec{W} = (W_1, \dots, W_k)$, another k-dimensional trinomial random vector $\vec{W}' = (W'_1, \dots, W'_k)$, which has the same distribution and is a nonrandom transformation of a random vector $\vec{U} = (U_1, \dots, U_k)$ uniformly distributed in the k-dimensional cube $\mathbb{U}_k = [0,1] \times \cdots \times [0,1]$.

Let a random vector $\vec{W} = (W_1, \dots, W_k)$ takes values $\bar{\delta}_{\bar{\jmath}_k} = (\delta_{1,\jmath_1}, \dots, \delta_{k,\jmath_k})$ with probability $p_{\bar{\jmath}_k}$ for $\bar{\jmath}_k = (\jmath_1, \dots, \jmath_k) \in \mathcal{J}_k$, where $\mathcal{J}_k = \{\bar{\jmath}_k : \jmath_i = +, \circ, -, i = 1, \dots, k\}$. The following conditions are also assumed: (a) $\delta_{i,-} < \delta_{i,\circ} < \delta_{i,+}$, $i = 1, \dots, k$, (b) $p_{\bar{\jmath}_k} \ge 0$, $\bar{\jmath}_k \in \mathcal{J}_k$ and $\sum_{\bar{\jmath}_k \in \mathcal{J}_k} p_{\bar{\jmath}_k} = 1$.

The distribution of the random vector $\vec{W}$ is determined by the pair of parameter families $\bar{\delta} = \langle \bar{\delta}_{\bar{\jmath}_k}, \bar{\jmath}_k \in \mathcal{J}_k \rangle$ and $\bar{p} = \langle p_{\bar{\jmath}_k}, \bar{\jmath}_k \in \mathcal{J}_k \rangle$.

Let us now define the joint distributions for the first $1 \le r \le k$ components of the random vector $\vec{W}$

$$p_{\bar{\jmath}_r} = p_{\jmath_1,\dots,\jmath_r} = \sum_{\bar{\jmath}_k : \jmath_{r+1},\dots,\jmath_N = +,\circ,-} p_{\bar{\jmath}_k}, \quad \bar{\jmath}_r = (\jmath_1, \dots, \jmath_r) \in \mathcal{J}_r . \tag{3.50}$$

Let us also define intervals $I_{\bar{\jmath}_{r+1}}[\bar{p}]$ for $\bar{\jmath}_{r+1} = (\jmath_1, \dots, \jmath_{r+1}), \jmath_1, \dots, \jmath_{r+1} = +, \circ, -$, for $r = 0$,

$$I_{\bar{\jmath}_1}[\bar{p}] = \begin{cases} [0, p_+) & \text{if } \jmath_1 = + , \\ [p_+, 1 - p_-) & \text{if } \jmath_1 = \circ , \\ [1 - p_-, 1] & \text{if } \jmath_1 = - , \end{cases} \tag{3.51}$$

and, for $r = 1, \dots, k-1$,

$$I_{\bar{\jmath}_{r+1}}[\bar{p}] = \begin{cases} \left[0, \frac{p_{\jmath_1,\dots,\jmath_r,+}}{p_{\jmath_1,\dots,\jmath_r}}\right) & \text{if } \jmath_{r+1} = + , \\ \left[\frac{p_{\jmath_1,\dots,\jmath_r,+}}{p_{\jmath_1,\dots,\jmath_r}}, 1 - \frac{p_{\jmath_1,\dots,\jmath_r,-}}{p_{\jmath_1,\dots,\jmath_r}}\right) & \text{if } \jmath_{r+1} = \circ , \\ \left[1 - \frac{p_{\jmath_1,\dots,\jmath_r,-}}{p_{\jmath_1,\dots,\jmath_r}}, 1\right] & \text{if } \jmath_{r+1} = - , \end{cases} \tag{3.52}$$

where one should count quotients $\frac{p_{\jmath_1,\dots,\jmath_r,\pm}}{p_{\jmath_1,\dots,\jmath_r}}$, for example, as $\frac{1}{3}$, if $p_{\jmath_1,\dots,\jmath_r} = 0$.

Finally, let us define disjoint cubes, whose union is $\mathbb{U}_k$

$$\bar{I}_{\bar{\jmath}_k}[\bar{p}] = I_{\bar{\jmath}_1}[\bar{p}] \times \cdots \times I_{\bar{\jmath}_k}[\bar{p}], \quad \bar{\jmath}_k \in \mathcal{J}_k . \tag{3.53}$$

By the definition, the length of the interval $I_{\bar{\jmath}_1}[\bar{p}]$ is $p_{\jmath_1}$ and the length $I_{\bar{\jmath}_r}[\bar{p}]$ is $\frac{p_{\jmath_1,\dots,\jmath_r}}{p_{\jmath_1,\dots,\jmath_{r-1}}}$ for $r = 2, \dots, k$. This holds for any $\bar{\jmath}_k \in \mathcal{J}_k$. Thus,

$$\mathsf{P}\{\vec{U} \in \bar{I}_{\bar{\jmath}_k}[\bar{p}]\} = p_{\jmath_1} \prod_{r=2}^{k} \frac{p_{\jmath_1,\dots,\jmath_r}}{p_{\jmath_1,\dots,\jmath_{r-1}}} = p_{\jmath_1,\dots,\jmath_k}, \quad \bar{\jmath}_k \in \mathcal{J}_k . \tag{3.54}$$

It follows from relation (3.54) that a random vector $\vec{W}$ has the same distribution with the following random vector:

$$\vec{W}' = \sum_{\bar{\jmath}_k \in \mathcal{J}_k} \bar{\delta}_{\bar{\jmath}_k} I(\vec{U} \in \bar{I}_{\bar{\jmath}_k}[\bar{p}]) . \tag{3.55}$$

Now we are prepared to define the model of the discrete time k-dimensional log-price process, represented by a trinomial inhomogeneous in space and time random walk modulated by an inhomogeneous Markov chain, in a dynamical form.

We restrict consideration by the model where values of jumps do not depend on the index argument, while the corresponding jump probabilities do depend on the index argument.

In this model, the log-price process is given by the following stochastic transition dynamic relation:

$$\begin{cases} \vec{Y}_n = \vec{Y}_{n-1} + \vec{W}_n\,, \\ \vec{W}_n = \sum\limits_{\bar{\jmath}_k \in \mathcal{J}_k} \bar{\delta}_{n,\bar{\jmath}_k}(\vec{Y}_{n-1}) I(\vec{U}_n \in \bar{I}_{\bar{\jmath}_k}[\bar{p}_n(\vec{Y}_{n-1}, X_{n-1})])\,, \\ n = 1, 2, \dots\,, \end{cases} \tag{3.56}$$

where (a) $\vec{Y}_0 = (Y_{0,1}, \dots, Y_{0,k})$ is a random vector taking value in $\mathbb{R}_k$ with a distribution $P_0(A)$, (b) $\vec{U}_n = (U_{n,1}, \dots, U_{n,k})$, $n = 1, 2, \dots$, is a sequence of i.i.d. random vectors uniformly distributed in the k-dimensional unit cube $\mathbb{U}_k$, (c) X_n, $n = 0, 1, \dots$, is an inhomogeneous in time Markov chain with the phase space $\mathbb{X}$, initial distribution $P(x) = \mathsf{P}\{X_0 = x\}, x \in \mathbb{X}$ and one-step transition probabilities $P_n(x', x'') = \mathsf{P}\{X_n = x''/X_{n-1} = x'\}, x', x'' \in \mathbb{X}, n = 1, 2, \dots$, (d) the random vector $\vec{Y}_0$, the random sequence $\vec{U}_n$, $n = 1, 2, \dots$, and the Markov chain X_n, $n = 0, 1, \dots$, are independent, (e) $\bar{\delta}_{n,\bar{\jmath}_k}(\vec{y}) = (\delta_{n,1,\jmath_1}(\vec{y}), \dots, \delta_{n,k,\jmath_k}(\vec{y}))$, $\bar{\jmath}_k = (\jmath_1, \dots, \jmath_k) \in \mathcal{J}_k$, $n = 1, 2, \dots$, where $\delta_{n,i,\jmath_i}(\vec{y}), \bar{\jmath}_k \in \mathcal{J}_k, i = 1, \dots, k, n = 1, 2, \dots$, are measurable functions acting from $\mathbb{R}_k$ to $\mathbb{R}_1$, and (f) $\bar{p}_n(\vec{y}, x) = \langle p_{n,\bar{\jmath}_k}(\vec{y}, x), \bar{\jmath} \in \mathcal{J}_k\rangle$, where $p_{n,\bar{\imath}_k}(\vec{y}, x), \bar{\jmath}_k \in \mathcal{J}_k, n = 1, 2, \dots, N$ are measurable functions acting from $\mathbb{Z} = \mathbb{R}_k \times \mathbb{X}$ to $[0, 1]$.

Also, usual assumptions are (g) $\delta_{n,i,-}(\vec{y}) < \delta_{n,i,\circ}(\vec{y}) < \delta_{n,i,+}(\vec{y}), \vec{y} \in \mathbb{R}_k, i = 1, \dots, k, n = 1, \dots$, and (h) $p_{n,\bar{\jmath}_k}(\vec{y}, x) \geq 0, (\vec{y}, x) \in \mathbb{Z}, \bar{\jmath}_k \in \mathcal{J}_k, n = 1, \dots$, and $\sum_{\bar{\jmath}_k \in \mathcal{J}_k} p_{n,\bar{\jmath}_k}(\vec{y}, x) = 1, (\vec{y}, x) \in \mathbb{Z}, n = 1, \dots$.

The Markov modulating index X_n is included in the model. Correspondingly, the pay-off function $g(n, e^{\vec{y}}, x)$ and the corresponding log-reward functions $\phi_n(\vec{y}, x)$ do depend on the index argument.

Let us assume that the initial state of the modulated log-price process $(\vec{Y}_0, X_0) = (\vec{y}_0, x_0)$ with probability 1, i.e. the initial distribution $P_0(A)P(x) = I(\vec{y}_0 \in A)I(x = x_0)$ is concentrated at a point $(\vec{y}_0, x_0) \in \mathbb{Z}$.

The tree structure of trajectories for the modulated log-price process $\vec{Z}_n$ is analogous to those described above for the simplest trinomial tree model. It is well visible from the recurrent algorithm that describes the sequential transitions of the modulated log-price process $\vec{Z}_n = (\vec{Y}_n, X_n)$. If $\vec{Z}_{n-1} = (\vec{y}, x)$ for some $n = 1, 2, \dots$, then $\vec{Z}_n = (\vec{y} + \bar{\delta}_{n,\bar{\jmath}_k}(\vec{y}), x')$ for some $\bar{\jmath}_k \in \mathcal{J}_k$ and $x' \in \mathbb{X}$.

The initial point $z_0 = (\vec{y}_0, x_0)$ determines the unique tree $T_{((\vec{y}_0,x_0),0)}$ with the root in the node $((\vec{y}_0, x_0), 0)$ in the space–time domain $\mathbb{Z} \times \{0, 1, \dots\}$.

According to the algorithm described above, the tree $T_{((\vec{y}_0,x_0),0)}$ includes one initial node and it may include $3^k m$ additional nodes after the first step, $3^{2k} m$ additional nodes after the second step, etc. So, it seems, that the tree may include $3^{nk} m$ additional nodes after the nth step.

If this is the case, i.e., the tree $T_{((\vec{y}_0,x_0),0)}$ possesses an exponential rate of growth for the number of nodes as a function of the number of steps n, any algorithm based on such a tree would not be effective from the computational point of view.

In order it would not be the case, the recombining condition should hold for the random walk $\vec{Y}_n = (Y_{n,1}, \dots, Y_{n,k})$. This condition means the following.

Let us assume that $\vec{Y}_{n-1} = \vec{y} = (y_1, \cdots, y_k)$ for some $n = 1, 2, \dots,$. Let us also assume that the random walk makes from this position two sequential jumps $\bar{\delta}_{n,\bar{j}'_k}(\vec{y})$ and then $\bar{\delta}_{n+1,\bar{j}''_k}(\vec{y} + \bar{\delta}_{n,\bar{i}'_k}(\vec{y}))$, where $\bar{j}'_k = (j'_1, \dots, j'_k), \bar{j}''_k = (j''_1, \dots, j''_k) \in \mathcal{J}_k$. Then, for every $j = 1, \dots, k$, the resulting position for the ith component of the random walk after these two jumps, $y_i + \delta_{n,i,j'_i}(\vec{y}) + \delta_{n+1,i,j''_i}(\vec{y} + \bar{\delta}_{n,\bar{j}'_k}(\vec{y}))$, should coincide for the following three cases: (a) $j'_i = +, j''_i = \circ$ or $j'_i = \circ, j''_i = +$, (b) $j'_i = -, j''_i = \circ$ or $j'_i = \circ, j''_i = -$; and (c) $j'_i = +, j''_i = -$ or $j'_i = -, j''_i = +$ or $j'_i = \circ, j''_i = \circ$. This should hold for any values of other components of vectors $\bar{j}'_k, \bar{j}''_k$, position $\vec{y} \in \mathbb{R}_1$, and $n = 1, 2, \dots$.

In the case of the model with values of jumps homogeneous in time and space, where there exists $\delta_{i,-} < \delta_{i,\circ} < \delta_{i,+}, i = 1, \dots, k$ such that $\delta_{n,i,j}(\vec{y}) = \delta_{i,j}, j = +, \circ, -$ for any $i = 1, \dots, k, \vec{y} \in \mathbb{R}_k, n = 1, 2 \dots$, the recombining condition take the form of the following relation:

$$\delta_{i,\circ} = \frac{\delta_{i,+} + \delta_{i,-}}{2}, \quad i = 1, \dots, k. \tag{3.57}$$

In the case of the model with values of jumps inhomogeneous in time but homogeneous in space, where there exists $\delta_{n,i,-} < \delta_{n,i,\circ} < \delta_{n,i,+}, i = 1, \dots, k, n = 1, 2, \dots,$ such that $\delta_{n,i,j}(\vec{y}) = \delta_{n,i,j}, j = +, \circ, -$ for $i = 1, \dots, k, \vec{y} \in \mathbb{R}_k, n = 1, 2 \dots$, the recombining condition holds if the following equalities take place for $i = 1, \dots, k$, $n = 1, 2, \dots$:

$$\begin{cases} \delta_{n,i,+} + \delta_{n+1,i,\circ} = \delta_{n,i,\circ} + \delta_{n+1,i,+}, \\ \delta_{n,i,-} + \delta_{n+1,i,\circ} = \delta_{n,i,\circ} + \delta_{n+1,i,-}, \\ \delta_{n,i,+} + \delta_{n+1,i,-} = \delta_{n,i,-} + \delta_{n+1,i,+} = \delta_{n,i,\circ} + \delta_{n+1,i,\circ}. \end{cases} \tag{3.58}$$

It follows from the third equality in (3.58) that there exists $\delta_i > 0, j = 1, \dots, k$ such that $\delta_{n,i,+} - \delta_{n,i,-} = 2\delta_i$, $n = 1, 2, \dots, i = 1, \dots, k$. Thus, $\delta_{n,i,+} = \delta_i + \delta_{n,i}, \delta_{n,i,-} = -\delta_i + \delta_{n,i}$, $n = 1, 2, \dots, i = 1, \dots, k$, where $\delta_{n,i} = \frac{\delta_{n,i,+} + \delta_{n,i,-}}{2}$, $n = 1, 2, \dots, i = 1, \dots, k$. By substituting the above $\delta_{n,i,\pm}$ in the expression on the right-hand side of the third equality in (3.58), we get the equalities $\delta_{n,i} + \delta_{n+1,i} = \delta_{n,i,\circ} + \delta_{n+1,i,\circ}$, $n = 1, 2, \dots,$ $i = 1, \dots, k$. Also, by adding and devising by 2 the first two equalities in (3.58), we get equalities $\delta_{n,i} - \delta_{n+1,i} = \delta_{n,i,\circ} - \delta_{n+1,i,\circ}$, $n = 1, 2, \dots, i = 1, \dots, k$. The above two series of equalities imply in an obvious way that $\delta_{n,i} = \delta_{n,i,\circ}$, $n = 1, 2, \dots, i = 1, \dots, k$.

Summarizing the above calculations, we can reformulate the recombining condition (3.58) in the equivalent form as the condition that require the existence of $\delta_i > 0$, $i = 1, \dots, k$ such that the following relation hold for $n = 1, 2, \dots, i = 1, \dots, k$:

$$\delta_{n,i,+} = \delta_i + \delta_{n,i,\circ}, \quad \delta_{n,i,-} = -\delta_i + \delta_{n,i,\circ}. \tag{3.59}$$

As an example of the model with values of jumps inhomogeneous in time and space, let us consider the model with values of jumps $\delta_{n,i,j}(\vec{y}) = \delta_{n,i,j} y_i, j = +, \circ, -$

are, for every $i = 1, \dots, k$, linear functions of y_i for $n = 1, 2, \dots$, such that $\delta_{n,i,-} < \delta_{n,i,\circ} < \delta_{n,i,+}$, $i = 1, \dots, k$, $n = 1, 2, \dots$,. In this case, the recombining condition takes the form of the following relations that should hold for any $n = 0, 1, \dots, i = 1, \dots, k$:

$$\begin{cases} \delta_{n,i,+} + \delta_{n+1,i,\circ} + \delta_{n,i,+}\delta_{n+1,i,\circ} \\ \quad = \delta_{n,i,\circ} + \delta_{n+1,i,+} + \delta_{n,i,\circ}\delta_{n+1,i,+}, \\ \delta_{n,i,-} + \delta_{n+1,i,\circ} + \delta_{n,i,-}\delta_{n+1,i,\circ} \\ \quad = \delta_{n,i,\circ} + \delta_{n+1,i,-} + \delta_{n,i,\circ}\delta_{n+1,i,-}, \\ \delta_{n,i,+} + \delta_{n+1,i,-} + \delta_{n,i,+}\delta_{n+1,i,-} \\ \quad = \delta_{n,i,-} + \delta_{n+1,i,+} + \delta_{n,i,-}\delta_{n+1,i,+} \\ \quad = \delta_{n,i,\circ} + \delta_{n+1,i,\circ} + \delta_{n,i,\circ}\delta_{n+1,i,\circ}. \end{cases} \tag{3.60}$$

For example, if the coefficients $\delta_{n,i,\jmath} = \delta_{i,\jmath}$, $\jmath = +, \circ, -$, $i = 1, \dots, k$ do not depend on n, then the first three equalities in (3.60) automatically hold, and, thus, relations (3.60) reduce to the following relations that should hold for $i = 1, \dots, k$:

$$\delta_{i,+} + \delta_{i,-} + \delta_{i,+}\delta_{i,-} = 2\delta_{i,\circ} + \delta_{i,\circ}^2. \tag{3.61}$$

Let us return back to the general case where the recombining condition holds. The tree $T_{((\vec{y}_0,x_0),0)}$ includes one initial node $((\vec{y}_0, x_0), 0)$. It includes $3^k m$ additional nodes $(\vec{y}_0 + \bar{\delta}_{1,\bar{\jmath}'_k}(\vec{y}_0), x_1), 1)$, $\bar{\jmath}'_k \in \mathcal{J}_k$, $x_1 \in \mathbb{X}$ after the first step. If the recombining condition holds, the tree contains $5^k m$ additional nodes $(\vec{y}_0 + \bar{\delta}_{1,\bar{\jmath}'_k}(\vec{y}_0) + \bar{\delta}_{2,\bar{\jmath}''_k}(\vec{y}_0 + \bar{\delta}_{1,\bar{\jmath}'_k}(\vec{y}_0)), x_2), 2)$, $\bar{\jmath}'_k, \bar{\jmath}''_k \in \mathcal{J}_k$, $x_2 \in \mathbb{X}$ after the second step. Indeed, the ith component of the vector $\bar{\delta}_{1,\bar{\jmath}'_k}(\vec{y}_0) + \bar{\delta}_{2,\bar{\jmath}''_k}(\vec{y}_0 + \bar{\delta}_{1,\bar{\jmath}'_k}(\vec{y}_0))$ can take five different values, $\delta_{1,i,+}(\vec{y}_0) + \delta_{2,i,+}(\vec{y}_0 + \bar{\delta}_{1,\bar{\jmath}'_k}(\vec{y}_0))$ if ith components of both vectors $\bar{\jmath}'_k$ and $\bar{\jmath}''_k$ equals $+$; $\delta_{1,i,+}(\vec{y}_0) + \delta_{2,i,\circ}(\vec{y}_0 + \bar{\delta}_{1,\bar{\jmath}'_k}(\vec{y}_0)) = \delta_{1,i,\circ}(\vec{y}_0) + \delta_{2,i,+}(\vec{y}_0 + \bar{\delta}_{1,\bar{\jmath}'_k}(\vec{y}_0))$ if the ith components of the vectors $\bar{\jmath}'_k$ and $\bar{\jmath}''_k$ equal, respectively $+$ and $\circ$, or vice versa; $\delta_{1,i,+}(\vec{y}_0) + \delta_{2,i,-}(\vec{y}_0 + \bar{\delta}_{1,\bar{\jmath}'_k}(\vec{y}_0)) = \delta_{1,i,-}(\vec{y}_0) + \delta_{2,i,+}(\vec{y}_0 + \bar{\delta}_{1,\bar{\jmath}'_k}(\vec{y}_0)) = \delta_{1,i,\circ}(\vec{y}_0) + \delta_{2,i,\circ}(\vec{y}_0 + \bar{\delta}_{1,\bar{\jmath}'_k}(\vec{y}_0))$ if the ith components of the vectors $\bar{\jmath}'_k$ and $\bar{\jmath}''_k$ equal, respectively, $+$ and $-$, or $-$ and $+$, or both equal $\circ$; $\delta_{1,i,-}(\vec{y}_0) + \delta_{2,i,\circ}(\vec{y}_0 + \bar{\delta}_{1,\bar{\jmath}'_k}(\vec{y}_0)) = \delta_{1,i,\circ}(\vec{y}_0) + \delta_{2,i,-}(\vec{y}_0 + \bar{\delta}_{1,\bar{\jmath}'_k}(\vec{y}_0))$ if the ith components of the vectors $\bar{\jmath}'_k$ and $\bar{\jmath}''_k$ equal, respectively $-$ and $\circ$, or vice versa; and, finally, $\delta_{1,i,-}(\vec{y}_0) + \delta_{2,i,-}(\vec{y}_0 + \bar{\delta}_{1,\bar{\jmath}'_k}(\vec{y}_0))$ if the ith components of both vectors $\bar{\jmath}'_k$ and $\bar{\jmath}''_k$ equals $-$. Subsequently, the tree includes $7^k m$ additional nodes after the third step, etc.

The tree $T_{(\vec{y}_0,x_0),0)}$ includes $(2n+1)^k m$ additional nodes after the nth step. The total number of nodes in the tree after n steps is $L_n = 1 + 3^k m + \cdots + (2n+1)^k m$.

Obviously,

$$1 - m + m\frac{(2n+1)^k}{2(k+1)} = 1 - m + m\int_{\frac{1}{2}}^{n+1}(2x-1)^k dx \le L_n$$
$$\le 1 - m + m\int_{1}^{n+2}(2x-1)^k dx = 1 - m + m\frac{(2n+3)^k - 1}{2(k+1)}. \quad (3.62)$$

Therefore, the tree possesses the polynomial of order $k+1$ rate of growth for the number of nodes in the tree after n steps as a function of n. This makes algorithms based on such a tree effective from the computational point of view.

Due to the recombining condition, states that can be reached by the ith component of the random walk after the nth step are determined by the number $l_{n,i}$, which is the difference between the numbers of jumps $\delta_{\cdot,i,+}(\cdot)$ and $\delta_{\cdot,i,-}(\cdot)$ made by the ith component of the random walk in n steps. By the definition, the parameters $l_{n,i}$, $i = 1, \dots, k$ can take values $-n, \dots, 0, \dots, n$. Correspondingly, states which can be reached by the random walk after n steps are determined by the vector $\bar{l}_n = (l_{n,1}, \dots, l_{n,k})$, which take values in the set $\mathbb{L}_n = \{\bar{l}_n : |l_{n,i}| \le n, i = 1, \dots, k\}$.

The states $\vec{y}_{n,\bar{l}_n} = (y_{n,1,l_{n,1}}, \dots, y_{n,k,l_{n,k}})$, $\bar{l}_n \in \mathbb{L}_n$, which can be reached by the log-price process $\vec{Y}_n = (Y_1, \dots, Y_n)$ with the initial value $\vec{Y}_0 = \vec{y}_0 = (y_{0,1}, \dots, y_{0,k})$, at moments $n = 0, 1, \dots,$ are given by the following recurrence relation, for $-n \le l_{n,i} \le n$, $i = 1, \dots, k$, $n = 0, 1, \dots$:

$$y_{n+1,i,l_{n+1,i}} = y_{n,i,l_{n,i}} + \begin{cases} \delta_{n+1,i,+}(\vec{y}_{n,\bar{l}_n}) & \text{if } l_{n+1,i} = l_{n,i} + 1, \\ \delta_{n+1,i,\circ}(\vec{y}_{n,\bar{l}_n}) & \text{if } l_{n+1,i} = l_{n,i}, \\ \delta_{n+1,i,-}(\vec{y}_{n,\bar{l}_n}) & \text{if } l_{n+1,i} = l_{n,i} - 1, \end{cases} \quad (3.63)$$

where $y_{0,i,0} = y_{0,i}$, $i = 1, \dots, k$.

Note that calculations in (3.63) should be made for $l_i = -n, \dots, n$, $i = 1, \dots, k$ sequentially for $n = 1, 2, \dots$. This procedure let one find vectors $\vec{y}_{n,\bar{l}_n}$, $\bar{l}_n \in \mathbb{L}_n$ sequentially for $n = 1, 2, \dots,$ for any given initial vector $\vec{y}_{0,\bar{l}_0} = \vec{y}_0$, $\bar{l}_0 = (0, \dots, 0)$.

Note that numbers $y_{n,i,l_{n,i}}$ satisfy the inequalities

$$y_{n,i,-n} < \cdots < y_{n,i,n}, \quad i = 1, \dots, k, \quad n = 0, 1, \dots. \quad (3.64)$$

According to the above remarks, the tree $T_{((\vec{y}_0,x_0),0)}$ is the following set of nodes:

$$T_{((\vec{y}_0,x_0),0)} = \langle ((\vec{y}_{n,\bar{l}_n}, x_n), n), \bar{l}_n \in \mathbb{L}_n, x_n \in \mathbb{X}, n = 0, 1, \dots \rangle. \quad (3.65)$$

The tree $T_{((\vec{y}_0,x_0),0)}$ includes a particular point $((\vec{y}, x), n)$, if $(\vec{y}_{n,\bar{l}_n}, x_n) = (\vec{y}, x)$ for some $\bar{l}_n \in \mathbb{L}_n$.

The condition $\mathbf{A}_2$ holds since, in this case, the following inequality take place for any $\vec{z} = (\vec{y}, x) \in \mathbb{Z} = \mathbb{R}_k \times \mathbb{X}$, $n = 0, 1, \dots, N$:

$$\begin{aligned} \mathsf{E}_{\vec{z},n} \max_{n \le r \le N} \left| g(r, e^{\vec{Y}_r}, X_r) \right| \\ \le \max_{n \le r \le N} \max_{m_i - r + n \le l_i \le m_i + r - n, i = 1, \dots, k, x_r \in \mathbb{X}} \left| g(r, e^{\vec{y}_{r,\bar{l}_r}}, x_r) \right| < \infty , \end{aligned} \tag{3.66}$$

where the points $\vec{y}_{r,\bar{l}}$ are given by formulas (3.63), with $\vec{y}_0$ and $\bar{m} = (m_1, \dots, m_k)$, $-n \le m_i \le n$, $i = 1, \dots, k$ chosen such that $\vec{y}_{n,\bar{m}} = \vec{y}$.

Theorem 2.2.3 takes in this case the following form.

Lemma 3.2.2. *Let the log-price process Y_n be represented by the trinomial random walk defined by the stochastic transition dynamic relation* (3.56). *Then, the log-reward functions $\phi_n(\vec{y}_{n,\bar{l}_n}, x_n)$, for points $(\vec{y}_{n,\bar{l}_n}, x_n)$, $\bar{l}_n \in \mathbb{L}_n$, $x_n \in \mathbb{X}$, $n = 0, 1, \dots, N$ given by formulas* (3.63), *are the unique solution for the following finite recurrence system of linear equations:*

$$\left\{ \begin{aligned} & \phi_N(\vec{y}_{N,\bar{l}_N}, x_N) = g(N, e^{\vec{y}_{N,\bar{l}_N}}, x_N) , \quad \bar{l}_N \in \mathbb{L}_N , \quad x_N \in \mathbb{X} , \\ & \phi_n(\vec{y}_{n,\bar{l}_n}, x_n) = \max \Bigg(g(n, e^{y_{n,\bar{l}_n}}, x_n) , \\ & \qquad \sum_{\bar{\jmath}_k \in \mathcal{J}_k, x_{n+1} \in \mathbb{X}} \phi_{n+1}(\vec{y}_{n+1,\bar{l}_n + \bar{e}_{\bar{\jmath}_k}}, x_{n+1}) \\ & \qquad \times p_{n+1,\bar{\jmath}_k}(\vec{y}_{n,\bar{l}_n}, x_n) P_{n+1}(x_n, x_{n+1}) \Bigg) , \quad \bar{l}_n \in \mathbb{L}_n , \quad x_n \in \mathbb{X} , \\ & \qquad n = N - 1, \dots, 0 , \end{aligned} \right. \tag{3.67}$$

where the points $\vec{y}_{n,\bar{l}_n}$ are given by formulas (3.63) *and vector $\bar{e}_{\bar{\jmath}_k} = (e_{\jmath_1}, \dots, e_{\jmath_k})$, where $e_{\jmath_i} = 1$ if $\jmath_i = +$, or $e_{\jmath_i} = 0$ if $\jmath_i = \circ$, or $e_{\jmath_i} = -1$ if $\jmath_i = -$, for $i = 1, \dots, k$.*

This system lets one compute the values of the log-reward function $\phi_n(\vec{y}, x)$ at points $(\vec{y}_{n,\bar{l}}, x_n)$, $\bar{l} \in \mathbb{L}_n$, $x_n \in \mathbb{X}$ given by formulas (3.63), recursively, for $n = N, \dots, 0$.

The particular value $\phi_n(\vec{y}, x)$ can be calculated using this system if the initial value $\vec{y}_0$ is chosen so that $y_{n,\bar{l}_n} = \vec{y}$, for some $\bar{l}_n \in \mathbb{L}_n$, and $x_n = x$.

The extension and reduction of the system of backward recurrence relations (3.67) can be realized in the same way as as described in Subsection 3.1.1, for the system of backward recurrence relations (3.8).

Also, the remarks concerning finding of optimal expected reward analogous to those made in Subsection 3.1.1 can be made.

3.2.3 Binomial and trinomial random walk models

In this subsection a recurrence algorithm for computing the transition probabilities for multivariate binomial and trinomial random walks is presented.

Let us introduce the sets

$$\mathbb{L}_r = \{\bar{l}_r = (l_{r,1}, \dots, l_{r,k}), l_{r,1}, \dots, l_{r,k} = -r, \dots, r\}, \quad r = 1, 2, \dots . \tag{3.68}$$

Consider the trinomial random vectors $\vec{W}_r = (W_{r,1}, \dots, W_{r,k})$, $r = 1, 2, \dots$, which can be represented in the following form:

$$\vec{W}_r = \vec{U}_1 + \cdots + \vec{U}_r, \tag{3.69}$$

where (a) $\vec{U}_n = (U_{n,1}, \dots, U_{n,k})$, $n = 1, 2, \dots$, are i.i.d. trinary random vectors taking values $\bar{l}_1 \in \mathbb{L}_1$ with probabilities $\mathsf{P}\{\vec{U}_1 = \bar{l}_1\} = p(1, \bar{l}_1)$ and (b) $p(1, \bar{l}_1) \geq 0$, $\bar{l}_1 \in \mathbb{L}_1$ and $\sum_{\bar{l}_1 \in \mathbb{L}_1} p(1, \bar{l}_1) = 1$.

The random vector $\vec{W}_r$ can be referred to as a trinomial random vector. In the particular case, where probabilities $p(1, \bar{l}_1) = 0$ for any $\bar{l}_1 = (l_{1,1}, \dots, l_{1,k})$ such that $\prod_{i=1}^{k} l_{1,i} = 0$, the random vector $\vec{W}_r$ can also be referred to as a binomial random vector.

The random vector $\vec{W}_r$ takes values $\bar{l}_r \in \mathbb{L}_r$, for every $r = 1, 2, \dots$. Let us denote the corresponding probabilities by

$$\mathsf{P}\{\vec{W}_r = \bar{l}_r\} = p(r, \bar{l}_r), \quad \bar{l}_r \in \mathbb{L}_r, \quad r = 1, 2, \dots . \tag{3.70}$$

By the definition, $p(r, \bar{l}_r) \geq 0$, $\bar{l}_r \in \mathbb{L}_r$ and $\sum_{\bar{l}_r \in \mathbb{L}_r} p(r, \bar{l}_r) = 1$.

The probabilities $p(r, \bar{l}_r)$, $\bar{l}_r \in \mathbb{L}_r$ satisfy the following convolution recurrence relations, which can be used for their sequential computing:

$$p(r+1, \bar{l}_{r+1}) = \sum_{\bar{l}_1 \in \mathbb{L}_1, \bar{l}_{r+1} - \bar{l}_1 \in \mathbb{L}_r} p(r, \bar{l}_{r+1} - \bar{l}_1) p(1, \bar{l}_1), \quad r = 1, 2, \dots . \tag{3.71}$$

It can be easily calculated that computing the probabilities $p(r, \bar{l}_r)$, $\bar{l}_r \in \mathbb{L}_r$ using these recurrence relations requires $N_r = \sum_{m=1}^{r-1} 3(2m+1)^k$ multiplication and $\frac{2}{3} N_r$ summation operations. Note that

$$\begin{aligned} N'_r &= \frac{3}{2(k+1)}((2r-1)^{k+1} - 1) = 3 \int_0^{r-1} (2x+1)^k dx \\ &\leq N_r \leq 3 \int_1^r (2x+1)^k dx = \frac{3}{2(k+1)}((2r+1)^{k+1} - 3^{k+1}) = N''_r . \end{aligned} \tag{3.72}$$

One can, however, improve the above recurrence algorithm.

The probabilities $p(r, \bar{l}_r)$, $\bar{l}_r \in \mathbb{L}_r$ also satisfy the following convolution recurrence relations, which can be used for their sequential computing:

$$p(2r, \bar{l}_{2r}) = \sum_{\bar{l}_r \in \mathbb{L}_r, \bar{l}_{2r} - \bar{l}_r \in \mathbb{L}_r} p(r, \bar{l}_{2r} - \bar{l}_r) p(r, \bar{l}_r), \quad r = 2^m, \quad m = 0, 1, \dots . \tag{3.73}$$

Any $r = 1, 2, \dots,$ can be uniquely decomposed in the binary sum $r = i_1 + \cdots + i_{n_r}$, where $1 \le i_1 = 2^{m_{r,1}} \cdots < i_{n_r} = 2^{m_{r,n_r}}$ and $0 \le m_{r,1} < \cdots < m_{r,n_r}$ are integer numbers. Note that $m_{r,n_r} = [\ln_2 r], m_{r,n_r-1} = [\ln_2(r - i_{n_r})], \dots, m_{r,1} = [\ln_2(r - i_{n_r} - \cdots - i_2)]$. Let us also denote $j_1 = i_1, j_2 = i_1 + i_2, \dots, j_{n_r} = i_1 + \cdots + i_{n_r} = r$.

The probabilities $p(r, \bar{l}_r),\ \bar{l}_r \in \mathbb{L}_r$ can be computed using the following recurrence relations:

$$p(j_{m+1}, \bar{l}_{j_{m+1}}) = \sum_{\bar{l}_{i_m} \in \mathbb{L}_{i_m}, \bar{l}_{j_{m+1}} - \bar{l}_{i_m} \in \mathbb{L}_{j_m}} p(j_m, \bar{l}_{j_{m+1}} - \bar{l}_{i_m}) \times p(i_m, \bar{l}_{i_m}), \quad m = 1, 2, \dots, n_r - 1. \tag{3.74}$$

The algorithm described above can be in obvious way translated to the general recombining case, where random variables U_n, $n = 1, 2, \dots,$ take values $\delta + \delta_\circ, \delta_\circ,$ and $-\delta + \delta_\circ$ for some $\delta > 0, \delta_\circ \in \mathbb{R}_1$, instead of values 1, 0 and −1.

3.3 Random walk reward algorithms

In this section, we describe backward tree algorithms for finding log-reward functions and optimal expected rewards for models, where log-price processes are represented by discrete random walks with bounded jumps. We consider classical homogeneous models, model inhomogeneous in time as well as models with Markov modulation. Random walk models can be effectively used for approximation of more general price processes represented by random walks with arbitrary distributions of jumps.

3.3.1 A classical homogeneous in time random walk model

This is the simplest model, where the log-price Y_n is a univariate random walk,

$$Y_n = Y_{n-1} + W_n\,, \quad n = 1, 2, \dots. \tag{3.75}$$

where (a) Y_0 is a real-valued random variable with the distribution $P_0(A)$, (b) W_n, $n = 1, 2, \dots,$ is a sequence of i.i.d. random variables which take values $l\delta$, $l = m_-, \dots, m_+$ with probabilities, respectively, $p(l)$, $l = m_-, \dots, m_+$, and (c) the random variable Y_0 and the random sequence W_n, $n = 1, 2, \dots,$ are independent.

Also, usual assumptions are (d) $\delta > 0$, (e) $m_- < m_+$ are integers, and (f) $0 \le p(l) \le 1$, $l = m_-, \dots, m_+$, and $p(m_-) + \cdots + p(m_+) = 1$.

In this case, the stochastic modulating index X_n is not included in the model. Correspondingly, a pay-off function $g(n, e^y)$ and the corresponding log-reward functions $\phi_n(y)$ do not depend on the index argument.

Let us assume that the initial state of the log-price process $Y_0 = y_0$ with probability 1, i.e. the initial distribution $P_0(A) = I(y_0 \in A)$ is concentrated in a point $y_0 \in \mathbb{R}_1$.

The tree structure of trajectories is well visible from the recurrent algorithm, which describes sequential transitions for the log-price process Y_n. If $Y_{n-1} = y$ for some $n = 1, 2, \ldots$, then $Y_n = y + l\delta$, where l takes one of the values $m_-, \ldots, m_+$.

The initial point y_0 determines the unique tree $T_{(y_0,0)}$ with the root in the node $(y_0, 0)$, in the space–time domain $\mathbb{R}_1 \times \{0, 1, \ldots\}$.

According to the algorithm described above the tree $T_{(y_0,0)}$ has one initial node $(y_0, 0)$. It may include $m_+ - m_- + 1$ additional nodes after the first step, then $(m_+ - m_- + 1)^2$ additional nodes after the second step, etc. The tree may include $(m_+ - m_- + 1)^n$ additional nodes after the nth step.

If this is the case, i.e. the tree $T_{(y_0,0)}$ possesses an exponential rate of growth for the number of nodes as a function of the number of steps n, any algorithms based on such a tree would not be effective from the computational point of view.

However, this is not the case. In fact, the tree includes only $m_+ - m_- + 1$ additional nodes after the first step, and $n(m_+ - m_-) + 1$ additional nodes after the nth step, for every $n \geq 2$.

This is because the tree $T_{(y_0,0)}$ satisfies the recombining condition. In this model, the recombining condition means that the random walk Y_n, which is in a position $Y_{n-1} = y$, for some $n = 1, 2, \ldots$, and makes from this position the first jump $l'\delta$ and then the second jump $l''\delta$ should end up at the same point as if it would make the first jump $l''\delta$ and then the second jump $l'\delta$. This should hold for any l' and l'' such that $l' + l'' = l$, for any $l = 2m_-, \ldots, 2m_+, y \in \mathbb{R}_1$, and $n = 1, 2, \ldots$.

Indeed, if $Y_{n-1} = y$ for some $n = 1, 2, \ldots$, then $Y_{n+1} = y + l'\delta + l''\delta = l\delta$ if $W_n = l'\delta, W_{n+1} = l''\delta$ or $Y_{n+1} = y + l''\delta + l'\delta = l\delta$ if $W_n = l''\delta, W_{n+1} = l'\delta$, for the case where $l' + l'' = l$. In all such cases, Y_{n+1} take the same value.

The tree $T_{(y_0,0)}$ obviously includes one initial node $(y_0, 0)$ and $m_+ - m_- + 1$ additional nodes $(y_0 + m_-\delta, 1), \ldots, (y_0 + m_+\delta, 1)$ after the first step. Due to the recombining property, the tree includes $2(m_+ - m_-) + 1$ additional nodes $(y_0 + 2m_-\delta, 2), \ldots, (y_0 + 2m_+\delta, 2)$ after the second step, etc.

The tree $T_{(y_0,0)}$ includes $n(m_+ - m_-) + 1$ additional nodes after the nth step. The total number of nodes in the tree after n steps is $L_n = 1 + (m_+ - m_-) + 1 + \cdots + n(m_+ - m_-) + 1 = n + 1 + (m_+ - m_-)\frac{n(n+1)}{2}$. This means that the tree possesses a quadratic rate of growth for the number of nodes in the tree after n steps. This makes algorithms based on such a tree effective from the computational point of view.

The nodes appearing in the tree after n steps are determined by the number $l = l_1 + \cdots + l_n$, where $l_1\delta, \ldots, l_n\delta$ are jumps at moments $1, \ldots, n$. The states that can be reached by the random walk at moments $n = 0, 1, \ldots$, are given by the following formula:

$$y_{n,l} = y_0 + l\delta\,, \quad l = nm_-, \ldots, nm_+\ n = 0, 1, \ldots\,. \tag{3.76}$$

The procedure for the construction of the tree $T_{(y_0,0)}$ described above does not take into account actual values of probabilities for nodes to appear in the tree generated by the random walk Y_n.

The probability that a node $y_{n,l}$ will be reached after n steps is given by the convolution recurrence relation, for $nm_- \leq l_n \leq nm_+$, $n = 1, 2, \dots$,

$$p_{n,l_n} = \begin{cases} p(l_1) & \text{if } l_1 = m_-, \dots, m_+, \quad n = 1, \\ \sum_{l=m_-}^{m_+} p_{n-1,l_n-l} p(l) & \text{if } l_n = nm_-, \dots, nm_+, n > 1, \end{cases} \tag{3.77}$$

where one should count $p_{n,l} = 0$ if $l < nm_-$ or $l > nm_+$, for $n = 1, 2, \dots$.

Note also that numbers $y_{n,l}$ satisfy the inequalities,

$$y_{n,nm_-} < y_{n,nm_-+1} < \cdots < y_{n,nm_+}, \quad n = 0, 1, \dots . \tag{3.78}$$

It is useful to note that these points are located on the grid of points $G[y_0, \delta] = \{y_0 + l\delta,\ l = 0, \pm 1, \pm 2, \dots\}$.

According to formulas (3.3), the tree $T_{(y_0,0)}$ is the following set of nodes:

$$T_{(y_0,0)} = \langle (y_{n,l}, n), l = nm_-, \dots, nm_+, n = 0, 1, \dots \rangle . \tag{3.79}$$

The tree $T_{(y_0,0)}$ includes a particular point (y, n), if $y_{n,l} = y_0 + l\delta = y$, for some $nm_- \leq l \leq nm_+$.

The condition $\mathbf{A_2}$ holds since, in this case, the following inequality takes place for any $y \in \mathbb{R}_1$, $n = 0, 1, \dots, N$:

$$\begin{aligned} \mathsf{E}_{y,n} \max_{n \leq r \leq N} \left| g(r, e^{Y_r}) \right| \\ \leq \max_{n \leq r \leq N} \max_{m+(r-n)m_- \leq l \leq m+(r-n)m_+} \left| g(r, e^{y_{r,l}}) \right| < \infty , \end{aligned} \tag{3.80}$$

where the points $y_{r,l}$ are given by formulas (3.76), with y_0 and $nm_- \leq m \leq nm_+$ chosen such that $y_{n,m} = y$.

Theorem 2.2.3 takes in this case the following form.

Lemma 3.3.1. *Let the log-price process Y_n be represented by the homogeneous random walk defined in the stochastic transition dynamic relation* (3.75). *Then the log-reward functions $\phi_n(y_{n,l})$, for points $y_{n,l} = y_0 + l\delta$, $l = nm_-, \dots, nm_+$, $n = 0, 1, \dots, N$ given by formulas* (3.76), *are the unique solution for the following finite recurrence system of linear equations:*

$$\begin{cases} \phi_N(y_{N,l}) = g(N, e^{y_{N,l}}), \quad l = Nm_-, \dots, Nm_+, \\ \phi_n(y_{n,l}) = \max \Bigg(g(n, e^{y_{n,l}}), \\ \qquad \sum_{r=m_-}^{m_+} \phi_{n+1}(y_{n+1,l+r}) p(r) \Bigg), \quad l = nm_-, \dots, nm_+, \\ n = N-1, \dots, 0 . \end{cases} \tag{3.81}$$

This system lets one find the values of the log-reward function $\phi_n(y)$ at points $y_{n,l}$, given by formulas (3.76), recursively, for $n = N, \dots, 0$.

The particular value $\phi_n(y)$ can be calculated using this system if the initial value y_0 is chosen so that $y_{n,l} = y_0 + l\delta = y$, for some $nm_- \leq l \leq nm_+$.

3.3.2 An inhomogeneous in time random walk model

This is the model, where the log-price Y_n is a univariate random walk,

$$Y_n = Y_{n-1} + W_n \,, \quad n = 1, 2, \dots . \tag{3.82}$$

where (a) Y_0 is a real-valued random variable with the distribution $P_0(A)$, (b) W_n, $n = 1, 2, \dots$, is a sequence of independent random variables which takes values $l\delta + \delta_{n,\circ}$, $l = m_{n,-}, \dots, m_{n,+}$ with probabilities, respectively, $p_n(l)$, $l = m_{n,-}, \dots, m_{n,+}$, and (c) the random variable Y_0 and the random sequence W_n, $n = 1, 2, \dots$, are independent.

Also usual assumptions are (d) $\delta > 0$; (e) $\delta_{n,\circ} \in \mathbb{R}_1$, $n = 1, \dots$; (f) $m_{n,-} < m_{n,+}$, $n = 1, 2, \dots$, are integers; (g) $0 \leq p_n(l) \leq 1$, $l = m_{n,-}, \dots, m_{n,+}$, $n = 1, 2, \dots$; (h) $p_n(m_{n,-}) + \cdots + p_n(m_{n,+}) = 1$, $n = 1, 2, \dots$.

Note that usually it is also assumed that (i) $\delta m_{n,-} + \delta_{n,\circ} < 0 < \delta m_{n,+} + \delta_{n,\circ}$, $n = 1, 2, \dots$.

In this case, the stochastic modulating index variable X_n is not included in the model. Correspondingly, a pay-off function $g(n, e^y)$ and the corresponding log-reward functions $\phi_n(y)$ do not depend on the index argument.

Let us assume that the initial state of the log-price process $Y_0 = y_0$ with probability 1, i.e. the initial distribution $P_0(A) = I(y_0 \in A)$ is concentrated in a point $y_0 \in \mathbb{R}_1$.

The tree structure of trajectories is well visible from the following recurrent algorithm, which describes sequential transitions of the log-price process Y_n. If $Y_{n-1} = y$ for some $n = 1, 2, \dots$, then $Y_n = y + l\delta + \delta_{n,\circ}$, where l takes one of the values $m_{n,-}, \dots, m_{n,+}$.

The initial point y_0 determines the unique tree $T_{(y_0,0)}$ with the root in the node $(y_0, 0)$, in the space–time domain $\mathbb{R}_1 \times \{0, 1, \dots\}$.

According to the algorithm described above, the tree $T_{(y_0,0)}$ has one initial node $(y_0, 0)$. It may include $m_{1,+} - m_{1,-} + 1$ additional nodes after the first step, then $(m_{1,+} - m_{1,-} + 1)(m_{2,+} - m_{2,-} + 1)$ additional nodes after the second step, etc. The tree may include $\prod_{k=1}^{n}(m_{k,+} - m_{k,-} + 1)$ additional nodes after the nth step.

If this is the case, i.e. the tree $T_{(y_0,0)}$ possesses an exponential rate of growth for the number of nodes as a function of the number of steps n, any algorithms based on such a tree would not be effective from the computational point of view.

However, this is not the case. In fact, the tree includes only $m_{1,+} - m_{1,-} + 1$ additional nodes after the first step, $m_{1,+} - m_{1,-} + m_{2,+} - m_{2,-} + 1$ additional nodes after the second step, etc.

This is because the tree $T_{(y_0,0)}$ satisfies the recombining condition. In this model, the recombining condition means that the random walk Y_n, which is in a position $Y_{n-1} = y$, for some $n = 1, 2, \dots$, and makes from this position the first jump $l'\delta + \delta_{n,\circ}$ and then the second jump $l''\delta + \delta_{n+1,\circ}$, ends up at the same point as if it would make the first jump $l''\delta + \delta_{n,\circ}$ and then the second jump $l'\delta + \delta_{n+1,\circ}$. This should hold for any l' and l'' such that $l' + l'' = l$. In this case, the end point is $l\delta + \delta_{n,\circ} + \delta_{n+1,\circ}$. This should hold for any $l = m_{n,-} + m_{n+1,-}, \dots, m_{n,+} + m_{n+1,+}$, $y \in \mathbb{R}_1$, and $n = 1, 2, \dots$.

The tree $T_{(y_0,0)}$ obviously includes one initial node $(y_0, 0)$ and $m_{1,+} - m_{1,-} + 1$ additional nodes $(y_0 + m_{1,-}\delta + \delta_{1,\circ}, 1), \dots, (y_0 + m_{1,+}\delta + \delta_{1,\circ}, 1)$ after the first step. Due to the recombining property, the tree includes $m_{1,+} - m_{1,-} + m_{2,+} - m_{2,-} + 1$ additional nodes $(y_0 + (m_{1,-} + m_{2,-})\delta + \delta_{1,\circ} + \delta_{2,\circ}, 2), \dots, (y_0 + (m_{1,+} + m_{2,+})\delta + \delta_{1,\circ} + \delta_{2,\circ}, 2)$ after the second step, etc.

The tree $T_{(y_0,0)}$ includes $\sum_{k=1}^{n}(m_{k,+} - m_{k,-}) + 1$ additional nodes after the nth step. The total number of nodes in the tree after n steps is $L_n = 1 + (m_{1,+} - m_{1,-}) + 1 + \cdots + \sum_{k=1}^{n}(m_{k,+} - m_{k,-}) + 1 = n + 1 + n(m_{1,+} - m_{1,-}) + \cdots + (m_{n,+} - m_{n,-})$.

This means that, in the most important case, where $m_{n,+} - m_{n,-} \leq L, n = 1, \dots, N$, the tree possesses not more than a quadratic rate of growth for the number of nodes in the tree after n steps. Indeed, in this case, $L_n \leq n + 1 + L\frac{n(n+1)}{2}$, $n = 1, \dots$. This makes algorithms based on such a tree effective from the computational point of view.

Nodes appeared in the tree after n steps are determined by the number $l = l_1 + \cdots + l_n$, where $l_1\delta + \delta_{1,\circ}, \dots, l_n\delta + \delta_{n,\circ}$ are jumps at moments $1, \dots, n$. The states that can be reached by the random walk at moments $n = 0, 1, \dots,$ are given by the formulas

$$y_{n,l} = y_0 + l\delta + \delta_{1,\circ} + \cdots + \delta_{n,\circ}, \quad l = M_{n,-}, \dots, M_{n,+}\ n = 0, 1, \dots. \tag{3.83}$$

where $M_{n,\pm} = \sum_{k=1}^{n} m_{k,\pm}$, $n = 1, 2, \dots$.

It should be mentioned that the procedure for the construction of the tree $T_{(y_0,0)}$ described above does not take into account the values of appearance probabilities for nodes in the tree generated by trajectories of the random walk Y_n.

The probability that a node $y_{n,l}$ will be reached after n steps is given by the convolution recurrence relation, for $M_{n,-} \leq l_n \leq M_{n,+}$, $n = 0, 1, \dots,$

$$p_{n,l_n} = \begin{cases} p_1(l_1) & \text{if } M_{1,-} \leq l_1 \leq M_{1,+}, \quad n = 1, \\ \sum_{l=m_{n,-}}^{m_{n,+}} p_{n-1,l_n-l}p_n(l) & \text{if } M_{n,-} \leq l_n \leq M_{n,-}, \quad n > 1, \end{cases} \tag{3.84}$$

where one should count $p_{n,l} = 0$ if $l < M_{n,-}$ or $l > M_{n,+}$, for $n = 1, 2, \dots$.

Note also that the states $y_{n,l}$ satisfy the inequalities,

$$y_{n,M_{n,-}} < y_{n,M_{n,-}+1} < \cdots < y_{n,M_{n,+}}, \quad n = 1, \dots. \tag{3.85}$$

According to formula (3.3), the tree $T_{(y_0,0)}$ is the following set of nodes:

$$T_{(y_0,0)} = \langle (y_{n,l}, n), l = M_{n,-}, \dots, M_{n,+}, n = 0, 1, \dots \rangle. \tag{3.86}$$

The tree $T_{(y_0,0)}$ includes a particular point (y, n), if $y_{n,l} = y_0 + l\delta + \delta_{1,\circ} + \cdots + \delta_{n,\circ} = y$ for some $M_{n,-} \leq l \leq M_{n,+}$.

The condition $\mathbf{A_2}$ holds since, in this case, the following inequality takes place for any $y \in \mathbb{R}_1$, $n = 0, 1, \dots, N$:

$$\begin{aligned} \mathsf{E}_{y,n} \max_{n \le r \le N} \left| g(r, e^{Y_r}) \right| \\ \le \max_{n \le r \le N} \max_{m + M_{r,-} - M_{n,-} \le l \le m + M_{r,+} - M_{n,+}} \left| g(r, e^{y_{r,l}}) \right| < \infty \,, \end{aligned} \tag{3.87}$$

where the points $y_{r,l}$ are given by formulas (3.83), with y_0 and $M_{n,-} \le m \le M_{n,+}$ chosen such that $y_{n,m} = y$.

Theorem 2.3.3 takes in this case the following form.

Lemma 3.3.2. *Let the log-price process Y_n be represented by the inhomogeneous in time random walk defined by the stochastic transition dynamic relation* (3.82). *Then the log-reward functions $\phi_n(y_{n,l})$, for points $y_{n,l}$, $l = M_{n,-}, \dots, M_{n,+}$, $n = 0, 1, \dots, N$ given by formulas* (3.83), *are the unique solution for the following finite recurrence system of linear equations:*

$$\begin{cases} \phi_N(y_{N,l}) = g(N, e^{y_{N,l}}) \,, \\ \qquad l = l = M_{N,-}, \dots, M_{N,+} \,, \\ \phi_n(y_{n,l}) = \max\left(g(n, e^{y_{n,l}}) \,, \displaystyle\sum_{r = m_{n+1,-}}^{m_{n+1,+}} \phi_{n+1}(y_{n+1,l+r}) p_{n+1}(r) \right) \,, \\ \qquad l = M_{n,-}, \dots, M_{n,+} \,, \\ \qquad n = N-1, \dots, 0 \,. \end{cases} \tag{3.88}$$

The system of equations (3.88) let one find values of the log-reward function $\phi_n(y)$ at points $y_{n,l}$, $l = 0, 1, \dots, n$ given by formulas (3.83), recursively, for $n = N, \dots, 0$.

The particular value $\phi_n(y)$ can be calculated using this system if the initial value y_0 is chosen so that $y_{n,l} = y_0 + l\delta + \delta_{1,\circ} + \cdots + \delta_{n,\circ} = y$, for some $M_{n,-} \le l \le M_{n,+}$.

Moreover, if $y_{n,l_y} = y$ for some $0 \le n \le N$, $y \in \mathbb{R}_1$, then the value $\phi_n(y)$ can be found by solving the reduced version of the system of equations (3.88),

$$\begin{cases} \phi_N(y_{N,l}) = g(N, e^{y_{N,l}}) \,, \\ \qquad l = l_y + M_{N,-} - M_{n,-}, \dots, l_y + M_{N,+} - M_{n,+} \,, \\ \phi_s(y_{s,l}) = \max\left(g(s, e^{y_{s,l}}), \displaystyle\sum_{r = m_{s+1,-}}^{m_{s+1,+}} \phi_{s+1}(y_{s+1,l+r}) p_{s+1}(r) \right) \,, \\ \qquad l = l_y + M_{s,-} - M_{n,-}, \dots, l_y + M_{s,+} - M_{n,+} \,, \\ \qquad s = N-1, \dots, n \,, \end{cases} \tag{3.89}$$

where $y_{s,l} = y + l\delta + \delta_{n+1,\circ} + \cdots + \delta_{s,\circ}$, $l = l_y + M_{s,-} - M_{n,-}, \dots, l_y + M_{s,+} - M_{n,+}$, $s = n, \dots, N$.

In the above model, it is also natural to assume that the initial distribution $P_0(A)$ is concentrated at a finite set of points $y_{0,l} = \delta l + \delta_{0,\circ}$, $l = m_{0,-}, \dots, m_{0,+}$, with the corresponding probabilities $p_{0,l}$, $l = m_{0,-}, \dots, m_{0,+}$. In this case, the optimal expected reward $\Phi(\mathcal{M}_{\max,N})$ can be found using the following summation formula:

$$\Phi(\mathcal{M}_{\max,N}) = \sum_{l=m_{0,-}}^{m_{0,+}} p_{0,l}\phi_0(y_{0,l}) \,. \tag{3.90}$$

3.4 Markov chain reward algorithms

In this section, we describe backward tree algorithms for finding log-reward functions and optimal expected rewards for models where modulated log-price processes are represented by Markov chains, with transition probabilities (distributions), which are concentrated at finite sets of points. Such log-price processes can be effectively used for approximation of more general modulated Markov-type log-price processes. Such so-called space-skeleton approximations are studied in Chapters 7 and 8.

3.4.1 An atomic Markov chain model

We assume that the modulated log-price process $\vec{Z}_n = (\vec{Y}_n, X_n)$ is an inhomogeneous in time Markov chain with a phase space $\mathbb{Z} = \mathbb{R}_k \times \mathbb{X}$, an initial distribution $P_0(A) = \mathsf{P}\{\vec{Z}_0 \in A\}$ and one-step transition probabilities $P_n(\vec{z}, A) = \mathsf{P}\{\vec{Z}_n \in A/\vec{Z}_{n-1} = \vec{z}\}$.

In this case, the stochastic modulating index X_n is included in the model. Correspondingly, a pay-off function $g(n, e^y, x)$ and the corresponding log-reward functions $\phi_n(\vec{y}, x)$ do depend on the index argument.

We assume that one-step transition probabilities $P_n(\vec{z}, A)$ are concentrated at finite sets $\mathbb{F}_{\vec{z},n} = \{f_l(\vec{z}, n),\ l = m^-(\vec{z}, n), \dots, m^+(\vec{z}, n)\} \in \mathcal{B}_{\mathbb{Z}}$, with the corresponding one-point transition probabilities $p_l(\vec{z}, n) = \mathsf{P}\{\vec{Z}_n = f_l(\vec{z}, n)\ /\vec{Z}_{n-1} = \vec{z}\}$, $l = m^-(\vec{z}, n), \dots, m^+(\vec{z}, n)$, i.e. for every $\vec{z} \in \mathbb{Z}$ and $n = 1, 2, \dots$,

$$P_n(\vec{z}, A) = \sum_{f_l(\vec{z},n)\in A} p_l(\vec{z}, n) \,, \quad A \in \mathcal{B}_{\mathbb{Z}} \,. \tag{3.91}$$

In this case, it is natural to assume that the initial distribution $P_0(A)$ also is concentrated at a finite set $\mathbb{F}_0 = \{\vec{z}_{0,l},\ l = m_0^-, \dots, m_0^+\} \in \mathcal{B}_{\mathbb{Z}}$ with the corresponding one-point probabilities $p_{0,l} = \mathsf{P}\{\vec{Z}_0 = \vec{z}_{0,l}\}$, $l = m_0^-, \dots, m_0^+$, i.e.

$$P_0(A) = \sum_{\vec{z}_{0,l}\in A} p_{0,l} \,, \quad A \in \mathcal{B}_{\mathbb{Z}} \,. \tag{3.92}$$

In the case, where the above conditions hold, $\vec{Z}_n$ can be referred to as an atomic Markov chain.

The above assumption (3.91) implies that the r-step transition probabilities $P(n, \vec{z}, n + r, A) = \mathsf{P}\{\vec{Z}_{n+r} \in A/\vec{Z}_n = \vec{z}\}$ are also concentrated on finite sets $\mathbb{F}_{\vec{z},n,n+r} =$

$\{f_l(\vec{z}, n, n+r), l = m^-(\vec{z}, n, n+r), \dots, m^+(\vec{z}, n, n+r)\}$, with the corresponding transition probabilities $p_l(\vec{z}, n, n+r) = \mathsf{P}\{\vec{Z}_{n+r} = f_l(\vec{z}, n, n+r)/\vec{Z}_n = \vec{z}\}$, $l = m^-(\vec{z}, n, n+r), \dots, m^+(\vec{z}, n, n+r)$, i.e. for every $\vec{z} \in \mathbb{Z}$ and $n = 0, 1, \dots,$ and $r = 1, 2, \dots,$

$$P(n, \vec{z}, n+r, A) = \sum_{f_l(\vec{z}, n, n+r) \in A} p_l(\vec{z}, n, n+r), \quad A \in \mathcal{B}_{\mathbb{Z}}. \tag{3.93}$$

The following recurrent relations link transition functions $f_l(\vec{z}, n, n+r)$ and the corresponding transition probabilities $p_l(\vec{z}, n, n+r)$ for $r = 1, 2, \dots$:

$$f_l(\vec{z}, n, n+r) = \begin{cases} f_l(\vec{z}, n+1) & \text{if } r = 1, \\ m^-(\vec{z}, n+1) \le l \le m^+(\vec{z}, n+1), & \\ f_j(f_i(\vec{z}, n, n+r-1), n+r) & \text{if } r > 1, \\ (i, j) \in \mathbb{F}_{\vec{z}, n, n+r, l}, & \\ m^-(\vec{z}, n, n+r) \le l \le m^+(\vec{z}, n, n+r). & \end{cases} \tag{3.94}$$

and

$$p_l(\vec{z}, n, n+r) = \begin{cases} p_l(\vec{z}, n+1) & \text{if } r = 1, \\ m^-(\vec{z}, n+1) \le l \le m^+(\vec{z}, n+1), & \\ \sum_{(i,j) \in \mathbb{F}_{\vec{z}, n, n+r, l}} p_i(\vec{z}, n, n+r-1) & \\ \times p_j(f_i(\vec{z}, n, n+r-1), n+r) & \text{if } r > 1, \\ m^-(\vec{z}, n, n+r) \le l \le m^+(\vec{z}, n, n+r), & \end{cases} \tag{3.95}$$

where

$$\begin{aligned} \mathbb{F}_{\vec{z}, n, n+r, l} = \big\{(i, j) : {} & f_j(f_i(\vec{z}, n, n+r-1), n+r) = f_l(\vec{z}, n, n+r) \\ & m^-(\vec{z}, n, n+r-1) \le i \le m^+(\vec{z}, n, n+r-1), \\ & m^-(f_i(\vec{z}, n, n+r-1), n+r) \le j \le m^+(f_i(\vec{z}, n, n+r-1), n+r)\big\}. \end{aligned} \tag{3.96}$$

Any node $(\vec{z}, n)$ generates the tree of nodes $T_{(\vec{z}, n)}$. This tree includes one initial node $(\vec{z}, n)$, additional nodes $(f_l(\vec{z}, n, n+1), n+1)$, $l = m^-(\vec{z}, n, n+1), \dots, m^+(\vec{z}, n, n+1)$, appearing in the tree after the first transition, additional nodes $(f_l(\vec{z}, n, n+2), n+2)$, $l = m^-(\vec{z}, n, n+2), \dots, m^+(\vec{z}, n, n+2)$, appearing in the tree after the second transition, etc.

According to the above remarks, the tree $T_{(\vec{z}, n)}$ is the following set of nodes:

$$\begin{aligned} T_{(\vec{z}, n)} = \langle & (f_l(\vec{z}, n, n+r), n+r), \\ & l = m^-(\vec{z}, n, n+r), \dots, \\ & m^+(\vec{z}, n, n+r), r = 0, 1, \dots \rangle. \end{aligned} \tag{3.97}$$

where $f_0(\vec{z}, n, n) = \vec{z}$, $m^{\pm}(\vec{z}, n, n) = 0$.

The total number of nodes in the tree after r transitions is

$$L_r(\vec{z}, n) = 1 + (m^+(\vec{z}, n, n+1) - m^-(\vec{z}, n, n+1) + 1) + \cdots + (m^+(\vec{z}, n, n+r) - m^-(\vec{z}, n, n+r) + 1) . \tag{3.98}$$

We are specially interested in trees, where the number $L_r(\vec{z}, n) \leq Lr^h$, $r = 1, 2, \ldots$, where $L, h > 0$, i.e. models where the trees have not more than polynomial rate of growth for $L_r(\vec{z}, n)$ as a function of n.

Examples of such binomial, trinomial and general discrete random walk trees have been presented in Sections 3.1–3.3. Below, we extend this list by presenting so-called skeleton-type models.

We should introduce additional notations explicitly separating log-price and index components for points $f_l(\vec{z}, n, n+r) \in \mathbb{Z}$, by presenting these points in the form

$$f_l(\vec{z}, n, n+r) = (f_l'(\vec{z}, n, n+r), f_l''(\vec{z}, n, n+r)) , \tag{3.99}$$

where $f_l'(\vec{z}, n, n+r) \in \mathbb{R}_k$ and $f_l''(\vec{z}, n, n+r) \in \mathbb{X}$.

The condition $\mathbf{A_2}$ holds since, in this case, the following inequality takes place for any $\vec{z} = (\vec{y}, x) \in \mathbb{Z}$, $n = 0, 1, \ldots, N$:

$$\begin{aligned} \mathsf{E}_{\vec{z},n} \max_{0 \leq r \leq N-n} \left| g(r, e^{\vec{Y}_{n+r}}, X_{n+r}) \right| &\leq \max_{0 \leq r \leq N-n} \max_{m^-(\vec{z},n,n+r) \leq l \leq m^+(\vec{z},n,n+r)} \left| g(n+r, e^{f_l'(\vec{z},n,n+r)}, f_l''(\vec{z}, n, n+r)) \right| \\ &< \infty . \end{aligned} \tag{3.100}$$

Theorem 2.3.3 takes in this case the following form.

Lemma 3.4.1. *Let the log-price process $\vec{Z}_n$ be an atomic Markov chain with transition probabilities and initial distribution defined, respectively, by relations* (3.91) *and* (3.92). *Then the log-reward functions $\phi_n(\vec{z})$ and $\phi_{n+r}(f_l(\vec{z}, n, n+r))$, for points $f_l(\vec{z}, n, n+r)$, $l = m^-(\vec{z}, n, n+r), \ldots, m^+(\vec{z}, n, n+r)$, $r = 1, \ldots N-n$ given by formulas* (3.94), *are, for every $\vec{z} \in \mathbb{Z}$, $n = 0, \ldots, N$, the unique solution for the following finite recurrence* *system of linear equations:*

$$\begin{cases} \phi_N(f_l(\vec{z}, n, N)) = g(N, e^{f_l'(\vec{z},n,N)}, f_l''(\vec{z}, n, N)) , \\ \qquad l = m^-(\vec{z}, n, N), \dots, m^+(\vec{z}, n, N) , \\ \phi_{n+r}(f_l(\vec{z}, n, n+r)) \\ \qquad = \max \Bigg(g\left(n + r, e^{f_l'(\vec{z},n,n+r)}, f_l''(\vec{z}, n, n+r)\right) , \\ \qquad \sum\limits_{l'=m^-(\vec{z},n,n+r+1)}^{m^+(\vec{z},n,n+r+1)} \phi_{n+r+1}(f_{l'}(f_l(\vec{z}, n, n+r), n+r+1)) \\ \qquad \times p_{l'}(f_l(\vec{z}, n, n+r), n+r+1) \Bigg) , \\ \qquad l = m^-(\vec{z}, n, n+r), \dots, m^+(\vec{z}, n, n+r) , \\ \qquad r = N - n - 1, \dots, 0 , \end{cases} \tag{3.101}$$

where $f_0(\vec{z}, n, n) = \vec{z}$, $m^\pm(\vec{z}, n, n) = 0$.

As far as the optimal expected reward $\Phi(\mathcal{M}_{\max,N})$ is concerned, it can be found using the following summation formula:

$$\Phi(\mathcal{M}_{\max,N}) = \sum_{l=m_0^-}^{m_0^+} p_{0,l}\phi_0(\vec{z}_{0,l}) . \tag{3.102}$$

3.4.2 An atomic Markov chain model with a fixed skeleton structure

This model is a particular case of the model introduced in Subsection 3.4.1.

Let us assume that the sets of points $\mathbb{F}_n = \{\vec{z}_{n,l}, l = m_n^-, \dots, m_n^+\} \in \mathcal{B}_\mathbb{Z}$ are defined for $n = 0, 1, \dots,$ and that points $f_l(\vec{z}, n) = \vec{z}_{n,l}$, $l = m^-(\vec{z}, n) = m_n^-, \dots,$ $m^+(\vec{z}, n) = m_n^+$, $n = 1, 2, \dots,$ and, respectively, sets $\mathbb{F}_{\vec{z},n} = \mathbb{F}_n$, $n = 1, 2, \dots,$ do not depend on $\vec{z} \in \mathbb{Z}$, while the probabilities $p_l(\vec{z}, n) = \mathsf{P}\{\vec{Z}_n = \vec{z}_{n,l}/\vec{Z}_{n-1} = \vec{z}\}$, $l = m_n^-, \dots, m_n^+$, $n = 1, 2, \dots,$ do depend on $\vec{z} \in \mathbb{Z}$.

In this case, measures $P_n(\vec{z}, A)$ have the following form, for $\vec{z} \in \mathbb{Z}$ and $n = 1, 2, \dots$:

$$P_n(\vec{z}, A) = \sum_{\vec{z}_{n,l} \in A} p_l(\vec{z}, n) , \quad A \in \mathcal{B}_\mathbb{Z} . \tag{3.103}$$

The log-price process $\vec{Z}_n$ can be referred to as a space-skeleton Markov chain.

The tree of nodes $T_{(\vec{z},n)}$ for the skeleton Markov chain $\vec{Z}_n$ includes one initial node $(\vec{z}, n)$, $m_{n+1}^+ - m_{n+1}^- + 1$, additional nodes $(\vec{z}_{n+1,l}, n+1)$, $l = m_{n+1}^-, \dots, m_{n+1}^+$, appearing in the tree after the first transition, $m^+_{n+2} - m^-_{n+2} + 1$ additional nodes $(\vec{z}_{n+2,l}, n+2)$, $l = m^-_{n+2}, \dots, m^+_{n+2}$, appearing in the tree after the second transition, etc.

According to the above remarks, the tree $T_{(\vec{z},n)}$ is the following set of nodes:

$$T_{(\vec{z},n)} = \langle (\vec{z}, n), (\vec{z}_{n+r,l}, n+r), l = m^-_{n+r}, \dots, m^+_{n+r}, r = 0, 1, \dots \rangle . \tag{3.104}$$

The total number of nodes in the tree after r transitions is given by the following formula:

$$L_r(n) = 1 + (m^+_{n+1} - m^-_{n+1} + 1) + \cdots + (m^+_{n+r} - m^-_{n+r} + 1) . \tag{3.105}$$

Theorem 2.3.3 takes in this case the following form.

Lemma 3.4.2. *Let the log-price process $\vec{Z}_n$ be a skeleton Markov chain with transition probabilities defined by relation* (3.103). *Then the log-reward functions $\phi_n(\vec{z})$ and $\phi_{n+r}(\vec{z}_{n+r,l})$, for points $\vec{z}_{n+r,l}$, $l = m^-_{n+r}, \dots, m^+_{n+r}$, $r = 1, \dots N-n$ are, for every $\vec{z} \in \mathbb{Z}$, $n = 0, \dots, N$, the unique solution for the following finite recurrence system of linear equations:*

$$\begin{cases} \phi_N(\vec{z}_{N,l}) = g(N, e^{\vec{y}_{N,l}}, x_{N,l}) , \quad l = m^-_N, \dots, m^+_N , \\ \phi_{n+r}(\vec{z}_{n+r,l}) = \max\Bigg(g(n+r, e^{\vec{y}_{n+r,l}}, x_{n+r,l}) , \\ \qquad \displaystyle\sum_{l'=m^-_{n+r+1}}^{m^+_{n+r+1}} \phi_{n+r+1}(\vec{z}_{n+r+1,l'}) p_{l'}(\vec{z}_{n+r,l}, n+r+1) \Bigg) , \\ \qquad l = m^-_{n+r}, \dots, m^+_{n+r} , \quad r = N-n-1, \dots, 1 , \\ \phi_n(\vec{z}) = \max\Bigg(g(n, e^{\vec{y}}, x) , \displaystyle\sum_{l'=m^-_{n+1}}^{m^+_{n+1}} \phi_{n+1}(\vec{z}_{n+1,l'}) p_{l'}(\vec{z}, n+1) \Bigg) . \end{cases} \tag{3.106}$$

3.4.3 A space-skeleton Markov chain model

This model is a particular case of the model introduced in Subsection 3.4.2.

Let us choose some $\delta_{n,i} > 0$, $i = 1, \dots, k$, $n = 0, 1, \dots$, integer $m^-_{n,i} \le m^+_{n,i}$, $i = 1, \dots, k$, $n = 0, 1, \dots$, and $\lambda_{n,i} \in \mathbb{R}_1$, $i = 1, \dots, k$, $n = 0, 1, \dots$.

Let us now define points $y_{n,i,l} = l\delta_{n,i} + \lambda_{n,i}$, $l = m^-_{n,i}, \dots, m^+_{n,i}$, $i = 1, \dots, k$, $n = 0, 1, \dots$, and also choose points $x_{n,l} \in \mathbb{X}$, $l = m^-_{n,0}, \dots, m^+_{n,0}$, $n = 0, 1, \dots$.

Now, let us define skeleton points for $\bar{l} = (l_0, l_1, \dots, l_k)$, $l_j = m^-_{n,j}, \dots, m^+_{n,j}$, $j = 0, \dots, k$, $n = 0, 1, \dots$,

$$\vec{z}_{n,\bar{l}} = (\vec{y}_{n,\bar{l}}, x_{n,\bar{l}}) = ((y_{n,1,l_1}, \dots, y_{n,k,l_k}), x_{n,l_0}) . \tag{3.107}$$

The difference in notations (for $\vec{z}_{n,l}$ and $\vec{z}_{n,\bar{l}}$ and between $p_l(\vec{z}, n)$ and $p_{\bar{l}}(\vec{z}, n)$) can be removed by a natural renumeration of indices, namely $(m^-_{n,0}, m^-_{n,1}, \dots, m^-_{n,k}) \leftrightarrow$

$m_n^-, \dots, (m_{n,0}^- + 1, m_{n,1}^-, \dots, m_{n,k}^+) \leftrightarrow m_n^- + 1, \dots, (m_{n,0}^+, m_{n,1}^+, \dots, m_{n,k}^+) \leftrightarrow m_n^+$. Obviously, the following relation holds, for $n = 0, 1, \dots$:

$$m_n^+ - m_n^- + 1 = \prod_{j=0}^{k} (m_{n,j}^+ - m_{n,j}^- + 1) . \tag{3.108}$$

The simplest variant is to choose integers $\pm m_n^{\pm} \geq 0$, $n = 0, 1, \dots$.

The tree of nodes $T_{(\vec{z},n)}$ includes one initial node $(\vec{z}, n)$, $m_{n+1}^+ - m_{n+1}^- + 1$ additional nodes $\vec{z}_{n+1,\bar{l}}, \bar{l} = (l_0, l_1, \dots, l_k), l_j = m_{n+1,j}^-, \dots, m_{n+1,j}^+, j = 0, \dots, k$ after the first jump, $m_{n+2}^+ - m_{n+2}^- + 1$ additional nodes $\vec{z}_{n+2,\bar{l}}, \bar{l} = (l_0, l_1, \dots, l_k)$, $l_j = m_{n+2,j}^-, \dots, m_{n+1,2}^+, j = 0, \dots, k$ after the second jump, etc.

According to the above remarks, the tree $T_{(\vec{z},n)}$ is the following set of nodes:

$$\begin{aligned} T_{(\vec{z},n)} = \langle (\vec{z}, n), (\vec{z}_{n+r,\bar{l}}, n + r), \quad & \bar{l} = (l_0, l_1, \dots, l_k), \\ & l_j = m_{n+r,j}^-, \dots, m_{n+r,j}^+, j = 0, \dots, k, r = 1, \dots \rangle . \end{aligned} \tag{3.109}$$

The total number of nodes in the tree after r transitions is given by the following formula:

$$L_r(n) = 1 + (m_{n+1}^+ - m_{n+1}^- + 1) + \cdots + (m_{n+r}^+ - m_{n+r}^- + 1) . \tag{3.110}$$

Theorem 2.2.3 takes in this case the following form.

Lemma 3.4.3. *Let the log-price process $\vec{Z}_n$ is a skeleton Markov chain with skeleton points defined in relation* (3.107). *Then the log-reward functions $\phi_n(\vec{z})$ and $\phi_{n+r}(\vec{z}_{n+r,\bar{l}})$, for points $\vec{z}_{n+r,\bar{l}} = (\vec{y}_{n+r,\bar{l}}, x_{n+r,\bar{l}}) = ((y_{n+r,1,l_1}, \dots, y_{n+r,k,l_k}), x_{n+r,l_0}), \bar{l} = (l_0, l_1, \dots, l_k)$, $l_j = m_{n+r,j}^-, \dots, m_{n+r,j}^+, j = 0, \dots, k, r = 1, \dots, N-n$, are, for every $\vec{z} \in \mathbb{Z}, n = 0, \dots, N$, the unique solution for the following recurrence finite system of linear equations:*

$$\begin{cases} \phi_N(\vec{z}_{N,\bar{l}}) = g(N, e^{\vec{y}_{N,\bar{l}}}, x_{N,\bar{l}}) , \\ \qquad \bar{l} = (l_0, l_1, \dots, l_k) , \quad l_j = m_{N,j}^-, \dots, m_{N,j}^+ , \quad j = 0, \dots, k , \\ \phi_{n+r}(\vec{z}_{n+r,\bar{l}}) = \max \Bigg(g(n + r, e^{\vec{y}_{n+r,\bar{l}}}, x_{n+r,\bar{l}}) , \\ \qquad \displaystyle\sum_{\vec{z}_{n+r+1,\bar{l}'} \in \mathbb{F}_{n+r+1}} \phi_{n+r+1}(\vec{z}_{n+r+1,\bar{l}'}) p_{\bar{l}'}(\vec{z}_{n+r,\bar{l}}, n + r + 1) \Bigg) , \\ \qquad \bar{l} = (l_0, l_1, \dots, l_k) , \quad l_j = m_{n+r,j}^-, \dots, m_{n+r,j}^+ , \quad j = 0, \dots, k , \\ \qquad r = N - n - 1, \dots, 1 , \\ \phi_n(\vec{z}) = \max \Bigg(g(n, e^{\vec{y}}, x) , \displaystyle\sum_{\vec{z}_{n+1,\bar{l}'} \in \mathbb{F}_{n+1}} \phi_{n+1}(\vec{z}_{n+1,\bar{l}'}) p_{\bar{l}'}(\vec{z}, n + 1) \Bigg) . \end{cases} \tag{3.111}$$

3.4.4 A space-skeleton Markov random walk model

A modulated Markov chain $\vec{Z}_n = (\vec{Y}_n, X_n)$, $n = 0, 1, \ldots$, can always be represented in the form of the modulated Markov random walk $\vec{Z}_n = (\vec{Y}_n, X_n) = (\vec{Y}_{n-1} + \vec{W}_n, X_n)$, $n = 1, 2, \ldots$, with an initial state $\vec{Z}_0 = (\vec{Y}_0, X_0)$ and random jumps $\vec{W}_n = \vec{Y}_n - \vec{Y}_{n-1}$, $n = 1, 2, \ldots$.

In this case, it is convenient to operate with the transition jump probabilities $\tilde{P}_n(\vec{y}, x, A) = \mathsf{P}\{(\vec{W}_n, X_n) \in A / \vec{Y}_{n-1} = \vec{y}, X_{n-1} = x\}$, which are connected with the transition probabilities of the Markov chain $\vec{Z}_n$ by the following relations, for $\vec{z} = (\vec{y}, x) \in \mathbb{Z}, A \in \mathcal{B}_{\mathbb{Z}}, n = 0, 1, \ldots$:

$$
\begin{aligned}
\tilde{P}_n(\vec{y}, x, A) &= \mathsf{P}\left\{(\vec{W}_n, X_n) \in A / \vec{Y}_{n-1} = \vec{y}, X_{n-1} = x\right\} \\
&= \mathsf{P}\left\{(\vec{Y}_n - \vec{y}, X_n) \in A / \vec{Y}_{n-1} = \vec{y}, X_{n-1} = x\right\} \\
&= P_n\left(\vec{z}, A_{[\vec{y}]}\right), \qquad (3.112)
\end{aligned}
$$

where $A_{[\vec{y}]} = \{\vec{z}\,' = (\vec{y}\,', x) : (\vec{y}\,' - \vec{y}, x) \in A\}$.

Let us consider particular cases of the model introduced above in Subsection 3.4.1, which can also be interpreted as a space-skeleton-type Markov random walk.

In this case, one can use the standard space-skeleton construction applied to jumps for the random walk.

Let us choose some $\delta_i > 0$, $i = 1, \ldots, k$, integers $m^-_{n,j} \le m^+_{n,j}$, $j = 0, \ldots, k$, $n = 0, 1, \ldots$, and $\lambda_{n,i} \in \mathbb{R}_1$, $i = 1, \ldots, k$, $n = 0, 1, \ldots$.

Let us now define points $y_{n,i,l_i} = l_i\delta_i + \lambda_{n,i}$, $l_i = m^-_{n,i} \ldots, m^+_{n,i}$, $i = 1, \ldots, k$, $n = 0, 1, \ldots$, and, also, choose points $x_{n,l_0} \in \mathbb{X}$, $l_0 = m^-_{n,0}, \ldots, m^+_{n,0}$, $n = 0, 1, \ldots$.

Finally let us define skeleton points $\vec{z}_{n,\bar{l}} = (\vec{y}_{n,\bar{l}}, x_{n,\bar{l}})$, $\bar{l} = (l_0, l_1, \ldots, l_k)$, $l_j = m^-_{n,j}, \ldots, m^+_{n,j}$, $j = 0, \ldots, k$, $n = 0, 1, \ldots$,

$$
\vec{z}_{n,\bar{l}} = (\vec{y}_{n,\bar{l}}, x_{n,\bar{l}}) = ((y_{n,1,l_1}, \ldots, y_{n,k,l_k}), x_{n,l_0}) \,. \qquad (3.113)
$$

Let us now introduce skeleton sets $\mathbb{F}_{\vec{z},n} = \{f_{\bar{l}}(\vec{z}, n) = (\vec{y} + \vec{y}_{n,\bar{l}}, x_{n,\bar{l}}) : (\vec{y}_{n,\bar{l}}, x_{n,\bar{l}}) \in \mathbb{F}_n\}$, where $\mathbb{F}_n = \{\vec{z}_{n,\bar{l}} : \bar{l} = (l_0, l_1, \ldots, l_k), l_j = m^-_{n,j}, \ldots, m^+_{n,j}, j = 0, \ldots, k\}$ for $\vec{z} = (\vec{y}, x) \in \mathbb{Z}$ and $n = 0, 1, \ldots$.

In this case, skeleton points $f_{\bar{l}}(\vec{z}, n)$ and sets $\mathbb{F}_{\vec{z},n}$ depend on $\vec{z} \in \mathbb{Z}$. Note that paranetrs δ_i, $i = 1, \ldots, k$, do not depend on n. This implies that the corresponding recombining condition holds.

The corresponding one-point transition probabilities $\tilde{p}_{\bar{l}}(\vec{y}, x, n) = \mathsf{P}\{(\vec{W}_n, X_n) = (\vec{y}_{n,\bar{l}}, x_{n,\bar{l}}) / \vec{Z}_{n-1} = \vec{z}\} = p_{\bar{l}}(\vec{z}, n) = \mathsf{P}\{\vec{Z}_n = f_{\bar{l}}(\vec{z}, n) / \vec{Z}_{n-1} = \vec{z}\}$, $\bar{l} = (l_0, l_1, \ldots, l_k)$, $l_j = m^-_{n,j}, \ldots, m^+_{n,j}$, $j = 0, \ldots, k$ also depend on $\vec{z} = (\vec{y}, x) \in \mathbb{Z}$.

In this case, the probability measures $\tilde{P}_n(\vec{z}, A)$ have the following form, for $\vec{z} = (\vec{y}, x) \in \mathbb{Z}$ and $n = 0, 1, \dots$:

$$\begin{aligned} \tilde{P}_n(\vec{y}, x, A) &= \sum_{(\vec{y}_{n,\bar{l}}, x_{n,\bar{l}}) \in A} \tilde{p}_{\bar{l}}(\vec{y}, x, n), \\ &= P_n(\vec{z}, A_{[\vec{y}]}) \\ &= \sum_{(\vec{y}+\vec{y}_{n,\bar{l}}, x_{n,\bar{l}}) \in A_{[\vec{y}]}} p_{\bar{l}}(\vec{z}, n), \quad A \in \mathcal{B}_{\mathbb{Z}}. \end{aligned} \tag{3.114}$$

The difference in notations (for $f_l(\vec{z}, n)$ and $f_{\bar{l}}(\vec{z}, n)$ and between $p_l(\vec{z}, n)$ and $p_{\bar{l}}(\vec{z}, n)$) can be removed, as above, by a natural renumeration of indices, namely $(m_{n,0}^-, m_{n,1}^-, \dots, m_{n,k}^-) \leftrightarrow m_n^-, \dots, (m_{n,0}^- + 1, m_{n,1}^-, \dots, m_{n,k}^+) \leftrightarrow m_n^- + 1, \dots, (m_{n,+}^-, m_{n,1}^+, \dots, m_{n,k}^+) \leftrightarrow m_n^+$, where

$$m_n^+ - m_n^- + 1 = \prod_{j=0}^{k} (m_{n,j}^+ - m_{n,j}^- + 1), \quad n = 0, 1, \dots. \tag{3.115}$$

The simplest variant is to choose integers $\pm m_n^\pm \geq 0$, $n = 0, 1, \dots$.

Let us denote $\vec{y}_{n,n+r,\bar{l}} = (\delta_1 l_1 + \lambda_{n,n+r,1}, \dots, \delta_k l_k + \lambda_{n,n+r,k})$, $x_{n,n+r,\bar{l}} = x_{n+r,l_0}$, $\bar{l} = (l_0, \dots, l_k)$, $l_j = m_{n,n+r,j}^-, \dots, m_{n,n+r,j}^+$, $j = 0, \dots, k$, where $m_{n,n+r,0}^\pm = m_{n+r,0}^\pm$ and $m_{n,n+r,j}^\pm = \sum_{l=n+1}^{n+r} m_{l,j}^\pm$, $\lambda_{n,n+r,j} = \sum_{l=n+1}^{n+r} \lambda_{l,j}$, $j = 1, \dots, k$, for $0 \leq n < n + r < \infty$.

If $(\vec{Y}_n, X_n) = (\vec{y}, x)$ for some $(\vec{y}, x) \in \mathbb{Z}$ and $n = 0, 1, \dots$, then the possible states for $(\vec{Y}_{n+r}, X_{n+r})$ for $r = 1, 2, \dots$, are $(\vec{y} + \vec{y}_{n,n+r,\bar{l}}, x_{n,n+r,\bar{l}})$, $\bar{l} = (l_0, \dots, l_k)$, $l_j = m_{n,n+r,j}^-, \dots, m_{n,n+r,j}^+$, $j = 0, \dots, k$.

Let us also define integer numbers $m_{n,n+r}^- \leq m_{n,n+r}^+$, $0 \leq n < n + r < \infty$ such that the following relation holds for $0 \leq n < n + r < \infty$:

$$m_{n,n+r}^+ - m_{n,n+r}^- + 1 = \prod_{j=0}^{k} (m_{n,n+r,j}^+ - m_{n,n+r,j}^- + 1). \tag{3.116}$$

The simplest variant is to choose integers $\pm m_{n,n+r}^\pm \geq 0$, $0 \leq n < n + r < \infty$.

The tree of nodes $T_{((\vec{y},x),n)}$ includes one initial node $((\vec{y}, x), n)$, $m_{n,n+1}^+ - m_{n,n+1}^- + 1$ additional nodes $((\vec{y} + \vec{y}_{n,n+1,\bar{l}}, \vec{x}_{n,n+1,\bar{l}}), n + 1)$, $\bar{l} = (l_0, l_1, \dots, l_k)$, $l_j = m_{n,n+1,j}^-, \dots, m_{n,n+1,j}^+$, $j = 0, \dots, k$ after the first jump, $m_{n,n+2}^+ - m_{n,n+2}^- + 1$ additional nodes $((\vec{y} + \vec{y}_{n,n+2,\bar{l}}, x_{n+2,\bar{l}}), n + 2)$, $\bar{l} = (l_0, l_1, \dots, l_k)$, $l_j = m_{n,n+2,j}^-, \dots, m_{n,n+2,j}^+$, $j = 0, \dots, k$ after the second jump, etc.

According to the above remarks, the skeleton Markov random walk tree $T_{((\vec{y},x),n)}$ is the following set of nodes:

$$\begin{aligned} T_{((\vec{y},x),n)} = \langle &((\vec{y}, x), n), ((\vec{y} + \vec{y}_{n,n+r,\bar{l}}, x_{n+r,\bar{l}}), n + r), \\ &\bar{l} = (l_0, l_1, \dots, l_k), \ l_j = m_{n,n+r,j}^-, \dots, m_{n,n+r,j}^+, \\ &j = 0, \dots, k, \ r = 1, \dots \rangle. \end{aligned} \tag{3.117}$$

The total number of nodes in the tree after r jumps is given by the following formula:

$$L_r(n) = 1 + m^+_{n,n+1} - m^-_{n,n+1} + 1 + \cdots + m^+_{n,n+r} - m^-_{n,n+r} + 1 . \tag{3.118}$$

Theorem 2.3.3 takes in this case the following form.

Lemma 3.4.4. *Let the log-price process $\vec{Z}_n$ be a skeleton Markov random walk with skeleton points and transition probabilities defined, respectively, by relations* (3.113) *and* (3.114). *Then the log-reward functions $\phi_n(\vec{y}, x)$ and $\phi_{n+r}(\vec{y} + \vec{y}_{n,n+r,\bar{l}}, x_{n+r,\bar{l}})$, for points $\bar{l} = (l_0, l_1, \ldots, l_k)$, $l_j = m^-_{n,n+r,j}, \ldots, m^+_{n,n+r,j}$, $j = 0, \ldots, k$, $r = 1, \ldots N - n$ are, for every $\vec{z} \in \mathbb{Z}$, $n = 0, \ldots, N$, the unique solution for the following finite recurrence system of linear equations:*

$$\left\{\begin{aligned}
&\phi_N(\vec{y} + \vec{y}_{n,N,\bar{l}}, x_{N,\bar{l}}) = g(N, e^{\vec{y}_{n,N,\bar{l}}}, x_{n,N,\bar{l}}) , \\
&\qquad \bar{l} = (l_0, l_1, \ldots, l_k) , \quad l_j = m^-_{n,N,j}, \ldots, m^+_{n,N,j} , \quad j = 0, \ldots, k , \\
&\phi_{n+r}(\vec{y} + \vec{y}_{n,n+r,\bar{l}}, x_{n+r,\bar{l}}) \\
&\qquad = \max\Bigg(g(n + r, e^{\vec{y} + \vec{y}_{n,n+r,\bar{l}}}, x_{n+r,\bar{l}}) , \\
&\qquad\qquad \sum_{(\vec{y}_{n+r+1,\bar{l}'}, x_{n+r+1,\bar{l}'}) \in \mathbb{F}_{n+r+1}} \phi_{n+r+1}\left(\vec{y} + \vec{y}_{n,n+r,\bar{l}} + \vec{y}_{n+r+1,\bar{l}'}, x_{n+r+1,\bar{l}'}\right) \\
&\qquad\qquad \times p_{\bar{l}'}\left(\vec{y} + \vec{y}_{n,n+r,\bar{l}}, x_{n+r,\bar{l}}, n + r + 1\right)\Bigg) , \\
&\qquad \bar{l} = (l_0, l_1, \ldots, l_k) , \quad l_j = m^-_{n+r,j}, \ldots, m^+_{n+r,j} , \quad j = 0, \ldots, k , \\
&\qquad r = N - n - 1, \ldots, 1 , \\
&\phi_n(\vec{y}, x) = \max\Bigg(g(n, e^{\vec{y}}, x) , \\
&\qquad\qquad \sum_{(\vec{y}_{n+1,\bar{l}'}, x_{n+1,\bar{l}'}) \in \mathbb{F}_{n+1}} \phi_{n+1}(\vec{y} + \vec{y}_{n+1,\bar{l}'}, x_{n+1,\bar{l}'}) \\
&\qquad\qquad \times p_{\bar{l}'}(\vec{y}, x, n + 1)\Bigg) .
\end{aligned}\right. \tag{3.119}$$

4 Upper bounds for option rewards

In this chapter, we give upper bounds for log-reward functions and optimal expected rewards for American-type options for multivariate modulated Markov log-price processes.

We consider models with pay-off functions, which do not have more than polynomial rate of growth in price arguments and multivariate modulated Markov log-price processes with bounded and unbounded characteristics.

In Section 4.1, we give upper bounds for log-reward functions and optimal expected rewards for American-type options for univariate and multivariate modulated Markov log-price processes with bounded exponential moments of increments.

In Section 4.2, we give upper bounds for log-reward functions and optimal expected rewards for American-type options for univariate and multivariate modulated log-price processes represented by standard and modulated random walks.

In Section 4.3, we present upper bounds for log-reward functions and optimal expected rewards for American-type options for univariate and multivariate modulated Markov log-price processes with unbounded exponential moments of increments.

In Section 4.4, we investigate upper bounds for log-reward functions and optimal expected rewards for American-type options for univariate Markov Gaussian log-price processes with bounded and unbounded drift and volatility coefficients.

In Section 4.5, we give upper bounds for log-reward functions and optimal expected rewards for American-type options for multivariate modulated Markov Gaussian log-price processes with bounded and unbounded drift and volatility coefficients.

The main results for multivariate modulated Markov log-price processes with bounded exponential moments of increments and for log-price processes represented by random walks are, respectively, given in Theorems 4.1.1–4.1.4 and 4.2.1–4.2.4.

These results generalize results obtained in papers Silvestrov, Jönsson, and Stenberg (2006, 2008, 2009), for univariate modulated Markov log-price processes and by Lundgren and Silvestrov (2009, 2011), and Silvetsrov and Lundgren (2011) for multivariate Markov log-price processes.

The results for multivariate modulated Markov log-price processes with unbounded exponential moments of increments given in Theorems 4.3.1–4.3.8, as well as results for univariate and multivariate modulated Markov Gaussian log-price processes given in Theorems 4.4.3–4.4.4 and 4.5.3–4.5.4, are new.

4.1 Markov LPP with bounded characteristics

In this section, we give explicit upper bounds for log-reward functions and optimal expected rewards for American-type options for discrete time-modulated Markov log-price processes, with conditional exponential moments of increments for log-price processes bounded as functions of log-price and index arguments.

4.1.1 Upper bounds for exponential moments of maxima for Markov log-price processes with bounded characteristics

We assume that a modulated log-price process $\vec{Z}_n = (\vec{Y}_n, X_n)$ is an inhomogeneous Markov chain with a phase space $\mathbb{Z} = \mathbb{R}_k \times \mathbb{X}$, an initial distribution $P_0(A) = \mathsf{P}\{\vec{Z}_0 \in A\}$, and one-step transition probabilities $P_n(\vec{z}, A) = \mathsf{P}\{\vec{Z}_n \in A / \vec{Z}_{n-1} = \vec{z}\}$.

Let us define the first-type modulus of exponential moment compactness for components of the log-price process $\vec{Y}_n = (Y_{n,1} \dots, Y_{n,k})$, for $\beta \geq 0$, $i = 1, \dots, k$,

$$\Delta_\beta(Y_{\cdot,i}, N) = \max_{0 \leq n \leq N-1} \sup_{\vec{z} \in \mathbb{Z}} \mathsf{E}_{\vec{z},n} e^{\beta |Y_{n+1,i} - Y_{n,i}|} . \tag{4.1}$$

Remind that we use the notations $\mathsf{P}_{\vec{z},n}$ and $\mathsf{E}_{\vec{z},n}$ for conditional probabilities and expectations under the condition that $\vec{Z}_n = \vec{z}$.

We use the following first-type condition of exponential moment compactness for the log-price process $\vec{Y}_n$, assumed to hold for some vector parameter $\bar{\beta} = (\beta_1, \dots, \beta_k)$ with nonnegative components:

$\mathbf{C_1}[\bar{\beta}]$: $\Delta_{\beta_i}(Y_{\cdot,i}, N) < K_{1,i}$, $i = 1, \dots, k$, for some $1 < K_{1,i} < \infty$, $i = 1, \dots, k$.

It is useful to note that the condition $\mathbf{C_1}[\bar{\beta}]$ implies that the condition $\mathbf{C_1}[\bar{\beta}']$ holds with the same constants $K_{1,i}$, $i = 1, \dots, k$, for any vector parameter $\bar{\beta}' = (\beta'_1, \dots, \beta'_k)$ such that $0 \leq \beta'_i \leq \beta_i$, $i = 1, \dots, k$.

The following lemma gives explicit upper bounds for conditional exponential moments of the maximal absolute values of components for the log-price process $\vec{Y}_n$.

Lemma 4.1.1. *Let the condition $\mathbf{C_1}[\bar{\beta}]$ holds. Then there exist constants $0 \leq M_{1,i} < \infty$, $i = 1, \dots, k$ such that the following inequalities take place for $\vec{z} = (\vec{y}, x) = ((y_1, \dots, y_k), x) \in \mathbb{Z}$, $0 \leq n \leq N$, $i = 1, \dots, k$:*

$$\mathsf{E}_{\vec{z},n} \exp\left\{\beta_i \max_{n \leq r \leq N} |Y_{r,i}|\right\} \leq M_{1,i} e^{\beta_i |y_i|} . \tag{4.2}$$

Proof. Note that $\Delta_0(Y_{i,\cdot}, N) \equiv 1$ and, thus, the relation penetrating condition $\mathbf{C_1}[\bar{\beta}]$ holds automatically for i such that $\beta_i = 0$. Also, the expectation on the left-hand side of inequality (4.2) equals 1, while the expression on the right-hand side of this inequality equals $M_{1,i}$. Thus, this inequality holds, in this case, with the constant $M_{1,i} = 1$.

Therefore, we should prove (4.2) only for i such that $\beta_i > 0$.

Let us use the following inequality:

$$\begin{aligned}
\exp\left\{\beta_i \max_{n\le r\le N} |Y_{r,i}|\right\}
&= \max_{n\le r\le N} \exp\left\{\beta_i |Y_{n,i} + \sum_{l=n}^{r-1}(Y_{l+1,i} - Y_{l,i})|\right\} \\
&\le \exp\left\{\beta_i |Y_{n,i}| + \beta_i \sum_{r=n}^{N-1} |Y_{r+1,i} - Y_{r,i}|\right\} . \qquad (4.3)
\end{aligned}$$

Using Markov property of the log-price process $\vec{Y}_n$ and then the recurrent condition $\mathbf{C}_1[\vec{\beta}]$, we get the following inequalities, for $\vec{z} \in \mathbb{Z}$, $0 \le n \le N$:

$$\begin{aligned}
&\mathsf{E}_{\vec{z},n} \exp\left\{\sum_{r=n}^{N-1} \beta_i |Y_{r+1,i} - Y_{r,i}|\right\} \\
&= \mathsf{E}_{\vec{z},n}\mathsf{E}\left\{\exp\left\{\sum_{r=n}^{N-1} \beta_i |Y_{r+1,i} - Y_{r,i}|\right\} / \mathcal{F}_{N-1}\right\} \\
&= \mathsf{E}_{\vec{z},n}\left\{\exp\left\{\beta_i \sum_{r=1}^{N-2} |Y_{r+1,i} - Y_{r,i}|\right\} \cdot \mathsf{E}\exp\left\{\beta_i |Y_{N,i} - Y_{N-1,i}| / \vec{Z}_{N-1}\right\}\right\} \\
&\le \mathsf{E}_{\vec{z},n} \exp\left\{\beta_i \sum_{r=n}^{N-2} |Y_{r,i}|\right\} \cdot K_{1,i} \le \cdots \le K_{1,i}^{N-n} . \qquad (4.4)
\end{aligned}$$

Finally, using relations (4.3) and (4.4) we get the following inequality, for $\vec{z} \in \mathbb{Z}$, $0 \le n \le N$:

$$\begin{aligned}
&\mathsf{E}_{\vec{z},n} \exp\left\{\beta_i \max_{n\le r\le N} |Y_{r,i}|\right\} \\
&\quad\le \mathsf{E}_{\vec{z},n} \exp\left\{\beta_i |Y_{n,i}| + \sum_{r=n}^{N-1} \beta_i |Y_{r+1,i} - Y_{r,i}|\right\} \\
&\quad= e^{\beta_i |y_i|} \mathsf{E}_{\vec{z},n} \exp\left\{\sum_{r=n}^{N-1} \beta_i |Y_{r+1,i} - Y_{r,i}|\right\} \le e^{\beta_i |y_i|} K_{1,i}^{N-n} . \qquad (4.5)
\end{aligned}$$

The remarks made at the beginning of the proof and inequality (4.5) imply that inequality (4.2) holds with the constant $K_{1,i}^N$. □

Remark 4.1.1. The constants $M_{1,i}$, $i = 1, \ldots, k$ are given by the following formulas:

$$M_{1,i} = K_{1,i}^N . \qquad (4.6)$$

Moreover, inequalities (4.5) give upper bounds for exponential moments of the maxima for the log-price processes $Y_{n,i}$ over discrete time intervals $[n, N]$, with the constants $K_{1,i}^{N-n}$ depending on n.

The first-type modulus of exponential moment compactness $\Delta_\beta(Y_{\cdot,i}, N)$ involves the supremum of conditional exponential moments for increments of the corresponding component for the log-price processes $\vec{Y}_n$ over the whole phase space $\mathbb{Z}$. In some cases, for example, in applications to space-skeleton approximation models, this can be too strong restriction. One can improve the definition of the above modulus in the following way.

Let $\mathcal{Z} = \langle \mathbb{Z}_0, \dots, \mathbb{Z}_N \rangle$ be a sequence of measurable sets and $\Delta_{\beta,\mathcal{Z}}(Y_{\cdot,i}, N)$ be a modified first-type modulus of exponential moment compactness defined for $\beta \geq 0$ by the following formula:

$$\Delta_{\beta,\mathcal{Z}}(Y_{\cdot,i}, N) = \max_{0 \leq n \leq N-1} \sup_{\vec{z} \in \mathbb{Z}_n} \mathsf{E}_{\vec{z},n} e^{\beta |Y_{n+1,i} - Y_{n,i}|} . \tag{4.7}$$

Obviously, if $\mathcal{Z} = \langle \mathbb{Z}, \dots, \mathbb{Z} \rangle$, then modulus $\Delta_{\beta,\mathcal{Z}}(Y_{\cdot,i}, N) = \Delta_\beta(Y_{\cdot,i}, N)$, $i = 1, \dots, k, \beta \geq 0$.

A sequence of measurable subsets $\mathcal{Z} = \langle \mathbb{Z}_0, \dots, \mathbb{Z}_N \rangle$ is called a complete sequence of phase sets for the multivariate modulated Markov log-price process $\vec{Z}_n$ with an initial distribution $P_0(A)$ and transition probabilities $P_n(\vec{z}, A)$, if $\mathsf{P}_n(\vec{z}, \mathbb{Z}_n) = 1, \vec{z} \in \mathbb{Z}_{n-1}$, $n = 1, \dots, N$, and an ultimately complete sequence of phase sets for this process, if also $\mathsf{P}(\mathbb{Z}_0) = 1$.

The following property of complete sequence of phase sets and ultimately complete sequence of phase sets follows in an obvious way from their definitions.

Lemma 4.1.2. *If $\mathcal{Z} = \langle \mathbb{Z}_0, \dots, \mathbb{Z}_N \rangle$ is a complete sequence of phase sets for the multivariate modulated Markov log-price process $\vec{Z}_n$, then $\mathsf{P}_{\vec{z},n}\{\vec{Z}_r \in \mathbb{Z}_r\} = 1$, $\vec{z} \in \mathbb{Z}_n$, $0 \leq n \leq r \leq N$. If $\mathcal{Z}$ is an ultimately complete sequence of sets for this process, then also $\mathsf{P}\{\vec{Z}_r \in \mathbb{Z}_r\} = 1, 0 \leq r \leq N$.*

Let us formulate a condition of exponential moment compactness for the log-price process $\vec{Y}_n$, which generalizes the condition $\mathbf{C}_1[\bar{\beta}]$. It should be assumed to hold for some vector parameter $\bar{\beta} = (\beta_1, \dots, \beta_k)$ with nonnegative components:

$\mathbf{C}_1'[\bar{\beta}]$: There exists a sequence of measurable sets $\mathcal{Z} = \langle \mathbb{Z}_0, \dots, \mathbb{Z}_N \rangle$ such that
- (a) $\Delta_{\beta_i,\mathcal{Z}}(Y_{\cdot,i}, N) < K'_{1,i}$, $i = 1, \dots, k$, for some $1 < K'_{1,i} < \infty$, $i = 1, \dots, k$;
- (b) $\mathcal{Z}$ is a complete sequence of phase sets, i.e., $\mathsf{P}_n(\vec{z}, \mathbb{Z}_n) = 1, \vec{z} \in \mathbb{Z}_{n-1}$, $n = 1, \dots, N$.

The following lemma gives alternative upper bounds for conditional exponential moments of the maximal absolute values of components for the log-price process $\vec{Y}_n$.

Lemma 4.1.3. *Let the condition $\mathbf{C}_1'[\bar{\beta}]$ holds. Then there exist constants $0 \leq M'_{1,i} < \infty$, $i = 1, \dots, k$ such that the following inequalities take place for $\vec{z} = (\vec{y}, x) = ((y_1, \dots, y_k), x) \in \mathbb{Z}_n$, $n = 0, \dots, N$, $i = 1, \dots, k$:*

$$\mathsf{E}_{\vec{z},n} \exp\left\{\beta_i \max_{n \leq r \leq N} Y_{r,i}\right\} \leq M'_{1,i} e^{\beta_i |y_i|} . \tag{4.8}$$

Proof. It repeats the proof of Lemma 4.1.1. Again, the case where $\beta_i = 0$ is trivial. As far as the case $\beta_i > 0$ is concerned, the only remark should be made that probability $\mathsf{P}_{\vec{z},n}\{\vec{Z}_r \in \mathbb{Z}_r\} = 1$, for $\vec{z} \in \mathbb{Z}_n, 0 \leq n \leq r \leq N$, by Lemma 4.1.2, and expectation $\mathsf{E}_{\vec{z},r} \exp\{\beta_i |Y_{r+1,i} - Y_{r,i}| \leq K'_{1,i}$, for $\vec{z} \in \mathbb{Z}_r, 0 \leq r \leq N$, by condition $\mathbf{C}'_1[\bar{\beta}]$.

Using these remarks one can write down a relations analogous to (4.4) and (4.5), but for $\vec{z} \in \mathbb{Z}_n, 0 \leq n \leq N$, and with the constants $K'_{1,i}$, $i = 1, \dots, k$ used instead of $K_{1,i}$, $i = 1, \dots, k$. These relations imply that inequality (4.8) holds for $\vec{z} \in \mathbb{Z}_n$, $0 \leq n \leq N$, $i = 1, \dots, k$. □

Remark 4.1.2. The constants $M'_{1,i}$, $i = 1, \dots, k$ are given by the following formulas:

$$M'_{1,i} = (K'_{1,i})^N . \tag{4.9}$$

Moreover, inequalities (4.5) give, in this case, upper bounds for exponential moments of the maxima for the log-price processes $Y_{n,i}$ over discrete time intervals $[n, N]$, with the constants $(K'_{1,i})^{N-n}$ depending on n.

4.1.2 Upper bounds for rewards of Markov log-price processes with bounded characteristics

In this book, we study the American-type options with pay-off functions $g(n, \vec{s}, x)$ that have not more than polynomial rate of growth in argument $\vec{s} = (s_1, \dots, s_k) \in \mathbb{R}^+_k$ and are bounded in arguments $0 \leq n \leq N$ and $x \in \mathbb{X}$. This condition can also be reformulated as the assumption that pay-off functions $g(n, e^{\vec{y}}, x)$ have not more than exponential rate of growth in argument $\vec{y} = (y_1, \dots, y_k) \in \mathbb{R}_k$ and are bounded in arguments $0 \leq n \leq N$ and $x \in \mathbb{X}$.

The following condition is assumed to hold for some vector parameter $\bar{\gamma} = (\gamma_1, \dots, \gamma_k)$ with nonnegative components:

$\mathbf{B}_1[\bar{\gamma}]$: $\max_{0 \leq n \leq N} \sup_{\vec{z}=(\vec{y},x) \in \mathbb{Z}} \frac{|g(n, e^{\vec{y}}, x)|}{1 + \sum_{i=1}^k L_{2,i} e^{\gamma_i |y_i|}} < L_1$, for some $0 < L_1 < \infty$ and $0 \leq L_{2,1}, \dots, L_{2,k} < \infty$.

The following lemma gives sufficient conditions for the condition $\mathbf{A}_2$ to hold and explicit upper bounds for conditional expectations of the maximal absolute values for the pay-off process $g(n, e^{\vec{Y}_n}, X_n)$.

Lemma 4.1.4. *Let the conditions $\mathbf{B}_1[\bar{\gamma}]$ and $\mathbf{C}_1[\bar{\beta}]$ hold and $0 \leq \gamma_i \leq \beta_i < \infty$, $i = 1, \dots, k$. Then, there exist constants $0 \leq M_2, M_{3,i} < \infty$, $i = 1, \dots, k$ such that the following inequalities take place for $0 \leq n \leq N$, $\vec{z} = ((y_1, \dots, y_k), x) \in \mathbb{Z}$:*

$$\mathsf{E}_{\vec{z},n} \max_{n \leq r \leq N} |g(r, e^{\vec{Y}_r}, X_r)| \leq M_2 + \sum_{i=1}^{k} M_{3,i} e^{\gamma_i |y_i|} . \tag{4.10}$$

Proof. Using the condition $\mathbf{B}_1[\bar{\gamma}]$ we get the following inequality for $0 \le n \le N$:

$$\begin{aligned}
\max_{n \le r \le N} \left| g(r, e^{\vec{Y}_r}, X_r) \right| &\le \max_{n \le r \le N} \left(L_1 + \sum_{i=1}^{k} L_1 L_{2,i} e^{\gamma_i |Y_{r,i}|} \right) \\
&\le L_1 + \sum_{i=1}^{k} L_1 L_{2,i} \max_{n \le r \le N} e^{\gamma_i |Y_{r,i}|} \\
&= L_1 + \sum_{i=1}^{k} L_1 L_{2,i} e^{\gamma_i \max_{n \le r \le N} |Y_{r,i}|} . \qquad (4.11)
\end{aligned}$$

Now, using well-known properties for moments of random variables and Lemma 4.1.2, we get the following inequality for $\vec{z} \in \mathbb{Z}$, $0 \le n \le N$:

$$\begin{aligned}
&\mathsf{E}_{\vec{z},n} \max_{n \le r \le N} \left| g\left(r, e^{\vec{Y}_r}, X_r\right) \right| \\
&\le L_1 + \sum_{i=1}^{k} L_1 L_{2,i} \mathsf{E}_{\vec{z},n} e^{\gamma_i \max_{n \le r \le N} |Y_{r,i}|} \\
&\le L_1 + \sum_{i:\gamma_i = 0} L_1 L_{2,i} + \sum_{i:\gamma_i > 0} L_1 L_{2,i} \left(\mathsf{E}_{\vec{z},n} \left(e^{\gamma_i \max_{n \le r \le N} |Y_{r,i}|} \right)^{\frac{\beta_i}{\gamma_i}} \right)^{\frac{\gamma_i}{\beta_i}} \\
&= L_1 + \sum_{i:\gamma_i = 0} L_1 L_{2,i} + \sum_{i:\gamma_i > 0} L_1 L_{2,i} \left(\mathsf{E}_{\vec{z},n} e^{\beta_i \max_{n \le r \le N} |Y_{r,i}|} \right)^{\frac{\gamma_i}{\beta_i}} \\
&\le L_1 + \sum_{i:\gamma_i = 0} L_1 L_{2,i} + \sum_{i:\gamma_i > 0} L_1 L_{2,i} \left(e^{\beta_i |y_i|} M_{1,i} \right)^{\frac{\gamma_i}{\beta_i}} \\
&= L_1 + \sum_{i:\gamma_i = 0} L_1 L_{2,i} + \sum_{i:\gamma_i > 0} L_1 L_{2,i} M_{1,i}^{\frac{\gamma_i}{\beta_i}} e^{\gamma_i |y_i|} . \qquad (4.12)
\end{aligned}$$

Thus, inequality (4.10) holds with the constants M_2 and $M_{3,i}$, $i = 1, \dots, k$, which are given by relation (4.12). This completes the proof. □

Remark 4.1.3. The explicit formulas for the constants M_2 and $M_{3,i}$, $i = 1, \dots, k$ follow from relations (4.6) and (4.12),

$$M_2 = L_1 , \quad M_{3,i} = L_1 L_{2,i} I(\gamma_i = 0) + L_1 L_{2,i} K_{1,i}^{N \frac{\gamma_i}{\beta_i}} I(\gamma_i > 0) . \qquad (4.13)$$

Lemma 4.1.5. *Let the conditions* $\mathbf{B}_1[\bar{\gamma}]$ *and* $\mathbf{C}'_1[\bar{\beta}]$ *hold and* $0 \le \gamma_i \le \beta_i < \infty$, $i = 1, \dots, k$. *Then, there exist constants* $0 \le M'_2, M'_{3,i} < \infty$, $i = 1, \dots, k$ *such that the following inequalities take place for* $\vec{z} = ((y_1, \dots, y_k), x) \in \mathbb{Z}_n$, $0 \le n \le N$:

$$\mathsf{E}_{\vec{z},n} \max_{n \le r \le N} \left| g(r, e^{\vec{Y}_r}, X_r) \right| \le M'_2 + \sum_{i=1}^{k} M'_{3,i} e^{\gamma_i |y_i|} . \qquad (4.14)$$

Proof. It repeats the proof of Lemma 4.1.4. The only difference is that the condition $\mathbf{C}'_1[\bar{\beta}]$ is assumed to hold instead the condition $\mathbf{C}_1[\bar{\beta}]$. In this case, inequality (4.12) takes place for $\vec{z} \in \mathbb{Z}_n, 0 \leq n \leq N$ with the same constants $L_1, L_{2,i}, i = 1, \dots, k$ and the constants $K'_{1,i}, i = 1, \dots, k$ used instead of the constants $K_{1,i}, i = 1, \dots, k$. □

Remark 4.1.4. The explicit formulas for the constants M'_2 and $M'_{3,i}, i = 1, \dots, k$ follow from relations (4.9) and (4.12), and the remarks made in the above proof,

$$M'_2 = L_1\,, \quad M'_{3,i} = L_1 L_{2,i} I(\gamma_i = 0) + L_1 L_{2,i} (K'_{1,i})^{N\frac{\gamma_i}{\beta_i}} I(\gamma_i > 0)\,. \tag{4.15}$$

According to Lemma 4.1.4, the conditions $\mathbf{B_1}[\bar{\gamma}]$ and $\mathbf{C_1}[\bar{\beta}]$, which are assumed to hold with parameters $0 \leq \gamma_i \leq \beta_i < \infty, i = 1, \dots, k$, imply that the condition $\mathbf{A_2}$ holds, and, therefore, the log-reward functions $\phi_n(\vec{y}, x)$ are well defined by the following relation, for any $\vec{z} = (\vec{y}, x) \in \mathbb{Z}, n = 0, \dots, N$:

$$\phi_n(\vec{y}, x) = \sup_{\tau_n \in \mathcal{M}_{\max,n,N}} \mathsf{E}_{\vec{z},n} g(\tau_n, e^{\vec{Y}_{\tau_n}}, X_{\tau_n})\,. \tag{4.16}$$

Moreover, for Lemma 4.1.4 let us obtain the following upper bounds for the log-reward functions of American-type options for the modulated log-price process $\vec{Z}_n$.

Theorem 4.1.1. *Let the conditions $\mathbf{B_1}[\bar{\gamma}]$ and $\mathbf{C_1}[\bar{\beta}]$ hold and $0 \leq \gamma_i \leq \beta_i < \infty, i = 1, \dots, k$. Then, the log-reward functions $\phi_n(\vec{y}, x)$ satisfy the following inequalities for $\vec{z} = (\vec{y}, x) = ((y_1, \dots, y_k), x) \in \mathbb{Z}, 0 \leq n \leq N$:*

$$|\phi_n(\vec{y}, x)| \leq M_2 + \sum_{i=1}^{k} M_{3,i} e^{\gamma_i |y_i|}\,. \tag{4.17}$$

Proof. Using Lemma 4.1.4 and the definition of the log-reward functions, we get the following inequalities for $\vec{z} = (\vec{y}, x) \in \mathbb{Z},\ 0 \leq n \leq N$:

$$\begin{aligned} |\phi_n(\vec{y}, x)| &= \left| \sup_{\tau_n \in \mathcal{M}_{\max,n,N}} \mathsf{E}_{\vec{z},n} g(\tau_n, e^{\vec{Y}_{\tau_n}}, X_{\tau_n}) \right| \\ &\leq \mathsf{E}_{\vec{z},n} \max_{n \leq r \leq N} \left| g(r, e^{\vec{Y}_r}, X_r) \right| \leq M_2 + \sum_{i=1}^{k} M_{3,i} e^{\gamma_i |y_i|}\,, \end{aligned} \tag{4.18}$$

which proves the theorem. □

Let us now assume that the conditions $\mathbf{B_1}[\bar{\gamma}]$ and $\mathbf{C'_1}[\bar{\beta}]$ hold with parameters $0 \leq \gamma_i \leq \beta_i < \infty, i = 1, \dots, k$. Lemma 4.1.5 guarantees in this case that the log-reward functions $\phi_n(\vec{y}, x)$ are defined by relation (4.16) only for $\vec{z} = (\vec{y}, x) \in \mathbb{Z}_n, n = 0, \dots, N$.

Moreover, for Lemma 4.1.5 let us get the following upper bounds for the log-reward functions of American-type options for the modulated log-price process $\vec{Z}_n$.

Theorem 4.1.2. *Let the conditions $\mathbf{B_1}[\bar{\gamma}]$ and $\mathbf{C'_1}[\bar{\beta}]$ hold and $0 \leq \gamma_i \leq \beta_i < \infty, i = 1, \dots, k$. Then, the log-reward functions $\phi_n(\vec{y}, x)$ satisfy the following inequalities for $\vec{z} = ((y_1, \dots, y_k), x) \in \mathbb{Z}_n,\ 0 \leq n \leq N$:*

$$|\phi_n(\vec{y}, x)| \leq M'_2 + \sum_{i=1}^{k} M'_{3,i} e^{\gamma_i |y_i|}\,. \tag{4.19}$$

Proof. It repeats the proof of Theorem 4.1.1. The only change is that Lemma 4.1.5 should be used instead of Lemma 4.1.4. This lets one write down inequality (4.18), for $\vec{z} = (\vec{y}, x) \in \mathbb{Z}_n$, $0 \le n \le N$ and with the constants $M'_2, M'_{3,i}$, $i = 1, \dots, k$ used instead of the constants $M_2, M_{3,i}$, $i = 1, \dots, k$. □

As far as points $\vec{z} = (\vec{y}, x) \in \overline{\mathbb{Z}}_n$, $n = 0, \dots, N$ are concerned, there is no guarantee that expectations on the right-hand side in formulas (4.16) do exist. Moreover, if these expectations are finite, there is no guarantee that suprema penetrating these formulas are finite. Thus, the log-reward functions $\phi_n(\vec{y}, x)$ may not be defined for such points.

There exist three alternatives.

First, some additional conditions should be imposed on transition probabilities $P_n(\vec{z}, A)$, in order to guarantee that formula (4.16) would well define the log-reward functions $\phi_n(\vec{y}, x)$ for points $\vec{z} = (\vec{y}, x) \in \overline{\mathbb{Z}}_n$, $n = 0, 1, \dots, N$.

For example, in the important case of a space-skeleton approximation model, distributions $P_n(\vec{z}, \cdot)$ are concentrated on finite sets of points $\mathbb{Z}^*_n(\vec{z}) \subseteq \mathbb{Z}_n$, for $\vec{z} = (\vec{y}, x) \in \mathbb{Z}$, $n = 1, \dots, N$. In this case, the log-reward functions $\phi_n(\vec{y}, x)$ are automatically finite and well defined by formula (4.16), for any $\vec{z} = (\vec{y}, x) \in \mathbb{Z}$, $n = 0, \dots, N$. As far as upper bounds for log-reward functions are concerned, they do require assuming conditions penetrating Theorems 4.1.1 or 4.1.2.

Second, one can define and operate with the log-reward functions only for points $\vec{z} = (\vec{y}, x) \in \mathbb{Z}_n$, $n = 0, \dots, N$.

Third, one can use the assumption that $\mathcal{Z} = \langle \mathbb{Z}_0, \dots, \mathbb{Z}_N \rangle$ is a complete sequence of phase sets such that values of the log-reward functions $\phi_n(\vec{y}, x)$, for points $\vec{z} = (\vec{y}, x) \in \mathbb{Z}_n$, $n = 0, \dots, N$, do not depend on transition probabilities $P_n(\vec{z}, A)$ for $\vec{z} = (\vec{y}, x) \in \overline{\mathbb{Z}}_{n-1}$, $n = 1, \dots, N$.

This assumption makes it possible to change transition probabilities $P_n(\vec{z}, A)$ for $\vec{z} = (\vec{y}, x) \in \overline{\mathbb{Z}}_{n-1}$, $n = 1, \dots, N$ that do not change the values of the log-reward functions $\phi_n(\vec{y}, x)$ for $\vec{z} = (\vec{y}, x) \in \mathbb{Z}_n$, $n = 0, \dots, N$. For example, one can define new transition probabilities $P_n(\vec{z}, A) = I(\vec{z} \in A)$ for $\vec{z} = (\vec{y}, x) \in \overline{\mathbb{Z}}_{n-1}$, $n = 1, \dots, N$. In this case, obviously, $\mathsf{P}_{\vec{z}}\{\vec{Z}_r = \vec{z}, n \le r \le N\} = 1$ for $\vec{z} = (\vec{y}, x) \in \overline{\mathbb{Z}}_n$, $n = 0, \dots, N$ and, therefore, $\phi_n(\vec{y}, x) = \max_{n \le r \le N} g(n, e^{\vec{y}}, x)$ for $\vec{z} = (\vec{y}, x) \in \overline{\mathbb{Z}}_n$, $n = 0, \dots, N$.

Let us now give upper bounds for optimal expected rewards using the above upper bounds for reward functions.

To do this, we should impose the following first-type condition of exponential moment boundedness on the initial distribution of the log-price process $\vec{Z}_n$. This condition should be assumed to hold for some vector parameter $\bar{\beta} = (\beta_1, \dots, \beta_k)$ with nonnegative components:

$\mathbf{D_1}[\bar{\beta}]$: $\mathsf{E}e^{\beta_i |Y_{0,i}|} < K_{2,i}$, $i = 1, \dots, k$, for some $1 < K_{2,i} < \infty$, $i = 1, \dots, k$.

It is appropriate to comment here, why conditions of type **C** are referred to as moment compactness conditions, while conditions of type **D** are referred to as moment boundedness conditions. As a matter of fact, conditions of type **C** are moment boundedness conditions imposed on increments of log-price processes, while conditions of type **D** are moment boundedness conditions imposed on values of log-price processes themselves.

The following lemma gives sufficient conditions for the condition $\mathbf{A}_1$ to hold and explicit upper bounds for the expectation of maximum for the pay-off process $g(n, e^{\vec{Y}_n}, X_n)$.

Lemma 4.1.6. *Let the conditions* $\mathbf{B}_1[\bar{\gamma}]$, $\mathbf{C}_1[\bar{\beta}]$, *and* $\mathbf{D}_1[\bar{\beta}]$ *hold and* $0 \leq \gamma_i \leq \beta_i < \infty, i = 1, \dots, k$. *Then, there exists a constant* $0 \leq M_4 < \infty$ *such that the following inequality takes place:*

$$\mathsf{E} \max_{0 \leq n \leq N} \left| g(n, e^{\vec{Y}_n}, X_n) \right| \leq M_4 . \tag{4.20}$$

Proof. Lemma 4.1.4, we get

$$\begin{aligned}
\mathsf{E} \max_{0 \leq n \leq N} \left| g(n, e^{\vec{Y}_n}, X_n) \right| &= \mathsf{E}\,\mathsf{E} \left\{ \max_{0 \leq n \leq N} \left| g\left(n, e^{\vec{Y}_n}, X_n\right) \right| / \vec{Z}_0 \right\} \\
&\leq \mathsf{E} \left(M_2 + \sum_{i=1}^{k} M_{3,i} e^{\gamma_i |Y_{0,i}|} \right) \\
&\leq M_2 + \sum_{i: \gamma_i = 0} M_{3,i} + \sum_{i: \gamma_i > 0} M_{3,i} \left(\mathsf{E} e^{\beta_i |Y_{0,i}|} \right)^{\frac{\gamma_i}{\beta_i}} \\
&\leq M_2 + \sum_{i: \gamma_i = 0} M_{3,i} + \sum_{i: \gamma_i > 0} M_{3,i} K_{2,i}^{\frac{\gamma_i}{\beta_i}} .
\end{aligned} \tag{4.21}$$

Relation (4.21) proves that inequality (4.20) holds with the constant M_4 given in formula (4.21). □

Remark 4.1.5. The explicit formula for the constant M_4 follows from the formulas given in Remark 4.1.3 and the last inequality in (4.21),

$$M_4 = L_1 + \sum_{i: \gamma_i = 0} L_1 L_{2,i} + \sum_{i: \gamma_i > 0} L_1 L_{2,i} K_{1,i}^{N \frac{\gamma_i}{\beta_i}} K_{2,i}^{\frac{\gamma_i}{\beta_i}} . \tag{4.22}$$

In the case, where the condition $\mathbf{C}'_1[\bar{\beta}]$ is used instead of the condition $\mathbf{C}_1[\bar{\beta}]$, the following modified first-type condition of exponential moment boundedness should be assumed to hold for some vector parameter $\bar{\beta} = (\beta_1, \dots, \beta_k)$ with nonnegative components:

$\mathbf{D}'_1[\bar{\beta}]$: (a) $\mathsf{E} e^{\beta_i |Y_{0,i}|} < K'_{2,i}$, $i = 1, \dots, k$, for some $1 \leq K'_{2,i} < \infty$, $i = 1, \dots, k$;
(b) $\mathsf{P}\{\vec{Z}_0 \in \mathbb{Z}_0\} = 1$, where $\mathbb{Z}_0$ is the set penetrating condition $\mathbf{C}'_1[\bar{\beta}]$.

Lemma 4.1.7. *Let the conditions* $\mathbf{B}_1[\bar{\gamma}]$, $\mathbf{C}'_1[\bar{\beta}]$, *and* $\mathbf{D}'_1[\bar{\beta}]$ *hold and* $0 \leq \gamma_i \leq \beta_i < \infty, i = 1, \dots, k$. *Then, there exists a constant* $0 \leq M'_4 < \infty$ *such that the following inequality takes place:*

$$\mathsf{E} \max_{0 \leq n \leq N} \left| g(n, e^{\vec{Y}_n}, X_n) \right| \leq M'_4 . \tag{4.23}$$

Proof. It repeats the proof of Lemma 4.1.6. The only difference is that Lemma 4.1.5 should be used instead of Lemma 4.1.4. The relation analogous to (4.21) can be written down, with the same constants $L_1, L_{2,i}$, $i = 1, \dots, k$ and the constants $K'_{1,i}, K'_{2,i}$, $i = 1, \dots, k$ used instead of the constants $K_{1,i}, K_{2,i}$, $i = 1, \dots, k$. □

Remark 4.1.6. The explicit formula for the constant M'_4 follows from the formulas given in Remark 4.1.4, the last inequality in (4.21) and the remarks made in the above proof,

$$M'_4 = L_1 + \sum_{i:\gamma_i = 0} L_1 L_{2,i} + \sum_{i:\gamma_i > 0} L_1 L_{2,i} (K'_{1,i})^{N\frac{\gamma_i}{\beta_i}} (K'_{2,i})^{\frac{\gamma_i}{\beta_i}} . \tag{4.24}$$

Let us assume that either conditions of Lemma 4.1.6 or Lemma 4.1.7 hold. In this case, the optimal expected reward $\Phi = \Phi(\mathcal{M}_{\max,N})$ for the log-price $\vec{Z}_n$ can be defined by the following relation:

$$\Phi = \sup_{\tau_0 \in \mathcal{M}_{\max,N}} \mathsf{E} g(\tau_0, e^{\vec{Y}_{\tau_0}}, X_{\tau_0}) . \tag{4.25}$$

The following two theorems give upper bound for the optimal expected reward Φ.

Theorem 4.1.3. *Let the conditions* $\mathbf{B}_1[\bar{\gamma}]$, $\mathbf{C}_1[\bar{\beta}]$, *and* $\mathbf{D}_1[\bar{\beta}]$ *hold, and* $0 \le \gamma_i \le \beta_i < \infty$, $i = 1, \dots, k$. *Then, the following inequality takes place:*

$$|\Phi| \le M_4 . \tag{4.26}$$

Proof. It follows from the definition of the functional Φ and Lemma 4.1.6 that

$$|\Phi| \le \sup_{\tau_0 \in \mathcal{M}_{\max,N}} \mathsf{E} \left| g(\tau_0, e^{\vec{Y}_{\tau_0}}, X_{\tau_0}) \right| \le \mathsf{E} \max_{0 \le n \le N} \left| g(n, e^{\vec{Y}_n}, X_n) \right| \le M_4 . \tag{4.27}$$

This inequality proves the theorem. □

Theorem 4.1.4. *Let the conditions* $\mathbf{B}_1[\bar{\gamma}]$, $\mathbf{C}'_1[\bar{\beta}]$, *and* $\mathbf{D}'_1[\bar{\beta}]$ *hold, and* $0 \le \gamma_i \le \beta_i < \infty$, $i = 1, \dots, k$. *Then, the following inequality takes place:*

$$|\Phi| \le M'_4 . \tag{4.28}$$

Proof. It repeats the proof of Theorem 4.1.3. The only difference is that Lemma 4.1.7 should be used instead of Lemma 4.1.6. The relation analogous to (4.27) can be written down, and the constant M'_4 used instead of the constant M_4. □

4.1.3 Modified conditions of exponential moment compactness for Markov log-price processes with bounded characteristics

The condition $\mathbf{C}_1[\bar{\beta}]$, used in Lemmas 4.1.1 and 4.1.2, is based on exponential moments for absolute values for increments of log-price process. It is useful for some models, for example, for log-price processes represented by random walks, to replace this condition by an alternative condition based on moment-generating functions for log-price processes.

Let us define the second-type modulus of exponential moment compactness for components of the log-price process $\vec{Y}_n = (Y_{n,1}\dots, Y_{n,k})$, for $\beta \geq 0$, $i = 1, \dots, k$,

$$\Xi^{\pm}_{\beta}(Y_{\cdot,i}, N) = \max_{0 \leq n \leq N-1} \sup_{\vec{z} \in \mathbb{Z}} \mathsf{E}_{\vec{z},n} e^{\pm\beta(Y_{n+1,i} - Y_{n,i})} . \tag{4.29}$$

The following second-type condition of exponential moment compactness based on moment-generating functions for increments of log-price processes $\vec{Y}_n$, is assumed to hold for some vector parameter $\bar{\beta} = (\beta_1, \dots, \beta_k)$ with nonnegative components:
$\mathbf{E}_1[\bar{\beta}]$: $\Xi^{\pm}_{\beta_i}(Y_{\cdot,i}, N) < K_{3,i}$, $i = 1, \dots, k$, for some $1 < K_{3,i} < \infty$, $i = 1, \dots, k$.

Lemma 4.1.8. *The condition $\mathbf{E}_1[\bar{\beta}]$ implies condition $\mathbf{C}_1[\bar{\beta}]$ to hold.*

Proof. The following inequality takes place, for every $0 \leq n \leq N-1$, $\vec{z} \in \mathbb{Z}$, $i = 1, \dots, k$,

$$\begin{aligned} &\mathsf{E}_{\vec{z},n} e^{\beta_i |Y_{n+1,i} - Y_{n,i}|} \\ &\quad = \mathsf{E}_{\vec{z},n} e^{\beta_i (Y_{n+1,i} - Y_{n,i})} I(Y_{n+1,i} \geq Y_{n,i}) \\ &\qquad + \mathsf{E}_{\vec{z},n} e^{-\beta_i (Y_{n+1,i} - Y_{n,i})} I(Y_{n+1,i} < Y_{n,i}) \\ &\quad \leq \mathsf{E}_{\vec{z},n} e^{\beta_i (Y_{n+1,i} - Y_{n,i})} + \mathsf{E}_{\vec{z},n} e^{-\beta_i (Y_{n+1,i} - Y_{n,i})} . \end{aligned} \tag{4.30}$$

Relation (4.30) obviously implies that the following inequality holds, for $i = 1, \dots, k$:

$$\Delta_{\beta_i}(Y_{\cdot,i}, N) \leq \Xi^{+}_{\beta_i}(Y_{\cdot,i}, N) + \Xi^{-}_{\beta_i}(Y_{\cdot,i}, N) . \tag{4.31}$$

Thus, the condition $\mathbf{E}_1[\bar{\beta}]$ implies that condition $\mathbf{C}_1[\bar{\beta}]$ holds. □

Remark 4.1.7. As follows from inequality (4.31), if condition $\mathbf{E}_1[\bar{\beta}]$ holds with the constants $K_{3,i}$, $i = 1, \dots, k$, then the condition $\mathbf{C}_1[\bar{\beta}]$ holds with the constants $K_{1,i}$, $i = 1, \dots, k$ given by the following formulas:

$$K_{1,i} = 2K_{3,i}, \ i = 1, \dots, k . \tag{4.32}$$

Let $\mathcal{Z} = \langle \mathbb{Z}_0, \dots, \mathbb{Z}_N \rangle$ be a sequence of measurable sets, and $\Xi^{\pm}_{\beta,\mathcal{Z}}(Y_{\cdot,i}, N)$ be a modified second-type exponential moment modulus of compactness defined for $\beta \geq 0$ by the following formula:

$$\Xi^{\pm}_{\beta,\mathcal{Z}}(Y_{\cdot,i}, N) = \max_{0 \leq n \leq N-1} \sup_{\vec{z} \in \mathbb{Z}_n} \mathsf{E}_{\vec{z},n} e^{\pm\beta(Y_{n+1,i} - Y_{n,i})} . \tag{4.33}$$

The following modified second-type condition of exponential moment compactness is based on moment-generating functions of increments for the log-price processes $\vec{Y}_n$. It generalizes the condition $\mathbf{E}_1[\bar{\beta}]$ and is assumed to hold for some vector parameter $\bar{\beta} = (\beta_1, \dots, \beta_k)$ with nonnegative components:

$\mathbf{E}_1'[\bar{\beta}]$: There exists a sequence of measurable sets $\mathcal{Z} = \langle \mathbb{Z}_0, \dots, \mathbb{Z}_N \rangle$ such that

(a) $\Xi^{\pm}_{\beta_i, \mathcal{Z}}(Y_{\cdot,i}, N) < K'_{3,i}$, $i = 1, \dots, k$, for some $1 < K'_{3,i} < \infty$, $i = 1, \dots, k$;

(b) $\mathcal{Z}$ is a complete sequence of phase sets, i.e., $\mathsf{P}_n(\vec{z}, \mathbb{Z}_n) = 1, \vec{z} \in \mathbb{Z}_{n-1}$, $n = 1, \dots, N$.

Lemma 4.1.9. *The condition $\mathbf{E}_1'[\bar{\beta}]$ implies the condition $\mathbf{C}_1'[\bar{\beta}]$ to hold.*

Proof. Relation (4.30) implies analogously to (4.31) the following inequality, for $i = 1, \dots, k$:

$$\Delta_{\beta_i, \mathcal{Z}}(Y_{\cdot,i}, N) \le \Xi^{+}_{\beta_i, \mathcal{Z}}(Y_{\cdot,i}, N) + \Xi^{-}_{\beta_i, \mathcal{Z}}(Y_{\cdot,i}, N) . \tag{4.34}$$

Thus, the condition $\mathbf{E}_1'[\bar{\beta}]$ implies that condition $\mathbf{C}_1'[\bar{\beta}]$ holds. □

Remark 4.1.8. As follows from inequality (4.31), if condition $\mathbf{E}_1'[\bar{\beta}]$ holds with the constants $K'_{3,i}$, $i = 1, \dots, k$, then the condition $\mathbf{C}_1'[\bar{\beta}]$ holds with the constants $K'_{1,i}$, $i = 1, \dots, k$ given by the following formulas:

$$K'_{1,i} = 2K'_{3,i}, \ i = 1, \dots, k . \tag{4.35}$$

One also can replace the first-type condition of exponential moment boundedness $\mathbf{D}_1[\bar{\beta}]$ by the following second-type condition of exponential moment boundedness based on moment-generating functions for the initial value of log-price processes $\vec{Y}_0$, assumed to hold for some vector parameter $\bar{\beta} = (\beta_1, \dots, \beta_k)$ with nonnegative components:

$\mathbf{F}_1[\bar{\beta}]$: $\mathsf{E} e^{\pm \beta_i Y_{0,i}} < K_{4,i}$, $i = 1, \dots, k$, for some $1 < K_{4,i} < \infty$, $i = 1, \dots, k$.

Lemma 4.1.10. *The condition $\mathbf{F}_1[\bar{\beta}]$ implies condition $\mathbf{D}_1[\bar{\beta}]$ to hold.*

Proof. The following inequality holds for any $\beta \ge 0$:

$$\mathsf{E} e^{\beta |Y_{0,i}|} \le \mathsf{E} e^{+\beta Y_{0,i}} + \mathsf{E} e^{-\beta Y_{0,i}} . \tag{4.36}$$

The proposition of Lemma 4.1.10 obviously follows from the above inequality. □

Remark 4.1.9. As follows from inequality (4.36), if condition $\mathbf{F}_1[\bar{\beta}]$ holds with the constants $K_{4,i}$, $i = 1, \dots, k$, then the condition $\mathbf{D}_1[\bar{\beta}]$ holds with the constants $K_{2,i}$, $i = 1, \dots, k$ given by the following formulas:

$$K_{2,i} = 2K_{4,i} , \quad i = 1, \dots, k . \tag{4.37}$$

Analogously, the condition $\mathbf{D}_1'[\bar{\beta}]$ can be replaced by the following second-type condition of exponential moment boundedness assumed to hold for some vector parameter $\bar{\beta} = (\beta_1, \dots, \beta_k)$ with nonnegative components:

$\mathbf{F}_1'[\bar{\beta}]$: (a) $\mathsf{E} e^{\pm \beta_i Y_{0,i}} < K_{4,i}$, $i = 1, \dots, k$, for some $1 < K_{4,i} < \infty$, $i = 1, \dots, k$;
(b) $\mathsf{P}\{\vec{Z}_0 \in \mathbb{Z}_0\} = 1$, where $\mathbb{Z}_0$ is the set penetrating condition $\mathbf{D}_1'[\bar{\beta}]$.

Lemma 4.1.11. *The condition $\mathbf{F}_1'[\bar{\beta}]$ implies condition $\mathbf{D}_1'[\bar{\beta}]$ to hold.*

Remark 4.1.10. As follows from inequality (5.26), if condition $\mathbf{F}_1'[\bar{\beta}]$ holds with the constants $K_{4,i}'$, $i = 1, \dots, k$, then the condition $\mathbf{D}_1'[\bar{\beta}]$ holds with the constants $K_{2,i}'$, $i = 1, \dots, k$ given by the following formulas:

$$K_{2,i}' = 2K_{4,i}', \ i = 1, \dots, k. \tag{4.38}$$

4.2 LPP represented by random walks

In this section, we get explicit upper bounds for log-reward functions and optimal expected rewards for American-type options for log-price processes represented by random walks.

4.2.1 Upper bounds for option rewards for log-price processes represented by standard random walks

Let us consider the simplest model, where the multivariate Markov log-price $\vec{Y}_n = (Y_{n,1} \dots, Y_{n,k})$ is a standard random walk,

$$\vec{Y}_n = \vec{Y}_{n-1} + \vec{W}_n, \quad n = 1, 2, \dots. \tag{4.39}$$

where (a) $\vec{Y}_0 = (Y_{0,1}, \dots, Y_{0,k})$ is a k-dimensional random vector, (b) $\vec{W}_n = (W_{n,1}, \dots, W_{n,k})$, $n = 1, 2, \dots$, is a sequence of k-dimensional independent random vectors, and (c) the random vector $\vec{Y}_0$ and the random sequence $\vec{W}_n$, $n = 1, 2, \dots$, are independent.

In this case, the model without index component is considered. In fact, one can always to define a "virtual" constant index component X_n which takes the constant value x_0 for every $n = 0, 1, \dots$.

In this model, a pay-off function $g(n, e^{\vec{y}}, x) = g(n, e^{\vec{y}})$ does not depend on the index argument x.

The first-type modulus of exponential moment compactness $\Delta_\beta(Y_{\cdot,i}, N)$, expressed in terms of components of jumps $\vec{W}_n$, takes the following form, for $\beta \geq 0$,

$i = 1, \dots, k$:

$$\Delta'_\beta(Y_{\cdot,i}, N) = \max_{0 \le n \le N-1} \mathsf{E} e^{\beta |W_{n+1,i}|} . \tag{4.40}$$

The condition $\mathbf{C_1}[\bar{\beta}]$, which is assumed to hold for some vector parameter $\bar{\beta} = (\beta_1, \dots, \beta_k)$ with nonnegative components, takes the following simpler form:
$\mathbf{C_2}[\bar{\beta}]$: $\Delta'_{\beta_i}(Y_{\cdot,i}, N) < K_{5,i}$, $i = 1, \dots, k$, for some $1 < K_{5,i} < \infty$, $i = 1, \dots, k$.

The condition $\mathbf{B_1}[\bar{\gamma}]$, which is assumed to hold for some vector parameter $\bar{\gamma} = (\gamma_1, \dots, \gamma_k)$ with nonnegative components, takes the following form:
$\mathbf{B_2}[\bar{\gamma}]$: $\max_{0 \le n \le N} \sup_{\vec{y} \in \mathbb{R}_k} \frac{|g(n, e^{\vec{y}})|}{1 + \sum_{i=1}^k L_{4,i} e^{\gamma_i |y_i|}} < L_3$, for some $0 < L_3 < \infty$ and $0 \le L_{4,1}, \dots, L_{4,k} < \infty$.

The condition $\mathbf{D_1}[\bar{\beta}]$ does not change.

The second-type modulus of exponential moment compactness for the components of the log-price process $\vec{Y}_n$, takes the following form for $\beta \ge 0$, $i = 1, \dots, k$:

$$\Xi'^{\pm}_\beta(Y_{\cdot,i}, N) = \max_{0 \le n \le N-1} \mathsf{E} e^{\pm \beta W_{n+1,i}} . \tag{4.41}$$

The condition $\mathbf{E_1}[\bar{\beta}]$, which is assumed to hold for some vector parameter $\bar{\beta} = (\beta_1, \dots, \beta_k)$ with nonnegative components, takes the following simpler form:
$\mathbf{E_2}[\bar{\beta}]$: $\Xi'^{\pm}_{\beta_i}(Y_{\cdot,i}, N) < K_{6,i}$, $i = 1, \dots, k$, for some $1 < K_{6,i} < \infty$, $i = 1, \dots, k$.

Note that, according to Lemma 4.1.8, the condition $\mathbf{E_2}[\bar{\beta}]$ implies the condition $\mathbf{C_2}[\bar{\beta}]$ to hold.

The conditions $\mathbf{F_1}[\bar{\beta}]$ do not change.

Lemmas 4.1.1, 4.1.3–4.1.11 and Theorems 4.1.1 and 4.1.3 can be reformulated with the replacement of the conditions $\mathbf{B_1}[\bar{\gamma}]$, $\mathbf{C_1}[\bar{\beta}]$, and $\mathbf{E_1}[\bar{\beta}]$, respectively, by the conditions $\mathbf{B_2}[\bar{\gamma}]$, $\mathbf{C_2}[\bar{\beta}]$ and $\mathbf{E_2}[\bar{\beta}]$ and the constants $L_1, L_{2,i}$, $i = 1, \dots, k$ and $K_{1,i}, K_{3,i}$, $i = 1, \dots, k$, respectively by the constants $L_3, L_{4,i}$, $i = 1, \dots, k$ and $K_{5,i}, K_{6,i}$, $i = 1, \dots, k$. As was mentioned above the conditions $\mathbf{D_1}[\bar{\beta}]$ and $\mathbf{F_1}[\bar{\beta}]$ do not change, and, thus, the same constants $K_{2,i}, K_{4,i}$, $i = 1, \dots, k$ should be used in the above listed lemmas and theorems.

Since the pay-off function $g(n, e^{\vec{y}}, x)$ does not depend on the index argument x, the log-reward functions $\phi_n(\vec{y}, x) = \phi_n(\vec{y})$ also do not depend on the index argument x.

Theorem 4.1.1 takes, according to the above remarks, the following form.

Theorem 4.2.1. *Let the conditions $\mathbf{B_2}[\bar{\gamma}]$ and $\mathbf{C_2}[\bar{\beta}]$ hold and $0 \le \gamma_i \le \beta_i < \infty$, $i = 1, \dots, k$. Then, the log-reward functions $\phi_n(\vec{y})$ satisfy the following inequalities for $\vec{y} = (y_1, \dots, y_k) \in \mathbb{R}_k$, $0 \le n \le N$:*

$$|\phi_n(\vec{y})| \le M_5 + \sum_{i=1}^k M_{6,i} e^{\gamma_i |y_i|} . \tag{4.42}$$

Remark 4.2.1. The explicit formulas for the constants M_5 and $M_{6,i}$, $i = 1, \dots, k$ follow from formula (4.13) given in Remark 4.1.3,

$$M_5 = L_3, \quad M_{6,i} = L_3 L_{4,i} I(\gamma_i = 0) + L_3 L_{4,i} K_{5,i}^{N\frac{\gamma_i}{\beta_i}} I(\gamma_i > 0)\,. \tag{4.43}$$

Theorem 4.1.3 takes, according to the above remarks, the following form.

Theorem 4.2.2. *Let the conditions* $\mathbf{B}_2[\bar{\gamma}]$, $\mathbf{C}_2[\bar{\beta}]$, *and* $\mathbf{D}_1[\bar{\beta}]$ *hold, and* $0 \le \gamma_i \le \beta_i < \infty$, $i = 1, \dots, k$. *Then, the following inequality takes place:*

$$|\Phi| \le M_7\,. \tag{4.44}$$

Remark 4.2.2. The explicit formula for the constant M_7 follows from formula (4.22) given in Remark 4.1.5,

$$M_7 = L_3 + \sum_{i:\gamma_i = 0} L_3 L_{4,i} + \sum_{i:\gamma_i > 0} L_3 L_{4,i} K_{5,i}^{N\frac{\gamma_i}{\beta_i}} (K_{2,i})^{\frac{\gamma_i}{\beta_i}}\,. \tag{4.45}$$

4.2.2 Upper bounds for option rewards for price processes represented by modulated random walks

In this model, a modulated price process $\vec{Z}_n = (\vec{Y}_n, X_n)$ can be introduced via the following stochastic transition dynamic relation:

$$\begin{cases} \vec{Y}_n = \vec{Y}_{n-1} + \vec{W}_n, & \text{where } \vec{W}_n = \tilde{B}_n(X_{n-1}, U_n)\,, \\ X_n = C_n(X_{n-1}, U_n), & n = 1, 2, \dots\,, \end{cases} \tag{4.46}$$

where (a) $(\vec{Y}_0, X_0) = ((Y_{0,1}, \dots, Y_{0,k}), X_0)$ is a random vector taking values in space $\mathbb{Z}$, (b) U_n, $n = 1, 2, \dots$, is a sequence of independent random variables taking values in some measurable space $\mathbb{U}$ with a σ-algebra of measurable subsets $\mathcal{B}_{\mathbb{U}}$, (c) the random vector $(\vec{Y}_0, X_0)$ and the random sequence U_n, $n = 1, 2, \dots$, are independent, (d) $\tilde{B}_n(x, u) = (\tilde{B}_{n,1}(x, u), \dots, \tilde{B}_{n,k}(x, u))$, $n = 1, 2, \dots$, are vector functions, which components are measurable functions acting from the space $\mathbb{X} \times \mathbb{U}$ to $\mathbb{R}_1$, (e) $C_n(x, u)$, $n = 1, 2, \dots$, are measurable functions acting from the space $\mathbb{X} \times \mathbb{U}$ to the space $\mathbb{X}$. In this model, (f) the sequence of jumps $\vec{W}_n = (W_{n,1}, \dots, W_{n,k})$, $n = 1, 2, \dots$, where $W_{n,i} = \tilde{B}_{n,i}(X_{n-1}, U_n)$, $i = 1, \dots, k$, $n = 1, 2, \dots$.

In this case, a pay-off function $g(n, e^{\vec{y}}, x)$ depends on the index argument x.

The first-type modulus of exponential moment compactness $\Delta''_\beta(Y_{\cdot,i}, N)$, expressed in terms of components of jumps $\vec{W}_n$, takes the following form, for $\beta \ge 0$, $i = 1, \dots, k$:

$$\Delta''_\beta(Y_{\cdot,i}, N) = \max_{0 \le n \le N-1} \sup_{x \in \mathbb{X}} \mathsf{E}_{x,n} e^{\beta |W_{n+1,i}|}\,, \tag{4.47}$$

where $\mathsf{E}_{x,n}$ is the conditional expectation under the condition $X_n = x$.

The condition $\mathbf{C}_1[\bar{\beta}]$, which is assumed to hold for some vector parameter $\bar{\beta} = (\beta_1, \dots, \beta_k)$ with nonnegative components, takes the following simpler form:
$\mathbf{C}_3[\bar{\beta}]$: $\Delta''_{\beta_i}(Y_{\cdot,i}, N) < K_{7,i}$, $i = 1, \dots, k$ for some $1 < K_{7,i} < \infty$, $i = 1, \dots, k$.

The conditions $\mathbf{B}_1[\bar{\gamma}]$ and $\mathbf{D}_1[\bar{\beta}]$ do not change.

The second-type modulus of exponential moment compactness for components of the log-price process $\vec{Y}_n$, takes the following form for $\beta \geq 0$, $i = 1, \dots, k$:

$$\Xi''^{\pm}_{\beta}(Y_{\cdot,i}, N) = \max_{0 \leq n \leq N-1} \sup_{x \in \mathbb{X}} \mathsf{E}_{x,n} e^{\pm \beta W_{n+1,i}} . \tag{4.48}$$

The condition $\mathbf{E}_1[\bar{\beta}]$, which is assumed to hold for some vector parameter $\bar{\beta} = (\beta_1, \dots, \beta_k)$ with nonnegative components, takes the following form:
$\mathbf{E}_3[\bar{\beta}]$: $\Xi''^{\pm}_{\beta_i}(Y_{\cdot,i}, N) < K_{8,i}$, $i = 1, \dots, k$, for some $1 < K_{8,i} < \infty$, $i = 1, \dots, k$.

The following lemma is a particular case of Lemma 4.1.8.

Lemma 4.2.1. *The condition $\mathbf{E}_3[\bar{\beta}]$ implies condition $\mathbf{C}_3[\bar{\beta}]$ to hold.*

The condition $\mathbf{F}_1[\bar{\beta}]$ does not change.

Lemmas 4.1.1, 4.1.3–4.1.11 and Theorems 4.1.1 and 4.1.3 can be reformulated with the replacement of the conditions $\mathbf{C}_1[\bar{\beta}]$, and $\mathbf{E}_1[\bar{\beta}]$, respectively, by the conditions $\mathbf{C}_3[\bar{\beta}]$ and $\mathbf{E}_3[\bar{\beta}]$, and the constants $K_{1,i}, K_{3,i}$, $i = 1. \dots, k$, respectively, by the constants $K_{7,i}, K_{8,i}$, $i = 1. \dots, k$. As was mentioned above the conditions $\mathbf{B}_1[\bar{\gamma}]$, $\mathbf{D}_1[\bar{\beta}]$ and $\mathbf{F}_1[\bar{\beta}]$ do not change, and, thus, the same constants $L_1, L_{2,i}$, $i = 1, \dots, k$ and $K_{2,i}, K_{4,i}$, $i = 1, \dots, k$ should be used in the above listed lemmas and theorems.

Theorem 4.1.1 takes, according to the above remarks, the following form.

Theorem 4.2.3. *Let the conditions $\mathbf{B}_1[\bar{\gamma}]$ and $\mathbf{C}_3[\bar{\beta}]$ hold and $0 \leq \gamma_i \leq \beta_i < \infty$, $i = 1, \dots, k$. Then, the log-reward functions $\phi_n(\vec{y}, x)$ satisfy the following inequalities for $\vec{z} = ((y_1, \dots, y_k), x) \in \mathbb{Z}$, $0 \leq n \leq N$:*

$$|\phi_n(\vec{y}, x)| \leq M_8 + \sum_{i=1}^{k} M_{9,i} e^{\gamma_i |y_i|} . \tag{4.49}$$

Remark 4.2.3. The explicit formulas for the constants M_8 and $M_{9,i}$, $i = 1, \dots, k$ follow from formula (4.13) given in Remark 1.4.1.3,

$$M_8 = L_1, \; M_{9,i} = L_1 L_{2,i} I(\gamma_i = 0) + L_1 L_{2,i} K_{7,i}^{N \frac{\gamma_i}{\beta_i}} I(\gamma_i > 0) . \tag{4.50}$$

Theorem 4.1.3 takes, according to the above remarks, the following form.

Theorem 4.2.4. *Let the conditions $\mathbf{B}_1[\bar{\gamma}]$, $\mathbf{C}_3[\bar{\beta}]$, and $\mathbf{D}_1[\bar{\beta}]$ hold, and $0 \leq \gamma_i \leq \beta_i < \infty$, $i = 1, \dots, k$. Then, the following inequality takes place:*

$$|\Phi| \leq M_{10} . \tag{4.51}$$

Remark 4.2.4. The explicit formula for the constant M_{10} follows from formula (4.22) given in Remark 4.1.5,

$$M_{10} = L_1 + \sum_{i:\gamma_i=0} L_1 L_{2,i} + \sum_{i:\gamma_i>0} L_1 L_{2,i} K_{7,i}^{N\frac{\gamma_i}{\beta_i}} K_{2,i}^{\frac{\gamma_i}{\beta_i}} . \tag{4.52}$$

4.2.3 Upper bounds for option rewards for price processes represented by Lévy random walks

Models of log-price processes represented by random walks often appear as the result of time discretization procedures applied to the continuous time log-price processes with independent increments.

In this case, a natural assumption can be such that the random variables $\vec{W}_n$, $n = 1, 2, \dots$, are infinitely divisible.

We can restrict consideration by univariate random variables and processes since the corresponding moment conditions used in this section are formulated in terms of one-dimensional components of jumps for the corresponding multivariate random walks.

As is known, the characteristic function for an infinitely divisible random variable W is given by Lévy–Khintchine representation formula,

$$\begin{aligned} \psi_W(s) &= \mathsf{E}\exp\{\mathrm{i}sW\} \\ &= \exp\Bigg\{\mathrm{i}\mu s - \frac{1}{2}\sigma^2 s^2 + \int\limits_{|y|<1} \left(e^{\mathrm{i}sy} - 1 - \mathrm{i}sy\right) \Pi(dy) \\ &\quad + \int\limits_{|y|\geq 1} \left(e^{\mathrm{i}sy} - 1\right) \Pi(dy)\Bigg\}, \ s \in \mathbb{R}_1 , \end{aligned} \tag{4.53}$$

where (a) μ is a real-valued constant, (b) σ^2 is a nonnegative constant, (c) $\Pi(A)$ is a measure on $\mathcal{B}_1$ without an atom in point 0 such that $\pi = \int_{\mathbb{R}_1} \frac{y^2}{1+y^2} \Pi(dy) < \infty$.

The triplet of characteristics $\langle \mu, \sigma^2, \Pi(\cdot) \rangle$ uniquely determines the characteristic function $\psi_W(s)$.

Let us introduce the following condition assumed to hold for some $\beta_0 \geq 0$:
$\mathbf{C}_2'[\beta_0]$: $\int_{|y|\geq 1} e^{\beta_0 |y|} \Pi(dy) < \infty$.

The following lemma gives the well-known explicit formula for the moment-generating function of the infinitely divisible random variable W.

Lemma 4.2.2. *Let W be an infinitely divisible random variable with the characteristic function given by relation (4.53). Then, condition $\mathbf{C}_2'[\beta_0]$ is necessary and sufficient for* *existence of exponential moments* $\mathsf{E}\exp\{\beta W\} < \infty$, $|\beta| \leq \beta_0$. *Moreover, the following formula takes place for* $|\beta| \leq \beta_0$*:*

$$\begin{aligned}\psi_W(\beta) &= \mathsf{E}\exp\{\beta W\} \\ &= \exp\left\{\mu\beta + \frac{1}{2}\sigma^2\beta^2 + \int\limits_{|y|<1}\left(e^{\beta y} - 1 - \beta y\right)\Pi(dy)\right. \\ &\quad \left. + \int\limits_{|y|\geq 1}\left(e^{\beta y} - 1\right)\Pi(dy)\right\} < \infty\,. \end{aligned} \tag{4.54}$$

Formula (4.54) implies the following inequality, which holds for any $0 \leq \beta \leq \beta_0$:

$$\begin{aligned}\psi_W(\pm\beta) \leq \exp&\left\{|\mu|\beta + \frac{1}{2}\sigma^2\beta^2 + \frac{1}{2}\beta^2 e^{\beta}\int\limits_{|y|<1} y^2\Pi(dy)\right. \\ &\left. +2\int\limits_{|y|\geq 1} e^{\beta|y|}\Pi(dy)\right\} < \infty\,. \end{aligned} \tag{4.55}$$

Now let us return to analysis of the model of log-price processes represented by a multivariate standard random walk, introduced in Subsection 4.2.1.

Let us additionally assume that the random jumps $\vec{W}_n = (W_{n,1}, \dots, W_{n,k})$, $n = 1, 2, \dots$, are infinitely divisible random vectors.

In this case, their components $W_{n,i}$, $i = 1, \dots, k$, $n = 1, 2, \dots$, are obviously infinitely divisible random variables. Let us assume that they have the corresponding triplets of characteristics $\langle \mu_{n,i}, \sigma^2_{n,i}, \Pi_{n,i}(\cdot)\rangle$, $i = 1, \dots, k$, $n = 1, 2, \dots$.

In this case, the second-type modulus of exponential moment compactness for the components of the log-price process $\vec{Y}_n$, takes the following form for $\beta \geq 0$, $i = 1, \dots, k$, $N = 0, 1, \dots$:

$$\Xi''^{\pm}_{\beta}(Y_{\cdot,i}, N) = \max_{0\leq n\leq N-1} \psi_{W_{n+1,i}}(\pm\beta)\,. \tag{4.56}$$

Let us now introduce the following condition which is assumed to hold for some vector parameter $\bar{\beta} = (\beta_1, \dots, \beta_k)$ with nonnegative components:

$\mathbf{C}_3'[\bar{\beta}]$: $\max_{0\leq n\leq N-1}\int_{|y|\geq 1} e^{\beta_i|y|}\Pi_{n+1,i}(dy) < K'_{8,i}(\beta_i)$, $i = 1, \dots, k$, for some $1 < K'_{8,i}(\beta_i) < \infty$, $i = 1, \dots, k$.

Lemma 4.2.3. *The condition* $\mathbf{C}_3'[\bar{\beta}]$ *implies condition* $\mathbf{E}_2[\bar{\beta}]$ *to hold.*

Proof. The condition $\mathbf{C}_3'[\bar{\beta}]$ and inequality (4.55) imply, in an obvious way, that condition $\mathbf{E}_2[\bar{\beta}]$ holds, for example, with constants defined for $i = 1, \dots, k$ by the following formula:

$$K_{8,i} = \exp\left\{ (\beta_i + \frac{1}{2}\beta_i^2 + \frac{1}{2}\beta_i^2 e^{\beta_i}) \max_{0 \le n \le N-1} (|\mu_{n+1,i}| + \sigma^2_{n+1,i} \right. \tag{4.57}$$

$$\left. + \int\limits_{|y| \le 1} y^2 \Pi_{n+1,i}(dy)) + 2K'_{8,i}(\beta_i) \right\} \in (1, \infty) . \qquad \square$$

4.3 Markov LPP with unbounded characteristics

Upper bounds for option rewards presented in Section 4.1 serve well for Markov log-price processes with conditional exponential moments of increments bounded as functions of log-price and index arguments. The corresponding results work well for the random walk models. They can also be effectively applied to Markov Gaussian random walks, with bounded drift and volatility coefficients. However, these results do not work well for the mean-reverse and other Markov-type models with unbounded characteristics. In this section, we present upper bounds for option rewards applicable for such models.

4.3.1 Upper bounds for exponential moments for maxima of multivariate modulated Markov log-price processes with unbounded characteristics

Let us consider a univariate modulated Markov log-price process $Z_n = (Y_n, X_n)$ with the phase space $\mathbb{Z} = \mathbb{R}_1 \times \mathbb{X}$.

Let us denote by $\mathcal{A}$, the class of measurable, nonnegative and nondecreasing functions $A(\beta)$ defined for $\beta \ge 0$ and such that $A(0) = 0$.

Any function $A(\beta) \in \mathcal{A}$ generates a nondecreasing sequence of functions $A_0(\beta) \le A_1(\beta) \le A_2(\beta) \le \cdots$ from the class $\mathcal{A}$ that are defined by the following recurrence relation, for every $\beta \ge 0$:

$$A_n(\beta) = \begin{cases} \beta & \text{for } n = 0 , \\ A_{n-1}(\beta) + A(A_{n-1}(\beta)) & \text{for } n = 1, 2, \dots . \end{cases} \tag{4.58}$$

According to relation (4.58), the first several functions $A_n(\beta)$ are,

$$A_0(\beta) = \beta , \quad A_1(\beta) = \beta + A(\beta) , \quad A_2(\beta) = \beta + A(\beta) + A(\beta + A(\beta)), \dots . \tag{4.59}$$

Let us define the first-type A-modulus of exponential moment compactness for the log-price process Y_n, for $\beta \geq 0$,

$$\Delta_{\beta,A}(Y_\cdot, N) = \max_{0 \leq n \leq N-1} \sup_{z=(y,x)\in \mathbb{Z}} \frac{\mathsf{E}_{z,n} e^{A_{N-n-1}(\beta)|Y_{n+1}-Y_n|}}{e^{A(A_{N-n-1}(\beta))|y|}} . \tag{4.60}$$

Remind that we use the notations $\mathsf{P}_{z,n}$ and $\mathsf{E}_{z,n}$ for conditional probabilities and expectations under the condition $Z_n = z$.

Note that the modulus $\Delta_{\beta,A}(Y_\cdot, N)$ coincides with the modulus $\Delta_\beta(Y_\cdot, N)$ introduced in Section 1.4.1, if function $A(\beta) \equiv 0$. Indeed, in this case all functions $A_n(\beta) \equiv \beta$, $n = 0, 1, \dots$, and functions $A(A_n(\beta)) \equiv 0$, $n = 0, 1, \dots$.

We use the following first-type condition of exponential moment compactness for the log-price process Y_n, which is assumed to hold for some $\beta \geq 0$:
$\mathbf{C_4}[\beta]$: $\Delta_{\beta,A}(Y_\cdot, N) < K_9$, for some $1 < K_9 < \infty$.

The following lemma gives explicit upper bounds for conditional exponential moments of the maximal absolute values for the log-price process Y_n.

Lemma 4.3.1. *Let the condition $\mathbf{C_4}[\beta]$ holds. Then there exists a constant $0 \leq M_{11} < \infty$ such that the following inequality takes place for $z = (y, x) \in \mathbb{Z}$, $0 \leq n \leq N$:*

$$\mathsf{E}_{z,n} \exp\{\beta \max_{n \leq r \leq N} |Y_r|\} \leq M_{11} \exp\{A_{N-n}(\beta)|y|\} . \tag{4.61}$$

Proof. Note that $\Delta_{0,A}(Y_\cdot, N) \equiv 1$ and, thus, the relation penetrating condition $\mathbf{C_4}[0]$ holds automatically. Also, the expectation on the left-hand side of inequality (4.61) equal 1, while the expression on the right-hand side of this inequality equal M_{11}. Thus, this inequality holds, in this case, with the constant $M_{11} = 1$.

Therefore, we should prove (4.61) only for the case $\beta > 0$.

Let us use the following inequality:

$$\exp\left\{\beta \max_{n \leq r \leq N} |Y_r|\right\} \leq \exp\left\{\beta(|Y_n| + \sum_{r=n}^{N-1} |Y_{r+1} - Y_r|)\right\} . \tag{4.62}$$

Using Markov property of the log-price process Y_n and then recurrently condition $\mathbf{C_4}[\beta]$, and relation (4.62), we get the following inequalities, for $z \in \mathbb{Z}$, $0 \leq n \leq N$:

$$\begin{aligned}
&\mathsf{E}_{z,n}\left\{\exp\left\{\beta\left(|Y_n| + \sum_{r=n}^{N-1}|Y_{r+1} - Y_r|\right)\right\}\right. \\
&\quad = \mathsf{E}_{z,n}\mathsf{E}\left\{\exp\left\{\beta\left(|Y_n| + \sum_{r=n}^{N-1}|Y_{r+1} - Y_r|\right)\right\}/\mathcal{F}_{N-1}\right\} \\
&\quad = \mathsf{E}_{z,n}\left(\exp\left\{\beta\left(|Y_n| + \sum_{r=1}^{N-2}|Y_{r+1} - Y_r|\right)\right\}\right. \\
&\qquad \left.\times\, \mathsf{E}\exp\{\beta|Y_N - Y_{N-1}|/Z_{N-1}\}\right) \\
&\quad \le \mathsf{E}_{z,n}\left(\exp\left\{\beta\left(|Y_n| + \sum_{r=n}^{N-2}|Y_{r+1} - Y_r|\right)\right\}\right. \\
&\qquad \left.\times \exp\{A(\beta)|Y_{N-1}|\}\right) K_9 \\
&\quad \le \mathsf{E}_{z,n}\exp\left\{A_1(\beta)\left(|Y_n| + \sum_{r=n}^{N-2}|Y_{r+1} - Y_r|\right)\right\}K_9 \\
&\quad \le \mathsf{E}_{z,n}\exp\left\{A_2(\beta)\left(|Y_n| + \sum_{r=n}^{N-3}|Y_{r+1} - Y_r|\right)\right\}K_9^2 \\
&\quad \le \cdots \le \mathsf{E}_{z,n}\exp\{A_{N-n}(\beta))|Y_n|\}\,K_9^{N-n} \\
&\quad = \exp\{A_{N-n}(\beta)|y|\}\,K_9^{N-n}\,.
\end{aligned} \tag{4.63}$$

Finally, using relations (4.62) and (4.63) we get the following inequality, for $0 \le n \le N$, $z \in \mathbb{Z}$:

$$\begin{aligned}
&\mathsf{E}_{z,n}\exp\left\{\beta\max_{n\le r\le N}|Y_r|\right\} \\
&\quad \le \mathsf{E}_{z,n}\left\{\exp\left\{\beta\left(|Y_n| + \sum_{r=n}^{N-1}|Y_{r+1,i} - Y_{r,i}|\right)\right\}\right. \\
&\quad \le \exp\{A_{N-n}(\beta)|y|\}\,K_9^{N-n}\,.
\end{aligned} \tag{4.64}$$

The remarks made at the beginning of the proof and inequality (4.64) implies that inequality (4.61) holds with the constant K_9^N. □

Remark 4.3.1. The constant M_{11} is given by the following formula:

$$M_{11} = K_9^N\,. \tag{4.65}$$

Moreover, inequalities (4.64) give upper bounds for exponential moments of the maxima for the log-price process Y_n over discrete time intervals $[n, N]$, with the constants K_9^{N-n} depending on n.

Let $\mathcal{Z} = \langle \mathbb{Z}_0, \dots, \mathbb{Z}_N \rangle$ be a sequence of measurable sets. Let us also introduce the first-type modified A-modulus of exponential moment compactness defined by the following formula, for $\beta \geq 0$:

$$\Delta_{\beta,A,\mathcal{Z}}(Y_\cdot, N) = \max_{0 \leq n \leq N-1} \sup_{z=(y,x)\in \mathbb{Z}_n} \frac{\mathsf{E}_{z,n} e^{A_{N-n-1}(\beta)|Y_{n+1}-Y_n|}}{e^{A(A_{N-n-1}(\beta))|y|}} . \tag{4.66}$$

Obviously, if $\mathcal{Z} = \langle \mathbb{Z}, \dots, \mathbb{Z} \rangle$, then modulus $\Delta_{\beta,A,\mathcal{Z}}(Y_\cdot, N) = \Delta_{\beta,A}(Y_\cdot, N)$, for $\beta \geq 0$.

The following first-type condition of exponential moment compactness for the log-price process Y_n generalizes the condition $\mathbf{C_4}[\beta]$. It is assumed to hold for some $\beta \geq 0$:

$\mathbf{C}_4'[\beta]$: There exists a sequence of sets $\mathcal{Z} = \langle \mathbb{Z}_0, \dots, \mathbb{Z}_N \rangle$ such that

(a) $\Delta_{\beta,A,\mathcal{Z}}(Y_\cdot, N) < K_9'$, for some $1 < K_9' < \infty$;

(b) $\mathcal{Z}$ is a complete sequence of phase sets, i.e., $\mathsf{P}_n(z, \mathbb{Z}_n) = 1, z \in \mathbb{Z}_{n-1}, n = 1, \dots, N$.

The following lemma gives explicit upper bounds for conditional exponential moments of the maximal absolute values for the log-price process Y_n.

Lemma 4.3.2. *Let the condition $\mathbf{C}_4'[\beta]$ holds. Then there exists the constant $0 \leq M_{11}' < \infty$ such that the following inequalities takes place for $z = (y, x) \in \mathbb{Z}_n$, $0 \leq n \leq N$:*

$$\mathsf{E}_{z,n} \exp\left\{\beta \max_{n \leq r \leq N} |Y_r|\right\} \leq M_{11}' \exp\{A_{N-n}(\beta)|y|\} . \tag{4.67}$$

Proof. In analogous to the proof of Lemma 4.3.1, Lemma 4.1.2 implies that probability $\mathsf{P}_{z,n}\{Z_r \in \mathbb{Z}_r\} = 1$, for $z \in \mathbb{Z}_n, 0 \leq n \leq r \leq N-1$. The condition $\mathbf{C}_4'[\beta]$ implies that expectation $\mathsf{E}_{z,r}\exp\{A_{N-r-1}(\beta)|Y_{r+1} - Y_r|\} \leq K_9' \exp\{A(A_{N-r-1}(\beta))|y|\}$, for $z \in \mathbb{Z}_r, 0 \leq r \leq N-1$. For these remarks let one write down relations analogous to (4.63) and (4.64), for $z \in \mathbb{Z}_n, 0 \leq n \leq N$ and with the constant K_9' used instead of the constant K_9. □

Remark 4.3.2. The constant M_{11}' is given by the following formula:

$$M_{11}' = (K_9')^N . \tag{4.68}$$

Moreover, inequalities (4.64), modified according to remarks made in the above proof, give upper bounds for exponential moments of the maxima for the log-price process Y_n over discrete time intervals $[n, N]$, with the constants $(K_9')^{N-n}$ depending on n.

4.3.2 Upper bounds for rewards of univariate modulated Markov log-price processes with unbounded characteristics

In the case of univariate log-price processes, the condition $\mathbf{B_1}[\bar{\gamma}]$ reduces to the following simpler condition, assumed to hold for some $\gamma \geq 0$:

$\mathbf{B}_3[\gamma]$: $\max_{0\le n\le N}\sup_{z=(y,x)\in\mathbb{Z}} \frac{|g(n,e^y,x)|}{1+L_6 e^{\gamma|y|}} < L_5$, for some $0 < L_5 < \infty$ and $0 \le L_6 < \infty$.

The following lemma gives sufficient conditions for the condition $\mathbf{A}_2$ to hold and explicit upper bound for conditional expectations of the maximal absolute values for the pay-off process $g(n, e^{Y_n}, X_n)$.

Lemma 4.3.3. *Let the conditions $\mathbf{B}_3[\gamma]$ and $\mathbf{C}_4[\beta]$ hold and $0 \le \gamma \le \beta < \infty$. Then, there exist the constants $0 \le M_{12}, M_{13} < \infty$ such that the following inequalities take place for $z = (y,x) \in \mathbb{Z}$, $0 \le n \le N$:*

$$\begin{aligned}
&\mathsf{E}_{z,n} \max_{n\le r\le N} \left| g(r, e^{Y_r}, X_r) \right| \\
&\quad \le M_{12} + M_{13} I(\gamma = 0) + M_{13} \exp\left\{ \frac{A_{N-n}(\beta)\gamma}{\beta} |y| \right\} I(\gamma > 0) . \qquad (4.69)
\end{aligned}$$

Proof. The case $\gamma = 0$ is trivial. Indeed, condition $\mathbf{B}_3[0]$ implies that $|g(n, e^y, x)| \le L_5 + L_5 L_6$ and the expression on the right-hand side of inequality (4.69) equals to $M_{12} + M_{13}$. Obviously, inequality (4.69) holds if one takes, for example, $M_{12} = L_5$ and $M_{13} = L_5 L_6$.

Let us consider the case, where $\gamma > 0$. Using condition $\mathbf{B}_3[\gamma]$, we get the following inequality for $0 \le n \le N$:

$$\begin{aligned}
\max_{n\le r\le N} \left| g(r, e^{Y_r}, X_r) \right| &\le \max_{n\le r\le N} (L_5 + L_5 L_6 e^{\gamma|Y_r|}) \\
&\le L_5 + L_5 L_6 \max_{n\le r\le N} e^{\gamma|Y_r|} = L_5 + L_5 L_6 e^{\gamma \max_{n\le r\le N} |Y_r|} . \qquad (4.70)
\end{aligned}$$

Now, using well-known properties of moments for random variables and Lemma 4.3.1, we get the following inequality for $z = (y,x) \in \mathbb{Z}$, $0 \le n \le N$:

$$\begin{aligned}
&\mathsf{E}_{z,n} \max_{n\le r\le N} \left| g(r, e^{Y_r}, X_r) \right| \\
&\quad \le L_5 + L_5 L_6 \mathsf{E}_{z,n} e^{\gamma \max_{n\le r\le N} |Y_r|} \\
&\quad \le L_5 + L_5 L_6 \left(\mathsf{E}_{z,n} \left(e^{\gamma \max_{n\le r\le N} |Y_r|} \right)^{\frac{\beta}{\gamma}} \right)^{\frac{\gamma}{\beta}} \\
&\quad = L_5 + L_5 L_6 \left(\mathsf{E}_{z,n} e^{\beta \max_{n\le r\le N} |Y_r|} \right)^{\frac{\gamma}{\beta}} \\
&\quad \le L_5 + L_5 L_6 \left(\exp\{A_{N-n}(\beta)|y|\} M_5 \right)^{\frac{\gamma}{\beta}} \\
&\quad = L_5 + L_5 L_6 (M_5)^{\frac{\gamma}{\beta}} \exp\left\{ \frac{A_{N-n}(\beta)\gamma}{\beta} |y| \right\} . \qquad (4.71)
\end{aligned}$$

Thus, inequality (4.69) holds with the constants M_{12} and M_{13}, which are given by relation (4.71). This completes the proof. □

Remark 4.3.3. The constants M_{12} and M_{13} are given by the following formulas:

$$M_{12} = L_5, \; M_{13} = L_5 L_6 I(\gamma = 0) + L_5 L_6 K_9^{N\frac{\gamma}{\beta}} I(\gamma > 0) . \qquad (4.72)$$

Lemma 4.3.4. *Let the conditions* $\mathbf{B}_3[\gamma]$ *and* $\mathbf{C}'_4[\beta]$ *hold and* $0 \leq \gamma \leq \beta < \infty$. *Then, there exist the constants* $0 \leq M'_{12}, M'_{13} < \infty$ *such that the following inequalities take place for* $z = (y, x) \in \mathbb{Z}_n, 0 \leq n \leq N$:

$$\begin{aligned} \mathsf{E}_{z,n} \max_{n \leq r \leq N} |g(r, e^{Y_r}, X_r)| \\ \leq M'_{12} + M'_{13} I(\gamma = 0) + M'_{13} \exp\left\{ \frac{A_{N-n}(\beta)\gamma}{\beta} |y| \right\} I(\gamma > 0)\,. \end{aligned} \tag{4.73}$$

Proof. It repeats the proof of Lemma 4.3.3. The only difference is that, in the case, where the condition $\mathbf{C}'_4[\beta]$ is assumed to hold instead of the condition $\mathbf{C}_4[\beta]$, inequality (4.71) takes place for $z \in \mathbb{Z}_n, 0 \leq n \leq N$, with the same constants L_5, L_6 and constant K'_9 replacing the constant K_9. □

Remark 4.3.4. The constants M'_{12} and M'_{13} are given by the following formulas:

$$M'_{12} = L_5,\; M'_{13} = L_5 L_6 I(\gamma = 0) + L_5 L_6 (K'_9)^{N\frac{\gamma}{\beta}} I(\gamma > 0)\,. \tag{4.74}$$

Lemma 4.3.3 implies that under conditions of this lemma, the log-reward functions $\phi_n(y, x)$ are well defined by the following relation, for $z = (y, x) \in \mathbb{Z}$, $n = 0, \dots, N$:

$$\phi_n(y, x) = \sup_{\tau_n \in \mathcal{M}_{\max,n,N}} \mathsf{E}_{\vec{z},n} g(\tau_n, e^{Y_{\tau_n}}, X_{\tau_n})\,. \tag{4.75}$$

Moreover, for Lemma 4.3.3 let us get the following upper bounds for the log-reward functions of American-type options for the log-price process Z_n.

Theorem 4.3.1. *Let the conditions* $\mathbf{B}_3[\gamma]$ *and* $\mathbf{C}_4[\beta]$ *hold and* $0 \leq \gamma \leq \beta < \infty$. *Then, the log-reward functions* $\phi_n(y, x)$ *satisfy the following inequalities for* $z = (y, x) \in \mathbb{Z}, 0 \leq n \leq N$:

$$\begin{aligned} |\phi_n(y, x)| &\leq M_{12} + M_{13} I(\gamma = 0) \\ &\quad + M_{13} \exp\left\{ \frac{A_{N-n}(\beta)\gamma}{\beta} |y| \right\} I(\gamma > 0)\,. \end{aligned} \tag{4.76}$$

Proof. Using Lemma 4.3.3 and the definition of the log-reward functions, we get the following inequalities for $z = (y, x) \in \mathbb{Z}, 0 \leq n \leq N$:

$$\begin{aligned} |\phi_n(y, x)| &= \left| \sup_{\tau_n \in \mathcal{M}_{\max,n,N}} \mathsf{E}_{z,n} g(\tau_n, e^{Y_{\tau_n}}, X_{\tau_n}) \right| \\ &\leq \mathsf{E}_{z,n} \max_{n \leq r \leq N} \left| g(r, e^{Y_r}, X_r) \right| \\ &\leq M_{12} + M_{13} I(\gamma = 0) + M_{13} \exp\left\{ \frac{A_{N-n}(\beta)\gamma}{\beta} |y| \right\} I(\gamma > 0)\,, \end{aligned} \tag{4.77}$$

which proves the theorem. □

Lemma 4.3.4 implies that under conditions of this lemma, the log-reward functions $\phi_n(y, x)$ are well defined by the relation (4.75), only for $z = (y, x) \in \mathbb{Z}_n$, $n = 0, \dots, N$.

Also, Lemma 4.3.4 let us get the following upper bounds for the log-reward functions of American-type options for the modulated log-price process Z_n.

Theorem 4.3.2. *Let the conditions* $\mathbf{B}_3[\gamma]$ *and* $\mathbf{C}_4'[\beta]$ *hold and* $0 \le \gamma \le \beta < \infty$. *Then, the log-reward functions* $\phi_n(y, x)$ *satisfy the following inequalities for* $z = (y, x) \in \mathbb{Z}_n$, $0 \le n \le N$:

$$\begin{aligned} |\phi_n(y, x)| &\le M_{12}' + M_{13}' I(\gamma = 0) \\ &\quad + M_{13}' \exp\left\{ \frac{A_{N-n}(\beta)\gamma}{\beta} |y| \right\} I(\gamma > 0) . \end{aligned} \tag{4.78}$$

Proof. It repeats the proof of Theorem 4.3.1. The only change is that Lemma 4.3.4 should be used instead of Lemma 4.3.3. This let one get inequalities analogous to (4.77), for $z = (y, x) \in \mathbb{Z}_n, 0 \le n \le N$, with the constants M_{12}', M_{13}' replacing the constants M_{12}, M_{13}. □

We can now get upper bounds for optimal expected rewards.

To do this, we should impose the following first-type condition of exponential moment boundedness (on the initial distribution of the log-price process), assumed to hold for some $\beta \ge 0$:

$\mathbf{D}_2[\beta]$: $\mathsf{E} \exp\{A_N(\beta)|Y_0|\} < K_{10}$, for some $1 < K_{10} < \infty$.

The following lemma gives sufficient conditions for the condition $\mathbf{A}_1$ to hold and explicit upper bound for the expectation of the maximum for the pay-off process $g(n, e^{Y_n}, X_n)$.

Lemma 4.3.5. *Let the conditions* $\mathbf{B}_3[\gamma]$, $\mathbf{C}_4[\beta]$, *and* $\mathbf{D}_2[\beta]$ *hold and* $0 \le \gamma \le \beta < \infty$. *Then, there exists a constant* $M_{14} < \infty$ *such that for the following inequality takes place:*

$$\mathsf{E} \max_{0 \le n \le N} \left| g(n, e^{Y_n}, X_n) \right| \le M_{14} . \tag{4.79}$$

Proof. If $\gamma = 0$ then inequality (4.79) holds with the constant $M_{14} = M_{12} + M_{13}$. Let $\gamma > 0$. Using Lemma 4.3.3, we get

$$\begin{aligned} \mathsf{E} \max_{0 \le n \le N} & \left| g(n, e^{Y_n}, X_n) \right| \\ &= \mathsf{E}\,\mathsf{E} \left\{ \max_{0 \le n \le N} \left| g(n, e^{Y_n}, X_n) \right| / Z_0 \right\} \\ &\le \mathsf{E} \left(M_{12} + M_{13} \exp\left\{ \frac{A_N(\beta)\gamma}{\beta} |Y_0| \right\} \right) \\ &\le M_{12} + M_{13} \left(\mathsf{E} \exp\left\{ A_N(\beta)|Y_0| \right\} \right)^{\frac{\gamma}{\beta}} \\ &\le M_{12} + M_{13} K_{10}^{\frac{\gamma}{\beta}} . \end{aligned} \tag{4.80}$$

Relation (4.80) proves that inequality (4.79) holds. □

Remark 4.3.5. The following formulas gives the explicit expression for constant M_{14}:

$$M_{14} = L_5 + L_5 L_6 I(\gamma = 0) + L_5 L_6 K_9^{N\frac{\gamma}{\beta}} K_{10}^{\frac{\gamma}{\beta}} I(\gamma > 0)\,. \tag{4.81}$$

In the case, where the condition $\mathbf{C}'_4[\beta]$ is used instead the condition $\mathbf{C}_4[\beta]$, the following condition should replace the condition $\mathbf{D}_2[\beta]$ and be assumed to hold for some $\beta \geq 0$:

$\mathbf{D}'_2[\beta]$: (a) $\mathsf{E}\exp\{A_N(\beta)|Y_0|\} < K'_{10}$, for some $1 < K'_{10} < \infty$;
(b) $\mathsf{P}\{Y_0 \in \mathbb{Z}_0\} = 1$, where $\mathbb{Z}_0$ is the set penetrating condition $\mathbf{C}'_4[\beta]$.

Lemma 4.3.6. *Let conditions the conditions $\mathbf{B}_3[\gamma]$, $\mathbf{C}'_4[\beta]$, and $\mathbf{D}'_2[\beta]$ hold and $0 \leq \gamma \leq \beta < \infty$. Then, there exists a constant $M'_{14} < \infty$ such that for the following inequality takes place:*

$$\mathsf{E} \max_{0 \leq n \leq N} |g(n, e^{Y_n}, X_n)| \leq M'_{14}\,. \tag{4.82}$$

Proof. It repeats the proof of Lemma 4.3.5. The only difference is that Lemma 4.3.4 should be use instead of Lemma 4.3.3. The relation analogous to (4.80) can be written down, with the same constants L_5, L_6 and the constants K'_9, K'_{10} used instead of the constants L_5, L_6 and the constants K_9, K_{10}. □

Remark 4.3.6. The following formula gives the explicit expression for constant M'_{14}:

$$M'_{14} = L_5 + L_5 L_6 I(\gamma = 0) + L_5 L_6 (K'_9)^{N\frac{\gamma}{\beta}} (K'_{10})^{\frac{\gamma}{\beta}} I(\gamma > 0)\,. \tag{4.83}$$

Let us assume that either conditions of Lemma 4.3.5 or Lemma 4.3.6 hold. In this case, the optimal expected reward $\Phi = \Phi(\mathcal{M}_{\max,N})$ for the log-price Z_n can be defined by the following relation:

$$\Phi = \sup_{\tau_0 \in \mathcal{M}_{\max,N}} \mathsf{E} g(\tau_0, e^{Y_{\tau_0}}, X_{\tau_0})\,. \tag{4.84}$$

The following two theorems give explicit upper bounds for the optimal expected reward Φ.

Theorem 4.3.3. *Let the conditions $\mathbf{B}_3[\gamma]$, $\mathbf{C}_4[\beta]$, and $\mathbf{D}_2[\beta]$ hold and $0 \leq \gamma \leq \beta < \infty$. Then, the following inequality takes place:*

$$|\Phi| \leq M_{14}\,. \tag{4.85}$$

Proof. The following inequality follows directly from the definition of the functional Φ and Lemma 4.3.5:

$$|\Phi| \leq \sup_{\tau_0 \in \mathcal{M}_{\max,N}} \mathsf{E}\left|g(\tau_0, e^{Y_{\tau_0}}, X_{\tau_0})\right| \leq \mathsf{E} \max_{0 \leq n \leq N} \left|g(n, e^{Y_n}, X_n)\right| \leq M_{14}\,. \tag{4.86}$$

This inequality proves the theorem. □

Theorem 4.3.4. *Let the conditions* $\mathbf{B}_3[\gamma]$, $\mathbf{C}'_4[\beta]$, *and* $\mathbf{D}'_2[\beta]$ *hold and* $0 \leq \gamma \leq \beta < \infty$. *Then, the following inequality takes place:*

$$|\Phi| \leq M'_{14}\,. \tag{4.87}$$

Proof. It repeats the proof of Theorem 4.3.3. The only difference is that Lemma 4.3.6 should be used instead of Lemma 4.3.5. The relation analogous to (4.86) can be written down, with and the constant M'_{14} used instead of the constant M_{14}. □

4.3.3 Modified conditions of exponential moment compactness for univariate modulated Markov log-price processes with unbounded characteristics

The condition $\mathbf{C}_4[\beta]$ used in Lemmas 4.3.1 and 4.3.2 is based on exponential moments for absolute values of increments for the log-price process Z_n. In some case, it is useful to replace this condition by an alternative condition based on moment-generating functions.

Let us define a second-type A-modulus of exponential moment compactness for the log-price process Y_n, for $\beta \geq 0$,

$$\Xi^{\pm}_{\beta,A}(Y_\cdot, N) = \max_{0 \leq n \leq N-1} \sup_{z=(y,x)\in\mathbb{Z}} \frac{\mathsf{E}_{z,n} e^{\pm A_{N-n-1}(\beta)(Y_{n+1}-Y_n)}}{e^{A(A_{N-n-1}(\beta))|y|}}\,. \tag{4.88}$$

The following second-type condition of exponential moment compactness based on moment-generating functions of increments for the log-price process Y_n, is assumed to hold for some $\beta \geq 0$:

$\mathbf{E}_4[\beta]$: $\Xi^{\pm}_{\beta,A}(Y_\cdot, N) < K_{11}$, for some $1 < K_{11} < \infty$.

Lemma 4.3.7. *The condition* $\mathbf{E}_4[\beta]$ *implies condition* $\mathbf{C}_4[\beta]$ *to hold.*

Proof. The following inequality takes place, for every $0 \leq n \leq N-1$, $z \in \mathbb{Z}$, $i = 1, \dots, k$, and $\beta \geq 0$:

$$\begin{aligned}
&\mathsf{E}_{z,n} e^{A_{N-n-1}(\beta)|Y_{n+1}-Y_n|} \\
&\quad = \mathsf{E}_{z,n} e^{A_{N-n-1}(\beta)(Y_{n+1}-Y_n)} I(Y_{n+1} \geq Y_n) \\
&\qquad + \mathsf{E}_{z,n} e^{-A_{N-n-1}(\beta)(Y_{n+1}-Y_n)} I(Y_{n+1} < Y_n) \\
&\quad \leq \mathsf{E}_{z,n} e^{A_{N-n-1}(\beta)(Y_{n+1}-Y_n)} + \mathsf{E}_{z,n} e^{-A_{N-n-1}(\beta)(Y_{n+1}-Y_n)}\,.
\end{aligned} \tag{4.89}$$

Relation (4.89) obviously implies the following inequality:

$$\Delta_{\beta,A}(Y_\cdot, N) \leq \Xi^{+}_{\beta,A}(Y_\cdot, N) + \Xi^{-}_{\beta,A}(Y_\cdot, N)\,. \tag{4.90}$$

Thus the condition $\mathbf{E}_4[\beta]$ implies that condition $\mathbf{C}_4[\beta]$ holds. □

Remark 4.3.7. As follows from inequality (4.31), if condition $\mathbf{E}_4[\beta]$ holds with the constant K_{11}, then condition $\mathbf{C}_4[\beta]$ holds with the constant K_9 given by the following formula:

$$K_9 = 2K_{11}\,. \tag{4.91}$$

Let $\mathcal{Z} = \langle \mathbb{Z}_0, \dots, \mathbb{Z}_N \rangle$ be a sequence of measurable sets and $\Xi^{\pm}_{\beta,A,\mathcal{Z}}(Y_\cdot, N)$ be the modified second-type A-modulus of exponential moment compactness, defined by the following formula for $\beta \geq 0$:

$$\Xi^{\pm}_{\beta,A,\mathcal{Z}}(Y_\cdot, N) = \max_{0 \leq n \leq N-1} \sup_{z=(y,x)\in\mathbb{Z}_n} \frac{\mathsf{E}_{z,n} e^{\pm A_{N-n-1}(\beta)(Y_{n+1}-Y_n)}}{e^{A(A_{N-n-1}(\beta))|y|}}\,. \tag{4.92}$$

The following modified second-type condition of exponential moment compactness is based on moment-generating functions of increments for the log-price process Y_n. It generalizes the condition $\mathbf{E}_4[\beta]$ and is assumed to hold for some $\beta \geq 0$:
$\mathbf{E}'_4[\beta]$: There exists a sequence of sets $\mathcal{Z} = \langle \mathbb{Z}_0, \dots, \mathbb{Z}_N \rangle$ such that
(a) $\Xi^{\pm}_{\beta,A,\mathcal{Z}}(Y_\cdot, N) < K'_{11}$, for some $1 < K'_{11} < \infty$;
(b) $\mathcal{Z}$ is a complete sequence of phase sets, i.e., $\mathsf{P}_n(z, \mathbb{Z}_n) = 1, z \in \mathbb{Z}_{n-1}, n = 1, \dots, N$.

Lemma 4.3.8. *The condition $\mathbf{E}'_4[\beta]$ implies condition $\mathbf{C}'_4[\beta]$ to hold.*

Proof. Relation (4.89) implies analogously to (4.90) the following inequality:

$$\Delta_{\beta,A,\mathcal{Z}}(Y_\cdot, N) \leq \Xi^{+}_{\beta,A,\mathcal{Z}}(Y_\cdot, N) + \Xi^{-}_{\beta,A,\mathcal{Z}}(Y_\cdot, N)\,. \tag{4.93}$$

Thus, the condition $\mathbf{E}'_4[\beta]$ implies that condition $\mathbf{C}'_4[\beta]$ holds. □

Remark 4.3.8. As follows from inequality (4.31), if condition $\mathbf{E}'_4[\beta]$ holds with the constant K'_{11}, then condition $\mathbf{C}'_4[\beta]$ holds with the constant K'_9 given by the following formula:

$$K'_9 = 2K'_{11}\,. \tag{4.94}$$

Similarly, one can replace condition $\mathbf{D}_2[\beta]$ by the following second-type condition of exponential moment boundedness based on moment-generating functions for the initial value of log-price processes $\vec{Y}_0$ and assumed to hold for some $\beta \geq 0$:
$\mathbf{F}_2[\beta]$: $\mathsf{E} e^{\pm A_N(\beta) Y_0} < K_{12}$, for some $1 < K_{12} < \infty$.

Lemma 4.3.9. *The condition $\mathbf{F}_2[\beta]$ implies condition $\mathbf{D}_2[\beta]$ to hold.*

Remark 4.3.9. If the condition $\mathbf{F}_2[\beta]$ holds with the constant K_{12}, then the condition $\mathbf{D}_2[\beta]$ holds with the constant K_{10} given by the following formula:

$$K_{10} = 2K_{12}\,. \tag{4.95}$$

Analogously, the condition $\mathbf{D}_2'[\beta]$ can be replaced by the following condition assumed to hold for some $\beta \geq 0$:

$\mathbf{F}_2'[\beta]$: (a) $\mathsf{E}e^{\pm A_N(\beta)Y_0} < K_{12}'$, for some $1 < K_{12}' < \infty$;
(b) $\mathsf{P}\{Y_0 \in \mathbb{Z}_0\} = 1$, where $\mathbb{Z}_0$ is the set penetrating condition $\mathbf{D}_2'[\beta]$.

Lemma 4.3.10. *The condition $\mathbf{F}_2'[\beta]$ implies condition $\mathbf{D}_2'[\beta]$ to hold.*

Remark 4.3.10. If the condition $\mathbf{F}_2'[\beta]$ holds with the constant K_{12}', then the condition $\mathbf{D}_2'[\beta]$ holds with the constant K_{10}' given by the following formula:

$$K_{10}' = 2K_{12}' . \tag{4.96}$$

4.3.4 Upper bounds for exponential moments of maxima for multivariate modulated Markov log-price processes with unbounded characteristics

We consider a modulated Markov log-price process $\vec{Z}_n = (\vec{Y}_n, X_n)$ with the phase space $\mathbb{Z} = \mathbb{R}_k \times \mathbb{X}$. Our goal is to generalise the condition $\mathbf{C}_1[\bar{\beta}]$.

Let us denote by $\mathcal{A}_k$, the class of measurable, vector functions $\vec{A}(\bar{\beta}) = (A_1(\bar{\beta}), \dots, A_k(\bar{\beta}))$ defined for $\bar{\beta} = (\beta_1, \dots, \beta_k), \beta_i \geq 0, i = 1, \dots, k$, whose components $A_i(\bar{\beta}), i = 1, \dots, k$ are real-valued functions, nondecreasing in every argument β_i, $i = 1, \dots, k$ and such that $A_i(\bar{0}) = 0, i = 1, \dots, k$.

Any function $\vec{A}(\bar{\beta}) \in \mathcal{A}_k$ generates a sequence of functions $\vec{A}_n(\bar{\beta}) = (A_{n,1}(\bar{\beta}), \dots, A_{n,k}(\bar{\beta})), n = 0, 1, \dots$ from the class $\mathcal{A}_k$ that is defined by the following recurrence relation, for any $\bar{\beta} = (\beta_1, \dots, \beta_k), \beta_i \geq 0, i = 1, \dots, k$:

$$\vec{A}_n(\bar{\beta}) = \begin{cases} \bar{\beta} & \text{for } n = 0, \\ \vec{A}_{n-1}(\bar{\beta}) + \vec{A}(\vec{A}_{n-1}(\bar{\beta})) & \text{for } n = 1, 2, \dots . \end{cases} \tag{4.97}$$

Note that, by the definition, $A_{0,j}(\bar{\beta}) \leq A_{1,j}(\bar{\beta}) \leq \cdots$ for every $\bar{\beta} = (\beta_1, \dots, \beta_k)$, $\beta_i \geq 0, i = 1, \dots, k$ and $j = 1, \dots, k$.

Let us define the first-type $\vec{A}$-modulus of exponential moment compactness for the log-price process $\vec{Y}_n$, for $\bar{\beta} = (\beta_1, \dots, \beta_k), \beta_i \geq 0, i = 1, \dots, k$,

$$\Delta_{\bar{\beta},\vec{A}}(\vec{Y}_\cdot, N) = \max_{0 \leq n \leq N-1} \sup_{\vec{z}=((y_1,\dots,y_k),x) \in \mathbb{Z}} \frac{\mathsf{E}_{\vec{z},n} e^{\sum_{j=1}^k A_{N-n-1,j}(\bar{\beta})|Y_{n+1,j} - Y_{n,j}|}}{e^{\sum_{j=1}^k A_j(\vec{A}_{N-n-1}(\bar{\beta}))|y_j|}} . \tag{4.98}$$

Remind that we use the notations $\mathsf{P}_{\vec{z},n}$ and $\mathsf{E}_{\vec{z},n}$ for conditional probabilities and expectations under the condition $\vec{Z}_n = \vec{z}$.

Let us introduce vectors $\bar{\beta}_i = (0, \dots, 0, \beta_i, 0, \dots, 0), \beta_i \geq 0, i = 1, \dots, k$ (the vector $\bar{\beta}_i$ has the ith component equals to β_i and other components equal to 0). If to denote $\beta_{i,j} = \beta_j I(j = i), j = 1, \dots, k$, then the vectors $\bar{\beta}_i = (0, \dots, 0, \beta_i, 0, \dots, 0) = (\beta_{i,1}, \dots, \beta_{i,k}), i = 1, \dots, k$.

We use the following first-type condition of exponential moment compactness for the log-price process $\vec{Y}_n$, which is assumed to hold for some vectors $\bar{\beta} = (\beta_1, \dots, \beta_k)$ with nonnegative components:

$\mathbf{C}_5[\bar{\beta}]$: $\Delta_{\bar{\beta}_i, \vec{A}}(\vec{Y}_\cdot, N) < K_{13,i}, i = 1, \dots, k$, for some $1 < K_{13,i} < \infty, i = 1, \dots, k$.

The following lemma gives explicit upper bounds for conditional moments of the maximum of absolute values for the log-price process $\vec{Y}_n$.

Lemma 4.3.11. *Let the condition* $\mathbf{C}_5[\bar{\beta}]$ *hold. Then there exist the constants* $0 \leq M_{15,i} < \infty, i = 1, \dots, k$ *such that the following inequality takes place for* $\vec{z} = ((y_1, \dots, y_k), x) \in \mathbb{Z}, 0 \leq n \leq N$ *and* $i = 1, \dots, k$:

$$\mathsf{E}_{\vec{z},n} \exp\left\{\beta_i \max_{n \leq r \leq N} |Y_{r,i}|\right\} \leq M_{15,i} \exp\left\{\sum_{j=1}^{k} A_{N-n,j}(\bar{\beta}_i)\,|y_j|\right\}. \tag{4.99}$$

Proof. Note that $\Delta_{\bar{0},\vec{A}}(Y_\cdot, N) \equiv 1$ and, thus, the relation penetrating condition $\mathbf{C}_5[\bar{\beta}]$ for a given $i = 1, \dots, k$ holds automatically if $\beta_i = 0$. Also, the expectation on the left hand side of inequality (4.99) equal 1, while the expression on the right-hand side of this inequality equal $M_{15,i}$, if $\beta_i = 0$. Thus, inequality (4.99) holds with the constant $M_{15,i} = 1$ for i such that $\beta_i = 0$.

Therefore, we should prove (4.99) only for i such that $\beta_i > 0$.

Let us use the following inequality, for $\beta_i \geq 0, i = 1, \dots, k$:

$$\exp\left\{\beta_i \max_{n \leq r \leq N} |Y_{r,i}|\right\} \leq \exp\left\{\beta_i \left(|Y_{n,i}| + \sum_{r=n}^{N-1} |Y_{r+1,i} - Y_{r,i}|\right)\right\}. \tag{4.100}$$

Using Markov property of the log-price process $\vec{Z}_n$ and then recurrently condition $\mathbf{C}_5[\bar{\beta}]$, we get the following inequalities, for $\vec{z} \in \mathbb{Z}, 0 \leq n \leq N, i = 1, \dots, k$:

$$
\begin{aligned}
&\mathsf{E}_{\vec{z},n} \exp\left\{\beta_i \left(|Y_{n,i}| + \sum_{r=n}^{N-1} |Y_{r+1,i} - Y_{r,i}| \right)\right\} \\
&= \mathsf{E}_{\vec{z},n} \exp\left\{\sum_{j=1}^{k} \beta_{i,j} \left(|Y_{n,j}| + \sum_{r=n}^{N-1} |Y_{r+1,j} - Y_{r,j}| \right)\right\} \\
&= \mathsf{E}_{\vec{z},n} \mathsf{E}\left\{\exp\left\{\sum_{j=1}^{k} \beta_{i,j} \left(|Y_{n,j}| + \sum_{r=n}^{N-1} |Y_{r+1,j} - Y_{r,j}| \right)\right\} \Big/ \mathcal{F}_{N-1}\right\} \\
&= \mathsf{E}_{\vec{z},n} \left(\exp\left\{\sum_{j=1}^{k} \beta_{i,j} \left(|Y_{n,j}| + \sum_{r=1}^{N-2} |Y_{r+1,j} - Y_{r,j}| \right)\right\} \right. \\
&\qquad \left. \times \mathsf{E} \exp\left\{\sum_{j=1}^{k} \beta_{i,j} |Y_{N,j} - Y_{N-1,j}| / \vec{Z}_{N-1}\right\} \right) \\
&\le \mathsf{E}_{\vec{z},n} \left(\exp\left\{\sum_{j=1}^{k} \beta_{i,j} \left(|Y_{n,j}| + \sum_{r=n}^{N-2} |Y_{r+1,j} - Y_{r,j}| \right)\right\} \right. \\
&\qquad \left. \times \exp\left\{\sum_{j=1}^{k} A_j(\bar{\beta}_i)\, |Y_{N-1,j}| \right\} \right) K_{13,i} \\
&\le \mathsf{E}_{\vec{z},n} \exp\left\{\sum_{j=1}^{k} \left(\beta_{i,j} + A_j(\bar{\beta}_i)\right) \left(|Y_{n,j}| + \sum_{r=n}^{N-2} |Y_{r+1,j} - Y_{r,j}| \right)\right\} K_{13,i} \\
&\le \mathsf{E}_{\vec{z},n} \exp\left\{\sum_{j=1}^{k} A_{1,j}(\bar{\beta}_i) \left(|Y_{n,j}| + \sum_{r=n}^{N-2} |Y_{r+1,j} - Y_{r,j}| \right)\right\} K_{13,i} \\
&\le \mathsf{E}_{\vec{z},n} \exp\left\{\sum_{j=1}^{k} A_{2,j}(\bar{\beta}_i) \left(|Y_{n,j}| + \sum_{r=n}^{N-3} |Y_{r+1,j} - Y_{r,j}| \right)\right\} K_{13,i}^{2} \\
&\le \cdots \le \mathsf{E}_{\vec{z},n} \exp\left\{\sum_{j=1}^{k} A_{N-n,j}(\bar{\beta}_i) |Y_{n,j}| \right\} K_{13,i}^{N-n} \\
&= \exp\left\{\sum_{j=1}^{k} A_{N-n,j}(\bar{\beta}_i) |y_j| \right\} K_{13,i}^{N-n} .
\end{aligned}
\tag{4.101}
$$

Finally, using relations (4.100) and (4.101) we get the following inequalities, for $\vec{z} \in \mathbb{Z}, 0 \leq n \leq N, i = 1, \dots, k$:

$$\begin{aligned} \mathsf{E}_{\vec{z},n} \exp\left\{\beta_i \max_{n \leq r \leq N} |Y_{r,i}|\right\} & \\ \leq \mathsf{E}_{\vec{z},n} \left\{\exp\left\{\beta_i \left(|Y_{n,i}| + \sum_{r=n}^{N-1} |Y_{r+1,i} - Y_{r,i}|\right)\right\}\right. & \\ \leq \exp\left\{\sum_{j=1}^{k} A_{N-n,j}(\bar{\beta}_i)|y_j|\right\} K_{13,i}^{N-n} . & \end{aligned} \tag{4.102}$$

The remarks made at the beginning of the proof and inequality (4.102) implies that inequality (4.99) holds with the constant $K_{13,i}^{N}$. □

Remark 4.3.11. The constants $M_{15,i}$, $i = 1, \dots, k$ are given by the following formulas:

$$M_{15,i} = K_{13,i}^{N} . \tag{4.103}$$

Moreover, inequalities (4.102) give upper bounds for exponential moments of the maxima for the log-price processes $Y_{n,i}$ over discrete time intervals $[n, N]$, with the constants $K_{13,i}^{N-n}$ depending on n.

Let $\mathcal{Z} = \langle \mathbb{Z}_0, \dots, \mathbb{Z}_N \rangle$ be a sequence of measurable sets. Let us define the first-type modified $\vec{A}$-modulus of exponential moment compactness $\Delta_{\bar{\beta},\vec{A},\mathcal{Z}}(\vec{Y}_\cdot, N)$ by the following formula, for $\bar{\beta} = (\beta_1, \dots, \beta_k), \beta_i \geq 0, i = 1, \dots, k$:

$$\begin{aligned} & \Delta_{\bar{\beta},\vec{A},\mathcal{Z}}(\vec{Y}_\cdot, N) \\ & = \max_{0 \leq n \leq N-1} \sup_{\vec{z}=((y_1,\dots,y_k),x) \in \mathbb{Z}_n} \frac{\mathsf{E}_{\vec{z},n} e^{\sum_{j=1}^{k} A_{N-n-1,j}(\bar{\beta})|Y_{n+1,j} - Y_{n,j}|}}{e^{\sum_{j=1}^{k} A_j(\vec{A}_{N-n-1}(\bar{\beta}))|y_j|}} . \end{aligned} \tag{4.104}$$

The following modified first-type condition of exponential moment compactness for the log-price process $\vec{Y}_n$ generalizes the condition $\mathbf{C}_5[\bar{\beta}]$, and should be assumed to hold for some vector parameter $\bar{\beta} = (\beta_1, \dots, \beta_k)$ with nonnegative components:

$\mathbf{C}_5'[\bar{\beta}]$: There exists a sequence of measurable sets $\mathcal{Z} = \langle \mathbb{Z}_0, \dots, \mathbb{Z}_N \rangle$ such that

(a) $\Delta_{\bar{\beta}_i,\vec{A},\mathcal{Z}}(\vec{Y}_\cdot, N) < K_{13,i}'$, $i = 1, \dots, k$, for some $1 < K_{13,i}' < \infty$, $i = 1, \dots, k$;

(b) $\mathcal{Z}$ is a complete sequence of phase sets, i.e., $\mathsf{P}_n(\vec{z}, \mathbb{Z}_n) = 1, \vec{z} \in \mathbb{Z}_{n-1}$, $n = 1, \dots, N$.

Lemma 4.3.12. *Let the condition $\mathbf{C}_5'[\bar{\beta}]$ holds. Then there exist the constants $0 \leq M_{15,i}' < \infty$, $i = 1, \dots, k$ such that the following inequalities take place for $\vec{z} = ((y_1, \dots, y_k), x) \in \mathbb{Z}_n$, $0 \leq n \leq N$ and $i = 1, \dots, k$:*

$$\mathsf{E}_{\vec{z},n} \exp\left\{\beta_i \max_{n \leq r \leq N} |Y_{r,i}|\right\} \leq M_{15,i}' \exp\left\{\sum_{j=1}^{k} A_{N-n,j}(\bar{\beta}_i)\, |y_j|\right\} . \tag{4.105}$$

Proof. Analogously to the proof of Lemma 4.3.10. Lemma 4.1.2 implies that probability $\mathsf{P}_{\vec{z},n}\{Z_r \in \mathbb{Z}_r\} = 1$, for $\vec{z} \in \mathbb{Z}_n, 0 \le n \le r \le N-1$. The condition $\mathbf{C}_5'[\bar{\beta}]$ implies that expectation $\mathsf{E}_{\vec{z},r} \exp\{\sum_{j=1}^k A_{N-r-1}(\bar{\beta}_i)|Y_{r+1,j} - Y_{r,j}|\} \le K_{13,i}' \exp\{\sum_{j=1}^k A_j(A_{N-r-1}(\bar{\beta}_i))|y_j|\}$, for $i = 1, \dots, k, \vec{z} = ((y-1, \dots, y_k), x) \in \mathbb{Z}_r$, $r = 0, \dots, N-1$. For these remarks let one write down relations analogous to (4.101) and (4.102), for $\vec{z} \in \mathbb{Z}_n, 0 \le n \le N$ and with the constants $K_{13,i}'$, $i = 1, \dots, k$ replacing the constants $K_{13,i}$, $i = 1, \dots, k$. This completes the proof. □

Remark 4.3.12. The constants $M_{15,i}'$, $i = 1, \dots, k$ are given by the following formulas:

$$M_{15,i}' = (K_{13,i}')^N . \tag{4.106}$$

Moreover, inequality (4.102), modified according to remarks made in the above proof, give upper bounds for exponential moments of the maxima for the log-price processes $Y_{n,i}$ over discrete time intervals $[n, N]$, with the constants $(K_{13,i}')^{N-n}$ depending on n.

4.3.5 Upper bounds for rewards of multivariate modulated Markov log-price processes with unbounded characteristics

The following lemma gives sufficient conditions for the condition $\mathbf{A}_2$ to hold and explicit upper bounds for conditional expectations of the maximal absolute values for the pay-off process $g(n, e^{\vec{Y}_n}, X_n)$.

Lemma 4.3.13. *Let the conditions $\mathbf{B}_1[\bar{\gamma}]$ and $\mathbf{C}_5[\bar{\beta}]$ hold and $0 \le \gamma_i \le \beta_i < \infty$, $i = 1, \dots, k$. Then, there exist the constants $0 \le M_{16}, M_{17,i} < \infty$, $i = 1, \dots, k$ such that the following inequalities take place for $\vec{z} = (\vec{y}, x) = ((y_1, \dots, y_k), x) \in \mathbb{Z}, 0 \le n \le N$:*

$$\begin{aligned} \mathsf{E}_{\vec{z},n} \max_{n \le r \le N} \left| g(r, e^{\vec{Y}_r}, X_r) \right| &\le M_{16} + \sum_{i:\gamma_i = 0} M_{17,i} \\ &\quad + \sum_{i:\gamma_i > 0} M_{17,i} \exp\left\{ \left(\sum_{j=1}^k A_{N-n,j}(\bar{\beta}_i)\, |y_j| \right) \frac{\gamma_i}{\beta_i} \right\} . \end{aligned} \tag{4.107}$$

Proof. Using the condition $\mathbf{B}_1[\bar{\gamma}]$ we get for $0 \le n \le N$,

$$\begin{aligned} \max_{n \le r \le N} \left| g(r, e^{\vec{Y}_r}, X_r) \right| &\le \max_{n \le r \le N} \left(L_1 + \sum_{i=1}^k L_1 L_{2,i} e^{\gamma_i |Y_{r,i}|} \right) \\ &\le L_1 + \sum_{i=1}^k L_1 L_{2,i} \max_{n \le r \le N} e^{\gamma_i |Y_{r,i}|} \\ &= L_1 + \sum_{i=1}^k L_1 L_{2,i} e^{\gamma_i \max_{n \le r \le N} |Y_{r,i}|} . \end{aligned} \tag{4.108}$$

Using relation (4.108) and well-known properties of moments for random variables and Lemma 4.3.11, we get the following inequality, for $\vec{z} \in \mathbb{Z}, 0 \leq n \leq N$:

$$
\begin{aligned}
\mathsf{E}_{\vec{z},n} \max_{n\leq r\leq N} \left| g(r, e^{\vec{Y}_r}, X_r) \right|
&\leq L_1 + \sum_{i=1}^{k} L_1 L_{2,i} \mathsf{E}_{\vec{z},n} e^{\gamma_i \max_{n\leq r\leq N} Y_{r,i}} \\
&\leq L_1 + \sum_{i:\gamma_i=0} L_1 L_{2,i} + \sum_{i:\gamma_i>0} L_1 L_{2,i} \left(\mathsf{E}_{\vec{z},n} \left(e^{\gamma_i \max_{n\leq r\leq N} Y_{r,i}} \right)^{\frac{\beta_i}{\gamma_i}} \right)^{\frac{\gamma_i}{\beta_i}} \\
&= L_1 + \sum_{i:\gamma_i=0} L_1 L_{2,i} + \sum_{i:\gamma_i>0} L_1 L_{2,i} \left(\mathsf{E}_{\vec{z},n} e^{\beta_i \max_{n\leq r\leq N} Y_{r,i}} \right)^{\frac{\gamma_i}{\beta_i}} \\
&\leq L_1 + \sum_{i:\gamma_i=0} L_1 L_{2,i} \\
&\quad + \sum_{i:\gamma_i>0} L_1 L_{2,i} \left(M_{15,i} \exp\left\{ \sum_{j=1}^{k} A_{N-n,j}(\bar{\beta}_i) |y_j| \right\} \right)^{\frac{\gamma_i}{\beta_i}} \\
&= L_1 + \sum_{i:\gamma_i=0} L_1 L_{2,i} \\
&\quad + \sum_{i:\gamma_i>0} L_1 L_{2,i} M_{15,i}^{\frac{\gamma_i}{\beta_i}} \exp\left\{ \left(\sum_{j=1}^{k} A_{N-n,j}(\bar{\beta}_i) |y_j| \right) \frac{\gamma_i}{\beta_i} \right\} .
\end{aligned} \tag{4.109}
$$

Thus, inequality (4.107) holds with the constants M_{16} and $M_{17,i}$, $i = 1, \dots, k$, which are given by relation (4.109). This completes the proof. □

Remark 4.3.13. The explicit formulas for the constants M_{16} and $M_{17,i}$, $i = 1, \dots, k$ follow from formulas (4.6) and (4.109),

$$
M_{16} = L_1 , \quad M_{17,i} = L_1 L_{2,i} I(\gamma_i = 0) + L_1 L_{2,i} K_{13,i}^{N\frac{\gamma_i}{\beta_i}} I(\gamma_i > 0) . \tag{4.110}
$$

Lemma 4.3.14. *Let the conditions* $\mathbf{B}_1[\bar{\gamma}]$ *and* $\mathbf{C}_5'[\bar{\beta}]$ *hold and* $0 \leq \gamma_i \leq \beta_i < \infty$, $i = 1, \dots, k$. *Then, there exist the constants* $0 \leq M'_{16}, M'_{17,i} < \infty$, $i = 1, \dots, k$ *such that the following inequalities take place for* $\vec{z} = (\vec{y}, x) = ((y_1, \dots, y_k), x) \in \mathbb{Z}_n, 0 \leq n \leq N$:

$$
\begin{aligned}
\mathsf{E}_{\vec{z},n} \max_{n\leq r\leq N} \left| g(r, e^{\vec{Y}_r}, X_r) \right| &\leq M'_{16} + \sum_{i:\gamma_i=0} M'_{17,i} \\
&\quad + \sum_{i:\gamma_i>0} M'_{17,i} \exp\left\{ \left(\sum_{j=1}^{k} A_{N-n,j}(\bar{\beta}_i) |y_j| \right) \frac{\gamma_i}{\beta_i} \right\} .
\end{aligned} \tag{4.111}
$$

Proof. It repeats the proof of Lemma 4.3.13. The only difference is that, in the case, where the condition $\mathbf{C}_5'[\bar{\beta}]$ is assumed to hold instead of the condition $\mathbf{C}_5[\bar{\beta}]$, inequality (4.109) takes place for $\vec{z} \in \mathbb{Z}_n, 0 \leq n \leq N$, with the same constants $L_1, L_{2,i}$, $i = 1, \dots, k$ and the constants $K'_{13,i}$, $i = 1, \dots, k$ replacing the constants $K_{13,i}$, $i = 1, \dots, k$. □

Remark 4.3.14. The explicit formulas for the constants M'_{16} and $M'_{17,i}$, $i = 1,\dots,k$ follow from formulas (4.110) and the remarks made in the above proof,

$$M'_{16} = L_1, \quad M'_{17,i} = L_1 L_{2,i} I(\gamma_i = 0) + L_1 L_{2,i} (K'_{13,i})^{N\frac{\gamma_i}{\beta_i}} I(\gamma_i > 0). \tag{4.112}$$

Lemma 4.3.12 implies that under conditions of this lemma, the log-reward functions $\phi_n(\vec{y}, x)$ are well defined by the following relation, for $\vec{z} = (\vec{y}, x) \in \mathbb{Z}$, $n = 0,\dots,N$:

$$\phi_n(\vec{y}, x) = \sup_{\tau_n \in \mathcal{M}_{\max,n,N}} \mathsf{E}_{\vec{z},n} g(\tau_n, e^{\vec{Y}_{\tau_n}}, X_{\tau_n}). \tag{4.113}$$

Moreover, for Lemma 4.3.13 let us obtain the following upper bounds for the log-reward functions of American-type options for the log-price process $\vec{Z}_n$.

Theorem 4.3.5. *Let the conditions* $\mathbf{B}_1[\bar{\gamma}]$ *and* $\mathbf{C}_5[\bar{\beta}]$ *hold and* $0 \le \gamma_i \le \beta_i < \infty$, $i = 1,\dots,k$. *Then, the log-reward functions* $\phi_n(\vec{y}, x)$ *satisfy the following inequalities for* $\vec{z} = (\vec{y}, x) = ((y_1,\dots,y_k), x) \in \mathbb{Z}, 0 \le n \le N$:

$$\begin{aligned} |\phi_n(\vec{y}, x)| &\le M_{16} + \sum_{i:\gamma_i=0} M_{17,i} \\ &\quad + \sum_{i:\gamma_i>0} M_{17,i} \exp\left\{ \left(\sum_{j=1}^{k} A_{N-n,j}(\bar{\beta}_i) |y_j| \right) \frac{\gamma_i}{\beta_i} \right\}. \end{aligned} \tag{4.114}$$

Proof. Using Lemma 4.3.13 and the definition of the log-reward functions, we get the following inequalities for $\vec{z} = (\vec{y}, x) \in \mathbb{Z}, 0 \le n \le N$:

$$\begin{aligned} |\phi_n(\vec{y}, x)| &= \left| \sup_{\tau_n \in \mathcal{M}_{\max,n,N}} \mathsf{E}_{\vec{z},n} g\left(\tau_n, e^{\vec{Y}_{\tau_n}}, X_{\tau_n}\right) \right| \\ &\le \mathsf{E}_{\vec{z},n} \max_{n \le r \le N} \left| g(r, e^{\vec{Y}_r}, X_r) \right| \\ &\le M_{16} + \sum_{i:\gamma_i=0} M_{17,i} + \sum_{i:\gamma_i>0} M_{17,i} \exp\left\{ \left(\sum_{j=1}^{k} A_{N-n,j}(\bar{\beta}_i) |y_j| \right) \frac{\gamma_i}{\beta_i} \right\}, \end{aligned} \tag{4.115}$$

which proves the theorem. □

Lemma 4.3.14 implies that, under conditions of this lemma, the log-reward functions $\phi_n(\vec{y}, x)$ are well defined by relation (4.113), only for $\vec{z} = (\vec{y}, x) \in \mathbb{Z}_n$, $n = 0,\dots,N$.

Also, for Lemma 4.3.14 let us get upper bounds for the log-reward functions of American-type options for the modulated log-price $\vec{Z}_n$.

Theorem 4.3.6. *Let the conditions* $\mathbf{B}_1[\bar{\gamma}]$ *and* $\mathbf{C}'_5[\bar{\beta}]$ *hold and* $0 \le \gamma_i \le \beta_i < \infty$, $i = 1,\dots,k$. *Then, the log-reward functions* $\phi_n(\vec{y}, x)$ *satisfy the following inequalities for* $\vec{z} = (\vec{y}, x) = ((y_1,\dots,y_k), x) \in \mathbb{Z}_n, 0 \le n \le N$:

$$\begin{aligned} |\phi_n(\vec{y}, x)| &\le M'_{16} + \sum_{i:\gamma_i=0} M'_{17,i} \\ &\quad + \sum_{i:\gamma_i>0} M'_{17,i} \exp\left\{ \left(\sum_{j=1}^{k} A_{N-n,j}(\bar{\beta}_i) |y_j| \right) \frac{\gamma_i}{\beta_i} \right\}. \end{aligned} \tag{4.116}$$

Proof. It repeats the proof of Theorem 4.3.5. The only difference is that Lemma 4.3.14 should be used instead of Lemma 4.3.13. This let one write down the relation analogous to (4.115), for $\vec{z} = (\vec{y}, x) \in \mathbb{Z}_n, 0 \leq n \leq N$ and with the constants $M'_{16}, M'_{17,i}, i = 1, \dots, k$ replacing the constants $M'_{16}, M'_{17,i}, i = 1, \dots, k$. □

We can now get upper bounds for optimal expected rewards of American-type options for the modulated log-price process $\vec{Z}_n$.

To do this, we should impose the following first-type condition of exponential moment boundedness on the initial distribution of the log-price process $\vec{Z}_n$, for some vector $\bar{\beta} = (\beta_1, \dots, \beta_k)$ with nonnegative components:

$\mathbf{D}_3[\bar{\beta}]$: $\mathsf{E}\exp\{\sum_{j=1}^{k} A_{N,j}(\bar{\beta}_i)|Y_{0,j}|\} < K_{14,i}$, $i = 1, \dots, k$, for some $1 < K_{14,i} < \infty$, $i = 1, \dots, k$.

The following lemma gives sufficient conditions for the condition $\mathbf{A}_1$ to hold and explicit upper bounds for the expectation of the maximum for the pay-off process $g(n, e^{\vec{Y}_n}, X_n)$.

Lemma 4.3.15. *Let the conditions $\mathbf{B}_1[\bar{\gamma}]$, $\mathbf{C}_5[\bar{\beta}]$, and $\mathbf{D}_3[\bar{\beta}]$ hold and $0 \leq \gamma_i \leq \beta_i < \infty$, $i = 1, \dots, k$. Then, there exists a constant $0 \leq M_{18} < \infty$ such that the following inequality takes place:*

$$\mathsf{E} \max_{0 \leq n \leq N} \left| g(n, e^{\vec{Y}_n}, X_n) \right| \leq M_{18} \,. \tag{4.117}$$

Proof. Using Lemma 4.3.13, we get

$$\begin{aligned}
&\mathsf{E} \max_{0 \leq n \leq N} \left| g(n, e^{\vec{Y}_n}, X_n) \right| \\
&\quad = \mathsf{E}\,\mathsf{E}\left\{ \max_{0 \leq n \leq N} \left| g\left(n, e^{\vec{Y}_n}, X_n\right) \right| / \vec{Z}_0 \right\} \\
&\quad \leq M_{16} + \sum_{i:\gamma_i = 0} M_{17,i} + \sum_{i:\gamma_i > 0} M_{17,i}\mathsf{E}\exp\left\{ \left(\sum_{j=1}^{k} A_{N-n,j}\left(\bar{\beta}_i\right)|Y_{0,j}| \right) \frac{\gamma_i}{\beta_i} \right\} \\
&\quad \leq M_{16} + \sum_{i:\gamma_i = 0} M_{17,i} + \sum_{i:\gamma_i > 0} M_{17,i}\left(\mathsf{E}\exp\left\{ \sum_{j=1}^{k} A_{N-n,j}(\bar{\beta}_i)|Y_{0,j}| \right\} \right)^{\frac{\gamma_i}{\beta_i}} \\
&\quad \leq M_{16} + \sum_{i:\gamma_i = 0} M_{17,i} + \sum_{i:\gamma_i > 0} M_{17,i}K_{14,i}^{\frac{\gamma_i}{\beta_i}} \,.
\end{aligned} \tag{4.118}$$

Relation (4.118) proves that inequality (4.117) holds with the constant M_{18} given in (4.118). □

Remark 4.3.15. The explicit formula for the constant M_{18} follows from the last inequality in (4.118),

$$M_{18} = L_1 + \sum_{i:\gamma_i = 0} L_1 L_{2,i} + \sum_{i:\gamma_i > 0} L_1 L_{2,i} K_{13,i}^{N\frac{\gamma_i}{\beta_i}} K_{14,i}^{\frac{\gamma_i}{\beta_i}} \,. \tag{4.119}$$

In the case, where the conditions $\mathbf{C}_5'[\bar{\beta}]$ and $\mathbf{D}_3'[\bar{\beta}]$ are used instead of the conditions $\mathbf{C}_5[\bar{\beta}]$ and $\mathbf{D}_3[\bar{\beta}]$, the following condition should be assumed to hold for some vector $\bar{\beta} = (\beta_1, \dots, \beta_k)$ with nonnegative components:

$\mathbf{D}_3'[\bar{\beta}]$: (a) $\mathsf{E}\exp\{\sum_{j=1}^{k} A_{N,j}(\bar{\beta}_i)|Y_{0,j}|\} < K'_{14,i}$, $i = 1, \dots, k$, for some $1 < K'_{14,i} < \infty$, $i = 1, \dots, k$;

(b) $\mathsf{P}\{\vec{Z}_0 \in \mathbb{Z}_0\} = 1$, where $\mathbb{Z}_0$ is the set penetrating condition $\mathbf{C}_5'[\beta]$.

Lemma 4.3.16. *Let the conditions $\mathbf{B}_1[\bar{\gamma}]$, $\mathbf{C}_5'[\bar{\beta}]$, and $\mathbf{D}_3'[\bar{\beta}]$ hold and $0 \leq \gamma_i \leq \beta_i < \infty$, $i = 1, \dots, k$. Then, there exists a constant $0 \leq M'_{18} < \infty$ such that the following inequality takes place:*

$$\mathsf{E} \max_{0 \leq n \leq N} \left| g(n, e^{\vec{Y}_n}, X_n) \right| \leq M'_{18} . \tag{4.120}$$

Proof. It repeats the proof of Lemma 4.3.15. The only difference is that Lemma 4.3.14 should be used instead of Lemma 4.3.13. This let one write down the relation analogous to (4.118), for $\vec{z} = (\vec{y}, x) \in \mathbb{Z}_n, 0 \leq n \leq N$ and with the constants $M'_{16}, M'_{17,i}$, $i = 1, \dots, k$ replacing the constants $M'_{16}, M'_{17,i}$, $i = 1, \dots, k$. □

Remark 4.3.16. The explicit formula for the constant M'_{18} follows from inequality (4.119) given in Remark 4.3.11 and the remarks made in the above proof,

$$M'_{18} = L_1 + \sum_{i:\gamma_i = 0} L_1 L_{2,i} + \sum_{i:\gamma_i > 0} L_1 L_{2,i} (K'_{13,i})^{N\frac{\gamma_i}{\beta_i}} (K'_{14,i})^{\frac{\gamma_i}{\beta_i}} . \tag{4.121}$$

Let us assume that either conditions of Lemma 4.3.15 or Lemma 4.3.16 hold. In this case, the optimal expected reward $\Phi = \Phi(\mathcal{M}_{\max,N})$ for the log-price $\vec{Z}_n$ can be defined by the following relation:

$$\Phi = \sup_{\tau_0 \in \mathcal{M}_{\max,N}} \mathsf{E} g(\tau_0, e^{\vec{Y}_{\tau_0}}, X_{\tau_0}) . \tag{4.122}$$

The following two theorems give upper bounds for the optimal expected reward $\Phi = \Phi(\mathcal{M}_{\max,N})$.

Theorem 4.3.7. *Let the conditions $\mathbf{B}_1[\bar{\gamma}]$, $\mathbf{C}_5[\bar{\beta}]$, and $\mathbf{D}_3[\bar{\beta}]$ hold and $0 \leq \gamma_i \leq \beta_i < \infty$, $i = 1, \dots, k$. Then, the following inequality takes place:*

$$|\Phi| \leq M_{18} . \tag{4.123}$$

Proof. As follows from the definition of the functional Φ and Lemma 4.3.15,

$$|\Phi| \leq \sup_{\tau_0 \in \mathcal{M}_{\max,N}} \mathsf{E} \left| g(\tau_0, e^{\vec{Y}_{\tau_0}}, X_{\tau_0}) \right| \leq \mathsf{E} \max_{0 \leq n \leq N} \left| g(n, e^{\vec{Y}_n}, X_n) \right| \leq M_{18} . \tag{4.124}$$

This inequality proves the theorem. □

Theorem 4.3.8. *Let the conditions $\mathbf{B}_1[\bar{\gamma}]$, $\mathbf{C}_5'[\bar{\beta}]$, and $\mathbf{D}_3'[\bar{\beta}]$ hold and $0 \leq \gamma_i \leq \beta_i < \infty$, $i = 1, \dots, k$. Then, the following inequality takes place:*

$$|\Phi| \leq M'_{18} . \tag{4.125}$$

Proof. It repeats the proof of Theorem 4.1.7. The only difference is that Lemma 4.3.16 is used instead of Lemma 4.3.15. The relation analogous to (4.124) can be written down, with the constant M'_{18} replacing the constant M_{18}. □

4.3.6 Modified conditions of exponential moment compactness for multivariate Markov log-price processes with unbounded characteristics

Let us define a second-type $\vec{A}$-modulus of exponential moment compactness the log-price process $\vec{Y}_n$, for vectors $\bar{\beta} = (\beta_1, \dots, \beta_k)$ with nonnegative components and vectors $\bar{\imath}_k = (\imath_1, \dots, \imath_k) \in \mathcal{I}_k$, where $\mathcal{I}_k = \{\bar{\imath}_k : \imath_j = +, -, j = 1, \dots, k\}$.

$$\Xi^{\bar{\imath}_k}_{\bar{\beta},\vec{A}}(\vec{Y}_\cdot, N) = \max_{0 \le n \le N-1} \sup_{\vec{z}=((y_1,\dots,y_k),x)\in\mathbb{Z}} \frac{\mathsf{E}_{\vec{z},n} e^{\sum_{j=1}^k \imath_j A_{N-n-1,j}(\bar{\beta})(Y_{n+1,j}-Y_{n,j})}}{e^{\sum_{j=1}^k A_j(\bar{A}_{N-n-1}(\bar{\beta}))|y_j|}} . \tag{4.126}$$

The following second-type condition of exponential moment boundedness based on moment-generating functions of increments for the multivariate log-price process $\vec{Y}_n$ is assumed to hold for some vector parameter $\bar{\beta} = (\beta_1, \dots, \beta_k)$ with nonnegative components:

$\mathbf{E}_5[\bar{\beta}]$: $\Xi^{\bar{\imath}_k}_{\bar{\beta}_i,\vec{A}}(\vec{Y}_\cdot, N) < K_{15,i}$, $\bar{\imath}_k = (\imath_1, \dots, \imath_k) \in \mathcal{I}_k$, $i = 1, \dots, k$, for some $1 < K_{15,i} < \infty$, $i = 1, \dots, k$.

Lemma 4.3.17. *The condition $\mathbf{E}_5[\bar{\beta}]$ implies condition $\mathbf{C}_5[\bar{\beta}]$ to hold.*

Proof. Let us introduce random indicator variables $I(\imath_j)$, $j = 1, \dots, k$, $n = 0, \dots, N-1$ in the following way:

$$I(\imath_j) = \begin{cases} I(Y_{n+1,j} - Y_{n,j} \ge 0) & \text{if } \imath_j = +, \\ I(Y_{n+1,j} - Y_{n,j} < 0) & \text{if } \imath_j = -. \end{cases} \tag{4.127}$$

The following inequality takes place, for every $0 \le n \le N-1$, $\vec{z} \in \mathbb{Z}$, $i = 1, \dots, k$:

$$\begin{aligned} &\mathsf{E}_{\vec{z},n} e^{\sum_{j=1}^k A_{N-n-1,j}(\bar{\beta}_i)|Y_{n+1,j}-Y_{n,j}|} \\ &\quad = \sum_{\bar{\imath}_k \in \mathcal{I}_k} \mathsf{E}_{\vec{z},n} e^{\sum_{j=1}^k \imath_j A_{N-n-1,j}(\bar{\beta}_i)(Y_{n+1,j}-Y_{n,j})} \prod_{j=1}^k I(\imath_j) \\ &\quad \le \sum_{\bar{\imath}_k \in \mathcal{I}_k} \mathsf{E}_{\vec{z},n} e^{\sum_{j=1}^k \imath_j A_{N-n-1,j}(\bar{\beta}_i)(Y_{n+1,j}-Y_{n,j})} . \end{aligned} \tag{4.128}$$

Relation (4.128) obviously implies the following inequalities, for $i = 1, \dots, k$:

$$\Delta_{\bar{\beta}_i,\vec{A}}(\vec{Y}_\cdot, N) \le \sum_{\bar{\imath}_k \in \mathcal{I}_k} \Xi^{\bar{\imath}_k}_{\bar{\beta}_i,\vec{A}}(\vec{Y}_\cdot, N) . \tag{4.129}$$

Thus, the condition $\mathbf{E}_5[\bar{\beta}]$ implies that condition $\mathbf{C}_5[\bar{\beta}]$ holds. □

Remark 4.3.17. As follows from inequality (4.129), if condition $\mathbf{E}_5[\bar{\beta}]$ holds with the constants $K_{15,i}$, $i = 1, \dots, k$, then the condition $\mathbf{C}_5[\bar{\beta}]$ holds with the constants $K_{13,i}$, $i = 1, \dots, k$ given by the following formulas:

$$K_{13,i} = 2^k K_{15,i}\,, \quad i = 1, \dots, k\,. \tag{4.130}$$

Let $\mathcal{Z} = \langle \mathbb{Z}_0, \dots, \mathbb{Z}_N \rangle$ be a sequence of measurable sets. Let us also introduce a second-type modified $\vec{A}$-modulus of exponential moment compactness, defined for vector $\bar{\beta} = (\beta_1, \dots, \beta_k)$ with nonnegative components and vectors $\bar{\imath}_k = (\imath_1, \dots, \imath_k) \in \mathcal{I}_k$, where $\mathcal{I}_k = \{\bar{\imath}_k : \imath_j = +, -, j = 1, \dots, k\}$.

$$\Xi^{\bar{\imath}_k}_{\bar{\beta},\vec{A},\mathcal{Z}}(\vec{Y}_\cdot, N) = \max_{0 \le n \le N-1} \sup_{\vec{z}=((y_1,\dots,y_k),x) \in \mathbb{Z}_n} \frac{\mathsf{E}_{\vec{z},n} e^{\sum_{j=1}^k \imath_j A_{N-n-1,j}(\bar{\beta})(Y_{n+1,j} - Y_{n,j})}}{e^{\sum_{j=1}^k A_j(\vec{A}_{N-n-1}(\bar{\beta}))|y_j|}}\,. \tag{4.131}$$

The following modified second-type condition of exponential moment compactness is based on moment-generating functions of increments of log-price processes Y_n. This condition generalizes the condition $\mathbf{E}_5[\bar{\beta}]$. It is assumed to hold for some vector parameter $\bar{\beta} = (\beta_1, \dots, \beta_k)$ with nonnegative components:

$\mathbf{E}'_5[\bar{\beta}]$: There exists a sequence of sets $\mathcal{Z} = \langle \mathbb{Z}_0, \dots, \mathbb{Z}_N \rangle$ such that

(a) $\Xi^{\bar{\imath}_k}_{\bar{\beta}_i,\vec{A},\mathcal{Z}}(Y_\cdot, N) < K'_{15,i}$, $\bar{\imath}_k = (\imath_1, \dots, \imath_k) \in \mathcal{I}_k$, $i = 1, \dots, k$, for some $1 < K'_{15,i} < \infty$, $i = 1, \dots, k$;

(b) $\mathcal{Z}$ is a complete sequence of phase sets, i.e., $\mathsf{P}_n(z, \mathbb{Z}_n) = 1, z \in \mathbb{Z}_{n-1}$, $n = 1, \dots, N$.

Lemma 4.3.18. *The condition $\mathbf{E}'_5[\bar{\beta}]$ implies condition $\mathbf{C}'_5[\bar{\beta}]$ to hold.*

Proof. Relation (4.128) implies analogously to (4.129), the following inequality, for $i = 1, \dots, k$:

$$\Delta_{\bar{\beta}_i,\vec{A},\mathcal{Z}}(\vec{Y}_\cdot, N) \le \sum_{\bar{\imath}_k \in \mathcal{I}_k} \Xi^{\bar{\imath}_k}_{\bar{\beta}_i,\vec{A},\mathcal{Z}}(\vec{Y}_\cdot, N)\,. \tag{4.132}$$

Thus, the condition $\mathbf{E}'_5[\bar{\beta}]$ implies that condition $\mathbf{C}'_5[\bar{\beta}]$ holds. □

Remark 4.3.18. As follows from inequality (4.129), if condition $\mathbf{E}'_5[\bar{\beta}]$ holds with the constants $K'_{15,i}$, $i = 1, \dots, k$, then the condition $\mathbf{C}'_5[\bar{\beta}]$ holds with the constants $K'_{13,i}$, $i = 1, \dots, k$ given by the following formulas:

$$K'_{13,i} = 2^k K'_{15,i}\,, \quad i = 1, \dots, k\,. \tag{4.133}$$

Similarly, one can replace condition $\mathbf{D}_3[\bar{\beta}]$ by the following second-type condition of exponential moment boundedness based on moment-generating functions for the initial value of log-price process. This condition should be assumed to hold for some vector parameter $\bar{\beta} = (\beta_1, \dots, \beta_k)$ with nonnegative components:

$\mathbf{F}_3[\bar{\beta}]$: $\mathsf{E}\exp\{\sum_{j=1}^{k} \imath_j A_{N,j}(\bar{\beta}_i) Y_{0,j}\} < K_{16,i}$, $\bar{\imath}_k = (\imath_1, \dots, \imath_k) \in \mathcal{I}_k$, $i = 1, \dots, k$ for some $1 < K_{16,i} < \infty$, $i = 1, \dots, k$.

Lemma 4.3.19. *The condition $\mathbf{F}_3[\bar{\beta}]$ implies condition $\mathbf{D}_3[\bar{\beta}]$ to hold.*

Proof. It follows from the following inequality:

$$\mathsf{E}e^{\sum_{j=1}^{k} A_{N,j}(\bar{\beta}_i)|Y_{0,j}|} \leq \sum_{\bar{\imath}_k \in \mathcal{I}_k} \mathsf{E}e^{\sum_{j=1}^{k} \imath_j A_{N,j}(\bar{\beta}_i) Y_{0,j}} . \tag{4.134}$$

□

Remark 4.3.19. If the condition $\mathbf{F}_3[\bar{\beta}]$ holds with constants $K_{16,i}$, $i = 1, \dots, k$, then condition $\mathbf{D}_3[\bar{\beta}]$ holds with the constants $K_{14,i}$, $i = 1, \dots, k$ given by the following formulas:

$$K_{14,i} = 2^k K_{16,i} , \quad i = 1, \dots, k . \tag{4.135}$$

Analogously, the condition $\mathbf{D}_3'[\bar{\beta}]$ can be replaced by the following condition assumed to hold for some some vector parameter $\bar{\beta} = (\beta_1, \dots, \beta_k)$ with nonnegative components:

$\mathbf{F}_3'[\bar{\beta}]$: (a) $\mathsf{E}\exp\{\sum_{j=1}^{k} \imath_j A_{N,j}(\bar{\beta}_i) Y_{0,j}\} < K'_{16,i}$, $\bar{\imath}_k = (\imath_1, \dots, \imath_k) \in \mathcal{I}_k$, $i = 1, \dots, k$ for some $1 < K'_{16,i} < \infty$, $i = 1, \dots, k$;

(b) $\mathsf{P}\{\vec{Z}_0 \in \mathbb{Z}_0\} = 1$, where $\mathbb{Z}_0$ is the set penetrating condition $\mathbf{D}_3'[\bar{\beta}]$.

Lemma 4.3.20. *The condition $\mathbf{F}_3'[\bar{\beta}]$ implies condition $\mathbf{D}_3'[\bar{\beta}]$ to hold.*

Remark 4.3.20. If the condition $\mathbf{F}_3'[\bar{\beta}]$ holds with constants $K'_{16,i}$, $i = 1, \dots, k$, then, respectively, condition $\mathbf{D}_3'[\bar{\beta}]$ holds with the constants $K'_{14,i}$, $i = 1, \dots, k$ given by the following formulas:

$$K'_{14,i} = 2^k K'_{16,i} , \quad i = 1, \dots, k . \tag{4.136}$$

4.4 Univariate Markov Gaussian LPP

In this section, we consider Markov Gaussian log-price processes. This model is one of the natural discrete time analogs for diffusion log-price processes. We get upper bounds for option rewards for two classes of such log-price processes. The first one includes processes with bounded drift and volatility coefficients. The second one includes models with drift and volatility coefficients with not more than linear rate of growth as a function of space arguments. The latter class includes, for example, mean-reverse modulated Markov Gaussian log-price processes.

4.4.1 Upper bounds for option rewards of univariate Markov Gaussian log-price processes with bounded drift and volatility coefficients

Let us consider the simplest univariate Markov Gaussian log-price process without modulation. In this case, the log-price process Y_n, $n = 0, 1, \ldots$, is given by the following stochastic transition dynamic relation:

$$Y_n = Y_{n-1} + \mu_n(Y_{n-1}) + \sigma_n(Y_{n-1}) W_n\,, \quad n = 1, 2, \ldots\,, \tag{4.137}$$

where (a) Y_0 is a real-valued random variable, (b) W_n, $n = 1, 2, \ldots$, is a sequence of standard i.i.d. normal random variables with mean 0 and variance 1, (c) random variable Y_0 and sequence of random variables W_n, $n = 1, 2, \ldots$, are independent, (d) $\mu_n(y)$ are real-valued Borel functions defined on $\mathbb{R}_1$, for $n = 1, 2, \ldots$, and (e) $\sigma_n(y)$ are real-valued Borel functions defined on $\mathbb{R}_1$, for $n = 1, 2, \ldots$.

This is a model without an index component, and, therefore, a pay-off function $g(n, e^y)$ is also assumed to be a function of the argument y. In the sequel, the log-reward functions $\phi_n(y)$ also depend only on argument y.

In this subsection, we give upper bounds for option rewards for the model, where the following condition holds:

G₁: $\max_{0 \le n \le N-1} \sup_{y \in \mathbb{R}_1} (|\mu_{n+1}(y)| + \sigma^2_{n+1}(y)) < K_{17}$, for some $0 < K_{17} < \infty$.

In this case, there exist the moment-generating functions $\mathsf{E}_{y,n} e^{\beta(Y_{n+1} - Y_n)}$, $y \in \mathbb{R}_1$, $n = 0, 1, \ldots, N-1$, defined for $\beta \in \mathbb{R}_1$,

$$\mathsf{E}_{y,n} e^{\beta(Y_{n+1} - Y_n)} = \mathsf{E} e^{\beta(\mu_{n+1}(y) + \sigma_{n+1}(y)\, W_{n+1})} = e^{\beta\mu_{n+1}(y) + \frac{1}{2}\beta^2\sigma^2_{n+1}(y)}\,. \tag{4.138}$$

The second-type modulus of exponential moment compactness $\Xi^{\pm}_{\beta}(Y_\cdot, N)$ takes, according to formula (4.138), the following form, for $\beta \ge 0$:

$$\Xi^{\pm}_{\beta}(Y_\cdot, N) = \max_{0 \le n \le N-1} \sup_{y \in \mathbb{R}_1} e^{\pm\beta\mu_{n+1}(y) + \frac{1}{2}\beta^2\sigma^2_{n+1}(y)}\,. \tag{4.139}$$

We are interested in the following one-dimensional analog of the condition $\mathbf{E}_1[\bar{\beta}]$ assumed to hold for the log-price process Y_n, for some $\beta \ge 0$:

$\mathbf{E}_6[\beta]$: $\Xi^{\pm}_{\beta}(Y_\cdot, N) < K_{18}$, for some $1 < K_{18} < \infty$.

Lemma 4.4.1. *The condition* $\mathbf{G}_1$ *implies that condition* $\mathbf{E}_6[\beta]$ *holds for any* $\beta \ge 0$.

Proof. The condition $\mathbf{G}_1$ and relations (4.138) and (4.139) imply that for any $\beta \ge 0$,

$$\begin{aligned}\Xi^{\pm}_{\beta}(Y_\cdot, N) &= \max_{0 \le n \le N-1} \sup_{y \in \mathbb{R}_1} e^{\pm\beta\mu_{n+1}(y) + \frac{1}{2}\beta^2\sigma^2_{n+1}(y)} \\ &< K_{18} = \frac{1}{2} + e^{K_{17}(\beta + \frac{1}{2}\beta^2)} \in (1, \infty)\,.\end{aligned} \tag{4.140}$$

Thus, the condition $\mathbf{G}_1$ implies that condition $\mathbf{E}_6[\beta]$ holds for any $\beta \ge 0$. □

Remark 4.4.1. The constant K_{18} is determined by the constant K_{17} via formula (4.140).

Lemma 4.4.1 makes it possible to get upper bounds for the log-reward functions and the optimal expected rewards, for the above model of log-price process using Theorems 4.1.1 and 4.1.2.

The one-dimensional analog of the condition $\mathbf{B}_1[\bar{\gamma}]$ takes the following form, for $\gamma \geq 0$:
$\mathbf{B}_4[\gamma]$: $\max_{0\leq n\leq N} \sup_{y\in\mathbb{R}_1} \frac{|g(n,e^y)|}{1+L_8 e^{\gamma|y|}} < L_7$, for some $0 < L_7 < \infty$ and $0 \leq L_8 < \infty$.

The following theorem is a corollary of Theorem 4.1.1.

Theorem 4.4.1. *Let the condition $\mathbf{B}_4[\gamma]$ holds for some $\gamma \geq 0$ and also the condition $\mathbf{G}_1$ holds. Then, there exist the constants $0 \leq M_{19}, M_{20} < \infty$ such that the log-reward functions $\phi_n(y)$ satisfy the following inequalities for $y \in \mathbb{R}_1, 0 \leq n \leq N$,*

$$|\phi_n(y)| \leq M_{19} + M_{20} e^{\gamma|y|}. \tag{4.141}$$

Proof. The case $\gamma = 0$ is trivial. In this case, inequality (4.141) obviously hold with the constants $M_{19} = L_7$, $M_{20} = L_7 L_8$. Let $\gamma > 0$. According to Lemma 4.4.1, the condition $\mathbf{E}_6[\beta]$ (which is a one-dimensional version of the condition $\mathbf{E}_1[\bar{\beta}]$) holds for any $\beta \geq \gamma$. Thus, by Lemma 4.1.8, the corresponding one-dimensional versions of conditions of Theorem 4.1.1 hold if the condition $\mathbf{B}_4[\gamma]$ holds. In this case, the inequality given in Theorem 4.1.1 takes the forms of inequality (4.141). □

Remark 4.4.2. The constants M_{19} and M_{20} take, according to Remarks 4.1.3 and 4.1.7, the following form:

$$\begin{aligned} M_{19} &= L_7, \\ M_{20} &= L_7 L_8 I(\gamma = 0) + L_7 L_8 (2K_{18})^{N\frac{\gamma}{\beta}} I(\gamma > 0) \\ &= L_7 L_8 I(\gamma = 0) + L_7 L_8 (1 + 2e^{K_{17}(\beta + \frac{1}{2}\beta^2)})^{N\frac{\gamma}{\beta}} I(\gamma > 0). \end{aligned} \tag{4.142}$$

One can take any $\beta \geq \gamma$, in formulas (4.142).

The condition $\mathbf{D}_1[\bar{\beta}]$ takes, in this case, the following forms, assumed to hold for some $\beta > 0$:
$\mathbf{D}_4[\beta]$: $\mathsf{E} e^{\beta|Y_0|} < K_{19}$, for some $1 < K_{19} < \infty$.

The following theorem, which is a corollary of Theorem 4.1.3, gives an upper bound for the optimal expected reward $\Phi = \Phi(\mathcal{M}_{\max,N})$.

Theorem 4.4.2. *Let the conditions $\mathbf{B}_4[\gamma]$, $\mathbf{G}_1$, and $\mathbf{D}_4[\beta]$ hold and $0 \leq \gamma \leq \beta < \infty$. Then, there exists a constant $0 \leq M_{21} < \infty$ such that the following inequality takes place:*

$$|\Phi| \leq M_{21}. \tag{4.143}$$

Remark 4.4.3. The explicit formula for the constant M_{21} take according to formulas given in Remarks 4.1.5, 4.1.7, and 4.1.9, the following form:

$$\begin{aligned} M_{21} &= L_7 + L_7 L_8 I(\gamma = 0) \\ &\quad + L_7 L_8 (1 + 2e^{K_{17}(\beta + \frac{1}{2}\beta^2)}) K_{19}^{\frac{\gamma}{\beta}} I(\gamma > 0) . \end{aligned} \tag{4.144}$$

4.4.2 Upper bounds for option rewards of univariate Markov Gaussian log-price processes with unbounded drift and volatility coefficients

In this subsection, we give upper bounds for option rewards in the case, where the following condition holds:

G$_2$: $\max_{0 \le n \le N-1} \sup_{y \in \mathbb{R}_1} \frac{|\mu_{n+1}(y)| + \sigma^2_{n+1}(y)}{1 + K_{21}|y|} < K_{20}$, for some $0 < K_{20} < \infty$ and $0 \le K_{21} < \infty$.

Formula (4.138) gives a hint to use function $A(\beta)$ given by the following formula:

$$A(\beta) = K_{20} K_{21} \left(\beta + \frac{1}{2}\beta^2\right) , \quad \beta \ge 0 , \tag{4.145}$$

In this case, the second-type modulus of exponential moment compactness $\Xi^{\pm}_{\beta,A}(Y_\cdot, N)$ takes the following form, for $\beta \ge 0$:

$$\begin{aligned} &\Xi^{\pm}_{\beta,A}(Y_\cdot, N) \\ &\quad = \max_{0 \le n \le N-1} \sup_{y \in \mathbb{R}_1} \frac{e^{\pm A_{N-n-1}(\beta)\mu_{n+1}(y) + \frac{1}{2}A^2_{N-n-1}(\beta)\sigma^2_{n+1}(y)}}{e^{A(A_{N-n-1}(\beta))|y|}} , \end{aligned} \tag{4.146}$$

where functions $A_n(\beta)$, $n = 0, 1, \dots$, are generated by the above function $A(\beta) = K_{20} K_{21}(\beta + \frac{1}{2}\beta^2)$ according to the following recurrence formulas, for every $\beta \ge 0$:

$$A_n(\beta) = \begin{cases} \beta & \text{for } n = 0 , \\ A_{n-1}(\beta) + A(A_{n-1}(\beta)) & \text{for } n = 1, 2, \dots . \end{cases} \tag{4.147}$$

We are interested in the following variant of the condition **E**$_4[\beta]$ for the log price process Y_n, assumed to hold for some $\beta \ge 0$:

E$_7[\beta]$: $\Xi^{\pm}_{\beta,A}(Y_\cdot, N) < K_{22}$, for some $1 < K_{22} < \infty$.

Lemma 4.4.2. *The condition* **G**$_2$ *implies that condition* **E**$_7[\beta]$ *holds for any* $\beta \ge 0$.

Proof. The condition $\mathbf{G_2}$ imply that, for any $y \in \mathbb{R}_1$, $n = 0, \dots, N-1$ and $\beta \geq 0$,

$$
\begin{aligned}
\pm\, & A_{N-n-1}(\beta)\mu_{n+1}(y) + \frac{1}{2}A^2_{N-n-1}(\beta)\sigma^2_{n+1}(y) \\
&\leq A_{N-n-1}(\beta)|\mu_{n+1}(y)| + \frac{1}{2}A^2_{N-n-1}(\beta)\sigma^2_{n+1}(y) \\
&< \left(A_{N-n-1}(\beta) + \frac{1}{2}A^2_{N-n-1}(\beta)\right)(K_{20} + K_{20}K_{21}|y|) \\
&= K_{20}\left(A_{N-n-1}(\beta) + \frac{1}{2}A^2_{N-n-1}(\beta)\right) \\
&\quad + K_{20}K_{21}\left(A_{N-n-1}(\beta) + \frac{1}{2}A^2_{N-n-1}(\beta)\right)|y| \\
&= K_{20}\left(A_{N-n-1}(\beta) + \frac{1}{2}A^2_{N-n-1}(\beta)\right) + A(A_{N-n-1}(\beta))|y|\,.
\end{aligned}
\tag{4.148}
$$

Relations (4.145)–(4.148) imply that, for $\beta \geq 0$,

$$
\begin{aligned}
\Xi^{\pm}_{\beta,A}&(Y_\cdot, N) \\
&= \max_{0\leq n\leq N-1}\sup_{y\in\mathbb{R}_1}\frac{e^{\pm A_{N-n-1}(\beta)\mu_{n+1}(y) + \frac{1}{2}A^2_{N-n-1}(\beta)\sigma^2_{n+1}(y)}}{e^{A(A_{N-n-1}(\beta))|y|}} \\
&\leq \max_{0\leq n\leq N-1} e^{K_{20}(A_{N-n-1}(\beta)+\frac{1}{2}A^2_{N-n-1}(\beta))} \\
&< \frac{1}{2} + e^{K_{20}(A_{N-1}(\beta)+\frac{1}{2}A^2_{N-1}(\beta))} = K_{22} \in (1, \infty)\,.
\end{aligned}
\tag{4.149}
$$

This inequality proves the lemma. □

Remark 4.4.4. The constant K_{22} is determined by the constant K_{20} via formula (4.149).

Lemma 4.4.2 makes it possible to get upper bounds for the log-reward functions $\phi_n(y)$.

Theorem 4.4.3. *Let the conditions $\mathbf{B_6}[\gamma]$ holds for some $\gamma \geq 0$ and also the condition $\mathbf{G_2}$ holds. Then, for every $\beta \geq \gamma$, there exist the constants $0 \leq M_{22}, M_{23} = M_{23}(\beta) < \infty$ such that the log-reward functions $\phi_n(y)$ satisfy the following inequalities for $y \in \mathbb{R}_1, 0 \leq n \leq N$:*

$$
|\phi_n(y)| \leq M_{22} + M_{23}I(\gamma = 0) + M_{23}\exp\{\frac{A_{N-n}(\beta)\gamma}{\beta}|y|\}I(\gamma > 0)\,. \tag{4.150}
$$

Proof. The case $\gamma = 0$ is trivial. Let $\gamma > 0$. According to Lemma 4.4.2, the condition $\mathbf{E_7}[\beta]$ holds for any $\beta \geq \gamma$. In this case inequality (4.76) given in Theorem 4.3.1 takes the form of inequality (4.150). □

Remark 4.4.5. The the constants M_{22} and $M_{23}(\beta)$ takes, according to Remarks 4.3.1, and 4.3.3, the following form:

$$
\begin{aligned}
M_{22} &= L_7, \\
M_{23}(\beta) &= L_7L_8I(\gamma = 0) + L_7L_8(2K_{22})^{N\frac{\gamma}{\beta}} \\
&= L_7L_8I(\gamma = 0) + L_7L_8\left(1 + 2e^{K_{20}\left(A_{N-1}(\beta)+\frac{1}{2}A^2_{N-1}(\beta)\right)}\right)^{N\frac{\gamma}{\beta}} I(\gamma > 0)\,.
\end{aligned}
\tag{4.151}
$$

One can take any $\beta \geq \gamma$, in formulas (4.151).

The condition $\mathbf{D}_2[\beta]$, where function $A(\beta) = K_{20}K_{21}(\beta + \frac{1}{2}\beta^2)$ should be used, takes the following form, for some $\beta \geq 0$:
$\mathbf{D}_5[\beta]$: $\mathsf{E}\exp\{A_N(\beta)|Y_0|\} < K_{23}$, for some $1 < K_{23} < \infty$.

The following theorem gives upper bound for the optimal expected reward $\Phi = \Phi(\mathcal{M}_{\max,N})$.

Theorem 4.4.4. *Let the conditions $\mathbf{B}_6[\gamma]$, $\mathbf{G}_2$, and $\mathbf{D}_5[\beta]$ hold and $0 \leq \gamma \leq \beta < \infty$. Then, there exists a constant $0 \leq M_{24} < \infty$ such that the following inequality takes place:*

$$|\Phi| \leq M_{24}. \tag{4.152}$$

Proof. The case $\gamma = 0$ is trivial. Let $\gamma > 0$. According to Lemma 4.4.2, the condition $\mathbf{E}_7[\beta]$ holds for any $\beta \geq \gamma$. In this case, inequality (4.85) given in Theorem 4.3.2 takes the form of inequality (4.152). □

Remark 4.4.6. The explicit formula for the constant M_{18} takes, according to formulas given in Remarks 4.31, 4.3.3, and 4.3.5, the following form:

$$\begin{aligned} M_{24} &= L_7 + L_7L_8I(\gamma = 0) \\ &\quad + L_7L_8\left(1 + 2e^{K_{20}(A_{N-1}(\beta) + \frac{1}{2}A^2_{N-1}(\beta))}\right)^{N\frac{\gamma}{\beta}} K_{23}^{\frac{\gamma}{\beta}} I(\gamma > 0). \end{aligned} \tag{4.153}$$

4.5 Multivariate modulated Markov Gaussian LPP

In this section, we consider multivariate modulated Markov Gaussian log-price processes. This model is one of natural discrete time analogs for multivariate modulated diffusion log-price processes. We get upper bounds for option rewards for two classes of such log-price processes. The first one includes log-price with bounded drift and volatility coefficients. The second one includes models with drift and volatility coefficients with not more than linear rate of growth as a function of space arguments. The latter class include various autoregressive and autoregressive stochastic volatility log-price processes with Gaussian noise terms.

4.5.1 Upper bounds for option rewards of multivariate modulated Markov Gaussian log-price processes with bounded characteristics

In this model, a log-price process is given by the following stochastic transition dynamic relation:

$$\begin{cases} \vec{Y}_n = \vec{Y}_{n-1} + \vec{\mu}_n(\vec{Y}_{n-1}, X_{n-1}) + \Sigma_n(\vec{Y}_{n-1}, X_{n-1})\vec{W}_n , \\ X_n = C_n(X_{n-1}, U_n) , \\ n = 1, 2, \dots , \end{cases} \tag{4.154}$$

where (a) $(\vec{Y}_0, X_0) = ((Y_{0,1}, \dots, Y_{0,k})), X_0)$ is a random vector taking values in the space $\mathbb{Z} = \mathbb{R}_k \times \mathbb{X}$, (b) $(\vec{W}_n, U_n) = ((W_{n,1}, \dots, W_{n,k}), U_n)$, $n = 1, 2, \dots$, is a sequence of independent random vectors taking values in the space $\mathbb{R}_k \times \mathbb{U}$, where $\mathbb{U}$ is a measurable space with a σ-algebra of measurable subsets $\mathcal{B}_{\mathbb{U}}$, (c) the random vector $(\vec{Y}_0, X_0)$ and the random sequence $(\vec{W}_n, U_n)$, $n = 1, 2, \dots$, are independent, (d) the random vector $\vec{W}_n = (W_{n,1}, \dots, W_{n,k})$ has a standard multivariate normal distribution with $\mathsf{E}W_{n,i} = 0, \mathsf{E}W_{n,i}W_{n,j} = I(i = j), i, j = 1, \dots, k$, for every $n = 1, 2, \dots$, (e) the random variables $U_1, U_2, \dots$, have a regular conditional distribution $G_{\vec{w}}(A) = \mathsf{P}\{U_n \in A / \vec{W}_n = \vec{w}\}$, $n = 1, 2, \dots$, (a probability measure in $B \in \mathcal{B}_{\mathbb{U}}$, for every $\vec{w} \in \mathbb{R}_k$, and a measurable function in $\vec{w} \in \mathbb{R}_k$, for every $B \in \mathcal{B}_{\mathbb{U}}$), for every $n = 1, 2, \dots$, (f) $\vec{\mu}_n(\vec{y}, x) = (\mu_{n,1}(\vec{y}, x), \dots, \mu_{n,k}(\vec{y}, x))$, $n = 1, 2, \dots$, are vector functions, which components are measurable functions acting from the space $\mathbb{Z} = \mathbb{R}_k \times \mathbb{X}$ to $\mathbb{R}_1$, (g) $\Sigma_n(\vec{y}, x) = \|\sigma_{n,i,j}(\vec{y}, x)\|$, $n = 1, 2, \dots$, are $k \times k$ matrix functions with elements $\sigma_{n,i,j}(\vec{y}, x), i, j = 1, \dots, k$, which are measurable functions acting from the space $\mathbb{Z} = \mathbb{R}_k \times \mathbb{X}$ to $\mathbb{R}_1$, and (h) $C_n(x, u)$ is a measurable function acting from the space $\mathbb{X} \times \mathbb{U}$ to the space $\mathbb{X}$.

The log-price process $\vec{Z}_n$ has an index component X_n, and, therefore, a pay-off function $g(n, e^{\vec{y}}, x)$ is also assumed to be a function of the argument $\vec{z} = (\vec{y}, x) \in \mathbb{Z}$. In sequel, the log-reward functions $\phi_n(\vec{y}, x)$ also are functions of $(\vec{y}, x) \in \mathbb{Z}$.

In this subsection, we give upper bounds for option rewards for the model, where the following condition holds:

G₃: $\max_{0 \leq n \leq N-1, i,j=1,\dots,k} \sup_{\vec{z}=(\vec{y},x) \in \mathbb{Z}} (|\mu_{n+1,i}(\vec{y}, x)| + \sigma^2_{n+1,i,j}(\vec{y}, x)) < K_{24}$, for some $0 < K_{24} < \infty$.

In this case, there exists, for every $\vec{z} = (\vec{y}, x) \in \mathbb{Z}$, $n = 0, 1, \dots, N-1$, $i = 1, \dots, k$, the moment generation function $\mathsf{E}_{\vec{z},n} e^{\beta(Y_{n+1,i} - Y_{n,i})}$, defined for $\beta \in \mathbb{R}_1$ by the following formula:

$$\begin{aligned} \mathsf{E}_{\vec{z},n} e^{\beta(Y_{n+1,i} - Y_{n,i})} &= \mathsf{E} e^{\beta(\mu_{n+1,i}(\vec{y},x) + \sum_{j=1}^k \sigma_{n+1,i,j}(\vec{y},x) W_{n+1,j})} \\ &= e^{\beta\mu_{n+1,i}(\vec{y},x) + \frac{1}{2}\beta^2 \sum_{j=1}^k \sigma^2_{n+1,i,j}(\vec{y},x)} . \end{aligned} \tag{4.155}$$

The second-type modulus of exponential moment compactness $\Xi^{\pm}_{\bar{\beta}}(Y_{\cdot,i}, N)$ takes, according to formula (4.155), the following form, for $\beta \geq 0$:

$$\Xi^{\pm}_{\bar{\beta}}(Y_{\cdot,i}, N) = \max_{0\leq n\leq N-1} \sup_{\vec{z}=(\vec{y},x)\in\mathbb{Z}} e^{\pm\beta\mu_{n+1,i}(\vec{y},x)+\frac{1}{2}\beta^2\sum_{j=1}^{k}\sigma^2_{n+1,i,j}(\vec{y},x)} . \tag{4.156}$$

We are interested in the following analog of the condition $\mathbf{E}_1[\bar{\beta}]$, assumed to hold for some vector parameter $\bar{\beta} = (\beta_1, \dots, \beta_k)$ with nonnegative components:
$\mathbf{E}_8[\bar{\beta}]$: $\Xi^{\pm}_{\beta_i}(Y_{\cdot,i}, N) < K_{25,i}$, $i = 1, \dots, k$, for some $1 < K_{25,i} < \infty$, $i = 1, \dots, k$.

The following lemma is a variant of Lemma 4.4.1.

Lemma 4.5.1. *The condition $\mathbf{G}_3$ implies that condition $\mathbf{E}_8[\bar{\beta}]$ holds for any vector parameter $\bar{\beta} = (\beta_1, \dots, \beta_k)$ with components $\beta_i \geq 0$, $i = 1, \dots, k$.*

Proof. It follows from the following relation, which holds for any $\beta \geq 0$:

$$\begin{aligned}\Xi^{\pm}_{\bar{\beta}}(Y_{i,\cdot}, N) &\leq \max_{0\leq n\leq N-1} \sup_{\vec{z}=(\vec{y},x)\in\mathbb{Z}} e^{\beta|\mu_{n+1,i}(\vec{y},x)|+\frac{1}{2}\beta^2\sum_{j=1}^{k}\sigma^2_{n+1,i,j}(\vec{y},x)} \\ &< \frac{1}{2} + e^{K_{24}(\beta+\frac{1}{2}k\beta^2)} = K_{25}(\beta) \in (1, \infty) .\end{aligned} \tag{4.157}$$

Relation (4.157) implies that condition $\mathbf{E}_8[\bar{\beta}]$ holds. □

Remark 4.5.1. The following formulas, analogous to formulas (4.140), determine the constants $K_{25,i} = K_{25}(\beta_i)$, $i = 1, \dots, k$, as functions of the constant K_{24}:

$$K_{25,i} = \frac{1}{2} + e^{K_{24}(\beta_i+\frac{1}{2}k\beta_i^2)}, \ i = 1, \dots, k . \tag{4.158}$$

The following theorems are corollaries, respectively, of Theorem 4.1.1 and 4.1.2.

Theorem 4.5.1. *Let the condition $\mathbf{B}_1[\bar{\gamma}]$ holds for some vector parameter $\bar{\gamma} = (\gamma_1, \dots, \gamma_k)$ with $\gamma_i \geq 0$, $i = 1, \dots, k$, and also condition $\mathbf{G}_3$ holds. Then, there exist the constants $0 \leq M_{25}, M_{26,1}, \dots, M_{26,k} < \infty$ such that the log-reward functions $\phi_n(\vec{y}, x)$ satisfy the following inequalities for $\vec{z} = (\vec{y}, x) = ((y_1, \dots, y_k), x) \in \mathbb{Z}, 0 \leq n \leq N$:*

$$|\phi_n(\vec{y}, x)| \leq M_{25} + \sum_{i=1}^{k} M_{26,i}e^{\gamma_i|y_i|} . \tag{4.159}$$

Proof. According to Lemma 4.5.1, the condition $\mathbf{E}_8[\bar{\beta}]$, which is the variant of the condition $\mathbf{E}_1[\bar{\beta}]$, holds for any $\beta_i \geq \gamma_i$, $i = 1, \dots, k$. In this case, by Lemma 4.1.8, the condition $\mathbf{C}_1[\bar{\beta}]$ holds. Thus condition of Theorem 4.1.1 hold if condition $\mathbf{B}_1[\bar{\gamma}]$ holds. In this case, the inequality given in Theorem 4.1.1 takes, respectively, the form of inequality (4.159). □

Remark 4.5.2. The constants M_{25} and $M_{26,i}$, $i = 1, \dots, k$ take, according to Remarks 4.1.3 and 4.1.7, the following form:

$$\begin{aligned} M_{25} &= L_1, \\ M_{26,i} &= L_1 L_{2,i} I(\gamma_i = 0) + L_1 L_{2,i} (2K_{25,i})^{N\frac{\gamma_i}{\beta_i}} I(\gamma_i > 0) \\ &= L_1 L_{2,i} I(\gamma_i = 0) + L_1 L_{2,i} (1 + 2e^{K_{24}(\beta_i + \frac{1}{2} k \beta_i^2)})^{N\frac{\gamma_i}{\beta_i}} I(\gamma_i > 0) . \end{aligned} \tag{4.160}$$

One can take any $\beta_i \geq \gamma_i$, $i = 1, \dots, k$, in formulas (4.160).

The following theorem, which is a corollary of Theorem 4.1.3, gives upper bounds for the optimal expected reward Φ.

Theorem 4.5.2. *Let the conditions* $\mathbf{B_1}[\bar{\gamma}]$, $\mathbf{G_3}$, *and* $\mathbf{D_1}[\bar{\beta}]$ *hold and* $0 \leq \gamma_i \leq \beta_i < \infty$, $i = 1, \dots, k$. *Then, there exists a constant* $0 \leq M_{27} < \infty$ *such that the following inequality takes place:*

$$|\Phi| \leq M_{27} . \tag{4.161}$$

Remark 4.5.3. The explicit formula for the constant M_{27} takes, according to formulas given in Remarks 4.1.3, 4.1.7, and 4.1.9, the following form:

$$M_{27} = L_1 + \sum_{i:\gamma_i = 0} L_1 L_{2,i} + \sum_{i:\gamma_i > 0} L_1 L_{2,i} (1 + 2e^{K_{24}(\beta_i + \frac{1}{2} k \beta_i^2)})^{N\frac{\gamma_i}{\beta_i}} K_{2,i}^{\frac{\gamma_i}{\beta_i}} . \tag{4.162}$$

4.5.2 Upper bounds for option rewards for multivariate modulated Markov Gaussian log-price processes with unbounded characteristics

In this subsection, we give upper bounds for option rewards for the model, where the following condition holds:

$\mathbf{G_4}$: $\max_{0 \leq n \leq N-1,\, i,j=1,\dots,k} \sup_{\vec{z}=(\vec{y},x)\in\mathbb{Z}} \frac{|\mu_{n+1,i}(\vec{y},x)| + \sigma^2_{n+1,i,j}(\vec{y},x)}{1+\sum_{l=1}^k K_{27,l}|y_l|} < K_{26}$, for some $0 < K_{26} < \infty$ and $0 \leq K_{27,1}, \dots, K_{27,k} < \infty$.

In this case, there exists, for every $\vec{z} = (\vec{y}, x) \in \mathbb{Z}$, $n = 0, 1, \dots, N-1$, $i = 1, \dots, k$, the multivariate moment-generating function $\mathsf{E}_{\vec{z},n} e^{\sum_{l=1}^k \beta_l (Y_{n+1,l} - Y_{n,l})}$, defined for $\bar{\beta} = (\beta_1, \dots, \beta_k) \in \mathbb{R}_k$,

$$\begin{aligned} f_{\vec{z},n}(\beta_1, \dots, \beta_k) &= \mathsf{E}_{\vec{z},n} e^{\sum_{l=1}^k \beta_l (Y_{n+1,l} - Y_{n,l})} \\ &= \mathsf{E} e^{\sum_{l=1}^k \beta_l (\mu_{n+1,l}(\vec{y},x) + \sum_{m=1}^k \sigma_{n+1,l,m}(\vec{y},x) W_{n+1,m})} \\ &= \mathsf{E} e^{\sum_{l=1}^k \beta_l \mu_{n+1,l}(\vec{y},x) + \sum_{m=1}^k (\sum_{l=1}^k \beta_l \sigma_{n+1,l,m}(\vec{y},x)) W_{n+1,m}} \\ &= e^{\sum_{l=1}^k \beta_l \mu_{n+1,l}(\vec{y},x) + \frac{1}{2} \sum_{m=1}^k (\sum_{l=1}^k \beta_l \sigma_{n+1,l,m}(\vec{y},x))^2} . \end{aligned} \tag{4.163}$$

Formula (4.163) gives a hint to use the function $\vec{A}(\bar{\beta}) = (A_1(\bar{\beta}), \dots, A_k(\bar{\beta}))$, with components given by the following formulas, for $\bar{\beta} = (\beta_1, \dots, \beta_k)$, $\beta_1, \dots, \beta_k \geq 0$:

$$A_j(\bar{\beta}) = K_{26} K_{27,j} \sum_{l=1}^{k} (\beta_l + \frac{1}{2} k^2 \beta_l^2), \quad j = 1, \dots, k, \tag{4.164}$$

Function $\vec{A}(\bar{\beta})$ generates the sequence of functions $\vec{A}_n(\bar{\beta}) = (A_{n,1}(\bar{\beta}), \dots, A_{n,k}(\bar{\beta}))$, $n = 0, 1, \dots$, from the class $\mathcal{A}_k$ that are defined by the following recurrence relation, for any $\bar{\beta} = (\beta_1, \dots, \beta_k)$, $\beta_i \geq 0$, $i = 1, \dots, k$:

$$\vec{A}_n(\bar{\beta}) = \begin{cases} \bar{\beta} & \text{for } n = 0, \\ \vec{A}_{n-1}(\bar{\beta}) + \vec{A}(\vec{A}_{n-1}(\bar{\beta})) & \text{for } n = 1, 2, \dots. \end{cases} \tag{4.165}$$

The modulus $\Xi^{\bar{\imath}_k}_{\bar{\beta}, \vec{A}}(\vec{Y}_\cdot, N)$ takes, according to formula (4.163), the following form for any vector $\bar{\beta} = (\beta_1, \dots, \beta_k)$ with nonnegative components and vectors $\bar{\imath}_k = (\imath_1, \dots, \imath_k) \in \mathcal{I}_k$, where $\mathcal{I}_k = \{\bar{\imath}_k : \imath_j = +, -, j = 1, \dots, k\}$:

$$\begin{aligned} &\Xi^{\bar{\imath}_k}_{\bar{\beta}, \vec{A}}(\vec{Y}_\cdot, N) \\ &= \sup_{0 \leq n \leq N-1, \vec{z} = ((y_1, \dots, y_k), x) \in \mathbb{Z}} \frac{\mathsf{E}_{\vec{z}, n} e^{\sum_{j=1}^{k} \imath_j A_{N-n-1,j}(\bar{\beta}))(Y_{n+1,j} - Y_{n,j})}}{e^{\sum_{j=1}^{k} A_j(\vec{A}_{N-n-1}(\bar{\beta}))|y_j|}} \\ &= \sup_{0 \leq n \leq N-1, \vec{z} \in \mathbb{Z}} \frac{f_{\vec{z}, n}(\imath_1 A_{N-n-1,1}(\bar{\beta}), \dots, \imath_k A_{N-n-1,k}(\bar{\beta}))}{e^{\sum_{j=1}^{k} A_j(\vec{A}_{N-n-1}(\bar{\beta}))|y_j|}}, \end{aligned} \tag{4.166}$$

where the vector functions $\vec{A}_n(\bar{\beta}) = (A_{n,1}(\bar{\beta}), \dots, A_{n,k}(\bar{\beta}))$, $n = 0, 1, \dots$, are generated by the vector function $\vec{A}(\bar{\beta}) = (A_1(\bar{\beta}), \dots, A_k(\bar{\beta}))$ (with components given by formulas (4.164)), according to the recurrence relations (4.165).

We are interested in the following variant of the condition $\mathbf{E}_5[\bar{\beta}]$, assumed to hold for some vector parameter $\bar{\beta} = (\beta_1, \dots, \beta_k)$ with nonnegative components:

$\mathbf{E}_9[\bar{\beta}]$: $\Xi^{\bar{\imath}_k}_{\bar{\beta}_i, \vec{A}}(\vec{Y}_\cdot, N) < K_{28,i}$, $\bar{\imath}_k = (\imath_1, \dots, \imath_k) \in \mathcal{I}_k$, $i = 1, \dots, k$, for some $1 < K_{28,i} < \infty$, $i = 1, \dots, k$.

Vectors $\bar{\beta}_i = (\beta_{i,1}, \dots, \beta_{i,k})$, $i = 1, \dots, k$, penetrating the above condition, have components $\beta_{i,j} = \beta_j I(j = i)$, $j = 1, \dots, k$ for $i = 1, \dots, k$.

Lemma 4.5.2. *The condition $\mathbf{G}_4$ implies that condition $\mathbf{E}_9[\bar{\beta}]$ holds for any vector $\bar{\beta}$ with nonnegative components.*

Proof. The condition $\mathbf{G}_4$ implies that, for any $\vec{z} = (\vec{y}, x) = ((y_1, \dots, y_k), x) \in \mathbb{Z}$, $n = 0, \dots, N-1$, vector $\bar{\imath}_k = (\imath_1, \dots, \imath_k) \in \mathcal{I}_k$ and vector $\bar{\beta}$ with nonnegative components,

$$\begin{aligned}
&\ln f_{\vec{z},n}(\imath_1 A_{N-n-1,1}(\bar{\beta}), \dots, \imath_k A_{N-n-1,k}(\bar{\beta})) \\
&\quad = \sum_{l=1}^{k} \imath_l A_{N-n-1,l}(\bar{\beta}) \mu_{n+1,l}(\vec{y}, x) \\
&\qquad + \frac{1}{2} \sum_{m=1}^{k} \left(\sum_{l=1}^{k} \imath_l A_{N-n-1,l}(\bar{\beta}) \sigma_{n+1,l,m}(\vec{y}, x) \right)^2 \\
&\quad \leq \sum_{l=1}^{k} A_{N-n-1,l}(\bar{\beta}) |\mu_{n+1,l}(\vec{y}, x)| \\
&\qquad + \frac{1}{2} \sum_{m=1}^{k} k \left(\sum_{l=1}^{k} A^2_{N-n-1,l}(\bar{\beta}) \sigma^2_{n+1,l,m}(\vec{y}, x) \right) \\
&\quad \leq \sum_{l=1}^{k} A_{N-n-1,l}(\bar{\beta}) (K_{26} + \sum_{j=1}^{k} K_{26} K_{27,j} |y_j|) \\
&\qquad + \frac{1}{2} \sum_{l=1}^{k} k^2 A^2_{N-n-1,l}(\bar{\beta}) (K_{26} + \sum_{j=1}^{k} K_{26} K_{27,j} |y_j|) \\
&\quad = K_{26} \sum_{l=1}^{k} \left(A_{N-n-1,l}(\bar{\beta}) + \frac{1}{2} k^2 A^2_{N-n-1,l}(\bar{\beta}) \right) \\
&\qquad + \sum_{j=1}^{k} K_{26} K_{27,j} \left(\sum_{l=1}^{k} A_{N-n-1,l}(\bar{\beta}) + \frac{1}{2} k^2 A^2_{N-n-1,l}(\bar{\beta})) \right) |y_j| \\
&\quad = K_{26} \sum_{l=1}^{k} (A_{N-n-1,l}(\bar{\beta}) + \frac{1}{2} k^2 A^2_{N-n-1,l}(\bar{\beta})) \\
&\qquad + \sum_{j=1}^{k} A_j(\vec{A}_{N-n-1}(\bar{\beta})) |y_j| \,.
\end{aligned} \tag{4.167}$$

Relations (4.163)–(4.167) imply that for any $\bar{\imath}_k = (\imath_1, \dots, \imath_k) \in \mathcal{I}_k$ and vector $\bar{\beta} = (\beta_1, \dots, \beta_k)$ with nonnegative components,

$$\begin{aligned}
&\Xi^{\bar{\imath}_k}_{\bar{\beta}, \vec{A}}(\vec{Y}_\cdot, N) \\
&\quad = \sup_{0 \leq n \leq N-1, \vec{z} \in \mathbb{Z}} \frac{f_{\vec{z},n}(\imath_1 A_{N-n-1,1}(\bar{\beta}), \dots, \imath_k A_{N-n-1,k}(\bar{\beta}))}{e^{\sum_{j=1}^{k} A_{N-n,j}(\bar{\beta}) |y_j|}} \\
&\quad < 2^{-k} + \max_{0 \leq n \leq N-1} e^{K_{26} \sum_{l=1}^{k} (A_{N-n-1,l}(\bar{\beta}) + \frac{1}{2} k^2 A^2_{N-n-1,l}(\bar{\beta}))} \\
&\quad = 2^{-k} + e^{K_{26} \sum_{l=1}^{k} (A_{N-1,l}(\bar{\beta}) + \frac{1}{2} k^2 A^2_{N-1,l}(\bar{\beta}))} = K_{28}(\bar{\beta}) \in (1, \infty) \,.
\end{aligned} \tag{4.168}$$

Relation (4.168) implies that condition $\mathbf{E}_9[\bar{\beta}]$ holds. □

Remark 4.5.4. The constants $K_{28,i}$, $i = 1, \dots, k$ are determined by constant K_{26}:

$$K_{28,i} = K_{28}(\bar{\beta}_i), \quad i = 1, \dots, k. \tag{4.169}$$

Lemma 4.5.2 makes it possible to get upper bounds for the log-reward functions $\phi_n(\vec{y}, x)$.

Theorem 4.5.3. *Let the conditions $\mathbf{B}_1[\bar{\gamma}]$ hold for some vector parameter $\bar{\gamma} = (\gamma_1, \dots, \gamma_k)$ with $\gamma_i \ge 0$, $i = 1, \dots, k$, and also condition $\mathbf{G}_4$ holds. Then, for any $\bar{\beta} = (\beta_1, \dots, \beta_k)$ such that $\beta_i \ge \gamma_i$, $i = 1, \dots, k$, there exist the constants $0 \le M_{28}$, $M_{29,1} = M_{29,1}(\beta_1), \dots, M_{29,k} = M_{29,k}(\beta_k) < \infty$ such that the reward functions $\phi_n(\vec{y}, x)$ satisfy the following inequalities for $\vec{z} = (\vec{y}, x) = ((y_1, \dots, y_k), x) \in \mathbb{Z}, 0 \le n \le N$:*

$$\begin{aligned} |\phi_n(\vec{y}, x)| \le \; & M_{28} + \sum_{i:\gamma_i=0} M_{29,i} \\ & + \sum_{i:\gamma_i>0} M_{29,i} \exp\left\{ \left(\sum_{j=1}^{k} A_{N-n,j}(\bar{\beta}_i)|y_j| \right) \frac{\gamma_i}{\beta_i} \right\}. \end{aligned} \tag{4.170}$$

Proof. According to Lemma 4.5.2, the condition $\mathbf{E}_9[\bar{\beta}]$, which is the variant of the condition $\mathbf{E}_5[\bar{\beta}]$, holds, for any $\beta_i \ge \gamma_i$, $i = 1, \dots, k$. In this case, the condition $\mathbf{C}_5[\bar{\beta}]$ holds by Lemma 4.3.16. Thus, conditions of Theorem 4.3.5 hold, if condition $\mathbf{B}_1[\bar{\gamma}]$ holds. In this case, the inequalities given in Theorem 4.3.5, take the form of inequalities (4.170). □

Remark 4.5.5. The explicit formulas for the constants M_{28} and $M_{29,i}$, $i = 1, \dots, k$ take, according to formulas given in Remarks 4.3.11, 4.3.13, 4.3.17, and 4.5.14, the following form:

$$\begin{aligned} M_{28} &= L_1, \\ M_{29,i}(\beta_i) &= L_1 L_{2,i} I(\gamma_i = 0) + L_1 L_{2,i} (2^k K_{28,i})^{N\frac{\gamma_i}{\beta_i}} I(\gamma_i > 0) \\ &= L_1 L_{2,i} I(\gamma_i = 0) \\ &\quad + L_1 L_{2,i} \left(1 + 2^k e^{K_{26} \sum_{l=1}^{k} (A_{N-1,l}(\bar{\beta}_i) + \frac{1}{2} k^2 A_{N-1,l}^2(\bar{\beta}_i))}\right)^{N\frac{\gamma_i}{\beta_i}} I(\gamma_i > 0). \end{aligned} \tag{4.171}$$

In the above formula, one can take any $\beta_i \ge \gamma_i$, $i = 1, \dots, k$.

The condition $\mathbf{D}_3[\bar{\beta}]$, where function $\vec{A}(\bar{\beta}) = (K_{26}K_{27,1} \cdot \sum_{l=1}^{k}(\beta_l + \frac{1}{2}k^2\beta_l^2), \dots, K_{26}K_{27,k} \cdot \sum_{l=1}^{k}(\beta_l + \frac{1}{2}k^2\beta_l^2)$ should be used, takes the following form, for some $\bar{\beta} = (\beta_1, \dots, \beta_k)$ with nonnegative components and the corresponding vectors $\bar{\beta}_i = (\beta_{i,1}, \dots, \beta_{i,k})$ with components $\beta_{i,j} = \beta_j I(i = j)$, $i, j = 1, \dots, k$:

$\mathbf{D}_6[\bar{\beta}]$: $\mathsf{E} \exp\{\sum_{j=1}^{k} A_{N,j}(\bar{\beta}_i)|Y_{0,j}|\} < K_{29,i}$, $i = 1, \dots, k$, for some $1 < K_{29,i} < \infty$, $i = 1, \dots, k$.

The following theorem gives conditions and the upper bound for the optimal expected reward $\Phi = \Phi(\mathcal{M}_{\max,N})$.

Theorem 4.5.4. *Let the conditions* $\mathbf{B}_1[\bar{\gamma}]$, $\mathbf{G}_4$, *and* $\mathbf{D}_6[\bar{\beta}]$ *hold and* $0 \leq \gamma_i \leq \beta_i < \infty$, $i = 1, \dots, k$. *Then, there exists a constant* $0 \leq M_{30} < \infty$ *such that the following inequality takes place:*

$$|\Phi| \leq M_{30} . \tag{4.172}$$

Proof. The case $\gamma_i = 0$ is trivial. Let $\gamma_i > 0$. According to Lemma 4.5.2, the condition $\mathbf{E}_9[\bar{\beta}]$ holds for any $\beta_i \geq \gamma_i$. In this case inequality given in Theorem 4.3.7 takes the form of inequality (4.172). □

Remark 4.5.6. The explicit formula for the constant M_{24} takes, according to formulas given in Remarks 4.3.11, 4.3.13, 4.3.15, 4.3.17, and 4.5.2, the following form:

$$\begin{aligned} M_{30} = {} & L_1 + \sum_{i:\gamma_i=0} L_1 L_{2,i} \\ & + \sum_{i:\gamma_i>0} L_1 L_{2,i} \left(1 + 2^k e^{K_{26} \sum_{l=1}^{k} (A_{N-1,l}(\bar{\beta}_i) + \frac{1}{2} k^2 A^2_{N-1,l}(\bar{\beta}_i))}\right)^{N \frac{\gamma_i}{\beta_i}} K_{29,i}^{\frac{\gamma_i}{\beta_i}} . \end{aligned} \tag{4.173}$$

5 Convergence of option rewards – I

In this chapter, we present our main convergence results for rewards of American-type options with general perturbed pay-off functions for perturbed multivariate modulated Markov log-price processes with bounded characteristics.

In Section 5.1, we give asymptotically uniform upper bounds for log-reward functions and optimal expected rewards for multivariate modulated Markov log-price processes with bounded characteristics (conditional exponential moments for increments) and with general perturbed pay-off functions.

In Section 5.2, we present our main convergence results for reward functions and optimal expected rewards for perturbed multivariate modulated Markov log-price processes with bounded characteristics.

In Section 5.3, we investigate convergence results for log-reward functions and optimal expected rewards for log-price processes represented by multivariate standard and modulated random walks, including Lévy random walks.

Our main results are given in Theorems 5.2.1–5.2.4, for perturbed multivariate modulated Markov log-price processes, and Theorems 5.3.1–5.3.6 for perturbed log-price processes represented by random walks. What is important is that we impose minimal conditions of smoothness on the limiting pay-off functions and transition probabilities. In the basic case, where the transition probabilities have densities with respect to some pivotal Lebesgue-type measure, it is usually required that the sets of weak discontinuity for the limiting transition probabilities and the sets of discontinuity for limiting pay-off functions are zero sets with respect to the above pivotal measure. In fact, such assumptions make it possible for the transition probabilities and pay-off functions to be very irregular. For example, the above discontinuity sets can be countable sets dense in the corresponding phase spaces.

The results presented in this chapter are based on and generalize in several aspects the results obtained in the literature: for univariate modulated Markov log-price processes the results of Silvestrov, Jönsson, and Stenberg (2006, 2008, 2009), for multivariate Markov log-price processes the results of Lundgren and Silvestrov (2009, 2011) and of Silvestrov and Lundgren (2011). First, we consider multivariate modulated log-price processes, i.e. combine together multivariate and modulation aspects together. Second, we consider pay-off functions, which also depend on an additional index component. Third, we improve moment compactness conditions (by taking the supremum of conditional exponential moments for increments in formulas defining the corresponding moduli of compactness over some special subsets of the phase space instead of the whole phase space). This generalization is important for some applications, for example, to skeleton approximation models. Fourth, we improve formulation of the corresponding conditions imposed on pay-off functions and the moment compactness condition on log-price processes by giving them in a natural asymptotic form.

5.1 Asymptotically uniform upper bounds for rewards – I

In this section, we introduce the model of perturbed modulated Markov log-price processes and give asymptotically uniform upper bounds for option rewards for such processes. These upper bounds play an important role in proofs of convergence for option rewards for perturbed log-price processes.

5.1.1 Perturbed log-price processes

We consider a discrete time multivariate modulated Markov log-price process $\vec{Z}_{\varepsilon,n} = (\vec{Y}_{\varepsilon,n}, X_{\varepsilon,n}) = ((Y_{\varepsilon,n,1}, \dots, Y_{\varepsilon,n,k}), X_{\varepsilon,n})$, $n = 0, 1, \dots$ with a phase space $\mathbb{Z} = \mathbb{R}_k \times \mathbb{X}$, an initial distribution $P_{\varepsilon,0}(A)$, and transition probabilities $P_{\varepsilon,n}(\vec{z}, A)$.

We assume that the process $\vec{Z}_{\varepsilon,n}$ and, therefore, its transition probabilities $P_{\varepsilon,n}(\vec{z}, A)$ and initial distribution $P_{\varepsilon,0}(A)$ depend on the perturbation parameter $\varepsilon \in [0, \varepsilon_0]$, where $\varepsilon_0 > 0$.

We consider American-type options for the log-price process $\vec{Z}_{\varepsilon,n}$ and a pay-off function $g_\varepsilon(n, e^{\vec{y}}, x)$ also depending on the perturbation parameter $\varepsilon \in [0, \varepsilon_0]$.

We are going to formulate conditions that would provide convergence for the corresponding log-reward functions $\phi_{\varepsilon,n}(\vec{y}, x)$ and optimal expected rewards $\Phi_\varepsilon = \Phi(\mathcal{M}_{\max,N}^{(\varepsilon)})$ for log-price processes $\vec{Z}_{\varepsilon,n}$, respectively, to the log-reward functions $\phi_{0,n}(\vec{y}, x)$ and optimal expected rewards $\Phi_0 = \Phi(\mathcal{M}_{\max,N}^{(0)})$ for log-price processes $\vec{Z}_{0,n}$, as $\varepsilon \to 0$.

The prelimiting log-price processes $\vec{Z}_{\varepsilon,n}$, for $\varepsilon > 0$, are usually simpler than the corresponding limiting log-price process $\vec{Z}_{0,n}$. For example, the prelimiting processes can be atomic Markov chains with initial distributions and transition probabilities concentrated at a finite set of points. This let one compute effectively values of log-reward functions and optimal expected rewards for prelimiting log-price processes using backward algorithms presented in Chapter 3. On the other hand, the limiting log-price process can be a nonatomic Markov chain with a continuous initial distribution and transition probabilities. In this case, the corresponding backward recurrence algorithms may not be effective. The convergence results may yield in such cases effective approximations for limiting option rewards.

The same approach can be applied to pay-off functions. Prelimiting pay-off functions can be constructed in such a way that they have a simpler structure than the corresponding limiting pay-off function. For example, prelimiting pay-off functions can be convex piecewise linear functions in price arguments. In such cases, one can describe effectively the structure of the corresponding optimal stopping domains. However, the corresponding limiting pay-off function can be a general convex function, with much more complex structure of the optimal stopping domain. The use of approximating pay-off functions may yield effective approximations for optimal stopping domains. We use this approach for designing of Monte Carlo-based approximation algorithms for option rewards. These results will be presented in the second volume of this book.

Conditions of convergence for option rewards, as usual, should be formulated in terms of "initial" characteristics of the model that are transition probabilities $P_{\varepsilon,n}(\vec{z}, A)$, initial distributions $P_\varepsilon(A)$, and pay-off functions $g_\varepsilon(n, e^{\vec{y}}, x)$. Moreover, these conditions should not involve these quantities in some complex combinations, in order to be really effective. The best variant is to formulate some separate and clearly formulated conditions for the above three quantities trying also to escape of making these conditions too restrictive. We systematically follow this approach.

Naturally, these conditions should include some moment assumptions for each of the above quantities and also assumptions about convergence in some sense for the transition probabilities $P_{\varepsilon,n}(\vec{z}, A)$, the initial distribution $P_\varepsilon(A)$, and the pay-off function $g_\varepsilon(n, e^{\vec{y}}, x)$, respectively, to the corresponding limit transition probabilities $P_{0,n}(\vec{z}, A)$, the initial distribution $P_0(A)$, and the pay-off functions $g_0(n, e^{\vec{y}}, x)$.

These convergence assumptions let us interpret the log-price process $\vec{Z}_{\varepsilon,n}$ and the pay-off function $g_\varepsilon(n, e^{\vec{y}}, x)$ for $\varepsilon > 0$ as the perturbed versions of the log-price process $\vec{Z}_{0,n}$ and the pay-off function $g_0(n, e^{\vec{y}}, x)$.

5.1.2 Asymptotically uniform upper bounds for rewards for log-price processes with bounded characteristics

In this subsection, we present upper bounds for option rewards analogous to those given in Chapter 4, but modified in such a way that they become asymptotically uniform with respect to parameter $\varepsilon \to 0$.

Recall that we use the notations $\mathsf{P}_{\vec{z},n}$ and $\mathsf{E}_{\vec{z},n}$ for conditional probabilities and expectations under the condition $\vec{Z}_{\varepsilon,n} = \vec{z}$.

Let us recall the first-type modulus of exponential moment compactness for the components of the log-price process $\vec{Y}_{\varepsilon,n} = (Y_{\varepsilon,n,1}, \dots, Y_{\varepsilon,n,k})$, for $\beta \geq 0$, $i = 1, \dots, k$,

$$\Delta_\beta(Y_{\varepsilon,\cdot,i}, N) = \max_{0 \leq n \leq N-1} \sup_{\vec{z} \in \mathbb{Z}} \mathsf{E}_{\vec{z},n} e^{\beta |Y_{\varepsilon,n+1,i} - Y_{\varepsilon,n,i}|} . \tag{5.1}$$

The condition $\mathbf{C}_1[\bar{\beta}]$ should be replaced by the following first-type condition of exponential moment compactness, which is assumed to hold for some vector parameter $\bar{\beta} = (\beta_1, \dots, \beta_k)$ with nonnegative components:

$\mathbf{C}_6[\bar{\beta}]$: $\overline{\lim}_{\varepsilon \to 0} \Delta_{\beta_i}(Y_{\varepsilon,\cdot,i}, N) < K_{30,i}$, $i = 1, \dots, k$, for some $1 < K_{30,i} < \infty$, $i = 1, \dots, k$.

The following lemmas give asymptotically uniform upper bounds for exponential moments of maximal absolute values for perturbed log-price processes.

Lemma 5.1.1. *Let the condition* $\mathbf{C}_6[\bar{\beta}]$ *holds. Then there exist* $0 < \varepsilon_1 \leq \varepsilon_0$ *and the constants* $0 \leq M_{31,i} < \infty$, $i = 1, \dots, k$ *such that the following inequality takes place for* $\vec{z} = (\vec{y}, x) = ((y_1, \dots, y_k), x) \in \mathbb{Z}$, $0 \leq n \leq N, 0 \leq \varepsilon \leq \varepsilon_1$, *and* $i = 1, \dots, k$:

$$\mathsf{E}_{\vec{z},n} \exp\left\{\beta_i \max_{n \leq r \leq N} |Y_{\varepsilon,r,i}|\right\} \leq M_{31,i} e^{\beta_i |y_i|} . \tag{5.2}$$

Proof. By the definition of the upper limit and the condition $\mathbf{C}_6[\bar{\beta}]$, one can choose $0 < \varepsilon_1 \leq \varepsilon_0$ such that for $0 \leq \varepsilon \leq \varepsilon_1$ and $i = 1, \dots, k$,

$$\Delta_{\beta_i}(Y_{\varepsilon,\cdot,i}(\cdot), N) < K_{30,i} . \tag{5.3}$$

Relation (5.3) means that, for every $\varepsilon \leq \varepsilon_1$, the condition $\mathbf{C}_1[\bar{\beta}]$ (with the constants $K_{30,i}$, $i = 1, \dots, k$, replacing the constants $K_{1,i}$, $i = 1, \dots, k$) holds for the process $\vec{Y}_{\varepsilon,n}$.

Therefore, Lemma 4.1.1 can be applied to the process $\vec{Y}_{\varepsilon,n}$, for every $\varepsilon \leq \varepsilon_1$.

This yields inequalities (5.2). For every $i = 1, \dots, k$, the constant $M_{1,i}$ penetrating the inequalities given in Lemma 4.1.1 is a function of the corresponding constant $K_{1,i}$. The explicit expression for this function is given in Remark 4.1.1. The same formula gives the expression for the constant $M_{31,i}$ as a function of the corresponding constant $K_{30,i}$. □

Remark 5.1.1. The constants $M_{31,i}$, $i = 1, \dots, k$ are given by the following formulas, which follows from relation (5.3), and the corresponding formulas given in Remark 4.1.1:

$$M_{31,i} = K_{30,i}^N . \tag{5.4}$$

Remark 5.1.2. The parameter ε_1 is determined by relation (5.3).

Let $\mathcal{Z}_\varepsilon = \langle \mathbb{Z}_{\varepsilon,0}, \dots, \mathbb{Z}_{\varepsilon,N} \rangle$ be the sequence of measurable subsets of space $\mathbb{Z}$ and $\Delta_{\beta,\mathcal{Z}_\varepsilon}(Y_{\varepsilon,\cdot,i}, N)$ be the modified first-type modulus of exponential moment compactness defined, for every $\varepsilon \in [0, \varepsilon_0]$ and $\beta \geq 0$, by the following formula:

$$\Delta_{\beta,\mathcal{Z}_\varepsilon}(Y_{\varepsilon,\cdot,i}, N) = \max_{0 \leq n \leq N-1} \sup_{\vec{z} \in \mathbb{Z}_{\varepsilon,n}} \mathsf{E}_{\vec{z},n} e^{\beta |Y_{\varepsilon,n+1,i} - Y_{\varepsilon,n,i}|} . \tag{5.5}$$

The following modified first-type condition of exponential moment compactness for the log-price process $\vec{Y}_n$ generalizes the condition $\mathbf{C}_6[\bar{\beta}]$. It should be assumed to hold for some vector parameter $\bar{\beta} = (\beta_1, \dots, \beta_k)$ with nonnegative components:

$\mathbf{C}_6'[\bar{\beta}]$: There exists a sequence of measurable sets $\mathcal{Z}_\varepsilon = \langle \mathbb{Z}_{\varepsilon,0}, \dots, \mathbb{Z}_{\varepsilon,N} \rangle$, for every $\varepsilon \in [0, \varepsilon_0]$, such that

(a) $\overline{\lim}_{\varepsilon \to 0} \Delta_{\beta_i,\mathcal{Z}_\varepsilon}(Y_{\varepsilon,\cdot,i}, N) < K_{30,i}'$, $i = 1, \dots, k$ for some $1 < K_{30,i}' < \infty$, $i = 1, \dots, k$;

(b) $\mathcal{Z}_\varepsilon$ is a complete sequence of phase sets for the process $\vec{Z}_{\varepsilon,n}$, i.e. $\mathsf{P}_{\varepsilon,n}(\vec{z}, \mathbb{Z}_{\varepsilon,n}) = 1, \vec{z} \in \mathbb{Z}_{\varepsilon,n-1}$, $n = 1, \dots, N$, for every $\varepsilon \in [0, \varepsilon_0]$.

Let us introduce, for every $\varepsilon \in [0, \varepsilon_0]$, sets $\mathbb{H}_{\varepsilon,n,k} = (h^-_{\varepsilon,n,1}, h^+_{\varepsilon,n,1}) \times \cdots \times (h^-_{\varepsilon,n,k}, h^+_{\varepsilon,n,k}) \times \mathbb{X}$, $n = 0, \dots, N$, where $-\infty \le h^-_{\varepsilon,n,i} \le h^+_{\varepsilon,n,i} \le +\infty$, $i = 1, \dots, k$, $n = 0, \dots, N$.

It is useful to note that in applications, for example to space skeleton models considered in Sections 1.7 and 1.8, the sets $\mathbb{Z}_{\varepsilon,n}$, $n = 0, \dots, N$ satisfy the following natural condition:

H₁: $\mathbb{Z}_{\varepsilon,n} \supseteq \mathbb{H}_{\varepsilon,n,k}$, $n = 0, \dots, N$, $\varepsilon \in [0, \varepsilon_0]$, where $h^\pm_{\varepsilon,n,i} \to h^\pm_{0,n,i} = \pm\infty$ as $\varepsilon \to 0$, for $i = 1, \dots, k$, $n = 0, \dots, N$.

Remark 5.1.3. According to the condition $\mathbf{H}_1$, for every $\vec{z} = (\vec{y}, x) = ((y_1, \dots, y_k), x)$, there exists $\varepsilon_1(\vec{z}) \in (0, \varepsilon_0]$ such that the following relation holds for every $\varepsilon \in [0, \varepsilon_1(\vec{z})]$,

$$h^-_{\varepsilon,n,i} < y_i < h^+_{\varepsilon,n,i}, \quad i = 1, \dots, k, \quad n = 0, \dots, N. \tag{5.6}$$

Obviously, $\vec{z} \in \mathbb{Z}_{\varepsilon,n}$, $n = 0, \dots, N$ for all $\varepsilon \in [0, \varepsilon_1(\vec{z})]$.

Note that, in the case, where the condition $\mathbf{H}_1$ holds, the sets $\mathbb{Z}_{0,n} = \mathbb{Z}$, $n = 0, \dots, N$.

Lemma 5.1.2. *Let the condition $\mathbf{C}'_6[\bar{\beta}]$ holds. Then there exists $0 < \varepsilon'_1 \le \varepsilon_0$ and the constants $0 \le M'_{31,i} < \infty$, $i = 1, \dots, k$ such that the following inequality takes place for $\vec{z} = (\vec{y}, x) = ((y_1, \dots, y_k), x) \in \mathbb{Z}_{\varepsilon,n}$, $0 \le n \le N$, $0 \le \varepsilon \le \varepsilon_1$, and $i = 1, \dots, k$,*

$$\mathsf{E}_{\vec{z},n} \exp\left\{\beta_i \max_{n \le r \le N} |Y_{\varepsilon,r,i}|\right\} \le M'_{31,i} e^{\beta_i |y_i|}. \tag{5.7}$$

Proof. By the definition of the upper limit and the condition $\mathbf{C}'_6[\bar{\beta}]$, one can choose $0 < \varepsilon'_1 \le \varepsilon_0$ such that for $0 \le \varepsilon \le \varepsilon'_1$ and $i = 1, \dots, k$,

$$\Delta_{\beta_i, Z_\varepsilon}(Y_{\varepsilon,\cdot,i}(\cdot), N) < K'_{30,i}. \tag{5.8}$$

The following part of the proof repeats the proof of Lemma 5.1.1. The only difference is that Lemma 4.1.3 should be applied to the process $\vec{Y}_{\varepsilon,n}$, for every $\varepsilon \le \varepsilon'_1$, instead of Lemma 4.1.1, which was applied to the process $\vec{Y}_{\varepsilon,n}$, for every $\varepsilon \le \varepsilon_1$, in the proof of Lemma 5.1.1.

This yields inequalities (5.7). For every $i = 1, \dots, k$, the constant $M'_{1,i}$ penetrating the inequalities given in Lemma 4.1.3 is a function of the corresponding constant $K'_{1,i}$. The explicit expression for this function is given in Remark 4.1.2. The same formula gives the expression for the constant $M'_{31,i}$ as a function of the corresponding constant $K'_{30,i}$. □

Remark 5.1.4. The constants $M'_{31,i}$, $i = 1, \dots, k$ are given by the following formulas, which follows from relation (5.8), and the corresponding formulas given in Remark 4.1.2,

$$M'_{31,i} = (K'_{30,i})^N. \tag{5.9}$$

Remark 5.1.5. The parameter ε_1' is determined by relation (5.8).

Let us now give asymptotically uniform upper bounds for log-reward functions for log-price processes $\vec{Z}_{\varepsilon,n}$.

We assume that pay-off functions $g_\varepsilon(n, e^{\vec{y}}, x)$ also depends on perturbation parameter ε.

The condition $\mathbf{B}_1[\bar{\gamma}]$ should be replaced by the following condition assumed to hold for some vector parameter $\bar{\gamma} = (\gamma_1, \dots, \gamma_k)$ with nonnegative components:

$\mathbf{B}_5[\bar{\gamma}]$: $\overline{\lim}_{\varepsilon\to 0} \max_{0\leq n\leq N} \sup_{\vec{z}=(\vec{y},x)\in\mathbb{Z}} \frac{|g_\varepsilon(n,e^{\vec{y}},x)|}{1+\sum_{i=1}^k L_{10,i}e^{\gamma_i|y_i|}} < L_9$, for some $0 < L_9 < \infty$ and $0 \leq L_{10,1}, \dots, L_{10,k} < \infty$.

The following two lemmas give asymptotically uniform upper bounds for conditional expectations of maximal absolute values for perturbed pay-off processes $g_\varepsilon(n, e^{\vec{Y}_{\varepsilon,n}}, X_{\varepsilon,n})$.

Lemma 5.1.3. *Let the conditions $\mathbf{B}_5[\bar{\gamma}]$ and $\mathbf{C}_6[\bar{\beta}]$ hold and $0 \leq \gamma_i \leq \beta_i$, $i = 1, \dots, k$. Then, there exist $0 < \varepsilon_3 \leq \varepsilon_0$ and the constants $0 \leq M_{32}, M_{33,i} < \infty$, $i = 1, \dots, k$ such that the following inequalities take place for $\vec{z} = ((y_1, \dots, y_k), x) \in \mathbb{Z}$, $0 \leq n \leq N$, $0 \leq \varepsilon \leq \varepsilon_3$,*

$$\mathsf{E}_{\vec{z},n} \max_{n\leq r\leq N} \left| g_\varepsilon\left(n, e^{\vec{Y}_{\varepsilon,n}}, X_{\varepsilon,n}\right)\right| \leq M_{32} + \sum_{i=1}^k M_{33,i}e^{\gamma_i|y_i|}. \tag{5.10}$$

Proof. The condition $\mathbf{B}_5[\bar{\gamma}]$ implies that there exists $0 < \varepsilon_2 \leq \varepsilon_0$ such that for any $0 \leq \varepsilon \leq \varepsilon_2$,

$$\max_{0\leq n\leq N} \sup_{\vec{z}=(\vec{y},x)\in\mathbb{Z}} \frac{|g_\varepsilon(n, e^{\vec{y}}, x)|}{1 + \sum_{i=1}^k L_{10,i}e^{\gamma_i|y_i|}} < L_9. \tag{5.11}$$

Relation (5.11) means that, for every $\varepsilon \leq \varepsilon_2$, the condition $\mathbf{B}_1[\bar{\gamma}]$ (with the constants $L_9, L_{10,i}$, $i = 1, \dots, k$, replacing the constants $L_1, L_{2,i}$, $i = 1, \dots, k$) holds for the process $\vec{Y}_{\varepsilon,n}$.

Let us now take $\varepsilon_3 = \min(\varepsilon_1, \varepsilon_2)$, where ε_1 was defined in relation (5.3). By the definition $\varepsilon_3 > 0$.

For every fixed $0 \leq \varepsilon \leq \varepsilon_3$, both the conditions $\mathbf{C}_1[\bar{\beta}]$ and $\mathbf{B}_1[\bar{\gamma}]$ (with the constants $K_{30,i}$, $i = 1, \dots, k$, and $L_9, L_{10,i}$, $i = 1, \dots, k$ replacing, respectively, the constants $K_{1,i}$, $i = 1, \dots, k$ and $L_1, L_{2,i}$, $i = 1, \dots, k$) hold for the process $\vec{Y}_{\varepsilon,n}$.

Therefore, Lemma 4.1.4 can be applied to the process $\vec{Y}_{\varepsilon,n}$, for every $0 \leq \varepsilon \leq \varepsilon_3$.

This yields inequality (5.10). For every $i = 1, \dots, k$, the constants M_2 and $M_{3,i}$ penetrating the inequalities given in Lemma 4.1.4 are functions of the corresponding constants $K_{1,i}$ and $L_1, L_{2,i}$. The explicit expressions for these functions are given in Remark 4.1.3. The same formulas give the expressions for the constant M_{32} and $M_{33,i}$ as functions of the corresponding constants $K_{30,i}$ and $L_9, L_{10,i}$. □

Remark 5.1.6. The explicit formula for the constants $M_{32}, M_{33,i}$, $i = 1, \dots, k$ are given by the following formulas, which follow from the corresponding formulas given in Remark 4.1.3:

$$M_{32} = L_9 \, , \quad M_{33,i} = L_9 L_{10,i} I(\gamma_i = 0) + L_9 L_{10,i} K_{30,i}^{N \frac{\gamma_i}{\beta_i}} I(\gamma_i > 0) \, . \tag{5.12}$$

Remark 5.1.7. The parameter $\varepsilon_3 = \min(\varepsilon_1, \varepsilon_2)$, where the parameters ε_1 and ε_2 are determined, respectively, by relations (5.3) and (5.11).

Lemma 5.1.4. *Let the conditions* $\mathbf{B}_5[\bar{\gamma}]$ *and* $\mathbf{C}_6'[\bar{\beta}]$ *hold and* $0 \leq \gamma_i \leq \beta_i$, $i = 1, \dots, k$. *Then, there exist* $0 < \varepsilon_3' \leq \varepsilon_0$ *and the constants* $0 \leq M_{32}', M_{33,i}' < \infty$, $i = 1, \dots, k$ *such that the following inequalities take place for* $\vec{z} = ((y_1, \dots, y_k), x) \in \mathbb{Z}_{\varepsilon,n}$, $0 \leq n \leq N$, $0 \leq \varepsilon \leq \varepsilon_3'$:

$$\mathsf{E}_{\vec{z},n} \max_{n \leq r \leq N} \left| g_\varepsilon \left(n, e^{\vec{Y}_{\varepsilon,n}}, X_{\varepsilon,n}\right) \right| \leq M_{32}' + \sum_{i=1}^{k} M_{33,i}' e^{\gamma_i |y_i|} \, . \tag{5.13}$$

Proof. It is analogous to the proof of Lemma 5.1.3. As was pointed out in the proof of this lemma, the condition $\mathbf{B}_5[\bar{\gamma}]$ implies that there exists $0 < \varepsilon_2 \leq \varepsilon_0$ such that inequality (5.11) holds for $0 \leq \varepsilon \leq \varepsilon_2$.

This means that, for every $\varepsilon \leq \varepsilon_2$, the condition $\mathbf{B}_1[\bar{\gamma}]$ (with the constants $L_9, L_{10,i}$, $i = 1, \dots, k$, replacing the constants $L_1, L_{2,i}$, $i = 1, \dots, k$) holds for the process $\vec{Y}_{\varepsilon,n}$.

Let us now take $\varepsilon_3' = \min(\varepsilon_1', \varepsilon_2)$, where ε_1' was defined in relation (5.8). By the definition $\varepsilon_3' > 0$.

For every fixed $0 \leq \varepsilon \leq \varepsilon_3'$ both the conditions $\mathbf{C}_1'[\bar{\beta}]$ and $\mathbf{B}_1[\bar{\gamma}]$ (with the constants $K_{30,i}'$, $i = 1, \dots, k$, and $L_9, L_{10,i}$, $i = 1, \dots, k$ replacing, respectively, the constants $K_{1,i}'$, $i = 1, \dots, k$ and $L_1, L_{2,i}$, $i = 1, \dots, k$) hold for the process $\vec{Y}_{\varepsilon,n}$.

Therefore, Lemma 4.1.5 can be applied to the process $\vec{Y}_{\varepsilon,n}$, for every $0 \leq \varepsilon \leq \varepsilon_3'$.

This yields inequality (5.10). For every $i = 1, \dots, k$, the constants M_2' and $M_{3,i}'$ penetrating the inequalities given in Lemma 4.1.5 are functions of the corresponding constants $K_{1,i}'$ and $L_1, L_{2,i}$. The explicit expressions for these functions are given in Remark 4.1.4. The same formulas give the expressions for the constant M_{32}' and $M_{33,i}'$ as functions of the corresponding constants $K_{30,i}'$ and $L_9, L_{10,i}$. □

Remark 5.1.8. The explicit formula for the constants $M_{32}', M_{33,i}'$, $i = 1, \dots, k$ are given by the following formulas, which follows from the corresponding formulas given in Remark 4.1.4:

$$M_{32}' = L_9 \, , \quad M_{33,i}' = L_9 L_{10,i} I(\gamma_i = 0) + L_9 L_{10,i} (K_{30,i}')^{N \frac{\gamma_i}{\beta_i}} I(\gamma_i > 0) \, . \tag{5.14}$$

Remark 5.1.9. The parameter $\varepsilon_3' = \min(\varepsilon_1', \varepsilon_2)$, where the parameters ε_1' and ε_2 are determined, respectively, by relations (5.8) and (5.11).

Let us recall the second-type modulus of exponential moment compactness for the components of the log-price process $\vec{Y}_{\varepsilon,n} = (Y_{\varepsilon,n,1}\dots, Y_{\varepsilon,n,k})$ defined for $\beta \geq 0$, $i = 1, \dots, k$,

$$\Xi^{\pm}_{\beta}(Y_{\varepsilon,\cdot,i}, N) = \max_{0 \leq n \leq N-1} \sup_{\vec{z} \in \mathbb{Z}} \mathsf{E}_{\vec{z},n} e^{\pm\beta(Y_{\varepsilon,n+1,i} - Y_{\varepsilon,n,i})} . \tag{5.15}$$

The following second-type condition of the exponential moment compactness, based on moment-generating functions of increments for one-dimensional Markov log-price processes $\vec{Y}_{\varepsilon,n}$, replaces the condition $\mathbf{E}_1[\bar{\beta}]$. It is assumed to hold for some vector parameter $\bar{\beta} = (\beta_1, \dots, \beta_k)$ with nonnegative components:

$\mathbf{E}_{10}[\bar{\beta}]$: $\overline{\lim}_{\varepsilon\to 0} \Xi^{\pm}_{\beta_i}(Y_{\varepsilon,\cdot,i}, N) < K_{31,i}$, $i = 1, \dots, k$, for some $1 < K_{31,i} < \infty$, $i = 1, \dots, k$.

Lemma 5.1.5. *The condition $\mathbf{E}_{10}[\bar{\beta}]$ implies the condition $\mathbf{C}_6[\bar{\beta}]$ to hold.*

Proof. It follows from Lemma 4.1.8 that should be applied to the log-price processes $Y_{\varepsilon,n,i}$. □

Remark 5.1.10. If the condition $\mathbf{E}_{10}[\bar{\beta}]$ holds with the constants $K_{31,i}$, $i = 1, \dots, k$, then the condition $\mathbf{C}_6[\bar{\beta}]$ holds with the constants $K_{30,i}$, $i = 1, \dots, k$ given by the following formulas:

$$K_{30,i} = 2K_{31,i} , \quad i = 1, \dots, k . \tag{5.16}$$

Let $\mathcal{Z}_\varepsilon = \langle \mathbb{Z}_{\varepsilon,0}, \dots, \mathbb{Z}_{\varepsilon,N} \rangle$ be a sequence of measurable sets. Recall the modified second-type modulus of the exponential moment compactness for the components defined, for every $\varepsilon \in [0, \varepsilon_0]$ and for $\beta \geq 0$, $i = 1, \dots, k$,

$$\Xi^{\pm}_{\beta,\mathcal{Z}_\varepsilon}(Y_{\varepsilon,\cdot,i}, N) = \max_{0 \leq n \leq N-1} \sup_{\vec{z} \in \mathbb{Z}_{\varepsilon,n}} \mathsf{E}_{\vec{z},n} e^{\pm\beta(Y_{\varepsilon,n+1,i} - Y_{\varepsilon,n,i})} . \tag{5.17}$$

The following condition is a modified version of the condition $\mathbf{E}_{10}[\bar{\beta}]$. It is assumed to hold for some vector parameter $\bar{\beta} = (\beta_1, \dots, \beta_k)$ with nonnegative components:

$\mathbf{E}'_{10}[\bar{\beta}]$: There exists a sequence of measurable sets $\mathcal{Z}_\varepsilon = \langle \mathbb{Z}_{\varepsilon,0}, \dots, \mathbb{Z}_{\varepsilon,N} \rangle$, for every $\varepsilon \in [0, \varepsilon_0]$, such that
- (a) $\overline{\lim}_{\varepsilon\to 0} \Xi^{\pm}_{\beta_i,\mathcal{Z}_\varepsilon}(Y_{\varepsilon,\cdot,i}, N) < K'_{31,i}$, $i = 1, \dots, k$, for some $1 < K'_{31,i} < \infty$, $i = 1, \dots, k$;
- (b) $\mathcal{Z}_\varepsilon$ is a complete sequence of phase sets for the process $\vec{Z}_{\varepsilon,n}$, i.e. $\mathsf{P}_{\varepsilon,n}(\vec{z}, \mathbb{Z}_{\varepsilon,n}) = 1$, $\vec{z} \in \mathbb{Z}_{\varepsilon,n-1}$, $n = 1, \dots, N$, for every $\varepsilon \in [0, \varepsilon_0]$.

Lemma 5.1.6. *The condition $\mathbf{E}'_{10}[\bar{\beta}]$ implies the condition $\mathbf{C}'_6[\bar{\beta}]$ to hold.*

Proof. It follows from Lemma 4.1.9 that should be applied to the log-price processes $Y_{\varepsilon,n,i}$. □

Remark 5.1.11. If the condition $\mathbf{E}_{10}'[\bar{\beta}]$ holds with the constants $K_{31,i}'$, $i = 1, \dots, k$, then the condition $\mathbf{C}_6'[\bar{\beta}]$ holds with the constants $K_{30,i}'$, $i = 1, \dots, k$ given by the following formulas:

$$K_{30,i}' = 2K_{31,i}', \quad i = 1, \dots, k. \tag{5.18}$$

In the model, where log-price processes $\vec{Y}_{\varepsilon,n}$ and pay-off functions $g_\varepsilon(n, e^{\vec{y}}, x)$ depend on a perturbation parameter $\varepsilon \in [0, \varepsilon_0]$, the corresponding log-reward functions also depend on this perturbation parameter.

Let us denote by $\mathcal{F}_{n,r}^{(\varepsilon)} = \sigma[\vec{Z}_{\varepsilon,l}, n \le l \le r]$ and $\mathcal{M}_{\max,n,N}^{(\varepsilon)}$ the class of all Markov moments $\tau_{\varepsilon,n}$ for the log-price process $\vec{Z}_{\varepsilon,n}$ such that (a) $n \le \tau_{\varepsilon,n} \le N$ and (b) event $\{\tau_{\varepsilon,n} = r\} \in \mathcal{F}_{n,r}^{(\varepsilon)}$, $n \le r \le N$.

Let us assume that the conditions $\mathbf{B}_5[\bar{\gamma}]$ and $\mathbf{C}_6[\bar{\beta}]$ hold and $0 \le \gamma_i \le \beta_i$, $i = 1, \dots, k$.

In this case, for Lemma 5.1.3 let us define log-reward functions for the American option for the log-price process $\vec{Z}_{\varepsilon,n}$, for every $\vec{z} = (\vec{y}, x) \in \mathbb{Z}, 0 \le n \le N$ and $0 \le \varepsilon \le \varepsilon_3$,

$$\phi_{\varepsilon,n}(\vec{z}) = \phi_{\varepsilon,n}(\vec{y}, x) = \sup_{\tau_{\varepsilon,n} \in \mathcal{M}_{\max,n,N}^{(\varepsilon)}} \mathsf{E}_{\vec{z},n} g_\varepsilon(\tau_{\varepsilon,n}, e^{\vec{Y}_{\tau_{\varepsilon,n}}}, X_{\tau_{\varepsilon n}}). \tag{5.19}$$

These functions are measurable functions of $\vec{z} \in \mathbb{Z}$ for every $0 \le n \le N$.

Moreover, for Lemma 5.1.3 let us obtain the following asymptotically uniform upper bounds for the log-reward functions $\phi_{\varepsilon,n}(\vec{y}, x)$.

Theorem 5.1.1. *Let the conditions $\mathbf{B}_{10}[\bar{\gamma}]$ and $\mathbf{C}_6[\bar{\beta}]$ hold and $0 \le \gamma_i \le \beta_i$, $i = 1, \dots, k$. Then, the log-reward functions $\phi_{\varepsilon,n}(\vec{y}, x)$ satisfy the following inequalities for $\vec{z} = (\vec{y}, x) \in \mathbb{Z}$, $0 \le n \le N, 0 \le \varepsilon \le \varepsilon_3$:*

$$|\phi_{\varepsilon,n}(\vec{y}, x)| \le M_{32} + \sum_{i=1}^{k} M_{33,i} e^{\gamma_i |y_i|}. \tag{5.20}$$

Proof. It is analogous to the proof of Theorem 4.1.1. Using Lemma 5.1.3 and the definition of the log-reward functions, we get the following inequalities for $\vec{z} = (\vec{y}, x) \in \mathbb{Z}, 0 \le n \le N$ and $0 \le \varepsilon \le \varepsilon_3$:

$$|\phi_{\varepsilon,n}(\vec{y}, x)| \le \mathsf{E}_{\vec{z},n} \max_{n \le r \le N} |g_\varepsilon(r, e^{\vec{Y}_{\varepsilon,r}}, X_{\varepsilon,r})| \le M_{32} + \sum_{i=1}^{k} M_{33,i} e^{\gamma_i |y_i|}, \tag{5.21}$$

which prove the above theorem. □

A more complicated is the case, where the conditions $\mathbf{B}_{10}[\bar{\gamma}]$ and $\mathbf{C}_6'[\bar{\beta}]$ hold and $0 \le \gamma_i \le \beta_i$, $i = 1, \dots, k$.

For relation (5.19) let us, according to Lemma 5.1.4, define log-reward functions for the American option for the log-price process $\vec{Z}_{\varepsilon,n}$, for every $0 \le \varepsilon \le \varepsilon_3'$ and $\vec{z} = (\vec{y}, x) \in \mathbb{Z}_{\varepsilon,n}$, $0 \le n \le N$, and get the following upper bounds.

Theorem 5.1.2. *Let the conditions $\mathbf{B}_5[\bar{\gamma}]$ and $\mathbf{C}'_6[\bar{\beta}]$ hold and $0 \leq \gamma_i \leq \beta_i$, $i = 1, \dots, k$. Then, the log-reward functions $\phi_{\varepsilon,n}(\vec{y}, x)$ satisfy the following inequalities for $\vec{z} = (\vec{y}, x) \in \mathbb{Z}_{\varepsilon,n}$, $0 \leq n \leq N$, $0 \leq \varepsilon \leq \varepsilon'_3$:*

$$|\phi_{\varepsilon,n}(\vec{y}, x)| \leq M'_{32} + \sum_{i=1}^{k} M'_{33,i} e^{\gamma_i |y_i|} . \tag{5.22}$$

Proof. It is analogous to the proof of Theorem 5.1.1. In this case Lemma 5.1.4 should be used instead of Lemma 5.1.3. An inequality analogous to (5.21) can be written down. The differences are that it holds for $\vec{z} = (\vec{y}, x) \in \mathbb{Z}_{\varepsilon,n}, 0 \leq n \leq N$, $0 \leq \varepsilon \leq \varepsilon'_3$, and with the constants $M'_{32}, M'_{33,i}$, $i = 1, \dots, k$. □

Remark 5.1.12. It is useful to note that the log-reward function $\phi_{\varepsilon,N}(\vec{y}, x) = g_\varepsilon(N, e^{\vec{y}}, x)$, $\vec{z} = (\vec{y}, x) \in \overline{\mathbb{Z}}_{\varepsilon,N}$ always exists. The corresponding upper bounds for this function are provided by the condition $\mathbf{B}_5[\bar{\gamma}]$. However, there is no guarantee that, under conditions of Theorem 5.1.2, the log-reward functions $\phi_{\varepsilon,n}(\vec{y}, x)$ exist for points $\vec{z} = (\vec{y}, x) \in \overline{\mathbb{Z}}_{\varepsilon,n}$, $0 \leq n \leq N - 1$. Even if these functions do exist for these points, Theorem 5.1.2 does not provide upper bounds for the log-reward functions in these points.

However, there is an important case where this problem is not essential. Let us assume that the condition $\mathbf{H}_1$ holds. Then, as was pointed out in Remark 5.1.3, for any point $\vec{z} = (\vec{y}, x) \in \mathbb{Z}$, there exists $\varepsilon_1(\vec{z}) \in (0, \varepsilon_0]$ such that $\vec{z} = (\vec{y}, x) \in \mathbb{H}_{\varepsilon,n,k} \subseteq \mathbb{Z}_{\varepsilon,n}$, $n = 0, \dots, N$ for $\varepsilon \in [0, \varepsilon_1(\vec{z})]$. The quantity $\varepsilon_1(\vec{z})$ is determined by relation (5.6) given in Remark 5.1.3. Thus, the upper bounds for $\phi_{\varepsilon,n}(\vec{y}, x)$ given in Theorem 5.1.2 take place for $\varepsilon \in [0, \varepsilon_1(\vec{z})]$.

Let us now get asymptotically uniform upper bounds for optimal expected rewards.

The condition $\mathbf{D}_1[\bar{\beta}]$ should be replaced, respectively, by the following condition assumed to hold for some vector parameter $\bar{\beta} = (\beta_1, \dots, \beta_k)$ with nonnegative components:

$\mathbf{D}_7[\bar{\beta}]$: $\overline{\lim}_{\varepsilon \to 0} \mathsf{E} e^{\beta_i |Y_{\varepsilon,0,i}|} < K_{32,i}$, $i = 1, \dots, k$, for some $1 < K_{32,i} < \infty$, $i = 1, \dots, k$.

The following lemma gives explicit upper bounds for the expectation of maximum of absolute values for the pay-off processes $g_\varepsilon(n, e^{\vec{Y}_{\varepsilon,n}}, X_{\varepsilon,n})$.

Lemma 5.1.7. *Let the conditions $\mathbf{B}_5[\bar{\gamma}]$, $\mathbf{C}_6[\bar{\beta}]$ and $\mathbf{D}_7[\bar{\beta}]$ hold, and $0 \leq \gamma_i \leq \beta_i$, $i = 1, \dots, k$. Then, there exist $0 < \varepsilon_5 \leq \varepsilon_0$ and a constant $0 \leq M_{34} < \infty$ such that the following inequality takes place, for $0 \leq \varepsilon \leq \varepsilon_5$:*

$$\mathsf{E} \max_{0 \leq n \leq N} |g_\varepsilon(n, e^{\vec{Y}_{\varepsilon,n}}, X_{\varepsilon,n})| \leq M_{34} . \tag{5.23}$$

Proof. The condition $\mathbf{D}_7[\bar{\beta}]$ implies that there exists $0 < \varepsilon_4 \leq \varepsilon_0$ such that for $0 \leq \varepsilon \leq \varepsilon_4$ and $i = 1, \dots, k$,

$$\mathsf{E} e^{\beta_i |Y_{\varepsilon,0,i}|} < K_{32,i} . \tag{5.24}$$

Relation (5.24) means that, for every $\varepsilon \le \varepsilon_4$, the condition $\mathbf{D}_1[\bar{\beta}]$ (with the constants $K_{32,i}$, $i = 1, \dots, k$, replacing the constants $K_{2,i}$, $i = 1, \dots, k$) holds for the process $\vec{Y}_{\varepsilon,n}$.

Let us now take $\varepsilon_5 = \min(\varepsilon_1, \varepsilon_2, \varepsilon_4)$, where ε_1 and ε_2 were defined in relations (5.3) and (5.11). By the definition $\varepsilon_5 > 0$.

For every fixed $0 \le \varepsilon \le \varepsilon_5$, the conditions $\mathbf{C}_1[\bar{\beta}]$, $\mathbf{B}_1[\bar{\gamma}]$, and $\mathbf{D}_1[\bar{\beta}]$ (with the constants $K_{30,i}, K_{32,i}$, $i = 1, \dots, k$, and $L_{17}, L_{18,i}$, $i = 1, \dots, k$ replacing, respectively, the constants $K_{1,i}, K_{2,i}$, $i = 1, \dots, k$ and $L_1, L_{2,i}$, $i = 1, \dots, k$) hold for the process $\vec{Y}_{\varepsilon,n}$.

Therefore, Lemma 4.1.6 can be applied to the process $\vec{Y}_{\varepsilon,n}$, for every $0 \le \varepsilon \le \varepsilon_5$.

This yields inequality (5.23). For every $i = 1, \dots, k$, the constant M_4 penetrating the inequality given in Lemma 4.1.6 is a function of the constants $K_{1,i}, K_{2,i}$, $i = 1, \dots, k$ and $L_1, L_{2,i}$, $i = 1, \dots, k$. The explicit expression for this function is given in Remark 4.1.5. The same formula gives the expression for the constant M_{58} as a function of the constants $K_{30,i}, K_{32,i}$, $i = 1, \dots, k$ and $L_{17}, L_{18,i}$, $i = 1, \dots, k$. □

Remark 5.1.13. The explicit formula for the constant M_{34} is given by the following formula, which follows from the corresponding formula given in Remark 4.1.5:

$$M_{34} = L_9 + \sum_{i:\gamma_i = 0} L_9 L_{10,i} + \sum_{i:\gamma_i > 0} L_9 L_{10,i} K_{30,i}^{N \frac{\gamma_i}{\beta_i}} K_{32,i}^{\frac{\gamma_i}{\beta_i}} . \tag{5.25}$$

Remark 5.1.14. The parameter $\varepsilon_5 = \min(\varepsilon_1, \varepsilon_2, \varepsilon_4)$, where the parameters $\varepsilon_1, \varepsilon_2$, and ε_4 are determined, respectively, by relations (5.3), (5.11), and (5.24).

Note also that one can replace the first-type condition of the exponential moment boundedness $\mathbf{D}_7[\bar{\beta}]$ by the following condition of the exponential moment boundedness based on moment-generating functions for the initial value of log-price processes $\vec{Y}_{\varepsilon,0}$, assumed to hold for some vector parameter $\bar{\beta} = (\beta_1, \dots, \beta_k)$ with nonnegative components:

$\mathbf{F}_4[\bar{\beta}]$: $\overline{\lim}_{\varepsilon \to 0} \mathsf{E} e^{\pm \beta_i Y_{\varepsilon,0,i}} < K_{33,i}$, $i = 1, \dots, k$, for some $1 \le K_{33,i} < \infty$, $i = 1, \dots, k$.

Lemma 5.1.8. *The condition* $\mathbf{F}_4[\bar{\beta}]$ *implies the condition* $\mathbf{D}_7[\bar{\beta}]$ *to hold.*

Proof. It follows from the following inequality that holds for any $\beta \ge 0$:

$$\mathsf{E} e^{\beta |Y_{\varepsilon,0,i}|} \le \mathsf{E} e^{+\beta Y_{\varepsilon,0,i}} + \mathsf{E} e^{-\beta Y_{\varepsilon,0,i}} . \tag{5.26}$$

□

Remark 5.1.15. As follows from inequality (5.26), if the condition $\mathbf{F}_4[\bar{\beta}]$ holds with the constants $K_{33,i}$, $i = 1, \dots, k$, then the condition $\mathbf{D}_7[\bar{\beta}]$ holds with the constants $K_{32,i}$, $i = 1, \dots, k$ given by the following formulas:

$$K_{32,i} = 2K_{33,i} , \quad i = 1, \dots, k . \tag{5.27}$$

The condition $\mathbf{D}_1'[\bar{\beta}]$ should be replaced by the following condition assumed to hold for some vector parameter $\bar{\beta} = (\beta_1, \dots, \beta_k)$ with nonnegative components:

$\mathbf{D}'_7[\bar{\beta}]$: (a) $\overline{\lim}_{\varepsilon\to 0} \mathsf{E}e^{\beta_i|Y_{\varepsilon,0,i}|} < K'_{32,i}$, $i = 1, \dots, k$, for some $1 < K'_{32,i} < \infty$, $i = 1, \dots, k$;
(b) $\mathsf{P}\{\vec{Z}_{\varepsilon,0} \in \mathbb{Z}_{\varepsilon,0}\} = 1$, for every $\varepsilon \in [0, \varepsilon_0]$, where $\mathbb{Z}_{\varepsilon,0}$ is the set penetrating the condition $\mathbf{C}'_6[\bar{\beta}]$.

The following lemma gives alternative explicit upper bounds for the expectation of maximal absolute values for the pay-off processes $g_\varepsilon(n, e^{\vec{Y}_{\varepsilon,n}}, X_{\varepsilon,n})$.

Lemma 5.1.9. *Let the conditions* $\mathbf{B}_5[\bar{\gamma}]$, $\mathbf{C}'_6[\bar{\beta}]$, *and* $\mathbf{D}'_7[\bar{\beta}]$ *hold, and* $0 \le \gamma_i \le \beta_i$, $i = 1, \dots, k$. *Then, there exist* $0 < \varepsilon'_5 \le \varepsilon_0$ *and a constant* $0 \le M'_{58} < \infty$ *such that the following inequality takes place for* $0 \le \varepsilon \le \varepsilon'_5$:

$$\mathsf{E} \max_{0\le n\le N} |g_\varepsilon(n, e^{\vec{Y}_{\varepsilon,n}}, X_{\varepsilon,n})| \le M'_{34}\,. \tag{5.28}$$

Proof. The condition $\mathbf{D}'_7[\bar{\beta}]$ implies that there exists $0 < \varepsilon'_4 \le \varepsilon_0$ such that $0 \le \varepsilon \le \varepsilon'_4$ and $i = 1, \dots, k$,

$$\mathsf{E}e^{\beta_i|Y_{\varepsilon,0,i}|} < K'_{32,i}\,. \tag{5.29}$$

Relation (5.29) means that, for every $\varepsilon \le \varepsilon'_4$, the condition $\mathbf{D}'_1[\bar{\beta}]$ (with the constants $K'_{32,i}$, $i = 1, \dots, k$, replacing the constants $K'_{2,i}$, $i = 1, \dots, k$) holds for the process $\vec{Y}_{\varepsilon,n}$.

Let us now take $\varepsilon'_5 = \min(\varepsilon'_1, \varepsilon_2, \varepsilon'_4)$, where ε'_1 and ε_2 were defined in relations (5.8) and (5.11). By the definition $\varepsilon'_5 > 0$.

For every fixed $0 \le \varepsilon \le \varepsilon'_5$, the conditions $\mathbf{C}'_1[\bar{\beta}]$, $\mathbf{B}_1[\bar{\gamma}]$, and $\mathbf{D}'_1[\bar{\beta}]$ (with the constants $K'_{30,i}, K'_{32,i} i = 1, \dots, k$, and $L_{17}, L_{18,i}$, $i = 1, \dots, k$ replacing, respectively, the constants $K'_{1,i}, K'_{2,i}$, $i = 1, \dots, k$ and $L_1, L_{2,i}$, $i = 1, \dots, k$) hold for the process $\vec{Y}_{\varepsilon,n}$.

Therefore, Lemma 4.1.7 can be applied to the process $\vec{Y}_{\varepsilon,n}$, for every $0 \le \varepsilon \le \varepsilon'_5$.

This yields inequality (5.28). For every $i = 1, \dots, k$, the constant M'_4 penetrating the inequality given in Lemma 4.1.7 is a function of the constants $K'_{1,i}, K'_{2,i}$, $i = 1, \dots, k$ and $L_1, L_{2,i}$, $i = 1, \dots, k$. The explicit expression for this function is given in Remark 4.1.6. The same formulas give the expression for the constant M'_{34} as a function of the constants $K'_{30,i}, K'_{32,i}$, $i = 1, \dots, k$ and $L_{17}, L_{18,i}$, $i = 1, \dots, k$. □

Remark 5.1.16. The explicit formula for the constant M'_{34} is given by the following formula, which follows from the corresponding formulas given in Remark 4.1.6:

$$M'_{34} = L_9 + \sum_{i:\gamma_i=0} L_9 L_{10,i} + \sum_{i:\gamma_i>0} L_9 L_{10,i} (K'_{30,i})^{N\frac{\gamma_i}{\beta_i}} (K'_{32,i})^{\frac{\gamma_i}{\beta_i}}\,. \tag{5.30}$$

Remark 5.1.17. The parameter $\varepsilon'_5 = \min(\varepsilon'_1, \varepsilon_2, \varepsilon'_4)$, where the parameters ε'_1, ε_2, and ε'_4 are determined, respectively, by relations (5.8), (5.11), and (5.29).

Note also that one can replace the second-type condition of the exponential moment boundedness $\mathbf{D}'_7[\bar{\beta}]$ by the following condition of the exponential moment boundedness based on moment-generating functions for the initial value of log-price processes $\vec{Y}_{\varepsilon,0}$, and assumed to hold for some vector parameter $\bar{\beta} = (\beta_1, \dots, \beta_k)$ with nonnegative components:

$\mathbf{F}'_4[\bar{\beta}]$: (a) $\overline{\lim}_{\varepsilon \to 0} \mathsf{E} e^{\pm \beta_i Y_{\varepsilon,0,i}} < K'_{33,i}$, $i = 1, \ldots, k$, for some $1 \leq K'_{33,i} < \infty$, $i = 1, \ldots, k$;
(b) $\mathsf{P}\{\vec{Z}_{\varepsilon,0} \in \mathbb{Z}_{\varepsilon,0}\} = 1$, for every $\varepsilon \in [0, \varepsilon_0]$, where $\mathbb{Z}_{\varepsilon,0}$ is the set penetrating the condition $\mathbf{C}'_6[\bar{\beta}]$.

Lemma 5.1.10. *The condition $\mathbf{F}'_4[\bar{\beta}]$ implies the condition $\mathbf{D}'_7[\bar{\beta}]$ to hold.*

Proof. It follows from inequality (5.26). □

Remark 5.1.18. If the condition $\mathbf{F}'_4[\bar{\beta}]$ holds with the constants $K'_{33,i}$, $i = 1, \ldots, k$, then the condition $\mathbf{D}'_7[\bar{\beta}]$ holds with the constants $K'_{32,i}$, $i = 1, \ldots, k$ given by the following formulas:

$$K'_{32,i} = 2K'_{33,i}, \quad i = 1, \ldots, k. \tag{5.31}$$

Let us assume that either conditions of Lemma 5.1.7 or Lemma 5.1.9 hold. In this case, the optimal expected rewards for the American option for the log-price process $\vec{Z}_{\varepsilon,n}$ can be defined, respectively, for every $0 \leq \varepsilon \leq \varepsilon_5$, or, for every $0 \leq \varepsilon \leq \varepsilon'_5$,

$$\Phi_\varepsilon = \Phi(\mathcal{M}^{(\varepsilon)}_{\max,N}) = \sup_{\tau_{\varepsilon,0} \in \mathcal{M}^{(\varepsilon)}_{\max,N}} \mathsf{E} g_\varepsilon(\tau_{\varepsilon,0}, e^{\vec{Y}_{\varepsilon,\tau_{\varepsilon,0}}}, X_{\varepsilon,\tau_{\varepsilon,0}}). \tag{5.32}$$

Also, as follows from Theorem 2.3.4, the following formula takes place, respectively, for every $0 \leq \varepsilon \leq \varepsilon_5$, or for every $0 \leq \varepsilon \leq \varepsilon'_5$:

$$\Phi_\varepsilon = \mathsf{E}\phi_{\varepsilon,0}(\vec{Y}_{\varepsilon,0}, X_{\varepsilon,0}). \tag{5.33}$$

The following two theorems give asymptotically uniform upper bounds for the optimal expected rewards Φ_ε.

Theorem 5.1.3. *Let the conditions $\mathbf{B}_{10}[\bar{\gamma}]$, $\mathbf{C}_6[\bar{\beta}]$, and $\mathbf{D}_7[\bar{\beta}]$ hold, and $0 \leq \gamma_i \leq \beta_i$, $i = 1, \ldots, k$. Then, the following inequality takes place, for $0 \leq \varepsilon \leq \varepsilon_5$:*

$$|\Phi_\varepsilon| \leq M_{34}. \tag{5.34}$$

Proof. It follows from the definition of Φ_ε and Lemma 5.1.7 that, for $0 \leq \varepsilon \leq \varepsilon_5$:

$$\begin{aligned} |\Phi_\varepsilon| &\leq \max_{\tau_{\varepsilon,0} \in \mathcal{M}^{(\varepsilon)}_{\max,N}} \mathsf{E}|g_\varepsilon(\tau_{\varepsilon,0}, e^{\vec{Y}_{\tau_{\varepsilon,0}}}, X_{\tau_{\varepsilon,0}})| \\ &\leq \mathsf{E} \max_{0 \leq n \leq N} |g_\varepsilon(n, e^{\vec{Y}_{\varepsilon,n}}, X_{\varepsilon,n})| \leq M_{34}. \end{aligned} \tag{5.35}$$

This inequality proves the theorem. □

Theorem 5.1.4. *Let the conditions $\mathbf{B}_5[\bar{\gamma}]$, $\mathbf{C}'_6[\bar{\beta}]$, and $\mathbf{D}'_{17}[\bar{\beta}]$ hold, and $0 \leq \gamma_i \leq \beta_i$, $i = 1, \ldots, k$. Then, the following inequality takes place, for $0 \leq \varepsilon \leq \varepsilon'_5$:*

$$|\Phi_\varepsilon| \leq M'_{34}. \tag{5.36}$$

Proof. It is analogous to the proof of Theorem 5.1.3. The only difference is that Lemma 5.1.9 is used instead of Lemma 5.1.7. In this case, the inequality analogous to (5.35) can be written for $0 \leq \varepsilon \leq \varepsilon'_5$. □

5.2 Modulated Markov LPP with bounded characteristics

In this section, we present results about the convergence of option rewards for perturbed multivariate modulated Markov log-price processes with bounded characteristics.

5.2.1 Locally uniform convergence for functions and probability measures

In order to formulate the convergence conditions, we should introduce notions of locally uniform convergence for functions and probability measures.

We assume that the phase space of the index component $\mathbb{X}$ is a Polish metric space, i.e. complete, separable, metric space, with a metric $d_{\mathbb{X}}(x', x'')$.

In this case, the phase space $\mathbb{Z} = \mathbb{R}_k \times \mathbb{X}$ is also a Polish space with the metric $d_{\mathbb{Z}}(\vec{z}\,', \vec{z}\,') = (|\vec{y}\,' - \vec{y}\,''|^2 + d^2_{\mathbb{X}}(x', x''))^{\frac{1}{2}}, \vec{z}\,' = (\vec{y}\,', x'), \vec{z}\,'' = (\vec{y}\,'', x'') \in \mathbb{Z}$, where $|\vec{y}\,' - \vec{y}\,''| = (|y'_1 - y''_1|^2 + \cdots + |y'_k - y''_k|^2)^{\frac{1}{2}}, \vec{y}\,' = (y'_1, \ldots, y'_k), \vec{y}\,'' = (y''_1, \ldots, y''_k) \in \mathbb{R}_k$ is the Euclidean metric in the space $\mathbb{R}_k$.

Let us denote by $\mathcal{B}_{\mathbb{X}}$ the corresponding Borel σ-algebra of subsets of $\mathbb{X}$ (the minimal σ-algebra containing all balls $R_d(x) = \{x' \in \mathbb{X} : d_{\mathbb{X}}(x, x') \leq d\}$ in the space $\mathbb{X}$) and let $\mathcal{B}_{\mathbb{Z}}$ be the corresponding Borel σ-algebra of subsets of $\mathbb{Z}$.

Let $f_\varepsilon(\vec{z})$ be real-valued functions defined on the space $\mathbb{Z}$ and depending on the parameter $\varepsilon \in [0, \varepsilon_0]$.

We use the symbol $f_\varepsilon(\vec{z}_0) \xrightarrow{U} f_0(\vec{z}_0)$ as $\varepsilon \to 0$ to indicate that the functions $f_\varepsilon(\vec{z})$ converge to a function $f_0(\vec{z})$ locally uniformly in point $\vec{z}_0$ as $\varepsilon \to 0$. This means that, for any $\vec{z}_\varepsilon \to \vec{z}_0$ as $\varepsilon \to 0$ (in the sense that $d_{\mathbb{Z}}(\vec{z}_\varepsilon, \vec{z}_0) \to 0$ as $\varepsilon \to 0$),

$$f_\varepsilon(\vec{z}_\varepsilon) \to f_0(\vec{z}_0) \quad \text{as} \quad \varepsilon \to 0\,. \tag{5.37}$$

The term used above for the locally uniform convergence is clarified by the following useful proposition.

Lemma 5.2.1. *Relation* (5.37) *holds if and only if the following relation holds:*

$$\lim_{0<d\to 0} \overline{\lim_{\varepsilon\to 0}} \sup_{d_{\mathbb{Z}}(\vec{z},\vec{z}_0)\leq d} |f_\varepsilon(\vec{z}) - f_0(\vec{z}_0)| = 0\,. \tag{5.38}$$

Proof. Let relation (5.37) holds but relation (5.38) does not. This assumption implies that (a) there exist $c > 0$ and a sequence $0 < d_n \to 0$ as $n \to \infty$ such that $\overline{\lim}_{\varepsilon\to 0} \sup_{d_{\mathbb{Z}}(\vec{z},\vec{z}_0)\leq d_n} |f_\varepsilon(\vec{z}) - f_0(\vec{z}_0)| \geq c$. Relation (a) implies that (b) there exists $0 < \varepsilon_n \to 0$, $n = 1, 2, \ldots$ such that $\sup_{d_{\mathbb{Z}}(\vec{z},\vec{z}_0)\leq d_n} |f_{\varepsilon_n}(\vec{z}) - f_0(\vec{z}_0)| \geq c/2$. Relations (b) imply that (c) there exists $\vec{z}_n$, $n = 1, 2, \ldots$ such that $d_{\mathbb{Z}}(\vec{z}_n, \vec{z}_0) \leq d_n$, $n = 1, 2, \ldots$ and $|f_{\varepsilon_n}(\vec{z}_n) - f_0(\vec{z}_0)| \geq c/4$. Finally, relations (c) imply that (d) $f_{\varepsilon_n}(\vec{z}_n) \not\to f_0(\vec{z}_0)$ as

$n \to \infty$. Obviously, (d) contradicts to the assumption that relation (5.37) holds. Thus, relation (5.37) implies relation (5.38).

Now, let us assume that relation (5.38) holds. Let us take an arbitrary $c > 0$. Relation (5.38) implies that (e) there exists $d = d(c) > 0$ such that $\overline{\lim}_{\varepsilon \to 0} \sup_{d_{\mathbb{Z}}(\vec{z}, \vec{z}_0) \le d(c)} |f_\varepsilon(\vec{z}) - f_0(\vec{z}_0)| \le c$. Let us take an arbitrary $\vec{z}_\varepsilon \to \vec{z}_0$ as $\varepsilon \to 0$. Obviously, there exists $\varepsilon(c) > 0$ such that (f) $d_{\mathbb{Z}}(\vec{z}, \vec{z}_0) \le d(c)$ for $\varepsilon \le \varepsilon(c)$. Relations (e) and (f) imply that (g) $|f_\varepsilon(\vec{z}_\varepsilon) - f_0(\vec{z}_0)| \le \sup_{d_{\mathbb{Z}}(\vec{z}, \vec{z}_0) \le d(c)} |f_\varepsilon(\vec{z}) - f_0(\vec{z}_0)| \le c$ for $\varepsilon \le \varepsilon(c)$. Since an arbitrary choice of $c > 0$, relations (g) imply that relation (5.37) holds. Thus, relation (5.38) implies relation (5.37). □

It is useful to mention two particular cases.

The locally uniform convergence claimed by relation (5.37) implies convergence of functions $f_\varepsilon(\vec{z})$ to the function $f_0(\vec{z})$ as $\varepsilon \to 0$, in point $\vec{z}_0$, i.e. $f_\varepsilon(\vec{z}_0) \to f_0(\vec{z}_0)$ as $\varepsilon \to 0$.

Indeed, it is enough to choose $\vec{z}_\varepsilon \equiv \vec{z}_0$ in relation (5.37) to see this.

The locally uniform convergence claimed by relation (5.37) implies continuity of function $f_0(\vec{z})$, in point $\vec{z}_0$.

Indeed, the relation $\varepsilon \to 0$ in (5.38) includes also the case, where $\varepsilon \equiv 0$.

For the set $A \in \mathcal{B}_{\mathbb{Z}}$, let us denote by A^-, the subset of internal points $\vec{z} \in A$, for which there exists $\delta_{\vec{z}} > 0$ such that the ball $R_{\delta_{\vec{z}}}(\vec{z}) \subseteq A$. Also, let us denote $A^+ = \mathbb{Z} \setminus \overline{A}^-$, the closure of set A (which contains all points from the set A plus all limits of convergent sequences of points from A). Finally, let us denote $\partial A = A^+ \setminus A^-$, the boundary of set A.

Let $P_\varepsilon(A)$ be the probability measures defined on the σ-algebra $\mathcal{B}_{\mathbb{Z}}$ and depending on the parameter $\varepsilon \in [0, \varepsilon_0]$.

We use the symbol $P_\varepsilon(\cdot) \Rightarrow P_0(\cdot)$ as $\varepsilon \to 0$ to indicate that the measures $P_\varepsilon(A)$ weakly converge to a measure $P_0(A)$ as $\varepsilon \to 0$. This means that $P_\varepsilon(A) \to P_0(A)$ as $\varepsilon \to 0$ for all the Borel sets A such that $P_0(\partial A) = 0$.

Let $P_\varepsilon(\vec{z}, A), \vec{z} \in \mathbb{Z}$ be a family of probability measures defined on the σ-algebra $\mathcal{B}_{\mathbb{Z}}$ and depending on the parameter $\varepsilon \in [0, \varepsilon_0]$.

We use the symbol $P_\varepsilon(\vec{z}_0, \cdot) \stackrel{U}{\Longrightarrow} P_0(\vec{z}_0, \cdot)$ as $\varepsilon \to 0$ to indicate that measures $P_\varepsilon(\vec{z}, A)$ weakly converge to a measure $P_0(\vec{z}, A)$ locally uniformly for point $\vec{z}_0$ as $\varepsilon \to 0$. This means that, for any $\vec{z}_\varepsilon \to \vec{z}_0$ as $\varepsilon \to 0$ (in the sense that $d_{\mathbb{Z}}(\vec{z}_\varepsilon, \vec{z}_0) \to 0$ as $\varepsilon \to 0$),

$$P_\varepsilon(\vec{z}_\varepsilon, \cdot) \Rightarrow P_0(\vec{z}_0, \cdot) \quad \text{as} \quad \varepsilon \to 0\,. \tag{5.39}$$

Again, it is useful to mention two particular cases.

The locally uniform weak convergence claimed by relation (5.39) implies the weak convergence of measures $P_\varepsilon(\vec{z}_0, \cdot) \Rightarrow P_0(\vec{z}_0, \cdot)$ as $\varepsilon \to 0$.

Indeed, it is enough to choose $\vec{z}_\varepsilon \equiv \vec{z}_0$ in (5.39) to see this.

The locally uniform convergence claimed by relation (5.39) implies weak continuity for the family of measures $P_0(\vec{z}, A)$, in point $\vec{z}_0$, i.e. that $P_0(\vec{z}_\varepsilon, \cdot) \Rightarrow P_0(\vec{z}_0, \cdot)$ as $\varepsilon \to 0$ for any $\vec{z}_\varepsilon \to \vec{z}_0$ as $\varepsilon \to 0$.

Indeed, relation (5.39) implies that, for any Borel set A such that $P_0(\partial A) = 0$ functions $P_0(\vec{z}_\varepsilon, A) \xrightarrow{U} P_0(\vec{z}_0, A)$ as $\varepsilon \to 0$. By the remarks above this relation implies that the function $P_0(\vec{z}, A)$ is continuous in point $\vec{z}_0$.

5.2.2 Convergence of log-reward functions for multivariate modulated Markov log-price processes with bounded characteristics

Let us now consider the log-reward functions $\phi_{\varepsilon,n}(\vec{y}, x)$ introduced in Subsection 5.1.2 and give conditions for their convergence.

We impose on the pay-off functions $g_\varepsilon(n, e^{\vec{y}}, x)$ the following condition of locally uniform convergence:

I$_1$: There exist measurable sets $\mathbb{Z}'_n \subseteq \mathbb{Z}$, $n = 0, \dots, N$, such that the pay-off functions $g_\varepsilon(n, e^{\vec{y}_0}, x_0) \xrightarrow{U} g_0(n, e^{\vec{y}_0}, x_0)$ as $\varepsilon \to 0$, i.e. $g_\varepsilon(n, e^{\vec{y}_\varepsilon}, x_\varepsilon) \to g_0(n, e^{\vec{y}_0}, x_0)$ as $\varepsilon \to 0$ for any $\vec{z}_\varepsilon = (\vec{y}_\varepsilon, x_\varepsilon) \to \vec{z}_0 = (\vec{y}_0, x_0) \in \mathbb{Z}'_n$ as $\varepsilon \to 0$, and $n = 0, \dots, N$.

We also impose on transition probabilities of log-price processes $P_{\varepsilon,n}(\vec{z}, A) = \mathsf{P}\{\vec{Z}_{\varepsilon,n} \in A / \vec{Z}_{\varepsilon,n-1} = \vec{z}\}$ the following condition of locally uniform weak convergence:

J$_1$: There exist measurable sets $\mathbb{Z}''_n \subseteq \mathbb{Z}$, $n = 0, \dots, N$ such that
(a) $P_{\varepsilon,n}(\vec{z}_0, \cdot) \stackrel{U}{\Longrightarrow} P_{0,n}(\vec{z}_0, \cdot)$ as $\varepsilon \to 0$, i.e. $P_{\varepsilon,n}(\vec{z}_\varepsilon, \cdot) \Rightarrow P_{0,n}(\vec{z}_0, \cdot)$ as $\varepsilon \to 0$, for any $\vec{z}_\varepsilon = (\vec{y}_\varepsilon, x_\varepsilon) \to \vec{z}_0 = (\vec{y}_0, x_0) \in \mathbb{Z}''_{n-1}$ as $\varepsilon \to 0$, and $n = 1, \dots, N$;
(b) $P_{0,n}(\vec{z}_0, \mathbb{Z}'_n \cap \mathbb{Z}''_n) = 1$, for every $\vec{z}_0 \in \mathbb{Z}'_{n-1} \cap \mathbb{Z}''_{n-1}$ and $n = 1, \dots, N$, where $\mathbb{Z}'_n$, $n = 0, \dots, N$ are sets introduced in the condition **I$_1$**.

A typical example is where the sets $\overline{\mathbb{Z}}'_n$, $\overline{\mathbb{Z}}''_n$, $n = 1, \dots, N$ are empty sets. Then the condition **J$_1$** (b) obviously holds.

Another typical example is where the sets $\overline{\mathbb{Z}}'_n$, $\overline{\mathbb{Z}}''_n$, $n = 1, \dots, N$ are at most finite or countable sets. Then the assumption that the measures $P_{0,n}(\vec{z}_0, A)$, $\vec{z}_0 \in \mathbb{Z}'_{n-1} \cap \mathbb{Z}''_{n-1}$ have no atoms at points from the sets $\overline{\mathbb{Z}}'_n$, $\overline{\mathbb{Z}}''_n$, for every $n = 1, \dots, N$, implies that the condition **J$_1$** (b) holds.

One more example is where measures $P_{0,n}(\vec{z}_0, A)$, $\vec{z}_0 \in \mathbb{Z}'_{n-1} \cap \mathbb{Z}''_{n-1}$, $n = 1, \dots, N$ are absolutely continuous with respect to some σ-finite measure $P(A)$ on $\mathcal{B}_{\mathbb{Z}}$ and $P(\overline{\mathbb{Z}}'_n), P(\overline{\mathbb{Z}}''_n) = 0$, $n = 1, \dots, N$. This assumption also implies that the condition **J$_1$** (b) holds.

Theorem 5.2.1. *Let the conditions* **B$_5$**$[\bar{\gamma}]$ *and* **C$_6$**$[\bar{\beta}]$ *hold with the vector parameters* $\bar{\gamma} = (\gamma_1, \dots, \gamma_k)$, $\bar{\beta} = (\beta_1, \dots, \beta_k)$ *such that, for every* $i = 1, \dots, k$, *either* $\beta_i > \gamma_i > 0$ *or* $\beta_i = \gamma_i = 0$, *and also the conditions* **I$_1$** *and* **J$_1$** *hold. Then, for every* $n = 0, 1, \dots, N$, *the following relation takes place for any* $\vec{z}_\varepsilon = (\vec{y}_\varepsilon, x_\varepsilon) \to \vec{z}_0 = (\vec{y}_0, x_0) \in \mathbb{Z}'_n \cap \mathbb{Z}''_n$:

$$\phi_{\varepsilon,n}(\vec{y}_\varepsilon, x_\varepsilon) \to \phi_{0,n}(\vec{y}_0, x_0) \quad as \quad \varepsilon \to 0. \tag{5.40}$$

Proof. Due to the above conditions imposed on the parameters $\gamma_i, \beta_i, i = 1, \dots, k$, one can always find $\alpha > 1$ such that $\gamma_i \alpha \leq \beta_i, i = 1, \dots, k$.

By Theorem 5.1.1, the conditions $\mathbf{B}_5[\bar{\gamma}]$ and $\mathbf{C}_6[\bar{\beta}]$ imply that there exist $\varepsilon_3 > 0$ and the constants $M_{32}, M_{33,i}, i = 1, \dots, k$ such that the following upper bounds take place for the log-reward functions $\phi_{\varepsilon,n}(\vec{y}, x)$ for $\vec{z} = (\vec{y}, x) \in \mathbb{Z}, 0 \leq n \leq N$ and $\varepsilon \leq \varepsilon_3$:

$$|\phi_{\varepsilon,n}(\vec{y}, x)| \leq M_{32} + \sum_{i=1}^{k} M_{33,i} e^{\gamma_i |y_i|} . \tag{5.41}$$

Using inequalities (5.41) and the condition $\mathbf{C}_6[\bar{\beta}]$ we get for any $\vec{z} = (\vec{y}, x) \in \mathbb{Z}, 0 \leq n \leq N - 1$ and $\varepsilon \leq \varepsilon_3$,

$$\begin{aligned}
&\mathsf{E}_{\vec{z},n} \left| \phi_{\varepsilon,n+1}\left(\vec{Y}_{\varepsilon,n+1}, X_{\varepsilon,n+1}\right) \right|^{\alpha} \\
&\quad \leq \mathsf{E}_{\vec{z},n} \left(M_{32} + \sum_{i=1}^{k} M_{33,i} e^{\gamma_i |Y_{\varepsilon,n+1,i}|} \right)^{\alpha} \\
&\quad \leq \mathsf{E}_{\vec{z},n} (k+1)^{\alpha-1} \left(M_{32}^{\alpha} + \sum_{i=1}^{k} M_{33,i}^{\alpha} e^{\gamma_i \alpha |Y_{\varepsilon,n+1,i}|} \right) \\
&\quad \leq (k+1)^{\alpha-1} \left(M_{32}^{\alpha} + \sum_{i=1}^{k} M_{33,i}^{\alpha} \mathsf{E}_{\vec{z},n} e^{\gamma_i \alpha |y_i|} e^{\gamma_i \alpha |Y_{\varepsilon,n+1,i} - y_i|} \right) \\
&\quad \leq (k+1)^{\alpha-1} \left(M_{32}^{\alpha} + \sum_{i=1}^{k} M_{33,i}^{\alpha} e^{\gamma_i \alpha |y_i|} \mathsf{E}_{\vec{z},n} e^{\beta_i |Y_{\varepsilon,n+1,i} - Y_{\varepsilon,n,i}|} \right) \\
&\quad \leq (k+1)^{\alpha-1} \left(M_{32}^{\alpha} + \sum_{i=1}^{k} M_{33,i}^{\alpha} e^{\gamma_i \alpha |y_i|} K_{30,i} \right) = Q(\vec{y}) < \infty .
\end{aligned} \tag{5.42}$$

According to Theorem 2.3.1, the reward functions $\phi_{\varepsilon,n}(\vec{y}, x)$ are defined by the following boundary relation:

$$\phi_{\varepsilon,N}(\vec{y}, x) = g_{\varepsilon}(N, e^{\vec{y}}, x) , \quad \vec{z} = (\vec{y}, x) \in \mathbb{Z} , \tag{5.43}$$

and backward recurrence relations

$$\begin{aligned}
\phi_{\varepsilon,n}(\vec{y}, x) &= \max(g(n, e^{\vec{y}}, x), \mathsf{E}_{\vec{z},n} \phi_{\varepsilon,n+1}(\vec{Y}_{\varepsilon,n+1}, X_{\varepsilon,n+1})), \\
&\qquad \vec{z} = (\vec{y}, x) \in \mathbb{Z} , \quad n = N-1, \dots, 0 .
\end{aligned} \tag{5.44}$$

We shall prove asymptotic relations (5.40) using recursive reasoning based on relations (5.42)–(5.44).

The condition $\mathbf{I}_1$ implies, due to the boundary relation (5.43) for the log-reward function $\phi_{\varepsilon,N}(\vec{y}, x)$, that the following relation holds for an arbitrary $\vec{z}_{\varepsilon} = (\vec{y}_{\varepsilon}, x_{\varepsilon}) \to \vec{z}_0 = (\vec{y}_0, x_0) \in \mathbb{Z}'_N \cap \mathbb{Z}''_N$ as $\varepsilon \to 0$:

$$\phi_{\varepsilon,N}(\vec{y}_{\varepsilon}, x_{\varepsilon}) \to \phi_{0,N}(\vec{y}_0, x_0) \quad \text{as} \quad \varepsilon \to 0 . \tag{5.45}$$

Let us prove that the following relation, similar to (5.45), holds for an arbitrary $\vec{z}_\varepsilon = (\vec{y}_\varepsilon, x_\varepsilon) \to \vec{z}_0 = (\vec{y}_0, x_0) \in \mathbb{Z}'_{N-1} \cap \mathbb{Z}''_{N-1}$ as $\varepsilon \to 0$:

$$\phi_{\varepsilon,N-1}(\vec{y}_\varepsilon, x_\varepsilon) \to \phi_{0,N-1}(\vec{y}_0, x_0) \quad \text{as} \quad \varepsilon \to 0 . \tag{5.46}$$

Let $\vec{Z}_{\varepsilon,n}(\vec{z}) = (\vec{Y}_{\varepsilon,n}(\vec{z}), X_{\varepsilon,n}(\vec{z}))$ be, for every $n = 1, \dots, N$ and $\vec{z} \in \mathbb{Z}$, a random vector such that $\mathsf{P}\{\vec{Z}_{\varepsilon,n}(\vec{z}) \in A\} = P_{\varepsilon,n}(\vec{z}, A)$, $A \in \mathcal{B}_{\mathbb{Z}}$.

Let us prove that the following relation takes place:

$$\phi_{\varepsilon,N}(\vec{Y}_{\varepsilon,N}(\vec{z}_\varepsilon), X_{\varepsilon,N}(\vec{z}_\varepsilon)) \xrightarrow{d} \phi_{0,N}(\vec{Y}_{0,N}(\vec{z}_0), X_{0,N}(\vec{z}_0)) \quad \text{as} \quad \varepsilon \to 0 . \tag{5.47}$$

Take an arbitrary sequence $\varepsilon_r \to \varepsilon_0 = 0$ as $r \to \infty$.

The condition $\mathbf{J}_1$ implies that

$$\vec{Z}_{\varepsilon_r,N}(\vec{z}_{\varepsilon_r}) \xrightarrow{d} \vec{Z}_{\varepsilon_0,N}(\vec{z}_{\varepsilon_0}) \quad \text{as} \quad r \to \infty , \tag{5.48}$$

and

$$\mathsf{P}\{\vec{Z}_{\varepsilon_0,N}(\vec{z}_{\varepsilon_0}) \in \mathbb{Z}'_N \cap \mathbb{Z}''_N\} = 1 . \tag{5.49}$$

According to the Skorokhod representation theorem, relation (5.48) let one construct the random variables $\vec{Z}'_{\varepsilon_r,N}(\vec{z}_{\varepsilon_r}) = (\vec{Y}'_{\varepsilon_r,N}(\vec{z}_{\varepsilon_r}), X'_{\varepsilon_r,N}(\vec{z}_{\varepsilon_r}))$, $r = 0, 1, \dots$ on some probability space $(\Omega, \mathcal{F}, \mathsf{P})$ such that, for every $r = 0, 1, \dots$,

$$\mathsf{P}\{\vec{Z}'_{\varepsilon_r,N}(\vec{z}_{\varepsilon_r}) \in A\} = \mathsf{P}\{\vec{Z}_{\varepsilon_r,N}(\vec{z}_{\varepsilon_r}) \in A\} , \quad A \in \mathcal{B}_{\mathbb{Z}} , \tag{5.50}$$

and such that random variables $\vec{Z}'_{\varepsilon_r,N}(\vec{z}_{\varepsilon_r})$ converge almost sure to the random variable $\vec{Z}'_{\varepsilon_0,N}(\vec{z}_{\varepsilon_0})$ as $r \to \infty$, i.e.

$$\vec{Z}'_{\varepsilon_r,N}(\vec{z}_{\varepsilon_r}) \xrightarrow{\text{a.s.}} \vec{Z}'_{\varepsilon_0,N}(\vec{z}_{\varepsilon_0}) \quad \text{as} \quad r \to \infty . \tag{5.51}$$

Let us denote

$$A_N = \{\omega \in \Omega : \vec{Z}'_{\varepsilon_r,N}(\vec{z}_{\varepsilon_r}, \omega) \to \vec{Z}'_{\varepsilon_0,N}(\vec{z}_{\varepsilon_0}, \omega) \text{ as } r \to \infty\}$$

and

$$B_N = \{\omega \in \Omega : \vec{Z}'_{\varepsilon_0,N}(\vec{z}_{\varepsilon_0}, \omega) \in \mathbb{Z}'_N \cap \mathbb{Z}''_N\} .$$

Relation (5.51) implies that $\mathsf{P}(A_N) = 1$. Also relations (5.49) and (5.50) imply that $\mathsf{P}(B_N) = 1$. Thus, $\mathsf{P}(A_N \cap B_N) = 1$.

By relation (5.45) and the definition of the sets A_N and B_N, the following relation takes place, for every $\omega \in A_N \cap B_N$:

$$\begin{aligned} &\phi_{\varepsilon_r,N}(\vec{Y}'_{\varepsilon_r,N}(\vec{z}_{\varepsilon_r}, \omega), X'_{\varepsilon_r,N}(\vec{z}_{\varepsilon_r}, \omega)) \\ &\quad \to \phi_{\varepsilon_0,N}(\vec{Y}'_{\varepsilon_0,N}(\vec{z}_{\varepsilon_0}, \omega), X'_{\varepsilon_0,N}(\vec{z}_{\varepsilon_0}, \omega)) \quad \text{as} \quad r \to \infty . \end{aligned} \tag{5.52}$$

Since $\mathsf{P}(A_N \cap B_N) = 1$, relation (5.52) implies that the following relation holds:

$$\begin{aligned} &\phi_{\varepsilon_r,N}(\vec{Y}'_{\varepsilon_r,N}(\vec{z}_{\varepsilon_r}), X'_{\varepsilon_r,N}(\vec{z}_{\varepsilon_r})) \\ &\quad \xrightarrow{a.s.} \phi_{\varepsilon_0,N}(\vec{Y}'_{\varepsilon_0,N}(\vec{z}_{\varepsilon_0}), X'_{\varepsilon_0,N}(\vec{z}_{\varepsilon_0})) \quad \text{as} \quad r \to \infty . \end{aligned} \tag{5.53}$$

Relation (5.50) implies that, for every $r = 0, 1, \dots,$

$$\begin{aligned} &\mathsf{P}\{\phi_{\varepsilon_r,N}(\vec{Y}'_{\varepsilon_r,N}(\vec{z}_{\varepsilon_r}), X'_{\varepsilon_r,N}(\vec{z}_{\varepsilon_r})) \in A\} \\ &\quad = \mathsf{P}\{\phi_{\varepsilon_r,N}(\vec{Y}_{\varepsilon_r,N}(\vec{z}_{\varepsilon_r}), X_{\varepsilon_r,N}(\vec{z}_{\varepsilon_r})) \in A\}, \quad A \in \mathcal{B}_{\mathbb{Z}}. \end{aligned} \tag{5.54}$$

Relations (5.53) and (5.54) imply that the following relation holds:

$$\begin{aligned} &\phi_{\varepsilon_r,N}(\vec{Y}_{\varepsilon_r,N}(\vec{z}_{\varepsilon_r}), X_{\varepsilon_r,N}(\vec{z}_{\varepsilon_r})) \\ &\quad \xrightarrow{d} \phi_{\varepsilon_0,N}(\vec{Y}_{\varepsilon_0,N}(\vec{z}_{\varepsilon_0}), X_{\varepsilon_0,N}(\vec{z}_{\varepsilon_0})) \quad \text{as} \quad r \to \infty. \end{aligned} \tag{5.55}$$

Since an arbitrary choice of a sequence $\varepsilon_r \to \varepsilon_0$, relation (5.55) implies the relation of weak convergence (5.47).

By the definition of the random vectors $\vec{Z}_{\varepsilon,n}(\vec{z}) = (\vec{Y}_{\varepsilon,n}(\vec{z}), X_{\varepsilon,n}(\vec{z}))$, the following relation take place for every $\vec{z} \in \mathbb{Z}$:

$$\mathsf{E}|\phi_{\varepsilon,N}(\vec{Y}_{\varepsilon,N}(\vec{z}), X_{\varepsilon,N}(\vec{z}))|^\alpha = \mathsf{E}_{\vec{z},N-1}|\phi_{\varepsilon,N}(\vec{Y}_{\varepsilon,N}, X_{\varepsilon,N})|^\alpha \tag{5.56}$$

The function $Q(\vec{y})$ defined in (5.42) is a continuous function of argument $\vec{y} \in \mathbb{R}_k$. Therefore, relation (5.42) implies that the following relation takes place:

$$\overline{\lim_{\varepsilon \to 0}} \mathsf{E}|\phi_{\varepsilon,N}(\vec{Y}_{\varepsilon,N}(\vec{z}_\varepsilon), X_{\varepsilon,N}(\vec{z}_\varepsilon))|^\alpha \le Q(\vec{y}_0) < \infty. \tag{5.57}$$

Relations (5.47) and (5.57) imply that the following relation holds:

$$\begin{aligned} \mathsf{E}\phi_{\varepsilon,N}(\vec{Y}_{\varepsilon,N}(\vec{z}_\varepsilon), X_{\varepsilon,N}(\vec{z}_\varepsilon)) &= \mathsf{E}_{\vec{z}_\varepsilon,N-1}\phi_{\varepsilon,N}(\vec{Y}_{\varepsilon,N}, X_{\varepsilon,N}) \\ &\to \mathsf{E}\phi_{\varepsilon,N}(\vec{Y}_{0,N}(\vec{z}_0), X_{0,N}(\vec{z}_0)) \\ &= \mathsf{E}_{\vec{z}_0,N-1}\phi_{0,N}(\vec{Y}_{0,N}, X_{0,N}) \quad \text{as} \quad \varepsilon \to 0. \end{aligned} \tag{5.58}$$

Finally, relation (5.58) and the condition $\mathbf{I}_1$ imply that the following relation, equivalent to (5.46), holds:

$$\begin{aligned} \phi_{\varepsilon,N-1}(\vec{y}_\varepsilon, x_\varepsilon) &= \max(g_\varepsilon(N-1, e^{\vec{y}_\varepsilon}, x_\varepsilon), \mathsf{E}_{\vec{z}_\varepsilon,N-1}\phi_{\varepsilon,N}(\vec{Y}_{\varepsilon,N}, X_{\varepsilon,N})) \\ &\to \phi_{0,N-1}(\vec{y}_0, x_0) = \max(g_0(N-1, e^{\vec{y}_0}, x_0), \\ &\quad \mathsf{E}_{\vec{z}_0,N-1}\phi_{0,N}(\vec{Y}_{0,N}, X_{0,N})) \quad \text{as} \quad \varepsilon \to 0. \end{aligned} \tag{5.59}$$

By repeating the recursive procedure described above we prove relations (5.40) for $n = N-2, \dots, 0$. □

Remark 5.2.1. According to Lemma 5.1.5, the conditions $\mathbf{C}_6[\bar{\beta}]$ can be replaced in Theorems 5.2.1 by the condition $\mathbf{E}_{10}[\bar{\beta}]$.

Remark 5.2.2. As follows from the remarks made in Subsection 5.2.1, the following point-wise convergence relation takes place for the log-reward functions $\phi_{\varepsilon,n}(\vec{y}, x)$, for every $\vec{z} = (\vec{y}, x) \in \mathbb{Z}'_n \cap \mathbb{Z}''_n$, $n = 0, 1, \dots, N$:

$$\phi_{\varepsilon,n}(\vec{y}, x) \to \phi_{0,n}(\vec{y}, x) \quad \text{as} \quad \varepsilon \to 0. \tag{5.60}$$

Also, the reward function $\phi_{0,n}(\vec{y}, x)$ is continuous in points $\vec{z} = (\vec{y}, x) \in \mathbb{Z}'_n \cap \mathbb{Z}''_n$, for every $n = 0, 1, \dots, N$.

The following theorem gives alternative conditions for the convergence of log-reward functions for the American-type options.

Theorem 5.2.2. *Let the conditions* $\mathbf{B}_5[\bar{\gamma}]$ *and* $\mathbf{C}'_6[\bar{\beta}]$ *hold with the vector parameters* $\bar{\gamma} = (\gamma_1, \dots, \gamma_k), \bar{\beta} = (\beta_1, \dots, \beta_k)$ *such that, for every* $i = 1, \dots, k$, *either* $\beta_i > \gamma_i > 0$ *or* $\beta_i = \gamma_i = 0$, *and also the conditions* $\mathbf{H}_1$, $\mathbf{I}_1$, *and* $\mathbf{J}_1$ *hold. Then, for every* $n = 0, 1, \dots, N$, *the following relation takes place for any* $\vec{z}_\varepsilon = (\vec{y}_\varepsilon, x_\varepsilon) \to \vec{z}_0 = (\vec{y}_0, x_0) \in \mathbb{Z}'_n \cap \mathbb{Z}''_n$:

$$\phi_{\varepsilon,n}(\vec{y}_\varepsilon, x_\varepsilon) \to \phi_{0,n}(\vec{y}_0, x_0) \quad \text{as} \quad \varepsilon \to 0\,. \tag{5.61}$$

Proof. It repeats the proof of Theorem 5.2.1, with some minor changes.

Due to the above conditions imposed on the parameters γ_i, β_i, $i = 1, \dots, k$, one can always find $\alpha > 1$ such that $\gamma_i \alpha \le \beta_i$, $i = 1, \dots, k$.

By Theorem 5.1.2, the conditions $\mathbf{B}_5[\bar{\gamma}]$ and $\mathbf{C}'_6[\bar{\beta}]$ imply that there exist $\varepsilon'_3 > 0$ and the constants $M'_{32}, M'_{33,i}$, $i = 1, \dots, k$ such that the following upper bounds take place for the log-reward functions $\phi_{\varepsilon,n}(\vec{y}, x)$, for $\vec{z} = (\vec{y}, x) \in \mathbb{Z}_{\varepsilon,n}$, $0 \le n \le N$, $\varepsilon \le \varepsilon'_3$:

$$|\phi_{\varepsilon,n}(\vec{y}, x)| \le M'_{32} + \sum_{i=1}^{k} M'_{33,i} e^{\gamma_i |y_i|}\,. \tag{5.62}$$

Recall that the condition $\mathbf{C}'_6[\bar{\beta}]$ (b) implies that $\mathsf{P}_{\vec{z},n}\{\vec{Z}_{\varepsilon,n+1} \in \mathbb{Z}_{\varepsilon,n+1}\} = 1$ for $\vec{z} \in \mathbb{Z}_{\varepsilon,n}$, $n = 0, \dots, N-1$. Using these relations, inequalities (5.62) and the condition $\mathbf{C}'_6[\bar{\beta}]$, we get for any $\vec{z} = (\vec{y}, x) \in \mathbb{Z}_{\varepsilon,n}$, $0 \le n \le N-1$, $\varepsilon \le \varepsilon'_3$,

$$\begin{aligned}
&\mathsf{E}_{\vec{z},n} \left| \phi_{\varepsilon,n+1}\left(\vec{Y}_{\varepsilon,n+1}, X_{\varepsilon,n+1}\right) \right|^{\alpha} \\
&\quad \le \mathsf{E}_{\vec{z},n} \left(M'_{32} + \sum_{i=1}^{k} M'_{33,i} e^{\gamma_i |Y_{\varepsilon,n+1,i}|} \right)^{\alpha} \\
&\quad \le \mathsf{E}_{\vec{z},n} (k+1)^{\alpha-1} \left((M'_{32})^{\alpha} + \sum_{i=1}^{k} (M'_{33,i})^{\alpha} e^{\gamma_i \alpha |Y_{\varepsilon,n+1,i}|} \right) \\
&\quad \le (k+1)^{\alpha-1} \left((M'_{32})^{\alpha} + \sum_{i=1}^{k} (M'_{33,i})^{\alpha} \mathsf{E}_{\vec{z},n} e^{\gamma_i \alpha |y_i|} e^{\gamma_i \alpha |Y_{\varepsilon,n+1,i} - y_i|} \right) \\
&\quad \le (k+1)^{\alpha-1} \left((M'_{32})^{\alpha} + \sum_{i=1}^{k} (M'_{33,i})^{\alpha} e^{\gamma_i \alpha |y_i|} \mathsf{E}_{\vec{z},n} e^{\beta_i |Y_{\varepsilon,n+1,i} - Y_{\varepsilon,n,i}|} \right) \\
&\quad \le (k+1)^{\alpha-1} \left((M'_{32})^{\alpha} + \sum_{i=1}^{k} (M'_{33,i})^{\alpha} e^{\gamma_i \alpha |y_i|} K'_{30,i} \right) = Q'(\vec{y}) < \infty\,.
\end{aligned} \tag{5.63}$$

The following part of the proof repeats the corresponding part of the proof for Theorem 5.2.1. The only difference is that additional reasoning is required for the relation, which should replace relation (5.57).

As in the proof of Theorem 5.2.1, we choose an arbitrary function $\vec{z}(\cdot) = \langle \vec{z}_\varepsilon, \varepsilon \in [0, \varepsilon_0] \rangle$ such that $\vec{z}_\varepsilon \to \vec{z}_0 \in \mathbb{Z}'_{N-1} \cap \mathbb{Z}''_{N-1}$ as $\varepsilon \to 0$, and should prove that

$$\overline{\lim_{\varepsilon \to 0}} \mathsf{E} |\phi_{\varepsilon,N}(\vec{Y}_{\varepsilon,N}(\vec{z}_\varepsilon), X_{\varepsilon,N}(\vec{z}_\varepsilon))|^\alpha \leq Q'(\vec{y}_0) < \infty . \tag{5.64}$$

In this case, we should use the condition $\mathbf{H}_1$, due to which the point $\vec{z}_0 = (\vec{y}_0, x_0) \in \mathbb{Z}_{\varepsilon,N-1}$, for every $\varepsilon \in [0, \varepsilon_1(\vec{z}_0)]$. The parameter $\varepsilon_1(\vec{z}_0) \in (0, \varepsilon_0]$ has been defined in relation (5.6) given in Remark 5.1.3.

Since $\vec{z}_\varepsilon \to \vec{z}_0$ as $\varepsilon \to 0$, this implies that there exists $\varepsilon'_1(\vec{z}(\cdot)) \in (0, \varepsilon_1(\vec{z}_0)]$ such that $\vec{z}_\varepsilon = (\vec{y}_\varepsilon, x_\varepsilon) \in \mathbb{Z}_{\varepsilon,n}$, $n = 0, \dots, N$, for every $\varepsilon \in [0, \varepsilon'_1(\vec{z}(\cdot))]$.

Therefore, the upper bounds for $\mathsf{E}_{\vec{z},N-1} |\phi_{\varepsilon,N}(\vec{Y}_{\varepsilon,N}, X_{\varepsilon,N})|^\alpha$ given in relation (5.63) hold for every $\varepsilon \in [0, \min(\varepsilon'_3, \varepsilon'_1(\vec{z}(\cdot)))]$.

Due to the above remarks and continuity of the function $Q'(\vec{y})$, relation (5.63) implies that relation (5.64), replacing relation (5.57), holds.

The rest of the proof repeats the corresponding part of the proof of Theorem 5.2.1. □

5.2.3 Convergence of optimal expected rewards for modulated Markov log-price processes with bounded characteristics

Let us now consider the optimal expected rewards $\Phi_\varepsilon = \Phi(\mathcal{M}^{(\varepsilon)}_{\max,N})$ introduced in Subsection 5.1.2 and give conditions for their convergence.

We should additionally impose a condition of weak convergence on the initial distributions $P_{\varepsilon,0}(A) = \mathsf{P}\{\vec{Z}_{\varepsilon,0} \in A\}$:

$\mathbf{K}_1$: (a) $P_{\varepsilon,0}(\cdot) \Rightarrow P_{0,0}(\cdot)$ as $\varepsilon \to 0$;
(b) $P_{0,0}(\mathbb{Z}'_0 \cap \mathbb{Z}''_0) = 1$, where $\mathbb{Z}'_0$ and $\mathbb{Z}''_0$ are the sets introduced in the conditions $\mathbf{I}_1$ and $\mathbf{J}_1$.

A typical example is where the sets $\overline{\mathbb{Z}}'_0, \overline{\mathbb{Z}}''_0$, $n = 1, \dots, N$ are empty sets. Then the condition $\mathbf{K}_1$ (b) obviously holds.

Another typical example is where sets $\overline{\mathbb{Z}}'_0, \overline{\mathbb{Z}}''_0$ are at most finite or countable sets. Then the assumption that measure $P_{0,0}(\cdot)$ has no atoms at points of sets $\overline{\mathbb{Z}}'_0, \overline{\mathbb{Z}}''_0$ implies that the condition $\mathbf{K}_1$ (b) holds.

One more example is where measure $P_{0,0}(\cdot)$ is absolutely continuous with respect to some σ-finite measure $P(A)$ on $\mathcal{B}_\mathbb{Z}$ and $P(\overline{\mathbb{Z}}'_0), P(\overline{\mathbb{Z}}''_0) = 0$. This assumption also implies that the condition $\mathbf{K}_1$ (b) holds.

Theorem 5.2.3. *Let the conditions $\mathbf{B}_5[\bar{\gamma}]$, $\mathbf{C}_6[\bar{\beta}]$, and $\mathbf{D}_7[\bar{\beta}]$ hold with the vector parameters $\bar{\gamma} = (\gamma_1, \dots, \gamma_k)$, $\bar{\beta} = (\beta_1, \dots, \beta_k)$ such that, for every $i = 1, \dots, k$, either $\beta_i > \gamma_i > 0$ or $\beta_i = \gamma_i = 0$, and also the conditions $\mathbf{I}_1$, $\mathbf{J}_1$, and $\mathbf{K}_1$ hold. Then the following relation takes place:*

$$\Phi_\varepsilon \to \Phi_0 \quad as \quad \varepsilon \to 0 . \tag{5.65}$$

Proof. Due to the above conditions imposed on the parameters $\gamma_i, \beta_i, i = 1, \dots, k$, one can always find $\alpha > 1$ such that $\gamma_i \alpha \le \beta_i, i = 1, \dots, k$.

According to Theorem 5.1.1, the conditions $\mathbf{B}_5[\bar{\gamma}]$ and $\mathbf{C}_6[\bar{\beta}]$ imply that there exist $\varepsilon_3 > 0$ and the constants $M_{32}, M_{33,i}, i = 1, \dots, k$ such that the following upper bounds take place for the reward functions $\phi_{\varepsilon,0}(\vec{y}, x)$ for $\vec{z} = (\vec{y}, x) \in \mathbb{Z}$ and $\varepsilon \le \varepsilon_3$:

$$|\phi_{\varepsilon,0}(\vec{y}, x)| \le M_{32} + \sum_{i=1}^{k} M_{33,i} e^{\gamma_i |y_i|} . \tag{5.66}$$

As was pointed out in the proof of Lemma 5.1.7, the condition $\mathbf{D}_7[\bar{\beta}]$ implies that there exists $0 < \varepsilon_4 \le \varepsilon_0$ such that $0 \le \varepsilon \le \varepsilon_4$ and $i = 1, \dots, k$,

$$\mathsf{E} e^{\beta_i |Y_{\varepsilon,0,i}|} < K_{32,i} . \tag{5.67}$$

Using relations (5.66), (5.67), and (5.69), we get that for any $\varepsilon \le \varepsilon_5 = \min(\varepsilon_3, \varepsilon_4)$,

$$\begin{aligned} \mathsf{E}|\phi_{\varepsilon,0}(\vec{Y}_{\varepsilon,0}, X_{\varepsilon,0})|^\alpha &\le \mathsf{E}\left(M_{32} + \sum_{i=1}^{k} M_{33,i} e^{\gamma_i |Y_{\varepsilon,0,i}|} \right)^\alpha \\ &\le \mathsf{E}(k+1)^{\alpha-1} \left(M_{32}^\alpha + \sum_{i=1}^{k} M_{33,i}^\alpha e^{\gamma_i \alpha |Y_{\varepsilon,0,i}|} \right) \\ &\le (k+1)^{\alpha-1} \left(M_{32}^\alpha + \sum_{i=1}^{k} M_{33,i}^\alpha \mathsf{E} e^{\beta_i |Y_{\varepsilon,0,i}|} \right) \\ &\le (k+1)^{\alpha-1} \left(M_{32}^\alpha + \sum_{i=1}^{k} M_{33,i}^\alpha K_{32,i} \right) = Q < \infty . \end{aligned} \tag{5.68}$$

Also, as follows from Theorem 2.3.4, the following formula takes place, for every $0 \le \varepsilon \le \varepsilon_5$:

$$\Phi_\varepsilon = \mathsf{E}\phi_{\varepsilon,0}(\vec{Y}_{\varepsilon,0}, X_{\varepsilon,0}) . \tag{5.69}$$

Let us prove that the following relation takes place:

$$\phi_{\varepsilon,0}(\vec{Y}_{\varepsilon,0}, X_{\varepsilon,0}) \xrightarrow{d} \phi_{0,0}(\vec{Y}_{0,0}, X_{0,0}) \quad \text{as} \quad \varepsilon \to 0 . \tag{5.70}$$

Take an arbitrary sequence $\varepsilon_r \to \varepsilon_0 = 0$ as $r \to \infty$.

The condition $\mathbf{K}_1$ implies that

$$\vec{Z}_{\varepsilon_r,0} \xrightarrow{d} \vec{Z}_{\varepsilon_0,0} \quad \text{as} \quad r \to \infty , \tag{5.71}$$

and

$$\mathsf{P}\{\vec{Z}_{\varepsilon_0,0} \in \mathbb{Z}'_0 \cap \mathbb{Z}''_0\} = 1 . \tag{5.72}$$

According to the Skorokhod representation theorem, for relation (5.71) let us construct the random variables $\vec{Z}'_{\varepsilon_r,0} = (\vec{Y}'_{\varepsilon_r,0}, X'_{\varepsilon_r,0}), r = 0, 1, \dots$ on some probability space $(\Omega, \mathcal{F}, \mathsf{P})$ such that, for every $r = 0, 1, \dots,$

$$\mathsf{P}\{\vec{Z}'_{\varepsilon_r,0} \in A\} = \mathsf{P}\{\vec{Z}_{\varepsilon_r,0} \in A\} , \quad A \in \mathcal{B}_{\mathbb{Z}} , \tag{5.73}$$

and

$$\vec{Z}'_{\varepsilon_r,0} \xrightarrow{a.s.} \vec{Z}'_{\varepsilon_0,0} \quad \text{as} \quad r \to \infty\,. \tag{5.74}$$

Let us denote

$$A = \{\omega \in \Omega : \vec{Z}'_{\varepsilon_r,0}(\omega) \to \vec{Z}'_{\varepsilon_0,0}(\omega) \text{ as } r \to \infty\}$$

and

$$B = \{\omega \in \Omega : \vec{Z}'_{\varepsilon_0,0}(\omega) \in \mathbb{Z}'_0 \cap \mathbb{Z}''_0\}\,.$$

Relation (5.74) implies that $\mathsf{P}(A) = 1$. Also relations (5.72) and (5.73) imply that $\mathsf{P}(B) = 1$. Thus, $\mathsf{P}(A \cap B) = 1$.

By relation (5.40) given in Theorem 5.2.1, which require the conditions $\mathbf{B}_5[\bar{\gamma}]$, $\mathbf{C}_6[\bar{\beta}]$, $\mathbf{I}_1$, and $\mathbf{J}_1$ to hold, and the definition of the sets A and B, for every $\omega \in A \cap B$,

$$\phi_{\varepsilon_r,0}(\vec{Y}'_{\varepsilon_r,0}(\omega), X'_{\varepsilon_r,0}(\omega)) \to \phi_{\varepsilon_0,0}(\vec{Y}'_{\varepsilon_0,0}(\omega), X'_{\varepsilon_0,0}(\omega)) \quad \text{as} \quad r \to \infty\,. \tag{5.75}$$

Since $\mathsf{P}(A \cap B) = 1$, relation (5.75) implies the following relation:

$$\phi_{\varepsilon_r,N}(\vec{Y}'_{\varepsilon_r,0}, X'_{\varepsilon_r,0}) \xrightarrow{a.s.} \phi_{\varepsilon_0,N}(\vec{Y}'_{\varepsilon_0,0}, X'_{\varepsilon_0,0}) \quad \text{as} \quad r \to \infty\,. \tag{5.76}$$

Relation (5.73) implies that, for every $r = 0, 1, \ldots,$

$$\mathsf{P}\{\phi_{\varepsilon_r,0}(\vec{Y}'_{\varepsilon_r,0}, X'_{\varepsilon_r,0}) \in A\} = \mathsf{P}\{\phi_{\varepsilon_r,0}(\vec{Y}_{\varepsilon_r,0}, X_{\varepsilon_r,0}) \in A\}\,, \quad A \in \mathcal{B}_{\mathbb{Z}}\,. \tag{5.77}$$

Relations (5.76) and (5.77) imply that the random variables

$$\phi_{\varepsilon_r,0}(\vec{Y}_{\varepsilon_r,0}, X_{\varepsilon_r,0}) \xrightarrow{d} \phi_{\varepsilon_0,0}(\vec{Y}_{\varepsilon_0,0}, X_{\varepsilon_0,0}) \quad \text{as} \quad r \to \infty\,. \tag{5.78}$$

Since an arbitrary choice of a sequence $\varepsilon_r \to \varepsilon_0$, relation (5.78) implies the relation of weak convergence (5.70).

Relations (5.68), (5.69), and (5.70) imply relation (5.65). □

Remark 5.2.3. According to Lemmas 5.1.5 and 5.1.8, the conditions $\mathbf{C}_6[\bar{\beta}]$ and $\mathbf{D}_7[\bar{\beta}]$ can be replaced in Theorem 5.2.3, respectively, by the conditions $\mathbf{E}_{10}[\bar{\beta}]$ and $\mathbf{F}_5[\bar{\beta}]$.

The following theorem gives alternative conditions of convergence for optimal expected rewards.

Theorem 5.2.4. *Let the conditions $\mathbf{B}_5[\bar{\gamma}]$, $\mathbf{C}'_6[\bar{\beta}]$, and $\mathbf{D}'_7[\bar{\beta}]$ hold with the vector parameters $\bar{\gamma} = (\gamma_1, \ldots, \gamma_k), \bar{\beta} = (\beta_1, \ldots, \beta_k)$ such that, for every $i = 1, \ldots, k$, either $\beta_i > \gamma_i > 0$ or $\beta_i = \gamma_i = 0$, and also the conditions $\mathbf{H}_1$, $\mathbf{I}_1$, $\mathbf{J}_1$, and $\mathbf{K}_1$ hold. Then the following relation takes place:*

$$\Phi_\varepsilon \to \Phi_0 \quad as \quad \varepsilon \to 0\,. \tag{5.79}$$

Proof. It repeats the proof of Theorem 5.2.3 with some minor changes.

Due to the above conditions imposed on the parameters γ_i, β_i, $i = 1, \dots, k$, one can always find $\alpha > 1$ such that $\gamma_i \alpha \leq \beta_i$, $i = 1, \dots, k$.

According to Theorem 5.1.2, the conditions $\mathbf{B}_5[\bar{\gamma}]$ and $\mathbf{C}'_6[\bar{\beta}]$ imply that there exist $\varepsilon'_3 > 0$ and the constants $M'_{32}, M'_{33,i}$, $i = 1, \dots, k$ such that the following upper bounds take place for the reward functions $\phi_{\varepsilon,0}(\vec{y}, x)$ for $\vec{z} = (\vec{y}, x) \in \mathbb{Z}_{\varepsilon,0}$, $\varepsilon \leq \varepsilon'_3$:

$$|\phi_{\varepsilon,0}(\vec{y}, x)| \leq M'_{32} + \sum_{i=1}^{k} M'_{33,i} e^{\gamma_i |y_i|} . \tag{5.80}$$

As was pointed out in the proof of Lemma 5.1.9, the condition $\mathbf{D}'_7[\bar{\beta}]$ implies that there exists $0 < \varepsilon'_4 \leq \varepsilon_0$ such that $0 \leq \varepsilon \leq \varepsilon'_4$ and $i = 1, \dots, k$,

$$\mathsf{E} e^{\beta_i |Y_{\varepsilon,0,i}|} < K'_{32,i} . \tag{5.81}$$

Recall that the condition $\mathbf{D}'_7[\bar{\beta}]$ (b) implies that $\mathsf{P}_{\vec{z},n}\{\vec{Z}_{\varepsilon,0} \in \mathbb{Z}_{\varepsilon,0}\} = 1$ for $\vec{z} \in \mathbb{Z}_{\varepsilon,0}$. Using this relation and relations (5.69), (5.80) and (5.81), we get that for any $\varepsilon \leq \varepsilon'_5 = \min(\varepsilon'_3, \varepsilon'_4)$,

$$\begin{aligned}
\mathsf{E}|\phi_{\varepsilon,0}(\vec{Y}_{\varepsilon,0}, X_{\varepsilon,0})|^{\alpha} &\leq \mathsf{E}\left(M'_{32} + \sum_{i=1}^{k} M'_{33,i} e^{\gamma_i |Y_{\varepsilon,0,i}|}\right)^{\alpha} \\
&\leq \mathsf{E}(k+1)^{\alpha-1}\left((M'_{32})^{\alpha} + \sum_{i=1}^{k} (M'_{33,i})^{\alpha} e^{\gamma_i \alpha |Y_{\varepsilon,0,i}|}\right) \\
&\leq (k+1)^{\alpha-1}\left((M'_{32})^{\alpha} + \sum_{i=1}^{k} (M'_{33,i})^{\alpha} \mathsf{E} e^{\beta_i |Y_{\varepsilon,0,i}|}\right) \\
&\leq (k+1)^{\alpha-1}\left((M'_{32})^{\alpha} + \sum_{i=1}^{k} (M'_{33,i})^{\alpha} K'_{32,i}\right) = Q' < \infty .
\end{aligned} \tag{5.82}$$

The following part of the proof repeats the corresponding part of the proof of Theorem 5.2.3. The only difference is that relation (5.82) should be used instead of relation (5.68). Also, relation analogous to (5.75) does require reference to relation (5.61) given in Theorem 5.2.2 instead of relation (5.40) given in Theorem 5.2.1. □

5.2.4 Convergence of log-rewards for modulated Markov log-price processes with a discrete phase space of the modulator

Some useful remarks can be made for the case, where the modulator $X_{\varepsilon,n}$ has the discrete phase space $\mathbb{X} = \{x_1, x_2, \dots\}$. In order to simplify notations one can assume that $x_i = i$, $i = 1, 2, \dots$.

The space $\mathbb{X}$ equipped with the metric $d_{\mathbb{X}}(i,j) = I(i \neq j)$ is a complete and separable metric space.

The moment conditions $\mathbf{B}_5[\bar{\gamma}]$, $\mathbf{C}_6[\bar{\beta}]$, and $\mathbf{D}_7[\bar{\beta}]$ as well as the conditions $\mathbf{E}_{10}[\bar{\beta}]$ and $\mathbf{F}_5[\bar{\beta}]$ do not change their formulations.

Convergence conditions can be simplified if we take into consideration that the convergence relation, $d_{\mathbb{X}}(i_\varepsilon, i_0) \to 0$ as $\varepsilon \to 0$, means that $i_\varepsilon \equiv i_0$ for all ε small enough.

The condition $\mathbf{I}_1$ can be replaced by the following simpler condition:

$\mathbf{I}_2$: There exist measurable sets $\mathbb{Y}'_{i,n} \subseteq \mathbb{Y}$, $i \in \mathbb{X}$, $n = 0, \dots, N$, such that the payoff function $g_\varepsilon(n, e^{\vec{y}_\varepsilon}, i) \to g_0(n, e^{\vec{y}_0}, i)$ as $\varepsilon \to 0$ for any $\vec{y}_\varepsilon \to \vec{y}_0 \in \mathbb{Y}'_{i,n}$, $i \in \mathbb{X}$ and $n = 0, \dots, N$.

The condition $\mathbf{I}_2$ implies that the condition $\mathbf{I}_1$ holds with the sets $\mathbb{Z}'_n = \cup_{i\in\mathbb{X}}(\mathbb{Y}'_{i,n} \times \{i\})$, $n = 0, \dots, N$.

The condition $\mathbf{J}_1$ can be replaced by the following simpler condition:

$\mathbf{J}_2$: There exist measurable sets $\mathbb{Y}''_{i,n} \subseteq \mathbb{Y}$, $i \in \mathbb{X}$, $n = 0, \dots, N$ such that
- (a) $P_{\varepsilon,n}((\vec{y}_\varepsilon, i), \cdot) \Rightarrow P_{0,n}((\vec{y}_0, i), \cdot)$ as $\varepsilon \to 0$, for any $\vec{y}_\varepsilon \to \vec{y}_0 \in \mathbb{Y}''_{i,n-1}$ as $\varepsilon \to 0$, $i \in \mathbb{X}$ and $n = 1, \dots, N$;
- (b) $\sum_{j\in\mathbb{X}} P_{0,n}((\vec{y}_0, i), (\mathbb{Y}'_{j,n} \cap \mathbb{Y}''_{j,n}) \times \{j\}) = 1$, for every $\vec{y}_0 \in \mathbb{Y}'_{i,n-1} \cap \mathbb{Y}''_{i,n-1}$, $i \in \mathbb{X}$, $n = 1, \dots, N$, where $\mathbb{Y}'_{i,n}$, $i \in \mathbb{X}$, $n = 0, \dots, N$ are sets introduced in the condition $\mathbf{I}_2$.

The condition $\mathbf{J}_2$ implies that the condition $\mathbf{J}_1$ holds with the sets $\mathbb{Z}'_n$, $n = 0, \dots, N$ defined above and sets $\mathbb{Z}''_n = \cup_{i\in\mathbb{X}}(\mathbb{Y}''_{i,n} \times \{i\})$, $n = 0, \dots, N$.

The condition $\mathbf{K}_1$ should be replaced by the following simpler condition:

$\mathbf{K}_2$:
- (a) $P_{\varepsilon,0}(\cdot) \Rightarrow P_{0,0}(\cdot)$ as $\varepsilon \to 0$;
- (b) $\sum_{i\in\mathbb{X}} P_{0,0}(\mathbb{Y}'_{i,0} \cap \mathbb{Y}''_{i,0}) = 1$, where $\mathbb{Y}'_{i,0}$, $i \in \mathbb{X}$ and $\mathbb{Y}''_{i,0}$, $i \in \mathbb{X}$ are sets introduced, respectively, in the conditions $\mathbf{I}_2$ and $\mathbf{J}_2$.

The condition $\mathbf{K}_2$ implies that the condition $\mathbf{K}_1$ holds with the sets $\mathbb{Z}'_0$, $\mathbb{Z}''_0$ introduced above.

5.2.5 Convergence of log-reward functions for modulated Markov log-price processes given in a dynamic form

Let us consider the case where a discrete time multivariate modulated Markov log-price process $\vec{Z}_{\varepsilon,n} = (\vec{Y}_{\varepsilon,n}, X_{\varepsilon,n})$, $n = 0, 1, \dots$ with a phase space $\mathbb{Z} = \mathbb{R}_k \times \mathbb{X}$, an initial distribution $P_\varepsilon(A)$, and transition probabilities $P_{\varepsilon,n}(\vec{z}, A)$, is given for every $\varepsilon \in [0, \varepsilon_0]$, by the following stochastic transition dynamic relation:

$$\vec{Z}_{\varepsilon,n} = A_{\varepsilon,n}(\vec{Z}_{\varepsilon,n-1}, U_{\varepsilon,n}), \quad n = 1, 2, \dots, \tag{5.83}$$

where (a) $\vec{Z}_{\varepsilon,0}$ is a random variable taking value in the space $\mathbb{Z}$, (b) $U_{\varepsilon,n}$, $n = 1, 2, \dots$, is a sequence of independent random variables taking values in the measurable space $\mathbb{U}$ (with a σ-algebra of measurable subsets $\mathcal{B}_{\mathbb{U}}$), with distributions $G_n(A) = \mathsf{P}\{U_{\varepsilon,n} \in A\}$, $n = 1, 2, \dots$, (c) the random variable $\vec{Z}_{\varepsilon,0}$ and the sequence of random variables $U_{\varepsilon,n}$, $n = 1, 2, \dots$ are independent, and (d) $A_{\varepsilon,n}(\vec{z}, u)$, $n = 1, 2, \dots$ are measurable functions acting from the space $\mathbb{Z} \times \mathbb{U}$ to the space $\mathbb{Z}$.

The random variable $\vec{Z}_{\varepsilon,0}$ can also be given in the dynamic form, $\vec{Z}_{\varepsilon,0} = A_{\varepsilon,0}(U_{\varepsilon,0})$, where: (e) (b) $U_{\varepsilon,0}$ is a random variable taking values in the space $\mathbb{U}$, with the distribution $G_0(A) = \mathsf{P}\{U_{\varepsilon,0} \in A\}$; (f) the random variable $\vec{U}_{\varepsilon,0}$ and the sequence of random variables $U_{\varepsilon,n}$, $n = 0, 1, \dots$ are independent; (g) $A_{\varepsilon,0}(u)$ is a measurable function acting from the space $\mathbb{U}$ to the space $\mathbb{Z}$.

Note that we can always assume that the distributions of random variables $U_{\varepsilon,n}$, $n = 0, 1, \dots$ do not depend on ε. This can be achieved by an appropriate modification of the functions $A_{\varepsilon,n}(\vec{z}, u)$ and $A_{\varepsilon,0}(u)$.

Moreover, according to our initial model assumptions, the phase space $\mathbb{Z}$ is a Polish metric space. In this case, for any transition probabilities $P_{\varepsilon,n}(\vec{z}, A)$ and initial distribution $P_\varepsilon(A)$, it is possible to construct a Markov chain $\vec{Z}_{\varepsilon,n}$ given in the dynamic form (5.83) with the space $\mathbb{U} = [0, 1]$ and i.i.d. random variables $U_{\varepsilon,n}$, $n = 0, 1, \dots$ uniformly distributed in the interval $\mathbb{U}$.

The moment and convergence conditions $\mathbf{B}_5[\bar{\gamma}]$ and $\mathbf{I}_1$ for the pay-off functions $g_\varepsilon(n, e^{\vec{y}}, x)$ do not change.

The moment conditions for $\mathbf{C}_6[\bar{\beta}]$, $\mathbf{C}'_6[\bar{\beta}]$, and $\mathbf{D}_7[\bar{\beta}]$, $\mathbf{D}'_7[\bar{\beta}]$ as well as the conditions $\mathbf{E}_{10}[\bar{\beta}]$, $\mathbf{E}'_{10}[\bar{\beta}]$ and $\mathbf{F}_4[\bar{\beta}]$, $\mathbf{F}'_4[\bar{\beta}]$ for log-price processes $\vec{Z}_{\varepsilon,n}$ do not change their formulations.

The convergence condition $\mathbf{J}_1$ for the transition probabilities of the log-price processes $\vec{Z}_{\varepsilon,n}$ can be replaced by the following condition imposed on the transition dynamic functions $A_{\varepsilon,n}(\vec{z}, u)$:

$\mathbf{J}_3$: There exist measurable sets $\mathbb{Z}''_n \subseteq \mathbb{Z}$, $n = 0, \dots, N$ and $\mathbb{U}_n(\vec{z}) \subseteq \mathbb{U}$, $\vec{z} \in \mathbb{Z}$, $n = 1, \dots, N$ such that (a) $A_{\varepsilon,n}(\vec{z}_\varepsilon, u) \to A_{0,n}(\vec{z}_0, u)$ as $\varepsilon \to 0$, for any $\vec{z}_\varepsilon \to \vec{z}_0 \in \mathbb{Z}''_{n-1}$ as $\varepsilon \to 0$ and $u \in \mathbb{U}_n(\vec{z}_0)$, $n = 1, \dots, N$, where (b) $\mathsf{P}\{U_{0,n} \in \mathbb{U}_n(\vec{z}_0)\} = 1$, $\vec{z}_0 \in \mathbb{Z}''_{n-1}$, $n = 1, \dots, N$; (c) $\mathsf{P}\{A_{0,n}(\vec{z}_0, U_{0,n}) \in \mathbb{Z}'_n \cap \mathbb{Z}''_n\} = 1$, for every $\vec{z}_0 \in \mathbb{Z}'_{n-1} \cap \mathbb{Z}''_{n-1}$, $n = 1, \dots, N$, where $\mathbb{Z}'_n$, $n = 0, \dots, N$ are sets introduced in the condition $\mathbf{I}_1$.

Lemma 5.2.2. *The condition $\mathbf{J}_3$ implies that the condition $\mathbf{J}_1$ holds for the log-price processes $\vec{Z}_{\varepsilon,n}$ defined by the dynamic relation* (5.83).

Proof. The condition $\mathbf{J}_3$ (a) implies that for any real-valued, bounded, and continuous function $f(\vec{z})$ defined on the space $\mathbb{Z}$, (a) functions $f(A_{\varepsilon,n}(\vec{z}_\varepsilon, u)) \to f(A_{0,n}(\vec{z}_0, u))$ as $\varepsilon \to 0$, for any $\vec{z}_\varepsilon \to \vec{z}_0 \in \mathbb{Z}''_{n-1}$ as $\varepsilon \to 0$ and $u \in \mathbb{U}_n(\vec{z}_0)$, $n = 1, \dots, N$.

Since the distributions $G_n(A) = \mathsf{P}\{U_{\varepsilon,n} \in A\}$ do not depend on ε, and by the condition $\mathbf{J}_3$ (b) $G_n(\mathbb{U}_n(\vec{z}_0)) = \mathsf{P}\{U_{\varepsilon,n} \in \mathbb{U}_n(\vec{z}_0)\} = 1$, $\vec{z}_0 \in \mathbb{Z}''_{n-1}$, $n = 1, \dots, N$. Thus, relation (a) implies, by the Lebesgue theorem, that (b) $\mathsf{E}f(A_{\varepsilon,n}(\vec{z}_\varepsilon, U_{\varepsilon,n})) \to \mathsf{E}f(A_{0,n}(\vec{z}_0, U_{0,n}))$ as $\varepsilon \to 0$, for any $\vec{z}_\varepsilon \to \vec{z}_0 \in \mathbb{Z}''_{n-1}$ as $\varepsilon \to 0$, and $n = 1, \dots, N$.

Since an arbitrary choice of real-valued bounded, and continuous function $f(\vec{z})$, relation (b) is equivalent to the following relation of weak convergence (c) $A_{\varepsilon,n}(\vec{z}_\varepsilon, U_{\varepsilon,n}) \Rightarrow A_{0,n}(\vec{z}_0, U_{0,n})$ as $\varepsilon \to 0$, for any $\vec{z}_\varepsilon \to \vec{z}_0 \in \mathbb{Z}''_{n-1}$ as $\varepsilon \to 0$, and $n = 1, \dots, N$. This relation is, in fact, a reformulation of the condition $\mathbf{J}_1$ (a).

It remain to note that the condition $\mathbf{J}_3$ (c) is a reformulation of the condition $\mathbf{J}_1$ (b).

□

Remark 5.2.4. The condition $\mathbf{J}_3$ (c) is equivalent to the assumption that there exist measurable sets $\mathbb{U}'_n \subseteq \mathbb{U}$, $n = 1, \dots, N$ such that $\mathsf{P}\{A_{0,n}(\vec{z}_0, u) \in \mathbb{Z}'_n \cap \mathbb{Z}''_n\} = 1$ for every $\vec{z}_0 \in \mathbb{Z}''_{n-1}$, $u \in \mathbb{U}'_n$, $n = 1, \dots, N$, and $\mathsf{P}\{U_{0,n} \in \mathbb{U}'_n\} = 1$, $n = 1, \dots, N$.

The convergence condition $\mathbf{K}_1$ for the initial distribution of the log-price processes can be replaced by the following condition imposed on the transformation function $A_{\varepsilon,0}(u)$, in the case where the random variable $\vec{Z}_{\varepsilon,0}$ is also given in the dynamic form:

$\mathbf{K}_3$: There exist a measurable set $\mathbb{U}_0 \subseteq \mathbb{U}$ such that
- (a) $A_{\varepsilon,0}(u) \to A_{0,0}(u)$ as $\varepsilon \to 0$, for $u \in \mathbb{U}_0$,
- (b) $\mathsf{P}\{A_{0,0}(U_{0,0}) \in \mathbb{Z}'_0 \cap \mathbb{Z}''_0\} = 1$.

Indeed, the condition $\mathbf{K}_3$ (a) implies that for any real-valued bounded, and continuous function $f(\vec{z})$ defined on the space $\mathbb{Z}$, (d) functions $f(A_{\varepsilon,0}(u)) \to f(A_{0,0}(u))$ as $\varepsilon \to 0$, for any $u \in \mathbb{U}_0$.

Since the distributions $G_0(A) = \mathsf{P}\{U_{\varepsilon,0} \in A\}$ do not depend on ε, and, by the condition $\mathbf{K}_3$ (b), $G_0(\mathbb{U}_0) = \mathsf{P}\{U_{\varepsilon,0} \in \mathbb{U}_0\} = 1$, relation (d) implies, by the Lebesgue theorem, that (e) $\mathsf{E}f(A_{\varepsilon,0}(U_{\varepsilon,0})) \to \mathsf{E}f(A_{0,0}(U_{0,0}))$ as $\varepsilon \to 0$.

Since an arbitrary choice of real-valued bounded, and continuous function $f(\vec{z})$, relation (e) is equivalent to the following relation of weak convergence (f) $A_{\varepsilon,0}(U_{\varepsilon,0}) \Rightarrow A_{0,0}(U_{0,0})$ as $\varepsilon \to 0$. This relation is, in fact, a reformulation of the condition $\mathbf{K}_1$ (a).

Remark 5.2.5. The condition $\mathbf{K}_3$ (b) is equivalent to the assumption that there exist measurable sets $\mathbb{U}'_0 \subseteq \mathbb{U}$ such that $\mathsf{P}\{A_{0,n}(\vec{z}_0, u) \in \mathbb{Z}'_0 \cap \mathbb{Z}''_0\} = 1$ for every $u \in \mathbb{U}'_0$ and $\mathsf{P}\{U_{0,0} \in \mathbb{U}'_0\} = 1$.

5.3 LPP represented by modulated random walks

In this section, we present results about the convergence of option rewards for log-price processes represented by standard and modulated random walks.

5.3.1 Convergence of rewards for price processes represented by standard random walks

Let us consider the model where the multivariate Markov log-price $\vec{Y}_{\varepsilon,n} = (Y_{\varepsilon,n,1}\dots, Y_{\varepsilon,n,k})$ is, for every $\varepsilon \in [0, \varepsilon_0]$, a standard random walk

$$\vec{Y}_{\varepsilon,n} = \vec{Y}_{\varepsilon,n-1} + \vec{W}_{\varepsilon,n}\,, \quad n = 1, 2, \dots . \tag{5.84}$$

where (a) $\vec{Y}_{\varepsilon,0} = (Y_{\varepsilon,0,1}, \dots, Y_{\varepsilon,0,k})$ is a k-dimensional random vector, (b) $\vec{W}_{\varepsilon,n} = (W_{\varepsilon,n,1}, \dots, W_{\varepsilon,n,k})$, $n = 1, 2, \dots$ is a sequence of k-dimensional independent random vectors, and (c) the random vector $\vec{Y}_{\varepsilon,0}$ and the random sequence $\vec{W}_{\varepsilon,n}$, $n = 1, 2, \dots$ are independent.

In this case, the model without index component is considered. In fact, one can always define a "virtual" constant index component $X_{\varepsilon,n}$ which takes the constant value x_0 for every $n = 0, 1, \dots$, and, the space $\mathbb{X} = \{x_0\}$ containing the only point x_0.

In this case, pay-off functions $g_\varepsilon(n, e^{\vec{y}}, x) = g_\varepsilon(n, e^{\vec{y}})$ do not depend on the index argument x, for every $\varepsilon \in [0, \varepsilon_0]$.

Correspondingly, the log-reward functions $\phi_{\varepsilon,n}(\vec{y}, x) = \phi_{\varepsilon,n}(\vec{y})$ also do not depend on the index argument x, for every $\varepsilon \in [0, \varepsilon_0]$.

The moment conditions $\mathbf{B}_5[\bar{\gamma}]$, $\mathbf{C}_6[\bar{\beta}]$ and $\mathbf{E}_{10}[\bar{\beta}]$, formulated in Subsection 1.6.1, are simplified for this model.

The first-type modulus of the exponential moment compactness $\Delta_\beta(Y_{\varepsilon,\cdot,i}, N)$ for the components of the log-price process $\vec{Y}_{\varepsilon,n}$, expressed in terms of the components of jumps $\vec{W}_n$, takes the following form, for $\beta \geq 0$:

$$\Delta'_\beta(Y_{\varepsilon,\cdot,i}, N) = \max_{0 \leq n \leq N-1} \mathsf{E} e^{\beta |W_{\varepsilon,n+1,i}|} . \tag{5.85}$$

The condition $\mathbf{C}_6[\bar{\beta}]$, assumed to hold for the vector parameter $\bar{\beta} = (\beta_1, \dots, \beta_k)$ with nonnegative components, takes the following form:
$\mathbf{C}_7[\bar{\beta}]$: $\overline{\lim}_{\varepsilon\to 0} \Delta'_{\beta_i}(Y_{\varepsilon,\cdot,i}, N) < K_{34,i}$, $i = 1, \dots, k$, for some $1 < K_{34,i} < \infty$, $i = 1, \dots, k$.

The second-type modulus of the exponential moment compactness $\Xi^\pm_\beta(Y_{\varepsilon,\cdot,i}, N)$ for components of the log-price process $\vec{Y}_{\varepsilon,n}$, expressed in terms of components of jumps $\vec{W}_n$, takes the following form for $\beta \geq 0$:

$$\Xi'^{\pm}_\beta(Y_{\varepsilon,\cdot,i}, N) = \max_{0 \leq n \leq N-1} \mathsf{E} e^{\pm\beta W_{\varepsilon,n+1,i}} . \tag{5.86}$$

The condition $\mathbf{E}_{10}[\bar{\beta}]$, assumed to hold for the vector parameter $\bar{\beta} = (\beta_1, \dots, \beta_k)$ with nonnegative components, takes the following form:
$\mathbf{E}_{11}[\bar{\beta}]$: $\overline{\lim}_{\varepsilon \to 0} \, \Xi'^{\pm}_{\beta_i}(Y_{\varepsilon,\cdot,i}, N) < K_{35,i}$, $i = 1, \dots, k$, for some $1 < K_{35,i} < \infty$, $i = 1, \dots, k$.

The following lemma is a direct corollary of Lemma 5.1.5.

Lemma 5.3.1. *The condition $\mathbf{E}_{11}[\bar{\beta}]$ implies the condition $\mathbf{C}_7[\bar{\beta}]$ to hold.*

The condition $\mathbf{B}_5[\bar{\gamma}]$, assumed to hold for the vector parameter $\bar{\gamma} = (\gamma_1, \dots, \gamma_k)$ with nonnegative components, takes the following form:
$\mathbf{B}_6[\bar{\gamma}]$: $\overline{\lim}_{\varepsilon \to 0} \max_{0 \le n \le N} \sup_{\vec{y} \in \mathbb{R}_k} \frac{|g_\varepsilon(n, e^{\vec{y}})|}{1 + \sum_{i=1}^{k} L_{12,i} e^{\gamma_i |y_i|}} < L_{11}$, for some $0 < L_{11} < \infty$ and $0 \le L_{12,1}, \dots, L_{12,k} < \infty$.

Conditions $\mathbf{D}_7[\bar{\beta}]$ and $\mathbf{F}_4[\bar{\beta}]$ do not change.

Convergence conditions $\mathbf{I}_1$ and $\mathbf{J}_1$ are also simplified in this model.

The condition $\mathbf{I}_1$ should be replaced by the following simpler condition:

$\mathbf{I}_3$: There exist measurable sets $\mathbb{Y}_n \subseteq \mathbb{R}_k$, $n = 0, \dots, N$, such that the pay-off function $g_\varepsilon(n, e^{\vec{y}_0}) \xrightarrow{U} g_0(n, e^{\vec{y}_0})$ as $\varepsilon \to 0$, i.e. $g_\varepsilon(n, e^{\vec{y}_\varepsilon}) \to g_0(n, e^{\vec{y}_0})$ as $\varepsilon \to 0$ for any $\vec{y}_\varepsilon \to \vec{y}_0 \in \mathbb{Y}_n$ as $\varepsilon \to 0$, and $n = 0, \dots, N$.

Obviously, the condition $\mathbf{I}_3$ implies that the condition $\mathbf{I}_1$ holds with the sets $\mathbb{Z}'_n = \mathbb{Y}_n \times \{x_0\}$, $n = 0, \dots, N$.

Let us define, for a Borel set $A \subseteq \mathbb{R}_k$ and a point $\vec{y} \in \mathbb{R}_k$, the set

$$A - \vec{y} = \{\vec{y}' : \vec{y}' + \vec{y} \in A\} . \tag{5.87}$$

The transition probabilities $P_{\varepsilon,n}(\vec{y}, A) = \mathsf{P}\{\vec{Y}_{\varepsilon,n} \in A / \vec{Y}_{\varepsilon,n-1} = \vec{y}\}$ of the Markov process $\vec{Y}_{\varepsilon,n}$ can be expressed in terms of distributions $\tilde{P}_{\varepsilon,n}(A) = \mathsf{P}\{\vec{W}_{\varepsilon,n} \in A\}$ of the jumps $\vec{W}_{\varepsilon,n}$, for $y \in \mathbb{R}_k$, $A \in \mathcal{B}_k$, $n = 1, \dots,$

$$P_{\varepsilon,n}(\vec{y}, A) = \mathsf{P}\{\vec{y} + \vec{W}_{\varepsilon,n} \in A\} = \tilde{P}_{\varepsilon,n}(A - \vec{y}) . \tag{5.88}$$

Let us assume that the following condition holds:

$\mathbf{J}_4$: (a) $\tilde{P}_{\varepsilon,n}(\cdot) \Rightarrow \tilde{P}_{0,n}(\cdot)$ as $\varepsilon \to 0$, for $n = 1, \dots, N$;
(b) $P_{0,n}(\vec{y}, \overline{\mathbb{Y}}_n) = \tilde{P}_{0,n}(\overline{\mathbb{Y}}_n - \vec{y}) = 1, \vec{y} \in \mathbb{Y}_{n-1}$, for $n = 1, \dots, N$, where $\mathbb{Y}_n$, $n = 0, \dots, N$ are the sets introduced in the conditions $\mathbf{I}_3$.

A typical example is where the sets $\overline{\mathbb{Y}}_n$, $n = 1, \dots, N$ are empty sets. Then the condition $\mathbf{J}_4$ (b) obviously holds.

Another typical example is where the sets $\overline{\mathbb{Y}}_n$ are at most finite or countable sets. Then the assumption that the measures $\tilde{P}_{0,n}(\cdot)$, $n = 1, \dots, N$ have no atoms implies that the condition $\mathbf{J}_4$ (b) holds.

One more example is where the sets $\overline{\mathbb{Y}}_n$ have zero Lebesgue measure. Then, the assumption that measures $\tilde{P}_{0,n}(\cdot)$, $n = 1, \dots, N$ are absolutely continuous with respect to the Lebesgue measure in $\mathbb{R}_k$, implies that the condition $\mathbf{J}_4$ (b) holds.

Lemma 5.3.2. *The condition* $\mathbf{J}_4$ *implies that* $P_{\varepsilon,n}(\vec{y}_0, \cdot) \overset{U}{\Rightarrow} P_{0,n}(\vec{y}_0, \cdot)$ *as* $\varepsilon \to 0$, *i.e.* $P_{\varepsilon,n}(\vec{y}_\varepsilon, \cdot) \Rightarrow P_{0,n}(\vec{y}_0, \cdot)$ *as* $\varepsilon \to 0$, *for any* $\vec{y}_\varepsilon \to \vec{y}_0 \in \mathbb{R}_k$ *as* $\varepsilon \to 0$, *and* $n = 1, \dots, N$.

Proof. The condition $\mathbf{J}_4$ implies that, for any $\vec{y}_\varepsilon \to \vec{y}_0 \in \mathbb{R}_k$ and $n = 1, \dots, N$, random vectors $\vec{y}_\varepsilon + \vec{W}_{\varepsilon,n} \xrightarrow{d} \vec{y}_0 + \vec{W}_{0,n}$ as $\varepsilon \to 0$. This is equivalent to the proposition of the lemma, due to relation (5.88). □

By Lemma 5.3.2, the condition $\mathbf{J}_4$ implies the condition $\mathbf{J}_1$ to hold. In this case, the role of sets $\mathbb{Z}'_n$ is played by sets $\mathbb{Y}_n \times \{x_0\}$ while the role of the set $\mathbb{Z}''_n$ is played by the set $\mathbb{R}_k \times \{x_0\}$, for every $n = 0, 1, \dots, N$.

The condition $\mathbf{K}_1$ imposed on the initial distributions $P_{\varepsilon,0}(A)$ should be replaced by the following condition:

$\mathbf{K}_4$: (a) $P_{\varepsilon,0}(\cdot) \Rightarrow P_{0,0}(\cdot)$ as $\varepsilon \to 0$;
(b) $P_{0,0}(\mathbb{Y}_0) = 1$, where $\mathbb{Y}_0$ is the set introduced in the condition $\mathbf{I}_3$.

Obviously, the condition $\mathbf{K}_4$ implies that the condition $\mathbf{K}_1$ holds with the sets $\mathbb{Z}'_0 = \mathbb{Y}_0 \times \{x_0\}$ and $\mathbb{Z}''_0 = \mathbb{R}_k \times \{x_0\}$.

Theorem 5.2.1 takes in this case the following form.

Theorem 5.3.1. *Let the log-price process* $\vec{Y}_{\varepsilon,n}$ *is represented, for every* $\varepsilon \in [0, \varepsilon_0]$, *by the standard random walk given by the stochastic transition dynamic relation* (5.84). *Let also the conditions* $\mathbf{B}_6[\bar{\gamma}]$ *and* $\mathbf{C}_7[\bar{\beta}]$ *hold with the vector parameters* $\bar{\gamma} = (\gamma_1, \dots, \gamma_k), \bar{\beta} = (\beta_1, \dots, \beta_k)$ *such that, for every* $i = 1, \dots, k$, *either* $\beta_i > \gamma_i > 0$ *or* $\beta_i = \gamma_i = 0$, *and also the conditions* $\mathbf{I}_3$ *and* $\mathbf{J}_4$ *hold. Then, for every* $n = 0, 1, \dots, N$, *the following relation takes place for any* $\vec{y}_\varepsilon \to \vec{y}_0 \in \mathbb{Y}_n$:

$$\phi_{\varepsilon,n}(\vec{y}_\varepsilon) \to \phi_{0,n}(\vec{y}_0) \quad \text{as} \quad \varepsilon \to 0. \tag{5.89}$$

Theorem 5.2.3 takes in this case the following form.

Theorem 5.3.2. *Let the log-price process* $\vec{Y}_{\varepsilon,n}$ *is represented, for every* $\varepsilon \in [0, \varepsilon_0]$, *by the standard random walk given by the stochastic transition dynamic relation* (5.84). *Let also the conditions* $\mathbf{B}_6[\bar{\gamma}]$, $\mathbf{C}_7[\bar{\beta}]$, *and* $\mathbf{D}_7[\bar{\beta}]$ *hold with the vector parameters* $\bar{\gamma} = (\gamma_1, \dots, \gamma_k), \bar{\beta} = (\beta_1, \dots, \beta_k)$ *such that, for every* $i = 1, \dots, k$, *either* $\beta_i > \gamma_i > 0$ *or* $\beta_i = \gamma_i = 0$ *for* $i = 1, \dots, k$, *and also the conditions* $\mathbf{I}_3$, $\mathbf{J}_4$, *and* $\mathbf{K}_4$ *hold. Then the following relation takes place:*

$$\Phi_\varepsilon \to \Phi_0 \quad \text{as} \quad \varepsilon \to 0. \tag{5.90}$$

5.3.2 Convergence of option rewards for log-price processes represented by Lévy random walks

Let us consider the model where, a log-price $Y_{\varepsilon,n}$ be, for every $\varepsilon \in [0, \varepsilon_0]$, a standard univariate random walk,

$$Y_{\varepsilon,n} = Y_{\varepsilon,n-1} + W_{\varepsilon,n}\,, \quad n = 1, 2, \dots . \tag{5.91}$$

where (a) $Y_{\varepsilon,0}$ is a real-valued random variable, (b) $W_{\varepsilon,n}$, $n = 1, 2, \dots$ is a sequence of independent random variables, (c) the random variable $Y_{\varepsilon,0}$ and the random sequence $W_{\varepsilon,n}$, $n = 1, 2, \dots$ are independent, and (d) random variable $W_{\varepsilon,n}$ has, for every $\varepsilon \in [0, \varepsilon_0]$ and $n = 1, 2, \dots$, an infinitely divisible distribution, i.e. its characteristic function has the following form:

$$\begin{aligned}\psi_{\varepsilon,n}(s) &= \mathsf{E}\exp\{\mathrm{i}sW_{\varepsilon,n}\} \\ &= \exp\Bigg\{\mathrm{i}\mu_{\varepsilon,n}s - \frac{1}{2}\sigma^2_{\varepsilon,n}s^2 + \int\limits_{|y|<1} \left(e^{\mathrm{i}sy} - 1 - \mathrm{i}sy\right)\Pi_{\varepsilon,n}(dy) \\ &\quad + \int\limits_{|y|\ge 1} \left(e^{\mathrm{i}sy} - 1\right)\Pi_{\varepsilon,n}(dy)\Bigg\}\,, \quad s \in \mathbb{R}_1\,,\end{aligned} \tag{5.92}$$

where (a) $\mu_{\varepsilon,n}$ is a real-valued constant, (b) $\sigma^2_{\varepsilon,n}$ is a nonnegative constant, and (c) $\Pi_{\varepsilon,n}(A)$ is a measure on $\mathcal{B}_1$ without an atom in 0 such that the integral $\pi_{\varepsilon,n} = \int_{\mathbb{R}_1} \frac{y^2}{1+y^2}\Pi_{\varepsilon,n}(dy) < \infty$.

The condition $\mathbf{C}_7[\bar{\beta}]$ (its univariate variant) can be replaced in this case by the following two conditions, the first one:

$\mathbf{C}_7'$: $\overline{\lim}_{\varepsilon\to 0} \max_{1\le n\le N}(|\mu_{\varepsilon,n}| + \sigma^2_{\varepsilon,n} + \int_{|y|<1} y^2\Pi_{\varepsilon,n}(dy)) < K_{36}$, for some $1 < K_{36} < \infty$,

and the second one, which should be assumed to hold for some $\beta \ge 0$:

$\mathbf{C}_7''[\beta]$: $\overline{\lim}_{\varepsilon\to 0} \max_{1\le n\le N} \int_{|y|\ge 1} e^{\beta|y|}\Pi_{\varepsilon,n}(dy) < K_{37}(\beta)$, for some $1 < K_{37}(\beta) < \infty$.

Lemma 5.3.3. *The conditions $\mathbf{C}_7'$ and $\mathbf{C}_7''[\beta]$ imply the univariate variant of the condition $\mathbf{E}_{11}[\bar{\beta}]$ to hold.*

Proof. The conditions $\mathbf{C}_7'$ and $\mathbf{C}_7''[\beta]$ imply that there exists $0 < \varepsilon_1'' \le \varepsilon_0$ such that for every $\varepsilon \in [0, \varepsilon_1'']$ and $n = 1, \dots, N$,

$$|\mu_{\varepsilon,n}| + \sigma^2_{\varepsilon,n} + \int\limits_{|y|<1} y^2\Pi_{\varepsilon,n}(dy) + \int\limits_{|y|\ge 1} e^{\beta|y|}\Pi_{\varepsilon,n}(dy) < K_{36} + K_{37}(\beta)\,. \tag{5.93}$$

Relation (5.93) implies that, for $0 \leq \varepsilon \leq \varepsilon_1''$, the moment-generating functions for the random variables $W_{\varepsilon,n}$, $n = 1, \dots, N$ exist, for $-\beta \leq \beta' \leq \beta$:

$$\begin{aligned}\psi_{\varepsilon,n}(\beta') &= \mathsf{E}\exp\{\beta' W_{\varepsilon,n}\} \\ &= \exp\Bigg\{\mu_{\varepsilon,n}\beta' + \frac{1}{2}\sigma^2_{\varepsilon,n}(\beta')^2 + \int\limits_{|y|<1}(e^{\beta' y} - 1 - \beta' y)\Pi_{\varepsilon,n}(dy) \\ &\quad + \int\limits_{|y|\geq 1}(e^{\beta' y} - 1)\Pi_{\varepsilon,n}(dy)\Bigg\} < \infty\,. \end{aligned} \tag{5.94}$$

Conditions $\mathbf{C}_7'$ and $\mathbf{C}_7''[\beta]$, and formula (5.94) imply that the following inequality holds for $\beta \geq 0$ penetrating condition $\mathbf{C}_7''[\beta]$, and every $\varepsilon \in [0, \varepsilon_1'']$, and $n = 1, \dots, N$:

$$\begin{aligned}\psi_{\varepsilon,n}(\pm\beta) &\leq \exp\Bigg\{|\mu_{\varepsilon,n}|\beta + \frac{1}{2}\sigma^2_{\varepsilon,n}\beta^2 + \frac{1}{2}\beta^2 e^{\beta}\int\limits_{|y|<1} y^2\Pi_{\varepsilon,n}(dy) + 2\int\limits_{|y|\geq 1} e^{\beta|y|}\Pi_{\varepsilon,n}(dy)\Bigg\} \\ &\leq \exp\left\{(\beta + \frac{1}{2}\beta^2 + \frac{1}{2}\beta^2 e^{\beta})K_{36} + 2K_{37}(\beta)\right\} \\ &= K_{38} \in (1, \infty)\,. \end{aligned} \tag{5.95}$$

Inequality (5.95) implies that the univariate variant of the condition $\mathbf{E}_{11}[\beta]$ holds, for example, with constant K_{38} given in the above formula. □

The condition $\mathbf{B}_{11}[\bar{\gamma}]$, can be replaced by its simpler univariate analog, assumed to hold for some $\gamma \geq 0$:
$\mathbf{B}_7[\gamma]$: $\overline{\lim}_{\varepsilon\to 0}\max_{0\leq n\leq N}\sup_{y\in\mathbb{R}_1}\frac{|g_\varepsilon(n, e^y)|}{1+L_{14}e^{\gamma|y|}} < L_{13}$, for some $0 < L_{13} < \infty$ and $0 \leq L_{14} < \infty$.

The condition $\mathbf{I}_3$ should be replaced by its simpler univariate analog:
$\mathbf{I}_4$: There exist measurable sets $\mathbb{Y}_n \subseteq \mathbb{R}_1$, $n = 0, \dots, N$, such that the payoff function $g_\varepsilon(n, e^{y_0}) \xrightarrow{U} g_0(n, e^{y_0})$ as $\varepsilon \to 0$, i.e. $g_\varepsilon(n, e^{y_\varepsilon}) \to g_0(n, e^{y_0})$ as $\varepsilon \to 0$ for any $y_\varepsilon \to y_0 \in \mathbb{Y}_n$ as $\varepsilon \to 0$, and $n = 0, \dots, N$.

Let us define the intervals $A^{\pm}(x) = \{y \in \mathbb{R}_1 : \pm y \geq x\}, x > 0$.

The condition $\mathbf{J}_4$ (a) can be replaced by the following condition which guarantees the weak convergence of infinitely divisible jump distributions:

$\mathbf{J}_5$: (a) $\mu_{\varepsilon,n} \to \mu_{0,n}$ as $\varepsilon \to 0$, for $n = 1, \dots, N$;
(b) $\sigma_{\varepsilon,n} \to \sigma_{0,n}$ as $\varepsilon \to 0$, for $n = 1, \dots, N$;
(c) $\Pi_{\varepsilon,n}(A^{\pm}(x)) \to \Pi_{0,n}(A^{\pm}(x))$ as $\varepsilon \to 0$, for all point $x > 0$, which are points of continuity for the limiting functions, for $n = 1, \dots, N$;
(d) $\lim_{0<\delta\to 0}\overline{\lim}_{\varepsilon\to 0}\int_{|y|<\delta} y^2\Pi_{\varepsilon,n}(dy) = 0$, for $n = 1, \dots, N$.

It is useful to note that the condition $\mathbf{J}_5$ implies the condition $\mathbf{C}_7'$ to hold.

As far as the condition $\mathbf{J}_4$ (b) is concerned, it takes the following form:

$\mathbf{J}_6$: $\mathsf{P}\{W_{0,n} + y \in \mathbb{Y}_n\} = 1$, for $n = 1, \ldots, N$, where $\mathbb{Y}_n$, $n = 1, \ldots, N$ are the sets introduced in the conditions $\mathbf{I}_4$.

Theorem 5.3.1 takes in this case the following form.

Theorem 5.3.3. *Let the log-price process $Y_{\varepsilon,n}$ is represented, for every $\varepsilon \in [0, \varepsilon_0]$, by the Lévy random walk given by the stochastic transition dynamic relation (5.91). Let also the conditions $\mathbf{B}_7[\gamma]$ and $\mathbf{C}_7''[\beta]$ hold with parameters γ and β such that, either $\beta > \gamma > 0$ or $\beta = \gamma = 0$, and also the conditions $\mathbf{I}_4$, $\mathbf{J}_5$ and $\mathbf{J}_6$ hold. Then, for every $n = 0, 1, \ldots, N$, the following relation takes place for any $\vec{y}_\varepsilon \to \vec{y}_0 \in \mathbb{Y}_n$:*

$$\phi_{\varepsilon,n}(\vec{y}_\varepsilon) \to \phi_{0,n}(\vec{y}_0) \quad \text{as} \quad \varepsilon \to 0\,. \tag{5.96}$$

The condition $\mathbf{K}_3$ imposed on the initial distributions $P_{\varepsilon,0}(A) = \mathsf{P}\{Y_{\varepsilon,n} \in A\}$ should be replaced by its univariate variant:

$\mathbf{K}_5$: (a) $P_{\varepsilon,0}(\cdot) \Rightarrow P_{0,0}(\cdot)$ as $\varepsilon \to 0$;

(b) $P_{0,0}(\mathbb{Y}_0) = 1$, where $\mathbb{Y}_0$ is the set introduced in the condition $\mathbf{I}_4$.

Obviously, the condition $\mathbf{K}_5$ implies that the condition $\mathbf{K}_1$ holds with the sets $\mathbb{Z}_0' = \mathbb{Y}_0 \times \{x_0\}$ and $\mathbb{Z}_0'' = \mathbb{R}_1 \times \{x_0\}$.

Theorem 5.2.3 takes in this case the following form.

Theorem 5.3.4. *Let the log-price process $Y_{\varepsilon,n}$ is represented, for every $\varepsilon \in [0, \varepsilon_0]$, by the Lévy random walk given by the stochastic transition dynamic relation (5.91). Let also the conditions $\mathbf{B}_7[\gamma]$, $\mathbf{C}_7''[\beta]$ and $\mathbf{D}_7[\beta]$ hold with parameters γ and β such that, either $\beta > \gamma > 0$ or $\beta = \gamma = 0$, and also the conditions $\mathbf{I}_4$, $\mathbf{J}_5$, $\mathbf{J}_6$, and $\mathbf{K}_5$ hold. Then the following relation takes place:*

$$\Phi_\varepsilon \to \Phi_0 \quad \text{as} \quad \varepsilon \to 0\,. \tag{5.97}$$

5.3.3 Convergence of rewards for log-price processes represented by modulated random walks

In this model, a multivariate modulated price processes $\vec{Z}_{\varepsilon,n} = (\vec{Y}_{\varepsilon,n}, X_{\varepsilon,n})$ can be introduced, for every $\varepsilon \in [0, \varepsilon_0]$, via the following stochastic transition dynamic relation:

$$\begin{cases} \vec{Y}_{\varepsilon,n} = \vec{Y}_{\varepsilon,n-1} + \vec{W}_{\varepsilon,n}\,, \quad X_{\varepsilon,n} = C_{\varepsilon,n}(X_{\varepsilon,n-1}, U_{\varepsilon,n})\,, \\ \text{where } \vec{W}_{\varepsilon,n} = \tilde{B}_{\varepsilon,n}(X_{\varepsilon,n-1}, U_{\varepsilon,n})\,, \\ n = 1, 2, \ldots\,, \end{cases} \tag{5.98}$$

where (a) $(\vec{Y}_{\varepsilon,0}, X_{\varepsilon,0}) = ((Y_{\varepsilon,0,1}, \ldots, Y_{\varepsilon,0,k}), X_{\varepsilon,0})$ is a random vector taking values in space $\mathbb{Z}$, (b) $U_{\varepsilon,n}$, $n = 1, 2, \ldots$ is a sequence of independent random variables taking value in some measurable space $\mathbb{U}$ (with a σ-algebra of measurable subsets $\mathcal{B}_{\mathbb{U}}$), with distributions $G_n(A) = \mathsf{P}\{U_{\varepsilon,n} \in A\}$, $n = 1, \dots,$ (c) the random vector $(\vec{Y}_{\varepsilon,0}, X_{\varepsilon,0})$ and the random sequence $U_{\varepsilon,n}$, $n = 1, 2, \dots$ are independent, (d) $\tilde{B}_{\varepsilon,n}(x, u) = (\tilde{B}_{\varepsilon,n,1}(x, u), \dots, \tilde{B}_{\varepsilon,n,k}(x, u))$, $n = 1, 2, \dots$ are the vector functions whose components are measurable functions acting from the space $\mathbb{X} \times \mathbb{U}$ to $\mathbb{R}_1$, (e) $C_{\varepsilon,n}(x, u)$, $n = 1, 2, \dots$ are measurable functions acting from the space $\mathbb{X} \times \mathbb{U}$ to the space $\mathbb{X}$, (f) the sequence of jumps $\vec{W}_{\varepsilon,n} = (W_{\varepsilon,n,1}, \dots, W_{\varepsilon,n,k})$, $n = 1, 2, \dots,$ where $W_{\varepsilon,n,i} = \tilde{B}_{\varepsilon,n,i}(X_{n-1}, U_{\varepsilon,n})$, $i = 1, \dots, k$, $n = 1, 2, \dots$.

In this case, the pay-off functions $g_\varepsilon(n, e^{\vec{y}}, x)$, which depend on the perturbation parameter $\varepsilon \in [0, \varepsilon_0]$, are considered.

The first-type modulus of the exponential moment compactness $\Delta''_\beta(Y_{\varepsilon,\cdot,i}, N)$, expressed in terms of jumps $\vec{W}_{\varepsilon,n}$, takes the following form, for $\beta \geq 0$, $i = 1, \dots, k$:

$$\Delta''_\beta(Y_{\varepsilon,\cdot,i}, N) = \max_{0 \leq n \leq N-1} \sup_{x \in \mathbb{X}} \mathsf{E}_{x,n} e^{\beta |W_{\varepsilon,n+1,i}|}, \tag{5.99}$$

where $\mathsf{E}_{x,n}$ is the conditional expectation under the condition $X_n = x$.

The condition $\mathbf{C_6}[\bar{\beta}]$, which is assumed to hold for some vector parameter $\bar{\beta} = (\beta_1, \dots, \beta_k)$ with nonnegative components, takes the following simpler form:

$\mathbf{C_8}[\bar{\beta}]$: $\overline{\lim}_{\varepsilon \to 0} \Delta''_{\beta_i}(Y_{\varepsilon,\cdot,i}, N) < K_{39,i}$, $i = 1, \dots, k$, for some $1 < K_{39,i} < \infty$, $i = 1, \dots, k$.

The second-type modulus of the exponential moment compactness $\Xi''^{\pm}_\beta(Y_{\varepsilon,\cdot,i}, N)$, expressed in terms of of jumps $\vec{W}_{\varepsilon,n}$, takes the following form for $\beta \geq 0$:

$$\Xi''^{\pm}_\beta(Y_{\varepsilon,\cdot,i}, N) = \max_{0 \leq n \leq N-1} \sup_{x \in \mathbb{X}} \mathsf{E}_{x,n} e^{\pm \beta W_{\varepsilon,n+1,i}}. \tag{5.100}$$

The condition $\mathbf{E_{10}}[\bar{\beta}]$, assumed to hold for some vector parameter $\bar{\beta} = (\beta_1, \dots, \beta_k)$ with nonnegative components, takes the following form:

$\mathbf{E_{12}}[\bar{\beta}]$: $\overline{\lim}_{\varepsilon \to 0} \Xi''^{\pm}_{\beta_i}(Y_{\varepsilon,\cdot,i}, N) < K_{40,i}$, $i = 1, \dots, k$, for some $1 < K_{40,i} < \infty$, $i = 1, \dots, k$.

The following lemma is a direct corollary of Lemma 5.1.5.

Lemma 5.3.4. *The condition $\mathbf{E_{12}}[\bar{\beta}]$ implies the condition $\mathbf{C_8}[\bar{\beta}]$ to hold.*

The conditions $\mathbf{B_5}[\bar{\gamma}]$, $\mathbf{D_7}[\bar{\beta}]$, and $\mathbf{F_5}[\bar{\beta}]$ do not change.

The convergence conditions $\mathbf{I_1}$ for pay-off functions do not change.

However, the convergence condition for transition probabilities $\mathbf{J_1}$ can be simplified.

Let us define, for a set $A \in \mathcal{B}_{\mathbb{Z}}$ and a point $\vec{y} \in \mathbb{R}_k$, the set

$$A_{[\vec{y}]} = \{(\vec{y}\,', x) : (\vec{y}\,' + \vec{y}, x) \in A\}. \tag{5.101}$$

The transition probabilities $P_{\varepsilon,n}(\vec{z}, A) = \mathsf{P}\{\vec{Z}_{\varepsilon,n} \in A / \vec{Z}_{\varepsilon,n-1} = \vec{z}\}$ for the Markov process $\vec{Z}_{\varepsilon,n} = (\vec{Y}_{\varepsilon,n}, X_{\varepsilon,n})$ are expressed via the transition probabilities $\tilde{P}_{\varepsilon,n}(x, A)$

$= \mathsf{P}\{(\vec{W}_{\varepsilon,n}, X_{\varepsilon,n}) \in A / X_{\varepsilon,n-1} = x\} = \mathsf{P}\{(\tilde{B}_{\varepsilon,n}(x, U_{\varepsilon,n}), C_{\varepsilon,n}(x, U_{\varepsilon,n})) \in A\}$ by the following formula, for $\vec{z} = (\vec{y}, x) \in \mathbb{Z} = \mathbb{R}_k \times \mathbb{X}$, $A \in \mathcal{B}_{\mathbb{Z}}$, $n = 1, 2, \dots$:

$$\begin{aligned} P_{\varepsilon,n}(\vec{z}, A) &= \mathsf{P}\{(\vec{y} + \vec{W}_{\varepsilon,n}, X_{\varepsilon,n}) \in A / \vec{Y}_{\varepsilon,n-1} = \vec{y}, X_{\varepsilon,n-1} = x\} \\ &= \mathsf{P}\{(\vec{y} + \tilde{B}_{\varepsilon,n+1}(x, U_{\varepsilon,n}), C_{\varepsilon,n}(x, U_{\varepsilon,n})) \in A\} \\ &= \tilde{P}_{\varepsilon,n}(x, A_{[\vec{y}]}) . \end{aligned} \tag{5.102}$$

Let us assume that the following condition holds:

J$_7$: There exist measurable sets $\mathbb{X}_n \subseteq \mathbb{X}$, $n = 0, \dots, N$ such that

(a) $\tilde{P}_{\varepsilon,n}(x_0, \cdot) \overset{U}{\Longrightarrow} \tilde{P}_{0,n}(x_0, \cdot)$ as $\varepsilon \to 0$, i.e. $\tilde{P}_{\varepsilon,n}(x_\varepsilon, \cdot) \Rightarrow \tilde{P}_{0,n}(x_0, \cdot)$ as $\varepsilon \to 0$, for any $x_\varepsilon \to x_0 \in \mathbb{X}_n$ as $\varepsilon \to 0$, and $n = 1, \dots, N$;

(b) $P_{0,n}((\vec{y}_0, x_0), \mathbb{Z}'_n \cap \mathbb{Z}''_n) = \tilde{P}_{0,n}(x_0, (\mathbb{Z}'_n \cap \mathbb{Z}''_n)_{[y_0]}) = 1$, for every $\vec{z}_0 = (\vec{y}_0, x_0) \in \mathbb{Z}'_{n-1} \cap \mathbb{Z}''_{n-1}$, $n = 1, \dots, N$, where $\mathbb{Z}'_n$, $n = 0, \dots, N$ are sets introduced in the condition **I**$_1$ and $\mathbb{Z}''_n = \mathbb{R}_k \times \mathbb{X}_n$, $n = 0, \dots, N$.

A typical example is where the sets $\overline{\mathbb{Z}}'_n, \overline{\mathbb{X}}_n$, $n = 1, \dots, N$ are empty sets. Then the condition **J**$_7$ (b) obviously holds.

Another typical example is where the sets $\overline{\mathbb{Z}}'_n, \overline{\mathbb{X}}_n$, $n = 1, \dots, N$ are at most finite or countable sets. Then, the assumption that the measures $\tilde{P}_{0,n}(x_0, A)$, $A \in \mathcal{B}_{\mathbb{Z}}$ and $\tilde{P}_{0,n}(x_0, \mathbb{R}_k \times B)$, $B \in \mathcal{B}_{\mathbb{X}}$ have no atoms, for every $x_0 \in \mathbb{X}_{n-1}$, $n = 1, \dots, N$, implies that the condition **J**$_7$ (b) holds.

One more example is where the measures $\tilde{P}_{0,n}(x_0, A)$, $x_0 \in \mathbb{X}_{n-1}$, $n = 1, \dots, N$ are absolutely continuous with respect to some σ-finite measure $P(A)$ on $\mathcal{B}_{\mathbb{Z}}$ and $P(\overline{\mathbb{Z}}'_n), P(\mathbb{R}_k \times \overline{\mathbb{X}}_n) = 0$, $n = 1, \dots, N$. This assumption implies that the condition **J**$_7$ (b) holds.

Lemma 5.3.5. *The condition* **J**$_7$ *implies that* $P_{\varepsilon,n}(\vec{z}_0, \cdot) \overset{U}{\Rightarrow} P_{0,n}(\vec{z}_0, \cdot)$ *as* $\varepsilon \to 0$, *i.e.* $P_{\varepsilon,n}(\vec{z}_\varepsilon, \cdot) \Rightarrow P_{0,n}(\vec{z}_0, \cdot)$ *as* $\varepsilon \to 0$, *for any* $\vec{z}_\varepsilon = (\vec{y}_\varepsilon, x_\varepsilon) \to \vec{z}_0 = (\vec{y}_0, x_0) \in \mathbb{R}_k \times \mathbb{X}_{n-1}$ *as* $\varepsilon \to 0$, *and* $n = 1, \dots, N$.

Proof. The condition **J**$_7$ implies that for any $n = 1, \dots, N$ and $\vec{z}_\varepsilon = (\vec{y}_\varepsilon, x_\varepsilon) \to \vec{z}_0 = (\vec{y}_0, x_0) \in \mathbb{R}_k \times \mathbb{X}_{n-1}$ as $\varepsilon \to 0$, the following relation of convergence in distribution holds:

$$\begin{aligned} &(\vec{y}_\varepsilon + \tilde{B}_{\varepsilon,n}(x_\varepsilon, U_{\varepsilon,n}), C_{\varepsilon,n}(x_\varepsilon, U_{\varepsilon,n})) \\ &\quad \xrightarrow{d} (\vec{y}_0 + \tilde{B}_{0,n}(x_0, U_{0,n}), C_{0,n+1}(x_0, U_{0,n})) \quad \text{as} \quad \varepsilon \to 0 . \end{aligned} \tag{5.103}$$

This relation is equivalent to the proposition of the lemma. □

By Lemma 5.3.5, the condition **J**$_7$ implies the condition **J**$_1$ to hold. In this case, the role of the set $\mathbb{Z}''_n$ is played by the set $\mathbb{R}_k \times \mathbb{X}_n$, for every $n = 0, 1, \dots, N$.

The modulated random walk $\vec{Z}_{\varepsilon,n}$ was introduced in the dynamic form. This makes it possible to replace the condition **J**$_7$ by the following sufficient condition, which by Lemma 5.2.1, implies the condition **J**$_7$:

$\mathbf{J}_7'$: There exist measurable sets $\mathbb{X}_n \subseteq \mathbb{X}$, $n = 0, \dots, N$ and $\mathbb{U}_n \in \mathbb{U}$, $n = 1, \dots, N$ such that

(a) $\tilde{B}_{\varepsilon,n}(x_\varepsilon, \cdot) \Rightarrow \tilde{B}_{0,n}(x_0, u)$ and $C_{\varepsilon,n}(x_\varepsilon, \cdot) \Rightarrow C_{0,n}(x_0, u)$ as $\varepsilon \to 0$, for any $x_\varepsilon \to x_0 \in \mathbb{X}_n$ as $\varepsilon \to 0$, $u \in \mathbb{U}_n$, $n = 1, \dots, N$;

(b) $G_n(\mathbb{U}_n) = \mathsf{P}\{U_{\varepsilon,n} \in \mathbb{U}_n\} = 1$, $n = 1, 2, \dots, N$;

(c) $P_{0,n}((\vec{y}_0, x_0), \mathbb{Z}_n' \cap \mathbb{Z}_n'') = \tilde{P}_{0,n}(x_0, (\mathbb{Z}_n' \cap \mathbb{Z}_n'')_{[y_0]}) = 1$, for every $\vec{z}_0 = (\vec{y}_0, x_0) \in \mathbb{Z}_{n-1}' \cap \mathbb{Z}_{n-1}''$, $n = 1, \dots, N$, where $\mathbb{Z}_n'$, $n = 0, \dots, N$ are sets introduced in the condition $\mathbf{I}_1$ and $\mathbb{Z}_n'' = \mathbb{R}_k \times \mathbb{X}_n$, $n = 0, \dots, N$.

The condition $\mathbf{K}_1$ imposed on the initial distributions $P_{\varepsilon,0}(A)$ should be replaced by the following condition:

$\mathbf{K}_6$: (a) $P_{\varepsilon,0}(\cdot) \Rightarrow P_{0,0}(\cdot)$ as $\varepsilon \to 0$;

(b) $P_{0,0}(\mathbb{Z}_0' \cap \mathbb{Z}_0'') = 1$, where $\mathbb{Z}_0'$ and $\mathbb{Z}_0''$ are the sets introduced in the conditions $\mathbf{I}_1$ and $\mathbf{J}_7$.

Obviously, the condition $\mathbf{K}_6$ implies that the condition $\mathbf{K}_1$ holds with with the sets $\mathbb{Z}_0'$ and $\mathbb{Z}_0'' = \mathbb{R}_k \times \mathbb{X}_0$.

Theorem 5.2.1 takes in this case the following form.

Theorem 5.3.5. *Let $\vec{Z}_{\varepsilon,n} = (\vec{Y}_{\varepsilon,n}, X_{\varepsilon,n})$ is the log-price process represented, for every $\varepsilon \in [0, \varepsilon_0]$, by the multivariate modulated random walk given by the stochastic transition dynamic relation* (5.98)*. Let also the conditions $\mathbf{B}_5[\bar{\gamma}]$ and $\mathbf{C}_8[\bar{\beta}]$ hold with the vector parameters $\bar{\gamma} = (\gamma_1, \dots, \gamma_k)$, $\bar{\beta} = (\beta_1, \dots, \beta_k)$ such that, for every $i = 1, \dots, k$, either $\beta_i > \gamma_i > 0$ or $\beta_i = \gamma_i = 0$, and also the conditions $\mathbf{I}_1$ and $\mathbf{J}_7$ hold. Then, for every $n = 0, 1, \dots, N$, the following relation takes place for any $\vec{z}_\varepsilon = (\vec{y}_\varepsilon, x_\varepsilon) \to \vec{z}_0 = (\vec{y}_0, x_0) \in \mathbb{Z}_n' \cap \mathbb{Z}_n''$:*

$$\phi_{\varepsilon,n}(\vec{y}_\varepsilon, x_\varepsilon) \to \phi_{0,n}(\vec{y}_0, x_0) \quad as \quad \varepsilon \to 0. \tag{5.104}$$

Theorem 5.2.3 takes in this case the following form.

Theorem 5.3.6. *Let $\vec{Z}_{\varepsilon,n} = (\vec{Y}_{\varepsilon,n}, X_{\varepsilon,n})$ is the log-price process represented, for every $\varepsilon \in [0, \varepsilon_0]$, by the multivariate modulated random walk given by the stochastic transition dynamic relation* (5.98)*. Let also the conditions $\mathbf{B}_5[\bar{\gamma}]$, $\mathbf{C}_8[\bar{\beta}]$ and $\mathbf{D}_7[\bar{\beta}]$ hold with the vector parameters $\bar{\gamma} = (\gamma_1, \dots, \gamma_k)$, $\bar{\beta} = (\beta_1, \dots, \beta_k)$ such that, for every $i = 1, \dots, k$, either $\beta_i > \gamma_i > 0$ or $\beta_i = \gamma_i = 0$, and also the conditions $\mathbf{I}_1$, $\mathbf{J}_7$, and $\mathbf{K}_6$, hold. Then the following relation takes place:*

$$\Phi_\varepsilon \to \Phi_0 \quad as \quad \varepsilon \to 0. \tag{5.105}$$

6 Convergence of option rewards – II

In this chapter, we present our main convergence results for rewards of American-type options with general perturbed pay-off functions for perturbed multivariate modulated Markov log-price processes with unbounded characteristics.

In Section 6.1, we give asymptotically uniform upper bounds for log-reward functions and optimal expected rewards for univariate modulated Markov log-price processes with unbounded characteristics (conditional exponential moments for increments).

In Section 6.2, we present convergence results for log-reward functions and optimal expected rewards for perturbed univariate modulated Markov log-price processes with unbounded characteristics.

In Section 6.3, we give asymptotically uniform upper bounds for log-reward functions and optimal expected rewards for multivariate modulated Markov log-price processes with unbounded characteristics.

In Section 6.4, we present convergence results for log-reward functions and optimal expected rewards for perturbed multivariate modulated Markov log-price processes with unbounded characteristics.

In Section 6.5, we show in which way conditions of convergence for rewards of American-type options can be reformulated in terms of modulated price Markov processes instead of modulated Markov log-price processes.

Our main results are given in Theorems 6.2.1–6.2.4 for perturbed univariate modulated Markov log-price processes with unbounded characteristics, and in Theorems 6.4.1–6.4.4 for perturbed multivariate modulated Markov log-price processes with unbounded characteristics.

What is important that we impose minimal conditions of smoothness on the limiting pay-off functions and transition probabilities of limiting log-price processes. For the basic case, where the transition probabilities have densities with respect to some pivotal Lebesgue-type measure, it is usually required that the sets of weak discontinuity for the limiting transition probabilities and the sets of discontinuity for pay-off functions are zero sets with respect to the above pivotal measure. In fact, such assumptions make it possible for the transition probabilities and pay-off functions to be very irregular.

The results presented in this chapter are new. They essentially improve results given in Chapter 5. The main improvement is that we consider much more general moduli of exponential moment compactness with normalizing coefficients for conditional exponential moments of increments. For example, this makes it possible to get the corresponding convergence results for Markov Gaussian log-price processes with drift and volatility coefficients with not more than linear rate of growth as a function of price arguments. The corresponding applications are presented in Chapters 7 and 8.

6.1 Asymptotically uniform upper bounds for rewards – II

In this section, we present asymptotically uniform upper bounds for option rewards for univariate modulated Markov log-price processes with unbounded characteristics.

6.1.1 Upper bounds for reward functions for univariate modulated Markov log-price processes with unbounded characteristics

We consider a one-dimensional modulated Markov log-price process $Z_{\varepsilon,n} = (Y_{\varepsilon,n}, X_{\varepsilon,n})$, $n = 0, 1, \dots$, with a phase space $\mathbb{Z} = \mathbb{R}_1 \times \mathbb{X}$, an initial distribution $P_\varepsilon(A)$, and transition probabilities $P_{\varepsilon,n}(z, A) = \mathsf{P}\{Z_{\varepsilon,n} \in A / Z_{\varepsilon,n-1} = z\}$, depending on the perturbation parameter $\varepsilon \in [0, \varepsilon_0]$.

We assume that the phase space of the index component $\mathbb{X}$ is a Polish metric space, i.e. complete, separable, metric space, with a metric $d_{\mathbb{X}}(x', x'')$.

Let us recall the class $\mathcal{A}$ of measurable functions $A(\beta)$ defined for $\beta \geq 0$ such that the function $A(\beta)$ is nondecreasing in β and $A(0) = 0$. As pointed out in Subsection 4.3.1, any function $A(\beta) \in \mathcal{A}$ generates the sequence of functions $A_n(\beta)$, $n = 0, 1, \dots$, from the class $\mathcal{A}$, which are defined by the following recurrence relations, for $\beta \geq 0$,

$$A_n(\beta) = \begin{cases} \beta & \text{for } n = 0, \\ A_{n-1}(\beta) + A(A_{n-1}(\beta)) & \text{for } n = 1, 2, \dots. \end{cases} \tag{6.1}$$

Note that by the definition $A_0(\beta) \leq A_1(\beta) \leq A_2(\beta) \leq \cdots$, for every $\beta \geq 0$.

Let us also recall the first-type A-modulus of exponential moment compactness for the log-price process $Y_{\varepsilon,n}$, defined for $\beta \geq 0$

$$\Delta_{\beta,A}(Y_{\varepsilon,\cdot}, N) = \max_{0 \leq n \leq N-1} \sup_{z=(y,x) \in \mathbb{Z}} \frac{\mathsf{E}_{z,n} e^{A_{N-n-1}(\beta)|Y_{\varepsilon,n+1} - Y_{\varepsilon,n}|}}{e^{A(A_{N-n-1}(\beta))|y|}}. \tag{6.2}$$

Recall that we use the notations $\mathsf{P}_{z,n}$ and $\mathsf{E}_{z,n}$ for conditional probabilities and expectations under the condition $Z_{\varepsilon,n} = z$.

The condition $\mathbf{C}_4[\beta]$ should be replaced by the following first-type condition of exponential moment compactness, assumed to hold for some $\beta \geq 0$:

$\mathbf{C}_9[\beta]$: $\overline{\lim}_{\varepsilon \to 0} \Delta_{\beta,A}(Y_{\varepsilon,\cdot}, N) < K_{41}$, for some $1 < K_{41} < \infty$.

The following lemma gives asymptotically uniform upper bounds for exponential moments of maxima for perturbed log-price processes.

Lemma 6.1.1. *Let the condition* $\mathbf{C}_9[\beta]$ *holds. Then there exist* $0 < \varepsilon_6 \leq \varepsilon_0$ *and the constants* $0 \leq M_{35} < \infty$ *such that the following inequality takes place for* $z = (y, x) \in \mathbb{Z}, 0 \leq n \leq N$ *and* $0 \leq \varepsilon \leq \varepsilon_6$*:*

$$\mathsf{E}_{z,n} \exp\left\{\beta \max_{n \leq r \leq N} |Y_{\varepsilon,r}|\right\} \leq M_{35} \exp\{A_{N-n}(\beta)|y|\} . \tag{6.3}$$

Proof. By the condition $\mathbf{C}_9[\beta]$, one can choose $0 < \varepsilon_6 \leq \varepsilon_0$ such that for $0 \leq \varepsilon \leq \varepsilon_6$

$$\Delta_{\beta,A}(Y_{\varepsilon,\cdot}, N) < K_{41} . \tag{6.4}$$

Relation (6.4) means that, for every $\varepsilon \leq \varepsilon_6$, the condition $\mathbf{C}_4[\beta]$ (with the constant K_{41} replacing the constants K_9) holds for the process $Y_{\varepsilon,n}$.

Therefore, Lemma 4.3.1 can be applied to the process $Y_{\varepsilon,n}$, for every $\varepsilon \leq \varepsilon_6$.

This yields inequalities (6.3). The constant M_{11} penetrating the inequality given in Lemma 4.3.1 is a function of the corresponding constant K_9. The explicit expression for this function is given in Remark 4.3.1. The same formula gives the expression for the constant M_{35} as a function of the corresponding constant K_{41}. □

Remark 6.1.1. The constant M_{35} is given by the following formula, which follows from the corresponding formula given in Remark 4.3.1:

$$M_{35} = K_{41}^N . \tag{6.5}$$

Remark 6.1.2. The parameter ε_6 is determined by relation (6.4).

Let $\mathcal{Z}_\varepsilon = \langle \mathbb{Z}_{\varepsilon,0}, \dots, \mathbb{Z}_{\varepsilon,N} \rangle$ be a sequence of measurable sets. Let us introduce a modified first-type A-modulus of exponential moment compactness defined, for every $\varepsilon \in [0, \varepsilon_0]$ and $\beta \geq 0$

$$\Delta_{\beta,A,\mathcal{Z}_\varepsilon}(Y_{\varepsilon,\cdot}, N) = \max_{0 \leq n \leq N-1} \sup_{z=(y,x) \in \mathbb{Z}_{\varepsilon,n}} \frac{\mathsf{E}_{z,n} e^{A_{N-n-1}(\beta)|Y_{\varepsilon,n+1} - Y_{\varepsilon,n}|}}{e^{A(A_{N-n-1}(\beta))|y|}} . \tag{6.6}$$

The following first-type condition of exponential moment compactness for the log-price process $Y_{\varepsilon,n}$ generalizes the condition $\mathbf{C}_9[\beta]$. It should be assumed to hold for some $\beta \geq 0$:

$\mathbf{C}'_9[\beta]$: There exists a sequence of measurable sets $\mathcal{Z}_\varepsilon = \langle \mathbb{Z}_{\varepsilon,0}, \dots, \mathbb{Z}_{\varepsilon,N} \rangle$, for every $\varepsilon \in [0, \varepsilon_0]$, such that

(a) $\overline{\lim}_{\varepsilon \to 0} \Delta_{\beta,A,\mathcal{Z}_\varepsilon}(Y_{\varepsilon,\cdot}, N) < K'_{41}$, for some $1 < K'_{41} < \infty$;

(b) $\mathcal{Z}_\varepsilon$ is a complete sequence of phase sets for the process $\vec{Z}_{\varepsilon,n}$, i.e. $\mathsf{P}_{\varepsilon,n}(\vec{z}, \mathbb{Z}_{\varepsilon,n}) = 1, \vec{z} \in \mathbb{Z}_{\varepsilon,n-1}, n = 1, \dots, N$, for every $\varepsilon \in [0, \varepsilon_0]$.

Let us introduce, for every $\varepsilon \in [0, \varepsilon_0]$, the sets $\mathbb{H}_{\varepsilon,n} = (h^-_{\varepsilon,n}, h^+_{\varepsilon,n}) \times \mathbb{X}$, $n = 0, \dots, N$, where $-\infty \leq h^-_{\varepsilon,n} \leq h^+_{\varepsilon,n} \leq +\infty$, $n = 0, \dots, N$.

It is useful to note that in applications, for example to space-skeleton models considered in Chapters 7 and 8, the sets $\mathbb{Z}_{\varepsilon,n}$, $n = 0, \dots, N$ satisfy the following condition analogous to $\mathbf{H}_1$:

H₂: $\mathbb{Z}_{\varepsilon,n} \supseteq \mathbb{H}_{\varepsilon,n}$, $n = 0, \dots, N$, $\varepsilon \in [0, \varepsilon_0]$, where $h^{\pm}_{\varepsilon,n} \to h^{\pm}_{0,n} = \pm\infty$ as $\varepsilon \to 0$, for $n = 0, \dots, N$.

Remark 6.1.3. According to the condition $\mathbf{H}_2$, for every $z = (y, x)$, there exists $\varepsilon_2(z) \in (0, \varepsilon_0]$ such that the following relation holds for every $\varepsilon \in [0, \varepsilon_2(z))$:

$$h^{-}_{\varepsilon,n} < y < h^{+}_{\varepsilon,n}\,, \quad n = 0, \dots, N\,. \tag{6.7}$$

Obviously, $z \in \mathbb{Z}_{\varepsilon,n}$, $n = 0, \dots, N$ for all $\varepsilon \in [0, \varepsilon_2(z))$.

Note that in the case where the condition $\mathbf{H}_2$ holds, the sets $\mathbb{Z}_{0,n} = \mathbb{Z}$, $n = 0, \dots, N$.

Lemma 6.1.2. *Let the condition $\mathbf{C}'_9[\beta]$ holds. Then there exist $0 < \varepsilon'_6 \le \varepsilon_0$ and a constant $0 \le M'_{35} < \infty$ such that the following inequalities take place for $z = (y, x) \in \mathbb{Z}_{\varepsilon,n}$, $0 \le n \le N$, $0 \le \varepsilon \le \varepsilon'_6$:*

$$\mathsf{E}_{\vec{z},n} \exp\left\{\beta \max_{n \le r \le N} |Y_{\varepsilon,r}|\right\} \le M'_{35} \exp\{A_{N-n}(\beta)|y|\}\,. \tag{6.8}$$

Proof. By the condition $\mathbf{C}'_9[\beta]$, one can choose $0 < \varepsilon'_6 \le \varepsilon_0$ such that for $0 \le \varepsilon \le \varepsilon'_6$,

$$\Delta_{\beta,A,\mathbb{Z}_\varepsilon}(Y_{\varepsilon,\cdot}(\cdot), N) < K'_{41}\,. \tag{6.9}$$

The following part of the proof repeats the proof of Lemma 6.1.1. The only difference is that Lemma 4.3.2 should be applied to the process $\vec{Y}_{\varepsilon,n}$, for every $\varepsilon \le \varepsilon'_6$, instead of Lemma 4.3.1 that was applied to the process $\vec{Y}_{\varepsilon,n}$, for every $\varepsilon \le \varepsilon_6$, in the proof of Lemma 6.1.1.

This yields inequalities (6.8). The constant M'_{11} penetrating the inequalities given in Lemma 4.3.2 is a function of the corresponding constant K'_9. The explicit expression for this function is given in Remark 4.3.2. The same formula gives the expression for the constant M'_{35} as a function of the corresponding constant K'_{41}. □

Remark 6.1.4. The constant M'_{35} is given by the following formula, which follows from the corresponding formula given in Remark 4.3.2:

$$M'_{35} = (K'_{41})^N\,. \tag{6.10}$$

Remark 6.1.5. The parameter ε'_6 is determined by relation (6.9).

6.1.2 Asymptotically uniform upper bounds for option rewards for univariate modulated Markov log-price processes

Let us now give asymptotically uniform upper bounds for log-reward functions for log-price processes $Z_{\varepsilon,n}$.

In this case, the pay-off functions $g_\varepsilon(n, e^y, x)$ are functions of the argument $z = (y, x) \in \mathbb{Z} = \mathbb{R}_1 \times \mathbb{X}$ depending on the perturbation parameter ε.

The condition $\mathbf{B}_3[\gamma]$ should be replaced by the following condition, assumed to hold for some $\gamma \geq 0$:

$\mathbf{B}_8[\gamma]$: $\overline{\lim}_{\varepsilon\to 0} \max_{0\leq n\leq N} \sup_{(y,x)\in\mathbb{Z}} \frac{|g_\varepsilon(n,e^y,x)|}{1+L_{16}e^{\gamma|y|}} < L_{15}$, for some $0 < L_{15} < \infty$ and $0 \leq L_{16} < \infty$.

The following lemma gives asymptotically uniform upper bounds for conditional expectations for maximal absolute values for the perturbed pay-off processes $g_\varepsilon(n, e^{Y_{\varepsilon,n}}, X_{\varepsilon,n})$.

Lemma 6.1.3. *Let the conditions* $\mathbf{B}_8[\gamma]$ *and* $\mathbf{C}_9[\beta]$ *hold and* $0 \leq \gamma \leq \beta$. *Then, there exist* $0 < \varepsilon_8 \leq \varepsilon_0$ *and the constants* $0 \leq M_{36}, M_{37} < \infty$ *such that the following inequalities take place for* $z = (y, x) \in \mathbb{Z}, 0 \leq n \leq N$ *and* $0 \leq \varepsilon \leq \varepsilon_8$*:*

$$\begin{aligned} &\mathsf{E}_{z,n} \max_{n\leq r\leq N} \left| g_\varepsilon\left(n, e^{Y_{\varepsilon,n}}, X_{\varepsilon,n}\right) \right| \\ &\quad \leq M_{36} + M_{37} I(\gamma = 0) + M_{61} \exp\left\{ \frac{A_{N-n}(\beta)\gamma}{\beta} |y| \right\} I(\gamma > 0). \end{aligned} \tag{6.11}$$

Proof. The condition $\mathbf{B}_8[\gamma]$ implies that there exists $0 < \varepsilon_7 \leq \varepsilon_0$ such that for any $0 \leq \varepsilon \leq \varepsilon_7$,

$$\max_{0\leq n\leq N} \sup_{z=(y,x)\in\mathbb{Z}} \frac{|g_\varepsilon(n, e^y, x)|}{1 + L_{16}e^{\gamma|y|}} < L_{15}. \tag{6.12}$$

Relation (6.12) means that for every $\varepsilon \leq \varepsilon_7$, the condition $\mathbf{B}_3[\gamma]$ (with the constants L_{15}, L_{16}, replacing the constants L_5, L_6) holds for the process $Y_{\varepsilon,n}$.

Let us now take $\varepsilon_8 = \min(\varepsilon_6, \varepsilon_7)$, where ε_6 was defined in relation (6.4). By the definition $\varepsilon_8 > 0$.

For every fixed $0 \leq \varepsilon \leq \varepsilon_8$ both the conditions $\mathbf{C}_4[\beta]$ and $\mathbf{B}_3[\gamma]$ (with the constants K_{41} and L_{15}, L_{16} replacing, respectively, the constants K_9 and L_5, L_6) hold for the process $Y_{\varepsilon,n}$.

Therefore, Lemma 4.3.3 can be applied to the process $Y_{\varepsilon,n}$, for every $0 \leq \varepsilon \leq \varepsilon_8$.

This yields inequality (6.11). The constants M_{12} and M_{13} penetrating the inequalities given in Lemma 4.3.3 are functions of the corresponding constants K_9 and L_5, L_6. The explicit expressions for these functions are given in Remark 4.3.3. The same formulas give the expressions for the constant M_{36} and M_{37} as functions of the corresponding constants K_{41} and L_{21}, L_{22}. □

Remark 6.1.6. The constants M_{36} and M_{37} are given by the following formulas:

$$M_{36} = L_{15}, \quad M_{37} = L_{15}L_{16}I(\gamma = 0) + L_{15}L_{16}K_{41}^{N\frac{\gamma}{\beta}} I(\gamma > 0). \tag{6.13}$$

Remark 6.1.7. The parameter $\varepsilon_8 = \min(\varepsilon_6, \varepsilon_7)$, where the parameters ε_6 and ε_7 are determined, respectively, by relations (6.4) and (6.12).

Lemma 6.1.4. *Let the conditions* $\mathbf{B}_8[\gamma]$ *and* $\mathbf{C}_9'[\beta]$ *hold and* $0 \le \gamma \le \beta$*. Then, there exist* $0 < \varepsilon_8' \le \varepsilon_0$ *and the constants* $0 \le M_{36}', M_{37}' < \infty$ *such that the following inequalities take place for* $z = (y, x) \in \mathbb{Z}_{\varepsilon,n}, 0 \le n \le N, 0 \le \varepsilon \le \varepsilon_8'$*:*

$$\mathsf{E}_{z,n} \max_{n \le r \le N} \left| g_\varepsilon\left(n, e^{Y_{\varepsilon,n}}, X_{\varepsilon,n}\right) \right|$$
$$\le M_{36}' + M_{37}' I(\gamma = 0) + M_{37}' \exp\left\{\frac{A_{N-n}(\beta)\gamma}{\beta}|y|\right\} I(\gamma > 0)\,. \tag{6.14}$$

Proof. It is analogous to the proof of Lemma 6.1.3. As was pointed out in the proof of this lemma, the condition $\mathbf{B}_8[\gamma]$ implies that there exists $0 < \varepsilon_7 \le \varepsilon_0$ such that inequality (6.12) holds for $0 \le \varepsilon \le \varepsilon_7$.

Relation (6.12) means that for every $\varepsilon \le \varepsilon_7$, the condition $\mathbf{B}_3[\gamma]$ (with the constants L_{15}, L_{16}, replacing the constants L_5, L_6) holds for the process $Y_{\varepsilon,n}$.

Let us now take $\varepsilon_8' = \min(\varepsilon_6', \varepsilon_7)$, where ε_6' was defined in relation (6.9). By the definition $\varepsilon_8' > 0$.

For every fixed $0 \le \varepsilon \le \varepsilon_8'$ both the conditions $\mathbf{C}_4'[\beta]$ and $\mathbf{B}_3[\gamma]$ (with the constants K_{41}' and L_{15}, L_{16} replacing, respectively, the constants K_9' and L_5, L_6) hold for the process $Y_{\varepsilon,n}$.

Therefore, Lemma 4.3.4 can be applied to the process $Y_{\varepsilon,n}$, for every $0 \le \varepsilon \le \varepsilon_8'$.

This yields inequality (6.14). The constants M_{12}' and M_{13}' penetrating the inequalities given in Lemma 4.3.4 are functions of the corresponding constants K_9' and L_5, L_6. The explicit expressions for these functions are given in Remark 4.3.4. The same formulas give the expressions for the constant M_{36}' and M_{37}' as functions of the corresponding constants K_{41}' and L_{15}, L_{16}. □

Remark 6.1.8. The constants M_{36}' and M_{37}' are given by the following formulas:

$$M_{36}' = L_{15}\,, \quad M_{37}' = L_{15}L_{16}I(\gamma = 0) + L_{15}L_{16}(K_{41}')^{N\frac{\gamma}{\beta}} I(\gamma > 0)\,. \tag{6.15}$$

Remark 6.1.9. The parameter $\varepsilon_8' = \min(\varepsilon_6', \varepsilon_7)$, where the parameters ε_6' and ε_7 are determined, respectively, by relations (6.9) and (6.12).

Let us recall the modified second-type A-modulus of exponential moment compactness for the log-price process $Y_{\varepsilon,n}$, defined for $\beta \ge 0$,

$$\Xi_{\beta,A}^{\pm}(Y_{\varepsilon,\cdot}, N) = \max_{0 \le n \le N-1} \sup_{z=(y,x)\in\mathbb{Z}} \frac{\mathsf{E}_{z,n} e^{\pm A_{N-n-1}(\beta)(Y_{\varepsilon,n+1} - Y_{\varepsilon,n})}}{e^{A(A_{N-n-1}(\beta))|y|}}\,. \tag{6.16}$$

The following second-type condition of exponential moment compactness, based on moment-generating functions for increments of log-price processes $Y_{\varepsilon,n}$, replaces the condition $\mathbf{E}_4[\beta]$. It is assumed to hold for some $\beta \ge 0$:
$\mathbf{E}_{13}[\beta]$: $\overline{\lim}_{\varepsilon\to 0}\, \Xi_{\beta,A}^{\pm}(Y_{\varepsilon,\cdot}, N) < K_{42}$, for some $1 < K_{42} < \infty$.

Lemma 6.1.5. *The condition* $\mathbf{E}_{13}[\beta]$ *implies that the condition* $\mathbf{C}_9[\beta]$ *holds.*

Proof. It follows from Lemma 4.3.7 that should be applied to the log-price processes $Y_{\varepsilon,n}$. □

Remark 6.1.10. If the condition $\mathbf{E}_{15}[\beta]$ holds with the constant K_{42}, then the condition $\mathbf{C}_9[\beta]$ holds with the constant K_{41} given by the following formula:

$$K_{41} = 2K_{42} \,. \tag{6.17}$$

Let $\mathcal{Z}_\varepsilon = \langle \mathbb{Z}_{\varepsilon,0}, \dots, \mathbb{Z}_{\varepsilon,N} \rangle$ be a sequence of measurable sets. Let us introduce and a modified second-type A-modulus of exponential moment compactness defined, for every $\varepsilon \in [0, \varepsilon_0]$ and $\beta \geq 0$,

$$\Xi^{\pm}_{\beta,A,\mathcal{Z}_\varepsilon}(Y_{\varepsilon,\cdot}, N) = \max_{0 \leq n \leq N-1} \sup_{z=(y,x)\in\mathbb{Z}_{\varepsilon,n}} \frac{\mathsf{E}_{z,n} e^{\pm A_{N-n-1}(\beta)(Y_{\varepsilon,n+1} - Y_{\varepsilon,n})}}{e^{A(A_{N-n-1}(\beta))|y|}} \,. \tag{6.18}$$

The following second-type condition of exponential moment compactness is based on moment-generating functions for increments of log-price processes $Y_{\varepsilon,n}$. It generalizes the condition $\mathbf{E}_{13}[\beta]$ and is assumed to hold for some $\beta \geq 0$:

$\mathbf{E}'_{13}[\beta]$: For every $\varepsilon \in [0, \varepsilon_0]$, there exists a sequence of measurable sets $\mathcal{Z}_\varepsilon = \langle \mathbb{Z}_{\varepsilon,0}, \dots, \mathbb{Z}_{\varepsilon,N} \rangle$ such that
- (a) $\overline{\lim}_{\varepsilon\to 0} \Xi^{\pm}_{\beta,A,\mathcal{Z}_\varepsilon}(Y_{\varepsilon,\cdot}, N) < K'_{42}$, for some $1 < K'_{42} < \infty$;
- (b) $\mathcal{Z}_\varepsilon$ is a complete sequence of phase sets for the process $Z_{\varepsilon,n}$, i.e. $\mathsf{P}_{z,n}\{Z_{\varepsilon,n} \in \mathbb{Z}_n\} = 1$, $z \in \mathbb{Z}_{\varepsilon,n-1}$, $n = 1, \dots, N$.

Lemma 6.1.6. *The condition* $\mathbf{E}'_{13}[\beta]$ *implies the condition* $\mathbf{C}'_9[\beta]$ *holds.*

Proof. It follows from inequality Lemma 4.3.8 that should be applied to the log-price processes $Y_{\varepsilon,n}$. □

Remark 6.1.11. If the condition $\mathbf{E}'_{13}[\beta]$ holds with constant K'_{42}, then the condition $\mathbf{C}'_9[\beta]$ holds with the constant K'_{41} given by the following formula:

$$K'_{41} = 2K'_{42} \,. \tag{6.19}$$

Let us assume that conditions $\mathbf{B}_{12}[\gamma]$ and $\mathbf{C}_9[\beta]$ hold and $0 \leq \gamma \leq \beta$.

In this case, for Lemma 6.1.3 let us define log-reward functions for American option for the modulated log-price process $Z_{\varepsilon,n}$, for every $0 \leq n \leq N$, and $0 \leq \varepsilon \leq \varepsilon_8$,

$$\begin{aligned} \phi_{\varepsilon,n}(z) &= \phi_{\varepsilon,n}(y, x) \\ &= \sup_{\tau_{\varepsilon,n} \in \mathcal{M}^{(\varepsilon)}_{\max,n,N}} \mathsf{E}_{z,n} g_\varepsilon(\tau_{\varepsilon,n}, e^{Y_{\varepsilon,\tau_{\varepsilon,n}}}, X_{\varepsilon,\tau_{\varepsilon,n}}) \,, \quad z = (y, x) \in \mathbb{Z} \,. \end{aligned} \tag{6.20}$$

These functions are the measurable functions of $z \in \mathbb{Z}$, for every $0 \leq n \leq N$.

Moreover, for Lemma 6.1.3 let us get the following asymptotically uniform upper bounds for the log-reward functions $\phi_{\varepsilon,n}(y, x)$.

Theorem 6.1.1. *Let the conditions* $\mathbf{B}_8[\gamma]$ *and* $\mathbf{C}_9[\beta]$ *hold and* $0 \leq \gamma \leq \beta$. *Then, the log-reward functions* $\phi_{\varepsilon,n}(y,x)$ *satisfy the following inequalities for* $z = (y,x) \in \mathbb{Z}, 0 \leq n \leq N, 0 \leq \varepsilon \leq \varepsilon_8$:

$$|\phi_{\varepsilon,n}(y,x)| \leq M_{36} + M_{37} I(\gamma = 0) + M_{37} \exp\left\{ \frac{A_{N-n}(\beta)\gamma}{\beta} |y| \right\} I(\gamma > 0) . \tag{6.21}$$

Proof. It is analogous to the proof of Theorem 5.1.1. Using Lemma 6.1.3 and the definition of the log-reward functions, we get the following inequalities for $z = (y,x) \in \mathbb{Z}, 0 \leq n \leq N, 0 \leq \varepsilon \leq \varepsilon_8$:

$$\begin{aligned} |\phi_{\varepsilon,n}(y,x)| &\leq \mathsf{E}_{z,n} \max_{n \leq r \leq N} |g_\varepsilon(r, e^{Y_{\varepsilon,r}}, X_{\varepsilon,r})| \\ &\leq M_{36} + M_{37} I(\gamma = 0) + M_{37} \exp\left\{ \frac{A_{N-n}(\beta)\gamma}{\beta} |y| \right\} I(\gamma > 0) , \end{aligned} \tag{6.22}$$

which proves the above theorem. □

Let us assume that the conditions $\mathbf{B}_8[\gamma]$ and $\mathbf{C}'_9[\beta]$ hold and $0 \leq \gamma \leq \beta$.

In this case, for Lemma 6.1.4 let us define the log-reward functions $\phi_{\varepsilon,n}(y,x)$ using relation (6.20), only for $z = (y,x) \in \mathbb{Z}_{\varepsilon,n}$, $0 \leq n \leq N$, $0 \leq \varepsilon \leq \varepsilon'_3$.

Also, for Lemma 6.1.4 let us get the following upper bounds for these log-reward functions.

Theorem 6.1.2. *Let the conditions* $\mathbf{B}_8[\gamma]$ *and* $\mathbf{C}'_9[\beta]$ *hold and* $0 \leq \gamma \leq \beta$. *Then, the log-reward functions* $\phi_{\varepsilon,n}(y,x)$ *satisfy the following inequalities for* $z = (y,x) \in \mathbb{Z}_{\varepsilon,n}, 0 \leq n \leq N, 0 \leq \varepsilon \leq \varepsilon'_8$:

$$|\phi_{\varepsilon,n}(y,x)| \leq M'_{36} + M'_{37} I(\gamma = 0) + M'_{37} \exp\left\{ \frac{(\beta + A_{N-n}(\beta))\gamma}{\beta} |y| \right\} I(\gamma > 0) . \tag{6.23}$$

Proof. It is analogous to the proof of Theorem 6.1.1. In this case, Lemma 6.1.4 should be used instead of Lemma 6.1.3. An inequality analogous to (6.22) can be written. The difference is that it holds for $z = (y,x) \in \mathbb{Z}_{\varepsilon,n}, 0 \leq n \leq N, 0 \leq \varepsilon \leq \varepsilon'_8$, and with the constants M'_{60}, M'_{61}. □

Remark 6.1.12. It is useful to note that the log-reward function $\phi_{\varepsilon,N}(y,x) = g_\varepsilon(N, e^y, x)$, $z = (y,x) \in \overline{\mathbb{Z}}_{\varepsilon,N}$ always exists. The corresponding upper bounds for this function are given by the condition $\mathbf{B}_8[\gamma]$. However, there is no guarantee that, under the conditions of Theorem 6.1.2, the log-reward functions $\phi_{\varepsilon,n}(y,x)$ exist for points $z = (y,x) \in \overline{\mathbb{Z}}_{\varepsilon,n}$, $0 \leq n \leq N-1$. Even, if these functions do exist for these points, Theorem 6.1.2 does not give upper bounds for the log-reward function in these points.

However, there is an important case, where this problem is not essential. Let us assume that the condition $\mathbf{H_2}$ holds. Then, as was pointed out in Remark 6.1.3, for any point $z = (y, x) \in \mathbb{Z}$, there exists $\varepsilon_2(z) \in (0, \varepsilon_0]$ such that $z = (y, x) \in \mathbb{H}_{\varepsilon,n} \subseteq \mathbb{Z}_{\varepsilon,n}$, $n = 0, \dots, N$, for $\varepsilon \in [0, \varepsilon_2(z)]$. The quantity $\varepsilon_2(z)$ is determined by relation (6.7) given in Remark 6.1.3. Thus, the upper bounds for $\phi_{\varepsilon,n}(y, x)$ given in Theorem 6.1.2 take place for $\varepsilon \in [0, \varepsilon_2(z)]$.

6.1.3 Upper bounds for optimal expected rewards for univariate modulated Markov log-price processes with unbounded characteristics

We can also get asymptotically uniform upper bounds for optimal expected rewards, for modulated Markov log-price processes $Z_{\varepsilon,n}$.

The condition $\mathbf{D_2}[\beta]$ should be replaced by the following condition, assumed to hold for some $\beta \geq 0$:

$\mathbf{D_8}[\beta]$: $\overline{\lim}_{\varepsilon\to 0} \mathsf{E} \exp\{A_N(\beta)\,|\,Y_{\varepsilon,0}|\} < K_{43}$, for some $1 < K_{43} < \infty$.

The following lemma gives asymptotically uniform upper bounds for the expectation of maximal absolute values for the pay-off processes $g_\varepsilon(n, e^{Y_{\varepsilon,n}}, X_{\varepsilon,n})$.

Lemma 6.1.7. *Let the conditions $\mathbf{B_8}[\gamma]$, $\mathbf{C_9}[\beta]$, and $\mathbf{D_8}[\beta]$ hold and $0 \leq \gamma \leq \beta$. Then, there exist $0 < \varepsilon_{10} \leq \varepsilon_0$ and the constant $0 \leq M_{38} < \infty$ such that for the following inequality takes place for $0 \leq \varepsilon \leq \varepsilon_{10}$:*

$$\mathsf{E} \max_{0\leq n\leq N} |g_\varepsilon(n, e^{Y_{\varepsilon,n}}, X_{\varepsilon,n})| \leq M_{38}\,. \tag{6.24}$$

Proof. The condition $\mathbf{D_8}[\beta]$ implies that there exists $0 < \varepsilon_9 \leq \varepsilon_0$ such that for $0 \leq \varepsilon \leq \varepsilon_9$,

$$\mathsf{E} e^{A_N(\beta)|Y_{\varepsilon,0}|} < K_{43}\,. \tag{6.25}$$

□

Relation (6.25) means that for every $\varepsilon \leq \varepsilon_9$, the condition $\mathbf{D_2}[\beta]$ (with the constant K_{86} replacing constant K_{10}) holds for the process $Y_{\varepsilon,n}$.

Let us now take $\varepsilon_{10} = \min(\varepsilon_6, \varepsilon_7, \varepsilon_9)$, where ε_6 and ε_7 were defined in relations (6.4) and (6.12). By the definition $\varepsilon_{10} > 0$.

For every fixed $0 \leq \varepsilon \leq \varepsilon_{10}$, the conditions $\mathbf{C_4}[\beta]$, $\mathbf{B_3}[\gamma]$, and $\mathbf{D_2}[\beta]$ (with the constants K_{41}, K_{43}, and L_{15}, L_{16} replacing, respectively, the constants K_9, K_{10} and L_5, L_6) hold for the process $Y_{\varepsilon,n}$.

Therefore, Lemma 4.3.5 can be applied to the process $Y_{\varepsilon,n}$, for every $0 \leq \varepsilon \leq \varepsilon_{10}$.

This yields inequality (6.24). The constant M_{14} penetrating the inequality given in Lemma 4.3.5 is a function of the constants K_9, K_{10}, and L_5, L_6. The explicit expressions for these functions are given in Remark 4.3.5. The same formulas give the expressions for constant M_{38} as a function of the corresponding constants K_{41}, K_{43} and L_{15}, L_{16}.

Remark 6.1.13. The constant M_{38} is given by the following formula, which follows from the corresponding formula given in Remark 4.1.5:

$$M_{38} = L_{15} + L_{15}L_{16}I(\gamma = 0) + L_{15}L_{16}K_{41}^{N\frac{\gamma}{\beta}} K_{43}^{\frac{\gamma}{\beta}} I(\gamma > 0)\,. \tag{6.26}$$

Remark 6.1.14. The parameter $\varepsilon_{10} = \min(\varepsilon_6, \varepsilon_7, \varepsilon_9)$, where the parameters $\varepsilon_6, \varepsilon_7$, and ε_9 are determined, respectively, by relations (6.4), (6.12), and (6.25).

Note also that one can replace the first-type condition of exponential moment boundedness $\mathbf{D}_8[\beta]$ by the following second-type condition of exponential moment boundedness based on moment-generating functions for the initial values of log-price processes $Y_{\varepsilon,0}$, assumed to hold for some $\beta \geq 0$:
$\mathbf{F}_5[\beta]$: $\overline{\lim}_{\varepsilon\to 0} \mathsf{E}\exp\{\pm A_N(\beta)Y_{\varepsilon,0}\} < K_{44}$, for some $1 < K_{44} < \infty$.

Lemma 6.1.8. *The condition $\mathbf{F}_5[\beta]$ implies that the condition $\mathbf{D}_8[\beta]$ holds.*

Proof. It follows from the following inequality that holds for any $\beta \geq 0$,

$$\mathsf{E}e^{A_N(\beta)|Y_{\varepsilon,0}|} \leq \mathsf{E}e^{+A_N(\beta)Y_{\varepsilon,0}} + \mathsf{E}e^{-A_N(\beta)Y_{\varepsilon,0}}\,. \tag{6.27}$$

□

Remark 6.1.15. As follows from inequality (6.27), if the condition $\mathbf{F}_5[\beta]$ holds with the constant K_{41}, then the condition $\mathbf{D}_8[\beta]$ holds with constant K_{43} given by the following formulas:

$$K_{43} = 2K_{44}\,. \tag{6.28}$$

We can also get alternative asymptotically uniform upper bounds for optimal expected rewards in the case, where the condition $\mathbf{C}_9'[\beta]$ is used instead of the condition $\mathbf{C}_9[\beta]$.

The condition $\mathbf{D}_2'[\bar{\beta}]$ should be replaced by the following condition, assumed to hold for some $\beta \geq 0$:
$\mathbf{D}_8'[\beta]$: (a) $\overline{\lim}_{\varepsilon\to 0} \mathsf{E}\exp\{A_N(\beta)|Y_{\varepsilon,0}|\} < K_{43}'$, for some $1 < K_{43}' < \infty$;
(b) $\mathsf{P}\{Z_{\varepsilon,0} \in \mathbb{Z}_{\varepsilon,0}\} = 1$, for every $\varepsilon \in [0, \varepsilon_0]$, where $\mathbb{Z}_{\varepsilon,0}$ is the set penetrating condition $\mathbf{C}_9'[\beta]$.

The following lemma gives alternative asymptotically uniform upper bounds for the expectation of maximum of absolute values for the pay-off processes.

Lemma 6.1.9. *Let the conditions $\mathbf{B}_8[\gamma]$, $\mathbf{C}_9'[\beta]$, and $\mathbf{D}_8'[\beta]$ hold and $0 \leq \gamma \leq \beta$. Then, there exist $0 < \varepsilon_{10}' \leq \varepsilon_0$ and a constant $0 \leq M_{38}' < \infty$ such that for the following inequality takes place for $0 \leq \varepsilon \leq \varepsilon_{10}'$,*

$$\mathsf{E}\max_{0\leq n\leq N} |g_\varepsilon(n, e^{Y_{\varepsilon,n}}, X_{\varepsilon,n})| \leq M_{38}'\,. \tag{6.29}$$

Proof. The condition $\mathbf{D}_8'[\beta]$ implies that there exists $0 < \varepsilon_9' \leq \varepsilon_0$ such that for $0 \leq \varepsilon \leq \varepsilon_9'$,

$$\mathsf{E}e^{A_N(\beta)|Y_{\varepsilon,0}|} < K_{43}'\,. \tag{6.30}$$

Relation (6.30) means that, for every $\varepsilon \leq \varepsilon'_9$, the condition $\mathbf{D}'_2[\beta]$ (with the constant K'_{43} replacing constant K'_{10}) holds for the process $Y_{\varepsilon,n}$.

Let us now take $\varepsilon'_{10} = \min(\varepsilon'_6, \varepsilon_7, \varepsilon'_9)$, where ε'_6 and ε_7 were defined in relations (6.9) and (6.12). By the definition $\varepsilon'_{10} > 0$.

For every fixed $0 \leq \varepsilon \leq \varepsilon'_{10}$, the conditions $\mathbf{C}'_4[\beta]$, $\mathbf{B}_3[\gamma]$, and $\mathbf{D}'_2[\beta]$ (with the constants K'_{41}, K'_{43}, and L_{15}, L_{16} replacing, respectively, the constants K_9, K_{10} and L_5, L_6) hold for the process $Y_{\varepsilon,n}$.

Therefore, Lemma 4.3.6 can be applied to the process $Y_{\varepsilon,n}$, for every $0 \leq \varepsilon \leq \varepsilon'_{10}$. This yields inequality (6.29). The constant M'_{14} penetrating the inequalities given in Lemma 4.3.6 is a function of the corresponding constants K'_9, K'_{10} and L_5, L_6. The explicit expressions for these functions are given in Remark 4.3.6. The same formulas give the expressions for constant M'_{38} as a function of the corresponding constants K'_{41}, K'_{43} and L_{15}, L_{16}. □

Remark 6.1.16. The constant M'_{38} is given by the following formula, which follows from the corresponding formula given in Remark 4.3.6:

$$M'_{38} = L_{15} + L_{15}L_{16}I(\gamma = 0) + L_{15}L_{16}(K'_{41})^{N\frac{\gamma}{\beta}}(K'_{43})^{\frac{\gamma}{\beta}}I(\gamma > 0)\,. \tag{6.31}$$

Remark 6.1.17. The parameter $\varepsilon'_{10} = \min(\varepsilon'_6, \varepsilon_7, \varepsilon'_9)$, where parameters $\varepsilon'_6, \varepsilon_7$, and ε'_9 are determined, respectively, by relations (6.9), (6.12), and (6.30).

Note also that one can replace the first-type condition of exponential moment boundedness $\mathbf{D}'_{16}[\beta]$ by the following second-type condition of exponential moment boundedness based on moment-generating functions for the initial value of log-price processes $Y_{\varepsilon,0}$, assumed to hold for some $\beta \geq 0$:

$\mathbf{F}'_5[\beta]$: (a) $\overline{\lim}_{\varepsilon\to 0} \mathsf{E}\exp\{\pm A_N(\beta)Y_{\varepsilon,0}\} < K'_{44}$, for some $1 \leq K'_{44} < \infty$;

(b) $\mathsf{P}\{Z_{\varepsilon,0} \in \mathbb{Z}_{\varepsilon,0}\} = 1$, for every $\varepsilon \in [0, \varepsilon_0]$, where $\mathbb{Z}_{\varepsilon,0}$ is the set penetrating condition $\mathbf{C}'_6[\bar{\beta}]$.

Lemma 6.1.10. *The condition* $\mathbf{F}'_5[\beta]$ *implies that the condition* $\mathbf{D}'_8[\beta]$ *holds.*

Proof. It follows from inequality (6.27). □

Remark 6.1.18. If the condition $\mathbf{F}'_5[\beta]$ holds with the constant K'_{44}, then the condition $\mathbf{D}'_8[\beta]$ holds with the constant K'_{43} given by the following formulas:

$$K'_{43} = 2K'_{44}\,. \tag{6.32}$$

Let us assume that either conditions of Lemma 6.1.7 or Lemma 6.1.9 hold. In this case, the optimal expected log-reward for American option for the log-price process $Z_{\varepsilon,n}$ can be defined, respectively, for every $0 \leq \varepsilon \leq \varepsilon_{10}$ or $0 \leq \varepsilon \leq \varepsilon'_{10}$,

$$\Phi_\varepsilon = \Phi(\mathcal{M}^{(\varepsilon)}_{\max,N}) = \sup_{\tau_{\varepsilon,0}\in\mathcal{M}^{(\varepsilon)}_{\max,N}} \mathsf{E}g_\varepsilon(\tau_{\varepsilon,0}, e^{Y_{\varepsilon,\tau_{\varepsilon,0}}}, X_{\varepsilon,\tau_{\varepsilon,0}})\,. \tag{6.33}$$

Also, as follows from Theorem 2.3.4, the following formula takes place, respectively, for every $0 \le \varepsilon \le \varepsilon_{10}$ or $0 \le \varepsilon \le \varepsilon_{10}$,

$$\Phi_\varepsilon = \mathsf{E}\phi_{\varepsilon,0}(Y_{\varepsilon,0}, X_{\varepsilon,0}) \,. \tag{6.34}$$

The following two theorems give asymptotically uniform upper bounds for the optimal expected rewards Φ_ε.

Theorem 6.1.3. *Let the conditions* $\mathbf{B_8}[\gamma]$, $\mathbf{C_9}[\beta]$, *and* $\mathbf{D_8}[\beta]$ *hold and* $0 \le \gamma \le \beta$. *Then, the following inequality takes place, for* $0 \le \varepsilon \le \varepsilon_{10}$*:*

$$|\Phi_\varepsilon| \le M_{38} \,. \tag{6.35}$$

Proof. It follows from Lemma 6.1.7 that for $0 \le \varepsilon \le \varepsilon_{10}$,

$$\begin{aligned} |\Phi_\varepsilon| &\le \max_{\tau_{\varepsilon,0} \in \mathcal{M}^{(\varepsilon)}_{\max,N}} \mathsf{E}|g_\varepsilon(\tau_{\varepsilon,0}, e^{Y_{\tau_{\varepsilon,0}}}, X_{\tau_{\varepsilon,0}})| \\ &\le \mathsf{E} \max_{0 \le n \le N} |g_\varepsilon(n, e^{Y_{\varepsilon,n}}, X_{\varepsilon,n})| \le M_{38} \,. \end{aligned} \tag{6.36}$$

This inequality proves the theorem. □

Theorem 6.1.4. *Let the conditions* $\mathbf{B_8}[\gamma]$, $\mathbf{C'_9}[\beta]$ *and* $\mathbf{D'_8}[\beta]$ *hold and* $0 \le \gamma \le \beta$. *Then, the following inequality takes place for* $0 \le \varepsilon \le \varepsilon'_{10}$*:*

$$|\Phi_\varepsilon| \le M'_{38} \,. \tag{6.37}$$

Proof. It is analogous to the proof of Theorem 6.1.3. The only difference that Lemma 6.1.9 is used instead of Lemma 6.1.7. In this case, the inequality analogous to (6.36) can be written for $0 \le \varepsilon \le \varepsilon'_{10}$. □

6.2 Univariate modulated LPP with unbounded characteristics

In this section, we present conditions of convergence of option rewards for univariate modulated Markov log-price processes with unbounded characteristics.

6.2.1 Convergence of option rewards for univariate modulated Markov log-price processes

Let us consider the log-reward functions $\phi_{\varepsilon,n}(y, x)$ introduced in Subsection 6.1.1 and give conditions for their convergence.

We impose the following condition of locally uniform convergence for the pay-off functions $g_\varepsilon(n, e^y, x)$:

$\mathbf{I}_5$: There exist measurable sets $\mathbb{Z}'_n \subseteq \mathbb{Z} = \mathbb{R}_1 \times \mathbb{X}$, $n = 0, \dots, N$, such that the pay-off function $g_\varepsilon(n, e^{y_0}, x_0) \stackrel{U}{\longrightarrow} g_0(n, e^{y_0}, x_0)$ as $\varepsilon \to 0$, i.e. $g_\varepsilon(n, e^{y_\varepsilon}, x_\varepsilon) \to g_0(n, e^{y_0}, x_0)$ as $\varepsilon \to 0$ for any $z_\varepsilon = (y_\varepsilon, x_\varepsilon) \to z_0 = (y_0, x_0) \in \mathbb{Z}'_n$ as $\varepsilon \to 0$, and $n = 0, \dots, N$.

We also impose the following condition of locally uniform weak convergence on the transition probabilities of log-price processes $P_{\varepsilon,n}(z, A) = \mathsf{P}\{Z_{\varepsilon,n} \in A / Z_{\varepsilon,n-1} = z\}$:

$\mathbf{J}_8$: There exist measurable sets $\mathbb{Z}''_n \subseteq \mathbb{Z}$, $n = 0, \dots, N$ such that:

(a) $P_{\varepsilon,n}(z_0, \cdot) \stackrel{U}{\Longrightarrow} P_{0,n}(z_0, \cdot)$ as $\varepsilon \to 0$, i.e. $P_{\varepsilon,n}(z_\varepsilon, \cdot) \Rightarrow P_{0,n}(z_0, \cdot)$ as $\varepsilon \to 0$, for any $z_\varepsilon = (y_\varepsilon, x_\varepsilon) \to z_0 = (y_0, x_0) \in \mathbb{Z}''_{n-1}$ as $\varepsilon \to 0$, and $n = 1, \dots, N$;

(b) $P_{0,n}(z_0, \mathbb{Z}'_n \cap \mathbb{Z}''_n) = 1$, for every $z_0 \in \mathbb{Z}'_{n-1} \cap \mathbb{Z}''_{n-1}$ and $n = 1, \dots, N$, where $\mathbb{Z}'_n$, $n = 0, \dots, N$ are sets introduced in the condition $\mathbf{I}_5$.

A typical example is where the sets $\overline{\mathbb{Z}}'_n, \overline{\mathbb{Z}}''_n$, $n = 1, \dots, N$ are empty sets. Then the condition $\mathbf{J}_8$ (b) obviously holds.

Another typical example is where the sets $\overline{\mathbb{Z}}'_n, \overline{\mathbb{Z}}''_n$, $n = 1, \dots, N$ are at most finite or countable sets. Then the assumption that measures $P_{0,n}(z_0, A)$, $z_0 \in \mathbb{Z}'_{n-1} \cap \mathbb{Z}''_{n-1}$, $n = 1, \dots, N$ have no atoms at points of the sets $\overline{\mathbb{Z}}'_n, \overline{\mathbb{Z}}''_n$, for every $n = 1, \dots, N$ implies that the condition $\mathbf{J}_8$ (b) holds.

One more example is, where the measures $P_{0,n}(z_0, A)$, $z_0 \in \mathbb{Z}'_{n-1} \cap \mathbb{Z}''_{n-1}$, $n = 1, \dots, N$ are absolutely continuous with respect to some σ-finite measure $P(A)$ on $\mathcal{B}_\mathbb{Z}$ and $P(\overline{\mathbb{Z}}'_n), P(\overline{\mathbb{Z}}''_n) = 0$, $n = 1, \dots, N$. This assumption implies that the condition $\mathbf{J}_8$ (b) holds.

Theorem 6.2.1. *Let the conditions $\mathbf{B}_8[\gamma]$ and $\mathbf{C}_9[\beta]$ hold with the parameters γ, β such that either $\beta > \gamma > 0$ or $\beta = \gamma = 0$, and also the conditions $\mathbf{I}_5$ and $\mathbf{J}_8$ hold. Then, for every $n = 0, 1, \dots, N$, the following relation takes place for any $z_\varepsilon = (y_\varepsilon, x_\varepsilon) \to z_0 = (y_0, x_0) \in \mathbb{Z}'_n \cap \mathbb{Z}''_n$:*

$$\phi_{\varepsilon,n}(y_\varepsilon, x_\varepsilon) \to \phi_{0,n}(y_0, x_0) \quad as \quad \varepsilon \to 0 . \tag{6.38}$$

Proof. It is analogous to the proof of Theorem 5.2.1.

Due to the above conditions imposed on the parameters γ and β, one can always find $\alpha > 1$ such that $\gamma\alpha \le \beta$.

According to Theorem 6.1.1, conditions $\mathbf{B}_8[\gamma]$ and $\mathbf{C}_9[\beta]$ imply that there exist $\varepsilon_8 > 0$ and the constant M_{36}, M_{37} such that the following upper bounds take place for the log-reward functions $\phi_{\varepsilon,n}(y, x)$, for $z = (y, x) \in \mathbb{Z}$, $n = 0, 1, \dots, N$ and $\varepsilon \le \varepsilon_8$,

$$\begin{aligned} |\phi_{\varepsilon,n}(y, x)| \le M_{36} &+ M_{37} I(\gamma = 0) \\ &+ M_{37} \exp\left\{\frac{A_{N-n}(\beta)\gamma}{\beta} |y|\right\} I(\gamma > 0) . \end{aligned} \tag{6.39}$$

Using inequalities (6.39) and the condition $\mathbf{C}_9[\beta]$, we get for any $z = (y, x) \in \mathbb{Z}$, $n = 0, 1, \dots, N-1$, $\varepsilon \le \varepsilon_8$,

$$\begin{aligned}
&\mathsf{E}_{z,n}|\phi_{\varepsilon,n+1}(Y_{\varepsilon,n+1}, X_{\varepsilon,n+1})|^\alpha \\
&\quad \le \mathsf{E}_{\vec{z},n}\left(M_{36} + M_{37}I(\gamma = 0) + M_{37}\exp\left\{\frac{A_{N-n-1}(\beta)\gamma}{\beta}|Y_{\varepsilon,n+1}|\right\}I(\gamma > 0)\right)^\alpha \\
&\quad \le \mathsf{E}_{z,n}2^{\alpha-1}\left(M_{36}^\alpha + M_{37}^\alpha I(\gamma = 0) + M_{37}^\alpha\exp\left\{\frac{A_{N-n-1}(\beta)\gamma\alpha}{\beta}|Y_{\varepsilon,n+1}|\right\}I(\gamma > 0)\right) \\
&\quad \le 2^{\alpha-1}\left(M_{36}^\alpha + M_{37}^\alpha I(\gamma = 0) + M_{37}^\alpha\exp\left\{\frac{A_{N-n-1}(\beta)\gamma\alpha}{\beta}|y|\right\}\right. \\
&\quad\quad \left.\times\, \mathsf{E}_{z,n}\exp\left\{\frac{A_{N-n-1}(\beta)\gamma\alpha}{\beta}|Y_{\varepsilon,n+1} - y|\right\}I(\gamma > 0)\right) \\
&\quad \le 2^{\alpha-1}\left(M_{36}^\alpha + M_{37}^\alpha I(\gamma = 0) + M_{37}^\alpha\exp\left\{\frac{A_{N-n-1}(\beta)\gamma\alpha}{\beta}|y|\right\}\right. \\
&\quad\quad \left.\times\, \mathsf{E}_{z,n}\exp\left\{A_{N-n-1}(\beta)|Y_{\varepsilon,n+1} - Y_{\varepsilon,n}|\right\}I(\gamma > 0)\right) \\
&\quad \le 2^{\alpha-1}\left(M_{36}^\alpha + M_{37}^\alpha I(\gamma = 0) + M_{37}^\alpha\exp\left\{\frac{A_{N-n-1}(\beta)\gamma\alpha}{\beta}|y|\right\}\right. \\
&\quad\quad \left.\times\, e^{A(A_{N-n-1}(\beta))|y|}K_{41}I(\gamma > 0)\right) = Q_A(y) < \infty\,.
\end{aligned} \tag{6.40}$$

The following part of the proof repeats the corresponding part in the proof of Theorem 5.2.1. The only difference is that relation (5.57) is replaced by the analogous relation

$$\varlimsup_{\varepsilon\to 0}\mathsf{E}|\phi_{\varepsilon,N}(Y_{\varepsilon,N}(z_\varepsilon), X_{\varepsilon,N}(z_\varepsilon))|^\alpha \le Q_A(y_0) < \infty\,, \tag{6.41}$$

where the function $Q(y)$ defined in relation (5.42) is replaced by the function $Q_A(y)$ defined in relation (6.40). □

Remark 6.2.1. According to Lemma 6.1.5, the condition $\mathbf{C}_9[\beta]$ can be replaced in Theorem 6.2.1 by the condition $\mathbf{E}_{13}[\beta]$.

Remark 6.2.2. As follows from the remarks made in Subsection 5.2.1, the following point-wise convergence relation takes place for the reward functions, for any $z = (y, x) \in \mathbb{Z}'_n \cap \mathbb{Z}''_n$, $n = 0, 1, \dots, N$,

$$\phi_{\varepsilon,n}(y, x) \to \phi_{0,n}(y, x) \quad \text{as} \quad \varepsilon \to 0\,. \tag{6.42}$$

Also, the reward function $\phi_{0,n}(y, x)$ is continuous in points $z = (y, x) \in \mathbb{Z}'_n \cap \mathbb{Z}''_n$, for every $n = 0, 1, \dots, N$.

The following theorem gives alternative conditions for the convergence of log-reward functions for American-type options, in the above model.

Theorem 6.2.2. *Let the conditions $\mathbf{B}_8[\gamma]$ and $\mathbf{C}'_9[\beta]$ hold with the parameters γ, β such that either $\beta > \gamma > 0$ or $\beta = \gamma = 0$, and also the conditions $\mathbf{H}_2$, $\mathbf{I}_5$, and $\mathbf{J}_8$ hold. Then, for every $n = 0, 1, \dots, N$, the following relation takes place for any $z_\varepsilon = (y_\varepsilon, x_\varepsilon) \to z_0 = (y_0, x_0) \in \mathbb{Z}'_n \cap \mathbb{Z}''_n$:*

$$\phi_{\varepsilon,n}(y_\varepsilon, x_\varepsilon) \to \phi_{0,n}(y_0, x_0) \quad as \quad \varepsilon \to 0\,. \tag{6.43}$$

Proof. It is analogous to the proof of Theorem 5.2.2. □

Due to the above conditions imposed on the parameters γ and β, one can always find $\alpha > 1$ such that $\gamma\alpha \leq \beta$.

By Theorem 6.1.2, conditions $\mathbf{B}_8[\gamma]$ and $\mathbf{C}'_9[\beta]$ imply that there exist $\varepsilon'_8 > 0$ and the constants M'_{36}, M'_{37} such that the following upper bounds take place for the reward functions $\phi_{\varepsilon,n}(y, x)$ for $z = (y, x) \in \mathbb{Z}_{\varepsilon,n}$, $n = 0, 1, \dots, N$, $\varepsilon \leq \varepsilon'_8$,

$$\begin{aligned}|\phi_{\varepsilon,n}(y, x)| &\leq M'_{36} + M'_{37} I(\gamma = 0) \\ &\quad + M'_{37} \exp\{\frac{A_{N-n}(\beta)\gamma}{\beta}|y|\} I(\gamma > 0)\,.\end{aligned} \tag{6.44}$$

Recall that the condition $\mathbf{C}'_9[\beta]$ (b) implies that $\mathsf{P}_{z,n}\{Z_{\varepsilon,n+1} \in \mathbb{Z}_{\varepsilon,n+1}\} = 1$, for $z \in \mathbb{Z}_{\varepsilon,n}$, $n = 0, \dots, N-1$. Using these relations, inequalities (6.44) and the condition $\mathbf{C}'_9[\beta]$, we get for any $z = (y, x) \in \mathbb{Z}_{\varepsilon,n}$, $0 \leq n \leq N-1$ and $\varepsilon \leq \varepsilon'_8$,

$$\begin{aligned}
&\mathsf{E}_{z,n}|\phi_{\varepsilon,n+1}(Y_{\varepsilon,n+1}, X_{\varepsilon,n+1})|^\alpha \\
&\quad \leq \mathsf{E}_{\vec{z},n}\left(M'_{36} + M'_{37} I(\gamma = 0) + M'_{37} \exp\left\{\frac{A_{N-n-1}(\beta)\gamma}{\beta}|Y_{\varepsilon,n+1}|\right\} I(\gamma > 0)\right)^\alpha \\
&\quad \leq \mathsf{E}_{z,n} 2^{\alpha-1}\left((M'_{36})^\alpha + (M'_{37})^\alpha I(\gamma = 0) + (M'_{37})^\alpha \exp\left\{\frac{A_{N-n-1}(\beta)\gamma\alpha}{\beta}|Y_{\varepsilon,n+1}|\right\} I(\gamma > 0)\right) \\
&\quad \leq 2^{\alpha-1}\Bigg((M'_{36})^\alpha + (M'_{37})^\alpha I(\gamma = 0) \\
&\qquad + (M'_{37})^\alpha \exp\left\{\frac{A_{N-n-1}(\beta)\gamma\alpha}{\beta}|y|\right\} \times \mathsf{E}_{z,n} \exp\left\{\frac{A_{N-n-1}(\beta)\gamma\alpha}{\beta}|Y_{\varepsilon,n+1} - y|\right\} I(\gamma > 0)\Bigg) \\
&\quad \leq 2^{\alpha-1}\Bigg((M'_{36})^\alpha + (M'_{37})^\alpha I(\gamma = 0) \\
&\qquad + (M'_{37})^\alpha \exp\left\{\frac{A_{N-n-1}(\beta)\gamma\alpha}{\beta}|y|\right\} \times \mathsf{E}_{z,n} \exp\left\{A_{N-n-1}(\beta)|Y_{\varepsilon,n+1} - Y_{\varepsilon,n}|\right\} I(\gamma > 0)\Bigg) \\
&\quad \leq 2^{\alpha-1}\Bigg((M'_{36})^\alpha + (M'_{37})^\alpha I(\gamma = 0) \\
&\qquad + (M'_{37})^\alpha \exp\left\{\frac{A_{N-n-1}(\beta)\gamma\alpha}{\beta}|y|\right\} \times e^{A(A_{N-n-1}(\beta))|y|} K'_{41} I(\gamma > 0)\Bigg) \\
&\quad = Q'_A(y) < \infty\,.
\end{aligned} \tag{6.45}$$

The following part of the proof repeats the corresponding part in the proof of Theorem 5.2.2. The only difference is that relation (5.64) should be slightly modified.

As in the proof of Theorem 5.2.2, we choose an arbitrary function $z(\cdot) = \langle z_\varepsilon, e \in [0, \varepsilon_0]\rangle$ such that $z_\varepsilon \to z_0 \in \mathbb{Z}'_{N-1} \cap \mathbb{Z}''_{N-1}$ as $\varepsilon \to 0$, and should prove that the following relation, which replaces relation (5.64), holds:

$$\varlimsup_{\varepsilon\to 0} \mathsf{E}|\phi_{\varepsilon,N}(Y_{\varepsilon,N}(z_\varepsilon), X_{\varepsilon,N}(z_\varepsilon))|^\alpha \le Q'_A(y_0) < \infty . \tag{6.46}$$

In this case, we should use the condition $\mathbf{H}_2$, due to which the point $z_0 = (y_0, x_0) \in \mathbb{Z}_{\varepsilon,N-1}$, for every $\varepsilon \in [0, \varepsilon_2(z_0)]$. The parameter $\varepsilon_2(z_0) \in (0, \varepsilon_0]$ has been defined in relation (6.7) given in Remark 6.1.3.

Since $z_\varepsilon \to z_0$ as $\varepsilon \to 0$, this implies that there exists $\varepsilon'_2(z(\cdot)) \in (0, \varepsilon_2(z_0)]$ such that $z_\varepsilon = (y_\varepsilon, x_\varepsilon) \in \mathbb{Z}_{\varepsilon,N-1}$ for every $\varepsilon \in [0, \varepsilon'_2(z(\cdot))]$.

Therefore, upper bounds for $\mathsf{E}_{z,N-1}|\phi_{\varepsilon,N}(Y_{\varepsilon,N}, X_{\varepsilon,N})|^\alpha$ given in relation (6.45) hold for every $\varepsilon \in [0, \min(\varepsilon'_8, \varepsilon'_2(z(\cdot)))]$.

Due to the above remarks and continuity of function $Q'_A(y)$, relation (6.45) implies relation (6.46), which replaces relation (5.64).

The remaining part of the proof repeats the corresponding part in the proof of Theorem 5.2.2. □

6.2.2 Convergence of optimal expected rewards for univariate modulated Markov log-price processes

Let us now consider optimal expected rewards $\Phi_\varepsilon = \Phi(\mathcal{M}^{(\varepsilon)}_{\max,N})$ introduced in Subsection 6.1.1 and give conditions for their convergence.

We should impose the following condition of weak convergence on the initial distributions $P_{\varepsilon,0}(A) = \mathsf{P}\{Z_{\varepsilon,0} \in A\}$:

$\mathbf{K}_7$: (a) $P_{\varepsilon,0}(\cdot) \Rightarrow P_{0,0}(\cdot)$ as $\varepsilon \to 0$;

(b) $P_{0,0}(\mathbb{Z}'_0 \cap \mathbb{Z}''_0) = 1$, where $\mathbb{Z}'_0$ and $\mathbb{Z}''_0$ are the sets introduced in conditions $\mathbf{I}_5$ and $\mathbf{J}_8$.

A typical example is where the sets $\overline{\mathbb{Z}}'_0, \overline{\mathbb{Z}}''_0$, $n = 1, \dots, N$ are empty sets. Then the condition $\mathbf{K}_7$ (b) obviously holds.

Another typical example is where the sets $\overline{\mathbb{Z}}'_0, \overline{\mathbb{Z}}''_0$ are at most finite or countable sets. Then the assumption that the measure $P_{0,0}(\cdot)$ has no atoms at points of the sets $\overline{\mathbb{Z}}'_0, \overline{\mathbb{Z}}''_0$ implies that the condition $\mathbf{K}_7$ (b) holds.

One more example is, where measure $P_{0,0}(\cdot)$ is absolutely continuous with respect to some σ-finite measure $P(A)$ on $\mathcal{B}_\mathbb{Z}$ and $P(\overline{\mathbb{Z}}'_0), P(\overline{\mathbb{Z}}''_0) = 0$. This assumption implies that the condition $\mathbf{K}_7$ (b) holds.

The following theorem gives conditions for convergence of optimal expected rewards Φ_ε.

Theorem 6.2.3. *Let the conditions $\mathbf{B}_8[\gamma]$, $\mathbf{C}_9[\beta]$, and $\mathbf{D}_8[\beta]$ hold with the vector parameters γ, β such that either $\beta > \gamma > 0$ or $\beta = \gamma = 0$, and also conditions $\mathbf{I}_5$, $\mathbf{J}_8$, and $\mathbf{K}_7$*

hold. Then the following relation takes place:

$$\Phi_\varepsilon \to \Phi_0 \quad as \quad \varepsilon \to 0\,. \tag{6.47}$$

Proof. It is analogous to the proof of Theorem 5.2.3.

Due to the above conditions imposed on the parameters γ and β, one can always find $\alpha > 1$ such that $\gamma\alpha \leq \beta$.

According to Theorem 6.1.1, conditions $\mathbf{B_8}[\gamma]$ and $\mathbf{C_9}[\beta]$ imply that there exist $\varepsilon_8 > 0$ and the constants M_{36}, M_{37} such that the following upper bounds take place for the log-reward functions $\phi_{\varepsilon,0}(y,x)$ for $z = (y,x) \in \mathbb{Z}$ and $\varepsilon \leq \varepsilon_8$,

$$\begin{aligned} |\phi_{\varepsilon,0}(y,x)| &\leq M_{36} + M_{37}I(\gamma = 0) \\ &\quad + M_{37}\exp\left\{\frac{A_N(\beta)\gamma}{\beta}|y|\right\} I(\gamma > 0)\,. \end{aligned} \tag{6.48}$$

Using inequalities (6.48) and the condition $\mathbf{D_8}[\beta]$, we get for any $\varepsilon \leq \varepsilon_{10}$,

$$\begin{aligned} &\mathsf{E}|\phi_{\varepsilon,0}(Y_{\varepsilon,0}, X_{\varepsilon,0})|^\alpha \\ &\leq \mathsf{E}\left(M_{36} + M_{37}I(\gamma = 0) + M_{37}\exp\left\{\frac{A_N(\beta)\gamma}{\beta}|Y_{\varepsilon,0}|\right\} I(\gamma > 0)\right)^\alpha \\ &\leq \mathsf{E}2^{\alpha-1}\left(M_{36}^\alpha + M_{37}^\alpha I(\gamma = 0) + M_{37}^\alpha \exp\left\{\frac{A_N(\beta)\gamma\alpha}{\beta}|Y_{\varepsilon,0}|\right\} I(\gamma > 0)\right) \\ &\leq 2^{\alpha-1}\left(M_{36}^\alpha + M_{37}^\alpha I(\gamma = 0) \quad + M_{37}^\alpha \mathsf{E}\exp\{A_N(\beta)|Y_{\varepsilon,0}|\}I(\gamma > 0)\right) \\ &\leq 2^{\alpha-1}\left(M_{36}^\alpha + M_{37}^\alpha I(\gamma = 0) + M_{37}^\alpha K_{43}I(\gamma > 0)\right) = Q_A < \infty\,. \end{aligned} \tag{6.49}$$

The following proof repeats the proof of Theorem 5.2.3. The only difference is that relation (6.49) should be used instead of relation (5.68) used in Theorem 5.2.3. □

Remark 6.2.3. According to Lemmas 6.1.5 and 6.1.8, the conditions $\mathbf{C_9}[\beta]$ and $\mathbf{D_8}[\beta]$ can be replaced in Theorem 6.2.3, respectively, by conditions $\mathbf{E_{13}}[\beta]$ and $\mathbf{F_6}[\beta]$.

The following theorem gives alternative conditions for convergence of optimal expected rewards Φ_ε.

Theorem 6.2.4. *Let the conditions $\mathbf{B_8}[\gamma]$, $\mathbf{C}'_9[\beta]$, and $\mathbf{D}'_8[\beta]$ hold with the vector parameters γ, β such that either $\beta > \gamma > 0$ or $\beta = \gamma = 0$, and also the conditions $\mathbf{H_2}$, $\mathbf{I_5}$, $\mathbf{J_8}$, and $\mathbf{K_7}$ hold. Then the following relation takes place:*

$$\Phi_\varepsilon \to \Phi_0 \quad as \quad \varepsilon \to 0\,. \tag{6.50}$$

Proof. It is analogous to the proof of Theorem 5.2.4.

Due to the above conditions imposed on the parameters γ, β, one can always find $\alpha > 1$ such that $\gamma\alpha \le \beta$.

According to Theorem 6.1.4, conditions $\mathbf{B_8}[\gamma]$ and $\mathbf{C}'_9[\beta]$ imply that there exist $\varepsilon'_8 > 0$ and the constants M'_{36}, M'_{37} such that the following upper bounds take place for the reward functions $\phi_{\varepsilon,0}(y,x)$ for $z = (y,x) \in \mathbb{Z}_{\varepsilon,0}$, $\varepsilon \le \varepsilon'_8$,

$$|\phi_{\varepsilon,0}(y,x)| \le M'_{36} + M'_{37}I(\gamma = 0) + M'_{37}\exp\{\frac{A_N(\beta)\gamma}{\beta}|y|\}I(\gamma > 0)\,. \tag{6.51}$$

Recall that the condition $\mathbf{D}'_8[\beta]$ (b) implies that $\mathsf{P}_{z,n}\{Z_{\varepsilon,0} \in \mathbb{Z}_{\varepsilon,0}\} = 1$ for $z \in \mathbb{Z}_{\varepsilon,0}$. Using this relation, inequalities (6.51) and the condition $\mathbf{D}'_{16}[\beta]$, we get for any $\varepsilon \le \varepsilon'_{10}$,

$$\begin{aligned}
&\mathsf{E}|\phi_{\varepsilon,0}(Y_{\varepsilon,0}, X_{\varepsilon,0})|^\alpha \\
&\le \mathsf{E}\left(M'_{36} + M'_{37}I(\gamma = 0) + M'_{37}\exp\left\{\frac{A_N(\beta)\gamma}{\beta}|Y_{\varepsilon,0}|\right\}I(\gamma > 0)\right)^\alpha \\
&\le \mathsf{E}2^{\alpha-1}\left((M'_{36})^\alpha + (M'_{37})^\alpha I(\gamma = 0) + (M'_{37})^\alpha\exp\left\{\frac{A_N(\beta)\gamma\alpha}{\beta}|Y_{\varepsilon,0}|\right\}I(\gamma > 0)\right) \\
&\le 2^{\alpha-1}\left((M'_{36})^\alpha + (M'_{37})^\alpha I(\gamma = 0) + (M'_{37})^\alpha \mathsf{E}\exp\{A_N(\beta)|Y_{\varepsilon,0}|\}I(\gamma > 0)\right) \\
&\le 2^{\alpha-1}\left((M'_{36})^\alpha + (M'_{37})^\alpha I(\gamma = 0) + (M'_{37})^\alpha K'_{43}I(\gamma > 0)\right) = Q'_A < \infty\,.
\end{aligned} \tag{6.52}$$

The following part of the proof repeats the corresponding part in the proof of Theorem 5.2.4. The only difference is that relation (6.52) should be used instead of relation (5.82). Also, relation analogous to (5.75) does require reference to relation (6.43) given in Theorem 6.2.2 instead of relation (6.38) given in Theorem 6.2.1. □

6.3 Asymptotically uniform upper bounds for rewards – III

In this section, we present asymptotically uniform upper bound for option rewards for multivariate modulated Markov log-price processes with unbounded characteristics.

6.3.1 Moduli of exponential moment compactness for multivariate modulated Markov log-price processes with unbounded characteristics

We consider a multivariate modulated Markov log-price process $\vec{Z}_{\varepsilon,n} = (\vec{Y}_{\varepsilon,n}, X_{\varepsilon,n}) = ((Y_{\varepsilon,n,1}, \dots, Y_{\varepsilon,n,k}), X_{\varepsilon,n})$, $n = 0, 1, \dots$, with a phase space $\mathbb{Z} = \mathbb{R}_k \times \mathbb{X}$, an initial distribution $P_\varepsilon(A)$, and the transition probabilities $P_{\varepsilon,n}(z, A) = \mathsf{P}\{Z_{\varepsilon,n} \in A/Z_{\varepsilon,n-1} = z\}$, which depend on a perturbation parameter $\varepsilon \in [0, \varepsilon_0]$.

Let us recall the notion of a class $\mathcal{A}_k$ of measurable, vector functions $\vec{A}(\bar{\beta}) = (A_1(\bar{\beta}), \dots, A_k(\bar{\beta}))$ defined for $\bar{\beta} = (\beta_1, \dots, \beta_k), \beta_i \geq 0, i = 1, \dots, k$ such that functions $A_i(\bar{\beta}), i = 1, \dots, k$ are nondecreasing in every argument $\beta_i, i = 1, \dots, k$ and such that $A_i(\bar{0}) = 0, i = 1, \dots, k$. As pointed out in Section 4.3 any function $\vec{A}(\bar{\beta}) \in \mathcal{A}_k$ generates the sequence of functions $\vec{A}_n(\bar{\beta}) = (A_{n,1}(\bar{\beta}), \dots, A_{n,k}(\bar{\beta})), n = 0, 1, \dots$ from the class $\mathcal{A}_k$ that are defined by the following recurrence relation, for any $\bar{\beta} = (\beta_1, \dots, \beta_k), \beta_i \geq 0, i = 1, \dots, k$:

$$\vec{A}_n(\bar{\beta}) = \begin{cases} \bar{\beta} & \text{for } n = 0\,, \\ \vec{A}_{n-1}(\bar{\beta}) + \vec{A}(\vec{A}_{n-1}(\bar{\beta})) & \text{for } n = 1, 2, \dots\,. \end{cases} \tag{6.53}$$

Note that by the definition $A_{0,j}(\bar{\beta}) \leq A_{1,j}(\bar{\beta}) \leq \cdots$ for every $\bar{\beta} = (\beta_1, \dots, \beta_k)$, $\beta_i \geq 0, i = 1, \dots, k$ and $j = 1, \dots, k$.

Let us also recall the first-type $\vec{A}$-modulus of exponential moment compactness for the log-price process $\vec{Y}_{\varepsilon,n}$. It is defined, for a vector parameter $\bar{\beta} = (\beta_1, \dots, \beta_k)$ with nonnegative components

$$\Delta_{\bar{\beta},\vec{A}}(\vec{Y}_{\varepsilon,\cdot}, N) = \max_{0 \leq n \leq N-1} \sup_{\vec{z}=((y_1,\dots,y_k),\,x)\in\mathbb{Z}} \frac{\mathsf{E}_{\vec{z},n} e^{\sum_{j=1}^k A_{N-n-1,j}(\bar{\beta})|Y_{\varepsilon,n+1,j} - Y_{\varepsilon,n,j}|}}{e^{\sum_{j=1}^k A_j(\vec{A}_{N-n-1}(\bar{\beta}))|y_j|}}\,. \tag{6.54}$$

Recall that we use the notations $\mathsf{P}_{\vec{z},n}$ and $\mathsf{E}_{\vec{z},n}$ for the conditional probabilities and expectations under the condition $\vec{Z}_{\varepsilon,n} = \vec{z}$.

The condition $\mathbf{C}_5[\bar{\beta}]$ should be replaced by the following condition, assumed to hold for some vector parameter $\bar{\beta} = (\beta_1, \dots, \beta_k)$ with nonnegative components and vectors $\bar{\beta}_i = (\beta_{i,1}, \dots, \beta_{i,k}) = (0, \dots, 0, \beta_i, 0, \dots, 0), i = 1, \dots, k$ (the vector $\bar{\beta}_i$ has the ith component equals β_i and other components equal to 0):

$\mathbf{C}_{10}[\bar{\beta}]$: $\overline{\lim}_{\varepsilon\to 0} \Delta_{\bar{\beta}_i,\vec{A}}(\vec{Y}_{\varepsilon,\cdot}, N) < K_{45,i}, i = 1, \dots, k$, for some $1 < K_{45,i} < \infty, i = 1, \dots, k$.

The following lemma gives asymptotically uniform upper bounds for moments of maxima for components of perturbed multivariate modulated Markov log-price processes.

Lemma 6.3.1. *Let the condition $\mathbf{C}_{10}[\bar{\beta}]$ holds. Then there exists $0 < \varepsilon_{11} \leq \varepsilon_0$ and the constants $0 \leq M_{39,i} < \infty, i = 1, \dots, k$ such that the following inequalities take place, for $\vec{z} = (\vec{y}, x) \in \mathbb{Z}, 0 \leq n \leq N, 0 \leq \varepsilon \leq \varepsilon_{11}$, and $i = 1, \dots, k$,*

$$\mathsf{E}_{\vec{z},n} \exp\left\{\beta_i \max_{n\leq r\leq N} |Y_{\varepsilon,r,i}|\right\} \leq M_{39,i} \exp\left\{\sum_{j=1}^k A_{N-n,j}(\bar{\beta}_i)|y_j|\right\}\,. \tag{6.55}$$

Proof. The condition $\mathbf{C}_{10}[\bar{\beta}]$ implies that there exists $0 < \varepsilon_{11} \leq \varepsilon_0$ such that, for $0 \leq \varepsilon \leq \varepsilon_{11}$ and $i = 1, \dots, k$,

$$\Delta_{\bar{\beta}_i,\vec{A}}(\vec{Y}_{\varepsilon,\cdot}, N) < K_{45,i}\,. \tag{6.56}$$

Relation (6.56) means that, for every $\varepsilon \leq \varepsilon_{11}$, the condition $\mathbf{C}_5[\bar{\beta}]$ (with the constants $K_{45,i}$, $i = 1, \dots, k$ replacing the constants $K_{13,i}$, $i = 1, \dots, k$) holds for the process $\vec{Y}_{\varepsilon,n}$.

Therefore, Lemma 4.3.11 can be applied to the process $\vec{Y}_{\varepsilon,n}$, for every $\varepsilon \leq \varepsilon_{11}$.

This yields inequalities (6.55). For every $i = 1, \dots, k$, the constant $M_{9,i}$ penetrating the inequality given in Lemma 4.3.11 is a function of the corresponding constant $K_{13,i}$. The explicit expression for this function is given in Remark 4.3.11. The same formula gives the expression for the constant $M_{39,i}$ as a function of the corresponding constant $K_{45,i}$. □

Remark 6.3.1. The constants $M_{39,i}$, $i = 1, \dots, k$ are given by the following formulas, which follows from the corresponding formulas given in Remark 4.3.11,

$$M_{39,i} = K_{45,i}^N \,. \tag{6.57}$$

Remark 6.3.2. The parameter ε_{11} is determined by relation (6.56).

Let $\mathcal{Z}_\varepsilon = \langle \mathbb{Z}_{\varepsilon,0}, \dots, \mathbb{Z}_{\varepsilon,N} \rangle$ be a sequence of measurable sets. Let us introduce a modified first-type $\vec{A}$-modulus of exponential moment compactness, defined for every $\varepsilon \in [0, \varepsilon_0]$ and vectors $\bar{\beta} = (\beta_1, \dots, \beta_k)$ with nonnegative components by the following formula:

$$\Delta_{\bar{\beta},\vec{A},\mathcal{Z}_\varepsilon}(\vec{Y}_{\varepsilon,\cdot}, N) = \max_{0 \leq n \leq N-1} \sup_{\vec{z}=((y_1,\dots,y_k),x) \in \mathbb{Z}_{\varepsilon,n}} \frac{\mathsf{E}_{\vec{z},n} e^{\sum_{j=1}^k A_{N-n-1,j}(\bar{\beta}) |Y_{\varepsilon,n+1,j} - Y_{\varepsilon,n,j}|}}{e^{\sum_{j=1}^k A_j(\vec{A}_{N-n-1}(\bar{\beta}))|y_j|}} \,. \tag{6.58}$$

The following first-type condition of exponential moment compactness for the log-price process $Y_{\varepsilon,n}$ generalizes the condition $\mathbf{C}_{10}[\bar{\beta}]$. It is assumed to hold for some vector parameter $\bar{\beta} = (\beta_1, \dots, \beta_k)$ with nonnegative components:

$\mathbf{C}_{10}'[\bar{\beta}]$: There exists a sequence of measurable sets $\mathcal{Z}_\varepsilon = \langle \mathbb{Z}_{\varepsilon,0}, \dots, \mathbb{Z}_{\varepsilon,N} \rangle$, for every $\varepsilon \in [0, \varepsilon_0]$, such that
- (a) $\overline{\lim}_{\varepsilon \to 0} \Delta_{\bar{\beta}_i,\vec{A},\mathcal{Z}_\varepsilon}(\vec{Y}_{\varepsilon,\cdot}, N) < K_{45,i}'$, $i = 1, \dots, k$, for some $1 < K_{45,i}' < \infty$, $i = 1, \dots, k$;
- (b) $\mathcal{Z}_\varepsilon$ is a complete sequence of phase sets for the process $\vec{Z}_{\varepsilon,n}$, i.e. $\mathsf{P}_{\varepsilon,n}(\vec{z}, \mathbb{Z}_{\varepsilon,n}) = 1$, $\vec{z} \in \mathbb{Z}_{\varepsilon,n-1}$, $n = 1, \dots, N$, for every $\varepsilon \in [0, \varepsilon_0]$.

Lemma 6.3.2. *Let the condition $\mathbf{C}_{10}'[\bar{\beta}]$ holds. Then there exists $0 < \varepsilon_{11}' \leq \varepsilon_0$ and the constants $0 \leq M_{39,i}' < \infty$, $i = 1, \dots, k$ such that the following inequalities take place for $\vec{z} = (\vec{y}, x) \in \mathbb{Z}_{\varepsilon,n}$, $0 \leq n \leq N$, $0 \leq \varepsilon \leq \varepsilon_{11}'$, and $i = 1, \dots, k$,*

$$\mathsf{E}_{\vec{z},n} \exp\left\{\beta_i \max_{n \leq r \leq N} |Y_{\varepsilon,r,i}|\right\} \leq M_{39,i}' \exp\left\{\sum_{j=1}^k A_{N-n,j}(\bar{\beta}_i)|y_j|\right\} \,. \tag{6.59}$$

Proof. By the condition $\mathbf{C}_{10}'[\bar{\beta}]$, one can choose $0 < \varepsilon_{11}' \leq \varepsilon_0$ such that, for $0 \leq \varepsilon \leq \varepsilon_{11}'$ and $i = 1, \dots, k$,

$$\Delta_{\bar{\beta}_i,\vec{A}}(\vec{Y}_{\varepsilon,\cdot}, N) < K_{45,i} \,. \tag{6.60}$$

The following part of the proof repeats the corresponding part in the proof of Lemma 6.3.1. The only difference is that Lemma 4.3.12 should be applied to the process $\vec{Y}_{\varepsilon,n}$, for every $\varepsilon \le \varepsilon'_{11}$, instead of Lemma 4.3.11 that was applied to the process $\vec{Y}_{\varepsilon,n}$, for every $\varepsilon \le \varepsilon_{11}$, in the proof of Lemma 6.3.1.

This yields inequalities (6.59). The constants $M'_{15,i}$, $i = 1, \dots, k$ penetrating the inequalities given in Lemma 4.3.12 are functions of the corresponding constant $K'_{13,i}$, $i = 1, \dots, k$. The explicit expression for these functions is given in Remark 4.3.8. The same formulas give expressions for the constant $M'_{39,i}$, $i = 1, \dots, k$ as functions of the corresponding constant $K'_{45,i}$, $i = 1, \dots, k$. □

Remark 6.3.3. The constants $M'_{39,i}$, $i = 1, \dots, k$ are given by the following formulas, which follows from the corresponding formula given in Remark 4.3.8:

$$M'_{39,i} = (K'_{45,i})^N . \tag{6.61}$$

Remark 6.3.4. The parameter ε'_{11} is determined by relation (6.60).

Let us now give asymptotically uniform upper bounds for log-reward functions for the multivariate modulated Markov log-price processes $\vec{Z}_{\varepsilon,n}$.

In this case, the pay-off functions $g_\varepsilon(n, e^{\vec{y}}, x)$ are functions of the argument $\vec{z} = (\vec{y}, x) \in \mathbb{Z} = \mathbb{R}_k \times \mathbb{X}$ and depend on perturbation parameter ε.

We impose on pay-off functions the condition $\mathbf{B}_5[\bar{\gamma}]$, which should be assumed to hold for some vector parameter $\bar{\gamma} = (\gamma_1, \dots, \gamma_k)$ with nonnegative components.

The following lemmas give asymptotically uniform upper bounds for the conditional expectations for the maximal absolute values for perturbed pay-off processes $g(n, e^{\vec{Y}_{\varepsilon,n}}, X_{\varepsilon,n})$.

Lemma 6.3.3. *Let the conditions $\mathbf{B}_5[\bar{\gamma}]$ and $\mathbf{C}_{10}[\bar{\beta}]$ hold and $0 \le \gamma_i \le \beta_i$, $i = 1, \dots, k$. Then, there exist $0 < \varepsilon_{12} \le \varepsilon_0$ and the constants $0 \le M_{40}, M_{41,i} < \infty$, $i = 1, \dots, k$ such that the following inequalities take place for $\vec{z} = (\vec{y}, x) \in \mathbb{Z}$, $0 \le n \le N$ and $0 \le \varepsilon \le \varepsilon_{12}$,*

$$\begin{aligned} &\mathsf{E}_{\vec{z},n} \max_{n \le r \le N} |g_\varepsilon(r, e^{\vec{Y}_{\varepsilon,r}}, X_{\varepsilon,r})| \\ &\quad \le M_{40} + \sum_{i:\gamma_i = 0} M_{41,i} + \sum_{i:\gamma_i > 0} M_{41,i} \exp\left\{ \left(\sum_{j=1}^k A_{N-n,j}(\bar{\beta}_i) |y_j| \right) \frac{\gamma_i}{\beta_i} \right\} . \end{aligned} \tag{6.62}$$

Proof. As pointed out in the proof of Lemma 5.1.3, the condition $\mathbf{B}_5[\bar{\gamma}]$ implies that there exists $0 < \varepsilon_2 \le \varepsilon_0$ such that, for every $\varepsilon \le \varepsilon_2$, the condition $\mathbf{B}_1[\bar{\gamma}]$ (with the constants $L_{19}, L_{20,i}$, $i = 1, \dots, k$, replacing the constants $L_1, L_{2,i}$, $I = 1, \dots, k$) holds for the process $\vec{Y}_{\varepsilon,n}$.

Let us now take $\varepsilon_{12} = \min(\varepsilon_{11}, \varepsilon_2)$, where ε_{11} was defined in relation (6.56). By the definition $\varepsilon_{12} > 0$.

For every fixed $0 \le \varepsilon \le \varepsilon_{12}$ both the conditions $\mathbf{C}_5[\bar{\beta}]$ and $\mathbf{B}_1[\bar{\gamma}]$ (with the constants $K_{45,i}$, $i = 1, \dots, k$, and $L_{19}, L_{20,i}$, $i = 1, \dots, k$ replacing, respectively, the constants $K_{13,i}$, $i = 1, \dots, k$ and $L_1, L_{2,i}$, $i = 1, \dots, k$) hold for the process $\vec{Y}_{\varepsilon,n}$.

Therefore, Lemma 4.3.13 can be applied to the process $\vec{Y}_{\varepsilon,n}$, for every $0 \le \varepsilon \le \varepsilon_{12}$.

This yields inequality (6.62). The constants M_{16} and $M_{17,i}$, $i = 1, \dots, k$ penetrating the inequalities given in Lemma 4.3.13 are functions of the corresponding constants $K_{13,i}$, $i = 1, \dots, k$ and $L_1, L_{2,i}$, $i = 1, \dots, k$. The explicit expressions for these functions are given in Remark 4.3.13. The same formulas give the expressions for the constant M_{40} and $M_{41,i}$, $i = 1, \dots, k$ as functions of the corresponding constants $K_{88,i}$, $i = 1, \dots, k$ and $L_{19}, L_{20,i}$, $i = 1, \dots, k$. □

Remark 6.3.5. The constants M_{40} and $M_{41,i}$, $i = 1, \dots, k$ are given by the following formulas:

$$
\begin{aligned}
M_{40} &= L_{19}, \\
M_{41,i} &= L_{19}L_{20,i}I(\gamma_i = 0) + L_{19}L_{20,i}K_{45,i}^{N\frac{\gamma_i}{\beta_i}} I(\gamma_i > 0) .
\end{aligned}
\tag{6.63}
$$

Remark 6.3.6. The parameter $\varepsilon_{12} = \min(\varepsilon_{11}, \varepsilon_2)$, where the parameters ε_{11} and ε_2 are determined, respectively, by relations (6.56) and (5.11).

Lemma 6.3.4. *Let the conditions* $\mathbf{B}_5[\bar{\gamma}]$ *and* $\mathbf{C}'_{10}[\bar{\beta}]$ *hold and* $0 \le \gamma_i \le \beta_i$, $i = 1, \dots, k$. *Then, there exist* $0 < \varepsilon'_{12} \le \varepsilon_0$ *and the constants* $0 \le M'_{40}, M'_{41,i} < \infty$, $i = 1, \dots, k$ *such that the following inequalities take place for* $\vec{z} = (\vec{y}, x) \in \mathbb{Z}_{\varepsilon,n}$, $0 \le n \le N$, $0 \le \varepsilon \le \varepsilon'_{12}$, *and* $i = 1, \dots, k$:

$$
\begin{aligned}
&\mathsf{E}_{\vec{z},n} \max_{n \le r \le N} \left| g_\varepsilon\left(r, e^{\vec{Y}_{\varepsilon,r}}, X_{\varepsilon,r}\right) \right| \\
&\quad \le M'_{40} + \sum_{i:\gamma_i=0} M'_{41,i} + \sum_{i:\gamma_i>0} M'_{41,i} \exp\left\{ \left(\sum_{j=1}^{k} A_{N-n,j}(\bar{\beta}_i)|y_j| \right) \frac{\gamma_i}{\beta_i} \right\} .
\end{aligned}
\tag{6.64}
$$

Proof. It is analogous to the proof of Lemma 6.3.3. As was pointed out in the proof of this lemma, the condition $\mathbf{B}_5[\bar{\gamma}]$ implies that there exists $0 < \varepsilon_2 \le \varepsilon_0$ such that, for every $\varepsilon \le \varepsilon_2$, the condition $\mathbf{B}_1[\bar{\gamma}]$ (with the constants $L_9, L_{10,i}$, $I = 1, \dots, k$, replacing constants L_1, L_2) holds for the process $\vec{Y}_{\varepsilon,n}$.

Let us now take $\varepsilon'_{12} = \min(\varepsilon'_{11}, \varepsilon_2)$, where ε'_{11} was defined in relation (6.60). By the definition $\varepsilon'_{12} > 0$.

For every fixed $0 \le \varepsilon \le \varepsilon'_{12}$ both the conditions $\mathbf{C}'_5[\bar{\beta}]$ and $\mathbf{B}_1[\bar{\gamma}]$ (with the constants $K'_{45,i}$, $i = 1, \dots, k$, and $L_9, L_{10,i}$, $i = 1, \dots, k$ replacing, respectively, the constants $K'_{13,i}$, $i = 1, \dots, k$ and $L_1, L_{2,i}$, $i = 1, \dots, k$) hold for the process $\vec{Y}_{\varepsilon,n}$.

Therefore, Lemma 4.3.14 can be applied to the process $\vec{Y}_{\varepsilon,n}$, for every $0 \le \varepsilon \le \varepsilon'_{12}$.

This yields inequality (6.64). The constants M'_{16} and $M'_{17,i}$, $i = 1, \dots, k$ penetrating the inequalities given in Lemma 4.3.14 are functions of the corresponding constants $K'_{13,i}$, $i = 1, \dots, k$ and $L_1, L_{2,i}$, $i = 1, \dots, k$. The explicit expressions for these functions are given in Remark 4.3.14. The same formulas give expressions for the constant M'_{40} and $M'_{41,i}$, $i = 1, \dots, k$ as functions of the corresponding constants $K'_{45,i}$, $i = 1, \dots, k$ and $L_9, L_{10,i}$, $i = 1, \dots, k$. □

Remark 6.3.7. The constants M'_{40} and $M'_{41,i}$, $i = 1, \dots, k$ are given by the following formulas:

$$\begin{aligned} M'_{40} &= L_9\,, \\ M'_{41,i} &= L_9 L_{10,i} I(\gamma_i = 0) + L_9 L_{10,i} (K'_{45,i})^{N\frac{\gamma_i}{\beta_i}} I(\gamma_i > 0)\,. \end{aligned} \tag{6.65}$$

Remark 6.3.8. The parameter $\varepsilon'_{12} = \min(\varepsilon'_{11}, \varepsilon_2)$, where the parameters ε'_{11} and ε_2 are determined, respectively, by relations (6.60) and (5.11).

Let us recall the second-type $\vec{A}$-modulus of exponential moment compactness for the components of the log-price process $\vec{Y}_{\varepsilon,n}$, defined for vector $\bar{\beta} = (\beta_1, \dots, \beta_k)$ with nonnegative components and vectors $\bar{\imath}_k = (\imath_1, \dots, \imath_k) \in \mathcal{I}_k$, where $\mathcal{I}_k = \{\bar{\imath}_k : \imath_j = +, -, j = 1, \dots, k\}$,

$$\Delta^{\bar{\imath}_k}_{\bar{\beta},\vec{A}}(\vec{Y}_{\varepsilon,\cdot}, N) = \max_{0 \le n \le N-1} \sup_{\vec{z}=((y_1,\dots,y_k),x)\in \mathbb{Z}} \frac{\mathsf{E}_{\vec{z},n} e^{\sum_{j=1}^k \imath_j A_{N-n-1,j}(\bar{\beta})(Y_{n+1,j} - Y_{n,j})}}{e^{\sum_{j=1}^k A_j(\vec{A}_{N-n-1}(\bar{\beta}))|y_j|}}\,. \tag{6.66}$$

The following second-type condition of exponential moment compactness based on moment-generating functions for increments of log-price processes $\vec{Y}_{\varepsilon,n}$ replaces the condition $\mathbf{E}_5[\bar{\beta}]$. It is assumed to hold for some vector parameter $\bar{\beta} = (\beta_1, \dots, \beta_k)$ with nonnegative components:

$\mathbf{E}_{14}[\bar{\beta}]$: $\overline{\lim}_{\varepsilon\to 0}\, \Xi^{\bar{\imath}_k}_{\bar{\beta}_i,\vec{A}}(\vec{Y}_{\varepsilon,\cdot}, N) < K_{46,i}$, $\bar{\imath}_k = (\imath_1, \dots, \imath_k) \in \mathcal{I}_k$, $i = 1, \dots, k$, for some $1 < K_{46,i} < \infty$, $i = 1, \dots, k$.

Lemma 6.3.5. *The condition $\mathbf{E}_{14}[\bar{\beta}]$ implies that the condition $\mathbf{C}_{10}[\bar{\beta}]$ holds.*

Proof. It follows from Lemma 4.3.17 that should be applied to the log-price processes $\vec{Y}_{\varepsilon,n}$. □

Remark 6.3.9. If the condition $\mathbf{E}_{14}[\bar{\beta}]$ holds with constant $K_{46,i}$, $i = 1, \dots, k$, then the condition $\mathbf{C}_{10}[\bar{\beta}]$ holds with the constant $K_{45,i}$, $i = 1, \dots, k$ given by the following formulas:

$$K_{45,i} = 2^k K_{46,i}\,, \quad i = 1, \dots, k\,. \tag{6.67}$$

Let $\mathcal{Z}_\varepsilon = \langle \mathbb{Z}_{\varepsilon,0}, \dots, \mathbb{Z}_{\varepsilon,N} \rangle$ be a sequence of measurable sets. Let us introduce a modified second-type $\vec{A}$-modulus of exponential moment compactness, defined for vector $\bar{\beta} = (\beta_1, \dots, \beta_k)$ with nonnegative components and vectors $\bar{\imath}_k = (\imath_1, \dots, \imath_k) \in \mathcal{I}_k$, where $\mathcal{I}_k = \{\bar{\imath}_k : \imath_j = +, -, j = 1, \dots, k\}$,

$$\Xi^{\bar{\imath}_k}_{\bar{\beta},\vec{A},\mathcal{Z}}(\vec{Y}_{\varepsilon,\cdot}, N) = \max_{0 \le n \le N-1} \sup_{\vec{z}=((y_1,\dots,y_k),x)\in \mathbb{Z}_{\varepsilon,n}} \frac{\mathsf{E}_{\vec{z},n} e^{\sum_{j=1}^k \imath_j A_{N-n-1,j}(\bar{\beta})(Y_{\varepsilon,n+1,j} - Y_{\varepsilon,n,j})}}{e^{\sum_{j=1}^k A_j(\vec{A}_{N-n-1}(\bar{\beta}))|y_j|}}\,. \tag{6.68}$$

The following modified second-type condition of exponential moment compactness is based on moment-generating functions for increments of log-price processes

$Y_{\varepsilon,n}$. It is assumed to hold for some vector parameter $\bar{\beta} = (\beta_1, \dots, \beta_k)$ with nonnegative components:

$\mathbf{E}'_{14}[\bar{\beta}]$: There exists, for every $\varepsilon \in [0, \varepsilon_0]$, a sequence of measurable sets $Z_\varepsilon = < \mathbb{Z}_{\varepsilon,0}, \dots, \mathbb{Z}_{\varepsilon,N} >$ such that
- (a) $\overline{\lim}_{\varepsilon \to 0} \Xi^{\bar{\imath}_k}_{\beta_i, \vec{A}, Z_\varepsilon}(Y_{\varepsilon,\cdot}, N) < K'_{46,i}$, $\bar{\imath}_k = (\imath_1, \dots, \imath_k) \in \mathcal{I}_k$, $i = 1, \dots, k$, for some $1 < K'_{46,i} < \infty$, $i = 1, \dots, k$;
- (b) Z_ε is a complete sequence of phase sets, i.e. $\mathsf{P}_n(z, \mathbb{Z}_{\varepsilon,n}) = 1$, $z \in \mathbb{Z}_{\varepsilon,n-1}$, $n = 1, \dots, N$.

Lemma 6.3.6. *The condition $\mathbf{E}'_{14}[\bar{\beta}]$ implies that the condition $\mathbf{C}'_{10}[\bar{\beta}]$ holds.*

Proof. It follows from Lemma 4.3.18 that should be applied to the log-price processes $\vec{Y}_{\varepsilon,n}$. □

Remark 6.3.10. If the condition $\mathbf{E}'_{14}[\bar{\beta}]$ holds with constant $K'_{46,i}$, $i = 1, \dots, k$, then condition $\mathbf{C}'_{10}[\bar{\beta}]$ holds with the constant $K'_{45,i}$, $i = 1, \dots, k$ given by the following formulas:

$$K'_{45,i} = 2^k K'_{46,i}, \quad i = 1, \dots, k. \tag{6.69}$$

6.3.2 Asymptotically uniform upper bounds for rewards for multivariate modulated Markov log-price processes with unbounded characteristics

Let us assume that the conditions $\mathbf{B}_5[\bar{\gamma}]$ and $\mathbf{C}_{10}[\bar{\beta}]$ hold and $0 \le \gamma_i \le \beta_i$, $i = 1, \dots, k$.

Lemma 6.3.3 lets us define log-reward functions $\phi_{\varepsilon,n}(\vec{z}) = \phi_{\varepsilon,n}(\vec{y}, x)$, $\vec{z} = (\vec{y}, x) \in \mathbb{Z}$ for the log-price process $\vec{Z}_{\varepsilon,n}$, for every $0 \le n \le N$ and $0 \le \varepsilon \le \varepsilon_{12}$,

$$\begin{aligned}\phi_{\varepsilon,n}(\vec{z}) &= \phi_{\varepsilon,n}(\vec{y}, x) \\ &= \sup_{\tau_{\varepsilon,n} \in \mathcal{M}^{(\varepsilon)}_{\max,n,N}} \mathsf{E}_{\vec{z},n} g_\varepsilon(\tau_{\varepsilon,n}, e^{\vec{Y}_{\varepsilon,\tau_{\varepsilon,n}}}, X_{\varepsilon,\tau_{\varepsilon,n}}), \quad \vec{z} = (\vec{y}, x) \in \mathbb{Z}.\end{aligned} \tag{6.70}$$

These functions are the measurable functions of $\vec{z} \in \mathbb{Z}$, for every $0 \le n \le N$.

Also, Lemma 6.3.3 lets us get the following asymptotically uniform upper bounds for the log-reward functions $\phi_{\varepsilon,n}(\vec{y}, x)$.

Theorem 6.3.1. *Let the conditions $\mathbf{B}_5[\bar{\gamma}]$ and $\mathbf{C}_{10}[\bar{\beta}]$ hold and $0 \le \gamma_i \le \beta_i$, $i = 1, \dots, k$. Then, the log-reward functions $\phi_{\varepsilon,n}(\vec{y}, x)$ satisfy the following inequalities for $\vec{z} = (\vec{y}, x) \in \mathbb{Z}$, $0 \le n \le N$, $0 \le \varepsilon \le \varepsilon_{12}$:*

$$|\phi_{\varepsilon,n}(\vec{y}, x)| \le M_{40} + \sum_{i:\gamma_i = 0} M_{41,i} + \sum_{i:\gamma_i > 0} M_{41,i} \exp\left\{\left(\sum_{j=1}^{k} A_{N-n,j}(\beta_i)|y_j|\right)\frac{\gamma_i}{\beta_i}\right\}. \tag{6.71}$$

Proof. It is analogous to the proof of Theorem 5.1.1. Using Lemma 6.3.3 and the definition of the log-reward functions, we get the following inequalities for $\vec{z} = (\vec{y}, x) \in \mathbb{Z}, 0 \leq n \leq N, 0 \leq \varepsilon \leq \varepsilon_{12}$:

$$|\phi_{\varepsilon,n}(\vec{y}, x)| \leq \mathsf{E}_{\vec{z},n} \max_{n \leq r \leq N} |g_\varepsilon(r, e^{\vec{Y}_{\varepsilon,r}}, X_{\varepsilon,r})|$$
$$\leq M_{40} + \sum_{i:\gamma_i=0} M_{41,i} + \sum_{i:\gamma_i>0} M_{41,i} \exp\left\{\left(\sum_{j=1}^{k} A_{N-n,j}(\bar{\beta}_i)|y_j|\right)\frac{\gamma_i}{\beta_i}\right\}, \quad (6.72)$$

which prove the above theorem. □

Let us now assume that the conditions $\mathbf{B}_5[\bar{\gamma}]$ and $\mathbf{C}'_{10}[\bar{\beta}]$ hold and $0 \leq \gamma_i \leq \beta_i, i = 1, \ldots, k$.

Also for Lemma 6.3.4, let us define log-reward functions $\phi_{\varepsilon,n}(\vec{y}, x)$ using relation (6.70), only for $\vec{z} = (\vec{y}, x) \in \mathbb{Z}_{\varepsilon,n}, \quad 0 \leq n \leq N, 0 \leq \varepsilon \leq \varepsilon'_{12}$.

Also for Lemma 6.3.4, let us obtain the following upper bounds for the log-reward functions.

Theorem 6.3.2. *Let the conditions $\mathbf{B}_5[\bar{\gamma}]$ and $\mathbf{C}'_{10}[\bar{\beta}]$ hold and $0 \leq \gamma_i \leq \beta_i, i = 1, \ldots, k$. Then, the log-reward functions $\phi_{\varepsilon,n}(\vec{y}, x)$ satisfy the following inequalities for $0 \leq n \leq N, \vec{z} = (\vec{y}, x) \in \mathbb{Z}_{\varepsilon,n}, 0 \leq \varepsilon \leq \varepsilon'_{12}$:*

$$|\phi_{\varepsilon,n}(\vec{y}, x)| \leq M'_{40} + \sum_{i:\gamma_i=0} M'_{41,i} + \sum_{i:\gamma_i>0} M'_{41,i} \exp\left\{\left(\sum_{j=1}^{k} A_{N-n,j}(\bar{\beta}_i)|y_j|\right)\frac{\gamma_i}{\beta_i}\right\}. \quad (6.73)$$

Proof. It is analogous to the proof of Theorem 6.3.1. In this case Lemma 6.3.4 should be used instead of Lemma 6.3.3. An inequality analogous to (6.72) can be written down. The differences are that it holds for $\vec{z} = (\vec{y}, x) \in \mathbb{Z}_{\varepsilon,n}, 0 \leq n \leq N, 0 \leq \varepsilon \leq \varepsilon'_{12}$, and with the constants $M'_{40}, M'_{41,i}, i = 1, \ldots, k$. □

Remark 6.3.11. It is useful to note that the log-reward function $\phi_{\varepsilon,N}(\vec{y}, x) = g_\varepsilon(N, e^{\vec{y}}, x), \vec{z} = (\vec{y}, x) \in \overline{\mathbb{Z}}_{\varepsilon,N}$ always exists. The corresponding upper bounds for this function are given by condition $\mathbf{B}_5[\bar{\gamma}]$. However, there is no guarantee that, under conditions of Theorem 6.3.2, the log-reward functions $\phi_{\varepsilon,n}(\vec{y}, x)$ exist for points $\vec{z} = (\vec{y}, x) \in \overline{\mathbb{Z}}_{\varepsilon,n}, 0 \leq n \leq N-1$. Even, if these functions do exist for these points, Theorem 6.3.2 does not give upper bounds for the log-reward functions in these points.

However, it is the important case where this problem is not essential. Let us assume that the condition $\mathbf{H}_1$ holds. Then, as was pointed out in Remark 5.1.3, for any point $\vec{z} = (\vec{y}, x) \in \mathbb{Z}$, there exists $\varepsilon_1(z) \in (0, \varepsilon_0]$ such that $\vec{z} = (\vec{y}, x) \in \mathbb{H}_{\varepsilon,n} \subseteq \mathbb{Z}_{\varepsilon,n}$, $n = 0, \ldots, N$ for $\varepsilon \in [0, \varepsilon_1(\vec{z})]$. The quantity $\varepsilon_1(\vec{z})$ is determined by relation (5.6) given in Remark 5.1.3. Thus, the upper bounds for $\phi_{\varepsilon,n}(\vec{y}, x)$ given in Theorem 6.3.2 take place for $\varepsilon \in [0, \varepsilon_1(\vec{z})]$.

Let us now give asymptotically uniform upper bounds for the optimal expected reward for the modulated Markov log-price processes $\vec{Z}_{\varepsilon,n}$.

The condition $\mathbf{D}_3[\bar{\beta}]$ should be replaced, respectively, by the following first-type condition of exponential moment boundedness, assumed to hold for some vector parameter $\bar{\beta} = (\beta_1, \dots, \beta_k)$ with nonnegative components:

$\mathbf{D}_9[\bar{\beta}]$: $\overline{\lim}_{\varepsilon\to 0} \mathsf{E} \exp\{\sum_{j=1}^{k} A_{N,j}(\bar{\beta}_i)|Y_{\varepsilon,0,j}|\} < K_{47,i}$, $i = 1, \dots, k$, for some $1 < K_{47,i} < \infty$, $i = 1, \dots, k$.

The following lemma gives explicit upper bounds for the expectation of maximal absolute values for the pay-off processes $g_\varepsilon(n, e^{\vec{Y}_{\varepsilon,n}}, X_{\varepsilon,n})$.

Lemma 6.3.7. *Let the conditions $\mathbf{B}_5[\bar{\gamma}]$, $\mathbf{C}_{10}[\bar{\beta}]$, and $\mathbf{D}_9[\bar{\beta}]$ hold and $0 \le \gamma_i \le \beta_i$, $i = 1, \dots, k$. Then, there exist $0 < \varepsilon_{14} \le \varepsilon_0$ and a constant $0 \le M_{42} < \infty$ such that for the following inequality takes place for $0 \le \varepsilon \le \varepsilon_{14}$:*

$$\mathsf{E} \max_{0\le n\le N} |g_\varepsilon(n, e^{\vec{Y}_{\varepsilon,n}}, X_{\varepsilon,n})| \le M_{42} . \tag{6.74}$$

Proof. The condition $\mathbf{D}_9[\bar{\beta}]$ implies that there exists $0 < \varepsilon_{13} \le \varepsilon_0$ such that for $0 \le \varepsilon \le \varepsilon_{13}$ and $i = 1, \dots, k$,

$$\mathsf{E} \exp\left\{\sum_{j=1}^{k} A_{N,j}(\bar{\beta}_i)|Y_{\varepsilon,0,j}|\right\} < K_{47,i} . \tag{6.75}$$

Relation (6.75) means that, for every $\varepsilon \le \varepsilon_{13}$, the condition $\mathbf{D}_3[\bar{\beta}]$ (with constants $K_{47,i}$, $i = 1, \dots, k$ replacing the constants $K_{14,i}$, $i = 1, \dots, k$) holds for the process $\vec{Y}_{\varepsilon,n}$.

Let us now take $\varepsilon_{14} = \min(\varepsilon_{11}, \varepsilon_2, \varepsilon_{13})$, where ε_{11} and ε_2 were defined in relations (6.56) and (5.11). By the definition $\varepsilon_{14} > 0$.

For every fixed $0 \le \varepsilon \le \varepsilon_{14}$, the conditions $\mathbf{C}_5[\bar{\beta}]$ and $\mathbf{B}_1[\bar{\gamma}]$, and $\mathbf{D}_3[\bar{\beta}]$ (with the constants $K_{45,i}, K_{47,i}$, $i = 1, \dots, k$, and $L_9, L_{10,i}$, $i = 1, \dots, k$ replacing, respectively, the constants $K_{13,i}, K_{14,i}$, $i = 1, \dots, k$ and $L_1, L_{2,i}$, $i = 1, \dots, k$) hold for the process $\vec{Y}_{\varepsilon,n}$.

Therefore, Lemma 4.3.15 can be applied to the process $\vec{Y}_{\varepsilon,n}$, for every $0 \le \varepsilon \le \varepsilon_{14}$.

This yields inequality (6.74). The constant M_{18} penetrating the inequalities given in Lemma 4.3.15 is a function of the corresponding constants $K_{13,i}$, $K_{14,i}$, $i = 1, \dots, k$ and $L_1, L_{2,i}$, $i = 1, \dots, k$. The explicit expression for these function is given in Remark 4.3.15. The same formula gives the expressions for constant M_{42} as a function of the corresponding constants $K_{45,i}, K_{47,i}$, $i = 1, \dots, k$, and $L_9, L_{10,i}$, $i = 1, \dots, k$. ☐

Remark 6.3.12. The constant M_{42} is given by the following formula, which follows from the corresponding formula given in Remark 4.3.15:

$$M_{42} = L_9 + \sum_{i:\gamma_i=0} L_9 L_{10,i} + \sum_{i:\gamma_i>0} L_9 L_{10,i} K_{45,i}^{N\frac{\gamma_i}{\beta_i}} K_{47,i}^{\frac{\gamma_i}{\beta_i}} . \tag{6.76}$$

Remark 6.3.13. The parameter $\varepsilon_{14} = \min(\varepsilon_{11}, \varepsilon_2, \varepsilon_{13})$, where ε_{11} and ε_2 were defined in relations (6.56), (5.11), and (6.75).

Note also that one can replace the following second-type condition of exponential moment boundedness $\mathbf{D}_9[\bar{\beta}]$ by the following condition of exponential moment boundedness based on moment-generating functions for the initial value of log-price processes $\vec{Y}_{\varepsilon,0}$. It is assumed to hold for some vector parameter $\bar{\beta} = (\beta_1, \dots, \beta_k)$ with nonnegative components:

$\mathbf{F}_6[\bar{\beta}]$: $\overline{\lim}_{\varepsilon\to 0} \mathsf{E}\exp\{\sum_{j=1}^k \imath_j A_{N,j}(\bar{\beta}_i) Y_{\varepsilon,0,j}\} < K_{48,i}$, $\bar{\imath}_k = (\imath_1, \dots, \imath_k) \in \mathcal{I}_k$, $i = 1, \dots, k$ for some $1 < K_{48,i} < \infty$, $i = 1, \dots, k$.

Lemma 6.3.8. *The condition $\mathbf{F}_6[\bar{\beta}]$ implies that the condition $\mathbf{D}_9[\bar{\beta}]$ holds.*

Proof. It follows from the following inequality that holds for any $\bar{\beta} = (\beta_1, \dots, \beta_k)$, $\beta_1, \dots, \beta_k \geq 0$,

$$\mathsf{E}e^{\sum_{j=1}^k A_{N,j}(\bar{\beta}_i)|Y_{0,j}|} \leq \sum_{\bar{\imath}_k \in \mathcal{I}_k} \mathsf{E}e^{\sum_{j=1}^k \imath_j A_{N,j}(\bar{\beta}_i) Y_{0,j}} . \tag{6.77}$$

□

Remark 6.3.14. If the condition $\mathbf{F}_6[\bar{\beta}]$ holds with the constant $K_{48,i}$, $i = 1, \dots, k$, then the condition $\mathbf{D}_9[\bar{\beta}]$ holds with the constant $K_{47,i}$, $i = 1, \dots, k$ given by the following formulas:

$$K_{47,i} = 2^k K_{48,i}\,, \quad i = 1, \dots, k. \tag{6.78}$$

The condition $\mathbf{D}'_3[\bar{\beta}]$ should be replaced, respectively, by the following condition, assumed to hold for some vector parameter $\bar{\beta} = (\beta_1, \dots, \beta_k)$ with nonnegative components:

$\mathbf{D}'_9[\bar{\beta}]$: (a) $\overline{\lim}_{\varepsilon\to 0} \mathsf{E}\exp\{\sum_{j=1}^k A_{N,j}(\bar{\beta}_i)|Y_{\varepsilon,0,j}|\} < K'_{47,i}$, $i = 1, \dots, k$, for some $1 < K'_{47,i} < \infty$, $i = 1, \dots, k$;
(b) $\mathsf{P}\{\vec{Z}_{\varepsilon,0} \in \mathbb{Z}_{\varepsilon,0}\} = 1$, for every $\varepsilon \in [0, \varepsilon_0]$, where $\mathbb{Z}_{\varepsilon,0}$ is the set penetrating condition $\mathbf{C}'_9[\beta]$.

The following lemma gives alternative explicit upper bounds for the expectation of maximal absolute values for the pay-off processes $g_\varepsilon(n, e^{\vec{Y}_{\varepsilon,n}}, X_{\varepsilon,n})$.

Lemma 6.3.9. *Let the conditions $\mathbf{B}_5[\bar{\gamma}]$, $\mathbf{C}'_{10}[\bar{\beta}]$ and $\mathbf{D}'_9[\bar{\beta}]$ hold and $0 \leq \gamma_i \leq \beta_i$, $i = 1, \dots, k$. Then, there exist $0 < \varepsilon'_{14} \leq \varepsilon_0$ and a constant $0 \leq M'_{42} < \infty$ such that for the following inequality takes place for $0 \leq \varepsilon \leq \varepsilon'_{14}$:*

$$\mathsf{E} \max_{0\leq n\leq N} |g_\varepsilon(n, e^{\vec{Y}_{\varepsilon,n}}, X_{\varepsilon,n})| \leq M'_{42}\,. \tag{6.79}$$

Proof. The condition $\mathbf{D}'_9[\bar{\beta}]$ implies that there exists $0 < \varepsilon'_{13} \leq \varepsilon_0$ such that for $0 \leq \varepsilon \leq \varepsilon'_{13}$,

$$\mathsf{E}\exp\left\{\sum_{j=1}^k A_{N,j}(\bar{\beta}_i)|Y_{\varepsilon,0,j}|\right\} < K'_{47,i}\,. \tag{6.80}$$

Relation (6.80) means that, for every $\varepsilon \leq \varepsilon'_{13}$, the condition $\mathbf{D}'_3[\bar{\beta}]$ (with the constants $K'_{47,i}$, $i = 1, \dots, k$ replacing the constants $K'_{14,i}$, $i = 1, \dots, k$) holds for the process $\vec{Y}_{\varepsilon,n}$.

Let us now take $\varepsilon'_{14} = \min(\varepsilon'_{11}, \varepsilon_2, \varepsilon'_{13})$, where ε'_{11} and ε_2 were defined in relations (6.60) and (5.11). By the definition $\varepsilon'_{14} > 0$.

For every fixed $0 \leq \varepsilon \leq \varepsilon'_{14}$, the conditions $\mathbf{C}'_5[\bar{\beta}]$ and $\mathbf{B}_1[\bar{\gamma}]$, and $\mathbf{D}'_3[\bar{\beta}]$ (with the constants $K'_{45,i}, K'_{47,i}$, $i = 1, \dots, k$, and $L_{19}, L_{20,i}$, $i = 1, \dots, k$ replacing, respectively, the constants $K'_{13,i}, K'_{14,i}$, $i = 1, \dots, k$ and $L_1, L_{2,i}$, $i = 1, \dots, k$) hold for the process $\vec{Y}_{\varepsilon,n}$.

Therefore, Lemma 4.3.16 can be applied to the process $\vec{Y}_{\varepsilon,n}$, for every $0 \leq \varepsilon \leq \varepsilon'_{14}$.

This yields inequality (6.79). The constant M'_{18} penetrating the inequalities given in Lemma 4.3.16 is a function of the corresponding constants $K'_{13,i}, K'_{14,i}$, $i = 1, \dots, k$ and $L_1, L_{2,i}$, $i = 1, \dots, k$. The explicit expression for these function is given in Remark 4.3.16. The same formula gives the expressions for the constant M'_{42} as a function of the corresponding constants $K'_{45,i}, K'_{47,i}$, $i = 1, \dots, k$, and $L_{19}, L_{20,i}$, $i = 1, \dots, k$. □

Remark 6.3.15. The constant M'_{42} is given by the following formula, which follows from the corresponding formula given in Remark 4.3.16:

$$M'_{42} = L_{19} + \sum_{i:\gamma_i = 0} L_{19} L_{20,i} + \sum_{i:\gamma_i > 0} L_{19} L_{20,i} (K'_{45,i})^{N \frac{\gamma_i}{\beta_i}} (K'_{47,i})^{\frac{\gamma_i}{\beta_i}} . \tag{6.81}$$

Remark 6.3.16. The parameter $\varepsilon'_{14} = \min(\varepsilon'_{11}, \varepsilon_2, \varepsilon'_{13})$, where ε_{11} and ε_2 were defined in relations (6.60), (5.11), and (6.80).

Note also that one can replace the following first-type condition of exponential moment boundedness $\mathbf{D}'_9[\bar{\beta}]$ by the following condition of exponential moment boundedness based on moment-generating functions for the initial value of log-price processes $\vec{Y}_{\varepsilon,0}$. It is assumed to hold for some vector parameter $\bar{\beta} = (\beta_1, \dots, \beta_k)$ with nonnegative components:

$\mathbf{F}'_6[\bar{\beta}]$: (a) $\overline{\lim}_{\varepsilon \to 0} \mathsf{E} \exp\{\sum_{j=1}^{k} \imath_j A_{N,j}(\bar{\beta}_i) Y_{\varepsilon,0,j}\} < K'_{48,i}$, $\bar{\imath}_k = (\imath_1, \dots, \imath_k) \in \mathcal{I}_k$, $i = 1, \dots, k$ for some $1 \leq K'_{48,i} < \infty$, $i = 1, \dots, k$;

(b) $\mathsf{P}\{\vec{Z}_{\varepsilon,0} \in \mathbb{Z}_{\varepsilon,0}\} = 1$, for every $\varepsilon \in [0, \varepsilon_0]$, where $\mathbb{Z}_{\varepsilon,0}$ is the set penetrating the condition $\mathbf{C}'_{10}[\bar{\beta}]$.

Lemma 6.3.10. *The condition $\mathbf{F}'_6[\bar{\beta}]$ implies that the condition $\mathbf{D}'_9[\bar{\beta}]$ holds.*

Proof. It follows from inequality (6.77). □

Remark 6.3.17. If the condition $\mathbf{F}'_6[\bar{\beta}]$ holds with the constant $K'_{48,i}$, $i = 1, \dots, k$, then the condition $\mathbf{D}'_{17}[\bar{\beta}]$ holds with the constant $K'_{47,i}$, $i = 1, \dots, k$ given by the following formulas:

$$K'_{47,i} = 2^k K'_{48,i} , \quad i = 1, \dots, k . \tag{6.82}$$

Let us now assume that either conditions of Lemma 6.3.7 or Lemma 6.3.9 hold. In this case, we can define the optimal expected reward $\Phi_\varepsilon = \Phi(\mathcal{M}^{(\varepsilon)}_{\max,N})$ for the log-price process $\vec{Z}_{\varepsilon,n}$, respectively, for every $0 \le \varepsilon \le \varepsilon_{14}$, or, for every $0 \le \varepsilon \le \varepsilon'_{14}$

$$\Phi_\varepsilon = \Phi(\mathcal{M}^{(\varepsilon)}_{\max,N}) = \sup_{\tau_{\varepsilon,0} \in \mathcal{M}^{(\varepsilon)}_{\max,N}} \mathsf{E} g_\varepsilon(\tau_{\varepsilon,0}, e^{\vec{Y}_{\varepsilon,\tau_{\varepsilon,0}}}, X_{\varepsilon,\tau_{\varepsilon,0}}) \,. \tag{6.83}$$

Also, as follows from Theorem 2.3.4, the following formula takes place, respectively, for every $0 \le \varepsilon \le \varepsilon_{14}$, or, for every $0 \le \varepsilon \le \varepsilon'_{14}$:

$$\Phi_\varepsilon = \mathsf{E}\phi_{\varepsilon,0}(\vec{Y}_{\varepsilon,0}, X_{\varepsilon,0}) \,. \tag{6.84}$$

The following two theorems give asymptotically uniform upper bounds for the optimal expected rewards for American-type options.

Theorem 6.3.3. *Let the conditions* $\mathbf{B}_5[\bar{\gamma}]$, $\mathbf{C}_{10}[\bar{\beta}]$, *and* $\mathbf{D}_9[\bar{\beta}]$ *hold and* $0 \le \gamma_i \le \beta_i$, $i = 1, \dots, k$. *Then, the following inequality takes place, for* $0 \le \varepsilon \le \varepsilon_{14}$,

$$|\Phi_\varepsilon| \le M_{42} \,. \tag{6.85}$$

Proof. It follows from the definition of the functional Φ_ε and Lemma 6.3.7 that, for $0 \le \varepsilon \le \varepsilon_{14}$,

$$\begin{aligned} |\Phi_\varepsilon| &\le \max_{\tau_{\varepsilon,0} \in \mathcal{M}^{(\varepsilon)}_{\max,N}} \mathsf{E}|g_\varepsilon(\tau_{\varepsilon,0}, e^{\vec{Y}_{\tau_{\varepsilon,0}}}, X_{\tau_{\varepsilon,0}})| \\ &\le \mathsf{E} \max_{0 \le n \le N} |g_\varepsilon(n, e^{\vec{Y}_{\varepsilon,n}}, X_{\varepsilon,n})| \le M_{42} \,. \end{aligned} \tag{6.86}$$

This inequality proves the theorem. □

Theorem 6.3.4. *Let the conditions* $\mathbf{B}_5[\bar{\gamma}]$, $\mathbf{C}'_{10}[\bar{\beta}]$, *and* $\mathbf{D}'_9[\bar{\beta}]$ *hold and* $0 \le \gamma_i \le \beta_i$, $i = 1, \dots, k$. *Then, the following inequality takes place, for* $0 \le \varepsilon \le \varepsilon'_{14}$,

$$|\Phi_\varepsilon| \le M'_{42} \,. \tag{6.87}$$

Proof. It is analogous to the proof of Theorem 6.3.3. The only difference is that Lemma 6.3.8 should be used instead of Lemma 6.3.7. In this case, the inequality analogous to (6.86) can be written down for $0 \le \varepsilon \le \varepsilon'_{14}$. □

6.4 Multivariate modulated LPP with unbounded characteristics

In this section, we present results about convergence for log-rewards for multivariate modulated Markov log-price processes with unbounded characteristics.

6.4.1 Convergence of option rewards for multivariate modulated Markov log-price processes

Let us now consider the log-reward functions $\phi_{\varepsilon,n}(\vec{y}, x)$ introduced in Subsection 6.3.2 and give conditions for their convergence.

Theorem 6.4.1. *Let the conditions* $\mathbf{B}_5[\bar{\gamma}]$ *and* $\mathbf{C}_{10}[\bar{\beta}]$ *hold with the vector parameters* $\bar{\gamma} = (\gamma_1, \dots, \gamma_k)$, $\bar{\beta} = (\beta_1, \dots, \beta_k)$ *such that, for every* $i = 1, \dots, k$, *either* $\beta_i > \gamma_i > 0$ *or* $\beta_i = \gamma_i = 0$, *and also conditions* $\mathbf{I}_1$ *and* $\mathbf{J}_1$ *hold. Then, for every* $n = 0, 1, \dots, N$, *the following relation takes place for any* $\vec{z}_\varepsilon = (\vec{y}_\varepsilon, x_\varepsilon) \to \vec{z}_0 = (\vec{y}_0, x_0) \in \mathbb{Z}'_n \cap \mathbb{Z}''_n$:

$$\phi_{\varepsilon,n}(\vec{y}_\varepsilon, x_\varepsilon) \to \phi_{0,n}(\vec{y}_0, x_0) \quad as \quad \varepsilon \to 0\,. \tag{6.88}$$

Proof. It is analogous to the proof of Theorem 5.2.1.

Due to the above conditions imposed on the parameters γ_i, β_i, $i = 1, \dots, k$, one can always find $\alpha > 1$ such that $\gamma_i \alpha \le \beta_i$, $i = 1, \dots, k$.

According to Theorem 6.3.1, the conditions $\mathbf{B}_5[\bar{\gamma}]$ and $\mathbf{C}_{10}[\bar{\beta}]$ imply that there exist $\varepsilon_{12} > 0$ and the constant $M_{40}, M_{41,i}$, $i = 1, \dots, k$ such that the following upper bounds take place for the reward functions $\phi_{\varepsilon,n}(\vec{y}, x)$ for $\vec{z} = (\vec{y}, x) \in \mathbb{Z}$, $n = 0, 1, \dots, N$, $\varepsilon \le \varepsilon_{12}$,

$$|\phi_{\varepsilon,n}(\vec{y}, x)| \le M_{40} + \sum_{i:\gamma_i=0} M_{41,i} + \sum_{i:\gamma_i>0} M_{41,i} \exp\left\{\left(\sum_{j=1}^{k} A_{N-n,j}(\bar{\beta}_i)|y_j|\right)\frac{\gamma_i}{\beta_i}\right\}. \tag{6.89}$$

Using inequalities (6.89) and the condition $\mathbf{C}_{10}[\bar{\beta}]$, we get for any $\vec{z} = (\vec{y}, x) \in \mathbb{Z}$, $n = 0, 1, \dots, N-1$, $\varepsilon \le \varepsilon_{12}$,

$$\begin{aligned}
\mathsf{E}_{\vec{z},n}|\phi_{\varepsilon,n+1}(\vec{Y}_{\varepsilon,n+1}, X_{\varepsilon,n+1})|^\alpha &\le \Bigg(M_{49} + \sum_{i:\gamma_i=0} M_{41,i} \\
&\quad + \sum_{i:\gamma_i>0} M_{41,i} \exp\left\{\left(\sum_{j=1}^{k} A_{N-n-1,j}(\bar{\beta}_i)|Y_{\varepsilon,n+1,j}|\right)\frac{\gamma_i}{\beta_i}\right\}\Bigg)^\alpha \\
&\le \mathsf{E}_{\vec{z},n}(k+1)^{\alpha-1}\Bigg(M_{40}^\alpha + \sum_{i:\gamma_i=0} M_{41,i}^\alpha \\
&\quad + \sum_{i:\gamma_i>0} M_{41,i}^\alpha \exp\left\{\left(\sum_{j=1}^{k} A_{N-n-1,j}(\bar{\beta}_i)|Y_{\varepsilon,n+1,j}|\right)\frac{\gamma_i\alpha}{\beta_i}\right\}\Bigg) \\
&\le (k+1)^{\alpha-1}\Bigg(M_{40}^\alpha + \sum_{i:\gamma_i=0} M_{41,i}^\alpha \\
&\quad + \sum_{i:\gamma_i>0} M_{41,i}^\alpha \exp\left\{\left(\sum_{j=1}^{k} A_{N-n-1,j}(\bar{\beta}_i)|y_j|\right)\frac{\gamma_i\alpha}{\beta_i}\right\}
\end{aligned}$$

$$
\begin{aligned}
&\times \mathsf{E}_{\vec{z},n} \exp\left\{\left(\sum_{j=1}^{k} A_{N-n-1,j}(\bar{\beta}_i)|Y_{\varepsilon,n+1,j} - y_j|\right)\frac{\gamma_i \alpha}{\beta_i}\right\}\Bigg) \\
&\leq (k+1)^{\alpha-1}\Bigg(M_{40}^{\alpha} + \sum_{i:\gamma_i=0} M_{41,i}^{\alpha} \\
&\quad + \sum_{i:\gamma_i>0} M_{41,i}^{\alpha} \exp\left\{\left(\sum_{j=1}^{k} A_{N-n-1,j}(\bar{\beta}_i)|y_j|\right)\frac{\gamma_i \alpha}{\beta_i}\right\} \\
&\quad \times \mathsf{E}_{\vec{z},n} \exp\left\{\left(\sum_{j=1}^{k} A_{N-n-1,j}(\bar{\beta}_i)|Y_{\varepsilon,n+1,j} - Y_{\varepsilon,n,j}|\right\}\right) \\
&\leq (k+1)^{\alpha-1}\Bigg(M_{40}^{\alpha} + \sum_{i:\gamma_i=0} M_{41,i}^{\alpha} \\
&\quad + \sum_{i:\gamma_i>0} M_{41,i}^{\alpha} \exp\left\{\left(\sum_{j=1}^{k} A_{N-n-1,j}(\bar{\beta}_i)|y_j|\right)\frac{\gamma_i \alpha}{\beta_i}\right\} \\
&\quad \times \exp\left\{\sum_{j=1}^{k} A_j(\vec{A}_{N-n-1}(\bar{\beta}_i))|y_j|\right\} K_{45,i}\Bigg) = Q_{\vec{A}}(\vec{y}) < \infty\,. \qquad (6.90)
\end{aligned}
$$

The following proof repeats the proof of Theorem 5.2.1. The only difference is that relation (5.57) is replaced by the analogous relation

$$\varlimsup_{\varepsilon\to 0} \mathsf{E}|\phi_{\varepsilon,N}(\vec{Y}_{\varepsilon,N}(\vec{z}_\varepsilon), X_{\varepsilon,N}(\vec{z}_\varepsilon))|^{\alpha} \leq Q_{\vec{A}}(\vec{y}_0) < \infty\,, \qquad (6.91)$$

where the function $Q(\vec{y})$ defined in relation (5.42) is replaced by the function $Q_{\vec{A}}(y)$ defined in relation (6.90). □

Remark 6.4.1. According to Lemma 6.3.5, the condition $\mathbf{C}_{10}[\bar{\beta}]$ can be replaced in Theorems 6.4.1 by the condition $\mathbf{E}_{14}[\bar{\beta}]$.

Remark 6.4.2. As follows from the remarks made in Subsection 5.2.1, the following pointwise convergence relation take place for the reward functions, for any $\vec{z} = (\vec{y}, x) \in \mathbb{Z}'_n \cap \mathbb{Z}''_n$, $n = 0, 1, \dots, N$,

$$\phi_{\varepsilon,n}(\vec{y}, x) \to \phi_{0,n}(\vec{y}, x) \quad \text{as} \quad \varepsilon \to 0\,. \qquad (6.92)$$

Also, the reward function $\phi_{0,n}(\vec{y}, x)$ is continuous in points $\vec{z} = (\vec{y}, x) \in \mathbb{Z}'_n \cap \mathbb{Z}''_n$, for every $n = 0, 1, \dots, N$.

The following theorem gives alternative conditions for convergence of log-reward functions for American-type options in the above model.

Theorem 6.4.2. *Let the conditions $\mathbf{B}_5[\bar{\gamma}]$ and $\mathbf{C}'_{10}[\bar{\beta}]$ hold with vector parameters $\bar{\gamma} = (\gamma_1, \dots, \gamma_k)$ and $\bar{\beta} = (\beta_1, \dots, \beta_k)$ such that, for every $i = 1, \dots, k$, either $\beta_i > \gamma_i > 0$ or $\beta_i = \gamma_i = 0$, and also conditions $\mathbf{H}_1$, $\mathbf{I}_1$ and $\mathbf{J}_1$ hold. Then, for every $n = 0, 1, \dots, N$, the* *following relation takes place for any* $\vec{z}_\varepsilon = (\vec{y}_\varepsilon, x_\varepsilon) \to \vec{z}_0 = (\vec{y}_0, x_0) \in \mathbb{Z}'_n \cap \mathbb{Z}''_n$,

$$\phi_{\varepsilon,n}(\vec{y}_\varepsilon, x_\varepsilon) \to \phi_{0,n}(\vec{y}_0, x_0) \quad as \quad \varepsilon \to 0\,. \tag{6.93}$$

Proof. It is analogous to the proof of Theorem 5.2.2.

Due to the above conditions imposed on the parameters γ_i, β_i, $i = 1, \dots, k$, one can always find $\alpha > 1$ such that $\gamma_i \alpha \leq \beta_i$, $i = 1, \dots, k$.

According to Theorem 6.3.2, the conditions $\mathbf{B}_5[\bar{\gamma}]$ and $\mathbf{C}'_{10}[\bar{\beta}]$ imply that there exist $\varepsilon'_{12} > 0$ and the constant $M'_{40}, M'_{41,i}$, $i = 1, \dots, k$ such that the following upper bounds take place for the reward functions $\phi_{\varepsilon,n}(\vec{y}, x)$ for $\vec{z} = (\vec{y}, x) \in \mathbb{Z}_{\varepsilon,n}$, $n = 0, 1, \dots, N$, $\varepsilon \leq \varepsilon'_{12}$,

$$|\phi_{\varepsilon,n}(\vec{y}, x)| \leq M'_{40} + \sum_{i:\gamma_i=0} M'_{41,i} + \sum_{i:\gamma_i>0} M'_{41,i} \exp\{(\sum_{j=1}^{k} A_{N-n,j}(\bar{\beta}_i)|y_j|)\frac{\gamma_i}{\beta_i}\}\,. \tag{6.94}$$

Recall that the condition $\mathbf{C}'_{10}[\bar{\beta}]$ (b) implies that $\mathsf{P}_{z,n}\{\vec{Z}_{\varepsilon,n+1} \in \mathbb{Z}_{\varepsilon,n+1}\} = 1$ for $z \in \mathbb{Z}_{\varepsilon,n}$, $n = 0, \dots, N-1$. Using inequalities (6.94) and the condition $\mathbf{C}'_{10}[\bar{\beta}]$, we get for any $\vec{z} = (\vec{y}, x) \in \mathbb{Z}_{\varepsilon,n}$, $n = 0, 1, \dots, N-1$, $\varepsilon \leq \varepsilon'_{12}$,

$$\begin{aligned}
\mathsf{E}_{\vec{z},n}|\phi_{\varepsilon,n+1}(\vec{Y}_{\varepsilon,n+1}, X_{\varepsilon,n+1})|^\alpha \leq \Bigg(M'_{40} + \sum_{i:\gamma_i=0} M'_{41,i} \\
+ \sum_{i:\gamma_i>0} M'_{41,i} \exp\left\{\left(\sum_{j=1}^{k} A_{N-n-1,j}(\bar{\beta}_i)|Y_{\varepsilon,n+1,j}|\right)\frac{\gamma_i}{\beta_i}\right\}\Bigg)^\alpha \\
\leq \mathsf{E}_{\vec{z},n}(k+1)^{\alpha-1}\Bigg((M'_{40})^\alpha + \sum_{i:\gamma_i=0} (M'_{41,i})^\alpha \\
+ \sum_{i:\gamma_i>0} (M'_{41,i})^\alpha \exp\left\{\left(\sum_{j=1}^{k} A_{N-n-1,j}(\bar{\beta}_i)|Y_{\varepsilon,n+1,j}|\right)\frac{\gamma_i\alpha}{\beta_i}\right\}\Bigg) \\
\leq (k+1)^{\alpha-1}\Bigg((M'_{40})^\alpha + \sum_{i:\gamma_i=0} (M'_{41,i})^\alpha \\
+ \sum_{i:\gamma_i>0} (M'_{41,i})^\alpha \exp\left\{\left(\sum_{j=1}^{k} A_{N-n-1,j}(\bar{\beta}_i)|y_j|\right)\frac{\gamma_i\alpha}{\beta_i}\right\} \\
\times \mathsf{E}_{\vec{z},n} \exp\left\{\left(\sum_{j=1}^{k} A_{N-n-1,j}(\bar{\beta}_i)|Y_{\varepsilon,n+1,j} - y_j|\right)\frac{\gamma_i\alpha}{\beta_i}\right\}\Bigg) \\
\leq (k+1)^{\alpha-1}\Bigg((M'_{40})^\alpha + \sum_{i:\gamma_i=0} (M'_{41,i})^\alpha \\
+ \sum_{i:\gamma_i>0} (M'_{41,i})^\alpha \exp\left\{\left(\sum_{j=1}^{k} A_{N-n-1,j}(\bar{\beta}_i)|y_j|\right)\frac{\gamma_i\alpha}{\beta_i}\right\}
\end{aligned}$$

$$\times\ \mathsf{E}_{\vec{z},n} \exp\left\{\left(\sum_{j=1}^{k} A_{N-n-1,j}(\bar{\beta}_i)\,|Y_{\varepsilon,n+1,j} - Y_{\varepsilon,n,j}|\right\}\right)$$

$$\le (k+1)^{\alpha-1}\left((M'_{40})^{\alpha} + \sum_{i:\gamma_i=0}(M'_{41,i})^{\alpha}\right.$$

$$+ \sum_{i:\gamma_i>0}(M'_{41,i})^{\alpha}\exp\left\{\left(\sum_{j=1}^{k} A_{N-n-1,j}(\bar{\beta}_i)\,|y_j|\right)\frac{\gamma_i\alpha}{\beta_i}\right\}$$

$$\left.\times\ \exp\left\{\sum_{j=1}^{k} A_j(\bar{A}_{N-n-1}(\bar{\beta}_i))\,|y_j|\right\}K'_{45,i}\right) = Q'_{\bar{A}}(\vec{y}) < \infty\,. \tag{6.95}$$

The following proof repeats the proof of Theorem 5.2.2. The only difference is that relation (5.64) should be slightly modified.

As in the proof of Theorem 5.2.2, we choose an arbitrary function $\vec{z}_\cdot = \langle \vec{z}_\varepsilon,\ e \in [0,\varepsilon_0]\rangle$ such that $\vec{z}_\varepsilon \to \vec{z}_0 \in \mathbb{Z}'_{N-1} \cap \mathbb{Z}''_{N-1}$ as $\varepsilon \to 0$, and one should prove that the following relation, which replaces relation (5.64), holds:

$$\overline{\lim_{\varepsilon\to 0}}\ \mathsf{E}|\phi_{\varepsilon,N}(\vec{Y}_{\varepsilon,N}(\vec{z}_\varepsilon), X_{\varepsilon,N}(\vec{z}_\varepsilon))|^{\alpha} \le Q'_{\bar{A}}(\vec{y}_0) < \infty\,. \tag{6.96}$$

In this case, we should use the condition $\mathbf{H}_1$, due to which the point $\vec{z}_0 = (\vec{y}_0, x_0) \in \mathbb{Z}_{\varepsilon,N-1}$, for every $\varepsilon \in [0,\varepsilon_1(\vec{z}_0)]$. The parameter $\varepsilon_1(\vec{z}_0) \in (0,\varepsilon_0]$ has been defined in relation (5.6) given in Remark 5.1.3.

Since $\vec{z}_\varepsilon \to \vec{z}_0$ as $\varepsilon \to 0$, this implies that there exists $\varepsilon'_1(\vec{z}(\cdot)) \in (0,\varepsilon_1(\vec{z}_0)]$ such that $\vec{z}_\varepsilon = (\vec{y}_\varepsilon, x_\varepsilon) \in \mathbb{Z}_{\varepsilon,N-1}$ for every $\varepsilon \in [0,\varepsilon'_1(\vec{z}(\cdot))]$.

Therefore, upper bounds for $\mathsf{E}_{z,N-1}|\phi_{\varepsilon,N}(\vec{Y}_{\varepsilon,N}, X_{\varepsilon,N})|^{\alpha}$ given in relation (6.95) hold for every $\varepsilon \in [0,\min(\varepsilon'_{12},\varepsilon'_1(\vec{z}(\cdot)))]$.

Due to the above remarks and continuity of the function $Q'_{\bar{A}}(\vec{y})$, relation (6.95) implies that relation (6.96), replacing relation (5.64), holds.

The remaining part of the proof repeats the corresponding part in the proof of Theorem 5.2.2. □

6.4.2 Convergence of optimal expected rewards for multivariate modulated Markov log-price processes

Let us now consider the optimal expected rewards $\Phi_\varepsilon = \Phi(\mathcal{M}^{(\varepsilon)}_{\max,N})$ introduced in Subsection 6.3.2 and give conditions for their convergence.

Theorem 6.4.3. *Let the conditions* $\mathbf{B}_5[\bar{\gamma}]$, $\mathbf{C}_{10}[\bar{\beta}]$, *and* $\mathbf{D}_9[\bar{\beta}]$ *hold with the vector parameters* $\bar{\gamma} = (\gamma_1,\dots,\gamma_k)$ *and* $\bar{\beta} = (\beta_1,\dots,\beta_k)$ *such that, for every* $i = 1,\dots,k$, *either* $\beta_i > \gamma_i > 0$ *or* $\beta_i = \gamma_i = 0$, *and also the conditions* $\mathbf{I}_1$, $\mathbf{J}_1$, *and* $\mathbf{K}_1$ *hold. Then the following relation takes place:*

$$\Phi_\varepsilon \to \Phi_0 \quad \text{as} \quad \varepsilon \to 0\,. \tag{6.97}$$

Proof. It is analogous to the proof of Theorem 5.2.3.

Due to the above conditions imposed on the parameters γ_i, β_i, $i = 1, \dots, k$, one can always find $\alpha > 1$ such that $\gamma_i \alpha \le \beta_i$, $i = 1, \dots, k$.

According to Theorem 6.3.1, the conditions $\mathbf{B}_5[\bar{\gamma}]$ and $\mathbf{C}_{10}[\bar{\beta}]$ imply that there exist $\varepsilon_{12} > 0$ and the constants $M_{40}, M_{41,i}$, $i = 1, \dots, k$ such that the following upper bounds take place for the reward functions $\phi_{\varepsilon,0}(\vec{y}, x)$ for $\vec{z} = (\vec{y}, x) \in \mathbb{Z}$ and $\varepsilon \le \varepsilon_{12}$,

$$|\phi_{\varepsilon,0}(\vec{y}, x)| \le M_{40} + \sum_{i:\gamma_i=0} M_{41,i} + \sum_{i:\gamma_i>0} M_{41,i} \exp\left\{\left(\sum_{j=1}^{k} A_{N,j}(\beta_i)|y_j|\right)\frac{\gamma_i}{\beta_i}\right\}. \tag{6.98}$$

Also, according to relation (6.75), the condition $\mathbf{D}_9[\bar{\beta}]$ implies that there exist $0 < \varepsilon_{13} \le \varepsilon_0$ and the constants $K_{47,i}$, $i = 1, \dots, k$ such that for $0 \le \varepsilon \le \varepsilon_{13}$ and $i = 1, \dots, k$,

$$\mathsf{E} \exp\left\{\sum_{j=1}^{k} A_{N,j}(\beta_i)\, |Y_{\varepsilon,0,j}|\right\} < K_{47,i}\,. \tag{6.99}$$

Using inequalities (6.98) and (6.99), we get for any $\varepsilon \le \varepsilon_{14} = \min(\varepsilon_{12}, \varepsilon_{13})$,

$$\begin{aligned}
&\mathsf{E}|\phi_{\varepsilon,0}(\vec{Y}_{\varepsilon,0}, X_{\varepsilon,0})|^{\alpha} \\
&\quad \le \left(M_{40} + \sum_{i:\gamma_i=0} M_{41,i} + \sum_{i:\gamma_i>0} M_{41,i} \exp\left\{\left(\sum_{j=1}^{k} A_{N,j}(\bar{\beta}_i)|Y_{\varepsilon,0,j}|\right)\frac{\gamma_i}{\beta_i}\right\}\right)^{\alpha} \\
&\quad \le \mathsf{E}(k+1)^{\alpha-1}\left(M_{40}^{\alpha} + \sum_{i:\gamma_i=0} M_{41,i}^{\alpha} + \sum_{i:\gamma_i>0} M_{41,i}^{\alpha} \exp\left\{\left(\sum_{j=1}^{k} A_{N,j}(\bar{\beta}_i)|Y_{\varepsilon,0,j}|\right)\frac{\gamma_i\alpha}{\beta_i}\right\}\right) \\
&\quad \le (k+1)^{\alpha-1}\left(M_{40}^{\alpha} + \sum_{i:\gamma_i=0} M_{41,i}^{\alpha} + \sum_{i:\gamma_i>0} M_{41,i}^{\alpha}\mathsf{E} \exp\left\{\sum_{j=1}^{k} A_{N,j}(\bar{\beta}_i)|Y_{\varepsilon,0,j}|\right\}\right) \\
&\quad \le (k+1)^{\alpha-1}\left(M_{40}^{\alpha} + \sum_{i:\gamma_i=0} M_{41,i}^{\alpha} + \sum_{i:\gamma_i>0} M_{41,i}^{\alpha} K_{47,i}\right) = Q_{\bar{A}} < \infty\,.
\end{aligned} \tag{6.100}$$

The following proof repeats the proof of Theorem 5.2.3. The only difference is that relation (6.100) should be used instead of relation (5.68) used in Theorem 5.2.3. □

Remark 6.4.3. According to Lemmas 6.3.5 and 6.3.8, the conditions $\mathbf{C}_{10}[\bar{\beta}]$ and $\mathbf{D}_9[\bar{\beta}]$ can be replaced in Theorem 6.4.3, respectively, by the conditions $\mathbf{E}_{14}[\bar{\beta}]$ and $\mathbf{F}_7[\bar{\beta}]$.

The following theorem gives alternative conditions of the convergence for optimal expected rewards.

Theorem 6.4.4. *Let the conditions $\mathbf{B}_5[\bar{\gamma}]$, $\mathbf{C}'_{10}[\bar{\beta}]$, and $\mathbf{D}'_9[\bar{\beta}]$ hold with the vector parameters $\bar{\gamma} = (\gamma_1, \dots, \gamma_k)$, $\bar{\beta} = (\beta_1, \dots, \beta_k)$ such that, for every $i = 1, \dots, k$, either $\beta_i > \gamma_i > 0$ or $\beta_i = \gamma_i = 0$, and also conditions $\mathbf{H}_1$, $\mathbf{I}_1$, $\mathbf{J}_1$, and $\mathbf{K}_1$ hold. Then the following relation takes place:*

$$\Phi_\varepsilon \to \Phi_0 \quad as \quad \varepsilon \to 0\,. \tag{6.101}$$

Proof. It is analogous to the proof of Theorem 5.2.4.

Due to the above conditions imposed on the parameters γ_i, β_i, one can always find $\alpha > 1$ such that $\gamma_i\alpha \leq \beta_i$, $i = 1, \dots, k$.

According to Theorem 6.3.2, the conditions $\mathbf{B}_5[\bar{\gamma}]$ and $\mathbf{C}'_{10}[\bar{\beta}]$ imply that there exist $\varepsilon'_{12} > 0$ and the constants $M'_{40}, M'_{41,i}$, $i = 1, \dots, k$ such that the following upper bounds take place for the reward functions $\phi_{\varepsilon,0}(\vec{y}, x)$ for $\vec{z} = (\vec{y}, x) \in \mathbb{Z}_{\varepsilon,0}$, $\varepsilon \leq \varepsilon'_{14} = \min(\varepsilon'_{12}, \varepsilon'_{13})$:

$$|\phi_{\varepsilon,0}(\vec{y}, x)| \leq M'_{40} + \sum_{i:\gamma_i=0} M'_{41,i} + \sum_{i:\gamma_i>0} M'_{41,i} \exp\left\{\left(\sum_{j=1}^{k} A_{N,j}(\bar{\beta}_i)|y_j|\right)\frac{\gamma_i}{\beta_i}\right\}. \tag{6.102}$$

Also, according to relation (6.80), the condition $\mathbf{D}'_9[\bar{\beta}]$ implies that there exist $0 < \varepsilon'_{13} \leq \varepsilon_0$ and the constants $K'_{47,i}$, $i = 1, \dots, k$ such that for $0 \leq \varepsilon \leq \varepsilon'_{13}$ and $i = 1, \dots, k$,

$$\mathsf{E}\exp\left\{\sum_{j=1}^{k} A_{N,j}(\bar{\beta}_i)|Y_{\varepsilon,0,j}|\right\} < K'_{47,i}. \tag{6.103}$$

Recall that the condition $\mathbf{D}'_9[\bar{\beta}]$ (b) implies that $\mathsf{P}_{z,n}\{\vec{Z}_{\varepsilon,0} \in \mathbb{Z}_{\varepsilon,0}\} = 1$ for $\vec{z} \in \mathbb{Z}_{\varepsilon,0}$. Using this relation, inequalities (6.102) and the condition $\mathbf{D}'_9[\bar{\beta}]$, we get for any $\varepsilon \leq \varepsilon'_{14} = \min(\varepsilon'_{12}, \varepsilon'_{13})$,

$$\begin{aligned}
&\mathsf{E}|\phi_{\varepsilon,0}(\vec{Y}_{\varepsilon,0}, X_{\varepsilon,0})|^{\alpha} \\
&\leq \left(M'_{40} + \sum_{i:\gamma_i=0} M'_{41,i} + \sum_{i:\gamma_i>0} M'_{41,i} \exp\left\{\left(\sum_{j=1}^{k} A_{N,j}(\bar{\beta}_i)|Y_{\varepsilon,0,j}|\right)\frac{\gamma_i}{\beta_i}\right\}\right)^{\alpha} \\
&\leq \mathsf{E}(k+1)^{\alpha-1}\left((M'_{40})^{\alpha} + \sum_{i:\gamma_i=0} (M'_{41,i})^{\alpha}\right. \\
&\quad \left. + \sum_{i:\gamma_i>0} (M'_{41,i})^{\alpha} \exp\left\{\left(\sum_{j=1}^{k} A_{N,j}(\bar{\beta}_i)|Y_{\varepsilon,0,j}|\right)\frac{\gamma_i\alpha}{\beta_i}\right\}\right) \\
&\leq (k+1)^{\alpha-1}\left((M'_{40})^{\alpha} + \sum_{i:\gamma_i=0} (M'_{41,i})^{\alpha} + \sum_{i:\gamma_i>0} (M'_{41,i})^{\alpha}\mathsf{E}\exp\left\{\sum_{j=1}^{k} A_{N,j}(\bar{\beta}_i)|Y_{\varepsilon,0,j}|\right\}\right) \\
&\leq (k+1)^{\alpha-1}\left((M'_{40})^{\alpha} + \sum_{i:\gamma_i=0} (M'_{41,i})^{\alpha} + \sum_{i:\gamma_i>0} (M'_{41,i})^{\alpha}K'_{47,i}\right) = Q'_{\bar{A}} < \infty.
\end{aligned} \tag{6.104}$$

The following part of the proof repeats the corresponding part of the proof of Theorem 5.2.4, with the only difference that relation (6.104) should be used instead of relation (5.82). Also, relation analogous to (5.75) does require reference to relation (6.93) given in Theorem 6.4.2 instead of relation (6.88) given in Theorem 6.4.1. □

6.5 Conditions of convergence for Markov price processes

In this section, we reformulate basic conditions and theorems given in Chapter 5 and in this chapter in terms of price processes.

6.5.1 Condition of asymptotic polynomial rate of growth for pay-off functions

In what follows the simple fact is used. If $e^y = s$ then $e^{|y|} = s \vee s^{-1}$, for any $y \in \mathbb{R}_1$.

The condition $\mathbf{B}_5[\bar{\gamma}]$, expressed in terms of argument $(\vec{s}, x)$, takes the following equivalent form:

$\dot{\mathbf{B}}_5[\bar{\gamma}]$: $\overline{\lim}_{\varepsilon\to 0} \max_{0\leq n\leq N} \sup_{\vec{v}=(\vec{s},x)\in\mathbb{V}} \frac{|g_\varepsilon(n,\vec{s},x)|}{1+\sum_{i=1}^k L_{10,i}(s_i\vee s_i^{-1})^{\gamma_i}} < L_9$, for some $0 < L_9 < \infty$ and $0 \leq L_{10,1}, \dots, L_{10,k} < \infty$.

6.5.2 Conditions of asymptotically uniform moment compactness for price processes

Recall that the price process $\vec{V}_{\varepsilon,n} = (\vec{S}_{\varepsilon,n}, X_{\varepsilon,n}) = ((S_{\varepsilon,n,1}, \dots, S_{\varepsilon,n,k}), X_{\varepsilon,n})$, is connected with the corresponding log-price process $\vec{Z}_{\varepsilon,n} = (\vec{Y}_{\varepsilon,n}, X_{\varepsilon,n}) = ((Y_{\varepsilon,n,1}, \dots, Y_{\varepsilon,n,k}), X_{\varepsilon,n})$ by the formulas $\vec{S}_{\varepsilon,n,i} = e^{\vec{Y}_{\varepsilon,n,i}}$, $i = 1, \dots, k$, $n = 0, 1, \dots$.

Recall that we use the notations $\dot{\mathsf{P}}_{\vec{v},n}$ and $\dot{\mathsf{E}}_{\vec{v},n}$ for conditional probabilities and expectations under the condition $\vec{V}_{\varepsilon,n} = \vec{v}$.

The modulus $\Delta_{\beta_i}(Y_{\varepsilon,\cdot,i}, N)$, expressed in terms of price processes, takes the following form:

$$\dot{\Delta}_\beta(S_{\varepsilon,\cdot,i}, N) = \max_{0\leq n\leq N-1} \sup_{\vec{v}\in\mathbb{V}} \dot{\mathsf{E}}_{\vec{v},n} \left(\frac{S_{\varepsilon,n+1,i}}{S_{\varepsilon,n,i}} \vee \frac{S_{\varepsilon,n,i}}{S_{\varepsilon,n+1,i}}\right)^\beta . \tag{6.105}$$

The condition $\mathbf{C}_6[\bar{\beta}]$, expressed in terms of price process, takes the following equivalent form:

$\dot{\mathbf{C}}_6[\bar{\beta}]$: $\overline{\lim}_{\varepsilon\to 0} \dot{\Delta}_{\beta_i}(S_{\varepsilon,\cdot,i}, c, T) = K_{30,i} < \infty$, $i = 1, \dots, k$, for some $1 < K_{30,i} < \infty$, $i = 1, \dots, k$.

The modulus $\Xi^\pm_\beta(Y_{\varepsilon,\cdot,i}, N)$, expressed in terms of price processes, takes the following equivalent form:

$$\dot{\Xi}^\pm_\beta(S_{\varepsilon,\cdot,i}, N) = \max_{0\leq n\leq N-1} \sup_{\vec{v}\in\mathbb{V}} \dot{\mathsf{E}}_{\vec{v},n}(\frac{S_{\varepsilon,n+1,i}}{S_{\varepsilon,n,i}})^{\pm\beta} . \tag{6.106}$$

The condition $\mathbf{E}_{10}[\bar{\beta}]$, expressed in terms of price processes, takes the following equivalent form:

$\dot{\mathbf{E}}_{10}[\bar{\beta}]$: $\overline{\lim}_{\varepsilon\to 0} \dot{\Xi}^{\pm}_{\beta_i}(S_{\varepsilon,\cdot,i}, N) < K_{31,i}$, $i = 1,\dots,k$, for some $1 < K_{31,i} < \infty$, $i = 1,\dots,k$.

The condition $\dot{\mathbf{E}}_{10}[\bar{\beta}]$ implies that the condition $\dot{\mathbf{C}}_6[\bar{\beta}]$ holds.

The condition $\mathbf{D}_7[\bar{\beta}]$, expressed in terms of price processes, takes the following equivalent form:

$\dot{\mathbf{D}}_7[\bar{\beta}]$: $\overline{\lim}_{\varepsilon\to 0} \mathsf{E}(S_{\varepsilon,0,i} \vee S^{-1}_{\varepsilon,0,i})^{\beta_i} < K_{32,i}$, $i = 1,\dots,k$, for some $1 < K_{32,i} < \infty$, $i = 1,\dots,k$.

The condition $\mathbf{F}_4[\bar{\beta}]$, expressed in terms of price processes, takes the following equivalent form:

$\dot{\mathbf{F}}_4[\bar{\beta}]$: $\overline{\lim}_{\varepsilon\to 0} \mathsf{E}(S_{\varepsilon,0,i})^{\pm\beta_i} < K_{33,i}$, $i = 1,\dots,k$, for some $1 < K_{33,i} < \infty$, $i = 1,\dots,k$.

The condition $\dot{\mathbf{F}}_4[\bar{\beta}]$ implies that the condition $\dot{\mathbf{D}}_7[\bar{\beta}]$ holds.

6.5.3 Conditions of convergence for pay-off functions

The condition $\mathbf{I}_1$, expressed in terms of argument $(\vec{s}, x)$, takes the following equivalent form:

$\dot{\mathbf{I}}_1$: There exists measurable sets $\mathbb{V}'_n \subseteq \mathbb{V}$, $n = 0,\dots,N$, such that the pay-off function $g_\varepsilon(n, \vec{s}_\varepsilon, x_\varepsilon) \to g_0(n, \vec{s}_0, x_0)$ as $\varepsilon \to 0$ for any $\vec{v}_\varepsilon = (\vec{s}_\varepsilon, x_\varepsilon) \to \vec{v}_0 = (\vec{s}_0, x_0) \in \mathbb{V}'_n$ as $\varepsilon \to 0$, and $n = 0,\dots,N$.

6.5.4 Conditions of convergence for price processes

The condition $\mathbf{J}_1$, expressed in terms of price processes, takes the following equivalent form:

$\dot{\mathbf{J}}_1$: There exist measurable sets $\mathbb{V}''_n \subseteq \mathbb{V}$, $n = 0,\dots,N$ such that:
(a) $\dot{P}_{\varepsilon,n}(\vec{v}_\varepsilon, \cdot) \Rightarrow \dot{P}_{0,n}(\vec{v}_0, \cdot)$ as $\varepsilon \to 0$, for any $\vec{v}_\varepsilon = (\vec{s}_\varepsilon, x_\varepsilon) \to \vec{v}_0 = (\vec{s}_0, x_0) \in \mathbb{V}''_{n-1}$ as $\varepsilon \to 0$, and $n = 1,\dots,N$;
(b) $\dot{P}_{0,n}(\vec{v}_0, \mathbb{V}'_n \cap \mathbb{V}''_n) = 1$, for every $\vec{v}_0 \in \mathbb{V}'_{n-1} \cap \mathbb{V}''_{n-1}$ and $n = 1,\dots,N$, where $\mathbb{V}'_n$, $n = 1,\dots,N$ are sets introduced in the condition $\dot{\mathbf{I}}_1$.

The condition $\dot{\mathbf{J}}_1$ (b) holds, if (a) the sets $\overline{\mathbb{V}}'_n, \overline{\mathbb{V}}''_n$, $n = 1,\dots,N$ are empty sets or (b) are at most finite or countable sets and measures $\dot{P}_{0,n}(\vec{v}_0, A)$, $\vec{v}_0 \in \mathbb{V}'_{n-1} \cap \mathbb{V}''_{n-1}$ have no atoms at points from the sets $\overline{\mathbb{V}}'_n, \overline{\mathbb{V}}''_n$, for every $n = 1,\dots,N$, or (c) measures $\dot{P}_{0,n}(\vec{v}_0, A)$, $\vec{v}_0 \in \mathbb{V}'_{n-1} \cap \mathbb{V}''_{n-1}$, $n = 1,\dots,N$ are absolutely continuous with respect to some σ-finite measure $\dot{P}(A)$ on $\mathcal{B}_{\mathbb{V}}$ and $\dot{P}(\overline{\mathbb{V}}'_n), \dot{P}(\overline{\mathbb{V}}''_n) = 0$, $n = 1,\dots,N$.

The condition $\mathbf{K}_1$, expressed in terms of price processes, takes the following equivalent form:

$\dot{\mathbf{K}}_1$: (a) $\dot{P}_{\varepsilon,0}(\cdot) \Rightarrow \dot{P}_{0,0}(\cdot)$ as $\varepsilon \to 0$;

(b) $\dot{P}_{0,0}(\mathbb{V}'_0 \cap \mathbb{V}''_0) = 1$, where $\mathbb{V}'_0$ and $\mathbb{V}''_0$ are the sets introduced in the conditions $\dot{\mathbf{I}}_1$ and $\dot{\mathbf{J}}_1$.

6.5.5 Examples of convergence results

As examples, we reformulate in terms of price processes just two basic Theorems 5.2.1 and 5.2.3 in terms of rewards for price processes $\vec{V}_{\varepsilon,n}$.

Theorem 6.5.1. *Let the conditions $\dot{\mathbf{B}}_5[\bar{\gamma}]$ and $\dot{\mathbf{C}}_6[\bar{\beta}]$ hold with vector parameters $\bar{\gamma} = (\gamma_1, \dots, \gamma_k), \bar{\beta} = (\beta_1, \dots, \beta_k)$ such that, for every $i = 1, \dots, k$, either $\beta_i > \gamma_i > 0$ or $\beta_i = \gamma_i = 0$, and also the conditions $\dot{\mathbf{I}}_1$ and $\dot{\mathbf{J}}_1$ hold. Then, for every $n = 0, 1, \dots, N$, the following relation takes place for any $\vec{v}_\varepsilon = (\vec{s}_\varepsilon, x_\varepsilon) \to \vec{v}_0 = (\vec{s}_0, x_0) \in \mathbb{V}'_n \cap \mathbb{V}''_n$:*

$$\dot{\phi}_{\varepsilon,n}(\vec{s}_\varepsilon, x_\varepsilon) \to \dot{\phi}_{0,n}(\vec{s}_0, x_0) \quad as \quad \varepsilon \to 0\,. \tag{6.107}$$

Theorem 6.5.2. *Let conditions $\dot{\mathbf{B}}_5[\bar{\gamma}]$, $\dot{\mathbf{C}}_6[\bar{\beta}]$, and $\dot{\mathbf{D}}_7[\bar{\beta}]$ hold with vector parameters $\bar{\gamma} = (\gamma_1, \dots, \gamma_k), \bar{\beta} = (\beta_1, \dots, \beta_k)$ such that, for every $i = 1, \dots, k$, either $\beta_i > \gamma_i > 0$ or $\beta_i = \gamma_i = 0$, and also conditions $\dot{\mathbf{I}}_1$, $\dot{\mathbf{J}}_1$, and $\dot{\mathbf{K}}_1$ hold. Then the following relation takes place,*

$$\dot{\Phi}_\varepsilon \to \dot{\Phi}_0 \quad as \quad \varepsilon \to 0\,. \tag{6.108}$$

The other conditions and theorems presented in Chapter 5 and in this chapter can be reformulated in terms of price processes in analogous way.

7 Space-skeleton reward approximations

In this chapter, we present results about space-skeleton approximations for rewards of American type options for discrete time multivariate modulated Markov log-price processes. This model is based on approximations of rewards for a multivariate modulated Markov log-price process by the corresponding rewards for American type options for multivariate modulated log-price processes represented by atomic Markov chains, whose transition probabilities and initial distributions are concentrated at finite sets of skeleton points. The rewards for approximating atomic Markov chains can be effectively computed using backward recurrence relations presented in Chapter 3. The space-skeleton approximations do also require special fitting of transition probabilities and initial distributions for approximating processes to the corresponding transition probabilities and initial distributions for approximated multivariate modulated Markov chains. Convergence of the approximating rewards can be proven using the general convergence results presented in Chapters 5 and 6.

In Section 7.1, we describe a general atomic approximation model for discrete time multivariate modulated Markov log-price processes, including the corresponding backward recurrence systems of linear equations for computing of reward functions for approximating atomic Markov log-price processes.

In Section 7.2, we give results about convergence of space-skeleton approximations for rewards of American type options for univariate modulated Markov log-price processes with bounded characteristics.

In Section 7.3, we give results about convergence of space-skeleton approximations for rewards of American type options for multivariate modulated Markov log-price processes with bounded characteristics.

In Section 7.4, we give results about convergence of space-skeleton approximations for rewards of American type options for log-price processes represented by multivariate modulated random walks.

In Section 7.5, we give results about convergence of space-skeleton approximations for rewards of American type options for multivariate modulated Markov log-price processes with unbounded characteristics.

The space-skeleton approximation results presented in this chapter are based on the combination of backward recurrence algorithms for computing of rewards for atomic Markov chains presented in Chapter 3 and general convergence results for rewards of American type options for multivariate modulated Markov log-price processes given in Chapters 5 and 6. This combination yields an effective stochastic approximation method for approximative computing of rewards for American type options for multivariate modulated Markov log-price processes.

The main convergence results given in the chapter are Theorems 7.4.1–7.4.6, and 7.5.1–7.5.4. These results are new.

7.1 Atomic approximation models

In this section, we describe general atomic approximation models for discrete time multivariate modulated Markov log-price processes. We also present the corresponding backward recurrence systems of linear equations for computing of log-reward functions for approximating atomic Markov log-price processes. Also, the two most important atomic approximation models, a space-skeleton approximation Markov chain model with fixed skeleton structure and a space-skeleton approximation random walk model with additive skeleton structure are described.

7.1.1 An atomic Markov chain approximation model

Let us consider a model, where a log-price process $\vec{Z}_{\varepsilon,n} = (\vec{Y}_{\varepsilon,n}, X_{\varepsilon,n})$, $n = 0, 1, \dots,$ is a discrete time multivariate modulated Markov process with a phase space $\mathbb{Z} = \mathbb{R}_k \times \mathbb{X}$, an initial distribution $P_{\varepsilon,0}(A) = \mathsf{P}\{\vec{Z}_{\varepsilon,0} \in A\}$, and the transition probabilities $P_{\varepsilon,n}(\vec{z}, A) = \mathsf{P}\{\vec{Z}_{\varepsilon,n} \in A/\vec{Z}_{\varepsilon,n-1} = \vec{z}\}$, which depend on a perturbation parameter $\varepsilon \in [0, \varepsilon_0]$.

We additionally assume that the limiting log-price process $\vec{Z}_{0,n}$ (for $\varepsilon = 0$) can be a nonatomic Markov chain, while the prelimiting log-price process $\vec{Z}_{\varepsilon,n}$ (for every $\varepsilon \in (0, \varepsilon_0]$) is an atomic Markov chain, i.e. its transition probabilities $P_{\varepsilon,n}(\vec{z}, A)$ and the initial distributions $P_{\varepsilon,0}(A)$ are discrete distributions concentrated at finite sets of points.

Thus, it is assumed that, for every $\varepsilon \in (0, \varepsilon_0]$, the one-step transition probabilities $P_{\varepsilon,n}(\vec{z}, A)$ are concentrated at the finite sets $\mathbb{F}_{\varepsilon,\vec{z},n} = \{f_{\varepsilon,l}(\vec{z}, n),\ l = m^-_\varepsilon(\vec{z}, n), \dots, m^+_\varepsilon(\vec{z}, n)\}$ and are connected with the corresponding one-point transition probabilities $p_{\varepsilon,l}(\vec{z}, n) = \mathsf{P}\{\vec{Z}_{\varepsilon,n} = f_{\varepsilon,l}(\vec{z}, n)/\vec{Z}_{\varepsilon,n-1} = \vec{z}\}$, $l = m^-_\varepsilon(\vec{z}, n), \dots, m^+_\varepsilon(\vec{z}, n)$, by the following relation, for every $\vec{z} \in \mathbb{Z}$ and $n = 1, 2, \dots$:

$$P_{\varepsilon,n}(\vec{z}, A) = \sum_{f_{\varepsilon,l}(\vec{z},n) \in A} p_{\varepsilon,l}(\vec{z}, n), \quad A \in \mathcal{B}_{\mathbb{Z}}. \tag{7.1}$$

Additionally, it is assumed that, for every $\varepsilon \in (0, \varepsilon_0]$, the initial distribution $P_{\varepsilon,0}(A)$ is also concentrated at a finite set $\mathbb{F}_{\varepsilon,0} = \{\vec{z}_{\varepsilon,0,l},\ l = m^-_{\varepsilon,0}, \dots, m^+_{\varepsilon,0}\}$ and is connected with the corresponding probabilities $p_{\varepsilon,l} = \mathsf{P}\{\vec{Z}_{\varepsilon,0} = \vec{z}_{\varepsilon,0,l}\}$, $l = m^-_{\varepsilon,0}, \dots, m^+_{\varepsilon,0}$, by the following relation:

$$P_{\varepsilon,0}(A) = \sum_{\vec{z}_{\varepsilon,0,l} \in A} p_{\varepsilon,l}, \quad A \in \mathcal{B}_{\mathbb{Z}}. \tag{7.2}$$

The above assumption (7.1) implies that the r-step transition probabilities $P_\varepsilon(n, \vec{z}, n + r, A) = \mathsf{P}\{\vec{Z}_{\varepsilon,n+r} \in A/\vec{Z}_{\varepsilon,n} = \vec{z}\}$ are also concentrated on the finite sets $\mathbb{F}_{\varepsilon,\vec{z},n,n+r} = \{f_{\varepsilon,l}(\vec{z}, n, n + r),\ l = m^-_\varepsilon(\vec{z}, n, n + r), \dots, m^+_\varepsilon(\vec{z}, n, n + r)\}$, with the corresponding one-point transition probabilities $p_{\varepsilon,l}(\vec{z}, n, n+r) = \mathsf{P}\{\vec{Z}_{\varepsilon,n+r} = f_{\varepsilon,l}(\vec{z}, n, n+r)/\vec{Z}_{\varepsilon,n} = \vec{z}\}$, $l = m^-_\varepsilon(\vec{z}, n, n+r), \dots, m^+_\varepsilon(\vec{z}, n, n+r)$, i.e. for every $\vec{z} \in \mathbb{Z}$ and $n = 0, 1, \dots,$ and $r = 1, 2, \dots,$

$$P_\varepsilon(n, \vec{z}, n+r, A) = \sum_{f_{\varepsilon,l}(\vec{z},n,n+r)\in A} p_{\varepsilon,l}(\vec{z}, n, n+r)\,, \quad A \in \mathcal{B}_{\mathbb{Z}}\,. \tag{7.3}$$

Also, the following recurrent relations link transition functions $f_{\varepsilon,l}(\vec{z}, n, n+r)$ and the corresponding transition probabilities $p_{\varepsilon,l}(\vec{z}, n, n+r)$, for $r = 1, 2, \dots,$

$$\begin{aligned} & f_{\varepsilon,l}(\vec{z}, n, n+r) \\ & = \begin{cases} f_{\varepsilon,l}(\vec{z}, n+1) & \text{if } r = 1, m^-_\varepsilon(\vec{z}, n+1) \\ & \quad \le l \le m^+_\varepsilon(\vec{z}, n+1)\,, \\ f_{\varepsilon,j}(f_{\varepsilon,i}(\vec{z}, n, n+r-1), n+r) & \text{if } r > 1, (i,j) \in \mathbb{F}_{\varepsilon,\vec{z},n,n+r,l}, \\ & \quad m^-_\varepsilon(\vec{z}, n, n+r) \\ & \quad \le l \le m^+_\varepsilon(\vec{z}, n, n+r)\,, \end{cases} \end{aligned} \tag{7.4}$$

and

$$\begin{aligned} & p_{\varepsilon,l}(\vec{z}, n, n+r) \\ & = \begin{cases} p_{\varepsilon,l}(\vec{z}, n+1) & \text{if } r = 1, m^-_\varepsilon(\vec{z}, n+1) \\ & \quad \le l \le m^+_\varepsilon(\vec{z}, n+1)\,, \\ \sum_{(i,j)\in\mathbb{F}_{\varepsilon,\vec{z},n,n+r,l}} p_{\varepsilon,i}(\vec{z}, n, n+r-1) & \\ \times p_{\varepsilon,j}(f_{\varepsilon,i}(\vec{z}, n, n+r-1), n+r) & \text{if } r > 1, m^-_\varepsilon(\vec{z}, n, n+r) \\ & \quad \le l \le m^+_\varepsilon(\vec{z}, n, n+r)\,, \end{cases} \end{aligned} \tag{7.5}$$

where

$$\begin{aligned} \mathbb{F}_{\varepsilon,\vec{z},n,n+r,l} = \Big\{ & (i,j) : f_{\varepsilon,j}(f_{\varepsilon,i}(\vec{z}, n, n+r-1), n+r) \\ & = f_{\varepsilon,l}(\vec{z}, n, n+r),\ m^-_\varepsilon(\vec{z}, n, n+r-1) \\ & \le i \le m^+_\varepsilon(\vec{z}, n, n+r-1), m^-_\varepsilon(f_{\varepsilon,i}(\vec{z}, n, n+r-1), n+r) \\ & \le j \le m^+_\varepsilon(f_{\varepsilon,i}(\vec{z}, n, n+r-1), n+r) \Big\}\,. \end{aligned} \tag{7.6}$$

Any node $(\vec{z}, n) \in \mathbb{Z} \times \mathbb{N}$ generates the tree of nodes $T_{\varepsilon,(\vec{z},n)}$. This tree includes one initial node $(\vec{z}, n)$, additional nodes $(f_{\varepsilon,l}(\vec{z}, n, n+1), n+1)$, $l = m^-_\varepsilon(\vec{z}, n, n+1), \dots, m^+_\varepsilon(\vec{z}, n, n+1)$, appearing in the tree after the first transition of the Markov chain, additional nodes $(f_{\varepsilon,l}(\vec{z}, n, n+2), n+2)$, $l = m^-_\varepsilon(\vec{z}, n, n+2), \dots, m^+_\varepsilon(\vec{z}, n, n+2)$, appearing in the tree after the second transition, etc.

According to the above remarks, the tree $T_{\varepsilon,(\vec{z},n)}$ is the following set of nodes:

$$T_{\varepsilon,(\vec{z},n)} = \langle (f_{\varepsilon,l}(\vec{z},n,n+r), n+r), \quad l = m_\varepsilon^-(\vec{z},n,n+r), \dots, m_\varepsilon^+(\vec{z},n,n+r), \quad r = 0,1,\dots \rangle , \tag{7.7}$$

where $f_{\varepsilon,0}(\vec{z},n,n) = \vec{z}$, $m_\varepsilon^\pm(\vec{z},n,n) = 0$.

The total number of nodes in the tree after r transitions is given by the following formula:

$$L_{\varepsilon,r}(\vec{z},n) = 1 + (m_\varepsilon^+(\vec{z},n,n+1) - m_\varepsilon^-(\vec{z},n,n+1) + 1) + \cdots + (m_\varepsilon^+(\vec{z},n,n+r) - m_\varepsilon^-(\vec{z},n,n+r) + 1) . \tag{7.8}$$

We are specially interested in trees, where the number of nodes after r steps $L_{\varepsilon,r}(\vec{z},n) \leq Lr^h$, $r = 1,2,\dots$, where $L, h > 0$, i.e. models where the tree has not more than polynomial rate of growth for $L_{\varepsilon,r}(\vec{z},n)$ as function of r.

The standard case is, where the parameters $m_\varepsilon^\pm(\vec{z},n,n+r) = \pm m_\varepsilon, 0 \leq n < n+r < \infty$, and m_ε are nonnegative integer numbers. In this case, $L_{\varepsilon,r}(\vec{z},n) = 1 + (2m_\varepsilon + 1) + \cdots + (2m_\varepsilon + 1) = r(2m_\varepsilon + 1) + 1$ is a linear function of r.

Another standard case is, where the parameters $m_\varepsilon^\pm(\vec{z},n,n+r) = \pm r m_\varepsilon, 0 \leq n < n+r < \infty$, and m_ε are nonnegative integer numbers. In this case, $L_{\varepsilon,r}(\vec{z},n) = 1 + (2m_\varepsilon + 1) + \cdots + (2rm_\varepsilon + 1) = r + 1 + r(r+1)m_\varepsilon = r^2 m_\varepsilon + r(m_\varepsilon + 1) + 1$ is a quadratic function of r.

In order to give a more convenient representation for the backward reward algorithm for atomic modulated Markov log-price processes, we should introduce additional notations, explicitly separating log-price and index components for transition points $f_{\varepsilon,l}(\vec{z},n,n+r) \in \mathbb{Z}$ in the following form:

$$f_{\varepsilon,l}(\vec{z},n,n+r) = (f'_{\varepsilon,l}(\vec{z},n,n+r), f''_{\varepsilon,l}(\vec{z},n,n+r)) , \tag{7.9}$$

where $f'_{\varepsilon,l}(\vec{z},n,n+r) \in \mathbb{R}_k$ while $f''_{\varepsilon,l}(\vec{z},n,n+r) \in \mathbb{X}$.

The condition $\mathbf{A_2}$ holds for every $\varepsilon \in (0, \varepsilon_0]$ since, in this case, the following inequality takes place for any $\vec{z} = (\vec{y}, x) \in \mathbb{Z}$, $n = 0,1,\dots$:

$$\begin{aligned} \mathsf{E}_{\vec{z},n} \max_{0 \leq r \leq N-n} & |g_\varepsilon(n+r, e^{\vec{Y}_{\varepsilon,n+r}}, X_{\varepsilon,n+r})| \\ & \leq \max_{0 \leq r \leq N-n} \max_{m_\varepsilon^-(\vec{z},n,n+r) \leq l \leq m_\varepsilon^+(\vec{z},n,n+r)} |g_\varepsilon(n, e^{f'_{\varepsilon,l}(\vec{z},n,n+r)}, \\ & \quad f''_{\varepsilon,l}(\vec{z},n,n+r))| < \infty . \end{aligned} \tag{7.10}$$

The following lemma is a variant of Lemma 3.4.1.

Lemma 7.1.1. *Let, for every $\varepsilon \in (0, \varepsilon_0]$, the prelimit log-price processes $\vec{Z}_{\varepsilon,n}$ is an atomic Markov chain, with transition probabilities and initial distribution defined, respectively, in relations (7.1) and (7.2). Then, the log-reward functions $\phi_{\varepsilon,n}(\vec{z})$ and $\phi_{\varepsilon,n+r}(f_{\varepsilon,l}(\vec{z},n,n+r))$, for points $f_{\varepsilon,l}(\vec{z},n,n+r)$, $l = m_\varepsilon^-(\vec{z},n,n+r), \dots, m_\varepsilon^+(\vec{z},n,n+$*

$r), r = 1, \dots N - n$ given by formulas (7.4), are, for every $\vec{z} \in \mathbb{Z}$, $n = 0, \dots, N$, the unique solution for the following finite recurrence system of linear equations:

$$\left\{\begin{array}{l} \phi_{\varepsilon,N}(f_{\varepsilon,l}(\vec{z}, n, N)) = g_\varepsilon\left(N, e^{f'_{\varepsilon,l}(\vec{z},n,N)}, f''_{\varepsilon,l}(\vec{z}, n, N)\right), \\ \qquad l = m^-_\varepsilon(\vec{z}, n, N), \dots, m^+_\varepsilon(\vec{z}, n, N), \\ \phi_{\varepsilon,n+r}(f_{\varepsilon,l}(\vec{z}, n, n + r)) \\ \qquad = \max\left(g_\varepsilon(n + r, e^{f'_{\varepsilon,l}(\vec{z},n,n+r)}, f''_{\varepsilon,l}(\vec{z}, n, n + r)),\right. \\ \qquad\qquad \left.\sum\limits_{l'=m^-_\varepsilon(\vec{z},n,n+r+1)}^{m^+_\varepsilon(\vec{z},n,n+r+1)} \phi_{\varepsilon,n+r+1}(f_{\varepsilon,l'}(f_{\varepsilon,l}(\vec{z}, n, n + r), n + r + 1))\right) \\ \qquad\quad \times p_{\varepsilon,l'}(f_{\varepsilon,l}(\vec{z}, n, n + r), n + r + 1))\,, \\ \qquad l = m^-_\varepsilon(\vec{z}, n, n + r), \dots, m^+_\varepsilon(\vec{z}, n, n + r), r = N - n - 1, \dots, 0. \end{array}\right. \tag{7.11}$$

where as above, $f_{\varepsilon,0}(\vec{z}, n, n) = \vec{z}$, $m^\pm_\varepsilon(\vec{z}, n, n) = 0$.

As far as the optimal expected reward $\Phi_\varepsilon = \Phi_\varepsilon(\mathcal{M}^{(\varepsilon)}_{\max,N})$ is concerned, it can be found using the following summation formula:

$$\Phi_\varepsilon = \sum_{l=m^-_{\varepsilon,0}}^{m^+_{\varepsilon,0}} p_{\varepsilon,l}\phi_{\varepsilon,0}(\vec{z}_{\varepsilon,0,l})\,. \tag{7.12}$$

7.1.2 An atomic Markov chain approximation model with a fixed skeleton structure

The atomic Markov chain approximation model with a fixed skeleton structure is an important particular case of the atomic Markov chain approximation model considered in Subsection 7.1.1.

Let us assume that the sets of points $\mathbb{F}_{\varepsilon,n} = \{\vec{z}_{\varepsilon,n,l} = (\vec{y}_{\varepsilon,n,l}, x_{\varepsilon,n,l}), l = m^-_{\varepsilon,n}, \dots, m^+_{\varepsilon,n}\} \in \mathcal{B}_\mathbb{Z}$ are defined for $n = 0, 1, \dots,$

In the space-skeleton Markov chain model: (a) the skeleton points $f_{\varepsilon,l}(\vec{z}, n) = \vec{z}_{\varepsilon,n,l}$, $m^-_\varepsilon(\vec{z}, n) = m^-_{\varepsilon,n} \leq l \leq m^+_\varepsilon(\vec{z}, n) = m^+_{\varepsilon,n}$, $n = 1, 2, \dots,$ and, in the sequel, the skeleton sets $\mathbb{F}_{\varepsilon,\vec{z},n} = \mathbb{F}_{\varepsilon,n}$, $n = 1, 2, \dots,$ do not depend on $\vec{z} \in \mathbb{Z}$, while (b) the one-point transition probabilities $p_{\varepsilon,l}(\vec{z}, n) = \mathsf{P}\{\vec{Z}_{\varepsilon,n} = \vec{z}_{\varepsilon,n,l}/\vec{Z}_{\varepsilon,n-1} = \vec{z}\}$, $l = m^-_{\varepsilon,n}, \dots, m^+_{\varepsilon,n}$, $n = 1, 2, \dots,$ do depend on $\vec{z} \in \mathbb{Z}$.

In this case, measures $P_{\varepsilon,n}(\vec{z}, A)$ have the following form, for $\vec{z} \in \mathbb{Z}$ and $n = 1, 2, \dots$:

$$P_{\varepsilon,n}(\vec{z}, A) = \sum_{\vec{z}_{\varepsilon,n,l} \in A} p_{\varepsilon,l}(\vec{z}, n)\,, \quad A \in \mathcal{B}_\mathbb{Z}\,. \tag{7.13}$$

As far as the initial distribution $P_{\varepsilon,0}(A)$ is concerned, it is defined, as in the general model of atomic Markov chain model, i.e. by the following formula, analogous to formula (7.2):

$$P_{\varepsilon,0}(A) = \sum_{\vec{z}_{\varepsilon,0,l} \in A} p_{\varepsilon,l}, \quad A \in \mathcal{B}_{\mathbb{Z}}. \tag{7.14}$$

The tree of nodes $T_{\varepsilon,(\vec{z},n)}$ includes one initial node $(\vec{z}, n)$, $m^+_{\varepsilon,n+1} - m^-_{\varepsilon,n+1} + 1$ additional nodes $(\vec{z}_{\varepsilon,n+1,l}, n+1)$, $l = m^-_{\varepsilon,n+1}, \dots, m^+_{\varepsilon,n+1}$, after the first transition, $m^+_{\varepsilon,n+2} - m^-_{\varepsilon,n+2} + 1$ additional nodes $(\vec{z}_{\varepsilon,n+2,l}, n+2)$, $l = m^-_{\varepsilon,n+2}, \dots, m^+_{\varepsilon,n+2}$, after the second transition, etc.

According to the above remarks, the tree $T_{\varepsilon,(\vec{z},n)}$ is the following set of nodes:

$$T_{\varepsilon,(\vec{z},n)} = \Big\langle (\vec{z}, n), (\vec{z}_{\varepsilon,n+r,l}, n+r), l = m^-_{\varepsilon,n+r}, \dots, m^+_{\varepsilon,n+r}, \quad r = 1, 2, \dots \Big\rangle . \tag{7.15}$$

The total number of nodes in the tree after r transitions is

$$L_{\varepsilon,r}(n) = 1 + (m^+_{\varepsilon,n+1} - m^-_{\varepsilon,n+1} + 1) + \cdots + (m^+_{\varepsilon,n+r} - m^-_{\varepsilon,n+r} + 1). \tag{7.16}$$

The standard case is where the parameters $m^{\pm}_{\varepsilon,n+r} = \pm m_\varepsilon, 0 \le n < n+r < \infty$ where m_ε are nonnegative integers. In this case, $L_{\varepsilon,r}(n) = 1 + (2m_\varepsilon + 1) + \cdots + (2m_\varepsilon + 1) = r(2m_\varepsilon + 1) + 1$ is a linear function of r.

The following lemma is a variant of Lemma 3.4.2.

Lemma 7.1.2. *Let, for every $\varepsilon \in (0, \varepsilon_0]$, the prelimit log-price processes $\vec{Z}_{\varepsilon,n}$ is a space-skeleton Markov chain, with transition probabilities and initial distribution defined, respectively, in relations (7.13) and (7.14). Then, the log-reward functions $\phi_{\varepsilon,n}(\vec{z})$ and $\phi_{\varepsilon,n+r}(\vec{z}_{\varepsilon,n+r,l})$, for points $\vec{z}_{\varepsilon,n+r,l}$, $l = m^-_{\varepsilon,n+r}, \dots, m^+_{\varepsilon,n+r}$, $r = 1, \dots N-n$, are, for every $\vec{z} \in \mathbb{Z}$, $n = 0, \dots, N$, the unique solution for the following finite recurrence system of linear equations:*

$$\begin{cases} \phi_{\varepsilon,N}(\vec{z}_{\varepsilon,N,l}) = g_\varepsilon(N, e^{\vec{y}_{\varepsilon,N,l}}, x_{\varepsilon,N,l}), \quad l = m^-_{\varepsilon,N}, \dots, m^+_{\varepsilon,N}, \\ \phi_{\varepsilon,n+r}(\vec{z}_{\varepsilon,n+r,l}) = \max\Bigg(g_\varepsilon(n+r, e^{\vec{y}_{\varepsilon,n+r,l}}, x_{\varepsilon,n+r,l}), \\ \qquad \sum\limits_{l'=m^-_{\varepsilon,n+r+1}}^{m^+_{\varepsilon,n+r+1}} \phi_{\varepsilon,n+r+1}(\vec{z}_{n+r+1,l'}) p_{\varepsilon,l'}(\vec{z}_{\varepsilon,n+r,l}, n+r+1) \Bigg), \\ \qquad l = m^-_{\varepsilon,n+r}, \dots, m^+_{\varepsilon,n+r}, \quad r = N-n-1, \dots, 1, \\ \phi_{\varepsilon,n}(\vec{z}) = \max\Bigg(g_\varepsilon(n, e^{\vec{y}}, x), \sum\limits_{l=m^-_{\varepsilon,n+1}}^{m^+_{\varepsilon,n+1}} \phi_{\varepsilon,n+1}(\vec{z}_{\varepsilon,n+1,l}) p_{\varepsilon,l}(\vec{z}, n+1) \Bigg). \end{cases} \tag{7.17}$$

7.1.3 A space-skeleton Markov chain approximation model

The space-skeleton Markov chain approximation model is an important particular case of the atomic Markov chain approximation model considered in Subsection 7.1.2.

Let us choose some $\delta_{\varepsilon,n,i} > 0$, $i = 1, \dots, k$, $n = 0, 1, \dots$, integers $m^-_{\varepsilon,n,j} \le m^+_{\varepsilon,n,j}$, $j = 0, \dots, k$, $n = 0, 1, \dots$, and $\lambda_{\varepsilon,n,i} \in \mathbb{R}_1$, $i = 1, \dots, k$, $n = 0, 1, \dots$.

Let us now define points $y_{\varepsilon,n,i,l} = l\delta_{\varepsilon,n,i} + \lambda_{\varepsilon,n,i}$, $l = m^-_{\varepsilon,n,i}, \dots, m^+_{\varepsilon,n,i}$, $i = 1, \dots, k$, $n = 0, 1, \dots$ and choose points $x_{\varepsilon,n,l} \in \mathbb{X}$, $l = m^-_{\varepsilon,n,0}, \dots, m^+_{\varepsilon,n,0}$, $n = 0, 1, \dots$.

Now, let us define the skeleton points for $\bar{l} = (l_0, l_1, \dots, l_k)$, $l_j = m^-_{\varepsilon,n,j}, \dots, m^+_{\varepsilon,n,j}$, $j = 0, \dots, k$, $n = 0, 1, \dots$,

$$\vec{z}_{\varepsilon,n,\bar{l}} = (\vec{y}_{\varepsilon,n,\bar{l}}, x_{\varepsilon,n,\bar{l}}) = ((y_{\varepsilon,n,1,l_1}, \dots, y_{\varepsilon,n,k,l_k}), x_{\varepsilon,n,l_0}) . \tag{7.18}$$

As in the general space-skeleton approximation model introduced above, the one-point transition probabilities $p_{\varepsilon,\bar{l}}(\vec{z}, n) = \mathsf{P}\{\vec{Z}_{\varepsilon,n} = \vec{z}_{\varepsilon,n,\bar{l}}/\vec{Z}_{\varepsilon,n-1} = \vec{z}\}$, $\bar{l} = (l_0, l_1, \dots, l_k)$, $l_j = m^-_{\varepsilon,n,j}, \dots, m^+_{\varepsilon,n,j}$, $j = 0, \dots, k$, $n = 1, 2, \dots$, do depend on $\vec{z} \in \mathbb{Z}$.

In this case, the transition probabilities $P_{\varepsilon,n}(\vec{z}, A)$ have the following form, for $\vec{z} \in \mathbb{Z}$ and $n = 1, 2, \dots$:

$$P_{\varepsilon,n}(\vec{z}, A) = \sum_{\vec{z}_{\varepsilon,n,\bar{l}} \in A} p_{\varepsilon,\bar{l}}(\vec{z}, n) , \quad A \in \mathcal{B}_{\mathbb{Z}} . \tag{7.19}$$

Also, let us $p_{\varepsilon,\bar{l}} = \mathsf{P}\{\vec{Z}_{\varepsilon,0} = \vec{z}_{\varepsilon,0,\bar{l}}\}$, $\bar{l} = (l_0, l_1, \dots, l_k)$, $l_j = m^-_{\varepsilon,0,j}, \dots, m^+_{\varepsilon,0,j}$, $j = 0, \dots, k$ be the corresponding one-point initial probabilities.

In this case, the initial distribution $P_{\varepsilon,0}(A)$ takes the following form:

$$P_{\varepsilon,0}(A) = \sum_{\vec{z}_{\varepsilon,0,\bar{l}} \in A} p_{\varepsilon,\bar{l}} , \quad A \in \mathcal{B}_{\mathbb{Z}} . \tag{7.20}$$

The difference in notations (between $\vec{z}_{\varepsilon,n,l}$ and $\vec{z}_{\varepsilon,n,\bar{l}}$ and between $p_{\varepsilon,l}(\vec{z}, n)$ and $p_{\varepsilon,\bar{l}}(\vec{z}, n)$) can be removed by a natural renumeration of indices, namely $(m^-_{\varepsilon,n,0}, m^-_{n,1}, \dots, m^-_{\varepsilon,n,k}) \leftrightarrow m^-_{\varepsilon,n}, \dots, (m^-_{\varepsilon,n,0}+1, m^-_{\varepsilon,n,1}, \dots, m^-_{\varepsilon,n,k}) \leftrightarrow m^-_{\varepsilon,n}+1, \dots, (m^+_{\varepsilon,n,0}, m^+_{\varepsilon,n,1}, \dots, m^+_{\varepsilon,n,k}) \leftrightarrow m^+_{\varepsilon,n}$, where

$$m^+_{\varepsilon,n} - m^-_{\varepsilon,n} + 1 = \prod_{j=0}^{k} (m^+_{\varepsilon,n,j} - m^-_{\varepsilon,n,j} + 1) , \quad n = 0, 1, \dots . \tag{7.21}$$

The simplest variant is to choose integerst $\pm m^\pm_{\varepsilon,n} \ge 0$, $n = 0, 1, \dots$.

The tree of nodes $T_{\varepsilon,(\vec{z},n)}$ includes one initial node $(\vec{z}, n)$, $m^+_{\varepsilon,n+1} - m^-_{\varepsilon,n+1} + 1$ additional nodes $\vec{z}_{\varepsilon,n+1,\bar{l}}$, $\bar{l} = (l_0, l_1, \dots, l_k)$, $l_j = m^-_{\varepsilon,n+1,j}, \dots, m^+_{\varepsilon,n+1,j}$, $j = 0, \dots, k$, appearing in the tree after the first jump, $m^+_{\varepsilon,n+2} - m^-_{\varepsilon,n+2} + 1$ additional nodes $\vec{z}_{\varepsilon,n+2,\bar{l}}$, $\bar{l} = (l_0, l_1, \dots, l_k)$, $l_j = m^-_{\varepsilon,n+2,j}, \dots, m^+_{\varepsilon,n+2,j}$, $j = 0, \dots, k$, appearing in the tree after the second jump, etc.

In this case, the tree $T_{\varepsilon,(\vec{z},n)}$ is the following set of nodes:

$$\begin{aligned} T_{\varepsilon,(\vec{z},n)} = \Big\langle (\vec{z}, n), (\vec{z}_{\varepsilon,n+r,\bar{l}}, n + r), \quad & \bar{l} = (l_0, l_1, \dots, l_k), \\ l_j = m^-_{\varepsilon,n+r,j}, \dots, m^+_{\varepsilon,n+r,j}, \quad & j = 0, \dots, k, \quad r = 1, 2, \dots \Big\rangle . \end{aligned} \tag{7.22}$$

The total number of nodes in the tree after r transitions is given by the following formula:

$$L_{\varepsilon,r}(n) = 1 + (m^+_{\varepsilon,n+1} - m^-_{\varepsilon,n+1} + 1) + \cdots + (m^+_{\varepsilon,n+r} - m^-_{\varepsilon,n+r} + 1)\,. \tag{7.23}$$

The following lemma is a variant of Lemma 3.4.3.

Lemma 7.1.3. *Let, for every $\varepsilon \in (0, \varepsilon_0]$, the prelimit log-price processes $\vec{Z}_{\varepsilon,n}$ is a space-skeleton Markov chain with transition probabilities and initial distribution defined, respectively, in relations (7.19) and (7.20). Then, the log-reward functions $\phi_{\varepsilon,n}(\vec{z})$ and $\phi_{\varepsilon,n+r}(\vec{z}_{n+r,\bar{l}})$, for points $\vec{z}_{n+r,\bar{l}}$, $\bar{l} = (l_0, l_1, \dots, l_k)$, $l_j = m^-_{\varepsilon,n,j}, \dots, m^+_{\varepsilon,n,j}$, $j = 0, \dots, k$, $r = 1, \dots, N-n$, given by formulas (7.18), are, for every $\vec{z} \in \mathbb{Z}$, $n = 0, \dots, N$, the unique solution for the following finite recurrence system of linear equations:*

$$\begin{cases} \phi_{\varepsilon,N}(\vec{z}_{\varepsilon,N,\bar{l}}) = g_\varepsilon(N, e^{\vec{y}_{\varepsilon,N,\bar{l}}}, x_{\varepsilon,N,\bar{l}})\,, \\ \qquad \bar{l} = (l_0, l_1, \dots, l_k),\ l_j = m^-_{\varepsilon,N,j}, \dots, m^+_{\varepsilon,N,j},\ j = 0, \dots, k\,, \\ \phi_{\varepsilon,n+r}(\vec{z}_{\varepsilon,n+r,\bar{l}}) = \max\Big(g_\varepsilon(n+r, e^{\vec{y}_{\varepsilon,n+r,\bar{l}}}, x_{\varepsilon,n+r,\bar{l}})\,, \\ \qquad \sum\limits_{\vec{z}_{\varepsilon,n+r+1,\bar{l}'} \in \mathbb{F}_{\varepsilon,n+r+1}} \phi_{\varepsilon,n+r+1}(\vec{z}_{\varepsilon,n+r+1,\bar{l}'}) p_{\varepsilon,\bar{l}'}(\vec{z}_{\varepsilon,n+r,\bar{l}}, n+r+1)\Big)\,, \\ \qquad \bar{l} = (l_0, l_1, \dots, l_k)\,, \quad l_j = m^-_{\varepsilon,n+r,j}, \dots, m^+_{\varepsilon,n+r,j}\,, \quad j = 0, \dots, k\,, \\ \qquad r = N - n - 1, \dots, 1\,, \\ \phi_{\varepsilon,n}(\vec{z}) = \max\Big(g_\varepsilon(n, e^{\vec{y}}, x)\,, \sum\limits_{\vec{z}_{\varepsilon,n+1,\bar{l}'} \in \mathbb{F}_{\varepsilon,n+1}} \phi_{\varepsilon,n+1}(\vec{z}_{\varepsilon,n+1,\bar{l}'}) \\ \qquad \times p_{\varepsilon,\bar{l}'}(\vec{z}, n+1)\Big)\,. \end{cases} \tag{7.24}$$

As far as the optimal expected reward $\Phi_\varepsilon = \Phi_\varepsilon(\mathcal{M}^{(\varepsilon)}_{max,N})$ is concerned, it is given by the following formula analogous of formula (3.102):

$$\Phi_\varepsilon = \sum_{\vec{z}_{\varepsilon,0,\bar{l}} \in \mathbb{F}_{\varepsilon,0}} p_{\varepsilon,\bar{l}}\, \phi_{\varepsilon,0}(\vec{z}_{\varepsilon,0,\bar{l}})\,. \tag{7.25}$$

7.1.4 A space-skeleton Markov chain approximation model with an additive skeleton structure

The space-skeleton Markov chain approximation model with an additive skeleton structure also is a particular case of the atomic Markov chain model, introduced in Subsection 7.1.1. It can also be referred to as a space-skeleton Markov random walk approximation model.

Note, first of all, that the modulated Markov chain $\vec{Z}_{\varepsilon,n} = (\vec{Y}_{\varepsilon,n}, X_{\varepsilon,n})$, $n = 0, 1, \dots$, can always be represented in the form of the modulated Markov random walk $\vec{Z}_{\varepsilon,n} = (\vec{Y}_{\varepsilon,n}, X_{\varepsilon,n}) = (\vec{Y}_{\varepsilon,n-1} + \vec{W}_{\varepsilon,n}, X_{\varepsilon,n})$, $n = 1, 2, \dots$, with an initial state $\vec{Z}_{\varepsilon,0} = (\vec{Y}_{\varepsilon,0}, X_{\varepsilon,0})$ and random jumps $\vec{W}_{\varepsilon,n} = \vec{Y}_{\varepsilon,n} - \vec{Y}_{\varepsilon,n-1}$, $n = 1, 2, \dots$.

In this case, it is convenient to operate with the transition jump probabilities $\tilde{P}_{\varepsilon,n}(\vec{y}, x, A) = \mathsf{P}\{(\vec{W}_{\varepsilon,n}, X_{\varepsilon,n}) \in A / \vec{Y}_{\varepsilon,n-1} = \vec{y}, X_{\varepsilon,n-1} = x\}$, which are connected with the transition probabilities of the Markov chain $\vec{Z}_{\varepsilon,n}$ by the following relations, for $\vec{z} = (\vec{y}, x) \in \mathbb{Z}$, $A \in \mathcal{B}_{\mathbb{Z}}$, $n = 1, 2, \dots$:

$$\begin{aligned} \tilde{P}_{\varepsilon,n}(\vec{y}, x, A) &= \mathsf{P}\{(\vec{W}_{\varepsilon,n}, X_{\varepsilon,n}) \in A / \vec{Y}_{\varepsilon,n-1} = \vec{y}, X_{\varepsilon,n-1} = x\} \\ &= \mathsf{P}\{(\vec{Y}_{\varepsilon,n} - \vec{y}, X_{\varepsilon,n}) \in A / \vec{Y}_{\varepsilon,n-1} = \vec{y}, X_{\varepsilon,n-1} = x\} \\ &= P_{\varepsilon,n}(\vec{z}, A_{[\vec{y}]}), \end{aligned} \tag{7.26}$$

where $A_{[\vec{y}]} = \{\vec{z}\,' = (\vec{y}\,', x) : (\vec{y}\,' - \vec{y}, x) \in A\}$.

In this case, we also use the standard space-skeleton construction, but applied to jumps of a random walk.

Let us choose, for every $\varepsilon \in (0, \varepsilon_0]$, some $\delta_{\varepsilon,i} > 0$, $i = 1, \dots, k$, integers $m^-_{\varepsilon,n,i} \leq m^+_{\varepsilon,n,i}$, $i = 1, \dots, k$, $n = 0, 1, \dots$, $\lambda_{\varepsilon,n,i} \in \mathbb{R}_1$, $i = 1, \dots, k$, $n = 0, 1, \dots$, and define points $y_{\varepsilon,n,i,l_i} = l_i\delta_{\varepsilon,i} + \lambda_{\varepsilon,n,i} \in \mathbb{R}_1$, $l_i = m^-_{\varepsilon,n,i} \dots, m^+_{\varepsilon,n,i}$, $i = 1, \dots, k$, $n = 0, 1, \dots$.

Then, let us choose some integers $m^-_{\varepsilon,n,0} \leq m^+_{\varepsilon,n,0}$, $n = 0, 1, \dots$, and define points $x_{\varepsilon,n,l_0} \in \mathbb{X}$, $l_0 = m^-_{\varepsilon,n,0}, \dots, m^+_{n,0}$, $n = 0, 1, \dots$.

Finally let us define skeleton points $\vec{z}_{\varepsilon,n,\bar{l}} = (\vec{y}_{\varepsilon,n,\bar{l}}, x_{\varepsilon,n,\bar{l}})$, $\bar{l} = (l_0, l_1, \dots, l_k)$, $l_j = m^-_{\varepsilon,n,j}, \dots, m^+_{\varepsilon,n,j}$, $j = 0, \dots, k$, $n = 0, 1, \dots$,

$$\vec{z}_{\varepsilon,n,\bar{l}} = (\vec{y}_{\varepsilon,n,\bar{l}}, x_{\varepsilon,n,\bar{l}}) = ((y_{\varepsilon,n,1,l_1}, \dots, y_{\varepsilon,n,k,l_k}), x_{\varepsilon,n,l_0}) . \tag{7.27}$$

Let us now introduce, for $\vec{z} = (\vec{y}, x) \in \mathbb{Z}$ and $n = 1, 2, \dots$, skeleton sets

$$\mathbb{F}_{\varepsilon,\vec{z},n} = \{f_{\varepsilon,\bar{l}}(\vec{z}, n) = (\vec{y} + \vec{y}_{\varepsilon,n,\bar{l}}, x_{\varepsilon,n,\bar{l}}) : (\vec{y}_{\varepsilon,n,\bar{l}}, x_{\varepsilon,n,\bar{l}}) \in \mathbb{F}_{\varepsilon,n}\} , \tag{7.28}$$

where the sets $\mathbb{F}_{\varepsilon,n} = \{\vec{z}_{\varepsilon,n,\bar{l}} : \bar{l} = (l_0, l_1, \dots, l_k), l_j = m^-_{\varepsilon,n,j}, \dots, m^+_{\varepsilon,n,j}, j = 0, \dots, k\}$, $n = 1, 2, \dots$.

In this case, the one-point transition probabilities $\tilde{p}_{\varepsilon,\bar{l}}(\vec{y}, x, n)$ are defined by the following formulas, for $\vec{z} = (\vec{y}, x) \in \mathbb{Z}, \bar{l} = (l_0, l_1, \dots, l_k)$, $l_j = m^-_{\varepsilon,n,j}, \dots, m^+_{\varepsilon,n,j}$, $j = 0, \dots, k$, $n = 1, 2, \dots$:

$$\begin{aligned} \tilde{p}_{\varepsilon,\bar{l}}(\vec{y}, x, n) &= \mathsf{P}\{(\vec{W}_{\varepsilon,n}, X_{\varepsilon,n}) = (\vec{y}_{\varepsilon,n,\bar{l}}, x_{\varepsilon,n,\bar{l}}) / \vec{Z}_{\varepsilon,n-1} = \vec{z}\} \\ &= \mathsf{P}\{\vec{Z}_{\varepsilon,n} = f_{\varepsilon,\bar{l}}(\vec{z}, n) / \vec{Z}_{\varepsilon,n-1} = \vec{z}\} \\ &= p_{\varepsilon,\bar{l}}(\vec{z}, n) . \end{aligned} \tag{7.29}$$

Sequentially, the transition probabilities $\tilde{P}_{\varepsilon,n}(\vec{y}, x, A)$ have the following form, for $\vec{z} \in \mathbb{Z}$, $A \in \mathcal{B}_{\mathbb{Z}}$, and $n = 1, 2, \dots$,

$$\begin{aligned} \tilde{P}_{\varepsilon,n}(\vec{y}, x, A) &= \sum_{(\vec{y}_{\varepsilon,n,\bar{l}}, x_{\varepsilon,n,\bar{l}}) \in A} \tilde{p}_{\varepsilon,\bar{l}}(\vec{y}, x, n), \\ &= \sum_{(\vec{y} + \vec{y}_{\varepsilon,n,\bar{l}}, x_{\varepsilon,n,\bar{l}}) \in A_{[\vec{y}]}} p_{\varepsilon,\bar{l}}(\vec{z}, n) \\ &= P_{\varepsilon,n}(\vec{z}, A_{[\vec{y}]}) . \end{aligned} \tag{7.30}$$

Also, let $p_{\varepsilon,\bar{l}} = \mathsf{P}\{\vec{Z}_{\varepsilon,0} = \vec{z}_{\varepsilon,0,\bar{l}}\}$, $\bar{l} = (l_0, l_1, \dots, l_k)$, $l_j = m^-_{\varepsilon,0,j}, \dots, m^+_{\varepsilon,0,j}$, $j = 0, \dots, k$ be the corresponding one-point initial probabilities.

In this case, the initial distribution $P_{\varepsilon,0}(A)$ takes the following form:

$$P_{\varepsilon,0}(A) = \sum_{\vec{z}_{\varepsilon,0,\bar{l}} \in A} p_{\varepsilon,\bar{l}}, \quad A \in \mathcal{B}_{\mathbb{Z}}. \tag{7.31}$$

The difference in notations (between $f_{\varepsilon,l}(\vec{z}, n)$ and $f_{\varepsilon,\bar{l}}(\vec{z}, n)$ and between $p_l(\vec{z}, n)$ and $p_{\varepsilon,\bar{l}}(\vec{z}, n) = p_{\varepsilon,\bar{l}}(\vec{y}, x, n)$) can be removed, as above, by a natural renumeration of indices, namely $(m^-_{\varepsilon,n,0}, m^-_{\varepsilon,n,1}, \dots, m^-_{\varepsilon,n,k}) \leftrightarrow m^-_{\varepsilon,n}, \dots, (m^-_{\varepsilon,n,0} + 1, m^-_{\varepsilon,n,1}, \dots, m^+_{\varepsilon,n,k}) \leftrightarrow m^-_{\varepsilon,n} + 1, \dots, (m^-_{\varepsilon,n,+}, m^+_{\varepsilon,n,1}, \dots, m^+_{\varepsilon,n,k}) \leftrightarrow m_{\varepsilon,n}$, where

$$m^+_{\varepsilon,n} - m^-_{\varepsilon,n} + 1 = \prod_{j=0}^{k} (m^+_{\varepsilon,n,j} - m^-_{\varepsilon,n,j} + 1), \quad n = 0, 1, \dots. \tag{7.32}$$

The simplest variant is to choose integers $\pm m^\pm_{\varepsilon,n} \geq 0$, $n = 0, 1, \dots$.

Let us denote $\vec{y}_{\varepsilon,n,n+r,\bar{l}} = (\delta_{\varepsilon,1} l_1 + \lambda_{\varepsilon,n,n+r,1}, \dots, \delta_{\varepsilon,k} l_k + \lambda_{\varepsilon,n,n+r,k})$, $x_{\varepsilon,n,n+r,\bar{l}} = x_{\varepsilon,n+r,l_0}$, $\bar{l} = (l_0, \dots, l_k)$, $l_j = m^-_{\varepsilon,n,n+r,j}, \dots, m^+_{\varepsilon,n,n+r,j}$, $j = 0, \dots, k$.

Here, $m^\pm_{\varepsilon,n,n+r,0} = m^\pm_{\varepsilon,n+r,0}$ and $m^\pm_{\varepsilon,n,n+r,j} = \sum_{l=n+1}^{n+r} m^\pm_{\varepsilon,l,j}$, $\lambda_{\varepsilon,n,n+r,j} = \sum_{l=n+1}^{n+r} \lambda_{\varepsilon,l,j}$, $j = 1, \dots, k$, for $0 \leq n < n + r < \infty$.

If $\vec{Z}_{\varepsilon,n} = (\vec{Y}_{\varepsilon,n}, X_{\varepsilon,n}) = (\vec{y}, x)$ for some $(\vec{y}, x) \in \mathbb{Z}$ and $n = 0, 1, \dots$, then, for $r = 1, 2, \dots$, possible states for $\vec{Z}_{\varepsilon,n+r} = (\vec{Y}_{\varepsilon,n+r}, X_{\varepsilon,n+r})$ are $(\vec{y} + \vec{y}_{\varepsilon,n,n+r,\bar{l}}, x_{\varepsilon,n+r,\bar{l}})$, $\bar{l} = (l_0, \dots, l_k)$, $l_j = m^-_{\varepsilon,n,n+r,j}, \dots, m^+_{\varepsilon,n,n+r,j}$, $j = 0, \dots, k$.

The total number of such states is given by the following formula, for $0 \leq n < n + r < \infty$:

$$m^+_{\varepsilon,n,n+r} - m^-_{\varepsilon,n,n+r} + 1 = \prod_{j=0}^{k} (m^+_{\varepsilon,n,n+r,j} - m^-_{\varepsilon,n,n+r,j} + 1). \tag{7.33}$$

The simplest variant is again to choose integers $\pm m^\pm_{\varepsilon,n,n+r} \geq 0$, $0 \leq n < n+r < \infty$.

The tree of nodes $T_{\varepsilon,((\vec{y},x),n)}$ includes one initial node $((\vec{y}, x), n)$, $m^+_{\varepsilon,n,n+1} - m^-_{\varepsilon,n,n+1} + 1$ additional nodes $((\vec{y} + \vec{y}_{\varepsilon,n+1,\bar{l}}, x_{\varepsilon,n+1,\bar{l}}), n + 1)$, $\bar{l} = (l_0, l_1, \dots, l_k)$, $l_j = m^-_{n,n+1,j}, \dots, m^+_{n,n+1,j}$, $j = 0, \dots, k$ after the first jump, $m^+_{\varepsilon,n,n+2} - m^-_{\varepsilon,n,n+2} + 1$ additional nodes $((\vec{y} + \vec{y}_{\varepsilon,n,n+2,\bar{l}}, x_{\varepsilon,n+2,\bar{l}}), n + 2)$, $\bar{l} = (l_0, l_1, \dots, l_k)$, $l_j = m^-_{\varepsilon,n,n+2,j}, \dots, m^+_{\varepsilon,n,n+2,j}$, $j = 0, \dots, k$ after the second jump, etc.

According to the above remarks, the tree $T_{\varepsilon,((\vec{y},x),n)}$ is the following set of nodes:

$$\begin{aligned} T_{\varepsilon,((\vec{y},x),n)} = \langle ((\vec{y}, x), n), & ((\vec{y} + \vec{y}_{\varepsilon,n,n+r,\bar{l}}, x_{\varepsilon,n+r,\bar{l}}), n + r), \\ & \bar{l} = (l_0, l_1, \dots, l_k), l_j = m^-_{\varepsilon,n,n+r,j}, \dots, m^+_{\varepsilon,n,n+r,j}, \\ & j = 0, \dots, k, r = 1, 2, \dots \rangle. \end{aligned} \tag{7.34}$$

The total number of nodes in the tree after r jumps is

$$L_{\varepsilon,r}(n) = 1 + m^+_{\varepsilon,n,n+1} - m^-_{\varepsilon,n,n+1} + 1 + \cdots + m^+_{\varepsilon,n,n+r} - m^-_{\varepsilon,n,n+r} + 1. \tag{7.35}$$

The following lemma is a variant of Lemma 3.4.4.

Lemma 7.1.4. *Let, for every $\varepsilon \in (0, \varepsilon_0]$, the prelimit log-price processes $\vec{Z}_{\varepsilon,n}$ is a space-skeleton random walk with transition probabilities and initial distribution defined, respectively, in relations (7.30) and (7.31). Then the log-reward functions $\phi_{\varepsilon,n}(\vec{y}, x)$ and $\phi_{\varepsilon,n+r}(\vec{y} + \vec{y}_{\varepsilon,n,n+r,\bar{l}}, x_{\varepsilon,n+r,\bar{l}})$ for points $(\vec{y} + \vec{y}_{\varepsilon,n,n+r,\bar{l}}, x_{\varepsilon,n+r,\bar{l}})$, $\bar{l} = (l_0, l_1, \dots, l_k)$, $l_j = m^-_{\varepsilon,n,n+r,j}, \dots, m^+_{\varepsilon,n,n+r,j}$, $j = 0, \dots, k$, $r = 1, \dots N - n$, given by formulas (7.27), are, for every $\vec{z} \in \mathbb{Z}$, $n = 0, \dots, N$, the unique solution for the following finite recurrence system of linear equations:*

$$
\left\{
\begin{aligned}
&\phi_{\varepsilon,N}(\vec{y} + \vec{y}_{\varepsilon,n,N,\bar{l}}, x_{\varepsilon,N,\bar{l}}) = g(N, e^{\vec{y}+\vec{y}_{\varepsilon,n,N,\bar{l}}}, x_{\varepsilon,N,\bar{l}}) \,, \\
&\qquad \bar{l} = (l_0, l_1, \dots, l_k)\,, \quad l_j = m^-_{\varepsilon,n,N,j}, \dots, m^+_{\varepsilon,n,N,j}\,, \quad j = 0, \dots, k\,, \\
&\phi_{\varepsilon,n+r}(\vec{y} + \vec{y}_{\varepsilon,n,n+r,\bar{l}}, x_{\varepsilon,n+r,\bar{l}}) \\
&\quad = \max\Bigg(g(n+r, e^{\vec{y}+\vec{y}_{\varepsilon,n,n+r,\bar{l}}}, x_{\varepsilon,n+r,\bar{l}}) \,, \\
&\qquad \sum_{(\vec{y}_{\varepsilon,n+r+1,\bar{l}'}, x_{\varepsilon,n+r+1,\bar{l}'}) \in \mathbb{F}_{\varepsilon,n+r+1}} \phi_{\varepsilon,n+r+1}(\vec{y} + \vec{y}_{\varepsilon,n,n+r,\bar{l}} + \vec{y}_{\varepsilon,n+r+1,\bar{l}'}, x_{\varepsilon,n+r+1,\bar{l}'}) \\
&\qquad \times p_{\varepsilon,\bar{l}'}(\vec{y} + \vec{y}_{\varepsilon,n,n+r,\bar{l}}, x_{\varepsilon,n+r,\bar{l}}, n+r+1) \Bigg) \,, \\
&\quad \bar{l} = (l_0, l_1, \dots, l_k)\,, \quad l_j = m^-_{\varepsilon,n+r,j}, \dots, m^+_{\varepsilon,n+r,j}\,, \quad j = 0, \dots, k\,, \\
&\quad r = N - n - 1, \dots, 1\,, \\
&\phi_{\varepsilon,n}(\vec{y}, x) = \max\Bigg(g(n, e^{\vec{y}}, x)\,, \\
&\qquad \sum_{(\vec{y}_{\varepsilon,n+1,\bar{l}'}, x_{\varepsilon,n+1,\bar{l}'}) \in \mathbb{F}_{\varepsilon,n+1}} \phi_{\varepsilon,n+1}(\vec{y} + \vec{y}_{\varepsilon,n+1,\bar{l}'}, x_{\varepsilon,n+1,\bar{l}'}) p_{\varepsilon,\bar{l}'}(\vec{y}, x, n+1) \Bigg) \,.
\end{aligned}
\right.
\tag{7.36}
$$

As far as the optimal expected reward $\Phi_\varepsilon = \Phi_\varepsilon(\mathcal{M}^{(\varepsilon)}_{max,N})$ is concerned, it is defined as in the general model of atomic modulated Markov log-price processes, i.e. by the following analogous of formula (3.102):

$$
\Phi_\varepsilon = \sum_{\vec{z}_{\varepsilon,0,\bar{l}} \in \mathbb{F}_{\varepsilon,0}} p_{\varepsilon,\bar{l}} \, \phi_{\varepsilon,0}(\vec{z}_{\varepsilon,0,\bar{l}}) \,. \tag{7.37}
$$

7.2 Univariate Markov LPP with bounded characteristics

In this section, we present space-skeleton approximation models for discrete time univariate Markov log-price processes with bounded characteristics.

7.2.1 Space-skeleton approximations for log-rewards functions for univariate Markov log-price processes

Let us assume that, for every $\varepsilon \in [0, \varepsilon_0]$, a modulated log-price process $Y_{\varepsilon,n}$, $n = 0, 1, \dots$, is an inhomogeneous Markov chain, with a phase space $\mathbb{R}_1$, an initial distribution $P_{\varepsilon,0}(A)$, and the one-step transition probabilities $P_{\varepsilon,n}(y, A)$.

We would like to approximate the log-price process $Y_{0,n}$ by the log-price processes $Y_{\varepsilon,n}$. We additionally assume that the log-price process $Y_{0,n}$ (for $\varepsilon = 0$) can be a nonatomic Markov chain while the log-price processes $Y_{\varepsilon,n}$ (for $\varepsilon \in (0, \varepsilon_0]$) are atomic Markov chains, i.e. their transition probabilities $P_{\varepsilon,n}(y, A)$ and initial distributions $P_{\varepsilon,0}(A)$ are concentrated at finite sets of points.

In this case, the model without index component is considered. In order to be able to interpret the model as modulated, one can always introduce a "virtual" index component $X_{\varepsilon,n}$ with the one-point phase space $\mathbb{X} = \{x_0\}$. Thus, we also assume that a pay-off function $g(n, e^y, x) = g(n, e^y)$ does not depend on argument x and is a real-valued Borel function of argument $y \in \mathbb{R}_1$, for every $n = 0, 1, \dots, N$.

Note also that, we consider the case where the pay-off function $g(n, e^y)$ does not depend on the parameter ε.

For every $\varepsilon \in (0, \varepsilon_0]$, let us choose nonempty Borel sets $\mathbb{A}_{\varepsilon,n,l} \subseteq \mathbb{R}_1$, $l = m^-_{\varepsilon,n}, \dots, m^+_{\varepsilon,n}$, $n = 0, 1, \dots$ such that,

$$\mathbb{A}_{\varepsilon,n,l'} \cap \mathbb{A}_{\varepsilon,n,l''} = \emptyset, l' \neq l''; \quad \mathbb{R}_1 = \cup_{l=m^-_{\varepsilon,n}}^{m^+_{\varepsilon,n}} \mathbb{A}_{\varepsilon,n,l}, \quad n = 0, 1, \dots . \tag{7.38}$$

and skeleton points,

$$y_{\varepsilon,n,l} \in \mathbb{A}_{\varepsilon,n,l}, \quad l = m^-_{\varepsilon,n}, \dots, m^+_{\varepsilon,n}, \quad n = 0, 1, \dots . \tag{7.39}$$

Let us now assume that the following fitting condition holds:

L₁: For every $\varepsilon \in (0, \varepsilon_0]$, transition probabilities $\mathsf{P}\{Y_{\varepsilon,n} = y_{\varepsilon,n,l}/Y_{\varepsilon,n-1} = y\} = p_{\varepsilon,l}(y, n) = P_{0,n}(y_{\varepsilon,n-1,l'}, \mathbb{A}_{\varepsilon,n,l})$, $y \in \mathbb{A}_{\varepsilon,n-1,l'}$, $l' = m^-_{\varepsilon,n-1}, \dots, m^+_{\varepsilon,n-1}$, $l = m^-_{\varepsilon,n}, \dots, m^+_{\varepsilon,n}$, $n = 1, 2, \dots$.

If we are also interested to fit initial distributions, the following fitting condition can be assumed:

M₁: For every $\varepsilon \in (0, \varepsilon_0]$, initial probabilities $\mathsf{P}\{Y_{\varepsilon,0} = y_{\varepsilon,0,l}\} = p_{\varepsilon,l} = P_{0,0}(\mathbb{A}_{\varepsilon,0,l})$, $l = m^-_{\varepsilon,0}, \dots, m^+_{\varepsilon,0}$.

A standard way is to chose as the skeleton sets $\mathbb{A}_{\varepsilon,n,l}$ and space-skeleton points $y_{\varepsilon,n,l}$ is the following.

Let us choose, for every $\varepsilon \in (0, \varepsilon_0]$, some $\delta_{\varepsilon,n} > 0$, $n = 0, 1, \dots$, integers $m^-_{\varepsilon,n} \leq m^+_{\varepsilon,n}$, $n = 0, 1, \dots$, and $\lambda_{\varepsilon,n} \in \mathbb{R}_1$, $n = 0, 1, \dots$.

First, the intervals $\mathbb{I}_{\varepsilon,n,l}$, which play the role of the skeleton sets $\mathbb{A}_{\varepsilon,n,l}$, are constructed for $l = m^-_{\varepsilon,n}, \dots, m^+_{\varepsilon,n}$, $n = 0, 1, \dots$,

$$\mathbb{I}_{\varepsilon,n,l} = \mathbb{A}_{\varepsilon,n,l} = \begin{cases} (-\infty,\ \delta_{\varepsilon,n}(m^-_{\varepsilon,n} + \frac{1}{2})] + \lambda_{\varepsilon,n} & \text{if } l = m^-_{\varepsilon,n}, \\ (\delta_{\varepsilon,n}(l - \frac{1}{2}),\ \delta_{\varepsilon,n}(l + \frac{1}{2})] + \lambda_{\varepsilon,n} & \text{if } m^-_{\varepsilon,n} < l < m^+_{\varepsilon,n}, \\ (\delta_{\varepsilon,n}(m^+_{\varepsilon,n} - \frac{1}{2}),\ \infty) + \lambda_{\varepsilon,n} & \text{if } l = m^+_{\varepsilon,n}. \end{cases} \tag{7.40}$$

Then the skeleton points $y_{\varepsilon,n,l} \in \mathbb{I}_{\varepsilon,n,l}$ are chosen in the following way:

$$y_{\varepsilon,n,l} = l\delta_{\varepsilon,n} + \lambda_{\varepsilon,n}, \quad l = m^-_{\varepsilon,n}, \dots, \quad m^+_{\varepsilon,n}, \quad n = 0, 1, \dots. \tag{7.41}$$

In this case, the fitting condition $\mathbf{L_1}$ takes the following form:

$\mathbf{L_2}$: For every $\varepsilon \in (0, \varepsilon_0]$, transition probabilities $\mathsf{P}\{Y_{\varepsilon,n} = y_{\varepsilon,n,l}/Y_{\varepsilon,n-1} = y\} = p_{\varepsilon,l}(y, n) = P_{0,n}(y_{\varepsilon,n-1,l'}, \mathbb{I}_{\varepsilon,n,l}), y \in \mathbb{I}_{\varepsilon,n-1,l'}, l' = m^-_{\varepsilon,n-1}, \dots, m^+_{\varepsilon,n-1}, l = m^-_{\varepsilon,n}, \dots, m^+_{\varepsilon,n}$, $n = 1, 2, \dots$,

The fitting condition $\mathbf{M_1}$ takes the following form:

$\mathbf{M_2}$: For every $\varepsilon \in (0, \varepsilon_0]$, initial probabilities $\mathsf{P}\{Y_{\varepsilon,0} = y_{\varepsilon,0,l}\} = p_{\varepsilon,l} = P_{0,0}(\mathbb{I}_{\varepsilon,0,l})$, $l = m^-_{\varepsilon,0}, \dots, m^+_{\varepsilon,0}$.

In this case, the corresponding skeleton sets $\mathbb{F}_{\varepsilon,n} = \{y_{\varepsilon,n,l}, l = m^-_{\varepsilon,n}, \dots, m^+_{\varepsilon,n}\}$, $n = 0, 1, \dots$.

Let us define, for $\varepsilon \in (0, \varepsilon_0]$ and $n = 0, 1, \dots$, skeleton functions,

$$h_{\varepsilon,n}(y) = \begin{cases} \delta_{\varepsilon,n} m^-_{\varepsilon,n} + \lambda_{\varepsilon,n} \\ \qquad \text{if } y \le \delta_{\varepsilon,n}(m^-_{\varepsilon,n} + \frac{1}{2}) + \lambda_{\varepsilon,n}, \\ \delta_{\varepsilon,n} l + \lambda_{\varepsilon,n} \\ \qquad \text{if } \delta_{\varepsilon,n}(l - \frac{1}{2}) + \lambda_{\varepsilon,n} < y \le \delta_{\varepsilon,n}(l + \frac{1}{2}) + \lambda_{\varepsilon,n}, \\ \qquad m^-_{\varepsilon,n} < l < m^+_{\varepsilon,n}, \\ \delta_{\varepsilon,n} m^+_{\varepsilon,n} + \lambda_{\varepsilon,n} \\ \qquad \text{if } y > \delta_{\varepsilon,n}(m^+_{\varepsilon,n} - \frac{1}{2}) + \lambda_{\varepsilon,n}. \end{cases} \tag{7.42}$$

In this case, the transition probabilities $P_{\varepsilon,n}(y, A) = \mathsf{P}\{Y_{\varepsilon,n} \in A/Y_{\varepsilon,n-1} = y\}$, $y \in \mathbb{R}_1, A \in \mathcal{B}_1, n = 1, \dots, N$ take the following form, for every $\varepsilon \in (0, \varepsilon_0]$:

$$\begin{aligned} P_{\varepsilon,n}(y, A) &= \sum_{y_{\varepsilon,n,l} \in A} P_{0,n}(h_{\varepsilon,n-1}(y), \mathbb{I}_{\varepsilon,n,l}) \\ &= \mathsf{P}\{h_{\varepsilon,n}(Y_{0,n}) \in A/Y_{0,n-1} = h_{\varepsilon,n-1}(y)\}. \end{aligned} \tag{7.43}$$

As far as the initial distribution $P_{\varepsilon,0}(A) = \mathsf{P}\{Y_{\varepsilon,0} \in A\}$, $A \in \mathcal{B}_1$ is concerned, it takes the following form, for every $\varepsilon \in (0, \varepsilon_0]$:

$$P_{\varepsilon,0}(A) = \sum_{y_{\varepsilon,0,l} \in A} P_{0,0}(\mathbb{I}_{\varepsilon,0,l}) = \mathsf{P}\{h_{\varepsilon,0}(Y_{0,0}) \in A\}. \tag{7.44}$$

The possibility to construct approximating Markov chains $Y_{\varepsilon,n}$ satisfying the fitting conditions $\mathbf{L_2}$ and $\mathbf{M_2}$ follows from the fact that the quantity $P_{\varepsilon,n}(y, A)$, defined in relation (7.43), satisfies conditions that should satisfy transition probabilities of a Markov chain. It is a measurable function in $y \in \mathbb{R}_1$, for every $A \in \mathcal{B}_1$, and a probability measure as a function of $A \in \mathcal{B}_1$, for every $y \in \mathbb{R}_1$. Also, the quantity $P_{\varepsilon,0}(A)$, defined in relation (7.44), is a probability measure.

It is worth to note that the above described space-skeleton approximation procedure realizes distributional fitting. The Markov chains $Y_{\varepsilon,n}$ can be defined for different ε on different probability spaces.

However, it is possible to realize the so-called strong space-skeleton approximation procedure, where Markov chains $Y_{\varepsilon,n}$ are defined for all ε on the same probability space.

The phase space of the Markov chain $Y_{0,n}$ is $\mathbb{R}_1$. In this case, we can assume, without loss of generality, that this Markov chain is defined in the dynamic form

$$\begin{cases} Y_{0,n} = A_{0,n}(Y_{0,n-1}, U_{0,n}) \,, \quad n = 1, 2, \dots \,, \\ Y_{0,0} = A_{0,0}(U_{0,0}) \end{cases} \tag{7.45}$$

where (a) $U_{0,n}$, $n = 0, 1, \dots$, are i.i.d. random variables uniformly distributed in the interval $[0, 1)$, (b) $A_{0,n}(y, u)$, $n = 1, 2, \dots$, are Borel functions acting from $\mathbb{R}_1 \times [0, 1)$ to $\mathbb{R}_1$, and (c) $A_{0,0}(u)$ is a Borel function acting from $[0, 1)$ to $\mathbb{R}_1$.

More precisely, as was pointed out in Subsection 1.1.4, we can always construct a Markov chain $Y'_{0,n}$ given in the described above dynamic form, with the same phase space, transition probabilities and initial distribution as the Markov chain $Y_{0,n}$. In this particular case, where the phase space is a real line, the corresponding construction is very simple one. Let us define $A_{0,n}(y, u) = \inf(v : P_{0,n}(y, (-\infty, v] > u)$, $u \in [0, 1)$, for every $y \in \mathbb{R}_1$, $n = 1, 2, \dots$, and $A_{0,0}(u) = \inf(v : P_{0,0}((-\infty, v] > u)$, $u \in [0, 1)$. Let also U_n, $n = 0, 1, \dots$, is a sequence of i.i.d. random variables uniformly distributed in interval $[0, 1)$. In this case, the random sequence $Y'_{0,n} = A_{0,n}(Y'_{0,n-1}, U_n)$, $n = 1, 2, \dots$, $Y'_{0,0} = A_{0,0}(U_0)$ is a Markov chain, which has the same phase space, transition probabilities and initial distribution as the Markov chain $Y_{0,n}$.

Thus, we can, indeed, assume that the Markov chain $Y_{0,n}$ is given in the dynamic form described above.

In this case, the Markov chain $Y_{\varepsilon,n}$ can also be defined in the dynamic form, for every $\varepsilon \in (0, \varepsilon_0]$, according to the following relation:

$$\begin{cases} Y_{\varepsilon,n} = h_{\varepsilon,n}(A_{0,n}(h_{\varepsilon,n-1}(Y_{\varepsilon,n-1}), U_n)) \,, \quad n = 1, 2, \dots \,, \\ Y_{\varepsilon,0} = h_{\varepsilon,0}(A_{0,0}(U_0)) \,. \end{cases} \tag{7.46}$$

It is readily seen that the random sequence $Y_{\varepsilon,n}$ is a Markov chain, and its transition probabilities and initial distribution satisfy the conditions $\mathbf{L_2}$ and $\mathbf{M_2}$ and, subsequently, relations (7.43) and (7.44).

It is worth to note that the conditions $\mathbf{L_2}$ and $\mathbf{M_2}$ imply that $\mathsf{P}\{Y_{\varepsilon,n} \in \mathbb{F}_{\varepsilon,n},\, n = 0, 1, \dots\} = 1$. Since $h_{\varepsilon,n}(y) = y$ for $y \in \mathbb{F}_{\varepsilon,n}$, $n = 0, 1, \dots$, it seems that one can use simpler dynamic transition function $h_{\varepsilon,n}(A_n(y,u))$ in relation (7.46). However, this is not possible.

The Markov chain $Y_{\varepsilon,n}$ has the phase space $\mathbb{R}_1$. The conditions $\mathbf{L_2}$ and $\mathbf{M_2}$ require that the transition probabilities and the initial distribution of this Markov chain are concentrated at the above finite sets. That is why, the corresponding dynamic transition functions in relation (7.46) are defined as $A_{\varepsilon,n}(y,u) = h_{\varepsilon,n}(A_n(h_{\varepsilon,n-1}(y),u))$ for $(y,u) \in \mathbb{R}_1 \times [0,1)$.

Since the pay-off function $g(n, e^y)$ does not depend on the index argument x, the log-reward functions $\phi_{\varepsilon,n}(y)$, $n = 0,1,\dots$, for the log-price processes $Y_{\varepsilon,n}$ are functions of y, for every $\varepsilon \in [0, \varepsilon_0]$.

The following lemma is a variant of Lemma 7.1.3.

Lemma 7.2.1. *Let, for every $\varepsilon \in (0, \varepsilon_0]$, the prelimiting log-price process $Y_{\varepsilon,n}$ is a space-skeleton Markov chain with transition probabilities and initial distribution defined, respectively, in relations* (7.43) *and* (7.44)*. Then, the log-reward functions $\phi_{\varepsilon,n}(y)$ and $\phi_{\varepsilon,n+r}(y_{\varepsilon,n+r,l})$, for points $y_{\varepsilon,n+r,l}$, $l = m^-_{\varepsilon,n+r},\dots,m^+_{\varepsilon,n+r}$, $r = 1,\dots N-n$, given by formulas* (7.41)*, are, for every $y \in \mathbb{I}_{\varepsilon,n,l''}$, $l'' = m^-_{\varepsilon,n},\dots,m^+_{\varepsilon,n}$, $n = 0,\dots,N$, the unique solution for the following finite recurrence system of linear equations:*

$$\begin{cases} \phi_{\varepsilon,N}(y_{\varepsilon,N,l}) = g(N, e^{y_{\varepsilon,N,l}}), \quad l = m^-_{\varepsilon,N},\dots,m^+_{\varepsilon,N}, \\ \phi_{\varepsilon,n+r}(y_{\varepsilon,n+r,l}) = \max\Bigg(g(n+r, e^{y_{\varepsilon,n+r,l}}), \\ \qquad \displaystyle\sum_{l'=m^-_{\varepsilon,n+r+1}}^{m^+_{\varepsilon,n+r+1}} \phi_{\varepsilon,n+r+1}(y_{\varepsilon,n+r+1,l'}) P_{0,n+r+1}(y_{\varepsilon,n+r,l''}, \mathbb{I}_{\varepsilon,n+r+1,l'})\Bigg), \\ \qquad l = m^-_{\varepsilon,n+r},\dots,m^+_{\varepsilon,n+r}, \quad r = N-n-1,\dots,1, \\ \phi_{\varepsilon,n}(y) = \max\Bigg(g(n, e^y), \displaystyle\sum_{l'=m^-_{\varepsilon,n+1}}^{m^+_{\varepsilon,n+1}} \phi_{\varepsilon,n+1}(y_{\varepsilon,n+1,l'}) \\ \qquad \times P_{0,n+1}(y_{\varepsilon,n,l''}, \mathbb{I}_{\varepsilon,n+1,l'})\Bigg). \end{cases} \tag{7.47}$$

7.2.2 Convergence of space-skeleton approximations for log-rewards functions for univariate Markov log-price processes

Let us formulate conditions, which should be imposed on the log-price process $Y_{0,n}$ and the pay-off function $g(n, e^y)$ in order the log-reward functions $\phi_{\varepsilon,n}(y)$ for the approximation processes $Y_{\varepsilon,n}$ would converge to the limiting log-reward functions $\phi_{0,n}(y)$ for the process $Y_{0,n}$, as $\varepsilon \to 0$.

Let us introduce special shorten notations for the maximal and the minimal skeleton points, for $\varepsilon \in (0, \varepsilon_0]$ and $n = 0,1,\dots$,

$$z^{\pm}_{\varepsilon,n} = \delta_{\varepsilon,n} m^{\pm}_{\varepsilon,n} + \lambda_{\varepsilon,n}. \tag{7.48}$$

We impose the following condition on parameters of the space-skeleton model:

N₁: (a) $\delta_{\varepsilon,n} \to 0$ as $\varepsilon \to 0$, for $n = 0, 1, \dots, N$;
(b) $\pm z^{\pm}_{\varepsilon,n} \to \infty$ as $\varepsilon \to 0$, for $n = 0, 1, \dots, N$;
(c) $\pm z^{\pm}_{\varepsilon,n}$, $n = 0, 1, \dots, N$ are nondecreasing sequences, for every $\varepsilon \in (0, \varepsilon_0]$.

Note that the condition $\mathbf{N}_1$ implies that $\delta_{\varepsilon,n}(m^{+}_{\varepsilon,n} - m^{-}_{\varepsilon,n}) \to \infty$ as $\varepsilon \to 0$, for $n = 0, 1, \dots, N$, and $m^{+}_{\varepsilon,n} - m^{-}_{\varepsilon,n} \to \infty$ as $\varepsilon \to 0$, for $n = 0, 1, \dots, N$.

Let us recall the first-type modulus of exponential moment compactness for the log-price process $Y_{0,n}$, defined for $\beta \geq 0$ and taken for the model without index component the following form:

$$\Delta_\beta(Y_{0,\cdot}, N) = \max_{0 \leq n \leq N-1} \sup_{y \in \mathbb{R}_1} \mathsf{E}_{y,n} e^{\beta |Y_{0,n+1} - Y_{0,n}|}. \tag{7.49}$$

We impose on the price process $Y_{0,n}$ the following first-type condition of exponential moment compactness, which is a one-dimensional variant of the condition $\mathbf{C}_1[\bar{\beta}]$ and should be assumed to hold for some $\beta \geq 0$:

$\mathbf{C}_{11}[\beta]$: $\Delta_\beta(Y_{0,\cdot}, N) < K_{49}$, for some $1 < K_{49} < \infty$.

Let us also recall the second-type modulus of exponential moment compactness for the log-price process Y_n, defined for $\beta \geq 0$,

$$\Xi^{\pm}_{\beta}(Y_{\cdot}, N) = \max_{0 \leq n \leq N-1} \sup_{y \in \mathbb{R}_1} \mathsf{E}_{y,n} e^{\pm\beta(Y_{n+1,i} - Y_{n,i})}. \tag{7.50}$$

As follows from Lemma 4.1.8, the following condition implies the condition $\mathbf{C}_{11}[\beta]$ to hold:

$\mathbf{E}_{15}[\beta]$: $\Xi^{\pm}_{\beta}(Y_{0,\cdot}, N) < K_{50}$, for some $1 < K_{50} < \infty$.

Let us introduce sequences of the sets $\mathcal{Z}_\varepsilon = \langle \mathbb{Z}_{\varepsilon,0}, \dots, \mathbb{Z}_{\varepsilon,N} \rangle$, where the sets $\mathbb{Z}_{\varepsilon,n}$ are intervals defined, for every $n = 0, \dots, N$, by the following relation:

$$\mathbb{Z}_{\varepsilon,n} = \begin{cases} [z^{-}_{\varepsilon,n}, z^{+}_{\varepsilon,n}] & \text{for } \varepsilon \in (0, \varepsilon_0], \\ \mathbb{R}_1 & \text{for } \varepsilon = 0. \end{cases} \tag{7.51}$$

Note that the condition $\mathbf{N}_1$ (c) implies the following imbedding relation for the intervals $\mathbb{Z}_{\varepsilon,n}$, for every $\varepsilon \in (0, \varepsilon_0]$:

$$\mathbb{Z}_{\varepsilon,0} \subseteq \cdots \subseteq \mathbb{Z}_{\varepsilon,N}. \tag{7.52}$$

Let us now recall the modified first-type modulus of exponential moment compactness, for every $\varepsilon \in [0, \varepsilon_0]$ and $\beta \geq 0$,

$$\Delta_{\beta,\mathcal{Z}_\varepsilon}(Y_{\varepsilon,\cdot}, N) = \max_{0 \leq n \leq N-1} \sup_{y \in \mathbb{Z}_{\varepsilon,n}} \mathsf{E}_{y,n} e^{\beta |Y_{\varepsilon,n+1} - Y_{\varepsilon,n}|}. \tag{7.53}$$

Note, that, by the definition, the modified modulus of exponential moment compactness $\Delta_{\beta,z_0}(Y_{0,\cdot}, N)$ coincides with the standard modulus of exponential moment compactness $\Delta_\beta(Y_{0,\cdot}, N)$.

We shall now prove below that, in the skeleton approximation model described above, the conditions $\mathbf{L_2}$, $\mathbf{N_1}$, and $\mathbf{C_{11}}[\beta]$ imply that the modified condition of exponential moment compactness $\mathbf{C'_6}[\bar{\beta}]$ (its one-dimensional variant) holds with the modified moduli of exponential moment compactness $\Delta_{\beta,z_\varepsilon}(Y_{\varepsilon,\cdot}, N)$. While the standard condition of exponential moment compactness $\mathbf{C_6}[\bar{\beta}]$ (also its one-dimensional variant) does not hold for the log-price processes $Y_{\varepsilon,n}$.

We assume that the pay-off function $g(n, e^y)$ satisfies the condition $\mathbf{B_4}[\gamma]$ holds for some $\gamma \geq 0$.

The condition $\mathbf{I_1}$ should be replaced by the following condition:

$\mathbf{I_6}$: There exist the Borel sets $\mathbb{Y}'_n \subseteq \mathbb{R}_1$, $n = 0, \dots, N$ such that function $g(n, e^y)$ is continuous in points $y \in \mathbb{Y}'_n$, for every $n = 0, \dots, N$.

The condition $\mathbf{I_6}$ obviously implies that $g(n, e^{y_\varepsilon}) \to g(n, e^{y_0})$ as $\varepsilon \to 0$ for any $y_\varepsilon \to y_0 \in \mathbb{Y}_n$ as $\varepsilon \to 0$ and $n = 0, \dots, N$. Thus, the condition $\mathbf{I_1}$ holds with the sets $\mathbb{Z}'_n = \mathbb{Y}'_n \times \{x_0\}$, $n = 0, \dots, N$.

The condition $\mathbf{J_1}$ should be replaced by the following condition:

$\mathbf{J_9}$: There exist the Borel sets $\mathbb{Y}''_n \subseteq \mathbb{R}_1$, $n = 0, \dots, N$ such that:
(a) $P_{0,n}(y_\varepsilon, \cdot) \Rightarrow P_{0,n}(y_0, \cdot)$ as $\varepsilon \to 0$, for any $y_\varepsilon \to y_0 \in \mathbb{Y}''_{n-1}$ as $\varepsilon \to 0$, and $n = 1, \dots, N$;
(b) $P_{0,n}(y_0, \mathbb{Y}'_n \cap \mathbb{Y}''_n) = 1$, for every $y_0 \in \mathbb{Y}'_{n-1} \cap \mathbb{Y}''_{n-1}$ and $n = 1, \dots, N$, where $\mathbb{Y}'_n$, $n = 0, \dots, N$ are sets introduced in the condition $\mathbf{I_6}$.

The condition $\mathbf{J_9}$ (a) means that the transition probabilities $P_{0,n}(y, \cdot)$ are weakly continuous at points $y \in \mathbb{Y}''_{n-1}$ for every $n = 1, \dots, N$.

Note also that the condition $\mathbf{J_9}$ (b) automatically holds, if the sets $\overline{Y}'_n, \overline{Y}''_n = \emptyset$, $n = 1, \dots, N$, or these sets are finite or countable and the measures $P_{0,n}(y, A)$, $y_0 \in \mathbb{Y}'_{n-1} \cap \mathbb{Y}''_{n-1}$ have no atoms in points of the sets $\overline{Y}'_n, \overline{Y}''_n$, for every $n = 1, \dots, N$, as well as in the case, where measures $P_{0,n}(y_0, \cdot)$, $y_0 \in \mathbb{Y}'_{n-1} \cap \mathbb{Y}''_{n-1}$, $n = 1, \dots, N$ are absolutely continuous with respect to some σ-finite measure $P(A)$ on the Borel σ-algebra $\mathcal{B}_1$ and $P(\overline{\mathbb{Y}}'_n), P(\overline{\mathbb{Y}}''_n) = 0$, $n = 1, \dots, N$.

We shall prove below that the conditions $\mathbf{L_2}$, $\mathbf{N_1}$ and $\mathbf{J_9}$ imply that the condition $\mathbf{J_1}$ holds with the sets $\mathbb{Z}'_n = \mathbb{Y}'_n \times \{x_0\}$ and $\mathbb{Z}''_n = \mathbb{Y}''_n \times \{x_0\}$, $n = 0, \dots, N$.

The following result can be obtained by application of Theorem 5.2.2 to the above skeleton model.

Theorem 7.2.1. *Let a Markov log-price process $Y_{0,n}$ and the corresponding space-skeleton approximation Markov processes $Y_{\varepsilon,n}$ are defined as described above. Let also the conditions $\mathbf{B_4}[\gamma]$ and $\mathbf{C_{11}}[\beta]$ hold with the parameters γ and β such that either $\beta > \gamma > 0$ or $\beta = \gamma = 0$, and also the conditions $\mathbf{L_2}$, $\mathbf{N_1}$, $\mathbf{I_6}$, and $\mathbf{J_9}$ hold. Then, for every $n =$*

$0, 1, \dots, N$, *the following relation takes place for any* $y_\varepsilon \to y_0 \in \mathbb{Y}'_n \cap \mathbb{Y}''_n$:

$$\phi_{\varepsilon,n}(y_\varepsilon) \to \phi_{0,n}(y_0) \quad as \quad \varepsilon \to 0\,. \tag{7.54}$$

Proof. We are going to check that all conditions of Theorem 5.2.2 hold and to apply this theorem.

The following inequality holds for $\varepsilon \in (0, \varepsilon_0]$ and $n = 0, \dots, N$, $y \in \mathbb{R}_1$,

$$\begin{aligned} |h_{\varepsilon,n}(y) \; - y| \le (z^-_{\varepsilon,n} - y) I(y \le z^-_{\varepsilon,n}) \\ + \; \delta_{\varepsilon,n} I(z^-_{\varepsilon,n} < y \le z^+_{\varepsilon,n}) + (y - z^+_{\varepsilon,n}) I(y > z^+_{\varepsilon,n})\,. \end{aligned} \tag{7.55}$$

Relation (7.43) and the condition $\mathbf{L_2}$ imply that the following inequality holds for every $\varepsilon \in (0, \varepsilon_0]$, $y \in \mathbb{R}_1$, $n = 0, \dots, N-1$, and $\beta \ge 0$,

$$\mathsf{E}_{y,n} e^{\beta |Y_{\varepsilon,n+1} - Y_{\varepsilon,n}|} = \mathsf{E}_{y,n} e^{\beta |Y_{\varepsilon,n+1} - y|} = \mathsf{E}_{h_{\varepsilon,n}(y),n} e^{\beta |h_{\varepsilon,n+1}(Y_{0,n+1}) - y|}\,. \tag{7.56}$$

Using relations (7.52), (7.55), and (7.56) we get the following inequalities for every $\varepsilon \in (0, \varepsilon_0]$, $n = 0, \dots, N-1$, and $\beta \ge 0$:

$$\begin{aligned} \sup_{y \in \mathbb{Z}_{\varepsilon,n}} \mathsf{E}_{y,n} e^{\beta |Y_{\varepsilon,n+1} - Y_{\varepsilon,n}|} &= \sup_{y \in \mathbb{Z}_{\varepsilon,n}} \mathsf{E}_{h_{\varepsilon,n}(y),n} e^{\beta |h_{\varepsilon,n+1}(Y_{0,n+1}) - y|} \\ &\le \sup_{y \in \mathbb{Z}_{\varepsilon,n}} \mathsf{E}_{h_{\varepsilon,n}(y),n} e^{\beta (y - Y_{0,n+1})} I(Y_{0,n+1} \le z^-_{\varepsilon,n+1}) \\ &\quad + e^{\beta \delta_{\varepsilon,n+1}} \sup_{y \in \mathbb{Z}_{\varepsilon,n}} \mathsf{E}_{h_{\varepsilon,n}(y)} e^{\beta |Y_{0,n+1} - y|} I(z^-_{\varepsilon,n+1} < Y_{0,n+1} \le z^+_{\varepsilon,n+1}) \\ &\quad + \sup_{y \in \mathbb{Z}_{\varepsilon,n}} \mathsf{E}_{h_{\varepsilon,n}(y),n} e^{\beta (Y_{0,n+1} - y)} I(Y_{0,n+1} > z^+_{\varepsilon,n+1}) \\ &\le e^{\beta \delta_{\varepsilon,n+1}} \sup_{y \in \mathbb{Z}_{\varepsilon,n}} \mathsf{E}_{h_{\varepsilon,n}(y),n} e^{\beta |Y_{0,n+1} - y|} \\ &\le e^{\beta \delta_{\varepsilon,n+1}} e^{\beta \delta_{\varepsilon,n}} \sup_{y \in \mathbb{Z}_{\varepsilon,n}} \mathsf{E}_{h_{\varepsilon,n}(y),n} e^{\beta |Y_{0,n+1} - h_{\varepsilon,n}(y)\,|} \\ &\le e^{\beta(\delta_{\varepsilon,n+1} + \delta_{\varepsilon,n})} \sup_{y \in \mathbb{R}_1} \mathsf{E}_{y,n} e^{\beta |Y_{0,n+1} - y|}\,. \end{aligned} \tag{7.57}$$

Relation (7.57) implies the following inequality for moduli $\Delta_{\beta, \mathbb{Z}_\varepsilon}(Y_{\varepsilon,\cdot}, N)$ and $\Delta_\beta(Y_{0,\cdot}, N)$ which holds for every $\varepsilon \in (0, \varepsilon_0]$ and $\beta \ge 0$,

$$\Delta_{\beta, \mathbb{Z}_\varepsilon}(Y_{\varepsilon,\cdot}, N) \le \max_{0 \le n \le N-1} e^{\beta(\delta_{\varepsilon,n+1} + \delta_{\varepsilon,n})} \Delta_\beta(Y_{0,\cdot}, N)\,. \tag{7.58}$$

By the condition $\mathbf{N_1}$, $\max_{0 \le n \le N-1} e^{\beta(\delta_{\varepsilon,n+1} + \delta_{\varepsilon,n})} \to 1$ as $\varepsilon \to 0$. This relation, inequality (7.58), and the condition $\mathbf{C_{11}}[\beta]$ imply the following relation:

$$\varlimsup_{\varepsilon \to 0} \Delta_{\beta, \mathbb{Z}_\varepsilon}(Y_{\varepsilon,\cdot}, N) \le \Delta_\beta(Y_{0,\cdot}, N) < K_{49}\,. \tag{7.59}$$

Therefore, the condition $\mathbf{C'_6}[\bar{\beta}]$ (its one-dimensional variant for the model without index component, and with the sets $\mathbb{Z}_{\varepsilon,n}$ given by relation (7.51)) holds for the modified moduli of exponential moment compactness $\Delta_{\beta, \mathbb{Z}_\varepsilon}(Y_{\varepsilon,\cdot}, N)$.

The condition $\mathbf{N}_1$ obviously implies that the condition $\mathbf{H}_1$, with the sets $\mathbb{H}_{\varepsilon,n} = \mathbb{Z}_{\varepsilon,n} \times \{x_0\}$, $n = 0, \dots, N$, holds.

As was mentioned above, the condition $\mathbf{I}_6$ implies that the condition $\mathbf{I}_1$ (its one-dimensional variant without modulating component) holds.

Let us choose an arbitrary $1 \leq n \leq N$ and $y_\varepsilon \to y_0 \in \mathbb{Y}'_{n-1} \cap \mathbb{Y}''_{n-1}$ as $\varepsilon \to 0$. It follows from relation (7.55) that, in this case, also

$$h_{\varepsilon,n-1}(y_\varepsilon) \to y_0 \quad \text{as} \quad \varepsilon \to 0\,. \tag{7.60}$$

Let us take arbitrary $\kappa, \varrho > 0$. One can find $T_\kappa > 0$ such that points $\pm T_\kappa$ are points of continuity for the distribution functions $\mathsf{P}_{y_0,n-1}\{Y_{0,n} \leq u\} = P_{0,n}(y_0, (-\infty, u])$, $n = 1, \dots, N$ and, for $n = 1, \dots, N$,

$$P_{0,n}(y_0, (-\infty, -T_\kappa]) + P_{0,n}(y_0, [T_\kappa, \infty)) \leq \kappa\,. \tag{7.61}$$

Using the above remarks, the conditions $\mathbf{N}_1$ and $\mathbf{J}_9$ (a) and relations (7.55), (7.60), and (7.61), we get the following relation, for $n = 1, \dots, N$:

$$\begin{aligned}
&\overline{\lim_{0<\varepsilon\to 0}}\, \mathsf{P}_{h_{\varepsilon,n-1}(y_\varepsilon),n-1}\{|h_{\varepsilon,n}(Y_{0,n}) - Y_{0,n}| \geq \varrho\} \\
&\quad \leq \overline{\lim_{0<\varepsilon\to 0}} \Big(\mathsf{P}_{h_{\varepsilon,n-1}(y_\varepsilon),n-1}\{Y_{0,n} \leq z^-_{\varepsilon,n}\} \\
&\qquad + I(\delta_{\varepsilon,n} \geq \varrho) + \mathsf{P}_{h_{\varepsilon,n-1}(y_\varepsilon),n-1}\{Y_{0,n} > z^+_{\varepsilon,n}\}\Big) \\
&\quad \leq \overline{\lim_{0<\varepsilon\to 0}} \Big(\mathsf{P}_{h_{\varepsilon,n-1}(y_\varepsilon),n-1}\{Y_{0,n} \leq -T_\kappa\} \\
&\qquad + \mathsf{P}_{h_{\varepsilon,n-1}(y_\varepsilon),n-1}\{Y_{0,n} \geq T_\kappa\}\Big) \\
&\quad = \mathsf{P}_{y_0,n-1}\{Y_{0,n} \leq -T_\kappa\} + \mathsf{P}_{y_0,n-1}\{Y_{0,n} \geq T_\kappa\} \leq \kappa\,.
\end{aligned} \tag{7.62}$$

Since an arbitrary choice of $\kappa, \varrho > 0$, relation (7.62) implies that for any for $n = 1, \dots, N$ and $\varrho > 0$,

$$\overline{\lim_{0<\varepsilon\to 0}}\, \mathsf{P}_{h_{\varepsilon,n-1}(y_\varepsilon),n-1}\{|h_{\varepsilon,n}(Y_{0,n}) - Y_{0,n}| \geq \varrho\} = 0\,. \tag{7.63}$$

Relation (7.63) implies that, for every $n = 1, \dots, N$, probability measures $\mathsf{P}_{h_{\varepsilon,n-1}(y_\varepsilon),n-1}\{h_{\varepsilon,n}(Y_{0,n}) \in \cdot\}$ and $\mathsf{P}_{h_{\varepsilon,n-1}(y_\varepsilon),n-1}\{Y_{0,n} \in \cdot\}$ have the same weak limits as $0 < \varepsilon \to 0$.

This proposition, relations (7.43), (7.60), and the condition $\mathbf{J}_9$ (a) imply that the following relation of weak convergence holds for any $y_\varepsilon \to y_0 \in \mathbb{Y}'_{n-1} \cap \mathbb{Y}''_{n-1}$ as $0 < \varepsilon \to 0$, for every $1 \leq n \leq N$,

$$\begin{aligned}
&P_{\varepsilon,n}(y_\varepsilon, \cdot) = \mathsf{P}_{h_{\varepsilon,n-1}(y_\varepsilon),n-1}\{h_{\varepsilon,n}(Y_{0,n}) \in \cdot\} \\
&\Rightarrow P_{0,n}(y_0, \cdot) = \mathsf{P}_{y_0,n-1}\{Y_{0,n} \in \cdot\} \quad \text{as} \quad 0 < \varepsilon \to 0\,.
\end{aligned} \tag{7.64}$$

The equality of the distributions on the right-hand side of relation (7.64) obviously implies that this asymptotic relation also holds as $0 \leq \varepsilon \to 0$. Thus, the condition $\mathbf{J}_1$ (a) holds with the sets $\mathbb{Z}'_n = \mathbb{Y}'_n \times \{x_0\}$ and $\mathbb{Z}''_n = \mathbb{Y}''_n \times \{x_0\}$, $n = 0, \dots, N$.

Since, the conditions $\mathbf{J}_9$ (b) and $\mathbf{J}_1$ (b) coincide, we can conclude that the condition $\mathbf{J}_1$ (its one-dimensional variant without modulating component) holds in this case, with sets $\mathbb{Z}'_n, \mathbb{Z}''_n$, $n = 0, \dots, N$.

It follows from the above remarks that all conditions of Theorem 5.2.2 hold, and, therefore, Theorem 7.2.1 is a corollary of Theorem 5.2.2. □

Remark 7.2.1. It is not out of picture to note that Theorem 5.2.1, based on the standard condition of exponential moment compactness $\mathbf{C}_6[\bar{\beta}]$ (its one-dimensional variant), cannot be applied in this case. Indeed, as follows from formula (7.56), the modulus $\Delta_\beta(Y_{\varepsilon,\cdot}, N) \equiv \infty$, for every $\varepsilon \in (0, \varepsilon_0]$ and, thus, the condition $\mathbf{C}_6[\bar{\beta}]$ (its one-dimensional variant) does not holds.

7.2.3 Convergence of space-skeleton approximations for optimal expected rewards for univariate Markov log-price processes

Let us now assume that we have found, using backward recurrence algorithm presented above in Subsection 7.2.1, the values of the reward functions $\phi_{\varepsilon,0}(y_{\varepsilon,0,l})$ for the following initial points:

$$y_{\varepsilon,0,l} = \delta_{\varepsilon,0} l + \lambda_{\varepsilon,0}\,, \quad m^-_{\varepsilon,0} \le l \le m^+_{\varepsilon,0}\,. \tag{7.65}$$

The optimal expected reward functional $\Phi_\varepsilon = \Phi(\mathcal{M}^{(\varepsilon)}_{max,N})$ for the approximation log-price processes $Y_{\varepsilon,n}$ can be found using the following formula:

$$\Phi_\varepsilon = \sum_{l=m^-_{\varepsilon,0}}^{m^+_{\varepsilon,0}} \phi_{\varepsilon,0}(y_{\varepsilon,0,l}) p_{\varepsilon,0}\,, \tag{7.66}$$

where $p_{\varepsilon,l} = P_{0,0}(\mathbb{I}_{\varepsilon,0,l}) = \mathsf{P}\{h_{\varepsilon,0}(Y_{0,0}) = y_{\varepsilon,0,l}\}$, $l = m^-_{\varepsilon,0}, \dots, m^+_{\varepsilon,0}$.

Let us formulate conditions that should be imposed on the log-price process $Y_{0,n}$ and the pay-off function $g(n, e^y)$ in order that the corresponding optimal expected rewards Φ_ε would converge to the limiting optimal expected reward Φ_0 as $\varepsilon \to 0$.

We impose (on the initial value of the log-price process $Y_{0,0}$) the following first-type condition of exponential moment boundedness, which is a one-dimensional variant of the condition $\mathbf{D}_1[\bar{\beta}]$ and should be assumed to hold for some $\beta \ge 0$:

$\mathbf{D}_{10}[\beta]$: $\mathsf{E}e^{\beta|Y_{0,0}|} < K_{51}$, for some $1 < K_{51} < \infty$.

The condition $\mathbf{K}_1$ should be replaced by the following condition imposed on the initial distribution $P_{0,0}(A)$:

$\mathbf{K}_8$: $P_{0,0}(\mathbb{Y}'_0 \cap \mathbb{Y}''_0) = 1$, where $\mathbb{Y}'_0$ and $\mathbb{Y}''_0$ are the sets introduced, respectively, in the conditions $\mathbf{I}_6$ and $\mathbf{J}_9$.

The following theorem is a variant of Theorem 5.2.4 for the above space-skeleton approximation model.

Theorem 7.2.2. *Let a Markov log-price process $Y_{0,n}$ and the corresponding space-skeleton approximation Markov processes $Y_{\varepsilon,n}$ are defined as described above. Also let conditions $\mathbf{B_4}[\gamma]$, $\mathbf{C_{11}}[\beta]$, and $\mathbf{D_{10}}[\beta]$ hold with the parameters γ and β such that either $\beta > \gamma > 0$ or $\beta = \gamma = 0$, and also the conditions $\mathbf{L_2}$, $\mathbf{M_2}$, $\mathbf{N_1}$, $\mathbf{I_6}$, $\mathbf{J_9}$, and $\mathbf{K_8}$ hold. Then, the following relation takes place:*

$$\Phi_\varepsilon \to \Phi_0 \quad as \quad \varepsilon \to 0 . \tag{7.67}$$

Proof. We are going to check that all conditions of Theorem 5.2.4 hold and to apply this theorem. In fact, we should only check that one-dimensional variants of the conditions $\mathbf{D}'_7[\bar{\beta}]$ (its one-dimensional variant for the model without index component, and the sets $\mathbb{Z}_{\varepsilon,0} = [z^-_{\varepsilon,0}, z^-_{\varepsilon,0}]$) and $\mathbf{K_1}$ hold. Indeed, this check was made for other conditions in the proof of Theorem 7.2.1.

By the condition $\mathbf{N_1}$, there exist $\varepsilon_{15} \in (0, \varepsilon_0]$ such that $\pm z^\pm_{\varepsilon,0} \geq 0$, for $\varepsilon \in (0, \varepsilon_{15}]$. The following inequality holds for $\varepsilon \in (0, \varepsilon_{15}]$:

$$\begin{aligned} |h_{\varepsilon,0}(y)| &\leq |y| I(y \notin [z^-_{\varepsilon,0}, z^+_{\varepsilon,0}]) + (|y| + \delta_{\varepsilon,0}) I(y \in [z^-_{\varepsilon,0}, z^+_{\varepsilon,0}]) \\ &\leq (|y| + \delta_{\varepsilon,0}) . \end{aligned} \tag{7.68}$$

Relation (7.44) and the condition $\mathbf{M_2}$ imply that the following relation holds for every $\varepsilon \in (0, \varepsilon_0]$ and $\beta \geq 0$:

$$\mathsf{E} e^{\beta |Y_{\varepsilon,0}|} = \mathsf{E} e^{\beta |h_{\varepsilon,0}(Y_{0,0})|} . \tag{7.69}$$

Using relations (7.68) and (7.69) we obtain the following inequality, for $\varepsilon \in (0, \varepsilon_{15}]$:

$$\mathsf{E} e^{\beta |Y_{\varepsilon,0}|} = \mathsf{E} e^{\beta |h_{\varepsilon,0}(Y_{0,0})|} \leq e^{\beta \delta_{\varepsilon,0}} \mathsf{E} e^{\beta |Y_{0,0}|} . \tag{7.70}$$

Since, by the condition $\mathbf{N_1}$, $e^{\beta \delta_{\varepsilon,0}} \to 1$ as $\varepsilon \to 0$, the condition $\mathbf{D_{10}}[\beta]$ and inequality (7.70) imply the following relation:

$$\overline{\lim_{\varepsilon \to 0}} \mathsf{E} e^{\beta |Y_{\varepsilon,0}|} \leq \overline{\lim_{\varepsilon \to 0}} e^{\beta \delta_{\varepsilon,0}} \mathsf{E} e^{\beta |Y_{0,0}|} < K_{51} . \tag{7.71}$$

Also, relation (7.44) implies that $\mathsf{P}\{Y_{\varepsilon,0} \in \mathbb{Z}_{\varepsilon,0}\} = \mathsf{P}\{h_{\varepsilon,0}(Y_{0,0}) \in \mathbb{Z}_{\varepsilon,0}\} = 1$, for every $\varepsilon \in (0, \varepsilon_0]$.

Thus, the one-dimensional variant of the condition $\mathbf{D}'_7[\bar{\beta}]$ (its one-dimensional variant for the model without index component and with the sets $\mathbb{Z}_{\varepsilon,0}$ given by relation (7.51)) holds.

Let us take an arbitrary $\varrho > 0$. Using the condition $\mathbf{N_1}$ and inequality (7.55), we get the following relation:

$$\begin{aligned} &\overline{\lim_{0<\varepsilon \to 0}} \mathsf{P}\{|h_{\varepsilon,0}(Y_{0,0}) - Y_{0,0}| \geq \varrho\} \\ &\quad \leq \overline{\lim_{0<\varepsilon \to 0}} \left(\mathsf{P}\{Y_{0,0} \leq z^-_{\varepsilon,0}\} + I(\delta_{\varepsilon,0} \geq \varrho) + \mathsf{P}\{Y_{0,0} > z^+_{\varepsilon,0}\} \right) = 0 . \end{aligned} \tag{7.72}$$

Relation (7.72) implies the following relation:

$$P_\varepsilon(\cdot) = \mathsf{P}\{h_{\varepsilon,0}(Y_{0,0}) \in \cdot\} \Rightarrow P_0(\cdot) = \mathsf{P}\{Y_{0,0} \in \cdot\} \quad \text{as} \quad \varepsilon \to 0\,. \tag{7.73}$$

Relation (7.73) and the condition $\mathbf{K_8}$ imply that the condition $\mathbf{K_1}$ (its univariate variant) holds with the sets $\mathbb{Z}'_0 = \mathbb{Y}'_0 \times \{x_0\}$ and $\mathbb{Z}''_0 = \mathbb{Y}''_0 \times \{x_0\}$.

It follows from the above remarks that all conditions of Theorem 5.2.4 hold, and, therefore, Theorem 7.2.2 is a corollary of Theorem 5.2.4. □

7.3 Multivariate Markov LPP with bounded characteristics

In this section, we present space-skeleton approximation models for discrete time multivariate modulated Markov log-price processes with bounded characteristics, describe the corresponding backward algorithms for computing reward functions and formulate convergence theorems for log-reward functions and optimal expected rewards for approximating space-skeleton Markov log-price processes.

7.3.1 Convergence of space-skeleton approximations for rewards functions for multivariate modulated Markov log-price processes

Let us now assume that the modulated log-price process $\vec{Z}_{\varepsilon,n} = (\vec{Y}_{\varepsilon,n}, X_{\varepsilon,n})$ is, for every $\varepsilon \in [0, \varepsilon_0]$, an inhomogeneous Markov chain with a phase space $\mathbb{Z} = \mathbb{R}_k \times \mathbb{X}$, an initial distribution $P_\varepsilon(A)$, and one-step transition probabilities $P_{\varepsilon,n}(\vec{z}, A)$.

We assume that the phase space of the index component $\mathbb{X}$ is a Polish metric space, i.e. complete, separable, metric space, with a metric $d_{\mathbb{X}}(x', x'')$.

We would like to approximate the log-price process $\vec{Z}_{0,n}$ by the log-price processes $\vec{Z}_{\varepsilon,n}$. We additionally assume that the log-price process $\vec{Z}_{0,n}$ (for $\varepsilon = 0$) can be a nonatomic Markov chain, while the log-price processes $\vec{Z}_{\varepsilon,n}$ (for $\varepsilon \in (0, \varepsilon_0]$) are atomic Markov chains, i.e. their transition probabilities $P_{\varepsilon,n}(\vec{z}, A)$ and initial distributions $P_\varepsilon(A)$ are distributions concentrated at finite sets of points.

We also assume that the pay-off function $g(n, e^{\vec{y}}, x)$ does not depend on the parameter ε.

In general, space-skeleton approximations can be constructed in the following way.

Let us choose nonempty sets $\mathbb{A}_{\varepsilon,n,l} \in \mathcal{B}_{\mathbb{Z}}$, $l = m^-_{\varepsilon,n}, \dots, m^+_{\varepsilon,n}$, $n = 0, 1, \dots$, such that

$$\mathbb{A}_{\varepsilon,n,l'} \cap \mathbb{A}_{\varepsilon,n,l''} = \emptyset,\, l' \neq l''\,; \quad \mathbb{Z} = \cup_{l=m^-_{\varepsilon,n}}^{m^+_{\varepsilon,n}} \mathbb{A}_{\varepsilon,n,l}\,, \quad n = 0, 1, \dots\,. \tag{7.74}$$

Let us also choose skeleton points,

$$\vec{z}_{\varepsilon,n,l} \in \mathbb{A}_{\varepsilon,n,l}, \quad l = m^-_{\varepsilon,n}, \dots, m^+_{\varepsilon,n}, \quad n = 0, 1, \dots . \tag{7.75}$$

Let us assume that the following fitting condition holds:

L$_3$: For every $\varepsilon \in (0, \varepsilon_0]$, transition probabilities $\mathsf{P}\{\vec{Z}_{\varepsilon,n} = \vec{z}_{\varepsilon,n,l}/\vec{Z}_{\varepsilon,n-1} = \vec{z}\} = p_{\varepsilon,l}(\vec{z}, n) = P_{0,n}(\vec{z}_{\varepsilon,n-1,l'}, \mathbb{A}_{\varepsilon,n,l})$, $\vec{z} \in \mathbb{A}_{\varepsilon,n-1,l'}$, $l' = m^-_{\varepsilon,n-1}, \dots, m^+_{\varepsilon,n-1}$, $l = m^-_{\varepsilon,n}, \dots, m^+_{\varepsilon,n}$, $n = 1, 2, \dots$.

If we are also interested to fit initial distributions, the following fitting condition should be assumed:

M$_3$: For every $\varepsilon \in (0, \varepsilon_0]$, initial probabilities $\mathsf{P}\{\vec{Z}_{\varepsilon,0} = \vec{z}_{\varepsilon,0,l}\} = p_{\varepsilon,l} = P_{0,0}(\mathbb{A}_{\varepsilon,0,l})$, $l = m^-_{\varepsilon,0}, \dots, m^+_{\varepsilon,0}$.

The most natural variant of the above space-skeleton approximation model is the following.

Let us choose $\delta_{\varepsilon,n,i} > 0$, $\lambda_{\varepsilon,n,i} \in \mathbb{R}_1$, $i = 1, \dots, k$, $n = 0, 1, \dots$ and integer numbers $m^-_{\varepsilon,n,j} \leq m^+_{\varepsilon,n,j}$, $j = 0, \dots, k$, $n = 0, 1, \dots$.

First, the intervals $\mathbb{I}_{\varepsilon,n,i,l}$ should be constructed for $l = m^-_{\varepsilon,n,i}, \dots, m^+_{\varepsilon,n,i}$, $i = 1, \dots, k$, $n = 0, 1, \dots$,

$$\mathbb{I}_{\varepsilon,n,i,l} = \begin{cases} (-\infty,\ \delta_{\varepsilon,n,i}(m^-_{\varepsilon,n,i} + \frac{1}{2})] + \lambda_{\varepsilon,n,i} & \text{if } l = m^-_{\varepsilon,n,i}, \\ \left(\delta_{\varepsilon,n,i}\left(l - \frac{1}{2}\right),\ \delta_{\varepsilon,n,i}\left(l + \frac{1}{2}\right)\right] + \lambda_{\varepsilon,n,i} & \text{if } m^-_{\varepsilon,n,i} < l < m^+_{\varepsilon,n,i}, \\ \left(\delta_{\varepsilon,n,i}(m^+_{\varepsilon,n,i} - \frac{1}{2}),\ \infty\right) + \lambda_{\varepsilon,n,i} & \text{if } l = m^+_{\varepsilon,n,i}. \end{cases} \tag{7.76}$$

Then cubes $\mathbb{I}_{\varepsilon,n,l_1,\dots,l_k} = \mathbb{I}_{\varepsilon,n,1,l_1} \times \cdots \times I_{\varepsilon,n,k,l_k}$, $l_i = m^-_{\varepsilon,n,i}, \dots, m^+_{\varepsilon,n,i}$, $i = 1, \dots, k$, $n = 0, 1, \dots$ should be defined.

By the definition, the points $y_{\varepsilon,n,i,l_i} = l_i\delta_{\varepsilon,n,i} + \lambda_{\varepsilon,n,i} \in \mathbb{I}_{\varepsilon,n,i,l_i}$ and, thus, the vector points $(y_{\varepsilon,n,1,l_1}, \dots, y_{\varepsilon,n,k,l_k}) \in \mathbb{I}_{\varepsilon,n,l_1,\dots,l_k}$, for $l_i = m^-_{\varepsilon,n,i}, \dots, m^+_{\varepsilon,n,i}$, $i = 1, \dots, k$, $n = 0, 1, \dots$.

Second, nonempty sets $\mathbb{J}_{\varepsilon,n,l} \subset \mathcal{B}_{\mathbb{X}}$, $l = m^-_{\varepsilon,n,0}, \dots, m^+_{\varepsilon,n,0}$, $n = 0, 1, \dots$, such that (a) $\mathbb{J}_{\varepsilon,n,l'} \cap \mathbb{J}_{\varepsilon,n,l''} = \emptyset$, $l' \neq l''$, $n = 0, 1, \dots$; (b) $\mathbb{X} = \cup_{m^-_{\varepsilon,n,0} \leq l \leq m^+_{\varepsilon,n,0}} \mathbb{J}_{\varepsilon,n,l}$, $n = 0, 1, \dots$, should be constructed.

Recall that one of our model assumption is that $\mathbb{X}$ is a Polish space, i.e. a complete, separable, metric space. In this case, it is natural to assume that there exist "large" sets $\mathbb{K}_{\varepsilon,n}$, $n = 0, 1, \dots$, and "small" nonempty sets $\mathbb{K}_{\varepsilon,n,l} \subseteq \mathcal{B}_{\mathbb{X}}$, $l = m^-_{\varepsilon,n,0}, \dots, m^+_{\varepsilon,n,0}$, $n = 0, 1, \dots$, such that (c) $\mathbb{K}_{\varepsilon,n,l'} \cap \mathbb{K}_{\varepsilon,n,l''} = \emptyset$, $l' \neq l''$, $n = 0, 1, \dots$; (d) $\cup_{m^-_{\varepsilon,n,0} \leq l \leq m^+_{\varepsilon,n,0}} \mathbb{K}_{\varepsilon,n,l} = \mathbb{K}_{\varepsilon,n}$, $n = 0, 1, \dots$. The exact sense of the epithets "large" and "small" used above is clarified in the condition **N**$_2$ formulated below.

Then, the sets $\mathbb{J}_{\varepsilon,n,l}$ can be defined in the following way, for $n = 0, 1, \dots$:

$$\mathbb{J}_{\varepsilon,n,l} = \begin{cases} \mathbb{K}_{\varepsilon,n,l} & \text{if } m^-_{\varepsilon,n,0} \leq l < m^+_{\varepsilon,n,0}, \\ \mathbb{K}_{\varepsilon,n,m^+_{\varepsilon,n,0}} \cup \overline{\mathbb{K}}_{\varepsilon,n} & \text{if } l = m^+_{\varepsilon,n,0}. \end{cases} \tag{7.77}$$

Finally, skeleton points $x_{\varepsilon,n,l} \in \mathbb{J}_{\varepsilon,n,l}$, $l = m^-_{\varepsilon,n,0}, \dots, m^+_{\varepsilon,n,0}$, $n = 0, 1, \dots$, should be chosen.

The particular important case is, where the space $\mathbb{X} = \{1, \dots, m\}$ is a finite set and metric $d_{\mathbb{X}}(x, y) = I(x \neq y)$. In this case, the standard choice is to take sets $\mathbb{K}_{\varepsilon,n} = \mathbb{X}$, $n = 0, 1, \dots$; one-point sets $\mathbb{J}_{\varepsilon,n,l} = \mathbb{K}_{\varepsilon,n,l} = \{l\}, 1 = m^-_{\varepsilon,n,0} \leq l \leq m^+_{\varepsilon,n,0} = m$, $n = 0, 1, \dots$; and points $x_{\varepsilon,n,l} = l$, $l = 1, \dots, m$, $n = 0, \dots$.

If the space $\mathbb{X} = \{1, 2, \dots\}$ is a countable set and again metric $d_{\mathbb{X}}(x, y) = I(x \neq y)$, the standard choice is to take sets $\mathbb{K}_{\varepsilon,n} = \{l : l \leq m_\varepsilon\}$, $n = 0, 1, \dots$, and one-point sets $\mathbb{K}_{\varepsilon,n,l} = \{l\}, 1 = m^-_{\varepsilon,n,0} \leq l \leq m^+_{\varepsilon,n,0} = m_\varepsilon$, $n = 0, 1, \dots$. In this case, the sets $\mathbb{J}_{\varepsilon,n,l} = \{l\}, 1 = m^-_{\varepsilon,n,0} \leq l < m^+_{\varepsilon,n,0} = m_\varepsilon$, $n = 0, 1, \dots$, while $\mathbb{J}_{\varepsilon,n,m_\varepsilon} = \{l : l \geq m_\varepsilon\}$, and points $x_{\varepsilon,n,l} = l$, $l = 1, \dots, m_\varepsilon$, $n = 0, 1, \dots$.

Another important case is, where $\mathbb{X} = \mathbb{R}_p$. In this case one can always choose as $K_{\varepsilon,n}$ and $\mathbb{K}_{\varepsilon,n,l}$, $m^-_{\varepsilon,n,0} \leq l \leq m^+_{\varepsilon,n,0}$ cubes satisfying the conditions imposed above on these sets.

Third, the sets $\mathbb{A}_{\varepsilon,n,\bar{l}}$ and skeleton points, $\vec{z}_{\varepsilon,n,\bar{l}} \in \mathbb{A}_{\varepsilon,n,\bar{l}}$ can be defined for $\bar{l} = (l_0, l_1, \dots, l_k)$, $l_j = m^-_{\varepsilon,n,j}, \dots, m^+_{\varepsilon,n,j}$, $j = 0, \dots, k$, $n = 0, 1, \dots$, in the following way:

$$\mathbb{A}_{\varepsilon,n,\bar{l}} = \mathbb{I}_{\varepsilon,n,l_1,\dots,l_k} \times \mathbb{J}_{\varepsilon,n,l_0}, \tag{7.78}$$

and

$$\vec{z}_{\varepsilon,n,\bar{l}} = (\vec{y}_{\varepsilon,n,\bar{l}}, x_{\varepsilon,n,\bar{l}}) = ((y_{\varepsilon,n,1,l_1}, \dots, y_{\varepsilon,n,k,l_k}), x_{\varepsilon,n,l_0}). \tag{7.79}$$

In this case, the corresponding skeleton sets $\mathbb{F}_{\varepsilon,n} = \{\vec{z}_{\varepsilon,n,\bar{l}} : \bar{l} \in \mathbb{L}_{\varepsilon,n}\}$, where sets of vector indices $\mathbb{L}_{\varepsilon,n}$ are defined in the following way, for $n = 0, 1, \dots$:

$$\mathbb{L}_{\varepsilon,n} = \{\bar{l} = (l_0, \dots, l_k), l_j = m^-_{\varepsilon,n,j}, \dots, m^+_{\varepsilon,n,j}, \quad j = 0, \dots, k\}. \tag{7.80}$$

In this case, the fitting condition $\mathbf{L_3}$ and $\mathbf{M_3}$ takes the following form:

$\mathbf{L_4}$: For every $\varepsilon \in (0, \varepsilon_0]$, transition probabilities $\mathsf{P}\{\vec{Z}_{\varepsilon,n} = \vec{z}_{\varepsilon,n,\bar{l}}/\vec{Z}_{\varepsilon,n-1} = \vec{z}\} = p_{\varepsilon,\bar{l}}(\vec{z}, n) = P_{0,n}(\vec{z}_{\varepsilon,n-1,\bar{l}'}, \mathbb{A}_{\varepsilon,n,\bar{l}})$, $\vec{z} \in \mathbb{A}_{\varepsilon,n-1,\bar{l}'}$, $\bar{l}' = (l'_0, \dots, l'_k) \in \mathbb{L}_{\varepsilon,n-1}$, $\bar{l} = (l_0, \dots, l_k) \in \mathbb{L}_{\varepsilon,n}$, $n = 1, 2, \dots$.

and

$\mathbf{M_4}$: For every $\varepsilon \in (0, \varepsilon_0]$, initial probabilities $\mathsf{P}\{\vec{Z}_{\varepsilon,0} = \vec{z}_{\varepsilon,0,\bar{l}}\} = p_{\varepsilon,\bar{l}} = P_{0,0}(\mathbb{A}_{\varepsilon,0,\bar{l}})$, $\bar{l} = (l_0, \dots, l_k) \in \mathbb{L}_{\varepsilon,0}$.

The difference in notations between the sets $\mathbb{A}_{\varepsilon,n,l}$ and $\mathbb{A}_{\varepsilon,n,\bar{l}}$ and between points $\vec{z}_{\varepsilon,n,l}$ and $\vec{z}_{\varepsilon,n,\bar{l}}$ can be removed by a natural renumeration of indices, namely $(m^-_{\varepsilon,n,0}, \dots, m^-_{\varepsilon,n,k}) \leftrightarrow m^-_{\varepsilon,n}, \dots, (m^-_{\varepsilon,n,0} + 1, \dots, m^-_{\varepsilon,n,k}) \leftrightarrow m^-_{\varepsilon,n} + 1, \dots, (m^+_{\varepsilon,n,0}, \dots, m^+_{\varepsilon,n,k}) \leftrightarrow m^+_{\varepsilon,n}$, where

$$m^+_{\varepsilon,n} - m^-_{\varepsilon,n} + 1 = \prod_{j=0}^{k} (m^+_{\varepsilon,n,j} - m^-_{\varepsilon,n,j} + 1), \quad n = 0, 1, \dots. \tag{7.81}$$

The simplest variant is to choose integers $\pm m^\pm_{\varepsilon,n} \geq 0$, $n = 0, 1, \dots$.

Let us define the skeleton functions, $h_{\varepsilon,n,i}(y)$, $y \in \mathbb{R}_1$, for $\varepsilon \in (0, \varepsilon_0]$ and $i = 1, \dots, k$, $n = 0, 1, \dots$,

$$h_{\varepsilon,n,i}(y) = \begin{cases} \delta_{\varepsilon,n,i} m^-_{\varepsilon,n,i} + \lambda_{\varepsilon,n,i} & \text{if } y \le \delta_{\varepsilon,n,j}(m^-_{\varepsilon,n,i} + \frac{1}{2}) + \lambda_{\varepsilon,n,i}, \\ \delta_{\varepsilon,n,i} l + \lambda_{\varepsilon,n,i} & \text{if } \delta_{\varepsilon,n,i}(l - \frac{1}{2}) + \lambda_{\varepsilon,n,i} < y \le \delta_{\varepsilon,n,i}(l + \frac{1}{2}) + \lambda_{\varepsilon,n,i}, \\ & \quad m^-_{\varepsilon,n,i} < l < m^+_{\varepsilon,n,i}, \\ \delta_{\varepsilon,n,i} m^+_{\varepsilon,n,i} + \lambda_{\varepsilon,n,i} & \text{if } y > \delta_{\varepsilon,n,i}(m^+_{\varepsilon,n,i} - \frac{1}{2}) + \lambda_{\varepsilon,n,i}. \end{cases} \tag{7.82}$$

and the skeleton functions $h_{\varepsilon,n,0}(x)$, $x \in \mathbb{X}$, for $\varepsilon \in (0, \varepsilon_0]$ and $n = 0, 1, \dots$,

$$h_{\varepsilon,n,0}(x) = \{x_{\varepsilon,n,l} \quad \text{if} \quad x \in \mathbb{J}_{\varepsilon,n,l},\, m^-_{\varepsilon,n,0} \le l \le m^+_{\varepsilon,n,0}\,. \tag{7.83}$$

Finally, let us define the vector skeleton functions $\hat{h}_{\varepsilon,n}(\vec{y})$, $\vec{y} = (y_1, \dots, y_k) \in \mathbb{Y}$, for $\varepsilon \in (0, \varepsilon_0]$ and $n = 0, 1, \dots$,

$$\hat{h}_{\varepsilon,n}(\vec{y}) = (h_{\varepsilon,n,1}(y_1), \dots, h_{\varepsilon,n,k}(y_k))\,. \tag{7.84}$$

and the vector skeleton functions $\bar{h}_{\varepsilon,n}(\vec{z})$, $\vec{z} = (\vec{y}, x) \in \mathbb{Z}$, for $\varepsilon \in (0, \varepsilon_0]$ and $n = 0, 1, \dots$,

$$\bar{h}_{\varepsilon,n}(\vec{z}) = (\hat{h}_{\varepsilon,n}(\vec{y}), h_{\varepsilon,n,0}(x))\,. \tag{7.85}$$

The transition probabilities $P_{\varepsilon,n}(\vec{z}, A) = \mathsf{P}\{\vec{Z}_{\varepsilon,n} \in A / \vec{Z}_{\varepsilon,n-1} = \vec{z}\}$, $\vec{z} \in \mathbb{Z}$, $n = 1, 2, \dots$, take the following form, for every $\varepsilon \in (0, \varepsilon_0]$:

$$\begin{aligned} P_{\varepsilon,n}(\vec{z}, A) &= \sum_{\vec{z}_{\varepsilon,n,\bar{l}} \in A} P_{0,n}(\bar{h}_{\varepsilon,n-1}(\vec{z}), \mathbb{A}_{\varepsilon,n,\bar{l}}) \\ &= \mathsf{P}\{\bar{h}_{\varepsilon,n}(\vec{Z}_{0,n}) \in A / \vec{Z}_{0,n-1} = \bar{h}_{\varepsilon,n-1}(\vec{z})\}\,, \quad A \in \mathcal{B}_{\mathbb{Z}}\,, \end{aligned} \tag{7.86}$$

As far as the initial distribution $P_{\varepsilon,0}(A) = \mathsf{P}\{\vec{Z}_{\varepsilon,0} \in A\}$ is concerned, it takes the following form, for every $\varepsilon \in (0, \varepsilon_0]$:

$$P_{\varepsilon,0}(A) = \sum_{\vec{z}_{\varepsilon,0,\bar{l}} \in A} P_{0,0}(\mathbb{A}_{\varepsilon,0,\bar{l}}) = \mathsf{P}\{\bar{h}_{\varepsilon,0}(\vec{Z}_{0,0}) \in A\}\,, \quad A \in \mathcal{B}_{\mathbb{Z}}\,. \tag{7.87}$$

The quantity $P_{\varepsilon,n}(\vec{z}, A)$, defined in relation (7.86), is a measurable function in $\vec{z}$ and a probability measure as a function of A and, thus, it can serve as transition probabilities for a Markov chain. Also the quantity $P_{\varepsilon,0}(A)$, defined in relation (7.87), is a probability measure and, thus, it can serve as an initial distribution for a Markov chain.

The role of approximating the log-price process $\vec{Z}_{\varepsilon,n}$ can play any Markov chain with the phase space $\mathbb{Z}$, the initial distribution $P_{\varepsilon,0}(A)$ defined in relation (7.87), and the transition probabilities $P_{\varepsilon,n}(\vec{z}, A)$ defined in relation (7.86).

Obviously, the transition probabilities $P_{\varepsilon,n}(\vec{z}, A)$ and the initial distribution $P_{\varepsilon,0}(A)$ of the Markov chain $\vec{Z}_{\varepsilon,n}$ satisfy the conditions $\mathbf{L}_4$ and $\mathbf{M}_4$ and, in sequel, relations (7.86) and (7.87).

As in the one-dimensional case, we can realize the strong fitting instead of the distributional one.

Let us assume that the Markov chain $\vec{Z}_{0,n}$ is given in the dynamic form

$$\begin{cases} \vec{Z}_{0,n} = A_{0,n}(\vec{Z}_{0,n-1}, U_{0,n}) , & n = 1, 2, \dots , \\ \vec{Z}_{0,0} = A_{0,0}(U_{0,0}) , \end{cases} \tag{7.88}$$

where (a) $U_{0,n}$, $n = 0, 1, \dots$, are independent random variables taking values in a measurable space $\mathbb{U}$ and with distributions $G_n(u) = \mathsf{P}\{U_{0,n} \in A\}$, $n = 0, 1, \dots$, (b) $A_{0,n}(\vec{z}, u)$, $n = 0, 1, \dots$, are the measurable functions acting from $\mathbb{Z} \times \mathbb{U}$ to $\mathbb{Z}$, and (c) $A_{0,0}(u)$ is a measurable function acting from $\mathbb{U}$ to $\mathbb{Z}$.

Since $\mathbb{Z}$ is a Polish metric space, the above construction can be realized with the space $\mathbb{U} = [0, 1]$ and random variables $U_{0,n}$, $n = 0, 1, \dots$, uniformly distributed in this interval.

The Markov chain $\vec{Z}_{\varepsilon,n}$ can be defined in the dynamic form

$$\begin{cases} \vec{Z}_{\varepsilon,n} = \bar{h}_{\varepsilon,n}(A_{0,n}(\bar{h}_{\varepsilon,n-1}(\vec{Z}_{\varepsilon,n-1}), U_{0,n})) , & n = 1, 2, \dots , \\ \vec{Z}_{\varepsilon,0} = \bar{h}_{\varepsilon,0}(A_0(U_{0,0})) . \end{cases} \tag{7.89}$$

The following lemma is a variant of Lemma 7.1.3.

Lemma 7.3.1. *Let, for every $\varepsilon \in (0, \varepsilon_0]$, the log-price process $\vec{Z}_{\varepsilon,n}$ is a space-skeleton Markov chain with transition probabilities and initial distribution defined, respectively, in relations (7.86) and (7.87). Then, the log-reward functions $\phi_{\varepsilon,n}(\vec{y}, x)$ and $\phi_{n+r}(\vec{y}_{\varepsilon,n,\bar{l}}, x_{\varepsilon,n,\bar{l}})$, for points $(\vec{y}_{\varepsilon,n,\bar{l}}, x_{\varepsilon,n,\bar{l}})$, $\bar{l} \in \mathbb{L}_{\varepsilon,n+r}$, $r = 1, \dots, N - n$, given by formulas (7.79), are, for every $\vec{z} = (\vec{y}, x) \in \mathbb{A}_{\varepsilon,n,\bar{l}''}$, $\bar{l}'' \in \mathbb{L}_{\varepsilon,n}$, $n = 0, \dots, N$, the unique solution for the finite recurrence system of linear equations,*

$$\begin{cases} \phi_{\varepsilon,N}(\vec{y}_{\varepsilon,N,\bar{l}}, x_{\varepsilon,N,\bar{l}}) = g(N, e^{\vec{y}_{\varepsilon,N,\bar{l}}}, x_{\varepsilon,N,\bar{l}}), \bar{l} \in \mathbb{L}_{\varepsilon,N} , \\ \phi_{\varepsilon,n+r}(\vec{y}_{\varepsilon,n+r,\bar{l}}, x_{\varepsilon,n+r,\bar{l}}) = \max\Bigg(g(n+r, e^{\vec{y}_{\varepsilon,n+r,\bar{l}}}, x_{\varepsilon,n+r,\bar{l}}) , \\ \qquad \sum\limits_{\bar{l}' \in \mathbb{L}_{\varepsilon,n+r+1}} \phi_{\varepsilon,n+r+1}(\vec{y}_{\varepsilon,n+r+1,\bar{l}'}, x_{\varepsilon,n+r+1,\bar{l}'}) \\ \qquad \times P_{0,n+r+1}(\vec{z}_{\varepsilon,n+r,\bar{l}}, \mathbb{A}_{\varepsilon,n+r+1,\bar{l}'}) \Bigg) , \\ \qquad \bar{l} \in \mathbb{L}_{\varepsilon,n+r}, \quad r = N - n - 1, \dots, 1 , \\ \phi_{\varepsilon,n}(\vec{y}, x) = \max\Bigg(g(n, e^{\vec{y}}, x) , \\ \qquad \sum\limits_{\bar{l}' \in \mathbb{L}_{\varepsilon,n+1}} \phi_{\varepsilon,n+1}(\vec{y}_{\varepsilon,n+1,\bar{l}'}, x_{\varepsilon,n+1,\bar{l}'}) \\ \qquad \times P_{0,n+1}(\vec{z}_{\varepsilon,n,\bar{l}''}, \mathbb{A}_{\varepsilon,n+1,\bar{l}'}) \Bigg) . \end{cases} \tag{7.90}$$

7.3.2 Convergence of space-skeleton approximations for log-rewards functions for multivariate modulated Markov log-price processes

The backward algorithm for computing of the reward functions $\phi_{\varepsilon,n}(\vec{y},x)$ given in Lemma 7.3.1 should be supplemented by the proposition that would show that the log-rewards functions $\phi_{\varepsilon,n}(\vec{y},x)$ converge to the log-reward functions $\phi_{0,n}(\vec{y},x)$ as $\varepsilon \to 0$, and, thus, they can serve as approximations for the log-reward functions $\phi_{0,n}(\vec{y},x)$.

Let us introduce special shorten notations for the maximal and the minimal skeleton points, for $j = 1,\dots,k$, $n = 0,\dots,N$, and $\varepsilon \in (0,\varepsilon_0]$,

$$z^{\pm}_{\varepsilon,n,j} = \delta_{\varepsilon,n,j} m^{\pm}_{\varepsilon,n,j} + \lambda_{\varepsilon,n,j}. \tag{7.91}$$

We impose the following on the parameters of the space-skeleton model defined in relations (7.74)–(7.87):

$\mathbf{N_2}$: (a) $\delta_{\varepsilon,n,j} \to 0$ as $\varepsilon \to 0$, for $j = 1,\dots,k$, $n = 0,1,\dots,N$;
(b) $\pm z^{\pm}_{\varepsilon,n,j} \to \infty$ as $\varepsilon \to 0$, for $j = 1,\dots,k$, $n = 0,1,\dots,N$;
(c) $\pm z^{\pm}_{\varepsilon,n,j}$, $n = 0,1,\dots,N$ are nondecreasing sequences, for every $j = 1,\dots,k$ and $\varepsilon \in (0,\varepsilon_0]$;
(d) for any $x \in \mathbb{X}$ and $d > 0$, there exists $\varepsilon_{x,d} \in (0,\varepsilon_0]$ such that the ball $R_d(x) \subseteq K_{\varepsilon,n}$, $n = 0,\dots,N$, for $\varepsilon \in (0,\varepsilon_{x,d}]$;
(e) the sets $\mathbb{K}_{\varepsilon,n,l}$ have diameters $d_{\varepsilon,n,l} = \sup_{x',x''\in\mathbb{K}_{\varepsilon,n,l}} d_{\mathbb{X}}(x',x'')$ such that $d_{\varepsilon,n} = \max_{m^{-}_{\varepsilon,n,0}\le l\le m^{+}_{\varepsilon,n,0}} d_{\varepsilon,n,l} \to 0$ as $\varepsilon \to 0$, $n = 0,\dots,N$.

Note that the condition $\mathbf{N_2}$ implies that $\delta_{\varepsilon,n,j}(m^{+}_{\varepsilon,n,j} - m^{-}_{\varepsilon,n,j}) \to \infty$ as $\varepsilon \to 0$, for $j = 1,\dots,k$, $n=0,1,\dots,N$, and $m^{+}_{\varepsilon,n,j} - m^{-}_{\varepsilon,n,j} \to \infty$ as $\varepsilon \to 0$, for $j = 1,\dots,k$, $n = 0,1,\dots,N$.

Note, however, that the condition $\mathbf{N_2}$ does not require that $m^{+}_{\varepsilon,n,0} - m^{-}_{\varepsilon,n,0} \to \infty$ as $\varepsilon \to 0$.

Note also that the sets $K_{\varepsilon,n}$ and $\mathbb{K}_{\varepsilon,n,l}$, $m^{-}_{\varepsilon,n,0} \le l \le m^{+}_{\varepsilon,n,0}$ satisfying the conditions $\mathbf{N_2}$ (d) and (e) can be always constructed for the most important cases, where $\mathbb{X}$ is a discrete space or a Euclidian space or a product of such spaces. The role of the sets $\mathbb{K}_{\varepsilon,n,l}$ can be played, respectively, by one-point sets, cubes, or products of one-point sets and cubes.

We impose on the price process $Y_{0,n}$ the following first-type condition of exponential moment compactness, which is a variant of the condition $\mathbf{C_1}[\bar{\beta}]$, and should be assumed to hold for some vector parameter $\bar{\beta} = (\beta_1,\dots,\beta_k)$ with nonnegative components:

$\mathbf{C_{12}}[\bar{\beta}]$: $\Delta_{\beta_i}(Y_{0,\cdot,i},N) < K_{52,i}$, $i = 1,\dots,k$, for some $1 < K_{52,i} < \infty$, $i = 1,\dots,k$.

Let us recall the second-type modulus of exponential moment compactness for the components of the log-price process,

$$\Xi^{\pm}_{\bar{\beta}}(Y_{\cdot,i},N) = \max_{0\le n\le N-1} \sup_{\vec{z}\in\mathbb{Z}} \mathsf{E}_{\vec{z},n} e^{\pm\beta(Y_{n+1,i}-Y_{n,i})}. \tag{7.92}$$

As follows from Lemma 4.1.8, the following condition implies the condition $\mathbf{C}_{12}[\bar{\beta}]$ to hold:
$\mathbf{E}_{16}[\bar{\beta}]$: $\Xi^{\pm}_{\bar{\beta}_i}(Y_{0,\cdot,i}, N) < K_{53,i}$, $i = 1, \ldots, k$, for some $1 < K_{53,i} < \infty$, $i = 1, \ldots, k$.

Let us introduce sequences of the sets $\mathcal{Z}_\varepsilon = \langle \mathbb{Z}_{\varepsilon,0}, \ldots, \mathbb{Z}_{\varepsilon,N} \rangle$, where the sets $\mathbb{Z}_{\varepsilon,n}$ are defined, for every $n = 0, \ldots, N$, by the following relation,:

$$\mathbb{Z}_{\varepsilon,n} = \begin{cases} [z^-_{\varepsilon,n,1}, z^+_{\varepsilon,n,1}] \times \cdots \times [z^-_{\varepsilon,n,k}, z^+_{\varepsilon,n,k}] \times \mathbb{X} & \text{for } \varepsilon \in (0, \varepsilon_0], \\ \mathbb{Z} = \mathbb{R}_k \times \mathbb{X} & \text{for } \varepsilon = 0. \end{cases} \tag{7.93}$$

Note that the condition $\mathbf{N}_2$ (c) implies the following imbedding relation for the sets $\mathbb{Z}_{\varepsilon,n}$, for every $\varepsilon \in (0, \varepsilon_0]$:

$$\mathbb{Z}_{\varepsilon,0} \subseteq \cdots \subseteq \mathbb{Z}_{\varepsilon,N}. \tag{7.94}$$

Let us also recall the modified first-type modulus of exponential moment compactness, defined for every $\varepsilon \in [0, \varepsilon_0]$ and $i = 1, \ldots, k$, $\beta \geq 0$,

$$\Delta_{\beta, \mathcal{Z}_\varepsilon}(Y_{\varepsilon,\cdot,i}, N) = \max_{0 \leq n \leq N-1} \sup_{\vec{z} \in \mathbb{Z}_{\varepsilon,n}} \mathsf{E}_{\vec{z},n} e^{\beta |Y_{\varepsilon,n+1,i} - Y_{\varepsilon,n,i}|}. \tag{7.95}$$

Note, that, by the definition, the modified first-type modulus of exponential moment compactness $\Delta_{\beta,\mathcal{Z}_0}(Y_{0,\cdot,i}, N)$ coincides with the standard first-type modulus of exponential moment compactness $\Delta_\beta(Y_{0,\cdot,i}, N)$.

It is worth to note that the conditions $\mathbf{L}_4$, $\mathbf{N}_2$, and $\mathbf{C}_{12}[\bar{\beta}]$ imply that the modified first-type condition of exponential moment compactness $\mathbf{C}'_6[\bar{\beta}]$ (with the modified moduli of exponential moment compactness $\Delta_{\beta_i,\mathcal{Z}_\varepsilon}(Y_{\varepsilon,\cdot,i}, N)$, $i = 1, \ldots, k$) holds, while the standard condition of exponential moment compactness $\mathbf{C}_6[\bar{\beta}]$ (with the standard moduli of exponential moment compactness $\Delta_{\beta_i}(Y_{\varepsilon,\cdot,i}, N)$, $i = 1, \ldots, k$) does not.

Let us recall the condition $\mathbf{B}_1[\bar{\gamma}]$ for the pay-off function $g(n, e^{\vec{y}}, x)$, which is assumed to hold for some vector parameter $\bar{\gamma} = (\gamma_1, \ldots, \gamma_k)$ with nonnegative components.

The condition $\mathbf{I}_1$ should be replaced by the following condition:

$\mathbf{I}_7$: There exist the sets $\mathbb{Z}'_n \in \mathcal{B}_{\mathbb{Z}}$, $n = 0, \ldots, N$ such that the function $g(n, e^{\vec{y}}, x)$ is continuous in points $\vec{z} = (\vec{y}, x) \in \mathbb{Z}'_n$, for $n = 0, \ldots, N$.

The condition $\mathbf{I}_7$ means that $g(n, e^{\vec{y}_\varepsilon}, x_\varepsilon) \to g(n, e^{\vec{y}_0}, x_0)$ as $\varepsilon \to 0$ for any $\vec{z}_\varepsilon = (\vec{y}_\varepsilon, x_\varepsilon) \to \vec{z}_0 = (\vec{y}_0, x_0) \in \mathbb{Z}'_n$ as $\varepsilon \to 0$, for $n = 0, \ldots, N$. Thus, the condition $\mathbf{I}_1$ holds with the sets $\mathbb{Z}'_n$, $n = 0, \ldots, N$.

The condition $\mathbf{J}_1$ should be replaced by the following condition:

$\mathbf{J}_{10}$: There exist the sets $\mathbb{Z}''_n \in \mathcal{B}_{\mathbb{Z}}$, $n = 0, \ldots, N$ such that:

(a) $P_{0,n}(\vec{z}_\varepsilon, \cdot) \Rightarrow P_{0,n}(\vec{z}_0, \cdot)$ as $\varepsilon \to 0$, for any $\vec{z}_\varepsilon \to \vec{z}_0 \in \mathbb{Z}''_{n-1}$ as $\varepsilon \to 0$, and $n = 1, \ldots, N$;

(b) $P_{0,n}(\vec{z}_0, \mathbb{Z}'_n \cap \mathbb{Z}''_n) = 1$, for every $\vec{z}_0 \in \mathbb{Z}'_{n-1} \cap \mathbb{Z}''_{n-1}$, $n = 1, \ldots, N$, where $\mathbb{Z}'_n$, $n = 0, \ldots, N$ are sets introduced in the condition $\mathbf{I}_7$.

The condition $\mathbf{J}_{10}$ (a) means that the transition probabilities $P_{0,n}(\vec{z}, \cdot)$ are weakly continuous at points $\vec{z} \in \mathbb{Z}''_{n-1}$ for every $n = 1, \ldots, N$.

Note also that the condition $\mathbf{J}_{10}$ (b) automatically holds if the sets $\overline{Z}'_n, \overline{Z}''_n = \emptyset$, $n = 1, \ldots, N$, or these sets are finite or countable and measures $P_{0,n}(\vec{z}_0, \cdot)$, $\vec{z}_0 \in \mathbb{Z}'_{n-1} \cap \mathbb{Z}''_{n-1}$ have no atoms in points of the sets $\overline{Z}'_n, \overline{Z}''_n$, for $n = 1, \ldots, N$, as well as in the case, where measures $P_{0,n}(y_0, \cdot), \vec{z}_0 \in \mathbb{Z}'_{n-1} \cap \mathbb{Z}''_{n-1}$, $n = 1, \ldots, N$ are absolutely continuous with respect to some σ-finite measure $P(A)$ on the σ-algebra $\mathcal{B}_{\mathbb{Z}}$ and $P(\overline{\mathbb{Z}}'_n), P(\overline{\mathbb{Z}}''_n) = 0$, for $n = 1, \ldots, N$.

We shall prove below that the conditions $\mathbf{L}_4$, $\mathbf{N}_2$, and $\mathbf{J}_{10}$ imply that the condition $\mathbf{J}_1$ holds with the sets $\mathbb{Z}'_n, \mathbb{Z}''_n$, $n = 0, \ldots, N$.

The following theorem is a variant of Theorem 5.2.2 for the space-skeleton approximation model for multivariate modulated Markov log-price processes.

Theorem 7.3.1. *Let the multivariate modulated Markov log-price process $\vec{Z}_{0,n}$ and the corresponding space-skeleton approximation Markov processes $\vec{Z}_{\varepsilon,n}$ are defined as described above. Also let the conditions $\mathbf{B}_1[\bar{\gamma}]$ and $\mathbf{C}_{12}[\bar{\beta}]$ hold with the vector parameters $\bar{\gamma} = (\gamma_1, \ldots, \gamma_k)$ and $\bar{\beta} = (\beta_1, \ldots, \beta_k)$ such that, for every $i = 1, \ldots, k$, either $\beta_i > \gamma_i > 0$ or $\beta_i = \gamma_i = 0$, and also the conditions $\mathbf{L}_4$, $\mathbf{N}_2$, $\mathbf{I}_7$, and $\mathbf{J}_{10}$ hold. Then, for every $n = 0, 1, \ldots, N$, the following relation takes place for any $\vec{z}_\varepsilon = (\vec{y}_\varepsilon, x_\varepsilon) \to \vec{z}_0 = (\vec{y}_0, x_0) \in \mathbb{Z}'_n \cap \mathbb{Z}''_n$,*

$$\phi_{\varepsilon,n}(\vec{y}_\varepsilon, x_\varepsilon) \to \phi_{0,n}(\vec{y}_0, x_0) \quad \text{as} \quad \varepsilon \to 0\,. \tag{7.96}$$

Proof. We are going to check that all conditions of Theorem 5.2.2 hold and then to apply this theorem.

It follows from the definition of functions $h_{\varepsilon,n,i}(y)$, $i = 1, \ldots, k$ that the following inequality holds for $\varepsilon \in (0, \varepsilon_0]$ and $i = 1, \ldots, k$, $n = 0, \ldots, N$, $y \in \mathbb{R}_1$,

$$\begin{aligned} |h_{\varepsilon,n,i}(y) - y| &\le (z^-_{\varepsilon,n,i} - y) I(y \le z^-_{\varepsilon,n,i}) \\ &\quad + \delta_{\varepsilon,n} I(z^-_{\varepsilon,n,i} < y \le z^+_{\varepsilon,n,i}) + (y - z^+_{\varepsilon,n,i}) I(y > z^+_{\varepsilon,n,i})\,. \end{aligned} \tag{7.97}$$

Relation (7.86) and the condition $\mathbf{L}_4$ imply that the following inequality holds for every $\varepsilon \in (0, \varepsilon_0]$ and $\vec{z} = (\vec{y}, x) = ((y_1, \ldots, y_k), x) \in \mathbb{Z}$, $i = 1, \ldots, k$, $n = 0, \ldots, N-1$, and $\beta \ge 0$,

$$\begin{aligned} \mathsf{E}_{\vec{z},n} e^{\beta |Y_{\varepsilon,n+1,i} - Y_{\varepsilon,n,i}|} &= \mathsf{E}_{\vec{z},n} e^{\beta |Y_{\varepsilon,n+1,i} - y|} \\ &= \mathsf{E}_{\vec{h}_{\varepsilon,n}(\vec{z}),n} e^{\beta |h_{\varepsilon,n+1,i}(Y_{0,n+1,i}) - y|}\,. \end{aligned} \tag{7.98}$$

Using relations (7.94), (7.97), and (7.98) we get the following inequalities, for every $\varepsilon \in (0, \varepsilon_0]$ and $i = 1, \dots, k$, $n = 0, \dots, N-1$, and $\beta \geq 0$,

$$\begin{aligned}
\sup_{\vec{z}=((y_1,\dots,y_k),x)\in\mathbb{Z}_{\varepsilon,n}} \mathsf{E}_{\vec{z},n} e^{\beta|Y_{\varepsilon,n+1,i}-Y_{\varepsilon,n,i}|} &= \sup_{\vec{z}\in\mathbb{Z}_{\varepsilon,n}} \mathsf{E}_{\bar{h}_{\varepsilon,n}(\vec{z}),n} e^{\beta|h_{\varepsilon,n+1,i}(Y_{0,n+1,i})-y_i|} \\
&\leq \sup_{\vec{z}\in\mathbb{Z}_{\varepsilon,n}} \mathsf{E}_{\bar{h}_{\varepsilon,n}(\vec{z}),n} e^{\beta(y_i - Y_{0,n+1,i})} I(Y_{0,n+1,i} \leq z^-_{\varepsilon,n+1,i}) \\
&\quad + e^{\beta\delta_{\varepsilon,n+1,i}} \sup_{\vec{z}\in\mathbb{Z}_{\varepsilon,n}} \mathsf{E}_{\bar{h}_{\varepsilon,n}(\vec{z})} e^{\beta|Y_{0,n+1,i}-y_i|} \\
&\quad \times I(z^-_{\varepsilon,n+1,i} < Y_{0,n+1,i} \leq z^+_{\varepsilon,n+1,i}) \\
&\quad + \sup_{\vec{z}\in\mathbb{Z}_{\varepsilon,n}} \mathsf{E}_{\bar{h}_{\varepsilon,n}(\vec{z}),n} e^{\beta(Y_{0,n+1,i}-y_i)} I(Y_{0,n+1,i} > z^+_{\varepsilon,n+1,i}) \\
&\leq e^{\beta\delta_{\varepsilon,n+1,i}} \sup_{\vec{z}\in\mathbb{Z}_{\varepsilon,n}} \mathsf{E}_{\bar{h}_{\varepsilon,n}(\vec{z}),n} e^{\beta_i|Y_{0,n+1,i}-y_i|} \\
&\leq e^{\beta\delta_{\varepsilon,n+1,i}} e^{\beta\delta_{\varepsilon,n,i}} \sup_{\vec{z}\in\mathbb{Z}_{\varepsilon,n}} \mathsf{E}_{\bar{h}_{\varepsilon,n}(\vec{z}),n} e^{\beta|Y_{0,n+1}-h_{\varepsilon,n,i}(y_i)|} \\
&\leq e^{\beta(\delta_{\varepsilon,n+1,i}+\delta_{\varepsilon,n,i})} \sup_{\vec{z}\in\mathbb{Z}} \mathsf{E}_{\vec{z},n} e^{\beta|Y_{0,n+1,i}-y_i|} \,. \qquad (7.99)
\end{aligned}$$

This relation implies that the following inequality holds, for every $\varepsilon \in (0, \varepsilon_0]$ and $i = 1, \dots, k$:

$$\Delta_{\beta,Z_\varepsilon}(Y_{\varepsilon,\cdot,i}, N) \leq \max_{0\leq n\leq N-1} e^{\beta(\delta_{\varepsilon,n+1,i}+\delta_{\varepsilon,n,i})} \Delta_\beta(Y_{0,\cdot,i}, N) \,. \qquad (7.100)$$

By the condition $\mathbf{N_2}$, $\max_{0\leq n\leq N-1} e^{\beta(\delta_{\varepsilon,n+1,i}+\delta_{\varepsilon,n,i})} \to 1$ as $\varepsilon \to 0$, for $i = 1, \dots, k$ and $\beta \geq 0$. This relation, inequality (7.100) and the condition $\mathbf{C_{12}}[\bar{\beta}]$ imply the following relation, for $i = 1, \dots, k$:

$$\overline{\lim_{\varepsilon\to 0}} \Delta_{\beta_i,Z_\varepsilon}(Y_{\varepsilon,\cdot,i}, N) \leq \Delta_{\beta_i}(Y_{0,\cdot,i}, N) < K_{52,i} \,. \qquad (7.101)$$

Therefore, the condition $\mathbf{C'_6}[\bar{\beta}]$ (with sets $\mathbb{Z}_{\varepsilon,n}$ given by relation (7.93)) holds.

The condition $\mathbf{N_2}$ obviously implies that the condition $\mathbf{H_1}$ (with the sets $\mathbb{H}_{\varepsilon,n} = \mathbb{Z}_{\varepsilon,n}$, $n = 0, \dots, N$) holds.

As was mentioned above, the condition $\mathbf{I_6}$ implies that the condition $\mathbf{I_1}$ holds.

Let us choose an arbitrary $1 \leq n \leq N$ and $\vec{z}_\varepsilon = (\vec{y}_\varepsilon, x_\varepsilon) = ((y_{\varepsilon,1}, \dots, y_{\varepsilon,k}), x_\varepsilon) \to \vec{z}_0 = (\vec{y}_0, x_0) = ((y_{0,1}, \dots, y_{0,k}), x_0) \in \mathbb{Z}'_{n-1} \cap \mathbb{Z}''_{n-1}$ as $\varepsilon \to 0$.

Inequality (7.97) and the condition $\mathbf{N_2}$ imply that, for $i = 1, \dots, k$, $n = 1, \dots, N$,

$$h_{\varepsilon,n-1,i}(y_{\varepsilon,i}) \to y_{0,i} \quad \text{as} \quad \varepsilon \to 0 \,. \qquad (7.102)$$

It follows from the definition of functions $h_{\varepsilon,n,0}(x)$ that the following inequality also takes place $\varepsilon \in (0, \varepsilon_0]$ and $n = 0, \dots, N$, $x \in \mathbb{X}$:

$$d_{\mathbb{X}}(h_{\varepsilon,n,0}(x)\,, x) \leq d_{\mathbb{X}}(x_{\varepsilon,n,m^+_{\varepsilon,n,0}}, x) I(x \in \overline{\mathbb{K}}_{\varepsilon,n}) + d_{\varepsilon,n} I(x \in \mathbb{K}_{\varepsilon,n}) \,. \qquad (7.103)$$

Inequality (7.103) and the condition $\mathbf{N_2}$ imply that, for $n = 1, \dots, N$,

$$h_{\varepsilon,n-1,0}(x_\varepsilon) \to x_0 \quad \text{as} \quad \varepsilon \to 0 \,. \qquad (7.104)$$

Relations (7.102) and (7.104) obviously imply that, for $n = 1, \dots, N$,

$$\bar{h}_{\varepsilon,n-1}(\vec{z}_\varepsilon) \to \vec{z}_0 \quad \text{as} \quad \varepsilon \to 0\,. \tag{7.105}$$

Let us take arbitrary $\kappa, \varrho > 0$. One can find $T_\kappa > 0$ such that points $\pm T_\kappa$ are points of continuity for the distribution functions $\mathsf{P}_{\vec{z}_0,n-1}\{Y_{0,n,i} \le u\}$, $i = 1, \dots, k$, $n = 1, \dots, N$ and, for $i = 1, \dots, k$, $n = 1, \dots, N$,

$$\mathsf{P}_{\vec{z}_0,n-1}\{Y_{0,n,i} \le -T_\kappa\} + \mathsf{P}_{\vec{z}_0,n-1}\{Y_{0,n,i} \ge T_\kappa\} \le \kappa\,. \tag{7.106}$$

Using the above remarks, the conditions $\mathbf{N_2}$ and $\mathbf{J_{10}}$ (a) and relations (7.97), (7.102), (7.104), and (7.106), we get the following relation, for $i = 1, \dots, k$, $n = 1, \dots, N$:

$$\begin{aligned}
&\overline{\lim_{0<\varepsilon\to 0}}\, \mathsf{P}_{\bar{h}_{\varepsilon,n-1}(\vec{z}_\varepsilon),n-1}\{|h_{\varepsilon,n,i}(Y_{0,n,i}) - Y_{0,n,i}| \ge \varrho\} \\
&\quad \le \overline{\lim_{0<\varepsilon\to 0}} \Big(\mathsf{P}_{\bar{h}_{\varepsilon,n-1}(\vec{z}_\varepsilon),n-1}\{Y_{0,n,i} \le z^-_{\varepsilon,n,i}\} \\
&\qquad + I(\delta_{\varepsilon,n,i} \ge \varrho) + \mathsf{P}_{\bar{h}_{\varepsilon,n-1}(\vec{z}_\varepsilon),n-1}\{Y_{0,n,i} > z^+_{\varepsilon,n,i}\}\Big) \\
&\quad \le \overline{\lim_{0<\varepsilon\to 0}} \Big(\mathsf{P}_{\bar{h}_{\varepsilon,n-1}(\vec{z}_\varepsilon),n-1}\{Y_{0,n,i} \le -T_\kappa\} \\
&\qquad + \mathsf{P}_{\bar{h}_{\varepsilon,n-1}(\vec{z}_\varepsilon),n-1}\{Y_{0,n,i} > T_\kappa\}\Big) \\
&\quad = \mathsf{P}_{\vec{z}_0,n-1}\{Y_{0,n,i} \le -T_\kappa\} + \mathsf{P}_{\vec{z}_0,n-1}\{Y_{0,n,i} \ge T_\kappa\} \le \kappa\,.
\end{aligned} \tag{7.107}$$

Since an arbitrary choice of $\kappa, \varrho > 0$, relation (7.107) implies that for any $\varrho > 0$ and $i = 1, \dots, k$, $n = 1, \dots, N$,

$$\overline{\lim_{0<\varepsilon\to 0}}\, \mathsf{P}_{\bar{h}_{\varepsilon,n-1}(\vec{z}_\varepsilon),n-1}\{|h_{\varepsilon,n,i}(Y_{0,n,i}) - Y_{0,n,i}| \ge \varrho\} = 0\,. \tag{7.108}$$

Let us choose an arbitrary sequence $\varepsilon_r \in (0, \varepsilon_0]$, $r = 1, 2, \dots$, such the $\varepsilon_r \to 0$ as $r \to \infty$.

The condition $\mathbf{J_{10}}$ and relation (7.105) imply that, for every $n = 1, \dots, N$, probability measures,

$$\mathsf{P}_{\bar{h}_{\varepsilon_r,n-1}(\vec{z}_{\varepsilon_r}),n-1}\{X_{0,n} \in \cdot\} \Rightarrow \mathsf{P}_{\vec{z}_0,n-1}\{X_{0,n} \in \cdot\} \quad \text{as} \quad r \to \infty\,. \tag{7.109}$$

Let us again take an arbitrary $\kappa > 0$. Relation (7.109) implies by the Prokhorov theorem that there exists a compact $C_\kappa \subseteq \mathbb{X}$ such that, for $n = 1, \dots, N$,

$$\max_{r\ge 1} \mathsf{P}_{\bar{h}_{\varepsilon_r,n-1}(\vec{z}_{\varepsilon_r}),n-1}\{X_{0,n} \in \overline{C}_\kappa\} \le \kappa. \tag{7.110}$$

Since $\mathbb{X}$ is a Polish metric space and C_κ is a compact, there exists a ball $R_\kappa = R_{d_\kappa}(x_\kappa)$ such that $C_\kappa \subseteq R_\kappa$. By conditions $\mathbf{N_2}$ (d) there exists a $\varepsilon_{x_\kappa,d_\kappa} \in (0, \varepsilon_0]$ such that $R_\kappa \subseteq K_{\varepsilon,n}$, $n = 0, \dots, N$, for $\varepsilon \in (0, \varepsilon_{x_\kappa,d_\kappa}]$. Thus,

$$\begin{aligned}
&\overline{\lim_{r\to\infty}}\, \mathsf{P}_{\bar{h}_{\varepsilon_r,n-1}(\vec{z}_{\varepsilon_r}),n-1}\{X_{0,n} \in \overline{\mathbb{K}}_{\varepsilon_r,n}\} \\
&\quad \le \overline{\lim_{r\to\infty}}\, \mathsf{P}_{\bar{h}_{\varepsilon_r,n-1}(\vec{z}_{\varepsilon_r}),n-1}\{X_{0,n} \in \overline{R}_\kappa\} \\
&\quad \le \overline{\lim_{r\to\infty}}\, \mathsf{P}_{\bar{h}_{\varepsilon_r,n-1}(\vec{z}_{\varepsilon_r}),n-1}\{X_{0,n} \in \overline{C}_\kappa\} \le \kappa\,.
\end{aligned} \tag{7.111}$$

Let us again take an arbitrary $\varrho > 0$. Using the condition $\mathbf{N_2}$ (e) and relations (7.103) and (7.111), we get the following relation, for $n = 1, \ldots, N$:

$$\begin{aligned}&\overline{\lim_{r\to\infty}} \mathsf{P}_{\bar{h}_{\varepsilon_r,n-1}(\vec{z}_{\varepsilon_r}),n-1}\{d_{\mathbb{X}}(h_{\varepsilon_r,n,0}(X_{0,n}), X_{0,n}) \geq \varrho\} \\ &\quad \leq \overline{\lim_{r\to\infty}} \left(\mathsf{P}_{\bar{h}_{\varepsilon_r,n-1}(\vec{z}_{\varepsilon_r}),n-1}\{X_{0,n} \in \overline{\mathbb{K}}_{\varepsilon_r,n}\} + I(d_{\varepsilon_r,n} \geq \varrho)\right) \leq \kappa\,. \end{aligned} \tag{7.112}$$

Since an arbitrary choice of $\kappa > 0$ and an arbitrary choice of sequence $0 < \varepsilon_r \to 0$ as $r \to \infty$, relation (7.112) implies that, for $\varrho > 0$ and $n = 1, \ldots, N$,

$$\overline{\lim_{0<\varepsilon\to 0}} \mathsf{P}_{\bar{h}_{\varepsilon,n-1}(\vec{z}_\varepsilon),n-1}\{d_{\mathbb{X}}(h_{\varepsilon,n,0}(X_{0,n}), X_{0,n}) \geq \varrho\} = 0\,. \tag{7.113}$$

Relations (7.108) and (7.113) imply that, for every $n = 1, \ldots, N$, probability measures $\mathsf{P}_{\bar{h}_{\varepsilon,n-1}(\vec{z}_\varepsilon),n-1}\{\bar{h}_{\varepsilon,n}(\vec{Z}_{0,n}) \in \cdot\}$ and $\mathsf{P}_{\bar{h}_{\varepsilon,n-1}(\vec{z}_\varepsilon),n-1}\{\vec{Z}_{0,n} \in \cdot\}$ have the same weak limits as $0 < \varepsilon \to 0$.

This proposition, relations (7.86), (7.105), and the condition $\mathbf{J_{10}}$ (a) imply that the following relation of weak convergence holds, for any $\vec{z}_\varepsilon \to \vec{z}_0 \in \mathbb{Z}'_{n-1} \cap \mathbb{Z}''_{n-1}$ as $0 < \varepsilon \to 0$, for every $1 \leq n \leq N$,

$$\begin{aligned}P_{\varepsilon,n}(\vec{z}_\varepsilon, \cdot) &= \mathsf{P}_{\bar{h}_{\varepsilon,n-1}(\vec{z}_\varepsilon),n-1}\{\bar{h}_{\varepsilon,n}(\vec{Z}_{0,n}) \in \cdot\} \\ \Rightarrow P_{0,n}(\vec{z}_0, \cdot) &= \mathsf{P}_{\vec{z}_0,n-1}\{\vec{Z}_{0,n} \in \cdot\} \quad \text{as} \quad \varepsilon \to 0\,. \end{aligned} \tag{7.114}$$

Thus, the condition $\mathbf{J_1}$ (a) holds with the sets $\mathbb{Z}'_n, \mathbb{Z}''_n$, $n = 0, \ldots, N$. Since, the conditions $\mathbf{J_{10}}$ (b) and $\mathbf{J_1}$ (b) coincide, we can conclude that the condition $\mathbf{J_1}$ holds in this case with the above sets $\mathbb{Z}'_n, \mathbb{Z}''_n$, $n = 0, \ldots, N$.

Therefore, all conditions of Theorem 5.2.2 hold, and, therefore, Theorem 7.2.3 is a corollary of Theorem 5.2.2. □

7.3.3 Convergence of space-skeleton approximations for optimal expected rewards for multivariate modulated Markov log-price processes

Let us now assume that we have found, using backward recurrence algorithm presented in Lemma 7.3.1, the values of log-reward functions $\phi_{\varepsilon,0}(\vec{y}_{\varepsilon,0,\bar{l}}, x_{\varepsilon,0,\bar{l}})$, $\bar{l} = (l_0, \ldots, l_k) \in \mathbb{L}_{\varepsilon,0}$ for the initial points,

$$\begin{aligned}\vec{z}_{\varepsilon,0,\bar{l}} &= (\vec{y}_{\varepsilon,0,\bar{l}}, x_{\varepsilon,0,\bar{l}}) = ((y_{\varepsilon,0,1,l_1}, \ldots, y_{\varepsilon,0,k,l_k}), x_{\varepsilon,0,l_0}) \\ &= ((\delta_{\varepsilon,0,1}l_1 + \lambda_{\varepsilon,0,1}, \ldots, \delta_{\varepsilon,0,k}l_k + \lambda_{\varepsilon,0,k}), x_{\varepsilon,0,l_0})\,. \end{aligned} \tag{7.115}$$

Then, the optimal expected reward functional $\Phi_\varepsilon = \Phi(\mathcal{M}^{(\varepsilon)}_{max,N})$ for the approximating log-price processes $\vec{Z}_{\varepsilon,n}$ can be found using the following formula:

$$\Phi_\varepsilon = \sum_{\bar{l}\in\mathbb{L}_{\varepsilon,0}} \phi_{\varepsilon,0}(\vec{z}_{\varepsilon,0,\bar{l}}) p_{\varepsilon,\bar{l}}\,, \tag{7.116}$$

where $p_{\varepsilon,\bar{l}} = P_{0,0}(\mathbb{A}_{\varepsilon,0,\bar{l}}) = \mathsf{P}\{\bar{h}_{\varepsilon,0}(\vec{Z}_{0,0}) = \vec{z}_{\varepsilon,0,\bar{l}}\}$, $\bar{l} = (l_0, \ldots, l_k) \in \mathbb{L}_{\varepsilon,0}$.

Let us formulate conditions that should be imposed on the log-price process $\vec{Z}_{0,n}$ and the pay-off function $g(n, e^{\vec{y}}, x)$ in order the optimal expected rewards $\Phi_\varepsilon = \Phi(\mathcal{M}^{(\varepsilon)}_{max,N})$ would converge to the limiting optimal expected reward $\Phi_0 = \Phi(\mathcal{M}^{(0)}_{max,N})$ as $\varepsilon \to 0$.

We impose on the initial value of the log-price process $\vec{Z}_{0,0}$ the following condition, which is a variant of the condition $\mathbf{D}_1[\bar{\beta}]$, and should be assumed to hold for some vector parameter $\bar{\beta} = (\beta_1, \dots, \beta_k)$ with nonnegative components:

$\mathbf{D}_{11}[\bar{\beta}]$: $\mathsf{E}e^{\beta_i |Y_{0,0,i}|} < K_{54,i}$, $i = 1, \dots, k$, for some $1 < K_{54,i} < \infty$, $i = 1, \dots, k$.

The condition $\mathbf{K}_1$ should be replaced by the following condition imposed on the initial distribution $P_{0,0}(A)$:

$\mathbf{K}_9$: $P_{0,0}(\mathbb{Z}'_0 \cap \mathbb{Z}''_0) = 1$, where $\mathbb{Z}'_0$ and $\mathbb{Z}''_0$ are the sets introduced, respectively, in the conditions $\mathbf{I}_7$ and $\mathbf{J}_{10}$.

The following theorem is a variant of Theorem 5.2.4 for the multivariate space-skeleton approximation model.

Theorem 7.3.2. *Let a Markov log-price process $\vec{Z}_{0,n}$ and the corresponding space space-skeleton approximation Markov processes $\vec{Z}_{\varepsilon,n}$ are defined as described above. Let also the conditions $\mathbf{B}_1[\bar{\gamma}]$, $\mathbf{C}_{12}[\bar{\beta}]$, and $\mathbf{D}_{11}[\bar{\beta}]$ hold with the vector parameters $\bar{\gamma} = (\gamma_1, \dots, \gamma_k)$ and $\bar{\beta} = (\beta_1, \dots, \beta_k)$ such that, for every $i = 1, \dots, k$, either $\beta_i > \gamma_i > 0$ or $\beta_i = \gamma_i = 0$, and also the conditions $\mathbf{L}_4$, $\mathbf{M}_4$, $\mathbf{N}_2$, $\mathbf{I}_7$, $\mathbf{J}_{10}$, and $\mathbf{K}_9$ hold. Then, the following relation takes place:*

$$\Phi_\varepsilon \to \Phi_0 \quad as \quad \varepsilon \to 0. \tag{7.117}$$

Proof. We are going to check that all conditions of Theorem 5.2.4 hold and to apply this theorem. In fact, we should only check that the condition $\mathbf{D}'_1[\bar{\beta}]$, with the sets $\mathbb{Z}_{\varepsilon,0}$ defined in relation (7.93), and the condition $\mathbf{K}_1$ hold. Indeed, this check was made for other conditions of Theorem 5.2.4 in the proof of Theorem 7.2.3.

By the condition $\mathbf{N}_2$, there exists $\varepsilon_{16} \in (0, \varepsilon_0]$ such that $\pm z^{\pm}_{\varepsilon,0,i} \geq 0$, $i = 1, \dots, k$, for $\varepsilon \in (0, \varepsilon_{16}]$. The following inequality holds for $i = 1, \dots, k$ and $\varepsilon \in (0, \varepsilon_{16}]$:

$$\begin{aligned} |h_{\varepsilon,0,i}(y)\,| &\leq |y| I(y \notin [z^-_{\varepsilon,0,i}, z^+_{\varepsilon,0,i}]) \\ &\quad + (|y| + \delta_{\varepsilon,0,i}) I(y \in [z^-_{\varepsilon,0,i}, z^+_{\varepsilon,0,i}]) \leq (|y| + \delta_{\varepsilon,0,i}) \,. \end{aligned} \tag{7.118}$$

Relation (7.87) and the condition $\mathbf{M}_2$ imply that the following relation holds for $\varepsilon \in (0, \varepsilon_0]$, $i = 1, \dots, k$, and $\beta \geq 0$:

$$\mathsf{E}e^{\beta |Y_{\varepsilon,0,i}|} = \mathsf{E}e^{\beta |h_{\varepsilon,0,i}(Y_{0,0,i})|} \,. \tag{7.119}$$

Using relations (7.118) and (7.119) we get the following inequality, for $i = 1, \dots, k$, $\varepsilon \in (0, \varepsilon_{16}]$, and $\beta \geq 0$:

$$\mathsf{E}e^{\beta |Y_{\varepsilon,0,i}|} = \mathsf{E}e^{\beta |h_{\varepsilon,0,i}(Y_{0,0,i})|} \leq e^{\beta \delta_{\varepsilon,0,i}} \mathsf{E}e^{\beta |Y_{0,0,i}|} \,. \tag{7.120}$$

Since, by the condition $\mathbf{N_2}$, $e^{\beta\delta_{\varepsilon,0,i}} \to 1$ as $\varepsilon \to 0$, for $i = 1, \dots, k$ and $\beta \geq 0$, the condition $\mathbf{D_{11}}[\bar{\beta}]$ and inequality (7.120) imply that, for $i = 1, \dots, k$,

$$\overline{\lim_{\varepsilon\to 0}} \mathsf{E}e^{\beta_i|Y_{\varepsilon,0,i}|} \leq \mathsf{E}e^{\beta_i|Y_{0,0,i}|} < K_{54,i}\,. \tag{7.121}$$

Also, relation (7.87) implies that, for every $\varepsilon \in (0, \varepsilon_0]$.

$$\mathsf{P}\{\vec{Z}_{\varepsilon,0} \in \mathbb{Z}_{\varepsilon,0}\} = \mathsf{P}\{\bar{h}_{\varepsilon,0}(\vec{Z}_{0,0}) \in \mathbb{Z}_{\varepsilon,0}\} = 1\,. \tag{7.122}$$

Thus, the condition $\mathbf{D}'_7[\bar{\beta}]$ (with sets $\mathbb{Z}_{\varepsilon,0}$ given by relation (7.93)) holds.

Let us take an arbitrary $\varrho > 0$. Using the condition $\mathbf{N_2}$ and inequality (7.97), we get the following relations, for $i = 1, \dots, k$:

$$\begin{aligned}
&\overline{\lim_{0<\varepsilon\to 0}} \mathsf{P}\{|h_{\varepsilon,0,i}(Y_{0,0,i}) - Y_{0,0,i}| \geq \varrho\} \\
&\quad \leq \overline{\lim_{0<\varepsilon\to 0}} \Big(\mathsf{P}\{Y_{0,0,i} \leq z^-_{\varepsilon,0,i}\} \\
&\qquad + I(\delta_{\varepsilon,0,i} \geq \varrho) + \mathsf{P}\{Y_{0,0,i} > z^+_{\varepsilon,0,i}\}\Big) = 0\,.
\end{aligned} \tag{7.123}$$

Let us take arbitrary $\kappa, \varrho > 0$. There exists a compact $C_\kappa \subseteq \mathbb{X}$ such that $\mathsf{P}\{X_{0,0} \in \overline{C}_\kappa\} \leq \kappa$. Since $\mathbb{X}$ is a Polish metric space and C_κ is a compact, there exists a ball $R_\kappa = R_{d_\kappa}(x_\kappa)$ such that $C_\kappa \subseteq R_\kappa$. By the conditions $\mathbf{N_2}$ (d) there exists a $\varepsilon_{x_\kappa,d_\kappa} \in (0, \varepsilon_0]$ such that $R_\kappa \subseteq K_{\varepsilon,0}$, for $\varepsilon \in (0, \varepsilon_{x_\kappa,d_\kappa}]$. The above remarks and relation (7.103) imply that,

$$\begin{aligned}
&\overline{\lim_{0<\varepsilon\to 0}} \mathsf{P}\{d_{\mathbb{X}}(h_{\varepsilon,0,0}(X_{0,0}), X_{0,0}) \geq \varrho\} \\
&\quad \leq \overline{\lim_{0<\varepsilon\to 0}} (\mathsf{P}\{X_{0,0} \in \overline{\mathbb{K}}_{\varepsilon,0}\} + I(d_{\varepsilon,0} \geq \varrho)) \leq \kappa\,.
\end{aligned} \tag{7.124}$$

Due to an arbitrary choice of $\kappa, \varrho > 0$, relations (7.123) and 7.124) imply the following relation:

$$P_{\varepsilon,0}(\cdot) = \mathsf{P}\{\bar{h}_{\varepsilon,0}(\vec{Z}_{0,0}) \in \cdot\} \Rightarrow P_{0,0}(\cdot) = \mathsf{P}\{\vec{Z}_{0,0} \in \cdot\} \quad \text{as} \quad \varepsilon \to 0\,. \tag{7.125}$$

Relation (7.125) and the condition $\mathbf{K_9}$ imply that the condition $\mathbf{K_1}$ holds with the sets $\mathbb{Z}'_0$ and $\mathbb{Z}''_0$.

It follows from the above remarks that all conditions of Theorem 5.2.4 hold, and, therefore, Theorem 7.1.4 is a corollary of Theorem 5.2.4. □

7.4 LPP represented by multivariate modulated random walks

In this section, we present space-skeleton approximation models for discrete time log-price processes represented by random walks. We describe the corresponding backward algorithms for computing log-reward functions and present convergence theorems, which give conditions of convergence for log-reward functions and optimal expected rewards for approximating space-skeleton approximation Markov log-price processes.

7.4.1 Space-skeleton approximations for log-price processes represented by univariate standard random walks

Let us consider the model where, for every $\varepsilon \in [0, \varepsilon_0]$, the log-price process $Y_{\varepsilon,n}$, $n = 0, 1, \dots$, is a one-dimensional random walk,

$$Y_{\varepsilon,n} = Y_{\varepsilon,n-1} + W_{\varepsilon,n}\,, \quad n = 1, 2, \dots\,, \tag{7.126}$$

where (a) $Y_{\varepsilon,0}$ is a real-valued random variable with distribution $P_{\varepsilon,0}(A)$, (b) $W_{\varepsilon,n}$, $n = 1, 2, \dots$, is a sequence of independent real-valued random variables with distributions $\tilde{P}_{\varepsilon,n}(A)$ for $n = 1, 2, \dots$, (c) the random variable $Y_{\varepsilon,0}$ and the random sequence $W_{\varepsilon,n}$, $n = 1, 2, \dots$, are independent.

We would like to approximate the log-price process $Y_{0,n}$ by the log-price processes $Y_{\varepsilon,n}$. We additionally assume that distributions of jumps and an initial distribution for the random walk $Y_{0,n}$ (for $\varepsilon = 0$) can be nondiscrete, while distributions of jumps and an initial distribution for the random walk $Y_{\varepsilon,n}$ (for every $\varepsilon \in (0, \varepsilon_0]$) should be discrete and finite, i.e. concentrated at finite sets of points.

In this case, the model without index component is considered. In order to be able to interpret the model as modulated, one can always introduce a "virtual" index component X_n with a one-point phase space $\mathbb{X} = \{x_0\}$.

Thus, we also assume that the pay-off function $g(n, e^y, x) = g(n, e^y)$ does not depend on argument x and is a real-valued Borel function of argument $y \in \mathbb{R}_1$ for every $n = 0, 1, \dots, N$.

Let us choose the space-skeleton parameters $\delta_\varepsilon > 0$, $\lambda_{\varepsilon,n} \in \mathbb{R}_1$, $n = 0, 1, \dots$, and integers $m^-_{\varepsilon,n} \leq m^+_{\varepsilon,n}$, $n = 0, 1, \dots$, as they have been defined in Subsection 7.1.4.

The intervals $\mathbb{I}_{\varepsilon,n,l}$ should be constructed, for $l = m^-_{\varepsilon,n}, \dots, m^+_{\varepsilon,n}$, $n = 0, 1, \dots$, using the following relation:

$$\mathbb{I}_{\varepsilon,n,l} = \begin{cases} (-\infty,\ \delta_\varepsilon(m^-_{\varepsilon,n} + \frac{1}{2})] + \lambda_{\varepsilon,n} & \text{if } l = m^-_{\varepsilon,n}\,, \\ (\delta_\varepsilon(l - \frac{1}{2}),\ \delta_\varepsilon(l + \frac{1}{2})] + \lambda_{\varepsilon,n} & \text{if } m^-_{\varepsilon,n} < l < m^+_{\varepsilon,n}\,, \\ (\delta_\varepsilon(m^+_{\varepsilon,n} - \frac{1}{2}),\ \infty) + \lambda_{\varepsilon,n} & \text{if } l = m^+_{\varepsilon,n}\,. \end{cases} \tag{7.127}$$

and the skeleton points $y_{\varepsilon,n,l} \in \mathbb{I}_{\varepsilon,n,l}$ should be defined according to the following formulas:

$$y_{\varepsilon,n,l} = l\delta_\varepsilon + \lambda_{\varepsilon,n}\,, \quad l = m^-_{\varepsilon,n}, \dots, m^+_{\varepsilon,n}\,, \quad n = 0, 1, \dots\,. \tag{7.128}$$

Let us now assume that the following fitting condition holds:

L₅: For every $\varepsilon \in (0, \varepsilon_0]$, probabilities $\mathsf{P}\{W_{\varepsilon,n} = y_{\varepsilon,n,l}\} = \tilde{p}_{\varepsilon,l}(n) = \mathsf{P}\{W_{0,n} \in \mathbb{I}_{\varepsilon,n,l}\} = \tilde{P}_{0,n}(\mathbb{I}_{\varepsilon,n,l})$, $l = m^-_{\varepsilon,n}, \dots, m^+_{\varepsilon,n}$, $n = 1, 2, \dots$.

If we also are interested to fit initial distributions, then the following fitting condition should also be assumed:

M₅: For every $\varepsilon \in (0, \varepsilon_0]$, initial probabilities $\mathsf{P}\{Y_{\varepsilon,0} = y_{\varepsilon,0,l}\} = p_{\varepsilon,l} = P_{0,0}(\mathbb{I}_{\varepsilon,0,l})$, $l = m^-_{\varepsilon,0}, \dots, m^+_{\varepsilon,0}$.

Let us define, for $\varepsilon \in (0, \varepsilon_0]$ and $n = 0, 1, \dots$, the skeleton functions,

$$h_{\varepsilon,n}(y) = \begin{cases} \delta_\varepsilon m^-_{\varepsilon,n} + \lambda_{\varepsilon,n} & \text{if } y \le \delta_\varepsilon(m^-_{\varepsilon,n} + \frac{1}{2}) + \lambda_{\varepsilon,n}\,, \\ \delta_\varepsilon l + \lambda_{\varepsilon,n} & \text{if } \delta_\varepsilon(l - \frac{1}{2}) + \lambda_{\varepsilon,n} < y \le \delta_\varepsilon(l + \frac{1}{2}) + \lambda_{\varepsilon,n}\,, \\ & \quad m^-_{\varepsilon,n} < l < m^+_{\varepsilon,n}\,, \\ \delta_\varepsilon m^+_{\varepsilon,n} + \lambda_{\varepsilon,n} & \text{if } y > \delta_\varepsilon(m^+_{\varepsilon,n} - \frac{1}{2}) + \lambda_{\varepsilon,n}\,. \end{cases} \tag{7.129}$$

In this case, the jump distributions $\tilde{P}_{\varepsilon,n}(A) = \mathsf{P}\{W_{\varepsilon,n} \in A\}$, $n = 1, \dots$, take the following form, for every $\varepsilon \in (0, \varepsilon_0]$:

$$\begin{aligned} \tilde{P}_{\varepsilon,n}(A) &= \sum_{y_{\varepsilon,n,l} \in A} \tilde{P}_{0,n}(\mathbb{I}_{\varepsilon,n,l}) \\ &= \mathsf{P}\{h_{\varepsilon,n}(W_{0,n}) \in A\}\,, \quad A \in \mathcal{B}_1\,. \end{aligned} \tag{7.130}$$

As far as the initial distribution $P_{\varepsilon,0}(A) = \mathsf{P}\{Y_{\varepsilon,0} \in A\}$ is concerned, it takes the following form, for every $\varepsilon \in (0, \varepsilon_0]$:

$$\begin{aligned} P_{\varepsilon,0}(A) &= \sum_{y_{\varepsilon,0,l} \in A} P_{0,0}(\mathbb{I}_{\varepsilon,0,l}) \\ &= \mathsf{P}\{h_{\varepsilon,n}(Y_{0,0}) \in A\}\,, \quad A \in \mathcal{B}_1\,. \end{aligned} \tag{7.131}$$

In order to realize the distributional fitting, we can use any standard random walk $Y_{\varepsilon,n} = Y_{\varepsilon,n-1} + W_{\varepsilon,n}$, $n = 1, 2, \dots$, defined by relation (7.126), with the jump distributions and the initial distribution defined, respectively, by relations (7.130) and (7.131). Any such random walk obviously will satisfy the fitting conditions **L₅** and **M₅**.

Moreover, we can easily realize the strong fitting defining the random walk $Y_{\varepsilon,n}$ in the following dynamic form:

$$Y_{\varepsilon,n} = Y_{\varepsilon,n-1} + h_{\varepsilon,n}(W_{0,n})\,, \quad n = 1, 2, \dots\,, \quad Y_{\varepsilon,0} = h_{\varepsilon,0}(Y_{0,0})\,. \tag{7.132}$$

In this case, relations (7.130) and (7.131) and, subsequently, the conditions **L₅** and **M₅** hold.

Let us denote

$$y_{\varepsilon,n,n+r,l} = \delta_\varepsilon l + \lambda_{\varepsilon,n,n+r}, \quad l = m^-_{\varepsilon,n,n+r}, \dots, m^+_{\varepsilon,n,n+r}, \tag{7.133}$$

where, for $0 \le n < n + r < \infty$,

$$m^\pm_{\varepsilon,n,n+r} = \sum_{l=n+1}^{n+r} m^\pm_{\varepsilon,l} \tag{7.134}$$

and

$$\lambda_{\varepsilon,n,n+r} = \sum_{l=n+1}^{n+r} \lambda_{\varepsilon,l}. \tag{7.135}$$

If $Y_{\varepsilon,n} = y$, for some $y \in \mathbb{R}_1$, and $n = 0, 1, \dots$, then possible states for $Y_{\varepsilon,n+r}$ are $y + y_{\varepsilon,n,n+r,l}$, $l = m^-_{\varepsilon,n,n+r}, \dots, m^+_{\varepsilon,n,n+r}$, $r = 1, 2, \dots$.

Since the pay-off function $g(n, e^y)$ does not depend on the index argument x, the reward functions $\phi_{\varepsilon,n}(y)$, $n = 0, 1, \dots$, for the log-price processes $Y_{\varepsilon,n}$ are functions of y, for every $\varepsilon \in [0, \varepsilon_0]$.

The following lemma is a variant of Lemma 7.1.4.

Lemma 7.4.1. *Let, for every $\varepsilon \in (0, \varepsilon_0]$, the log-price process $Y_{\varepsilon,n}$ is a space-skeleton random walk with distributions of jumps and the initial distribution defined, respectively, in relations (7.130) and (7.131). Then, the log-reward functions $\phi_{\varepsilon,n}(y)$ and $\phi_{\varepsilon,n+r}(y + y_{\varepsilon,n,n+r,l})$, for points $y + y_{\varepsilon,n,n+r,l}$, $l = m^-_{\varepsilon,n,n+r}, \dots, m^+_{\varepsilon,n,n+r}$, $r = 1, \dots N - n$ given by formulas (7.128), are, for every $y \in \mathbb{R}_1$, $n = 0, \dots, N$, the unique solution for the following finite recurrence system of linear equations:*

$$\begin{cases} \phi_{\varepsilon,N}(y + y_{\varepsilon,n,N,l}) = g(N, e^{y+y_{\varepsilon,n,N,l}}), \\ \qquad l = m^-_{\varepsilon,n,N}, \dots, m^+_{\varepsilon,n,N}, \\ \phi_{\varepsilon,n+r}(y + y_{\varepsilon,n,n+r,l}) = \max\Bigg(g(n+r, e^{y+y_{\varepsilon,n,n+r,l}}), \\ \qquad \sum\limits_{l'=m^-_{\varepsilon,n+r+1}}^{m^+_{\varepsilon,n+r+1}} \phi_{\varepsilon,n+r+1}(y + y_{\varepsilon,n,n+r,l} + y_{\varepsilon,n+r+1,l'}) \\ \qquad \times \tilde{P}_{0,n+r+1}(\mathbb{I}_{\varepsilon,n+r+1,l'}) \Bigg), \\ \qquad l = m^-_{\varepsilon,n,n+r}, \dots, m^+_{\varepsilon,n,n+r}, \quad r = N - n - 1, \dots, 1, \\ \phi_{\varepsilon,n}(y) = \max\Bigg(g(n, e^y), \\ \qquad \sum\limits_{l'=m^-_{\varepsilon,n+1}}^{m^+_{\varepsilon,n+1}} \phi_{\varepsilon,n+1}(y + y_{\varepsilon,n+1,l'}) \tilde{P}_{0,n+1}(\mathbb{I}_{\varepsilon,n+1,l'}) \Bigg). \end{cases} \tag{7.136}$$

7.4.2 Convergence of space-skeleton approximations for log-price processes represented by univariate standard random walks

The above backward algorithm for computing the log-reward functions $\phi_{\varepsilon,n}(y)$ should be supplemented by the proposition that would prove that the log-rewards functions $\phi_{\varepsilon,n}(y)$ converge to the log-reward functions $\phi_{0,n}(y)$ as $\varepsilon \to 0$.

We impose the following condition on the parameters of the space-skeleton model defined in relations (7.126)–(7.131):

N₃: (a) $\delta_\varepsilon \to 0$ as $\varepsilon \to 0$;
(b) $z^{\pm}_{\varepsilon,n} = \delta_\varepsilon m^{\pm}_{\varepsilon,n} + \lambda_{\varepsilon,n} \to \pm\infty$ as $\varepsilon \to 0$, for $n = 0, 1, \dots, N$.

Note that the condition $\mathbf{N}_3$ implies that $\delta_\varepsilon(m^{+}_{\varepsilon,n} - m^{-}_{\varepsilon,n}) \to \infty$ as $\varepsilon \to 0$, for $n = 0, 1, \dots, N$, and $m^{+}_{\varepsilon,n} - m^{-}_{\varepsilon,n} \to \infty$ as $\varepsilon \to 0$, for $n = 0, 1, \dots, N$.

In this case, we are going to apply Theorem 5.3.1, which give conditions of convergence for reward functions for log-price processes represented by random walks.

Let us recall the first-type modulus of exponential moment compactness for the log-price process $Y_{0,n}$, defined for $\beta \geq 0$,

$$\Delta'_\beta(Y_{0,\cdot}, N) = \max_{0 \leq n \leq N-1} \mathsf{E} e^{\beta |W_{0,n+1}|} . \tag{7.137}$$

We impose on the price process $Y_{0,n}$ the following first-type condition of exponential moment compactness, which is a one-dimensional variant of the condition $\mathbf{C}_7[\bar{\beta}]$ and should be assumed to hold for some $\beta \geq 0$:

$\mathbf{C}_{13}[\beta]$: $\Delta'_\beta(Y_{0,\cdot}, N) < K_{55}$, for some $1 < K_{55} < \infty$.

Let us recall the second-type modulus of exponential moment compactness for the components of the log-price process $Y_{0,n}$, takes the following form for $\beta \geq 0$:

$$\Xi'^{\pm}_\beta(Y_{0,\cdot}, N) = \max_{0 \leq n \leq N-1} \mathsf{E} e^{\pm\beta W_{0,n+1}} . \tag{7.138}$$

As follows from Lemma 4.1.8, the condition $\mathbf{C}_{13}[\beta]$ is implied by the following second-type condition of exponential moment compactness, which should be assumed to hold for the same parameter $\beta \geq 0$:

$\mathbf{E}_{17}[\beta]$: $\Xi'^{\pm}_\beta(Y_{0,\cdot}, N) < K_{56}$, for some $1 < K_{56} < \infty$.

We shall prove below that the conditions $\mathbf{L}_5$, $\mathbf{N}_3$, and $\mathbf{C}_{13}[\beta]$ imply that the first-type condition of exponential moment compactness $\mathbf{C}_7[\bar{\beta}]$ (its one-dimensional variant) holds for the log-price processes $Y_{\varepsilon,n}$.

As in Subsection 7.2.1, we impose the conditions $\mathbf{B}_4[\gamma]$ and $\mathbf{I}_6$ on the pay-off function $g(n, e^y)$.

As was pointed out in Subsection 7.2.1, the condition $\mathbf{I}_6$ implies that $g(n, e^{y_\varepsilon}) \to g(n, e^{y_0})$ as $\varepsilon \to 0$ for any $y_\varepsilon \to y_0 \in \mathbb{Y}'_n$ as $\varepsilon \to 0$ and $n = 0, \dots, N$. Thus, the condition $\mathbf{I}_3$ holds with the sets $\mathbb{Z}'_n = \mathbb{Y}'_n \times \{x_0\}$, $n = 0, \dots, N$.

The condition $\mathbf{J}_4$ should be replaced by the following condition:

$\mathbf{J}_{11}$: $\tilde{P}_{0,n}(\mathbb{Y}'_n - y) = \mathsf{P}\{W_{0,n} + y \in \mathbb{Y}'_n\} = 1, y \in \mathbb{Y}'_{n-1}$, for $n = 1, \dots, N$, where $\mathbb{Y}'_n$, $n = 0, \dots, N$ are the sets introduced in conditions $\mathbf{I}_6$.

We shall prove below that the conditions $\mathbf{L}_5$ and $\mathbf{N}_3$ imply that probability measures $\tilde{P}_{\varepsilon,n}(\cdot) \Rightarrow \tilde{P}_{0,n}(\cdot)$ as $\varepsilon \to 0$, for $n = 1, \dots, N$, and, thus, the condition $\mathbf{J}_4$ (a) holds.

Also, the condition $\mathbf{J}_{11}$ coincides with the condition $\mathbf{J}_4$ (b).

Thus the conditions $\mathbf{L}_5$, $\mathbf{N}_3$, and $\mathbf{J}_{11}$ imply that the condition $\mathbf{J}_4$ holds with the sets $\mathbb{Z}'_n = \mathbb{Y}'_n \times \{x_0\}$ and $\mathbb{Z}''_n = \mathbb{R}_1 \times \{x_0\}$, $n = 0, \dots, N$.

Note also that the condition $\mathbf{J}_{11}$ (b) automatically holds if the sets $\overline{Y}'_n = \emptyset$, $n = 1, \dots, N$, or if these sets are finite or countable and the measures $P_{0,n}(A - y)$, $y \in \mathbb{Y}'_{n-1}$ have no atoms in points of the set $\overline{Y}'_n$, for every $n = 1, \dots, N$, as well as in the case, where measures $P_{0,n}(A - y)$, $y \in \mathbb{Y}'_{n-1}$, $n = 1, \dots, N$ are absolutely continuous with respect to some σ-finite measure $P(A)$ on the Borel σ-algebra $\mathcal{B}_1$ and $P(\overline{\mathbb{Y}}'_n) = 0$, $n = 1, \dots, N$.

The following result can be obtained by application of Theorem 5.3.1 to the above space-skeleton approximation model.

Theorem 7.4.1. *Let a log-price process $Y_{0,n}$ and the corresponding space skeleton approximation processes $Y_{\varepsilon,n}$ represented by random walks are defined as described above. Let also the conditions $\mathbf{B}_4[\gamma]$ and $\mathbf{C}_{13}[\beta]$ hold with the parameters γ and β such that either $\beta > \gamma > 0$ or $\beta = \gamma = 0$, and also the conditions $\mathbf{L}_5$, $\mathbf{N}_3$, $\mathbf{I}_6$, and $\mathbf{J}_{11}$ hold. Then, for every $n = 0, 1, \dots, N$, the following relation takes place for any $y_\varepsilon \to y_0 \in \mathbb{Y}'_n$:*

$$\phi_{\varepsilon,n}(y_\varepsilon) \to \phi_{0,n}(y_0) \quad as \quad \varepsilon \to 0. \tag{7.139}$$

Proof. By the condition $\mathbf{N}_3$, there exists $\varepsilon_{17} \in (0, \varepsilon_0]$ such that $\pm z^{\pm}_{\varepsilon,n} \geq 0$, $n = 0, \dots, N$, for $\varepsilon \in (0, \varepsilon_{17}]$. The following inequality holds for every $\varepsilon \in (0, \varepsilon_{17}]$ and $y \in \mathbb{R}_1$, $n = 0, \dots, N$:

$$\begin{aligned} |h_{\varepsilon,n}(y)\,| &\leq |y| I(y \notin [z^-_{\varepsilon,n}, z^+_{\varepsilon,n}]) + (|y| + \delta_\varepsilon) I(y \in [z^-_{\varepsilon,n}, z^+_{\varepsilon,n}]) \\ &\leq (|y| + \delta_\varepsilon)\,. \end{aligned} \tag{7.140}$$

Relation (7.130) and the condition $\mathbf{L}_5$ imply that the following inequality holds for every $n = 1, \dots, N$, $\varepsilon \in (0, \varepsilon_{17}]$, and $\beta \geq 0$:

$$\mathsf{E} e^{\beta |W_{\varepsilon,n}|} = \mathsf{E} e^{\beta |h_{\varepsilon,n}(W_{0,n})|}\,. \tag{7.141}$$

Using relations (7.140) and (7.141) we get the following inequality, for $\varepsilon \in (0, \varepsilon_{17}]$, and $\beta \geq 0$:

$$\begin{aligned} \Delta'_\beta(Y_{\varepsilon,\cdot}, N) &= \max_{0 \leq n \leq N-1} \mathsf{E} e^{\beta |W_{\varepsilon,n+1}|} = \max_{0 \leq n \leq N-1} \mathsf{E} e^{\beta |h_{\varepsilon,n+1}(W_{0,n+1})|} \\ &\leq e^{\beta \delta_\varepsilon} \max_{0 \leq n \leq N-1} \mathsf{E} e^{\beta |W_{0,n+1}|} = e^{\beta \delta_\varepsilon} \Delta'_\beta(Y_{0,\cdot}, N)\,. \end{aligned} \tag{7.142}$$

Since, by the condition $\mathbf{N}_3$, $e^{\beta\delta_\varepsilon} \to 1$ as $\varepsilon \to 0$, the condition $\mathbf{C}_{13}[\beta]$ and inequality (7.142) imply the following relation:

$$\varlimsup_{\varepsilon\to 0} \Delta'_\beta(Y_{\varepsilon,\cdot}, N) \le \Delta'_\beta(Y_{0,\cdot}, N) < K_{55} . \tag{7.143}$$

Thus, the condition $\mathbf{C}_7[\bar{\beta}]$ (its one-dimensional variant) holds.

Also, the following inequality holds for every $\varepsilon \in (0, \varepsilon_0]$ and $n = 0, \dots, N$, $y \in \mathbb{R}_1$:

$$\begin{aligned} |h_{\varepsilon,n}(y) - y| &\le (z^-_{\varepsilon,n} - y) I(y \le z^-_{\varepsilon,n}) \\ &\quad + \delta_\varepsilon I(z^-_{\varepsilon,n} < y \le z^+_{\varepsilon,n}) + (y - z^+_{\varepsilon,n}) I(y > z^+_{\varepsilon,n}) . \end{aligned} \tag{7.144}$$

The condition $\mathbf{N}_3$ and relation (7.144) imply that, for every $y \in \mathbb{R}_1$ and $n = 0, 1, \dots, N$,

$$h_{\varepsilon,n}(y) \to y \quad \text{as} \quad \varepsilon \to 0 . \tag{7.145}$$

Relation (7.145) obviously implies that, for every $y \in \mathbb{R}_1$ and $n = 0, 1, \dots, N$,

$$h_{\varepsilon,n}(W_{0,n}) \xrightarrow{a.s.} W_{0,n} \quad \text{as} \quad \varepsilon \to 0 . \tag{7.146}$$

Since a.s. convergence of random variables implies their weak convergence, relation (7.146) implies that, for every $y \in \mathbb{R}_1$ and $n = 0, 1, \dots, N$,

$$\begin{aligned} \tilde{P}_{\varepsilon,n}(\cdot) &= \mathsf{P}\{h_{\varepsilon,n}(W_{0,n}) \in \cdot\} \\ \Rightarrow \tilde{P}_{0,n}(\cdot) &= \mathsf{P}\{W_{0,n} \in \cdot\} \quad \text{as} \quad 0 < \varepsilon \to 0 . \end{aligned} \tag{7.147}$$

As follows from the remarks made above, other conditions of Theorem 5.3.1 also holds. Therefore, Theorem 7.4.1 is a corollary of Theorem 5.3.1. □

7.4.3 Convergence of space-skeleton approximations for optimal expected rewards of log-price processes represented by univariate standard random walks

Let us now assume that we have found, using backward recurrence algorithm presented above in Subsection 7.3.1, the values of the reward functions $\phi_{\varepsilon,0}(y_{\varepsilon,0,l})$, for the following initial points:

$$y_{\varepsilon,0,l} = \delta_{\varepsilon,0} l + \lambda_{\varepsilon,0} , \quad m^-_{\varepsilon,0} \le l \le m^+_{\varepsilon,0} . \tag{7.148}$$

Then the optimal expected reward functional $\Phi_\varepsilon = \Phi(\mathcal{M}^{(\varepsilon)}_{max,N})$ for the approximation log-price processes $Y_{\varepsilon,n}$ can be found using the following formula:

$$\Phi_\varepsilon = \sum_{l=m^-_{\varepsilon,0}}^{m^+_{\varepsilon,0}} \phi_{\varepsilon,0}(y_{\varepsilon,0,l}) p_{\varepsilon,0} , \tag{7.149}$$

where $p_{\varepsilon,0} = P_{0,0}(\mathbb{I}_{\varepsilon,0,l}) = \mathsf{P}\{h_{\varepsilon,0}(Y_{0,0}) = y_{\varepsilon,0,l}\}$, $l = m^-_{\varepsilon,0}, \dots, m^+_{\varepsilon,0}$.

We impose on the initial value of the price process $Y_{0,n}$ the condition $\mathbf{D}_{10}[\beta]$ and the following variant of the condition $\mathbf{K}_6$:
$\mathbf{K}_{10}$: $P_{0,0}(\mathbb{Y}'_0) = 1$, where $\mathbb{Y}'_0$ is the set introduced in the condition $\mathbf{I}_6$.

The following theorem is a corollary of Theorem 5.3.2.

Theorem 7.4.2. *Let a log-price process $Y_{0,n}$ and the corresponding space skeleton approximation processes $Y_{\varepsilon,n}$ represented by random walks are defined as described above. Also let the conditions $\mathbf{B}_4[\gamma]$, $\mathbf{C}_{13}[\beta]$, and $\mathbf{D}_{10}[\beta]$ hold with the parameters γ and β such that either $\beta > \gamma > 0$ or $\beta = \gamma = 0$, and also the conditions $\mathbf{L}_5$, $\mathbf{M}_5$, $\mathbf{N}_3$, $\mathbf{I}_6$, $\mathbf{J}_{11}$, and $\mathbf{K}_{10}$ hold. Then, for every $n = 0, 1, \dots, N$, the following relation takes place:*

$$\Phi_\varepsilon \to \Phi_0 \quad as \quad \varepsilon \to 0. \tag{7.150}$$

7.4.4 Space-skeleton approximations for option rewards of log-price processes represented by Lévy random walks

Let us consider the case, where the $Y_{0,n}$ is a Lévy random walk. This means that the random jumps $W_{0,n}$, $n = 1, 2, \dots$, have infinitely divisible distributions with characteristic functions,

$$\begin{aligned} \psi_{0,n}(s) &= \mathsf{E}\exp\{\mathrm{i}sW_{0,n}\} \\ &= \exp\Big\{\mathrm{i}\mu_n s - \frac{1}{2}\sigma_n^2 s^2 + \int\limits_{|y|<1} (e^{\mathrm{i}sy} - 1 - \mathrm{i}sy)\Pi_n(dy) \\ &\quad + \int\limits_{|y|\ge 1} (e^{\mathrm{i}sy} - 1)\Pi_n(dy)\Big\}, \quad s \in \mathbb{R}_1, \end{aligned} \tag{7.151}$$

where (a) μ_n, $n = 1, \dots$ are real-valued constants, (b) σ_n^2, $n = 1, \dots$ are nonnegative constants, and (c) $\Pi_n(A)$, $n = 1, \dots$ are measures on $\mathcal{B}_1$ without atoms in point 0 such that $\pi_n = \int_{\mathbb{R}_1} \frac{y^2}{1+y^2}\Pi_n(dy) < \infty$, $n = 1, \dots$.

As follows from Lemma 5.3.3, the condition $\mathbf{E}_{17}[\beta]$ is implied by the following condition, which should be assumed to hold for some $\beta \ge 0$:
$\mathbf{C}'_{13}[\beta]$: $\max_{1\le n\le N} \int_{|y|\ge 1} e^{\beta|y|}\Pi_n(dy) < K_{57}$, for some $1 < K_{57} < \infty$.

The conditions $\mathbf{I}_6$ and $\mathbf{J}_{11}$ do not change.

Theorem 7.4.1 takes the following form.

Theorem 7.4.3. *Let a log-price process $Y_{0,n}$ represented by a Lévy random walk and the corresponding space skeleton approximation processes $Y_{\varepsilon,n}$ are defined as described above. Also let the conditions $\mathbf{B}_4[\gamma]$ and $\mathbf{C}'_{13}[\beta]$ hold with the parameters γ and β such* *that either $\beta > \gamma > 0$ or $\beta = \gamma = 0$, and also the conditions* $\mathbf{L_5}$, $\mathbf{N_3}$, $\mathbf{I_6}$, *and* $\mathbf{J_{11}}$ *hold. Then, for every $n = 0, 1, \ldots, N$, the following relation takes place for any $y_\varepsilon \to y_0 \in \mathbb{Y}'_n$:*

$$\phi_{\varepsilon,n}(y_\varepsilon) \to \phi_{0,n}(y_0) \quad as \quad \varepsilon \to 0\,. \tag{7.152}$$

The conditions $\mathbf{D_{10}}[\beta]$ and $\mathbf{K_{10}}$ do not change.

Theorem 7.3.2 takes the following form.

Theorem 7.4.4. *Let a log-price process $Y_{0,n}$ represented by a Lévy random walk and the corresponding space skeleton approximation processes $Y_{\varepsilon,n}$ are defined as described above. Let also the conditions $\mathbf{B_4}[\gamma]$, $\mathbf{C'_{13}}[\beta]$, and $\mathbf{D_{10}}[\beta]$ hold with the parameters γ and β such that either $\beta > \gamma > 0$ or $\beta = \gamma = 0$, and also the conditions $\mathbf{L_5}$, $\mathbf{M_5}$, $\mathbf{N_3}$, $\mathbf{I_6}$, $\mathbf{J_{11}}$, and $\mathbf{K_{10}}$ hold. Then, for every $n = 0, 1, \ldots, N$, the following relation takes place:*

$$\Phi_\varepsilon \to \Phi_0 \quad as \quad \varepsilon \to 0\,. \tag{7.153}$$

7.4.5 Space-skeleton approximations for log-price processes represented by multivariate modulated random walks

In this model, a modulated price process $\vec{Z}_{\varepsilon,n} = (\vec{Y}_{\varepsilon,n}, X_{\varepsilon,n})$ can be introduced, for every $\varepsilon \in [0, \varepsilon_0]$, via the following stochastic transition dynamic relation:

$$\begin{cases} & \vec{Y}_{\varepsilon,n} = \vec{Y}_{\varepsilon,n-1} + \vec{W}_{\varepsilon,n}\,, \quad X_{\varepsilon,n} = C_{\varepsilon,n}(X_{\varepsilon,n-1}, U_{\varepsilon,n})\,, \\ \text{where} & \vec{W}_{\varepsilon,n} = \tilde{B}_{\varepsilon,n}(X_{\varepsilon,n-1}, U_{\varepsilon,n}), n = 1, 2, \ldots\,, \\ & \vec{Y}_{\varepsilon,0} = B_{\varepsilon,0}(U_{\varepsilon,0})\,, \quad X_{\varepsilon,0} = C_{\varepsilon,0}(U_{\varepsilon,0})\,. \end{cases} \tag{7.154}$$

where (a) $U_{\varepsilon,n}$, $n = 0, 1, \ldots$, is a sequence of independent random variables taking value in some measurable space $\mathbb{U}$ with a σ-algebra of measurable subsets $\mathcal{B}_\mathbb{U}$, (b) $\tilde{B}_{\varepsilon,n}(\vec{y}, u) = (\tilde{B}_{\varepsilon,n,1}(y_1, u), \ldots, \tilde{B}_{\varepsilon,n,k}(y_k, u))$, $n = 1, 2, \ldots$, are vector measurable functions acting from the space $\mathbb{R}_k \times \mathbb{U}$ to $\mathbb{R}_k$, (c) $C_{\varepsilon,n}(x, u)$, $n = 1, 2, \ldots$, are measurable functions acting from the space $\mathbb{X} \times \mathbb{U}$ to the space $\mathbb{X}$, (d) $B_{\varepsilon,0}(u) = (B_{\varepsilon,0,1}(u), \ldots, B_{\varepsilon,0,k}(u))$ is a vector measurable function acting from the space $\mathbb{U}$ to $\mathbb{R}_k$, (e) $C_{\varepsilon,0}(u)$ is a measurable function acting from the space $\mathbb{U}$ to the space $\mathbb{X}$, and (f) the sequence of jumps $\vec{W}_{\varepsilon,n} = (W_{\varepsilon,n,1}, \ldots, W_{\varepsilon,n,k})$, $n = 1, 2, \ldots$, where $W_{\varepsilon,n,i} = \tilde{B}_{\varepsilon,n,i}(X_{\varepsilon,n-1}, U_{\varepsilon,n})$, $i = 1, \ldots, k$, $n = 1, 2, \ldots$.

In this case, the transition probabilities $P_{\varepsilon,n}(\vec{z}, A) = \mathsf{P}\{\vec{Z}_{\varepsilon,n} \in A/\vec{Z}_{\varepsilon,n-1} = \vec{z}\}$ of the Markov process $\vec{Z}_{\varepsilon,n} = (\vec{Y}_{\varepsilon,n}, X_{\varepsilon,n})$ can be expressed via the transition probabilities $\tilde{P}_{\varepsilon,n}(x, A) = \mathsf{P}\{(\vec{W}_{\varepsilon,n}, X_{\varepsilon,n}) \in A/X_{\varepsilon,n-1} = x\} = \mathsf{P}\{(\tilde{B}_{\varepsilon,n}(x, U_{\varepsilon,n}), C_{\varepsilon,n}(x, U_{\varepsilon,n})) \in A\}$ by the following formulas, for $\vec{z} = (\vec{y}, x) \in \mathbb{Z} = \mathbb{R}_k \times \mathbb{X}$, $A \in \mathcal{B}_\mathbb{Z}$, $n = 1, 2, \ldots$:

$$\begin{aligned} P_{\varepsilon,n}(\vec{z}, A) &= \mathsf{P}\{(\vec{y} + \vec{W}_{\varepsilon,n}, X_{\varepsilon,n}) \in A/\vec{Y}_{\varepsilon,n-1} = \vec{y}, X_{\varepsilon,n-1} = x\} \\ &= \mathsf{P}\{(\vec{y} + \tilde{B}_{\varepsilon,n}(x, U_{\varepsilon,n}), C_{\varepsilon,n}(x, U_{\varepsilon,n})) \in A\} \\ &= \tilde{P}_{\varepsilon,n}(x, A_{[\vec{y}]})\,, \end{aligned} \tag{7.155}$$

where $A_{[\vec{y}]} = A_{[\vec{y}]} = \{(\vec{y}\,', x) : (\vec{y}\,' + \vec{y}, x) \in A\}$.

As far as the initial distribution $P_{\varepsilon,0}(A) = \mathsf{P}\{\vec{Z}_{\varepsilon,0} \in A\}$ is concerned, it has the following form, for $A \in \mathcal{B}_{\mathbb{Z}}$:

$$P_{\varepsilon,0}(A) = \mathsf{P}\{(\vec{Y}_{\varepsilon,0}, X_{\varepsilon,0}) \in A\} = \mathsf{P}\{(B_{\varepsilon,0}(U_{\varepsilon,0}), C_{\varepsilon,0}(U_{\varepsilon,0})) \in A\}\,. \tag{7.156}$$

A standard way for constructing skeleton sets and space-skeleton approximation for the modulated random walk $\vec{Z}_{0,n}$ is the following.

Let us choose $\delta_{\varepsilon,i} > 0$ and $\lambda_{\varepsilon,n,i} \in \mathbb{R}_1$, $i = 1, \dots, k$, $n = 0, 1, \dots$, and integers $m^-_{\varepsilon,n,j} \le m^+_{\varepsilon,n,j}$, $j = 0, \dots, k$, $n = 0, 1, \dots$.

First, the intervals $\mathbb{I}_{\varepsilon,n,i,l}$ should be constructed for $l_i = m^-_{\varepsilon,n,i}, \dots, m^+_{\varepsilon,n,i}$, $i = 1, \dots, k$, $n = 0, 1, \dots$,

$$\mathbb{I}_{\varepsilon,n,i,l} = \begin{cases} (-\infty,\ \delta_{\varepsilon,i}(m^-_{\varepsilon,n,i} + \frac{1}{2})] + \lambda_{\varepsilon,n,i} & \text{if } l = m^-_{\varepsilon,n,i}\,, \\ \left(\delta_{\varepsilon,i}(l - \frac{1}{2}),\ \delta_{\varepsilon,i}(l + \frac{1}{2})\right] + \lambda_{\varepsilon,n,i} & \text{if } m^-_{\varepsilon,n,i} < l < m^+_{\varepsilon,n,i}\,, \\ (\delta_{\varepsilon,i}(m^+_{\varepsilon,n,i} - \frac{1}{2}),\ \infty) + \lambda_{\varepsilon,n,i} & \text{if } l = m^+_{\varepsilon,n,i}\,. \end{cases} \tag{7.157}$$

Then the cubes $\mathbb{I}_{\varepsilon,n,l_1,\dots,l_k} = \mathbb{I}_{\varepsilon,n,1,l_1} \times \cdots \times I_{\varepsilon,n,k,l_k}$, $l_i = m^-_{\varepsilon,n,i}, \dots, m^+_{\varepsilon,n,i}$, $i = 1, \dots, k$, $n = 0, 1, \dots$, should be defined.

By the definition, the points $y_{\varepsilon,n,i,l_i} = l_i\delta_{\varepsilon,i} + \lambda_{\varepsilon,n,i} \in \mathbb{I}_{\varepsilon,n,i,l_i}$ and, thus, the vector points $(y_{\varepsilon,n,1,l_1}, \dots, y_{\varepsilon,n,k,l_k}) \in \mathbb{I}_{\varepsilon,n,l_1,\dots,l_k}$, for $l_i = m^-_{\varepsilon,n,i}, \dots, m^+_{\varepsilon,n,i}$, $i = 1, \dots, k$, $n = 0, 1, \dots$.

Second, nonempty sets $\mathbb{J}_{\varepsilon,n,l} \in \mathcal{B}_{\mathbb{X}}$, $l = m^-_{\varepsilon,n,0}, \dots, m^+_{\varepsilon,n,0}$, $n = 0, 1, \dots$, such that (a) $\mathbb{J}_{\varepsilon,n,l'} \cap \mathbb{J}_{\varepsilon,n,l''} = \varnothing$, $l' \ne l''$, $n = 0, 1, \dots$; (b) $\mathbb{X} = \cup_{m^-_{\varepsilon,n,0} \le l \le m^+_{\varepsilon,n,0}} \mathbb{J}_{\varepsilon,n,l}$, $n = 0, 1, \dots$, should be constructed.

Recall that one of our model assumption is the $\mathbb{X}$ is a Polish space, i.e. a complete, separable, metric space. In this case, it is natural to assume that there exist some "large" sets $\mathbb{K}_{\varepsilon,n}$, $n = 0, 1, \dots$, "small" nonempty sets $\mathbb{K}_{\varepsilon,n,l} \subseteq \mathcal{B}_{\mathbb{X}}$, $l = m^-_{\varepsilon,n,0}, \dots, m^+_{\varepsilon,n,0}$, $n = 0, 1, \dots$, such that (c) $\mathbb{K}_{\varepsilon,n,l'} \cap \mathbb{K}_{\varepsilon,n,l''} = \varnothing$, $l' \ne l''$, $n = 0, 1, \dots$; (d) $\cup_{m^-_{\varepsilon,n,0} \le l \le m^+_{\varepsilon,n,0}} \mathbb{K}_{\varepsilon,n,l} = \mathbb{K}_{\varepsilon,n}$, $n = 0, 1, \dots$.

Then, the sets $\mathbb{J}_{\varepsilon,n,l}$ can be defined in the following way, for $n = 0, 1, \dots$:

$$\mathbb{J}_{\varepsilon,n,l} = \begin{cases} \mathbb{K}_{\varepsilon,n,l} & \text{if } m^-_{\varepsilon,n,0} \le l < m^+_{\varepsilon,n,0}\,, \\ \mathbb{K}_{\varepsilon,n,m^+_{\varepsilon,n,0}} \cup \overline{\mathbb{K}}_{\varepsilon,n} & \text{if } l = m^+_{\varepsilon,n,0}\,. \end{cases} \tag{7.158}$$

Then skeleton points $x_{\varepsilon,n,l} \in \mathbb{K}_{\varepsilon,n,l}$, $l = m^-_{\varepsilon,n,0}, \dots, m^+_{\varepsilon,n,0}$, $n = 0, 1, \dots$, should be chosen.

A particular important case is, where the space $\mathbb{X} = \{1, \dots, m\}$ is a finite set and metric $d_{\mathbb{X}}(x, y) = I(x \ne y)$. In this case, the standard choice is to take $\mathbb{K}_{\varepsilon,n} = \mathbb{X}$, $n = 0, 1, \dots$, one-point sets $\mathbb{J}_{\varepsilon,n,l} = \mathbb{K}_{\varepsilon,n,l} = \{l\}, 1 = m^-_{\varepsilon,n,0} \le l \le m^+_{\varepsilon,n,0} = m$, $n = 0, 1, \dots, N$, and points $x_{\varepsilon,n,l} = l$, $l = 1, \dots, m$, $n = 0, \dots$.

If the space $\mathbb{X} = \{1, 2, \dots\}$ is a countable set and again metric $d_{\mathbb{X}}(x, y) = I(x \ne y)$, the standard choice is to take one-point sets $\mathbb{K}_{\varepsilon,n,l} = \{l\}, 1 = m^-_{\varepsilon,n,0} \le l \le m^+_{\varepsilon,n,0} =$

m_ε, $n = 0, 1, \ldots$, sets $\mathbb{K}_{\varepsilon,n} = \{l : l \leq m_\varepsilon\}$, $n = 0, 1, \ldots$. In this case, the sets $\mathbb{J}_{\varepsilon,n,l} = \{l\}, 1 = m^-_{\varepsilon,n,0} \leq l < m^+_{\varepsilon,n,0} = m_\varepsilon$, $n = 0, 1, \ldots$, while $\mathbb{J}_{\varepsilon,n,m_\varepsilon} = \{l : l \geq m_\varepsilon\}$, and points $x_{\varepsilon,n,l} = l$, $l = 1, \ldots, m_\varepsilon$, $n = 0, 1, \ldots$.

Third, the sets $\mathbb{A}_{\varepsilon,n,\bar{l}}$ and skeleton points, $\vec{z}_{\varepsilon,n,\bar{l}} \in \mathbb{A}_{\varepsilon,n,\bar{l}}$ can be defined for $\bar{l} = (l_0, l_1, \ldots, l_k)$, $l_j = m^-_{\varepsilon,n,j}, \ldots, m^+_{\varepsilon,n,j}$, $j = 0, \ldots, k$, $n = 0, 1, \ldots$, in the following way:

$$\mathbb{A}_{\varepsilon,n,\bar{l}} = \mathbb{I}_{\varepsilon,n,l_1,\ldots,l_k} \times \mathbb{J}_{\varepsilon,n,l_0}, \tag{7.159}$$

and

$$\vec{z}_{\varepsilon,n,\bar{l}} = (\vec{y}_{\varepsilon,n,\bar{l}}, x_{\varepsilon,n,\bar{l}}) = ((y_{\varepsilon,n,1,l_1}, \ldots, y_{\varepsilon,n,k,l_k}), x_{\varepsilon,n,l_0}). \tag{7.160}$$

Let us introduce sets of vector indices $\mathbb{L}_{\varepsilon,n}$ which are defined in the following way, for $n = 0, 1, \ldots$:

$$\mathbb{L}_{\varepsilon,n} = \{\bar{l} = (l_0, \ldots, l_k), l_j = m^-_{\varepsilon,n,j}, \ldots, m^+_{\varepsilon,n,j}, j = 0, \ldots, k\}. \tag{7.161}$$

Let us now assume that the following fitting condition holds:

L₆: For every $\varepsilon \in (0, \varepsilon_0]$, transition probabilities $\mathsf{P}\{\vec{W}_{\varepsilon,n} = \vec{y}_{\varepsilon,n,\bar{l}}, X_{\varepsilon,n} = x_{\varepsilon,n,\bar{l}}/X_{\varepsilon,n-1} = x\} = \tilde{p}_{\varepsilon,\bar{l}}(x, n) = \mathsf{P}\{(\vec{W}_{0,n}, X_{0,n}) \in \mathbb{A}_{\varepsilon,n,\bar{l}}/X_{0,n-1} = x\} = \tilde{P}_{0,n}(x_{\varepsilon,n-1,l'_0}, \mathbb{A}_{\varepsilon,n,\bar{l}})$, $x \in \mathbb{J}_{\varepsilon,n-1,l'_0}$, $l'_0 = m^-_{\varepsilon,n-1,0}, \ldots, m^+_{\varepsilon,n-1,0}$, $\bar{l} = (l_0, \ldots, l_k) \in \mathbb{L}_{\varepsilon,n}$, $n = 1, \ldots$.

If we are also interested to fit initial distributions, the following fitting condition should be assumed:

M₆: For every $\varepsilon \in (0, \varepsilon_0]$, initial probabilities $\mathsf{P}\{\vec{Y}_{\varepsilon,0} = \vec{y}_{\varepsilon,0,\bar{l}}, X_{\varepsilon,0} = x_{\varepsilon,0,\bar{l}}\} = p_{\varepsilon,\bar{l}} = \mathsf{P}\{(\vec{Y}_{0,0}, X_{0,0}) \in \mathbb{A}_{\varepsilon,0,\bar{l}})\} = P_{0,0}(\mathbb{A}_{\varepsilon,0,\bar{l}})$, $\bar{l} = (l_0, \ldots, l_k) \in \mathbb{L}_{\varepsilon,0}$.

Let us define the skeleton functions, $h_{\varepsilon,n,i}(y)$, $y \in \mathbb{R}_1$, for $\varepsilon \in (0, \varepsilon_0]$ and $i = 1, \ldots, k$, $n = 0, 1, \ldots$,

$$h_{\varepsilon,n,i}(y) = \begin{cases} \delta_{\varepsilon,i} m^-_{\varepsilon,n,i} + \lambda_{\varepsilon,n,i} & \text{if } y \leq \delta_{\varepsilon,i}(m^-_{\varepsilon,n,i} + \frac{1}{2}) + \lambda_{\varepsilon,n,i}, \\ \delta_{\varepsilon,i} l + \lambda_{\varepsilon,n,i} & \text{if } \delta_{\varepsilon,i}(l - \frac{1}{2}) + \lambda_{\varepsilon,n,i} < y \leq \delta_{\varepsilon,i}(l + \frac{1}{2}) + \lambda_{\varepsilon,n,i}, \\ & \quad m^-_{\varepsilon,n,i} < l < m^+_{\varepsilon,n,i}, \\ \delta_{\varepsilon,i} m^+_{\varepsilon,n,i} + \lambda_{\varepsilon,n,i} & \text{if } y > \delta_{\varepsilon,i}(m^+_{\varepsilon,n,i} - \frac{1}{2}) + \lambda_{\varepsilon,n,i}. \end{cases} \tag{7.162}$$

and the vector skeleton functions $\hat{h}_{\varepsilon,n}(\vec{y})$, $\vec{y} = (y_1, \ldots, y_k) \in \mathbb{R}_k$, for $\varepsilon \in (0, \varepsilon_0]$ and

$n = 0, 1, \ldots$,

$$\hat{h}_{\varepsilon,n}(\vec{y}) = (h_{\varepsilon,n,1}(y_1), \ldots, h_{\varepsilon,n,k}(y_k)). \tag{7.163}$$

Let us also define skeleton functions $h_{\varepsilon,n,0}(x)$, $x \in \mathbb{X}$, for $\varepsilon \in (0, \varepsilon_0]$ and $n = 0, 1, \ldots$,

$$h_{\varepsilon,n,0}(x) = \{x_{\varepsilon,n,l} \quad \text{if} \quad x \in \mathbb{J}_{\varepsilon,n,l}, m^-_{\varepsilon,n,0} \leq l \leq m^+_{\varepsilon,n,0}. \tag{7.164}$$

In this case, the transition probabilities $\tilde{P}_{\varepsilon,n}(x, A) = \mathsf{P}\{(\vec{W}_{\varepsilon,n}, X_{\varepsilon,n}) \in A / X_{\varepsilon,n-1} = x\}$, $x \in X$, $A \in \mathcal{B}_{\mathbb{Z}}$, $n = 1, \dots$, take the following form, for every $\varepsilon \in (0, \varepsilon_0]$:

$$\begin{aligned}\tilde{P}_{\varepsilon,n}(x, A) &= \sum_{(\vec{y}_{\varepsilon,n,\bar{l}}, x_{\varepsilon,n,\bar{l}}) \in A} \tilde{P}_{0,n}(h_{\varepsilon,n-1,0}(x), \mathbb{A}_{\varepsilon,n,\bar{l}}) \\ &= \mathsf{P}\left\{(\hat{h}_{\varepsilon,n}(\vec{W}_{0,n}), h_{\varepsilon,n,0}(X_{0,n})) \in A / X_{0,n-1} = h_{\varepsilon,n-1,0}(x)\right\} . \end{aligned} \tag{7.165}$$

As far as the initial distribution $P_{\varepsilon,0}(A) = \mathsf{P}\{(\vec{Y}_{\varepsilon,0}, X_{\varepsilon,0}) \in A\}$, $A \in \mathcal{B}_{\mathbb{Z}}$ is concerned, it takes the following form, for every $\varepsilon \in (0, \varepsilon_0]$:

$$\begin{aligned}P_{\varepsilon,0}(A) &= \sum_{(\vec{y}_{\varepsilon,0,\bar{l}}, x_{\varepsilon,0,\bar{l}}) \in A} P_{0,0}(\mathbb{A}_{\varepsilon,0,\bar{l}}) \\ &= \mathsf{P}\left\{(\hat{h}_{\varepsilon,0}(\vec{Y}_{0,0}), h_{\varepsilon,0,0}(X_{0,0})) \in A\right\} . \end{aligned} \tag{7.166}$$

In this case, the dynamic representation (7.154) for the modulated log-price process $\vec{Z}_{\varepsilon,n}$ takes, for every $\varepsilon \in (0, \varepsilon_0]$, the following form:

$$\begin{cases} \quad\quad\;\; \vec{Y}_{\varepsilon,n} = \vec{Y}_{\varepsilon,n-1} + \vec{W}_{\varepsilon,n} , \\ \text{where} \;\; \vec{W}_{\varepsilon,n} = \hat{h}_{\varepsilon,n}(\tilde{B}_{0,n}(h_{\varepsilon,n-1,0}(X_{\varepsilon,n-1}), U_{0,n})) , \\ \quad\quad\;\; X_{\varepsilon,n} = h_{\varepsilon,n,0}(C_{0,n}(h_{\varepsilon,n-1,0}(X_{\varepsilon,n-1}), U_{0,n})) , \quad n = 1, 2, \dots , \\ \quad\quad\;\; \vec{Y}_{\varepsilon,0} = \hat{h}_{\varepsilon,0}(B_{0,0}(U_{0,0})) , \\ \quad\quad\;\; X_{\varepsilon,0} = h_{\varepsilon,0,0}(C_{0,0}(U_{0,0})) . \end{cases} \tag{7.167}$$

Obviously, the random sequence $\vec{Z}_{\varepsilon,n}$ given by the dynamic relations (7.167) is a modulated random walk with transition probabilities and initial distribution satisfying relations (7.165) and (7.166) and, therefore, satisfying fitting conditions $\mathbf{L_6}$ and $\mathbf{M_6}$.

Let us denote $\vec{y}_{\varepsilon,n,n+r,\bar{l}} = (\delta_{\varepsilon,1} l_1 + \lambda_{\varepsilon,n,n+r,1}, \dots, \delta_{\varepsilon,k} l_k + \lambda_{\varepsilon,n,n+r,k})$, $x_{\varepsilon,n,n+r,\bar{l}} = x_{\varepsilon,n+r,l_0}$, $\bar{l} = (l_0, \dots, l_k)$, $l_j = m^-_{\varepsilon,n,n+r,j}, \dots, m^+_{\varepsilon,n,n+r,j}$, $j = 0, \dots, k$.

Here, for $i = 1, \dots, k$, $0 \le n < n + r < \infty$,

$$m^{\pm}_{\varepsilon,n,n+r,0} = m^{\pm}_{\varepsilon,n+r,0} , \quad m^{\pm}_{\varepsilon,n,n+r,i} = \sum_{l=n+1}^{n+r} m^{\pm}_{\varepsilon,l,i} , \tag{7.168}$$

and

$$\lambda_{\varepsilon,n,n+r,i} = \sum_{l=n+1}^{n+r} \lambda_{\varepsilon,l,i} . \tag{7.169}$$

If $\vec{Z}_{\varepsilon,n} = (\vec{Y}_{\varepsilon,n}, X_{\varepsilon,n}) = (\vec{y}, x)$ for some $(\vec{y}, x) \in \mathbb{Z}$ and $n = 0, 1, \dots$, then, for $r = 1, 2, \dots$, possible states for $\vec{Z}_{\varepsilon,n+r} = (\vec{Y}_{\varepsilon,n+r}, X_{\varepsilon,n+r})$ are $(\vec{y} + \vec{y}_{\varepsilon,n,n+r,\bar{l}}, x_{\varepsilon,n,n+r,\bar{l}})$, $\bar{l} = (l_0, \dots, l_k)$, $l_j = m^-_{\varepsilon,n,n+r,j}, \dots, m^+_{\varepsilon,n,n+r,j}$, $j = 0, \dots, k$.

The following lemma is a variant of Lemma 7.1.4.

Lemma 7.4.2. *Let, for every $\varepsilon \in (0, \varepsilon_0]$, the log-price process $\vec{Z}_{\varepsilon,n}$ is a space-skeleton random walk with jump distributions probabilities and the initial distribution defined,* *respectively, in relations* (7.165) *and* (7.166). *Then, the log-reward functions* $\phi_{\varepsilon,n}(\vec{y},x)$ *and* $\phi_{\varepsilon,n+r}(\vec{y}+\vec{y}_{\varepsilon,n,n+r,\bar{l}},x_{\varepsilon,n+r,\bar{l}})$ *for points* $\vec{z}=(\vec{y},x)=((y_1,\dots,y_k),x)$ *and* $(\vec{y}+\vec{y}_{\varepsilon,n,n+r,\bar{l}},x_{\varepsilon,n+r,\bar{l}})$, $\bar{l}=(l_0,l_1,\dots,l_k)$, $l_j=m^-_{\varepsilon,n,n+r,j},\dots,m^+_{\varepsilon,n,n+r,j}$, $j=0,\dots,k$, $r=1,\dots N-n$, *are, for every* $\vec{z}=(\vec{y},x)\in\mathbb{R}_k\times\mathbb{J}_{\varepsilon,n,l''_0}$, $l''_0=m^-_{\varepsilon,n,0},\dots,m^+_{\varepsilon,n,0}$, $n=0,\dots,N$, *the unique solution for the following finite recurrence system of linear equations:*

$$\begin{cases}
\phi_{\varepsilon,N}(\vec{y}+\vec{y}_{\varepsilon,n,N,\bar{l}},x_{\varepsilon,N,\bar{l}}) = g(N,e^{\vec{y}+\vec{y}_{\varepsilon,n,N,\bar{l}}},x_{\varepsilon,N,\bar{l}})\,, \\
\qquad \bar{l}=(l_0,l_1,\dots,l_k)\,,\quad l_j=m^-_{\varepsilon,n,N,j},\dots,m^+_{\varepsilon,n,N,j}\,, \\
\qquad j=0,\dots,k\,, \\
\phi_{\varepsilon,n+r}(\vec{y}+\vec{y}_{\varepsilon,n,n+r,\bar{l}},x_{\varepsilon,n+r,\bar{l}}) = \max\Big(g(n+r,e^{\vec{y}+\vec{y}_{\varepsilon,n,n+r,\bar{l}}},x_{\varepsilon,n+r,\bar{l}})\,, \\
\qquad \sum\limits_{(\vec{y}_{\varepsilon,n+r+1,\bar{l}'},x_{\varepsilon,n+r+1,\bar{l}'})\in\mathbb{F}_{\varepsilon,n+r+1}} \phi_{\varepsilon,n+r+1}\big(\vec{y}+\vec{y}_{\varepsilon,n,n+r,\bar{l}} \\
\qquad\quad +\vec{y}_{\varepsilon,n+r+1,\bar{l}'},x_{\varepsilon,n+r+1,\bar{l}'}\big) \\
\qquad\quad \times\tilde{P}_{0,n+r+1}(x_{\varepsilon,n+r,\bar{l}},\mathbb{A}_{\varepsilon,n+r+1,\bar{l}'})\Big)\,, \\
\qquad \bar{l}=(l_0,l_1,\dots,l_k)\,,\quad l_j=m^-_{\varepsilon,n,n+r,j},\dots,m^+_{\varepsilon,n,n+r,j}\,, \\
\qquad j=0,\dots,k\,,\quad r=N-n-1,\dots,1\,, \\
\phi_{\varepsilon,n}(\vec{y},x) = \max\Big(g(n,e^{\vec{y}},x), \\
\qquad \sum\limits_{(\vec{y}_{\varepsilon,n+1,\bar{l}'},x_{\varepsilon,n+1,\bar{l}'})\in\mathbb{F}_{\varepsilon,n+1}} \phi_{\varepsilon,n+1}\big(\vec{y}+\vec{y}_{\varepsilon,n+1,\bar{l}'},x_{\varepsilon,n+1,\bar{l}'}\big) \\
\qquad\quad \times\tilde{P}_{0,n+1}(x_{\varepsilon,n,l''_0},\mathbb{A}_{\varepsilon,n+1,\bar{l}'})\Big)\,.
\end{cases} \tag{7.170}$$

7.4.6 Convergence of space-skeleton approximations for log-price processes represented by multivariate modulated random walks

The above backward algorithm for computing of the log-reward functions $\phi_{\varepsilon,n}(\vec{y},x)$ given in Lemma 7.4.2 should be supplemented by the proposition that would prove that the log-rewards functions $\phi_{\varepsilon,n}(\vec{y},x)$ converge to the log-reward functions $\phi_{0,n}(\vec{y},x)$ as $\varepsilon\to 0$, and, thus, they can serve as approximations for the log-reward functions $\phi_{0,n}(\vec{y},x)$.

Let us introduce special shorten notations for the maximal and the minimal skeleton points, for $i = 1, \dots, k$, $n = 0, \dots, N$ and $\varepsilon \in (0, \varepsilon_0]$,

$$z^{\pm}_{\varepsilon,n,i} = \delta_{\varepsilon,i} m^{\pm}_{\varepsilon,n,i} + \lambda_{\varepsilon,n,i} . \tag{7.171}$$

We impose the following condition on the parameters of the space-skeleton model defined in relations (7.154)–(7.166):

N₄: (a) $\delta_{\varepsilon,i} \to 0$ as $\varepsilon \to 0$, for $i = 1, \dots, k$;
(b) $\pm z^{\pm}_{\varepsilon,n,i} \to \infty$ as $\varepsilon \to 0$, for $i = 1, \dots, k$, $n = 0, 1, \dots, N$;
(c) for any $x \in \mathbb{X}$ and $d > 0$, there exists $\varepsilon_{x,d} \in (0, \varepsilon_0]$ such that the ball $R_d(x) \subseteq K_{\varepsilon,n}$, $n = 0, \dots, N$, for $\varepsilon \in (0, \varepsilon_{x,d}]$;
(d) the sets $\mathbb{K}_{\varepsilon,n,l}$ have diameters $d_{\varepsilon,n,l} = \sup_{x',x'' \in \mathbb{K}_{\varepsilon,n,l}} d_{\mathbb{X}}(x', x'')$ such that $d_{\varepsilon,n} = \max_{m^-_{\varepsilon,n,0} \le l \le m^+_{\varepsilon,n,0}} d_{\varepsilon,n,l} \to 0$ as $\varepsilon \to 0$, $n = 0, \dots, N$.

Note that the condition **N₄** implies that $\delta_{\varepsilon,i}(m^+_{\varepsilon,n,i} - m^-_{\varepsilon,n,i}) \to \infty$ as $\varepsilon \to 0$, for $i = 1, \dots, k$, $n = 0, 1, \dots, N$, and $m^+_{\varepsilon,n,i} - m^-_{\varepsilon,n,i} \to \infty$ as $\varepsilon \to 0$, for $i = 1, \dots, k$, $n = 0, 1, \dots, N$.

Note, however, that the condition **N₄** does not require that $m^+_{\varepsilon,n,0} - m^-_{\varepsilon,n,0} \to \infty$ as $\varepsilon \to 0$.

Let us recall the first-type modulus of exponential moment compactness for the components log-price process $Y_{0,n,i}$, for $i = 1, \dots, k$, defined for $\beta \ge 0$,

$$\Delta''_{\beta}(Y_{0,\cdot,i}, N) = \max_{0 \le n \le N-1} \sup_{x \in \mathbb{X}} \mathsf{E}_{x,n} e^{\beta |Y_{0,n+1,i} - Y_{0,n,i}|} , \tag{7.172}$$

where $\mathsf{E}_{x,n}$ and $\mathsf{P}_{x,n}$ denote, respectively, conditional expectation and conditional probability under condition that $X_n = x$.

We impose on the price process $\vec{Z}_{0,n} = (\vec{Y}_{0,n}, X_{0,n})$ the following first-type condition of exponential moment compactness, which is a variant of the condition $\mathbf{C_2}[\bar{\beta}]$, and should be assumed to hold for some vector parameter $\bar{\beta} = (\beta_1, \dots, \beta_k)$ with nonnegative components:

$\mathbf{C_{14}}[\bar{\beta}]$: $\Delta''_{\beta_i}(Y_{0,\cdot,i}, N) < K_{58,i}$, $i = 1, \dots, k$ for some $1 < K_{58,i} < \infty$, $i = 1, \dots, k$.

Let us recall the second-type modulus of exponential moment compactness for components of the log-price process $\vec{Z}_{0,n}$, defined for $\beta \ge 0$,

$$\Xi''^{\pm}_{\beta}(Y_{0,\cdot,i}, N) = \max_{0 \le n \le N-1} \sup_{x \in \mathbb{X}} \mathsf{E}_{x,n} e^{\pm \beta W_{0,n+1}} . \tag{7.173}$$

As follows from Lemma 4.2.1, the condition $\mathbf{C_{14}}[\bar{\beta}]$ is implied by the following condition, which should be assumed to hold for the same vector parameter $\bar{\beta} = (\beta_1, \dots, \beta_k)$ with nonnegative components:

$\mathbf{E_{18}}[\bar{\beta}]$: $\Xi''^{\pm}_{\beta_i}(Y_{0,\cdot,i}, N) < K_{59,i}$, $i = 1, \dots, k$, for some $1 < K_{59,i} < \infty$, $i = 1, \dots, k$.

We are going to apply Theorem 5.3.5, which give conditions of convergence for reward functions for log-price processes represented by multivariate modulated random walks.

We shall prove below that, in the skeleton approximation model described above, the conditions $\mathbf{L_6}$, $\mathbf{N_4}$, and $\mathbf{C_{14}}[\bar{\beta}]$ imply that the first-type condition of exponential moment compactness $\mathbf{C_8}[\bar{\beta}]$ holds for the log-price processes $\vec{Z}_{\varepsilon,n}$.

In this case, we impose on pay-off function $g(n, e^{\vec{y}}, x)$ the condition $\mathbf{B_1}[\bar{\gamma}]$ assumed to hold for some vector parameter $\bar{\gamma} = (\gamma_1, \dots, \gamma_k)$ with nonnegative components.

Also, the condition $\mathbf{I_1}$ should be replaced by the condition $\mathbf{I_7}$. As was pointed out in Subsection 7.2.3, the condition $\mathbf{I_7}$ means that $g(n, e^{\vec{y}_\varepsilon}, x_\varepsilon) \to g(n, e^{\vec{y}_0}, x_0)$ as $\varepsilon \to 0$ for any $\vec{z}_\varepsilon = (\vec{y}_\varepsilon, x_\varepsilon) \to \vec{z}_0 = (\vec{y}_0, x_0) \in \mathbb{Z}'_n$ as $\varepsilon \to 0$ and $n = 0, \dots, N$. Thus, the condition $\mathbf{I_1}$ holds with the sets $\mathbb{Z}'_n$, $n = 0, \dots, N$.

The condition $\mathbf{J_7}$ should be replaced by the following condition:

$\mathbf{J_{12}}$: There exist measurable sets $\mathbb{X}_n \subseteq \mathbb{X}$, $n = 0, \dots, N$ such that:
(a) $\tilde{P}_{0,n}(x_\varepsilon, \cdot) \Rightarrow \tilde{P}_{0,n}(x_0, \cdot)$ as $\varepsilon \to 0$, for any $x_\varepsilon \to x_0 \in \mathbb{X}_{n-1}$ as $\varepsilon \to 0$, and $n = 1, \dots, N$;
(b) $\tilde{P}_{0,n}(x_0, (\mathbb{Z}'_n \cap \mathbb{Z}''_n)_{[\vec{y}_0]}) = 1$, for every $\vec{z}_0 = (\vec{y}_0, x_0) \in \mathbb{Z}'_{n-1} \cap \mathbb{Z}''_{n-1}$, $n = 1, \dots, N$, where $\mathbb{Z}'_n$, $n = 0, \dots, N$ are sets introduced in the condition $\mathbf{I_7}$ and $\mathbb{Z}''_n = \mathbb{R}_k \times \mathbb{X}_n$, $n = 0, \dots, N$.

The condition $\mathbf{J_{12}}$ (a) means that the transition probabilities $\tilde{P}_{0,n}(x, \cdot)$ are weakly continuous at points $x_0 \in \mathbb{X}_{n-1}$ for every $n = 1, \dots, N$.

A typical example is where the sets $\overline{\mathbb{Z}}'_n$, $\overline{\mathbb{X}}_n$, $n = 1, \dots, N$ are empty sets. Then the condition $\mathbf{J_{12}}$ (b) obviously holds.

Another typical example is where the sets $\overline{\mathbb{Z}}'_n$, $\overline{\mathbb{X}}_n$, $n = 1, \dots, N$ are at most finite or countable sets. Then, the assumption that the measures $\tilde{P}_{0,n}(x_0, A)$, $A \in \mathcal{B}_{\mathbb{Z}}$ and $\tilde{P}_{0,n}(x_0, \mathbb{R}_k \times B)$, $B \in \mathcal{B}_{\mathbb{X}}$ have no atoms, for every $x_0 \in \mathbb{X}_{n-1}$, $n = 1, \dots, N$, implies that the condition $\mathbf{J_{12}}$ (b) holds.

One more example is, where the measures $\tilde{P}_{0,n}(x_0, A)$, $x_0 \in \mathbb{X}_{n-1}$, $n = 1, \dots, N$ are absolutely continuous with respect to some σ-finite measure $P(A)$ on $\mathcal{B}_{\mathbb{Z}}$ and $P(\overline{\mathbb{Z}}'_n), P(\mathbb{R}_k \times \overline{\mathbb{X}}_n) = 0$, $n = 1, \dots, N$. This assumption implies that the condition $\mathbf{J_{12}}$ (b) holds.

We shall prove below that the conditions $\mathbf{L_6}$, $\mathbf{N_4}$, and $\mathbf{J_{12}}$ imply that the condition $\mathbf{J_{12}}$ holds with the sets $\mathbb{Z}'_n, \mathbb{Z}''_n$, $n = 0, \dots, N$.

The modulated random walk $\vec{Z}_{0,n}$ was introduced in the dynamic form. This makes it possible to replace the condition $\mathbf{J_{12}}$ by the following sufficient condition, which, by Lemma 5.2.1, implies the condition $\mathbf{J_{12}}$ to hold:

$\mathbf{J'_{12}}$: There exist measurable sets $\mathbb{X}_n \subseteq \mathbb{X}$, $n = 0, \dots, N$ and $\mathbb{U}_n \in \mathbb{U}$, $n = 1, \dots, N$ such that:
(a) $\tilde{B}_{0,n}(x_\varepsilon, u) \to \tilde{B}_{0,n}(x_0, u)$ and $C_{0,n}(x_\varepsilon, u) \to C_{0,n}(x_0, u)$ as $\varepsilon \to 0$, for any $x_\varepsilon \to x_0 \in \mathbb{X}_{n-1}$ as $\varepsilon \to 0$, $u \in \mathbb{U}_n$, $n = 1, \dots, N$;

(b) $\mathsf{P}\{U_{0,n} \in \mathbb{U}_n\} = 1,\ n = 1, \dots, N$;

(c) $\tilde{P}_{0,n}(x_0, (\mathbb{Z}'_n \cap \mathbb{Z}''_n)_{[y_0]}) = 1$, for every $\vec{z}_0 = (\vec{y}_0, x_0) \in \mathbb{Z}'_{n-1} \cap \mathbb{Z}''_{n-1}$, $n = 1, \dots, N$, where $\mathbb{Z}'_n$, $n = 0, \dots, N$ are sets introduced in the condition $\mathbf{I}_7$ and $\mathbb{Z}''_n = \mathbb{R}_k \times \mathbb{X}_n$, $n = 0, \dots, N$.

The following result can be obtained by application of Theorem 5.3.5. This theorem should be applied to the above skeleton model log-price processes represented by multivariate modulated random walks.

Theorem 7.4.5. *Let a log-price process $\vec{Z}_{0,n}$ and the corresponding space skeleton approximation processes $\vec{Z}_{\varepsilon,n}$ represented by multivariate modulated random walks are defined as described above. Let also the conditions $\mathbf{B}_1[\bar{\gamma}]$ and $\mathbf{C}_{14}[\bar{\beta}]$ hold with the vector parameters $\bar{\gamma} = (\gamma_1, \dots, \gamma_k)$, $\bar{\beta} = (\beta_1, \dots, \beta_k)$ such that, for every $i = 1, \dots, k$, either $\beta_i > \gamma_i > 0$ or $\beta_i = \gamma_i = 0$, and also the conditions $\mathbf{L}_6$, $\mathbf{N}_4$, $\mathbf{I}_7$, and $\mathbf{J}_{12}$ hold. Then, for every $n = 0, 1, \dots, N$, the following relation takes place for any $\vec{z}_\varepsilon = (\vec{y}_\varepsilon, x_\varepsilon) \to \vec{z}_0 = (\vec{y}_0, x_0) \in \mathbb{Z}'_{n-1} \cap \mathbb{Z}''_{n-1}$,*

$$\phi_{\varepsilon,n}(y_\varepsilon, x_\varepsilon) \to \phi_{0,n}(y_0, x_0) \quad \text{as} \quad \varepsilon \to 0. \tag{7.174}$$

Proof. By the condition $\mathbf{N}_4$, there exists $\varepsilon_{18} \in (0, \varepsilon_0]$ such that $\pm z^{\pm}_{\varepsilon,n,i} \geq 0$, $i = 1, \dots, k$, $n = 0, \dots, N$, for $\varepsilon \in (0, \varepsilon_{18}]$. The following inequality holds for every $\varepsilon \in (0, \varepsilon_{18}]$ and $y \in \mathbb{R}_1$, $i = 1, \dots, k$, $n = 0, \dots, N$,

$$\begin{aligned} &|h_{\varepsilon,n,i}(y)| \\ &\leq |y| I(y \notin [z^-_{\varepsilon,n,i}, z^+_{\varepsilon,n,i}]) + (|y| + \delta_{\varepsilon,i}) I(y \in [z^-_{\varepsilon,n,i}, z^+_{\varepsilon,n,i}]) \leq (|y| + \delta_{\varepsilon,i}). \end{aligned} \tag{7.175}$$

Relation (7.165) and the condition $\mathbf{L}_6$ imply that the following inequality holds for every $x \in \mathbb{X}$, $i = 1, \dots, k$, $n = 0, \dots, N-1$, $\varepsilon \in (0, \varepsilon_{18}]$, and $\beta \geq 0$:

$$\mathsf{E}_{x,n} e^{\beta |W_{\varepsilon,n+1,i}|} = \mathsf{E}_{h_{\varepsilon,n,0}(x),n} e^{\beta |h_{\varepsilon,n+1,i}(W_{0,n+1,i})|}. \tag{7.176}$$

Using relations (7.175) and (7.176) we get the following inequality, for $i = 1, \dots, k$, $e \in (0, \varepsilon_{18}]$, and $\beta \geq 0$:

$$\begin{aligned} \Delta''_\beta(Y_{\varepsilon,\cdot,i}, N) &= \max_{0 \leq n \leq N-1} \sup_{x \in \mathbb{X}} \mathsf{E}_{x,n} e^{\beta |W_{\varepsilon,n+1,i}|} \\ &= \max_{0 \leq n \leq N-1} \sup_{x \in \mathbb{X}} \mathsf{E}_{h_{\varepsilon,n,0}(x),n} e^{\beta |h_{\varepsilon,n+1,i}(W_{0,n+1,i})|} \\ &\leq e^{\beta \delta_{\varepsilon,i}} \max_{0 \leq n \leq N-1} \sup_{x \in \mathbb{X}} \mathsf{E}_{h_{\varepsilon,n,0}(x),n} e^{\beta |W_{0,n+1,i}|} \\ &\leq e^{\beta \delta_{\varepsilon,i}} \max_{0 \leq n \leq N-1} \sup_{x \in \mathbb{X}} \mathsf{E}_{x,n} e^{\beta_i |W_{0,n+1,i}|} \\ &= e^{\beta \delta_{\varepsilon,i}} \Delta''_\beta(Y_{0,\cdot,i}, N). \end{aligned} \tag{7.177}$$

Since, by the condition $\mathbf{N}_4$, $e^{\beta \delta_{\varepsilon,i}} \to 1$ as $\varepsilon \to 0$, for $i = 1, \dots, k$ and $\beta \geq 0$, the condition $\mathbf{C}_{14}[\bar{\beta}]$ and inequality (7.177) imply that, for $i = 1, \dots, k$,

$$\overline{\lim_{\varepsilon \to 0}} \Delta''_{\beta_i}(Y_{\varepsilon,\cdot,i}, N) \leq \Delta''_{\beta_i}(Y_{0,\cdot,i}, N) < K_{58,i}. \tag{7.178}$$

Thus, the condition $\mathbf{C}_8[\bar{\beta}]$ holds.

Let us choose an arbitrary $1 \leq n \leq N$ and $x_\varepsilon \to x_0 \in \mathbb{X}_{n-1}$ as $\varepsilon \to 0$.

It follows from the definition of functions $h_{\varepsilon,n,0}(x)$ that the following inequality also takes place $\varepsilon \in (0, \varepsilon_0]$ and $x \in \mathbb{X}$, $n = 0, \ldots, N$:

$$d_{\mathbb{X}}(h_{\varepsilon,n,0}(x), x) \leq d_{\mathbb{X}}(x_{\varepsilon,n,m^+_{\varepsilon,n,0}}, x) I(x \in \overline{\mathbb{K}}_{\varepsilon,n}) + d_{\varepsilon,n} I(x \in \mathbb{K}_{\varepsilon,n}) . \tag{7.179}$$

Inequality (7.179) and the condition $\mathbf{N}_4$ imply that, for $n = 1, \ldots, N$,

$$h_{\varepsilon,n-1,0}(x_\varepsilon) \to x_0 \quad \text{as} \quad \varepsilon \to 0 . \tag{7.180}$$

The following inequality holds for every $\varepsilon \in (0, \varepsilon_0]$ and $y \in \mathbb{R}_1$, $i = 1, \ldots, k$, $n = 0, \ldots, N$,

$$\begin{aligned} |h_{\varepsilon,n,i}(y) - y| &\leq (z^-_{\varepsilon,n,i} - y) I(y \leq z^-_{\varepsilon,n,i}) \\ &\quad + \delta_{\varepsilon,i} I(z^-_{\varepsilon,n,i} < y \leq z^+_{\varepsilon,n,i}) + (y - z^+_{\varepsilon,n,i}) I(y > z^+_{\varepsilon,n,i}) . \end{aligned} \tag{7.181}$$

Let us take arbitrary $\kappa, \varrho > 0$. One can find $T_\kappa > 0$ such that points $\pm T_\kappa$ are points of continuity for the distribution functions $\mathsf{P}_{x_0,n-1}\{W_{0,n,i} \leq u\}$, $i = 1, \ldots, k$, $n = 1, \ldots, N$ and, for $i = 1, \ldots, k$, $n = 1, \ldots, N$,

$$\mathsf{P}_{x_0,n-1}\{W_{0,n,i} \leq -T_\kappa\} + \mathsf{P}_{x_0,n-1}\{W_{0,n,i} \geq T_\kappa\} \leq \kappa . \tag{7.182}$$

Using the above remarks, the conditions $\mathbf{N}_4$ and $\mathbf{J}_{12}$ (a) and relations (7.180) and (7.181), we get the following relation, for $i = 1, \ldots, k$, $n = 1, \ldots, N$:

$$\begin{aligned} &\overline{\lim_{0<\varepsilon\to 0}} \mathsf{P}_{h_{\varepsilon,n-1,0}(x_\varepsilon),n-1}\{|h_{\varepsilon,n,i}(W_{0,n,i}) - W_{0,n,i}| \geq \varrho\} \\ &\quad \leq \overline{\lim_{0<\varepsilon\to 0}} \Big(\mathsf{P}_{h_{\varepsilon,n-1,0}(x_\varepsilon),n-1}\{W_{0,n,i} \leq z^-_{\varepsilon,n,i}\} \\ &\qquad + I(\delta_{\varepsilon,i} \geq \varrho) + \mathsf{P}_{h_{\varepsilon,n-1,0}(x_\varepsilon),n-1}\{W_{0,n,i} > z^+_{\varepsilon,n,i}\}\Big) \\ &\quad \leq \overline{\lim_{0<\varepsilon\to 0}} \Big(\mathsf{P}_{h_{\varepsilon,n-1,0}(x_\varepsilon),n-1}\{W_{0,n,i} \leq -T_\kappa\} \\ &\qquad + \mathsf{P}_{h_{\varepsilon,n-1,0}(x_\varepsilon),n-1}\{W_{0,n,i} > T_\kappa\}\Big) \\ &\quad = \mathsf{P}_{x_0,n-1}\{W_{0,n,i} \leq -T_\kappa\} + \mathsf{P}_{x_0,n-1}\{W_{0,n,i} \geq T_\kappa\} \leq \kappa . \end{aligned} \tag{7.183}$$

Since an arbitrary choice of $\kappa, \varrho > 0$, relation (7.183) implies that for any $\varrho > 0$ and $i = 1, \ldots, k$, $n = 1, \ldots, N$,

$$\overline{\lim_{0<\varepsilon\to 0}} \mathsf{P}_{h_{\varepsilon,n-1,0}(x_\varepsilon),n-1}\{|h_{\varepsilon,n,i}(W_{0,n,i}) - W_{0,n,i}| \geq \varrho\} = 0 . \tag{7.184}$$

Let us choose an arbitrary sequence $\varepsilon_r \in (0, \varepsilon_0]$, $r = 1, 2, \ldots$, such the $\varepsilon_r \to 0$ as $r \to \infty$.

The condition $\mathbf{J}_{12}$ and relation (7.180) imply that, for every $n = 1, \ldots, N$, probability measures,

$$\mathsf{P}_{h_{\varepsilon_r,n-1,0}(x_{\varepsilon_r}),n-1}\{X_{0,n} \in \cdot\} \Rightarrow \mathsf{P}_{x_0,n-1}\{X_{0,n} \in \cdot\} \quad \text{as} \quad r \to \infty . \tag{7.185}$$

Let us take again an arbitrary $\kappa > 0$. Relation (7.185) implies by the Prokhorov theorem that there exists a compact $C_\kappa \subseteq \mathbb{X}$ such that, for $n = 1, \dots, N$,

$$\max_{r\geq 1} \mathsf{P}_{h_{\varepsilon_r,n-1,0}(x_{\varepsilon_r}),n-1}\{X_{0,n} \in \overline{C}_\kappa\} \leq \kappa\,. \tag{7.186}$$

Since $\mathbb{X}$ is a Polish metric space and C_κ is a compact, there exists a ball $R_\kappa = R_{d_\kappa}(x_\kappa)$ such that $C_\kappa \subseteq R_\kappa$. By conditions $\mathbf{N_4}$ (c), there exists a $\varepsilon_{x_\kappa,d_\kappa} \in (0, \varepsilon_0]$ such that $R_\kappa \subseteq K_{\varepsilon,n}$, $n = 0, \dots, N$, for $\varepsilon \in (0, \varepsilon_{x_\kappa,d_\kappa}]$. Thus,

$$\begin{aligned}
&\overline{\lim_{r\to\infty}} \mathsf{P}_{h_{\varepsilon_r,n-1,0}(x_{\varepsilon_r}),n-1}\{X_{0,n} \in \overline{\mathbb{K}}_{\varepsilon_r,n}\}\\
&\quad\leq \overline{\lim_{r\to\infty}} \mathsf{P}_{h_{\varepsilon_r,n-1,0}(x_{\varepsilon_r}),n-1}\{X_{0,n} \in \overline{R}_\kappa\}\\
&\quad\leq \overline{\lim_{r\to\infty}} \mathsf{P}_{h_{\varepsilon_r,n-1,0}(x_{\varepsilon_r}),n-1}\{X_{0,n} \in \overline{C}_\kappa\} \leq \kappa\,.
\end{aligned} \tag{7.187}$$

Let us again take an arbitrary $\varrho > 0$. Using the condition $\mathbf{N_4}$ (d) and relations (7.179) and (7.187), we get the following relation, for $n = 1, \dots, N$:

$$\begin{aligned}
&\overline{\lim_{r\to\infty}} \mathsf{P}_{h_{\varepsilon_r,n-1,0}(x_{\varepsilon_r}),n-1}\{d_{\mathbb{X}}(h_{\varepsilon_r,n,0}(X_{0,n}), X_{0,n}) \geq \varrho\}\\
&\quad\leq \overline{\lim_{r\to\infty}} \left(\mathsf{P}_{h_{\varepsilon_r,n-1,0}(x_{\varepsilon_r}),n-1}\{X_{0,n} \in \overline{\mathbb{K}}_{\varepsilon_r,n}\} + I(d_{\varepsilon_r,n} \geq \varrho)\right) \leq \kappa\,.
\end{aligned} \tag{7.188}$$

Since an arbitrary choice of $\kappa > 0$ and an arbitrary choice of sequence $0 < \varepsilon_r \to 0$ as $r \to \infty$, relation (7.188) implies that, for $\varrho > 0$ and $n = 1, \dots, N$,

$$\overline{\lim_{0<\varepsilon\to 0}} \mathsf{P}_{h_{\varepsilon,n-1,0}(x_\varepsilon),n-1}\{d_{\mathbb{X}}(h_{\varepsilon,n,0}(X_{0,n}), X_{0,n}) \geq \varrho\} = 0\,. \tag{7.189}$$

Relations (7.184) and (7.189) obviously imply that, for every $n = 1, \dots, N$, the probability measures $\mathsf{P}_{h_{\varepsilon,n-1,0}(x_\varepsilon),n-1}\{(\hat{h}_{\varepsilon,n}(\vec{W}_{0,n}), h_{\varepsilon,n,0}(X_{0,n})) \in \cdot\}$ and $\mathsf{P}_{h_{\varepsilon,n-1,0}(x_\varepsilon),n-1}$ $\{(\vec{W}_{0,n}, X_{0,n}) \in \cdot\}$ have the same weak limits as $0 < \varepsilon \to 0$.

This proposition, relations (7.165), (7.180), and the condition $\mathbf{J_{12}}$ (a) imply that the following relation of weak convergence holds, for any $x_\varepsilon \to x_0 \in \mathbb{X}_{n-1}$ as $0 < \varepsilon \to 0$, for every $n = 1, \dots, N$:

$$\begin{aligned}
&\tilde{P}_{\varepsilon,n}(x_\varepsilon, \cdot) = \mathsf{P}_{h_{\varepsilon,n-1,0}(x_\varepsilon),n-1}\{(\hat{h}_{\varepsilon,n}(\vec{W}_{0,n}), h_{\varepsilon,n,0}(X_{0,n})) \in \cdot\}\\
&\Rightarrow \tilde{P}_{0,n}(x_0, \cdot) = \mathsf{P}_{x_0,n-1}\{(\vec{W}_{0,n}, X_{0,n}) \in \cdot\} \quad \text{as} \quad \varepsilon \to 0\,.
\end{aligned} \tag{7.190}$$

Thus, the condition $\mathbf{J_7}$ (a) holds with the sets $\mathbb{Z}'_n, \mathbb{Z}''_n$, $n = 0, \dots, N$. Since, the conditions $\mathbf{J_{12}}$ (b) and $\mathbf{J_7}$ (b) coincide, we can conclude that the condition $\mathbf{J_7}$ holds, in this case, with the above sets $\mathbb{Z}'_n, \mathbb{Z}''_n$, $n = 0, \dots, N$.

As follows from the remarks made above, other conditions of Theorem 5.3.5 also holds. Therefore, Theorem 7.3.5 is a corollary of Theorem 5.3.5. □

7.4.7 Convergence of space-skeleton approximations for optimal expected rewards of log-price processes represented by multivariate modulated random walks

Let us now assume that we have found, using backward recurrence algorithm presented in Lemma 7.4.2, the values of log-reward functions $\phi_{\varepsilon,0}(\vec{y}_{\varepsilon,0,\bar{l}}, x_{\varepsilon,0,\bar{l}})$, $\bar{l} = (l_0, \dots, l_k) \in \mathbb{L}_{\varepsilon,0}$ for the initial points,

$$\begin{aligned}\vec{z}_{\varepsilon,0,\bar{l}} &= (\vec{y}_{\varepsilon,0,\bar{l}}, x_{\varepsilon,0,\bar{l}}) = ((y_{\varepsilon,0,1,l_1}, \dots, y_{\varepsilon,0,k,l_k}), x_{\varepsilon,0,l_0}) \\ &= ((\delta_{\varepsilon,0,1}l_1 + \lambda_{\varepsilon,0,1}, \dots, \delta_{\varepsilon,0,k}l_k + \lambda_{\varepsilon,0,k}), x_{\varepsilon,0,l_0}) . \end{aligned} \tag{7.191}$$

Then, the optimal expected reward functional Φ_ε can be found using the following formula:

$$\Phi_\varepsilon = \sum_{\bar{l} \in \mathbb{L}_{\varepsilon,0}} \phi_{\varepsilon,0}(\vec{z}_{\varepsilon,0,\bar{l}}) p_{\varepsilon,\bar{l}} , \tag{7.192}$$

where $p_{\varepsilon,\bar{l}} = P_{0,0}(\mathbb{A}_{\varepsilon,0,\bar{l}}) = \mathsf{P}\{\hat{h}_{\varepsilon,0}(\vec{Y}_{0,0}) = \vec{Y}_{\varepsilon,0,\bar{l}}, h_{\varepsilon,0,0}(X_{0,0}) = x_{\varepsilon,0,\bar{l}}\}$, $\bar{l} = (l_0, \dots, l_k) \in \mathbb{L}_{\varepsilon,0}$.

Let us formulate conditions of convergence for optimal expected rewards.

We impose on the initial value of the log-price process $\vec{Z}_{0,0}$ the condition $\mathbf{D}_{11}[\bar{\beta}]$, which should be assumed to hold for some vector parameter $\bar{\beta} = (\beta_1, \dots, \beta_k)$ with nonnegative components.

Also, the following variant of the condition $\mathbf{K}_5$ is assumed to hold:

$\mathbf{K}_{11}$: x $P_{0,0}(\mathbb{Z}'_0 \cap \mathbb{Z}''_0) = 1$, where $\mathbb{Z}'_0$ and $\mathbb{Z}''_0$ are the sets introduced, respectively, in the conditions $\mathbf{I}_7$ and $\mathbf{J}_{12}$.

The following result can be obtained by application of Theorem 5.3.6.

Theorem 7.4.6. *Let a Markov log-price process $\vec{Z}_{0,n}$ and the corresponding space space-skeleton approximating Markov processes $\vec{Z}_{\varepsilon,n}$ represented by multivariate modulated random walks are defined as described above. Let also the conditions $\mathbf{B}_1[\bar{\gamma}]$, $\mathbf{C}_{14}[\bar{\beta}]$, and $\mathbf{D}_{11}[\bar{\beta}]$ hold with the vector parameters $\bar{\gamma} = (\gamma_1, \dots, \gamma_k)$ and $\bar{\beta} = (\beta_1, \dots, \beta_k)$ such that, for every $i = 1, \dots, k$, either $\beta_i > \gamma_i > 0$ or $\beta_i = \gamma_i = 0$, and also the conditions $\mathbf{L}_6$, $\mathbf{M}_6$, $\mathbf{N}_4$, $\mathbf{I}_7$, $\mathbf{J}_{12}$, and $\mathbf{K}_{11}$ hold. Then, the following relation takes place:*

$$\Phi_\varepsilon \to \Phi_0 \quad \text{as} \quad \varepsilon \to 0 . \tag{7.193}$$

Proof. We are going to check that all conditions of Theorem 5.3.6 hold and then to apply this theorem. In fact, we should only check that the condition $\mathbf{D}_7[\bar{\beta}]$ and the condition $\mathbf{K}_6$ hold. Indeed, this check was made for other conditions of Theorem 5.3.6 in the proof of Theorem 7.3.5.

Relation (7.166) and the condition $\mathbf{M}_6$ imply that the following relation holds for $i = 1, \dots, k$ and $\varepsilon \in (0, \varepsilon_0]$:

$$\mathsf{E}e^{\beta|Y_{\varepsilon,0,i}|} = \mathsf{E}e^{\beta|h_{\varepsilon,0,i}(Y_{0,0,i})|} . \tag{7.194}$$

Using inequality (7.175) and relation (7.194) we get the following inequality, for $i = 1, \dots, k$, $\varepsilon \in (0, \varepsilon_{18}]$, and $\beta \geq 0$:

$$\mathsf{E}e^{\beta|Y_{\varepsilon,0,i}|} = \mathsf{E}e^{\beta|h_{\varepsilon,0,i}(Y_{0,0,i})|} \leq e^{\beta\delta_{\varepsilon,i}}\mathsf{E}e^{\beta|Y_{0,0,i}|}\,. \tag{7.195}$$

Since, by the condition $\mathbf{N}_4$, $e^{\beta\delta_{\varepsilon,i}} \to 1$ as $\varepsilon \to 0$, for $i = 1, \dots, k$, the condition $\mathbf{D}_{11}[\bar{\beta}]$ and inequality (7.195) imply that, for $i = 1, \dots, k$,

$$\varlimsup_{\varepsilon\to 0} \mathsf{E}e^{\beta_i|Y_{\varepsilon,0,i}|} \leq \mathsf{E}e^{\beta_i|Y_{0,0,i}|} < K_{54,i}\,. \tag{7.196}$$

Thus, the condition $\mathbf{D}_7[\bar{\beta}]$ holds.

Let us take an arbitrary $\varrho > 0$. Using the condition $\mathbf{N}_4$ and inequality (7.181), we get the following relations, for $i = 1, \dots, k$:

$$\begin{aligned}
&\varlimsup_{0<\varepsilon\to 0} \mathsf{P}\{|h_{\varepsilon,0,i}(Y_{0,0,i}) - Y_{0,0,i}| \geq \varrho\} \\
&\quad \leq \varlimsup_{0<\varepsilon\to 0} \left(\mathsf{P}\{Y_{0,0,i} \leq z^-_{\varepsilon,0,i}\} + I(\delta_{\varepsilon,i} \geq \varrho) + \mathsf{P}\{Y_{0,0,i} > z^+_{\varepsilon,0,i}\}\right) = 0\,.
\end{aligned} \tag{7.197}$$

Let us take arbitrary $\kappa, \varrho > 0$. There exists a compact $C_\kappa \subseteq \mathbb{X}$ such that $\mathsf{P}\{X_{0,0} \in \overline{C}_\kappa\} \leq \kappa$. Since $\mathbb{X}$ is a Polish metric space and C_κ is a compact, there exists a ball $R_\kappa = R_{d_\kappa}(x_\kappa)$ such that $C_\kappa \subseteq R_\kappa$. By the conditions $\mathbf{N}_2$ (d) there exists a $\varepsilon_{x_\kappa,d_\kappa} \in (0, \varepsilon_0]$ such that $R_\kappa \subseteq K_{\varepsilon,0}$, for $\varepsilon \in (0, \varepsilon_{x_\kappa,d_\kappa}]$. The above remarks and relation (7.103) imply that,

$$\begin{aligned}
&\varlimsup_{0<\varepsilon\to 0} \mathsf{P}\{d_{\mathbb{X}}(h_{\varepsilon,0,0}(X_{0,0}), X_{0,0}) \geq \varrho\} \\
&\quad \leq \varlimsup_{0<\varepsilon\to 0} (\mathsf{P}\{X_{0,0} \in \overline{\mathbb{K}}_{\varepsilon,0}\} + I(d_{\varepsilon,0} \geq \varrho)) \leq \kappa\,.
\end{aligned} \tag{7.198}$$

Due to an arbitrary choice of $\kappa, \varrho > 0$, relations (7.197) and (7.198) imply the following relation:

$$\begin{aligned}
&P_{\varepsilon,0}(\cdot) = \mathsf{P}\{(\hat{h}_{\varepsilon,0}(\vec{Y}_{0,0}), h_{\varepsilon,0,0}(X_{0,0})) \in \cdot\} \\
&\Rightarrow P_{0,0}(\cdot) = \mathsf{P}\{(\vec{Y}_{0,0}, X_{0,0}) \in \cdot\} \quad \text{as} \quad \varepsilon \to 0\,.
\end{aligned} \tag{7.199}$$

Relation (7.199) and the condition $\mathbf{K}_{11}$ imply that the condition $\mathbf{K}_6$ holds with the sets $\mathbb{Z}'_0$ and $\mathbb{Z}''_0$.

It follows from the above remarks that all conditions of Theorem 5.3.6 hold, and, therefore, Theorem 7.3.6 is a corollary of Theorem 5.3.6. □

7.5 Multivariate Markov LPP with unbounded characteristics

In this section, we present space-skeleton approximation models for discrete time multivariate modulated Markov log-price processes with unbounded characteristics.

7.5.1 Space-skeleton approximations for univariate Markov log-price processes with unbounded characteristics

In this subsection, we consider the same model of log-price processes $Y_{\varepsilon,n}$, $n = 0, 1, \dots$, and the same skeleton approximation model as in Subsection 7.2.1. The only difference is that we assume that the corresponding characteristics of the process $Y_{0,n}$ can be unbounded.

In this case, we are going to use Theorems 6.2.2 and 6.2.4, which give conditions of convergence for, respectively, log-reward functions and optimal expected rewards for univariate modulated Markov log-price processes with unbounded characteristics. In fact, we shall use reduced variants of these theorems for the model without index component.

Let us recall the class $\mathcal{A}$ of measurable, nonnegative, and nondecreasing functions $A(\beta)$ defined for $\beta \geq 0$ and such that $A(0) = 0$.

Any function $A(\beta) \in \mathcal{A}$ generates a nondecreasing sequence of functions $A_0(\beta) \leq A_1(\beta) \leq A_2(\beta) \leq \cdots$ from the class $\mathcal{A}$ that are defined by the following recurrence relation, for every $\beta \geq 0$:

$$A_n(\beta) = \begin{cases} \beta & \text{for } n = 0, \\ A_{n-1}(\beta) + A(A_{n-1}(\beta)) & \text{for } n = 1, 2, \dots . \end{cases} \tag{7.200}$$

Let us also recall the first-type A-modulus of exponential moment compactness for the log-price process $Y_{0,n}$, defined for $\beta \geq 0$,

$$\Delta_{\beta,A}(Y_{0,\cdot}, N) = \max_{0 \leq n \leq N-1} \sup_{y \in \mathbb{R}_1} \frac{\mathsf{E}_{y,n} e^{A_{N-n-1}(\beta)|Y_{0,n+1} - Y_{0,n}|}}{e^{A(A_{N-n-1}(\beta))|y|}} . \tag{7.201}$$

Recall that we use the notations $\mathsf{P}_{y,n}$ and $\mathsf{E}_{y,n}$ for conditional probabilities and expectations under the condition $Y_n = y$.

Note that modulus $\Delta_{\beta,A}(Y_{0,\cdot}, N)$ coincides with modulus $\Delta_\beta(Y_{0,\cdot}, N)$, if function $A(\beta) = 0$. Indeed, in this case all functions $A(A_n(\beta)) \equiv 0$ and $A_n(\beta) \equiv \beta$ for $n = 0, 1, \dots$.

Instead of the condition $\mathbf{C_{11}}[\beta]$ based on the modulus $\Delta_\beta(Y_{0,\cdot}, N)$, we use the following first-type condition of exponential moment compactness for the log-price process $Y_{0,n}$, which is assumed to hold for some $\beta \geq 0$:

$\mathbf{C}_{15}[\beta]$: $\Delta_{\beta,A}(Y_{0,\cdot}, N) < K_{60}$, for some $1 < K_{60} < \infty$.

Let us also recall the second-type A-modulus of exponential moment compactness for the log-price process $Y_{0,n}$, for $\beta \geq 0$,

$$\Xi^{\pm}_{\beta,A}(Y_{0,\cdot}, N) = \max_{0\leq n\leq N-1} \sup_{y\in\mathbb{R}_1} \frac{\mathsf{E}_{y,n} e^{\pm A_{N-n-1}(\beta)(Y_{0,n+1}-Y_{0,n})}}{e^{A(A_{N-n-1}(\beta))|y|}}. \tag{7.202}$$

The following second-type condition of exponential moment compactness is assumed to hold for some $\beta \geq 0$:
$\mathbf{E}_{19}[\beta]$: $\Xi^{\pm}_{\beta,A}(Y_{0,\cdot}, N) < K_{61}$, for some $1 < K_{61} < \infty$.

The following lemma is a corollary of Lemma 4.3.7.

Lemma 7.5.1. *The condition $\mathbf{E}_{19}[\beta]$ implies the condition $\mathbf{C}_{15}[\beta]$ to hold.*

Let us also recall the sequences of the intervals $\mathcal{Z}_\varepsilon = \langle \mathbb{Z}_{\varepsilon,0}, \ldots, \mathbb{Z}_{\varepsilon,N} >$ introduced in relation (7.51).

Instead of the modulus of exponential moment compactness $\Delta_{\beta,\mathcal{Z}_\varepsilon}(Y_{\varepsilon,\cdot}, N)$ we recall the modified first-type A-modulus of exponential moment compactness, defined for every $\varepsilon \in [0, \varepsilon_0]$ and $\beta \geq 0$ by the following formula:

$$\Delta_{\beta,A,\mathcal{Z}_\varepsilon}(Y_{\varepsilon,\cdot}, N) = \max_{0\leq n\leq N-1} \sup_{y\in\mathbb{Z}_{\varepsilon,n}} \frac{\mathsf{E}_{y,n} e^{A_{N-n-1}(\beta)|Y_{\varepsilon,n+1}-Y_{\varepsilon,n}|}}{e^{A(A_{N-n-1}(\beta))|y|}}. \tag{7.203}$$

Theorem 7.5.1. *Let a Markov log-price process $Y_{0,n}$ and the corresponding space-skeleton approximation Markov processes $Y_{\varepsilon,n}$ be defined as described in Subsection 7.2.1. Also let the conditions $\mathbf{B}_4[\gamma]$ and $\mathbf{C}_{15}[\beta]$ hold with the parameters γ and β such that either $\beta > \gamma > 0$ or $\beta = \gamma = 0$, and also the conditions $\mathbf{L}_2$, $\mathbf{N}_1$, $\mathbf{I}_6$, and $\mathbf{J}_9$ hold. Then, for every $n = 0, 1, \ldots, N$, the following relation takes place for any $y_\varepsilon \to y_0 \in \mathbb{Y}'_n \cap \mathbb{Y}''_n$,*

$$\phi_{\varepsilon,n}(y_\varepsilon) \to \phi_{0,n}(y_0) \quad \text{as} \quad \varepsilon \to 0. \tag{7.204}$$

Proof. We are going to check that all conditions of Theorem 6.2.2 hold and to apply this theorem.

The first part of the proof is similar with the corresponding part in the proof of Theorem 7.2.1. The only difference that we use the modulus $\Delta_{\beta,A,\mathcal{Z}_\varepsilon}(Y_{\varepsilon,\cdot}, N)$ instead on the modulus $\Delta_{\beta,\mathcal{Z}_\varepsilon}(Y_{\varepsilon,\cdot}, N)$,

Using relations (7.52), (7.55), and (7.56) we get the following inequalities (replacing inequalities (7.57) given in the proof of Theorem 7.2.1), for every $\varepsilon \in (0, \varepsilon_0]$, $n = 0, \ldots, N-1$, and $\beta \geq 0$,

$$\sup_{y\in\mathbb{Z}_{\varepsilon,n}} \frac{\mathsf{E}_{y,n} e^{A_{N-n-1}(\beta)|Y_{\varepsilon,n+1}-Y_{\varepsilon,n}|}}{e^{A(A_{N-n-1}(\beta))|y|}}$$

$$= \sup_{y\in\mathbb{Z}_{\varepsilon,n}} \mathsf{E}_{h_{\varepsilon,n}(y),n} \frac{e^{A_{N-n-1}(\beta)|h_{\varepsilon,n+1}(Y_{0,n+1})-y|}}{e^{A(A_{N-n-1}(\beta))|y|}}$$

$$= \sup_{y \in \mathbb{Z}_{\varepsilon,n}} \mathsf{E}_{h_{\varepsilon,n}(y),n} \Bigg(\frac{e^{A_{N-n-1}(\beta)(y - Y_{0,n+1})}}{e^{A(A_{N-n-1}(\beta))|y|}} I^-_{\varepsilon,n}$$
$$+ e^{A_{N-n-1}(\beta)\delta_{\varepsilon,n+1}} \frac{e^{A_{N-n-1}(\beta)|Y_{0,n+1} - y|}}{e^{A(A_{N-n-1}(\beta))|y|}} I_{\varepsilon,n}$$
$$+ \frac{e^{A_{N-n-1}(\beta)(Y_{0,n+1} - y)}}{e^{A(A_{N-n-1}(\beta))|y|}} I^+_{\varepsilon,n} \Bigg)$$
$$\leq e^{A_{N-n-1}(\beta)\delta_{\varepsilon,n+1}} \sup_{y \in \mathbb{Z}_{\varepsilon,n}} \mathsf{E}_{h_{\varepsilon,n}(y),n} \frac{e^{A_{N-n-1}(\beta)|Y_{0,n+1} - y|}}{e^{A(A_{N-n-1}(\beta))|y|}}$$
$$\leq e^{A_{N-n-1}(\beta)\delta_{\varepsilon,n+1}} e^{A_{N-n-1}(\beta)\delta_{\varepsilon,n}}$$
$$\times \sup_{y \in \mathbb{Z}_{\varepsilon,n}} \mathsf{E}_{h_{\varepsilon,n}(y),n} \frac{e^{A_{N-n-1}(\beta)|Y_{0,n+1} - h_{\varepsilon,n}(y)|}}{e^{A(A_{N-n-1}(\beta))|y|}}$$
$$\leq e^{A_{N-n-1}(\beta)\delta_{\varepsilon,n+1}} e^{A_{N-n-1}(\beta)\delta_{\varepsilon,n}} e^{A(A_{N-n-1}(\beta))\delta_{\varepsilon,n}}$$
$$\times \sup_{y \in \mathbb{R}_1} \mathsf{E}_{h_{\varepsilon,n}(y),n} \frac{e^{A_{N-n-1}(\beta)|Y_{0,n+1} - h_{\varepsilon,n}(y)|}}{e^{A(A_{N-n-1}(\beta))|h_{\varepsilon,n}(y)|}}$$
$$\leq e^{A_{N-n-1}(\beta)\delta_{\varepsilon,n+1} + A_{N-n-1}(\beta)\delta_{\varepsilon,n} + A(A_{N-n-1}(\beta))\delta_{\varepsilon,n}}$$
$$\times \sup_{y \in \mathbb{R}_1} \frac{\mathsf{E}_{y,n} e^{A_{N-n-1}(\beta)|Y_{0,n+1} - y|}}{e^{A(A_{N-n-1}(\beta))|y|}}, \tag{7.205}$$

where, for $n = 0, \dots, N-1$,

$$\begin{aligned} I^-_{\varepsilon,n} &= I(Y_{0,n+1} \leq z^-_{\varepsilon,n+1}), \\ I_{\varepsilon,n} &= I(z^-_{\varepsilon,n+1} < Y_{0,n+1} \leq z^+_{\varepsilon,n+1}), \\ I^+_{\varepsilon,n} &= I(Y_{0,n+1} > z^+_{\varepsilon,n+1}). \end{aligned} \tag{7.206}$$

This relation implies the following inequality, for the moduli $\Delta_{\beta,A,Z_\varepsilon}(Y_{\varepsilon,\cdot}, N)$ and $\Delta_{\beta,A}(Y_{0,\cdot}, N)$, which holds for every $\varepsilon \in (0, \varepsilon_0]$ and $\beta \geq 0$,

$$\Delta_{\beta,A,Z_\varepsilon}(Y_{\varepsilon,\cdot}, N) \leq \max_{0 \leq n \leq N-1} e^{A_{N-n-1}(\beta)\delta_{\varepsilon,n+1} + A_{N-n-1}(\beta)\delta_{\varepsilon,n} + A(A_{N-n-1}(\beta))\delta_{\varepsilon,n}} \times \Delta_{\beta,A}(Y_{0,\cdot}, N). \tag{7.207}$$

By the condition $\mathbf{N_1}$,

$$\max_{0 \leq n \leq N-1} e^{A_{N-n-1}(\beta)\delta_{\varepsilon,n+1} + A_{N-n-1}(\beta)\delta_{\varepsilon,n} + A(A_{N-n-1}(\beta))\delta_{\varepsilon,n}} \to 1 \quad \text{as} \quad \varepsilon \to 0. \tag{7.208}$$

Inequality (7.207), relation (7.208), and the condition $\mathbf{C_{15}}[\beta]$ imply the following relation:

$$\overline{\lim_{\varepsilon \to 0}} \Delta_{\beta,A,Z_\varepsilon}(Y_{\varepsilon,\cdot}, N) \leq \Delta_{\beta,A}(Y_{0,\cdot}, N) < K_{60}. \tag{7.209}$$

Therefore, the condition $\mathbf{C}'_9[\beta]$ (its variant for the model without modulating index component) holds for the modified moduli of exponential moment compactness $\Delta_{\beta,A,Z_\varepsilon}(Y_{\varepsilon,\cdot}, N)$.

The following part of the proof repeats the corresponding part of the proof for Theorem 7.2.1.

In particular, it was shown in the proof of Theorem 7.2.1 that the condition $\mathbf{I}_6$ implies that the condition $\mathbf{I}_1$ (its univariate variant without modulating component) holds. But, this reduced variant of the condition $\mathbf{I}_1$ just coincides with the condition $\mathbf{I}_5$.

Analogously, it was shown in the proof of Theorem 7.2.1 that the condition $\mathbf{J}_9$ implies that the condition $\mathbf{J}_1$ (its univariate variant without modulating component) holds. But, this reduced variant of the condition $\mathbf{J}_1$ just coincides with the condition $\mathbf{J}_8$.

It follows from the above remarks that all conditions of Theorem 6.2.2 hold, and, therefore, Theorem 7.4.1 is a corollary of Theorem 6.2.2. □

7.5.2 Convergence of space-skeleton approximations for optimal expected rewards of univariate Markov log-price processes with unbounded characteristics

We continue consider the model of one-dimensional Markov log-price processes introduced in Subsection 7.2.1. In this case, the condition $\mathbf{D}_{10}[\beta]$ should be replaced by the following condition assumed to hold for some $\beta \geq 0$:
$\mathbf{D}_{12}[\beta]$: $\mathsf{E}\exp\{A_N(\beta)|Y_{0,0}|\} < K_{62}$, for some $1 < K_{62} < \infty$.

The following result can be obtained by application Theorem 6.2.4.

Theorem 7.5.2. *Let a Markov log-price process $Y_{0,n}$ and the corresponding space-skeleton approximation Markov processes $Y_{\varepsilon,n}$ are defined in Subsection 7.2.1. Let also the conditions $\mathbf{B}_4[\gamma]$, $\mathbf{C}_{15}[\beta]$, and $\mathbf{D}_{12}[\beta]$ hold with the parameters γ and β such that either $\beta > \gamma > 0$ or $\beta = \gamma = 0$, and also the conditions $\mathbf{L}_2$, $\mathbf{M}_2$, $\mathbf{N}_1$, $\mathbf{I}_6$, $\mathbf{J}_9$, and $\mathbf{K}_8$ hold. Then, the following relation takes place:*

$$\Phi_\varepsilon \to \Phi_0 \quad as \quad \varepsilon \to 0\,. \tag{7.210}$$

Proof. It repeats the proof of Theorem 7.2.2. The only difference is that the condition $\mathbf{D}_{12}[\beta]$ should be used instead of the condition $\mathbf{D}_{10}[\beta]$ that let one get the inequality analogous to (7.71),

$$\overline{\lim_{\varepsilon\to 0}}\, \mathsf{E}e^{A_N(\beta)|Y_{\varepsilon,0}|} \leq \overline{\lim_{\varepsilon\to 0}}\, e^{A_N(\beta)\delta_{\varepsilon,0}}\mathsf{E}e^{A_N(\beta)|Y_{0,0}|} < K_{62}\,. \tag{7.211}$$

In the above relation, we used the relation $e^{A_N(\beta)\delta_{\varepsilon,0}} \to 1$ as $\varepsilon \to 0$ implied by the condition $\mathbf{N}_1$.

As was pointed out in the proof of Theorem 7.2.2, relation (7.44) implies that, for every $\varepsilon \in (0, \varepsilon_0]$,

$$\mathsf{P}\{Y_{\varepsilon,0} \in \mathbb{Z}_{\varepsilon,0}\} = \mathsf{P}\{h_{\varepsilon,0}(Y_{0,0}) \in \mathbb{Z}_{\varepsilon,0}\} = 1\,. \tag{7.212}$$

Relations (7.211) and (7.212) imply that the condition $\mathbf{D}'_8[\beta]$ (its variant for the model without modulating index component) holds.

The remaining part of the proof repeats the corresponding part of the proof for Theorem 7.2.2. □

7.5.3 Space-skeleton approximations for multivariate modulated Markov log-price processes with unbounded characteristics

In this subsection, we consider the same model of log-price processes $\vec{Z}_{\varepsilon,n} = (\vec{Y}_{\varepsilon,n}, X_{\varepsilon,n})$, $n = 0, 1, \dots$, and the same skeleton approximation model as those considered in Subsection 7.3.1. The only difference is that we assume that the corresponding characteristics of the process $\vec{Z}_{\varepsilon,n}$ can be unbounded.

In this case, we are going to use Theorem 6.4.2 instead of Theorem 6.2.2.

Let us recall the class $\mathcal{A}_k$ of measurable, vector functions $\vec{A}(\bar{\beta}) = (A_1(\bar{\beta}), \dots, A_k(\bar{\beta}))$ defined for $\bar{\beta} = (\beta_1, \dots, \beta_k), \beta_i \geq 0$, $i = 1, \dots, k$ such that functions $A_i(\bar{\beta})$, $i = 1, \dots, k$ are nondecreasing in every argument β_i, $i = 1, \dots, k$ and $A_i(\bar{0}) = 0$, $i = 1, \dots, k$.

Any function $\vec{A}(\bar{\beta}) \in \mathcal{A}_k$ generates a sequence of functions $\vec{A}_n(\bar{\beta}) = (A_{n,1}(\bar{\beta}), \dots, A_{n,k}(\bar{\beta}))$, $n = 0, 1, \dots$ from the class $\mathcal{A}_k$ that are defined by the following recurrence relation, for any $\bar{\beta} = (\beta_1, \dots, \beta_k), \beta_i \geq 0$, $i = 1, \dots, k$,

$$\vec{A}_n(\bar{\beta}) = \begin{cases} \bar{\beta} & \text{for } n = 0\,, \\ \vec{A}_{n-1}(\bar{\beta}) + \vec{A}(\vec{A}_{n-1}(\bar{\beta})) & \text{for } n = 1, 2, \dots\,. \end{cases} \tag{7.213}$$

Note that, by the definition, $0 = A_{0,j}(\bar{\beta}) \leq A_{1,j}(\bar{\beta}) \leq \cdots$ for every $\bar{\beta} = (\beta_1, \dots, \beta_k), \beta_i \geq 0$, $i = 1, \dots, k$ and $j = 1, \dots, k$.

Let us recall the first-type $\vec{A}$-modulus of exponential moment compactness for the log-price process $\vec{Y}_{0,n}$, for $\bar{\beta} = (\beta_1, \dots, \beta_k), \beta_i \geq 0$, $i = 1, \dots, k$,

$$\Delta_{\bar{\beta},\vec{A}}(\vec{Y}_{0,\cdot}, N) = \max_{0 \leq n \leq N-1} \sup_{\vec{z}=((y_1,\dots,y_k),\,x)\in\mathbb{Z}} \frac{\mathsf{E}_{\vec{z},n} e^{\sum_{j=1}^k A_{N-n-1,j}(\bar{\beta})|Y_{0,n+1,j} - Y_{0,n,j}|}}{e^{\sum_{j=1}^k A_j(\vec{A}_{N-n-1}(\bar{\beta}))|y_j|}}\,. \tag{7.214}$$

Recall that we use the notations $\mathsf{P}_{\vec{z},n}$ and $\mathsf{E}_{\vec{z},n}$ for conditional probabilities and expectations under the condition $\vec{Z}_n = \vec{z}$.

Instead of the condition $\mathbf{C}_{12}[\bar{\beta}]$ based on the modulus $\Delta_\beta(\vec{Y}_{0,\cdot}, N)$, we use the following first-type condition of exponential moment compactness for the log-price process $\vec{Y}_{0,n}$, which is assumed to hold for some vector parameter $\bar{\beta} = (\beta_1, \dots, \beta_k)$ with nonnegative components and vectors $\bar{\beta}_i = (0, \dots, 0, \beta_i, 0, \dots, 0)$, $i = 1, \dots, k$:

$\mathbf{C}_{16}[\bar{\beta}]$: $\Delta_{\bar{\beta}_i,\vec{A}}(\vec{Y}_{0,\cdot}, N) < K_{63,i}$, $i = 1, \dots, k$, for some $1 < K_{63,i} < \infty$, $i = 1, \dots, k$.

Let us recall the second-type $\vec{A}$-modulus of exponential moment compactness for the components of the log-price process $\vec{Y}_{0,n}$, defined for vector $\bar{\beta} = (\beta_1, \dots, \beta_k)$ with nonnegative components and vectors $\bar{\imath}_k = (\imath_1, \dots, \imath_k) \in \mathcal{I}_k$, where $\mathcal{I}_k = \{\bar{\imath}_k : \imath_j = +, -, j = 1, \dots, k\}$.

$$\Xi^{\bar{\imath}_k}_{\bar{\beta},\vec{A}}(\vec{Y}_{0,\cdot}, N) = \max_{0\le n\le N-1} \sup_{\vec{z}=((y_1,\dots,y_k),x)\in\mathbb{Z}} \frac{\mathsf{E}_{\vec{z},n} e^{\sum_{j=1}^k \imath_j A_{N-n-1,j}(\bar{\beta})(Y_{n+1,j}-Y_{n,j})}}{e^{\sum_{j=1}^k A_j(\vec{A}_{N-n-1}(\bar{\beta}))|y_j|}} . \tag{7.215}$$

The following second-type condition of exponential moment compactness is assumed to hold for some vector parameter $\bar{\beta}$ with nonnegative components:

$\mathbf{E_{20}}[\bar{\beta}]$: $\Xi^{\bar{\imath}_k}_{\bar{\beta}_i,\vec{A}}(\vec{Y}_{0,\cdot}, N) < K_{64,i}$, $\bar{\imath}_k = (\imath_1, \dots, \imath_k) \in \mathcal{I}_k$, $i = 1, \dots, k$, for some $1 < K_{64,i} < \infty$, $i = 1, \dots, k$.

The following lemma is a corollary of Lemma 4.3.17.

Lemma 7.5.2. *The condition $\mathbf{E_{20}}[\beta]$ implies the condition $\mathbf{C_{16}}[\beta]$ to hold.*

Let us recall the sequences of the sets $\mathcal{Z}_\varepsilon = \langle \mathbb{Z}_{\varepsilon,0}, \dots, \mathbb{Z}_{\varepsilon,N}\rangle$ introduced in relation (7.93).

Let us also recall the modified first-type $\vec{A}$-modulus of exponential moment compactness, defined for every $\varepsilon \in [0, \varepsilon_0]$ and a vector parameter $\bar{\beta} = (\beta_1, \dots, \beta_k)$ with nonnegative components by the following formula:

$$\Delta_{\bar{\beta},\vec{A},\mathcal{Z}_\varepsilon}(\vec{Y}_{\varepsilon,\cdot}, N) = \max_{0\le n\le N-1} \sup_{\vec{z}=((y_1,\dots,y_k),x)\in\mathbb{Z}_{\varepsilon,n}} \frac{\mathsf{E}_{\vec{z},n} e^{\sum_{j=1}^k A_{N-n-1,j}(\bar{\beta})|Y_{\varepsilon,n+1,j}-Y_{\varepsilon,n,j}|}}{e^{\sum_{j=1}^k A_j(\vec{A}_{N-n-1}(\bar{\beta}))|y_j|}} . \tag{7.216}$$

Theorem 7.5.3. *Let a Markov multivariate modulated log-price process $\vec{Z}_{0,n}$ and the corresponding space-skeleton approximation Markov processes $\vec{Z}_{\varepsilon,n}$ be defined as in Subsection 7.3.1. Let also the conditions $\mathbf{B_1}[\bar{\gamma}]$ and $\mathbf{C_{16}}[\bar{\beta}]$ hold with the vector parameters $\bar{\gamma} = (\gamma_1, \dots, \gamma_k)$ and $\bar{\beta} = (\beta_1, \dots, \beta_k)$ such that, for every $i = 1, \dots, k$, either $\beta_i > \gamma_i > 0$ or $\beta_i = \gamma_i = 0$, and also the conditions $\mathbf{L_4}$, $\mathbf{N_2}$, $\mathbf{I_7}$, and $\mathbf{J_{10}}$ hold. Then, for every $n = 0, 1, \dots, N$, the following relation takes place for any $\vec{z}_\varepsilon = (\vec{y}_\varepsilon, x_\varepsilon) \to \vec{z}_0 = (\vec{y}_0, x_0) \in \mathbb{Z}'_n \cap \mathbb{Z}''_n$,*

$$\phi_{\varepsilon,n}(\vec{y}_\varepsilon, x_\varepsilon) \to \phi_{0,n}(\vec{y}_0, x_0) \quad as \quad \varepsilon \to 0 . \tag{7.217}$$

Proof. We are going to check that all conditions of Theorem 6.4.2 hold and then to apply this theorem.

The first part of the proof is similar with the corresponding part in the proof of Theorem 7.3.2. The only difference is that we use the modulus $\Delta_{\bar{\beta},\vec{A},\mathcal{Z}_\varepsilon}(\vec{Y}_{\varepsilon,\cdot}, N)$ instead on the modulus $\Delta_{\bar{\beta},\mathcal{Z}_\varepsilon}(\vec{Y}_{\varepsilon,\cdot}, N)$.

Using relations (7.94), (7.97), and (7.98), we get the following inequalities for every $\varepsilon \in (0, \varepsilon_0]$ and $i = 1, \dots, k$, $n = 0, \dots, N-1$,

$$
\begin{aligned}
\sup_{\vec{z}=((y_1,\dots,y_k),x)\in\mathbb{Z}_{\varepsilon,n}} & \frac{\mathsf{E}_{\vec{z},n} e^{\sum_{j=1}^{k} A_{N-n-1,j}(\bar{\beta})|Y_{\varepsilon,n+1,j}-Y_{\varepsilon,n,j}|}}{e^{\sum_{j=1}^{k} A_j(\vec{A}_{N-n-1}(\bar{\beta}))|y_j|}} \\
&= \sup_{\vec{z}\in\mathbb{Z}_{\varepsilon,n}} \frac{\mathsf{E}_{\bar{h}_{\varepsilon,n}(\vec{z}),n} e^{\sum_{j=1}^{k} A_{N-n-1,j}(\bar{\beta})|h_{\varepsilon,n+1,j}(Y_{0,n+1,j})-y_j|}}{e^{\sum_{j=1}^{k} A_j(\vec{A}_{N-n-1}(\bar{\beta}))|y_j|}} \\
&= \sup_{\vec{z}\in\mathbb{Z}_{\varepsilon,n}} \frac{1}{e^{\sum_{j=1}^{k} A_j(\vec{A}_{N-n-1}(\bar{\beta}))|y_j|}} \\
&\quad \times \mathsf{E}_{\bar{h}_{\varepsilon,n}(\vec{z}),n} \prod_{j=1}^{k} \Big(e^{A_{N-n-1,j}(\bar{\beta})(y_j-Y_{0,n+1,j})} I^{-}_{\varepsilon,n,j} \\
&\quad + e^{A_{N-n-1,j}(\bar{\beta})\delta_{\varepsilon,n+1,j}} e^{A_{N-n-1,j}(\bar{\beta})|Y_{0,n+1,j}-y_j|} I_{\varepsilon,n,j} \\
&\quad + e^{A_{N-n-1,j}(\bar{\beta})(Y_{0,n+1,j}-y_j)} I^{+}_{\varepsilon,n,j} \Big) \\
&\le \prod_{j=1}^{k} e^{A_{N-n-1,j}(\bar{\beta})\delta_{\varepsilon,n+1,j}} \sup_{\vec{z}\in\mathbb{Z}_{\varepsilon,n}} \frac{1}{e^{\sum_{j=1}^{k} A_j(\vec{A}_{N-n-1}(\bar{\beta}))|y_j|}} \\
&\quad \times \mathsf{E}_{\bar{h}_{\varepsilon,n}(\vec{z}),n} \prod_{j=1}^{k} e^{A_{N-n-1,j}(\bar{\beta})|Y_{0,n+1,j}-y_j|} \\
&\le \prod_{j=1}^{k} e^{A_{N-n-1,j}(\bar{\beta})\delta_{\varepsilon,n+1,j}} e^{A_{N-n-1,j}(\bar{\beta})\delta_{\varepsilon,n,j}} e^{A_j(\vec{A}_{N-n-1}(\bar{\beta}))\delta_{\varepsilon,n,j}} \\
&\quad \times \sup_{\vec{z}\in\mathbb{Z}_{\varepsilon,n}} \frac{1}{e^{\sum_{j=1}^{k} A_j(\vec{A}_{N-n-1}(\bar{\beta}))|h_{\varepsilon,n,j}(y_j)|}} \\
&\quad \times \mathsf{E}_{\bar{h}_{\varepsilon,n}(\vec{z}),n} \prod_{j=1}^{k} e^{A_{N-n-1,j}(\bar{\beta})|Y_{0,n+1,j}-h_{\varepsilon,n,j}(y_j)|} \\
&\le \prod_{j=1}^{k} e^{A_{N-n-1,j}(\bar{\beta})\delta_{\varepsilon,n+1,j}} e^{A_{N-n-1,j}(\bar{\beta})\delta_{\varepsilon,n,j}} e^{A_j(\vec{A}_{N-n-1}(\bar{\beta}))\delta_{\varepsilon,n,j}} \\
&\quad \times \sup_{\vec{z}\in\mathbb{Z}_{\varepsilon,n}} \frac{\mathsf{E}_{\vec{z},n} e^{\sum_{j=1}^{k} A_{N-n-1,j}(\bar{\beta})|Y_{0,n+1,j}-y_j|}}{e^{\sum_{j=1}^{k} A_j(\vec{A}_{N-n-1}(\bar{\beta}))|y_j|}} ,
\end{aligned}
\tag{7.218}
$$

where, for $i = 1, \dots, k$, $n = 0, \dots, N-1$,

$$
\begin{aligned}
I^{-}_{\varepsilon,n,i} &= I(Y_{0,n+1,i} \le z^{-}_{\varepsilon,n+1,i}) , \\
I_{\varepsilon,n,i} &= I(z^{-}_{\varepsilon,n+1,i} < Y_{0,n+1,i} \le z^{+}_{\varepsilon,n+1,i}) , \\
I^{+}_{\varepsilon,n,i} &= I(Y_{0,n+1,i} > z^{+}_{\varepsilon,n+1,i}) .
\end{aligned}
\tag{7.219}
$$

Let us denote

$$\delta_\varepsilon(\bar{\beta}) = \max_{0\le n\le N-1} \big(e^{\sum_{j=1}^k A_{N-n-1,j}(\bar{\beta})\delta_{\varepsilon,n+1,j}} \times e^{\sum_{j=1}^k A_{N-n-1,j}(\bar{\beta})\delta_{\varepsilon,n,j}} e^{\sum_{j=1}^k A_j(\vec{A}_{N-n-1}(\bar{\beta}))\delta_{\varepsilon,n,j}}\big) . \tag{7.220}$$

Relation (7.218) implies the following inequality, which holds for every $\varepsilon \in (0, \varepsilon_0]$ and vectors $\bar{\beta}$ with nonnegative components,

$$\Delta_{\bar{\beta},\vec{A},Z_\varepsilon}(\vec{Y}_{\varepsilon,\cdot}, N) \le \delta_\varepsilon(\bar{\beta})\Delta_{\bar{\beta},\vec{A}}(\vec{Y}_{0,\cdot}, N) . \tag{7.221}$$

Since $\delta_\varepsilon(\bar{\beta}_i) \to 0$ as $\varepsilon \to 0$, for $i = 1, \dots, k$, by the condition $\mathbf{N_4}$, inequality (7.221) and the condition $\mathbf{C_{16}}[\bar{\beta}]$ imply the following relation, for $i = 1, \dots, k$:

$$\overline{\lim_{\varepsilon\to 0}} \Delta_{\bar{\beta}_i,\vec{A},Z_\varepsilon}(\vec{Y}_{\varepsilon,\cdot}, N) \le \Delta_{\bar{\beta}_i,\vec{A}}(\vec{Y}_{0,\cdot}, N) < K_{63,i} . \tag{7.222}$$

Therefore, the condition $\mathbf{C'_{10}}[\bar{\beta}]$ holds for the modified modula of exponential moment compactness $\Delta_{\bar{\beta}_i,\vec{A},Z_\varepsilon}(Y_{\varepsilon,\cdot}, N)$, $i = 1, \dots, k$.

The following part of the proof repeats the corresponding part of the proof for Theorem 7.2.3.

In particular, it was shown in the proof of Theorem 7.2.3 that the condition $\mathbf{I_7}$ implies that the condition $\mathbf{I_1}$ holds.

Analogously, it was shown in the proof of Theorem 7.2.3 that the condition $\mathbf{J_{10}}$ implies that the condition $\mathbf{J_1}$.

It follows from the above remarks that all conditions of Theorem 6.4.2 hold, and, therefore, Theorem 7.4.1 is a corollary of Theorem 6.4.2. □

7.5.4 Convergence of space-skeleton approximations for optimal expected rewards for multivariate modulated Markov log-price processes with unbounded characteristics

We continue consider the model of multidimensional modulated Markov log-price processes introduced in Subsection 7.3.1.

In this case, the condition $\mathbf{D_{11}}[\bar{\beta}]$ should be replaced by the following condition assumed to hold for some vector $\bar{\beta} = (\beta_1, \dots, \beta_k)$ with nonnegative components and vectors $\bar{\beta}_i = (0, \dots, 0, \beta_i, 0, \dots, 0) = (\beta_{i,1}, \dots, \beta_{i,k})$, $i = 1, \dots, k$:

$\mathbf{D_{13}}[\bar{\beta}]$ $\mathsf{E}e^{\sum_{j=1}^k A_{N,j}(\bar{\beta}_i)|Y_{0,0,j}|} < K_{65,i}$, $i = 1, \dots, k$, for some $1 < K_{65,i} < \infty$, $i = 1, \dots, k$.

Theorem 7.5.4. *Let a Markov multivariate modulated log-price process $\vec{Z}_{0,n}$ and the corresponding space-skeleton approximation Markov processes $\vec{Z}_{\varepsilon,n}$ are defined as in Subsection 7.3.1. Let also the conditions $\mathbf{B_1}[\bar{\gamma}]$ $\mathbf{C_{16}}[\bar{\beta}]$ and $\mathbf{D_{13}}[\bar{\beta}]$ hold with the vector parameters $\bar{\gamma} = (\gamma_1, \dots, \gamma_k)$ and $\bar{\beta} = (\beta_1, \dots, \beta_k)$ such that, for every $i = 1, \dots, k$, either*

$\beta_i > \gamma_i > 0$ or $\beta_i = \gamma_i = 0$, and also the conditions $\mathbf{L_4}, \mathbf{M_4}, \mathbf{N_2}, \mathbf{I_7}, \mathbf{J_{10}}$, *and* $\mathbf{K_9}$ *hold. Then, the following relation takes place:*

$$\Phi_\varepsilon \to \Phi_0 \quad as \quad \varepsilon \to 0\,. \tag{7.223}$$

Proof. This theorem is a corollary of Theorem 6.4.4. The proof is analogous to the proof of Theorem 7.3.2. The only difference is that the condition $\mathbf{D_{13}}[\bar\beta]$ should be used instead of the condition $\mathbf{D_{11}}[\bar\beta]$ that let one get the inequality analogous to (7.120), for $i = 1, \dots, k$,

$$\overline{\lim_{\varepsilon\to 0}} \, \mathsf{E} e^{\sum_{j=1}^k A_{N,j}(\bar\beta_i)|Y_{\varepsilon,0,j}|} \le \overline{\lim_{\varepsilon\to 0}} \, e^{\sum_{j=1}^k A_{N,j}(\bar\beta_i)\delta_{\varepsilon,0,j}} \mathsf{E} e^{\sum_{j=1}^k A_{N,j}(\bar\beta_i)|Y_{0,0,j}|} < K_{65,i}\,. \tag{7.224}$$

In the above relation, we used the relation $e^{\sum_{j=1}^k A_{N,j}(\bar\beta_i)\delta_{\varepsilon,0,j}} \to 1$ as $\varepsilon \to 0$ implied by the condition $\mathbf{N_2}$.

As was pointed out in the proof of Theorem 7.3.2, relation (7.87) implies that, for every $\varepsilon \in (0, \varepsilon_0]$,

$$\mathsf{P}\{\vec Z_{\varepsilon,0} \in \mathbb{Z}_{\varepsilon,0}\} = \mathsf{P}\{\bar h_{\varepsilon,0}(\vec Z_{0,0}) \in \mathbb{Z}_{\varepsilon,0}\} = 1\,. \tag{7.225}$$

Relations (7.224) and (7.225) imply that the condition $\mathbf{D}'_9[\bar\beta]$ holds.

The remaining part of the proof repeats the corresponding part of the proof for Theorem 7.3.2. □

8 Convergence of rewards for Markov Gaussian LPP

In this chapter, we present results about convergence of rewards for American-type options for discrete time modulated Markov Gaussian log-price processes. These processes play, in the class of discrete time Markov log-price processes, the role, analogous to those played by diffusion processes in the class of continuous time Markov log-price processes. They can also be effectively used for constructing of stochastic approximations for rewards of diffusion-type log-price processes.

In Section 8.1, we discuss results about convergence of rewards for perturbed, univariate modulated Markov Gaussian log-price processes with bounded and unbounded (linear bounded) drift and volatility coefficients. These results are applied to Markov Gaussian log-price processes with estimated parameters, in Section 8.3.

In Section 8.2, we present results about convergence of rewards for perturbed, multivariate modulated Markov Gaussian log-price processes with bounded and unbounded (linear bounded) drift and volatility coefficients.

In Section 8.3, we consider Markov Gaussian log-price processes with random drift and volatility coefficients given in the form of consistent statistical estimates and present results about consistency of the corresponding estimates for option rewards.

In Section 8.4, we present results about convergence of space-skeleton reward approximations for multivariate modulated Markov Gaussian log-price processes with bounded and unbounded (linear bounded) drift and volatility coefficients.

In Section 8.5, we present results about convergence of space-skeleton reward approximations for log-price processes represented by multivariate modulated Markov Gaussian random walks.

The main results are given in Theorems 8.3.1–8.3.8 and 8.4.1–8.4.8. These results are new.

8.1 Univariate Markov Gaussian LPP

In this section, we presents general results about convergence of rewards for American-type options for Markov Gaussian log-price processes.

8.1.1 Asymptotically uniform upper bounds for option rewards for univariate Markov Gaussian log-price processes with bounded characteristics

Let us assume that a modulated log-price process $Y_{\varepsilon,n}$, $n = 0, 1, \dots$ is given, for every $\varepsilon \in [0, \varepsilon_0]$, by the following stochastic transition dynamic relation

$$Y_{\varepsilon,n} = Y_{\varepsilon,n-1} + \mu_{\varepsilon,n}(Y_{\varepsilon,n-1}) + \sigma_{\varepsilon,n}(Y_{\varepsilon,n-1}) W_{\varepsilon,n}, \quad n = 1, 2, \dots, \tag{8.1}$$

where (a) $Y_{\varepsilon,0}$ is a real-valued random variable, (b) $W_{\varepsilon,n}$, $n = 1, 2, \dots$ is a sequence of standard i.i.d. normal random variables with mean 0 and variance 1, (c) the random variable $Y_{\varepsilon,0}$ and the sequence of random variables $W_{\varepsilon,n}$, $n = 1, 2, \dots$ are independent, (d) $\mu_{\varepsilon,n}(y)$ are real-valued Borel functions defined on $\mathbb{R}_1$, for $n = 1, 2, \dots,$ and (e) $\sigma_{\varepsilon,n}(y)$ are real-valued Borel functions defined on $\mathbb{R}_1$, for $n = 1, 2, \dots$.

This is a model without an index component, and, therefore, the pay-off function $g_\varepsilon(n, e^y)$ is also assumed to be, for every $\varepsilon \in [0, \varepsilon_0]$, a function of the argument y. In sequel, the log-reward functions $\phi_{\varepsilon,n}(y)$ are also functions argument y.

Let us give upper bounds for option rewards for the model, where the following condition holds:

$\mathbf{G_5}$: $\overline{\lim}_{\varepsilon\to 0} \max_{1\le n\le N} \sup_{y\in\mathbb{R}_1}(|\mu_{\varepsilon,n}(y)| + \sigma^2_{\varepsilon,n}(y)) < K_{66}$, for some $0 < K_{66} < \infty$.

In this case, there exist the moment-generating functions $\mathsf{E}_{y,n}e^{\beta(Y_{\varepsilon,n+1}-Y_n)}$ defined, for every $\varepsilon \in [0, \varepsilon_0]$ and $y \in \mathbb{R}_1$, $n = 0, 1, \dots, N-1$,

$$\begin{aligned} \mathsf{E}_{y,n}e^{\beta(Y_{\varepsilon,n+1}-Y_{\varepsilon,n})} &= \mathsf{E}e^{\beta(\mu_{\varepsilon,n+1}(y)\ +\sigma_{\varepsilon,n+1}(y)\ W_{\varepsilon,n+1})} \\ &= e^{\beta\mu_{\varepsilon,n+1}(y)\ +\frac{1}{2}\beta^2\sigma^2_{\varepsilon,n+1}(y)}, \quad \beta \in \mathbb{R}_1 . \end{aligned} \tag{8.2}$$

The second-type modulus of exponential moment compactness $\Xi^{\pm}_{\beta}(Y_{\varepsilon,\cdot}, N)$ takes, according to formula (8.2), the following form, for $\beta \ge 0$:

$$\Xi^{\pm}_{\beta}(Y_{\varepsilon,\cdot}, N) = \max_{0\le n\le N-1} \sup_{y\in\mathbb{R}_1} e^{\pm\beta\mu_{\varepsilon,n+1}(y)\ +\frac{1}{2}\beta^2\sigma^2_{\varepsilon,n+1}(y)} . \tag{8.3}$$

We are interested in the following condition, assumed to hold for some $\beta \ge 0$:

$\mathbf{E_{21}}[\beta]$: $\overline{\lim}_{\varepsilon\to 0}\, \Xi^{\pm}_{\beta}(Y_{\varepsilon,\cdot}, N) < K_{67}$, for some $1 < K_{67} < \infty$.

Lemma 8.1.1. *The condition* $\mathbf{G_5}$ *implies that the condition* $\mathbf{E_{21}}[\beta]$ *holds for any* $\beta \ge 0$.

Proof. It is analogous to the proof of Lemma 4.4.1.

The condition $\mathbf{G_5}$ and relations (8.2) and (8.3) imply that for any $\beta \ge 0$,

$$\begin{aligned} \overline{\lim_{\varepsilon\to 0}}\, \Xi^{\pm}_{\beta}(Y_{\varepsilon,\cdot}, N) &= \overline{\lim_{\varepsilon\to 0}} \max_{0\le n\le N-1} \sup_{y\in\mathbb{R}_1} e^{\pm\beta\mu_{\varepsilon,n+1}(y)\ +\frac{1}{2}\beta^2\sigma^2_{\varepsilon,n+1}(y)} \\ &< K_{67}(\beta) = \frac{1}{2} + e^{K_{66}(\beta+\frac{1}{2}\beta^2)} < \infty . \end{aligned} \tag{8.4}$$

Thus, the condition $\mathbf{G_5}$ implies that the condition $\mathbf{E_{21}}[\beta]$ holds for any $\beta \ge 0$. □

Remark 8.1.1. The constant $K_{67} = K_{67}(\beta)$ is determined by the constant K_{66} via formula (8.4).

We assume that the condition $\mathbf{B_7}[\gamma]$ holds for pay-off functions.

Theorem 8.1.1. *Let the condition* $\mathbf{G}_5$ *holds and, also, the condition* $\mathbf{B}_7[\gamma]$ *holds for some* $\gamma \geq 0$. *Then, there exist* $0 < \varepsilon_{19} \leq \varepsilon_0$ *and the constants* $0 \leq M_{43}, M_{44} < \infty$ *such that the log-reward functions* $\phi_{\varepsilon,n}(y)$ *satisfy the following inequality, for* $y \in \mathbb{R}_1, \quad 0 \leq n \leq N$ *and* $0 \leq \varepsilon \leq \varepsilon_{19}$*:*

$$|\phi_{\varepsilon,n}(y)| \leq M_{43} + M_{44}e^{\gamma|y|}. \tag{8.5}$$

Proof. This theorem is a corollary of Theorem 5.1.1.

Indeed, according to Lemma 8.1.1, the condition $\mathbf{G}_5$ implies that the condition $\mathbf{E}_{21}[\beta]$ holds for any $\beta \geq 0$. This condition is a variant of the condition $\mathbf{E}_{10}[\bar{\beta}]$ (its univariate variant for the model without index component).

According to Lemma 5.1.6, this condition implies that a reduced variant of the condition $\mathbf{C}_6[\bar{\beta}]$ (its univariate variant for the model without index component) holds. Moreover, the condition $\mathbf{G}_5$ implies that there exists $0 < \varepsilon_{20} \leq \varepsilon_0$ such that for $0 \leq \varepsilon \leq \varepsilon_{20}$,

$$\max_{1 \leq n \leq N} \sup_{y \in \mathbb{R}_1} (|\mu_{\varepsilon,n}(y)| + \sigma^2_{\varepsilon,n}(y)) < K_{66}. \tag{8.6}$$

By Lemmas 4.1.8, Remark 4.1.7, and Lemma 8.1.1, the following inequality holds for any $\beta \geq 0$ and $0 \leq \varepsilon \leq \varepsilon_{20}$:

$$\Delta_\beta(Y_{\varepsilon,\cdot,i}(\cdot), N) < K_{68} = 2K_{67}(\beta) = 1 + 2e^{K_{66}(\beta + \frac{1}{2}\beta^2)}. \tag{8.7}$$

Also, the condition $\mathbf{B}_7[\gamma]$ is a reduced variant of the condition $\mathbf{B}_5[\bar{\gamma}]$ (its univariate variant for the model without index component). This condition implies that there exists $0 < \varepsilon_{21} \leq \varepsilon_0$ such that for $0 \leq \varepsilon \leq \varepsilon_{21}$,

$$\max_{0 \leq n \leq N} \sup_{y \in \mathbb{R}_1} \frac{|g_\varepsilon(n, e^y)|}{1 + L_{14}e^{\gamma|y|}} < L_{13}. \tag{8.8}$$

Note that we can always choose $\beta \geq \gamma$ that is required in Theorem 5.1.1.

Thus, Theorem 5.1.1 can be applied that yields inequalities (8.5). □

Remark 8.1.2. The explicit formulas for constants M_{43}, M_{44} can be obtained by using formulas given in Remark 5.1.6, for constants M_{32} and $M_{33,1}$, in the univariate case $k = 1$. They take the following form:

$$M_{43} = L_{13}, \quad M_{44} = L_{13}L_{14}I(\gamma = 0) + L_{13}L_{14}(2K_{67}(\beta))^{N\frac{\gamma}{\beta}}I(\gamma > 0). \tag{8.9}$$

One can choose any $\beta \geq \gamma$ in formula (8.9).

Remark 8.1.3. The parameter ε_{19} can be obtained by using formulas given in Remark 5.1.7 for the parameter ε_3. It takes the following form: $\varepsilon_{19} = \min(\varepsilon_{20}, \varepsilon_{21})$, where the parameters ε_{20} and ε_{21} replace the parameters ε_1 and ε_2 penetrating Remark 5.1.7 and are determined, respectively, by relations (8.6) and (8.8).

Let us now impose the following condition on the distribution of random variables $Y_{\varepsilon,0}$, assumed to hold for some $\beta \geq 0$:
$\mathbf{D_{14}}[\beta]$: $\overline{\lim}_{\varepsilon \to 0} \mathsf{E} e^{\beta |Y_{\varepsilon,0}|} < K_{69}$, for some $1 < K_{69} < \infty$.

The following theorem gives asymptotically uniform upper bounds for optimal expected rewards Φ_ε.

Theorem 8.1.2. *Let the condition* $\mathbf{G_5}$ *holds and, also, the conditions* $\mathbf{B_7}[\gamma]$ *and* $\mathbf{D_{14}}[\beta]$ *hold for some* $\beta \geq \gamma \geq 0$. *Then, there exist* $0 < \varepsilon_{22} \leq \varepsilon_0$ *and constants* $0 \leq M_{45} < \infty$ *such that the following inequality holds for* $0 \leq \varepsilon \leq \varepsilon_{22}$:

$$|\Phi_\varepsilon| \leq M_{45} \,. \tag{8.10}$$

Proof. This theorem follows from Theorem 5.1.3.

Indeed, the condition $\mathbf{D_{14}}[\beta]$ is a reduced variant of the condition $\mathbf{D_7}[\bar{\beta}]$ (its univariate variant for the model without index component). Moreover, the condition $\mathbf{D_{14}}[\beta]$ implies that there exists $0 < \varepsilon_{23} \leq \varepsilon_0$ such that for $0 \leq \varepsilon \leq \varepsilon_{23}$,

$$\mathsf{E} e^{\beta |Y_{\varepsilon,0,i}|} < K_{69} \,. \tag{8.11}$$

Other conditions of this theorem also holds that was pointed out in the proof of Theorem 8.1.1.

Thus, Theorem 5.1.3 can be applied that yields inequality (8.10). □

Remark 8.1.4. The explicit formulas for the constant M_{45} can be obtained with the use of formula given in Remark 5.1.13, for the constant M_{34}, in the univariate case where $k = 1$. It takes the following form:

$$M_{45} = L_{13} + L_{13} L_{14} I(\gamma = 0) + L_{13} L_{14} (2K_{67}(\beta))^{N \frac{\gamma}{\beta}} K_{69}^{\frac{\gamma}{\beta}} I(\gamma > 0) \,. \tag{8.12}$$

Remark 8.1.5. The parameter ε_{22} can be obtained with the use of formulas given in Remark 5.1.14 for the parameter ε_5. It takes the following form, $\varepsilon_{22} = \min(\varepsilon_{19}, \varepsilon_{23})$, where the parameter ε_{23}, determined by relation (8.11), replaces the parameter ε_4 penetrating Remark 5.1.14.

8.1.2 Convergence of option rewards for univariate Markov Gaussian log-price processes with bounded characteristics

We impose the following convergence condition on pay-off functions:
$\mathbf{I_8}$: There exist sets $\mathbb{Y}'_n \in \mathcal{B}_1$, $n = 0, \dots, N$, such that the pay-off function $g_\varepsilon(n, e^{y_\varepsilon}) \to g_0(n, e^{y_0})$ as $\varepsilon \to 0$ for any $y_\varepsilon \to y_0 \in \mathbb{Y}'_n$ as $\varepsilon \to 0$, $n = 0, \dots, N$.

It is useful to note that, in this case, the log-price process $Y_{\varepsilon,n}$ has Gaussian one-step transition probabilities, i.e. for $y \in \mathbb{R}_1$, $A \in \mathcal{B}_1$, $n = 1, 2, \ldots$,

$$P_{\varepsilon,n}(y, A) = \mathsf{P}\{y + \mu_{\varepsilon,n}(y) + \sigma_{\varepsilon,n}(y)\, W_{\varepsilon,n} \in A\}\,. \tag{8.13}$$

We impose on parameters of the log-price processes $Y_{\varepsilon,n}$ the following convergence condition:

$\mathbf{O}_1$: There exist sets $\mathbb{Y}''_n \in \mathcal{B}_1$, $n = 0, \ldots, N$, such that
  (a) $\mu_{\varepsilon,n}(y_\varepsilon) \to \mu_{0,n}(y_0)$ and $\sigma_{\varepsilon,n}(y_\varepsilon) \to \sigma_{0,n}(y_0)$ as $\varepsilon \to 0$ for any $y_\varepsilon \to y_0 \in \mathbb{Y}''_{n-1}$ as $\varepsilon \to 0$, and $n = 1, \ldots, N$;
  (b) $\mathsf{P}\{y_0 + \mu_{0,n}(y_0) + \sigma_{0,n}(y_0) W_{0,n} \in \mathbb{Y}'_n \cap \mathbb{Y}''_n\} = 1$, for every $y_0 \in \mathbb{Y}'_{n-1} \cap \mathbb{Y}''_{n-1}$, $n = 1, \ldots, N$, where $\mathbb{Y}'_n$, $n = 0, \ldots, N$ are sets introduced in the condition $\mathbf{I}_8$.

Let us consider two important examples, where the condition $\mathbf{O}_1$ (a) holds.

The first model is, where the drift and volatility coefficients $\mu_{\varepsilon,n}(y) = \mu_{\varepsilon,n}$ and $\sigma_{\varepsilon,n}(y) = \sigma_{\varepsilon,n}$, for $y \in \mathbb{R}_1$, $n = 1, \ldots, N$, i.e. these coefficients do not depend on y. In this case, the condition $\mathbf{O}_1$ (a) reduces to the relations $\mu_{\varepsilon,n} \to \mu_{0,n}$ and $\sigma_{\varepsilon,n} \to \sigma_{0,n}$ as $\varepsilon \to 0$, for $n = 1, \ldots, N$.

Note that, in this case, the condition $\mathbf{G}_5$ holds.

Another model is, where the parameters $\mu_{\varepsilon,n}(y) = \mu'_{\varepsilon,n} + \mu''_{\varepsilon,n} y$ and $\sigma_{\varepsilon,n}(y) = \sigma'_{\varepsilon,n} + \sigma''_{\varepsilon,n}\sigma_n(y)$, where $|\sigma_n(y)\,| \leq \sqrt{|y|}$ for $y \in \mathbb{R}_1$, $n = 1, \ldots, N$. In this case, the condition $\mathbf{O}_1$ (a) reduces to the relations $\mu'_{\varepsilon,n} \to \mu'_{0,n}$, $\mu''_{\varepsilon,n} \to \mu''_{0,n}$ and $\sigma'_{\varepsilon,n} \to \sigma'_{0,n}$, $\sigma''_{\varepsilon,n} \to \sigma''_{0,n}$ as $\varepsilon \to 0$, for $n = 1, \ldots, N$.

In this case, the condition $\mathbf{G}_5$ does not holds, but the weaker condition $\mathbf{G}_6$ formulated in the next subsection, does hold.

As far as the condition $\mathbf{O}_1$ (b) is concerned, it obviously holds, if, for every $y_0 \in \mathbb{Y}'_{n-1} \cap \mathbb{Y}''_{n-1}$, $n = 1, \ldots, N$, either $\sigma_{0,n}(y_0) > 0$ and the sets $\overline{\mathbb{Y}}'_n$, $\overline{\mathbb{Y}}''_n$ have zero Lebesgue measure, or $\sigma_{0,n}(y_0) = 0$ and $y_0 + \mu_{0,n}(y_0) \in \mathbb{Y}'_n \cap \mathbb{Y}''_n$.

The following theorem gives conditions of convergence for the log-reward functions $\phi_{\varepsilon,n}(y)$.

Theorem 8.1.3. *Let the condition $\mathbf{B}_7[\gamma]$ holds for some $\gamma \geq 0$ and also the conditions $\mathbf{G}_5$, $\mathbf{I}_8$, and $\mathbf{O}_1$ hold. Then, for every $n = 0, 1, \ldots, N$, the following relation takes place, for any $y_\varepsilon \to y_0 \in \mathbb{Y}'_n \cap \mathbb{Y}''_n$:*

$$\phi_{\varepsilon,n}(y_\varepsilon) \to \phi_{0,n}(y_0) \quad \text{as} \quad \varepsilon \to 0\,. \tag{8.14}$$

Proof. This theorem follows from Theorem 5.2.1.

As was pointed out in the proof of Theorem 8.1.1, in this case, the conditions $\mathbf{C}_6[\bar{\beta}]$ and $\mathbf{B}_5[\bar{\gamma}]$ (their univariate variants for the model without index component) hold.

Note also that we can always choose $\beta \geq \gamma$ if $\gamma > 0$ or $\beta = 0$ if $\gamma = 0$ that is required in Theorem 5.2.1.

The condition $\mathbf{I}_8$ is a reduced version of the condition $\mathbf{I}_1$ (its univariate variant for the model without index component).

Also, the condition $\mathbf{O}_1$ (a) is equivalent to the condition $\mathbf{J}_1$ (a) (its univariate variant for the model without index component) since convergence of expectations and standard deviations for Gaussian random variables is equivalent to their convergence in distribution. The condition $\mathbf{O}_1$ (b) is a reformulation of the condition $\mathbf{J}_1$ (b) (its univariate variant for the model without index component).

Thus, all conditions of Theorem 5.2.1 hold. By applying this theorem we get relation (8.14). □

Let us now impose the following convergence condition for the initial distributions $P_{\varepsilon,0}(A) = \mathsf{P}\{Y_{\varepsilon,0} \in A\}$:

$\mathbf{K}_{12}$: (a) $P_{\varepsilon,0}(\cdot) \Rightarrow P_{0,0}(\cdot)$ as $\varepsilon \to 0$;

(b) $P_{0,0}(\mathbb{Y}'_0 \cap \mathbb{Y}''_0) = 1$, where $\mathbb{Y}'_0$ and $\mathbb{Y}''_0$ are the sets introduced in the conditions $\mathbf{I}_8$ and $\mathbf{O}_1$.

The following theorem gives conditions of convergence for the optimal expected rewards Φ_ε.

Theorem 8.1.4. *Let the conditions $\mathbf{B}_7[\gamma]$ and $\mathbf{D}_{14}[\beta]$ hold for some $\beta > \gamma > 0$ or $\beta = \gamma = 0$. Also let the conditions $\mathbf{G}_5$, $\mathbf{I}_8$, $\mathbf{O}_1$, and $\mathbf{K}_{12}$ hold. Then, the following relation takes place:*

$$\Phi_\varepsilon \to \Phi_0 \quad as \quad \varepsilon \to 0 \,. \tag{8.15}$$

Proof. This theorem follows from Theorem 5.2.3.

Indeed, the condition $\mathbf{K}_{12}$ is a variant of the condition $\mathbf{K}_1$ (its one-dimensional variant for the model without index component).

Other conditions of Theorem 5.2.3 also hold that was pointed out in the proofs of Theorems 8.1.1–8.1.3.

Thus, Theorem 5.2.3 can be applied that yields relation (8.15). □

8.1.3 Asymptotically uniform upper bounds for option rewards of univariate Markov Gaussian log-price processes with unbounded characteristics

In this subsection, we give asymptotically uniform upper bounds for option rewards for the model, where the following condition holds:

$\mathbf{G}_6$: $\overline{\lim}_{\varepsilon\to 0} \max_{1\le n\le N} \sup_{y\in\mathbb{R}_1} \frac{|\mu_{\varepsilon,n}(y)|+\sigma^2_{\varepsilon,n}(y)}{1+K_{71}|y|} < K_{70}$, for some $0 < K_{70} < \infty$ and $0 \le K_{71} < \infty$.

It is worth to note that the condition $\mathbf{G}_6$ is weaker than the condition $\mathbf{G}_5$ and reduces to this condition in the case where the constant $K_{71} = 0$.

In this case, according to the results presented in Subsection 4.4.2, one can use the second-type modulus of exponential moment compactness $\Xi^{\pm}_{\beta,A}(Y_{\varepsilon,\cdot}, N)$, defined for $\beta \geq 0$

$$\Xi^{\pm}_{\beta,A}(Y_{\varepsilon,\cdot}, N) = \max_{0\leq n\leq N-1} \sup_{y\in\mathbb{R}_1} \frac{e^{\pm A_{N-n-1}(\beta)\mu_{\varepsilon,n+1}(y) + \frac{1}{2}A^2_{N-n-1}(\beta)\sigma^2_{\varepsilon,n+1}(y)}}{e^{A(A_{N-n-1}(\beta))|y|}}, \tag{8.16}$$

where $A_0(\beta) \leq A_1(\beta) \leq A_2(\beta) \leq \cdots$ is a nondecreasing sequence of functions generated by function $A(\beta) = K_{70}K_{71}(\beta + \frac{1}{2}\beta^2)$, $\beta \geq 0$ with the use of the following recurrence relation, for every $\beta \geq 0$:

$$A_n(\beta) = \begin{cases} \beta, & \text{for } n = 0, \\ A_{n-1}(\beta) + A(A_{n-1}(\beta)), & \text{for } n = 1, 2, \dots. \end{cases} \tag{8.17}$$

We are interested in the following condition, assumed to hold for some $\beta \geq 0$:
$\mathbf{E}_{22}[\beta]$: $\Xi^{\pm}_{\beta,A}(Y_{\varepsilon,\cdot}, N) < K_{72}$, for some $1 < K_{72} < \infty$.

Lemma 8.1.2. *Condition $\mathbf{G}_6$ implies that the condition $\mathbf{E}_{22}[\beta]$ holds for any $\beta \geq 0$.*

Proof. It is analogous to the proof of Lemma 4.4.2.

The condition $\mathbf{G}_6$ implies that there exists $\varepsilon_{24} \in (0, \varepsilon_0]$ such that, for any $y \in \mathbb{R}_1$ and $\varepsilon \in [0, \varepsilon_{24}]$,

$$\max_{0\leq n\leq N-1} (|\mu_{\varepsilon,n+1}(y)| + \sigma^2_{\varepsilon,n+1}(y)) \leq K_{70} + K_{70}K_{71}|y|. \tag{8.18}$$

Relation (8.18) implies that, for any $y \in \mathbb{R}_1$, $n = 0, \dots, N-1$, $\varepsilon \in [0, \varepsilon_{24}]$ and $\beta \geq 0$,

$$\begin{aligned} \pm A_{N-n-1}&(\beta)\mu_{\varepsilon,n+1}(y) + \tfrac{1}{2}A^2_{N-n-1}(\beta)\sigma^2_{\varepsilon,n+1}(y) \\ &\leq \left(A_{N-n-1}(\beta) + \tfrac{1}{2}A^2_{N-n-1}(\beta)\right)(K_{70} + K_{70}K_{71}|y|) \\ &= K_{70}\left(A_{N-n-1}(\beta) + \tfrac{1}{2}A^2_{N-n-1}(\beta)\right) \\ &\quad + K_{70}K_{71}\left(A_{N-n-1}(\beta) + \tfrac{1}{2}A^2_{N-n-1}(\beta)\right)|y| \\ &= K_{70}\left(A_{N-n-1}(\beta) + \tfrac{1}{2}A^2_{N-n-1}(\beta)\right) \\ &\quad + A(A_{N-n-1}(\beta))|y|. \end{aligned} \tag{8.19}$$

Relations (8.16) and (8.19) imply that for any $y \in \mathbb{R}_1$, $n = 0, \dots, N-1$, $\varepsilon \in [0, \varepsilon_{24}]$ and $\beta \geq 0$,

$$\begin{aligned} \overline{\lim_{\varepsilon\to 0}}\, \Xi^{\pm}_{\beta,A}(Y_{\cdot}, N) &\leq \overline{\lim_{\varepsilon\to 0}} \max_{0\leq n\leq N-1} e^{K_{70}(A_{N-n-1}(\beta)+\frac{1}{2}A^2_{N-n-1}(\beta))} \\ &< \frac{1}{2} + e^{K_{70}(A_{N-1}(\beta)+\frac{1}{2}A^2_{N-1}(\beta))} = K_{72}(\beta). \end{aligned} \tag{8.20}$$

This inequality proves the lemma. □

Remark 8.1.6. The constant $K_{72} = K_{72}(\beta)$ is determined by the constant K_{70} via formula (8.20).

The following theorem gives asymptotically uniform upper bounds for the log-reward functions $\phi_{\varepsilon,n}(y)$.

Theorem 8.1.5. *Let the condition* $\mathbf{G}_6$ *holds and, also, the condition* $\mathbf{B}_7[\gamma]$ *holds for some* $\gamma \geq 0$. *Then, for any* $\beta \geq \gamma$, *there exist* $0 < \varepsilon_{25} \leq \varepsilon_0$ *and constants* $0 \leq M_{46}, M_{47} = M_{47}(\beta) < \infty$ *such that the log-reward functions* $\phi_{\varepsilon,n}(y)$ *satisfy the following inequalities, for* $y \in \mathbb{R}_1, \quad 0 \leq n \leq N, \quad 0 \leq \varepsilon \leq \varepsilon_{25}$:

$$|\phi_{\varepsilon,n}(y)| \leq M_{46} + M_{47}I(\gamma = 0) + M_{47}\exp\left\{\frac{A_{N-n}(\beta)\gamma}{\beta}|y|\right\}I(\gamma > 0)\,. \tag{8.21}$$

Proof. This theorem follows from Theorem 6.1.1.

Indeed, according to Lemma 8.1.2, the condition $\mathbf{G}_6$ implies that the condition $\mathbf{E}_{23}[\beta]$ holds for any $\beta \geq 0$. This is a variant of the condition $\mathbf{E}_{13}[\beta]$ (its variant for the model without index component) for the univariate Markov Gaussian log-price process $Y_{\varepsilon,n}$. According to Lemma 6.1.5 this condition implies that a reduced variant of the condition $\mathbf{C}_9[\beta]$ (its variant for the model without index component) holds.

Also, the condition $\mathbf{B}_7[\gamma]$ is a reduced variant of the condition $\mathbf{B}_8[\gamma]$ (its variant for the model without index component).

Note that we can always choose $\beta \geq \gamma$ that is required in Theorem 6.1.1.

Thus, Theorem 6.1.1 can be applied that yields inequalities (8.21). □

Remark 8.1.7. The explicit formulas for constants $M_{46}, M_{47}(\beta)$ can be obtained with the use of formulas given in Remarks 6.1.6 and 6.1.7. These formulas are analogous to formulas (8.12) and take the following form:

$$M_{46} = L_{13}\,, \quad M_{47}(\beta) = L_{13}L_{14}I(\gamma = 0) + L_{13}L_{14}(2K_{72}(\beta))^{N\frac{\gamma}{\beta}}I(\gamma > 0)\,. \tag{8.22}$$

One can choose any $\beta \geq \gamma$ in formula (8.22).

Remark 8.1.8. The parameter ε_{25} can be obtained with the use of formulas given in Remarks 6.1.6 and 6.1.7. This parameter is given by the formula $\varepsilon_{25} = \min(\varepsilon_{21}, \varepsilon_{24})$ analogous to those given in Remark 8.1.5 for the parameter ε_{22}.

The following theorem gives asymptotically uniform upper bounds for optimal expected rewards Φ_ε, for the model with unbounded characteristics.

Theorem 8.1.6. *Let the condition* $\mathbf{G}_6$ *holds and, also, the conditions* $\mathbf{B}_7[\gamma]$ *and* $\mathbf{D}_{14}[\beta]$ *hold for some* $\beta > \gamma \geq 0$. *Then, there exist* $0 < \varepsilon_{26} \leq \varepsilon_0$ *and a constant* $0 \leq M_{48} < \infty$ *such that the following inequality holds for* $0 \leq \varepsilon \leq \varepsilon_{26}$:

$$|\Phi_\varepsilon| \leq M_{48}\,. \tag{8.23}$$

Proof. This theorem follows from Theorem 6.1.3.

All conditions of this theorems holds that was pointed out in the proof of Theorems 8.1.2 and 8.1.5.

Thus, Theorem 6.1.3 can be applied that yields inequality (8.23). □

Remark 8.1.9. The explicit formulas for the constant M_{48} is given by the following formula analogous to (8.12):

$$M_{48} = L_{13} + L_{13}L_{14}I(\gamma = 0) + L_{13}L_{14}(2K_{72}(\beta))^{N\frac{\gamma}{\beta}} K_{69}^{\frac{\gamma}{\beta}} I(\gamma > 0) . \tag{8.24}$$

Remark 8.1.10. Parameter ε_{26} can be obtained with the use of formulas given in Remarks 6.1.13 and 6.1.14. This parameter is given by the formula $\varepsilon_{26} = \min(\varepsilon_{25}, \varepsilon_{23})$ analogous to those given in Remark 8.1.5 for the parameter ε_{22}.

8.1.4 Convergence of option rewards for univariate Markov Gaussian log-price processes with unbounded characteristics

The following theorem gives conditions of convergence for log-reward functions $\phi_{\varepsilon,n}(y)$ for univariate Markov Gaussian log-price processes with unbounded characteristics.

Theorem 8.1.7. *Let the condition* $\mathbf{B}_7[\gamma]$ *holds for some* $\gamma \geq 0$ *and also the conditions* $\mathbf{G}_6$, $\mathbf{I}_8$ *and* $\mathbf{O}_1$ *hold. Then, for every* $n = 0, 1, \dots, N$, *the following relation takes place for any* $y_\varepsilon \to y_0 \in \mathbb{Y}'_n \cap \mathbb{Y}''_n$:

$$\phi_{\varepsilon,n}(y_\varepsilon) \to \phi_{0,n}(y_0) \quad \text{as} \quad \varepsilon \to 0 . \tag{8.25}$$

Proof. This theorem follows from Theorem 6.2.1.

As was pointed out in the proof of Theorem 8.1.5, in this case, the conditions $\mathbf{C}_9[\beta]$ and $\mathbf{B}_8[\gamma]$ (their variants for the model without index component) hold.

The condition $\mathbf{I}_8$ is a variant of the condition $\mathbf{I}_5$ (its variant for the model without index component).

The condition $\mathbf{O}_1$ (a) is equivalent to the condition $\mathbf{J}_8$ (a) (its variant for the model without index component). Also, the condition $\mathbf{O}_1$ (b) is, just, reformulation of the condition $\mathbf{J}_8$ (b) (its variant for the model without index component).

Note also that we can always choose $\beta \geq \gamma$ if $\gamma > 0$ or $\beta = 0$ if $\gamma = 0$ that is required in Theorem 6.2.1.

Thus, Theorem 6.2.1 can be applied that yields relation (8.25). □

The following theorem gives conditions of convergence for optimal expected rewards Φ_ε, for the model with unbounded characteristics.

Theorem 8.1.8. *Let the conditions* $\mathbf{B}_7[\gamma]$ *and* $\mathbf{D}_{14}[\beta]$ *hold for some* $\beta > \gamma > 0$ *or* $\beta = \gamma = 0$. *Also let the conditions* $\mathbf{G}_6$, $\mathbf{I}_8$, $\mathbf{O}_1$, *and* $\mathbf{K}_{12}$ *hold. Then, the following relation takes place:*

$$\Phi_\varepsilon \to \Phi_0 \quad as \quad \varepsilon \to 0\,. \tag{8.26}$$

Proof. This theorem follows from Theorem 6.2.3.

All conditions of this theorem hold that was pointed out in the proofs of Theorems 8.1.2, 8.1.4, 8.1.5, and 8.1.7.

Thus, Theorem 6.2.3 can be applied that yields relation (8.26). □

8.2 Multivariate modulated Markov Gaussian LPP

In this subsection, we presents general results about convergence of rewards for American-type options for multivariate modulated Markov Gaussian log-price processes.

8.2.1 Asymptotically uniform upper bounds for option rewards of multivariate modulated Markov Gaussian log-price processes with bounded characteristics

Let us assume that a modulated log-price process $\vec{Z}_{\varepsilon,n} = (\vec{Y}_{\varepsilon,n}, X_{\varepsilon,n})$, $n = 0, 1, \dots$ is, for every $\varepsilon \in [0, \varepsilon_0]$, a multivariate modulated Markov Gaussian log-price process given by the following stochastic transition dynamic relation:

$$\begin{cases} \vec{Y}_{\varepsilon,n} = \vec{Y}_{\varepsilon,n-1} + \vec{\mu}_{\varepsilon,n}(\vec{Y}_{\varepsilon,n-1}, X_{\varepsilon,n-1}) + \Sigma_{\varepsilon,n}(\vec{Y}_{\varepsilon,n-1}, X_{\varepsilon,n-1})\vec{W}_{\varepsilon,n}\,, \\ X_{\varepsilon,n} = C_{\varepsilon,n}(X_{\varepsilon,n-1}, U_{\varepsilon,n})\,, \\ \quad n = 1, 2, \dots\,, \end{cases} \tag{8.27}$$

where (a) $\vec{Z}_{\varepsilon,0} = (\vec{Y}_{\varepsilon,0}, X_{\varepsilon,0}) = ((Y_{\varepsilon,0,1}, \dots, Y_{\varepsilon,0,k})), X_{\varepsilon,0})$ is a random vector taking values in the space $\mathbb{Z} = \mathbb{R}_k \times \mathbb{X}$, (b) $(\vec{W}_{\varepsilon,n}, U_{\varepsilon,n}) = ((W_{\varepsilon,n,1}, \dots, W_{\varepsilon,n,k}), U_{\varepsilon,n})$, $n = 1, 2, \dots$ is a sequence of independent random vectors taking values in the space $\mathbb{R}_k \times \mathbb{U}$, where $\mathbb{U}$ is a measurable space with a σ-algebra of measurable subsets $\mathcal{B}_\mathbb{U}$, (c) the random vector $(\vec{Y}_{\varepsilon,0}, X_{\varepsilon,0})$ and the random sequence $(\vec{W}_{\varepsilon,n}, U_{\varepsilon,n})$, $n = 1, 2, \dots$ are independent; (d) the random vector $\vec{W}_{\varepsilon,n} = (W_{\varepsilon,n,1}, \dots, W_{\varepsilon,n,k})$ has a standard multivariate normal distribution with $\mathsf{E}W_{\varepsilon,n,i} = 0$, $\mathsf{E}W_{\varepsilon,n,i}W_{\varepsilon,n,j} = I(i = j)$, $i, j = 1, \dots, k$, for every $n = 1, 2, \dots$, (e) the random variables $U_{\varepsilon,1}, U_{\varepsilon,2}, \dots$ have a regular conditional distribution $G_{\vec{w}}(B) = \mathsf{P}\{U_{\varepsilon,n} \in B / \vec{W}_{\varepsilon,n} = \vec{w}\}$ (a probability measure in $B \in \mathcal{B}_\mathbb{U}$ for every $\vec{w} \in \mathbb{R}_k$ and a measurable function in $\vec{w} \in \mathbb{R}_k$ for every $B \in \mathcal{B}_\mathbb{U}$), for every $n = 1, 2, \dots$; (e) $\vec{\mu}_{\varepsilon,n}(\vec{y}, x) = (\mu_{\varepsilon,n,1}(\vec{y}, x), \dots, \mu_{\varepsilon,n,k}(\vec{y}, x))$, $n = 1, 2, \dots$ are vector functions, which components are measurable functions acting from the space $\mathbb{Z} = \mathbb{R}_k \times \mathbb{X}$ to $\mathbb{R}_1$, (f) $\Sigma_{\varepsilon,n}(\vec{y}, x) = \|\sigma_{\varepsilon,n,i,j}(\vec{y}, x)\|$, $n = 1, 2, \dots$ are $k \times k$ matrix functions with elements $\sigma_{\varepsilon,n,i,j}(\vec{y}, x), i, j = 1, \dots, k$, which are measurable functions acting from the space $\mathbb{Z} = \mathbb{R}_k \times \mathbb{X}$ to $\mathbb{R}_1$, and (g) $C_{\varepsilon,n}(x, u)$ is a measurable function acting from the space $\mathbb{X} \times \mathbb{U}$ to the space $\mathbb{X}$.

The process $\vec{Z}_{\varepsilon,n}$ is a multivariate modulated Markov log-price process with an index component. Therefore, a pay-off function $g_\varepsilon(n, e^{\vec{y}}, x)$ is also assumed to be a function of the argument $\vec{z} = (\vec{y}, x) \in \mathbb{Z}$. In sequel, the log-reward functions $\phi_{\varepsilon,n}(\vec{y}, x)$ also are functions of the argument $(\vec{y}, x) \in \mathbb{Z}$.

In this subsection, we give upper bounds for option rewards for the model, where the following condition holds:

G₇: $\overline{\lim}_{\varepsilon \to 0} \max_{1 \le n \le N, i,j=1,\dots,k} \sup_{\vec{z}=(\vec{y},x)\in\mathbb{Z}} (|\mu_{\varepsilon,n,i}(\vec{y}, x)| + \sigma^2_{\varepsilon,n,i,j}(\vec{y}, x)) < K_{73}$, for some $0 < K_{73} < \infty$.

In this case, for every $\varepsilon \in [0, \varepsilon_0]$ and $\vec{z} = (\vec{y}, x) \in \mathbb{Z}$, $n = 0, 1, \dots, N-1$, $i = 1, \dots, k$, there exists the moment-generating function $\mathsf{E}_{\vec{z},n} e^{\beta(Y_{\varepsilon,n+1,i} - Y_{\varepsilon,n,i})}$, defined for $\beta \in \mathbb{R}_1$,

$$\begin{aligned} \mathsf{E}_{\vec{z},n} e^{\beta(Y_{\varepsilon,n+1,i} - Y_{\varepsilon,n,i})} &= \mathsf{E} e^{\beta(\mu_{\varepsilon,n+1,i}(\vec{y},x) + \sum_{j=1}^k \sigma_{\varepsilon,n+1,i,j}(\vec{y},x) W_{\varepsilon,n+1,j})} \\ &= e^{\beta\mu_{\varepsilon,n+1,i}(\vec{y},x) + \frac{1}{2}\beta^2 \sum_{j=1}^k \sigma^2_{\varepsilon,n+1,i,j}(\vec{y},x)} . \end{aligned} \tag{8.28}$$

The second-type modulus of exponential moment compactness $\Xi^{\pm}_{\beta}(Y_{\varepsilon,i,\cdot}, N)$ takes, according to formula (8.28), the following form, for $\beta \ge 0$:

$$\Xi^{\pm}_{\beta}(Y_{\varepsilon,i,\cdot}, N) = \max_{0 \le n \le N-1} \sup_{\vec{z}=(\vec{y},x)\in\mathbb{Z}} e^{\pm\beta\mu_{\varepsilon,n+1,i}(\vec{y},x) + \frac{1}{2}\beta^2 \sum_{j=1}^k \sigma^2_{\varepsilon,n+1,i,j}(\vec{y},x)} . \tag{8.29}$$

We are interested in the following condition, assumed to hold for some vector parameter $\bar{\beta} = (\beta_1, \dots, \beta_k)$ with nonnegative components:

E₂₃$[\bar{\beta}]$: $\overline{\lim}_{\varepsilon \to 0} \Xi^{\pm}_{\beta_i}(Y_{\varepsilon,\cdot,i}, N) < K_{74,i}$, $i = 1, \dots, k$, for some $1 < K_{74,i} < \infty$, $i = 1, \dots, k$.

Lemma 8.2.1. *The condition* **G₇** *implies that the condition* **E₂₃**$[\bar{\beta}]$ *holds for any vector parameter* $\bar{\beta} = (\beta_1, \dots, \beta_k)$ *with components* $\beta_i \ge 0$, $i = 1, \dots, k$.

Proof. It is analogous to the proof of Lemma 8.1.1.

The condition **G₇** and relation (8.28) and (8.29) imply that for any $\beta_i \ge 0$, $i = 1, \dots, k$,

$$\begin{aligned} \overline{\lim_{\varepsilon\to 0}} \Xi^{\pm}_{\beta_i}(Y_{\varepsilon,\cdot,i}, N) &= \overline{\lim_{\varepsilon\to 0}} \max_{0 \le n \le N-1} \sup_{\vec{z}=(\vec{y},x)\in\mathbb{Z}} e^{\pm\beta_i\mu_{\varepsilon,n+1,i}(\vec{y},x) + \frac{1}{2}\beta_i^2 \sum_{j=1}^k \sigma^2_{\varepsilon,n+1,i,j}(\vec{y},x)} \\ &< \tfrac{1}{2} + e^{K_{73}(\beta_i + \frac{1}{2}k\beta_i^2)} = K_{74,i}(\beta_i) < \infty . \end{aligned} \tag{8.30}$$

This inequality proves the lemma. □

Remark 8.2.1. The constants $K_{74,i} = K_{74,i}(\beta_i)$, $i = 1, \dots, k$ are determined by the constant K_{73} via formula (8.30).

We impose on pay-off functions the condition **B₅**$[\bar{\gamma}]$, assumed to hold for some vector parameter $\bar{\gamma} = (\gamma_1, \dots, \gamma_k)$ with nonnegative components.

The following theorem gives asymptotically uniform upper bounds for the log-reward functions $\phi_{\varepsilon,n}(\vec{y}, x)$.

Theorem 8.2.1. *Let the condition* $\mathbf{G}_7$ *holds and, also, the condition* $\mathbf{B}_5[\bar{\gamma}]$ *holds for some vector parameter* $\bar{\gamma} = (\gamma_1, \dots, \gamma_k)$ *with the components* $\gamma_i \geq 0$, $i = 1, \dots, k$. *Then, there exist* $0 < \varepsilon_{27} \leq \varepsilon_0$ *and constants* $0 \leq M_{49}, M_{50,i} < \infty$, $i = 1, \dots, k$ *such that the log-reward functions* $\phi_{\varepsilon,n}(\vec{y}, x)$ *satisfy the following inequalities for* $\vec{z} = (\vec{y}, x) \in \mathbb{Z}$, $0 \leq n \leq N$, $0 \leq \varepsilon \leq \varepsilon_{27}$:

$$|\phi_{\varepsilon,n}(\vec{y}, x)| \leq M_{49} + \sum_{i=1}^{k} M_{50,i} e^{\gamma_i |y_i|} . \tag{8.31}$$

Proof. This theorem is a corollary of Theorem 5.1.1.

Indeed, according to Lemma 8.2.1, the condition $\mathbf{G}_7$ implies that the condition $\mathbf{E}_{23}[\bar{\beta}]$ holds for any vector parameter $\bar{\beta} = (\beta_1, \dots, \beta_k)$ with the components $\beta_i \geq 0$, $i = 1, \dots, k$. This condition is a variant of the condition $\mathbf{E}_{10}[\bar{\beta}]$, for the model of multivariate modulated Markov Gaussian log-price processes $\vec{Z}_{\varepsilon,n}$. According to Lemma 5.1.6, this condition implies that the condition $\mathbf{C}_6[\bar{\beta}]$ holds.

Note that we can always choose the parameter $\bar{\beta} = (\beta_1, \dots, \beta_k)$ such that $\beta_i \geq \gamma_i$, $i = 1, \dots, k$, where $\bar{\gamma} = (\gamma_1, \dots, \gamma_k)$ is the parameter penetrating condition $\mathbf{B}_5[\bar{\gamma}]$, which also is required in Theorem 5.1.1.

Thus, Theorem 5.1.1 can be applied that yields inequalities (8.31). □

Remark 8.2.2. The explicit formulas for the constants $M_{49}, M_{50,i}$, $i = 1, \dots, k$ and the parameter ε_{27} can be obtained by using formulas given in Remarks 5.1.6 and 5.1.7.

Let us now impose the following condition on the distribution of random variables $\vec{Y}_{\varepsilon,0}$, assumed to hold for some vector parameter $\bar{\beta} = (\beta_1, \dots, \beta_k)$ with nonnegative components:

$\mathbf{D}_{15}[\bar{\beta}]$: $\overline{\lim}_{\varepsilon \to 0} \mathsf{E} e^{\beta_i |Y_{\varepsilon,0,i}|} < K_{75,i}$, $i = 1, \dots, k$, for some $1 < K_{75,i} < \infty$, $i = 1, \dots, k$.

The following theorem gives asymptotically uniform upper bounds for the optimal expected rewards Φ_ε.

Theorem 8.2.2. *Let the condition* $\mathbf{G}_7$ *holds and, also, the conditions* $\mathbf{B}_5[\bar{\gamma}]$ *and* $\mathbf{D}_{15}[\bar{\beta}]$ *hold, and* $0 \leq \gamma_i \leq \beta_i$, $i = 1, \dots, k$. *Then, there exist* $0 < \varepsilon_{28} \leq \varepsilon_0$ *and a constant* $0 \leq M_{51} < \infty$ *such that the following inequality holds for* $0 \leq \varepsilon \leq \varepsilon_{28}$:

$$|\Phi_\varepsilon| \leq M_{51} . \tag{8.32}$$

Proof. This theorem follows from Theorem 5.1.3.

The condition $\mathbf{D}_{15}[\bar{\beta}]$ is a variant of the condition $\mathbf{D}_7[\bar{\beta}]$. Other conditions of Theorem 5.1.3 also hold that was pointed out in the proof of Theorem 8.2.1.

Thus, Theorem 5.1.3 can be applied that yields inequality (8.32). □

Remark 8.2.3. The explicit formulas for the constant M_{51} and the parameter ε_{28} can be obtained with the use of formulas given in Remarks 6.3.12 and 6.3.13.

8.2.2 Convergence of option rewards of modulated multivariate Markov Gaussian log-price processes with bounded characteristics

Let us formulate convergence conditions that should be imposed on the pay-off functions $g_\varepsilon(n, e^{\vec{y}}, x)$ and Markov Gaussian log-price processes $\vec{Z}_{\varepsilon,n}$.

We impose the convergence condition $\mathbf{I}_1$ on pay-off functions.

In this case, the log-price $\vec{Z}_{\varepsilon,n}$ has one-step transition probabilities given for $\vec{z} = (\vec{y}, x) \in \mathbb{Z} = \mathbb{R}_k \times \mathbb{X}$, $A \in \mathcal{B}_{\mathbb{Z}}$, $n = 1, 2, \dots$ by the following relation:

$$\begin{aligned} P_{\varepsilon,n}(\vec{z}, A) &= \mathsf{P}\Big\{\big(\vec{y} + \vec{\mu}_{\varepsilon,n}(\vec{y}, x) + \Sigma_{\varepsilon,n}(\vec{y}, x)\vec{W}_{\varepsilon,n}, C_{\varepsilon,n}(x, U_{\varepsilon,n})\big) \in A\Big\} \\ &= \int\limits_{\mathbb{R}_k}\int\limits_{\mathbb{U}} I\big((\vec{y} + \vec{\mu}_{\varepsilon,n}(\vec{y}, x) + \Sigma_{\varepsilon,n}(\vec{y}, x)\vec{w}, C_{\varepsilon,n}(x, u)) \in A\big) \\ &\quad \times G_{\vec{w}}(du)\frac{1}{(\sqrt{2\pi})^k} e^{-\sum_{i=1}^k \frac{w_i^2}{2}}\, d\vec{w}\,. \end{aligned} \tag{8.33}$$

We impose on parameters of the log-price processes $\vec{Z}_{\varepsilon,n}$ the following convergence condition:

$\mathbf{O}_2$: There exist sets $\mathbb{Z}''_n \in \mathcal{B}_{\mathbb{Z}}$, $n = 0, \dots, N$ and $\mathbb{U}_n \in \mathcal{B}_{\mathbb{U}}$, $n = 1, \dots, N$ such that

(a) $\vec{\mu}_{\varepsilon,n}(\vec{y}_\varepsilon, x_\varepsilon) \to \vec{\mu}_{0,n}(\vec{y}_0, x_0)$, $\Sigma_{\varepsilon,n}(\vec{y}_\varepsilon, x_\varepsilon) \to \Sigma_{0,n}(\vec{y}_0, x_0)$, and $C_{\varepsilon,n}(x_\varepsilon, u) \to C_{0,n}(x_0, u)$ as $\varepsilon \to 0$ for any $\vec{z}_\varepsilon = (\vec{y}_\varepsilon, x_\varepsilon) \to \vec{z}_0 = (\vec{y}_0, x_0) \in \mathbb{Z}''_{n-1}$, $u \in \mathbb{U}_n$ and $n = 1, \dots, N$;

(b) $\mathsf{P}\{U_{0,n} \in \mathbb{U}_n\} = 1$, $n = 1, \dots, N$;

(c) $\mathsf{P}\{(\vec{y}_0 + \vec{\mu}_{0,n}(\vec{y}_0, x_0) + \Sigma_{0,n}(\vec{y}_0, x_0)\vec{W}_{0,n}, C_{0,n}(x_0, U_{0,n})) \in \mathbb{Z}'_n \cap \mathbb{Z}''_n\} = 1$ for every $\vec{z}_0 = (\vec{y}_0, x_0) \in \mathbb{Z}'_{n-1} \cap \mathbb{Z}''_{n-1}$ and $n = 1, \dots, N$, where $\mathbb{Z}'_n$, $n = 0, \dots, N$ are sets introduced in the condition $\mathbf{I}_1$.

The following theorem gives conditions of convergence for the log-reward functions $\phi_{\varepsilon,n}(\vec{y}, x)$.

Theorem 8.2.3. *Let the condition $\mathbf{B}_5[\bar{\gamma}]$ holds for some vector parameter $\bar{\gamma} = (\gamma_1, \dots, \gamma_k)$ with nonnegative components, and also the conditions $\mathbf{G}_7$, $\mathbf{I}_1$ and $\mathbf{O}_2$ hold. Then, for every $n = 0, 1, \dots, N$, the following relation takes place for any $\vec{z}_\varepsilon = (\vec{y}_\varepsilon, x_\varepsilon) \to \vec{z}_0 = (\vec{y}_0, x_0) \in \mathbb{Z}'_n \cap \mathbb{Z}''_n$:*

$$\phi_{\varepsilon,n}(\vec{y}_\varepsilon, x_\varepsilon) \to \phi_{0,n}(\vec{y}_0, x_0) \quad \text{as} \quad \varepsilon \to 0\,. \tag{8.34}$$

Proof. This theorem follows from Theorem 5.2.1.

As was pointed out in the proof of Theorem 8.2.1, in this case, the condition $\mathbf{C}_6[\bar{\beta}]$ holds.

Note also that we can always choose the vector parameter $\bar{\beta} = (\beta_1, \dots, \beta_k)$ such that, for every $i = 1, \dots, k$, either $\beta_i > \gamma_i > 0$ or $\beta_i = \gamma_i = 0$ that is required in Theorem 5.2.1.

The condition $\mathbf{O}_2$ (a) and (b) imply due to relation (8.33) that the following relation holds any $\vec{z}_\varepsilon = (\vec{y}_\varepsilon, x_\varepsilon) \to \vec{z}_0 = (\vec{y}_0, x_0) \in \mathbb{Z}''_{n-1}$ and $n = 1, \dots, N$:

$$\Big(\vec{y_\varepsilon} + \vec{\mu}_{\varepsilon,n}(\vec{y}_\varepsilon, x_\varepsilon) + \Sigma_{\varepsilon,n}(\vec{y}_\varepsilon, x_\varepsilon)\vec{W}_{\varepsilon,n}, C_{\varepsilon,n}(x_\varepsilon, U_{\varepsilon,n})\Big)$$
$$\xrightarrow{d} \Big(\vec{y_0} + \vec{\mu}_{0,n}(\vec{y}_0, x_0) + \Sigma_{0,n}(\vec{y}_0, x_0)\vec{W}_{0,n}, C_{0,n}(x_0, U_{0,n})\Big) \quad \text{as} \quad \varepsilon \to 0\,. \tag{8.35}$$

Relation (8.35) is equivalent to the condition $\mathbf{J}_3$ (a). The conditions $\mathbf{O}_2$ (b) and (c) are reformulations, respectively, for the conditions $\mathbf{J}_3$ (b) and (c).

Therefore, the condition $\mathbf{J}_3$ holds and, thus, by Lemma 5.2.1, the condition $\mathbf{J}_1$ holds for the log-price processes $\vec{Z}_{\varepsilon,n}$.

All conditions of Theorem 5.2.1 hold. By applying this theorem we get relation (8.34). □

Let us now impose the following convergence condition on the initial distributions $P_{\varepsilon,0}(A) = \mathsf{P}\{\vec{Z}_{\varepsilon,0} \in A\}$:

$\mathbf{K}_{13}$: (a) $P_{\varepsilon,0}(\cdot) \Rightarrow P_{0,0}(\cdot)$ as $\varepsilon \to 0$;

(b) $P_{0,0}(\mathbb{Z}'_0 \cap \mathbb{Z}''_0) = 1$, where $\mathbb{Z}'_0$ and $\mathbb{Z}''_0$ are the sets introduced in the conditions $\mathbf{I}_1$ and $\mathbf{O}_2$.

The following theorem gives conditions of convergence for optimal expected rewards Φ_ε.

Theorem 8.2.4. *Let the conditions $\mathbf{B}_5[\bar{\gamma}]$ and $\mathbf{D}_{15}[\bar{\beta}]$ hold for some vector parameters $\bar{\beta} = (\beta_1, \dots, \beta_k)$ and $\bar{\gamma} = (\gamma_1, \dots, \gamma_k)$ such that, for every $i = 1, \dots, k$, either $\beta_i > \gamma_i > 0$ or $\beta_i = \gamma_i = 0$. Also let the conditions $\mathbf{G}_7$, $\mathbf{I}_1$, $\mathbf{O}_2$, and $\mathbf{K}_{13}$ hold. Then, the following relation takes place:*

$$\Phi_\varepsilon \to \Phi_0 \quad as \quad \varepsilon \to 0\,. \tag{8.36}$$

Proof. This theorem follows from Theorem 5.2.3.

Indeed, the condition $\mathbf{K}_{13}$ is a variant of the condition $\mathbf{K}_1$.

Other conditions of Theorem 5.2.3 also hold that was pointed out in the proof of Theorem 8.2.3.

Thus, Theorem 5.2.3 can be applied that yields relation (8.36). □

8.2.3 Asymptotically uniform upper bounds for option rewards for modulated multivariate Markov Gaussian log-price processes with unbounded characteristics

In this subsection, we give asymptotically uniform upper bounds for option rewards for modulated multivariate Markov Gaussian log-price processes in the case, where the following condition holds:

G$_8$: $\overline{\lim}_{\varepsilon\to 0} \max_{1\le n\le N,\, i,j=1,\dots,k} \sup_{\vec{z}=(\vec{y},x)\in\mathbb{Z}} \frac{|\mu_{\varepsilon,n,i}(\vec{y},x)|+\sigma^2_{\varepsilon,n,i,j}(\vec{y},x)}{1+\sum_{l=1}^k K_{77,l}|y_l|} < K_{76}$, for some $0 < K_{76} < \infty$ and $0 \le K_{77,1},\dots,K_{77,k} < \infty$.

It is worth to note that the condition **G**$_8$ is weaker than the condition **G**$_7$ and reduces to this condition in the case where the constants $K_{77,l} = 0$, $l = 1,\dots,k$.

In this case, there exists, for every $\varepsilon \in [0,\varepsilon_0]$ and $\vec{z} = (\vec{y},x) \in \mathbb{Z}$, $n = 0,1,\dots,N-1$, $i = 1,\dots,k$, the following multivariate moment-generating function: defined for $\bar{\beta} = (\beta_1,\dots,\beta_k) \in \mathbb{R}_k$,

$$
\begin{aligned}
f_{\varepsilon,\vec{z},n}(\beta_1,\dots,\beta_k) &= \mathsf{E}_{\vec{z},n} e^{\sum_{p=1}^k \beta_p (Y_{\varepsilon,n+1,p}-Y_{\varepsilon,n,p})} \\
&= \mathsf{E} e^{\sum_{p=1}^k \beta_p (\mu_{n+1,p}(\vec{y},x)+\sum_{q=1}^k \sigma_{n+1,p,q}(\vec{y},x) W_{\varepsilon,n+1,q})} \\
&= \mathsf{E} e^{\sum_{p=1}^k \beta_p \mu_{\varepsilon,n+1,p}(\vec{y},x)+\sum_{q=1}^k (\sum_{p=1}^k \beta_p \sigma_{\varepsilon,n+1,p,q}(\vec{y},x)) W_{\varepsilon,n+1,q}} \\
&= e^{\sum_{p=1}^k \beta_p \mu_{\varepsilon,n+1,p}(\vec{y},x)+\frac{1}{2}\sum_{q=1}^k (\sum_{p=1}^k \beta_p \sigma_{\varepsilon,n+1,p,q}(\vec{y},x))^2} .
\end{aligned}
\tag{8.37}
$$

In this case, according to the results presented in Subsection 4.4.4, one can use the second-type modulus of exponential moment compactness $\Xi^{\pm}_{\bar{\beta},\vec{A}}(\vec{Y}_{\varepsilon,\cdot},N)$ defined for any vector $\bar{\beta} = (\beta_1,\dots,\beta_k)$ with nonnegative components and vectors $\bar{\imath}_k = (\imath_1,\dots,\imath_k) \in \mathcal{I}_k$, where $\mathcal{I}_k = \{\bar{\imath}_k : \imath_j = +,-, j = 1,\dots,k\}$,

$$
\begin{aligned}
\Delta^{\bar{\imath}_k}_{\bar{\beta},\vec{A}}(\vec{Y}_{\varepsilon,\cdot},N) &= \sup_{0\le n\le N-1, \vec{z}=((y_1,\dots,y_k),x)\in\mathbb{Z}} \frac{\mathsf{E}_{\vec{z},n} e^{\sum_{j=1}^k \imath_j A_{N-n-1,j}(\bar{\beta})(Y_{\varepsilon,n+1,j}-Y_{\varepsilon,n,j})}}{e^{\sum_{j=1}^k A_j(A_{N-n-1}(\bar{\beta}))|y_j|}} \\
&= \sup_{0\le n\le N-1, \vec{z}\in\mathbb{Z}} \frac{f_{\varepsilon,\vec{z},n}(\imath_1 A_{N-n-1,1}(\bar{\beta}),\dots,\imath_k A_{N-n-1,k}(\bar{\beta}))}{e^{\sum_{j=1}^k A_j(\vec{A}_{N-n-1}(\bar{\beta}))|y_j|}} ,
\end{aligned}
\tag{8.38}
$$

where the vector functions $\vec{A}_n(\bar{\beta}) = (A_{n,1}(\bar{\beta}),\dots,A_{n,k}(\bar{\beta}))$, $n = 0,1,\dots$ are generated by the vector function

$$
\begin{aligned}
\vec{A}(\bar{\beta}) &= (A_1(\bar{\beta}),\dots,A_k(\bar{\beta})) \\
&= \left(K_{76}K_{77,1} \sum_{p=1}^k \left(\beta_p + \tfrac{1}{2}k^2\beta_p^2\right),\dots, K_{76}K_{77,k} \sum_{p=1}^k \left(\beta_p + \tfrac{1}{2}k^2\beta_p^2\right) \right) ,
\end{aligned}
\tag{8.39}
$$

according to the following recurrence formulas, for every vector $\bar{\beta} = (\beta_1,\dots,\beta_k)$ with nonnegative components:

$$
\vec{A}_n(\bar{\beta}) = \begin{cases} \bar{\beta} & \text{for } n = 0, \\ \vec{A}_{n-1}(\bar{\beta}) + \vec{A}(\vec{A}_{n-1}(\bar{\beta})) & \text{for } n = 1,2,\dots . \end{cases}
\tag{8.40}
$$

Recall the vectors $\bar{\beta}_i = (\beta_{i,1},\dots,\beta_{i,k})$, $i = 1,\dots,k$, which have the components $\beta_{i,j} = \beta_i I(j = i)$, $j = 1,\dots,k$ for $i = 1,\dots,k$.

We are interested in the following variant of condition, assumed to hold for some vector $\bar{\beta} = (\beta_1, \dots, \beta_k)$ with nonnegative components:

$\mathbf{E_{24}}[\bar{\beta}]$: $\overline{\lim}_{\varepsilon\to 0} \Xi^{\bar{\imath}_k}_{\bar{\beta}_i, \vec{A}}(\vec{Y}_{\varepsilon,\cdot}, N) < K_{78,i}$, $\bar{\imath}_k = (\imath_1, \dots, \imath_k) \in \mathcal{I}_k$, $i = 1, \dots, k$, for some $1 < K_{78,i} < \infty$, $i = 1, \dots, k$.

Lemma 8.2.2. *The condition $\mathbf{G_8}$ implies that the condition $\mathbf{E_{24}}[\bar{\beta}]$ holds for any vector $\bar{\beta}$ with nonnegative components.*

Proof. It is analogous to the proof of Lemma 4.4.4.

The condition $\mathbf{G_8}$ implies that there exists $\varepsilon_{29} \in (0, \varepsilon_0]$ such that, for any $\vec{z} = (\vec{y}, x) = ((y_1, \dots, y_k), x) \in \mathbb{Z}$ and $\varepsilon \in [0, \varepsilon_{29}]$,

$$\max_{0\le n\le N-1,\, i,j=1,\dots,k} \Big(|\mu_{\varepsilon,n+1,i}(\vec{y}, x)| + \sigma^2_{\varepsilon,n+1,i,j}(\vec{y}, x) \Big) \le K_{76} + \sum_{l=1}^{k} K_{76} K_{77,l} |y_l| \,. \tag{8.41}$$

Relation (8.41) implies that, for any $\vec{z} = (\vec{y}, x) = ((y_1, \dots, y_k), x) \in \mathbb{Z}$, $n = 0, \dots, N-1$, $\varepsilon \in [0, \varepsilon_{29}]$, and any vector $\bar{\beta} = (\beta_1, \dots, \beta_k)$ with nonnegative components,

$$\begin{aligned}
&\ln f_{\varepsilon,\vec{z},n}(\imath_1 A_{N-n-1,1}(\bar{\beta}), \dots, \imath_k A_{N-n-1,k}(\bar{\beta})) \\
&\quad = \sum_{p=1}^{k} \imath_p A_{N-n-1,p}(\bar{\beta}) \mu_{\varepsilon,n+1,p}(\vec{y}, x) + \frac{1}{2} \sum_{q=1}^{k} \Bigg(\sum_{p=1}^{k} \imath_p A_{N-n-1,p}(\bar{\beta}) \sigma_{\varepsilon,n+1,p,q}(\vec{y}, x) \Bigg)^2 \\
&\quad \le \sum_{p=1}^{k} A_{N-n-1,p}(\bar{\beta}) |\mu_{\varepsilon,n+1,p}(\vec{y}, x)| + \frac{1}{2} \sum_{q=1}^{k} k \Bigg(\sum_{p=1}^{k} A^2_{N-n-1,p}(\bar{\beta}) \sigma^2_{\varepsilon,n+1,p,q}(\vec{y}, x) \Bigg) \\
&\quad \le \sum_{p=1}^{k} A_{N-n-1,p}(\bar{\beta}) \Bigg(K_{76} + \sum_{j=1}^{k} K_{76} K_{77,j} |y_j| \Bigg) \\
&\qquad + \frac{1}{2} \sum_{p=1}^{k} A^2_{N-n-1,p}(\bar{\beta}) k^2 \Bigg(K_{76} + \sum_{j=1}^{k} K_{76} K_{77,j} |y_j| \Bigg) \\
&\quad = K_{76} \sum_{p=1}^{k} (A_{N-n-1,p}(\bar{\beta}) + \frac{1}{2} k^2 A^2_{N-n-1,p}(\bar{\beta})) \\
&\qquad + \sum_{j=1}^{k} K_{76} K_{77,j} \Bigg(\sum_{p=1}^{k} A_{N-n-1,p}(\bar{\beta}) + \frac{1}{2} k^2 A^2_{N-n-1,p}(\bar{\beta})) \Bigg) |y_j| \\
&\quad = K_{76} \sum_{p=1}^{k} \Bigg(A_{N-n-1,p}(\bar{\beta}) + \frac{1}{2} k^2 A^2_{N-n-1,p}(\bar{\beta}) \Bigg) + \sum_{j=1}^{k} A_j(\vec{A}_{N-n-1}(\bar{\beta})) |y_j| \,.
\end{aligned} \tag{8.42}$$

Relations (8.38) and (8.42) imply that for $\bar{\imath}_k = (\imath_1, \dots, \imath_k) \in \mathcal{I}_k$ and any vector $\bar{\beta} = (\beta_1, \dots, \beta_k)$ with nonnegative components,

$$\begin{aligned}\overline{\lim_{\varepsilon \to 0}} \Delta^{\bar{\imath}_k}_{\bar{\beta}, \vec{A}}(\vec{Y}_{\cdot}, N) &= \overline{\lim_{\varepsilon \to 0}} \sup_{0 \le n \le N-1, \vec{z} \in \mathbb{Z}} \frac{f_{\varepsilon, \vec{z}, n}(\imath_1 A_{N-n-1,1}(\bar{\beta}), \dots, \imath_k A_{N-n-1,k}(\bar{\beta}))}{e^{\sum_{j=1}^k A_j(\vec{A}_{N-n-1}(\bar{\beta}))|y_j|}} \\ &< \tfrac{1}{2} + \max_{0 \le n \le N-1} e^{K_{76} \sum_{p=1}^k (A_{N-n-1,p}(\bar{\beta}) + \frac{1}{2} k^2 A^2_{N-n-1,p}(\bar{\beta}))} \\ &= \tfrac{1}{2} + e^{K_{76} \sum_{p=1}^k (A_{N-1,p}(\bar{\beta}) + \frac{1}{2} k^2 A^2_{N-1,p}(\bar{\beta}))} = K_{78}(\bar{\beta}) . \end{aligned} \tag{8.43}$$

Relation (8.43) implies that for any vector $\bar{\beta} = (\beta_1, \dots, \beta_k)$ with nonnegative components and vector $\bar{\imath}_k = (\imath_1, \dots, \imath_k) \in \mathcal{I}_k$, and $i = 1, \dots, k$,

$$\Xi^{\bar{\imath}_k}_{\bar{\beta}_i, \vec{A}}(\vec{Y}_{\varepsilon, \cdot}, N) < K_{78,i} = K_{78}(\bar{\beta}_i) . \tag{8.44}$$

This inequality proves the lemma. □

Remark 8.2.4. The constants $K_{78,i}$, $i = 1, \dots, k$ are determined by the constant K_{76} via formulas (8.44).

The following theorem gives asymptotically uniform upper bounds for the log-reward functions $\phi_{\varepsilon,n}(\vec{y}, x)$.

Theorem 8.2.5. *Let the condition* $\mathbf{G_8}$ *holds and, also, the condition* $\mathbf{B_5}[\bar{\gamma}]$ *holds for some vector parameter* $\bar{\gamma} = (\gamma_1, \dots, \gamma_k)$ *with nonnegative components. Then, for every* $\bar{\beta} = (\beta_1, \dots, \beta_k)$ *such that* $\beta_i \ge \gamma_i$, $i = 1, \dots, k$, *there exist* $0 < \varepsilon_{30} \le \varepsilon_0$ *and the constants* $0 \le M_{52}, M_{53,i} = M_{53,i}(\bar{\beta}) < \infty, I = 1, \dots, k$ *such that the log-reward functions* $\phi_n(\vec{y}, x)$ *satisfy the following inequalities for* $\vec{z} = (\vec{y}, x) = ((y_1, \dots, y_k), x) \in \mathbb{Z}$, $0 \le n \le N$ $0 \le \varepsilon \le \varepsilon_{30}$*:*

$$|\phi_{\varepsilon,n}(\vec{y}, x)| \le M_{52} + \sum_{i:\gamma_i = 0} M_{53,i} + \sum_{i:\gamma_i > 0} M_{53,i} \exp\left\{ \left(\sum_{j=1}^k A_{N-n,j}(\bar{\beta}_i)|y_j| \right) \frac{\gamma_i}{\beta_i} \right\} . \tag{8.45}$$

Proof. This theorem follows from Theorem 6.3.1.

Indeed, according to Lemma 8.2.2 the condition $\mathbf{G_8}$ implies that the condition $\mathbf{E_{24}}[\bar{\beta}]$ holds for any vector $\bar{\beta} = (\beta_1, \dots, \beta_k)$ with nonnegative components. This condition is a reformulation of the condition $\mathbf{E_{14}}[\bar{\beta}]$, which, according to Lemma 6.3.5, implies that the condition $\mathbf{C_{10}}[\bar{\beta}]$ holds.

The condition $\mathbf{B_5}[\bar{\gamma}]$ is imposed on the pay-off functions. Note that we can always chose $\beta_i \ge \gamma_i$, $i = 1, \dots, k$ that is required in Theorem 6.3.1.

Thus, Theorem 6.3.1 can be applied that yields inequalities (8.45). □

Remark 8.2.5. The explicit formulas for the constants $M_{52}, M_{53,i}(\bar{\beta})$, $i = 1, \dots, k$ and the parameter ε_{30} can be obtained with the use of formulas given in Remarks 6.3.5 and 6.3.6.

The following theorem gives asymptotically uniform upper bounds for optimal expected rewards Φ_ε.

Theorem 8.2.6. *Let the condition* $\mathbf{G_8}$ *holds and, also, the conditions* $\mathbf{B_5}[\bar{\gamma}]$ *and* $\mathbf{D_{15}}[\bar{\beta}]$ *hold for some vector parameters* $\bar{\gamma} = (\gamma_1, \dots, \gamma_k)$ *and* $\bar{\beta} = (\beta_1, \dots, \beta_k)$ *such that* $\beta_i \geq \gamma_i \geq 0$, $i = 1, \dots, k$. *Then there exist* $0 < \varepsilon_{31} \leq \varepsilon_0$ *and a constant* $0 \leq M_{54} < \infty$ *such that the following inequality holds for* $0 \leq \varepsilon \leq \varepsilon_{31}$:

$$|\Phi_\varepsilon| \leq M_{54} . \tag{8.46}$$

Proof. This theorem is a corollary of Theorem 6.3.3.

The condition $\mathbf{D_{15}}[\bar{\beta}]$ is a reformulation of the condition $\mathbf{D_9}[\bar{\beta}]$. Other conditions of Theorem 6.3.3 also hold that was checked in the proof of Theorem 8.2.5.

Thus, Theorem 6.3.3 can be applied that yields inequalities (8.46). □

Remark 8.2.6. The explicit formulas for the constants M_{54} and the parameter ε_{31} can be obtained with the use of formulas given in Remarks 6.3.5 and 6.3.6.

8.2.4 Convergence of option rewards for multivariate modulated Markov Gaussian log-price processes with unbounded characteristics

The following theorem gives conditions of convergence for the log-reward functions for the multivariate modulated Markov Gaussian log-price processes with unbounded characteristics.

Theorem 8.2.7. *Let the condition* $\mathbf{B_5}[\bar{\gamma}]$ *holds for some vector parameter* $\bar{\gamma} = (\gamma_1, \dots, \gamma_k)$ *with nonnegative components, and also the conditions* $\mathbf{G_8}$, $\mathbf{I_1}$ *and* $\mathbf{O_2}$ *hold. Then, for every* $n = 0, 1, \dots, N$, *the following relation takes place for any* $\vec{z}_\varepsilon = (\vec{y}_\varepsilon, x_\varepsilon) \to \vec{z}_0 = (\vec{y}_0, x_0) \in \mathbb{Z}'_n \cap \mathbb{Z}''_n$:

$$\phi_{\varepsilon,n}(\vec{y}_\varepsilon, x_\varepsilon) \to \phi_{0,n}(\vec{y}_0, x_0) \quad \text{as} \quad \varepsilon \to 0 . \tag{8.47}$$

Proof. This theorem follows from Theorem 6.4.1.

As was pointed out in the proof of Theorem 8.2.5, in this case, the condition $\mathbf{C_{10}}[\bar{\beta}]$ holds for any vector parameter $\bar{\beta} = (\beta_1, \dots, \beta_k)$ with nonnegative components.

The condition $\mathbf{B_5}[\bar{\gamma}]$ is required in both Theorems 8.2.7 and 6.4.1. Note also that we can always choose the vector parameter $\bar{\beta} = (\beta_1, \dots, \beta_k)$ such that, for every $i = 1, \dots, k$, either $\beta_i > \gamma_i > 0$ or $\beta_i = \gamma_i = 0$ that is required in Theorem 6.4.1.

The condition $\mathbf{I_1}$ is required in both Theorems 8.2.7 and 6.4.1.

As was shown in Theorem 8.2.3, the condition $\mathbf{O_2}$ implies that the condition $\mathbf{J_1}$ holds.

Thus, all conditions of Theorem 6.4.1 hold. By applying this theorem we get relation (8.47). □

The following theorem gives conditions of convergence for optimal expected rewards Φ_ε.

Theorem 8.2.8. *Let the conditions* $\mathbf{B}_5[\bar{\gamma}]$ *and* $\mathbf{D}_{15}[\bar{\beta}]$ *hold for some vector parameters* $\bar{\beta} = (\beta_1, \dots, \beta_k)$ *and* $\bar{\gamma} = (\gamma_1, \dots, \gamma_k)$ *such that, for every* $i = 1, \dots, k$, *either* $\beta_i > \gamma_i > 0$ *or* $\beta_i = \gamma_i = 0$. *Also let the conditions* $\mathbf{G}_8$, $\mathbf{I}_1$, $\mathbf{O}_2$, *and* $\mathbf{K}_{13}$ *hold. Then, the following relation takes place:*

$$\Phi_\varepsilon \to \Phi_0 \quad as \quad \varepsilon \to 0\,. \tag{8.48}$$

Proof. This theorem follows from Theorem 6.4.3. All conditions of this theorem hold that was pointed out in the proofs of Theorems 8.2.3, 8.2.4, and 8.2.7.

Thus, Theorem 6.4.3 can be applied that yields relation (8.48). □

8.3 Markov Gaussian LPP with estimated characteristics

In this section, we consider the model, where drift and volatility functional coefficients of a Markov Gaussian log-price process depend on unknown parameter, for which consistent statistical estimates are given. In this case, we show that the log-reward functions and the optimal expected rewards for the log-price processes with estimated coefficients are consistent estimates for the corresponding log-reward functions and the optimal expected rewards for the log-price processes with the true coefficients. We restrict our consideration by univariate models. However, analogous results can also be obtained for the multivariate modulated Markov Gaussian log-price processes.

8.3.1 Option rewards for univariate Markov Gaussian log-price processes with estimated bounded parameters

Let Θ be a Polish metric space with a metric $\rho_\Theta(\theta', \theta'')$. In this case the space $\Theta \times \mathbb{R}_1$ is also a Polish metric space with the metric $\rho((\theta', y'), (\theta'', y'')) = (\rho^2_\Theta(\theta', \theta'') + |y' - y''|^2)^{\frac{1}{2}}$, $(\theta', y'), (\theta'', y'') \in \Theta \times \mathbb{R}_1$. Let also $\mathcal{B}_\Theta$ be the Borel σ-algebra of subsets for space Θ.

Let us consider the model, where the univariate Markov Gaussian log-price processes $Y_{\varepsilon,n}(\theta)$, $n = 0, 1, \dots$ is given, for every $\theta \in \Theta$ and $\varepsilon \in [0, \varepsilon_0]$, by the following stochastic transition dynamic relation:

$$\begin{aligned} Y_{\varepsilon,n}(\theta) &= Y_{\varepsilon,n-1}(\theta) + \mu_{\varepsilon,n}(\theta, Y_{\varepsilon,n-1}(\theta)) \\ &\quad + \sigma_{\varepsilon,n}(\theta, Y_{\varepsilon,n-1}(\theta)) W_{\varepsilon,n}, \quad n = 1, 2, \dots, \end{aligned} \tag{8.49}$$

where (a) $Y_{\varepsilon,0}(\theta) \equiv Y_{\varepsilon,0}, \theta \in \Theta$ is a real-valued random variable, (b) $W_{\varepsilon,n}, n = 1, 2, \dots$ is a sequence of standard i.i.d. normal random variables with mean 0 and variance 1, (c) the random variable $Y_{\varepsilon,0}$ and the sequence of random variables $W_{\varepsilon,n}, n = 1, 2, \dots$ are independent, (d) $\mu_{\varepsilon,n}(\theta, y)$ are real-valued Borel functions acting from the space $\Theta \times \mathbb{R}_1$ to $\mathbb{R}_1$, for $n = 1, 2, \dots$, and (e) $\sigma_{\varepsilon,n}(\theta, y)$ are real-valued Borel functions acting from the space $\Theta \times \mathbb{R}_1$ to $\mathbb{R}_1$, for $n = 1, 2, \dots$.

Let us assume that the following condition of boundedness for pay-off functions holds:

G₉: $\overline{\lim}_{\varepsilon \to 0} \sup_{\theta \in \Theta} \max_{1 \le n \le N} \sup_{y \in \mathbb{R}_1} \frac{|\mu_{\varepsilon,n}(\theta,y)| + \sigma^2_{\varepsilon,n}(\theta,y)}{K'(\theta)} < 1$, where $K'(\theta)$ is a continuous function acting from the space Θ to interval $(0, \infty)$.

The condition $\mathbf{G}_9$ implies that there exists $0 < \varepsilon' \le \varepsilon_0$ such that the following relation holds for $0 \le \varepsilon \le \varepsilon'$:

$$\max_{1 \le n \le N} \sup_{y \in \mathbb{R}_1} (|\mu_{\varepsilon,n}(\theta, y)| + \sigma^2_{\varepsilon,n}(\theta, y)) < K'(\theta)\,, \quad \theta \in \Theta\,. \tag{8.50}$$

Relation (8.50) implies that the condition $\mathbf{G}_5$ holds for trend and volatility coefficients $\mu_{\varepsilon,n}(\theta, y)$ and $\sigma_{\varepsilon,n}(\theta, y)$, for every $\theta \in \Theta$. Moreover, in this case the constant $K'(\theta)$, replacing the constant K_{66}, depends on the parameter θ, while the the parameter ε', which guarantees holding of inequality (8.50), for $0 \le \varepsilon \le \varepsilon'$, does not depend on the parameter θ.

Let us also assume that the condition $\mathbf{B}_7[\gamma]$ holds for the pay-off functions $g_\varepsilon(n, e^y)$.

Theorem 8.1.1 implies that the log-reward functions $\phi_{\varepsilon,n}(\theta, y)$, $y \in \mathbb{R}_1$, $n = 0, 1, \dots, N$ are well defined for the log-price process $Y_{\varepsilon,n}(\theta)$, for every $\theta \in \Theta$. Moreover, it follows from relation (8.50) that, in this case, parameters ε_{19} and ε_{22} penetrating Theorem 8.1.1 does not depend on parameter θ.

The following theorem follows from Theorem 8.1.1 and the above remarks.

Theorem 8.3.1. *Let the condition $\mathbf{G}_9$ holds and, also, the condition $\mathbf{B}_7[\gamma]$ holds for some $\gamma \ge 0$. Then, for every $\theta \in \Theta$, there exist $0 < \varepsilon_{32} \le \varepsilon_0$ and constants $0 \le M_{55}, M_{56} = M_{56}(\theta) < \infty$ such that the log-reward functions $\phi_{\varepsilon,n}(\theta, y)$ satisfy the following inequalities for $y \in \mathbb{R}_1$, $0 \le n \le N$ and $0 \le \varepsilon \le \varepsilon_{32}$:*

$$|\phi_{\varepsilon,n}(\theta, y)| \le M_{55} + M_{56}(\theta) e^{\gamma|y|}\,. \tag{8.51}$$

Remark 8.3.1. The constants $M_{55}, M_{56} = M_{56}(\theta)$ are given by the following formulas, which follows from formulas (8.9):

$$M_{55} = L_{13}\,, \quad M_{56}(\theta) = L_{13}L_{14}I(\gamma = 0) + L_{13}L_{14}\left(1 + 2e^{K'(\theta)(\beta + \frac{1}{2}\beta^2)}\right)^{N\frac{\gamma}{\beta}} I(\gamma > 0)\,. \tag{8.52}$$

One can choose any $\beta \ge \gamma$ in formulas (8.52).

Remark 8.3.2. The parameter ε_{32} is given by the following formula $\varepsilon_{32} = \min(\varepsilon', \varepsilon_{21})$, where the parameters ε_{21} and ε' are determined by relations (8.8) and (8.50).

Let us now assume that the condition $\mathbf{D}_{14}[\beta]$ holds.

Theorem 8.1.2 implies that the optimal expected rewards $\Phi_\varepsilon(\theta)$ is well defined for the log-price process $Y_{\varepsilon,n}(\theta)$, for every $\theta \in \Theta$.

Moreover, it follows from relation (8.50) that, in this case, the parameter ε_{22} penetrating Theorem 8.1.2 does not depend on parameter θ.

The following theorem follows from Theorem 8.1.2 and the above remarks.

Theorem 8.3.2. *Let the condition $\mathbf{G}_9$ holds and, also, the conditions $\mathbf{B}_7[\gamma]$ and $\mathbf{D}_{14}[\beta]$ hold for some $\beta \geq \gamma \geq 0$. Then, for every $\theta \in \Theta$, there exist $0 < \varepsilon_{33} \leq \varepsilon_0$ and a constant $0 \leq M_{57} = M_{57}(\theta) < \infty$ such that the following inequality holds for $0 \leq \varepsilon \leq \varepsilon_{33}$:*

$$|\Phi_\varepsilon(\theta)| \leq M_{57}(\theta) . \tag{8.53}$$

Remark 8.3.3. The constant $M_{57}(\theta)$ is given by the following formula:

$$M_{57}(\theta) = L_{13} + L_{13}L_{14}I(\gamma = 0) + L_{13}L_{14}(1 + 2e^{K'(\theta)(\beta + \frac{1}{2}\beta^2)})^{N\frac{\gamma}{\beta}} K_{69}^{\frac{\gamma}{\beta}} I(\gamma > 0) . \tag{8.54}$$

Remark 8.3.4. The parameter ε_{33} is given by the following formula $\varepsilon_{33} = \min(\varepsilon', \varepsilon_{21}, \varepsilon_{23})$, where the parameters ε_{21}, ε_{23}, and ε' are determined by relations (8.8), (8.11), and (8.50).

8.3.2 Strong consistency of option rewards for univariate Markov Gaussian log-price processes with estimated bounded parameters

Let us also assume the following convergence condition for drift and volatility coefficients of the log-price processes $Y_{\varepsilon,n}(\theta)$ holds:

$\mathbf{O}_3$: There exist sets $\Theta_0 \in \mathcal{B}_\Theta$ and $\mathbb{Y}''_n \in \mathcal{B}_1$, $n = 0, \ldots, N$ such that

(a) $\mu_{\varepsilon,n}(\theta_\varepsilon, y_\varepsilon) \to \mu_{0,n}(\theta_0, y_0)$ and $\sigma_{\varepsilon,n}(\theta_\varepsilon, y_\varepsilon) \to \sigma_{0,n}(\theta_0, y_0)$ as $\varepsilon \to 0$, for any $\theta_\varepsilon \to \theta_0 \in \Theta_0$ and $y_\varepsilon \to y_0 \in \mathbb{Y}''_{n-1}$ as $\varepsilon \to 0$, $n = 1, \ldots, N$;

(b) $\mathsf{P}\{y_0 + \mu_{0,n}(\theta_0, y_0) + \sigma_{0,n}(\theta_0, y_0) W_{0,n} \in \mathbb{Y}'_n \cap \mathbb{Y}''_n\} = 1$, for every $\theta_0 \in \Theta_0$, $y_0 \in \mathbb{Y}'_{n-1} \cap \mathbb{Y}''_{n-1}$ and $n = 1, \ldots, N$, where $\mathbb{Y}'_n$, $n = 0, \ldots, N$ are sets introduced in the condition $\mathbf{I}_8$.

Let $\langle \hat{\theta}_\varepsilon, \varepsilon \in [0, \varepsilon_0] > $ be a family of random variables defined on some probability space $\langle \Omega', \mathcal{F}', P' \rangle$.

We interpret random variables $\hat{\theta}_\varepsilon$, for $\varepsilon \in (0, \varepsilon_0]$ as estimates of the random variable $\hat{\theta}_0$ and assume that the following consistency condition, coherent with convergence condition $\mathbf{O}_3$, for drift and volatility coefficients, holds:

$\mathbf{P_1}$: (a) $\hat{\theta}_\varepsilon \xrightarrow{a.s.} \hat{\theta}_0$ as $\varepsilon \to 0$;
(b) $\mathsf{P}\{\hat{\theta}_0 \in \Theta_0\} = 1$, where Θ_0 is the set penetrating condition $\mathbf{O_3}$.

In applications it can be shown that $\hat{\theta}_0 = const$ with probability 1. In this case, the condition $\mathbf{P_1}$ means that $\hat{\theta}_\varepsilon$ are strongly consistent estimates for the parameter $\hat{\theta}_0$.

Let us mention two important particular models.

The first model, where the functions $\mu_{\varepsilon,n}(y) = \mu_{\varepsilon,n}$ and $\sigma_{\varepsilon,n}(y) = \sigma_{\varepsilon,n}$, for $y \in \mathbb{R}_1$, $n = 1, \dots, N$, where $\mu_{\varepsilon,n}$, $n = 1, \dots, N$ and $\sigma_{\varepsilon,n}$, $n = 1, \dots, N$ are real-valued random variable defined on a probability space $\langle \Omega', \mathcal{F}', P' \rangle$. In this case, the random vectors $\hat{\theta}_\varepsilon = (\mu_{\varepsilon,1}, \sigma_{\varepsilon,1}, \dots, \mu_{\varepsilon,n}, \sigma_{\varepsilon,n})$, $\varepsilon \in [0, \varepsilon_0]$. Note also that the trend and volatility coefficients can be defined as $\mu_{\varepsilon,n}(y) = \mu_{\varepsilon,N}, \sigma_{\varepsilon,n}(y) = \sigma_{\varepsilon,N}$, for $y \in \mathbb{R}_1, n > N$. The condition $\mathbf{P_1}$ reduces to the relations $\mu_{\varepsilon,n} \xrightarrow{a.s.} \mu_{0,n}$ and $\sigma_{\varepsilon,n} \xrightarrow{a.s.} \sigma_{0,n}$ as $\varepsilon \to 0$, for $n = 1, \dots, N$.

The second model is where the parameters $\mu_{\varepsilon,n}(y) = \mu'_{\varepsilon,n} + \mu''_{\varepsilon,n} y$ and $\sigma_{\varepsilon,n}(y) = \sigma'_{\varepsilon,n} + \sigma''_{\varepsilon,n}\sigma_n(y)$, for $y \in \mathbb{R}_1$, $n = 1, \dots, N$. In this model, $\mu'_{\varepsilon,n}, \mu''_{\varepsilon,n}$, $n = 1, \dots, N$ and $\sigma'_{\varepsilon,n}, \sigma''_{\varepsilon,n}$, $n = 1, \dots, N$ are real-valued random variable defined on a probability space $\langle \Omega', \mathcal{F}', P' \rangle$, while $|\sigma_n(y)\ | \le \sqrt{|y|}$ are nonrandom real-valued measurable functions. In this case, the random vectors $\hat{\theta}_\varepsilon = (\mu'_{\varepsilon,1}, \mu''_{\varepsilon,1}, \sigma'_{\varepsilon,1}, \sigma''_{\varepsilon,1}, \dots, \mu'_{\varepsilon,n}, \mu''_{\varepsilon,n}, \sigma'_{\varepsilon,n}, \sigma''_{\varepsilon,n})$, $\varepsilon \in [0, \varepsilon_0]$. Note also that the trend and volatility coefficients can be defined as $\mu'_{\varepsilon,n}(y) = \mu'_{\varepsilon,N}, \mu''_{\varepsilon,n}(y) = \mu''_{\varepsilon,N}, \sigma'_{\varepsilon,n}(y) = \sigma_{\varepsilon,N}, \sigma''_{\varepsilon,n}(y) = \sigma''_{\varepsilon,N}$, for $y \in \mathbb{R}_1, n > N$. The condition $\mathbf{O_1}$ (a) reduces to the relations $\mu'_{\varepsilon,n} \to \mu'_{0,n}, \mu''_{\varepsilon,n} \xrightarrow{a.s.} \mu''_{0,n}$ and $\sigma'_{\varepsilon,n} \xrightarrow{a.s.} \sigma'_{0,n}$, $\sigma''_{\varepsilon,n} \xrightarrow{a.s.} \sigma''_{0,n}$ as $\varepsilon \to 0$, for $n = 1, \dots, N$.

The object of our interest are random variables $\phi_{\varepsilon,n}(\hat{\theta}_\varepsilon, y), y \in \mathbb{R}_1, n = 0, 1, \dots, N$ and $\Phi_\varepsilon(\hat{\theta}_\varepsilon)$.

However, we would also like to analyze the stochastic processes $Y_{\varepsilon,n}(\hat{\theta}_\varepsilon)$, $n = 0, 1, \dots$. In this case, some additional assumptions about joint distributions of random variables $\hat{\theta}_\varepsilon$ and $Y_{\varepsilon,0}$, $W_{\varepsilon,n}$, $n = 1, 2, \dots$ should be made.

It is clear that the superposition form $\phi_{\varepsilon,n}(\hat{\theta}_\varepsilon, y)$, $y \in \mathbb{R}_1$, $n = 0, 1, \dots, N$ and $\Phi_\varepsilon(\hat{\theta}_\varepsilon)$ is based on some independence assumptions for the random variables $\hat{\theta}_\varepsilon$ and the sequence of random variables $Y_{\varepsilon,0}$, $W_{\varepsilon,n}$, $n = 1, 2, \dots$. A natural variant to formalize the above-mentioned independence assumption is to assume that the following condition holds:

$\mathbf{Q_1}$: (a) Random variables $\hat{\theta}_\varepsilon$, $\varepsilon \in [0, \varepsilon_0]$ and $Y_{\varepsilon,0}$, $W_{\varepsilon,n}$, $n = 1, 2, \dots$, $\varepsilon \in (0, \varepsilon_0]$ are defined on the same probability space $\langle \Omega, \mathcal{F}, P \rangle$;
(b) families of random variables $\langle \hat{\theta}_\varepsilon, \varepsilon \in [0, \varepsilon_0] \rangle$, $\langle Y_{\varepsilon,0}, \varepsilon \in [0, \varepsilon_0] \rangle$, $\langle W_{\varepsilon,n}, \varepsilon \in [0, \varepsilon_0] \rangle$, $n = 1, 2, \dots$ are mutually independent.

If the condition $\mathbf{Q_1}$ holds, then the log-price process $Y_{\varepsilon,n}(\hat{\theta}_\varepsilon)$, $n = 0, 1, \dots$ is well defined, for every $\varepsilon \in [0, \varepsilon_0]$, by the following stochastic transition dynamic relation:

$$\begin{aligned} Y_{\varepsilon,n}(\hat{\theta}_\varepsilon) = {} & Y_{\varepsilon,n-1}(\hat{\theta}_\varepsilon) + \mu_{\varepsilon,n}(\hat{\theta}_\varepsilon, Y_{\varepsilon,n-1}(\hat{\theta}_\varepsilon)) \\ & + \sigma_{\varepsilon,n}(\hat{\theta}_\varepsilon, Y_{\varepsilon,n-1}(\hat{\theta}_\varepsilon)) W_{\varepsilon,n}, \quad n = 1, 2, \dots, \end{aligned} \tag{8.55}$$

In this case, the random variables $\phi_{\varepsilon,n}(\hat{\theta}_\varepsilon, y)$, $y \in \mathbb{R}_1$, $n = 0, 1, \ldots, N$ and $\Phi_\varepsilon(\hat{\theta}_\varepsilon)$ can be interpreted as estimates, respectively, for the log-reward functions and the optimal expected rewards for the log-price processes $Y_{0,n}(\theta)$, $n = 0, 1, \ldots$, with the drift and volatility coefficients $\mu_{0,n}(\theta, y)$ and $\sigma_{0,n}(\theta, y)$ depending on parameter θ, for which information about the true value $\hat{\theta}_0$ is given in the form of consistent estimates $\hat{\theta}_\varepsilon$.

Let us describe a natural variant of model construction, which provides holding of the condition $\mathbf{Q}_1$.

Let us assume that the sequences of independent random variables $Y_{\varepsilon,0}$, $W_{\varepsilon,n}$, $n = 1, 2, \ldots$ are defined on different probability spaces $\langle \Omega''_\varepsilon, \mathcal{F}''_\varepsilon, P''_\varepsilon \rangle$. In this case, one can construct random sequences $Y'_{\varepsilon,0}$, $W'_{\varepsilon,n}$, $n = 1, 2, \ldots$ defined on some probability space $\langle \Omega'', \mathcal{F}'', P'' \rangle$, for all $\varepsilon \in [0, \varepsilon_0]$, and such that the random sequence $Y'_{\varepsilon,0}$, $W'_{\varepsilon,n}$, $n = 1, 2, \ldots$ is equivalent in distributional sense to the random sequence $Y_{\varepsilon,0}$, $W_{\varepsilon,n}$, $n = 1, 2, \ldots$, for every $\varepsilon \in [0, \varepsilon_0]$.

Let us assume that a sequence of random variables ρ_n, $n = 0, 1, \ldots$, independent and uniformly distributed on interval $[0, 1]$, is defined on some probability space $\langle \Omega'', \mathcal{F}'', P'' \rangle$.

Let $G_{\varepsilon,0}(u) = \mathsf{P}\{Y_{\varepsilon,0} \leq u\}$ and $G_{\varepsilon,n}(u) = \mathsf{P}\{W_{\varepsilon,n} \leq u\}$, $n = 1, 2, \ldots$ be distribution functions of these random variables and $G^{-1}_{\varepsilon,n}(v) = \inf(u : G_{\varepsilon,n}(u) > v)$, $n = 0, 1, \ldots$ be the corresponding inverse functions. In this case, the random variables $Y'_{\varepsilon,0} = G^{-1}_{\varepsilon,0}(\rho_0) \stackrel{d}{=} Y_{\varepsilon,0}$ (have the same distribution), and the random variables $W'_{\varepsilon,n} = G^{-1}_{\varepsilon,n}(\rho_n) \stackrel{d}{=} W_{\varepsilon,n}$, $n = 1, 2, \ldots$. By the definition, the random variables $Y'_{\varepsilon,0}$, $W'_{\varepsilon,n}$, $n = 1, 2, \ldots$ are defined on the same probability space $\langle \Omega'', \mathcal{F}'', P'' \rangle$, for all $\varepsilon \in (0, \varepsilon_0]$. Also the families of random variables $\langle Y'_{\varepsilon,0}, \varepsilon \in (0, \varepsilon_0] \rangle$ and $\langle W'_{\varepsilon,n}, \varepsilon \in [0, \varepsilon_0] \rangle$, $n = 1, 2, \ldots$ are mutually independent.

It follows from the above remarks that the log-reward functions $\phi_{\varepsilon,n}(\theta, y)$ and the optimal expected rewards $\Phi_\varepsilon(\theta)$ are the same for the log-price process $Y_{\varepsilon,n}(\theta)$, $n = 0, 1, \ldots$ and the log-price process $Y'_{\varepsilon,n}(\theta)$, $n = 0, 1, \ldots$ constructed with the use of stochastic transition relation (8.49), in which the random variables $Y_{\varepsilon,0}$, $W_{\varepsilon,n}$, $n = 1, 2, \ldots$ should be replaced by the corresponding random variables $Y'_{\varepsilon,0}$, $W'_{\varepsilon,n}$, $n = 1, 2, \ldots$.

Now, let us define the probability space $\langle \Omega, \mathcal{F}, P \rangle$, where $\Omega = \Omega' \times \Omega''$, $\mathcal{F} = \sigma(\mathcal{F}' \times \mathcal{F}'')$ (that is the minimal σ-algebra of subsets of Ω containing all rectangles $B \times C$, $B \in \mathcal{F}'$, $C \in \mathcal{F}''$), and the probability measure $P(A)$ is a product measure on $\mathcal{F}$, which is uniquely defined by its values on rectangles $P(B \times C) = P'(A) \cdot P''(B)$, $B \in \mathcal{F}'$, $C \in \mathcal{F}''$, via the measure continuation theorem.

The random variable $\hat{\theta}_\varepsilon = \hat{\theta}_\varepsilon(\omega')$, $\omega' \in \Omega'$ can be considered as a function of $\omega = (\omega', \omega) \in \Omega = \Omega' \times \Omega''$, which, in fact, are functions of $\omega' \in \Omega'$, for every $\varepsilon \in [0, \varepsilon_0]$. Analogously, the random variables $Y_{\varepsilon,0} = Y_{\varepsilon,0}(\omega'')$, $\omega'' \in \Omega''$ and $W_{\varepsilon,n} = W_{\varepsilon,n}(\omega'')$, $\omega'' \in \Omega''$ can be considered as functions of $\omega = (\omega', \omega'') \in \Omega = \Omega' \times \Omega''$, which, in fact, are functions of $\omega'' \in \Omega''$, for every $n = 1, 2, \ldots$ and $\varepsilon \in [0, \varepsilon_0]$.

In this case, the condition $\mathbf{Q}_1$ holds.

The conditions $\mathbf{O}_3$ and $\mathbf{P}_1$ are also consistent with the construction described above.

In this construction let us consider $\phi_{\varepsilon,n}(\hat{\theta}_\varepsilon, y)$ and $\Phi_\varepsilon(\hat{\theta}_\varepsilon)$ as random variables defined either on the probability space $\langle \Omega, \mathcal{F}, P\rangle$ or on the probability space $\langle \Omega', \mathcal{F}', P'\rangle$.

In fact, the quantities $\phi_{\varepsilon,n}(\hat{\theta}_\varepsilon(\omega'), y)$ and $\Phi_\varepsilon(\hat{\theta}_\varepsilon(\omega'))$ for given outcome $\omega' \in \Omega'$ can be computed as the values of log-reward functions and optimal expected rewards for the log-price process $Y_{\varepsilon,n}(\hat{\theta}_\varepsilon(\omega'))$ given by the following stochastic transition relation:

$$\begin{aligned} Y_{\varepsilon,n}(\hat{\theta}_\varepsilon(\omega')) &= Y_{\varepsilon,n-1}(\hat{\theta}_\varepsilon(\omega')) + \mu_{\varepsilon,n}(\hat{\theta}_\varepsilon(\omega'), Y_{\varepsilon,n-1}(\hat{\theta}_\varepsilon(\omega'))) \\ &\quad + \sigma_{\varepsilon,n}(\hat{\theta}_\varepsilon(\omega'), Y_{\varepsilon,n-1}(\hat{\theta}_\varepsilon(\omega')))W_{\varepsilon,n}, \quad n = 1, 2, \dots, \end{aligned} \tag{8.56}$$

The following theorem takes place.

Theorem 8.3.3. *Let the condition $\mathbf{B}_7[\gamma]$ holds for some $\gamma \geq 0$ and also the conditions $\mathbf{Q}_1$, $\mathbf{G}_9$, $\mathbf{I}_8$, $\mathbf{O}_3$, and $\mathbf{P}_1$ hold. Then, for every $n = 0, 1, \dots, N$, the following relation takes place for any $y_\varepsilon \to y_0 \in \mathbb{Y}'_n \cap \mathbb{Y}''_n$:*

$$\phi_{\varepsilon,n}(\hat{\theta}_\varepsilon, y_\varepsilon) \xrightarrow{\text{a.s.}} \phi_{0,n}(\hat{\theta}_0, y_0) \quad \text{as} \quad \varepsilon \to 0\,. \tag{8.57}$$

Proof. Let us choose an arbitrary $n = 1, \dots, N$ and $y_\varepsilon \to y_0 \in \mathbb{Y}''_{n-1}$ as $\varepsilon \to 0$.

Denote by A the set of those $\omega' \in \Omega'$, for which,

$$\hat{\theta}_\varepsilon(\omega') \to \hat{\theta}_0(\omega') \quad \text{as} \quad \varepsilon \to 0\,. \tag{8.58}$$

Also, denote by B the set of those $\omega' \in \Omega'$, for which

$$\hat{\theta}_0(\omega') \in \Theta_0\,. \tag{8.59}$$

Relations (8.58) – (8.59), and conditions $\mathbf{O}_3$ and $\mathbf{P}_1$ imply that for $\omega' \in A \cap B$,

$$\mu_{\varepsilon,n}(\hat{\theta}_\varepsilon(\omega'), y_\varepsilon) \to \mu_{0,n}(\hat{\theta}_0(\omega'), y_0) \quad \text{as} \quad \varepsilon \to 0\,, \tag{8.60}$$

and

$$\sigma_{\varepsilon,n}(\hat{\theta}_\varepsilon(\omega'), y_\varepsilon) \to \sigma_{0,n}(\hat{\theta}_0(\omega'), y_0) \quad \text{as} \quad \varepsilon \to 0\,, \tag{8.61}$$

Since an arbitrary choice of $n = 1, \dots, N$ and $y_\varepsilon \to y_0 \in \mathbb{Y}''_{n-1}$ as $\varepsilon \to 0$, relations (8.60) and (8.61) imply that the condition $\mathbf{O}_1$ (a) holds for the log-price processes $Y_{\varepsilon,n}(\hat{\theta}_\varepsilon(\omega'))$ with trend and volatility coefficients $\mu_{\varepsilon,n}(\hat{\theta}_\varepsilon(\omega'), y)$ and $\sigma_{\varepsilon,n}(\hat{\theta}_\varepsilon(\omega'), y)$, for every $\omega' \in A \cap B$.

Relations (8.59) and the condition $\mathbf{P}_1$ (b) imply that for $\omega' \in B$,

$$\mathsf{P}''\{y_0 + \mu_{0,n}(\hat{\theta}_0(\omega'), y_0) + \sigma_{0,n}(\hat{\theta}_0(\omega'), y_0)W_{0,n} \in \mathbb{Y}'_n \cap \mathbb{Y}''_n\} = 1\,. \tag{8.62}$$

Since an arbitrary choice of $y_0 \in \mathbb{Y}''_{n-1}$ as $\varepsilon \to 0$ and $n = 1, \dots, N$, relation (8.62) implies that the condition $\mathbf{O_1}$ (b) holds for the log-price processes $Y_{\varepsilon,n}(\hat{\theta}_\varepsilon(\omega'))$, with drift and volatility coefficients $\mu_{\varepsilon,n}(\hat{\theta}_\varepsilon(\omega'), y)$ and $\sigma_{\varepsilon,n}(\hat{\theta}_\varepsilon(\omega'), y)$, for every $\omega' \in B$.

The above remarks imply that the condition $\mathbf{O_1}$ holds for the log-price processes $Y_{\varepsilon,n}(\hat{\theta}_\varepsilon(\omega'))$, with the drift and volatility coefficients $\mu_{\varepsilon,n}(\hat{\theta}_\varepsilon(\omega'), y)$ and $\sigma_{\varepsilon,n}(\hat{\theta}_\varepsilon(\omega'), y)$, for every $\omega' \in C = A \cap B$.

The condition $\mathbf{G_9}$ and relations (8.50) and (8.58) imply that the following relation holds, for any $\omega' \in A$:

$$\begin{aligned} &\overline{\lim_{\varepsilon \to 0}} \max_{1 \le n \le N} \sup_{y \in \mathbb{R}_1} (|\mu_{\varepsilon,n}(\hat{\theta}_\varepsilon(\omega'), y)| + \sigma^2_{\varepsilon,n}(\hat{\theta}_\varepsilon(\omega'), y)) \\ &\qquad \le \overline{\lim_{\varepsilon \to 0}} K'(\hat{\theta}_\varepsilon(\omega')) = K'(\hat{\theta}_0(\omega')) < \infty . \end{aligned} \tag{8.63}$$

Relation (8.63) implies that the condition $\mathbf{G_5}$ holds for the log-price processes $Y_{\varepsilon,n}(\hat{\theta}_\varepsilon(\omega'))$, with the drift and volatility coefficients $\mu_{\varepsilon,n}(\hat{\theta}_\varepsilon(\omega'), y)$ and $\sigma_{\varepsilon,n}(\hat{\theta}_\varepsilon(\omega'), y)$, for every $\omega' \in A$.

We can now conclude that conditions in Theorem 8.1.3 hold for the log-price processes $Y_{\varepsilon,n}(\hat{\theta}_\varepsilon(\omega'))$, for every $\omega' \in C = A \cap B$.

Therefore, for every $n = 0, 1, \dots, N$, the following relations holds for the log-reward functions $\phi_{\varepsilon,n}(\hat{\theta}_\varepsilon(\omega'), y)$, for any $y_\varepsilon \to y_0 \in \mathbb{Y}'_n \cap \mathbb{Y}''_n$:

$$\phi_{\varepsilon,n}(\hat{\theta}_\varepsilon(\omega'), y_\varepsilon) \to \phi_{0,n}(\hat{\theta}_0(\omega'), y_0) \quad \text{as} \quad \varepsilon \to 0, \quad \omega' \in C . \tag{8.64}$$

The condition $\mathbf{P_1}$ implies that $P'(A) = 1$ and $P'(B) = 1$. Thus, $P'(C) = 1$. Therefore, relation (8.64) implies relation (8.57) to hold. □

Let us assume that the following condition holds:

$\mathbf{K_{14}}$: (a) $P_{\varepsilon,0}(\cdot) \Rightarrow P_{0,0}(\cdot)$ as $\varepsilon \to 0$;

(b) $P_{0,0}(\mathbb{Y}'_0 \cap \mathbb{Y}''_0) = 1$, where $\mathbb{Y}'_0$ and $\mathbb{Y}''_0$ are the sets introduced in the conditions $\mathbf{I_8}$ and $\mathbf{O_3}$.

Theorem 8.3.4. *Let the conditions $\mathbf{B_7}[\gamma]$ and $\mathbf{D_{14}}[\beta]$ hold for some $\beta > \gamma > 0$ or $\beta = \gamma = 0$, and, also, the conditions $\mathbf{Q_1}$, $\mathbf{G_9}$, $\mathbf{I_8}$, $\mathbf{O_3}$, $\mathbf{P_1}$, and $\mathbf{K_{14}}$ hold. Then, the following relation takes place:*

$$\Phi_\varepsilon(\hat{\theta}_\varepsilon) \xrightarrow{\text{a.s.}} \Phi_0(\hat{\theta}_0) \quad \text{as} \quad \varepsilon \to 0 . \tag{8.65}$$

Proof. All conditions of Theorem 8.1.4 hold for the log-price processes $Y_{\varepsilon,n}(\hat{\theta}_\varepsilon(\omega'))$ with the trend and volatility coefficients $\mu_{\varepsilon,n}(\hat{\theta}_\varepsilon(\omega'), y)$ and $\sigma_{\varepsilon,n}(\hat{\theta}_\varepsilon(\omega'), y)$, for every $\omega' \in A \cap B$.

This was checked for the conditions $\mathbf{G_5}$ and $\mathbf{O_1}$ in the proof of Theorem 8.3.3.

The conditions $\mathbf{D_{14}}[\beta]$ and $\mathbf{I_8}$ are assumed in Theorem 8.1.4. The condition $\mathbf{K_{14}}$ is a variant of the condition $\mathbf{K_{12}}$ used in Theorem 8.1.4.

Thus, conditions of Theorem 8.1.4 hold for the log-price processes $Y_{\varepsilon,n}(\hat{\theta}_\varepsilon(\omega'))$, for every $\omega' \in C = A \cap B$.

By applying this theorem, we get the following relation, for $\omega' \in C$:

$$\Phi_\varepsilon(\hat{\theta}_\varepsilon(\omega')) \to \Phi_0(\hat{\theta}_0(\omega')) \quad \text{as} \quad \varepsilon \to 0 . \tag{8.66}$$

Since $P'(C) = 1$, relation (8.66) implies relation (8.65) to hold. □

Let us now assume that the object of our interest is just the asymptotic relations (8.57) and (8.65).

The random variables $\phi_{\varepsilon,n}(\hat{\theta}_\varepsilon, y)$, $y \in \mathbb{R}_1$, $n = 0, 1, \dots, N$ and $\Phi_\varepsilon(\hat{\theta}_\varepsilon)$ are superpositions of the random variables $\hat{\theta}_\varepsilon$ and nonrandom functions $\phi_{\varepsilon,n}(\theta, y)$, $(\theta, y) \in \Theta \times \mathbb{R}_1$, $n = 0, 1, \dots, N$ and $\Phi_\varepsilon(\theta)$, $\theta \in \Theta$. This makes it possible to analyze the random variables $\phi_{\varepsilon,n}(\hat{\theta}_\varepsilon, y)$, $y \in \mathbb{R}_1$, $n = 0, 1, \dots, N$ and $\Phi_\varepsilon(\hat{\theta}_\varepsilon)$ without any assumptions about joint distributions of the random variables $\hat{\theta}_\varepsilon$ and $Y_{\varepsilon,0}$, $W_{\varepsilon,n}$, $n = 1, 2, \dots$.

Recall that the condition $\mathbf{P}_1$ requires that $\langle \hat{\theta}_\varepsilon, \varepsilon \in [0, \varepsilon_0] \rangle$ be a family of random variables defined on some probability space $\langle \Omega', \mathcal{F}', P' \rangle$. As far as sequences of random variables $Y_{\varepsilon,0}$, $W_{\varepsilon,n}$, $n = 1, 2, \dots$ are concerned, they can be defined on different probability spaces $\langle \Omega''_\varepsilon, \mathcal{F}''_\varepsilon, P''_\varepsilon \rangle$, for different values of $\varepsilon \in [0, \varepsilon_0]$.

In this case, the random variables $\phi_{\varepsilon,n}(\hat{\theta}_\varepsilon, y)$, $y \in \mathbb{R}_1$, $n = 0, 1, \dots, N$ and $\Phi_\varepsilon(\hat{\theta}_\varepsilon)$ can be interpreted as estimates of the functions $\phi_{0,n}(\theta, y)$, $y \in \mathbb{R}_1$, $n = 0, 1, \dots, N$ and $\Phi_0(\theta)$ depending on the parameter θ, whose true value $\hat{\theta}_0$ is given in the form of consistent estimates $\hat{\theta}_\varepsilon$.

The following two theorems show that the condition $\mathbf{Q}_1$ can be omitted in Theorem 8.3.3 and the asymptotic relation (8.57) still can be proved.

Theorem 8.3.5. *Let the condition $\mathbf{B}_7[\gamma]$ holds for some $\gamma \geq 0$ and also the conditions $\mathbf{G}_9$, $\mathbf{I}_8$, $\mathbf{O}_3$, and $\mathbf{P}_1$ hold. Then, for every $n = 0, 1, \dots, N$, the following relation takes place for any $y_\varepsilon \to y_0 \in \mathbb{Y}'_n \cap \mathbb{Y}''_n$:*

$$\phi_{\varepsilon,n}(\hat{\theta}_\varepsilon, y_\varepsilon) \xrightarrow{\text{a.s.}} \phi_{0,n}(\hat{\theta}_0, y_0) \quad \text{as} \quad \varepsilon \to 0 . \tag{8.67}$$

Proof. Let us take an arbitrary nonrandom points $\theta_\varepsilon \in \Theta$, $\varepsilon \in [0, \varepsilon_0]$ such that

$$\theta_\varepsilon \to \theta_0 \in \Theta_0 \quad \text{as} \quad \varepsilon \to 0 . \tag{8.68}$$

Let us consider the log-price processes $Y_{\varepsilon,n}(\theta_\varepsilon)$, with the trend and volatility coefficients $\mu_{\varepsilon,n}(\theta_\varepsilon, y)$ and $\sigma_{\varepsilon,n}(\theta_\varepsilon, y)$.

The conditions $\mathbf{G}_9$ and $\mathbf{O}_3$ imply that the following relation holds:

$$\varlimsup_{\varepsilon \to 0} \max_{1 \leq n \leq N} \sup_{y \in \mathbb{R}_1} (|\mu_{\varepsilon,n}(\theta_\varepsilon, y)| + \sigma^2_{\varepsilon,n}(\theta_\varepsilon, y)) \leq \varlimsup_{\varepsilon \to 0} K'(\theta_\varepsilon) = K'(\theta_0) < \infty . \tag{8.69}$$

Relation (8.69) implies that the condition $\mathbf{G}_5$ holds for the log-price processes $Y_{\varepsilon,n}(\theta_\varepsilon)$.

The condition $\mathbf{O}_3$ implies that the condition $\mathbf{O}_1$ holds for the log-price processes $Y_{\varepsilon,n}(\theta_\varepsilon)$.

Therefore, all conditions of Theorem 8.1.3 hold for the log-price processes $Y_{\varepsilon,n}(\theta_\varepsilon)$. This theorem yields, in this case, that, for every $n = 0, 1, \dots, N$, the following asymptotic relation holds, for any $y_\varepsilon \to y_0 \in \mathbb{Y}'_n \cap \mathbb{Y}''_n$:

$$\phi_{\varepsilon,n}(\theta_\varepsilon, y_\varepsilon) \to \phi_{0,n}(\theta_0, y_0) \quad \text{as} \quad \varepsilon \to 0\,. \tag{8.70}$$

The following part of the proof reminds the corresponding part in the proof of Theorem 8.3.3.

According to the condition $\mathbf{P}_1$, the random variables $\hat{\theta}_\varepsilon$, $\varepsilon \in [0, \varepsilon_0]$ are defined on the same probability space $\langle \Omega'', \mathcal{F}'', P'' \rangle$. Let us denote by C the set of elementary events ω' such that

$$\hat{\theta}_\varepsilon(\omega') \to \hat{\theta}_0(\omega') \quad \text{as} \quad \varepsilon \to 0\,. \tag{8.71}$$

and

$$\hat{\theta}_0(\omega') \in \Theta_0\,. \tag{8.72}$$

The condition $\mathbf{P}_1$ implies that $P'(C) = 1$.

Relation (8.70) holds for any nonrandom points $\theta_\varepsilon \in \Theta$, $\varepsilon \in [0, \varepsilon_0]$ satisfying assumption (8.68).

Using this, we get, substituting points $\hat{\theta}_\varepsilon(\omega')$ in relation (8.70), that the following relation holds, for any $\omega' \in C$:

$$\phi_{\varepsilon,n}(\hat{\theta}_\varepsilon(\omega'), y_\varepsilon) \to \phi_{0,n}(\hat{\theta}_0(\omega'), y_0) \quad \text{as} \quad \varepsilon \to 0\,. \tag{8.73}$$

Since $P'(C) = 1$, relation (8.73) implies relation (8.67) to hold. □

Theorem 8.3.6. *Let the conditions $\mathbf{B}_7[\gamma]$ and $\mathbf{D}_{14}[\beta]$ hold for some $\beta > \gamma > 0$ or $\beta = \gamma = 0$, and, also, the conditions $\mathbf{Q}_1$, $\mathbf{G}_9$, $\mathbf{I}_8$, $\mathbf{O}_3$, $\mathbf{P}_1$, and $\mathbf{K}_{14}$ hold. Then, the following relation takes place:*

$$\Phi_\varepsilon(\hat{\theta}_\varepsilon) \xrightarrow{\text{a.s.}} \Phi_0(\hat{\theta}_0) \quad \textit{as} \quad \varepsilon \to 0\,. \tag{8.74}$$

Proof. Let us take an arbitrary nonrandom points $\theta_\varepsilon \in \Theta$, $\varepsilon \in [0, \varepsilon_0]$ such that $\theta_\varepsilon \to \theta_0 \in \Theta_0$ as $\varepsilon \to 0$.

As was shown in the proof of Theorem 8.3.5, the conditions $\mathbf{G}_5$ and $\mathbf{O}_1$ hold for the log-price processes $Y_{\varepsilon,n}(\theta_\varepsilon)$ with trend and volatility coefficients $\mu_{\varepsilon,n}(\theta_\varepsilon, y)$ and $\sigma_{\varepsilon,n}(\theta_\varepsilon, y)$.

The conditions $\mathbf{D}_{14}[\beta]$ and $\mathbf{I}_8$ are assumed in Theorem 8.1.4. The condition $\mathbf{K}_{14}$ is a variant of the condition $\mathbf{K}_{12}$ used in Theorem 8.1.4.

Thus, all conditions of Theorem 8.1.4 hold for the log-price processes $Y_{\varepsilon,n}(\theta_\varepsilon)$, for any nonrandom points $\theta_\varepsilon \in \Theta$, $\varepsilon \in [0, \varepsilon_0]$ such that $\theta_\varepsilon \to \theta_0 \in \Theta_0$ as $\varepsilon \to 0$.

By applying Theorem 8.1.4, we get the following relation, for any $\theta_\varepsilon \to \theta_0 \in \Theta_0$ as $\varepsilon \to 0$,

$$\Phi_\varepsilon(\theta_\varepsilon) \to \Phi_0(\theta_0) \quad \text{as} \quad \varepsilon \to 0\,. \tag{8.75}$$

By substituting points $\hat{\theta}_\varepsilon(\omega')$, defined in relations (8.71) and (8.72), in relation (8.75), we get the following relation, which holds for any $\omega' \in C$:

$$\Phi_\varepsilon(\hat{\theta}_\varepsilon(\omega')) \to \Phi_0(\hat{\theta}_0(\omega')) \quad \text{as} \quad \varepsilon \to 0\,. \tag{8.76}$$

Since $P'(C) = 1$, relation (8.76) implies relation (8.74) to hold. □

8.3.3 Option rewards for univariate Markov Gaussian log-price processes with estimated unbounded parameters

Analogous results can be obtained for the model where the condition $\mathbf{G}_9$ is replace by the weaker condition

$\mathbf{G}_{10}$: $\overline{\lim}_{\varepsilon\to 0}\sup_{\theta\in\Theta}\max_{1\leq n\leq N}\sup_{y\in\mathbb{R}_1}\frac{|\mu_{\varepsilon,n}(\theta,y)|+\sigma^2_{\varepsilon,n}(\theta,y)}{K''(\theta)(1+K_{79}|y|)} < 1$, where $K''(\theta)$ is a continuous function acting from the space Θ to interval $(0,\infty)$ and the constant $0 \leq K_{79} < \infty$.

The condition $\mathbf{G}_{10}$ implies that there exists $0 < \varepsilon'' \leq \varepsilon_0$ such that, for the following relation holds for $0 \leq \varepsilon \leq \varepsilon''$:

$$\max_{1\leq n\leq N}\sup_{y\in\mathbb{R}_1}\frac{|\mu_{\varepsilon,n}(\theta,y)|+\sigma^2_{\varepsilon,n}(\theta,y)}{1+K_{79}|y|} < K''(\theta)\,, \quad \theta\in\Theta\,. \tag{8.77}$$

Relation (8.77) implies that the condition $\mathbf{G}_6$ holds for trend and volatility coefficients $\mu_{\varepsilon,n}(\theta,y)$ and $\sigma_{\varepsilon,n}(\theta,y)$, for every $\theta\in\Theta$. Moreover, in this case the constant K_{79} replaces the constant K_{71}, the constant $K''(\theta)$, replacing the constant K_{70}, depends on parameter θ, while the parameter ε'', which guarantees holding of inequality (8.77), for $0 \leq \varepsilon \leq \varepsilon''$, does not depend on parameter θ.

Let us also assume that the condition $\mathbf{B}_7[\gamma]$ holds for pay-off functions $g_\varepsilon(n, e^y)$.

Theorem 8.1.5 implies that the log-reward functions $\phi_{\varepsilon,n}(\theta,y)$, $y\in\mathbb{R}_1$, $n = 0,1,\dots,N$ and the optimal expected rewards $\Phi_\varepsilon(\theta)$ are well defined for the log-price process $Y_{\varepsilon,n}(\theta)$, for every $\theta\in\Theta$. Moreover, it follows from relation (8.77) that, in this case, the parameters ε_{25} and ε_{26} penetrating, respectively, Theorems 8.1.5 and 8.1.6 do not depend on parameter θ.

The following theorem follows from Theorem 8.1.5 and the above remarks.

Theorem 8.3.7. *Let the condition $\mathbf{G}_{10}$ holds and, also, the condition $\mathbf{B}_7[\gamma]$ holds for some $\gamma \geq 0$. Then, for every $\theta\in\Theta$ and $\beta\geq\gamma$, there exist $0 < \varepsilon_{34} \leq \varepsilon_0$ and constants $0 \leq M_{58}, M_{59} = M_{59}(\theta,\beta) < \infty$ such that the log-reward functions $\phi_{\varepsilon,n}(\theta,y)$ satisfy the following inequalities for $y\in\mathbb{R}_1$, $0\leq n\leq N$ and $0\leq\varepsilon\leq\varepsilon_{34}$:*

$$|\phi_{\varepsilon,n}(y)\,| \leq M_{58} + M_{59}(\theta,\beta)I(\gamma=0) + M_{59}(\theta,\beta)\exp\left\{\frac{A_{N-n}(\beta)\gamma}{\beta}|y|\right\}I(\gamma>0)\,. \tag{8.78}$$

Remark 8.3.5. The explicit formulas for the constants $M_{58}, M_{59}(\theta,\beta)$ are given by the following formulas, which follows from formulas (8.22):

$$\begin{aligned} M_{58} = L_{13}\,, \quad M_{59}(\theta,\beta) &= L_{13}L_{14}I(\gamma=0) \\ &+ L_{13}L_{14}\left(1+2e^{K''(\theta)((\beta+A_{N-1}(\beta))+\frac{1}{2}(\beta+A_{N-1}(\beta))^2)}\right)^{N\frac{\gamma}{\beta}} I(\gamma>0)\,. \end{aligned} \tag{8.79}$$

One can choose any $\beta\geq\gamma$ in formula (8.79).

Remark 8.3.6. The parameter ε_{34} is given by the following formula $\varepsilon_{34} = \min(\varepsilon'', \varepsilon_{21})$, where the parameters ε_{21} and ε'' are determined by relations (8.8) and (8.77).

Theorem 8.3.8. *Let the condition* $\mathbf{G}_{10}$ *holds and, also, the conditions* $\mathbf{B}_7[\gamma]$ *and* $\mathbf{D}_{14}[\beta]$ *hold for some* $\beta \geq \gamma \geq 0$. *Then, for every* $\theta \in \Theta$, *there exist* $0 < \varepsilon_{35} \leq \varepsilon_0$ *and the constant* $0 \leq M_{60} = M_{60}(\theta) < \infty$ *such that the following inequality holds for* $0 \leq \varepsilon \leq \varepsilon_{35}$*:*

$$|\Phi_\varepsilon(\theta)| \leq M_{60}(\theta)\,. \tag{8.80}$$

Remark 8.3.7. The constant $M_{60}(\theta)$ is given by the following formula:

$$M_{60}(\theta) = L_{13} + L_{13}L_{14}I(\gamma = 0) + L_{13}L_{14}\left(1 + 2e^{K''(\theta)(\beta + \frac{1}{2}\beta^2)}\right)^{N\frac{\gamma}{\beta}} K_{69}^{\frac{\gamma}{\beta}} I(\gamma > 0)\,. \tag{8.81}$$

Remark 8.3.8. The parameter ε_{35} is given by the following formula $\varepsilon_{35} = \min(\varepsilon'', \varepsilon_{21}, \varepsilon_{23})$, where the parameters $\varepsilon_{21}, \varepsilon_{23}$, and ε'' are determined by relations (8.8), (8.11), and (8.77).

8.3.4 Strong consistency of option rewards for univariate Markov Gaussian log-price processes with estimated unbounded parameters

As in the case of the model with estimated bounded parameters, two variants of convergence results can be obtained.

In the case, where the condition $\mathbf{Q}_1$ is assumed to hold, one can analyze of random log-reward functions $\phi_{\varepsilon,n}(\hat{\theta}_\varepsilon, y)$, $y \in \mathbb{R}_1$, $n = 0, 1, \dots, N$ and random optimal expected rewards $\Phi_\varepsilon(\hat{\theta}_\varepsilon)$ for the log-reward processes $Y_{\varepsilon,n}(\hat{\theta}_\varepsilon)$, $n = 0, 1, \dots$.

The following theorem takes place.

Theorem 8.3.9. *Let the condition* $\mathbf{B}_7[\gamma]$ *holds for some* $\gamma \geq 0$ *and also the conditions* $\mathbf{Q}_1$, $\mathbf{G}_{10}$, $\mathbf{I}_8$, $\mathbf{O}_3$, *and* $\mathbf{P}_1$ *hold. Then, for every* $n = 0, 1, \dots, N$, *the following relation takes place for any* $y_\varepsilon \to y_0 \in \mathbb{Y}'_n \cap \mathbb{Y}''_n$*:*

$$\phi_{\varepsilon,n}(\hat{\theta}_\varepsilon, y_\varepsilon) \xrightarrow{\text{a.s.}} \phi_{0,n}(\hat{\theta}_0, y_0) \quad as \quad \varepsilon \to 0\,. \tag{8.82}$$

Proof. It is analogous to the proof of Theorem 8.3.3. The only difference is that the condition $\mathbf{G}_{10}$ is used instead of the condition $\mathbf{G}_9$. Thus let one check that all conditions of Theorem 8.1.7 hold, for the log-price process $Y_{\varepsilon,n}(\hat{\theta}_\varepsilon(\omega'))$, for every $\omega' \in C$ defined in relations (8.71) and (8.72). It can be made in the same way, as holding of all conditions of Theorem 8.1.3 was checked in the proof of Theorem 8.3.3. So, Theorem 8.1.7 can be applied to the processes $Y_{\varepsilon,n}(\hat{\theta}_\varepsilon(\omega'))$, for $\omega' \in C$. Since, $P'(C) = 1$, one get relation (8.82). □

Theorem 8.3.10. *Let the conditions $\mathbf{B}_7[\gamma]$ and $\mathbf{D}_{14}[\beta]$ hold for some $\beta > \gamma > 0$ or $\beta = \gamma = 0$, and, also, the conditions $\mathbf{Q}_1$, $\mathbf{G}_{10}$, $\mathbf{I}_8$, $\mathbf{O}_3$, $\mathbf{P}_1$, and $\mathbf{K}_{14}$ hold. Then, the following relation takes place:*

$$\Phi_\varepsilon(\hat{\theta}_\varepsilon) \xrightarrow{\text{a.s.}} \Phi_0(\hat{\theta}_0) \quad as \quad \varepsilon \to 0\,. \tag{8.83}$$

Proof. It is analogous to the proof of Theorem 8.3.4. The only difference is that the condition $\mathbf{G}_{10}$ is used instead of the condition $\mathbf{G}_{10}$. Thus let one check that all conditions of Theorem 8.1.8 hold for log-price processes $Y_{\varepsilon,n}(\hat{\theta}_\varepsilon(\omega'))$, for any $\omega' \in C$ such that $P'(C) = 1$. It can be made in the same way, as holding of all conditions of Theorem 8.1.4 was checked in the proof of Theorem 8.3.4. By applying Theorem 8.1.8 to the processes $Y_{\varepsilon,n}(\hat{\theta}_\varepsilon(\omega'))$, for $\omega' \in C$, and taking into account that $P'(C) = 1$, one get relation (8.83). □

In the case, where the condition $\mathbf{Q}_1$ does not holds, the asymptotic relation (8.82) can also be proved by the direct analysis of superpositions $\phi_{\varepsilon,n}(\hat{\theta}_\varepsilon, y)$, $y \in \mathbb{R}_1$, $n = 0, 1, \dots, N$.

Theorem 8.3.11. *Let the condition $\mathbf{B}_7[\gamma]$ holds for some $\gamma \geq 0$ and also the conditions $\mathbf{G}_{10}$, $\mathbf{I}_8$, $\mathbf{O}_3$, and $\mathbf{P}_1$ hold. Then, for every $n = 0, 1, \dots, N$, the following relation takes place for any $y_\varepsilon \to y_0 \in \mathbb{Y}'_n \cap \mathbb{Y}''_n$,*

$$\phi_{\varepsilon,n}(\hat{\theta}_\varepsilon, y_\varepsilon) \xrightarrow{\text{a.s.}} \phi_{0,n}(\hat{\theta}_0, y_0) \quad as \quad \varepsilon \to 0\,. \tag{8.84}$$

Proof. It is analogous to the proof of Theorem 8.3.5. The only difference is that the condition $\mathbf{G}_{10}$ is used instead of the condition $\mathbf{G}_9$. Thus let one check that all conditions of Theorem 8.1.7 hold, for log-price processes $Y_{\varepsilon,n}(\theta_\varepsilon)$, for any nonrandom points $\theta_\varepsilon \to \theta_0 \in \Theta_0$, where Θ_0 is the set penetrating condition $\mathbf{O}_3$. It can be made in the same way, as holding of all conditions of Theorem 8.1.7 was checked in the proof of Theorem 8.3.5. By applying Theorem 8.1.7 to the log-price processes $Y_{\varepsilon,n}(\theta_\varepsilon)$ one can get relation $\phi_{\varepsilon,n}(\theta_\varepsilon, y_\varepsilon) \to \phi_{0,n}(\theta_0, y_0)$ as $\varepsilon \to 0$, for the corresponding nonrandom points θ_ε and y_ε. Then, using the condition $\mathbf{P}_1$, one can get relation (8.67). □

Theorem 8.3.12. *Let the conditions $\mathbf{B}_7[\gamma]$ and $\mathbf{D}_{14}[\beta]$ hold for some $\beta > \gamma > 0$ or $\beta = \gamma = 0$, and, also, the conditions $\mathbf{G}_{10}$, $\mathbf{I}_8$, $\mathbf{O}_3$, $\mathbf{P}_1$, and $\mathbf{K}_{14}$ hold. Then, the following relation takes place:*

$$\Phi_\varepsilon(\hat{\theta}_\varepsilon) \xrightarrow{\text{a.s.}} \Phi_0(\hat{\theta}_0) \quad as \quad \varepsilon \to 0\,. \tag{8.85}$$

Proof. It is analogous to the proof of Theorem 8.3.6. The only difference is that the condition $\mathbf{G}_{10}$ is used instead of the condition $\mathbf{G}_9$. Thus let us check that all conditions of Theorem 8.1.8 hold, for log-price processes $Y_{\varepsilon,n}(\theta_\varepsilon)$, for any nonrandom points $\theta_\varepsilon \to \theta_0 \in \Theta_0$, where Θ_0 is the set penetrating condition $\mathbf{O}_3$. It can be made in the same way, as holding of all conditions of Theorem 8.1.8 was checked in the proof of Theorem 8.3.6. By applying Theorem 8.1.8 to the log-price processes $Y_{\varepsilon,n}(\theta_\varepsilon)$ one can get relation $\phi_{\varepsilon,n}(\theta_\varepsilon, y_\varepsilon) \to \phi_{0,n}(\theta_0, y_0)$ as $\varepsilon \to 0$, for the corresponding nonrandom points θ_ε and y_ε. Then, using the condition $\mathbf{P}_1$, one can get relation (8.85). □

8.3.5 Weak consistency of option rewards for univariate Markov Gaussian log-price processes with estimated parameters

Let us now assume that random variables $\hat{\theta}_\varepsilon$ is defined on different probability spaces $\langle \Omega'_\varepsilon, \mathcal{F}'_\varepsilon, P'_\varepsilon \rangle$, for $\varepsilon \in [0, \varepsilon_0]$.

We interpret the random variables $\hat{\theta}_\varepsilon$, for $\varepsilon \in (0, \varepsilon_0]$ as estimates for the random variable $\hat{\theta}_0$ and assume that the following weak consistency condition, coherent with convergence condition $\mathbf{O}_3$, for trend and volatility coefficients, holds:

$\mathbf{P}_2$: (a) $\hat{\theta}_\varepsilon \xrightarrow{d} \hat{\theta}_0$ as $\varepsilon \to 0$;

(b) $\mathsf{P}\{\hat{\theta}_0 \in \Theta_0\} = 1$, where Θ_0 is the set penetrating condition $\mathbf{O}_3$.

In applications, it can be seen that $\hat{\theta}_0 = const$ with probability 1. In this case, the condition $\mathbf{P}_2$ means that $\hat{\theta}_\varepsilon \xrightarrow{P} \hat{\theta}_0$ as $\varepsilon \to 0$, i.e. $\hat{\theta}_\varepsilon$ are weakly consistent estimates for the parameter $\hat{\theta}_0$.

The object of our interest are again the random variables $\phi_{\varepsilon,n}(\hat{\theta}_\varepsilon, y)$, $y \in \mathbb{R}_1$, $n = 0, 1, \dots, N$ and $\Phi_\varepsilon(\hat{\theta}_\varepsilon)$.

However, we would also like to analyze stochastic processes $Y_{\varepsilon,n}(\hat{\theta}_\varepsilon)$, $n = 0, 1, \dots$. In this case, some additional independence assumptions for random variables $\hat{\theta}_\varepsilon$ and $Y_{\varepsilon,0}$, $W_{\varepsilon,n}$, $n = 1, 2, \dots$ should be made.

A natural variant to formalize the above mentioned independence assumptions is to assume that the following condition holds:

$\mathbf{Q}_2$: For every $\varepsilon \in [0, \varepsilon_0]$:

(a) random variables $\hat{\theta}_\varepsilon$, $Y_{\varepsilon,0}$, $W_{\varepsilon,n}$, $n = 1, 2, \dots$ are defined on the same probability space $\langle \Omega_\varepsilon, \mathcal{F}_\varepsilon, P_\varepsilon \rangle$;

(b) random variables $\hat{\theta}_\varepsilon$, $Y_{\varepsilon,0}$, and $W_{\varepsilon,n}$, $n = 1, 2, \dots$ are mutually independent.

If the condition $\mathbf{Q}_2$ holds, then the log-price process $Y_{\varepsilon,n}(\hat{\theta}_\varepsilon)$ is well defined, for every $\varepsilon \in [0, \varepsilon_0]$, by the following stochastic dynamic relation:

$$
\begin{aligned}
Y_{\varepsilon,n}(\hat{\theta}_\varepsilon) &= Y_{\varepsilon,n-1}(\hat{\theta}_\varepsilon) + \mu_{\varepsilon,n}(\hat{\theta}_\varepsilon, Y_{\varepsilon,n-1}(\hat{\theta}_\varepsilon)) \\
&\quad + \sigma_{\varepsilon,n}(\hat{\theta}_\varepsilon, Y_{\varepsilon,n-1}(\hat{\theta}_\varepsilon)) W_{\varepsilon,n}, \quad n = 1, 2, \dots, \qquad (8.86)
\end{aligned}
$$

Let us describe a natural variant of model construction, which provides holding of the condition $\mathbf{Q}_2$.

Let us assume that the sequence of independent random variables $Y_{\varepsilon,0}$, $W_{\varepsilon,n}$, $n = 1, 2, \dots$ is defined on a probability space $\langle \Omega''_\varepsilon, \mathcal{F}''_\varepsilon, P''_\varepsilon \rangle$, for $\varepsilon \in [0, \varepsilon_0]$.

Now, let us define, for every $\varepsilon \in [0, \varepsilon_0]$, the probability space $\langle \Omega_\varepsilon, \mathcal{F}_\varepsilon, P_\varepsilon \rangle$, where $\Omega_\varepsilon = \Omega'_\varepsilon \times \Omega''_\varepsilon$, $\mathcal{F}_\varepsilon = \sigma(\mathcal{F}'_\varepsilon \times \mathcal{F}''_\varepsilon)$ (that is the minimal σ-algebra of subsets of Ω containing all rectangles $B \times C$, $B \in \mathcal{F}'_\varepsilon$, $C \in \mathcal{F}''_\varepsilon$), and the probability measure $P_\varepsilon(A)$ is a product measure on $\mathcal{F}_\varepsilon$, which is uniquely defined by its values on rectangles $P_\varepsilon(B \times C) = P'_\varepsilon(B) P''_\varepsilon(C)$, $B \in \mathcal{F}'_\varepsilon$, $C \in \mathcal{F}''_\varepsilon$, via the measure continuation theorem.

Let $\varepsilon \in [0, \varepsilon_0]$. The random variable $\hat{\theta}_\varepsilon = \hat{\theta}_\varepsilon(\omega'), \omega' \in \Omega'_\varepsilon$ can be considered as a function of $\omega = (\omega', \omega) \in \Omega_\varepsilon = \Omega'_\varepsilon \times \Omega''_\varepsilon$, which, in fact, is a function of $\omega' \in \Omega'_\varepsilon$. Analogously, the random variables $Y_{\varepsilon,0} = Y_{\varepsilon,0}(\omega''), \omega'' \in \Omega''_\varepsilon$ and $W_{\varepsilon,n} = W_{\varepsilon,n}(\omega''), \omega'' \in \Omega''_\varepsilon$ can be considered as functions of $\omega = (\omega', \omega'') \in \Omega_\varepsilon = \Omega'_\varepsilon \times \Omega''_\varepsilon$, which, in fact, are functions of $\omega'' \in \Omega''_\varepsilon$, for all $n = 1, 2, \ldots$.

In this case, the condition $\mathbf{Q}_2$ holds.

The conditions $\mathbf{O}_3$ and $\mathbf{P}_2$ are also consistent with the construction described above.

This construction enables one to consider $\phi_{\varepsilon,n}(\hat{\theta}_\varepsilon, y)$ and $\Phi_\varepsilon(\hat{\theta}_\varepsilon)$ as random variables defined either on the probability space $\langle \Omega_\varepsilon, \mathcal{F}_\varepsilon, P_\varepsilon \rangle$ or on the probability space $\langle \Omega'_\varepsilon, \mathcal{F}'_\varepsilon, P'_\varepsilon \rangle$.

The following theorem takes place.

Theorem 8.3.13. *Let the condition $\mathbf{B}_7[\gamma]$ holds for some $\gamma \geq 0$ and also the conditions $\mathbf{Q}_2$, $\mathbf{G}_9$ or $\mathbf{G}_{10}$, $\mathbf{I}_8$, $\mathbf{O}_3$, and $\mathbf{P}_2$ hold. Then, for every $n = 0, 1, \ldots, N$, the following relation takes place for any $y_\varepsilon \to y_0 \in \mathbb{Y}'_n \cap \mathbb{Y}''_n$:*

$$\phi_{\varepsilon,n}(\hat{\theta}_\varepsilon, y_\varepsilon) \xrightarrow{d} \phi_{0,n}(\hat{\theta}_0, y_0) \quad \text{as} \quad \varepsilon \to 0\,. \tag{8.87}$$

Theorem 8.3.14. *Let the conditions $\mathbf{B}_7[\gamma]$ and $\mathbf{D}_{14}[\beta]$ hold for some $\beta > \gamma > 0$ or $\beta = \gamma = 0$, and, also, the conditions $\mathbf{Q}_2$, $\mathbf{G}_9$ or $\mathbf{G}_{10}$, $\mathbf{I}_8$, $\mathbf{O}_3$, $\mathbf{P}_2$, and $\mathbf{K}_{14}$ hold. Then, the following relation takes place:*

$$\Phi_\varepsilon(\hat{\theta}_\varepsilon) \xrightarrow{d} \Phi_0(\hat{\theta}_0) \quad \text{as} \quad \varepsilon \to 0\,. \tag{8.88}$$

Proof of Theorems 8.3.13–8.3.14. Since (8.87) is a relation of convergence in distribution, we can assume that the parameter $\varepsilon \to 0$ runs only over some countable sequence of values. If we prove relation (8.87) under this additional assumption, then we will be able to prove relation (8.87) for any subsequence of values of the parameter $\varepsilon \to 0$. This would prove relation (8.87) for any $\varepsilon \to 0$.

Let us assume that $\rho, \rho_0, \rho_1, \ldots$ is a sequence of independent uniformly distributed random variables defined of some probability space $\langle \Omega, \mathcal{F}, P \rangle$.

According to the Skorokhod representation theorem, if the condition $\mathbf{P}_2$ holds, one can construct measurable functions $f_\varepsilon(v)$ acting from interval $[0, 1]$ to the Polish metric space Θ such that random variables $\hat{\theta}'_\varepsilon = f_\varepsilon(\rho) \stackrel{d}{=} \hat{\theta}_\varepsilon$, for every $\varepsilon \in [0, \varepsilon_0]$, and $\hat{\theta}'_\varepsilon \xrightarrow{\text{a.s.}} \hat{\theta}'_0$ as $\varepsilon \to 0$.

We again employ random variables $Y'_{\varepsilon,0} = G^{-1}_{\varepsilon,0}(\rho_0)$ and $W'_{\varepsilon,n} = G^{-1}_n(\rho_n)$, $n = 1, 2, \ldots$, defined in Subsection 8.3.2. By the definition, $Y'_{\varepsilon,0} \stackrel{d}{=} Y_{\varepsilon,0}$ and $W'_{\varepsilon,n} \stackrel{d}{=} W_{\varepsilon,n}$, $n = 1, 2, \ldots$, for every $\varepsilon \in [0, \varepsilon_0]$.

Obviously, random variables $\hat{\theta}'_\varepsilon, Y'_{\varepsilon,0}, W'_{\varepsilon,n}, n = 1, 2, \ldots, \varepsilon \in [0, \varepsilon_0]$ are defined on the same probability space $\langle \Omega, \mathcal{F}, P \rangle$ and satisfy the condition $\mathbf{Q}_1$ stronger than the condition $\mathbf{Q}_2$.

It follows from the above remarks that the log-reward functions $\phi_{\varepsilon,n}(\theta, y)$ and the optimal expected rewards $\Phi_\varepsilon(\theta)$ are the same for the log-price process $Y_{\varepsilon,n}(\theta)$, $n = 0, 1, \dots$ and the log-price process $Y'_{\varepsilon,n}(\theta)$, $n = 0, 1, \dots$ constructed with the use of stochastic transition relation (8.49), in which the random variables $Y_{\varepsilon,0}$, $W_{\varepsilon,n}$, $n = 1, 2, \dots$ should be replaced by the corresponding random variables $Y'_{\varepsilon,0}$, $W'_{\varepsilon,n}$, $n = 1, 2, \dots$.

Also the random variables $\phi_{\varepsilon,n}(\hat{\theta}_\varepsilon, y) \overset{d}{=} \phi_{\varepsilon,n}(\hat{\theta}'_\varepsilon, y)$, for every $y \in \mathbb{R}_1$, $n = 0, 1, \dots, N$ and $\varepsilon \in [0, \varepsilon_0]$ and the random variables $\Phi_\varepsilon(\hat{\theta}_\varepsilon) \overset{d}{=} \Phi_\varepsilon(\hat{\theta}'_\varepsilon)$.

Obviously, the condition $\mathbf{P_1}$ holds for random variables $\hat{\theta}'_\varepsilon$, $\varepsilon \in [0, \varepsilon_0]$.

The above remarks imply that Theorem 8.3.3 or 8.3.9 (respectively, in the case, where the condition $\mathbf{G_9}$ or $\mathbf{G_{10}}$ is assumed to hold) can be applied to the log-price processes $Y'_{\varepsilon,n}(\theta)$ and random variables $\hat{\theta}'_\varepsilon$ that yields the relation,

$$\phi_{\varepsilon,n}(\hat{\theta}'_\varepsilon, y_\varepsilon) \xrightarrow{\text{a.s.}} \phi_{0,n}(\hat{\theta}'_0, y_0) \quad \text{as} \quad \varepsilon \to 0\,. \tag{8.89}$$

Since a.s. convergence is stronger than convergence in distribution and random variables $\phi_{\varepsilon,n}(\hat{\theta}'_\varepsilon, y_\varepsilon) \overset{d}{=} \phi_{\varepsilon,n}(\hat{\theta}_\varepsilon, y_\varepsilon)$, relation (8.89) implies relation (8.87) to hold.

The above remarks imply that Theorem 8.3.4 or 8.3.10 (respectively, in the case, where the condition $\mathbf{G_9}$ or $\mathbf{G_{10}}$ is assumed to hold) can be applied to the log-price processes $Y'_{\varepsilon,n}(\theta)$ and random variables $\hat{\theta}'_\varepsilon$ that yields the relation,

$$\Phi_\varepsilon(\hat{\theta}'_\varepsilon) \xrightarrow{\text{a.s.}} \Phi_0(\hat{\theta}'_0) \quad \text{as} \quad \varepsilon \to 0\,. \tag{8.90}$$

Since, random variables $\Phi_\varepsilon(\hat{\theta}'_\varepsilon) \overset{d}{=} \Phi_\varepsilon(\hat{\theta}_\varepsilon)$, relation (8.90) implies relation (8.88) to hold. □

Remark 8.3.9. In the same way, as it was made for Theorems 8.3.5, 8.3.6, 8.3.11, and 8.3.12, one can prove that the condition $\mathbf{Q_2}$ can be omitted in these theorems. The corresponding proofs will only require to use the Skorokhod representation theorem for constructing random variables $\hat{\theta}'_\varepsilon \overset{d}{=} \hat{\theta}_\varepsilon$, for every $\varepsilon \in [0, \varepsilon_0]$, and such that $\hat{\theta}'_\varepsilon \xrightarrow{\text{a.s.}} \hat{\theta}'_0$ as $\varepsilon \to 0$. Then, the theorems mentioned above should be applied, in order, to yield relations $\phi_{\varepsilon,n}(\hat{\theta}'_\varepsilon, y_\varepsilon) \xrightarrow{\text{a.s.}} \phi_{0,n}(\hat{\theta}'_0, y_0)$ as $\varepsilon \to 0$ and $\Phi_\varepsilon(\hat{\theta}'_\varepsilon) \xrightarrow{\text{a.s.}} \Phi_0(\hat{\theta}'_0)$ as $\varepsilon \to 0$.

8.4 Skeleton reward approximations for Markov Gaussian LPP

In this section, we present results about space-skeleton approximations for reward of American-type options for univariate and multivariate modulated Markov Gaussian log-price processes.

8.4.1 Space-skeleton approximations for option rewards of univariate Markov Gaussian log-price processes

Let the univariate Markov Gaussian log-price process $Y_{0,n}$, $n = 0, 1, \ldots$ be defined by the following stochastic transition dynamic relation:

$$Y_{0,n} = Y_{0,n-1} + \mu_{0,n}(Y_{0,n-1}) + \sigma_{0,n}(Y_{0,n-1})W_{0,n}\,, \quad n = 1, 2, \ldots\,, \tag{8.91}$$

where (a) $Y_{0,0}$ is a real-valued random variable, (b) $W_{0,n}$, $n = 1, 2, \ldots$ is a sequence of standard i.i.d. normal random variables with mean 0 and variance 1, (c) the random variable $Y_{0,0}$ and sequence of random variables $W_{0,n}$, $n = 1, 2, \ldots$ are independent, (d) $\mu_{0,n}(y)$, $n = 1, 2, \ldots$ are real-valued Borel functions defined on $\mathbb{R}_1$, and (e) $\sigma_{0,n}(y)$, $n = 1, 2, \ldots$ are real-valued Borel functions defined on $\mathbb{R}_1$.

Let us construct the corresponding space-skeleton approximating processes $Y_{\varepsilon,n}$, $n = 0, 1, \ldots$, for $\varepsilon \in (0, \varepsilon_0]$, according to the algorithm described in Subsection 7.2.1.

Let us choose, for every $\varepsilon \in (0, \varepsilon_0]$, the parameters $\delta_{\varepsilon,n} > 0$, $\lambda_{\varepsilon,n} \in \mathbb{R}_1$, $n = 0, 1, \ldots$, and integers $m^-_{\varepsilon,n} \leq m^+_{\varepsilon,n}$, $n = 0, 1, \ldots$.

First, the corresponding intervals $\mathbb{I}_{\varepsilon,n,l}$, which play the role of skeleton sets $\mathbb{A}_{\varepsilon,n,l}$, are constructed for $l = m^-_{\varepsilon,n}, \ldots, m^+_{\varepsilon,n}$, $n = 0, 1, \ldots, N$,

$$\mathbb{I}_{\varepsilon,n,l} = \mathbb{A}_{\varepsilon,n,l} = \begin{cases} (-\infty,\ \delta_{\varepsilon,n}(m^-_{\varepsilon,n} + \frac{1}{2})] + \lambda_{\varepsilon,n} & \text{if } l = m^-_{\varepsilon,n}\,, \\ (\delta_{\varepsilon,n}(l - \frac{1}{2}),\ \delta_{\varepsilon,n}(l + \frac{1}{2})] + \lambda_{\varepsilon,n} & \text{if } m^-_{\varepsilon,n} < l < m^+_{\varepsilon,n}\,, \\ (\delta_{\varepsilon,n}(m^+_{\varepsilon,n} - \frac{1}{2}),\ \infty) + \lambda_{\varepsilon,n} & \text{if } l = m^+_{\varepsilon,n}\,. \end{cases} \tag{8.92}$$

Then, the skeleton points $y_{\varepsilon,n,l} \in \mathbb{I}_{\varepsilon,n,l}$ are chosen in the following way:

$$y_{\varepsilon,n,l} = l\delta_{\varepsilon,n} + \lambda_{\varepsilon,n}\,, \quad l = m^-_{\varepsilon,n}, \ldots, m^+_{\varepsilon,n}\,, \quad n = 0, 1, \ldots, N\,. \tag{8.93}$$

Finally, skeleton functions $h_{\varepsilon,n}(y)$, $n = 0, 1, \ldots$, are defined by the following formulas:

$$h_{\varepsilon,n}(y) = \Big\{ y_{\varepsilon,n,l} \quad \text{if } y \in \mathbb{I}_{\varepsilon,n,l}\,, \quad l = m^-_{\varepsilon,n}, \ldots, m^+_{\varepsilon,n}\,, \tag{8.94}$$

The corresponding space-skeleton approximating Markov processes $Y_{\varepsilon,n}$ are defined by the following stochastic transition dynamic relations:

$$\begin{cases} Y_{\varepsilon,n} = h_{\varepsilon,n}\big(h_{\varepsilon,n-1}(Y_{\varepsilon,n-1}) + \mu_{0,n}(h_{\varepsilon,n-1}(Y_{\varepsilon,n-1})) \\ \qquad\quad + \sigma_{0,n}(h_{\varepsilon,n-1}(Y_{\varepsilon,n-1}))W_{0,n}\big)\,, \quad n = 1, 2, \ldots\,, \\ Y_{\varepsilon,0} = h_{\varepsilon,0}(Y_{0,0})\,. \end{cases} \tag{8.95}$$

The approximating log-price process $Y_{\varepsilon,n}$ is, for every $\varepsilon \in (0, \varepsilon_0]$, the skeleton Markov chain with one-step transition probabilities,

$$
\begin{aligned}
P_{\varepsilon,n}(y, A) &= \sum_{y_{\varepsilon,n,l} \in A} P_{0,n}(h_{\varepsilon,n-1}(y), \mathbb{I}_{\varepsilon,n,l}) \\
&= \sum_{y_{\varepsilon,n,l} \in A} \mathsf{P}\Big\{h_{\varepsilon,n-1}(y) \\
&\qquad + \mu_{0,n}(h_{\varepsilon,n-1}(y)) + \sigma_{0,n}(h_{\varepsilon,n-1}(y))W_{0,n} \in \mathbb{I}_{\varepsilon,n,l}\Big\}\,, \qquad (8.96)
\end{aligned}
$$

and the initial distribution,

$$
P_{\varepsilon,0}(A) = \sum_{y_{\varepsilon,0,l} \in A} P_{0,0}(\mathbb{I}_{\varepsilon,0,l}) \qquad (8.97)
$$

Note that the one-point transition probabilities $P_{0,n}(y, \mathbb{I}_{\varepsilon,n,l}) = \mathsf{P}\{y + \mu_{0,n}(y) + \sigma_{0,n}(y)\, W_{0,n} \in \mathbb{I}_{\varepsilon,n,l}\}$ are interval Gaussian probabilities.

Let us also assume that a pay-off function $g_\varepsilon(n, e^y) = g(n, e^y)$ does not depend on the parameter ε.

Let us $\phi_{\varepsilon,n}(y)$ be the corresponding log-reward functions for the log-price processes $Y_{\varepsilon,n}$.

The following lemma is the direct corollary of Lemma 7.2.1.

Lemma 8.4.1. *Let, for every $\varepsilon \in (0, \varepsilon_0]$, the log-price process $Y_{\varepsilon,n}$ be a space-skeleton Markov chain with the transition probabilities and the initial distribution defined, respectively, in relations (8.96) and (8.97). Then the reward functions $\phi_{\varepsilon,n}(y)$ and $\phi_{\varepsilon,n+r}(y_{\varepsilon,n+r,l})$, for points $y_{\varepsilon,n+r,l}$, $l = m^-_{\varepsilon,n+r}, \dots, m^+_{\varepsilon,n+r}$, $r = 1, \dots N - n$, given by formulas (8.93), are, for every $y \in \mathbb{I}_{\varepsilon,n,l''}$, $l'' = m^-_{\varepsilon,n}, \dots, m^+_{\varepsilon,n}$, $n = 0, \dots, N$, the unique solution for the following recurrence finite system of linear equations:*

$$
\begin{cases}
\phi_{\varepsilon,N}(y_{\varepsilon,N,l}) = g(N, e^{y_{\varepsilon,N,l}})\,, l = m^-_{\varepsilon,N}, \dots, m^+_{\varepsilon,N}\,, \\
\phi_{\varepsilon,n+r}(y_{\varepsilon,n+r,l}) = \max\Bigg(g(n+r, e^{y_{\varepsilon,n+r,l}}), \\
\qquad \displaystyle\sum_{l' = m^-_{\varepsilon,n+r+1}}^{m^+_{\varepsilon,n+r+1}} \phi_{\varepsilon,n+r+1}(y_{\varepsilon,n+r+1,l'}) P_{0,n+r}(y_{\varepsilon,n+r,l}, \mathbb{I}_{\varepsilon,n+r+1,l'})\Bigg)\,, \\
\qquad l = m^-_{\varepsilon,n+r}, \dots, m^+_{\varepsilon,n+r}\,, \quad r = N-n-1, \dots, 1\,, \\
\phi_{\varepsilon,n}(y) = \max\Bigg(g(n, e^y), \displaystyle\sum_{l' = m^-_{\varepsilon,n+1}}^{m^+_{\varepsilon,n+1}} \phi_{\varepsilon,n+1}(y_{\varepsilon,n+1,l'}) \\
\qquad \times P_{0,n}(y_{\varepsilon,n,l''}, \mathbb{I}_{\varepsilon,n+1,l'})\Bigg)\,.
\end{cases} \qquad (8.98)
$$

Let us now formulate conditions of convergence for the above log-reward functions.

We are going to apply Theorem 7.2.1.

We assume that the condition $\mathbf{I}_6$ holds for the pay-off function $g(n, e^y)$.

The condition $\mathbf{J}_7$ should be replaced in this case by the following condition:

$\mathbf{O}_4$: There exist sets $\mathbb{Y}''_n \in \mathcal{B}_1$, $n = 0, \dots, N$ such that:
(a) $\mu_{0,n}(y_\varepsilon) \to \mu_{0,n}(y_0)$ and $\sigma_{0,n}(y_\varepsilon) \to \sigma_{0,n}(y_0)$ as $\varepsilon \to 0$, for any $y_\varepsilon \to y_0 \in \mathbb{Y}''_{n-1}$ as $\varepsilon \to 0$, and $n = 1, \dots, N$;
(b) $\mathsf{P}\{y_0 + \mu_{0,n}(y_0) + \sigma_{0,n}(y_0)W_{0,n} \in \mathbb{Y}'_n \cap \mathbb{Y}''_n\} = 1$, for every $y_0 \in \mathbb{Y}'_{n-1} \cap \mathbb{Y}''_{n-1}$ and $n = 1, \dots, N$, where $\mathbb{Y}'_n$, $n = 0, \dots, N$ are sets introduced in the condition $\mathbf{I}_6$.

The condition $\mathbf{O}_4$ (b) holds if, for every $y_0 \in \mathbb{Y}'_{n-1} \cap \mathbb{Y}''_{n-1}$, $n = 1, \dots, N$, either $\sigma_{0,n}(y_0) > 0$ and the sets $\overline{\mathbb{Y}}'_n, \overline{\mathbb{Y}}''_n$ have zero Lebesgue measure, or $\sigma_{0,n}(y_0) = 0$ and $y_0 + \mu_{0,n}(y_0) \in \mathbb{Y}'_n \cap \mathbb{Y}''_n$.

The following theorem is a corollary of Theorems 7.2.1.

Theorem 8.4.1. *Let a Markov Gaussian log-price process $Y_{0,n}$ and the corresponding space-skeleton approximating Markov log-price processes $Y_{\varepsilon,n}$ are defined, respectively, by stochastic transition dynamic relations (8.91) and (8.95). Let also the conditions $\mathbf{B}_4[\gamma]$ holds for some $\gamma \geq 0$, and, also, the conditions $\mathbf{G}_1$, $\mathbf{N}_1$, $\mathbf{I}_6$, and $\mathbf{O}_4$ hold. Then, for every $n = 0, 1, \dots, N$, the following relation takes place for any $y_\varepsilon \to y_0 \in \mathbb{Y}'_n \cap \mathbb{Y}''_n$:*

$$\phi_{\varepsilon,n}(y_\varepsilon) \to \phi_{0,n}(y_0) \quad as \quad \varepsilon \to 0\,. \tag{8.99}$$

Proof. The proof can be obtained by application of Theorem 7.2.1.

The fitting condition $\mathbf{L}_2$ holds since the corresponding transition probabilities are defined according to formula (8.96).

The condition $\mathbf{C}_{11}[\beta]$ holds for any $\beta \geq 0$ that follows from Lemmas 4.1.8 and 4.4.1. We can always choose some $\beta > \gamma$ if $\gamma > 0$ of $\beta = 0$ if $\gamma = 0$.

The condition $\mathbf{O}_4$ (a) implies that the condition $\mathbf{J}_9$ (a) holds since convergence of parameters for normal random variables is equivalent to weak convergence for these random variables. The condition $\mathbf{O}_4$ (b) is just a variant of the condition $\mathbf{J}_9$ (b) .

Thus, all conditions of Theorem 7.2.1 hold. By applying this theorem we get the convergence relation (8.99). □

Let us now formulate conditions of convergence for the corresponding optimal expected rewards.

In this case the optimal expected reward Φ_ε for the approximation log-price processes $Y_{\varepsilon,n}$ can be found, for every $\varepsilon \in (0, \varepsilon_0]$, using the following formula:

$$\Phi_\varepsilon = \sum_{l=m^-_{\varepsilon,0}}^{m^+_{\varepsilon,0}} \phi_{\varepsilon,0}(y_{\varepsilon,0,l})p_{\varepsilon,0}\,, \tag{8.100}$$

where $p_{\varepsilon,l} = P_{0,0}(\mathbb{I}_{\varepsilon,0,l}) = \mathsf{P}\{h_{\varepsilon,0}(Y_{0,0}) = y_{\varepsilon,0,l}\}$, $l = m^-_{\varepsilon,0}, \dots, m^+_{\varepsilon,0}$.

We are going to use Theorem 7.2.2.

The condition $\mathbf{D}_{10}[\beta]$ used in Theorem 7.2.2 should be replaced by an analog of the condition $\mathbf{D}_4[\beta]$, assumed to hold for some $\beta \geq 0$:

$\mathbf{D}_{16}[\beta]$: $\mathsf{E}e^{\beta|Y_{0,0}|} < K_{80}$, for some $1 < K_{80} < \infty$.

The condition $\mathbf{K}_7$ used in Theorem 7.2.2 should be replaced by the following condition imposed on the initial distribution $\mathsf{P}\{Y_{0,0} \in A\} = P_{0,0}(A)$:
$\mathbf{K}_{14}$: $P_{0,0}(\mathbb{Y}'_0 \cap \mathbb{Y}''_0) = 1$, where $\mathbb{Y}'_0$ and $\mathbb{Y}''_0$ are the sets introduced, respectively, in the conditions $\mathbf{I}_6$ and $\mathbf{O}_4$.

Theorem 8.4.2. *Let a Markov log-price process $Y_{0,n}$ and the corresponding space-skeleton approximating Markov log-price processes $Y_{\varepsilon,n}$ are defined, respectively, by stochastic transition dynamic relations* (8.91) *and* (8.95). *Let also the conditions $\mathbf{B}_4[\gamma]$ and $\mathbf{D}_{16}[\beta]$ hold with parameters γ and β such that either $\beta > \gamma > 0$ or $\beta = \gamma = 0$, and also the conditions $\mathbf{G}_1$, $\mathbf{N}_1$, $\mathbf{I}_6$, $\mathbf{O}_4$, and $\mathbf{K}_{14}$ hold. Then, the following relation takes place,*

$$\Phi_\varepsilon \to \Phi_0 \quad as \quad \varepsilon \to 0 \,. \tag{8.101}$$

Proof. The fitting condition $\mathbf{M}_2$ holds since the corresponding initial distributions are defined according to formula (8.97).

The condition $\mathbf{K}_{14}$ is just reformulation of the condition $\mathbf{K}_8$.

Other conditions of Theorem 7.2.2 also hold that were shown in the proof of Theorem 8.4.1. By applying Theorem 7.2.2 we get convergence relations (8.125). □

Analogous results can be formulated for univariate Markov Gaussian log-price processes with unbounded characteristics. In this case, Theorems 7.4.1 and 7.4.2 should be used instead Theorems 7.1.1 and 7.1.2.

Theorem 8.4.3. *Let a Markov Gaussian log-price process $Y_{0,n}$ and the corresponding space-skeleton approximating Markov log-price processes $Y_{\varepsilon,n}$ are defined, respectively, by stochastic transition dynamic relations* (8.91) *and* (8.95). *Let also the conditions $\mathbf{B}_4[\gamma]$ holds for some $\gamma \geq 0$, and, also, the conditions $\mathbf{G}_2$, $\mathbf{N}_1$, $\mathbf{I}_6$, and $\mathbf{O}_4$ hold. Then, for every $n = 0, 1, \dots, N$, the following relation takes place for any $y_\varepsilon \to y_0 \in \mathbb{Y}'_n \cap \mathbb{Y}''_n$,*

$$\phi_{\varepsilon,n}(y_\varepsilon) \to \phi_{0,n}(y_0) \quad as \quad \varepsilon \to 0 \,. \tag{8.102}$$

Proof. The proof repeats the proof of Theorem 8.4.1. The only change is that Theorem 7.4.1 should be used instead of Theorem 7.2.1.

In this case, the condition $\mathbf{C}_{15}[\beta]$ holds for any $\beta \geq 0$ according to Lemmas 4.3.6 and 4.4.2. We can always choose some $\beta > \gamma$ if $\gamma > 0$ of $\beta = 0$ if $\gamma = 0$.

Other conditions of Theorem 7.4.1 hold as it was shown in the proof of Theorem 8.4.1.

By applying Theorem 7.4.1 we get the convergence relation (8.102). □

The condition $\mathbf{D}_{12}[\beta]$ used in Theorem 7.4.2 should be replaced by analog of the condition $\mathbf{D}_5[\beta]$ with the function $A(\beta) = K_{16}K_{17}(\beta + \frac{1}{2}\beta^2)$, for some $\beta \geq 0$:
$\mathbf{D}_{17}[\beta]$: $\mathsf{E}\exp\{A_N(\beta)|Y_{0,0}|\} < K_{81}$, for some $1 < K_{81} < \infty$.

Theorem 8.4.4. *Let a Markov log-price process $Y_{0,n}$ and the corresponding space-skeleton approximating Markov log-price processes $Y_{\varepsilon,n}$ are defined, respectively, by stochastic transition dynamic relations* (8.91) *and* (8.95). *Let also the conditions* $\mathbf{B}_4[\gamma]$ *and* $\mathbf{D}_{17}[\beta]$ *hold with parameters γ and β such that either $\beta > \gamma > 0$ or $\beta = \gamma = 0$, and also the conditions* $\mathbf{G}_2$, $\mathbf{N}_1$, $\mathbf{I}_6$, $\mathbf{O}_4$, *and* $\mathbf{K}_{14}$ *hold. Then, the following relation takes place:*

$$\Phi_\varepsilon \to \Phi_0 \quad as \quad \varepsilon \to 0\,. \tag{8.103}$$

Proof. Theorem 8.4.4 is a corollary of Theorem 7.4.2. All conditions of the latter theorem hold as was shown in the proofs of Theorems 8.4.1–8.4.3.

By applying Theorem 7.4.2, we get the convergence relation (8.103). □

8.4.2 Space-skeleton approximations for option rewards of multivariate modulated Markov Gaussian log-price processes

Let us assume that a log-price process $\vec{Z}_{0,n} = (\vec{Y}_{0,n}, X_{0,n})$, $n = 0, 1, \dots$ is a multivariate modulated Markov Gaussian log-price process given by the following stochastic transition dynamic relation:

$$\begin{cases} \vec{Y}_{0,n} = \vec{Y}_{0,n-1} + \vec{\mu}_{0,n}(\vec{Y}_{0,n-1}, X_{0,n-1}) + \Sigma_{0,n}(\vec{Y}_{0,n-1}, X_{0,n-1})\vec{W}_{0,n}\,, \\ X_{0,n} = C_{0,n}(X_{0,n-1}, U_{0,n})\,, \\ \quad n = 1, 2, \dots\,, \end{cases} \tag{8.104}$$

where (a) $\vec{Z}_{0,0} = (\vec{Y}_{0,0}, X_{0,0}) = ((Y_{0,0,1}, \dots, Y_{0,0,k}), X_{0,0})$ is a random vector taking values in the space $\mathbb{Z} = \mathbb{R}_k \times \mathbb{X}$, (b) $(\vec{W}_{0,n}, U_{0,n}) = ((W_{0,n,1}, \dots, W_{0,n,k}), U_{0,n})$, $n = 1, 2, \dots$ is a sequence of independent random vectors taking values in the space $\mathbb{R}_k \times \mathbb{U}$, where $\mathbb{U}$ is a measurable space with a σ-algebra of measurable subsets $\mathcal{B}_{\mathbb{U}}$, (c) the random vector $(\vec{Y}_{0,0}, X_{0,0})$ and the random sequence $(\vec{W}_{0,n}, U_{0,n})$, $n = 1, 2, \dots$ are independent, (d) the random vector $\vec{W}_{0,n} = (W_{0,n,1}, \dots, W_{0,n,k})$ has a standard multivariate normal distribution with $\mathsf{E}W_{0,n,i} = 0, \mathsf{E}W_{0,n,i}W_{0,n,j} = I(i = j)$, i, $j = 1, \dots, k$, for every $n = 1, 2, \dots$, (e) the random variables $U_{0,n}$ have a regular conditional distribution $G_{\vec{w}}(B) = \mathsf{P}\{U_{0,n} \in B / \vec{W}_{0,n} = \vec{w}\}$, for every $n = 1, 2, \dots$, (f) $\vec{\mu}_{0,n}(\vec{y}, x) = (\mu_{0,n,1}(\vec{y}, x), \dots, \mu_{0,n,k}(\vec{y}, x))$, $n = 1, 2, \dots$ are vector functions, which components are measurable functions acting from the space $\mathbb{Z} = \mathbb{R}_k \times \mathbb{X}$ to $\mathbb{R}_1$, (g) $\Sigma_{0,n}(\vec{y}, x) = \|\sigma_{0,n,i,j}(\vec{y}, x)\|$, $n = 1, 2, \dots$ are $k \times k$ matrix functions with elements $\sigma_{0,n,i,j}(\vec{y}, x)$, i, $j = 1, \dots, k$, which are measurable functions acting from the space $\mathbb{Z} = \mathbb{R}_k \times \mathbb{X}$ to $\mathbb{R}_1$, and (h) $C_{0,n}(x, u)$ is a measurable function acting from the space $\mathbb{X} \times \mathbb{U}$ to the space $\mathbb{X}$.

Let us construct the corresponding space-skeleton approximating processes $\vec{Z}_{\varepsilon,n}$, $n = 0, 1, \dots$, for $\varepsilon \in (0, \varepsilon_0]$, according to the algorithm described in Subsection 7.2.3.

Let $\delta_{\varepsilon,n,i} > 0$, $\lambda_{\varepsilon,n,i} \in \mathbb{R}_1$, $i = 1, \dots, k$, $n = 0, 1, \dots$ and $m^-_{\varepsilon,n,j} \leq m^+_{\varepsilon,n,j}$, $j = 0, \dots, k$, $n = 0, 1, \dots$ be integer numbers.

First, the intervals $\mathbb{I}_{\varepsilon,n,i,l}$ should be constructed for $l_i = m^-_{\varepsilon,n,i}, \dots, m^+_{\varepsilon,n,i}$, $i = 1, \dots, k$, $n = 0, 1, \dots$,

$$\mathbb{I}_{\varepsilon,n,i,l} = \begin{cases} (-\infty, \delta_{\varepsilon,n,i}(m^-_{\varepsilon,n,i} + \frac{1}{2})] + \lambda_{\varepsilon,n,i} & \text{if } l = m^-_{\varepsilon,n,i}, \\ (\delta_{\varepsilon,n,i}(l - \frac{1}{2}), \delta_{\varepsilon,n,i}(l + \frac{1}{2})] + \lambda_{\varepsilon,n,i} & \text{if } m^-_{\varepsilon,n,i} < l < m^+_{\varepsilon,n,i}, \\ (\delta_{\varepsilon,n,i}(m^+_{\varepsilon,n,i} - \frac{1}{2}), \infty) + \lambda_{\varepsilon,n,i} & \text{if } l = m^+_{\varepsilon,n,i}. \end{cases} \tag{8.105}$$

Then cubes $\mathbb{I}_{\varepsilon,n,l_1,\dots,l_k} = \mathbb{I}_{\varepsilon,n,1,l_1} \times \dots \times I_{\varepsilon,n,k,l_k}$, $l_i = m^-_{\varepsilon,n,i}, \dots, m^+_{\varepsilon,n,i}$, $i = 1, \dots, k$, $n = 0, 1, \dots$ should be defined.

By the definition, the skeleton points $y_{\varepsilon,n,i,l_i} = l_i\delta_{\varepsilon,n,i} + \lambda_{\varepsilon,n,i} \in \mathbb{I}_{\varepsilon,n,i,l_i}$ and, thus, vector points $(y_{\varepsilon,n,1,l_1}, \dots, y_{\varepsilon,n,k,l_k}) \in \mathbb{I}_{\varepsilon,n,l_1,\dots,l_k}$, for $l_i = m^-_{\varepsilon,n,i}, \dots, m^+_{\varepsilon,n,i}$, $i = 1, \dots, k$, $n = 0, 1, \dots$.

Second, nonempty sets $\mathbb{J}_{\varepsilon,n,l} \subset \mathcal{B}_{\mathbb{X}}$, $l = m^-_{\varepsilon,n,0}, \dots, m^+_{\varepsilon,n,0}$, $n = 0, 1, \dots$, such that (a) $\mathbb{J}_{\varepsilon,n,l'} \cap \mathbb{J}_{\varepsilon,n,l''} = \emptyset$, $l' \neq l''$, $n = 0, 1, \dots, N$; (b) $\mathbb{X} = \cup_{m^-_{\varepsilon,n,0} \leq l \leq m^+_{\varepsilon,n,0}} \mathbb{J}_{\varepsilon,n,l}$, $n = 0, 1, \dots$, should be constructed.

Remind that one of our model assumption is the $\mathbb{X}$ which is a Polish metric space. The sets $\mathbb{K}_{\varepsilon,n}$, $n = 0, 1, \dots$, should be chosen and then the nonempty sets $\mathbb{K}_{\varepsilon,n,l} \subseteq \mathcal{B}_{\mathbb{X}}$, $l = m^-_{\varepsilon,n,0}, \dots, m^+_{\varepsilon,n,0}$, $n = 0, 1, \dots$, such that (c) $\mathbb{K}_{\varepsilon,n,l'} \cap \mathbb{K}_{\varepsilon,n,l''} = \emptyset$, $l' \neq l''$, $n = 0, 1, \dots$; (d) $\cup_{m^-_{\varepsilon,n,0} \leq l \leq m^+_{\varepsilon,n,0}} \mathbb{K}_{\varepsilon,n,l} = \mathbb{K}_{\varepsilon,n}$, $n = 0, 1, \dots$.

The sets $\mathbb{J}_{\varepsilon,n,l}$ can be defined in the following way, for $n = 0, 1, \dots$,

$$\mathbb{J}_{\varepsilon,n,l} = \begin{cases} \mathbb{K}_{\varepsilon,n,l} & \text{if } m^-_{\varepsilon,n,0} \leq l < m^+_{\varepsilon,n,0}, \\ \mathbb{K}_{\varepsilon,n,m^+_{\varepsilon,n,0}} \cup \overline{\mathbb{K}}_{\varepsilon,n} & \text{if } l = m^+_{\varepsilon,n,0}. \end{cases} \tag{8.106}$$

Then, skeleton points $x_{\varepsilon,n,l} \in \mathbb{J}_{\varepsilon,n,l}$, $l = m^-_{\varepsilon,n,0}, \dots, m^+_{\varepsilon,n,0}$, $n = 0, 1, \dots, N$ should be chosen.

Third, skeleton sets $\mathbb{A}_{\varepsilon,n,\bar{l}}$ and skeleton points, $\vec{z}_{\varepsilon,n,\bar{l}} \in \mathbb{A}_{\varepsilon,n,\bar{l}}$ can be defined for $\bar{l} \in \mathbb{L}_{\varepsilon,n} = \{\bar{l} = (l_0, \dots, l_k), l_j = m^-_{\varepsilon,n,j}, \dots, m^+_{\varepsilon,n,j}, j = 0, \dots, k\}$, $n = 0, 1, \dots$, in the following way:

$$\mathbb{A}_{\varepsilon,n,\bar{l}} = \mathbb{I}_{\varepsilon,n,l_1,\dots,l_k} \times \mathbb{J}_{\varepsilon,n,l_0}, \tag{8.107}$$

and

$$\vec{z}_{\varepsilon,n,\bar{l}} = (\vec{y}_{\varepsilon,n,\bar{l}}, x_{\varepsilon,n,\bar{l}}) = ((y_{\varepsilon,n,1,l_1}, \dots, y_{\varepsilon,n,k,l_k}), x_{\varepsilon,n,l_0}). \tag{8.108}$$

Fourth, skeleton functions, $h_{\varepsilon,n,i}(y)$, $y \in \mathbb{R}_1$, $i = 1, \dots, k$, $n = 0, 1, \dots$, should be defined by the following formulas:

$$h_{\varepsilon,n,i}(y) = \big\{ y_{\varepsilon,n,i,l} \quad \text{if } y \in \mathbb{I}_{\varepsilon,n,i,l}, \quad m^-_{\varepsilon,n,i} \leq l \leq m^+_{\varepsilon,n,i}, \tag{8.109}$$

and skeleton functions $h_{\varepsilon,n,0}(x)$, $x \in \mathbb{X}$, $n = 0, 1, \dots$, should be defined by the following formula:

$$h_{\varepsilon,n,0}(x) = \big\{ x_{\varepsilon,n,l} \quad \text{if } x \in \mathbb{J}_{\varepsilon,n,l}, \quad m^-_{\varepsilon,n,0} \leq l \leq m^+_{\varepsilon,n,0}. \tag{8.110}$$

Finally, vector skeleton functions $\hat{h}_{\varepsilon,n}(\vec{y})$, $\vec{y} = (y_1, \dots, y_k) \in \mathbb{R}_k$, $n = 0, 1, \dots, N$ should be defined by the following formula:

$$\hat{h}_{\varepsilon,n}(\vec{y}) = (h_{\varepsilon,n,1}(y_1), \dots, h_{\varepsilon,n,k}(y_k)), \tag{8.111}$$

and then vector skeleton functions $\bar{h}_{\varepsilon,n}(\vec{z})$, $\vec{z} = (\vec{y}, x) \in \mathbb{Z}$, $n = 0, 1, \dots, N$ can be defined by the following formula:

$$\bar{h}_{\varepsilon,n}(\vec{z}) = (\hat{h}_{\varepsilon,n}(\vec{y}), h_{\varepsilon,n,0}(x)). \tag{8.112}$$

The corresponding space-skeleton approximating modulated Markov processes $\vec{Z}_{\varepsilon,n} = (\vec{Y}_{\varepsilon,n}, X_{\varepsilon,n})$ are defined by the following stochastic transition dynamic relations:

$$\begin{cases} \vec{Y}_{\varepsilon,n} = \hat{h}_{\varepsilon,n}\Big(\hat{h}_{\varepsilon,n-1}(\vec{Y}_{\varepsilon,n-1}) \\ \qquad + \vec{\mu}_{0,n}(\hat{h}_{\varepsilon,n-1}(\vec{Y}_{\varepsilon,n-1}), h_{\varepsilon,n-1,0}(X_{\varepsilon,n-1})) \\ \qquad + \Sigma_{0,n}(\hat{h}_{\varepsilon,n-1}(\vec{Y}_{\varepsilon,n-1}), h_{\varepsilon,n-1,0}(X_{\varepsilon,n-1}))\, \vec{W}_{0,n}\Big), \\ X_{\varepsilon,n} = h_{\varepsilon,n,0}\left(C_{0,n}(h_{\varepsilon,n-1,0}(X_{\varepsilon,n-1}), U_{0,n})\right), \\ \quad n = 1, 2, \dots, \\ \vec{Y}_{\varepsilon,0} = \hat{h}_{\varepsilon,0}(\vec{Y}_{0,0}), \quad X_{\varepsilon,0} = h_{\varepsilon,0,0}(X_{0,0}). \end{cases} \tag{8.113}$$

The approximating log-price process $\vec{Z}_{\varepsilon,n}$ is, for every $\varepsilon \in (0, \varepsilon_0]$, a skeleton Markov chain with one-step transition probabilities $P_{\varepsilon,n}(\vec{z}, A)$, $\vec{z} = (\vec{y}, x) \in \mathbb{Z}$, $A \in \mathcal{B}_{\mathbb{Z}}$, $n = 1, 2, \dots$ given by the following formula:

$$\begin{aligned} P_{\varepsilon,n}(\vec{z}, A) &= \sum_{\vec{z}_{\varepsilon,n,\bar{l}} \in A} P_{0,n}(\bar{h}_{\varepsilon,n-1}(\vec{z}), \mathbb{A}_{\varepsilon,n,\bar{l}}) \\ &= \sum_{\vec{z}_{\varepsilon,n,\bar{l}} \in A} \mathsf{P}\Big\{\Big(\hat{h}_{\varepsilon,n-1}(\vec{y}) + \vec{\mu}_{0,n}(\hat{h}_{\varepsilon,n-1}(\vec{y}), h_{\varepsilon,n-1,0}(x)) \\ &\qquad + \Sigma_{0,n}(\hat{h}_{\varepsilon,n-1}(\vec{y}), h_{\varepsilon,n-1,0}(x))\, \vec{W}_{0,n}, \\ &\qquad C_{0,n}(h_{\varepsilon,n-1,0}(x), U_{0,n})\Big) \in \mathbb{A}_{\varepsilon,n,\bar{l}}\Big\}. \end{aligned} \tag{8.114}$$

As far as the initial distribution $P_{\varepsilon,0}(A) = \mathsf{P}\{\vec{Z}_{\varepsilon,0} \in A\}$, $A \in \mathcal{B}_{\mathbb{Z}}$ is concerned, it takes, for every $\varepsilon \in (0, \varepsilon_0]$, the following form:

$$\begin{aligned} P_{\varepsilon,0}(A) &= \sum_{\vec{z}_{\varepsilon,0,\bar{l}} \in A} P_{0,0}(\mathbb{A}_{\varepsilon,0,\bar{l}}) \\ &= \sum_{\vec{z}_{\varepsilon,0,\bar{l}} \in A} \mathsf{P}\{(\vec{Y}_{0,0}, X_{0,0}) \in \mathbb{A}_{\varepsilon,0,\bar{l}}\}. \end{aligned} \tag{8.115}$$

In this case, a pay-off function $g_\varepsilon(n, e^{\vec{y}}, x) = g(n, e^{\vec{y}}, x)$ does not depend on parameter ε.

Let $\phi_{\varepsilon,n}(\vec{y}, x)$ be the corresponding log-reward functions for the log-price processes $\vec{Z}_{\varepsilon,n}$.

The following lemma is the direct corollary of Lemma 7.2.2.

Lemma 8.4.2. *Let, for every $\varepsilon \in (0, \varepsilon_0]$, the log-price process $\vec{Z}_{\varepsilon,n}$ is a space-skeleton Markov chain with the transition probabilities and the initial distribution defined, respectively, in relations (8.114) and (8.115). Then the reward functions $\phi_{\varepsilon,n}(\vec{y}, x)$ and $\phi_{n+r}(\vec{y}_{\varepsilon,n,\bar{l}}, x_{\varepsilon,n,\bar{l}})$, for the points $(\vec{y}_{\varepsilon,n,\bar{l}}, x_{\varepsilon,n,\bar{l}})$, $\bar{l} \in \mathbb{L}_{\varepsilon,n+r}$, $r = 1, \dots N - n$, given by formulas (8.108), are, for every $\vec{z} = (\vec{y}, x) \in \mathbb{A}_{\varepsilon,n,\bar{l}''}$, $\bar{l}'' \in \mathbb{L}_{\varepsilon,n}$, $n = 0, \dots, N$, the unique solution for the following recurrence finite system of linear equations:*

$$\begin{cases} \phi_{\varepsilon,N}(\vec{y}_{\varepsilon,N,\bar{l}}, x_{\varepsilon,N,\bar{l}}) = g(N, e^{\vec{y}_{\varepsilon,N,l}}, x_{\varepsilon,N,\bar{l}}) \,, \quad \bar{l} \in \mathbb{L}_{\varepsilon,N} \,, \\ \phi_{\varepsilon,n+r}(\vec{y}_{\varepsilon,n+r,\bar{l}}, x_{\varepsilon,n+r,\bar{l}}) = \max\Bigg(g(n+r, e^{\vec{y}_{\varepsilon,n+r,\bar{l}}}, x_{\varepsilon,n+r,\bar{l}}) \,, \\ \qquad \sum\limits_{\bar{l}' \in \mathbb{L}_{\varepsilon,n+r+1}} \phi_{\varepsilon,n+r+1}(\vec{y}_{\varepsilon,n+r+1,\bar{l}'}, x_{\varepsilon,n+r+1,\bar{l}'}) \\ \qquad \times P_{0,n+r}(\vec{z}_{\varepsilon,n+r,l}, \mathbb{A}_{\varepsilon,n+r+1,\bar{l}'}) \Bigg) \,, \\ \qquad \bar{l} \in \mathbb{L}_{\varepsilon,n+r} \,, \quad r = N - n - 1, \dots, 1 \,, \\ \phi_{\varepsilon,n}(\vec{y}, x) = \max\Bigg(g(n, e^{\vec{y}}, x) \,, \\ \qquad \sum\limits_{\bar{l}' \in \mathbb{L}_{\varepsilon,n+1}} \phi_{\varepsilon,n+1}(\vec{y}_{\varepsilon,n+1,\bar{l}'}, x_{\varepsilon,n+1,\bar{l}'}) P_{0,n}(\vec{z}_{\varepsilon,n,\bar{l}''}, \mathbb{A}_{\varepsilon,n+1,\bar{l}'}) \Bigg) \,. \end{cases} \tag{8.116}$$

Let us now formulate conditions of convergence for the above log-reward functions.

We are going to apply Theorem 7.2.3.

We assume that the condition $\mathbf{I}_7$ holds for the pay-off function $g(n, e^y, x)$.

The condition $\mathbf{J}_8$ should be replaced in this case by the following condition:

$\mathbf{O}_5$: There exist sets $\mathbb{Z}''_n \in \mathcal{B}_{\mathbb{Z}}$, $n = 0, \dots, N$ and $\mathbb{U}_n \in \mathcal{B}_{\mathbb{U}}$, $n = 1, \dots, N$ such that
- (a) $\vec{\mu}_{0,n}(\vec{y}_\varepsilon, x_\varepsilon) \to \vec{\mu}_{0,n}(\vec{y}_0, x_0)$, $\Sigma_{0,n}(\vec{y}_\varepsilon, x_\varepsilon) \to \Sigma_{0,n}(\vec{y}_0, x_0)$, and $C_{0,n}(x_\varepsilon, u) \to C_{0,n}(x_0, u)$ as $\varepsilon \to 0$ for any $\vec{z}_\varepsilon = (\vec{y}_\varepsilon, x_\varepsilon) \to \vec{z}_0 = (\vec{y}_0, x_0) \in \mathbb{Z}''_{n-1}$ as $\varepsilon \to 0$, and $u \in \mathbb{U}_n$ and $n = 1, \dots, N$;
- (b) $\mathsf{P}\{U_{0,n} \in \mathbb{U}_n\} = 1$, $n = 1, \dots, N$;
- (c) $\mathsf{P}\{\big(\vec{y}_0 + \vec{\mu}_{0,n}(\vec{y}_0, x_0) + \Sigma_{0,n}(\vec{y}_0, x_0)\vec{W}_{0,n}, C_{0,n}(x_0, U_{0,n})\big) \in \mathbb{Z}'_n \cap \mathbb{Z}''_n\} = 1$, for every $\vec{z}_0 = (\vec{y}_0, x_0) \in \mathbb{Z}'_{n-1} \cap \mathbb{Z}''_{n-1}$ and $n = 1, \dots, N$, where $\mathbb{Z}'_n$, $n = 0, \dots, N$ are sets introduced in the condition $\mathbf{I}_7$.

Using the definition of the model, we can write the following formula:

$$\mathsf{P}\left\{\big(\vec{y_0} + \vec{\mu}_{0,n}(\vec{y}_0, x_0) + \Sigma_{0,n}(\vec{y}_0, x_0)\vec{W}_{0,n}, C_{0,n}(x_0, U_{0,n})\big) \in \overline{\mathbb{Z}}'_n \cup \overline{\mathbb{Z}}''_n\right\} \\ = \int\limits_{\mathbb{R}_k} g_n(\vec{z_0}, \vec{w}) \frac{1}{(\sqrt{2\pi})^k} e^{-\sum_{i=1}^k \frac{w_i^2}{2}} \, d\vec{w} \,, \tag{8.117}$$

where

$$g_n(\vec{z}_0, \vec{w}) = \int_{\mathbb{U}} I\Big((\vec{y}_0 + \vec{\mu}_{0,n}(\vec{y}_0, x_0) + \Sigma_{0,n}(\vec{y}_0, x_0)\vec{w}, \\ C_{0,n}(x_0, u)) \in \overline{\mathbb{Z}}'_n \cup \overline{\mathbb{Z}}''_n \Big) G_{\vec{w}}(du). \tag{8.118}$$

Let us also introduce sets

$$\mathbb{G}_{n,\vec{z}_0} = \{\vec{w} : g_n(\vec{z}_0, \vec{w}) > 0\}. \tag{8.119}$$

Then, condition $\mathbf{O}_5$ (c) can be rewritten in the form of the following relation, which should be required to hold for $\vec{z}_0 = (\vec{y}_0, x_0) \in \mathbb{Z}'_{n-1} \cap \mathbb{Z}''_{n-1}$ and $n = 1, \dots, N$:

$$L_k(\mathbb{G}_{n,\vec{z}_0}) = 0. \tag{8.120}$$

This form of the condition $\mathbf{O}_5$ (c) reduces to the condition $\mathbf{O}_4$ (b) for the univariate model without index component considered in Subsection 8.4.1.

The following theorem is a corollary of Theorems 7.3.1.

Theorem 8.4.5. *Let the multivariate modulated Markov Gaussian log-price process $\vec{Z}_{0,n}$ and the corresponding space-skeleton approximating Markov log-price processes $\vec{Z}_{\varepsilon,n}$ are defined, respectively, by stochastic transition dynamic relations* (8.104) *and* (8.113). *Let also the conditions $\mathbf{B}_1[\bar{\gamma}]$ hold for some vector parameter $\bar{\gamma} = (\gamma_1, \dots, \gamma_k)$ with non-negative components, and, also, the conditions $\mathbf{G}_3$, $\mathbf{N}_2$, $\mathbf{I}_7$, and $\mathbf{O}_5$ hold. Then, for every $n = 0, 1, \dots, N$, the following relation takes place for any $\vec{z}_\varepsilon = (\vec{y}_\varepsilon, x_\varepsilon) \to \vec{z}_0 = (\vec{y}_0, x_0) \in \mathbb{Z}'_n \cap \mathbb{Z}''_n$,*

$$\phi_{\varepsilon,n}(\vec{y}_\varepsilon, x_\varepsilon) \to \phi_{0,n}(\vec{y}_0, x_0) \quad as \quad \varepsilon \to 0. \tag{8.121}$$

Proof. The fitting condition $\mathbf{L}_4$ holds since the corresponding transition probabilities are defined according to formula (8.114).

The condition $\mathbf{C}_{12}[\bar{\beta}]$ holds for any vector parameter $\bar{\beta} = (\beta_1, \dots, \beta_k)$ with non-negative components that follows from Lemmas 4.1.8 and 4.5.1. We can always choose a vector $\bar{\beta}$ such that, for every $i = 1, \dots, k$, either $\beta_i > \gamma_i$ if $\gamma_i > 0$ of $\beta_i = 0$ if $\gamma_i = 0$.

The condition $\mathbf{O}_5$ (a) and (b) imply that the following relation holds any $\vec{z}_\varepsilon = (\vec{y}_\varepsilon, x_\varepsilon) \to \vec{z}_0 = (\vec{y}_0, x_0) \in \mathbb{Z}''_n$ as $\varepsilon \to 0$, and $n = 1, \dots, N$:

$$\Big(\vec{y_\varepsilon} + \vec{\mu}_{0,n}(\vec{y}_\varepsilon, x_\varepsilon) + \Sigma_{0,n}(\vec{y}_\varepsilon, x_\varepsilon)\vec{W}_{0,n}, C_{0,n}(x_\varepsilon, U_{0,n})\Big) \\ \xrightarrow{\text{a.s.}} \Big(\vec{y_0} + \vec{\mu}_{0,n}(\vec{y}_0, x_0) + \Sigma_{0,n}(\vec{y}_0, x_0)\vec{W}_{0,n}, C_{0,n}(x_0, U_{0,n})\Big) \quad \text{as} \quad \varepsilon \to 0. \tag{8.122}$$

Relation (8.122) obviously implies that the following relation holds any $\vec{z}_\varepsilon = (\vec{y}_\varepsilon, x_\varepsilon) \to \vec{z}_0 = (\vec{y}_0, x_0) \in \mathbb{Z}''_n$ and $n = 1, \dots, N$:

$$\Big(\vec{y_\varepsilon} + \vec{\mu}_{0,n}(\vec{y}_\varepsilon, x_\varepsilon) + \Sigma_{0,n}(\vec{y}_\varepsilon, x_\varepsilon)\vec{W}_{0,n}, C_{0,n}(x_\varepsilon, U_{0,n})\Big) \\ \xrightarrow{d} \Big(\vec{y_0} + \vec{\mu}_{0,n}(\vec{y}_0, x_0) + \Sigma_{0,n}(\vec{y}_0, x_0)\vec{W}_{0,n}, C_{0,n}(x_0, U_{0,n})\Big) \quad \text{as} \quad \varepsilon \to 0. \tag{8.123}$$

Relation (8.123) is a variant of the condition $\mathbf{J}_{10}$ (a) . The condition $\mathbf{O}_5$ (b) is just a variant of the condition $\mathbf{J}_{10}$ (b) .

Thus, all conditions of Theorem 7.3.1 hold. By applying this theorem we get the convergence relation (8.121). □

Let us now formulate conditions of convergence for the corresponding optimal expected rewards.

In this case, the optimal expected reward functional Φ_ε for the approximation log-price processes $\vec{Z}_{\varepsilon,n}$ can be found using the following formula:

$$\Phi_\varepsilon = \sum_{\bar{l} \in \mathbb{L}_{\varepsilon,0}} \phi_{\varepsilon,0}(\vec{z}_{\varepsilon,0,\bar{l}}) p_{\varepsilon,\bar{l}}, \tag{8.124}$$

where $p_{\varepsilon,\bar{l}} = P_{0,0}(\mathbb{A}_{\varepsilon,0,\bar{l}}) = \mathsf{P}\{\vec{h}_{\varepsilon,0}(\vec{Z}_{0,0}) = \vec{z}_{\varepsilon,0,\bar{l}}\}, \bar{l} = (l_0, \dots, l_k) \in \mathbb{L}_{\varepsilon,0}$.

We are going to apply Theorem 7.3.2.

The condition $\mathbf{D}_{11}[\bar{\beta}]$ should be replaced by the following condition, assumed to hold for some vector parameter $\bar{\beta} = (\beta_1, \dots, \beta_k)$ with nonnegative components:

$\mathbf{D}_{18}[\bar{\beta}]$: $\mathsf{E}e^{\beta_i |Y_{0,0,i}|} < K_{82,i}$, $i = 1, \dots, k$, for some $1 < K_{82,i} < \infty$, $i = 1, \dots, k$.

The condition $\mathbf{K}_9$ should be replaced by the following condition imposed on the initial distribution $\mathsf{P}\{Z_{0,0} \in A\} = P_{0,0}(A)$:

$\mathbf{K}_{15}$: $P_{0,0}(\mathbb{Z}'_0 \cap \mathbb{Z}''_0) = 1$, where $\mathbb{Z}'_0$ and $\mathbb{Z}''_0$ are the sets introduced, respectively, in the conditions $\mathbf{I}_7$ and $\mathbf{O}_5$.

Theorem 8.4.6. *Let the multivariate modulated Markov Gaussian log-price process $\vec{Z}_{0,n}$ and the corresponding space-skeleton approximating Markov log-price processes $\vec{Z}_{\varepsilon,n}$ are defined, respectively, by stochastic transition dynamic relations* (8.104) *and* (8.113). *Let also the conditions $\mathbf{B}_1[\bar{\gamma}]$ and $\mathbf{D}_{18}[\bar{\beta}]$ hold for some vector parameters $\bar{\gamma} = (\gamma_1, \dots, \gamma_k)$ and $\bar{\beta} = (\beta_1, \dots, \beta_k)$ such that, for every $i = 1, \dots, k$ either $\beta_i > \gamma_i > 0$ or $\beta_i = \gamma_i = 0$, and also the conditions $\mathbf{G}_3$, $\mathbf{N}_2$, $\mathbf{I}_7$, $\mathbf{O}_5$, and $\mathbf{K}_{15}$ hold. Then, the following relation takes place:*

$$\Phi_\varepsilon \to \Phi_0 \quad \text{as} \quad \varepsilon \to 0\,. \tag{8.125}$$

Proof. The fitting condition $\mathbf{M}_4$ holds since the corresponding initial distributions are defined according to formula (8.115).

The condition $\mathbf{K}_{15}$ is just reformulation of the condition $\mathbf{K}_9$.

Other conditions of Theorem 7.3.2 also hold that was shown in the proof of Theorem 8.4.5. By applying Theorem 7.3.2 we get the convergence relations (8.125). □

Analogous results can be formulated for multidimensional modulated Markov Gaussian log-price processes with unbounded characteristics.

Theorem 8.4.7. *Let the multivariate modulated Markov Gaussian log-price process $\vec{Z}_{0,n}$ and the corresponding space-skeleton approximating Markov log-price processes $\vec{Z}_{\varepsilon,n}$*

are defined, respectively, by stochastic transition dynamic relations (8.104) *and* (8.113). *Let also the conditions* $\mathbf{B}_1[\bar{\gamma}]$ *hold for some vector parameter* $\bar{\gamma} = (\gamma_1, \dots, \gamma_k)$ *with nonnegative components, and, also, the conditions* $\mathbf{G}_4$, $\mathbf{N}_2$, $\mathbf{I}_7$, *and* $\mathbf{O}_5$ *hold. Then, for every* $n = 0, 1, \dots, N$, *the following relation takes place for any* $\vec{z}_\varepsilon = (\vec{y}_\varepsilon, x_\varepsilon) \to \vec{z}_0 = (\vec{y}_0, x_0) \in \mathbb{Z}'_n \cap \mathbb{Z}''_n$:

$$\phi_{\varepsilon,n}(\vec{y}_\varepsilon, x_\varepsilon) \to \phi_{0,n}(\vec{y}_0, x_0) \quad \text{as} \quad \varepsilon \to 0\,. \tag{8.126}$$

Proof. The proof repeats the proof of Theorem 8.4.5. The only change is that Theorem 7.5.3 should be used instead of Theorem 7.3.1.

In this case, the condition $\mathbf{C}_{16}[\bar{\beta}]$ holds for any vector parameter $\bar{\beta} = (\beta_1, \dots, \beta_k)$ with nonnegative components that follows from Lemmas 4.3.17 and 4.5.2. We can always choose a vector $\bar{\beta}$ such that, for every $i = 1, \dots, k$, either $\beta_i > \gamma_i$ if $\gamma_i > 0$ of $\beta_i = 0$ if $\gamma_i = 0$.

Other conditions of Theorem 7.5.3 hold as it was shown in the proof of Theorem 8.4.5.

By applying Theorem 7.5.3 we get the convergence relation (8.126). □

Let us also give conditions of convergence for optimal expected rewards Φ_ε.

In this case, the corresponding functions $\vec{A}(\bar{\beta}) = (A_1(\bar{\beta}), \dots, A_k(\bar{\beta}))$ given by the following formula, for $\bar{\beta} = (\beta_1, \dots, \beta_k)$, $\beta_1, \dots, \beta_k \geq 0$,

$$A_j(\bar{\beta}) = K_{26} K_{27,j} \sum_{l=1}^{k} (\beta_l + \tfrac{1}{2} k^2 \beta_l^2)\,, \quad j = 1, \dots, k\,, \tag{8.127}$$

where $K_{26}, K_{27,j}$, $j = 1, \dots, k$ are constants penetrating condition $\mathbf{G}_4$.

The condition $\mathbf{D}_6[\bar{\beta}]$, with the above functions $\vec{A}(\bar{\beta})$, should assumed to hold for some vector parameter $\bar{\beta} = (\beta_1, \dots, \beta_k)$ with nonnegative components and the corresponding vectors $\bar{\beta}_i = (\beta_{i,1}, \dots, \beta_{i,k})$ with components $\beta_{i,j} = I(i = j)$, $i, j = 1, \dots, k$. It takes, in this case, the following form:

$\mathbf{D}_{19}[\bar{\beta}]$: $\mathsf{E} \exp\{\sum_{j=1}^{k} A_{N,j}(\bar{\beta}_i) |Y_{0,0,j}|\} < K_{83,i}$, $i = 1, \dots, k$, for some $1 < K_{83,i} < \infty$, $i = 1, \dots, k$.

Theorem 8.4.8. *Let the multivariate modulated Markov Gaussian log-price process* $\vec{Z}_{0,n}$ *and the corresponding space-skeleton approximating Markov log-price processes* $\vec{Z}_{\varepsilon,n}$ *are defined, respectively, by stochastic transition dynamic relations* (8.104) *and* (8.113). *Let also the conditions* $\mathbf{B}_1[\bar{\gamma}]$ *and* $\mathbf{D}_{19}[\bar{\beta}]$ *hold for some vector parameters* $\bar{\gamma} = (\gamma_1, \dots, \gamma_k)$ *and* $\bar{\beta} = (\beta_1, \dots, \beta_k)$ *such that, for every* $i = 1, \dots, k$ *either* $\beta_i > \gamma_i > 0$ *or* $\beta_i = \gamma_i = 0$, *and also the conditions* $\mathbf{G}_4$, $\mathbf{N}_2$, $\mathbf{I}_7$, $\mathbf{O}_5$, *and* $\mathbf{K}_{15}$ *hold. Then, the following relation takes place:*

$$\Phi_\varepsilon \to \Phi_0 \quad \text{as} \quad \varepsilon \to 0\,. \tag{8.128}$$

Proof. Theorem 8.4.8 is a corollary of Theorem 7.5.4. All conditions of the latter theorem hold as was shown in the proofs of Theorems 8.4.5–8.4.7.

By applying Theorem 7.5.4, we get the convergence relation (8.128). □

8.4.3 Space-skeleton approximations for option rewards of multivariate modulated autoregressive type log-price processes with Gaussian noise terms

All models of multivariate modulated autoregressive type and autoregressive stochastic volatility type log-price processes with Gaussian noise terms introduced in Sections 1.3 and 1.4 can be considered as particular cases of multivariate modulated Markov Gaussian log-price processes.

Thus let one apply space-skeleton reward approximations presented in Subsections 8.4.1 and 8.4.2 to such autoregressive-type log-price processes. The corresponding results will be given in the second volume of the present book.

8.5 LPP represented by Gaussian random walks

In this section, we present results concerning space-skeleton approximations for reward of American-type options for log-price processes represented by univariate and multivariate modulated Gaussian random walks.

8.5.1 Space-skeleton approximations for option rewards of log-price processes represented by univariate Gaussian random walks

Let the univariate Markov Gaussian log-price process $Y_{0,n}$, $n = 0, 1, \dots$ be defined by the following stochastic transition dynamic relation:

$$Y_{0,n} = Y_{0,n-1} + \mu_{0,n} + \sigma_{0,n} W_{0,n}, \quad n = 1, 2, \dots, \tag{8.129}$$

where (a) $Y_{0,0}$ is a real-valued random variable, (b) $W_{0,n}$, $n = 1, 2, \dots$ is a sequence of standard i.i.d. normal random variables with mean 0 and variance 1, (c) the random variable $Y_{0,0}$ and sequence of random variables $W_{0,n}$, $n = 1, 2, \dots$ are independent, and (d) $\mu_{0,n}, \sigma_{0,n}$, $n = 1, 2, \dots$ are real numbers.

In this case, an alternative space-skeleton approximation scheme with an additive skeleton structure can also be used.

Let us construct the corresponding space-skeleton approximating processes $Y_{\varepsilon,n}$, $n = 0, 1, \dots$, for $\varepsilon \in (0, \varepsilon_0]$, according to the algorithm described in Subsection 7.3.1.

Let us take the space-skeleton parameters $\delta_\varepsilon > 0$, $\lambda_{\varepsilon,n} \in \mathbb{R}_1$, $n = 0, 1, \dots$, and integers $m^-_{\varepsilon,n} \leq m^+_{\varepsilon,n}$, $n = 0, 1, \dots$.

First, the intervals $\mathbb{I}_{\varepsilon,n,l}$, should be constructed for $l = m^-_{\varepsilon,n}, \dots, m^+_{\varepsilon,n}$, $n = 0, 1, \dots, N$,

$$\mathbb{I}_{\varepsilon,n,l} = \begin{cases} (-\infty,\ \delta_\varepsilon(m^-_{\varepsilon,n} + \frac{1}{2})] + \lambda_{\varepsilon,n} & \text{if } l = m^-_{\varepsilon,n}\,, \\ (\delta_\varepsilon(l - \frac{1}{2}),\ \delta_\varepsilon(l + \frac{1}{2})] + \lambda_{\varepsilon,n} & \text{if } m^-_{\varepsilon,n} < l < m^+_{\varepsilon,n}\,, \\ (\delta_\varepsilon(m^+_{\varepsilon,n} - \frac{1}{2}),\ \infty) + \lambda_{\varepsilon,n} & \text{if } l = m^+_{\varepsilon,n}\,. \end{cases} \tag{8.130}$$

and the skeleton points $y_{\varepsilon,n,l} \in \mathbb{I}_{\varepsilon,n,l}$ should be chosen in the following way:

$$y_{\varepsilon,n,l} = l\delta_\varepsilon + \lambda_{\varepsilon,n}\,, \quad l = m^-_{\varepsilon,n}, \dots, m^+_{\varepsilon,n}\,, \quad n = 0, 1, \dots, N\,. \tag{8.131}$$

Second, the skeleton functions $h_{\varepsilon,n}(y)$, $n = 0, 1, \dots$ should be defined,

$$h_{\varepsilon,n}(y) = \Big\{ y_{\varepsilon,n,l} \quad \text{if } y \in \mathbb{I}_{\varepsilon,n,l}\,, \quad m^-_{\varepsilon,n} \le l \le m^+_{\varepsilon,n}\,. \tag{8.132}$$

The corresponding space-skeleton approximating Markov process $Y_{\varepsilon,n}$ is defined, for every $\varepsilon \in (0, \varepsilon_0]$, by the following stochastic transition dynamic relation:

$$\begin{cases} Y_{\varepsilon,n} = Y_{\varepsilon,n-1} + h_{\varepsilon,n}(\mu_{0,n} + \sigma_{0,n}W_{0,n}),\ n = 1, 2, \dots\,, \\ Y_{\varepsilon,0} = h_{\varepsilon,0}(Y_{0,0})\,. \end{cases} \tag{8.133}$$

The approximating log-price process $Y_{\varepsilon,n}$ is a skeleton random walk with independent random jumps $W_{\varepsilon,n} = h_{\varepsilon,n}(\mu_{0,n} + \sigma_{0,n}W_{0,n})$, $n = 1, 2, \dots$, which have distributions defined by the following formula, for $A \in \mathcal{B}_1$, $n = 1, 2, \dots$:

$$\begin{aligned} \tilde{P}_{\varepsilon,n}(A) &= \mathsf{P}\{W_{\varepsilon,n} \in A\} = \mathsf{P}\{h_{\varepsilon,n}(\mu_{0,n} + \sigma_{0,n}W_{0,n}) \in A\} \\ &= \sum_{y_{\varepsilon,n,l} \in A} \tilde{P}_{0,n}(\mathbb{I}_{\varepsilon,n,l}) = \sum_{y_{\varepsilon,n,l} \in A} \mathsf{P}\{\mu_{0,n} + \sigma_{0,n}W_{0,n} \in \mathbb{I}_{\varepsilon,n,l}\}\,. \end{aligned} \tag{8.134}$$

Note that one-point probabilities $\mathsf{P}\{W_{\varepsilon,n} = y_{\varepsilon,n,l}\} = \mathsf{P}\{\mu_{0,n} + \sigma_{0,n}W_{0,n} \in \mathbb{I}_{\varepsilon,n,l}\}$ are interval Gaussian probabilities.

As far as the initial random value $Y_{\varepsilon,0} = h_{\varepsilon,0}(Y_{0,0})$ is concerned, it has the distribution defined by the following formula, for $A \in \mathcal{B}_1$:

$$\begin{aligned} P_{\varepsilon,0}(A) &= \mathsf{P}\{Y_{\varepsilon,0} \in A\} = \mathsf{P}\{h_{\varepsilon,0}(Y_{0,0}) \in A\} \\ &= \sum_{y_{\varepsilon,0,l} \in A} P_{0,0}(\mathbb{I}_{\varepsilon,n,l}) = \sum_{y_{\varepsilon,0,l} \in A} \mathsf{P}\{Y_{0,0} \in \mathbb{I}_{\varepsilon,n,l}\}\,. \end{aligned} \tag{8.135}$$

In this case, a pay-off function $g_\varepsilon(n, e^y) = g(n, e^y)$ does not depend on parameter ε.

Let $\phi_{\varepsilon,n}(y)$ be the corresponding log-reward functions for the log-price processes $Y_{\varepsilon,n}$.

Let us denote,

$$y_{\varepsilon,n,n+r,l} = \delta_\varepsilon l + \lambda_{\varepsilon,n,n+r}\,, \quad l = m^-_{\varepsilon,n,n+r}, \dots, m^+_{\varepsilon,n,n+r}\,, \tag{8.136}$$

where, for $0 \leq n \leq n + r < \infty$,

$$m^{\pm}_{\varepsilon,n,n+r} = \sum_{l=n+1}^{n+r} m^{\pm}_{\varepsilon,l}, \quad \lambda_{\varepsilon,n,n+r} = \sum_{l=n+1}^{n+r} \lambda_{\varepsilon,l}. \tag{8.137}$$

The following lemma is a variant of Lemma 7.4.1.

Lemma 8.5.1. *Let, for every $\varepsilon \in (0, \varepsilon_0]$, the log-price process $Y_{\varepsilon,n}$ is a random walk with the distributions of jumps and the initial distribution defined, respectively, in relations (8.134) and (8.135). Then the reward functions $\phi_{\varepsilon,n}(y)$ and $\phi_{\varepsilon,n+r}(y + y_{\varepsilon,n,n+r,l})$, for points $y + y_{\varepsilon,n,n+r,l}$, $l = m^-_{\varepsilon,n,n+r}, \dots, m^+_{\varepsilon,n,n+r}$, $r = 1, \dots N - n$, are, for every $y \in \mathbb{R}_1$, $n = 0, \dots, N$, the unique solution for the following recurrence finite system of linear equations:*

$$\left\{\begin{aligned} \phi_{\varepsilon,N}(y + y_{\varepsilon,n,N,l}) &= g(N, e^{y + y_{\varepsilon,n,N,l}}), \\ & \qquad l = m^-_{\varepsilon,n,N}, \dots, m^+_{\varepsilon,n,N}, \\ \phi_{\varepsilon,n+r}(y + y_{\varepsilon,n,n+r,l}) &= \max\Bigg(g(n + r, e^{y + y_{\varepsilon,n,n+r,l}}), \\ & \qquad \sum_{l' = m^-_{\varepsilon,n+r+1}}^{m^+_{\varepsilon,n+r+1}} \phi_{\varepsilon,n+r+1}(y + y_{\varepsilon,n,n+r,l} + y_{\varepsilon,n+r+1,l'}) \\ & \qquad \times \mathsf{P}\{\mu_{0,n+r+1} + \sigma_{0,n+r+1} W_{0,n+r+1} \in \mathbb{I}_{\varepsilon,n+r+1,l'}\}\Bigg), \\ & \qquad l = m^-_{\varepsilon,n,n+r}, \dots, m^+_{\varepsilon,n,n+r}, \quad r = N - n - 1, \dots, 1, \\ \phi_{\varepsilon,n}(y) &= \max\Bigg(g(n, e^y), \sum_{l' = m^-_{\varepsilon,n+1}}^{m^+_{\varepsilon,n+1}} \phi_{\varepsilon,n+1}(y + y_{\varepsilon,n+1,l'}) \\ & \qquad \times \mathsf{P}\{\mu_{0,n+1} + \sigma_{0,n+1} W_{0,n+1} \in \mathbb{I}_{\varepsilon,n+1,l'}\}\Bigg). \end{aligned}\right. \tag{8.138}$$

Let us now formulate conditions of convergence for the above log-reward functions. We are going to apply Theorem 7.4.1.

We assume that the condition $\mathbf{I}_6$ holds for the pay-off functions $g(n, e^y)$.

The condition $\mathbf{J}_9$ takes in this case the following form:

$\mathbf{O}_6$: $\mathsf{P}\{y + \mu_{0,n} + \sigma_{0,n} W_{0,n} \in \mathbb{Y}'_n\} = 1$, $y \in \mathbb{Y}'_{n-1}$, for $n = 1, \dots, N$, where $\mathbb{Y}'_n$, $n = 0, \dots, N$ are the sets introduced in the conditions $\mathbf{I}_6$.

The condition $\mathbf{O}_6$ obviously holds if, for every $y \in \mathbb{Y}'_{n-1}$, $n = 1, \dots, N$, either $\sigma_{0,n} > 0$ and the set $\overline{\mathbb{Y}}'_n$ has zero Lebesgue measure or $\sigma_{0,n} = 0$ and $y + \mu_{0,n} \in \mathbb{Y}'_n$.

The following theorem is a variant of Theorem 7.4.1 for log-price processes represented by Gaussian random walks.

Theorem 8.5.1. *Let the log-price process $Y_{0,n}$ is a Gaussian random walk defined by the stochastic transition dynamic relation* (8.129) *and the corresponding space-skeleton approximating processes $Y_{\varepsilon,n}$ is the random walk defined by stochastic transition dynamic relations* (8.133). *Let also the condition* $\mathbf{B}_4[\gamma]$ *holds for some $\gamma \geq 0$, and also the conditions* $\mathbf{N}_3$, $\mathbf{I}_6$, *and* $\mathbf{O}_6$ *hold. Then, for every $n = 0, 1, \dots, N$, the following relation takes place for any $y_\varepsilon \to y_0 \in \mathbb{Y}'_n$:*

$$\phi_{\varepsilon,n}(y_\varepsilon) \to \phi_{0,n}(y_0) \quad as \quad \varepsilon \to 0\,. \tag{8.139}$$

Proof. The condition $\mathbf{L}_5$ holds in this case that follows from relation (8.134).

In this case, the condition $\mathbf{G}_1$ automatically holds. Thus, by Lemmas 4.1.8 and 4.5.1, the condition $\mathbf{C}_{13}[\beta]$ holds for any $\beta \geq 0$. We can always choose some $\beta > \gamma$ if $\gamma > 0$ of $\beta = 0$ if $\gamma = 0$.

As was mentioned above, the condition $\mathbf{J}_9$ takes in this case the form of the condition $\mathbf{O}_6$.

Thus, all conditions of Theorem 7.4.1 hold. By applying this theorem we get the convergence relation (8.139). □

Let us now formulate conditions of convergence for the corresponding optimal expected rewards.

In this case, the optimal expected reward functional Φ_ε for the approximation log-price processes $Y_{\varepsilon,n}$ can be found, for every $\varepsilon \in (0, \varepsilon_0]$, using the following formula:

$$\Phi_\varepsilon = \sum_{l=m^-_{\varepsilon,0}}^{m^+_{\varepsilon,0}} \phi_{\varepsilon,0}(y_{\varepsilon,0,l}) p_{\varepsilon,0}\,, \tag{8.140}$$

where $p_{\varepsilon,l} = P_{0,0}(\mathbb{I}_{\varepsilon,0,l}) = \mathsf{P}\{h_{\varepsilon,0}(Y_{0,0}) = y_{\varepsilon,0,l}\}$, $l = m^-_{\varepsilon,0}, \dots, m^+_{\varepsilon,0}$.

We are going to apply Theorem 7.4.2.

We assume that the condition $\mathbf{D}_{16}[\beta]$ holds for some appropriately chosen parameter $\beta \geq 0$.

The condition $\mathbf{K}_9$ should be replaced by the following condition imposed on the initial distribution $\mathsf{P}\{Y_{0,0} \in A\} = P_{0,0}(A)$:

$\mathbf{K}_{16}$: $P_{0,0}(\mathbb{Y}'_0) = 1$, where $\mathbb{Y}'_0$ is the set introduced in the condition $\mathbf{I}_6$.

Theorem 8.5.2. *Let the log-price process $Y_{0,n}$ is a Gaussian random walk defined by the stochastic transition dynamic relation* (8.129) *and the corresponding space-skeleton approximating log-price processes $Y_{\varepsilon,n}$ are random walks defined by stochastic transition dynamic relations* (8.133). *Let also the conditions* $\mathbf{B}_4[\gamma]$ *and* $\mathbf{D}_{16}[\beta]$ *hold with parameters γ and β such that either $\beta > \gamma > 0$ or $\beta = \gamma = 0$, and also the conditions* $\mathbf{N}_3$, $\mathbf{I}_6$, $\mathbf{O}_6$, *and* $\mathbf{K}_{16}$ *hold. Then, the following relation takes place:*

$$\Phi_\varepsilon \to \Phi_0 \quad as \quad \varepsilon \to 0\,. \tag{8.141}$$

Proof. The fitting condition $\mathbf{M}_5$ holds since the corresponding initial distributions are defined according to formula (8.135).

The condition $\mathbf{K}_{16}$ is just reformulation of the condition $\mathbf{K}_9$.

Other conditions of Theorem 7.4.2 also hold that was shown in the proof of Theorem 8.4.9. By applying Theorem 7.4.2 we get the convergence relations (8.141). □

8.5.2 Space-skeleton approximations for option rewards for log-price processes represented by multivariate modulated Gaussian random walks

Let us assume that a log-price process $\vec{Z}_{0,n} = (\vec{Y}_{0,n}, X_{0,n})$, $n = 0, 1, \dots$ is a multivariate modulated Markov Gaussian log-price process given by the following stochastic transition dynamic relation:

$$\begin{cases} \vec{Y}_{0,n} = \vec{Y}_{0,n-1} + \vec{\mu}_{0,n}(X_{0,n-1}) + \Sigma_{0,n}(X_{0,n-1})\vec{W}_{0,n}, \\ X_{0,n} = C_{0,n}(X_{0,n-1}, U_{0,n}), \ n = 1, 2, \dots, \end{cases} \tag{8.142}$$

where (a) $\vec{Z}_{0,0} = (\vec{Y}_{0,0}, X_{0,0}) = ((Y_{0,0,1}, \dots, Y_{0,0,k})), X_{0,0})$ is a random vector taking values in space $\mathbb{Z} = \mathbb{R}_k \times \mathbb{X}$, (b) $(\vec{W}_{0,n}, U_{0,n}) = ((W_{0,n,1}, \dots, W_{0,n,k}), U_{0,n})$, $n = 1, 2, \dots$ is a sequence of independent random vectors taking values in space $\mathbb{R}_k \times \mathbb{U}$, where $\mathbb{U}$ is a measurable space with a σ-algebra of measurable subsets $\mathcal{B}_{\mathbb{U}}$, (c) the random vector $(\vec{Y}_{0,0}, X_{0,0})$ and the random sequence $(\vec{W}_{0,n}, U_{0,n})$, $n = 1, 2, \dots$ are independent, (d) the random vector $\vec{W}_{0,n} = (W_{0,n,1}, \dots, W_{0,n,k})$ has a standard multivariate normal distribution with $\mathsf{E}W_{0,n,i} = 0$, $\mathsf{E}W_{0,n,i}W_{0,n,j} = I(i = j)$, $i, j = 1, \dots, k$, for every $n = 1, 2, \dots$, (e) the random variables $U_{0,n}$ have a regular conditional distribution $G_{\vec{w}}(B) = \mathsf{P}\{U_{0,n} \in B / \vec{W}_{0,n} = \vec{w}\}$ (a probability measure in $B \in \mathcal{B}_{\mathbb{U}}$ for every $\vec{w} \in \mathbb{R}_k$ and a measurable function in $\vec{w} \in \mathbb{R}_k$ for every $B \in \mathcal{B}_{\mathbb{U}}$), for every $n = 1, 2, \dots$, (f) $\vec{\mu}_{0,n}(x) = (\mu_{0,n,1}(x), \dots, \mu_{0,n,k}(x))$, $n = 1, 2, \dots$ are vector functions, which components are measurable functions acting from the space $\mathbb{X}$ to $\mathbb{R}_1$, (g) $\Sigma_{0,n}(x) = \|\sigma_{0,n,i,j}(x)\|$, $n = 1, 2, \dots$ are $k \times k$ matrix functions with elements $\sigma_{0,n,i,j}(x)$, $i, j = 1, \dots, k$, which are measurable functions acting from the space $\mathbb{X}$ to $\mathbb{R}_1$; (h) $C_{0,n}(x, u)$ is a measurable function acting from the space $\mathbb{X} \times \mathbb{U}$ to the space $\mathbb{X}$.

In this case, a space-skeleton approximation scheme with the additive skeleton structure can also be used.

Let us construct the corresponding space-skeleton approximating processes $\vec{Z}_{\varepsilon,n}$, $n = 0, 1, \dots$, for $\varepsilon \in (0, \varepsilon_0]$, according to the algorithm described in Subsection 7.4.5.

Let $\delta_{\varepsilon,i} > 0$, $\lambda_{\varepsilon,n,i} \in \mathbb{R}_1$, $i = 1, \dots, k$, $n = 0, 1, \dots$. and $m^-_{\varepsilon,n,j} \le m^+_{\varepsilon,n,j}$, $j = 0, \dots, k$, $n = 0, 1, \dots$ be integer numbers.

First, the intervals $\mathbb{I}_{\varepsilon,n,i,l}$ should be constructed for $l_i = m^-_{\varepsilon,n,i}, \dots, m^+_{\varepsilon,n,i}$, $i = 1, \dots, k$, $n = 0, 1, \dots$,

$$\mathbb{I}_{\varepsilon,n,i,l} = \begin{cases} (-\infty,\ \delta_{\varepsilon,i}(m^-_{\varepsilon,n,i} + \frac{1}{2})] + \lambda_{\varepsilon,n,i} & \text{if } l = m^-_{\varepsilon,n,i}\,, \\ (\delta_{\varepsilon,i}(l - \frac{1}{2}),\ \delta_{\varepsilon,i}(l + \frac{1}{2})] + \lambda_{\varepsilon,n,i} & \text{if } m^-_{\varepsilon,n,i} < l < m^+_{\varepsilon,n,i}\,, \\ (\delta_{\varepsilon,i}(m^+_{\varepsilon,n,i} - \frac{1}{2}),\ \infty) + \lambda_{\varepsilon,n,i} & \text{if } l = m^+_{\varepsilon,n,i}\,. \end{cases} \tag{8.143}$$

and cubes $\mathbb{I}_{\varepsilon,n,l_1,\dots,l_k}$, $l_i = m^-_{\varepsilon,n,i}, \dots, m^+_{\varepsilon,n,i}$, $i = 1, \dots, k$, $n = 0, 1, \dots$ should be defined,

$$\mathbb{I}_{\varepsilon,n,l_1,\dots,l_k} = \mathbb{I}_{\varepsilon,n,1,l_1} \times \cdots \times I_{\varepsilon,n,k,l_k}, \tag{8.144}$$

Then, the skeleton points $y_{\varepsilon,n,i,l_i} \in \mathbb{I}_{\varepsilon,n,i,l_i}$, $l_i = m^-_{\varepsilon,n,i}, \dots, m^+_{\varepsilon,n,i}$, $i = 1, \dots, k$, $n = 0, 1, \dots$ should be defined,

$$y_{\varepsilon,n,i,l_i} = l_i \delta_{\varepsilon,i} + \lambda_{\varepsilon,n,i}\,. \tag{8.145}$$

Second, nonempty sets $\mathbb{J}_{\varepsilon,n,l} \subset \mathcal{B}_{\mathbb{X}}$, $l = m^-_{\varepsilon,n,0}, \dots, m^+_{\varepsilon,n,0}$, $n = 0, 1, \dots$, such that (a) $\mathbb{J}_{\varepsilon,n,l'} \cap \mathbb{J}_{\varepsilon,n,l''} = \emptyset$, $l' \neq l''$, $n = 0, 1, \dots$, (b) $\mathbb{X} = \cup_{m^-_{\varepsilon,n,0} \leq l \leq m^+_{\varepsilon,n,0}} \mathbb{J}_{\varepsilon,n,l}$, $n = 0, 1, \dots$, should be constructed.

Recall that our model assumption is the $\mathbb{X}$ which is a Polish metric space. In this case, it is natural to assume that there exist some "large" sets $\mathbb{K}_{\varepsilon,n}$, $n = 0, 1, \dots$, "small" nonempty sets $\mathbb{K}_{\varepsilon,n,l} \subseteq \mathcal{B}_{\mathbb{X}}$, $l = m^-_{\varepsilon,n,0}, \dots, m^+_{\varepsilon,n,0}$, $n = 0, 1, \dots$ such that (c) $\mathbb{K}_{\varepsilon,n,l'} \cap \mathbb{K}_{\varepsilon,n,l''} = \emptyset$, $l' \neq l''$, $n = 0, 1, \dots$, (d) $\cup_{m^-_{\varepsilon,n,0} \leq l \leq m^+_{\varepsilon,n,0}} \mathbb{K}_{\varepsilon,n,l} = \mathbb{K}_{\varepsilon,n}$, $n = 0, 1, \dots$.

The sets $\mathbb{J}_{\varepsilon,n,l}$ can be defined in the following way, for $n = 0, 1, \dots$:

$$\mathbb{J}_{\varepsilon,n,l} = \begin{cases} \mathbb{K}_{\varepsilon,n,l} & \text{if } m^-_{\varepsilon,n,0} \leq l < m^+_{\varepsilon,n,0}\,, \\ \mathbb{K}_{\varepsilon,n,m^+_{\varepsilon,n,0}} \cup \overline{\mathbb{K}}_{\varepsilon,n} & \text{if } l = m^+_{\varepsilon,n,0}\,. \end{cases} \tag{8.146}$$

Then the skeleton points x_{ε,n,l_0}, $l_0 = m^-_{\varepsilon,n,0}, \dots, m^+_{\varepsilon,n,0}$, $n = 0, 1, \dots$ should be chosen so that the following inclusion relation would hold, for every $l_0 = m^-_{\varepsilon,n,0}, \dots, m^+_{\varepsilon,n,0}$, $n = 0, 1, \dots$:

$$x_{\varepsilon,n,l_0} \in \mathbb{J}_{\varepsilon,n,l_0}\,. \tag{8.147}$$

Let us introduce index sets, for $n = 0, 1, \dots$,

$$\mathbb{L}_{\varepsilon,n} = \{\bar{l} = (l_0, \dots, l_k), l_j = m^-_{\varepsilon,n.j}, \dots, m^+_{\varepsilon,n,j}, j = 0, \dots, k\} \tag{8.148}$$

Third, the skeleton sets $\mathbb{A}_{\varepsilon,n,\bar{l}}$ and skeleton points, $\vec{z}_{\varepsilon,n,\bar{l}} \in \mathbb{A}_{\varepsilon,n,\bar{l}}$ can be defined for $\bar{l} = (l_0, l_1, \dots, l_k) \in \mathbb{L}_{\varepsilon,n}$, $n = 1, 2, \dots$, in the following way:

$$\mathbb{A}_{\varepsilon,n,\bar{l}} = \mathbb{I}_{\varepsilon,n,l_1,\dots,l_k} \times \mathbb{J}_{\varepsilon,n,l_0}\,, \tag{8.149}$$

and

$$\vec{z}_{\varepsilon,n,\bar{l}} = (\vec{y}_{\varepsilon,n,\bar{l}}, x_{\varepsilon,n,\bar{l}}) = ((y_{\varepsilon,n,1,l_1}, \dots, y_{\varepsilon,n,k,l_k}), x_{\varepsilon,n,l_0})\,. \tag{8.150}$$

Fourth, the skeleton functions, $h_{\varepsilon,n,i}(y)$, $y \in \mathbb{R}_1$, for $\varepsilon \in (0, \varepsilon_0]$ and $i = 1, \dots, k$, $n = 0, 1, \dots$ should be defined,

$$h_{\varepsilon,n,i}(y) = \Big\{ y_{\varepsilon,n,l} \quad \text{if } y \in \mathbb{I}_{\varepsilon,n,i,l}, \quad m^-_{\varepsilon,n,i} \le l \le m^+_{\varepsilon,n,i}, \tag{8.151}$$

and the vector skeleton functions $\hat{h}_{\varepsilon,n}(\vec{y}), \vec{y} = (y_1, \dots, y_k) \in \mathbb{R}_k$, for $\varepsilon \in (0, \varepsilon_0]$ and $n = 0, 1, \dots$ should be defined,

$$\hat{h}_{\varepsilon,n}(\vec{y}) = (h_{\varepsilon,n,1}(y_1), \dots, h_{\varepsilon,n,k}(y_k)). \tag{8.152}$$

Also, the skeleton functions $h_{\varepsilon,n,0}(x)$, $x \in \mathbb{X}$, for $\varepsilon \in (0, \varepsilon_0]$ and $n = 0, 1, \dots,$ should be defined

$$h_{\varepsilon,n,0}(x) = \Big\{ x_{\varepsilon,n,l} \quad \text{if } x \in \mathbb{J}_{\varepsilon,n,l}, \quad m^-_{\varepsilon,n,0} \le l \le m^+_{\varepsilon,n,0}. \tag{8.153}$$

The corresponding space-skeleton approximating modulated Markov processes $Y_{\varepsilon,n}$ are defined by the following stochastic transition dynamic relations:

$$\begin{cases} \vec{Y}_{\varepsilon,n} = \vec{Y}_{\varepsilon,n-1} + \hat{h}_{\varepsilon,n}\Big(\vec{\mu}_{0,n}(h_{\varepsilon,n-1,0}(X_{\varepsilon,n-1})) + \Sigma_{0,n}(h_{\varepsilon,n-1,0}(X_{\varepsilon,n-1}))\vec{W}_{0,n}\Big), \\ X_{\varepsilon,n} = h_{\varepsilon,n,0}(C_{0,n}(h_{\varepsilon,n-1,0}(X_{\varepsilon,n-1}), U_{0,n})), \ n = 1, 2, \dots, \\ \vec{Y}_{\varepsilon,0} = \hat{h}_{\varepsilon,0}(\vec{Y}_{0,0}), \quad X_{\varepsilon,0} = h_{\varepsilon,0,0}(X_{0,0}). \end{cases} \tag{8.154}$$

In this case, the Markov log-price process $\vec{Z}_{\varepsilon,n} = (\vec{Y}_{\varepsilon,n}, X_{\varepsilon,n})$ has one-step transition probabilities $\tilde{P}_{\varepsilon,n}(x, A) = \mathsf{P}\{(\vec{W}_{\varepsilon,n}, X_{\varepsilon,n}) \in A / X_{\varepsilon,n-1} = x\}$ given for $x \in X$, $A \in \mathcal{B}_{\mathbb{Z}}$, $n = 1, \dots$ by the following relation:

$$\begin{aligned} \tilde{P}_{\varepsilon,n}(x, A) &= \sum_{(\vec{y}_{\varepsilon,n,\bar{l}}, x_{\varepsilon,n,\bar{l}}) \in A} \tilde{P}_{0,n}(h_{\varepsilon,n-1,0}(x), \mathbb{A}_{\varepsilon,n,\bar{l}}) \\ &= \sum_{(\vec{y}_{\varepsilon,n,\bar{l}}, x_{\varepsilon,n,\bar{l}}) \in A} \mathsf{P}\Big\{ \Big(\vec{\mu}_{0,n}(h_{\varepsilon,n-1,0}(x)) + \Sigma_{0,n}(h_{\varepsilon,n-1,0}(x))\vec{W}_{0,n}, \\ &\qquad C_{0,n}(h_{\varepsilon,n-1,0}(x), U_{0,n})\Big) \in A_{\varepsilon,n,\bar{l}} \Big\} \\ &= \sum_{(\vec{y}_{\varepsilon,n,\bar{l}}, x_{\varepsilon,n,\bar{l}}) \in A} \int_{\mathbb{R}_k} \int_{\mathbb{U}} I\Big(\Big(\vec{\mu}_{0,n}(h_{\varepsilon,n-1,0}(x)) + \Sigma_{0,n}(h_{\varepsilon,n-1,0}(x))\vec{w}, \\ &\qquad C_{0,n}(x, u)\Big) \in A_{\varepsilon,n,\bar{l}}\Big) G_{\vec{w}}(du) \frac{1}{(\sqrt{2\pi})^k} e^{-\sum_{i=1}^k \frac{w_i^2}{2}} d\vec{w}. \end{aligned} \tag{8.155}$$

As far as the initial distribution $P_{\varepsilon,0}(A) = \mathsf{P}\{(\vec{Y}_{\varepsilon,0}, X_{\varepsilon,0}) \in A\}$, $A \in \mathcal{B}_{\mathbb{Z}}$ is concerned, it takes the following form:

$$\begin{aligned} P_{\varepsilon,0}(A) &= \sum_{(\vec{y}_{\varepsilon,0,\bar{l}}, x_{\varepsilon,0,\bar{l}}) \in A} P_{0,0}(\mathbb{A}_{\varepsilon,0,\bar{l}}) \\ &= \sum_{(\vec{y}_{\varepsilon,0,\bar{l}}, x_{\varepsilon,0,\bar{l}}) \in A} \mathsf{P}\{(\vec{Y}_{0,0}, X_{0,0}) \in A_{\varepsilon,0,\bar{l}}\}. \end{aligned} \tag{8.156}$$

In this case, the pay-off function $g_\varepsilon(n, e^{\vec{y}}, x) = g(n, e^{\vec{y}}, x)$ does not depend on the parameter ε.

Let us $\phi_{\varepsilon,n}(\vec{y}, x)$ be the corresponding log-reward functions for the log-price processes $\vec{Z}_{\varepsilon,n}$.

Let us denote, for $\bar{l} = (l_0, \dots, l_k)$, $l_j = m^-_{\varepsilon,n,n+r,j}, \dots, m^+_{\varepsilon,n,n+r,j}$, $j = 0, \dots, k$,

$$\vec{y}_{\varepsilon,n,n+r,\bar{l}} = (\delta_{\varepsilon,1} l_1 + \lambda_{\varepsilon,n,n+r,1}, \dots, \delta_{\varepsilon,k} l_k + \lambda_{\varepsilon,n,n+r,k}), \tag{8.157}$$

and

$$x_{\varepsilon,n,n+r,\bar{l}} = x_{\varepsilon,n+r,\bar{l}} = x_{\varepsilon,n+r,l_0}, \tag{8.158}$$

where, for $j = 1, \dots, k$, $0 \le n < n + r < \infty$,

$$m^\pm_{\varepsilon,n,n+r,j} = \sum_{l=n+1}^{n+r} m^\pm_{\varepsilon,l,j}, \quad \lambda_{\varepsilon,n,n+r,j} = \sum_{l=n+1}^{n+r} \lambda_{\varepsilon,l,j}. \tag{8.159}$$

and, for $0 \le n < n + r < \infty$,

$$m^\pm_{\varepsilon,n,n+r,0} = m^\pm_{\varepsilon,n+r,0}. \tag{8.160}$$

The following lemma is a variant of Lemma 7.4.2.

Lemma 8.5.2. *Let, for every $\varepsilon \in (0, \varepsilon_0]$, the space-skeleton approximating process $\vec{Z}_{\varepsilon,n}$ is represented modulated random walk with jump distributions probabilities and the initial distribution defined, respectively, in relations* (8.155) *and* (8.156). *Then the reward functions $\phi_{\varepsilon,n}(\vec{y}, x)$ and $\phi_{\varepsilon,n+r}(\vec{y} + \vec{y}_{\varepsilon,n,n+r,\bar{l}}, x_{\varepsilon,n+r,\bar{l}})$ for points $\vec{z} = (\vec{y}, x) = ((y_1, \dots, y_k), x)$ and $(\vec{y} + \vec{y}_{\varepsilon,n,n+r,\bar{l}}, x_{\varepsilon,n+r,\bar{l}})$, $\bar{l} = (l_0, l_1, \dots, l_k)$, $l_j = m^-_{\varepsilon,n,n+r,j}, \dots, m^+_{\varepsilon,n,n+r,j}$, $j = 0, \dots, k$, $r = 1, \dots N - n$, are, for every $\vec{z} \in \mathbb{R}_k \times \mathbb{J}_{\varepsilon,n,l''_0}$, $l''_0 = m^-_{\varepsilon,n,0}, \dots, m^+_{\varepsilon,n,0}$, $n = 0, \dots, N$, the unique solution for the following recurrence finite system of linear equations:*

$$\begin{cases} \phi_{\varepsilon,N}(\vec{y} + \vec{y}_{\varepsilon,n,N,\bar{l}}, x_{\varepsilon,N,\bar{l}}) = g(N, e^{\vec{y}+\vec{y}_{\varepsilon,n,N,\bar{l}}}, x_{\varepsilon,N,\bar{l}}), \\ \qquad \bar{l} = (l_0, l_1, \dots, l_k), \quad l_j = m^-_{\varepsilon,n,N,j}, \dots, m^+_{\varepsilon,n,N,j}, \quad j = 0, \dots, k, \\ \phi_{\varepsilon,n+r}(\vec{y} + \vec{y}_{\varepsilon,n,n+r,\bar{l}}, x_{\varepsilon,n+r,\bar{l}}) \\ \qquad = \max\Big(g(n + r, e^{\vec{y}+\vec{y}_{\varepsilon,n,n+r,\bar{l}}}, x_{\varepsilon,n+r,\bar{l}}), \sum_{\bar{l}' \in \mathbb{L}_{\varepsilon,n+r+1}} \phi_{\varepsilon,n+r+1}(\vec{y} + \vec{y}_{\varepsilon,n,n+r,\bar{l}} \\ \qquad\qquad + \vec{y}_{\varepsilon,n+r+1,\bar{l}'}, x_{\varepsilon,n+r+1,\bar{l}'})\tilde{P}_{0,n+r+1}(x_{\varepsilon,n+r,\bar{l}}, \mathbb{A}_{\varepsilon,n+r+1,\bar{l}'})\Big), \\ \qquad \bar{l} = (l_0, l_1, \dots, l_k), \quad l_j = m^-_{\varepsilon,n,n+r,j}, \dots, m^+_{\varepsilon,n,n+r,j}, \\ \qquad j = 0, \dots, k, \quad r = N - n - 1, \dots, 1, \\ \phi_{\varepsilon,n}(\vec{y}, x) = \max\Big(g(n, e^{\vec{y}}, x), \\ \qquad \sum_{\bar{l}' \in \mathbb{L}_{\varepsilon,n+1}} \phi_{\varepsilon,n+1}(\vec{y} + \vec{y}_{\varepsilon,n+1,\bar{l}'}, x_{\varepsilon,n+1,\bar{l}'})\tilde{P}_{0,n+1}(x_{\varepsilon,n,l''_0}, \mathbb{A}_{\varepsilon,n+1,\bar{l}'})\Big). \end{cases} \tag{8.161}$$

Let us now formulate conditions of convergence for the above log-reward functions. We are going to apply Theorem 7.4.5.

The condition $\mathbf{C_{14}}[\bar{\beta}]$ should be replaced in this case by the following reduced version of the condition $\mathbf{G_3}$:

$\mathbf{G_{11}}$: $\max_{1 \leq n \leq N,\, i,j=1,\dots,k} \sup_{x \in \mathbb{X}} (|\mu_{0,n,i}(x)| + \sigma^2_{0,n,i,j}(x)) < K_{84}$, for some $0 < K_{84} < \infty$.

We assume that the condition $\mathbf{I_7}$ holds for pay-off functions $g(n, e^{\vec{y}}, x)$.

The condition $\mathbf{J'_{12}}$ takes in this case the following form:

$\mathbf{O_7}$: There exist sets $\mathbb{X}_n \in \mathcal{B}_{\mathbb{X}}$, $n = 0, \dots, N$ and $\mathbb{U}_n \in \mathcal{B}_{\mathbb{U}}$, $n = 1, \dots, N$ such that:
- (a) $\vec{\mu}_{0,n}(x_\varepsilon) \to \vec{\mu}_{0,n}(x_0)$, $\Sigma_{0,n}(x_\varepsilon) \to \Sigma_{0,n}(x_0)$ and $C_{0,n}(x_\varepsilon, u) \to C_{0,n}(x_0, u)$ as $\varepsilon \to 0$, for any $x_\varepsilon \to x_0 \in \mathbb{X}_{n-1}$ as $\varepsilon \to 0$, $u \in \mathbb{U}_n$, $n = 1, \dots, N$;
- (b) $\mathsf{P}\{U_{0,n} \in \mathbb{U}_n\} = 1$, $n = 1, \dots, N$;
- (c) $\mathsf{P}\{(\vec{y}_0 + \vec{\mu}_{0,n}(x_0) + \Sigma_{0,n}(x_0)\vec{W}_{0,n}, C_{0,n}(x_0, U_{0,n})) \in \mathbb{Z}'_n \cap \mathbb{Z}''_n\} = 1$ for every $\vec{z}_0 = (\vec{y}_0, x_0) \in \mathbb{Z}'_{n-1} \cap \mathbb{Z}''_{n-1}$ and $n = 1, \dots, N$, where $\mathbb{Z}'_n$, $n = 0, \dots, N$ are sets introduced in the condition $\mathbf{I_7}$ and $\mathbb{Z}''_n = \mathbb{R}_k \times \mathbb{X}_n$, $n = 0, \dots, N$.

Theorem 8.5.3. *Let the log-price process $\vec{Z}_{0,n}$ is a multivariate modulated Gaussian random walk defined by the stochastic transition dynamic relation* (8.142) *and the corresponding space skeleton approximating Markov log-price process processes $\vec{Z}_{\varepsilon,n}$ are defined by the stochastic transition dynamic relations* (8.154). *Let also the condition $\mathbf{B_1}[\bar{\gamma}]$ holds for some vector parameter $\bar{\gamma} = (\gamma_1, \dots, \gamma_k)$ with nonnegative components, and also the conditions $\mathbf{G_{11}}$, $\mathbf{N_4}$, $\mathbf{I_7}$, and $\mathbf{O_7}$ hold. Then, for every $n = 0, 1, \dots, N$, the following relation takes place for any $\vec{z}_\varepsilon = (\vec{y}_\varepsilon, x_\varepsilon) \to \vec{z}_0 = (\vec{y}_0, x_0) \in \mathbb{Z}'_{n-1} \cap \mathbb{Z}''_{n-1}$:*

$$\phi_{\varepsilon,n}(\vec{y}_\varepsilon, x_\varepsilon) \to \phi_{0,n}(\vec{y}_0, x_0) \quad as \quad \varepsilon \to 0\,. \tag{8.162}$$

Proof. The condition $\mathbf{L_6}$ holds in this case that follows from relation (8.155).

The condition $\mathbf{G_{11}}$ implies, by Lemmas 4.2.1 and 4.5.1, that the condition $\mathbf{C_{14}}[\bar{\beta}]$ holds for any vector parameter $\bar{\beta} = (\beta_1, \dots, \beta_k)$ with nonnegative components. We can always choose parameter $\bar{\beta} = (\beta_1, \dots, \beta_k)$ such that, for every $i = 1, \dots, k$, either $\beta_i > \gamma_i > 0$ or $\beta_i = \gamma_i = 0$.

As was mentioned above, the condition $\mathbf{J'_{12}}$ takes in this case the form of the condition $\mathbf{O_7}$.

Thus, all conditions of Theorem 7.4.5 hold. By applying this theorem we get the convergence relation (8.162). □

Let us now formulate conditions of convergence for the corresponding optimal expected rewards. We are going to apply Theorem 7.4.6.

In this case, the optimal expected reward Φ_ε for the approximation log-price processes $\vec{Z}_{\varepsilon,n}$ can be found using the following formula:

$$\Phi_\varepsilon = \sum_{\bar{l} \in \mathbb{L}_{\varepsilon,0}} \phi_{\varepsilon,0}(\vec{z}_{\varepsilon,0,\bar{l}}) p_{\varepsilon,\bar{l}}\,, \tag{8.163}$$

where $p_{\varepsilon,\bar{l}} = \mathsf{P}\{\hat{h}_{\varepsilon,0}(\vec{Y}_{0,0}) = \vec{y}_{\varepsilon,0,\bar{l}}, h_{\varepsilon,0}(\vec{X}_{0,0}) = x_{\varepsilon,0,\bar{l}}\} = P_{0,0}(\mathbb{A}_{\varepsilon,0,\bar{l}})$, $\bar{l} = (l_0, \dots, l_k) \in \mathbb{L}_{\varepsilon,0}$.

We assume that the condition $\mathbf{D_{18}}[\bar{\beta}]$, replacing in this case the condition $\mathbf{D_{11}}[\bar{\beta}]$, holds for some appropriately chosen vector parameter $\bar{\beta} = (\beta_1, \dots, \beta_k)$ with nonnegative components.

The condition $\mathbf{K_1}$ should be replaced by the following condition imposed on the initial distribution $\mathsf{P}\{\vec{Z}_{0,0}\} = P_{0,0}(A)$:

$\mathbf{K_{17}}$: $P_{0,0}(\mathbb{Z}'_0 \cap \mathbb{Z}''_0) = 1$, where $\mathbb{Z}'_0$ and $\mathbb{Z}''_0$ are the sets introduced, respectively, in the conditions $\mathbf{I_7}$ and $\mathbf{O_7}$.

Theorem 8.5.4. *Let the conditions $\mathbf{B_1}[\bar{\gamma}]$ and $\mathbf{D_{18}}[\bar{\beta}]$ hold for some vector parameters $\bar{\beta} = (\beta_1, \dots, \beta_k)$ and $\bar{\gamma} = (\gamma_1, \dots, \gamma_k)$ such that, for every $i = 1, \dots, k$, either $\beta_i > \gamma_i > 0$ or $\beta_i = \gamma_i = 0$. Also let the conditions $\mathbf{G_{11}}$, $\mathbf{N_4}$, $\mathbf{I_7}$, $\mathbf{O_7}$, and $\mathbf{K_{17}}$ hold. Then, the following relation takes place:*

$$\Phi_\varepsilon \to \Phi_0 \quad \text{as} \quad \varepsilon \to 0 . \tag{8.164}$$

Proof. The fitting condition $\mathbf{M_6}$ holds since the corresponding initial distributions are defined according to formula (8.156).

The condition $\mathbf{K_{17}}$ is just reformulation of the condition $\mathbf{K_{10}}$.

Other conditions of Theorem 7.4.6 also hold that was shown in the proof of Theorem 8.5.3. By applying Theorem 7.4.6, we get the convergence relations (8.164). □

9 Tree-type approximations for Markov Gaussian LPP

In this chapter, we investigate problems of fitting parameters for discrete time Markov log-price processes with binomial or trinomial transition probabilities to parameters of a discrete time Markov log-price processes with Gaussian transition probabilities. The corresponding fitting procedures are based on exact or asymptotic equalizing of moments for the corresponding processes (first two moments in the univariate case or vectors of expectations and covariance matrices, in the multivariate case). It should also be mentioned that this moment fitting should be realized under additional recombining conditions that should be imposed on parameters of the corresponding binomial or trinomial models, in order the backward recurrence tree reward algorithms would be effective for these models due to polynomial rate of growth for the number of tree nodes as functions of the number of steps in corresponding trees.

In Section 9.1, we present well-known exact and asymptotic fitting formulas for univariate log-price processes represented by homogeneous and inhomogeneous in time binomial random walks.

In Section 9.2, we present fitting formulas for multivariate log-price processes represented by homogeneous in time binomial random walks. In the case of multivariate models with dimension $k \geq 3$, the special attention should also be given to the problem of construction of multivariate binomial distributions of jumps with given expectations and covariances.

In Section 9.3, we present fitting formulas for multivariate log-price processes represented by inhomogeneous in time trinomial random walks. It should be mentioned that in the case of multivariate inhomogeneous in time models the binomial random walks may possess not enough free parameters for exact or even asymptotic fitting. However, such fitting can be realized by using trinomial models. As in the case of binomial models, if dimension $k \geq 3$, the special attention should also be given to the problem of construction of multivariate trinomial distributions of jumps with given expectations and covariances.

In Section 9.4, we present fitting formulas for univariate log-price processes represented by homogeneous in time but inhomogeneous in space binomial random walks.

In Section 9.5, we obtain fitting formulas for multivariate log-price processes represented by inhomogeneous in time and space trinomial random walks.

In the case of Markov Gaussian-type log-price processes, the large number of binomial, trinomial and multinomial tree approximations has been proposed in the works related to reward approximation models for both discrete and continuous time processes. It should be noted that discrete time models and the corresponding approximations have their own important value. At the same time, they can be effectively used as approximations for the corresponding continuous time models.

We think that the fitting procedures for multivariate models, inhomogeneous in time or time and space, given in Lemmas 9.3.1 and 9.5.2, are the most interesting in this chapter, and to our knowledge, these lammas contain new results.

Results about the convergence of the corresponding reward binomial and trinomial approximations are given in Chapter 10.

9.1 Univariate binomial tree approximations

In this section, we present fitting formulas for univariate binomial tree approximation models inhomogeneous in time but homogeneous is space.

9.1.1 Exact fitting of parameters for a classical binomial tree model

Let us consider the model of homogeneous in time Gaussian random walk, which can be considered as a discrete time analog of the Wienner process (Brownian motion). In this case, a log-price process $Y_{0,n}$ is given by the following stochastic transition dynamic relation:

$$Y_{0,n} = Y_{0,n-1} + W_{0,n}\,, \quad n = 1, 2, \dots . \tag{9.1}$$

where (a) $Y_{0,0}$ is a real-valued random variable, (b) $W_{0,n}$, $n = 1, 2, \dots$ is a sequence of i.i.d. normal random variables with a mean $\mathsf{E} W_{0,1} = \mu$ and a variance $\mathrm{Var} W_{0,1} = \sigma^2$, and (c) the random variable $Y_{0,0}$ and the random sequence $W_{0,n}$, $n = 1, 2, \dots$ are independent.

Also, usual assumptions are (d) $\mu \in \mathbb{R}_1$ and (e) $\sigma > 0$.

Note that the assumption (e) excludes the trivial case of degenerated log-price process $Y_{0,n}$. Indeed, if $\sigma = 0$, then $Y_{0,n} = Y_{0,0} + \mu n$, $n = 0, 1, \dots$.

We shall approximate the log-price process $Y_{0,n}$ by log-price processes $Y_{\varepsilon,n}$ represented by homogeneous in time binomial random walk. In this case, the log-price process $Y_{\varepsilon,n}$ is given, for every $\varepsilon \in (0, \varepsilon_0]$, by the following dynamic transition relation:

$$Y_{\varepsilon,n} = Y_{\varepsilon,n-1} + W_{\varepsilon,n}\,, \quad n = 1, 2, \dots . \tag{9.2}$$

where (a) $Y_{\varepsilon,0}$ is a real-valued random variable, (b) $W_{\varepsilon,n} = W_{\varepsilon,n,1} + \cdots + W_{\varepsilon,n,r_\varepsilon}$, $n = 1, 2, \dots$, where $W_{\varepsilon,n,l}$, $l, n = 1, 2, \dots$ is a family of i.i.d. binary random binomial variables, which take values $\delta_{\varepsilon,+}$ and $\delta_{\varepsilon,-}$ with probabilities, respectively, $p_{\varepsilon,+}$ and $p_{\varepsilon,-}$, and (c) the random variable $Y_{\varepsilon,0}$ and the family of random variables $W_{\varepsilon,n,l}$, l, $n = 1, 2, \dots$ are independent.

Also, usual assumptions are (d) $\delta_{\varepsilon,-} < \delta_{\varepsilon,+}$, (e) $0 \le p_{\varepsilon,+} = 1 - p_{\varepsilon,-} \le 1$, and (f) r_ε is a positive integer number.

It is useful to note that $W_{\varepsilon,n}$, $n = 1, 2, \dots$ are binomial random variables taking values $w_{\varepsilon,r_\varepsilon,l} = l\delta_{\varepsilon,+} + (r_\varepsilon - l)\delta_{\varepsilon,-}$, $l = 0, \dots, r_\varepsilon$ with the corresponding probabilities $p_{\varepsilon,l} = C_{r_\varepsilon}^l p_{\varepsilon,+}^l p_{\varepsilon,-}^{r_\varepsilon - l}$, $l = 0, \dots, r_\varepsilon$.

We use the process $Y_{\varepsilon,n}$ for constructing approximation algorithms for rewards of American-type options for the process $Y_{0,n}$. As a matter of fact, rewards for the log-price processes $Y_{\varepsilon,n}$ well approximate the rewards for log-price process $Y_{0,n}$, if we fit in a natural way parameters of these processes.

A usual approach is to choose the parameters $\delta_{\varepsilon,\pm}$, $p_{\varepsilon,\pm}$, and r_ε in such a way that the following exact moment-fitting equalities hold, for every $\varepsilon \in (0, \varepsilon_0]$:

$$\{ \mathsf{E} W_{\varepsilon,1} = \mathsf{E} W_{0,1} , \quad \mathrm{Var} W_{\varepsilon,1} = \mathrm{Var} W_{0,1} . \tag{9.3}$$

Using the sum-representation for binomial random variables $W_{\varepsilon,n}$ used in the defining transition relation (9.2), one can easily transform equalities (9.3) to the equivalent form of the following system of equations, for every $\varepsilon \in (0, \varepsilon_0]$:

$$\begin{cases} \mathsf{E} W_{\varepsilon,1} = r_\varepsilon(\delta_{\varepsilon,+} p_{\varepsilon,+} + \delta_{\varepsilon,-} p_{\varepsilon,-}) = \mu , \\ \mathrm{Var} W_{\varepsilon,1} = r_\varepsilon(\delta_{\varepsilon,+}^2 p_{\varepsilon,+} + \delta_{\varepsilon,-}^2 p_{\varepsilon,-} - (\frac{\mu}{r_\varepsilon})^2) = \sigma^2 . \end{cases} \tag{9.4}$$

It is convenient to symmetrize the values of jumps by representing binary random variables $W_{\varepsilon,n,l}$, $l, n = 1, 2, \dots$ in the following form:

$$W_{\varepsilon,n,l} = \lambda_\varepsilon + W'_{\varepsilon,n,l} = \lambda_\varepsilon + \begin{cases} \delta_\varepsilon & \text{with probability} \quad p_{\varepsilon,+} , \\ -\delta_\varepsilon & \text{with probability} \quad p_{\varepsilon,-} , \end{cases} \tag{9.5}$$

where

$$\delta_\varepsilon = \frac{\delta_{\varepsilon,+} - \delta_{\varepsilon,-}}{2} , \quad \lambda_\varepsilon = \frac{\delta_{\varepsilon,+} + \delta_{\varepsilon,-}}{2} . \tag{9.6}$$

It follows from relation (9.5) that the transition dynamic relation (9.2) can be rewritten in the following form:

$$Y_{\varepsilon,n} = Y_{\varepsilon,n-1} + \lambda'_\varepsilon + W'_{\varepsilon,n} , \quad n = 1, 2, \dots . \tag{9.7}$$

where (a) $Y_{\varepsilon,0}$ is a real-valued random variable, (b) $W'_{\varepsilon,n} = W'_{\varepsilon,n,1} + \cdots + W'_{\varepsilon,n,r_\varepsilon}$, $n = 1, 2, \dots$, where $W'_{\varepsilon,n,l}$, $l, n = 1, 2, \dots$ is a family of i.i.d. binary random variables, which takes values δ_ε and $-\delta_\varepsilon$ with probabilities, respectively, $p_{\varepsilon,+}$ and $p_{\varepsilon,-}$, (c) the random variable $Y_{\varepsilon,0}$ and the family of random variables $W'_{\varepsilon,n,l}$, $l, n = 1, 2, \dots$ are independent, and (d) $\lambda'_\varepsilon = r_\varepsilon \lambda_\varepsilon$.

Indeed, it follows from the following representation that takes place for the random variable $W_{\varepsilon,n}$, for every $n = 1, 2, \dots$:

$$W_{\varepsilon,n} = W_{\varepsilon,n,1} + \cdots + W_{\varepsilon,n,r_\varepsilon} = r_\varepsilon \lambda_\varepsilon + W'_{\varepsilon,n,1} + \cdots + W'_{\varepsilon,n,r_\varepsilon} = \lambda'_\varepsilon + W'_{\varepsilon,n} . \tag{9.8}$$

The system (9.4) can be rewritten in following form:

$$\begin{cases} \delta_\varepsilon(p_{\varepsilon,+} - p_{\varepsilon,-}) + \lambda_\varepsilon = \frac{\mu}{r_\varepsilon} , \\ \delta_\varepsilon^2 - (\frac{\mu}{r_\varepsilon} - \lambda_\varepsilon)^2 = \frac{\sigma^2}{r_\varepsilon} . \end{cases} \tag{9.9}$$

The system (9.9) should be supplemented by the following consistency relations that should hold for admissible values of parameters:

$$p_{\varepsilon,\pm} \geq 0\,, \quad p_{\varepsilon,+} + p_{\varepsilon,-} = 1\,, \quad \delta_\varepsilon > 0\,, \quad \lambda_\varepsilon \in \mathbb{R}_1\,, \quad r_\varepsilon = 1, 2, \dots\,. \tag{9.10}$$

As was pointed out in Chapter 3, the recombining condition should hold for the process $Y_{\varepsilon,n}$, in order the corresponding algorithms for computing of option rewards for the log-price processes $Y_{\varepsilon,n}$ would be effective.

As was pointed in Subsection 3.3.1, the recombining condition holds for any homogeneous in time random walk with bounded jumps taking values on a grid of points $\lambda_\varepsilon r_\varepsilon \pm \delta_\varepsilon l$, $l = 0, 1, \dots$. The binomial random walk $Y_{\varepsilon,n}$ belongs to this class, for any parameters $\delta_\varepsilon > 0, \lambda_\varepsilon \in \mathbb{R}_1$, and $r_\varepsilon = 1, 2, \dots$.

The system (9.9) has infinitely many solutions given by the following formulas:

$$\begin{cases} \delta_\varepsilon = \sqrt{\frac{\sigma^2}{r_\varepsilon} + (\frac{\mu}{r_\varepsilon} - \lambda_\varepsilon)^2}\,, \\ p_{\varepsilon,\pm} = \frac{1}{2}\left(1 \pm \frac{\frac{\mu}{r_\varepsilon} - \lambda_\varepsilon}{\delta_\varepsilon}\right)\,, \end{cases} \tag{9.11}$$

where solutions are parameterized by the parameters $\lambda_\varepsilon \in \mathbb{R}_1$ and $r_\varepsilon = 1, 2, \dots$.

Note that relations (9.10) hold for solutions given by formulas (9.11).

Let now take $\imath = +, -\,, \quad \delta \geq \sigma > 0$ and choose,

$$\lambda_\varepsilon = \frac{\imath\sqrt{\delta^2 - \sigma^2}}{\sqrt{r_\varepsilon}} + \frac{\mu}{r_\varepsilon}\,. \tag{9.12}$$

Substituting λ_ε defined by relation (9.12) in relation (9.11), we get the following formulas:

$$\begin{cases} \delta_\varepsilon = \frac{\delta}{\sqrt{r_\varepsilon}}\,, \\ p_{\varepsilon,\pm} = \frac{1}{2}\left(1 \mp \frac{\imath\sqrt{\delta^2 - \sigma^2}}{\delta}\right)\,, \end{cases} \tag{9.13}$$

Thus, the following formulas give solutions of the system (9.9):

$$\lambda_\varepsilon = \frac{\imath\sqrt{\delta^2 - \sigma^2}}{\sqrt{r_\varepsilon}} + \frac{\mu}{r_\varepsilon}\,, \quad \delta_\varepsilon = \frac{\delta}{\sqrt{r_\varepsilon}}\,, \quad p_{\varepsilon,\pm} = \frac{1}{2}\left(1 \mp \frac{\imath\sqrt{\delta^2 - \sigma^2}}{\delta}\right)\,, \tag{9.14}$$

parameterized by the parameters $\imath = +, -\,, \ \delta \geq \sigma$ and $r_\varepsilon = 1, 2, \dots$.

In this case, $\delta_{\varepsilon,\pm} = \pm\frac{\delta}{\sqrt{r_\varepsilon}} + \frac{\imath\sqrt{\delta^2-\sigma^2}}{\sqrt{r_\varepsilon}} + \frac{\mu}{r_\varepsilon}$. Thus, $\delta_{\varepsilon,-} < \delta_{\varepsilon,+}$.

Note also that the consistency relations (9.10) hold for solutions given by formulas (9.14) for any $\imath = +, -\,, \quad \delta \geq \sigma$ and $r_\varepsilon = 1, 2, \dots$.

Formulas (9.14) take the simplest form in the case, where the parameter $\delta = \sigma$. In this case, we get the following formulas for the corresponding solutions of the system (9.9):

$$\lambda_\varepsilon = \frac{\mu}{r_\varepsilon}\,, \quad \delta_\varepsilon = \frac{\sigma}{\sqrt{r_\varepsilon}}\,, \quad p_{\varepsilon,\pm} = \frac{1}{2}\,, \tag{9.15}$$

parameterized by the parameter $r_\varepsilon = 1, 2, \dots$.

In this case, $\delta_{\varepsilon,\pm} = \pm\frac{\sigma}{\sqrt{r_\varepsilon}}$. Thus, $\delta_{\varepsilon,-} < 0 < \delta_{\varepsilon,+}$.

There are also alternative variants of solutions with simplified parametrization for the system (9.9). For example, we can choose the parameter $\lambda_\varepsilon = 0$. In this case we get the following formulas for solutions of the system (9.9):

$$\lambda_\varepsilon = 0\,, \quad \delta_\varepsilon = \frac{\sigma}{\sqrt{r_\varepsilon}}\sqrt{1 + \frac{1}{r_\varepsilon}\frac{\mu^2}{\sigma^2}}\,, \quad p_{\varepsilon,\pm} = \frac{1}{2}\left(1 \pm \frac{\mu}{\sigma\sqrt{r_\varepsilon}}\frac{1}{\sqrt{1+\frac{1}{r_\varepsilon}\frac{\mu^2}{\sigma^2}}}\right)\,, \tag{9.16}$$

where solutions are parameterized by the parameter $r_\varepsilon = 1, 2, \dots$.

In this case, $\delta_{\varepsilon,\pm} = \pm\frac{\sigma}{\sqrt{r_\varepsilon}}\sqrt{1 + \frac{1}{r_\varepsilon}\frac{\mu^2}{\sigma^2}}$. Thus, $\delta_{\varepsilon,-} < 0 < \delta_{\varepsilon,+}$.

Note also that the consistency relations (9.10) hold for solutions given by formulas (9.16) for any $r_\varepsilon = 1, 2, \dots$.

The above remarks can be summarized in the following lemma.

Lemma 9.1.1. *The log-price process $Y_{\varepsilon,n}$ represented, for every $\varepsilon \in (0, \varepsilon_0]$, by binomial random walk, which is defined by the stochastic transition dynamic relation* (9.2), *realizes the exact moment fitting given by relation* (9.3) *for the log-price processes $Y_{0,n}$ represented by Gaussian random walk, which is defined by the stochastic transition dynamic relation* (9.1). *This holds if the corresponding parameters for the process $Y_{\varepsilon,n}$ are given by one of relations* (9.11), (9.14), (9.15), *or* (9.16).

It should be pointed out that the important recombining condition (which guarantees not more than the polynomial rate of growth for number of nodes in the corresponding binomial tree representation for the above model) discussed in Chapter 3 holds automatically for any values of the parameters δ_ε and λ_ε as well as the parameters $p_{\varepsilon,\pm}$ and r_ε, in particular, those given by relations (9.11), (9.14), (9.15), or (9.16).

The corresponding tree $T_{\varepsilon,(y,n)}$ generated by the trajectories of the log-price process $Y_{\varepsilon,n+m}$, $m = 0, 1, \dots$ with the value $Y_{\varepsilon,n} = y$ has the total number of points after r steps,

$$L_{\varepsilon,n,n+r} \le 1 + 2r_\varepsilon + 1 + \cdots + 2r_\varepsilon r + 1 = r_\varepsilon r^2 + (r_\varepsilon + 1)r + 1\,. \tag{9.17}$$

The expression on the right-hand side in (9.17) is a polynomial (in this case quadratic) function of r.

9.1.2 Exact fitting of parameters for an univariate inhomogeneous in time binomial tree model

Let us consider the model of inhomogeneous in time Gaussian random walk, which can be considered as a discrete time analog of a Gaussian process with independent increments. In this case, a log-price process $Y_{0,n}$ is given by the following stochastic transition dynamic relation:

$$Y_{0,n} = Y_{0,n-1} + W_{0,n}\,, \quad n = 1, 2, \dots\,. \tag{9.18}$$

where (a) $Y_{0,0}$ is a real-valued random variable, (b) $W_{0,n}$, $n = 1, 2, \dots$ is a sequence of independent normal random variables with means $\mathsf{E}W_{0,n} = \mu_n$, $n = 1, 2, \dots$ and variances $\mathrm{Var}W_{0,n} = \sigma_n^2$, $n = 1, 2, \dots$, and (c) the random variable $Y_{0,0}$ and the random sequence $W_{0,n}$, $n = 1, 2, \dots$ are independent.

Also, usual assumptions are (d) $\mu_n \in \mathbb{R}_1$, $n = 1, 2, \dots$ and (e) $\sigma_n > 0$, $n = 1, 2, \dots$.

We approximate the log-price process $Y_{0,n}$ by log-price processes $Y_{\varepsilon,n}$ represented by inhomogeneous in time binomial random walks. In this case, the log-price process $Y_{\varepsilon,n}$ is given, for every $\varepsilon \in (0, \varepsilon_0]$, by the following stochastic transition dynamic relation:

$$Y_{\varepsilon,n} = Y_{\varepsilon,n-1} + W_{\varepsilon,n}\,, \quad n = 1, 2, \dots . \tag{9.19}$$

where (a) $Y_{\varepsilon,0}$ is a real-valued random variable, (b) $W_{\varepsilon,n} = W_{\varepsilon,n,1} + \cdots + W_{\varepsilon,n,r_{\varepsilon,n}}$, $n = 1, 2, \dots$, where $W_{\varepsilon,n,l}$, $l, n = 1, 2, \dots$ is a family of independent binary random variables, which take, for every $n = 1, 2, \dots$, values $\delta_{\varepsilon,n,+}$ and $\delta_{\varepsilon,n,-}$ with probabilities, respectively, $p_{\varepsilon,n,+}$ and $p_{\varepsilon,n,-}$, and (c) the random variable $Y_{\varepsilon,0}$ and the family of random variables $W_{\varepsilon,n,l}$, $l, n = 1, 2, \dots$ are independent.

Also, usual assumptions are (d) $\delta_{\varepsilon,n,-} < \delta_{\varepsilon,n,+}$, $n = 1, 2, \dots$, (e) $0 \leq p_{\varepsilon,n,+} = 1 - p_{\varepsilon,n,-} \leq 1$, $n = 1, 2, \dots$, and (f) $r_{\varepsilon,n}$, $n = 1, 2, \dots$ are positive integer numbers.

It is useful to note that $W_{\varepsilon,n}$, $n = 1, 2, \dots$ are binomial random variables taking, for every $n = 1, 2, \dots$, values $w_{\varepsilon,r_{\varepsilon,n},l} = l\delta_{\varepsilon,n,+} + (r_{\varepsilon,n} - l)\delta_{\varepsilon,n,-}$, $l = 0, \dots, r_{\varepsilon,n}$ with the corresponding probabilities $p_{\varepsilon,n,l} = C_{r_{\varepsilon,n}}^l p_{\varepsilon,n,+}^l p_{\varepsilon,n,-}^{r_{\varepsilon,n}-l}$, $l = 0, \dots, r_{\varepsilon,n}$.

The fitting should usually be made for a finite discrete time interval $[1, \dots, N]$.

We can try to choose parameters $\delta_{\varepsilon,n,\pm}$, $p_{\varepsilon,n,\pm}$, $r_{\varepsilon,n}$, $n = 1, \dots, N$ in such a way that the following exact moment-fitting equalities hold, for every $\varepsilon \in (0, \varepsilon_0]$:

$$\Big\{ \mathsf{E}W_{\varepsilon,n} = \mathsf{E}W_{0,n}\,, \quad \mathrm{Var}W_{\varepsilon,n} = \mathrm{Var}W_{0,n},\, n = 1, \dots, N\,. \tag{9.20}$$

Using the sum-representation for binomial random variables $W_{\varepsilon,n}$ used in the defining transition relation (9.19), one can easily transform equalities (9.20) in the equivalent form of the following system of equations, for every $\varepsilon \in (0, \varepsilon_0]$:

$$\begin{cases} \mathsf{E}W_{\varepsilon,n} = r_{\varepsilon,n}(\delta_{\varepsilon,n,+}p_{\varepsilon,n,+} + \delta_{\varepsilon,n,-}p_{\varepsilon,n,-}) = \mu_n\,, \\ \mathrm{Var}W_{\varepsilon,n} = r_{\varepsilon,n}\left(\delta_{\varepsilon,n,+}^2 p_{\varepsilon,n,+} + \delta_{\varepsilon,n,-}^2 p_{\varepsilon,n,-} - \left(\dfrac{\mu_n}{r_{\varepsilon,n}}\right)^2\right) = \sigma_n^2\,, \\ n = 1, \dots, N\,. \end{cases} \tag{9.21}$$

As in the case of homogeneous in time model, it is convenient to symmetrize the values of jumps by representing binary random variables $W_{\varepsilon,n,l}$ in the following form, for $l, n = 1, 2, \dots$:

$$W_{\varepsilon,n,l} = \lambda_{\varepsilon,n} + W'_{\varepsilon,n,l} = \lambda_{\varepsilon,n} + \begin{cases} +\delta_{\varepsilon,n} & \text{with probability} \quad p_{\varepsilon,n,+}\,, \\ -\delta_{\varepsilon,n} & \text{with probability} \quad p_{\varepsilon,n,-}\,, \end{cases} \tag{9.22}$$

where

$$\delta_{\varepsilon,n} = \frac{\delta_{\varepsilon,n,+} - \delta_{\varepsilon,n,-}}{2}\,, \quad \lambda_{\varepsilon,n} = \frac{\delta_{\varepsilon,n,+} + \delta_{\varepsilon,n,-}}{2}\,. \tag{9.23}$$

It follows from relation (9.22) that the transition dynamic relation (9.19) can be rewritten in the following form:

$$Y_{\varepsilon,n} = Y_{\varepsilon,n-1} + \lambda'_{\varepsilon,n} + W'_{\varepsilon,n}, \quad n = 1, 2, \dots . \tag{9.24}$$

where (a) $Y_{\varepsilon,0}$ is a real-valued random variable, (b) $W'_{\varepsilon,n} = W'_{\varepsilon,n,1} + \cdots + W'_{\varepsilon,n,r_{\varepsilon,n}}$, $n = 1, 2, \dots$, where $W'_{\varepsilon,n,l}$, $l, n = 1, 2, \dots$ is a family of i.i.d. binary random variables, which take values $\delta_{\varepsilon,n}$ and $-\delta_{\varepsilon,n}$ with probabilities, respectively, $p_{\varepsilon,n,+}$ and $p_{\varepsilon,n,-}$, (c) the random variable $Y_{\varepsilon,0}$ and the family of random variables $W'_{\varepsilon,n,l}$, $l, n = 1, 2, \dots$ are independent; (d) $\lambda'_{\varepsilon,n} = r_{\varepsilon,n}\lambda_{\varepsilon,n}$, $n = 1, 2, \dots$.

The system (9.21) can be rewritten, for every $\varepsilon \in (0, \varepsilon_0]$, in the following form:

$$\begin{cases} \delta_{\varepsilon,n}(p_{\varepsilon,n,+} - p_{\varepsilon,n,-}) + \lambda_{\varepsilon,n} = \dfrac{\mu_n}{r_{\varepsilon,n}}, \\[2ex] \delta^2_{\varepsilon,n} - \left(\dfrac{\mu_n}{r_{\varepsilon,n}} - \lambda_{\varepsilon,n}\right)^2 = \dfrac{\sigma_n^2}{r_{\varepsilon,n}} \end{cases} \tag{9.25}$$

The system (9.25) should be supplemented by the following consistency relations that should hold for admissible values of parameters, for every $\varepsilon \in (0, \varepsilon_0]$:

$$p_{\varepsilon,n,\pm} \geq 0, \quad p_{\varepsilon,n,+} + p_{\varepsilon,n,-} = 1, \delta_{\varepsilon,n} > 0, \quad \lambda_{\varepsilon,n} \in \mathbb{R}_1, r_{\varepsilon,n} = 1, 2, \dots, \\ n = 1, \dots, N. \tag{9.26}$$

In the case of inhomogeneous in time model, the recombining condition does not hold automatically. As was shown in Subsection 3.3.2, the recombining condition holds for random walk $Y_{\varepsilon,n}$ if the parameter $\delta_{\varepsilon,n}$ does not depend on n, i.e. the following additional condition holds:

$$\delta_{\varepsilon,n} = \delta_\varepsilon > 0, \quad n = 1, 2, \dots . \tag{9.27}$$

If relation (9.27) holds then the recombining condition holds for any values of the parameters δ_ε and $\lambda_{\varepsilon,n}, p_{\varepsilon,n,\pm}, r_{\varepsilon,n}$, $n = 1, \dots$.

The system (9.25) takes in this case the following form:

$$\begin{cases} \delta_\varepsilon\,(p_{\varepsilon,n,+} - p_{\varepsilon,n,-}) + \lambda_{\varepsilon,n} = \dfrac{\mu_n}{r_{\varepsilon,n}}, \\[2ex] \delta^2_\varepsilon - \left(\dfrac{\mu_n}{r_{\varepsilon,n}} - \lambda_{\varepsilon,n}\right)^2 = \dfrac{\sigma_n^2}{r_{\varepsilon,n}}, n = 1, \dots, N. \end{cases} \tag{9.28}$$

The system of equations (9.28) can be transformed to the following equivalent form:

$$\begin{cases} \delta_\varepsilon = \sqrt{\dfrac{\sigma_n^2}{r_{\varepsilon,n}} + (\dfrac{\mu_n}{r_{\varepsilon,n}} - \lambda_{\varepsilon,n})^2}, \\[2ex] p_{\varepsilon,n,\pm} = \dfrac{1}{2}\left(1 \pm \dfrac{\frac{\mu_n}{r_{\varepsilon,n}} - \lambda_{\varepsilon,n}}{\delta_\varepsilon}\right), \\[2ex] n = 1, \dots, N. \end{cases} \tag{9.29}$$

Let us denote

$$\sigma_+ = \max_{1 \le n \le N} \sigma_n . \tag{9.30}$$

Let us now take $\imath_n = +, -, n = 1, \dots, N, \delta \ge \sigma_+ > 0$ and choose,

$$\lambda_{\varepsilon,n} = \frac{\imath_n \sqrt{\delta^2 - \sigma_n^2}}{\sqrt{r_{\varepsilon,n}}} + \frac{\mu_n}{r_{\varepsilon,n}}, \quad n = 1, \dots, N . \tag{9.31}$$

Substituting $\lambda_{\varepsilon,n}$, $n = 1, \dots, N$ defined by relation (9.31) in relations (9.29), we get the following formulas:

$$\begin{cases} \delta_\varepsilon = \dfrac{\delta}{\sqrt{r_{\varepsilon,n}}}, \\ p_{\varepsilon,n,\pm} = \dfrac{1}{2}\left(1 \mp \dfrac{\imath_n \sqrt{\delta^2 - \sigma_n^2}}{\delta}\right), \\ n = 1, \dots, N . \end{cases} \tag{9.32}$$

This relation implies that we should choose, for every $\varepsilon \in (0, \varepsilon_0]$, the parameters $r_{\varepsilon,n}$ independent on $n = 1, \dots, N$, i.e.

$$r_{\varepsilon,n} = r_\varepsilon , \quad n = 1, \dots, N . \tag{9.33}$$

In this case, the following formulas give solutions of the system (9.25):

$$\delta_\varepsilon = \frac{\delta}{\sqrt{r_\varepsilon}}, \quad \lambda_{\varepsilon,n} = \frac{\imath_n \sqrt{\delta^2 - \sigma_n^2}}{\sqrt{r_\varepsilon}} + \frac{\mu_n}{r_\varepsilon}, \quad p_{\varepsilon,n,\pm} = \frac{1}{2}\left(1 \mp \frac{\imath_n \sqrt{\delta^2 - \sigma_n^2}}{\delta}\right), \quad n = 1, \dots, N, \tag{9.34}$$

parameterized by the parameters $\imath_n = +, -, n = 1, \dots, N, \delta \ge \sigma_+$ and $r_\varepsilon = 1, 2, \dots$.

Note that the consistency relations (9.26) hold for any solution given by formulas (9.34).

The above remarks can be summarized in the following lemma.

Lemma 9.1.2. *The log-price process $Y_{\varepsilon,n}$ represented, for every $\varepsilon \in (0, \varepsilon_0]$, by the binomial random walk, which is defined by the stochastic transition dynamic relation* (9.19), *realizes the exact moment fitting given by relation* (9.20) *for the log-price processes $Y_{0,n}$ represented by the Gaussian random walk, which is defined by the stochastic transition dynamic relation* (9.18). *This holds if the corresponding parameters of the process $Y_{\varepsilon,n}$ are given by relations* (9.34).

It should be pointed out that the recombining condition (which guarantees not more than polynomial rate of growth for the number of nodes in the corresponding binomial tree representation) holds automatically for any values of parameters of the log-price process $Y_{\varepsilon,n}$ given by relation (9.34).

As in the homogeneous case, the corresponding tree $T_{\varepsilon,(y,n)}$ generated by the trajectories of the log-price process $Y_{\varepsilon,n+r}$, $r = 0, 1, \ldots$ with the value $Y_{\varepsilon,n} = y$ has the total number of points after r steps,

$$L_{\varepsilon,n,n+r} \leq 1 + 2r_\varepsilon + 1 + \cdots + 2r_\varepsilon r + 1 = r_\varepsilon r^2 + (r_\varepsilon + 1)r + 1 . \tag{9.35}$$

The expression on the right-hand side in (9.35) is a polynomial (in this case quadratic) function of r.

9.1.3 Asymptotic fitting of parameters for a univariate inhomogeneous in time binomial tree model

We use the log-price processes $Y_{\varepsilon,n}$ for approximation of the log-price process $Y_{0,n}$ asymptotically, as $\varepsilon \to 0$, in the sense of weak convergence and convergence for the corresponding reward functions.

In order to realize the above approximation, we can weaken the condition of exact fitting replacing it by condition of asymptotic moment fitting for the log-price processes $Y_{\varepsilon,n}$ and $Y_{0,n}$.

In this case, the exact fitting relations (9.20) are replaced by the following weaker asymptotic relations:

$$\mathsf{E}W_{\varepsilon,n} \to \mathsf{E}W_{0,n}\,, \quad \operatorname{Var}W_{\varepsilon,n} \to \operatorname{Var}W_{0,n} \quad \text{as } \varepsilon \to 0\,, \quad n = 1, \ldots, N\,. \tag{9.36}$$

Let us return back to the system of equations (9.29). Let us choose $\lambda_{\varepsilon,n} = \frac{\mu_n}{r_{\varepsilon,n}}$, $n = 1, \ldots, N$ and represent $\delta_\varepsilon > 0$ in the form $\delta_\varepsilon = \frac{\delta}{\sqrt{r_\varepsilon}}$, where $\delta, r_\varepsilon > 0$. In this case, system (9.29) takes the following form:

$$\begin{cases} \dfrac{\delta}{\sqrt{r_\varepsilon}} = \sqrt{\dfrac{\sigma_n^2}{r_{\varepsilon,n}}}\,, \\ p_{\varepsilon,n,\pm} = \dfrac{1}{2}\,, \\ \qquad n = 1, \ldots, N\,. \end{cases} \tag{9.37}$$

The second equation in (9.37) yields the following expressions for the parameters $r_{\varepsilon,n}$:

$$r_{\varepsilon,n} = r_\varepsilon \frac{\sigma_n^2}{\delta^2}\,, \quad n = 1, \ldots, N\,. \tag{9.38}$$

The problem is that quantities $r_{\varepsilon,n}$, $n = 1, \ldots, N$ given by relation (9.38) may not take positive integer values, except some special cases for values of the parameters σ_n^2, $n = 1, \ldots, N$.

Let us show that the above asymptotic moment fitting can be achieved by computing real-valued parameters $r^*_{\varepsilon,n}$, $n = 1, \ldots, N$ using relation (9.38) and then taking as

$r_{\varepsilon,n}$, $n = 1, \dots, N$ the corresponding nearest positive integer numbers that are larger or equal than $r^*_{\varepsilon,n}$, $n = 1, \dots, N$.

Summarizing the above remarks, let us choose parameters of the approximating process $Y_{\varepsilon,n}$ in the following way, for every $\varepsilon \in (0, \varepsilon_0]$:

$$\delta_\varepsilon = \frac{\delta}{\sqrt{r_\varepsilon}}, \quad \lambda_{\varepsilon,n} = \frac{\mu_n}{r_{\varepsilon,n}}, \quad p_{\varepsilon,n,\pm} = \frac{1}{2}, \quad r_{\varepsilon,n} = [r_\varepsilon \frac{\sigma_n^2}{\delta^2}] + 1, \quad n = 1, \dots, N, \tag{9.39}$$

where one can take any positive real values for parameters δ and r_ε and then pass the parameter $r_\varepsilon \to \infty$ as $\varepsilon \to 0$.

The exact fitting equalities take place, for every $\varepsilon \in (0, \varepsilon_0]$:

$$\mathsf{E} W_{\varepsilon,n} = r_{\varepsilon,n} \left(\frac{\delta}{\sqrt{r_\varepsilon}} \frac{1}{2} - \frac{\delta}{\sqrt{r_\varepsilon}} \frac{1}{2} + \frac{\mu_n}{r_{\varepsilon,n}} \right) = \mu_n, \quad n = 1, \dots, N, \tag{9.40}$$

and the following asymptotic fitting relations take place if $r_\varepsilon \to \infty$ as $\varepsilon \to 0$:

$$\mathrm{Var} W_{\varepsilon,n} = r_{\varepsilon,n} \frac{\delta^2}{r_\varepsilon} = \left([r_\varepsilon \frac{\sigma_n^2}{\delta^2}] + 1 \right) \frac{\delta^2}{r_\varepsilon} \to \sigma_n^2 \quad \text{as} \quad \varepsilon \to 0, \quad n = 1, \dots, N. \tag{9.41}$$

The above remarks can be summarized in the following lemma.

Lemma 9.1.3. *The log-price process $Y_{\varepsilon,n}$ represented, for every $\varepsilon \in (0, \varepsilon_0]$, by the binomial random walk, which is defined by the stochastic transition dynamic relation* (9.19), *realizes the asymptotic moment fitting given by relation* (9.36) *for the log-price processes $Y_{0,n}$ represented by the Gaussian random walk, which is defined by the stochastic transition dynamic relation* (9.18). *This holds if the corresponding parameters of the process $Y_{\varepsilon,n}$ are given by relation* (9.39) *and the parameter $r_\varepsilon \to 0$ as $\varepsilon \to 0$.*

It should be pointed out that the recombining condition (which guarantees not more than polynomial rate of growth for the number of nodes in the corresponding binomial tree representation for the above model) holds automatically for any values of parameters of the log-price process $Y_{\varepsilon,n}$ given by relation (9.39).

The corresponding tree $T_{\varepsilon,(y,n)}$ generated by the trajectories of the log-price process $Y_{\varepsilon,n+r}$, $r = 0, 1, \dots$ with the value $Y_{\varepsilon,n} = y$ has the total number of points after r steps,

$$\begin{aligned} L_{\varepsilon,n,n+r} &\leq 1 + 2r_{\varepsilon,n+1} + 1 + \cdots + 2\left(r_{\varepsilon,n+1} + \cdots + r_{\varepsilon,n+r}\right) + 1 \\ &\leq 1 + 2\left(r_\varepsilon \frac{\sigma_{n+1}^2}{\delta^2} + 1 \right) + 1 + \cdots + 2\left(r_\varepsilon \frac{\sigma_{n+1}^2}{\delta^2} + 1 + \cdots + r_\varepsilon \frac{\sigma_{n+r}^2}{\delta^2} + 1 \right) + 1 \\ &= \left(r_\varepsilon \frac{\sigma^2}{\delta^2} + 1 \right) r^2 + \left(r_\varepsilon \frac{\sigma^2}{\delta^2} + 2 \right) r + 1, \end{aligned} \tag{9.42}$$

where $\sigma^2 = \max_{1 \leq n \leq N} \sigma_n^2$.

The expression on the right-hand side in (9.42) is a polynomial (in this case quadratic) function of r.

9.2 Multivariate binomial tree approximations

In this section, we present fitting formulas for multivariate binomial tree approximation models homogeneous in time and space.

9.2.1 Exact fitting of parameters for a multivariate homogeneous in time binomial tree model

Let us consider the model of multivariate homogeneous in time Gaussian random walk, which can be considered as a discrete time analog of the multivariate Wiener process (Brownian motion). In this case, a log-price $\vec{Y}_{0,n}$ is given by the following stochastic transition dynamic relation:

$$\vec{Y}_{0,n} = \vec{Y}_{0,n-1} + \vec{W}_{0,n}\,, \quad n = 1, 2, \dots . \tag{9.43}$$

where (a) $\vec{Y}_{0,0} = (Y_{0,1}, \dots, Y_{0,k})$ is a random vector taking values in the space $\mathbb{R}_k$, (b) $\vec{W}_{0,n} = (W_{0,n,1}, \dots, W_{0,n,k})$, $n = 1, 2, \dots$ is a sequence of i.i.d. k-dimensional normal random vectors with mean values $\mathsf{E}W_{0,n,i} = \mu_i$, $i = 1, \dots, k$, variances $\mathrm{Var}W_{0,n,i} = \sigma_i^2$, $i = 1, \dots, k$, and covariances $\mathsf{E}(W_{0,n,i} - \mu_i)(W_{0,n,i} - \mu_i) = \rho_{i,j}\sigma_i\sigma_j$, $i, j = 1, \dots, k$, and (c) the random vector $\vec{Y}_{0,0}$ and the family of the random vectors $\vec{W}_{0,n}$, $n = 1, 2, \dots$ are independent.

Usual assumptions are: (d) $\mu_i \in \mathbb{R}_1$, $i = 1, \dots$; (d) $\sigma_i > 0$, $i = 1, \dots, k$; (e) $|\rho_{i,j}| \leq 1$ and $\rho_{i,i} = 1$, $i, j = 1, \dots, k$; (f) $\Sigma = \|\sigma_{i,j}\|$, where $\sigma_{i,j} = \rho_{i,j}\sigma_i\sigma_j$, $i, j = 1, \dots, k$, is a nonnegatively defined $k \times k$ matrix.

We shall approximate the log-price process $Y_{0,n}$ by multivariate homogeneous in time binomial random walks. In this case, the log-price process $Y_{\varepsilon,n}$ is given, for every $\varepsilon \in (0, \varepsilon_0]$, by the following stochastic transition dynamic relation:

$$\vec{Y}_{\varepsilon,n} = \vec{Y}_{\varepsilon,n-1} + \vec{W}_{\varepsilon,n}\,, \quad n = 1, 2, \dots . \tag{9.44}$$

where (a) $\vec{Y}_{\varepsilon,0} = (Y_{\varepsilon,1}, \dots, Y_{\varepsilon,k})$ is a random vector taking values in the space $\mathbb{R}_k$, (b) $\vec{W}_{\varepsilon,n} = (W_{\varepsilon,n,1}, \dots, W_{\varepsilon,n,k}) = \vec{W}_{\varepsilon,n,1} + \cdots + \vec{W}_{\varepsilon,n,r_\varepsilon}$, $n = 1, 2, \dots$, where $\vec{W}_{\varepsilon,n,l} = (W_{\varepsilon,n,l,1}, \dots, W_{\varepsilon,n,l,k})$, $l, n = 1, 2, \dots$ is a family of i.i.d. binary random vectors taking values $\bar{\delta}_{\varepsilon,\bar{\imath}} = (\delta_{\varepsilon,1,\imath_1}, \dots, \delta_{\varepsilon,k,\imath_k})$ with the corresponding probabilities $p_{\varepsilon,\bar{\imath}} = p_\varepsilon(\imath_1, \dots \imath_k)$ for $\bar{\imath} \in \mathcal{I}_k = \{\bar{\imath} = \langle \imath_1, \dots, \imath_k \rangle : \imath_1, \dots \imath_k = \pm\}$, (c) the random vector $\vec{Y}_{\varepsilon,0}$ and the family of the random vectors $\vec{W}_{\varepsilon,n,l}$, $l, n = 1, 2, \dots$ are independent.

Also, usual assumptions are (d) $-\infty < \delta_{\varepsilon,i,-} < \delta_{\varepsilon,i,+} < \infty$, $i = 1, \dots, k$, (e) $p_{\varepsilon,\bar{\imath}} \geq 0$, $\bar{\imath} \in \mathcal{I}_k$, $\sum_{\bar{\imath},\in\mathcal{I}_k} p_{\varepsilon,\bar{\imath}} = 1$, and (f) r_ε is a positive integer number.

We use the log-price processes $\vec{Y}_{\varepsilon,n}$ for approximation of the log-price process $\vec{Y}_{0,n}$ as $\varepsilon \to 0$. In this case, a usual approach is to fit the corresponding parameters, i.e. to try choosing the parameters $\bar{\delta}_{\varepsilon,\bar{\imath}}$, $\bar{\imath}, \in \mathcal{I}_k$, probabilities $p_{\varepsilon,\bar{\imath}}$, $\bar{\imath}, \in \mathcal{I}_k$ and r_ε in such a way that the following exact fitting equalities holds, for every $\varepsilon \in (0, \varepsilon_0]$:

$$\begin{cases} \mathsf{E} W_{\varepsilon,1,i} = \mathsf{E} W_{0,1,i}, \quad i = 1, \dots, k, \\ \mathrm{Var} W_{\varepsilon,1,i} = \mathrm{Var} W_{0,1,i}, \quad i = 1, \dots, k, \\ \mathsf{E}(W_{\varepsilon,1,i} - \mathsf{E} W_{\varepsilon,1,i})(W_{\varepsilon,1,i} - \mathsf{E} W_{\varepsilon,1,j}) \\ \qquad = \mathsf{E}(W_{0,1,i} - \mathsf{E} W_{0,1,i})(W_{0,1,i} - \mathsf{E} W_{0,1,j}), \quad 1 \le i < j \le k. \end{cases} \tag{9.45}$$

Using the sum-representation for binomial random vectors $\vec{W}_{\varepsilon,n}$ used in the defining dynamic relation (9.44), one can easily transform, for every $\varepsilon \in (0, \varepsilon_0]$, equalities (9.45) to the equivalent form of the following system of equations:

$$\begin{cases} \mathsf{E} W_{\varepsilon,1,i} = r_\varepsilon \left(\delta_{\varepsilon,i,+} p_{\varepsilon,i,+} + \delta_{\varepsilon,i,-} p_{\varepsilon,i,-}\right) = \mu_i, i = 1, \dots, k \\ \mathrm{Var} W_{\varepsilon,1,i} = r_\varepsilon \left(\delta^2_{\varepsilon,i,+} p_{\varepsilon,i,+} + \delta^2_{\varepsilon,i,-} p_{\varepsilon,i,-} - (\frac{\mu_i}{r_\varepsilon})^2\right) = \sigma_i^2, i = 1, \dots, k, \\ \mathsf{E}(W_{\varepsilon,1,i} - \mathsf{E} W_{\varepsilon,1,i})(W_{\varepsilon,1,j} - \mathsf{E} W_{\varepsilon,1,j}) \\ \qquad = r_\varepsilon \delta_{\varepsilon,i,+} \delta_{\varepsilon,j,+} (p_{\varepsilon,i,j,+,+} + p_{\varepsilon,i,j,-,-} - p_{\varepsilon,i,j,+,-} - p_{\varepsilon,i,j,-,+}) - \mu_i \mu_j \\ \qquad = \rho_{i,j} \sigma_i \sigma_j, 1 \le i < j \le k, \end{cases} \tag{9.46}$$

where

$$p_{\varepsilon,i,j,\pm,\pm} = \mathsf{P}\{W_{\varepsilon,1,1,i} = \delta_{\varepsilon,i,\pm}, W_{\varepsilon,1,1,j} = \delta_{\varepsilon,j,\pm}\}, \quad 1 \le i < j \le k, \tag{9.47}$$

and

$$p_{\varepsilon,i,\pm} = \mathsf{P}\{W_{\varepsilon,1,1,i} = \delta_{\varepsilon,i,\pm}\}, \quad 1 \le i \le k. \tag{9.48}$$

The system of equations (9.46) should be supplemented by the following probability inequalities:

$$0 \le p_{\varepsilon,i,j,+,+}, p_{\varepsilon,i,j,+,-}, p_{\varepsilon,i,j,-,+}, p_{\varepsilon,i,j,-,-} \le 1, \quad 1 \le i < j \le k, \tag{9.49}$$

and the following system of consistency relations for probabilities $p_{\varepsilon,i,j,\pm,\pm}, 1 \le i < j \le k$ and $p_{\varepsilon,i,\pm}, 1 \le i \le k$,

$$\begin{cases} p_{\varepsilon,i,j,\pm,+} + p_{\varepsilon,i,j,\pm,-} = p_{\varepsilon,i,\pm}, \quad 1 \le i < j \le k, \\ p_{\varepsilon,i,j,+,\pm} + p_{\varepsilon,i,j,-,\pm} = p_{\varepsilon,j,\pm}, \quad 1 \le i < j \le k, \\ p_{\varepsilon,i,+} + p_{\varepsilon,i-} = 1, \quad 1 \le i \le k. \end{cases} \tag{9.50}$$

The multivariate random walk $\vec{Y}_{\varepsilon,n} = (Y_{\varepsilon,n,1}, \dots, Y_{\varepsilon,n,k})$ possesses recombining property since every component of this random walk $Y_{\varepsilon,n,i}$, which is a univariate homogeneous in time binomial random walk, possesses this property. Thus, the recombining condition do not impose any additional assumptions on parameters of the process $\vec{Y}_{\varepsilon,n}$.

As in the one-dimensional case, it is convenient to symmetrize the values of jumps by representing binary random vectors $\vec{W}_{\varepsilon,n,l}$ in the following form, for $l, n = 1, 2, \dots$:

$$\vec{W}_{\varepsilon,n,l} = \bar{\lambda}_\varepsilon + \vec{W}'_{\varepsilon,n,l} = \bar{\lambda}_\varepsilon + \left\{ \bar{\delta}'_{\varepsilon,\bar{\imath}} \text{ with probability } p_{\varepsilon,\bar{\imath}}, \quad \bar{\imath} \in \mathcal{I}_k \right. \tag{9.51}$$

where

$$\bar{\lambda}_\varepsilon = (\lambda_{\varepsilon,1}, \dots, \lambda_{\varepsilon,k}), \quad \bar{\delta}'_{\varepsilon,\bar{\imath}} = (\imath_1 \delta_{\varepsilon,1}, \dots, \imath_k \delta_{\varepsilon,k}), \quad \bar{\imath} = \langle \imath_1, \dots, \imath_k \rangle \in \mathcal{I}_k, \tag{9.52}$$

and

$$\delta_{\varepsilon,i} = \frac{\delta_{\varepsilon,i,+} - \delta_{\varepsilon,i,-}}{2}, \quad \lambda_{\varepsilon,i} = \frac{\delta_{\varepsilon,i,+} + \delta_{\varepsilon,i,-}}{2}, \quad i = 1, \dots, k. \tag{9.53}$$

It follows from relation (9.52) that the transition dynamic relation (9.44) can be rewritten in the following form:

$$\vec{Y}_{\varepsilon,n} = \vec{Y}_{\varepsilon,n-1} + \bar{\lambda}'_\varepsilon + \vec{W}'_{\varepsilon,n}, \quad n = 1, 2, \dots. \tag{9.54}$$

where (a) $\vec{Y}_{\varepsilon,0} = (Y_{\varepsilon,1}, \dots, Y_{\varepsilon,k})$ is a random vector taking values in the space $\mathbb{R}_k$, (b) $\vec{W}'_{\varepsilon,n} = (W'_{\varepsilon,n,1}, \dots, W'_{\varepsilon,n,k}) = \vec{W}'_{\varepsilon,n,1} + \cdots + \vec{W}'_{\varepsilon,n,r_\varepsilon}$, $n = 1, 2, \dots$, where $\vec{W}'_{\varepsilon,n,l} = (W'_{\varepsilon,n,l,1}, \dots, W'_{\varepsilon,n,l,k})$, $l, n = 1, 2, \dots$ is a family of i.i.d. binary random vectors taking values $\bar{\delta}'_{\varepsilon,\bar{\imath}}$ with the corresponding probabilities $p_{\varepsilon,\bar{\imath}}$, for $\bar{\imath} \in \mathcal{I}_k$, (c) the random vector $\vec{Y}_{\varepsilon,0}$ and the family of the random vectors $\vec{W}'_{\varepsilon,n,l}$, $l, n = 1, 2, \dots$ are independent; (d) $\bar{\lambda}'_\varepsilon = r_\varepsilon \bar{\lambda}_\varepsilon$.

According to the first equation in the system (9.46), for every $i = 1, \dots, k$,

$$\mathsf{E} W_{\varepsilon,1,1,i} = \mathsf{E} W'_{\varepsilon,1,1,i} + \lambda_{\varepsilon,i} = \delta_{\varepsilon,i}(p_{\varepsilon,i,+} - p_{\varepsilon,i,-}) + \lambda_{\varepsilon,i} = \frac{\mu_j}{r_\varepsilon}. \tag{9.55}$$

Also, $p_{\varepsilon,i,+} + p_{\varepsilon,i,-} = 1$, $i = 1, \dots, k$. Thus, for every $i = 1, \dots, k$,

$$\begin{aligned} \mathrm{Var} W_{\varepsilon,1,1,i} &= \mathrm{Var} W'_{\varepsilon,1,1,i} \\ &= \delta^2_{\varepsilon,i}(p_{\varepsilon,i,+} + p_{\varepsilon,i,-}) - (\frac{\mu_i}{r_\varepsilon} - \lambda_{\varepsilon,i})^2 = \delta^2_{\varepsilon,i} - (\frac{\mu_i}{r_\varepsilon} - \lambda_{\varepsilon,i})^2. \end{aligned} \tag{9.56}$$

and, for every $1 \leq i < j \leq k$,

$$\begin{aligned} \mathsf{E}\left(W_{\varepsilon,1,1,i} - \frac{\mu_i}{r_\varepsilon}\right)\left(W_{\varepsilon,1,1,j} - \frac{\mu_j}{r_\varepsilon}\right) &= \mathsf{E}\left(W'_{\varepsilon,1,1,i} - \left(\frac{\mu_i}{r_\varepsilon} - \lambda_{\varepsilon,i}\right)\right)\left(W'_{\varepsilon,1,1,j} - \left(\frac{\mu_j}{r_\varepsilon} - \lambda_{\varepsilon,j}\right)\right) \\ &= \mathsf{E} W q'_{\varepsilon,1,1,i} W'_{\varepsilon,1,1,j} - \left(\frac{\mu_i}{r_\varepsilon} - \lambda_{\varepsilon,i}\right)\left(\frac{\mu_j}{r_\varepsilon} - \lambda_{\varepsilon,j}\right), \\ &= \delta_{\varepsilon,i}\delta_{\varepsilon,j}\Big(p_{\varepsilon,i,j,+,+} \cdot p_{\varepsilon,i,j,+,+} + p_{\varepsilon,i,j,-,-} \cdot p_{\varepsilon,i,j,-,-} \\ &\quad - p_{\varepsilon,i,j,+,-} \cdot p_{\varepsilon,i,j,-,+} - p_{\varepsilon,i,j,-,+} \cdot p_{\varepsilon,i,j,+,-}\Big) \\ &\quad - \left(\frac{\mu_i}{r_\varepsilon} - \lambda_{\varepsilon,i}\right)\left(\frac{\mu_j}{r_\varepsilon} - \lambda_{\varepsilon,j}\right). \end{aligned} \tag{9.57}$$

Therefore, the system (9.46) takes the following form:

$$\begin{cases} \delta_{\varepsilon,i}\,(p_{\varepsilon,i,+} - p_{\varepsilon,i,-}) + \lambda_{\varepsilon,i} = \dfrac{\mu_i}{r_\varepsilon}, \quad i = 1,\dots,k, \\ \delta^2_{\varepsilon,i} - \left(\dfrac{\mu_i}{r_\varepsilon} - \lambda_{\varepsilon,i}\right)^2 = \dfrac{\sigma_i^2}{r_\varepsilon}, \quad i = 1,\dots,k, \\ \delta_{\varepsilon,i}\delta_{\varepsilon,j}(p_{\varepsilon,i,j,+,+} + p_{\varepsilon,i,j,-,-} - p_{\varepsilon,i,j,+,-} - p_{\varepsilon,i,j,-,+}) - \left(\dfrac{\mu_i}{r_\varepsilon} - \lambda_{\varepsilon,i}\right)\left(\dfrac{\mu_j}{r_\varepsilon} - \lambda_{\varepsilon,j}\right) \\ \qquad = \dfrac{\rho_{i,j}\sigma_i\sigma_j}{r_\varepsilon}, \quad 1 \le i < j \le k. \end{cases} \tag{9.58}$$

The second equation in (9.58) yields the following formulas for quantities $\delta_{\varepsilon,i}$, $i = 1,\dots,k$:

$$\delta^2_{\varepsilon,i} = \sqrt{\frac{\sigma_i^2}{r_\varepsilon} + (\frac{\mu_i}{r_\varepsilon} - \lambda_{\varepsilon,i})^2}\,. \tag{9.59}$$

Also, the first series of equations in (9.58) plus the last equations in (9.58) and (9.50) yield the following system of equations:

$$\begin{cases} p_{\varepsilon,i,j,+,+} + p_{\varepsilon,i,j,+,-} - p_{\varepsilon,i,j,-,+} - p_{\varepsilon,i,j,-,-} = \dfrac{\frac{\mu_i}{r_\varepsilon} - \lambda_{\varepsilon,i}}{\delta_{\varepsilon,i}}, \\ p_{\varepsilon,i,j,+,+} + p_{\varepsilon,i,j,-,+} - p_{\varepsilon,i,j,+,-} - p_{\varepsilon,i,j,-,-} = \dfrac{\frac{\mu_j}{r_\varepsilon} - \lambda_{\varepsilon,j}}{\delta_{\varepsilon,j}}, \\ p_{\varepsilon,i,j,+,+} + p_{\varepsilon,i,j,-,-} - p_{\varepsilon,i,j,+,-} - p_{\varepsilon,i,j,-,+} = \dfrac{\frac{\rho_{i,j}\sigma_i\sigma_j}{r_\varepsilon} + (\frac{\mu_i}{r_\varepsilon} - \lambda_{\varepsilon,i})(\frac{\mu_j}{r_\varepsilon} - \lambda_{\varepsilon,j})}{\delta_{\varepsilon,i}\delta_{\varepsilon,j}} \\ p_{\varepsilon,i,j,+,+} + p_{\varepsilon,i,j,-,-} + p_{\varepsilon,i,j,+,-} + p_{\varepsilon,i,j,-,+} = 1, \quad 1 \le i < j \le k. \end{cases} \tag{9.60}$$

The system of equations (9.60) has the unique solution for any fixed values of free terms.

The explicit formulas for $p_{\varepsilon,i,j,+,+}$ can be obtained by adding of all equations in (9.60), dividing the result by 4, and taking into account formula (9.59) for the parameters $\delta_{\varepsilon,i}$, $i = 1,\dots,k$.

The explicit formulas for $p_{\varepsilon,i,j,+,-}$ can be obtained by adding the first and the fourth and subtracting the second and the third equation in (9.60), dividing the result by 4, and taking into account formula (9.59) for the parameters $\delta_{\varepsilon,i}$, $i = 1,\dots,k$.

The explicit formulas for $p_{\varepsilon,i,j,-,+}$ can be obtained by adding the second and the fourth and subtracting the first and the third equations in (9.60), dividing the result by 4, and taking into account formula (9.59) for the parameters $\delta_{\varepsilon,i}$, $i = 1,\dots,k$.

The explicit formulas for $p_{\varepsilon,i,j,-,-}$ can be obtained by adding the third and the fourth and subtracting the first and the second equations in (9.60), dividing the result by 4, and taking into account formula (9.59) for the parameters $\delta_{\varepsilon,i}$, $i = 1,\dots,k$.

Taking into account the above remarks, we can conclude that the system of equations (9.58) has infinitely many solutions given by the following formulas:

$$\begin{cases} \delta_{\varepsilon,i} = \sqrt{\frac{\sigma_i^2}{r_\varepsilon} + (\frac{\mu_i}{r_\varepsilon} - \lambda_{\varepsilon,i})^2}, \quad i = 1, \dots, k, \\ p_{\varepsilon,i,j,+,+} = \frac{1}{4} + \frac{1}{4}\left(\frac{\frac{\mu_i}{r_\varepsilon} - \lambda_{\varepsilon,i}}{\sqrt{\frac{\sigma_i^2}{r_\varepsilon} + (\frac{\mu_i}{r_\varepsilon} - \lambda_{\varepsilon,i})^2}} + \frac{\frac{\mu_j}{r_\varepsilon} - \lambda_{\varepsilon,j}}{\sqrt{\frac{\sigma_j^2}{r_\varepsilon} + (\frac{\mu_j}{r_\varepsilon} - \lambda_{\varepsilon,j})^2}} \right) \\ \quad + \frac{1}{4} \frac{\frac{\rho_{i,j}\sigma_i\sigma_j}{r_\varepsilon} + (\frac{\mu_i}{r_\varepsilon} - \lambda_{\varepsilon,i})(\frac{\mu_j}{r_\varepsilon} - \lambda_{\varepsilon,j})}{\sqrt{\frac{\sigma_i^2}{r_\varepsilon} + (\frac{\mu_i}{r_\varepsilon} - \lambda_{\varepsilon,i})^2}\sqrt{\frac{\sigma_j^2}{r_\varepsilon} + (\frac{\mu_j}{r_\varepsilon} - \lambda_{\varepsilon,j})^2}}, \\ p_{\varepsilon,i,j,+,-} = \frac{1}{4} + \frac{1}{4}\left(\frac{\frac{\mu_i}{r_\varepsilon} - \lambda_{\varepsilon,i}}{\sqrt{\frac{\sigma_i^2}{r_\varepsilon} + (\frac{\mu_i}{r_\varepsilon} - \lambda_{\varepsilon,i})^2}} - \frac{\frac{\mu_j}{r_\varepsilon} - \lambda_{\varepsilon,j}}{\sqrt{\frac{\sigma_j^2}{r_\varepsilon} + (\frac{\mu_j}{r_\varepsilon} - \lambda_{\varepsilon,j})^2}} \right) \\ \quad - \frac{1}{4} \frac{\frac{\rho_{i,j}\sigma_i\sigma_j}{r_\varepsilon} + (\frac{\mu_i}{r_\varepsilon} - \lambda_{\varepsilon,i})(\frac{\mu_j}{r_\varepsilon} - \lambda_{\varepsilon,j})}{\sqrt{\frac{\sigma_i^2}{r_\varepsilon} + (\frac{\mu_i}{r_\varepsilon} - \lambda_{\varepsilon,i})^2}\sqrt{\frac{\sigma_j^2}{r_\varepsilon} + (\frac{\mu_j}{r_\varepsilon} - \lambda_{\varepsilon,j})^2}}, \\ p_{\varepsilon,i,j,-,+} = \frac{1}{4} + \frac{1}{4}\left(-\frac{\frac{\mu_i}{r_\varepsilon} - \lambda_{\varepsilon,i}}{\sqrt{\frac{\sigma_i^2}{r_\varepsilon} + (\frac{\mu_i}{r_\varepsilon} - \lambda_{\varepsilon,i})^2}} + \frac{\frac{\mu_j}{r_\varepsilon} - \lambda_{\varepsilon,j}}{\sqrt{\frac{\sigma_j^2}{r_\varepsilon} + (\frac{\mu_j}{r_\varepsilon} - \lambda_{\varepsilon,j})^2}} \right) \\ \quad - \frac{1}{4} \frac{\frac{\rho_{i,j}\sigma_i\sigma_j}{r_\varepsilon} + (\frac{\mu_i}{r_\varepsilon} - \lambda_{\varepsilon,i})(\frac{\mu_j}{r_\varepsilon} - \lambda_{\varepsilon,j})}{\sqrt{\frac{\sigma_i^2}{r_\varepsilon} + (\frac{\mu_i}{r_\varepsilon} - \lambda_{\varepsilon,i})^2}\sqrt{\frac{\sigma_j^2}{r_\varepsilon} + (\frac{\mu_j}{r_\varepsilon} - \lambda_{\varepsilon,j})^2}}, \\ p_{\varepsilon,i,j,-,-} = \frac{1}{4} + \frac{1}{4}\left(-\frac{\frac{\mu_i}{r_\varepsilon} - \lambda_{\varepsilon,i}}{\sqrt{\frac{\sigma_i^2}{r_\varepsilon} + (\frac{\mu_i}{r_\varepsilon} - \lambda_{\varepsilon,i})^2}} - \frac{\frac{\mu_j}{r_\varepsilon} - \lambda_{\varepsilon,j}}{\sqrt{\frac{\sigma_j^2}{r_\varepsilon} + (\frac{\mu_j}{r_\varepsilon} - \lambda_{\varepsilon,j})^2}} \right) \\ \quad + \frac{1}{4} \frac{\frac{\rho_{i,j}\sigma_i\sigma_j}{r_\varepsilon} + (\frac{\mu_i}{r_\varepsilon} - \lambda_{\varepsilon,i})(\frac{\mu_j}{r_\varepsilon} - \lambda_{\varepsilon,j})}{\sqrt{\frac{\sigma_i^2}{r_\varepsilon} + (\frac{\mu_i}{r_\varepsilon} - \lambda_{\varepsilon,i})^2}\sqrt{\frac{\sigma_j^2}{r_\varepsilon} + (\frac{\mu_j}{r_\varepsilon} - \lambda_{\varepsilon,j})^2}}, \\ 1 \leq i < j \leq k, \end{cases} \tag{9.61}$$

where solutions are parameterized by the parameters $\lambda_{\varepsilon,1}, \dots, \lambda_{\varepsilon,k} \in \mathbb{R}_1$ and $r_\varepsilon = 1, 2, \dots$.

It is easy to check that solutions given by formulas (9.61) satisfy the consistency relations (9.50).

Solutions of the system (9.58) given by relation (9.61) take especially simple form, if one to choose the parameters $\lambda_{\varepsilon,i} = \frac{\mu_i}{r_\varepsilon}$, $i = 1, \dots, k$.

In this case, parameters of the process $\vec{Y}_{\varepsilon,n}$ take the following very simple form:

$$\begin{cases} \lambda_{\varepsilon,i} = \frac{\mu_i}{r_\varepsilon}, \quad \delta_{\varepsilon,i} = \frac{\sigma_i}{\sqrt{r_\varepsilon}}, i = 1, \dots, k, \\ p_{\varepsilon,i,j,+,+} = p_{\varepsilon,i,j,-,-} = \frac{1}{4}(1 + \rho_{i,j}), 1 \leq i < j \leq k, \\ p_{\varepsilon,i,j,+,-} = p_{\varepsilon,i,j,-,+} = \frac{1}{4}(1 - \rho_{i,j}), 1 \leq i < j \leq k, \end{cases} \tag{9.62}$$

parameterized by the parameter $r_\varepsilon = 1, 2, \dots$.

These solutions satisfy both probability inequalities (9.49) and the consistency relations (9.50), for every $\varepsilon \in (0, \varepsilon_0]$.

In this case, quantities $p_{\varepsilon,i,\pm}$ take the following form:

$$p_{\varepsilon,i,\pm} = \frac{1}{2}, \quad i = 1, \dots, k. \tag{9.63}$$

Let us now take $\imath_i = +, -$, $\delta_i \geq \sigma_i > 0$, $i = 1, \dots, k$ and choose quantities $\lambda_{\varepsilon,i}$ given by the following formulas:

$$\lambda_{\varepsilon,i} = \frac{\imath_i \sqrt{\delta_i^2 - \sigma_i^2}}{\sqrt{r_\varepsilon}} + \frac{\mu_i}{r_\varepsilon}, \quad i = 1, \dots, k. \tag{9.64}$$

Substituting $\lambda_{\varepsilon,i}$, $i = 1, \dots, k$ given by formulas (9.64) in relations (9.61), we get solutions of system of equations (9.58) in the following form:

$$\begin{cases} \delta_{\varepsilon,i} = \dfrac{\delta_i}{\sqrt{r_\varepsilon}}, \ \lambda_{\varepsilon,i} = \dfrac{\imath_i \sqrt{\delta_i^2 - \sigma_i^2}}{\sqrt{r_\varepsilon}} + \dfrac{\mu_i}{r_\varepsilon}, \ i = 1, \dots, k. \\ p_{\varepsilon,i,j,+,+} = \dfrac{1}{4} + \dfrac{1}{4}\rho_{i,j}\dfrac{\sigma_i\sigma_j}{\delta_i\delta_j} + \dfrac{1}{4}\left(-\dfrac{\imath_i\sqrt{\delta_i^2 - \sigma_i^2}}{\delta_i} - \dfrac{\imath_j\sqrt{\delta_j^2 - \sigma_j^2}}{\delta_j} \right) \\ \qquad + \dfrac{1}{4}\dfrac{\imath_i\sqrt{\delta_i^2 - \sigma_i^2}\,\imath_j\sqrt{\delta_j^2 - \sigma_j^2}}{\delta_i\delta_j}, \quad 1 \leq i < j \leq k, \\ p_{\varepsilon,i,j,+,-} = \dfrac{1}{4} - \dfrac{1}{4}\rho_{i,j}\dfrac{\sigma_i\sigma_j}{\delta_i\delta_j} + \dfrac{1}{4}\left(-\dfrac{\imath_i\sqrt{\delta_i^2 - \sigma_i^2}}{\delta_i} + \dfrac{\imath_j\sqrt{\delta_j^2 - \sigma_j^2}}{\delta_j} \right) \\ \qquad - \dfrac{1}{4}\dfrac{\imath_i\sqrt{\delta_i^2 - \sigma_i^2}\,\imath_j\sqrt{\delta_j^2 - \sigma_j^2}}{\delta_i\delta_j}, \quad 1 \leq i < j \leq k, \\ p_{\varepsilon,i,j,-,+} = \dfrac{1}{4} - \dfrac{1}{4}\rho_{i,j}\dfrac{\sigma_i\sigma_j}{\delta_i\delta_j} + \dfrac{1}{4}\left(\dfrac{\imath_i\sqrt{\delta_i^2 - \sigma_i^2}}{\delta_i} - \dfrac{\imath_j\sqrt{\delta_j^2 - \sigma_j^2}}{\delta_j} \right) \\ \qquad - \dfrac{1}{4}\dfrac{\imath_i\sqrt{\delta_i^2 - \sigma_i^2}\,\imath_j\sqrt{\delta_j^2 - \sigma_j^2}}{\delta_i\delta_j}, \quad 1 \leq i < j \leq k, \\ p_{\varepsilon,i,j,-,-} = \dfrac{1}{4} + \dfrac{1}{4}\rho_{i,j}\dfrac{\sigma_i\sigma_j}{\delta_i\delta_j} + \dfrac{1}{4}\left(\dfrac{\imath_i\sqrt{\delta_i^2 - \sigma_i^2}}{\delta_i} + \dfrac{\imath_j\sqrt{\delta_j^2 - \sigma_j^2}}{\delta_j} \right) \\ \qquad + \dfrac{1}{4}\dfrac{\imath_i\sqrt{\delta_i^2 - \sigma_i^2}\,\imath_j\sqrt{\delta_j^2 - \sigma_j^2}}{\delta_i\delta_j}, \quad 1 \leq i < j \leq k, \end{cases} \tag{9.65}$$

parameterized by the parameters $\imath_i = +, -$, $\delta_i \geq \sigma_i$, $i = 1, \dots, k$ and $r_\varepsilon = 1, 2, \dots$.

It is easy to check that solutions given by the formulas (9.65) satisfy the consistency relations (9.50).

For example, the following formulas can be obtained for quantities $p_{\varepsilon,i,\pm}$ by applying the corresponding summation formulas given in consistency relations (9.50) to quantities $p_{\varepsilon,i,j,\imath_i,\imath_j}$, $\imath_i, \imath_j = +, -$ given by formulas (9.65)

$$p_{\varepsilon,i,\pm} = \frac{1}{2}(1 \mp \frac{\imath_i\sqrt{\delta_i^2 - \sigma_i^2}}{\delta_i}), \quad i = 1, \dots, k. \tag{9.66}$$

The question about holding the probability inequalities (9.49) for quantities $p_{\varepsilon,i,j,\pm,\pm}$ is more complicated. These quantities, given by formulas (9.65), can take negative values for large values of parameters δ_i, $i = 1, \dots, k$

Let us introduce the parameter

$$\rho = \max_{1 \le i < j \le k} |\rho_{i,j}| \in [0, 1]. \tag{9.67}$$

It follows from formulas (9.65) that inequalities (9.49) for quantities $p_{\varepsilon,i,j,\imath_i,\imath_j}$, $\imath_i, \imath_j = +, -$ hold if the following inequalities hold, for every $1 \le i < j \le k$:

$$\frac{1}{4}(1 - \rho\frac{\sigma_i\sigma_j}{\delta_i\delta_j} - \frac{\sqrt{\delta_i^2 - \sigma_i^2}}{\delta_i} - \frac{\sqrt{\delta_j^2 - \sigma_j^2}}{\delta_j} + \frac{\sqrt{\delta_i^2 - \sigma_i^2}\sqrt{\delta_j^2 - \sigma_j^2}}{\delta_i\delta_j}) \ge 0. \tag{9.68}$$

For the moment, let us use parameters

$$\kappa_i = \frac{\delta_i}{\sigma_i} \ge 1, \quad i = 1, \dots, k. \tag{9.69}$$

Inequalities (9.68) can be rewritten in the following equivalent form:

$$(\kappa_i - \sqrt{\kappa_i^2 - 1})(\kappa_j - \sqrt{\kappa_j^2 - 1}) \ge \rho. \tag{9.70}$$

A simple calculus calculations show that functions $f(\kappa) = \kappa - \sqrt{\kappa^2 - 1}, \kappa \ge 1$ takes value $f(1) = 1$, strictly decrease, and $f(\kappa) \to 0$ as $\kappa \to \infty$.

That is why, $f(\kappa) \ge \sqrt{\rho}$ if and only if $1 \le \kappa \le \kappa(\rho)$, where $\kappa(\rho)$ is the root of equation,

$$\kappa(\rho) - \sqrt{\kappa^2(\rho) - 1} = \sqrt{\rho}. \tag{9.71}$$

This equation has the root

$$\kappa(\rho) = \frac{1 + \rho}{2\sqrt{\rho}} \ge 1. \tag{9.72}$$

Inequality (9.70) and the above remarks imply that inequalities (9.49) hold if the following condition holds for the parameters δ_i, $i = 1, \dots, k$:

$$\sigma_i \le \delta_i \le \kappa(\rho)\sigma_i, \quad i = 1, \dots, k. \tag{9.73}$$

At the same time, the above remarks imply that inequalities (9.49) do not hold if there exist a pair of indices $1 \le i < j \le k$ such that $\rho_{i,j} = -\rho$ and $\delta_i > \kappa(\rho)\sigma_i$,

$\delta_j > \kappa(\rho)\sigma_j$. In this case, at least one of the quantities $p_{\varepsilon,i,j,\imath_i,\imath_j}$, $\imath_i, \imath_j = +, -$, given by formulas (9.65), takes a negative value.

It is useful to note that condition (9.73) let one choose parameters $\delta_i = \sigma_i$, $i = 1, \dots, k$. In this case, formulas (9.65) reduce to simpler formulas (9.62).

Let us introduce the condition:

$\mathbf{R}_1$: $\rho < 1$.

If the condition $\mathbf{R}_1$ holds, then $\kappa(\rho) > 1$, and, thus, all intervals $[\sigma_i, \kappa(\rho)\sigma_i]$, $i = 1, \dots, k$, for admissible values of the parameters δ_i, $i = 1, \dots, k$ have nonzero lengths.

If the condition $\mathbf{R}_1$ does not hold, i.e. $\rho = 1$, then $\kappa(\rho) = 1$. In this case the only values $\delta_i = \sigma_i$, $i = 1, \dots, k$ are admissible.

The fitting procedure described above, yields the one-dimensional and two-dimensional distributions for components of the random vectors $\vec{W}_{\varepsilon,1,1} = (W_{\varepsilon,1,1,1}, \dots, W_{\varepsilon,1,1,k})$, respectively, $p_{\varepsilon,i,\pm}$, $i = 1, \dots, k$ and $p_{\varepsilon,i,j,\pm,\pm}$, $1 \leq i < j \leq k$. These distributions are consistent in the sense of relations (9.47), (9.48), and (9.50).

The analysis of the fitting problem presented above is complete in the case of two-dimensional binomial random walks.

However, in the case of k-dimensional binomial random walk, with $k \geq 3$, the additional question, about existence of a multivariate distribution of the random vector $\vec{W}_{\varepsilon,1,1}$ (with the above one- and two-dimensional distributions $p_{\varepsilon,i,\pm}$, $1 \leq i \leq k$ and $p_{\varepsilon,i,j,\pm,\pm}$, $1 \leq i < j \leq k$) arises.

This question has an affirmative answer. Moreover, the algorithm for constructing of the corresponding k-dimensional distributions is given below in Lemma 9.2.2, which should be applied to the above one- and two-dimensional distributions, for every $\varepsilon \in (0, \varepsilon_0]$.

The above remarks can be summarized in the following lemma.

Lemma 9.2.1. *The log-price process $\vec{Y}_{\varepsilon,n}$ represented, for $\varepsilon \in (0, \varepsilon_0]$, by the multivariate binomial random walk, which is defined by the stochastic transition dynamic relation* (9.44), *realizes the exact moment fitting given by relation* (9.45) *for the log-price process $Y_{0,n}$ represented by the Gaussian random walk defined by the stochastic transition dynamic relation* (9.43). *Parameters of the process $\vec{Y}_{\varepsilon,n}$ namely, $\delta_{\varepsilon,i}$, $\lambda_{\varepsilon,i}$, $p_{\varepsilon,i,\pm}$, $1 \leq i \leq k$, $p_{\varepsilon,i,j,\pm,\pm}$, $1 \leq i < i \leq k$, r_ε, can be chosen, for every $\varepsilon \in (0, \varepsilon_0]$, according to relations* (9.62) *and* (9.63) *and an arbitrary choice of $r_\varepsilon = 1, 2, \dots$, or according to relations* (9.65) *and* (9.66), *the choice of the parameters δ_i, $i = 1, \dots, k$ satisfying condition* (9.73) *and an arbitrary choice of $r_\varepsilon = 1, 2, \dots$. The corresponding high-order distributions for components of the random vectors $\vec{W}_{\varepsilon,1,1}$ can be constructed with the use of algorithm described in Lemma 9.2.2, which should be applied to the above one- and two-dimensional distributions $p_{\varepsilon,i,\pm}$, $i = 1, \dots, k$ and $p_{\varepsilon,i,j,\pm,\pm}$, $1 \leq i < j \leq k$, for every $\varepsilon \in (0, \varepsilon_0]$.*

It is worth noting that the fitting procedure based on expectations and covariances involves only one- and two-dimensional distributions for the random vectors $\vec{W}_{\varepsilon,1,1}$.

In fact, we need only know that the random vectors $\vec{W}_{\varepsilon,1,1}$, with the corresponding fitted one- and two-dimensional distributions, do exist.

The same is valid for the corresponding convergence results for option rewards presented in the next chapter. The only expectations and covariances and, thus, only one- and two-dimensional distributions for the random vectors $\vec{W}_{\varepsilon,1,1}$ are involved in the corresponding convergence conditions, which provide convergence of the reward functions for the binomial log-price processes $\vec{Y}_{\varepsilon,n}$ to the reward functions for the Gaussian log-price process $\vec{Y}_{0,n}$.

However, the high-dimensional distributions for the random vectors $\vec{W}_{\varepsilon,1,1}$ (for the case, $k \geq 3$) are explicitly used in the backward recurrence relations for computing of the reward functions for the approximating binomial log-price processes $\vec{Y}_{\varepsilon,n}$.

It should be pointed out that the recombining condition (which guarantees not more than polynomial rate of growth for number of nodes in the corresponding binomial tree representation for the above model) discussed in Chapter 3 holds automatically for any values of parameters for the log-price process $\vec{Y}_{\varepsilon,n}$ given Lemma 9.2.1.

The corresponding tree $T_{\varepsilon,(\vec{y},n)}$ generated by the trajectories of the log-price process $Y_{\varepsilon,n+l}$, $l = 0, 1, \dots$ with the value $\vec{Y}_{\varepsilon,n} = \vec{y}$ has the total number of points after r steps,

$$
\begin{aligned}
L_{\varepsilon,n,n+r} &\leq 1 + (2r_\varepsilon + 1)^k + \cdots + (2r_\varepsilon r + 1)^k \\
&\leq \int_0^{r+1} (2r_\varepsilon x + 1)^k dx = \frac{(2r_\varepsilon(r+1) + 1)^{k+1} - 1}{2r_\varepsilon(k+1)} .
\end{aligned} \tag{9.74}
$$

The expression on the right-hand side in (9.74) is a polynomial function of r of order $k + 1$.

9.2.2 Multivariate binary distributions with given one- and two-dimensional distributions

Let $k \geq 3$ and $\vec{W} = (W_1, \dots, W_k)$ be a k-dimensional binary random vector, which takes values $\bar{\delta}_{\bar{\imath}} = (\delta_{1,\imath_1}, \dots, \delta_{k,\imath_k})$, where $\delta_{i,\imath_i}, \imath_i = +, -, i = 1, \dots, k$ are real numbers and $\bar{\imath} = \langle \imath_1, \dots, \imath_k \rangle \in \mathcal{I}_k = \{\bar{\imath} : \imath_1, \dots, \imath_k = +, -\}$.

Also, let $1 \leq i_1 < \cdots < i_l \leq k$ and $p_{i_1,\dots,i_l,\imath_{i_1},\dots,\imath_{i_l}} = \mathsf{P}\{W_{i_1} = \delta_{i_1,\imath_{i_1}}, \dots, W_{i_l} = \delta_{i_l,\imath_{i_l}}\}$, $\imath_{i_1}, \dots, \imath_{i_l} = +, -$ be the corresponding joint l-dimensional distributions for components of the random vector $\vec{W}$.

The distributions $p_{i_1,\dots,i_l,\imath_{i_1},\dots,\imath_{i_l}}$ should satisfy the following consistency relations:

$$
\begin{cases}
p_{i_1,\dots,i_l,\imath_{i_1},\dots,\imath_{i_l}} \geq 0 , \\
\sum\limits_{\imath_{i_j}=+,-} p_{i_1,\dots,i_l,\imath_{i_1},\dots,\imath_{i_l}} = p_{i_1,\dots,i_{j-1},i_{j+1},\dots,i_l,\imath_{i_1},\dots,\imath_{i_{j-1}},\imath_{i_{j+1}},\dots,\imath_{i_l}} \\
\qquad \imath_{i_1}, \dots, \imath_{i_l} = +, - , \quad 1 \leq i_1 < \cdots < i_l \leq k, \quad 1 \leq j \leq l \leq k ,
\end{cases} \tag{9.75}
$$

where the quantity $p = p_{i_1,\dots,i_l,\imath_{i_1},\dots,\imath_{i_l}}$ should be counted as 1 for the case $l = 0$.

By the Kolmogorov theorem about consistent finite-dimensional distributions, there exists a random vector $\vec{W}$ with any consistent family of joint distributions of components satisfying relations (9.75).

The random vector $\vec{W} = (W_1, \dots, W_k)$ can be constructed in the following way.

One can order indices + and − (using the ordering symbol $\succ$) as $+ \succ -$ and then vectors $\bar{\imath} = \langle \imath_1, \dots, \imath_k \rangle \in \mathcal{I}_k$ (using the ordering symbol $\succ$) in a natural way, by counting $\bar{\imath}' \succ \bar{\imath}''$, if there exists $1 \le j \le k$ such that $\imath'_i = \imath''_i$, for $i < j$, but $\imath'_j = +, \imath''_j = -$.

Then, one can define disjoint subintervals $\mathbb{I}_{\bar{\imath}} = [q^-_{\bar{\imath}}, q^+_{\bar{\imath}}), \bar{\imath} \in \mathcal{I}_k$ of interval $[0, 1)$ such that: (a) $q^+_{\bar{\imath}} - q^-_{\bar{\imath}} = p_{1,\dots,k,\imath_1,\dots,\imath_k}, \bar{\imath} \in \mathcal{I}_k$, (b) interval $\mathbb{I}_{\bar{\imath}'}$ is located to the right of interval $\mathbb{I}_{\bar{\imath}''}$, if $\bar{\imath}' \succ \bar{\imath}''$.

Finally, the random vector $\vec{W} = (W_1, \dots, W_k)$ can be defined as

$$\vec{W} = \sum_{\bar{\imath} \in \mathcal{I}_k} \bar{\delta}_{\bar{\imath}} \mathrm{I}(\rho \in \mathbb{I}_{\bar{\imath}}) , \tag{9.76}$$

where ρ is the random variable uniformly distributed in the interval $[0, 1)$.

The fitting algorithm presented in Subsection 9.2.1 does require to answer a slightly different question.

Let us assume that some one-dimensional and two-dimensional distributions, respectively $p_{i,\imath_i}, \imath_i = +, -, 1 \le i \le k$ and $p_{i,j,\imath_i,\imath_j}, \imath_i, \imath_j = +, -, 1 \le i < j \le k$ are given and satisfy the consistency relations, which are the corresponding relations in (9.75) related to one- and two-dimensional distributions,

$$\begin{cases} p_{i,j,\imath_i,\imath_j} \ge 0 , \\ p_{i,j,\imath_i,+} + p_{i,j,\imath_i,-} = p_{i,\imath_i} , \quad p_{i,j,+,\imath_j} + p_{i,j,-,\imath_j} = p_{j,\imath_j} , \\ \imath_i, \imath_j = +, - , \quad 1 \le i < j \le k , \\ p_{i,+} + p_{i,-} = 1 , \quad 1 \le i \le k . \end{cases} \tag{9.77}$$

The following question arises. Does a random vector $\vec{W} = (W_1, \dots, W_k)$, which has the prescribed above consistent one- and two-dimensional distributions of components always exists, for any for $k \ge 3$?

Taking into account the above remarks, one can reduce this question to the following equivalent question. Do multivariate distributions $p_{i_1,\dots,i_l,\imath_{i_1},\dots,\imath_{i_l}}, \imath_{i_1}, \dots, \imath_{i_l} = +, - , 1 \le i_1 < \dots < i_l \le k, 1 \le l \le k$, which have prescribed one- and two-dimensional distributions consistent in the sense of relations (9.77), always exist, for any for $k \ge 3$?

In the following, we describe the algorithm for constructing high-order (three-, four-, etc.) distributions with the above one- and two-dimensional distributions consistent in the sense of relations (9.75).

Let us describe how the corresponding three-dimensional distributions can be constructed.

It is enough to show how to choose, for any prescribed indices $1 \le i < j < r \le k$, values for eight probabilities $p_{i,j,r,\imath_i,\imath_j,\imath_r}, \imath_i, \imath_j, \imath_r = +, -$ such that the following consistency relation hold:

$$\begin{cases} p_{i,j,r,\imath_i,\imath_j,\imath_r} \geq 0\,, & \imath_i, \imath_j, \imath_r = \pm\,, \\ p_{i,j,r,+,\imath_j,\imath_r} + p_{i,j,r,-,\imath_j,\imath_r} = p_{j,r,\imath_j,\imath_r}\,, & \imath_j, \imath_r = \pm\,, \\ p_{i,j,r,\imath_i,+,\imath_r} + p_{i,j,r,\imath_i,-,\imath_r} = p_{i,r,\imath_i,\imath_r}\,, & \imath_i, \imath_r = \pm\,, \\ p_{i,j,r,\imath_i,\imath_j,+} + p_{\varepsilon,i,j,r,\imath_i,\imath_j,-} = p_{i,j,\imath_i,\imath_j}\,, & \imath_i, \imath_j = \pm\,. \end{cases} \tag{9.78}$$

Let us order triplets of indices $\langle \imath_i, \imath_j, \imath_r \rangle$ (using the ordering symbol $\succ$) in the way described above, i.e.

$$\langle +,+,+\rangle \succ \langle +,+,-\rangle \succ \langle +,-,+\rangle \succ \cdots \succ \langle -,-,-\rangle\,. \tag{9.79}$$

One should begin by choosing a value for probability $p_{i,j,r,+,+,+}$. It follows from relations (9.78) that $p_{\varepsilon,i,j,r,+,+,+} \leq p_{i,j,+,+}, p_{i,r,+,+}, p_{j,r,+,+}$. Thus, one can choose for this probability any values satisfying inequality,

$$0 \leq p_{i,j,r,+,+,+} \leq p_{i,j,+,+} \wedge p_{i,r,+,+} \wedge p_{j,r,+,+}\,. \tag{9.80}$$

Then, one should define probabilities $p_{i,j,r,\imath_i,\imath_j,\imath_r}$, sequentially, in three groups.

First, the group of probabilities $p_{i,j,r,+,+,-}, p_{i,j,r,+,-,+}, p_{\varepsilon,i,j,r,-,+,+}$, which index triplets differ just by one index from triplet $\langle +,+,+\rangle$, should be defined.

Second, the group of probabilities $p_{i,j,r,+,-,-}, p_{i,j,r,-,+,-}, p_{i,j,r,-,-,+}$, which index triplets differ by two indices from triplet $\langle +,+,+\rangle$, should be defined.

Third, the group includeing one probability $p_{i,j,r,-,-,-}$, which index triplet differs by three indices from triplet $\langle +,+,+\rangle$, should be defined.

The order established by relation (9.79) is used within the above groups. Relation (9.78) imply the following formulas:

$$\begin{cases} p_{i,j,r,+,+,-} = p_{i,j,+,+} - p_{i,j,r,+,+,+}\,, \\ p_{i,j,r,+,-,+} = p_{i,r,+,+} - p_{i,j,r,+,+,+}\,, \\ p_{i,j,r,-,+,+} = p_{j,r,+,+} - p_{i,j,r,+,+,+}\,, \\ p_{i,j,r,+,-,-} = p_{i,j,+,-} - p_{i,j,r,+,-,+} = p_{i,r,+,-} - p_{i,j,r,+,+,-}\,, \\ p_{i,j,r,-,+,-} = p_{i,j,-,+} - p_{i,j,r,-,+,+} = p_{j,r,+,-} - p_{i,j,r,+,+,-}\,, \\ p_{i,j,r,-,-,+} = p_{i,r,-,+} - p_{i,j,r,-,+,+} = p_{j,r,-,+} - p_{i,j,r,+,-,+}\,, \\ p_{i,j,r,-,-,-} = p_{i,j,-,-} - p_{i,j,r,-,-,+} = p_{i,r,-,-} - p_{i,j,r,-,+,-} = p_{j,r,-,-} - p_{i,j,r,+,-,-}\,. \end{cases} \tag{9.81}$$

One can check that alternative formulas for the above probabilities follows from consistency of the corresponding one- and two-dimensional probabilities.

For example, the equality $p_{i,j,+,-} - p_{i,j,r,+,-,+} = p_{i,r,+,-} - p_{i,j,r,+,+,-}$ is equivalent, due to first two equalities in (9.81), to the equality $p_{i,j,+,-} - (p_{i,r,+,+} - p_{i,j,r,+,+,+}) = p_{i,r,+,-} - (p_{i,j,+,+} - p_{i,j,r,+,+,+})$ that is equivalent to the equality $p_{i,j,+,-} + p_{i,j,+,+} = p_{i,r,+,-} + p_{i,r,+,+} = p_{i,+}$.

Also, one can easily check that probabilities $p_{\varepsilon,i,j,r,+,+,+}\,\ldots, p_{\varepsilon,i,j,r,-,-,-}$ defined by relations (9.80) and (9.81) satisfy all consistency inequalities and equations in (9.78).

It follows from relations (9.80) and (9.81) that there is one semifree parameter $p_{i,j,r,+,+,+} \in [0, p_{i,j,+,+} \wedge p_{i,r,+,+} \wedge p_{j,r,+,+}]$ in these relations.

The algorithm described above should be repeated for every $1 \leq i < j < r \leq k$.

In total, there are C_k^3 semifree parameters $p_{i,j,r,+,+,+}, 1 \leq i < j < r \leq k$ for the family of three-dimensional distributions.

As soon as the three-dimensional distributions consistent with given one- and two-dimensional distributions are constructed, one can construct, in the similar way, the corresponding four-dimensional distributions, consistent with the corresponding one-, two- and three-dimensional distributions.

Note only that, again, there are C_k^4 semifree parameters $p_{i,j,r,s,+,+,+,+} \in [0, p_{i,j,r,+,++} \wedge p_{i,j,s,+,+,+} \wedge p_{i,r,s,+,+,+} \wedge p_{j,r,s,+,+,+}]$, for the family of four-dimensional distributions, for every chosen indices $1 \leq i < j < r < s \leq k$.

In total, there are $C_k^3 + C_k^4$ semifree parameters $p_{i,j,r,+,+,+}, 1 \leq i < j < r \leq k$ and $p_{i,j,r,s,+,+,+,+}, 1 \leq i < j < r < s \leq k$ for the family of three- and four-dimensional distributions.

The procedure can be repeated for five-dimensional distributions, etc. It will be completed when the corresponding k-dimensional distributions will be constructed.

As soon as k-dimensional distributions $p_{1,\dots,k,\imath_1,\dots,\imath_k}, \imath_1, \dots, \imath_k = +, -$ are constructed, one can construct a random vector $\vec{W} = (W_1, \dots, W_k)$ with the above distributions, as this was described above.

Let us summarize the above remarks in the following lemma.

Lemma 9.2.2. *A k-dimensional binary random vector $\vec{W} = (W_1, \dots, W_k)$, with any prescribed consistent one- and two-dimensional distributions, exists. The corresponding high-order distributions for this vector can be constructed according to algorithm described in the above relations* (9.78)–(9.81) *and related remarks.*

Now, one can construct, for every $\varepsilon \in (0, \varepsilon_0]$, the random walk $Y_{\varepsilon,n}$ defined by the transition dynamic relation (9.44).

First, one should choose jump values $\bar{\delta}_{\varepsilon,\bar{\imath}} = (\delta_{\varepsilon,\imath_1}, \dots, \delta_{\varepsilon,\imath_k}), \bar{\imath} = \langle \imath_{i_1}, \dots, \imath_{i_k} \rangle \in \mathrm{I}_k \in \mathrm{I}_k$ and one- and two-dimensional distributions, respectively, $p_{\varepsilon,i.\imath_i}, \imath_i = +, -, 1 \leq i \leq k$ and $p_{\varepsilon,i,j,\imath_i.\imath_j}, \imath_i, \imath_j = +, -, 1 \leq i < j \leq k$ for random vector $\vec{W}_{\varepsilon,1,1}$, according to relations pointed in Lemma 9.2.1.

Second, one should construct the corresponding consistent k-dimensional distributions $p_{\varepsilon,1,\dots,k,\imath_1,\dots,\imath_k}, \bar{\imath} \in \mathrm{I}_k$, with the above one- and two dimensional distributions, using the algorithm described in Lemma 9.2.2, and then to define the corresponding intervals $\mathbb{I}_{\bar{\imath}}, = \mathbb{I}_{\varepsilon,\bar{\imath}}, \bar{\imath} \in \mathrm{I}_k$, which, in this case, also depend on parameter ε.

Third, one can define the corresponding sequence of i.i.d. random variables $\vec{W}_{\varepsilon,n,l}, n, l = 1, 2, \dots$ used in the transition dynamic relation (9.44) as

$$\vec{W}_{\varepsilon,n,l} = \sum_{\bar{\imath} \in \mathrm{I}_k} \bar{\delta}_{\varepsilon,\bar{\imath}} \mathrm{I}(\rho_{n,l} \in \mathbb{I}_{\varepsilon,\bar{\imath}}), \quad n, l = 1, 2, \dots, \tag{9.82}$$

where $\rho_{n,l}, \ n, l = 1, 2, \dots$ is a family of i.i.d. random variables uniformly distributed in the interval $[0, 1)$.

9.3 Multivariate trinomial tree approximations

In this section, we present fitting formulas for multivariate trinomial tree approximation models inhomogeneous in time but homogeneous is space.

9.3.1 Exact fitting of parameters for a multivariate inhomogeneous in time trinomial tree model

Let us consider the model of multivariate inhomogeneous in time Gaussian random walk, which can be considered as a discrete time analog of a multivariate Gaussian process with independent increments. In this case, a log-price $\vec{Y}_{0,n}$ is given by the following stochastic transition dynamic relation:

$$\vec{Y}_{0,n} = \vec{Y}_{0,n-1} + \vec{W}_{0,n}, \quad n = 1, 2, \ldots . \tag{9.83}$$

where (a) $\vec{Y}_{0,0} = (Y_{0,1}, \ldots, Y_{0,k})$ is a random vector taking values in the space $\mathbb{R}_k$, (b) $\vec{W}_{0,n} = (W_{0,n,1}, \ldots, W_{0,n,k})$, $n = 1, 2, \ldots$ is a sequence of independent k-dimensional normal random vectors with mean values $\mathsf{E}W_{0,n,i} = \mu_{n,i}$, $i = 1, \ldots, k$, $n = 1, 2, \ldots$, variances $\mathrm{Var}W_{0,n,i} = \sigma^2_{n,i}$, $i = 1, \ldots, k$, and covariances $\mathsf{E}(W_{0,n,i} - \mu_{n,i})(W_{0,n,i} - \mu_{n,i}) = \rho_{n,i,j}\sigma_{n,i}\sigma_{n,j}$, $i, j = 1, \ldots, k$, $n = 1, 2, \ldots$, and (c) the random vector $\vec{Y}_{0,0}$ and the family of the random vectors $\vec{W}_{0,n,l}$, $l, n = 1, 2, \ldots$ are independent.

Usual assumptions are: (d) $\mu_{n,i} \in \mathbb{R}_1$, $i = 1, \ldots, n = 1, 2, \ldots$, (d) $\sigma_{n,i} > 0$, $i = 1, \ldots, k$, $n = 1, 2, \ldots$, (e) $|\rho_{n,i,j}| \leq 1, \rho_{n,i,i} = 1, i, j = 1, \ldots, k$, $n = 1, 2, \ldots$, (f) $\Sigma_n = \|\sigma_{n,i,j}\|$, $n = 1, 2, \ldots$, where $\sigma_{n,i,j} = \rho_{n,i,j}\sigma_{n,i}\sigma_{n,j}$, $i, j = 1, \ldots, k$, are nonnegatively defined $k \times k$ matrices.

We shall approximate the log-price process $Y_{0,n}$ by multivariate inhomogeneous in time trinomial random walks.

In this case, the log-price process $Y_{\varepsilon,n}$ is given, for every $\varepsilon \in (0, \varepsilon_0]$, by the following stochastic transition dynamic relation:

$$\vec{Y}_{\varepsilon,n} = \vec{Y}_{\varepsilon,n-1} + \vec{W}_{\varepsilon,n}, \quad n = 1, 2, \ldots . \tag{9.84}$$

where (a) $\vec{Y}_{\varepsilon,0} = (Y_{0,1}, \ldots, Y_{0,k})$ is a random vector taking values in the space $\mathbb{R}_k$, (b) $\vec{W}_{\varepsilon,n} = (W_{\varepsilon,n,1}, \ldots, W_{\varepsilon,n,k}) = \vec{W}_{\varepsilon,n,1} + \cdots + \vec{W}_{\varepsilon,n,r_{\varepsilon,n}}$, $n = 1, 2, \ldots$, where $\vec{W}_{\varepsilon,n,l} = (W_{\varepsilon,n,l,1}, \ldots, W_{\varepsilon,n,l,k})$, $l, n = 1, 2, \ldots$ is a family of independent random vectors taking values $\bar{\delta}_{\varepsilon,n,\bar{\jmath}} = (\delta_{\varepsilon,n,1,\jmath_1}, \ldots, \delta_{\varepsilon,n,k,\jmath_k})$ with the corresponding probabilities $p_{\varepsilon,n,\bar{\jmath}} = p_{\varepsilon,n}(\jmath_1, \ldots \jmath_k)$ for $\bar{\jmath} \in \mathcal{J}_k = \{\bar{\jmath} = \langle \jmath_1, \ldots, \jmath_k\rangle : \jmath_j, \ldots \jmath_k = +, \circ, -\}$, (c) the random vector $\vec{Y}_{\varepsilon,0}$ and the family of the random vectors $\vec{W}_{\varepsilon,n,l}$, $l, n = 1, 2, \ldots$ are independent.

Usual assumptions are: (d) $-\infty < \delta_{\varepsilon,n,i,-} < \delta_{\varepsilon,n,i,\circ} < \delta_{\varepsilon,n,i,+} < \infty$, $\quad i = 1, \ldots, k$, $n = 1, 2, \ldots$; (e) $p_{\varepsilon,n,\bar{\jmath}} \geq 0, \bar{\jmath} \in \mathcal{J}_k$, $\sum_{\bar{\jmath} \in \mathcal{J}_k} p_{\varepsilon,n,\bar{\jmath}} = 1$, $n = 1, 2, \ldots$; and (f) $r_{\varepsilon,n}$, $n = 1, 2, \ldots$ are positive integer numbers.

We use the log-price processes $\vec{Y}_{\varepsilon,n}$ for approximation of the log-price process $\vec{Y}_{0,n}$ as $\varepsilon \to 0$.

In this case, a usual approach is to fit the corresponding parameters, i.e. to try to choose parameters $\bar{\delta}_{\varepsilon,n,\bar{\jmath}}, \bar{\jmath} \in \mathcal{J}_k$, $n = 1, \dots, N$, probabilities $p_{\varepsilon,n,\bar{\jmath}}, \bar{\jmath} \in \mathcal{J}_k$, $n = 1, \dots, N$, and $r_{\varepsilon,n}$, $n = 1, \dots, N$ in such a way that the following exact fitting equalities holds, for every $\varepsilon \in (0, \varepsilon_0]$:

$$\begin{cases} \mathsf{E}W_{\varepsilon,n,i} = \mathsf{E}W_{0,n,i}, \ i = 1, \dots, k, \ n = 1, \dots, N, \\ \mathrm{Var}W_{\varepsilon,n,i} = \mathrm{Var}W_{0,1,i}, \ i = 1, \dots, k, \ n = 1, \dots, N, \\ \mathsf{E}(W_{\varepsilon,n,i} - \mathsf{E}W_{\varepsilon,n,i})(W_{\varepsilon,n,i} - \mathsf{E}W_{\varepsilon,n,j}) = \mathsf{E}(W_{0,n,i} - \mathsf{E}W_{0,n,i}) \\ \qquad \times (W_{0,n,i} - \mathsf{E}W_{0,n,j}), \ 1 \le i < j \le k, \ n = 1, \dots, N. \end{cases} \tag{9.85}$$

As pointed out in Subsection 3.3.2, an inhomogeneous in time univariate random walk with bounded jumps possess the recombining property if nth jump of this random walk take values on a grid of points of the following form $\delta l + \delta_{n,\circ}$, $l = 0, \pm 1, \pm 2, \dots$ for every $n = 1, 2, \dots$, where $\delta > 0$, $\delta_{n,\circ} \in \mathbb{R}_1$, $n = 1, 2, \dots$.

The multivariate random walk possesses the recombining property if and only if every its component, which is an inhomogeneous in time univariate random walk, possesses this property. Thus, the above condition should hold for every component of the multivariate random walk, with parameters δ and $\delta_{n,\circ}$, $n = 1, 2, \dots$ specific for every component.

As follows from the above remarks, in the case, of multivariate inhomogeneous in time trinomial random walk $Y_{\varepsilon,n}$ considered in this section, the above recombining condition reduces to the following assumption that the parameters $\delta_{\varepsilon,n,i,+}, \delta_{\varepsilon,n,i,\circ}$, $\delta_{\varepsilon,n,i,-}$, $n = 1, 2, \dots$ have, for every $\varepsilon \in (0, \varepsilon_0]$, the following form:

$$\delta_{\varepsilon,n,i,\pm} = \pm\delta_{\varepsilon,i} + \lambda_{\varepsilon,n,i}, \quad \delta_{\varepsilon,n,i,\circ} = \lambda_{\varepsilon,n,i}, \quad i = 1, \dots, k, \quad n = 1, 2, \dots, \tag{9.86}$$

where $\delta_{\varepsilon,i} > 0$, $\lambda_{\varepsilon,n,i} \in \mathbb{R}_1$, $i = 1, \dots, k$, $n = 1, 2, \dots$.

As in the one-dimensional case, it is convenient to symmetrize the values of jumps by representing binary random vectors $\vec{W}_{\varepsilon,n,l}$ in the following form, for $l, n = 1, 2, \dots$:

$$\vec{W}_{\varepsilon,n,l} = \bar{\lambda}_{\varepsilon,n} + \vec{W}'_{\varepsilon,n,l} = \bar{\lambda}_{\varepsilon,n} + \Big\{ \bar{\delta}'_{\varepsilon,\bar{\jmath}} \ \text{ with probability } \ p_{\varepsilon,n,\bar{\jmath}}, \quad \bar{\jmath} \in \mathcal{J}_k, \tag{9.87}$$

where

$$\bar{\lambda}_{\varepsilon,n} = (\lambda_{\varepsilon,n,1}, \dots, \lambda_{\varepsilon,n,k}), \quad n = 1, 2, \dots, \tag{9.88}$$

and

$$\bar{\delta}'_{\varepsilon,\bar{\jmath}} = (\delta'_{\varepsilon,1,\jmath_1}, \dots, \delta'_{\varepsilon,k,\jmath_k}), \quad \bar{\jmath} = \langle \jmath_1, \dots, \jmath_k \rangle \in \mathcal{J}_k, \tag{9.89}$$

where

$$\delta'_{\varepsilon,i,+} = \delta_{\varepsilon,i}, \quad \delta'_{\varepsilon,i,\circ} = 0, \quad \delta'_{\varepsilon,i,-} = -\delta_{\varepsilon,i}, \quad i = 1, \dots, k. \tag{9.90}$$

It follows from relation (9.87) that the transition dynamic relation (9.83) can be rewritten in the following form:

$$\vec{Y}_{\varepsilon,n} = \vec{Y}_{\varepsilon,n-1} + \vec{\lambda}'_{\varepsilon,n} + \vec{W}'_{\varepsilon,n}, \quad n = 1, 2, \dots . \tag{9.91}$$

where (a) $\vec{Y}_{\varepsilon,0} = (Y_{0,1}, \dots, Y_{0,k})$ is a random vector taking values in the space $\mathbb{R}_k$, (b) $\vec{W}'_{\varepsilon,n} = (W'_{\varepsilon,n,1}, \dots, W'_{\varepsilon,n,k}) = \vec{W}'_{\varepsilon,n,1} + \cdots + \vec{W}'_{\varepsilon,n,r_{\varepsilon,n}}$, $n = 1, 2, \dots$, where $\vec{W}'_{\varepsilon,n,l} = (W'_{\varepsilon,n,l,1}, \dots, W'_{\varepsilon,n,l,k})$, $l, n = 1, 2, \dots$ is a family of i.i.d. trinary random vectors taking values $\bar{\delta}'_{\varepsilon,n,\bar{\jmath}}$ with the corresponding probabilities $p_{\varepsilon,n,\bar{\jmath}}$, for $\bar{\jmath} \in \mathcal{J}_k$, $n = 1, \dots$, and (c) the random vector $\vec{Y}_{\varepsilon,0}$ and the family of the random vectors $\vec{W}'_{\varepsilon,n,l}$, $l, n = 1, 2, \dots$ are independent; (d) $\vec{\lambda}'_{\varepsilon,n} = r_{\varepsilon,n}\vec{\lambda}_{\varepsilon,n}$, $n = 1, 2, \dots$.

Calculations analogous to those made in relations (9.55)–(9.57) yield the following formulas:

$$\begin{aligned}
\mathsf{E}W_{\varepsilon,n,1,i} &= \mathsf{E}W'_{\varepsilon,n,1,i} + \lambda_{\varepsilon,n,i} \\
&= \delta_{\varepsilon,i}(p_{\varepsilon,n,i,+} - p_{\varepsilon,n,i,-}) + \lambda_{\varepsilon,n,i} = \frac{\mu_{n,i}}{r_{\varepsilon,n}}, \quad i = 1, \dots, k, \quad n = 1, \dots, \\
\mathrm{Var}\, W_{\varepsilon,n,1,i} &= \mathrm{Var}\, W'_{\varepsilon,n,1,i} = \delta^2_{\varepsilon,i}(1 - p_{\varepsilon,n,i,\circ}) \\
&\quad - \left(\frac{\mu_{n,i}}{r_{\varepsilon,n}} - \lambda_{\varepsilon,n,i}\right)^2, \quad i = 1, \dots, k, \quad n = 1, \dots, \\
\mathsf{E}\left(W_{\varepsilon,n,1,i} - \frac{\mu_{n,i}}{r_\varepsilon}\right)&\left(W_{\varepsilon,n,1,j} - \frac{\mu_{n,j}}{r_{\varepsilon,n}}\right) \\
&= \mathsf{E}\left(W'_{\varepsilon,n,1,i} - \left(\frac{\mu_{n,i}}{r_{\varepsilon,n}} - \lambda_{\varepsilon,n,i}\right)\right)\left(W'_{\varepsilon,n,1,j} - \left(\frac{\mu_{n,j}}{r_{\varepsilon,n}} - \lambda_{\varepsilon,n,j}\right)\right) \\
&= \delta_{\varepsilon,i}\delta_{\varepsilon,j}(p_{\varepsilon,n,i,j,+,+} + p_{\varepsilon,n,i,j,-,-} - p_{\varepsilon,n,i,j,+,-} - p_{\varepsilon,n,i,j,-,+}) \\
&\quad - \left(\frac{\mu_{n,i}}{r_{\varepsilon,n}} - \lambda_{\varepsilon,n,i}\right)\left(\frac{\mu_{n,j}}{r_{\varepsilon,n}} - \lambda_{\varepsilon,n,j}\right), \quad 1 \le i < j \le k, \quad n = 1, \dots,
\end{aligned} \tag{9.92}$$

where

$$\begin{aligned}
p_{\varepsilon,n,i,j,\jmath_i,\jmath_j} &= \mathsf{P}\{W'_{\varepsilon,n,1,i} = \delta'_{\varepsilon,i,\jmath_i}, W'_{\varepsilon,n,1,j} = \delta'_{\varepsilon,n,j,\jmath_j}\}, \\
&\quad \jmath_i, \jmath_j = +, \circ, -, 1 \le i < j \le k, \quad n = 1, \dots,
\end{aligned} \tag{9.93}$$

and

$$\begin{aligned}
p_{\varepsilon,n,i,\jmath} &= \mathsf{P}\{W'_{\varepsilon,n,1,i} = \delta'_{\varepsilon,i,\jmath}\}, \\
&\quad \jmath = +, \circ, -, \quad i = 1, \dots, k, \quad n = 1, \dots .
\end{aligned} \tag{9.94}$$

Relations (9.92) let us rewrite the fitting relations (9.85) in the following equivalent form:

$$\begin{cases} \delta_{\varepsilon,i}(p_{\varepsilon,n,i,+} - p_{\varepsilon,n,i,-}) = \dfrac{\mu_{n,i}}{r_{\varepsilon,n}} - \lambda_{\varepsilon,n,i}\,, \\ \quad i = 1,\dots,k,\ n = 1,\dots,N\,, \\ \delta^2_{\varepsilon,i}(1 - p_{\varepsilon,n,i,\circ}) = \dfrac{\sigma^2_{n,i}}{r_{\varepsilon,n}} + \left(\dfrac{\mu_{n,i}}{r_{\varepsilon,n}} - \lambda_{\varepsilon,n,i}\right)^2, \\ \quad i = 1,\dots,k, \quad n = 1,\dots,N\,, \\ \delta_{\varepsilon,i}\delta_{\varepsilon,j}\left(p_{\varepsilon,n,i,j,+,+} + p_{\varepsilon,n,i,j,-,-} - p_{\varepsilon,n,i,j,+,-} - p_{\varepsilon,n,i,j,-,+}\right) \\ \quad = \dfrac{\rho_{n,i,j}\sigma_{n,i}\sigma_{n,j}}{r_{\varepsilon,n}} + \left(\dfrac{\mu_{n,i}}{r_{\varepsilon,n}} - \lambda_{\varepsilon,n,i}\right)\left(\dfrac{\mu_{n,j}}{r_{\varepsilon,n}} - \lambda_{\varepsilon,n,j}\right), \\ \quad 1 \le i < j \le k, \quad n = 1,\dots,N\,. \end{cases} \tag{9.95}$$

The system of equations (9.95) should be supplemented by the following probability inequalities:

$$0 \le p_{\varepsilon,n,i,j,\jmath_i,\jmath_j} \le 1, \quad \jmath_i, \jmath_j = +, \circ, -, \quad 1 \le i < j \le k, \quad n = 1,\dots,N\,, \tag{9.96}$$

and the following system of consistency relations for probabilities $p_{\varepsilon,i,j,\jmath_i,\jmath_j}$ and $p_{\varepsilon,i,\jmath_i}$, $\jmath_i, \jmath_j = +, \circ, -, 1 \le i < j \le k, n = 1,\dots,N$:

$$\begin{cases} p_{\varepsilon,n,i,j,\jmath_i,+} + p_{\varepsilon,n,i,j,\jmath_i,\circ} + p_{\varepsilon,n,i,j,\jmath_i,-} = p_{\varepsilon,n,i,\jmath_i}, \\ \quad \jmath_i = +, \circ, -,\ 1 \le i < j \le k,\ n = 1,\dots,N\,, \\ p_{\varepsilon,n,i,j,+,\jmath_j} + p_{\varepsilon,n,i,j,\circ,\jmath_j} + p_{\varepsilon,n,i,j,-,\jmath_j} = p_{\varepsilon,n,j,\jmath_j}, \\ \quad \jmath_j = +, \circ, -,\ 1 \le i < j \le k, \quad n = 1,\dots,N\,, \\ p_{\varepsilon,n,i,+} + p_{\varepsilon,n,i,\circ} + p_{\varepsilon,n,i,-} = 1, \quad 1 \le i \le k, \quad n = 1,\dots,N\,. \end{cases} \tag{9.97}$$

In order to simplify the fitting problem, let us choose the parameters $r_{\varepsilon,n}$ independent on n, i.e. assume that they satisfy the following relation:

$$r_{\varepsilon,n} = r_\varepsilon\,, \quad n = 1, 2, \dots, N\,, \tag{9.98}$$

and choose the parameters $\lambda_{\varepsilon,n,i}$ such that they compensate the drift of the log-price process $Y_{\varepsilon,n}$, i.e. these parameters take the following form:

$$\lambda_{\varepsilon,n,i} = \frac{\mu_{n,i}}{r_\varepsilon}, \quad i = 1,\dots,k, \quad n = 1, 2, \dots, N\,. \tag{9.99}$$

In this case, the system of equations (9.95) is essentially simplified. It takes the following form:

$$\begin{cases} p_{\varepsilon,n,i,+} - p_{\varepsilon,n,i,-} = 0\,, \quad i = 1,\dots,k, \quad n = 1,\dots,N\,, \\ \delta^2_{\varepsilon,i}(1 - p_{\varepsilon,n,i,\circ}) = \dfrac{\sigma^2_{n,i}}{r_\varepsilon}, \quad i = 1,\dots,k, \quad n = 1,\dots,N\,, \\ \delta_{\varepsilon,i}\delta_{\varepsilon,j}(p_{\varepsilon,n,i,j,+,+} + p_{\varepsilon,n,i,j,-,-} - p_{\varepsilon,n,i,j,+,-} - p_{\varepsilon,n,i,j,-,+}) \\ \quad = \dfrac{\rho_{n,i,j}\sigma_{n,i}\sigma_{n,j}}{r_\varepsilon}, \quad 1 \le i < j \le k, \quad n = 1,\dots,N\,. \end{cases} \tag{9.100}$$

Let us introduce the parameters

$$\sigma_{i,+} = \max_{1 \le n \le N} \sigma_{n,i} > 0, \quad i = 1, \ldots, k. \tag{9.101}$$

For the second equation in (9.100) let us choose the parameters $\delta_{\varepsilon,i}$ in the following form:

$$\delta_{\varepsilon,i} = \frac{\delta_i}{\sqrt{r_\varepsilon}}, \quad \text{where } \delta_i \ge \sigma_{i,+}, \quad i = 1, \ldots, k, \tag{9.102}$$

and then get the following explicit formulas for quantities $p_{\varepsilon,n,i,\circ}$, which, in this case, do not depend on the parameter $\varepsilon \in (0, \varepsilon_0]$:

$$p_{\varepsilon,n,i,\circ} = p_{n,i,\circ} = 1 - \frac{\sigma_{n,i}^2}{\delta_i^2}, \quad i = 1, \ldots, k, \quad n = 1, \ldots, N. \tag{9.103}$$

Note that the quantities $p_{\varepsilon,n,i,\circ} \in [0, 1)$, $i = 1, \ldots, k$, $n = 1, \ldots, N$, i.e. they take values consistent with relations (9.96) and (9.97).

Formula (9.103) gives us the hint to try to find values of quantities $p_{\varepsilon,n,i,\jmath} = p_{n,i,\jmath}$, $\jmath = +, \circ, -$, $i = 1, \ldots, k$ and $p_{\varepsilon,n,i,j,\jmath_i,\jmath_j} = p_{n,i,j,\jmath_i,\jmath_j}$, $\jmath_i, \jmath_j = +, \circ, -$, $1 \le i < j \le k$ in the form independent on the parameter $\varepsilon \in (0, \varepsilon_0]$.

The first equation in (9.100) yields the following formulas for quantities $p_{\varepsilon,n,i,\jmath}$:

$$p_{\varepsilon,n,i,\pm} = p_{n,i,\pm} = \frac{1 - p_{n,i,\circ}}{2} = \frac{\sigma_{n,i}^2}{2\delta_i^2}, \quad i = 1, \ldots, k, \quad n = 1, \ldots, N. \tag{9.104}$$

Relations (9.97) and (9.104) imply the following relation, for every $1 \le i < j \le k$, $n = 1, \ldots, N$:

$$\begin{aligned} p_{n,i,+} + p_{n,i,-} &= p_{n,i,j,+,+} + p_{n,i,j,+,\circ} + p_{n,i,j,+,-} + p_{n,i,j,-,+} + p_{n,i,j,-,\circ} + p_{n,i,j,-,-} \\ &= p_{n,i,j,+,+} + p_{n,i,j,+,-} + p_{n,i,j,-,+} + p_{n,i,j,-,-} + p_{n,j,\circ} - p_{n,i,j,\circ,\circ} \\ &= 1 - p_{n,i,\circ} \end{aligned} \tag{9.105}$$

which can be rewritten in the following form:

$$p_{n,i,j,+,+} + p_{n,i,j,+,-} + p_{n,i,j,-,+} + p_{n,i,j,-,-} = 1 - p_{n,i,\circ} - p_{n,j,\circ} + p_{n,i,j,\circ,\circ} \tag{9.106}$$

By substituting quantities $\delta_{\varepsilon,i}$ given by relation (9.102) in the last equation in (9.100), we get the following equation, for every $1 \le i < j \le k$, $n = 1, \ldots, N$:

$$p_{n,i,j,+,+} + p_{n,i,j,-,-} - p_{n,i,j,+,-} - p_{n,i,j,-,+} = \rho_{n,i,j}\sqrt{1 - p_{n,i,\circ}}\sqrt{1 - p_{n,j,\circ}}\,. \tag{9.107}$$

Let us denote, for the moment, $P_{n,i,j} = p_{n,i,j,+,+} + p_{n,i,j,-,-}$ and $Q_{n,i,j} = p_{n,i,j,+,-} + p_{n,i,j,-,+}$. For relations (9.106) and (9.107), let us write the following system of linear equations for quantities $P_{n,i,j}$, $Q_{n,i,j}$, for every $1 \le i < j \le k$, $n = 1, \ldots, N$:

$$\begin{cases} P_{n,i,j} + Q_{n,i,j} = 1 - p_{n,i,\circ} - p_{n,j,\circ} + p_{n,i,j,\circ,\circ}\,, \\ P_{n,i,j} - Q_{n,i,j} = \rho_{n,i,j}\sqrt{1 - p_{n,i,\circ}}\sqrt{1 - p_{n,j,\circ}}\,. \end{cases} \tag{9.108}$$

By solving this system of equation, we get the following formulas, for every $1 \leq i < j \leq k$, $n = 1, \dots, N$:

$$\begin{cases} P_{n,i,j} = \frac{1}{2}\Big(\rho_{n,i,j}\sqrt{1 - p_{n,i,\circ}}\sqrt{1 - p_{n,j,\circ}} + 1 - p_{n,i,\circ} - p_{n,j,\circ} + p_{n,i,j,\circ,\circ}\Big)\,, \\ Q_{n,i,j} = \frac{1}{2}\Big(- \rho_{n,i,j}\sqrt{1 - p_{n,i,\circ}}\sqrt{1 - p_{n,j,\circ}} + 1 - p_{n,i,\circ} - p_{n,j,\circ} + p_{n,i,j,\circ,\circ}\Big)\,. \end{cases} \tag{9.109}$$

Relations (9.96) and (9.97) imply that quantities $P_{n,i,j}$ and $Q_{n,i,j}$ should satisfy inequalities (a) $P_{n,i,j} \geq 0$, (b) $Q_{n,i,j} \geq 0$ and (c) $P_{n,i,j} + Q_{n,i,j} \leq 1$, for every $1 \leq i < j \leq k$, $n = 1, \dots, N$.

In order to improve chances for holding inequalities (a) – (c), let us choose for probabilities $p_{n,i,j,\circ,\circ}$ the maximal admissible value, which is obviously given by the following formula:

$$p_{n,i,j,\circ,\circ} = p_{n,i,\circ} \wedge p_{n,j,\circ}\,, \quad 1 \leq i < j \leq k\,, \quad n = 1, \dots, N\,. \tag{9.110}$$

In this case inequality (c) obviously holds since $1 - p_{n,i,\circ} - p_{n,j,\circ} + p_{n,i,\circ} \wedge p_{n,j,\circ} = (1 - p_{n,i,\circ}) \wedge (1 - p_{n,j,\circ}) \leq 1$. As far as inequalities (a) and (b) are concerned, they hold if and only if the following inequality holds, for every $1 \leq i < j \leq k$, $n = 1, \dots, N$:

$$\begin{aligned} |\rho_{n,i,j}|\sqrt{1 - p_{n,i,\circ}}\sqrt{1 - p_{n,j,\circ}} &\leq 1 - p_{n,i,\circ} - p_{n,j,\circ} + p_{n,i,\circ} \wedge p_{n,j,\circ} \\ &= (1 - p_{n,i,\circ}) \wedge (1 - p_{n,j,\circ})\,. \end{aligned} \tag{9.111}$$

By substituting quantities $\sqrt{1 - p_{n,i,\circ}} = \frac{\sigma_{n,i}}{\delta_i}$, $i = 1, \dots, k$, $n = 1, \dots, N$, given by formulas (9.103), in relation (9.111), we can transform the latter relation to the following equivalent form, for every $1 \leq i < j \leq k$, $n = 1, \dots, N$:

$$|\rho_{n,i,j}| \frac{\sigma_{n,i}}{\delta_i} \frac{\sigma_{n,j}}{\delta_j} \leq \frac{\sigma^2_{n,i}}{\delta^2_i} \wedge \frac{\sigma^2_{n,j}}{\delta^2_j}\,. \tag{9.112}$$

Relation (9.112) is a condition, under which the exact fitting represented by relation (9.85) can be realized.

Unfortunately, relation (9.112) does not separate parameters of the Gaussian log-price process $Y_{0,n}$ and parameters of the approximating log-price process $Y_{\varepsilon,n}$. Such separation can be realized in the following way.

Let us assume that the following condition holds:

R$_2$: $\max_{1 \leq i < j \leq k, 1 \leq n \leq N} |\rho_{n,i,j}| \frac{\sigma_{n,i} \vee \sigma_{n,j}}{\sigma_{n,i} \wedge \sigma_{n,j}} = \rho \leq 1$.

Let us also introduce the parameter

$$\kappa = \min_{1 \leq i,j \leq k} \frac{\delta_i}{\delta_j} \leq 1\,. \tag{9.113}$$

It is easily seen that, in the case, where the condition **R**$_2$ holds, the following condition imposed on parameter κ implies relation (9.112) to hold:

$$\rho \leq \kappa \leq 1\,. \tag{9.114}$$

In particular, if we choose all the parameters $\delta_i = \delta, i = 1, \dots, k$, then $\kappa = 1$, and, therefore, condition (9.113) automatically holds if the condition $\mathbf{R_2}$ holds. Also note that, in the case, where the condition $\mathbf{R_2}$ holds and $\rho = 1$, then condition (9.113) holds if and only if $\kappa = 1$ and, thus, all the parameters $\delta_i = \delta, i = 1, \dots, k$.

Now, we are going to use symmetry proposition and to choose quantities $p_{n,i,j,\pm,\pm}$ in the following form, for $1 \le i < j \le k, n = 1, \dots, N$:

$$p_{n,i,j,+,+} = p_{n,i,j,-,-} = \tfrac{1}{2} P_{n,i,j}, \tag{9.115}$$

and

$$p_{n,i,j,+,-} = p_{n,i,j,-,+} = \tfrac{1}{2} Q_{n,i,j}. \tag{9.116}$$

Obviously, these quantities satisfy probability inequalities (9.96) since quantities $P_{n,i,j}$ and $Q_{n,i,j}$ satisfy these inequalities.

Finally, we can define quantities $p_{n,i,j,\circ,\pm}$ using formulas, for $1 \le i < j \le k, n = 1, \dots, N$,

$$p_{n,i,j,\circ,+} = p_{n,i,j,\circ,-} = \tfrac{1}{2}(p_{n,i,\circ} - p_{n,i,j,\circ,\circ}), \tag{9.117}$$

and

$$p_{n,i,j,+,\circ} = p_{n,i,j,-,\circ} = \tfrac{1}{2}(p_{n,j,\circ} - p_{n,i,j,\circ,\circ}). \tag{9.118}$$

These quantities obviously satisfy probability inequalities (9.96).

By the definition, quantities $p_{n,i,j_i}, p_{n,i,j,j_i,j_j}, \imath_i, \jmath_j = +, \circ, -, 1 \le i < j \le k, n = 1, \dots, N$ defined in formulas (9.103), (9.104), (9.110), and (9.115)–(9.118) satisfy relations (9.97) as well.

Summarizing all the above remarks, we get (under assumption that the condition $\mathbf{R_2}$ holds) solutions of the fitting problem, for every $\varepsilon \in (0, \varepsilon_0]$, in the following form:

$$\begin{cases} \delta_{\varepsilon,n,i,\pm} = \pm\delta_{\varepsilon,i} + \lambda_{\varepsilon,n,i}, \quad \delta_{\varepsilon,n,i,\circ} = \lambda_{\varepsilon,n,i}, \\ \lambda_{\varepsilon,n,i} = \dfrac{\mu_{n,i}}{r_\varepsilon}, \quad \delta_{\varepsilon,i} = \dfrac{\delta_i}{\sqrt{r_\varepsilon}}, \\ p_{\varepsilon,n,i,\circ} = p_{n,i,\circ} = 1 - \dfrac{\sigma^2_{n,i}}{\delta^2_i}, \quad p_{\varepsilon,n,i,\pm} = p_{n,i,\pm} = \dfrac{\sigma^2_{n,i}}{2\delta^2_i}, \\ \qquad i = 1, \dots, k, \quad n = 1, \dots, N, \\ p_{\varepsilon,n,i,j,\pm,\pm} = p_{n,i,j,\pm,\pm} = \dfrac{1}{4}\left(\rho_{n,i,j}\dfrac{\sigma_{n,i}}{\delta_i}\dfrac{\sigma_{n,j}}{\delta_j} + \dfrac{\sigma^2_{n,i}}{\delta^2_i} \wedge \dfrac{\sigma^2_{n,j}}{\delta^2_j}\right), \\ p_{\varepsilon,n,i,j,\pm,\mp} = p_{n,i,j,\pm,\mp} = \dfrac{1}{4}\left(-\rho_{n,i,j}\dfrac{\sigma_{n,i}}{\delta_i}\dfrac{\sigma_{n,i}}{\delta_j} + \dfrac{\sigma^2_{n,i}}{\delta^2_i} \wedge \dfrac{\sigma^2_{n,j}}{\delta^2_j}\right), \\ p_{\varepsilon,n,i,j,\circ,\pm} = p_{n,i,j,\circ,\pm} = \dfrac{1}{2}\left(1 - \dfrac{\sigma^2_{n,i}}{\delta^2_i} - \left(1 - \dfrac{\sigma^2_{n,i}}{\delta^2_i}\right) \wedge \left(1 - \dfrac{\sigma^2_{n,j}}{\delta^2_j}\right)\right), \\ p_{\varepsilon,n,i,j,\pm,\circ} = p_{n,i,j,\pm,\circ} = \dfrac{1}{2}\left(1 - \dfrac{\sigma^2_{n,j}}{\delta^2_j} - \left(1 - \dfrac{\sigma^2_{n,i}}{\delta^2_i}\right) \wedge \left(1 - \dfrac{\sigma^2_{n,j}}{\delta^2_j}\right)\right), \\ p_{\varepsilon,n,i,j,\circ,\circ} = p_{n,i,j,\circ,\circ} = \left(1 - \dfrac{\sigma^2_{n,i}}{\delta^2_i}\right) \wedge \left(1 - \dfrac{\sigma^2_{n,j}}{\delta^2_j}\right), \quad 1 \le i < j \le k, \quad n = 1, \dots, N, \end{cases} \tag{9.119}$$

which are parameterized by the parameters $\delta_i \geq \sigma_{i,+}$, $i = 1, \dots, k$ and $r_\varepsilon = 1, 2, \dots$.

If the conditions $\mathbf{R_2}$ and (9.114) hold, then the solutions of the system (9.95) given by relation (9.119) satisfy both probability inequality (9.96) and the consistency relations (9.97), for every $\varepsilon \in (0, \varepsilon_0]$.

The fitting procedure described above, yields the one-dimensional and the two-dimensional distributions for components of the random vectors $\vec{W}_{\varepsilon,1,1} = (W_{\varepsilon,1,1,1}, \dots, W_{\varepsilon,1,1,k})$. These distributions are consistent in the sense of relations (9.47), (9.48), and (9.50).

The analysis of the fitting problem presented above is complete in the case of two-dimensional trinomial random walks.

However, in the case of k-dimensional binomial random walk, with $k \geq 3$, some additional question about existence of a multivariate distribution of the random vector $\vec{W}_{\varepsilon,1,1}$ (with the above one- and two-dimensional distributions $p_{\varepsilon,i,\jmath_i}$, $\jmath_i = +, \circ, -, 1 \leq i \leq k$ and $p_{\varepsilon,i,j,\jmath_i,\jmath_j}$, $\jmath_i, \jmath_j = +, \circ, -, 1 \leq i < j \leq k$) arises.

This question has an affirmative answer. Moreover, the algorithm for constructing of the corresponding k-dimensional distributions is given below in Lemma 9.3.2, which should be applied to the above one- and two-dimensional distributions.

The above remarks can be summarized in the following lemma.

Lemma 9.3.1. *The vector log-price process $\vec{Y}_{\varepsilon,n}$ represented, for $\varepsilon \in (0, \varepsilon_0]$, by the multivariate binomial random walk, which is defined by the stochastic transition dynamic relation* (9.84)*, realizes the exact moment fitting given by relation* (9.85) *for the log-price processes $Y_{0,n}$ represented by the Gaussian random walk defined by the stochastic transition dynamic relation* (9.83)*, under assumption that the condition* $\mathbf{R_2}$ *holds. Parameters of the process $\vec{Y}_{\varepsilon,n}$ namely, $\delta_{\varepsilon,i}$, $\lambda_{\varepsilon,n,i}$, $p_{\varepsilon,n,i,\jmath_i}$, $\jmath_i = +, \circ, -, 1 \leq i \leq k$, $p_{\varepsilon,n,i,j,\jmath_i,\jmath_j}$, $\jmath_i, \jmath_j = +, \circ, -, 1 \leq i < j \leq k$, $n = 1, 2, \dots, N$, r_ε, can be chosen, for every $\varepsilon \in (0, \varepsilon_0]$, according to relations* (9.119)*, where the parameters $\delta_i \geq \sigma_{i,+}$, $i = 1, \dots, k$ satisfy condition* (9.114) *and $r_\varepsilon = 1, 2, \dots$. The corresponding high-order distributions for components of random vectors $\vec{W}_{\varepsilon,n,1}$, $n = 1, \dots, N$ can be constructed with the use of algorithm described in Lemma 9.3.2, which should be applied to the above one- and two-dimensional distributions $p_{\varepsilon,n,i,\jmath_i}$, $\jmath_i = +, \circ, -, 1 \leq i \leq k$ and $p_{\varepsilon,n,i,j,\jmath_i,\jmath_j}$, $\jmath_i, \jmath_j = +, \circ, -, 1 \leq i < j \leq k$, for every $n = 1, 2, \dots, N$ and $\varepsilon \in (0, \varepsilon_0]$.*

It is worth to note that the fitting procedure based on expectations and covariances does involve only one- and two-dimensional distributions for the random vectors $\vec{W}_{\varepsilon,1,1}$. In fact, we need only know that the random vectors $\vec{W}_{\varepsilon,1,1}$ with the corresponding fitted one- and two-dimensional distributions do exist.

The same is valid for the corresponding convergence results for option rewards presented in the next chapter. The only expectations and covariances and, thus, only one- and two-dimensional distributions for the random vectors $\vec{W}_{\varepsilon,1,1}$ are involved in the corresponding convergence conditions, which provide convergence of reward functions for the trinomial log-price processes $\vec{Y}_{\varepsilon,n}$ to the reward functions for the Gaussian log-price process $\vec{Y}_{0,n}$.

However, the high-dimensional distributions for the random vectors $\vec{W}_{\varepsilon,n,1}$, $n = 1, \dots, N$ are explicitly used in the backward recurrence relations for computing the reward functions for the approximating trinomial log-price processes $\vec{Y}_{\varepsilon,n}$.

It should be pointed out that the recombining condition (which guarantees not more than polynomial rate of growth for the number of nodes in the corresponding trinomial tree representation for the above model) discussed in Chapter 3 holds automatically for any values of parameters of the log-price process $\vec{Y}_{\varepsilon,n}$ given by relation (9.119).

The corresponding tree $T_{\varepsilon,(\vec{y},n)}$ generated by the trajectories of the log-price process $\vec{Y}_{\varepsilon,n+r}$, $r = 0, 1, \dots$ with the value $\vec{Y}_{\varepsilon,n} = \vec{y}$ has the total number of points after r steps,

$$
\begin{aligned}
L_{\varepsilon,n,n+r} &\le 1 + (2r_\varepsilon + 1)^k + \cdots + (2r_\varepsilon r + 1)^k \\
&\le \int_0^{r+1} (2r_\varepsilon x + 1)^k dx = \frac{(2r_\varepsilon(r+1) + 1)^{k+1} - 1}{2r_\varepsilon(k+1)} .
\end{aligned}
\tag{9.120}
$$

The expression on the right-hand side in (9.120) is a polynomial function of r of order $k + 1$.

9.3.2 Multivariate trinary distributions with given one- and two-dimensional distributions

Let $\vec{W} = (W_1, \dots, W_k)$ be a k-dimensional trinary random vector (with $k \ge 3$), which take values $\bar{\delta}_{\bar{\jmath}} = (\delta_{1,\jmath_1}, \dots, \delta_{k,\jmath_k})$, where $\delta_{i,\jmath_i}, \jmath_i = +, \circ, -$, $i = 1, \dots, k$ are real numbers and $\bar{\jmath} = \langle \jmath_1, \dots, \jmath_k \rangle \in \mathcal{J}_k = \{\bar{\jmath} : \jmath_1, \dots, \jmath_k = +, \circ, -\}$.

Let $1 \le i_1 < \cdots < i_l \le k$ and $p_{i_1,\dots,i_l,\jmath_{i_1},\dots,\jmath_{i_l}} = \mathsf{P}\{W_{i_1} = \delta_{i_1,\jmath_{i_1}}, \dots, W_{i_l} = \delta_{i_l,\jmath_{i_l}}\}$, $\jmath_{i_1}, \dots, \jmath_{i_l} = +, \circ, -$ be the corresponding joint l-dimensional distributions for components of the random vector $\vec{W}$.

The distributions $p_{i_1,\dots,i_l,\jmath_{i_1},\dots,\jmath_{i_l}}$ should satisfy the following consistency relations:

$$
\begin{cases}
p_{i_1,\dots,i_l,\jmath_{i_1},\dots,\jmath_{i_l}} \ge 0, \\
\sum\limits_{\jmath_{i_j}=+,\circ,-} p_{i_1,\dots,i_l,\jmath_{i_1},\dots,\jmath_{i_l}} = p_{i_1,\dots,i_{j-1},i_{j+1},\dots,i_l,\jmath_{i_1},\dots,\jmath_{i_{j-1}},\jmath_{i_{j+1}},\dots,\jmath_{i_l}} \\
\jmath_{i_1}, \dots, \jmath_{i_l} = +, \circ, -, \quad 1 \le i_1 < \cdots < i_l \le k, \quad 1 \le j \le l \le k,
\end{cases}
\tag{9.121}
$$

where the quantity $p = p_{i_1,\dots,i_l,\jmath_{i_1},\dots,\jmath_{i_l}}$ should be counted as 1, for the case $l = 0$.

By the Kolmogorov theorem about consistent finite-dimensional distributions, there exists a random vector $\vec{W}$ with any consistent family of joint distributions of components satisfying relations (9.75).

The random vector $\vec{W} = (W_1, \dots, W_k)$ can be, in fact, constructed in the following way.

One can order indices $+, \circ$ and $-$ (using the ordering symbol $\succ$) as $+ \succ \circ \succ -$ and, then, vectors $\bar{\jmath} = \langle \jmath_1, \dots, \jmath_k \rangle \in \mathcal{J}_k$ (using the ordering symbol $\succ$) in a natural way, by counting $\bar{\jmath}' \succ \bar{\jmath}''$, if there exist $1 \leq j \leq k$ such that $\jmath_i' = \jmath_i''$, for $i < j$, but $\jmath_j' \succ \jmath_j''$.

Then one can, in the obvious way, define disjoint subintervals $\mathbb{J}_{\bar{\jmath}} = [q_{\bar{\jmath}}^-, q_{\bar{\jmath}}^+), \bar{\jmath} \in \mathcal{J}_k$ of interval $[0, 1)$ such that: (a) $q_{\bar{\jmath}}^+ - q_{\bar{\jmath}}^- = p_{1,\dots,k,\jmath_1,\dots,\jmath_k}, \bar{\jmath} \in \mathcal{J}_k$, (b) interval $\mathbb{J}_{\bar{\jmath}'}$ is located to the right of interval $\mathbb{J}_{\bar{\jmath}''}$ if $\bar{\jmath}' \succ \bar{\jmath}''$.

Finally, the random vector $\vec{W} = (W_1, \dots, W_k)$ can be defined as

$$\vec{W} = \sum_{\bar{\jmath} \in \mathcal{J}_k} \bar{\delta}_{\bar{\jmath}} \mathrm{I}(\rho \in \mathbb{J}_{\bar{\jmath}}), \tag{9.122}$$

where ρ is a random variable uniformly distributed in the interval $[0, 1)$.

The fitting algorithm presented in Subsection 9.3.1 does require to answer slightly different question.

Let us assume that some one-dimensional and two-dimensional distributions, respectively $p_{i,\jmath_i}, \jmath_i = +, \circ, -, 1 \leq i \leq k$ and $p_{i,j,\jmath_i,\jmath_j}, \jmath_i, \jmath_j = +, \circ, -, 1 \leq i < j \leq k$ are given and satisfy the consistency relations, which are, in fact, corresponding relations in (9.121) related to one- and two-dimensional distributions

$$\begin{cases} p_{i,j,\jmath_i,\jmath_j} \geq 0\,, \\ p_{i,j,\jmath_i,+} + p_{i,j,\jmath_i,\circ} + p_{i,j,\jmath_i,-} = p_{i,\jmath_i}\,, \\ p_{i,j,+,\jmath_j} + p_{i,j,\circ,\jmath_j} + p_{i,j,-,\jmath_j} = p_{j,\jmath_j}\,, \\ \jmath_i, \jmath_j = +, \circ, -, \quad 1 \leq i < j \leq k\,, \\ p_{i,+} + p_{i,\circ} + p_{i,-} = 1, \quad 1 \leq i \leq k\,. \end{cases} \tag{9.123}$$

The following question arises. Does a random vector $\vec{W} = (W_1, \dots, W_k)$ with the prescribed above consistent one- and two-dimensional distributions of components exist, for any for $k \geq 3$?

Taking into account the above remarks, one can reduce this question to the following equivalent question. Do multivariate distributions $p_{i_1,\dots,i_l,\jmath_{i_1},\dots,\jmath_{i_l}}, \jmath_{i_1}, \dots, \jmath_{i_l} = +, \circ, -, \quad 1 \leq i_1 < \cdots < i_l \leq k, 1 \leq l \leq k$, which have prescribed one- and two-dimesional distributions consistent in the sense of relations (9.123), always exist, for any $k \geq 3$?

Below, we describe algorithm for constructing high order (three-, four-, etc.) distributions with the above one- and two-dimensional distributions consistent in the sense of relations (9.121).

Let us describe how the corresponding three-dimensional distributions can be constructed.

It is enough to show how to choose, for any prescribed indices $1 \leq i < j < r \leq k$, values for 27 probabilities $p_{i,j,r,\jmath_i,\jmath_j,\jmath_r}, \jmath_i, \jmath_j, \jmath_r = +, \circ, -$ such that the following consistency relation would hold:

$$\begin{cases} p_{n,i,j,r,\jmath_i,\jmath_j,\jmath_r} \geq 0\,, \\ p_{n,i,j,r,+,\jmath_j,\jmath_r} + p_{n,i,j,r,\circ,\jmath_j,\jmath_r} + p_{n,i,j,r,-,\jmath_j,\jmath_r} = p_{n,j,r,\jmath_j,\jmath_r}\,, \\ p_{n,i,j,r,\jmath_i,+,\jmath_r} + p_{n,i,j,r,\jmath_i,\circ,\jmath_r} + p_{n,i,j,r,\jmath_i,-,\jmath_r} = p_{n,i,r,\jmath_i,\jmath_r}\,, \\ p_{n,i,j,r,\jmath_i,\jmath_j,+} + p_{n,i,j,r,\jmath_i,\jmath_j,\circ} + p_{n,i,j,r,\jmath_i,\jmath_j,-} = p_{n,i,j,\jmath_i,\jmath_j}\,, \\ \jmath_i, \jmath_j, \jmath_r = +, \circ, -\,. \end{cases} \tag{9.124}$$

Let us order indices $+, \circ, -$ and then triplets of indices $\langle \imath_i, \imath_j, \imath_r \rangle$ (using the same ordering symbol $\succ$) in the way described above, i.e. as $+ \succ \circ \succ -$ and then

$$\langle +, +, + \rangle \succ \langle +, +, \circ \rangle \succ \langle +, +, - \rangle \succ \cdots \succ \langle -, -, - \rangle\,. \tag{9.125}$$

One can define probabilities $p_{i,j,r,\jmath_i,\jmath_j,\jmath_r}$ sequentially, in groups.

First, one should should choose values for eight probabilities $p_{i,j,r,\jmath_i,\jmath_j,\jmath_r}$, $\jmath_i, \jmath_j, \jmath_r = +, \circ, -$, which are semifree in the sense that they are defined by a sequence of inequalities.

To begin, one can choose a value for probability $p_{\varepsilon,i,j,r,+,+,+}$. It follows from relations (9.124) that $p_{\varepsilon,i,j,r,+,+,+} \leq p_{i,j,+,+}, p_{i,r,+,+}, p_{j,r,+,+}$. One can choose for this probability any value satisfying inequality

$$0 \leq p_{i,j,r,+,+,+} \leq p_{i,j,+,+} \wedge p_{\varepsilon,i,r,+,+} \wedge p_{j,r,+,+}\,. \tag{9.126}$$

Then, one can choose values for probabilities $p_{n,i,j,r,+,+,\circ}, p_{n,i,j,r,+,\circ,+}$, and $p_{n,i,j,r,\circ,+,+}$, which index triplets can be obtained from the triplet $\langle +, +, + \rangle$ just by replacing one index $+$ by index $\circ$. The order established above for triplets $\langle \imath_i, \imath_j, \imath_r \rangle$ is used within the group.

Relations (9.124) imply that $p_{i,j,r,+,+,\circ} \leq p_{i,r,+,\circ}, p_{j,r,+,\circ}, p_{i,j,+,+} - p_{i,j,r,+,+,+}$. Analogously, $p_{i,j,r,+,\circ,+} \leq p_{i,j,+,\circ}, p_{i,r,+,+} - p_{i,j,r,+,+,+}, p_{j,r,\circ,+}$, and $p_{i,j,r,\circ,+,+} \leq p_{j,r,+,+} - p_{i,j,r,+,+,+}, p_{i,j,\circ,+}, p_{\varepsilon,i,r,+,+}$. One can choose, for these probabilities any values satisfying inequalities

$$\begin{cases} 0 \leq p_{i,j,r,+,+,\circ} \leq p_{i,r,+,\circ} \wedge p_{j,r,+,\circ} \wedge (p_{i,j,+,+} - p_{i,j,r,+,+,+})\,, \\ 0 \leq p_{i,j,r,+,\circ,+} \leq p_{i,j,+,\circ} \wedge (p_{i,r,+,+} - p_{i,j,r,+,+,+}) \wedge p_{j,r,\circ,+}\,, \\ 0 \leq p_{i,j,r,\circ,+,+} \leq (p_{j,r,+,+} - p_{i,j,r,+,+,+}) \wedge p_{i,j,\circ,+} \wedge p_{i,r,+,+}\,. \end{cases} \tag{9.127}$$

Then, one can choose values for probabilities $p_{i,j,r,+,\circ,\circ}, p_{i,j,r,\circ,+,\circ}$, and $p_{i,j,r,\circ,\circ,+}$, which index triplets can be obtained from the triplet $\langle +, +, + \rangle$ by replacing two indices $+$ by index $\circ$.

The order established above for triplets $\langle \imath_i, \imath_j, \imath_r \rangle$ is used within the group.

By relations (9.124), $p_{i,j,r,+,\circ,\circ} \leq p_{i,j,+,\circ} - p_{i,j,r,+,\circ,+}, p_{i,r,+,\circ} - p_{i,j,r,+,+,\circ}, p_{j,r,\circ,\circ}$. Analogously, $p_{i,j,r,\circ,+,\circ} \leq p_{i,j,\circ,+} - p_{i,j,r,\circ,+,+}, p_{j,r,+,\circ} - p_{i,j,r,+,+,\circ}, p_{i,r,\circ,\circ}$ and $p_{i,j,r,\circ,\circ,+} \leq p_{i,r,\circ,+} - p_{i,j,r,\circ,+,+}, p_{j,r,\circ,+} - p_{i,j,r,+,\circ,+}, p_{i,j,\circ,\circ}$. One can choose, for these probabilities any values satisfying inequalities

$$\begin{cases} 0 \le p_{i,j,r,+,\circ,\circ} \le (p_{i,j,+,\circ} - p_{i,j,r,+,\circ,+}) \wedge (p_{i,r,+,\circ} - p_{\varepsilon,i,j,r,+,+,\circ}) \wedge p_{j,r,\circ,\circ}\,, \\ 0 \le p_{i,j,r,\circ,+,\circ} \le (p_{i,j,\circ,+} - p_{\varepsilon,i,j,r,\circ,+,+}) \wedge (p_{j,r,+,\circ} - p_{i,j,r,+,+,\circ}) \wedge p_{i,r,\circ,\circ}\,, \\ 0 \le p_{i,j,r,\circ,\circ,+} \le (p_{i,r,\circ,+} - p_{i,j,r,\circ,+,+}) \wedge (p_{j,r,\circ,+} - p_{i,j,r,+,\circ,+}) \wedge p_{i,j,\circ,\circ}\,. \end{cases} \tag{9.128}$$

Finally, one should choose a value for probability $p_{i,j,r,\circ,\circ,\circ}$, which index triplets can be obtained from the triplet $\langle +, +, + \rangle$ by replacing all three indices $+$ by index $\circ$.

Relations (9.124) imply that $p_{i,j,r,\circ,\circ,\circ} \le p_{i,j,\circ,\circ} - p_{i,j,r,\circ,\circ,+}, p_{i,r,\circ,\circ} - p_{i,j,r,\circ,+,\circ}, p_{j,r,\circ,\circ} - p_{\varepsilon,i,j,r,+,\circ,\circ}$. One can choose, for this probability any value satisfying inequality

$$\begin{cases} 0 \le p_{i,j,r,\circ,\circ,\circ} \le (p_{i,j,\circ,\circ} - p_{i,j,r,\circ,\circ,+}) \wedge (p_{i,r,\circ,\circ} - p_{i,j,r,\circ,+,\circ}) \\ \qquad \wedge\, (p_{j,r,\circ,\circ} - p_{i,j,r,+,\circ,\circ})\,. \end{cases} \tag{9.129}$$

Second, one can find values for 12 probabilities $p_{\varepsilon,i,j,r,\jmath_i,\imath_j,\jmath_r}$ in three groups, where one index in triplet $\langle \jmath_i, \jmath_j, \jmath_r \rangle$ is $-$, respectively, the third, second, or first, while the other two indices are $+$ or $\circ$.

The order established above for triplets $\langle \imath_i, \imath_j, \imath_r \rangle$ is used within the group.

Using relations (9.124) and values of probabilities $p_{i,j,r,\jmath_i,\imath_j,\jmath_r}, \jmath_i, \jmath_j, \jmath_r = +, \circ$ chosen according to relations (9.126)–(9.129), one can get the following formula, for probabilities on the left-hand side in the equalities as follows:

$$\begin{cases} p_{i,j,r,+,+,-} = p_{i,j,+,+} - p_{i,j,r,+,+,+} - p_{\varepsilon,i,j,r,+,+,\circ}\,, \\ p_{i,j,r,+,\circ,-} = p_{i,j,+,\circ} - p_{i,j,r,+,\circ,+} - p_{i,j,r,+,\circ,\circ}\,, \\ p_{i,j,r,\circ,+,-} = p_{i,j,\circ,+} - p_{i,j,r,\circ,+,+} - p_{i,j,r,\circ,+,\circ}\,, \\ p_{i,j,r,\circ,\circ,-} = p_{i,j,\circ,\circ} - p_{i,j,r,\circ,\circ,+} - p_{i,j,r,\circ,\circ,\circ}\,, \\ p_{i,j,r,+,-,+} = p_{i,r,+,+} - p_{i,j,r,+,+,+} - p_{i,j,r,+,\circ,+}\,, \\ p_{i,j,r,+,-,\circ} = p_{i,r,+,\circ} - p_{i,j,r,+,+,\circ} - p_{i,j,r,+,\circ,\circ}\,, \\ p_{i,j,r,\circ,-,+} = p_{i,r,\circ,+} - p_{i,j,r,\circ,+,+} - p_{i,j,r,\circ,\circ,+}\,, \\ p_{i,j,r,\circ,-,\circ} = p_{i,r,\circ,\circ} - p_{i,j,r,\circ,+,\circ} - p_{i,j,r,\circ,\circ,\circ}\,, \\ p_{i,j,r,-+,+} = p_{j,r,+,+} - p_{i,j,r,+,+,+} - p_{i,j,r,\circ,+,+}\,, \\ p_{i,j,r,-,+,\circ} = p_{j,r,+,\circ} - p_{i,j,r,+,+,\circ} - p_{i,j,r,\circ,+,\circ}\,, \\ p_{i,j,r,-,\circ,+} = p_{j,r,\circ,+} - p_{i,j,r,+,\circ,+} - p_{i,j,r,\circ,\circ,+}\,, \\ p_{i,j,r,-,\circ,\circ} = p_{j,r,\circ,\circ} - p_{i,j,r,+,\circ,\circ} - p_{i,j,r,\circ,\circ,\circ}\,. \end{cases} \tag{9.130}$$

Third, one can find values of six probabilities $p_{i,j,r,\jmath_i,\imath_j,\jmath_r}$, in three groups, where two indices in triplet $\langle \jmath_i, \jmath_j, \jmath_r \rangle$ are, respectively, second and third, or first and third, or first and second, while the remaining index is $+$ or $\circ$, as well as a value for probability $p_{i,j,r,-,-,-}$.

The order established above for triplets $\langle \imath_i, \imath_j, \imath_r \rangle$ is used within the group.

Using relations (9.124) and values of the probabilities $p_{\varepsilon,i,j,r,\jmath_i,\imath_j,\jmath_r}$, whose triplets $\langle \jmath_i, \imath_j, \jmath_r \rangle$ have not more than one index $-$, one can get the following formulas:

$$\begin{cases} p_{i,j,r,+,-,-} = p_{i,j,+,-} - p_{i,j,r,+,-,+} - p_{i,j,r,+,-,\circ} = p_{i,r,+,-} - p_{i,j,r,+,+,-} - p_{i,j,r,+,\circ,-}\,, \\ p_{i,j,r,\circ,-,-} = p_{i,j,\circ,-} - p_{i,j,r,\circ,-,+} - p_{i,j,r,\circ,-,\circ} = p_{i,r,\circ,-} - p_{i,j,r,\circ,+,-} - p_{i,j,r,\circ,\circ,-}\,, \\ p_{i,j,r,-,+,-} = p_{i,j,-,+} - p_{i,j,r,-,+,+} - p_{i,j,r,-,+,\circ} = p_{j,r,+,-} - p_{i,j,r,+,+,-} - p_{i,j,r,\circ,+,-}\,, \\ p_{i,j,r,-,\circ,-} = p_{i,j,-,\circ} - p_{i,j,r,-,\circ,+} - p_{i,j,r,-,\circ,\circ} = p_{j,r,\circ,-} - p_{i,j,r,+,\circ,-} - p_{i,j,r,\circ,\circ,-}\,, \\ p_{i,j,r,-,-,+} = p_{i,r,-,+} - p_{i,j,r,-,+,+} - p_{i,j,r,-,\circ,+} = p_{j,r,-,+} - p_{i,j,r,+,-,+} - p_{i,j,r,\circ,-,+}\,, \\ p_{i,j,r,-,-,\circ} = p_{i,r,-,\circ} - p_{i,j,r,-,+,\circ} - p_{i,j,r,-,\circ,\circ} = p_{j,r,-,\circ} - p_{i,j,r,+,-,\circ} - p_{i,j,r,\circ,-,\circ}\,, \\ p_{i,j,r,-,-,-} = p_{i,j,-,-} - p_{i,j,r,-,-,+} - p_{i,j,r,-,-,\circ} = p_{i,r,-,-} - p_{i,j,r,-,+,-} - p_{i,j,r,-,\circ,-} \\ \qquad = p_{j,r,-,-} - p_{i,j,r,+,-,-} - p_{i,j,r,\circ,-,-}\,. \end{cases} \tag{9.131}$$

One can check that alternative formulas for some of the above probabilities follows from consistency of the corresponding one- and two-dimensional probabilities.

For example, the equality $p_{i,j,+,-} - p_{i,j,r,+,-,+} - p_{i,j,r,+,-,\circ} = p_{i,r,+,-} - p_{i,j,r,+,+,-} - p_{i,j,r,+,\circ,-}$ is equivalent, due to four first equalities in (9.130), to the equality $p_{i,j,+,-} - (p_{i,r,+,+} - p_{i,j,r,+,+,+} - p_{i,j,r,+,\circ,+}) - (p_{i,r,+,\circ} - p_{i,j,r,+,+,\circ} - p_{i,j,r,+,\circ,\circ}) = p_{i,r,+,-} - (p_{i,j,+,+} - p_{i,j,r,+,+,+} - p_{i,j,r,+,+,\circ}) - (p_{i,j,+,\circ} - p_{i,j,r,+,\circ,+} - p_{i,j,r,+,\circ,\circ})$ that is equivalent to the equality $p_{i,j,+,+} + p_{i,j,+,\circ} + p_{i,j,+,-} = p_{i,r,+,+} + p_{i,r,+,\circ} + p_{i,r,+,-} = p_{i,+}$.

Also, one can easily check that probabilities $p_{i,j,r,+,+,+}, \ldots, p_{i,j,r,-,-,-}$ defined by relations (9.126)–(9.131) satisfy all consistency inequalities and equations in (9.124).

Also note that as it follows from relations (9.126)–(9.131) , there is 8 semifree parameters in these relations.

The algorithm described above should be repeated for every $1 \leq i < j < r \leq k$. In total, there are $8C_k^3$ semifree parameters $p_{i,j,r,\jmath_i,\jmath_j,\imath_r}, \jmath_i, \jmath_j, \jmath_r = +, \circ, 1 \leq i < j < r \leq k$ for the family of three-dimensional distributions.

As soon as the three-dimensional distributions consistent with given one- and two-dimensional distributions are constructed, in the similar way one can construct the corresponding four-dimensional distributions consistent with one-, two- and three-dimensional distributions.

Note that there is $8 \times 4 = 32$ semifree parameters $p_{j,r,s,\jmath_j,\jmath_r,\jmath_s}, p_{i,r,s,\jmath_i,\jmath_r,\jmath_s}, p_{i,j,s,\jmath_i,\jmath_j,\jmath_s}, p_{i,j,r,\jmath_i,\jmath_j,\jmath_r}, \jmath_i, \jmath_j, \jmath_r, \jmath_s = +, \circ$, plus 16 semifree parameters $p_{i,j,r,s,\jmath_i,\jmath_j,\jmath_r,\jmath_s}, \jmath_i, \jmath_j, \jmath_r, \jmath_s = +, \circ$, for the family of three- and four-dimensional distributions, for every chosen indices $1 \leq i < j < r < s \leq k$.

In total, there are $8C_k^3 + 16C_k^4$ semifree parameters $p_{i,r,s,\jmath_i,\jmath_j,\jmath_r}, \jmath_i, \jmath_j, \jmath_r = +, \circ, 1 \leq i < j < r \leq k$ and $p_{i,j,r,s,\jmath_i,\jmath_j,\jmath_r,\jmath_s}, \jmath_i, \jmath_j, \jmath_r, \jmath_s = +, \circ, \ 1 \leq i < j < r < s \leq k$, for the family of three- and four-dimensional distributions.

The procedure can be repeated for five-dimensional distributions, etc. It will be completed when the corresponding k-dimensional distributions are constructed.

As soon as k-dimensional distributions $p_{1,\ldots,k,\jmath_1,\ldots,\jmath_k}, \jmath_1, \ldots, \jmath_k = +, \circ, -$ are constructed, one can construct a random vector $\vec{W} = (W_1, \ldots, W_k)$ with the above distributions, as this was described above.

Let us summarize the above remarks in the following lemma.

Lemma 9.3.2. *A k-dimensional trinary random vector $\vec{W} = (W_1, \dots, W_k)$ with any given consistent one- and two-dimensional distributions exists. The corresponding high--order distributions for this vector can be constructed according to the algorithm described in relations* (9.124)–(9.131) *and related remarks.*

Now, one can construct, for every $\varepsilon \in (0, \varepsilon_0]$, the random walk $Y_{\varepsilon,n}$ defined by the transition dynamic relation (9.84).

First, one should choose, for every $n = 1, \dots, N$, jump values $\bar{\delta}_{\varepsilon,n,\bar{\jmath}} = (\delta_{\varepsilon,n,\jmath_1}, \dots, \delta_{\varepsilon,n,\jmath_k}), \bar{\jmath} = \langle \jmath_{i_1}, \dots, \jmath_{i_k} \rangle \in \mathcal{J}_k$ and one- and two-dimensional distributions, respectively, $p_{\varepsilon,n,i,\jmath_i}, \jmath_i = +, \circ, -, \ 1 \le i \le k$ and $p_{\varepsilon,n,i,j,\jmath_i,\jmath_j}, \jmath_i, \jmath_j = +, \circ, -, \ 1 \le i < j \le k$, for random vector $\vec{W}_{\varepsilon,n,1}$, according to relations pointed in Lemma 9.2.3.

Second, one should construct the corresponding consistent k-dimensional distributions $p_{\varepsilon,n,1,\dots,k,\jmath_1,\dots,\jmath_k}, \bar{\jmath} \in \mathcal{J}_k$, with the above one- and two dimensional distributions, for every $n = 1, \dots, N$, using the algorithm described in Lemma 9.3.2, and then to define the corresponding intervals $\mathbb{I}_{\bar{\jmath}}, = \mathbb{I}_{\varepsilon,n,\bar{\jmath}}, \bar{\jmath} \in \mathcal{J}_k$, which, in this case, also depend on parameters ε and n.

Third, one can define the corresponding sequence of i.i.d. random vectors $\vec{W}_{\varepsilon,n,l}$, $n, l = 1, 2, \dots$ used in the transition dynamic relation (9.84) as

$$\vec{W}_{\varepsilon,n,l} = \sum_{\bar{\jmath} \in \mathcal{J}_k} \bar{\delta}_{\varepsilon,n,\bar{\jmath}} \mathrm{I}(\rho_{n,l} \in \mathbb{I}_{\varepsilon,n,\bar{\jmath}}), \tag{9.132}$$

where $\rho_{n,l}, n, l = 1, 2, \dots$ is the family of i.i.d. random variables uniformly distributed in the interval $[0, 1)$.

9.3.3 Comparison of fitting algorithms for binomial and trinomial tree models

Let us explain the cases in which the binomial fitting is possible and those in which the binomial model has not enough free parameters, while the trinomial fitting is possible due to the presence of some additional free parameters in the trinomial model.

Let us return back to the system of equations (9.95),

$$\begin{cases} \delta_{\varepsilon,i}(p_{\varepsilon,n,i,+} - p_{\varepsilon,n,i,-}) = \dfrac{\mu_{n,i}}{r_{\varepsilon,n}} - \lambda_{\varepsilon,n,i}, \quad i = 1, \dots, k, \quad n = 1, \dots, N, \\[2ex] \delta^2_{\varepsilon,i}(1 - p_{\varepsilon,n,i,\circ}) = \dfrac{\sigma^2_{n,i}}{r_{\varepsilon,n}} + \left(\dfrac{\mu_{n,i}}{r_{\varepsilon,n}} - \lambda_{\varepsilon,n,i} \right)^2, \quad i = 1, \dots, k, \quad n = 1, \dots, N, \\[2ex] \delta_{\varepsilon,i}\delta_{\varepsilon,j}(p_{\varepsilon,n,i,j,+,+} + p_{\varepsilon,n,i,j,-,-} - p_{\varepsilon,n,i,j,+,-} - p_{\varepsilon,n,i,j,-,+}) \\[1ex] = \dfrac{\rho_{n,i,j}\sigma_{n,i}\sigma_{n,j}}{r_{\varepsilon,n}} + \left(\dfrac{\mu_{n,i}}{r_{\varepsilon,n}} - \lambda_{\varepsilon,n,i} \right) \left(\dfrac{\mu_{n,j}}{r_{\varepsilon,n}} - \lambda_{\varepsilon,n,j} \right), \\[1ex] 1 \le i < j \le k, \quad n = 1, \dots, N. \end{cases} \tag{9.133}$$

Let us restrict consideration by the basic case, where the drift of the corresponding Gaussian log-price process $Y_{0,n}$ is compensated by shift parameters for the corresponding approximating log-price processes $Y_{\varepsilon,n}$, i.e.

$$\lambda_{\varepsilon,n,i} = \frac{\mu_{n,i}}{r_{\varepsilon,n}}, \quad i = 1,\dots,k, \quad n = 1,\dots,N. \tag{9.134}$$

In this case, the system (9.133) takes the following form:

$$\begin{cases} \delta_{\varepsilon,i}(p_{\varepsilon,n,i,+} - p_{\varepsilon,n,i,-}) = 0, \quad i = 1,\dots,k, \quad n = 1,\dots,N, \\ \delta^2_{\varepsilon,i}(1 - p_{\varepsilon,n,i,\circ}) = \dfrac{\sigma^2_{n,i}}{r_{\varepsilon,n}}, \quad i = 1,\dots,k, \quad n = 1,\dots,N, \\ \delta_{\varepsilon,i}\delta_{\varepsilon,j}(p_{\varepsilon,n,i,j,+,+} + p_{\varepsilon,n,i,j,-,-} - p_{\varepsilon,n,i,j,+,-} - p_{\varepsilon,n,i,j,-,+}) = \dfrac{\rho_{n,i,j}\sigma_{n,i}\sigma_{n,j}}{r_{\varepsilon,n}}, \\ \qquad 1 \le i < j \le k, \quad n = 1,\dots,N. \end{cases} \tag{9.135}$$

Let us first consider a binomial tree model, which is, in fact, a particular case of the trinomial model, where quantities,

$$p_{\varepsilon,n,i,\circ} = 0, \quad i = 1,\dots,k, \quad n = 1,\dots,N. \tag{9.136}$$

As was shown in Subsections 9.1.1 and 9.1.2, the exact fitting of parameters can be realized with the use of binomial tree models in the univariate case, where $k = 1$, for both homogeneous and inhomogeneous in time models.

Also, as was shown in Section 9.2, the exact fitting of parameters can be realized with the use of binomial tree models in the multivariate case, where $k > 1$, for homogeneous in time models.

Let us consider the multivariate case, where $k > 1$, for inhomogeneous in time model.

In this case, the second series of equations in (9.135) takes the form,

$$\delta^2_{\varepsilon,i} r_{\varepsilon,n} = \sigma^2_{n,i}, \quad i = 1,\dots,k, \quad n = 1,\dots,N. \tag{9.137}$$

These equalities or even their asymptotic analogs $\delta^2_{\varepsilon,i} r_{\varepsilon,n} \sim \sigma^2_{n,i}$ as $\varepsilon \to 0$, $i = 1,\dots,k$, $n = 1,\dots,N$ cannot achieved to hold by any choice of parameters $\delta_{\varepsilon,i}$, $i = 1,\dots,k$ and $r_{\varepsilon,n}$, $n = 1,\dots,N$, except the case, where the parameters $\sigma^2_{n,i}$ have a specific multiplicative form $\sigma^2_{n,i} = \delta^2_i \sigma^2_{n}$,, $i = 1,\dots,k$, $n = 1,\dots,N$.

In the case of trinomial model, the second series of equations in (9.135) takes the form,

$$\delta^2_{\varepsilon,i}(1 - p_{\varepsilon,n,i,\circ}) r_{\varepsilon,n} = \sigma^2_{n,i}, \quad i = 1,\dots,k, \quad n = 1,\dots,N. \tag{9.138}$$

As was shown in Subsections 9.3.1 and 9.3.2, this equation and the whole system of equations (9.135) have appropriate solutions, which realize the exact fitting of parameters for the multivariate case, where $k > 1$ for inhomogeneous in time trinomial model.

This is due to the presence of additional parameters $p_{\varepsilon,n,i,\circ}$ and $p_{\varepsilon,n,i,j,\pm,\circ}$, $p_{\varepsilon,n,i,j,\circ,\pm}$, $p_{\varepsilon,n,i,j,\circ,\circ}$ in the trinomial models, which are absent in the binomial models.

9.4 Inhomogeneous in space binomial approximations

In this section, we present results about fitting of parameters for homogeneous in time but inhomogeneous in space binomial tree approximation models.

9.4.1 Fitting of parameters for univariate homogeneous in time and inhomogeneous in space binomial tree models

Let us assume that a log-price process $Y_{0,n}$ is a homogeneous in time Gaussian Markov chain, with the phase space $\mathbb{R}_1$, an initial distribution $P_0(A)$, and transition probabilities given for $y \in \mathbb{R}_1$, $A \in \mathcal{B}_1$, $n = 1, 2, \dots$ by the following relation:

$$P_0(y, A) = \mathsf{P}\{Y_{0,n} \in A / Y_{0,n-1} = y\} = \mathsf{P}\{y + W_0(y) \in A\}, \tag{9.139}$$

where (a) $W_0(y)$ is, for every $y \in \mathbb{R}_1$, a normal random variable with the mean value $\mathsf{E}W_0(y) = \mu(y)$ and the variance $\mathrm{Var}W_0(y) = \sigma^2(y)$.

Usual assumptions are: (b) $\mu(y)$ is a measurable function acting from $\mathbb{R}_1$ to $\mathbb{R}_1$, (c) $\sigma^2(y)$ is a measurable function acting from $\mathbb{R}_1$ to $(0, \infty)$.

We shall approximate the log-price process $Y_{0,n}$ by the log-price process $Y_{\varepsilon,n}$ which is, for every $\varepsilon \in (0, \varepsilon_0]$, a homogeneous binomial Markov chain with the phase space $\mathbb{R}_1$, an initial distribution $P_\varepsilon(A)$, and transition probabilities given for $y \in \mathbb{R}_1$, $A \in \mathcal{B}_1$, $n = 1, 2, \dots$ by the following relation:

$$P_\varepsilon(y, A) = \mathsf{P}\{Y_{\varepsilon,n} \in A / Y_{\varepsilon,n-1} = y\} = \mathsf{P}\{y + W_\varepsilon(y) \in A\}, \tag{9.140}$$

where (a) $W_\varepsilon(y) = W_{\varepsilon,1}(y) + \cdots + W_{\varepsilon,r_\varepsilon(y)}(y)$, where $W_{\varepsilon,n}(y)$, $n = 1, \dots$ is, for every $y \in \mathbb{R}_1$, a sequence of i.i.d. binary random variables, which take values $\pm\delta_\varepsilon$ with probabilities $p_{\varepsilon,\pm}(y)$.

Usual assumptions are: (b) $\delta_\varepsilon > 0$, (c) $r_\varepsilon(y)$ is a measurable function defined on space $\mathbb{R}_1$ and taking positive integer values; (d) $p_{\varepsilon,\pm}(y)$ are the measurable functions acting from $\mathbb{R}_1$ to $[0, 1]$, such that $p_{\varepsilon,-}(y) = 1 - p_{\varepsilon,+}(y)$.

In this case, the use of asymmetric jumps $\delta_{\varepsilon,\pm} = \pm\delta_\varepsilon + \lambda_\varepsilon$ has no advantage, since any constant compensator $\lambda_\varepsilon \in \mathbb{R}_1$ cannot compensate an inhomogeneous in space drift $\mu(y)$.

It is useful to note that $W_\varepsilon(y)$ are binomial random variables taking values $w_{\varepsilon,l} = (2l - r_\varepsilon(y))\delta_\varepsilon$, $l = 0, \dots, r_\varepsilon(y)$ with the corresponding probabilities $p_{\varepsilon,l}(y) = C^l_{r_\varepsilon(y)} p_{\varepsilon,+}(y)^l p_{\varepsilon,-}(y)^{r_\varepsilon(y)-l}$, $l = 0, \dots, r_\varepsilon(y)$.

We use the log-price processes $Y_{\varepsilon,n}$ for approximation of the log-price process $Y_{0,n}$, as $\varepsilon \to 0$. In this case, a usual approach is to fit the corresponding parameters, i.e. to try choosing the parameters δ_ε, $p_{\varepsilon,\pm}$, and $r_\varepsilon(y)$ in such a way that the following exact moment-fitting equalities hold, for every $\varepsilon \in (0, \varepsilon_0]$:

$$\Big\{ \mathsf{E}W_\varepsilon(y) = \mathsf{E}W_{0,1}(y)\,, \quad \mathrm{Var}W_{\varepsilon,1}(y) = \mathrm{Var}W_{0,1}(y)\,, \quad y \in \mathbb{R}_1\,. \tag{9.141}$$

The system (9.141) can be rewritten in the following form:

$$\begin{cases} r_\varepsilon(y)\,(\delta_\varepsilon p_{\varepsilon,+}(y) - \delta_\varepsilon p_{\varepsilon,-}(y)) = \mu(y)\,, \\ r_\varepsilon(y)\left(\delta_\varepsilon^2 - \left(\dfrac{\mu(y)}{r_\varepsilon(y)}\right)^2\right) = \sigma^2(y)\,, \quad y \in \mathbb{R}_1\,. \end{cases} \tag{9.142}$$

It is convenient to parameterize $\delta_\varepsilon > 0$ representing it in the form $\delta_\varepsilon = \frac{\delta}{\sqrt{r_\varepsilon}}$, where δ, r_ε are the positive real numbers.

Substituting $\delta_\varepsilon = \frac{\delta}{\sqrt{r_\varepsilon}}$ in relation (9.142) and taking into account that $p_{\varepsilon,+}(y) + p_{\varepsilon,-}(y) = 1$ we can transform relation (9.142) to the following equivalent form:

$$\begin{cases} \delta_\varepsilon = \dfrac{\delta}{\sqrt{r_\varepsilon}}\,, \\ p_{\varepsilon,+}(y) = \dfrac{1}{2}\left(1 + \dfrac{\mu(y)\,\sqrt{r_\varepsilon}}{r_\varepsilon(y)\,\delta}\right), \\ \dfrac{\delta^2}{r_\varepsilon} r_\varepsilon^2(y) - \sigma^2(y) r_\varepsilon(y) - \mu^2(y) = 0\,, \quad y \in \mathbb{R}_1\,. \end{cases} \tag{9.143}$$

The third equality in (9.143) is a quadratic equation with respect to the unknown $r_\varepsilon(y)$. Solving this equation and taking into account that $r_\varepsilon(y) > 0$, we get the following formula for function $r_\varepsilon(y)$:

$$r_\varepsilon(y) = \frac{r_\varepsilon\left(\sigma^2(y) + \sqrt{\sigma^4(y) + \frac{4\delta^2\mu^2(y)}{r_\varepsilon}}\right)}{2\delta^2}\,, \quad y \in \mathbb{R}_1\,. \tag{9.144}$$

The problem is that quantity $r_\varepsilon(y)$ may not take positive integer values for all $y \in \mathbb{R}_1$, except some special cases for functions $\mu(y)$ and $\sigma^2(y)$.

We use the log-price processes $Y_{\varepsilon,n}$ for approximation of the log-price process $Y_{0,n}$ in the sense of weak convergence and convergence of the corresponding reward functions.

We shall show in Chapter 10 that in order to realize the above approximation, one should assume that the parameter $r_\varepsilon \to \infty$ as $\varepsilon \to 0$ and to use asymptotic moment fitting for log-price processes $Y_{\varepsilon,n}$ and $Y_{0,n}$, where the exact fitting relations (9.141) are replaced by the following weaker asymptotic relations:

$$\begin{cases} \mathsf{E}W_\varepsilon(y) \to \mathsf{E}W_{0,1}(y) \quad \text{as} \quad \varepsilon \to 0\,, \quad y \in \mathbb{R}_1\,, \\ \mathrm{Var}W_{\varepsilon,1}(y) \to \mathrm{Var}W_{0,1}(y) \quad \text{as} \quad \varepsilon \to 0\,, \quad y \in \mathbb{R}_1\,. \end{cases} \tag{9.145}$$

Let us show that the above asymptotic moment fitting can be achieved, first, by computing real-valued $r_\varepsilon^*(y)$ using relation (9.144), second, by computing probability

$p_{\varepsilon,+}(y)$ as function of $r^*_\varepsilon(y)$ given by the second equality in relation (9.143), third, by taking as $r_\varepsilon(y)$ the nearest positive integer number that is larger or equal than $r^*_\varepsilon(y)$, and, fourth, by passing the parameter $r_\varepsilon \to \infty$ as $\varepsilon \to 0$.

Thus, according to the above remarks, we define the parameters of the approximating log-price process $Y_{\varepsilon,n}$ according to the following formulas:

$$\begin{cases} \delta_\varepsilon = \dfrac{\delta}{\sqrt{r_\varepsilon}} \\ p_{\varepsilon,+}(y) = \dfrac{1}{2}\left(1 + \dfrac{\mu(y)\sqrt{r_\varepsilon}}{r^*_{\varepsilon,y}\delta}\right), \\ r^*_\varepsilon(y) = \dfrac{r_\varepsilon\left(\sigma^2(y) + \sqrt{\sigma^4(y) + \frac{4\delta^2\mu^2(y)}{r_\varepsilon}}\right)}{2\delta^2}, \\ r_\varepsilon(y) = [r^*_\varepsilon(y)] + 1, \quad y \in \mathbb{R}_1 . \end{cases} \tag{9.146}$$

where $\delta, r_\varepsilon > 0$, and finally passing the parameter $r_\varepsilon \to \infty$ as $\varepsilon \to 0$.

It should also be noted that, according to relation (9.146), $r^*_\varepsilon(y) \geq \frac{\mu(y)\sqrt{r_\varepsilon}}{\delta}$ and, thus, the second equality in (9.146) yields that $0 \leq p_{\varepsilon,+}(y) \leq 1$, for all $y \in \mathbb{R}_1$.

Let us show that the asymptotic moment-fitting relation (9.145) takes place in the case, where parameters of the approximating log-price process $Y_{\varepsilon,n}$ are chosen according to relation (9.146).

The second equality in relation (9.146) can be rewritten in the following equivalent form, for every $\varepsilon \in (0, \varepsilon_0]$ and $y \in \mathbb{R}_1$:

$$r^*_{\varepsilon,y}\left(\frac{\delta}{\sqrt{r_\varepsilon}}p_{\varepsilon,+}(y) - \frac{\delta}{\sqrt{r_\varepsilon}}p_{\varepsilon,-}(y)\right) = \mu(y) . \tag{9.147}$$

Taking into account relation (9.147), inequality $0 \leq r_\varepsilon(y) - r^*_\varepsilon(y) \leq 1$, relation $r^*_\varepsilon(y) \to \infty$ as $\varepsilon \to 0$, which follows from the third equality in (9.146), and the assumption $r_\varepsilon \to \infty$ as $\varepsilon \to 0$, we get the following relation, for every $y \in \mathbb{R}_1$:

$$\begin{aligned} \mathsf{E}W_\varepsilon(y) &= r_\varepsilon(y)\left(\frac{\delta}{\sqrt{r_\varepsilon}}p_{\varepsilon,+}(y) - \frac{\delta}{\sqrt{r_\varepsilon}}p_{\varepsilon,-}(y)\right) \\ &= \mu(y) + (r_\varepsilon(y) - r^*_\varepsilon(y))\left(\frac{\delta}{\sqrt{r_\varepsilon}}\frac{1}{2}\left(1 + \frac{\mu(y)\sqrt{r_\varepsilon}}{r^*_\varepsilon(y)\delta}\right) - \frac{\delta}{\sqrt{r_\varepsilon}}\frac{1}{2}\left(1 - \frac{\mu(y)\sqrt{r_\varepsilon}}{r^*_\varepsilon(y)\,\delta}\right)\right) \\ &= \mu(y) + \mu(y)\frac{r_\varepsilon(y) - r^*_\varepsilon(y)}{r^*_\varepsilon(y)} \to \mu(y) \quad \text{as} \quad \varepsilon \to 0 . \end{aligned} \tag{9.148}$$

Thus, the first asymptotic moment-fitting relation in (9.145) holds.

The function $r^*_\varepsilon(y)$, given by the third equality in relation (9.146), is a root of the quadratic equation. It is given by the third equality in relation (9.143), or, equivalently, in the form of the second equality in relation (9.142). Thus, $r^*_\varepsilon(y)$ satisfies the following relation, for every $\varepsilon \in (0, \varepsilon_0]$ and $y \in \mathbb{R}_1$:

$$r^*_\varepsilon(y)\left(\delta^2_\varepsilon - \left(\frac{\mu(y)}{r^*_\varepsilon(y)}\right)^2\right) = \sigma^2(y) . \tag{9.149}$$

Taking into account relation (9.149), inequality $0 \leq r_\varepsilon(y) - r^*_\varepsilon(y) \leq 1$, relation $r^*_\varepsilon(y), r_\varepsilon(y) \to \infty$ as $\varepsilon \to 0$, which follows from the third equality in (9.146), and the assumption that $r_\varepsilon \to \infty$ as $\varepsilon \to 0$, we get the following relation, for every $y \in \mathbb{R}_1$:

$$\begin{aligned}\mathrm{Var} W_\varepsilon(y) &= r_\varepsilon(y)\left(\frac{\delta^2}{r_\varepsilon} - \left(\frac{\mu(y)}{r_\varepsilon(y)}\right)^2\right) \\ &= \sigma^2(y) + (r_\varepsilon(y) - r^*_\varepsilon(y))\frac{\delta^2}{r_\varepsilon} \\ &\quad + (r_\varepsilon(y) - r^*_\varepsilon(y))\frac{\mu^2(y)}{r_\varepsilon(y)\, r^*_\varepsilon(y)} \to \sigma^2(y) \quad \text{as} \quad \varepsilon \to 0\,. \end{aligned} \tag{9.150}$$

Relation (9.150) is equivalent to the second asymptotic moment-fitting relation in (9.145).

The above remarks can be summarized in the following lemma.

Lemma 9.4.1. *The log-price process $Y_{\varepsilon,n}$ represented, for every $\varepsilon \in (0, \varepsilon_0]$, by the homogeneous binomial Markov chain with transition probabilities defined by relation* (9.140)*, realizes the asymptotic moment fitting given by relation* (9.145)*, for the log-price processes $Y_{0,n}$ represented by the Gaussian Markov chain with transition probabilities defined by relation* (9.139)*. This holds, if the corresponding parameters of the process $Y_{\varepsilon,n}$ are given by relation* (9.146)*, and under the assumption that the parameter $r_\varepsilon \to \infty$ as $\varepsilon \to 0$.*

9.4.2 The recombining condition

The question about recombining property of the binomial Markov chain $Y_{\varepsilon,n}$ does require a special analysis.

Let us first consider the case, where the following condition holds:
$\mathbf{G_{12}}$: $\sup_{y \in \mathbb{R}_1}(|\mu(y)| + \sigma^2(y)) < K_{85}$, for some $0 < K_{85} < \infty$.

The condition $\mathbf{G_{12}}$ and relation (9.146) implies the following relation:

$$\begin{aligned}\sup_{y \in \mathbb{R}_1} r_\varepsilon(y) &\leq \sup_{y \in \mathbb{R}_1} \frac{r_\varepsilon\left(\sigma^2(y) + \sqrt{\sigma^4(y) + \frac{4\delta^2\mu^2(y)}{r_\varepsilon}}\right)}{2\delta^2} + 1 \\ &\leq \sup_{y \in \mathbb{R}_1} \frac{r_\varepsilon\left(\sigma^2(y) + \frac{\delta|\mu(y)|}{\sqrt{r_\varepsilon}}\right)}{\delta^2} + 1 \\ &\leq \frac{K_{85} + K_{85}\delta + \delta^2}{\delta^2} r_\varepsilon = K_{86}(\delta) r_\varepsilon\,. \end{aligned} \tag{9.151}$$

As follows from inequality (9.151), the corresponding tree $T_{\varepsilon,(y,n)}$ generated by the trajectories of the log-price process $Y_{\varepsilon,n+m}$, $m = 0, 1, \dots$, with the initial value $Y_{\varepsilon,n} =$

y, has the total number of points after r steps,

$$L_{\varepsilon,n,n+r} \leq 1 + 2K_{86}(\delta)r_\varepsilon + 1 + \cdots + 2K_{86}(\delta)r_\varepsilon r + 1$$
$$= K_{86}(\delta)r_\varepsilon r^2 + (K_{86}(\delta)r_\varepsilon + 1)r + 1, \quad (9.152)$$

The expression on the right-hand side in 9.152) is a polynomial (in this case quadratic) function of r.

Let us now assume that the condition $\mathbf{G}_{12}$ does not hold. In this case, the number of nodes, which appear in the corresponding tree $T_{\varepsilon,(y,n)}$ generated by the trajectories of the log-price process $Y_{\varepsilon,n+r}$, $r = 0, 1, \ldots$, with the initial value $\vec{Y}_{\varepsilon,n} = \vec{y}$, may possess an exponential rate of growth as a function of the number of steps.

For example, let us assume that the coefficients $\mu(y)$ and $\sigma^2(y)$ have not less than linear rate of growth as functions of y, i.e. that following condition holds:

$\mathbf{G}_{13}$: $\inf_{y\in\mathbb{R}_1} \frac{|\mu(y)|+\sigma^2(y)}{1+K_{88}|y|} \geq K_{87}$, for some $0 < K_{87}, K_{88} < \infty$.

Let us assume for simplicity that $(y, n) = (0, 0)$. Let us denote by $y_{\varepsilon,r}$ the maximal value that can take the random variable $|Y_{\varepsilon,r}|$. Formula (9.146) and the condition $\mathbf{G}_{13}$ imply that $y_{\varepsilon,1} \geq \frac{r_\varepsilon K_{88}}{\delta^2}\frac{\delta}{\sqrt{r_\varepsilon}} = \frac{\sqrt{r_\varepsilon}K_{88}}{\delta}$. Also, $y_{\varepsilon,2} \geq y_{\varepsilon,1} + \frac{r_\varepsilon K_{87}K_{88}y_{\varepsilon,1}}{\delta^2}\frac{\delta}{\sqrt{r_\varepsilon}} = y_{\varepsilon,1}(1+\frac{\sqrt{r_\varepsilon}K_{87}K_{88}}{\delta})$. Since, the parameter $r_\varepsilon \to \infty$ as $\varepsilon \to 0$, we can always assume that $r_\varepsilon \geq 1$. In this case, the above inequalities imply that $y_{\varepsilon,2} \geq y_{\varepsilon,1}(1 + L)$, where $L = \frac{K_{87}K_{88}}{\delta} > 0$. The repetition of the above calculations yields inequalities $y_{\varepsilon,r} \geq y_{\varepsilon,1}(1 + L)^{r-1}$, $r = 1, 2, \ldots$. These inequalities imply that the number of nodes appearing in the tree $T_{\varepsilon,(0,0)}$, as the result of rth step, is larger than or equal to $\frac{2y_{\varepsilon,r}\sqrt{r_\varepsilon}}{\delta} = \frac{2r_\varepsilon(1+L)^{r-1}}{\delta^2}$, and, thus, this tree has the total number of nodes after r steps,

$$L_{\varepsilon,n,n+r} \geq \frac{2r_\varepsilon}{\delta^2} + \cdots \frac{2r_\varepsilon(1+L)^{r-1}}{\delta^2} = \frac{2r_\varepsilon((1+L)^r - 1)}{\delta^2 L}, \quad (9.153)$$

Therefore, the tree has an exponential rate of growth for the number of nodes as a function of r.

9.5 Inhomogeneous in time and space trinomial approximations

In this section, we present results about fitting of parameters for inhomogeneous in time and space trinomial tree approximation models.

9.5.1 Fitting of parameters for univariate inhomogeneous in time and space trinomial tree models

Let us assume that the log-price process $Y_{0,n}$ is an inhomogeneous in time Gaussian Markov chain, with the phase space $\mathbb{R}_1$, an initial distribution $P_0(A)$, and transition probabilities, given for $y \in \mathbb{R}_1$, $A \in \mathcal{B}_1$, $n = 1, 2, \dots$ by the following relation:

$$P_{0,n}(y, A) = \mathsf{P}\{Y_{0,n} \in A / Y_{0,n-1} = y\} = \mathsf{P}\{y + W_{0,n}(y) \in A\}, \tag{9.154}$$

where (a) $W_{0,n}(y)$ is, for every $y \in \mathbb{R}_1$, $n = 1, 2, \dots$, a normal random variable with mean value $\mathsf{E}W_{0,n}(y) = \mu_n(y)$ and variance $\mathrm{Var}W_{0,n}(y) = \sigma_n^2(y)$.

Usual assumptions are: (b) $\mu_n(y)$ is, for every $n = 1, 2, \dots$, a measurable function acting from $\mathbb{R}_1$ to $\mathbb{R}_1$, (c) $\sigma_n^2(y)$ is, for every $n = 1, 2, \dots$,a measurable function acting from $\mathbb{R}_1$ to $(0, \infty)$.

We shall approximate the log-price process $Y_{0,n}$ by the log-price process $Y_{\varepsilon,n}$, which is, for every $\varepsilon \in (0, \varepsilon_0]$, an inhomogeneous trinomial Markov chain, with the phase space $\mathbb{R}_1$, an initial distribution $P_\varepsilon(A)$, and transition probabilities, given for $y \in \mathbb{R}_1$, $A \in \mathcal{B}_1$, $n = 1, 2, \dots$ by the following relation:

$$P_{\varepsilon,n}(y, A) = \mathsf{P}\{Y_{\varepsilon,n} \in A / Y_{\varepsilon,n-1} = y\} = \mathsf{P}\{y + W_{\varepsilon,n}(y) \in A\}, \tag{9.155}$$

where (a) $W_{\varepsilon,n}(y)$ is, for every $y \in \mathbb{R}_1$, $n = 1, 2, \dots$, a trinomial random variable, which can be represented in the form $W_{\varepsilon,n}(y) = W_{\varepsilon,n,1}(y) + \cdots + W_{\varepsilon,n,r_{\varepsilon,n}(y)}(y)$, where $W_{\varepsilon,n,l}(y)$, $l = 1, 2, \dots$ are the i.i.d. random variables taking values $\delta_{\varepsilon,\jmath}$ with probabilities $p_{\varepsilon,n,\jmath}(y)$, for $\jmath = +, \circ, -\}$.

Usual assumptions are: (b) $\delta_{\varepsilon,-} = -\delta_\varepsilon < \delta_{\varepsilon,\circ} = 0 < \delta_{\varepsilon,+} = +\delta_\varepsilon$, (c) $p_{\varepsilon,n,\jmath}(y)$, $\jmath = +, \circ, -$, $n = 1, 2, \dots$ are the measurable functions acting from $\mathbb{R}_1$ to $[0, 1]$ such that $\sum_{\jmath=+,\circ,-} p_{\varepsilon,n,\jmath}(y) = 1$, $y \in \mathbb{R}_1$, $n = 1, 2, \dots$; (d) $r_{\varepsilon,n}(y)$, $n = 1, 2, \dots$ are the measurable functions defined on the space $\mathbb{R}_1$ and taking positive integer values.

Note that, in this case, the use of asymmetric jumps $\delta_{\varepsilon,n,\jmath} = \delta_{\varepsilon,\jmath} + \lambda_{\varepsilon,n}$, $n = 1, 2, \dots$ has no advantage, since any compensators $\lambda_{\varepsilon,n} \in \mathbb{R}_1$ inhomogeneous in time cannot compensate an inhomogeneous in time and space drift $\mu_n(y)$.

Trinomial models have larger number of the parameters $p_{\varepsilon,n,\jmath}(y)$ than binomial models. This makes it possible to simplify the choice of the parameters $r_{\varepsilon,n}(y)$.

We simplify the model by choosing the parameters $r_{\varepsilon,n}(y)$ of the simplest form

$$r_{\varepsilon,n}(y) = r_\varepsilon, \quad y \in \mathbb{R}_1, \quad n = 1, 2, \dots. \tag{9.156}$$

In this case, the Markov random walk $Y_{\varepsilon,n}$ satisfies the recombining condition. Indeed, the random variable $W_{\varepsilon,n}(y)$ can, for every $y \in \mathbb{R}_1$, $n = 1, 2, \dots$, take with positive probabilities only values $l\delta_\varepsilon$, $-r_\varepsilon \leq l \leq r_\varepsilon$ and, therefore, the random variable $Y_{\varepsilon,n}$ can, under condition that $Y_{\varepsilon,0} = y$, take with positive probabilities only values $y + l\delta_\varepsilon$, $-nr_\varepsilon \leq l \leq nr_\varepsilon$. The number of such values is $2nr_\varepsilon + 1$.

We use the log-price processes $Y_{\varepsilon,n}$ for approximation of the log-price process $Y_{0,n}$ asymptotically, as $\varepsilon \to 0$. In this case, the usual approach is to fit the corresponding parameters, i.e. to try choosing the parameters δ_ε, probabilities $p_{\varepsilon,n,\jmath}(y)$, and the parameters r_ε in such a way that the following exact fitting equalities holds, for every $\varepsilon \in (0, \varepsilon_0]$:

$$\begin{cases} \mathsf{E}W_{\varepsilon,n}(y) = \mathsf{E}W_{0,n}(y), \\ \mathrm{Var}W_{\varepsilon,n}(y) = \mathrm{Var}W_{0,n}(y), \quad y \in \mathbb{R}_1, \quad n = 1, \dots, N. \end{cases} \tag{9.157}$$

Calculations analogous to those made in relations (9.55)–(9.57) yield the following formulas, for every $y \in \mathbb{R}_1$, $n = 1, \dots$:

$$\mathsf{E}W_{\varepsilon,n,1}(y) = \delta_\varepsilon(p_{\varepsilon,n,+}(y) - p_{\varepsilon,n,-}(y)) = \frac{\mu_n(y)}{r_\varepsilon}, \tag{9.158}$$

and

$$\mathrm{Var}W_{\varepsilon,n,1}(y) = \delta_\varepsilon^2(1 - p_{\varepsilon,n,\circ}(y)) - \left(\frac{\mu_n(y)}{r_\varepsilon}\right)^2 = \frac{\sigma_n^2(y)}{r_\varepsilon}. \tag{9.159}$$

By analogy with the homogeneous in space case, we shall choose the parameter δ_ε in the following form:

$$\delta_\varepsilon = \frac{\delta}{\sqrt{r_\varepsilon}}. \tag{9.160}$$

For relations (9.158) – (9.160), let us rewrite the fitting relations (9.157) in the following equivalent form:

$$\begin{cases} p_{\varepsilon,n,+}(y) - p_{\varepsilon,n,-}(y) = \dfrac{\mu_n(y)}{\delta\sqrt{r_\varepsilon}}, \\ 1 - p_{\varepsilon,n,\circ}(y) = \dfrac{\sigma_n^2(y)}{\delta^2} + \dfrac{\mu_n^2(y)}{\delta^2 r_\varepsilon}, \quad y \in \mathbb{R}_1, \quad n = 1, \dots, N. \end{cases} \tag{9.161}$$

The system of equations (9.161) should be supplemented by the following consistency relations:

$$\begin{cases} p_{\varepsilon,n,\jmath}(y) \geq 0, \quad \jmath = +, \circ, -, \\ p_{\varepsilon,n,+}(y) + p_{\varepsilon,n,\circ}(y) + p_{\varepsilon,n,-}(y) = 1, \quad y \in \mathbb{R}_1, \quad n = 1, \dots, N. \end{cases} \tag{9.162}$$

The system of equations (9.161) supplemented by the consistency equalities from relations (9.162) has the following solution:

$$\begin{cases} p_{\varepsilon,n,\pm}(y) = \dfrac{1}{2}\left(\dfrac{\sigma_n^2(y)}{\delta^2} + \dfrac{\mu_n^2(y)}{\delta^2 r_\varepsilon} \pm \dfrac{\mu_n(y)}{\delta\sqrt{r_\varepsilon}}\right), \\ p_{\varepsilon,n,\circ}(y) = 1 - \dfrac{\sigma_n^2(y)}{\delta^2} - \dfrac{\mu_n^2(y)}{\delta^2 r_\varepsilon}, \quad y \in \mathbb{R}_1, \quad n = 1, \dots, N. \end{cases} \tag{9.163}$$

Let us assume that the following conditions hold:
G₁₄: $\min_{1\leq n\leq N} \inf_{y\in\mathbb{R}_1} \sigma_n^2(y) \geq \sigma^2$, for some $\sigma^2 > 0$.

and
G₁₅: $\max_{1\leq n\leq N} \sup_{y\in\mathbb{R}_1} (|\mu_n(y)| + \sigma_n^2(y)) < K_{89}$, for some $0 < K_{89} < \infty$.

In this case, the consistency inequalities given in relation (9.162) hold for probabilities $p_{\varepsilon,n,\jmath}(y)$, $\jmath = +, \circ, -$, for every $y \in \mathbb{R}_1$, $n = 1, \dots, N$, if to choose parameter δ satisfying the following inequality:

$$\delta \geq \sqrt{K_{89}^2 + K_{89}}, \tag{9.164}$$

and then to choose the parameter r_ε satisfying the following inequality:

$$r_\varepsilon \geq \frac{K_{89}^2 \delta^2}{\sigma^4}. \tag{9.165}$$

Indeed, the condition $\mathbf{G}_{15}$ and inequality (9.164) imply that, for every $y \in \mathbb{R}_1$, $n = 1, \dots, N$,

$$p_{\varepsilon,n,\circ}(\vec{y}) = 1 - \frac{\sigma_n^2(\vec{y})}{\delta^2} - \frac{\mu_n^2(y)}{\delta^2 r_\varepsilon} \geq 1 - \frac{K_{89}}{\delta^2} - \frac{K_{89}^2}{\delta^2} \geq 0, \tag{9.166}$$

Also, the conditions $\mathbf{G}_{14}$ and $\mathbf{G}_{15}$, and inequality (9.165) imply that, for every $y \in \mathbb{R}_1$, $n = 1, \dots, N$,

$$p_{\varepsilon,n,\pm}(y) = \frac{1}{2}\left(\frac{\sigma_n^2(y)}{\delta^2} + \frac{\mu_n^2(y)}{\delta^2 r_\varepsilon} \pm \frac{\mu_n(y)}{\delta\sqrt{r_\varepsilon}}\right) \geq \frac{1}{2}\left(\frac{\sigma^2}{\delta^2} - \frac{K_{89}}{\delta\sqrt{r_\varepsilon}}\right) \geq 0. \tag{9.167}$$

The above remarks can be summarized in the following lemma.

Lemma 9.5.1. *The log-price process $Y_{\varepsilon,n}$ represented, for every $\varepsilon \in (0, \varepsilon_0]$, by the inhomogeneous trinomial Markov chain with transition probabilities defined by relation* (9.155), *realizes the exact moment fitting given by relation* (9.157) *for the log-price process $Y_{0,n}$ represented by the Gaussian–Markov chain with transition probabilities defined by relation* (9.154). *This holds, if the corresponding parameters of the process $Y_{\varepsilon,n}$ are given by relations* (9.160) *and* (9.163), *the conditions $\mathbf{G}_{14}$ and $\mathbf{G}_{15}$ hold, and parameters δ and r_ε satisfy conditions* (9.164) *and* (9.165).

The corresponding tree $T_{\varepsilon,(y,n)}$ generated by the trajectories of the log-price process $Y_{\varepsilon,n+r}$, $r = 0, 1, \dots$, with the initial value $Y_{\varepsilon,n} = y$, has the total number of points after r steps,

$$L_{\varepsilon,n,n+r} \leq 1 + 2r_\varepsilon + 1 + \cdots + 2r_\varepsilon r + 1 = r_\varepsilon r^2 + (r_\varepsilon + 1)r + 1. \tag{9.168}$$

The expression on the right-hand side in (9.168) is a polynomial (in this case quadratic) function of r.

9.5.2 Fitting of parameters for multivariate inhomogeneous in time and space trinomial tree models

Let us assume that a log-price process $\vec{Y}_{0,n}$ is an inhomogeneous in time Gaussian Markov chain with the phase space $\mathbb{R}_k$, an initial distribution $P_0(A)$, and transition probabilities, given for $\vec{y} \in \mathbb{R}_k$, $A \in \mathcal{B}_k$, $n = 1, 2, \dots$ by the following relation:

$$P_{0,n}(\vec{y}, A) = \mathsf{P}\{\vec{Y}_{0,n} \in A / \vec{Y}_{0,n-1} = \vec{y}\} = \mathsf{P}\{\vec{y} + \vec{W}_{0,n}(\vec{y}) \in A\}, \tag{9.169}$$

where (a) $\vec{W}_{0,n}(\vec{y}) = (W_{0,n,1}(\vec{y}), \dots, W_{0,n,k}(\vec{y}))$ is, for every $n = 1, 2, \dots, \vec{y} \in \mathbb{R}_k$, a normal random variable with means $\mathsf{E}W_{0,n,i}(\vec{y}) = \mu_{n,i}(\vec{y})$, $i = 1, \dots, k$, variances $\mathrm{Var}W_{0,n,i}(\vec{y}) = \sigma_{n,i}^2(\vec{y})$, $i = 1, \dots, k$, and covariances $\mathsf{E}(W_{0,n,i}(\vec{y}) - \mu_{n,i}(\vec{y})) \cdot (W_{0,n,j}(\vec{y}) - \mu_{n,j}(\vec{y}))) = \rho_{n,i,j}(\vec{y})\sigma_{n,i}(\vec{y})\sigma_{n,j}(\vec{y})$, $i, j = 1, \dots, k$.

Usual assumptions are: (b) $\mu_{n,i}(\vec{y})$, $i = 1, \dots, k$, $n = 1, 2, \dots$ are the measurable functions acting from $\mathbb{R}_k$ to $\mathbb{R}_1$, (c) $\sigma^2_{n,i}(\vec{y})$, $i = 1, \dots, k$, $n = 1, 2, \dots$ are the measurable function acting from $\mathbb{R}_1$ to $(0, \infty)$; (d) $\rho_{n,i,j}(\vec{y}), i, j = 1, \dots, k$, $n = 1, 2, \dots$ are the measurable functions acting from $\mathbb{R}_k$ to $[-1, 1]$ and $\rho_{n,i,i}(\vec{y}) = 1, \vec{y} \in \mathbb{R}_k$, $i = 1, \dots, k$, $n = 1, 2, \dots$; (e) $\Sigma(\vec{y}) = \|\sigma_{n,i,j}(\vec{y})\|, \vec{y} \in \mathbb{R}_k$, $n = 1, 2, \dots$, where $\sigma_{n,i,j}(\vec{y}) = \rho_{n,i,j}(\vec{y})\sigma_{n,i}(\vec{y})\sigma_{n,j}(\vec{y}), i, j = 1, \dots, k$, are nonnegatively defined $k \times k$ matrices.

We shall approximate the log-price process $\vec{Y}_{0,n}$ by the log-price process $\vec{Y}_{\varepsilon,n}$ which is, for every $\varepsilon \in (0, \varepsilon_0]$, an inhomogeneous trinomial Markov chain with the phase space $\mathbb{R}_k$, an initial distribution $P_\varepsilon(A)$, and transition probabilities, given for $\vec{y} \in \mathbb{R}_k$, $A \in \mathcal{B}_k$, $n = 1, 2, \dots$ by the following relation:

$$P_{\varepsilon,n}(\vec{y}, A) = \mathsf{P}\{\vec{Y}_{\varepsilon,n} \in A / \vec{Y}_{\varepsilon,n-1} = \vec{y}\} = \mathsf{P}\{\vec{y} + \vec{W}_{\varepsilon,n}(\vec{y}) \in A\}, \tag{9.170}$$

where (a) $\vec{W}_{\varepsilon,n}(\vec{y}) = (W_{\varepsilon,n,1}(\vec{y}), \dots, W_{\varepsilon,n,k}(\vec{y}))$ is, for every $\vec{y} \in \mathbb{R}_k$, $n = 1, 2, \dots$, a trinomial random vector, which can be represented in the form $\vec{W}_{\varepsilon,n}(\vec{y}) = \vec{W}_{\varepsilon,n,1}(\vec{y}) + \dots + \vec{W}_{\varepsilon,n,r_{\varepsilon,n}(\vec{y})}(\vec{y})$, where $\vec{W}_{\varepsilon,n,l}(\vec{y}) = (W_{\varepsilon,n,l,1}(\vec{y}), \dots, W_{\varepsilon,n,l,k}(\vec{y}))$, $l = 1, 2, \dots$ are the i.i.d. random vectors taking, for every $\vec{y} \in \mathbb{R}_k$ and $n = 1, 2, \dots$, values $\bar{\delta}_{\varepsilon,\bar{\jmath}} = (\delta_{\varepsilon,1,\jmath_1}, \dots, \delta_{\varepsilon,k,\jmath_k})$ with probabilities $p_{\varepsilon,n,\bar{\jmath}}(\vec{y}) = p_{\varepsilon,n,\jmath_1,\dots\jmath_k}(\vec{y})$, for $\bar{\jmath} \in \mathcal{J}_k = \{\bar{\jmath} = \langle \jmath_1, \dots, \jmath_k \rangle :$ $\jmath_j, \dots \jmath_k = +, \circ, -\}$;

Usual assumptions are: (b) $\delta_{\varepsilon,i,-} = -\delta_{\varepsilon,i} < \delta_{\varepsilon,i,\circ} = 0 < \delta_{\varepsilon,i,+} = +\delta_{\varepsilon,i}$, $i = 1, \dots, k$; (c) $p_{\varepsilon,n,\bar{\jmath}}(\vec{y}), \bar{\jmath} \in \mathcal{J}_k$, $n = 1, 2, \dots$ are measurable function acting from $\mathbb{R}_k$ to $[0, 1]$ such that $\sum_{\bar{\jmath} \in \mathcal{J}_k} p_{\varepsilon,n,\bar{\jmath}}(\vec{y}) = 1, \vec{y} \in \mathbb{R}_k$, $n = 1, 2, \dots$; (d) $r_{\varepsilon,n}(\vec{y})$, $n = 1, 2, \dots$ are measurable functions defined on the space $\mathbb{R}_k$ and taking positive integer values.

Note that, in this case, the use of asymmetric jumps $\delta_{\varepsilon,n,i,\jmath} = \delta_{\varepsilon,i,\jmath} + \lambda_{\varepsilon,n,i}$, $i = 1, \dots, k$, $n = 1, 2, \dots$ has no advantage, since any compensators $\lambda_{\varepsilon,n,i} \in \mathbb{R}_1$ inhomogeneous in time cannot compensate an inhomogeneous in time and space drifts $\mu_{n,i}(\vec{y})$.

We also simplify the model by choosing positive integer parameters $r_{\varepsilon,n}(\vec{y})$ of the simplest form,

$$r_{\varepsilon,n}(\vec{y}) = r_\varepsilon, \quad \vec{y} \in \mathbb{R}_k, \quad n = 1, 2, \dots. \tag{9.171}$$

As a matter of fact, trinomial models have larger number of parameters $p_{\varepsilon,n,\bar{\jmath}}(\vec{y})$ than binomial models. This makes it possible to simplify the choice of the parameters $r_{\varepsilon,n}(\vec{y})$.

In this case, the Markov random walk $\vec{Y}_{\varepsilon,n}$ satisfies the recombining condition. Indeed, the random vector $\vec{W}_{\varepsilon,n}(\vec{y})$ can, for every $\vec{y} \in \mathbb{R}_k$, $n = 1, 2, \dots$, take with positive probabilities only values $(l_1\delta_{\varepsilon,1}, \dots, l_k\delta_{\varepsilon,k})$, $-r_\varepsilon \leq l_1, \dots, l_k \leq r_\varepsilon$ and, therefore, the random vector $\vec{Y}_{\varepsilon,n}$ can, under condition that $\vec{Y}_{\varepsilon,0} = \vec{y}$, take with positive probabilities only values $\vec{y} + (l_1\delta_{\varepsilon,1}, \dots, l_k\delta_{\varepsilon,k})$, $-nr_\varepsilon \leq l_1, \dots, l_k \leq nr_\varepsilon$. The number of such values is $(2nr_\varepsilon + 1)^k$.

We use the log-price processes $\vec{Y}_{\varepsilon,n}$ for approximation the log-price process $\vec{Y}_{0,n}$ asymptotically, as $\varepsilon \to 0$. In this case, a usual approach is to fit the corresponding parameters, i.e. to try to choose parameters $\delta_{\varepsilon,i}$, probabilities $p_{\varepsilon,n,\bar{j}}(\vec{y})$ and a parameters r_ε in such a way that the following exact fitting equalities hold, for every $\varepsilon \in (0, \varepsilon_0]$:

$$\begin{cases} \mathsf{E}W_{\varepsilon,n,i}(\vec{y}) = \mathsf{E}W_{0,n,i}(\vec{y}), \\ \qquad i = 1, \dots, k, \quad \vec{y} \in \mathbb{R}_k, \quad n = 1, \dots, N, \\ \mathrm{Var}W_{\varepsilon,n,i}(\vec{y}) = \mathrm{Var}W_{0,n,i}(\vec{y}), \\ \qquad i = 1, \dots, k, \quad \vec{y} \in \mathbb{R}_k, \quad n = 1, \dots, N, \\ \mathsf{E}(W_{\varepsilon,n,i}(\vec{y}) - \mathsf{E}W_{\varepsilon,n,i}(\vec{y}))(W_{\varepsilon,n,i}(\vec{y}) - \mathsf{E}W_{\varepsilon,n,j}(\vec{y})) \\ \qquad = \mathsf{E}(W_{0,n,i}(\vec{y}) - \mathsf{E}W_{0,n,i}(\vec{y}))(W_{0,n,i}(\vec{y}) - \mathsf{E}W_{0,n,j}(\vec{y})), \\ \qquad 1 \leq i < j \leq k, \quad \vec{y} \in \mathbb{R}_k, \quad n = 1, \dots, N. \end{cases} \tag{9.172}$$

Calculations analogous to those made in relations (9.55)–(9.57) yield the following formulas:

$$\begin{aligned} \mathsf{E}W_{\varepsilon,n,1,i}(\vec{y}) &= \delta_{\varepsilon,i}(p_{\varepsilon,n,i,+}(\vec{y}) - p_{\varepsilon,n,i,-}(\vec{y})) = \frac{\mu_{n,i}(\vec{y})}{r_\varepsilon}, \\ & i = 1, \dots, k,\ \vec{y} \in \mathbb{R}_k, \quad n = 1, \dots, \\ \mathrm{Var}W_{\varepsilon,n,1,i}(\vec{y}) &= \delta^2_{\varepsilon,i}(1 - p_{\varepsilon,n,i,\circ}(\vec{y})) - (\frac{\mu_{n,i}(\vec{y})}{r_\varepsilon})^2 = \frac{\sigma^2_{n,i}(\vec{y})}{r_\varepsilon}, \\ & i = 1, \dots, k,\ \vec{y} \in \mathbb{R}_k, \quad n = 1, \dots, \\ \mathsf{E}\left(W_{\varepsilon,n,1,i}(\vec{y}) - \frac{\mu_{n,i}(\vec{y})}{r_\varepsilon}\right)&\left(W_{\varepsilon,n,1,j}(\vec{y}) - \frac{\mu_{n,j}(\vec{y})}{r_\varepsilon}\right) \\ &= \delta_{\varepsilon,i}\delta_{\varepsilon,j}\left(p_{\varepsilon,n,i,j,+,+}(\vec{y}) + p_{\varepsilon,n,i,j,-,-}(\vec{y}) - p_{\varepsilon,n,i,j,+,-}(\vec{y}) - p_{\varepsilon,n,i,j,-,+}(\vec{y})\right) \\ &\quad - \frac{\mu_{n,i}(\vec{y})\mu_{n,j}(\vec{y})}{r^2_\varepsilon} \\ &= \frac{\rho_{n,i,j}(\vec{y})\sigma_{n,i}(\vec{y})\sigma_{n,j}(\vec{y})}{r_\varepsilon}, \\ & 1 \leq i < j \leq k, \vec{y} \in \mathbb{R}_k, \quad n = 1, \dots, \end{aligned} \tag{9.173}$$

where

$$p_{\varepsilon,n,i,j,\jmath_i,\jmath_j}(\vec{y}) = \mathsf{P}\{W_{\varepsilon,n,1,i}(\vec{y}) = \delta_{\varepsilon,i,\jmath_i}, W_{\varepsilon,n,1,j}(\vec{y}) = \delta_{\varepsilon,j,\jmath_j}\},$$
$$\jmath_i, \jmath_j = +, \circ, -, \quad 1 \leq i < j \leq k, \vec{y} \in \mathbb{R}_k, \quad n = 1, \dots, \tag{9.174}$$

and

$$p_{\varepsilon,n,i,\jmath_i} = \mathsf{P}\{W_{\varepsilon,n,1,i} = \delta_{\varepsilon,i,\jmath_i}\}, \quad \jmath_i = +, \circ, -, \quad i = 1, \dots, k, \quad n = 1, \dots. \tag{9.175}$$

By analogy with the homogeneous in space case, we shall search the parameters $\delta_{\varepsilon,i}$ in the following form:

$$\delta_{\varepsilon,i} = \frac{\delta_i}{\sqrt{r_\varepsilon}}, \quad i = 1, \dots, k. \tag{9.176}$$

For relations (9.173) and (9.176), let us rewrite the fitting relations (9.172) in the following equivalent form:

$$\begin{cases} p_{\varepsilon,n,i,+}(\vec{y}) - p_{\varepsilon,n,i,-}(\vec{y}) = \dfrac{\mu_{n,i}(\vec{y})}{\delta_i\sqrt{r_\varepsilon}}\,, \\ \qquad i = 1,\dots,k, \quad \vec{y} \in \mathbb{R}_k, \quad n = 1,\dots,N, \\ p_{\varepsilon,n,i,\circ}(\vec{y}) = 1 - \dfrac{\sigma^2_{n,i}(\vec{y})}{\delta_i^2} - \dfrac{\mu^2_{n,i}(\vec{y})}{\delta_i^2 r_\varepsilon}\,, \\ \qquad i = 1,\dots,k, \quad \vec{y} \in \mathbb{R}_k, \quad n = 1,\dots,N, \\ p_{\varepsilon,n,i,j,+,+}(\vec{y}) + p_{\varepsilon,n,i,j,-,-}(\vec{y}) - p_{\varepsilon,n,i,j,+,-}(\vec{y}) - p_{\varepsilon,n,i,j,-,+}(\vec{y}) \\ \qquad = \dfrac{\rho_{n,i,j}(\vec{y})\sigma_{n,i}(\vec{y})\sigma_{n,j}(\vec{y})}{\delta_i\delta_j} + \dfrac{\mu_{n,i}(\vec{y})\mu_{n,j}(\vec{y})}{\delta_i\delta_j r_\varepsilon}\,, \\ \qquad 1 \le i < j \le k, \quad \vec{y} \in \mathbb{R}_k, \quad n = 1,\dots,N. \end{cases} \tag{9.177}$$

The system of equations (9.95) should be supplemented by the following probability inequalities:

$$\begin{aligned} &0 \le p_{\varepsilon,n,i,j,\jmath_i,\jmath_j}(\vec{y}) \le 1, \quad \jmath_i, \jmath_j = +, \circ, -\,, \\ &\qquad 1 \le i < j \le k, \quad \vec{y} \in \mathbb{R}_k, \quad n = 1,\dots,N, \end{aligned} \tag{9.178}$$

and the following system of consistency relations:

$$\begin{cases} p_{\varepsilon,n,i,j,\jmath_i,+}(\vec{y}) + p_{\varepsilon,n,i,j,\jmath_i,\circ}(\vec{y}) + p_{\varepsilon,n,i,j,\jmath_i,-}(\vec{y}) = p_{\varepsilon,n,i,\jmath_i}(\vec{y})\,, \\ p_{\varepsilon,n,i,j,+,\jmath_j}(\vec{y}) + p_{\varepsilon,n,i,j,\circ,\jmath_j}(\vec{y}) + p_{\varepsilon,n,i,j,-,\jmath_j}(\vec{y}) = p_{\varepsilon,n,j,\jmath_j}(\vec{y})\,, \\ \jmath_i, \jmath_j = +, \circ, -, \quad 1 \le i < j \le k, \quad \vec{y} \in \mathbb{R}_k, \quad n = 1,\dots,N, \\ p_{\varepsilon,n,i,+}(\vec{y}) + p_{\varepsilon,n,i,\circ}(\vec{y}) + p_{\varepsilon,n,i-}(\vec{y}) = 1\,, \\ \qquad 1 \le i \le k, \quad \vec{y} \in \mathbb{R}_k, \quad n = 1,\dots,N. \end{cases} \tag{9.179}$$

The first two series of equalities in (9.177) and the last series of equalities in (9.179) yield the following system of equations:

$$\begin{cases} p_{\varepsilon,n,i,+}(\vec{y}) - p_{\varepsilon,n,i,-}(\vec{y}) = \dfrac{\mu_{n,i}(\vec{y})}{\delta_i\sqrt{r_\varepsilon}}\,, \\ p_{\varepsilon,n,i,\circ}(\vec{y}) = 1 - \dfrac{\sigma^2_{n,i}(\vec{y})}{\delta_i^2} - \dfrac{\mu^2_{n,i}(\vec{y})}{\delta_i^2 r_\varepsilon}\,, \\ p_{\varepsilon,n,i,+}(\vec{y}) + p_{\varepsilon,n,i-}(\vec{y}) = 1 - p_{\varepsilon,n,i,\circ}(\vec{y}) \\ \qquad i = 1,\dots,k, \quad \vec{y} \in \mathbb{R}_k, \quad n = 1,\dots,N, \end{cases} \tag{9.180}$$

which yields the following formulas:

$$\begin{cases} p_{\varepsilon,n,i,\pm}(\vec{y}) = \dfrac{1}{2}\left(\dfrac{\sigma^2_{n,i}(\vec{y})}{\delta_i^2} + \dfrac{\mu^2_{n,i}(\vec{y})}{\delta_i^2 r_\varepsilon} \pm \dfrac{\mu_{n,i}(\vec{y})}{\delta_i\sqrt{r_\varepsilon}}\right), \\ p_{\varepsilon,n,i,\circ}(\vec{y}) = 1 - \dfrac{\sigma^2_{n,i}(\vec{y})}{\delta_i^2} - \dfrac{\mu^2_{n,i}(\vec{y})}{\delta_i^2 r_\varepsilon}\,, \\ \qquad i = 1,\dots,k, \quad \vec{y} \in \mathbb{R}_k, \quad n = 1,\dots,N, \end{cases} \tag{9.181}$$

Let us assume that the following conditions hold:
$\mathbf{G_{16}}$: $\min_{1\le i\le k, 1\le n\le N}\inf_{\vec{y}\in\mathbb{R}_k}\sigma^2_{n,i}(\vec{y}) \ge \sigma^2$, for some $\sigma^2 > 0$,

and
$\mathbf{G_{17}}$: $\max_{1\le i\le k, 1\le n\le N}\sup_{\vec{y}\in\mathbb{R}_k}(|\mu_{n,i}(\vec{y})| + \sigma^2_{n,i}(\vec{y})) < K_{90}$, for some $0 < K_{90} < \infty$.

In this case, the following consistency relations hold:

$$\begin{cases} p_{\varepsilon,n,i,+}(\vec{y}), p_{\varepsilon,n,i,\circ}(\vec{y}), p_{\varepsilon,n,i,-}(\vec{y}) \ge 0\,, \\ p_{\varepsilon,n,i,+}(\vec{y}) + p_{\varepsilon,n,i,\circ}(\vec{y}) + p_{\varepsilon,n,i,-}(\vec{y}) = 1\,, \\ 1 \le i \le k\,, \quad \vec{y}\in\mathbb{R}_k\,, \quad n = 1,\dots,N\,, \end{cases} \tag{9.182}$$

if we choose the parameters δ_i, $i = 1,\dots,k$ satisfying the following inequalities:

$$\delta_i \ge \sqrt{K^2_{90} + K_{90}}\,, \quad i = 1,\dots,k\,. \tag{9.183}$$

and then choose the parameter r_ε satisfying the following inequality:

$$r_\varepsilon \ge \frac{K^2_{90}\delta^2}{\sigma^4}\,, \tag{9.184}$$

where

$$\delta = \max_{1\le i\le k}\delta_i\,. \tag{9.185}$$

Indeed, the condition $\mathbf{G_{17}}$ and inequality (9.183) imply that, for every $i = 1,\dots,k$, $\vec{y}\in\mathbb{R}_k$, $n = 1,\dots,N$,

$$p_{\varepsilon,n,i,\circ}(\vec{y}) = 1 - \frac{\sigma^2_{n,i}(\vec{y})}{\delta^2_i} - \frac{\mu^2_{n,i}(\vec{y})}{\delta^2_i r_\varepsilon} \ge 1 - \frac{K_{90}}{\delta^2_i} - \frac{K^2_{90}}{\delta^2_i} \ge 0\,, \tag{9.186}$$

Also, the conditions $\mathbf{G_{16}}$, $\mathbf{G_{17}}$, and inequality (9.184) imply that, for every $i = 1,\dots,k$, $\vec{y}\in\mathbb{R}_k$, $n = 1,\dots,N$,

$$\begin{aligned} p_{\varepsilon,n,i,\pm}(\vec{y}) &= \frac{1}{2}\left(\frac{\sigma^2_{n,i}(\vec{y})}{\delta^2_i} + \frac{\mu^2_{n,i}(\vec{y})}{\delta^2_i r_\varepsilon} \pm \frac{\mu_{n,i}(\vec{y})}{\delta_i\sqrt{r_\varepsilon}}\right) \\ &\ge \frac{1}{2}\left(\frac{\sigma^2}{\delta^2_i} - \frac{K_{90}}{\delta_i\sqrt{r_\varepsilon}}\right) \ge 0\,. \end{aligned} \tag{9.187}$$

Relations (9.179) imply the following relation, for every $1 \le i < j \le k$, $\vec{y}\in\mathbb{R}_k$, $n = 1,\dots,N$:

$$\begin{aligned} p_{\varepsilon,n,i,+}(\vec{y}) + p_{\varepsilon,n,i,-}(\vec{y}) &= p_{\varepsilon,n,i,j,+,+}(\vec{y}) + p_{\varepsilon,n,i,j,+,\circ}(\vec{y}) + p_{\varepsilon,n,i,j,+,-}(\vec{y}) \\ &\quad + p_{\varepsilon,n,i,j,-,+}(\vec{y}) + p_{\varepsilon,n,i,j,-,\circ}(\vec{y}) + p_{\varepsilon,n,i,j,-,-}(\vec{y}) \\ &= p_{\varepsilon,n,i,j,+,+}(\vec{y}) + p_{\varepsilon,n,i,j,+,-}(\vec{y}) + p_{\varepsilon,n,i,j,-,+}(\vec{y}) \\ &\quad + p_{\varepsilon,n,i,j,-,-}(\vec{y}) + p_{\varepsilon,n,j,\circ}(\vec{y}) - p_{\varepsilon,n,i,j,\circ,\circ}(\vec{y}) = 1 - p_{\varepsilon,n,i,\circ}(\vec{y})\,, \end{aligned} \tag{9.188}$$

which can be rewritten in the following form:

$$\begin{aligned}
& p_{\varepsilon,n,i,j,+,+}(\vec{y}) + p_{\varepsilon,n,i,j,+,-}(\vec{y}) + p_{\varepsilon,n,i,j,-,+}(\vec{y}) + p_{\varepsilon,n,i,j,-,-}(\vec{y}) \\
& \quad = 1 - p_{\varepsilon,n,i,\circ}(\vec{y}) - p_{\varepsilon,n,j,\circ}(\vec{y}) + p_{\varepsilon,n,i,j,\circ,\circ}(\vec{y}) \\
& \quad = \frac{\sigma^2_{n,i}(\vec{y})}{\delta_i^2} + \frac{\mu^2_{n,i}(\vec{y})}{\delta_i^2 r_\varepsilon} + \frac{\sigma^2_{n,j}(\vec{y})}{\delta_j^2} + \frac{\mu^2_{n,j}(\vec{y})}{\delta_j^2 r_\varepsilon} - 1 + p_{\varepsilon,n,i,j,\circ,\circ}(\vec{y})
\end{aligned} \tag{9.189}$$

Let us denote, for $1 \le i < j \le k, \vec{y} \in \mathbb{R}_k, n = 1, \dots, N$,

$$\begin{aligned}
P_{\varepsilon,n,i,j}(\vec{y}) &= p_{\varepsilon,n,i,j,+,+}(\vec{y}) + p_{\varepsilon,n,i,j,-,-}(\vec{y}), \\
Q_{\varepsilon,n,i,j}(\vec{y}) &= p_{\varepsilon,n,i,j,+,-}(\vec{y}) + p_{\varepsilon,n,i,j,-,+}(\vec{y}),
\end{aligned} \tag{9.190}$$

The last series of equations in (9.177) and relations (9.189) constitute the following system of equations:

$$\begin{cases}
P_{\varepsilon,n,i,j}(\vec{y}) + Q_{\varepsilon,n,i,j}(\vec{y}) = \dfrac{\sigma^2_{n,i}(\vec{y})}{\delta_i^2} + \dfrac{\mu^2_{n,i}(\vec{y})}{\delta_i^2 r_\varepsilon} \\
\qquad + \dfrac{\sigma^2_{n,j}(\vec{y})}{\delta_j^2} + \dfrac{\mu^2_{n,j}(\vec{y})}{\delta_j^2 r_\varepsilon} - 1 + p_{\varepsilon,n,i,j,\circ,\circ}(\vec{y}), \\
P_{\varepsilon,n,i,j}(\vec{y}) - Q_{\varepsilon,n,i,j}(\vec{y}) = \dfrac{\rho_{n,i,j}(\vec{y})\sigma_{n,i}(\vec{y})\sigma_{n,j}(\vec{y})}{\delta_i\delta_j} + \dfrac{\mu_{n,i}(\vec{y})\mu_{n,j}(\vec{y})}{\delta_i\delta_j r_\varepsilon}, \\
\qquad 1 \le i < j \le k, \quad \vec{y} \in \mathbb{R}_k, \quad n = 1, \dots, N
\end{cases} \tag{9.191}$$

By solving this system of equation, we get the following formulas:

$$\begin{cases}
P_{\varepsilon,n,i,j}(\vec{y}) = \dfrac{1}{2}\Bigg(\dfrac{\rho_{n,i,j}(\vec{y})\sigma_{n,i}(\vec{y})\sigma_{n,j}(\vec{y})}{\delta_i\delta_j} + \dfrac{\mu_{n,i}(\vec{y})\mu_{n,j}(\vec{y})}{\delta_i\delta_j r_\varepsilon} + \dfrac{\sigma^2_{n,i}(\vec{y})}{\delta_i^2} \\
\qquad + \dfrac{\mu^2_{n,i}(\vec{y})}{\delta_i^2 r_\varepsilon} + \dfrac{\sigma^2_{n,j}(\vec{y})}{\delta_j^2} + \dfrac{\mu^2_{n,j}(\vec{y})}{\delta_j^2 r_\varepsilon} - 1 + p_{\varepsilon,n,i,j,\circ,\circ}(\vec{y})\Bigg), \\
Q_{\varepsilon,n,i,j}(\vec{y}) = \dfrac{1}{2}\Bigg(-\dfrac{\rho_{n,i,j}(\vec{y})\sigma_{n,i}(\vec{y})\sigma_{n,j}(\vec{y})}{\delta_i\delta_j} - \dfrac{\mu_{n,i}(\vec{y})\mu_{n,j}(\vec{y})}{\delta_i\delta_j r_\varepsilon} + \dfrac{\sigma^2_{n,i}(\vec{y})}{\delta_i^2} \\
\qquad + \dfrac{\mu^2_{n,i}(\vec{y})}{\delta_i^2 r_\varepsilon} + \dfrac{\sigma^2_{n,j}(\vec{y})}{\delta_j^2} + \dfrac{\mu^2_{n,j}(\vec{y})}{\delta_j^2 r_\varepsilon} - 1 + p_{\varepsilon,n,i,j,\circ,\circ}(\vec{y})\Bigg), \\
\qquad 1 \le i < j \le k, \quad \vec{y} \in \mathbb{R}_k, \quad n = 1, \dots, N.
\end{cases} \tag{9.192}$$

Consistency relations (9.178) and (9.179) imply that the quantities $P_{\varepsilon,n,i,j}(\vec{y})$ and $Q_{\varepsilon,n,i,j}(\vec{y})$ should satisfy inequalities (a) $P_{\varepsilon,n,i,j}(\vec{y}) \ge 0$, (b) $Q_{\varepsilon,n,i,j}(\vec{y}) \ge 0$; and (c) $P_{\varepsilon,n,i,j}(\vec{y}) + Q_{\varepsilon,n,i,j}(\vec{y}) \le 1$, for every $1 \le i < j \le k, \quad \vec{y} \in \mathbb{R}_k, \quad n = 1, \dots, N$.

In order to improve chances for holding inequalities (a)–(c), let us choose for probabilities $p_{\varepsilon,n,i,j,\circ,\circ}(\vec{y})$ the maximal admissible values, which are given by the following formula:

$$\begin{aligned} p_{\varepsilon,n,i,j,\circ,\circ}(\vec{y}) &= p_{\varepsilon,n,i,\circ}(\vec{y}) \wedge p_{\varepsilon,n,j,\circ}(\vec{y}) \\ &= \left(1 - \frac{\sigma^2_{n,i}(\vec{y})}{\delta_i^2} - \frac{\mu^2_{n,i}(\vec{y})}{\delta_i^2 r_\varepsilon}\right) \wedge \left(1 - \frac{\sigma^2_{n,j}(\vec{y})}{\delta_j^2} - \frac{\mu^2_{n,j}(\vec{y})}{\delta_j^2 r_\varepsilon}\right), \\ &1 \le i < j \le k, \quad \vec{y} \in \mathbb{R}_k, \quad n = 1, \dots, N. \end{aligned} \tag{9.193}$$

In this case, inequality (c) obviously holds since,

$$\begin{aligned} 1 - p_{\varepsilon,n,i,\circ}(\vec{y}) - p_{\varepsilon,n,j,\circ}(\vec{y}) + p_{n,i,\circ}(\vec{y}) \wedge p_{\varepsilon,n,j,\circ}(\vec{y}) \\ = \left(1 - p_{\varepsilon,n,i,\circ}(\vec{y})\right) \wedge \left(1 - p_{\varepsilon,n,j,\circ}(\vec{y})\right) \le 1. \end{aligned} \tag{9.194}$$

In this case, formulas (9.192) take the following form:

$$\begin{cases} P_{\varepsilon,n,i,j}(\vec{y}) = \frac{1}{2}\Bigg(\frac{\rho_{n,i,j}(\vec{y})\sigma_{n,i}(\vec{y})\sigma_{n,j}(\vec{y})}{\delta_i\delta_j} + \frac{\mu_{n,i}(\vec{y})\mu_{n,j}(\vec{y})}{\delta_i\delta_j r_\varepsilon} \\ \qquad + \left(\frac{\sigma^2_{n,i}(\vec{y})}{\delta_i^2} + \frac{\mu^2_{n,i}(\vec{y})}{\delta_i^2 r_\varepsilon}\right) \wedge \left(\frac{\sigma^2_{n,j}(\vec{y})}{\delta_j^2} + \frac{\mu^2_{n,j}(\vec{y})}{\delta_i^2 r_\varepsilon}\right)\Bigg), \\ Q_{\varepsilon,n,i,j}(\vec{y}) = \frac{1}{2}\Bigg(- \frac{\rho_{n,i,j}(\vec{y})\sigma_{n,i}(\vec{y})\sigma_{n,j}(\vec{y})}{\delta_i\delta_j} - \frac{\mu_{n,i}(\vec{y})\mu_{n,j}(\vec{y})}{\delta_i\delta_j r_\varepsilon} \\ \qquad + \left(\frac{\sigma^2_{n,i}(\vec{y})}{\delta_i^2} + \frac{\mu^2_{n,i}(\vec{y})}{\delta_i^2 r_\varepsilon}\right) \wedge \left(\frac{\sigma^2_{n,j}(\vec{y})}{\delta_j^2} + \frac{\mu^2_{n,j}(\vec{y})}{\delta_i^2 r_\varepsilon}\right)\Bigg), \\ \qquad 1 \le i < j \le k, \quad \vec{y} \in \mathbb{R}_k, \quad n = 1, \dots, N. \end{cases} \tag{9.195}$$

As far as inequalities (a) and (b) are concerned, they hold if the following inequality holds, for every $1 \le i < j \le k, \vec{y} \in \mathbb{R}_k, n = 1, \dots, N$:

$$\begin{aligned} |\rho_{n,i,j}(\vec{y})|\frac{\sigma_{n,i}(\vec{y})\sigma_{n,j}(\vec{y})}{\delta_i\delta_j} + \frac{|\mu_{n,i}(\vec{y})\mu_{n,j}(\vec{y})|}{\delta_i\delta_j r_\varepsilon} \\ \le \left(\frac{\sigma^2_{n,i}(\vec{y})}{\delta_i^2} + \frac{\mu^2_{n,i}(\vec{y})}{\delta_i^2 r_\varepsilon}\right) \wedge \left(\frac{\sigma^2_{n,j}(\vec{y})}{\delta_j^2} + \frac{\mu^2_{n,j}(\vec{y})}{\delta_j^2 r_\varepsilon}\right). \end{aligned} \tag{9.196}$$

Relation (9.112) is a condition, under which the exact fitting represented by relation (9.85) can be realized.

Unfortunately, relation (9.112) do not separate parameters of the Gaussian log-price process $Y_{0,n}$ and parameters of the approximating log-price process $Y_{\varepsilon,n}$. Such separation can be realized in the following way.

Let us assume that the following condition holds:

R$_3$: $\max_{1\le i<j\le k, 1\le n\le N} \sup_{\vec{y}\in\mathbb{R}_k} |\rho_{n,i,j}(\vec{y})| \frac{\sigma_{n,i}(\vec{y})\vee\sigma_{n,j}(\vec{y})}{\sigma_{n,i}(\vec{y})\wedge\sigma_{n,j}(\vec{y})} = \rho < 1$.

Let us also introduce the parameter

$$\kappa = \min_{1\le i,j\le k} \frac{\delta_i}{\delta_j} \le 1. \tag{9.197}$$

In the case, where the conditions $\mathbf{G}_{16}$, $\mathbf{G}_{17}$, and $\mathbf{R}_3$ hold, inequalities (9.196) also hold, if we choose the parameters δ_i, $i = 1, \dots, k$ such that the following inequality hold:

$$\rho < \kappa \leq 1\,. \tag{9.198}$$

and then we choose the parameter r_ε satisfying the following inequality:

$$r_\varepsilon \geq \frac{K_{90}^2}{\sigma^2(\kappa - \rho)}\,. \tag{9.199}$$

Indeed, let us, for example, $\frac{\sigma^2_{n,i}(\vec{y})}{\delta_i^2} \leq \frac{\sigma^2_{n,j}(\vec{y})}{\delta_j^2}$ (in the case of opposite inequality, the calculations are analogous). Then, the conditions $\mathbf{G}_{16}$, $\mathbf{G}_{17}$, and $\mathbf{R}_3$ and inequalities (9.114) and (9.199) implies that

$$\begin{aligned}
&\left(\frac{\sigma^2_{n,i}(\vec{y})}{\delta_i^2} + \frac{\mu^2_{n,i}(\vec{y})}{\delta_i^2 r_\varepsilon}\right) \wedge \left(\frac{\sigma^2_{n,j}(\vec{y})}{\delta_j^2} + \frac{\mu^2_{n,j}(\vec{y})}{\delta_j^2 r_\varepsilon}\right) \\
&\quad - |\rho_{n,i,j}(\vec{y})| \frac{\sigma_{n,i}(\vec{y})\sigma_{n,j}(\vec{y})}{\delta_i \delta_j} - \frac{|\mu_{n,i}(\vec{y})\mu_{n,j}(\vec{y})|}{\delta_i \delta_j r_\varepsilon} \\
&\quad \geq \frac{\sigma^2_{n,i}(\vec{y})}{\delta_i \delta_j}\left(\frac{\delta_j}{\delta_i} - |\rho_{n,i,j}(\vec{y})| \frac{\sigma_{n,j}(\vec{y})}{\sigma_{n,i}(\vec{y})} - \frac{|\mu_{n,i}(\vec{y})\mu_{n,j}(\vec{y})|}{\sigma^2_{n,i}(\vec{y}) r_\varepsilon}\right) \\
&\quad \geq \frac{\sigma^2}{\delta_i \delta_j}\left(\kappa - \rho - \frac{K_{90}^2}{\sigma^2 r_\varepsilon}\right) \geq 0\,.
\end{aligned} \tag{9.200}$$

Now, let us show how to choose quantities $p_{\varepsilon,n,i,j,\jmath_i,\jmath_j}(\vec{y}), \jmath_i, \jmath_j = +, \circ, -, 1 \leq i < j \leq k, \vec{y} \in \mathbb{R}_k, n = 1, \dots, N$, which satisfy the system of equations (9.177) and consistency relations (9.178) and (9.179) (where quantities $p_{\varepsilon,n,i,\jmath_i}(\vec{y}), \jmath_i, = +, \circ, -, 1 \leq i \leq k$, $\vec{y} \in \mathbb{R}_k$, $n = 1, \dots, N$ are given by relations (9.181)) as well as the derivative consistency relations (9.190).

We achieve this by constructing, for every $1 \leq i < j \leq k, \vec{y} \in \mathbb{R}_k$, $n = 1, \dots, N$, a 4×4 matrix with nonnegative elements such that the sum of vectors given by three first rows equal to the vector given in the fourth row and the sum of vectors given by three first columns equal to the vector given in the fourth column, and with elements of $p_{\varepsilon,n,i,j,\pm,\pm}(\vec{y})$ satisfying the consistency relations (9.190),

$$\mathbf{P}_{\varepsilon,i,j,\vec{y}} = \begin{bmatrix}
p_{\varepsilon,n,i,j,+,+}(\vec{y}) & p_{\varepsilon,n,i,j,+,-}(\vec{y}) & p_{\varepsilon,n,i,j,+,\circ}(\vec{y}) & p_{\varepsilon,n,i,+}(\vec{y}) \\
p_{\varepsilon,n,i,j,-,+}(\vec{y}) & p_{\varepsilon,n,i,j,-,-}(\vec{y}) & p_{\varepsilon,n,i,j,\circ,\circ}(\vec{y}) & p_{\varepsilon,n,i,-}(\vec{y}) \\
p_{\varepsilon,n,i,j,\circ,+}(\vec{y}) & p_{\varepsilon,n,i,j,\circ,-}(\vec{y}) & p_{\varepsilon,n,i,j,\circ,\circ}(\vec{y}) & p_{\varepsilon,n,i,\circ}(\vec{y}) \\
p_{\varepsilon,n,j,+}(\vec{y}) & p_{\varepsilon,n,j,-}(\vec{y}) & p_{\varepsilon,n,j,\circ}(\vec{y}) & 1
\end{bmatrix} \tag{9.201}$$

Let us denote, for $1 \leq i < j \leq k, \vec{y} \in \mathbb{R}_k, n = 1, \dots, N$,

$$\begin{aligned}
R_{\varepsilon,n,i,j}(\vec{y}) &= p_{\varepsilon,n,i,\circ}(\vec{y}) - p_{\varepsilon,n,i,j,\circ,\circ}(\vec{y}) \\
&= 1 - \frac{\sigma^2_{n,i}(\vec{y})}{\delta_i^2} - \frac{\mu^2_{n,i}(\vec{y})}{\delta_i^2 r_\varepsilon} \\
&\quad - \left(1 - \frac{\sigma^2_{n,i}(\vec{y})}{\delta_i^2} - \frac{\mu^2_{n,i}(\vec{y})}{\delta_i^2 r_\varepsilon}\right) \wedge \left(1 - \frac{\sigma^2_{n,j}(\vec{y})}{\delta_j^2} - \frac{\mu^2_{n,j}(\vec{y})}{\delta_j^2 r_\varepsilon}\right),
\end{aligned} \tag{9.202}$$

and

$$\begin{aligned} R_{\varepsilon,n,j,i}(\vec{y}) &= p_{\varepsilon,n,j,\circ}(\vec{y}) - p_{\varepsilon,n,i,j,\circ,\circ}(\vec{y}) \\ &= 1 - \frac{\sigma^2_{n,j}(\vec{y})}{\delta^2_j} - \frac{\mu^2_{n,j}(\vec{y})}{\delta^2_j r_\varepsilon} \\ &\quad - \left(1 - \frac{\sigma^2_{n,i}(\vec{y})}{\delta^2_i} - \frac{\mu^2_{n,i}(\vec{y})}{\delta^2_i r_\varepsilon}\right) \wedge \left(1 - \frac{\sigma^2_{n,j}(\vec{y})}{\delta^2_j} - \frac{\mu^2_{n,j}(\vec{y})}{\delta^2_j r_\varepsilon}\right), \end{aligned} \tag{9.203}$$

Let us also denote, for $1 \leq i \leq k$, $\vec{y} \in \mathbb{R}_k$, $n = 1, \ldots, N$,

$$S_i = S_{\varepsilon,n,i}(\vec{y}) = \frac{\mu_{n,i}(\vec{y})}{\delta_i \sqrt{r_\varepsilon}}. \tag{9.204}$$

The matrix $\mathbf{P}_{\varepsilon,i,j,\vec{y}}$ can be chosen in the following way:

$$\mathbf{P}_{\varepsilon,i,j,\vec{y}} = \begin{bmatrix} \frac{P_{\varepsilon,n,i,j}(\vec{y})}{2} + \frac{S_i+S_j}{2} & \frac{Q_{\varepsilon,n,i,j}(\vec{y})}{2} + \frac{S_i-S_j}{2} & \frac{R_{\varepsilon,n,j,i}(\vec{y})}{2} & \frac{1-p_{\varepsilon,n,i,\circ}(\vec{y})}{2} + S_i \\ \frac{Q_{\varepsilon,n,i,j}(\vec{y})}{2} + \frac{S_j-S_i}{2} & \frac{P_{\varepsilon,n,i,j}(\vec{y})}{2} + \frac{-S_i-S_j}{2} & \frac{R_{\varepsilon,n,j,i}(\vec{y})}{2} & \frac{1-p_{\varepsilon,n,i,\circ}(\vec{y})}{2} - S_i \\ \frac{R_{\varepsilon,n,i,j}(\vec{y})}{2} & \frac{R_{\varepsilon,n,i,j}(\vec{y})}{2} & p_{\varepsilon,n,i,j,\circ,\circ}(\vec{y}) & p_{\varepsilon,n,i,\circ}(\vec{y}) \\ \frac{1-p_{\varepsilon,n,j,\circ}(\vec{y})}{2} + S_j & \frac{1-p_{\varepsilon,n,j,\circ}(\vec{y})}{2} - S_j & p_{\varepsilon,n,j,\circ}(\vec{y}) & 1 \end{bmatrix} \tag{9.205}$$

The direct simple calculations show that the formulated above consistency relations connected with summations of rows and columns hold.

Finally, conditions that provide nonnegativity of elements $p_{\varepsilon,n,i,j,\pm,\pm}(\vec{y})$ and $p_{\varepsilon,n,i,j,\pm,\mp}(\vec{y})$, should be investigated.

Other elements of the above matrix are obviously nonnegative.

Nonnegativity conditions for quantities $P_{\varepsilon,n,i,j}(\vec{y})$ and $Q_{\varepsilon,n,i,j}(\vec{y})$ have been given in relations (9.195)–(9.199).

Nonnegativity conditions for quantities $p_{\varepsilon,n,i,j,+,+}(\vec{y})$ can be obtained analogously. The difference is that one should take into account the presence of the additional term $\frac{S_{\varepsilon,n,i}(\vec{y})+S_{\varepsilon,n,j}(\vec{y})}{2}$.

The calculations in the following show that $p_{\varepsilon,n,i,j,+,+}(\vec{y}) \geq 0$ if the conditions $\mathbf{G}_{16}$, $\mathbf{G}_{17}$, and $\mathbf{R}_3$ hold, and the parameters δ_i, $i = 1, \ldots, k$ satisfying inequality (9.198) are chosen, and then the parameter r_ε satisfying the following inequality, slightly stronger than (9.199), is chosen:

$$r_\varepsilon \geq \frac{K^2_{91}}{\sigma^2(\kappa - \rho)} + \frac{16 K^2_{90} \delta^2}{\sigma^4(\kappa - \rho)^2}. \tag{9.206}$$

Indeed, let us assume, for example, $\frac{\sigma^2_{n,i}(\vec{y})}{\delta^2_i} \leq \frac{\sigma^2_{n,j}(\vec{y})}{\delta^2_j}$ (in the case of the opposite inequality, the calculations are analogous). We get using formulas (9.181), (9.195), and (9.204),

$$\begin{aligned}
p_{\varepsilon,n,i,j,+,+}(\vec{y}) &= \frac{P_{\varepsilon,n,i,j}(\vec{y})}{2} + \frac{S_{\varepsilon,n,i}(\vec{y}) + S_{\varepsilon,n,j}(\vec{y})}{2} \\
&= \frac{1}{4}\Bigg(\frac{\rho_{n,i,j}(\vec{y})\sigma_{n,i}(\vec{y})\sigma_{n,j}(\vec{y})}{\delta_i\delta_j} + \frac{\mu_{n,i}(\vec{y})\mu_{n,j}(\vec{y})}{\delta_i\delta_j r_\varepsilon} \\
&\quad + \left(\frac{\sigma^2_{n,i}(\vec{y})}{\delta_i^2} + \frac{\mu^2_{n,i}(\vec{y})}{\delta_i^2 r_\varepsilon}\right) \wedge \left(\frac{\sigma^2_{n,j}(\vec{y})}{\delta_j^2} + \frac{\mu^2_{n,j}(\vec{y})}{\delta_j^2 r_\varepsilon}\right)\Bigg) \\
&\quad + \frac{1}{2}\left(\frac{\mu_{n,i}(\vec{y})}{\delta_i\sqrt{r_\varepsilon}} + \frac{\mu_{n,j}(\vec{y})}{\delta_j\sqrt{r_\varepsilon}}\right) \\
&\geq \frac{\sigma^2_{n,i}(\vec{y})}{4\delta_i\delta_j}\Bigg(\frac{\delta_j}{\delta_i} - |\rho_{n,i,j}(\vec{y})|\frac{\sigma_{n,j}(\vec{y})}{\sigma_{n,i}(\vec{y})} - \frac{|\mu_{n,i}(\vec{y})\mu_{n,j}(\vec{y})|}{\sigma^2_{n,i}(\vec{y})r_\varepsilon} \\
&\quad - \frac{2|\mu_{n,i}(\vec{y})|\delta_j}{\sigma^2_{n,i}(\vec{y})\sqrt{r_\varepsilon}} - \frac{2|\mu_{n,j}(\vec{y})|\delta_i}{\sigma^2_{n,i}(\vec{y})\sqrt{r_\varepsilon}}\Bigg) \\
&\geq \frac{\sigma^2}{4\delta_i\delta_j}\left(\kappa - \rho - \frac{K^2_{90}}{\sigma^2 r_\varepsilon} - \frac{4K_{90}\delta}{\sigma^2\sqrt{r_\varepsilon}}\right) \geq 0\,. \qquad (9.207)
\end{aligned}$$

Similar calculations made for probabilities $p_{\varepsilon,n,i,j,-,-}(\vec{y})$, $p_{\varepsilon,n,i,j,+,-}(\vec{y})$, and $p_{\varepsilon,n,i,j,-,+}(\vec{y})$ yield the same conditions for their nonnegativity.

Conditions imposed on the parameters δ_i, $i = 1, \ldots, k$ and r_ε by relations (9.183), (9.184), (9.198), (9.199), and (9.206) do not contradict each other and can be satisfied simultaneously by choosing values for the parameters δ_i, $i = 1, \ldots, k$ large enough and close to each other, according to relations, respectively, (9.183) and (9.198), and then by choosing a value of the parameter r_ε large enough, according to relation (9.206). Note that inequalities (9.184) and (9.199) are weaker than inequality (9.206) and, thus, can be omitted.

The analysis of the fitting problem presented above is complete in the case of two-dimensional random binomial random walks.

However, in the case of k-dimensional trinomial random walk, which has dimension $k \geq 3$, some additional question arises, if, for every $\varepsilon \in (0, \varepsilon_0]$, $\vec{y} \in \mathbb{R}_k$ and $n = 1, \ldots, N$, there exists a multivariate distribution of the random vector $\vec{W}_{\varepsilon,n,1}(\vec{y}) = (W_{\varepsilon,n,1,1}(\vec{y}), \ldots, W_{\varepsilon,n,1,k}(\vec{y}))$, with the above one- and two-dimensional distributions $p_{\varepsilon,n,i,j_i}(\vec{y}), j_i = +, \circ, -, 1 \leq i \leq k$ and $p_{\varepsilon,i,j,j_i,j_j}(\vec{y}), j_i, j_j = +, \circ, -, 1 \leq i < j \leq k$.

This question has an affirmative answer. Moreover, the algorithm for constructing of the corresponding k-dimensional distributions is given in Lemma 9.3.2, which should be applied to the random vectors $\vec{W}_{\varepsilon,n,1}(\vec{y})$ with the above one- and two-dimensional distributions, for every $\varepsilon \in (0, \varepsilon_0]$, $\vec{y} \in \mathbb{R}_k$ and $n = 1, \ldots, N$.

The following lemma summarizes the above remarks.

Lemma 9.5.2. *The log-price process $Y_{\varepsilon,n}$ represented, for every $\varepsilon \in (0, \varepsilon_0]$, by the multivariate inhomogeneous in time and space trinomial random walk, which is defined by the transition relation* (9.170), *realizes the exact moment fitting given by re-*

lation (9.172) *for the log-price processes* $Y_{0,n}$ *represented by the multivariate inhomogeneous in time and space Gaussian random walk, which is defined by the transition relation* (9.169), *if the conditions* $\mathbf{G}_{16}$, $\mathbf{G}_{17}$ *and* $\mathbf{R}_3$ *hold. Parameters of the process* $\vec{Y}_{\varepsilon,n}$ $(\delta_i, p_{\varepsilon,n,i,j,\jmath_i,\jmath_j}(\vec{y}), \jmath_i, \jmath_j = +, \circ, -, i, j = 1, \dots, k, \vec{y} \in \mathbb{R}_k, n = 1, \dots, N, r_\varepsilon)$ *can be chosen, for every* $\varepsilon \in (0, \varepsilon_0]$, *according to relations* (9.181), (9.193), (9.195), *and* (9.201)–(9.205), *and the choice of the parameters* δ_i, $i = 1, \dots, k$ *and* r_ε *satisfying conditions* (9.183), (9.198), *and* (9.206). *The corresponding high-order distributions for components of the random vectors* $\vec{W}_{\varepsilon,n,1}(\vec{y}), \vec{y} \in \mathbb{R}_k$, $n = 1, \dots, N$ *can be constructed with the use of algorithm described in Lemma 9.3.2, which should be applied to the above one- and two-dimensional distributions* $p_{\varepsilon,n,i,\jmath_i}(\vec{y}), \jmath_i = +, \circ, -, 1 \le i \le k$ *and* $p_{\varepsilon,i,j,\jmath_i,\jmath_j}(\vec{y}), \jmath_i, \jmath_j = +, \circ, -, 1 \le i < j \le k$, *for every* $\vec{y} \in \mathbb{R}_k$, $n = 1, 2, \dots, N$, *and* $\varepsilon \in (0, \varepsilon_0]$.

It is worth noting that, in fact, the fitting procedure based on expectations and covariances do involve only one- and two-dimensional distributions for the random vectors $\vec{W}_{\varepsilon,n,1}(\vec{y})$. In fact, we need to know that the random vectors $\vec{W}_{\varepsilon,n,1}(\vec{y})$, with the corresponding fitted one- and two-dimensional distributions do exist.

The same is valid for the corresponding convergence results for option rewards presented in Chapter 10. The only expectations and covariances and, thus, only one- and two-dimensional distributions for the random vectors $\vec{W}_{\varepsilon,n,1}(\vec{y})$ are involved in the corresponding convergence conditions.

However, the high-dimensional distributions for the random vectors $\vec{W}_{\varepsilon,n,1}(\vec{y})$ are explicitly used in the backward recurrence relations for computing the reward functions for the approximating trinomial log-price processes $\vec{Y}_{\varepsilon,n}$.

One can construct, for every $\varepsilon \in (0, \varepsilon_0]$ and $\vec{y} \in \mathbb{R}_k$, $n = 1, \dots, N$, i.i.d. random vectors $\vec{W}_{\varepsilon,n,l}(\vec{y})$, $l = 1, 2, \dots$, used in relation (9.170) defining the transition probabilities of the Markov chain $\vec{Y}_{\varepsilon,n}$.

First, one should choose jump values $\bar{\delta}_{\varepsilon,\bar{\jmath}} = (\delta_{\varepsilon,n,\jmath_1}, \dots, \delta_{\varepsilon,n,\jmath_k}), \bar{\jmath} = \langle \jmath_{i_1}, \dots, \jmath_{i_k} \rangle \in \mathcal{J}_k$ and one- and two-dimensional distributions, respectively, $p_{\varepsilon,n,i,\jmath_i}(\vec{y}), \jmath_i = +, \circ, -, 1 \le i \le k$ and $p_{\varepsilon,n,i,j,\jmath_i,\jmath_j}(\vec{y}), \jmath_i, \jmath_j = +, \circ, -, 1 \le i < j \le k$ for random vector $\vec{W}_{\varepsilon,n,1}(\vec{y})$, according to relations pointed in Lemma 9.5.2.

Second, one should construct the corresponding consistent k-dimensional distributions $p_{\varepsilon,n,1,\dots,k,\jmath_1,\dots,\jmath_k}(\vec{y}), \bar{\jmath} \in \mathcal{J}_k$, with the above one- and two dimensional distributions, using the algorithm described in Lemma 9.3.2, and then to define the corresponding intervals $\mathbb{I}_{\bar{\jmath}}, = \mathbb{I}_{\varepsilon,n,\bar{\jmath}}(\vec{y}), \bar{\jmath} \in \mathcal{J}_k$, which, in this case, also depend on the parameters ε, n and $\vec{y}$.

Third, one can define the corresponding sequence of i.i.d. random vectors $\vec{W}_{\varepsilon,n,l}(\vec{y})$, $l = 1, 2, \dots$ used in relation (9.170), defining the transition probabilities of the Markov chain $\vec{Y}_{\varepsilon,n}$,

$$\vec{W}_{\varepsilon,n,l}(\vec{y}) = \sum_{\bar{\jmath} \in \mathcal{J}_k} \bar{\delta}_{\varepsilon,\bar{\jmath}} \mathrm{I}(\rho_l \in \mathbb{I}_{\varepsilon,n,\bar{\jmath}}(\vec{y})), \quad l = 1, 2, \dots, \tag{9.208}$$

where ρ_l, $l = 1, 2, \dots$ is a sequence of i.i.d. random variables uniformly distributed in the interval $[0, 1)$.

It should be pointed out that the recombining condition (which guarantees not more than polynomial rate of growth for number of nodes in the corresponding trinomial tree representation for the above model) holds automatically for any values of parameters of the log-price process $\vec{Y}_{\varepsilon,n}$ given by relation (9.170).

The corresponding tree $T_{\varepsilon,(\vec{y},n)}$ generated by the trajectories of the log-price process $\vec{Y}_{\varepsilon,n+r}$, $r = 0, 1, \dots$ with the value $\vec{Y}_{\varepsilon,n} = \vec{y}$ has the total number of points after r steps,

$$L_{\varepsilon,n,n+r} \leq 1 + (2r_\varepsilon + 1)^k + \cdots + (2r_\varepsilon r + 1)^k \leq \int_0^{r+1} (2r_\varepsilon x + 1)^k dx = \frac{(2r_\varepsilon(r+1) + 1)^{k+1} - 1}{2r_\varepsilon(k+1)} . \tag{9.209}$$

The expression on the right-hand side in (9.209) is a polynomial function of r of order $k + 1$.

10 Convergence of tree-type reward approximations

In this chapter, we present results about convergence of rewards for American-type options for log-price processes represented by discrete time Markov binomial and trinomial processes approximating discrete time univariate and multivariate Markov Gaussian log-price processes homogeneous and inhomogeneous in time and space.

In Section 10.1, the results about convergence of rewards for log-price processes represented by univariate binomial random walks to the rewards for log-price processes represented by univariate Gaussian random walks are given.

In Section 10.2, the results about convergence of rewards for log-price processes represented by multivariate binomial and trinomial random walks to the rewards for log-price processes represented by multivariate Gaussian random walks are given.

In Section 10.3, we present results about convergence of rewards for univariate Markov binomial and trinomial log-price processes to the rewards for univariate Markov Gaussian log-price processes inhomogeneous in space.

In Section 10.4, we present results about convergence of rewards for multivariate Markov trinomial log-price processes to the rewards for multivariate Markov Gaussian log-price processes inhomogeneous in time and space.

In all convergence theorems given in the chapter, we impose minimal regularity conditions on pay-off functions and drift and volatility functional coefficients (for inhomogeneous in space models). The main new results are given in Theorems 10.2.1–10.2.4 and 10.4.1–10.4.2.

10.1 Univariate binomial tree approximation models

In this section, we present results about convergence of rewards for American-type options for univariate binomial tree approximations of log-price processes represented by Gaussian random walks. We give the corresponding results for inhomogeneous in time binomial tree approximations for two variants of the model, respectively, with exactly and asymptotically fitted parameters.

10.1.1 Convergence of option rewards for a univariate binomial tree model with exactly fitted parameters

Let us consider the model of log-price processes represented by inhomogeneous in time Gaussian random walks. These processes can be considered as discrete time analogs of inhomogeneous in time Gaussian processes with independent increments.

In this case, a log-price $Y_{0,n}$ is given by the following stochastic transition dynamic relation:

$$Y_{0,n} = Y_{0,n-1} + W_{0,n}\,, \quad n = 1, 2, \dots . \tag{10.1}$$

where (a) $Y_{0,0}$ is a real-valued random variable with a distribution $P_0(A)$, (b) $W_{0,n}$, $n = 1, 2, \dots$ is a sequence of independent normal random variables with mean values $\mathsf{E}W_{0,n} = \mu_n$, $n = 1, 2, \dots$ and variances $\mathrm{Var}W_{0,n} = \sigma_n^2$, $n = 1, 2, \dots$, and (c) the random variable $Y_{0,0}$ and the random sequence $W_{0,n}$, $n = 1, 2, \dots$ are independent.

Also, usual assumptions are (d) $\mu_n \in \mathbb{R}_1$, $n = 1, 2, \dots$ and (e) $\sigma_n > 0$, $n = 1, 2, \dots$.

We approximate the log-price process $Y_{0,n}$ by the log-price $Y_{\varepsilon,n}$ represented by an inhomogeneous in time binomial random walk. In this case, the log-price $Y_{\varepsilon,n}$ is given, for every $\varepsilon \in (0, \varepsilon_0]$, by the following stochastic transition dynamic relation:

$$Y_{\varepsilon,n} = Y_{\varepsilon,n-1} + W_{\varepsilon,n}\,, \quad n = 1, 2, \dots . \tag{10.2}$$

where (a) $Y_{\varepsilon,0}$ is a real-valued random variable with a distribution $P_\varepsilon(A)$, (b) $W_{\varepsilon,n} = W_{\varepsilon,n,1} + \dots + W_{\varepsilon,n,r_\varepsilon}$, $n = 1, 2, \dots$, where $W_{\varepsilon,n,l}$, $n, l = 1, 2, \dots$ is a family of independent binary random variables, which take values $\delta_{\varepsilon,n,+}$ and $\delta_{\varepsilon,n,-}$ with probabilities, respectively, $p_{\varepsilon,n,+}$ and $p_{\varepsilon,n,-}$, and (c) the random variable $Y_{\varepsilon,0}$ and the family of random variables $W_{\varepsilon,n,l}$, $n, l = 1, 2, \dots$ are independent.

Also, usual assumptions are (d) $\delta_{\varepsilon,n,-} < \delta_{\varepsilon,n,+}$, $n = 1, 2, \dots$, (e) $0 \leq p_{\varepsilon,n,+} = 1 - p_{\varepsilon,n,-} \leq 1$, $n = 1, 2, \dots$, and (f) r_ε is a positive integer number.

We consider a model, where a pay-off function $g(n, e^y)$ does not depend on the parameter ε and assume that $g(n, e^y)$ are real-valued Borel functions of argument $y \in \mathbb{R}_1$, for every $n = 0, 1, \dots, N$.

In this case, the corresponding reward function $\phi_{\varepsilon,n}(y)$, for the log-price process $Y_{\varepsilon,n}$, is also a Borel function of the argument $y \in \mathbb{R}_1$, for every $n = 0, 1, \dots, N$.

We fit the parameters of approximating log-price processes $Y_{\varepsilon,n}$ in such a way that the following exact moment fitting: relations hold, for every $\varepsilon \in (0, \varepsilon_0]$,

$$\mathsf{E}W_{\varepsilon,n} = \mu_n\,, \quad \mathrm{Var}W_{\varepsilon,n} = \sigma_n^2\,, \quad n = 1, \dots, N\,. \tag{10.3}$$

The solution of this fitting problem is given in Lemma 9.1.2. According to this lemma, the fitting relations (10.3) hold, if parameters of the log-price process $Y_{\varepsilon,n}$ are given, for every $\varepsilon \in (0, \varepsilon_0]$, by the following formulas:

$$\begin{cases} \delta_{\varepsilon,n,\pm} = \pm \dfrac{\delta}{\sqrt{r_\varepsilon}} + \lambda_{\varepsilon,n}\,, \quad \lambda_{\varepsilon,n} = \dfrac{\imath_n\sqrt{\delta^2 - \sigma_n^2}}{\sqrt{r_\varepsilon}} + \dfrac{\mu_n}{r_\varepsilon}\,, \\[2ex] p_{\varepsilon,n,\pm} = \dfrac{1}{2}\left(1 \mp \dfrac{\imath_n\sqrt{\delta^2 - \sigma_n^2}}{\delta}\right)\,, \quad n = 1, \dots, N\,, \end{cases} \tag{10.4}$$

where (i) one can choose for the parameters $\imath_n$, $n = 1, \dots, N$, δ, r_ε any values $\imath_n = +, -$, $n = 1, \dots, N$, $\delta \geq \sigma$ and $r_\varepsilon = 1, 2, \dots$, (ii) $\sigma^2 = \max_{1\leq n\leq N} \sigma_n^2 > 0$ (according to the initial model assumptions).

The random variables $W_{\varepsilon,n}$, $n = 1, \dots, N$ are binomial random variables. Namely, the random variable $W_{\varepsilon,n}$ takes values

$$w_{\varepsilon,n,l} = l\frac{\delta}{\sqrt{r_\varepsilon}} + r_\varepsilon \lambda_{\varepsilon,n}, \quad l = -r_\varepsilon, \dots, r_\varepsilon, \tag{10.5}$$

with the corresponding probabilities,

$$p_{\varepsilon,n,l} = \begin{cases} C_{r_\varepsilon}^{\frac{r_\varepsilon+l}{2}} (p_{\varepsilon,n,+})^{\frac{r_\varepsilon+l}{2}} (p_{\varepsilon,n,-})^{\frac{r_\varepsilon-l}{2}} & \text{if } \frac{r_\varepsilon+l}{2} \text{ is integer}, \\ 0 & \text{if } \frac{r_\varepsilon+l}{2} \text{ is not integer}, \\ & l = -r_\varepsilon, \dots, r_\varepsilon. \end{cases} \tag{10.6}$$

Note that we slightly modified the above standard definition of binomial distribution, by including in the range of possible values all integer numbers $-r_\varepsilon \leq l \leq r_e$ including the values with the corresponding probabilities equal to zero.

Let us assume that $Y_{\varepsilon,n} = y = \text{const} \in \mathbb{R}_1$. In this case, trajectories of the log-price process $Y_{\varepsilon,n+r}$, $r = 0, 1, \dots$ generate, for every $\varepsilon \in (0, \varepsilon_0]$, the unique tree of nodes

$$T_{\varepsilon,(y,n)} = \{(y_{\varepsilon,n,n+r,l}, n + r), \quad l = -r_\varepsilon r, \dots, r_\varepsilon r, \quad r = 0, 1, \dots\}, \tag{10.7}$$

where

$$y_{\varepsilon,n,n+r,l} = y + l\frac{\delta}{\sqrt{r_\varepsilon}} + r_\varepsilon \lambda_{\varepsilon,n,n+r}, \quad l = -r_\varepsilon r, \dots, r_\varepsilon r, \quad r = 0, 1, \dots. \tag{10.8}$$

and

$$\lambda_{\varepsilon,n,n+r} = \sum_{m=n+1}^{n+r} \lambda_{\varepsilon,m}, \quad r = 0, 1, \dots. \tag{10.9}$$

The tree $T_{\varepsilon,y,n}$ contains the total number of nodes after r steps

$$L_{\varepsilon,n,n+r} \leq 1 + 2r_\varepsilon + 1 + \cdots + 2r_\varepsilon r + 1 = r_\varepsilon r^2 + (r_\varepsilon + 1)r + 1. \tag{10.10}$$

The following lemma is a direct corollary of Lemma 3.3.3.

Lemma 10.1.1. *Let, for every $\varepsilon \in (0, \varepsilon_0]$, the approximating log-price process $Y_{\varepsilon,n}$ be the binomial random walk given by the stochastic transition dynamic relation* (10.2), *with parameters defined in relation* (10.4). *Then the reward functions $\phi_{\varepsilon,n+r}(y_{\varepsilon,n,n+r,l})$, for points $y_{\varepsilon,n,n,0} = y$ and $y_{\varepsilon,n,n+r,l}$, $l = -r_\varepsilon r, \dots, r_\varepsilon r$, $r = 1, \dots N - n$, are, for every $\varepsilon \in (0, \varepsilon_0]$, and $y \in \mathbb{R}_1$, $n = 0, \dots, N$, the unique solution for the following recurrence finite system of linear equations:*

$$\begin{cases} \phi_{\varepsilon,N}(y_{\varepsilon,n,N,l}) = g_\varepsilon(N, e^{y_{\varepsilon,n,N,l}}), \quad l = -r_\varepsilon(N-n), \dots, r_\varepsilon(N-n), \\ \phi_{\varepsilon,n+r}(y_{\varepsilon,n,n+r,l}) = \max\left(g_\varepsilon(n+r, e^{y_{\varepsilon,n,n+r,l}}), \sum_{l'=-r_\varepsilon}^{r_\varepsilon} \phi_{\varepsilon,n+r+1}(y_{n,n+r+1,l+l'}) p_{\varepsilon,n+r+1,l'}\right), \\ \qquad l = -r_\varepsilon r, \dots, r_\varepsilon r, \quad r = N-n-1, \dots, 1, \\ \phi_{\varepsilon,n}(y) = \max\left(g_\varepsilon(n, e^y), \sum_{l'=-r_\varepsilon}^{r_\varepsilon} \phi_{\varepsilon,n}(y_{\varepsilon,n,n+1,l'}) p_{\varepsilon,n+1,l'}\right). \end{cases} \tag{10.11}$$

We are going to apply Theorem 5.3.1, in order to prove convergence of the reward functions for log-price processes $Y_{\varepsilon,n}$ to the corresponding reward functions for the log-price process $Y_{0,n}$ as $\varepsilon \to 0$.

First, we assume that the condition $\mathbf{B}_4[\gamma]$ holds for the pay-off function $g(n, e^y)$, for some $\gamma \geq 0$.

We also assume that the following condition, analogous to the condition $\mathbf{I}_6$, holds:

$\mathbf{I}_9$: There exist sets $\mathbb{Y}_n \in \mathcal{B}_1$, $n = 0, \dots, N$ such that function $g(n, e^y)$ is continuous in points $y \in \mathbb{Y}_n$ for every $n = 0, \dots, N$.

We shall see that, in this case, the condition $\mathbf{J}_4$ (a) holds for the log-prices $Y_{\varepsilon,n}$, with parameters given by relation (10.4), if $r_\varepsilon \to \infty$ as $\varepsilon \to 0$.

Also the random variables $W_{0,n}$, $n = 1, \dots, N$ have normal distributions that make it possible to replace the condition $\mathbf{J}_4$ (b) by the following condition:

$\mathbf{J}_{13}$: $L_1(\overline{\mathbb{Y}}_n) = 0$, $n = 1, \dots, N$, where $\mathbb{Y}_n$, $n = 1, \dots, N$ are sets penetrating the condition $\mathbf{I}_9$.

Recall that the notation $L_k(A)$ is used for the Lebesgue measure in $\mathbb{R}_k$.

The above backward recurrence algorithm for finding reward functions is supported by the following convergence result.

Theorem 10.1.1. *Let the log-price process $Y_{0,n}$ be represented by the Gaussian random walk given by the stochastic transition dynamic relation* (10.1), *and the approximating log-price process is represented, for every $\varepsilon \in (0, \varepsilon_0]$, by the binomial random walk $Y_{\varepsilon,n}$ given by the stochastic transition dynamic relation* (10.2), *with parameters given by relation* (10.4). *Let also the conditions $\mathbf{B}_4[\gamma]$ holds for some $\gamma \geq 0$, the conditions $\mathbf{I}_9$ and $\mathbf{J}_{13}$ hold, and $r_\varepsilon \to \infty$ as $\varepsilon \to 0$. Then the following relation takes place for any $y_\varepsilon \to y_0 \in \mathbb{Y}_n$, $n = 0, 1, \dots, N$:*

$$\phi_{\varepsilon,n}(y_\varepsilon) \to \phi_{0,n}(y_0) \quad \text{as } \varepsilon \to 0\,. \tag{10.12}$$

Proof. The moment-generating functions for the random variables $W_{\varepsilon,n}$, $n = 1, \dots, N$ have, for every $\varepsilon \in (0, \varepsilon_0]$, the following form:

$$\begin{aligned} \mathsf{E}e^{\beta W_{\varepsilon,n}} = e^{\beta\mu_n} \Bigg(& \exp\left\{\beta \frac{\delta + \imath_n\sqrt{\delta^2 - \sigma_n^2}}{\sqrt{r_\varepsilon}}\right\} \frac{1}{2}\left(1 - \frac{\imath_n\sqrt{\delta^2 - \sigma_n^2}}{\delta}\right) \\ & + \exp\left\{\beta \frac{-\delta + \imath_n\sqrt{\delta^2 - \sigma_n^2}}{\sqrt{r_\varepsilon}}\right\} \frac{1}{2}\left(1 + \frac{\imath_n\sqrt{\delta^2 - \sigma_n^2}}{\delta}\right)\Bigg)^{r_\varepsilon}, \quad \beta \in \mathbb{R}_1\,, \end{aligned} \tag{10.13}$$

while, for $\varepsilon = 0$, these moment-generating functions have the following form:

$$\mathsf{E}e^{\beta W_{0,n}} = e^{\beta\mu_n + \frac{1}{2}\beta^2\sigma_n^2}, \quad \beta \in \mathbb{R}_1\,. \tag{10.14}$$

It follows from formula (10.13) and from the Taylor asymptotic expansion for the exponential function that the following asymptotic relation takes place, for every $n = 1, \dots, N$ and $\beta \in \mathbb{R}_1$:

$$
\begin{aligned}
\mathsf{E}e^{\beta W_{\varepsilon,n}} &= e^{\beta\mu_n}\Bigg(\Bigg(1 + \beta\frac{\delta + \imath_n\sqrt{\delta^2 - \sigma_n^2}}{\sqrt{r_\varepsilon}} + \frac{1}{2}\beta^2\left(\frac{\delta + \imath_n\sqrt{\delta^2 - \sigma_n^2}}{\sqrt{r_\varepsilon}}\right)^2 \\
&\quad + o_2\left(\frac{\beta(\delta + \imath_n\sqrt{\delta^2 - \sigma_n^2})}{\sqrt{r_\varepsilon}}\right)\Bigg)\frac{1}{2}\left(1 - \frac{\imath_n\sqrt{\delta^2 - \sigma_n^2}}{\delta}\right) \\
&\quad + \left(1 + \beta\frac{-\delta + \imath_n\sqrt{\delta^2 - \sigma_n^2}}{\sqrt{r_\varepsilon}} + \frac{1}{2}\beta^2\left(\frac{-\delta + \imath_n\sqrt{\delta^2 - \sigma_n^2}}{\sqrt{r_\varepsilon}}\right)^2\right. \\
&\quad \left. + o_2\left(\frac{\beta(-\delta + \imath_n\sqrt{\delta^2 - \sigma_n^2})}{\sqrt{r_\varepsilon}}\right)\right)\frac{1}{2}\left(1 + \frac{\imath_n\sqrt{\delta^2 - \sigma_n^2}}{\delta}\right)\Bigg)^{r_\varepsilon} \\
&= e^{\beta\mu_n}\Bigg(1 + \frac{\beta}{2}\Bigg(\left(\frac{\delta + \imath_n\sqrt{\delta^2 - \sigma_n^2}}{\sqrt{r_\varepsilon}}\right)\left(\frac{\delta - \imath_n\sqrt{\delta^2 - \sigma_n^2}}{\delta}\right) \\
&\quad + \left(\frac{-\delta + \imath_n\sqrt{\delta^2 - \sigma_n^2}}{\sqrt{r_\varepsilon}}\right)\left(\frac{\delta + \imath_n\sqrt{\delta^2 - \sigma_n^2}}{\delta}\right)\Bigg) \\
&\quad + \frac{\beta^2}{4}\Bigg(\left(\frac{(\delta + \imath_n\sqrt{\delta^2 - \sigma_n^2})^2}{r_\varepsilon}\right)\left(\frac{\delta - \imath_n\sqrt{\delta^2 - \sigma_n^2}}{\delta}\right) \\
&\quad + \left(\frac{(-\delta + \imath_n\sqrt{\delta^2 - \sigma_n^2})^2}{r_\varepsilon}\right)\left(\frac{\delta + \imath_n\sqrt{\delta^2 - \sigma_n^2}}{\delta}\right)\Bigg) \\
&\quad + o_2\left(\frac{\beta(\delta + \imath_n\sqrt{\delta^2 - \sigma_n^2})}{\sqrt{r_\varepsilon}}\right)\frac{1}{2}\left(1 - \frac{\imath_n\sqrt{\delta^2 - \sigma_n^2}}{\delta}\right) \\
&\quad + o_2\left(\frac{\beta(-\delta + \imath_n\sqrt{\delta^2 - \sigma_n^2})}{\sqrt{r_\varepsilon}}\right)\frac{1}{2}\left(1 + \frac{\imath_n\sqrt{\delta^2 - \sigma_n^2}}{\delta}\right)\Bigg)^{r_\varepsilon} \\
&= e^{\beta\mu_n}\Bigg(1 + \frac{\beta^2\sigma_n^2}{2r_\varepsilon} \\
&\quad + o_2\left(\frac{\beta\left(\delta + \imath_n\sqrt{\delta^2 - \sigma_n^2}\right)}{\sqrt{r_\varepsilon}}\right)\frac{1}{2}\left(1 - \frac{\imath_n\sqrt{\delta^2 - \sigma_n^2}}{\delta}\right) \\
&\quad + o_2\left(\frac{\beta(-\delta + \imath_n\sqrt{\delta^2 - \sigma_n^2})}{\sqrt{r_\varepsilon}}\right)\frac{1}{2}\left(1 + \frac{\imath_n\sqrt{\delta^2 - \sigma_n^2}}{\delta}\right)\Bigg)^{r_\varepsilon} \\
&\to e^{\beta\mu_n + \frac{1}{2}\beta^2\sigma_n^2} = \mathsf{E}e^{\beta W_{0,n}} \quad \text{as } \varepsilon \to 0\,.
\end{aligned}
\tag{10.15}
$$

Here and henceforth, we use the notation $o_k(x)$ for a function of $x \in \mathbb{R}_1$ if $o_k(x)$ is $o(x^k)$ as $x \to 0$, and write the second-order Taylor asymptotic expansion for the exponential function in the form $e^x = 1 + x + \frac{1}{2}x^2 + o_2(x)$ instead of the traditional form $e^x = 1 + x + \frac{1}{2}x^2 + o(x^2)$ as $x \to 0$, in order to distinguish the corresponding residual terms for arguments x and $-x$.

Relation (10.15) implies that for any $\beta \geq 0$,

$$\begin{aligned} \Xi'^{\pm}_{\beta}(Y_{\varepsilon,\cdot}, N) &= \max_{0 \leq n \leq N-1} \mathsf{E} e^{\pm \beta W_{\varepsilon,n+1}} \\ \to \Xi'^{\pm}_{\beta}(Y_{0,\cdot}, N) &= \max_{0 \leq n \leq N-1} e^{\pm \beta \mu_{n+1} + \frac{1}{2}\beta^2 \sigma^2_{n+1}} < \infty \, . \end{aligned} \tag{10.16}$$

Relation (10.16) implies in an obvious way that the condition $\mathbf{E_{10}}[\bar{\beta}]$ (its univariate version) and, therefore, by Lemma 5.1.6 (its variant for one-dimensional model without modulating component) the condition $\mathbf{C_6}[\bar{\beta}]$ (its univariate version) holds, for any parameter $\beta \geq 0$.

The condition $\mathbf{B_4}[\gamma]$ is a one-dimensional variant of the condition $\mathbf{B_6}[\bar{\gamma}]$.

Note that one can always take parameter $\beta > \gamma$ if $\gamma > 0$ or $\beta = \gamma$ if $\gamma = 0$.

The condition $\mathbf{I_9}$ is a particular case of the condition $\mathbf{I_3}$.

Relation (10.16) also implies, by the corresponding variant of the continuous theorem, the following relation, for every $n = 1, \dots, N$:

$$W_{\varepsilon,n} \xrightarrow{d} W_{0,n} \quad \text{as } \varepsilon \to 0 \, . \tag{10.17}$$

Relation (10.17) implies that the condition $\mathbf{J_4}$ (a) (its univariate variant) holds.

The condition $\mathbf{J_{13}}$ implies that the condition $\mathbf{J_4}$ (b) (its univariate variant) holds, since random variables $W_{0,n}$, $n = 1, \dots, N$ have normal distributions with positive variances and, therefore, possess positive everywhere probability density functions.

Thus, all conditions of Theorem 5.3.1 hold. By applying this theorem to the above approximation model, we get the asymptotic relation (10.12). □

It is worth to note one more time that Theorem 10.1.1 also covers the case of univariate homogeneous in time binomial tree approximation for log-price processes represented by univariate homogeneous in time Gaussian random walk.

10.1.2 Convergence of option rewards for a univariate binomial tree model with asymptotically fitted parameters

Alternatively, we can approximate the log-price process $Y_{0,n}$ by the log-price $Y_{\varepsilon,n}$ represented by inhomogeneous in time binomial random walk given, for every $\varepsilon \in (0, \varepsilon_0]$, by the following stochastic transition dynamic relation:

$$Y_{\varepsilon,n} = Y_{\varepsilon,n-1} + W_{\varepsilon,n} \, , \quad n = 1, 2, \dots \, . \tag{10.18}$$

where (a) $Y_{\varepsilon,0}$ is a real-valued random variable with the distribution $P_\varepsilon(A)$, (b) $W_{\varepsilon,n} = W_{\varepsilon,n,1} + \cdots + W_{\varepsilon,n,r_{\varepsilon,n}}$, $n = 1, 2, \dots$, where $W_{\varepsilon,n,l}$, $n, l = 1, 2, \dots$ is a family of independent binary random binomial variables, which take values $\delta_{\varepsilon,n,+}$ and $\delta_{\varepsilon,n,-}$ with probabilities, respectively, $p_{\varepsilon,n,+}$ and $p_{\varepsilon,n,-}$, and (c) the random variable $Y_{\varepsilon,0}$ and the family of random variables $W_{\varepsilon,n,l}$, $n, l = 1, 2, \dots$ are independent.

Also, usual assumptions are (d) $\delta_{\varepsilon,n,-} < \delta_{\varepsilon,n,+}$, $n = 1, 2, \dots$, (e) $0 \leq p_{\varepsilon,n,+} = 1 - p_{\varepsilon,n,-} \leq 1$, $n = 1, 2, \dots$, and (f) $r_{\varepsilon,n}$, $n = 1, 2, \dots$ are positive integer numbers.

The difference with the previous approximation model is that we admit possibility for the corresponding binomial random variables $W_{\varepsilon,n}$ to have the parameters $r_{\varepsilon,n}$ depending on n.

In this case, we fit parameters of approximating log-price processes $Y_{\varepsilon,n}$ in such a way that the following asymptotic moment fitting relation holds:

$$\mathsf{E}W_{\varepsilon,n} \to \mu_n\,, \quad \mathrm{Var}W_{\varepsilon,n} \to \sigma_n^2\,, \quad \text{as } \varepsilon \to 0\,, \quad n = 1, \dots, N\,. \tag{10.19}$$

The solution of this fitting problem is presented in Lemma 9.1.3. According to this lemma relation (10.19) holds, if one can choose the parameters of the log-price process $Y_{\varepsilon,n}$ given by the following formulas, for every $\varepsilon \in (0, \varepsilon_0]$ and $n = 1, \dots, N$:

$$\delta_{\varepsilon,n,\pm} = \frac{\delta}{\sqrt{r_\varepsilon}} + \lambda_{\varepsilon,n}\,, \quad \lambda_{\varepsilon,n} = \frac{\mu_n}{r_{\varepsilon,n}}\,, \quad p_{\varepsilon,n,\pm} = \frac{1}{2}\,, \quad r_{\varepsilon,n} = \left[r_\varepsilon \frac{\sigma_n^2}{\delta^2} \right] + 1\,, \tag{10.20}$$

where (i) one can choose for the parameters δ, r_ε any real values $\delta, r_\varepsilon > 0$, (ii) then, to pass parameter $r_\varepsilon \to \infty$ as $\varepsilon \to 0$.

The random variables $W_{\varepsilon,n}$, $n = 1, \dots, N$ are binomial random variables. Namely, the random variable $W_{\varepsilon,n}$ takes values,

$$w_{\varepsilon,n,l} = l\frac{\delta}{\sqrt{r_\varepsilon}} + r_{\varepsilon,n}\lambda_{\varepsilon,n} = l\frac{\delta}{\sqrt{r_\varepsilon}} + \mu_n\,, \quad l = -r_{\varepsilon,n}, \dots, r_{\varepsilon,n} \tag{10.21}$$

with the corresponding probabilities,

$$p_{\varepsilon,n,l} = \begin{cases} C_{r_{\varepsilon,n}}^{\frac{r_{\varepsilon,n}+l}{2}} \left(\frac{1}{2}\right)^{r_{\varepsilon,n}} & \text{if } \frac{r_{\varepsilon,n}+l}{2} \text{ is integer}\,, \\ 0 & \text{if } \frac{r_{\varepsilon,n}+l}{2} \text{ is not integer}\,, \\ l = -r_{\varepsilon,n}, \dots, r_{\varepsilon,n}\,. \end{cases} \tag{10.22}$$

Note that we slightly modified the above standard definition of binomial distribution by including the range of all possible values integer numbers $-r_{\varepsilon,n} \leq l \leq r_{\varepsilon,n}$, including the values with the corresponding probabilities equal to zero.

Let us assume that $Y_{\varepsilon,n} = y = \text{const} \in \mathbb{R}_1$. In this case, trajectories of the log-price process $Y_{\varepsilon,n+r}$, $r = 0, 1, \dots$ generate, for every $\varepsilon \in (0, \varepsilon_0]$, the unique tree of nodes

$$T_{\varepsilon,(y,n)} = \{(y_{\varepsilon,n,n+r,l}, n + r)\,, l = -r_{\varepsilon,n,n+r}, \dots, r_{\varepsilon,n,n+r}, r = 0, 1, \dots\}\,, \tag{10.23}$$

where

$$y_{\varepsilon,n,n+r,l} = y + l\frac{\delta}{\sqrt{r_\varepsilon}} + \mu_{n,n+r}\,, \quad -r_{\varepsilon,n,n+r} \leq l \leq r_{\varepsilon,n,n+r}\,, \quad r = 0, 1, \dots\,. \tag{10.24}$$

and

$$\mu_{n,n+r} = \sum_{m=n+1}^{n+r} \mu_m \,, \quad r_{\varepsilon,n,n+r} = \sum_{m=n+1}^{n+r} r_{\varepsilon,m} \,, \quad r = 0, 1, \dots . \tag{10.25}$$

The tree $T_{\varepsilon,(y,n)}$ contains the total number of nodes after r steps,

$$L_{\varepsilon,n,n+r} \leq \left(r_\varepsilon \frac{\sigma^2}{\delta^2} + 1 \right) r^2 + \left(r_\varepsilon \frac{\sigma^2}{\delta^2} + 2 \right) r + 1 \,, \tag{10.26}$$

where $\sigma^2 = \max_{1 \leq n \leq N} \sigma_n^2$.

The following lemma is a direct corollary of Lemma 3.3.3.

Lemma 10.1.2. *Let, for every $\varepsilon \in (0, \varepsilon_0]$, the approximating log-price process $Y_{\varepsilon,n}$ is the binomial random walk given by the stochastic transition dynamic relation* (10.18), *with parameters defined in relation* (10.20). *Then the reward functions $\phi_{\varepsilon,n+r}(y_{\varepsilon,n,n+r,l})$, for points $y_{\varepsilon,n,n,0} = y$ and $y_{\varepsilon,n,n+r,l}$, $l = -r_{\varepsilon,n,n+r}, \dots, r_{\varepsilon,n,n+r}$, $r = 1, \dots N-n$, are, for every $\varepsilon \in (0, \varepsilon_0]$, and $y \in \mathbb{R}_1$, $n = 0, \dots, N$, the unique solution for the following recurrence finite system of linear equations:*

$$\left\{ \begin{aligned} \phi_{\varepsilon,N}(y_{\varepsilon,n,N,l}) &= g_\varepsilon(N, e^{y_{\varepsilon,n,N,l}}) \,, \\ & \quad l = -r_{\varepsilon,n,N}, \dots, r_{\varepsilon,n,N} \,, \\ \phi_{\varepsilon,n+r}(y_{\varepsilon,n,n+r,l}) &= \max \Bigg(g_\varepsilon(n+r, e^{y_{\varepsilon,n,n+r,l}}) \,, \\ & \qquad \sum_{l'=-r_{\varepsilon,n+r+1}}^{r_{\varepsilon,n+r+1}} \phi_{\varepsilon,n+r+1}(y_{\varepsilon,n,n+r+1,l+l'}) p_{\varepsilon,n+r+1,l'} \Bigg) \,, \\ & \quad l = -r_{\varepsilon,n,n+r}, \dots, r_{\varepsilon,n,n+r} \,, \quad r = N-n-1, \dots, 1 \,, \\ \phi_{\varepsilon,n}(y) &= \max \left(g_\varepsilon(n, e^y), \sum_{l'=-r_{\varepsilon,n+1}}^{r_{\varepsilon,n+1}} \phi_{\varepsilon,n}(y_{\varepsilon,n,n+1,l'}) p_{\varepsilon,n+1,l'} \right) . \end{aligned} \right. \tag{10.27}$$

The above backward recurrence algorithm for finding reward functions is supported by the following convergence result.

Theorem 10.1.2. *Let the log-price process $Y_{0,n}$ be represented by the Gaussian random walk given by the stochastic transition dynamic relation* (10.1) *and the approximating log-price process is represented, for every $\varepsilon \in (0, \varepsilon_0]$, by the binomial random walks $Y_{\varepsilon,n}$ given by the stochastic transition dynamic relation* (10.18), *with parameters given by relation* (10.20). *Let also the conditions $\mathbf{B}_4[\gamma]$ hold for some $\gamma \geq 0$, the conditions $\mathbf{I}_9$ and $\mathbf{J}_{13}$ hold, and $r_\varepsilon \to \infty$ as $\varepsilon \to 0$. Then the following relation takes place, for any $y_\varepsilon \to y_0 \in \mathbb{Y}_n$, $n = 0, 1, \dots, N$:*

$$\phi_{\varepsilon,n}(y_\varepsilon) \to \phi_{0,n}(y_0) \quad \text{as } \varepsilon \to 0 \,. \tag{10.28}$$

Proof. It is analogous to the proof of Theorem 10.1.1. The moment-generating functions for the random variable $W_{\varepsilon,n}$ have for $n = 1, \dots, N$ and $\varepsilon \in (0, \varepsilon_0]$ the following form:

$$\mathsf{E}e^{\beta W_{\varepsilon,n}} = e^{\beta\mu_n}\left(e^{\beta\frac{\delta}{\sqrt{r_\varepsilon}}}\frac{1}{2} + e^{-\beta\frac{\delta}{\sqrt{r_\varepsilon}}}\frac{1}{2}\right)^{r_{\varepsilon,n}}, \quad \beta \in \mathbb{R}_1, \tag{10.29}$$

while, for $\varepsilon = 0$, it has the following form:

$$\mathsf{E}e^{\beta W_{0,n}} = e^{\beta\mu_n + \frac{1}{2}\beta^2\sigma_n^2}, \quad \beta \in \mathbb{R}_1. \tag{10.30}$$

It follows from formula (10.29) and the Taylor asymptotic expansion for the exponential function that the following asymptotic relation takes place, for every $n = 1, \dots, N$ and $\beta \in \mathbb{R}_1$:

$$\begin{aligned}\mathsf{E}e^{\beta W_{\varepsilon,n}} &= e^{\beta\mu_n}\left(e^{\beta\frac{\delta}{\sqrt{r_\varepsilon}}}\frac{1}{2} + e^{-\beta\frac{\delta}{\sqrt{r_\varepsilon}}}\frac{1}{2}\right)^{r_{\varepsilon,n}} \\ &= e^{\beta\mu_n}\left(1 + \frac{1}{2}\frac{\beta^2\delta^2}{r_\varepsilon} + o_2\left(\frac{\beta\delta}{\sqrt{r_\varepsilon}}\right) + o_2\left(-\frac{\beta\delta}{\sqrt{r_\varepsilon}}\right)\right)^{[r_\varepsilon\frac{\sigma_n^2}{\delta^2}]+1} \\ \to e^{\beta\mu_n + \frac{1}{2}\beta^2\sigma_n^2} &= \mathsf{E}e^{\beta W_{0,n}} \quad \text{as } \varepsilon \to 0.\end{aligned} \tag{10.31}$$

The continuation of the proof repeats the corresponding part of the proof for Theorem 10.1.1. □

10.1.3 Convergence of optimal expected rewards for a univariate binomial tree model

The optimal expected reward $\Phi_0 = \Phi_0(\mathcal{M}^{(0)}_{\max,N})$ for the log-price process $Y_{0,n}$, defined by the stochastic transition dynamic relation (10.1), is given by the following formula:

$$\Phi_0 = \mathsf{E}\phi_{0,0}(Y_{0,0}) = \int_{\mathbb{R}_1} \phi_{0,0}(y)P_0(dy). \tag{10.32}$$

It is natural to choose, for every $\varepsilon \in (0, \varepsilon_0]$, the initial distribution $P_\varepsilon(A)$ for the approximating log-price process $Y_{\varepsilon,n}$ as the corresponding skeleton approximation for the distribution $P_0(A)$

$$P_\varepsilon(A) = \sum_{y_{\varepsilon,0,0,l} \in A} P_0(\mathbb{I}_{\varepsilon,l}), \quad A \in \mathcal{B}_1, \tag{10.33}$$

where

$$y_{\varepsilon,0,0,l} = l\frac{\delta}{\sqrt{r_{\varepsilon,0}}} + \lambda_{\varepsilon,0}, \quad l = -r_{\varepsilon,0}, \dots, r_{\varepsilon,0}, \tag{10.34}$$

and

$$\mathbb{I}_{\varepsilon,l} = \begin{cases} \left(-\infty, \frac{\delta}{\sqrt{r_{\varepsilon,0}}}(-r_{\varepsilon,0} + \frac{1}{2})\right] + \lambda_{\varepsilon,0} & \text{if } l = -r_{\varepsilon,0}, \\ \left(\frac{\delta}{\sqrt{r_{\varepsilon,0}}}(l - \frac{1}{2}), \frac{\delta}{\sqrt{r_{\varepsilon,0}}}(l + \frac{1}{2})\right] + \lambda_{\varepsilon,0} & \text{if } -r_{\varepsilon,0} < l < r_{\varepsilon,0}, \\ \frac{\delta}{\sqrt{r_{\varepsilon,0}}}(r_{\varepsilon,0} - \frac{1}{2}), \infty) + \lambda_{\varepsilon,0} & \text{if } l = r_{\varepsilon,0}, \end{cases} \tag{10.35}$$

where real numbers $\lambda_{\varepsilon,0}$ are positive integer numbers $r_{\varepsilon,0}$ chosen in such a way that the following condition holds:

N$_5$: (a) $r_{\varepsilon,0} \to \infty$ as $\varepsilon \to 0$,
(b) $\pm\delta\sqrt{r_{\varepsilon,0}} + \lambda_{\varepsilon,0} \to \pm\infty$ as $\varepsilon \to 0$.

Correspondingly, the optimal expected reward $\Phi_\varepsilon = \Phi_\varepsilon(\mathcal{M}^{(\varepsilon)}_{max,N})$ for the log-price process $Y_{\varepsilon,n}$, defined by the transition dynamic relation (10.2), is given, for every $\varepsilon \in (0, \varepsilon_0]$, by the following formula:

$$\Phi_\varepsilon = \sum_{l=-r_{\varepsilon,0}}^{r_{\varepsilon,0}} \phi_{\varepsilon,0}(y_{\varepsilon,0,0,l}) P_0(\mathbb{I}_{\varepsilon,l}) . \tag{10.36}$$

We are going to apply Theorem 5.3.2, in order to prove convergence of the optimal expected rewards for log-price processes $Y_{\varepsilon,n}$, to the corresponding reward functions for log-price processes $Y_{0,n}$ as $\varepsilon \to 0$.

Let us assume that the following condition holds for some $\beta \geq 0$:

D$_{20}$ $[\beta]$: $\mathsf{E}e^{\beta|Y_{0,0}|} < K_{91}$, for some $1 < K_{91} < \infty$,

and also that the following condition holds:

K$_{18}$: $P_0(\overline{\mathbb{Y}}_0) = 0$, where $\mathbb{Y}_0$ is the set defined in the condition **I$_9$**.

Theorem 10.1.3. *Let the log-price process $Y_{0,n}$ be represented by the Gaussian random walk given by the stochastic transition dynamic relation* (10.1) *and the approximating log-price process is represented, for every $\varepsilon \in (0, \varepsilon_0]$, by the binomial random walks $Y_{\varepsilon,n}$ given by the stochastic transition dynamic relation* (10.2), *with parameters given by relation* (10.4). *Let also the conditions* **B$_4$**$[\gamma]$ *and* **D$_{20}$**$[\beta]$ *hold for parameters γ and β such that, either $\beta > \gamma > 0$ or $\beta = \gamma = 0$, and also the conditions* **N$_5$**, **I$_9$**, **J$_{13}$** *and* **K$_{18}$** *hold, and $r_\varepsilon \to \infty$ as $\varepsilon \to 0$. Then the following relation takes place:*

$$\Phi_\varepsilon \to \Phi_0 \quad as \ \varepsilon \to 0 . \tag{10.37}$$

Proof. We shall use Theorem 5.3.2.

Let us define skeleton functions

$$h_\varepsilon(y) = \sum_{l=-r_{\varepsilon,0}}^{r_{\varepsilon,0}} \left(l\frac{\delta}{\sqrt{r_\varepsilon}} + \lambda_{\varepsilon,0} \right) I(y \in \mathbb{I}_{\varepsilon,l}) \tag{10.38}$$

Obviously, for every $\varepsilon \in (0, \varepsilon_0]$,

$$P_\varepsilon(A) = \mathsf{P}\{h_\varepsilon(Y_{0,n}) \in A\} , \quad A \in \mathcal{B}_1 . \tag{10.39}$$

The skeleton functions $h_\varepsilon(y)$ satisfy the following inequalities for $y \in \mathbb{R}_1$:

$$\begin{aligned}|h_\varepsilon(y)| &\le |y| I\left(y \notin \left[-\delta\sqrt{r_{\varepsilon,0}} + \lambda_{\varepsilon,0}, \delta\sqrt{r_{\varepsilon,0}} + \lambda_{\varepsilon,0}\right]\right) \\ &\quad + \left(|y| + \frac{\delta}{\sqrt{r_{\varepsilon,0}}}\right) I\left(y \in [-\delta\sqrt{r_{\varepsilon,0}} + \lambda_{\varepsilon,0}, \delta\sqrt{r_{\varepsilon,0}} + \lambda_{\varepsilon,0}]\right) \\ &\le |y| + \frac{\delta}{\sqrt{r_{\varepsilon,0}}} .\end{aligned} \tag{10.40}$$

This inequality and the condition $\mathbf{D}_{20}[\beta]$ imply the following relation:

$$\mathsf{E}e^{\beta|Y_{\varepsilon,0}|} = \mathsf{E}e^{\beta|h_\varepsilon(Y_{0,0})|} \le e^{\frac{\beta\delta}{\sqrt{r_{\varepsilon,0}}}} \mathsf{E}e^{\beta|Y_{0,0}|} \to \mathsf{E}e^{\beta|Y_{0,0}|} < \infty \quad \text{as } \varepsilon \to 0 . \tag{10.41}$$

Relation (10.41) implies that the condition $\mathbf{D}_7[\bar{\beta}]$ (its one-dimensional variant) holds.

The skeleton functions $h_\varepsilon(y)$ also satisfy the following inequalities for $y \in \mathbb{R}_1$:

$$\begin{aligned}|h_\varepsilon(y) - y| &\le (-\delta\sqrt{r_{\varepsilon,0}} + \lambda_{\varepsilon,0} - y) I(y \le -\delta\sqrt{r_{\varepsilon,0}} + \lambda_{\varepsilon,0}) \\ &\quad + \frac{\delta}{\sqrt{r_{\varepsilon,0}}} I(\delta\sqrt{r_{\varepsilon,0}} + \lambda_{\varepsilon,0} < y \le \delta\sqrt{r_{\varepsilon,0}} + \lambda_{\varepsilon,0}) \\ &\quad + (y - \delta\sqrt{r_{\varepsilon,0}} - \lambda_{\varepsilon,0}) I(y > \delta\sqrt{r_{\varepsilon,0}} + \lambda_{\varepsilon,0}) .\end{aligned} \tag{10.42}$$

This inequalities and the condition $\mathbf{N}_{10}$ obviously imply that $h_\varepsilon(y) \to y$ as $\varepsilon \to 0$, for every $y \in \mathbb{R}_1$, and, therefore, $h_\varepsilon(Y_{0,0}) \xrightarrow{\text{a.s.}} Y_{0,0}$ as $\varepsilon \to 0$. Therefore,

$$P_\varepsilon(\cdot) \Rightarrow P_0(\cdot) \quad \text{as } \varepsilon \to 0 . \tag{10.43}$$

Relation (10.43) implies that the condition $\mathbf{K}_4$ (a) (its one-dimensional variant) holds. The condition $\mathbf{K}_{18}$ is a variant of the condition $\mathbf{K}_4$ (b).

Other conditions of Theorem 5.3.2 also hold that was shown in the proof of Theorem 10.1.1. By applying Theorem 5.3.2 to the log-price processes $Y_{\varepsilon,n}$ we get the asymptotic relation (10.37). □

The following theorem give analogous result for the binomial tree approximation model with asymptotically fitted parameters.

Theorem 10.1.4. *Let the log-price process $Y_{0,n}$ be represented by the Gaussian random walk given by the stochastic transition dynamic relation* (10.1) *and the approximating log-price process is represented, for every $\varepsilon \in (0, \varepsilon_0]$, by the binomial random walks $Y_{\varepsilon,n}$ given by the stochastic transition dynamic relation* (10.18), *with parameters given by relation* (10.20). *Let also the conditions $\mathbf{B}_4[\gamma]$ and $\mathbf{D}_{20}[\beta]$ hold for parameters γ and β such that, either $\beta > \gamma > 0$ or $\beta = \gamma = 0$, and also the conditions $\mathbf{N}_5$, $\mathbf{I}_9$, $\mathbf{J}_{13}$ and $\mathbf{K}_{18}$ hold, and $r_\varepsilon \to \infty$ as $\varepsilon \to 0$. Then the following relation takes place:*

$$\Phi_\varepsilon \to \Phi_0 \quad \text{as } \varepsilon \to 0 . \tag{10.44}$$

The proof of this theorem is analogous to the proof of Theorem 10.1.3.

Remark 10.1.1. If the model assumption (e) $\sigma_n^2 > 0$, $n = 1, 2, \dots$ may not hold, the condition $\mathbf{J}_{13}$ in Theorems 10.1.1–10.1.4 can be replaced by the assumption that $\mathsf{P}\{y_0 + W_{0,n} \in \overline{\mathbb{Y}}_n\} = 0, y_0 \in \mathbb{Y}_{n-1}, n = 1, \dots, N$.

10.2 Multivariate homogeneous in space tree models

In this section, we present results about convergence of rewards for American-type options for multivariate binomial and trinomial tree approximations for log-price processes represented by multivariate Gaussian random walks. We give the corresponding results for homogeneous in time binomial tree approximations and inhomogeneous in time trinomial tree approximations.

10.2.1 Convergence of option rewards for multivariate homogeneous in time binomial tree models

Let us consider the model of multivariate homogeneous in time Gaussian random walk, which can be considered as a discrete time analog of the multivariate Wienner process (Brownian motion). In this case, the log-price $\vec{Y}_{0,n}$ is given by the following stochastic transition dynamic relation:

$$\vec{Y}_{0,n} = \vec{Y}_{0,n-1} + \vec{W}_{0,n}, \quad n = 1, 2, \dots . \tag{10.45}$$

where (a) $\vec{Y}_{0,0} = (Y_{0,0,1}, \dots, Y_{0,0,k})$ is a random vector taking values in the space $\mathbb{R}_k$, and with distribution $P_0(A)$, (b) $\vec{W}_{0,n} = (W_{0,n,1}, \dots, W_{0,n,k})$, $n = 1, 2, \dots$ is a sequence of i.i.d. k-dimensional normal random vectors with means $\mathsf{E}W_{0,n,i} = \mu_i$, $i = 1, \dots, k$, variances $\mathrm{Var}W_{0,n,i} = \sigma_i^2$, $i = 1, \dots, k$, and covariances $\mathsf{E}(W_{0,n,i} - \mu_i)(W_{0,n,i} - \mu_i) = \rho_{i,j}\sigma_i\sigma_j$, $i, j = 1, \dots, k$, and (c) the random vector $\vec{Y}_{0,0}$ and the sequence of random vectors $\vec{W}_{0,n}$, $n = 1, 2, \dots$ are independent.

Also, usual assumptions are (d) $\mu_i \in \mathbb{R}_1$, $i = 1, \dots,$ (e) $\sigma_i > 0$, $i = 1, \dots, k$; (f) $|\rho_{i,j}| \leq 1, \rho_{i,i} = 1, i, j = 1, \dots, k$, (g) $\Sigma = \|\sigma_{i,j}\|$, where $\sigma_{i,j} = \rho_{i,j}\sigma_i\sigma_j$, $i, j = 1, \dots, k$, is a nonnegatively defined $k \times k$ matrix, and (h) $\det(\Sigma) \neq 0$.

We shall approximate the log-price process $\vec{Y}_{0,n}$ by the multivariate homogeneous in time binomial random walks $\vec{Y}_{\varepsilon,n}$. In this case, the log-price $\vec{Y}_{\varepsilon,n}$ is given, for every $\varepsilon \in (0, \varepsilon_0]$, by the following stochastic transition dynamic relation:

$$\vec{Y}_{\varepsilon,n} = \vec{Y}_{\varepsilon,n-1} + \vec{W}_{\varepsilon,n}, \quad n = 1, 2, \dots . \tag{10.46}$$

where (a) $\vec{Y}_{\varepsilon,0} = (Y_{\varepsilon,0,1}, \dots, Y_{\varepsilon,0,k})$ is a random vector taking values in the space $\mathbb{R}_k$, and with distribution $P_\varepsilon(A)$, (b) $\vec{W}_{\varepsilon,n} = (W_{\varepsilon,n,1}, \dots, W_{\varepsilon,n,k}) = \vec{W}_{\varepsilon,n,1} + \dots + \vec{W}_{\varepsilon,n,r_\varepsilon}$,

$n = 1, 2, \ldots$, where $\vec{W}_{\varepsilon,n,l} = (W_{\varepsilon,n,l,1}, \ldots, W_{\varepsilon,n,l,k})$, $l, n = 1, 2, \ldots$ is a family of i.i.d. binary random vectors taking values $\bar{\delta}_{\varepsilon,\bar{\imath}} = (\delta_{\varepsilon,1,\imath_1}, \ldots, \delta_{\varepsilon,k,\imath_k})$ with the corresponding probabilities $p_{\varepsilon,\bar{\imath}}$, for $\bar{\imath} \in \mathcal{I}_k = \{\bar{\imath} = (\imath_1, \ldots \imath_k) : \imath_1, \ldots \imath_k = \pm\}$, and (c) the random vector $\vec{Y}_{\varepsilon,0}$ and the family of random vectors $\vec{W}_{\varepsilon,n,l}$, $l, n = 1, 2, \ldots$ are independent.

Also, usual assumptions are (d) $-\infty < \delta_{\varepsilon,i,-} < \delta_{\varepsilon,i,+} < \infty$, $i = 1, \ldots, k$, (e) $p_{\varepsilon,\bar{\imath}} \geq 0$, $\sum_{\bar{\imath},\in \mathcal{I}_k} p_{\varepsilon,\bar{\imath}} = 1$, (f) r_ε is a positive integer number.

We assume that a pay-off function $g(n, e^{\vec{y}})$ does not depend on the parameter ε and also $g(n, e^{\vec{y}})$ is a real-valued Borel function of the argument $\vec{y} \in \mathbb{R}_k$, for every $n = 0, 1, \ldots, N$.

In this case, the corresponding reward function $\phi_{\varepsilon,n}(\vec{y})$ for the log-price process $\vec{Y}_{\varepsilon,n}$ is also a Borel function of argument $\vec{y} \in \mathbb{R}_k$, for every $n = 0, 1, \ldots, N$.

We fit parameters of approximating log-price processes $\vec{Y}_{\varepsilon,n}$ in such a way that the following exact moment-fitting relations hold, for every $\varepsilon \in (0, \varepsilon_0]$:

$$\begin{cases} \mathsf{E}W_{\varepsilon,1,i} = \mu_i, \quad \mathrm{Var}W_{\varepsilon,1,i} = \sigma_i^2, \quad i = 1, \ldots, k, \\ \mathsf{E}(W_{\varepsilon,1,i} - \mathsf{E}W_{\varepsilon,1,i})(W_{\varepsilon,1,i} - \mathsf{E}W_{\varepsilon,1,j}) = \rho_{i,j}\sigma_i\sigma_j, \\ 1 \leq i < j \leq k. \end{cases} \tag{10.47}$$

Lemma 9.2.1 gives two variants for solution of this fitting problem.

According to the first variant presented in Lemma 9.2.1, the fitting relations (10.47) hold, if parameters for the log-price process $Y_{\varepsilon,n}$ are given, for every $\varepsilon \in (0, \varepsilon_0]$, by the following formulas:

$$\begin{cases} \delta_{\varepsilon,i,\pm} = \pm\delta_{\varepsilon,i} + \lambda_{\varepsilon,i}, \quad \delta_{\varepsilon,i} = \dfrac{\sigma_i}{\sqrt{r_\varepsilon}}, \quad i = 1, \ldots, k, \\ \lambda_{\varepsilon,i} = \dfrac{\mu_i}{r_\varepsilon}, \quad p_{\varepsilon,i,\pm} = \dfrac{1}{2}, \quad i = 1, \ldots, k, \\ p_{\varepsilon,i,j,+,\pm} = p_{\varepsilon,i,j,-,\mp} = \frac{1}{4}(1 \pm \rho_{i,j}), \quad 1 \leq i < j \leq k. \end{cases} \tag{10.48}$$

where (i) one can choose for the parameter r_ε any value $r_\varepsilon = 1, 2, \ldots$, and (ii) the corresponding higher order distributions for components of random vectors $\vec{W}_{\varepsilon,1,1}$ consistent with the above one- and two-dimensional distributions should be defined using algorithm described in Lemma 9.2.2.

According to the second variant also presented in Lemma 9.2.1, the fitting relations (10.47) hold, if parameters of the log-price process $Y_{\varepsilon,n}$ are given, for every $\varepsilon \in (0, \varepsilon_0]$, by the following formulas:

$$\begin{cases} \delta_{\varepsilon,i,\pm} = \pm\delta_{\varepsilon,i} + \lambda_{\varepsilon,i}, \quad \delta_{\varepsilon,i} = \dfrac{\delta_i}{\sqrt{r_\varepsilon}}, \quad \lambda_{\varepsilon,i} = \dfrac{\imath_i\sqrt{\delta_i^2 - \sigma_i^2}}{\sqrt{r_\varepsilon}} + \dfrac{\mu_i}{r_\varepsilon}, \\ p_{\varepsilon,i,\pm} = \dfrac{1}{2}\left(1 \mp \dfrac{\imath_i\sqrt{\delta_i^2 - \sigma_i^2}}{\delta_i}\right), \quad i = 1, \ldots, k, \end{cases} \tag{10.49}$$

and

$$\begin{cases} p_{\varepsilon,i,j,+,+} = \frac{1}{4} + \frac{1}{4}\rho_{i,j}\frac{\sigma_i\sigma_j}{\delta_i\delta_j} + \frac{1}{4}\left(-\frac{\imath_i\sqrt{\delta_i^2-\sigma_i^2}}{\delta_i} - \frac{\imath_j\sqrt{\delta_j^2-\sigma_j^2}}{\delta_j}\right) \\ \quad + \frac{1}{4}\frac{\imath_i\sqrt{\delta_i^2-\sigma_i^2}\,\imath_j\sqrt{\delta_j^2-\sigma_j^2}}{\delta_i\delta_j}, \quad 1 \le i < j \le k, \\ p_{\varepsilon,i,j,+,-} = \frac{1}{4} - \frac{1}{4}\rho_{i,j}\frac{\sigma_i\sigma_j}{\delta_i\delta_j} + \frac{1}{4}\left(-\frac{\imath_i\sqrt{\delta_i^2-\sigma_i^2}}{\delta_i} + \frac{\imath_j\sqrt{\delta_j^2-\sigma_j^2}}{\delta_j}\right) \\ \quad - \frac{1}{4}\frac{\imath_i\sqrt{\delta_i^2-\sigma_i^2}\,\imath_j\sqrt{\delta_j^2-\sigma_j^2}}{\delta_i\delta_j}, \quad 1 \le i < j \le k, \\ p_{\varepsilon,i,j,-,+} = \frac{1}{4} - \frac{1}{4}\rho_{i,j}\frac{\sigma_i\sigma_j}{\delta_i\delta_j} + \frac{1}{4}\left(\frac{\imath_i\sqrt{\delta_i^2-\sigma_i^2}}{\delta_i} - \frac{\imath_j\sqrt{\delta_j^2-\sigma_j^2}}{\delta_j}\right) \\ \quad - \frac{1}{4}\frac{\imath_i\sqrt{\delta_i^2-\sigma_i^2}\,\imath_j\sqrt{\delta_j^2-\sigma_j^2}}{\delta_i\delta_j}, \quad 1 \le i < j \le k, \\ p_{\varepsilon,i,j,-,-} = \frac{1}{4} + \frac{1}{4}\rho_{i,j}\frac{\sigma_i\sigma_j}{\delta_i\delta_j} + \frac{1}{4}\left(\frac{\imath_i\sqrt{\delta_i^2-\sigma_i^2}}{\delta_i} + \frac{\imath_j\sqrt{\delta_j^2-\sigma_j^2}}{\delta_j}\right) \\ \quad + \frac{1}{4}\frac{\imath_i\sqrt{\delta_i^2-\sigma_i^2}\,\imath_j\sqrt{\delta_j^2-\sigma_j^2}}{\delta_i\delta_j}, \quad 1 \le i < j \le k, \end{cases} \tag{10.50}$$

where (i) one can choose for the parameters $\imath_i, \delta_i$, $i = 1, \dots, k, r_\varepsilon$ any values such that (a) $\imath_i = +, -$, $i = 1, \dots, k$; (b) $\sigma_i \le \delta_i \le \kappa(\rho)\sigma_i$, $i = 1, \dots, k$, where $\kappa(\rho) = \frac{1+\rho}{2\sqrt{\rho}}$ and $\rho = \max_{1\le i<j\le k}|\rho_{i,j}|$, and (c) $r_\varepsilon = 1, 2, \dots$; (ii) the corresponding higher order distributions for components of random vectors $\vec{W}_{\varepsilon,1,1}$ consistent with the above one- and two-dimensional distributions can be defined using algorithm described in Lemma 9.2.2.

Let us denote, $\mathcal{L}_{k,r} = \{\bar{l} = (l_1, \dots, l_k) : -r \le l_1, \dots, l_k \le r\}$, for $r = 1, 2, \dots$,

The random vectors $\vec{W}_{\varepsilon,n}$, $n = 1, \dots, N$ are the i.i.d. binomial random vectors. Namely, the random vector $\vec{W}_{\varepsilon,n}$ takes values $\vec{w}_{\varepsilon,r_\varepsilon,\bar{l}}$ with the corresponding probabilities $p_{\varepsilon,r_\varepsilon}(\bar{l})$, for $\bar{l} = (l_1, \dots, l_k) \in \mathcal{L}_{k,r_\varepsilon}$.

Here, for every $\bar{l} = (l_1, \dots, l_k) \in \mathcal{L}_{k,r}$, $r = 1, 2, \dots$,

$$\vec{w}_{\varepsilon,r,\bar{l}} = (l_1\delta_{\varepsilon,1} + r\lambda_{\varepsilon,1}, \dots, l_k\delta_{\varepsilon,k} + r\lambda_{\varepsilon,k}), \tag{10.51}$$

and

$$p_{\varepsilon,r}(\bar{l}) = \mathsf{P}\{\vec{W}_{\varepsilon,1} = \vec{W}_{\varepsilon,1,1} + \cdots + \vec{W}_{\varepsilon,1,r} = \vec{w}_{\varepsilon,r,\bar{l}}\} I(\bar{l} \in \mathcal{L}_{k,r}). \tag{10.52}$$

The probabilities $p_{\varepsilon,r}(\bar{l})$ can be found using the following recurrent convolution formulas:

$$p_{\varepsilon,r}(\bar{l}) = \begin{cases} p_{\varepsilon,\bar{\imath}} I(\bar{l} = \bar{e}_{\bar{\imath}}) & \text{if } r = 1, \bar{l} \in \mathcal{L}_{k,1}, \\ \sum_{\bar{l}' \in \mathcal{L}_{k,1}} p_{\varepsilon,r-1}(\bar{l} - \bar{l}') p_{\varepsilon,1}(\bar{l}') & \text{if } r > 1, \bar{l} \in \mathcal{L}_{k,r}. \end{cases} \tag{10.53}$$

where the vectors $\bar{e}_{\bar{\imath}} = (e_{\imath_1}, \dots e_{\imath_k}), \bar{\imath} = (\imath_1, \dots, \imath_k) \in \mathcal{I}_k$, $e_\imath = \pm 1$ if $\imath = \pm$.

The additional comments concerning algorithms for computing of probabilities $p_{\varepsilon,r}(\bar{l})$ can be found in Subsection 3.2.3.

Let us assume that $\vec{Y}_{\varepsilon,n} = \vec{y} = (y_1, \dots, y_k) = \text{const} \in \mathbb{R}_k$. In this case, trajectories of the log-price process $\vec{Y}_{\varepsilon,n+r}$, $r = 0, 1, \dots$ generate, for every $\varepsilon \in (0, \varepsilon_0]$, the unique tree of nodes,

$$T_{\varepsilon(y,n)} = \{(\vec{y}_{\varepsilon,n,n+r,\bar{l}}, n + r), \bar{l} = (l_1, \dots, l_k) \in \mathcal{L}_{k,r_\varepsilon r}, r = 0, 1, \dots\}, \tag{10.54}$$

where

$$\begin{aligned}\vec{y}_{\varepsilon,n,n+r,\bar{l}} &= (y_{\varepsilon,n,n+r,1,l_1}, \dots, y_{\varepsilon,n,n+r,k,l_k}) \\ &= (y_1 + l_1\delta_{\varepsilon,1} + r_\varepsilon r\lambda_{\varepsilon,1}, \dots, y_k + l_k\delta_{\varepsilon,k} + r_\varepsilon r\lambda_{\varepsilon,k}), \\ &\quad \bar{l} = (l_1, \dots, l_k) \in \mathcal{L}_{k,r_\varepsilon r}, \quad r = 0, 1, \dots .\end{aligned} \tag{10.55}$$

The tree $T_{\varepsilon,(\vec{y},n)}$ contains the total number of nodes after r steps,

$$\begin{aligned}L_{\varepsilon,n,n+r} &\le 1 + (2r_\varepsilon + 1)^k + \cdots + (2r_\varepsilon r + 1)^k \\ &\le \int_0^{r+1} (2r_\varepsilon x + 1)^k dx = \frac{(2r_\varepsilon(r+1) + 1)^{k+1} - 1}{2r_\varepsilon(k+1)}.\end{aligned} \tag{10.56}$$

The expression on the right-hand side in (10.56) is a polynomial function of r of the order $k + 1$.

The following lemma is a direct corollary of Lemma 3.3.3.

Lemma 10.2.1. *Let, for every $\varepsilon \in (0, \varepsilon_0]$, the approximating log-price process $\vec{Y}_{\varepsilon,n}$ is a multivariate binomial random walk given by the stochastic transition dynamic relation (10.46), with parameters defined in relation (10.48) or (10.49) – (10.50). Then the reward functions $\phi_{\varepsilon,n+r}(\vec{y}_{\varepsilon,n,n+r,\bar{l}})$, for points $\vec{y}_{\varepsilon,n,n,0} = \vec{y}$ and $\vec{y}_{\varepsilon,n,n+r,\bar{l}}$, $\bar{l} \in \mathcal{L}_{k,r_\varepsilon r}$, $r = 1, \dots N - n$, are, for every $\varepsilon \in (0, \varepsilon_0]$, and $\vec{y} \in \mathbb{R}_k$, $n = 0, \dots, N$, the unique solution for the following recurrence finite system of linear equations:*

$$\left\{\begin{aligned} \phi_{\varepsilon,N}(\vec{y}_{\varepsilon,n,N,l}) &= g_\varepsilon(N, e^{\vec{y}_{\varepsilon,n,N,l}}), \quad \bar{l} \in \mathcal{L}_{k,r_\varepsilon(N-n)}, \\ \phi_{\varepsilon,n+r}(\vec{y}_{\varepsilon,n,n+r,l}) &= \max\Bigg(g_\varepsilon(n + r, e^{\vec{y}_{\varepsilon,n,n+r,l}}), \\ &\qquad \sum_{\bar{l}' \in \mathcal{L}_{k,r_\varepsilon}} \phi_{\varepsilon,n+r+1}(\vec{y}_{\varepsilon,n,n+r+1,\bar{l}+\bar{l}'}) p_{\varepsilon,r_\varepsilon}(\bar{l}')\Bigg), \\ &\qquad \bar{l} \in \mathcal{L}_{k,r_\varepsilon r}, \quad r = N - n - 1, \dots, 1, \\ \phi_{\varepsilon,n}(\vec{y}) &= \max\Bigg(g_\varepsilon(n, e^{\vec{y}}), \sum_{\bar{l}' \in \mathcal{L}_{k,r_\varepsilon}} \phi_{\varepsilon,n}(\vec{y}_{\varepsilon,n,n+1,\bar{l}'}) p_{\varepsilon,r_\varepsilon}(\bar{l}')\Bigg). \end{aligned}\right. \tag{10.57}$$

We are going to apply Theorem 5.3.1, in order to prove convergence of the reward functions for the log-price processes $\vec{Y}_{\varepsilon,n}$ to the corresponding reward functions for the log-price processes $\vec{Y}_{0,n}$ as $\varepsilon \to 0$.

First, we assume that the condition $\mathbf{B}_6[\bar{\gamma}]$ holds for the pay-off function $g(n, e^{\vec{y}})$, for some vector parameter $\bar{\gamma} = (\gamma_1, \dots, \gamma_k)$ with nonnegative components.

The condition $\mathbf{I}_3$ reduces in this case to the following condition:

$\mathbf{I}_{10}$: There exists sets $\mathbb{Y}_n \in \mathcal{B}_k$, $n = 0, \dots, N$ such that function $g(n, e^{\vec{y}})$ is continuous in points $\vec{y} \in \mathbb{Y}_n$ for every $n = 0, \dots, N$.

We shall see that, in this case, the condition $\mathbf{J}_4$ (a) holds for log-prices $\vec{Y}_{\varepsilon,n}$ with parameters given by relation (10.48) if $r_\varepsilon \to \infty$ as $\varepsilon \to 0$. Also, the random vector $\vec{W}_{0,1}$ has normal distribution that makes it possible to replace the condition $\mathbf{J}_4$ (b) by the following condition:

$\mathbf{J}_{14}$: $L_k(\overline{\mathbb{Y}}_n) = 0$, $n = 1, \dots, N$, where $\mathbb{Y}_n$, $n = 1, \dots, N$ are sets penetrating condition $\mathbf{I}_{10}$.

The above backward recurrence algorithm for finding reward functions is supported by the following convergence result.

Theorem 10.2.1. *Let the log-price process $\vec{Y}_{0,n}$ be represented by the multivariate Gaussian random walk given by the stochastic transition dynamic relation* (10.45) *and the approximating log-price process is represented, for every $\varepsilon \in (0, \varepsilon_0]$, by the binomial random walks $\vec{Y}_{\varepsilon,n}$ given by the stochastic transition dynamic relation* (10.46), *with parameters given by relation* (10.48). *Let also the conditions $\mathbf{B}_6[\bar{\gamma}]$ hold for some vector parameter $\bar{\gamma} = (\gamma_1, \dots, \gamma_k)$ with nonnegative components, the conditions $\mathbf{I}_{10}$ and $\mathbf{J}_{14}$ hold, and $r_\varepsilon \to \infty$ as $\varepsilon \to 0$. Then the following relation takes place for any $\vec{y}_\varepsilon \to \vec{y}_0 \in \mathbb{Y}_n$, $n = 0, 1, \dots, N$:*

$$\phi_{\varepsilon,n}(\vec{y}_\varepsilon) \to \phi_{0,n}(\vec{y}_0) \quad \text{as } \varepsilon \to 0. \tag{10.58}$$

Proof. Recall that random vectors $\vec{W}_{\varepsilon,n} = \vec{W}_{\varepsilon,n,1} + \cdots + \vec{W}_{\varepsilon,n,r_\varepsilon}$, $n = 1, 2, \dots$, where $\vec{W}_{\varepsilon,n,l}$, $n, l = 1, 2, \dots$ are the i.i.d. random vectors.

In the case, where parameters of the log-price process $Y_{\varepsilon,n}$ are given by relations (10.48), the moment-generating function for the random vector $\vec{W}_{\varepsilon,1} = (W_{\varepsilon,1,1}, \dots, W_{\varepsilon,1,k})$ has, for every $\varepsilon \in (0, \varepsilon_0]$, the following form, for $\bar{\beta} = (\beta_1, \dots, \beta_k) \in \mathbb{R}_k$:

$$\begin{aligned} \mathsf{E} e^{\sum_{i=1}^k \beta_i W_{\varepsilon,1,i}} &= \left(\sum_{\bar{\imath} \in \mathcal{I}_k} e^{\sum_{i=1}^k \beta_i (\frac{\imath_i \sigma_i}{\sqrt{r_\varepsilon}} + \frac{\mu_i}{r_\varepsilon})} p_{\varepsilon,\bar{\imath}} \right)^{r_\varepsilon} \\ &= e^{\sum_{i=1}^k \beta_i \mu_i} \left(\sum_{\bar{\imath} \in \mathcal{I}_k} e^{\sum_{i=1}^k \beta_i \frac{\imath_i \sigma_i}{\sqrt{r_\varepsilon}}} p_{\varepsilon,\bar{\imath}} \right)^{r_\varepsilon} . \end{aligned} \tag{10.59}$$

For $\varepsilon = 0$, this moment-generating function has the following form, for $\bar{\beta} = (\beta_1, \dots, \beta_k) \in \mathbb{R}_k$:

$$\mathsf{E}e^{\sum_{i=1}^k \beta_i W_{0,1,i}} = \exp\left\{ \sum_{i=1}^k \beta_i \mu_i + \frac{1}{2} \sum_{i=1}^k \beta_i^2 \sigma_i^2 + \sum_{1 \le i < j \le k} \beta_i \beta_j \rho_{i,j} \sigma_i \sigma_j \right\}. \tag{10.60}$$

Relations (10.48), formula (10.59), and Taylor expansion for exponential function imply that the following asymptotic relation takes place, for every $\bar{\beta} = (\beta_1, \dots, \beta_k) \in \mathbb{R}_k$:

$$\begin{aligned}
\mathsf{E}e^{\sum_{i=1}^k \beta_i W_{\varepsilon,1,i}} &= e^{\sum_{i=1}^k \beta_i \mu_i} \left(\sum_{\bar{\imath} \in \mathbb{I}_k} \left(1 + \sum_{i=1}^k \beta_i \frac{\imath_i \sigma_i}{\sqrt{r_\varepsilon}} + \frac{1}{2} \left(\sum_{i=1}^k \beta_i \frac{\imath_i \sigma_i}{\sqrt{r_\varepsilon}} \right)^2 + o_2 \left(\sum_{i=1}^k \beta_i \frac{\imath_i \sigma_i}{\sqrt{r_\varepsilon}} \right) \right) p_{\varepsilon,\bar{\imath}} \right)^{r_\varepsilon} \\
&= e^{\sum_{i=1}^k \beta_i \mu_i} \left(1 + \sum_{i=1}^k \beta_i \frac{\sigma_i}{\sqrt{r_\varepsilon}} (p_{\varepsilon,i,+} - p_{\varepsilon,i,-}) + \frac{1}{2} \sum_{i=1}^k \beta_i^2 \frac{\sigma_i^2}{r_\varepsilon} \right. \\
&\quad + \sum_{1 \le i < j \le k} \beta_i \beta_j \frac{\sigma_i \sigma_j}{r_\varepsilon} (p_{\varepsilon,i,j,+,+} + p_{\varepsilon,i,j,-,-} - p_{\varepsilon,i,j,+,-} - p_{\varepsilon,i,j,-,+}) \\
&\quad \left. + \sum_{\bar{\imath} \in \mathbb{I}_k} o_2 \left(\sum_{i=1}^k \beta_i \frac{\imath_i \sigma_i}{\sqrt{r_\varepsilon}} \right) p_{\varepsilon,\bar{\imath}} \right)^{r_\varepsilon} \\
&= e^{\sum_{i=1}^k \beta_i \mu_i} \left(1 + \frac{1}{r_\varepsilon} \left(\frac{1}{2} \sum_{i=1}^k \beta_i^2 \sigma_i^2 + \sum_{1 \le i < j \le k} \beta_i \beta_j \sigma_i \sigma_j \rho_{i,j} \right) \right. \\
&\quad \left. + \sum_{\bar{\imath} \in \mathbb{I}_k} o_2 \left(\sum_{i=1}^k \beta_i \frac{\imath_i \sigma_i}{\sqrt{r_\varepsilon}} \right) p_{\varepsilon,\bar{\imath}} \right)^{r_\varepsilon} \\
&\to \exp\left\{ \sum_{i=1}^k \beta_i \mu_i + \frac{1}{2} \sum_{i=1}^k \beta_i^2 \sigma_i^2 + \sum_{1 \le i < j \le k} \beta_i \beta_j \rho_{i,j} \sigma_i \sigma_j \right\} \\
&= \mathsf{E}e^{\sum_{i=1}^k \beta_i W_{0,1,i}} \quad \text{as } \varepsilon \to 0,
\end{aligned} \tag{10.61}$$

since,

$$\left| \sum_{\bar{\imath} \in \mathbb{I}_k} o_2 \left(\sum_{i=1}^k \beta_i \frac{\imath_i \sigma_i}{\sqrt{r_\varepsilon}} \right) p_{\varepsilon,\bar{\imath}} \right| r_\varepsilon \le \sum_{\bar{\imath} \in \mathbb{I}_k} \left| o_2 \left(\sum_{i=1}^k \beta_i \frac{\imath_i \sigma_i}{\sqrt{r_\varepsilon}} \right) r_\varepsilon \right| \to 0 \quad \text{as } \varepsilon \to 0. \tag{10.62}$$

Relation (10.61) implies that for any $i = 1, \dots, k$ and $\beta \ge 0$,

$$\Xi_\beta'^{\pm}(Y_{\varepsilon,i,\cdot}, N) = \mathsf{E}e^{\pm \beta W_{\varepsilon,1,i}} \to \Xi_\beta'^{\pm}(Y_{0,i,\cdot}, N) = e^{\pm \beta \mu_i + \frac{1}{2} \beta^2 \sigma_i^2} < \infty. \tag{10.63}$$

Relation (10.63) implies, in an obvious way, that the condition $\mathbf{E}_{11}[\bar{\beta}]$ holds and, therefore, by Lemma 5.3.1 the condition $\mathbf{C}_7[\bar{\beta}]$ holds, for any parameter $\bar{\beta} = (\beta_1, \dots, \beta_k)$ with nonnegative components.

The condition $\mathbf{B}_6[\bar{\gamma}]$ is used in Theorem 5.3.1.

Note that one can always take the parameter $\bar{\beta}$ in such a way that, for every $i = 1, \dots, k$, either $\beta_i > \gamma_i$ if $\gamma_i > 0$ or $\beta_i = \gamma_i$ if $\gamma_i = 0$.

As was pointed above, the condition $\mathbf{I}_1$ reduces in this case to the form of the condition $\mathbf{I}_{10}$.

Relation (10.61) also implies, by the corresponding variant of the continuous theorem for moment-generating functions, the following relation:

$$\vec{W}_{\varepsilon,1} \xrightarrow{d} \vec{W}_{0,1} \quad \text{as } \varepsilon \to 0\,. \tag{10.64}$$

Relation (10.64) implies that the condition $\mathbf{J}_4$ (a) holds.

Condition $\mathbf{J}_{14}$ implies that the condition $\mathbf{J}_4$ (b) holds, since the random vector $\vec{W}_{0,1}$ has, according to initial model assumptions, a k-dimensional normal distribution with positive everywhere in $\mathbb{R}_k$ probability density function.

Thus, all conditions of Theorem 5.3.1 hold. By applying this theorem to the above approximation model, we get the asymptotic relation (10.58). □

The following theorem presents an analogous convergence results for the case, where parameters of the approximating log-price processes $Y_{\varepsilon,n}$ are given by relations (10.49) – (10.50) instead of relation (10.48).

Theorem 10.2.2. *Let the log-price process $\vec{Y}_{0,n}$ be represented by the multivariate Gaussian random walk given by the stochastic transition dynamic relation* (10.45) *and the approximating log-price process is represented, for every $\varepsilon \in (0, \varepsilon_0]$, by the binomial random walks $\vec{Y}_{\varepsilon,n}$ given by the stochastic transition dynamic relation* (10.46), *with parameters given by relation* (10.49) – (10.50). *Let also conditions $\mathbf{B}_6[\bar{\gamma}]$ hold for some vector parameter $\bar{\gamma} = (\gamma_1, \dots, \gamma_k)$ with nonnegative components, the conditions $\mathbf{I}_{10}$ and $\mathbf{J}_{14}$ hold, and $r_\varepsilon \to \infty$ as $\varepsilon \to 0$. Then the following relation takes place for any $\vec{y}_\varepsilon \to \vec{y}_0 \in \mathbb{Y}_n$, $n = 0, 1, \dots, N$:*

$$\phi_{\varepsilon,n}(\vec{y}_\varepsilon) \to \phi_{0,n}(\vec{y}_0) \quad \text{as } \varepsilon \to 0\,. \tag{10.65}$$

Proof. In the case, where parameters of the log-price process $Y_{\varepsilon,n}$ are given by relations (10.49), the moment-generating function for the random vector $\vec{W}_{\varepsilon,1} = (W_{\varepsilon,1,1}, \dots, W_{\varepsilon,1,k})$ has, for every $\varepsilon \in (0, \varepsilon_0]$, the following form, for $\bar{\beta} = (\beta_1, \dots, \beta_k) \in \mathbb{R}_k$:

$$\begin{aligned} \mathsf{E}e^{\sum_{i=1}^k \beta_i W_{\varepsilon,1,i}} &= \Bigg(\sum_{\bar{\imath}' = \langle \imath_1', \dots, \imath_k' \rangle \in \mathcal{I}_k} e^{\sum_{i=1}^k \beta_i \left(\frac{\imath_i' \delta_i}{\sqrt{r_\varepsilon}} + \frac{\imath_i \sqrt{\delta_i^2 - \sigma_i^2}}{\sqrt{r_\varepsilon}} + \frac{\mu_i}{r_\varepsilon} \right)} p_{\varepsilon,\bar{\imath}'} \Bigg)^{r_\varepsilon} \\ &= e^{\sum_{i=1}^k \beta_i \mu_i} \Bigg(\sum_{\bar{\imath}' \in \mathcal{I}_k} e^{\sum_{i=1}^k \beta_i \left(\frac{\imath_i' \delta_i}{\sqrt{r_\varepsilon}} + \frac{\imath_i \sqrt{\delta_i^2 - \sigma_i^2}}{\sqrt{r_\varepsilon}} \right)} p_{\varepsilon,\bar{\imath}'} \Bigg)^{r_\varepsilon}\,. \end{aligned} \tag{10.66}$$

Relations (10.49), formula (10.66), and Taylor expansion for the exponential function imply that the following asymptotic relation takes place, for every $\bar{\beta} = (\beta_1, \dots, \beta_k) \in \mathbb{R}_k$:

$$\begin{aligned}
\mathsf{E}e^{\sum_{i=1}^k \beta_i W_{\varepsilon,1,i}} &= e^{\sum_{i=1}^k \beta_i\mu_i} \Bigg(\sum_{\bar{\imath}' \in \mathcal{I}_k} \Bigg(1 + \sum_{i=1}^k \beta_i \Bigg(\frac{\imath_i' \delta_i}{\sqrt{r_\varepsilon}} + \frac{\imath_i \sqrt{\delta_i^2 - \sigma_i^2}}{\sqrt{r_\varepsilon}} \Bigg) \\
&\quad + \frac{1}{2} \Bigg(\sum_{i=1}^k \beta_i \Bigg(\frac{\imath_i' \delta_i}{\sqrt{r_\varepsilon}} + \frac{\imath_i \sqrt{\delta_i^2 - \sigma_i^2}}{\sqrt{r_\varepsilon}} \Bigg) \Bigg)^2 + o_2 \Bigg(\sum_{i=1}^k \beta_i \Bigg(\frac{\imath_i' \delta_i}{\sqrt{r_\varepsilon}} + \frac{\imath_i \sqrt{\delta_i^2 - \sigma_i^2}}{\sqrt{r_\varepsilon}} \Bigg) \Bigg) \Bigg) p_{\varepsilon,\bar{\imath}'} \Bigg)^{r_\varepsilon} \\
&= e^{\sum_{i=1}^k \beta_i\mu_i} \Bigg(1 + \sum_{i=1}^k \beta_i \Bigg(\frac{\delta_i}{\sqrt{r_\varepsilon}} (p_{\varepsilon,i,+} - p_{\varepsilon,i,-}) + \frac{\imath_i \sqrt{\delta_i^2 - \sigma_i^2}}{\sqrt{r_\varepsilon}} \Bigg) \\
&\quad + \frac{1}{2} \sum_{i=1}^k \beta_i^2 \Bigg(\frac{\delta_i^2}{r_\varepsilon} + 2 \frac{\delta_i \imath_i \sqrt{\delta_i^2 - \sigma_i^2}}{r_\varepsilon} (p_{\varepsilon,i,+} - p_{\varepsilon,i,-}) + \frac{\delta_i^2 - \sigma_i^2}{r_\varepsilon} \Bigg) \\
&\quad + \sum_{1 \le i < j \le k} \beta_i \beta_j \Bigg(\frac{\delta_i \delta_j}{r_\varepsilon} \big(p_{\varepsilon,i,j,+,+} + p_{\varepsilon,i,j,-,-} - p_{\varepsilon,i,j,+,-} - p_{\varepsilon,i,j,-,+} \big) \\
&\quad + \frac{\delta_i \imath_j \sqrt{\delta_j^2 - \sigma_j^2}}{r_\varepsilon} (p_{\varepsilon,i,+} - p_{\varepsilon,i,-}) + \frac{\delta_i \imath_i \sqrt{\delta_i^2 - \sigma_i^2}}{r_\varepsilon} \big(p_{\varepsilon,j,+} - p_{\varepsilon,j,-} \big) \\
&\quad + \frac{\imath_i \sqrt{\delta_i^2 - \sigma_i^2}\, \imath_j \sqrt{\delta_j^2 - \sigma_j^2}}{r_\varepsilon} \Bigg) + \sum_{\bar{\imath}' \in \mathcal{I}_k} o_2 \Bigg(\sum_{i=1}^k \beta_i \Bigg(\frac{\imath_i' \delta_i}{\sqrt{r_\varepsilon}} + \frac{\imath_i \sqrt{\delta_i^2 - \sigma_i^2}}{\sqrt{r_\varepsilon}} \Bigg) \Bigg) p_{\varepsilon,\bar{\imath}'} \Bigg)^{r_\varepsilon} \\
&= e^{\sum_{i=1}^k \beta_i\mu_i} \Bigg(1 + \sum_{i=1}^k \beta_i \Bigg(\frac{\delta_i}{\sqrt{r_\varepsilon}} \Bigg(- \frac{\imath_i \sqrt{\delta_i^2 - \sigma_i^2}}{\delta_i} \Bigg) + \frac{\imath_i \sqrt{\delta_i^2 - \sigma_i^2}}{\sqrt{r_\varepsilon}} \Bigg) \\
&\quad + \frac{1}{2} \sum_{i=1}^k \beta_i^2 \Bigg(\frac{\delta_i^2}{r_\varepsilon} + 2 \frac{\delta_i \imath_i \sqrt{\delta_i^2 - \sigma_i^2}}{r_\varepsilon} \Bigg(- \frac{\imath_i \sqrt{\delta_i^2 - \sigma_i^2}}{\delta_i} \Bigg) + \frac{\delta_i^2 - \sigma_i^2}{r_\varepsilon} \Bigg) \\
&\quad + \sum_{1 \le i < j \le k} \beta_i \beta_j \Bigg(\frac{\delta_i \delta_j}{r_\varepsilon} \Bigg(\rho_{i,j} \frac{\sigma_i \sigma_j}{\delta_i \delta_j} + \frac{\imath_i \sqrt{\delta_i^2 - \sigma_i^2} \imath_j \sqrt{\delta_j^2 - \sigma_j^2}}{\delta_i \delta_j} \Bigg) \\
&\quad + \frac{\delta_i \imath_j \sqrt{\delta_j^2 - \sigma_j^2}}{r_\varepsilon} \Bigg(- \frac{\imath_i \sqrt{\delta_i^2 - \sigma_i^2}}{\delta_i} \Bigg) + \frac{\delta_j \imath_i \sqrt{\delta_i^2 - \sigma_i^2}}{r_\varepsilon} \Bigg(- \frac{\imath_j \sqrt{\delta_j^2 - \sigma_j^2}}{\delta_j} \Bigg) \\
&\quad + \frac{\imath_i \sqrt{\delta_i^2 - \sigma_i^2}\, \imath_j \sqrt{\delta_j^2 - \sigma_j^2}}{r_\varepsilon} \Bigg) + \sum_{\bar{\imath}' \in \mathcal{I}_k} o_2 \Bigg(\sum_{i=1}^k \beta_i \Bigg(\frac{\imath_i' \delta_i}{\sqrt{r_\varepsilon}} + \frac{\imath_i \sqrt{\delta_i^2 - \sigma_i^2}}{\sqrt{r_\varepsilon}} \Bigg) \Bigg) p_{\varepsilon,\bar{\imath}'} \Bigg)^{r_\varepsilon} \\
&= e^{\sum_{i=1}^k \beta_i\mu_i} \Bigg(1 + \frac{1}{r_\varepsilon} \Bigg(\frac{1}{2} \sum_{i=1}^k \beta_i^2 \sigma_i^2 + \sum_{1 \le i < j \le k} \beta_i \beta_j \sigma_i \sigma_j \rho_{i,j} \Bigg) \\
&\quad + \sum_{\bar{\imath}' \in \mathcal{I}_k} o_2 \Bigg(\sum_{i=1}^k \beta_i \Bigg(\frac{\imath_i' \delta_i}{\sqrt{r_\varepsilon}} + \frac{\jmath_i \sqrt{\delta_i^2 - \sigma_i^2}}{\sqrt{r_\varepsilon}} \Bigg) \Bigg) p_{\varepsilon,\bar{\imath}'} \Bigg)^{r_\varepsilon} \\
&\to \exp\Bigg\{ \sum_{i=1}^k \beta_i \mu_i + \frac{1}{2} \sum_{i=1}^k \beta_i^2 \sigma_i^2 + \sum_{1 \le i < j \le k} \beta_i \beta_j \rho_{i,j} \sigma_i \sigma_j \Bigg\} = \mathsf{E}e^{\sum_{i=1}^k \beta_i W_{0,1,i}} \quad \text{as } \varepsilon \to 0, \qquad (10.67)
\end{aligned}$$

since,

$$\left| \sum_{\bar{\imath}' \in \mathcal{I}_k} o_2 \left(\sum_{i=1}^{k} \beta_i \left(\frac{\imath_i' \delta_i}{\sqrt{r_\varepsilon}} + \frac{\imath_i \sqrt{\delta_i^2 - \sigma_i^2}}{\sqrt{r_\varepsilon}} \right) \right) p_{\varepsilon, \bar{\imath}'} \right| r_\varepsilon$$

$$\leq \sum_{\bar{\imath}' \in \mathcal{I}_k} \left| o_2 \left(\sum_{i=1}^{k} \beta_i \left(\frac{\imath_i' \delta_i}{\sqrt{r_\varepsilon}} + \frac{\imath_i \sqrt{\delta_i^2 - \sigma_i^2}}{\sqrt{r_\varepsilon}} \right) \right) r_\varepsilon \right| \to 0 \quad \text{as } \varepsilon \to 0 \,. \tag{10.68}$$

The following part of the proof repeats the proof of Theorem 10.2.1. □

10.2.2 Convergence of option rewards for multivariate inhomogeneous in time trinomial tree models

Let us consider the model of multivariate inhomogeneous in time Gaussian random walk, which can be considered as a discrete time analog of a multivariate inhomogeneous in time Gaussian process with independent increments. In this case, a log-price $\vec{Y}_{0,n}$ is given by the following stochastic transition dynamic relation:

$$\vec{Y}_{0,n} = \vec{Y}_{0,n-1} + \vec{W}_{0,n} \,, \quad n = 1, 2, \dots \,. \tag{10.69}$$

where (a) $\vec{Y}_{0,0} = (Y_{0,0,1}, \dots, Y_{0,0,k})$ is a random vector taking values in the space $\mathbb{R}_k$ and with distribution $P_0(A)$, (b) $\vec{W}_{0,n} = (W_{0,n,1}, \dots, W_{0,n,k})$, $n = 1, 2, \dots$ is a sequence of independent k-dimensional normal random vectors with mean values $\mathsf{E} W_{0,n,i} = \mu_{n,i}$, $i = 1, \dots, k$, $n = 1, 2, \dots$, variances $\mathrm{Var} W_{0,n,i} = \sigma_{n,i}^2$, $i = 1, \dots, k$, and covariances $\mathsf{E}(W_{0,n,i} - \mu_{n,i})(W_{0,n,i} - \mu_{n,i}) = \rho_{n,i,j}\sigma_{n,i}\sigma_{n,j}$, $i, j = 1, \dots, k$, $n = 1, 2, \dots$, and (c) the random vector $\vec{Y}_{0,0}$ and the sequence of random vectors $\vec{W}_{0,n}$, $n = 1, 2, \dots$ are independent.

Also, usual assumptions are (d) $\mu_{n,i} \in \mathbb{R}_1$, $i = 1, \dots, n = 1, 2, \dots$, (e) $\sigma_{n,i} > 0$, $i = 1, \dots, k$, $n = 1, 2, \dots$, (f) $|\rho_{n,i,j}| \leq 1$, $i, j = 1, \dots, k$, $n = 1, 2, \dots$ and $\rho_{n,i,i} = 1$, $i = 1, \dots, k$, $n = 1, 2, \dots$, (g) $\Sigma_n = \|\sigma_{n,i,j}\|$, $n = 1, 2, \dots$, where $\sigma_{n,i,j} = \rho_{n,i,j}\sigma_{n,i}\sigma_{n,j}$, i, $j = 1, \dots, k$, $n = 1, 2, \dots$, are nonnegatively defined $k \times k$ matrices, and (h) $\det(\Sigma_n) \neq 0$, $n = 1, 2, \dots$.

We approximate the log-price process $\vec{Y}_{0,n}$ by the log-price process $\vec{Y}_{\varepsilon,n}$ represented by a multivariate inhomogeneous in time trinomial random walk. In this case, the log-price process $\vec{Y}_{\varepsilon,n}$ is given, for every $\varepsilon \in (0, \varepsilon_0]$, by the following stochastic transition dynamic relation:

$$\vec{Y}_{\varepsilon,n} = \vec{Y}_{\varepsilon,n-1} + \vec{W}_{\varepsilon,n} \,, \quad n = 1, 2, \dots \,. \tag{10.70}$$

where (a) $\vec{Y}_{\varepsilon,0}$ is a random vector taking values in the space $\mathbb{R}_k$ and with the distribution $P_\varepsilon(A)$, (b) $\vec{W}_{\varepsilon,n} = (W_{\varepsilon,n,1}, \dots, W_{\varepsilon,n,k}) = \vec{W}_{\varepsilon,n,1} + \cdots + \vec{W}_{\varepsilon,n,r_\varepsilon}$, where $\vec{W}_{\varepsilon,n,l} = (W_{\varepsilon,n,l,1}, \dots, W_{\varepsilon,n,l,k})$, $l, n = 1, 2, \dots$ is a family of independent random vectors taking values $\bar{\delta}_{\varepsilon,n,\bar{\jmath}} = (\delta_{\varepsilon,n,1,\jmath_1}, \dots, \delta_{\varepsilon,n,k,\jmath_k})$ with the corresponding probabilities

$p_{\varepsilon,n,\bar{\jmath}} = p_{\varepsilon,n}(\jmath_1,\dots\jmath_k)$, for $\bar{\jmath} \in \mathcal{J}_k = \{\bar{\jmath} = \langle \jmath_1,\dots\jmath_k\rangle : \imath_j,\dots\jmath_k = +,\circ,-\}$, and (c) the random vector $\vec{Y}_{\varepsilon,0}$ and the family of random vectors $\vec{W}_{\varepsilon,n,l}$, $l, n = 1,2,\dots$ are independent.

Also, usual assumptions are (d) $-\infty < \delta_{\varepsilon,n,i,-} < \delta_{\varepsilon,n,i,\circ} < \delta_{\varepsilon,n,i,+} < \infty$, $i = 1,\dots,k$, $n = 1,2,\dots$, (e) $p_{\varepsilon,n,\bar{\jmath}} \geq 0$, $\sum_{\bar{\jmath},\in \mathcal{J}_k} p_{\varepsilon,n,\bar{\jmath}} = 1$, $n = 1,2,\dots$, and (f) r_ε is a positive integer number.

We assume that the pay-off function $g(n, e^{\vec{y}})$ does not depend on the parameter ε, and also assume that $g(n, e^{\vec{y}})$ are real-valued Borel functions of the argument $\vec{y} \in \mathbb{R}_k$ for every $n = 0,1,\dots,N$.

In this case, the corresponding reward functions $\phi_{\varepsilon,n}(\vec{y})$ for the log-price process $\vec{Y}_{\varepsilon,n}$ are also Borel functions of argument $\vec{y} \in \mathbb{R}_k$, for every $n = 0,1,\dots,N$.

We fit parameters of approximating log-price processes $\vec{Y}_{\varepsilon,n}$ in such a way that the following exact moment-fitting relations hold, for every $\varepsilon \in (0, \varepsilon_0]$:

$$\begin{cases} \mathsf{E}W_{\varepsilon,n,i} = \mathsf{E}W_{0,n,i}, \quad i = 1,\dots,k, \quad n = 1,\dots,N, \\ \mathrm{Var}W_{\varepsilon,n,i} = \mathrm{Var}W_{0,1,i}, \quad i = 1,\dots,k, \quad n = 1,\dots,N, \\ \mathsf{E}(W_{\varepsilon,n,i} - \mathsf{E}W_{\varepsilon,n,i})(W_{\varepsilon,n,i} - \mathsf{E}W_{\varepsilon,n,j}) \\ \qquad = \mathsf{E}(W_{0,n,i} - \mathsf{E}W_{0,n,i})(W_{0,n,i} - \mathsf{E}W_{0,n,j}), \\ \qquad 1 \leq i < j \leq k, \quad n = 1,\dots,N. \end{cases} \tag{10.71}$$

The solution of this fitting problem is presented in Lemma 9.3.1. According to this lemma, the fitting relations (10.71) hold, if to assume that the condition $\mathbf{R}_2$ holds and to choose parameters of the log-price process $\vec{Y}_{\varepsilon,n}$ given by the following formulas, for every $\varepsilon \in (0, \varepsilon_0]$ and $n = 1,\dots,N$:

$$\begin{cases} \delta_{\varepsilon,n,i,\pm} = \pm\delta_{\varepsilon,i} + \lambda_{\varepsilon,n,i}, \quad \delta_{\varepsilon,n,i,\circ} = \lambda_{\varepsilon,n,i}, \quad \delta_{\varepsilon,i} = \dfrac{\delta_i}{\sqrt{r_\varepsilon}}, \quad \lambda_{\varepsilon,n,i} = \dfrac{\mu_{n,i}}{r_\varepsilon}, \\ p_{\varepsilon,n,i,\pm} = \dfrac{\sigma_{n,i}^2}{2\delta_i^2}, \quad p_{\varepsilon,n,i,\circ} = 1 - \dfrac{\sigma_{n,i}^2}{\delta_i^2} \\ \qquad i = 1,\dots,k, \quad n = 1,\dots,N, \\ p_{\varepsilon,n,i,j,\pm,\pm} = \dfrac{1}{4}\left(\rho_{n,i,j}\dfrac{\sigma_{n,i}}{\delta_i}\dfrac{\sigma_{n,j}}{\delta_j} + \dfrac{\sigma_{n,i}^2}{\delta_i^2} \wedge \dfrac{\sigma_{n,j}^2}{\delta_j^2}\right), \\ p_{\varepsilon,n,i,j,\pm,\mp} = \dfrac{1}{4}\left(-\rho_{n,i,j}\dfrac{\sigma_{n,i}}{\delta_i}\dfrac{\sigma_{n,i}}{\delta_j} + \dfrac{\sigma_{n,i}^2}{\delta_i^2} \wedge \dfrac{\sigma_{n,j}^2}{\delta_j^2}\right), \\ p_{\varepsilon,n,i,j,\circ,\pm} = \dfrac{1}{2}\left(1 - \dfrac{\sigma_{n,i}^2}{\delta_i^2} - (1 - \dfrac{\sigma_{n,i}^2}{\delta_i^2}) \wedge (1 - \dfrac{\sigma_{n,j}^2}{\delta_j^2})\right), \\ p_{\varepsilon,n,i,j,\pm,\circ} = \dfrac{1}{2}\left(1 - \dfrac{\sigma_{n,j}^2}{\delta_j^2} - \left(1 - \dfrac{\sigma_{n,i}^2}{\delta_i^2}\right) \wedge \left(1 - \dfrac{\sigma_{n,j}^2}{\delta_j^2}\right)\right), \\ p_{\varepsilon,n,i,j,\circ,\circ} = \left(1 - \dfrac{\sigma_{n,i}^2}{\delta_i^2}\right) \wedge \left(1 - \dfrac{\sigma_{n,j}^2}{\delta_j^2}\right), \\ \qquad 1 \leq i < j \leq k, \quad n = 1,\dots,N, \end{cases} \tag{10.72}$$

where (i) one can choose for the parameters δ_i, $i = 1, \dots, k$, r_ε any values such that (a) $\delta_i \geq \sigma_{i,+}$, $i = 1, \dots, k$, (b) $\rho \leq \kappa \leq 1$, where $\kappa = \min_{1 \leq i,j \leq k} \frac{\delta_i}{\delta_j} \leq 1$, while $\sigma_{i,+} = \max_{1 \leq n \leq N} \sigma_{n,i} > 0$, $i = 1, \dots, k$ (according to the initial model assumptions), and $0 \leq \rho < 1$ is the parameter penetrating the condition $\mathbf{R_2}$, (c) $r_\varepsilon = 1, 2, \dots$ (ii) the corresponding higher order distributions for components of random vectors $\vec{W}_{\varepsilon,n,1}$ consistent with the above one- and two-dimensional distributions can be defined using algorithm described in Lemma 9.3.2.

Recall sets $\mathcal{L}_{k,r} = \{\bar{l} = (l_1, \dots, l_k) : -r \leq l_1, \dots, l_k \leq r\}$, for $r = 1, 2, \dots$.

The random vectors $\vec{W}_{\varepsilon,n}$, $n = 1, \dots, N$ are independent trinomial random vectors. Namely, the random vector $\vec{W}_{\varepsilon,n}$ takes values $\vec{w}_{\varepsilon,n,r_\varepsilon,\bar{l}}$ with the corresponding probabilities $p_{\varepsilon,n,r_\varepsilon}(\bar{l})$, for $\bar{l} = (l_1, \dots, l_k) \in \mathcal{L}_{k,r_\varepsilon}$.

Here, for every $\bar{l} = (l_1, \dots, l_k) \in \mathcal{L}_{k,r}$, $r = 1, 2, \dots$,

$$\vec{w}_{\varepsilon,n,r,\bar{l}} = (l_1\delta_{\varepsilon,1} + r\lambda_{\varepsilon,n,1}, \dots, l_k\delta_{\varepsilon,k} + r\lambda_{\varepsilon,n,k}), \tag{10.73}$$

and

$$p_{\varepsilon,n,r}(\bar{l}) = \mathsf{P}\{\vec{W}_{\varepsilon,n} = \vec{W}_{\varepsilon,n,1} + \cdots + \vec{W}_{\varepsilon,n,r} = \vec{w}_{\varepsilon,n,r,\bar{l}}\} I(\bar{l} \in \mathcal{L}_{k,r}). \tag{10.74}$$

The probabilities $p_{\varepsilon,n,r}(\bar{l})$ can be found using the following recurrent convolution formulas:

$$p_{\varepsilon,n,r}(\bar{l}) = \begin{cases} p_{\varepsilon,n,\bar{\jmath}} I(\bar{l} = \bar{e}_{\bar{\jmath}}) & \text{if } r = 1, \bar{l} \in \mathcal{L}_{k,1}, \\ \sum_{\bar{l}' \in \mathcal{L}_{k,1}} p_{\varepsilon,n,r-1}(\bar{l} - \bar{l}') p_{\varepsilon,n,1}(\bar{l}') & \text{if } r > 1, \bar{l} \in \mathcal{L}_{k,r}, \end{cases} \tag{10.75}$$

where the vectors $\bar{e}_{\bar{\jmath}} = (e_{\jmath_1}, \dots, e_{\jmath_k})$, $\bar{\jmath} = \langle \jmath_1, \dots, \jmath_k \rangle \in \mathcal{J}_k$, $e_\jmath = 1, 0, -1$ if, respectively, $\jmath = +, \circ, -$.

The additional comments concerning algorithms for computing of probabilities $p_{\varepsilon,n,r}(\bar{l})$ can be found in Subsection 3.2.3.

Let us assume that $\vec{Y}_{\varepsilon,n} = \vec{y} = (y_1, \dots, y_k) = \text{const} \in \mathbb{R}_k$. In this case, trajectories of the log-price process $\vec{Y}_{\varepsilon,n+r}$, $r = 0, 1, \dots$ generate, for every $\varepsilon \in (0, \varepsilon_0]$, the unique tree of nodes,

$$T_{\varepsilon,(y,n)} = \{(\vec{y}_{\varepsilon,n,n+r,\bar{l}}, n + r), \quad \bar{l} = (l_1, \dots, l_k) \in \mathcal{L}_{k,r_\varepsilon r}, \quad r = 0, 1, \dots\}, \tag{10.76}$$

where

$$\begin{aligned} \vec{y}_{\varepsilon,n,n+r,\bar{l}} &= (y_{\varepsilon,n,n+r,1,l_1}, \dots, y_{\varepsilon,n,n+r,k,l_k}) \\ &= (y_1 + l_1\delta_{\varepsilon,1} + r_\varepsilon\lambda_{\varepsilon,n,n+r,1}, \dots, y_k + l_k\delta_{\varepsilon,k} + r_\varepsilon\lambda_{\varepsilon,n,n+r,k}), \\ &= (y_1 + l_1\delta_{\varepsilon,1} + \mu_{n,n+r,1}, \dots, y_k + l_k\delta_{\varepsilon,k} + \mu_{n,n+r,k}), \\ \bar{l} &= (l_1, \dots, l_k) \in \mathcal{L}_{k,r_\varepsilon r}, \quad r = 0, 1, \dots. \end{aligned} \tag{10.77}$$

and, for $i = 1, \dots, k$, $r = 0, 1, \dots$,

$$\lambda_{\varepsilon,n,n+r,i} = \sum_{m=n+1}^{n+r} \lambda_{\varepsilon,m,i}, \quad \mu_{n,n+r,i} = \sum_{m=n+1}^{n+r} \mu_{m,i}. \tag{10.78}$$

The tree $T_{\varepsilon,(\vec{y},n)}$ contains the total number of nodes after r steps,

$$L_{\varepsilon,n,n+r} \leq 1 + (2r_\varepsilon + 1)^k + \cdots + (2r_\varepsilon r + 1)^k$$
$$\leq \int_0^{r+1} (2r_\varepsilon x + 1)^k dx = \frac{(2r_\varepsilon(r+1)+1)^{k+1} - 1}{2r_\varepsilon(k+1)} . \tag{10.79}$$

The expression on the right-hand side in (10.79) is a polynomial function of r of the order $k+1$.

The following lemma is a direct corollary of Lemma 3.3.4.

Lemma 10.2.2. *Let, for every $\varepsilon \in (0, \varepsilon_0]$, the approximating log-price process $\vec{Y}_{\varepsilon,n}$ is the multivariate trinomial random walk given by the stochastic transition dynamic relation* (10.70), *with parameters defined in relation* (10.72), *under the assumption that the condition* $\mathbf{R}_2$ *holds. Then the reward functions $\phi_{\varepsilon,n+r}(\vec{y}_{\varepsilon,n,n+r,\bar{l}})$, for points $\vec{y}_{\varepsilon,n,n,0} = \vec{y}$ and $\vec{y}_{\varepsilon,n,n+r,\bar{l}}$, $\bar{l} \in \mathcal{L}_{k,r_\varepsilon r}$, $r = 1,\dots N-n$, are, for every $\varepsilon \in (0,\varepsilon_0]$, and $\vec{y} \in \mathbb{R}_k$, $n = 0,\dots,N$, the unique solution for the following recurrence finite system of linear equations:*

$$\begin{cases} \phi_{\varepsilon,N}(\vec{y}_{\varepsilon,n,N,l}) = g_\varepsilon(N, e^{\vec{y}_{\varepsilon,n,N,l}}) , \quad \bar{l} \in \mathcal{L}_{k,r_\varepsilon(N-n)} , \\ \phi_{\varepsilon,n+r}(\vec{y}_{\varepsilon,n,n+r,l}) = \max\Bigg(g_\varepsilon(n+r, e^{\vec{y}_{\varepsilon,n,n+r,l}}) , \\ \qquad \sum\limits_{\bar{l}'\in\mathcal{L}_{k,r_\varepsilon}} \phi_{\varepsilon,n+r+1}(\vec{y}_{\varepsilon,n,n+r+1,\bar{l}+\bar{l}'}) p_{\varepsilon,n+r+1,r_\varepsilon}(\bar{l}') \Bigg) , \\ \qquad \bar{l} \in \mathcal{L}_{k,r_\varepsilon r} , \quad r = N-n-1,\dots,1 , \\ \phi_{\varepsilon,n}(\vec{y}) = \max\Bigg(g_\varepsilon(n, e^{\vec{y}}), \sum\limits_{\bar{l}'\in\mathcal{L}_{k,r_\varepsilon}} \phi_{\varepsilon,n}(\vec{y}_{\varepsilon,n,n+1,\bar{l}'}) p_{\varepsilon,n+1,r_\varepsilon}(\bar{l}') \Bigg) . \end{cases} \tag{10.80}$$

We are going to apply Theorem 5.3.5 in order to prove convergence of the reward functions for log-price processes $\vec{Y}_{\varepsilon,n}$ to the corresponding reward functions for log-price processes $\vec{Y}_{0,n}$ as $\varepsilon \to 0$.

We assume that the condition $\mathbf{B}_6[\bar{\gamma}]$ holds for the pay-off function $g(n, e^{\vec{y}})$, for some vector parameter $\bar{\gamma} = (\gamma_1,\dots,\gamma_k)$ with nonnegative components.

The condition $\mathbf{I}_3$ reduces in this case to the condition $\mathbf{I}_{10}$.

We shall see that, in this case, the condition $\mathbf{J}_4$ (a) holds for log-prices $\vec{Y}_{\varepsilon,n}$ with parameters given by relation (10.72), if $r_\varepsilon \to \infty$ as $\varepsilon \to 0$. Also, the random vectors $\vec{W}_{0,n}$, $n = 1,\dots,N$ have normal distributions that make it possible to replace the condition $\mathbf{J}_4$ (b) by the condition $\mathbf{J}_{14}$.

The above backward recurrence algorithm for finding reward functions is supported by the following convergence result.

Theorem 10.2.3. *Let the log-price process $\vec{Y}_{0,n}$ be represented by the multivariate Gaussian random walk given by the stochastic transition dynamic relation* (10.69) *and the approximating log-price process is represented, for every $\varepsilon \in (0, \varepsilon_0]$, by the trinomial random walks $Y_{\varepsilon,n}$ given by the stochastic transition dynamic relation* (10.70), *with parameters satisfying the condition* $\mathbf{R}_2$ *and given by relation* (10.72). *Let also the conditions $\mathbf{B}_6[\bar{\gamma}]$ hold for some vector parameter $\bar{\gamma} = (\gamma_1, \dots, \gamma_k)$ with nonnegative components, conditions $\mathbf{I}_{10}$ and $\mathbf{J}_{14}$ hold, and $r_\varepsilon \to \infty$ as $\varepsilon \to 0$. Then the following relation takes place for any $\vec{y}_\varepsilon \to \vec{y}_0 \in \mathbb{Y}_n$, $n = 0, 1, \dots, N$:*

$$\phi_{\varepsilon,n}(\vec{y}_\varepsilon) \to \phi_{0,n}(\vec{y}_0) \quad \text{as } \varepsilon \to 0 . \tag{10.81}$$

Proof. Recall that random vectors $\vec{W}_{\varepsilon,n} = \vec{W}_{\varepsilon,n,1} + \cdots + \vec{W}_{\varepsilon,n,r_\varepsilon}$, $n = 1, 2, \dots$, where $\vec{W}_{\varepsilon,n,l}$, $n, l = 1, 2, \dots$ are independent random vectors.

In the case, where parameters of the log-price process $Y_{\varepsilon,n}$ are given by relations (10.72), the moment-generating functions for the random vectors $\vec{W}_{\varepsilon,n} = (W_{\varepsilon,n,1}, \dots, W_{\varepsilon,n,k})$, $n = 1, \dots, N$ have, for every $\varepsilon \in (0, \varepsilon_0]$, the following form, for $\bar{\beta} = (\beta_1, \dots, \beta_k) \in \mathbb{R}_k$:

$$\begin{aligned} \mathsf{E} e^{\sum_{i=1}^k \beta_i W_{\varepsilon,n,i}} &= \Bigg(\sum_{\bar{j} \in \mathcal{J}_k} e^{\sum_{i=1}^k \beta_i (\frac{\delta_i e_{\jmath_i}}{\sqrt{r_\varepsilon}} + \frac{\mu_{n,i}}{r_\varepsilon})} p_{\varepsilon,n,\bar{j}} \Bigg)^{r_\varepsilon} \\ &= e^{\sum_{i=1}^k \beta_i \mu_{n,i}} \Bigg(\sum_{\bar{j} \in \mathcal{J}_k} e^{\sum_{i=1}^k \beta_i \frac{\delta_i e_{\jmath_i}}{\sqrt{r_\varepsilon}}} p_{\varepsilon,n,\bar{j}} \Bigg)^{r_\varepsilon} . \end{aligned} \tag{10.82}$$

For $\varepsilon = 0$, the moment-generating functions for the random vectors $\vec{W}_{0,n} = (W_{0,n,1}, \dots, W_{0,n,k})$, $n = 1, \dots, N$ take the following form, for $\bar{\beta} = (\beta_1, \dots, \beta_k) \in \mathbb{R}_k$:

$$\begin{aligned} \mathsf{E} e^{\sum_{i=1}^k \beta_i W_{0,n,i}} = \exp \Bigg\{ & \sum_{i=1}^k \beta_i \mu_{n,i} + \frac{1}{2} \sum_{i=1}^k \beta_i \sigma_{n,i}^2 \\ & + \sum_{1 \le i < j \le k} \beta_i \beta_j \rho_{n,i,j} \sigma_{n,i} \sigma_{n,j} \Bigg\} . \end{aligned} \tag{10.83}$$

Relations (10.48), formula (10.59), and Taylor expansion for exponential function imply that the following asymptotic relation takes place, for every $n = 1, \dots, N$ and $\bar{\beta} = (\beta_1, \dots, \beta_k) \in \mathbb{R}_k$:

$$\begin{aligned} \mathsf{E} e^{\sum_{i=1}^k \beta_i W_{\varepsilon,n,i}} = e^{\sum_{i=1}^k \beta_i \mu_{n,i}} \Bigg(\sum_{\bar{j} \in \mathcal{J}_k} \Bigg(1 + \sum_{i=1}^k \beta_i \frac{\delta_i e_{\jmath_i}}{\sqrt{r_\varepsilon}} \\ + \frac{1}{2} \Bigg(\sum_{i=1}^k \beta_i \frac{\delta_i e_{\jmath_i}}{\sqrt{r_\varepsilon}} \Bigg)^2 + o_2 \Bigg(\sum_{i=1}^k \beta_i \frac{\delta_i e_{\jmath_i}}{\sqrt{r_\varepsilon}} \Bigg) \Bigg) p_{\varepsilon,n,\bar{j}} \Bigg)^{r_\varepsilon} \end{aligned}$$

$$
\begin{aligned}
&= e^{\sum_{i=1}^k \beta_i \mu_{n,i}} \Bigg(1 + \sum_{i=1}^k \beta_i \frac{\delta_i}{\sqrt{r_\varepsilon}}(p_{\varepsilon,n,i,+} - p_{\varepsilon,n,i,-}) + \frac{1}{2}\sum_{i=1}^k \beta_i^2 \frac{\delta_i^2}{r_\varepsilon}(p_{\varepsilon,n,i,+} + p_{\varepsilon,n,i,-}) \\
&\quad + \sum_{1 \le i < j \le k} \beta_i \beta_j \frac{\delta_i \delta_j}{r_\varepsilon}(p_{\varepsilon,n,i,j,+,+} + p_{\varepsilon,n,i,j,-,-} \\
&\quad - p_{\varepsilon,n,i,j,+,-} - p_{\varepsilon,n,i,j,-,+}) + \sum_{\bar{\jmath} \in \mathcal{J}_k} o_2\left(\sum_{i=1}^k \beta_i \frac{\delta_i e_{\jmath_i}}{\sqrt{r_\varepsilon}}\right) p_{\varepsilon,n,\bar{\jmath}}\Bigg)^{r_\varepsilon} \\
&= e^{\sum_{i=1}^k \beta_i \mu_{n,i}} \Bigg(1 + \frac{1}{r_\varepsilon}\left(\frac{1}{2}\sum_{i=1}^k \beta_i^2 \sigma_{n,i}^2 + \sum_{1 \le i < j \le k} \beta_i \beta_j \sigma_{n,i}\sigma_{n,j}\rho_{n,i,j}\right) \\
&\quad + \sum_{\bar{\jmath} \in \mathcal{J}_k} o_2\left(\sum_{i=1}^k \beta_i \frac{\delta_i e_{\jmath_i}}{\sqrt{r_\varepsilon}}\right) p_{\varepsilon,n,\bar{\jmath}}\Bigg)^{r_\varepsilon} \\
&\to \exp\left\{\sum_{i=1}^k \beta_i \mu_{n,i} + \frac{1}{2}\sum_{i=1}^k \beta_i^2 \sigma_{n,i}^2 + \sum_{1 \le i < j \le k} \beta_i \beta_j \rho_{n,i,j}\sigma_{n,i}\sigma_{n,j}\right\} \\
&= \mathsf{E} e^{\sum_{i=1}^k \beta_i W_{0,n,i}} \quad \text{as } \varepsilon \to 0\,, \qquad (10.84)
\end{aligned}
$$

since,

$$
\left|\sum_{\bar{\jmath} \in \mathcal{J}_k} o_2\left(\sum_{i=1}^k \beta_i \frac{\delta_i e_{\jmath_i}}{\sqrt{r_\varepsilon}}\right) p_{\varepsilon,n,\bar{\jmath}}\right| r_\varepsilon \le \sum_{\bar{\jmath} \in \mathcal{J}_k} \left| o_2\left(\sum_{i=1}^k \beta_i \frac{\delta_i e_{\jmath_i}}{\sqrt{r_\varepsilon}}\right) r_\varepsilon \right| \to 0 \quad \text{as } \varepsilon \to 0\,. \qquad (10.85)
$$

Relation (10.84) implies that for any $i = 1, \dots, k$ and $\beta \ge 0$,

$$
\begin{aligned}
\Xi'^{\pm}_{\beta}(Y_{\varepsilon,i,\cdot}, N) &= \max_{1 \le n \le N} \mathsf{E} e^{\pm\beta W_{\varepsilon,n,i}} \\
\to \Xi'^{\pm}_{\beta}(Y_{0,i,\cdot}, N) &= \max_{1 \le n \le N} e^{\pm\beta\mu_{n,i} + \frac{1}{2}\beta^2\sigma_{n,i}^2} < \infty\,.
\end{aligned} \qquad (10.86)
$$

Relation (10.86) implies in an obvious way that the condition $\mathbf{E}_{11}[\bar{\beta}]$ holds and, therefore, by Lemma 5.3.1, the condition $\mathbf{C}_7[\bar{\beta}]$ holds, for any parameter $\bar{\beta} = (\beta_1, \dots, \beta_k)$ with nonnegative components.

The condition $\mathbf{B}_6[\bar{\gamma}]$ is used in Theorem 5.3.1.

Note that one can always take the parameter $\bar{\beta}$ in such a way that, for every $i = 1, \dots, k$, either $\beta_i > \gamma_i$ if $\gamma_i > 0$ or $\beta_i = \gamma_i$ if $\gamma_i = 0$.

As was pointed above, the condition $\mathbf{I}_3$ reduces in this case to the form of the condition $\mathbf{I}_{10}$.

Relation (10.84) also implies by the corresponding variant of the continuous theorem for moment-generating functions, the following relation, for every $n = 1, \dots, N$:

$$
\vec{W}_{\varepsilon,n} \xrightarrow{d} \vec{W}_{0,n} \quad \text{as } \varepsilon \to 0\,. \qquad (10.87)
$$

Relation (10.87) implies that the condition $\mathbf{J}_4$ (a) holds.

The condition $\mathbf{J}_{14}$ implies the condition $\mathbf{J}_4$ (b) to hold, since the random vectors $\vec{W}_{0,n}$, $n = 1, \ldots, N$ have, according initial model assumptions, k-dimensional normal distributions with positive everywhere in $\mathbb{R}_k$ probability density functions.

Thus, all conditions of Theorem 5.3.1 hold. By applying this theorem to the above approximation model, we get the asymptotic relation (10.81). □

10.2.3 Convergence of optimal expected rewards for a multivariate binomial and trinomial models

Let us first study the corresponding results for the multivariate binomial tree model considered in Subsection 10.2.1.

In this case, the optimal expected reward $\Phi_0 = \Phi_0(\mathcal{M}^{(0)}_{max,N})$ for the log-price process $\vec{Y}_{0,n}$ is defined by the following formula:

$$\Phi_0 = \mathsf{E}\phi_{0,0}(\vec{Y}_{0,0}) = \int_{\mathbb{R}_k} \phi_{0,0}(\vec{y})P_0(d\vec{y}) \,. \tag{10.88}$$

It is natural to choose, for every $\varepsilon \in (0, \varepsilon_0]$, the initial distribution $P_\varepsilon(A)$ for the approximating log-price process $\vec{Y}_{\varepsilon,n}$ as the corresponding skeleton approximation for the distribution $P_0(A)$,

$$P_\varepsilon(A) = \sum_{\vec{y}_{\varepsilon,0,0,\bar{l}} \in A} P_0(\bar{\mathbb{I}}_{\varepsilon,\bar{l}}) \,, \quad A \in \mathcal{B}_k \,, \tag{10.89}$$

where

$$\begin{aligned} \vec{y}_{\varepsilon,0,0,\bar{l}} &= \left(l_1 \frac{\delta_1}{\sqrt{r_{\varepsilon,0}}} + \lambda_{\varepsilon,0,1}, \ldots, l_k \frac{\delta_k}{\sqrt{r_{\varepsilon,0}}} + \lambda_{\varepsilon,0,k} \right) , \\ \bar{l} &= (l_1, \ldots, l_k) \in \mathcal{L}_{k,r_{\varepsilon,0}} \,, \end{aligned} \tag{10.90}$$

and

$$\bar{\mathbb{I}}_{\varepsilon,\bar{l}} = \mathbb{I}_{\varepsilon,1,l_1} \times \cdots \times \mathbb{I}_{\varepsilon,k,l_k} \,, \quad \bar{l} = (l_1, \ldots, l_k) \in \mathcal{L}_{k,r_{\varepsilon,0}} \,, \tag{10.91}$$

where, for $i = 1, \ldots, k$,

$$\mathbb{I}_{\varepsilon,i,l} = \begin{cases} \left(-\infty, \frac{\delta_i}{\sqrt{r_{\varepsilon,0}}}\left(-r_{\varepsilon,0} + \frac{1}{2}\right)\right] + \lambda_{\varepsilon,0,i} & \text{if } l = -r_{\varepsilon,0} \,, \\ \left(\frac{\delta_i}{\sqrt{r_{\varepsilon,0}}}(l - \frac{1}{2}), \frac{\delta_i}{\sqrt{r_{\varepsilon,0}}}(l + \frac{1}{2})\right] + \lambda_{\varepsilon,0,i} & \text{if } -r_{\varepsilon,0} < l < r_{\varepsilon,0} \,, \\ \frac{\delta_i}{\sqrt{r_{\varepsilon,0}}}(r_{\varepsilon,0} - \frac{1}{2}), \infty) + \lambda_{\varepsilon,0,i} & \text{if } l = r_{\varepsilon,0} \,. \end{cases} \tag{10.92}$$

where real numbers $\lambda_{\varepsilon,0,i}$, $i = 1, \ldots, k$ and positive integer numbers $r_{\varepsilon,0}$ are chosen in such a way that the following condition holds:

$\mathbf{N}_6$: (a) $r_{\varepsilon,0} \to \infty$ as $\varepsilon \to 0$,

(b) $\pm\delta_i\sqrt{r_{\varepsilon,0}} + \lambda_{\varepsilon,0,i} \to \pm\infty$ as $\varepsilon \to 0$, $i = 1, \ldots, k$.

The optimal expected reward $\Phi_\varepsilon = \Phi_\varepsilon(\mathcal{M}^{(\varepsilon)}_{max,N})$ for the log-price process $\vec{Y}_{\varepsilon,n}$ is concerned, it is defined, for every $\varepsilon \in (0, \varepsilon_0]$, by the following formula:

$$\Phi_\varepsilon = \sum_{\bar{l} \in \mathcal{L}_{k,r_{\varepsilon,0}}} \phi_{\varepsilon,0}(\vec{y}_{\varepsilon,0,0,\bar{l}}) P_0(\bar{\mathbb{I}}_{\varepsilon,\bar{l}}) . \tag{10.93}$$

We are going to apply Theorem 5.3.2, in order to prove convergence of the optimal expected rewards for log-price processes $\vec{Y}_{\varepsilon,n}$ to the corresponding reward functions for log-price processes $\vec{Y}_{0,n}$ as $\varepsilon \to 0$.

Let us assume that the following condition holds for some vector parameter $\bar{\beta} = (\beta_1, \dots, \beta_k)$ with nonnegative components:

D$_{21}[\bar{\beta}]$: $\mathsf{E}e^{\beta_i |Y_{0,0,i}|} < K_{92}$, for some $1 < K_{92} < \infty$,

and also that the following condition holds:

K$_{19}$: $P_0(\overline{\mathbb{Y}}_0) = 0$, where $\mathbb{Y}_0$ is the set defined in condition **I**$_{10}$.

Theorem 10.2.4. *Let the log-price process $\vec{Y}_{0,n}$ be represented by the multivariate Gaussian random walk given by the stochastic transition dynamic relation* (10.45) *and the approximating log-price process is represented, for every $\varepsilon \in (0, \varepsilon_0]$, by the binomial random walks $Y_{\varepsilon,n}$ given by the stochastic transition dynamic relation* (10.46), *with parameters given by relation* (10.48) *or* (10.49) – (10.50). *Let also conditions* **B**$_6[\bar{\gamma}]$ *and* **D**$_{21}[\bar{\beta}]$ *hold for the parameters $\bar{\gamma}$ and $\bar{\beta}$ such that, for every $i = 1, \dots, k$, either $\beta_i > \gamma_i > 0$ or $\beta_i = \gamma_i = 0$, and also the conditions* **N**$_6$, **I**$_{10}$, **J**$_{14}$ *and* **K**$_{19}$ *hold, and $r_\varepsilon \to \infty$ as $\varepsilon \to 0$. Then the following relation takes place:*

$$\Phi_\varepsilon \to \Phi_0 \quad \text{as } \varepsilon \to 0 . \tag{10.94}$$

Proof. We are going to use Theorem 5.3.2.

Let us define vector skeleton functions,

$$\bar{h}_\varepsilon(\vec{y}) = (h_{\varepsilon,1}(y_1), \dots, h_{\varepsilon,k}(y_k)) , \quad \vec{y} = (y_1, \dots, y_k) \in \mathbb{R}_k , \tag{10.95}$$

where

$$h_{\varepsilon,i}(y) = \sum_{l=-r_{\varepsilon,0}}^{r_{\varepsilon,0}} \left(l\frac{\delta_i}{\sqrt{r_{\varepsilon,0}}} + \lambda_{\varepsilon,0,i} \right) I(y \in \mathbb{I}_{\varepsilon,i,l}) , \quad y \in \mathbb{R}_1 , \quad i = 1, \dots, k . \tag{10.96}$$

Obviously, for every $\varepsilon \in (0, \varepsilon_0]$,

$$P_\varepsilon(A) = \mathsf{P}\{\bar{h}_\varepsilon(\vec{Y}_{0,n}) \in A\} , \quad A \in \mathcal{B}_k . \tag{10.97}$$

The skeleton functions $h_{\varepsilon,i}(y)$ satisfy the following inequalities for $y \in \mathbb{R}_1$, $i = 1, \dots, k$:

$$\begin{aligned} |h_{\varepsilon,i}(y)| &\leq |y| I(y \notin [-\delta_i\sqrt{r_{\varepsilon,0}} + \lambda_{\varepsilon,0,i}, \delta_i\sqrt{r_{\varepsilon,0}} + \lambda_{\varepsilon,0,i}]) \\ &\quad + \left(|y| + \frac{\delta_i}{\sqrt{r_{\varepsilon,0}}}\right) I\left(y \in \left[-\delta_i\sqrt{r_{\varepsilon,0}} + \lambda_{\varepsilon,0_i}, \delta_i\sqrt{r_{\varepsilon,0}} + \lambda_{\varepsilon,0,i}\right]\right) \\ &\leq |y| + \frac{\delta_i}{\sqrt{r_{\varepsilon,0}}} . \end{aligned} \tag{10.98}$$

This inequality and the condition $\mathbf{D}_{21}[\bar{\beta}]$ imply the following relation, for every $i = 1, \dots, k$:

$$\mathsf{E}e^{\beta_i|Y_{\varepsilon,0,i}|} = \mathsf{E}e^{\beta_i|h_{\varepsilon,i}(Y_{0,0,i})|} \leq e^{\frac{\beta_i\delta_i}{\sqrt{r_{\varepsilon,0}}}} \mathsf{E}e^{\beta_i|Y_{0,0,i}|} \to \mathsf{E}e^{\beta|Y_{0,0,i}|} < \infty \quad \text{as } \varepsilon \to 0 . \tag{10.99}$$

Thus, the condition $\mathbf{D}_7[\bar{\beta}]$ holds.

The skeleton functions $h_{\varepsilon,i}(y)$ also satisfy the following inequalities for $y \in \mathbb{R}_1$, $i = 1, \dots, k$:

$$\begin{aligned} |h_{\varepsilon,i}(y) - y| &\leq (-\delta_i\sqrt{r_{\varepsilon,0}} + \lambda_{\varepsilon,0,i} - y) I(y \leq -\delta_i\sqrt{r_{\varepsilon,0}} + \lambda_{\varepsilon,0,i}) \\ &\quad + \frac{\delta_i}{\sqrt{r_{\varepsilon,0}}} I(\delta_i\sqrt{r_{\varepsilon,0}} + \lambda_{\varepsilon,0,i} < y \leq \delta_i\sqrt{r_{\varepsilon,0}} + \lambda_{\varepsilon,0,i}) \\ &\quad + (y - \delta_i\sqrt{r_{\varepsilon,0}} - \lambda_{\varepsilon,0,i}) I(y > \delta_i\sqrt{r_{\varepsilon,0}} + \lambda_{\varepsilon,0,i}) . \end{aligned} \tag{10.100}$$

This inequalities and the condition $\mathbf{N}_6$ obviously imply that $h_{\varepsilon,i}(y) \to y$ as $\varepsilon \to 0$, for every $y \in \mathbb{R}_1$, $i = 1, \dots, k$, and, thus, $\bar{h}_\varepsilon(\vec{Y}_{0,0}) \xrightarrow{a.s.} \vec{Y}_{0,0}$ as $\varepsilon \to 0$. Therefore,

$$P_\varepsilon(\cdot) \Rightarrow P_0(\cdot) \quad \text{as } \varepsilon \to 0 . \tag{10.101}$$

Relation (10.101) implies that the condition $\mathbf{K}_4$ (a) holds. The condition $\mathbf{K}_{19}$ is a variant of the condition $\mathbf{K}_4$ (b) .

Other conditions of Theorem 5.3.2 also hold that was shown in the proof of Theorems 10.2.1 and 10.2.2. By applying Theorem 5.3.2 to the log-price processes $\vec{Y}_{\varepsilon,n}$ we get the asymptotic relation (10.94). □

Let us now study the corresponding results for the multivariate trinomial tree model considered in Subsection 10.2.2.

In fact, formulas (10.88) and (10.93) defining the optimal expected reward as well as the conditions $\mathbf{D}_{21}[\bar{\beta}]$ and $\mathbf{K}_{19}$ remain the same. The only difference is that the reward functions $\phi_{0,0}(\vec{y})$ for the trinomial model considered in Subsection 10.2.2 should be used in formulas (10.88) and (10.93) instead of the reward functions $\phi_{0,0}(\vec{y})$ for the binomial model considered in Subsection 10.2.1.

The following theorem is an analog of Theorem 10.2.3.

Theorem 10.2.5. *Let the log-price process $\vec{Y}_{0,n}$ be represented by the multivariate Gaussian random walk given by stochastic transition dynamic relation* (10.69) *and the approximating log-price processes are represented, for every $\varepsilon \in (0, \varepsilon_0]$, by the multivariate trinomial random walks $\vec{Y}_{\varepsilon,n}$ given by stochastic transition dynamic relation* (10.70), *with parameters satisfying the condition* $\mathbf{R}_2$ *and given by relations* (10.72). *Let also conditions* $\mathbf{B}_6[\bar{\gamma}]$ *and* $\mathbf{D}_{21}[\bar{\beta}]$ *hold for the parameters* $\bar{\gamma}$ *and* $\bar{\beta}$ *such that, for every* $i = 1, \dots, k$, *either* $\beta_i > \gamma_i > 0$ *or* $\beta_i = \gamma_i = 0$, *and also the conditions* $\mathbf{N}_6$, $\mathbf{I}_{10}$, $\mathbf{J}_{14}$ *and* $\mathbf{K}_{19}$ *hold, and* $r_\varepsilon \to \infty$ *as* $\varepsilon \to 0$. *Then the following relation takes place:*

$$\Phi_\varepsilon \to \Phi_0 \quad as\ \varepsilon \to 0\,. \tag{10.102}$$

The proof of this theorem is analogous to the proof of Theorem 10.2.3.

Remark 10.2.1. If the model assumption (h) $\det(\Sigma_n) \neq 0$, $n = 1, 2, \dots$ (reduced the assumption $\det(\Sigma) \neq 0$ in the case of homogeneous in time model, where matrices $\Sigma_n = \Sigma$, $n = 1, 2, \dots$) may not hold, the condition $\mathbf{J}_{14}$ in Theorems 10.2.1–10.2.5 can be replaced by the assumption that $\mathsf{P}\{\vec{y}_0 + \vec{W}_{0,n} \in \overline{\mathbb{Y}}_n\} = 0, \vec{y} \in \mathbb{Y}_{n-1}$, $n = 1, \dots, N$.

10.3 Univariate inhomogeneous in space tree models

In this section, we present results about convergence of rewards for American-type options for binomial and trinomial tree approximations for log-price processes represented by univariate inhomogeneous in time and space Gaussian random walks.

10.3.1 Convergence of reward functions for univariate inhomogeneous in space binomial tree approximations

Let us consider the model of univariate homogeneous in time but inhomogeneous in space Gaussian random walk, which can be considered as a discrete time analog of a diffusion process. In this case, a log-price $Y_{0,n}$ is a Markov chain with the phase space $\mathbb{R}_1$, an initial distribution $P_0(A)$, and transition probabilities given for $y \in \mathbb{R}_1$, $A \in \mathcal{B}_1$, $n = 1, 2, \dots$ by the following relation:

$$P_0(y, A) = \mathsf{P}\{Y_{0,n} \in A / Y_{0,n-1} = y\} = \mathsf{P}\{y + W_0(y) \in A\}\,, \tag{10.103}$$

where (a) $W_0(y)$ is, for every $y \in \mathbb{R}_1$, a normal random variable with a mean $\mathsf{E}W_0(y) = \mu(y)$ and a variance $\mathrm{Var}W_0(y) = \sigma^2(y)$.

Also, usual assumptions are (b) $\mu(y)$ is a measurable function acting from $\mathbb{R}_1$ to $\mathbb{R}_1$ and (c) $\sigma^2(y)$ is a measurable function acting from $\mathbb{R}_1$ to $(0, \infty)$.

Note that we exclude the case, where $\sigma^2(y)$ can take zero value in some points.

We shall approximate the log-price process $Y_{0,n}$ by a log-price process $Y_{\varepsilon,n}$, which is, for every $\varepsilon \in (0, \varepsilon_0]$, a univariate inhomogeneous in space binomial random walk. In this case, the log-price process $Y_{\varepsilon,n}$ is a Markov chain with the phase space $\mathbb{R}_1$, an initial distribution $P_\varepsilon(A)$, and transition probabilities given for $y \in \mathbb{R}_1$, $A \in \mathcal{B}_1$, $n = 1, 2, \dots$ by the following relation:

$$P_\varepsilon(y, A) = \mathsf{P}\{Y_{\varepsilon,n} \in A / Y_{\varepsilon,n-1} = y\} = \mathsf{P}\{y + W_\varepsilon(y) \in A\}, \tag{10.104}$$

where (a) $W_\varepsilon(y) = W_{\varepsilon,1}(y) + \cdots + W_{\varepsilon,r_\varepsilon(y)}(y)$, where $W_{\varepsilon,n}(y)$, $n = 1, 2, \dots$ is, for every $y \in \mathbb{R}_1$, a sequence of i.i.d. binary random variables, which take values δ_ε and $-\delta_\varepsilon$ with probabilities, respectively, $p_{\varepsilon,+}(y)$ and $p_{\varepsilon,-}(y)$.

Also, usual assumptions are (b) $\delta_\varepsilon > 0$, (c) $r_\varepsilon(y)$ is a measurable function defined on space $\mathbb{R}_1$ and taking positive integer values, and (d) $p_{\varepsilon,\pm}(y)$ are the measurable function acting from $\mathbb{R}_1$ to $[0, 1]$ such that $p_{\varepsilon,-}(y) = 1 - p_{\varepsilon,+}(y)$.

We assume that a pay-off function $g(n, e^y)$ does not depend on the parameter ε and also assume that $g(n, e^y)$ is a real-valued Borel function of argument $y \in \mathbb{R}_1$, for every $n = 0, 1, \dots, N$.

In this case, the corresponding reward function $\phi_{\varepsilon,n}(y)$ for the log price process $Y_{\varepsilon,n}$ is also a Borel function of argument $y \in \mathbb{R}_1$, for every $n = 0, 1, \dots, N$.

We fit parameters for the log-price processes $Y_{\varepsilon,n}$ and $Y_{0,n}$ in such a way that the following asymptotic moment fitting relations hold:

$$\mathsf{E}W_\varepsilon(y) \to \mu(y), \quad \mathrm{Var}W_\varepsilon(y) \to \sigma^2(y), \quad \text{as } \varepsilon \to 0, \quad y \in \mathbb{R}_1. \tag{10.105}$$

The solution of this fitting problem is presented in Lemma 9.4.1. According to this lemma, relations (10.105) hold if one choose the parameters of the log-price process $Y_{\varepsilon,n}$ given by the following formulas, for every $\varepsilon \in (0, \varepsilon_0]$ and $y \in \mathbb{R}_1$:

$$\begin{cases} \delta_\varepsilon = \dfrac{\delta}{\sqrt{r_\varepsilon}} \\ p_{\varepsilon,\pm}(y) = \dfrac{1}{2}\left(1 \pm \dfrac{\mu(y)\sqrt{r_\varepsilon}}{r^*_\varepsilon(y)\,\delta}\right), \\ r^*_\varepsilon(y) = \dfrac{r_\varepsilon\left(\sigma^2(y) + \sqrt{\sigma^4(y) + \frac{4\delta^2\mu^2(y)}{r_\varepsilon}}\right)}{2\delta^2}, \\ r_\varepsilon(y) = [r^*_\varepsilon(y)] + 1, \end{cases} \tag{10.106}$$

where (i) $\delta, r_\varepsilon > 0$; (ii) $r_\varepsilon \to \infty$ as $\varepsilon \to 0$.

The random variables $W_\varepsilon(y)$, $y \in \mathbb{R}_1$ are binomial random variables. Namely, the random variable $W_\varepsilon(y)$ takes values,

$$w_{\varepsilon,l}(y) = l\frac{\delta}{\sqrt{r_\varepsilon}}, \quad l = -r_\varepsilon(y), \dots, r_\varepsilon(y), \tag{10.107}$$

with the corresponding probabilities,

$$p_{\varepsilon,l}(y) = \begin{cases} C_{r_\varepsilon(y)}^{\frac{r_\varepsilon(y)+l}{2}}(p_{\varepsilon,+}(y))^{\frac{r_\varepsilon(y)+l}{2}}(p_{\varepsilon,-}(y))^{\frac{r_\varepsilon(y)+l}{2}} & \text{if } \frac{r_\varepsilon(y)+l}{2} \text{ is integer}, \\ 0 & \text{if } \frac{r_\varepsilon(y)-l}{2} \text{ is not integer}, \\ l = -r_\varepsilon(y), \dots, r_\varepsilon(y). \end{cases} \tag{10.108}$$

Note that we slightly modified above the standard definition of binomial distribution by including in the range of possible values all integer numbers $-r_\varepsilon(y) \le l \le r_\varepsilon(y)$ including the values with the corresponding probabilities equal to zero.

Let us assume that $Y_{\varepsilon,n} = y = \text{const} \in \mathbb{R}_1$, for some $0 \le n \le N$. In this case, trajectories of the log-price process $Y_{\varepsilon,n+r}$, $r = 0, 1, \dots$ generate, for every $\varepsilon \in (0, \varepsilon_0]$, the unique tree of nodes,

$$T_{\varepsilon,(y,n)} = \{(y_{\varepsilon,n,n+r,l}, n + r), r^-_{\varepsilon,n,n+r}(y) \le l \le r^+_{\varepsilon,n,n+r}(y), r = 0, 1, \dots\}, \tag{10.109}$$

where

$$y_{\varepsilon,n,n+r,l} = y + l\frac{\delta}{\sqrt{r_\varepsilon}}, r^-_{\varepsilon,n,n+r}(y) \le l \le r^+_{\varepsilon,n,n+r}(y), r = 0, 1, \dots, \tag{10.110}$$

and

$$r^\pm_{\varepsilon,n,n+r}(y) = \begin{cases} 0 & \text{if } r = 0, \\ \pm r_\varepsilon(y) & \text{if } r = 1, \\ \pm \max_{r^-_{\varepsilon,n,n+r-1}(y) \le l \le r^+_{\varepsilon,n,n+r-1}(y)} \pm \left(l \pm r_\varepsilon\left(y + \frac{\delta}{\sqrt{r_\varepsilon}}l\right)\right) & \text{if } r > 1. \end{cases} \tag{10.111}$$

The following lemma is a direct corollary of Lemma 3.4.1.

Lemma 10.3.1. *Let, for every $\varepsilon \in (0, \varepsilon_0]$, the approximating log-price process $Y_{\varepsilon,n}$ is a homogeneous binomial Markov chain with transition probabilities given by relation (10.104) and parameters defined in relation (10.106). Then the reward functions $\phi_{\varepsilon,n+r}(y_{\varepsilon,n,n+r,l})$, for points $y_{\varepsilon,n,n,0} = y$ and $y_{\varepsilon,n,n+r,l}$, $l = r^-_{\varepsilon,n,n+r}(y), \dots, r^+_{\varepsilon,n,n+r}(y)$, $r = 1, \dots N - n$, are, for every $\varepsilon \in (0, \varepsilon_0]$, and $y \in \mathbb{R}_1$, $n = 0, \dots, N$, the unique solution for the following recurrence finite system of linear equations:*

$$\begin{cases} \phi_{\varepsilon,N}(y_{\varepsilon,n,N,l}) = g_\varepsilon(N, e^{y_{\varepsilon,n,N,l}}), \\ \qquad l = r^-_{\varepsilon,n,N}(y), \dots, r^+_{\varepsilon,n,N}(y), \\ \phi_{\varepsilon,n+r}(y_{\varepsilon,n,n+r,l}) = \max\Bigg(g_\varepsilon(n + r, e^{y_{\varepsilon,n,n+r,l}}), \\ \qquad \sum\limits_{l'=-r_\varepsilon(y)}^{r_\varepsilon(y)} \phi_{\varepsilon,n+r+1}(y_{\varepsilon,n,n+r+1,l+l'})p_{\varepsilon,l'}(y_{\varepsilon,n,n+r,l})\Bigg), \\ \qquad l = r^-_{\varepsilon,n,n+r}(y), \dots, r^+_{\varepsilon,n,n+r}(y), \quad r = N - n - 1, \dots, 1, \\ \phi_{\varepsilon,n}(y) = \max\Bigg(g_\varepsilon(n, e^y), \sum\limits_{l'=-r_\varepsilon(y)}^{r_\varepsilon(y)} \phi_{\varepsilon,n}(y_{\varepsilon,n,n+1,l'})p_{\varepsilon,l'}(y)\Bigg). \end{cases} \tag{10.112}$$

Let us assume that the functions $\mu(y)$ and $\sigma^2(y)$ are bounded, i.e. the condition $\mathbf{G_{12}}$ holds.

This condition implies that the corresponding tree $T_{\varepsilon,(\vec{y},n)}$ generated by the trajectories of the log-price process $Y_{\varepsilon,n+r}$, $r = 0, 1, \dots$ with the value $\vec{Y}_{\varepsilon,n} = \vec{y}$ has the total number of nodes after r steps,

$$L_{\varepsilon,n,n+r} \le 1 + 2K_{86}(\delta)r_\varepsilon + 1 + \cdots + 2K_{86}(\delta)r_\varepsilon r + 1$$
$$= K_{86}(\delta)r_\varepsilon r^2 + (K_{86}(\delta)r_\varepsilon + 1)r + 1\,. \quad (10.113)$$

The explicit formula for $K_{86}(\delta)$ was given in Subsection 9.4.1. The expression on the right-hand side in (10.113) is a polynomial (quadratic) function of r.

We are going to apply Theorem 5.2.1, in order to prove convergence of the reward functions for log-price processes $Y_{\varepsilon,n}$ to the corresponding reward functions for log-price processes $Y_{0,n}$ as $\varepsilon \to 0$.

We assume that the condition $\mathbf{B}_4[\gamma]$ holds for the pay-off function $g(n, e^y)$, for some $\gamma \ge 0$.

The condition $\mathbf{I}_1$ reduces, in this case, to the form of the condition $\mathbf{I}_9$.

Let us also assume that the following condition holds:

$\mathbf{J}_{15}$: (a) There exists a set $\mathbb{Y} \in \mathcal{B}_1$ such that functions $\mu(y)$ and $\sigma^2(y)$ are continuous in points $y \in \mathbb{Y}$;

(b) $L_1(\overline{\mathbb{Y}}) = 0$ and $L_1(\overline{\mathbb{Y}}_n) = 0$, $n = 1, \dots, N$, where $\mathbb{Y}_n$, $n = 1, \dots, N$ are sets penetrating the condition $\mathbf{I}_9$.

We shall see that, in this case, the condition $\mathbf{J}_1$ (a) (its one-dimensional variant for the model without index component) holds for log-prices $Y_{\varepsilon,n}$ with parameters given by relation (10.106), if $r_\varepsilon \to \infty$ as $\varepsilon \to 0$. Also the random variables $W_0(y)$, $y \in \mathbb{R}_1$ have normal distributions that make it possible to replace the condition $\mathbf{J}_1$ (b) (its one-dimensional variant for the model without index component) by the condition $\mathbf{J}_{15}$.

The above backward recurrence algorithm for finding reward functions can be now supported by the following convergence result.

Theorem 10.3.1. *Let the log-price process $Y_{0,n}$ be represented by the inhomogeneous in space Gaussian random walk with transition probabilities given by relation* (10.103) *and the approximating log-price process is represented, for every $\varepsilon \in (0, \varepsilon_0]$, by the inhomogeneous binomial random walk $Y_{\varepsilon,n}$ with transition probabilities given by relation* (10.104) *and parameters satisfying the condition $\mathbf{G}_{12}$ and given by relation* (10.106). *Let also conditions $\mathbf{B}_4[\gamma]$ holds for some $\gamma \ge 0$, conditions $\mathbf{I}_9$ and $\mathbf{J}_{15}$ hold, and $r_\varepsilon \to \infty$ as $\varepsilon \to 0$. Then the following relations take place for any $y_\varepsilon \to y_0 \in \mathbb{Y}_n \cap \mathbb{Y}$, $n = 0, 1, \dots, N$:*

$$\phi_{\varepsilon,n}(y_\varepsilon) \to \phi_{0,n}(y_0) \quad \text{as } \varepsilon \to 0\,. \quad (10.114)$$

Proof. We are going to use Theorem 5.2.1.

The moment-generating function for the random variables $W_\varepsilon(y)$ has, for every $\varepsilon \in (0, \varepsilon_0]$ and $y \in \mathbb{R}_1$, the following form, for $\beta \in \mathbb{R}_1$:

$$\mathsf{E}e^{\beta W_\varepsilon(y)} = \left(e^{\beta \frac{\delta}{\sqrt{r_\varepsilon}}} \frac{1}{2} \left(1 + \frac{\mu(y)}{r^*_\varepsilon(y)} \frac{\sqrt{r_\varepsilon}}{\delta}\right) + e^{-\beta \frac{\delta}{\sqrt{r_\varepsilon}}} \frac{1}{2} \left(1 - \frac{\mu(y)}{r^*_\varepsilon(y)} \frac{\sqrt{r_\varepsilon}}{\delta}\right)\right)^{r_\varepsilon(y)}, \quad (10.115)$$

while, for $\varepsilon = 0$, these moment-generating functions have the following form:

$$\mathsf{E}e^{\beta W_0(y)} = e^{\beta\mu(y) + \frac{1}{2}\beta^2\sigma^2(y)}. \tag{10.116}$$

Using the condition $\mathbf{G_{12}}$, we get the following inequalities, for every $\varepsilon \in (0, \varepsilon_0]$, $y \in \mathbb{R}_1$, and any $\beta \in \mathbb{R}_1$:

$$\begin{aligned}
0 &\le e^{\beta\frac{\delta}{\sqrt{r_\varepsilon}}}\frac{1}{2}\left(1 + \frac{\mu(y)\sqrt{r_\varepsilon}}{r_\varepsilon^*(y)\,\delta}\right) + e^{-\beta\frac{\delta}{\sqrt{r_\varepsilon}}}\frac{1}{2}\left(1 - \frac{\mu(y)\sqrt{r_\varepsilon}}{r_\varepsilon^*(y)\,\delta}\right) \\
&= \left(1 + \frac{\beta\delta}{\sqrt{r_\varepsilon}} + \frac{\beta^2\delta^2}{2!r_\varepsilon} + \cdots\right)\frac{1}{2}\left(1 + \frac{\mu(y)\sqrt{r_\varepsilon}}{r_\varepsilon^*(y)\,\delta}\right) \\
&\quad + \left(1 - \frac{\beta\delta}{\sqrt{r_\varepsilon}} + \frac{\beta^2\delta^2}{2!r_\varepsilon} + \cdots\right)\frac{1}{2}\left(1 - \frac{\mu(y)\sqrt{r_\varepsilon}}{r_\varepsilon^*(y)\,\delta}\right) \\
&= \left(1 + \frac{\beta^2\delta^2}{2!r_\varepsilon} + \frac{\beta^4\delta^4}{4!r_\varepsilon^2} + \cdots\right) + \left(\frac{\beta\delta}{\sqrt{r_\varepsilon}} + \frac{\beta^3\delta^3}{3!(\sqrt{r_\varepsilon})^2} + \cdots\right)\frac{\mu(y)\sqrt{r_\varepsilon}}{r_\varepsilon^*(y)\,\delta} \\
&= 1 + \left(1 + \frac{\beta^2\delta^2}{\frac{4!}{2}r_\varepsilon} + \cdots\right)\frac{\beta^2\delta^2}{2r_\varepsilon} + \left(1 + \frac{\beta^2\delta^2}{r_\varepsilon} + \cdots\right)\frac{\beta\mu(y)}{r_\varepsilon^*(y)} \\
&\le 1 + e^{\frac{\beta^2\delta^2}{r_\varepsilon}}\left(\frac{\beta^2\delta^2}{2r_\varepsilon} + \frac{|\beta||\mu(y)|}{r_\varepsilon^*(y)}\right).
\end{aligned} \tag{10.117}$$

The condition $\mathbf{G_{12}}$ also implies the following inequality, for every $\varepsilon \in (0, \varepsilon_0]$, and $y \in \mathbb{R}_1$:

$$r_\varepsilon^*(y) = \frac{r_\varepsilon\left(\sigma^2(y) + \sqrt{\sigma^4(y) + \frac{4\delta^2\mu^2(y)}{r_\varepsilon}}\right)}{2\delta^2} \le r_\varepsilon K_{85}\frac{1 + \sqrt{1 + \frac{4\delta^2}{r_\varepsilon}}}{2\delta^2}, \tag{10.118}$$

where K_{85} is the constant penetrating condition $\mathbf{G_{12}}$.

Using formula (10.115), inequalities (10.117) and (10.118), and taking into account inequalities $r_\varepsilon(y) \le r_\varepsilon^*(y) + 1$ and $\frac{|\mu(y)|}{r_\varepsilon^*(y)} \le \frac{\delta}{\sqrt{r_\varepsilon}}$, we get the following inequality, for every $\varepsilon \in (0, \varepsilon_0]$, $y \in \mathbb{R}_1$, and $\beta \in \mathbb{R}_1$:

$$\begin{aligned}
\mathsf{E}e^{\beta W_\varepsilon(y)} &\le \left(1 + e^{\frac{\beta^2\delta^2}{r_\varepsilon}}\left(\frac{\beta^2\delta^2}{2r_\varepsilon} + \frac{|\beta||\mu(y)|}{r_\varepsilon^*(y)}\right)\right)^{r_\varepsilon^*(y)+1} \\
&\le \left(1 + e^{\frac{\beta^2\delta^2}{r_\varepsilon}}\left(\frac{\beta^2\delta^2}{2r_\varepsilon} + \frac{|\beta|\delta}{\sqrt{r_\varepsilon}}\right)\right) \\
&\quad \times \exp\left\{e^{\frac{\beta^2\delta^2}{r_\varepsilon}}\left(\frac{\beta^2\delta^2 r_\varepsilon^*(y)}{2r_\varepsilon} + |\beta||\mu(y)|\right)\right\} \\
&\le \left(1 + e^{\frac{\beta^2\delta^2}{r_\varepsilon}}\left(\frac{\beta^2\delta^2}{2r_\varepsilon} + \frac{|\beta|\delta}{\sqrt{r_\varepsilon}}\right)\right) \\
&\quad \times \exp\left\{e^{\frac{\beta^2\delta^2}{r_\varepsilon}}\left(\frac{1}{2}\beta^2K_{85}\frac{1 + \sqrt{1 + \frac{4\delta^2}{r_\varepsilon}}}{2} + |\beta|K_{85}\right)\right\}.
\end{aligned} \tag{10.119}$$

Also, the condition $\mathbf{G}_{12}$ implies that the following inequality holds, for every $y \in \mathbb{R}_1$, and any $\beta \in \mathbb{R}_1$:

$$\mathsf{E}e^{\beta W_0(y)} = e^{\beta\mu(y) + \frac{1}{2}\beta^2\sigma^2(y)} \leq \exp\{(|\beta| + \tfrac{1}{2}\beta^2)K_{85}\} . \tag{10.120}$$

The modulus $\Delta^{\pm}_{\beta}(Y_{\varepsilon,\cdot}, N)$ takes for the above log-price processes $Y_{\varepsilon,n}$ the following form, for every $\varepsilon \in [0, \varepsilon_0]$ and $\beta \geq 0$:

$$\Delta^{\pm}_{\beta}(Y_{\varepsilon,\cdot}, N) = \sup_{y\in\mathbb{R}_1} e^{\pm\beta W_\varepsilon(y)} . \tag{10.121}$$

Relations (10.119) and (10.120) imply the following asymptotic relation, which holds for any $\beta \geq 0$:

$$\begin{aligned}\overline{\lim_{\varepsilon\to 0}}\, \Delta^{\pm}_{\beta}(Y_{\varepsilon,\cdot}, N) &\leq \overline{\lim_{\varepsilon\to 0}} \left(\left(1 + e^{\frac{\beta^2\delta^2}{r_\varepsilon}} \left(\frac{\beta^2\delta^2}{2r_\varepsilon} + \frac{|\beta|\delta}{\sqrt{r_\varepsilon}} \right)\right)\right.\\ &\quad \left.\times \exp\left\{ e^{\frac{\beta^2\delta^2}{r_\varepsilon}} \left(\frac{1}{2}\beta^2 K_{85} \frac{1 + \sqrt{1 + \frac{4\delta^2}{r_\varepsilon}}}{2} + \beta K_{85} \right)\right\}\right)\\ &= \exp\left\{ \left(\beta + \tfrac{1}{2}\beta^2\right) K_{85} \right\} < \infty .\end{aligned} \tag{10.122}$$

Relations (10.120) and (10.122) imply that the condition $\mathbf{E}_{10}[\bar{\beta}]$ (its univariate variant for the model without index component) holds. Therefore, by Lemma 5.1.5, the condition $\mathbf{C}_6[\bar{\beta}]$ (its one-dimensional variant for the model without index component) holds for any $\beta \geq 0$, for example with the constant $K_{93} = 1 + 2e^{(|\beta|+\beta^2)K_{86}}$ replacing the constant $K_{30,1}$ penetrating this condition.

The condition $\mathbf{B}_4[\gamma]$ is a one-dimensional variant of the condition $\mathbf{B}_5[\bar{\gamma}]$ (its univariate variant for the model without index component and a pay-off function, which does not depend on parameter ε).

Note that one can always take the parameter $\beta > \gamma$ if $\gamma > 0$ or $\beta = \gamma$ if $\gamma = 0$.

The condition $\mathbf{I}_1$ (its univariate l variant for the model without index component and a pay-off function, which does not depend on the parameter ε) reduces in this case to the form of the condition $\mathbf{I}_9$.

Let us take an arbitrary $n = 0, 1, \ldots, N$ and $y_\varepsilon \to y_0 \in \mathbb{Y}_n \cap \mathbb{Y}$ as $\varepsilon \to 0$. By the initial model assumptions, $\sigma^2(y_0) > 0$ and, by the conditions $\mathbf{I}_9$ and $\mathbf{J}_{15}$, $\mu(y_\varepsilon) \to \mu(y_0)$ and $\sigma^2(y_\varepsilon) \to \sigma^2(y_0)$ as $\varepsilon \to 0$, and, therefore,

$$\begin{aligned} r_\varepsilon(y_\varepsilon) &= [r^*_\varepsilon(y_\varepsilon)] + 1\\ &\sim r^*_\varepsilon(y_\varepsilon) = \frac{r_\varepsilon\left(\sigma^2(y_\varepsilon) + \sqrt{\sigma^4(y_\varepsilon) + \frac{4\delta^2\mu^2(y_\varepsilon)}{r_\varepsilon}}\right)}{2\delta^2}\\ &\sim r_\varepsilon \frac{\sigma^2(y_0)}{\delta^2} \quad \text{as } \varepsilon \to 0 . \end{aligned} \tag{10.123}$$

and, for every $\beta \in \mathbb{R}_1$,

$$\begin{aligned}
\mathsf{E}e^{\beta W_\varepsilon(y_\varepsilon)} &= \Bigg(\Bigg(1 + \frac{\beta\delta}{\sqrt{r_\varepsilon}} + \frac{\beta^2\delta^2}{2r_\varepsilon} + o_2\left(\frac{\beta\delta}{\sqrt{r_\varepsilon}}\right)\Bigg)\frac{1}{2}\left(1 + \frac{\mu(y_\varepsilon)\sqrt{r_\varepsilon}}{r^*_\varepsilon(y_\varepsilon)\delta}\right) \\
&\quad + \left(1 - \frac{\beta\delta}{\sqrt{r_\varepsilon}} + \frac{\beta^2\delta^2}{2r_\varepsilon} + o_2\left(-\frac{\beta\delta}{\sqrt{r_\varepsilon}}\right)\right)\frac{1}{2}\left(1 - \frac{\mu(y_\varepsilon)\sqrt{r_\varepsilon}}{r^*_\varepsilon(y_\varepsilon)\delta}\right)\Bigg)^{r_\varepsilon(y_\varepsilon)} \\
&= \Bigg(1 + \frac{\beta\mu(y_\varepsilon)}{r^*_\varepsilon(y_\varepsilon)} + \frac{1}{2}\frac{\beta^2\delta^2}{r_\varepsilon} + o_2\left(\frac{\beta\delta}{\sqrt{r_\varepsilon}}\right)\frac{1}{2}\left(1 + \frac{\mu(y_\varepsilon)\sqrt{r_\varepsilon}}{r^*_\varepsilon(y_\varepsilon)\delta}\right) \\
&\quad + o_2\left(-\frac{\beta\delta}{\sqrt{r_\varepsilon}}\right)\frac{1}{2}\left(1 - \frac{\mu(y_\varepsilon)\sqrt{r_\varepsilon}}{r^*_\varepsilon(y_\varepsilon)\delta}\right)\Bigg)^{r_\varepsilon(y_\varepsilon)} \\
&\to \exp\left\{\beta\mu(y_0) + \tfrac{1}{2}\beta^2\sigma^2(y_0)\right\} = \mathsf{E}e^{\beta W_0(y_0)} \quad \text{as } \varepsilon \to 0,
\end{aligned} \tag{10.124}$$

since by relation (10.123),

$$\begin{aligned}
&\left| o_2\left(\frac{\beta\delta}{\sqrt{r_\varepsilon}}\right)\frac{1}{2}\left(1 + \frac{\mu(y_\varepsilon)\sqrt{r_\varepsilon}}{r^*_\varepsilon(y_\varepsilon)\delta}\right) + o_2\left(-\frac{\beta\delta}{\sqrt{r_\varepsilon}}\right)\frac{1}{2}\left(1 - \frac{\mu(y_\varepsilon)\sqrt{r_\varepsilon}}{r^*_\varepsilon(y_\varepsilon)\delta}\right)\right| r_\varepsilon(y_\varepsilon) \\
&\quad \le \left| o_2\left(\frac{\beta\delta}{\sqrt{r_\varepsilon}}\right) r_\varepsilon(y_\varepsilon)\right| + \left| o_2\left(-\frac{\beta\delta}{\sqrt{r_\varepsilon}}\right) r_\varepsilon(y_\varepsilon)\right| \to 0 \quad \text{as } \varepsilon \to 0.
\end{aligned} \tag{10.125}$$

Relation (10.124) implies, by the corresponding variant of the continuous theorem, the following relation, for an arbitrary $n = 1, \dots, N$ and $y_\varepsilon \to y_0 \in \mathbb{Y}_{n-1} \cap \mathbb{Y}$ as $\varepsilon \to 0$:

$$y_\varepsilon + W_\varepsilon(y_\varepsilon) \xrightarrow{d} y_0 + W_0(y_0) \quad \text{as } \varepsilon \to 0. \tag{10.126}$$

Relation (10.126) is equivalent to the condition $\mathbf{J}_1$ (a) (its univariate variant for the model without index component).

The condition $\mathbf{J}_{15}$ (b) implies that the condition $\mathbf{J}_1$ (b) (its univariate variant for the model without index component) holds, since random variables $W_0(y)$, $y \in \mathbb{R}_1$ have normal distributions with positive variances, and, thus, possess positive probability density functions with respect to the Lebesgue measure.

Thus, all conditions of Theorem 5.2.1 hold. By applying this theorem to the above approximation model, we get the asymptotic relation (10.114). □

10.3.2 Convergence of reward functions for univariate inhomogeneous in time and space trinomial tree approximations

Let us consider the model, where a log-price process $Y_{0,n}$ is a univariate inhomogeneous in time and space Gaussian random walk, with the phase space $\mathbb{R}_1$, an initial distribution $P_0(A)$, and transition probabilities given for $y \in \mathbb{R}_1$, $A \in \mathcal{B}_1$, $n = 1, 2, \dots$

by the following relation:

$$P_{0,n}(y,A) = \mathsf{P}\{Y_{0,n} \in A / Y_{0,n-1} = y\} = \mathsf{P}\{y + W_{0,n}(y) \in A\}, \tag{10.127}$$

where (a) $W_{0,n}(y)$ is, for every $y \in \mathbb{R}_1$, $n = 1, 2, \ldots$, a normal random variable with a mean value $\mathsf{E}W_{0,n}(y) = \mu_n(y)$ and a variance $\mathrm{Var}W_{0,n}(y) = \sigma_n^2(y)$.

Also, usual assumptions are (b) $\mu_n(y)$ is, for every $n = 1, 2, \ldots$, a measurable function acting from $\mathbb{R}_1$ to $\mathbb{R}_1$, (c) $\sigma_n^2(y)$ is, for every $n = 1, 2, \ldots$, a measurable function acting from $\mathbb{R}_1$ to $(0, \infty)$.

We shall approximate the log-price process $Y_{0,n}$ by a log-price process $Y_{\varepsilon,n}$ which is, for every $\varepsilon \in (0, \varepsilon_0]$, a univariate inhomogeneous in time and space trinomial random walk, with the phase space $\mathbb{R}_1$, an initial distribution $P_\varepsilon(A)$, and transition probabilities given for $y \in \mathbb{R}_1$, $A \in \mathcal{B}_1$, $n = 1, 2, \ldots$ by the following relation:

$$P_{\varepsilon,n}(y,A) = \mathsf{P}\{Y_{\varepsilon,n} \in A / Y_{\varepsilon,n-1} = y\} = \mathsf{P}\{y + W_{\varepsilon,n}(y) \in A\}, \tag{10.128}$$

where (a) $W_{\varepsilon,n}(y)$ is, for every $y \in \mathbb{R}_1$, $n = 1, 2, \ldots$, a trinomial random variable, which can be represented in the form $W_{\varepsilon,n}(\vec{y}) = W_{\varepsilon,n,1}(\vec{y}) + \cdots + W_{\varepsilon,n,r_\varepsilon}(\vec{y})$, where $\vec{W}_{\varepsilon,n,l}(\vec{y})$, $l = 1, 2, \ldots$ are the i.i.d. random variables taking values $\delta_{\varepsilon,\jmath}$ with the probabilities $p_{\varepsilon,n,\jmath}(\vec{y})$, for $\jmath = +, \circ, -$.

Also, usual assumptions are (b) $\delta_{\varepsilon,-} = -\delta_\varepsilon < \delta_{\varepsilon,\circ} = 0 < \delta_{\varepsilon,+} = +\delta_\varepsilon$, (c) $p_{\varepsilon,n,\jmath}(y)$, $\jmath = +, \circ, -$, $n = 1, 2, \ldots$ are the measurable functions acting from $\mathbb{R}_1$ to $[0, 1]$ such that $\sum_{\jmath=+,\circ,-} p_{\varepsilon,n,\jmath}(y) = 1$, $y \in \mathbb{R}_1$, $n = 1, 2, \ldots$; (d) r_ε is a positive integer number.

We use the log-prices $Y_{\varepsilon,n}$ for approximation of the log-price $Y_{0,n}$ asymptotically, as $\varepsilon \to 0$. In this case, a usual approach is to fit the corresponding parameters, i.e. to try to choose parameter δ_ε, probabilities $p_{\varepsilon,n,\jmath}(y)$, and parameter r_ε in such a way that the following exact fitting relations hold, for every $\varepsilon \in (0, \varepsilon_0]$:

$$\begin{cases} \mathsf{E}W_{\varepsilon,n}(y) = \mathsf{E}W_{0,n}(y), \\ \mathrm{Var}W_{\varepsilon,n}(y) = \mathrm{Var}W_{0,n}(y), \quad y \in \mathbb{R}_1, \quad n = 1, \ldots, N. \end{cases} \tag{10.129}$$

The solution of this fitting problem is presented in Lemma 9.5.1. According to this lemma, the fitting relations (10.129) hold, if to assume that the conditions $\mathbf{G_{14}}$, $\mathbf{G_{15}}$ and to choose parameters of the log-price process $Y_{\varepsilon,n}$ given by the following formulas, for every $\varepsilon \in (0, \varepsilon_0]$ and $n = 1, \ldots, N$:

$$\begin{cases} \delta_\varepsilon = \dfrac{\delta}{\sqrt{r_\varepsilon}}, \\ p_{\varepsilon,n,\pm}(y) = \dfrac{1}{2}\left(\dfrac{\sigma_n^2(y)}{\delta^2} + \dfrac{\mu_n^2(y)}{\delta^2 r_\varepsilon} \pm \dfrac{\mu_n(y)}{\delta\sqrt{r_\varepsilon}}\right), \\ p_{\varepsilon,n,\circ}(y) = 1 - \dfrac{\sigma_n^2(y)}{\delta^2} - \dfrac{\mu_n^2(y)}{\delta^2 r_\varepsilon}, \quad y \in \mathbb{R}_1, \quad n = 1, \ldots, N, \end{cases} \tag{10.130}$$

where (i) one can choose for parameters δ and r_ε any values such that (a) $\delta \geq \sqrt{K_{89}^2 + K_{89}}$, (b) $r_\varepsilon \geq \frac{K_{89}^2 \delta^2}{\sigma^4}$, where $\sigma^2 > 0$ and $K_{89} < \infty$ are parameters penetrating, respectively the conditions $\mathbf{G_{14}}$ and $\mathbf{G_{15}}$.

The random variable $W_{\varepsilon,n}(y)$ is for every $y \in \mathbb{R}_1$, $n = 1, \dots, N$ a trinomial random variable. Namely, the random variable $W_{\varepsilon,n}(y)$ takes values $w_{\varepsilon,r_\varepsilon,l}$ with the corresponding probabilities $p_{\varepsilon,n,r_\varepsilon,l}(y)$, for $-r_\varepsilon \le l \le r_\varepsilon$.

Here, for every $y \in \mathbb{R}_1$, $n = 1, \dots, N$, $r = 1, 2, \dots$,

$$w_{\varepsilon,r,l} = l\delta_\varepsilon = l\frac{\delta}{\sqrt{r_\varepsilon}}, \ -r \le l \le r, \tag{10.131}$$

and, for $-r \le l \le r$,

$$p_{\varepsilon,n,r,l}(y) = \mathsf{P}\{W_{\varepsilon,n}(y) = W_{\varepsilon,n,1}(y) + \cdots + W_{\varepsilon,n,r}(y) = w_{\varepsilon,r,l}\} I(|l| \le r). \tag{10.132}$$

The probabilities $p_{\varepsilon,n,r,l}(y)$ can be found using the following recurrent convolution formulas:

$$p_{\varepsilon,n,r,l}(y) = \begin{cases} p_{\varepsilon,n,\jmath}(y)\, I(l = e_\jmath) & \text{if } r = 1, -1 \le l \le 1, \\ \sum_{l'=-1}^{1} p_{\varepsilon,n,r-1,l-l'}(y) p_{\varepsilon,n,1,l'}(y) & \text{if } r > 1, -r \le l \le r. \end{cases} \tag{10.133}$$

where $e_\jmath = 1, 0, -1$ if $\jmath = +, \circ, -$.

The additional comments concerning algorithms for computing of the probabilities $p_{\varepsilon,n,r,l}(y)$ can be found in Subsection 3.2.3.

Let us assume that $Y_{\varepsilon,n} = y = \text{const} \in \mathbb{R}_1$. In this case, trajectories of the log-price process $Y_{\varepsilon,n+r}$, $r = 0, 1, \dots$ generate, for every $\varepsilon \in (0, \varepsilon_0]$, the unique tree of nodes

$$T_{\varepsilon,(y,n)} = \{(y_{\varepsilon,n,n+r,l}, n + r), -r_\varepsilon r \le l \le r_\varepsilon r, r = 0, 1, \dots\}, \tag{10.134}$$

where

$$y_{\varepsilon,n,n+r,l} = y + l\delta_\varepsilon = y + l\frac{\delta}{\sqrt{r_\varepsilon}}, \quad -r_\varepsilon r \le l \le r_\varepsilon r. \tag{10.135}$$

Note that actual values of points $y_{\varepsilon,n,n+r,l}$ do not depend on the parameter $n = 0, 1, \dots$.

The tree $T_{\varepsilon,(y,n)}$ contains the total number of nodes after the r steps,

$$L_{\varepsilon,n,n+r} \le 1 + (2r_\varepsilon + 1) + \cdots + (2r_\varepsilon r + 1) = r_\varepsilon r^2 + (r_\varepsilon + 1)r + 1. \tag{10.136}$$

The expression on the right-hand side in (10.136) is a polynomial (quadratic) function of r.

The following lemma is a direct corollary of Lemma 3.4.1.

Lemma 10.3.2. *Let, for every $\varepsilon \in (0, \varepsilon_0]$, the approximating log-price process $Y_{\varepsilon,n}$ is the multivariate trinomial Markov chain given by the stochastic transition dynamic relation* (10.128), *with parameters defined in relation* (10.130), *under the assumption that the conditions* $\mathbf{G_{14}}$ *and* $\mathbf{G_{15}}$ *hold. Then the reward functions $\phi_{\varepsilon,n+r}(y_{\varepsilon,n+r,l})$, for points $y_{\varepsilon,n,n,0} = y$ and $y_{\varepsilon,n,n+r,l}$, $-r_\varepsilon r \le l \le r_\varepsilon r$, $r = 1, \dots N - n$, are, for every $\varepsilon \in (0, \varepsilon_0]$,* *and $y \in \mathbb{R}_1$, $n = 0, \ldots, N$, the unique solution for the following recurrence finite system of linear equations:*

$$
\begin{cases}
\phi_{\varepsilon,N}(y_{\varepsilon,n,N,l}) = g_\varepsilon(N, e^{y_{\varepsilon,n,N,l}}) \,, \\
\qquad\qquad - r_\varepsilon(N-n) \le l \le r_\varepsilon(N-n) \,, \\
\phi_{\varepsilon,n+r}(y_{\varepsilon,n,n+r,l}) = \max\Big(g_\varepsilon(n+r, e^{y_{\varepsilon,n,n+r,l}}), \\
\qquad\qquad \sum\limits_{l'=-r_\varepsilon}^{r_\varepsilon} \phi_{\varepsilon,n+r+1}(y_{\varepsilon,n,n+r+1,l+l'}) p_{\varepsilon,n+r+1,r_\varepsilon,l'}(y_{\varepsilon,n,n+r,l}) \Big) \,, \\
\qquad\qquad - r_\varepsilon r \le l \le r_\varepsilon r \,, \quad r = N-n-1, \ldots, 1 \,, \\
\phi_{\varepsilon,n}(y) = \max\Big(g_\varepsilon(n, e^{\vec{y}}), \sum\limits_{l'=-r_\varepsilon}^{r_\varepsilon} \phi_{\varepsilon,n}(y_{\varepsilon,n,n+1,l'}) p_{\varepsilon,n+1,r_\varepsilon,l'}(y) \Big) \,.
\end{cases}
\tag{10.137}
$$

We are going to apply Theorem 5.2.1, in order to prove convergence of the reward functions for log-price processes $Y_{\varepsilon,n}$ to the corresponding reward functions for log-price processes $Y_{0,n}$ as $\varepsilon \to 0$.

First, we assume that the condition $\mathbf{B}_4[\gamma]$ holds for the pay-off function $g(n, e^y)$, for some $\gamma \ge 0$.

The condition $\mathbf{I}_1$ reduces in this case to the form of the condition $\mathbf{I}_9$.

Let us also assume that the following condition holds:

$\mathbf{J}_{16}$: (a) There exist sets $\mathbb{Y}'_n \in \mathcal{B}_1$, $n = 0, 1, \ldots, N$ such that functions $\mu_n(y)$ and $\sigma_n^2(y)$ are continuous in points $y \in \mathbb{Y}'_{n-1}$, for $n = 1, \ldots, N$;
(b) $L_1(\overline{\mathbb{Y}}'_n) = 0$, $L_1(\overline{\mathbb{Y}}_n) = 0$, $n = 1, \ldots, N$, where $\mathbb{Y}_n$, $n = 1, \ldots, N$ are sets penetrating the condition $\mathbf{I}_9$.

We shall see that, in this case, the condition $\mathbf{J}_1$ (a) (its one-dimensional variant for the model without index component) holds for log-prices $Y_{\varepsilon,n}$ with parameters given by relation (10.106), if $r_\varepsilon \to \infty$ as $\varepsilon \to 0$. Also the random variables $W_0(y)$, $y \in \mathbb{R}_1$ have normal distributions that make it possible to replace the condition $\mathbf{J}_1$ (b) (its one-dimensional variant for the model without index component) by the condition $\mathbf{J}_{15}$.

The above backward recurrence algorithm for finding reward functions is supported by the following convergence result.

Theorem 10.3.2. *Let the log-price process $Y_{0,n}$ be represented by the inhomogeneous in time and space Gaussian random walk, with transition probabilities given by relation* (10.127) *and the approximating log-price process is represented, for every $\varepsilon \in (0, \varepsilon_0]$, by the inhomogeneous trinomial random walk $Y_{\varepsilon,n}$, with transition probabilities given by relation* (10.128) *and parameters satisfying the conditions $\mathbf{G}_{14}$, $\mathbf{G}_{15}$ and given by relation* (10.130). *Let also the conditions $\mathbf{B}_4[\gamma]$ holds for some $\gamma \ge 0$, the conditions $\mathbf{I}_9$ and $\mathbf{J}_{16}$ hold, and $r_\varepsilon \to \infty$ as $\varepsilon \to 0$. Then the following relations take place for any $y_\varepsilon \to y_0 \in \mathbb{Y}_n \cap \mathbb{Y}'_n$, $n = 0, 1, \ldots, N$:*

$$\phi_{\varepsilon,n}(y_\varepsilon) \to \phi_{0,n}(y_0) \quad \text{as } \varepsilon \to 0 \,. \tag{10.138}$$

Proof. We are going to use Theorem 5.2.1.

Let the conditions $\mathbf{G}_{14}$ and $\mathbf{G}_{15}$ hold and parameters of the process $Y_{\varepsilon,n}$ are given by relation (10.130).

In this case, the moment-generating function for the random variables $W_{\varepsilon,n}(y)$ has, for every $\varepsilon \in (0, \varepsilon_0]$ and $y \in \mathbb{R}_1$, $n = 1, \dots, N$, the following form:

$$\begin{aligned}\mathsf{E}e^{\beta W_{\varepsilon,n}(y)} &= \left(e^{\beta\frac{\delta}{\sqrt{r_\varepsilon}}} p_{\varepsilon,n,+}(y) + e^{-\beta\frac{\delta}{\sqrt{r_\varepsilon}}} p_{\varepsilon,n,-}(y) + p_{\varepsilon,n,\circ}(y)\right)^{r_\varepsilon} \\ &= \left(e^{\beta\frac{\delta}{\sqrt{r_\varepsilon}}} \frac{1}{2}\left(\frac{\sigma_n^2(y)}{\delta^2} + \frac{\mu_n^2(y)}{\delta^2 r_\varepsilon} + \frac{\mu_n(y)}{\delta\sqrt{r_\varepsilon}}\right)\right. \\ &\quad + e^{-\beta\frac{\delta}{\sqrt{r_\varepsilon}}} \frac{1}{2}\left(\frac{\sigma_n^2(y)}{\delta^2} + \frac{\mu_n^2(y)}{\delta^2 r_\varepsilon} - \frac{\mu_n(y)}{\delta\sqrt{r_\varepsilon}}\right) \\ &\quad \left. + \left(1 - \frac{\sigma_n^2(y)}{\delta^2} - \frac{\mu_n^2(y)}{\delta^2 r_\varepsilon}\right)\right)^{r_\varepsilon}, \quad \beta \in \mathbb{R}_1, \end{aligned} \tag{10.139}$$

while, for $\varepsilon = 0$, these moment-generating functions have the following form:

$$\mathsf{E}e^{\beta W_{0,n}(y)} = e^{\beta\mu_n(y) + \frac{1}{2}\beta^2\sigma_n^2(y)}, \quad \beta \in \mathbb{R}_1. \tag{10.140}$$

Using the condition $\mathbf{G}_{15}$, we get the following inequalities, for every $\varepsilon \in (0, \varepsilon_0]$, $y \in \mathbb{R}_1$, $n = 1, \dots, N$, and any $\beta \in \mathbb{R}_1$:

$$\begin{aligned} 0 &\le \exp\left\{\beta\frac{\delta}{\sqrt{r_\varepsilon}}\right\}\frac{1}{2}\left(\frac{\sigma_n^2(y)}{\delta^2} + \frac{\mu_n^2(y)}{\delta^2 r_\varepsilon} + \frac{\mu_n(y)}{\delta\sqrt{r_\varepsilon}}\right) \\ &\quad + \exp\left\{-\beta\frac{\delta}{\sqrt{r_\varepsilon}}\right\}\frac{1}{2}\left(\frac{\sigma_n^2(y)}{\delta^2} + \frac{\mu_n^2(y)}{\delta^2 r_\varepsilon} - \frac{\mu_n(y)}{\delta\sqrt{r_\varepsilon}}\right) \\ &\quad + \left(1 - \frac{\sigma_n^2(y)}{\delta^2} - \frac{\mu_n^2(y)}{\delta^2 r_\varepsilon}\right) \\ &= \left(1 + \frac{\beta\delta}{\sqrt{r_\varepsilon}} + \frac{\beta^2\delta^2}{2! r_\varepsilon} + \cdots\right)\frac{1}{2}\left(\frac{\sigma_n^2(y)}{\delta^2} + \frac{\mu_n^2(y)}{\delta^2 r_\varepsilon} + \frac{\mu_n(y)}{\delta\sqrt{r_\varepsilon}}\right) \\ &\quad + \left(1 - \frac{\beta\delta}{\sqrt{r_\varepsilon}} + \frac{\beta^2\delta^2}{2! r_\varepsilon} + \cdots\right)\frac{1}{2}\left(\frac{\sigma_n^2(y)}{\delta^2} + \frac{\mu_n^2(y)}{\delta^2 r_\varepsilon} - \frac{\mu_n(y)}{\delta\sqrt{r_\varepsilon}}\right) \\ &\quad + \left(1 - \frac{\sigma_n^2(y)}{\delta^2} - \frac{\mu_n^2(y)}{\delta^2 r_\varepsilon}\right) \\ &= 1 + \left(\frac{\beta^2\delta^2}{2! r_\varepsilon} + \frac{\beta^4\delta^4}{4! r_\varepsilon^2} + \cdots\right)\left(\frac{\sigma_n^2(y)}{\delta^2} + \frac{\mu_n^2(y)}{\delta^2 r_\varepsilon}\right) \\ &\quad + \left(\frac{\beta\delta}{\sqrt{r_\varepsilon}} + \frac{\beta^3\delta^3}{3!(\sqrt{r_\varepsilon})^3} + \cdots\right)\frac{\mu_n(y)}{\delta\sqrt{r_\varepsilon}} \\ &\le 1 + \frac{\beta^2\delta^2}{2r_\varepsilon}\left(1 + \frac{\beta^2\delta^2}{r_\varepsilon} + \cdots\right)\left(\frac{K_{89}}{\delta^2} + \frac{K_{89}^2}{\delta^2 r_\varepsilon}\right) + \frac{|\beta|\delta}{\sqrt{r_\varepsilon}}\left(1 + \frac{\beta^2\delta^2}{r_\varepsilon} + \cdots\right)\frac{K_{89}}{\delta\sqrt{r_\varepsilon}} \\ &= 1 + \frac{1}{r_\varepsilon}e^{\frac{\beta^2\delta^2}{r_\varepsilon}}\left(\frac{1}{2}\beta^2\left(K_{89} + \frac{K_{89}^2}{r_\varepsilon}\right) + |\beta| K_{89}\right). \end{aligned} \tag{10.141}$$

The relation (10.141) implies that for every $\varepsilon \in (0, \varepsilon_0]$, $y \in \mathbb{R}_1$, $n = 1, \ldots, N$, and any $\beta \in \mathbb{R}_1$,

$$\mathsf{E}e^{\beta W_{\varepsilon,n}(y)} \leq \left(1 + \frac{1}{r_\varepsilon} e^{\frac{\beta^2\delta^2}{r_\varepsilon}} \left(\frac{1}{2}\beta^2 \left(K_{89} + \frac{K_{89}^2}{r_\varepsilon}\right) + |\beta| K_{89}\right)\right)^{r_\varepsilon}$$
$$\leq \exp\left\{e^{\frac{\beta^2\delta^2}{r_\varepsilon}} \left(\frac{1}{2}\beta^2 \left(K_{89} + \frac{K_{89}^2}{r_\varepsilon}\right) + |\beta| K_{89}\right)\right\}. \tag{10.142}$$

Also, relation (10.140) implies that for every $y \in \mathbb{R}_1$, $n = 1, \ldots, N$, and any $\beta \in \mathbb{R}_1$,

$$\mathsf{E}e^{\beta W_{0,n}(y)} \leq e^{(|\beta| + \frac{1}{2}\beta^2)K_{89}}. \tag{10.143}$$

The modulus $\Delta_{\bar{\beta}}^{\pm}(Y_{\varepsilon,\cdot}, N)$ takes for the above log-price processes $Y_{\varepsilon,n}$ the following form, for every $\varepsilon \in [0, \varepsilon_0]$ and $\beta \geq 0$:

$$\Delta_{\bar{\beta}}^{\pm}(Y_{\varepsilon,\cdot}, N) = \max_{1 \leq n \leq N} \sup_{y \in \mathbb{R}_1} e^{\pm\beta W_{\varepsilon,n}(y)}. \tag{10.144}$$

Relations (10.142) and (10.143) imply the following asymptotic relation, which holds for any $\beta \geq 0$:

$$\overline{\lim_{\varepsilon\to 0}} \Delta_{\bar{\beta}}^{\pm}(Y_{\varepsilon,\cdot}, N) \leq \overline{\lim_{\varepsilon\to 0}} \exp\left\{e^{\frac{\beta^2\delta^2}{r_\varepsilon}} \left(\frac{1}{2}\beta^2 \left(K_{89} + \frac{K_{89}^2}{r_\varepsilon}\right) + \beta K_{89}\right)\right\}$$
$$= \exp\left\{\left(\beta + \frac{1}{2}\beta^2\right) K_{89}\right\} < \infty. \tag{10.145}$$

Relations (10.142) and (10.145) imply that the condition $\mathbf{E}_{10}[\bar{\beta}]$ (its one-dimensional variant for the model without index component) holds.

Therefore, by Lemma 5.1.5, the condition $\mathbf{C}_6[\bar{\beta}]$ (its one-dimensional variant for the model without index component) holds for any $\beta \geq 0$, for example with the constant $K_{94} = 1 + 2e^{(|\beta| + \beta^2)K_{89}}$ replacing the constant $K_{30,1}$ penetrating this condition.

The condition $\mathbf{B}_4[\gamma]$ is a one-dimensional variant of the condition $\mathbf{B}_5[\bar{\gamma}]$ (its one-dimensional variant for the model without index component and a pay-off function, which does not depend on the parameter ε). Note that one can always take parameter $\beta > \gamma$, if $\gamma > 0$, or $\beta = \gamma$, if $\gamma = 0$.

The condition $\mathbf{I}_1$ (its one-dimensional variant for the model without index component and a pay-off function, which does not depend on the parameter ε) reduces in this case to the form of condition $\mathbf{I}_9$.

Let us take an arbitrary $n = 1, \ldots, N$ and $y_\varepsilon \to y_0 \in \mathbb{Y}_{n-1} \cap \mathbb{Y}'_{n-1}$ as $\varepsilon \to 0$. This relation and the conditions $\mathbf{I}_9$ and $\mathbf{J}_{16}$ imply that

$$\mu_n(y_\varepsilon) \to \mu_n(y_0) \quad \text{and} \quad \sigma_n^2(y_\varepsilon) \to \sigma_n(y_0) \quad \text{as } \varepsilon \to 0. \tag{10.146}$$

Therefore, for every $n = 1, \dots, N$ and $\beta \in \mathbb{R}_1$

$$
\begin{aligned}
\mathsf{E}e^{\beta W_{\varepsilon,n}(y_\varepsilon)} &= \Bigg(\Bigg(1 + \frac{\beta\delta}{\sqrt{r_\varepsilon}} + \frac{\beta^2\delta^2}{2r_\varepsilon} + o_2\Big(\frac{\beta\delta}{\sqrt{r_\varepsilon}}\Big)\Bigg) \\
&\quad \times \frac{1}{2}\Bigg(\frac{\sigma_n^2(y_\varepsilon)}{\delta^2} + \frac{\mu_n^2(y_\varepsilon)}{\delta^2 r_\varepsilon} + \frac{\mu_n(y_\varepsilon)}{\delta\sqrt{r_\varepsilon}}\Bigg) \\
&\quad + \Bigg(1 - \frac{\beta\delta}{\sqrt{r_\varepsilon}} + \frac{\beta^2\delta^2}{2r_\varepsilon} + o_2\Big(-\frac{\beta\delta}{\sqrt{r_\varepsilon}}\Big)\Bigg) \\
&\quad \times \frac{1}{2}\Bigg(\frac{\sigma_n^2(y_\varepsilon)}{\delta^2} + \frac{\mu_n^2(y_\varepsilon)}{\delta^2 r_\varepsilon} - \frac{\mu_n(y_\varepsilon)}{\delta\sqrt{r_\varepsilon}}\Bigg) \\
&\quad + \Bigg(1 - \frac{\sigma_n^2(y_\varepsilon)}{\delta^2} - \frac{\mu_n^2(y_\varepsilon)}{\delta^2 r_\varepsilon}\Bigg)\Bigg)^{r_\varepsilon} \\
&= \Bigg(1 + \beta\frac{\mu_n(y_\varepsilon)}{r_\varepsilon} + \beta^2\frac{\sigma_n^2(y_\varepsilon)}{2r_\varepsilon} + \beta^2\frac{\mu_n^2(y_\varepsilon)}{2r_\varepsilon^2} \\
&\quad + o_2\Big(\frac{\beta\delta}{\sqrt{r_\varepsilon}}\Big)\frac{1}{2}\Bigg(\frac{\sigma_n^2(y_\varepsilon)}{\delta^2} + \frac{\mu_n^2(y_\varepsilon)}{\delta^2 r_\varepsilon} + \frac{\mu_n(y_\varepsilon)}{\delta\sqrt{r_\varepsilon}}\Bigg) \\
&\quad + o_2\Big(-\frac{\beta\delta}{\sqrt{r_\varepsilon}}\Big)\Bigg)\frac{1}{2}\Bigg(\frac{\sigma_n^2(y_\varepsilon)}{\delta^2} + \frac{\mu_n^2(y_\varepsilon)}{\delta^2 r_\varepsilon} - \frac{\mu_n(y_\varepsilon)}{\delta\sqrt{r_\varepsilon}}\Bigg)\Bigg)^{r_\varepsilon} \\
&\to \exp\Big\{\beta\mu(y_0) + \frac{1}{2}\beta^2\sigma^2(y_0)\Big\} = \mathsf{E}e^{\beta W_0(y_0)} \quad \text{as } \varepsilon \to 0,
\end{aligned} \tag{10.147}
$$

since,

$$
\begin{aligned}
&\Bigg|\beta^2\frac{\mu_n^2(y_\varepsilon)}{2r_\varepsilon^2} + o_2\Big(\frac{\beta\delta}{\sqrt{r_\varepsilon}}\Big)\frac{1}{2}\Bigg(\frac{\sigma_n^2(y_\varepsilon)}{\delta^2} + \frac{\mu_n^2(y_\varepsilon)}{\delta^2 r_\varepsilon} + \frac{\mu_n(y_\varepsilon)}{\delta\sqrt{r_\varepsilon}}\Bigg) \\
&\quad + o_2\Big(-\frac{\beta\delta}{\sqrt{r_\varepsilon}}\Big)\frac{1}{2}\Bigg(\frac{\sigma_n^2(y_\varepsilon)}{\delta^2} + \frac{\mu_n^2(y_\varepsilon)}{\delta^2 r_\varepsilon} - \frac{\mu_n(y_\varepsilon)}{\delta\sqrt{r_\varepsilon}}\Bigg)\Bigg| r_\varepsilon \\
&\le \beta^2\frac{\mu_n^2(y_\varepsilon)}{2r_\varepsilon} + \Big|o_2\Big(\frac{\beta\delta}{\sqrt{r_\varepsilon}}\Big) r_\varepsilon\Big| + \Big|o_2\Big(-\frac{\beta\delta}{\sqrt{r_\varepsilon}}\Big) r_\varepsilon\Big| \to 0 \quad \text{as } \varepsilon \to 0.
\end{aligned} \tag{10.148}
$$

Relation (10.147) implies, by the corresponding variant of the continuous theorem, the following relation, for an arbitrary $n = 1, \dots, N$ and $y_\varepsilon \to y_0 \in \mathbb{Y}_{n-1} \cap \mathbb{Y}'_{n-1}$ as $\varepsilon \to 0$:

$$
y_\varepsilon + W_{\varepsilon,n}(y_\varepsilon) \xrightarrow{d} y_0 + W_{0,n}(y_0) \quad \text{as } \varepsilon \to 0. \tag{10.149}
$$

Relation (10.149) is equivalent to the condition $\mathbf{J}_1$ (a) (its univariate variant for the model without index component).

The condition $\mathbf{J}_{16}$ (b) implies condition $\mathbf{J}_1$ (b) (its univariate variant for the model without index component) to hold, since random variables $W_{0,n}(y)$, $y \in \mathbb{R}_1$, $n = 1, \dots, N$ have normal distributions with positive variances, and, thus, possess positive probability density functions with respect to Lebesgue measure.

Thus, all conditions of Theorem 5.2.1 hold. By applying this theorem to the above approximation model, we get the asymptotic relation (10.138). □

10.3.3 Convergence of optimal expected rewards for inhomogeneous in space binomial and trinomial tree models

Let us first present the corresponding results for the univariate inhomogeneous in space binomial tree model considered in Subsection 10.3.1.

In this case, the optimal expected reward $\Phi_0 = \Phi_0(\mathcal{M}_{max,N}^{(0)})$ for the log-price process $Y_{0,n}$ is given by the following formula:

$$\Phi_0 = \mathsf{E}\phi_{0,0}(Y_{0,0}) = \int_{\mathbb{R}_1} \phi_{0,0}(y) P_0(dy) \,. \tag{10.150}$$

It is natural to choose, for every $\varepsilon \in (0, \varepsilon_0]$, the initial distribution $P_\varepsilon(A)$ for the approximating log-price process $Y_{\varepsilon,n}$ as the corresponding skeleton approximation for the distribution $P_0(A)$,

$$P_\varepsilon(A) = \sum_{y_{\varepsilon,0,0,l} \in A} P_0(\mathbb{I}_{\varepsilon,l}) \,, \quad A \in \mathcal{B}_1 \,, \tag{10.151}$$

where

$$y_{\varepsilon,0,0,l} = l \frac{\delta}{\sqrt{r_{\varepsilon,0}}} \,, \qquad l = -r_{\varepsilon,0}, \dots, r_{\varepsilon,0} \,, \tag{10.152}$$

and

$$\mathbb{I}_{\varepsilon,l} = \begin{cases} \left(-\infty, \frac{\delta}{\sqrt{r_{\varepsilon,0}}}\left(-r_{\varepsilon,0} + \frac{1}{2}\right)\right] & \text{if } l = -r_{\varepsilon,0} \,, \\ \left(\frac{\delta}{\sqrt{r_{\varepsilon,0}}}\left(l - \frac{1}{2}\right), \frac{\delta}{\sqrt{r_{\varepsilon,0}}}\left(l + \frac{1}{2}\right)\right] & \text{if } -r_{\varepsilon,0} < l < r_{\varepsilon,0} \,, \\ \frac{\delta}{\sqrt{r_\varepsilon}}(r_{\varepsilon,0} - \frac{1}{2}), \infty) & \text{if } l = r_{\varepsilon,0} \,, \end{cases} \tag{10.153}$$

where $r_{\varepsilon,0}$ is a positive integer number.

Correspondingly, the optimal expected reward $\Phi_\varepsilon = \Phi_\varepsilon(\mathcal{M}_{\mathrm{max},N}^{(\varepsilon)})$ for the log-price process $Y_{\varepsilon,n}$ is defined, for every $\varepsilon \in (0, \varepsilon_0]$, by the following formula:

$$\Phi_\varepsilon = \sum_{l=-r_{\varepsilon,0}}^{r_{\varepsilon,0}} \phi_{\varepsilon,0}(y_{\varepsilon,0,0,l}) P_0(\mathbb{I}_{\varepsilon,l}) \,. \tag{10.154}$$

Let us assume that the following condition holds for some $\beta \geq 0$:

$\mathbf{D}_{22}[\beta]$: $\mathsf{E}e^{\beta|Y_{0,0}|} < K_{95}$, for some $1 < K_{95} < \infty$,

and also that the following condition holds:

$\mathbf{K}_{20}$: $P_0(\overline{\mathbb{Y}}_0) = 0$, where $\mathbb{Y}_0$ is the set penetrating condition $\mathbf{I}_9$.

Theorem 10.3.3. *Let the log-price process $Y_{0,n}$ be represented by the inhomogeneous in space Gaussian random walk with transition probabilities given by relation* (10.103) *and the approximating log-price process is represented, for every $\varepsilon \in (0, \varepsilon_0]$, by the inhomogeneous binomial random walk $Y_{\varepsilon,n}$ with transition probabilities given by relation*

(10.104) *and parameters satisfying the condition* $\mathbf{G}_{12}$, *and given by relation* (10.106). *Let also conditions* $\mathbf{B}_4[\gamma]$ *and* $\mathbf{D}_{22}[\beta]$ *hold for parameters* γ *and* β *such that, either* $\beta > \gamma > 0$ *or* $\beta = \gamma = 0$, *and also the conditions* $\mathbf{I}_9$, $\mathbf{J}_{15}$, *and* $\mathbf{K}_{20}$ *hold, and* $r_{\varepsilon,0}, r_\varepsilon \to \infty$ *as* $\varepsilon \to 0$. *Then the following relation takes place:*

$$\Phi_\varepsilon \to \Phi_0 \quad as\ \varepsilon \to 0\,. \tag{10.155}$$

The proof of this theorem is analogous to the proof of Theorem 10.1.3.

Let us now present the corresponding results for the univariate inhomogeneous in space and time trinomial tree model considered in Subsection 10.3.2

In fact, formulas (10.150)–(10.154) defining the optimal expected reward as well as the condition $\mathbf{D}_{22}[\beta]$ and $\mathbf{K}_{20}$ remain the same. The only difference is that the reward functions $\phi_{0,0}(\vec{y})$ for the trinomial model considered in Subsection 10.3.2 should be used in formulas (10.150) and (10.154) instead of the reward functions $\phi_{0,0}(\vec{y})$ for the binomial model considered in Subsection 10.3.1.

Theorem 10.3.4. *Let the log-price process* $Y_{0,n}$ *be represented by the inhomogeneous in time and space Gaussian random walk with transition probabilities given by relation* (10.127) *and the approximating log-price process is represented, for every* $\varepsilon \in (0, \varepsilon_0]$, *by the inhomogeneous trinomial random walk* $Y_{\varepsilon,n}$ *with transition probabilities given by relation* (10.128) *and parameters satisfying the conditions* $\mathbf{G}_{14}$, $\mathbf{G}_{15}$ *and given by relation* (10.130). *Let also the conditions* $\mathbf{B}_4[\gamma]$ *and* $\mathbf{D}_{22}[\beta]$ *hold for the parameters* γ *and* β *such that, either* $\beta > \gamma > 0$ *or* $\beta = \gamma = 0$, *and also the conditions* $\mathbf{I}_9$, $\mathbf{J}_{16}$, *and* $\mathbf{K}_{20}$ *hold, and* $r_{\varepsilon,0}, r_\varepsilon \to \infty$ *as* $\varepsilon \to 0$. *Then the following relation takes place:*

$$\Phi_\varepsilon \to \Phi_0 \quad as\ \varepsilon \to 0\,. \tag{10.156}$$

The proof of this theorem is analogous to the proof of Theorem 10.1.3.

Remark 10.3.1. If the model assumption (c) $\sigma^2(y) > 0$, $y \in \mathbb{R}_1$ (in the case of homogeneous in time model) may not hold, the condition $\mathbf{J}_{15}$ (a) can be replaced in Theorems 10.3.1 and 10.3.3 by the assumption that there exists a set $\mathbb{Y} \in \mathcal{B}_1$ such that the functions $\mu(y)$ and $\sigma^2(y)$ are continuous in points $y \in \mathbb{Y}$ and $\sigma^2(y) > 0$ for $y \in \mathbb{Y}$ (see relation (10.123)) and the condition $\mathbf{J}_{15}$ (b) by the assumption that $\mathsf{P}\{y_0 + W_{0,n}(y_0) \in \overline{\mathbb{Y}} \cup \overline{\mathbb{Y}}_n\} = 0$, for $y_0 \in \mathbb{Y} \cap \mathbb{Y}_{n-1}$, $n = 1, \dots, N$.

Remark 10.3.2. If the model assumption (c) $\sigma_n^2(y) > 0$, $y \in \mathbb{R}_1$, $n = 1, 2, \dots$ (in the case of inhomogeneous in time model) may not hold, the condition $\mathbf{J}_{16}$ (b) can be replaced in Theorems 10.3.2 and 10.3.4 by the assumption that $\mathsf{P}\{y_0 + W_{0,n}(y_0) \in \overline{\mathbb{Y}}'_n \cup \overline{\mathbb{Y}}_n\} = 0$, for $y_0 \in \mathbb{Y}'_{n-1} \cap \mathbb{Y}_{n-1}$, $n = 1, \dots, N$.

10.4 Multivariate inhomogeneous in space tree models

In this section, we present results about convergence of rewards for American-type options for multivariate trinomial tree approximations for log-price processes represented by multivariate inhomogeneous in time and space Gaussian random walks.

10.4.1 Convergence of reward functions for multivariate inhomogeneous in time and space trinomial tree approximations

Let us assume that a log-price process $\vec{Y}_{0,n}$ is an inhomogeneous in time and space Gaussian Markov chain with the phase space $\mathbb{R}_k$, an initial distribution $P_0(A)$, and transition probabilities given for $\vec{y} \in \mathbb{R}_k$, $A \in \mathcal{B}_k$, $n = 1, 2, \dots$ by the following relation:

$$P_{0,n}(\vec{y}, A) = \mathsf{P}\left\{\vec{Y}_{0,n} \in A / \vec{Y}_{0,n-1} = \vec{y}\right\} = \mathsf{P}\left\{\vec{y} + \vec{W}_{0,n}(\vec{y}) \in A\right\}, \tag{10.157}$$

where (a) $\vec{W}_{0,n}(\vec{y}) = (W_{0,n,1}(\vec{y}), \dots, W_{0,n,k}(\vec{y}))$ is, for every $n = 1, 2, \dots, \vec{y} \in \mathbb{R}_k$, a normal random variable with mean values $\mathsf{E}W_{0,n,i}(\vec{y}) = \mu_{n,i}(\vec{y})$, $i = 1, \dots, k$, variances $\mathrm{Var}W_{0,n,i}(\vec{y}) = \sigma^2_{n,i}(\vec{y})$, $i = 1, \dots, k$, and covariances $\mathsf{E}(W_{0,n,i}(\vec{y}) - \mu_{n,i}(\vec{y}))(W_{0,n,i}(\vec{y}) - \mu_{n,j}(\vec{y}))) = \rho_{n,i,j}(\vec{y})\sigma_{n,i}(\vec{y})\sigma_{n,j}(\vec{y})$, $i, j = 1, \dots, k$.

Also, usual assumptions are (b) $\mu_{n,i}(\vec{y})$, $i = 1, \dots, k$, $n = 1, 2, \dots$ are the measurable functions acting from $\mathbb{R}_k$ to $\mathbb{R}_1$, (c) $\sigma^2_{n,i}(\vec{y})$, $i = 1, \dots, k$, $n = 1, 2, \dots$ is a measurable function acting from $\mathbb{R}_1$ to $(0, \infty)$, (d) $\rho_{n,i,j}(\vec{y})$, $i, j = 1, \dots, k$, $n = 1, 2, \dots$ are the measurable functions acting from $\mathbb{R}_k$ to $[-1, 1]$, and $\rho_{n,i,i}(\vec{y}) = 1, \vec{y} \in \mathbb{R}_k$, $i = 1, \dots, k$, $n = 1, 2, \dots$, and (e) $\Sigma_n(\vec{y}) = \|\sigma_{n,i,j}(\vec{y}),\|, \vec{y} \in \mathbb{R}_k$, $n = 1, 2, \dots$, where $\sigma_{n,i,j}(\vec{y}) = \rho_{n,i,j}(\vec{y})\sigma_{n,i}(\vec{y})\sigma_{n,j}(\vec{y})$, $\vec{y} \in \mathbb{R}_k$, $i, j = 1, \dots, k$, $n = 1, 2, \dots$, are nonnegatively defined $k \times k$ matrices; (f) $\det(\Sigma_n(\vec{y})) \neq 0$, $\vec{y} \in \mathbb{R}_k$, $n = 1, 2, \dots$.

We shall approximate the log-price process $\vec{Y}_{0,n}$ by the log-price processes $\vec{Y}_{\varepsilon,n}$, which is, for every $\varepsilon \in (0, \varepsilon_0]$, an inhomogeneous in time and space trinomial Markov chain, with the phase space $\mathbb{R}_1$, an initial distribution $P_\varepsilon(A)$, and transition probabilities given for $\vec{y} \in \mathbb{R}_k$, $A \in \mathcal{B}_k$, $n = 1, 2, \dots$ by the following relation:

$$P_{\varepsilon,n}(\vec{y}, A) = \mathsf{P}\left\{\vec{Y}_{\varepsilon,n} \in A / \vec{Y}_{\varepsilon,n-1} = \vec{y}\right\} = \mathsf{P}\left\{\vec{y} + \vec{W}_{\varepsilon,n}(\vec{y}) \in A\right\}, \tag{10.158}$$

where (a) $\vec{W}_{\varepsilon,n}(\vec{y}) = (W_{\varepsilon,n,1}(\vec{y}), \dots, W_{\varepsilon,n,k}(\vec{y}))$ is, for every $\vec{y} \in \mathbb{R}_k$, $n = 1, 2, \dots$, a trinomial random vector, which can be represented in the form $\vec{W}_{\varepsilon,n}(\vec{y}) = \vec{W}_{\varepsilon,n,1}(\vec{y}) + \dots + \vec{W}_{\varepsilon,n,r_\varepsilon}(\vec{y})$, where $\vec{W}_{\varepsilon,n,l}(\vec{y}) = (W_{\varepsilon,n,l,1}(\vec{y}), \dots, W_{\varepsilon,n,l,k}(\vec{y}))$, $l = 1, 2, \dots$ are the i.i.d. random vectors taking values $\bar{\delta}_{\varepsilon,\bar{\jmath}} = (\delta_{\varepsilon,1,\jmath_1}, \dots, \delta_{\varepsilon,k,\jmath_k})$ with the probabilities $p_{\varepsilon,n,\bar{\jmath}}(\vec{y}) = p_{\varepsilon,n,\jmath_1,\dots\jmath_k}(\vec{y})$, for $\bar{\jmath} \in \mathcal{J}_k = \{\bar{\jmath} = \langle \jmath_1, \dots \jmath_k\rangle : \jmath_j, \dots \jmath_k = +, \circ, -\}$.

Also, usual assumptions are (b) $\delta_{\varepsilon i,-} = -\delta_{\varepsilon i} < \delta_{\varepsilon i,\circ} = 0 < \delta_{\varepsilon i,+} = +\delta_{\varepsilon i}$, $i = 1, \dots, k$, (c) $p_{\varepsilon,n,\bar{\jmath}}(\vec{y})$, $\bar{\jmath} \in \mathcal{J}_k$, $n = 1, 2, \dots$ are the measurable functions acting from $\mathbb{R}_k$

to $[0, 1]$ such that $\sum_{\bar{j} \in \mathcal{J}_k} p_{\varepsilon,n,\bar{j}}(\vec{y}) = 1, \vec{y} \in \mathbb{R}_k, n = 1, 2, \dots$; (d) r_ε is a positive integer number.

We use the log-price processes $\vec{Y}_{\varepsilon,n}$ for approximation of the log-price process $\vec{Y}_{0,n}$ as $\varepsilon \to 0$. In this case, a usual approach is to fit the corresponding parameters, i.e. to try to choose parameter $\delta_{\varepsilon,i}$, probabilities $p_{\varepsilon,n,\bar{j}}(\vec{y})$, and parameter r_ε in such a way that the following exact fitting equalities hold, for every $\varepsilon \in (0, \varepsilon_0]$:

$$\begin{cases} \mathsf{E}W_{\varepsilon,n,i}(\vec{y}) = \mathsf{E}W_{0,n,i}(\vec{y}), \quad i = 1, \dots, k, \quad \vec{y} \in \mathbb{R}_k, \quad n = 1, \dots, N, \\ \mathrm{Var}W_{\varepsilon,n,i}(\vec{y}) = \mathrm{Var}W_{0,1,i}(\vec{y}), \quad i = 1, \dots, k, \quad \vec{y} \in \mathbb{R}_k, \quad n = 1, \dots, N, \\ \mathsf{E}(W_{\varepsilon,n,i}(\vec{y}) - \mathsf{E}W_{\varepsilon,n,i}(\vec{y}))(W_{\varepsilon,n,i}(\vec{y}) - \mathsf{E}W_{\varepsilon,n,j}(\vec{y})) \\ \qquad = \mathsf{E}(W_{0,n,i}(\vec{y}) - \mathsf{E}W_{0,n,i}(\vec{y}))(W_{0,n,i}(\vec{y}) - \mathsf{E}W_{0,n,j}(\vec{y})), \\ \qquad 1 \le i < j \le k, \quad \vec{y} \in \mathbb{R}_k, \quad n = 1, \dots, N. \end{cases} \tag{10.159}$$

The solution of this fitting problem is presented in Lemma 9.5.2. According to this lemma, the fitting relations (10.159) hold, if to assume that the conditions $\mathbf{G}_{16}$, $\mathbf{G}_{17}$, and $\mathbf{R}_3$ hold, and to choose parameters of the log-price process $\vec{Y}_{\varepsilon,n}$ given by the following formulas, for every $\varepsilon \in (0, \varepsilon_0]$ and $n = 1, \dots, N$:

$$\begin{cases} \delta_{\varepsilon,i,\pm} = \pm \dfrac{\delta_i}{\sqrt{r_\varepsilon}}, \quad \delta_{\varepsilon,i,\circ} = 0, \\ p_{\varepsilon,n,i,\pm}(\vec{y}) = \dfrac{1}{2}\left(\dfrac{\sigma^2_{n,i}(\vec{y})}{\delta_i^2} + \dfrac{\mu^2_{n,i}(\vec{y})}{\delta_i^2 r_\varepsilon} \pm \dfrac{\mu_{n,i}(\vec{y})}{\delta_i \sqrt{r_\varepsilon}} \right), \\ p_{\varepsilon,n,i,\circ}(\vec{y}) = 1 - \dfrac{\sigma^2_{n,i}(\vec{y})}{\delta_i^2} - \dfrac{\mu^2_{n,i}(\vec{y})}{\delta_i^2 r_\varepsilon}, \quad i = 1, \dots, k, \quad \vec{y} \in \mathbb{R}_k, \quad n = 1, \dots, N, \end{cases} \tag{10.160}$$

and

$$\begin{cases} p_{\varepsilon,n,i,j,\pm,\pm}(\vec{y}) = \dfrac{1}{4}\left(\dfrac{\rho_{n,i,j}(\vec{y})\sigma_{n,i}(\vec{y})\sigma_{n,j}(\vec{y})}{\delta_i\delta_j} + \dfrac{\mu_{n,i}(\vec{y})\mu_{n,j}(\vec{y})}{\delta_i\delta_j r_\varepsilon} \right. \\ \qquad \left. + \left(\dfrac{\sigma^2_{n,i}(\vec{y})}{\delta_i^2} + \dfrac{\mu^2_{n,i}(\vec{y})}{\delta_i^2 r_\varepsilon} \right) \wedge \left(\dfrac{\sigma^2_{n,j}(\vec{y})}{\delta_j^2} + \dfrac{\mu^2_{n,j}(\vec{y})}{\delta_j^2 r_\varepsilon} \right) \right) \pm \dfrac{1}{2}\left(\dfrac{\mu_{n,i}(\vec{y})}{\delta_i\sqrt{r_\varepsilon}} + \dfrac{\mu_{n,j}(\vec{y})}{\delta_j\sqrt{r_\varepsilon}} \right) \\ p_{\varepsilon,n,i,j,\pm,\mp}(\vec{y}) = \dfrac{1}{4}\left(-\dfrac{\rho_{n,i,j}(\vec{y})\sigma_{n,i}(\vec{y})\sigma_{n,j}(\vec{y})}{\delta_i\delta_j} - \dfrac{\mu_{n,i}(\vec{y})\mu_{n,j}(\vec{y})}{\delta_i\delta_j r_\varepsilon} \right. \\ \qquad \left. + \left(\dfrac{\sigma^2_{n,i}(\vec{y})}{\delta_i^2} + \dfrac{\mu^2_{n,i}(\vec{y})}{\delta_i^2 r_\varepsilon} \right) \wedge \left(\dfrac{\sigma^2_{n,j}(\vec{y})}{\delta_j^2} + \dfrac{\mu^2_{n,j}(\vec{y})}{\delta_j^2 r_\varepsilon} \right) \right) \pm \dfrac{1}{2}\left(\dfrac{\mu_{n,i}(\vec{y})}{\delta_i\sqrt{r_\varepsilon}} - \dfrac{\mu_{n,j}(\vec{y})}{\delta_j\sqrt{r_\varepsilon}} \right) \\ p_{\varepsilon,n,i,j,\circ,\pm}(\vec{y}) = 1 - \dfrac{\sigma^2_{n,i}(\vec{y})}{\delta_i^2} - \dfrac{\mu^2_{n,i}(\vec{y})}{\delta_i^2 r_\varepsilon} - \left(1 - \dfrac{\sigma^2_{n,i}(\vec{y})}{\delta_i^2} - \dfrac{\mu^2_{n,i}(\vec{y})}{\delta_i^2 r_\varepsilon}\right) \wedge \left(1 - \dfrac{\sigma^2_{n,j}(\vec{y})}{\delta_j^2} - \dfrac{\mu^2_{n,j}(\vec{y})}{\delta_j^2 r_\varepsilon}\right), \\ p_{\varepsilon,n,i,j,\pm,\circ}(\vec{y}) = 1 - \dfrac{\sigma^2_{n,j}(\vec{y})}{\delta_j^2} - \dfrac{\mu^2_{n,j}(\vec{y})}{\delta_j^2 r_\varepsilon} - \left(1 - \dfrac{\sigma^2_{n,i}(\vec{y})}{\delta_i^2} - \dfrac{\mu^2_{n,i}(\vec{y})}{\delta_i^2 r_\varepsilon}\right) \wedge \left(1 - \dfrac{\sigma^2_{n,j}(\vec{y})}{\delta_j^2} - \dfrac{\mu^2_{n,j}(\vec{y})}{\delta_j^2 r_\varepsilon}\right), \\ p_{\varepsilon,n,i,j,\circ,\circ}(\vec{y}) = \left(1 - \dfrac{\sigma^2_{n,i}(\vec{y})}{\delta_i^2} - \dfrac{\mu^2_{n,i}(\vec{y})}{\delta_i^2 r_\varepsilon}\right) \wedge \left(1 - \dfrac{\sigma^2_{n,j}(\vec{y})}{\delta_j^2} - \dfrac{\mu^2_{n,j}(\vec{y})}{\delta_j^2 r_\varepsilon}\right), \\ 1 \le i < j \le k, \quad \vec{y} \in \mathbb{R}_k, \quad n = 1, \dots, N, \end{cases} \tag{10.161}$$

where (i) one can choose for the parameters δ_i, $i = 1, \dots, k$ and r_ε any values such that (a) $\delta_i \geq \sqrt{K_{90}^2 + K_{90}}$, $i = 1, \dots, k$, (b) $\rho < \kappa \leq 1$, (c) $r_\varepsilon \geq \frac{K_{90}^2}{\sigma^2(\kappa-\rho)} + \frac{16K_{90}^2\delta^2}{\sigma^4(\kappa-\rho)^2}$, where $\delta = \max_{1\leq i\leq k}\delta_i$, $\kappa = \min_{1\leq i,j\leq k}\frac{\delta_i}{\delta_j}$, while $\sigma^2 > 0$, $K_{90} < \infty$, and $0 \leq \rho < 1$ are parameters penetrating, respectively, the conditions $\mathbf{G}_{16}$, $\mathbf{G}_{17}$, and $\mathbf{R}_3$; (ii) the corresponding higher order distributions for components of random vectors $\vec{W}_{\varepsilon,n,1}$ consistent with the above one- and two-dimensional distributions can be defined using algorithm described in Lemma 9.3.2.

Let us denote,

$$\mathcal{L}_{k,r} = \{\bar{l} = (l_1, \dots, l_k) : -r \leq l_1, \dots, l_k \leq r\},\ r = 1, 2, \dots . \tag{10.162}$$

The random vector $\vec{W}_{\varepsilon,n}(\vec{y})$ is for every $\vec{y} \in \mathbb{R}_k$, $n = 1, \dots, N$ a trinomial random vector. Namely, the random vector $\vec{W}_{\varepsilon,n}(\vec{y})$ takes values $\vec{w}_{\varepsilon,n,r_\varepsilon,\bar{l}}$ with the corresponding probabilities $p_{\varepsilon,n,r_\varepsilon}(\bar{l})$, for $\bar{l} = (l_1, \dots, l_k) \in \mathcal{L}_{k,r_\varepsilon}$.

Here, for every $\bar{l} = (l_1, \dots, l_k) \in \mathcal{L}_{k,r}$, $r = 1, 2, \dots$,

$$\vec{w}_{\varepsilon,n,r,\bar{l}} = (l_1\delta_{\varepsilon,1}, \dots, l_k\delta_{\varepsilon,k})\,, \tag{10.163}$$

and

$$\begin{aligned} p_{\varepsilon,n,r,\bar{l}}(\vec{y}) = \mathsf{P}\Big\{\vec{W}_{\varepsilon,n}(\vec{y}) &= \vec{W}_{\varepsilon,n,1}(\vec{y}) \\ &+ \cdots + \vec{W}_{\varepsilon,n,r}(\vec{y}) = \vec{w}_{\varepsilon,n,r,\bar{l}}\Big\} I(\bar{l} \in \mathcal{L}_{k,r})\,. \end{aligned} \tag{10.164}$$

The probabilities $p_{\varepsilon,n,r,\bar{l}}(\vec{y})$ can be found using the following recurrent convolution formulas:

$$p_{\varepsilon,n,r,\bar{l}}(\vec{y}) = \begin{cases} p_{\varepsilon,n,\bar{\jmath}}(\vec{y}) I(\bar{l} = \bar{e}_{\bar{\jmath}}) & \text{if } r = 1, \bar{l} \in \mathcal{L}_{k,1}\,, \\ \sum_{\bar{l}' \in \mathcal{L}_{k,1}} p_{\varepsilon,n,r-1,\bar{l}-\bar{l}'}(\vec{y}) p_{\varepsilon,n,1,\bar{l}'}(\vec{y}) & \text{if } r > 1, \bar{l} \in \mathcal{L}_{k,r}\,. \end{cases} \tag{10.165}$$

where the vectors $\bar{e}_{\bar{\jmath}} = (e_{\jmath_1}, \dots, e_{\jmath_k})$, $\bar{\jmath} = (\jmath_1, \dots, \jmath_k) \in \mathcal{J}_k$, $e_\jmath = 1, 0, -1$ if, respectively, $\jmath = +, \circ, -$.

The additional comments concerning algorithms for computing of probabilities $p_{\varepsilon,n,r,\bar{l}}(\vec{y})$ can be found in Subsection 3.2.3.

Let us assume that $\vec{Y}_{\varepsilon,n} = \vec{y} = (y_1, \dots, y_k) = \text{const} \in \mathbb{R}_k$. In this case, trajectories of the log-price process $\vec{Y}_{\varepsilon,n+r}$, $r = 0, 1, \dots$ generate, for every $\varepsilon \in (0, \varepsilon_0]$, the unique tree of nodes

$$T_{\varepsilon,(\vec{y},n)} = \{(\vec{y}_{\varepsilon,n,n+r,\bar{l}}, n + r),\ \bar{l} = (l_1, \dots, l_k) \in \mathcal{L}_{k,r_\varepsilon r}\,,\ r = 0, 1, \dots\}\,, \tag{10.166}$$

where

$$\begin{aligned} \vec{y}_{\varepsilon,n,n+r,\bar{l}} &= (y_{\varepsilon,n,n+r,1,l_1}, \dots, y_{\varepsilon,n,n+r,k,l_k}) \\ &= (y_1 + l_1\delta_{\varepsilon,1}, \dots, y_k + l_k\delta_{\varepsilon,k}), \\ &\quad \bar{l} = (l_1, \dots, l_k) \in \mathcal{L}_{k,r_\varepsilon r}\,,\ r = 0, 1, \dots . \end{aligned} \tag{10.167}$$

The tree $T_{\varepsilon,(\vec{y},n)}$ contains the total number of nodes after r steps,

$$L_{\varepsilon,n,n+r} \leq 1 + (2r_\varepsilon + 1)^k + \cdots + (2r_\varepsilon r + 1)^k$$
$$\leq \int_0^{r+1} (2r_\varepsilon x + 1)^k dx = \frac{(2r_\varepsilon(r+1)+1)^{k+1} - 1}{2r_\varepsilon(k+1)} . \tag{10.168}$$

The expression on the right-hand side in (10.168) is a polynomial function of r of the order $k + 1$.

The following lemma is a direct corollary of Lemma 3.4.1.

Lemma 10.4.1. *Let, for every $\varepsilon \in (0, \varepsilon_0]$, the approximating log-price process $\vec{Y}_{\varepsilon,n}$ is the multivariate trinomial random walk given by the stochastic transition dynamic relation* (10.158), *with parameters defined in relations* (10.160) *and* (10.161), *under the assumption that conditions* $\mathbf{G}_{16}$, $\mathbf{G}_{17}$, *and* $\mathbf{R}_3$ *hold. Then the reward functions $\phi_{\varepsilon,n+r}(\vec{y}_{\varepsilon,n,n+r,\bar{l}})$, for the points $\vec{y}_{\varepsilon,n,n,0} = \vec{y}$ and $\vec{y}_{\varepsilon,n,n+r,\bar{l}}, \bar{l} \in \mathcal{L}_{k,r_\varepsilon r}$, $r = 1, \ldots N - n$, are, for every $\varepsilon \in (0, \varepsilon_0]$, and $\vec{y} \in \mathbb{R}_k$, $n = 0, \ldots, N$, the unique solution for the following recurrence finite system of linear equations:*

$$\begin{cases} \phi_{\varepsilon,N}(\vec{y}_{\varepsilon,n,N,\bar{l}}) = g_\varepsilon(N, e^{\vec{y}_{\varepsilon,n,N,\bar{l}}}) , \\ \qquad \bar{l} \in \mathcal{L}_{k,r_\varepsilon(N-n)} , \\ \phi_{\varepsilon,n+r}(\vec{y}_{\varepsilon,n,n+r,\bar{l}}) = \max\Big(g_\varepsilon(n+r, e^{\vec{y}_{\varepsilon,n,n+r,\bar{l}}}), \\ \qquad \sum\limits_{\bar{l}' \in \mathcal{L}_{k,r_\varepsilon}} \phi_{\varepsilon,n+r+1}(\vec{y}_{\varepsilon,n,n+r+1,\bar{l}+\bar{l}'}) p_{\varepsilon,n+r+1,r_\varepsilon,\bar{l}'}(\vec{y}_{\varepsilon,n,n+r,\bar{l}}) \Big) , \\ \qquad \bar{l} \in \mathcal{L}_{k,r_\varepsilon r} , \quad r = N-n-1, \ldots, 1 , \\ \phi_{\varepsilon,n}(\vec{y}) = \max\Big(g_\varepsilon(n, e^{\vec{y}}), \sum\limits_{\bar{l}' \in \mathcal{L}_{k,r_\varepsilon}} \phi_{\varepsilon,n}(\vec{y}_{\varepsilon,n,n+1,\bar{l}'}) p_{\varepsilon,n+1,r_\varepsilon,\bar{l}'}(\vec{y}) \Big) . \end{cases} \tag{10.169}$$

We are going to apply Theorem 5.2.1 in order to prove convergence of the reward functions for the log-price processes $\vec{Y}_{\varepsilon,n}$ to the corresponding reward functions for the log-price process $\vec{Y}_{0,n}$ as $\varepsilon \to 0$.

We assume that the condition $\mathbf{B}_2[\bar{\gamma}]$ holds for the pay-off function $g(n, e^{\vec{y}})$, for some vector parameter $\bar{\gamma} = (\gamma_1, \ldots, \gamma_k)$ with nonnegative components.

The condition $\mathbf{I}_1$ reduces in this case to the condition $\mathbf{I}_{10}$.

Let us also assume that the following condition holds:

$\mathbf{J}_{17}$: (a) There exists set $\mathbb{Y}'_n \in \mathcal{B}_k$, $n = 0, 1, \ldots, N$ such that functions $\mu_{n,i}(\vec{y})$, $\sigma^2_{n,i}(\vec{y}), \rho^2_{n,i,j}(\vec{y}), i, j = 1, \ldots, k$ are continuous in points $y \in \mathbb{Y}'_{n-1}$, for every $n = 1, \ldots, N$;

(b) $L_k(\overline{\mathbb{Y}}_n) = 0, L_k(\overline{\mathbb{Y}}'_n) = 0$, $n = 1, \ldots, N$, where $\mathbb{Y}_n$, $n = 1, \ldots, N$ are sets penetrating condition $\mathbf{I}_{10}$.

We shall see that, in this case, the condition $\mathbf{J}_1$ (a) holds for log-prices $\vec{Y}_{\varepsilon,n}$ with parameters given by relation (10.160), if $r_\varepsilon \to \infty$ as $\varepsilon \to 0$. Also, the random vectors $\vec{W}_{0,n}(\vec{y}), \vec{y} \in \mathbb{R}_k, n = 1, \dots, N$ have normal distributions that make it possible to replace the condition $\mathbf{J}_1$ (b) by the condition $\mathbf{J}_{17}$.

The above backward recurrence algorithm for finding reward functions is supported by the following convergence result.

Theorem 10.4.1. *Let the log-price process $\vec{Y}_{0,n}$ be represented by the multivariate inhomogeneous in space and time Gaussian random walk with transition probabilities given by relation* (10.157) *and the approximating log-price processes are represented, for every $\varepsilon \in (0, \varepsilon_0]$, by the multivariate inhomogeneous in space and time binomial random walks $Y_{\varepsilon,n}$ with transition probabilities given by relation* (10.158) *and parameters satisfying the conditions $\mathbf{G}_{16}$, $\mathbf{G}_{17}$, $\mathbf{R}_3$ and given by relation* (10.160) *and* (10.161). *Let also conditions $\mathbf{B}_2[\bar{\gamma}]$ holds for some vector parameter $\bar{\gamma} = (\gamma_1, \dots, \gamma_k)$ with nonnegative components, the conditions $\mathbf{I}_{10}$ and $\mathbf{J}_{17}$ hold, and $r_\varepsilon \to \infty$ as $\varepsilon \to 0$. Then the following relations take place for any $\vec{y}_\varepsilon \to \vec{y}_0 \in \mathbb{Y}_n \cap \mathbb{Y}'_n$, $n = 0, 1, \dots, N$:*

$$\phi_{\varepsilon,n}(\vec{y}_\varepsilon) \to \phi_{0,n}(\vec{y}_0) \quad \text{as } \varepsilon \to 0. \tag{10.170}$$

Proof. We are going to use Theorem 5.2.1.

Let the conditions $\mathbf{G}_{16}$, $\mathbf{G}_{17}$ hold and parameters of the process $Y_{\varepsilon,n}$ are given by relations (10.160) and (10.161).

In this case, for $\varepsilon \in (0, \varepsilon_0]$, the moment-generating functions for the random vectors $\vec{W}_{\varepsilon,n}(\vec{y}) = (W_{\varepsilon,n,1}(\vec{y}), \dots, W_{\varepsilon,n,k}(\vec{y})), \vec{y} \in \mathbb{R}_k, n = 1, \dots, N$ have the following form, for $\bar{\beta} = (\beta_1, \dots, \beta_k) \in \mathbb{R}_k$:

$$\mathsf{E} e^{\sum_{i=1}^k \beta_i W_{\varepsilon,n,i}(\vec{y})} = \left(\sum_{\bar{j} \in \mathcal{J}_k} e^{\sum_{i=1}^k \beta_i \frac{\delta_i e_{ji}}{\sqrt{r_\varepsilon}}} p_{\varepsilon,n,\bar{j}}(\vec{y}) \right)^{r_\varepsilon}, \tag{10.171}$$

while, for $\varepsilon = 0$, the moment-generating functions for the random vectors $\vec{W}_{0,n}(\vec{y}) = (W_{0,n,1}(\vec{y}), \dots, W_{0,n,k}(\vec{y})), \vec{y} \in \mathbb{R}_k$, $n = 1, \dots, N$ have the following form, for $\bar{\beta} = (\beta_1, \dots, \beta_k) \in \mathbb{R}_k$:

$$\begin{aligned} \mathsf{E} e^{\sum_{i=1}^k \beta_i W_{0,n,i}(\vec{y})} &= \exp\Bigg\{ \sum_{i=1}^k \beta_i \mu_{n,i}(\vec{y}) + \frac{1}{2} \sum_{i=1}^k \beta_i^2 \sigma_{n,i}^2(\vec{y}) \\ &\quad + \sum_{1 \le i < j \le k} \beta_i \beta_j \rho_{n,i,j}(\vec{y}) \sigma_{n,i}(\vec{y}) \sigma_{n,j}(\vec{y}) \Bigg\}. \end{aligned} \tag{10.172}$$

Relations (10.160) and (10.171) yield, for $\varepsilon \in (0, \varepsilon_0]$, the following formulas for the marginal moment-generating functions for the random variables $W_{\varepsilon,n,i}(\vec{y})$,

$i = 1, \dots, k, \vec{y} \in \mathbb{R}_k$, $n = 1, \dots, N$, for $\beta \in \mathbb{R}_1$:

$$\begin{aligned} \mathsf{E}e^{\beta W_{\varepsilon,n,i}(\vec{y})} &= \left(e^{\beta \frac{\delta_i}{\sqrt{r_\varepsilon}}} p_{\varepsilon,n,i,+}(\vec{y}) + e^{-\beta \frac{\delta_i}{\sqrt{r_\varepsilon}}} p_{\varepsilon,n,i,-}(\vec{y}) + p_{\varepsilon,n,i,\circ}(\vec{y})\right)^{r_\varepsilon} \\ &= \left(e^{\beta \frac{\delta_i}{\sqrt{r_\varepsilon}}} \frac{1}{2}\left(\frac{\sigma^2_{n,i}(\vec{y})}{\delta_i^2} + \frac{\mu^2_{n,i}(\vec{y})}{\delta_i^2 r_\varepsilon} + \frac{\mu_{n,i}(\vec{y})}{\delta_i \sqrt{r_\varepsilon}}\right)\right. \\ &\quad + e^{-\beta \frac{\delta_i}{\sqrt{r_\varepsilon}}} \frac{1}{2}\left(\frac{\sigma^2_{n,i}(\vec{y})}{\delta_i^2} + \frac{\mu^2_{n,i}(\vec{y})}{\delta_i^2 r_\varepsilon} - \frac{\mu_{n,i}(\vec{y})}{\delta_i \sqrt{r_\varepsilon}}\right) \\ &\quad \left. + \left(1 - \frac{\sigma^2_{n,i}(\vec{y})}{\delta_i^2} - \frac{\mu^2_{n,i}(\vec{y})}{\delta_i^2 r_\varepsilon}\right)\right)^{r_\varepsilon}, \end{aligned} \tag{10.173}$$

while, for $\varepsilon = 0$, these moment-generating functions have the following form:

$$\mathsf{E}e^{\beta W_{0,n,i}(\vec{y})} = e^{\beta \mu_{n,i}(\vec{y}) + \frac{1}{2}\beta^2 \sigma^2_{n,i}(\vec{y})} . \tag{10.174}$$

Relations (10.173)–(10.174) are analogous, respectively, to relations (10.139)–(10.140) used in the proof of Theorem 10.3.2. This makes it possible to repeat calculations given in the proof of this theorem and to get a relation analogous to relation (10.142), i.e. to get, under the assumption that the following condition holds, for every $\varepsilon \in (0, \varepsilon_0]$, $i = 1, \dots, k, \vec{y} \in \mathbb{R}_k$, $n = 1, \dots, N$, and any $\beta \in \mathbb{R}_1$:

$$\begin{aligned} \mathsf{E}e^{\beta W_{\varepsilon,n,i}(\vec{y})} &\le \left(1 + \frac{1}{r_\varepsilon} e^{\frac{\beta^2 \delta_i^2}{r_\varepsilon}} \left(\frac{1}{2}\beta^2 \left(K_{90} + \frac{K_{90}^2}{r_\varepsilon}\right) + |\beta| K_{90}\right)\right)^{r_\varepsilon} \\ &\le \exp\left\{e^{\frac{\beta^2 \delta_i^2}{r_\varepsilon}} \left(\frac{1}{2}\beta^2 \left(K_{90} + \frac{K_{90}^2}{r_\varepsilon}\right) + |\beta| K_{90}\right)\right\} , \end{aligned} \tag{10.175}$$

where K_{90} is the constant penetrating condition $\mathbf{G_{17}}$.

Also the condition $\mathbf{G_{17}}$ and relation (10.174) imply that, for every $\varepsilon \in (0, \varepsilon_0]$, $i = 1, \dots, k, \vec{y} \in \mathbb{R}_k$, $n = 1, \dots, N$, and any $\beta \in \mathbb{R}_1$,

$$\mathsf{E}e^{\beta W_{0,n,i}(\vec{y})} \le e^{(|\beta| + \frac{1}{2}\beta^2) K_{90}} . \tag{10.176}$$

The modulus $\Delta^{\pm}_{\bar{\beta}}(Y_{\varepsilon,i,\cdot}, N)$ takes for the above log-price processes $Y_{\varepsilon,n}$ the following form, for every $\varepsilon \in [0, \varepsilon_0]$, $i = 1, \dots, k$ and $\beta \ge 0$:

$$\Delta^{\pm}_{\bar{\beta}}(Y_{\varepsilon,i,\cdot}, N) = \max_{1 \le n \le N} \sup_{\vec{y} \in \mathbb{R}_k} e^{\pm \beta W_{\varepsilon,n,i}(\vec{y})} . \tag{10.177}$$

Relations (10.175) and (10.176) imply the following asymptotic relation, which holds for any $i = 1, \dots, k$ and $\beta \ge 0$:

$$\begin{aligned} \varlimsup_{\varepsilon \to 0} \Delta^{\pm}_{\bar{\beta}}(Y_{\varepsilon,i,\cdot}, N) &\le \varlimsup_{\varepsilon \to 0} \exp\left\{e^{\frac{\beta^2 \delta_i^2}{r_\varepsilon}} \left(\frac{1}{2}\beta^2 \left(K_{90} + \frac{K_{90}^2}{r_\varepsilon}\right) + \beta K_{90}\right)\right\} \\ &= \exp\left\{\left(\beta + \frac{1}{2}\beta^2\right) K_{90}\right\} < \infty . \end{aligned} \tag{10.178}$$

Relations (10.176) and (10.178) imply that the condition $\mathbf{E_{10}}[\bar{\beta}]$ (its variant for the model without index component) holds. Therefore, by Lemma 5.1.5, the condition $\mathbf{C_6}[\bar{\beta}]$ (its variant for the model without index component) holds for any vector parameter $\bar{\beta} = (\beta_1, \dots, \beta_k)$ with nonnegative components, for example, with the constant $K_{96} = 1 + 2e^{(|\beta| + \beta^2)K_{90}}$ replacing the constants $K_{30,i}$, $i = 1, \dots, k$ penetrating this condition.

The condition $\mathbf{B}_2[\bar{\gamma}]$ is a variant of the condition $\mathbf{B}_5[\bar{\gamma}]$ (its variant for the model without index component and a pay-off function, which does not depend on the parameter ε).

Note that one can always choose the parameter $\bar{\beta}$ such that, for every $i = 1, \dots, k$, either $\beta_i > \gamma_i$, if $\gamma_i > 0$, or $\beta_i = \gamma_i$, if $\gamma_i = 0$.

The condition $\mathbf{I}_1$ (its variant for the model without index component and a pay-off function, which does not depend on the parameter ε) reduces in this case to the form of the condition $\mathbf{I}_{10}$.

Let us take an arbitrary $n = 0, 1, \dots, N$ and $\vec{y}_\varepsilon \to \vec{y}_0 \in \mathbb{Y}_n \cap \mathbb{Y}'_n$ as $\varepsilon \to 0$. By the conditions $\mathbf{I}_{10}$ and $\mathbf{J}_{17}$, $\mu_{n,i}(\vec{y}_\varepsilon) \to \mu_{n,i}(\vec{y}_0)$, $\sigma^2_{n,i}(\vec{y}_\varepsilon) \to \sigma^2_{n,i}(\vec{y}_0)$ and $\rho_{n,i,j}(\vec{y}_\varepsilon) \to \rho_{n,i,j}(\vec{y}_0)$ as $\varepsilon \to 0$, for $i, j = 1, \dots, k$, and, therefore, for every $n = 1, \dots, N$ and $\bar{\beta} = (\beta_1, \dots, \beta_k) \in \mathbb{R}_k$,

$$
\begin{aligned}
\mathsf{E} e^{\sum_{i=1}^k \beta_i W_{\varepsilon,n,i}(\vec{y}_\varepsilon)} &= \left(\sum_{\bar{j} \in \mathcal{J}_k} \left(1 + \sum_{i=1}^k \beta_i \frac{\delta_i e_{j_i}}{\sqrt{r_\varepsilon}} + \frac{1}{2} \left(\sum_{i=1}^k \beta_i \frac{\delta_i e_{j_i}}{\sqrt{r_\varepsilon}} \right)^2 + o_2 \left(\sum_{i=1}^k \beta_i \frac{\delta_i e_{j_i}}{\sqrt{r_\varepsilon}} \right) \right) p_{\varepsilon,n,\bar{j}}(\vec{y}_\varepsilon) \right)^{r_\varepsilon} \\
&= \Bigg(1 + \sum_{i=1}^k \beta_i \frac{\delta_i}{\sqrt{r_\varepsilon}} \left(p_{\varepsilon,n,i,+}(\vec{y}_\varepsilon) - p_{\varepsilon,n,i,-}(\vec{y}_\varepsilon) \right) \\
&\quad + \frac{1}{2} \sum_{i=1}^k \beta_i^2 \frac{\delta_i^2}{r_\varepsilon} \left(p_{\varepsilon,n,i,+}(\vec{y}_\varepsilon) + p_{\varepsilon,n,i,-}(\vec{y}_\varepsilon) \right) \\
&\quad + \sum_{1 \le i < j \le k} \beta_i \beta_j \frac{\delta_i \delta_j}{r_\varepsilon} \Big(p_{\varepsilon,n,i,j,+,+}(\vec{y}_\varepsilon) + p_{\varepsilon,n,i,j,-,-}(\vec{y}_\varepsilon) \\
&\quad - p_{\varepsilon,n,i,j,+,-}(\vec{y}_\varepsilon) - p_{\varepsilon,n,i,j,-,+}(\vec{y}_\varepsilon) \Big) + \sum_{\bar{j} \in \mathcal{J}_k} o_2 \left(\sum_{i=1}^k \beta_i \frac{\delta_i e_{j_i}}{\sqrt{r_\varepsilon}} \right) p_{\varepsilon,n,\bar{j}}(\vec{y}_\varepsilon) \Bigg)^{r_\varepsilon} \\
&= \Bigg(1 + \frac{1}{r_\varepsilon} \Bigg(\sum_{i=1}^k \beta_i \mu_{n,i}(\vec{y}_\varepsilon) + \frac{1}{2} \sum_{i=1}^k \beta_i^2 \left(\sigma^2_{n,i}(\vec{y}_\varepsilon) + \frac{\mu^2_{n,i}(\vec{y}_\varepsilon)}{r_\varepsilon} \right) \\
&\quad + \sum_{1 \le i < j \le k} \beta_i \beta_j \Big(\rho_{n,i,j}(\vec{y}_\varepsilon) \sigma_{n,i}(\vec{y}_\varepsilon) \sigma_{n,j}(\vec{y}_\varepsilon) \\
&\quad + \frac{\mu_{n,i}(\vec{y}_\varepsilon) \mu_{n,j}(\vec{y}_\varepsilon)}{r_\varepsilon} \Big) \Bigg) + \sum_{\bar{j} \in \mathcal{J}_k} o_2 \left(\sum_{i=1}^k \beta_i \frac{\delta_i e_{j_i}}{\sqrt{r_\varepsilon}} \right) p_{\varepsilon,n,\bar{j}}(\vec{y}_\varepsilon) \Bigg)^{r_\varepsilon} \\
&\to \exp\left\{ \sum_{i=1}^k \beta_i \mu_{n,i}(\vec{y}_0) + \frac{1}{2} \sum_{i=1}^k \beta_i^2 \sigma^2_{n,i}(\vec{y}_0) + \sum_{1 \le i < j \le k} \beta_i \beta_j \rho_{n,i,j}(\vec{y}_0) \sigma_{n,i}(\vec{y}_0) \sigma_{n,j}(\vec{y}_0)) \right\} \\
&= \mathsf{E} e^{\sum_{i=1}^k \beta_i W_{0,n,i}(\vec{y}_0)} \quad \text{as } \varepsilon \to 0,
\end{aligned}
\tag{10.179}
$$

since,

$$\left| \frac{1}{r_\varepsilon} \left(\frac{1}{2} \sum_{i=1}^{k} \beta_i^2 \frac{\mu_{n,i}^2(\vec{y}_\varepsilon)}{r_\varepsilon} + \sum_{1 \le i < j \le k} \beta_i \beta_j \frac{\mu_{n,i}(\vec{y}_\varepsilon)\mu_{n,j}(\vec{y}_\varepsilon)}{r_\varepsilon} \right) + \sum_{\bar{\jmath} \in \mathcal{J}_k} o_2 \left(\sum_{i=1}^{k} \beta_i \frac{\delta_i e_{j_i}}{\sqrt{r_\varepsilon}} \right) p_{\varepsilon,n,\bar{\jmath}}(\vec{y}_\varepsilon) \right| r_\varepsilon$$
$$\le \frac{1}{2} \sum_{i=1}^{k} \beta_i^2 \frac{K_{90}^2}{r_\varepsilon} + \sum_{1 \le i < j \le k} |\beta_i \beta_j| \frac{K_{90}^2}{r_\varepsilon} + \sum_{\bar{\jmath} \in \mathcal{J}_k} \left| o_2 \left(\sum_{i=1}^{k} \beta_i \frac{\delta_i e_{j_i}}{\sqrt{r_\varepsilon}} \right) r_\varepsilon \right| \to 0 \quad \text{as } \varepsilon \to 0 . \tag{10.180}$$

Relation (10.179) implies, by the corresponding variant of the continuous theorem, the following relation, for an arbitrary $n = 1, \dots, N$ and $y_\varepsilon \to y_0 \in \mathbb{Y}_{n-1} \cap \mathbb{Y}'_{n-1}$ as $\varepsilon \to 0$:

$$y_\varepsilon + W_{\varepsilon,n}(y_\varepsilon) \xrightarrow{d} y_0 + W_{0,n}(y_0) \quad \text{as } \varepsilon \to 0 . \tag{10.181}$$

Relation (10.181) is equivalent to the condition $\mathbf{J}_1$ (a) (its variant for the model without index component).

The condition $\mathbf{J}_{17}$ (b) implies the condition $\mathbf{J}_1$ (b) (its variant for the model without index component) to hold, since the random vectors $\vec{W}_{0,n,}(y)$, $y \in \mathbb{R}_k$, $n = 1, \dots, N$ have, according to initial model assumptions, k-dimensional normal distributions with positive everywhere in $\mathbb{R}_k$ probability density functions.

Thus, all conditions of Theorem 5.2.1 hold. By applying this theorem to the above approximation model, we get the asymptotic relation (10.170). □

10.4.2 Convergence of optimal expected rewards for multivariate inhomogeneous in time and space trinomial tree models

Let us study the corresponding results for the multivariate trinomial tree model considered in Subsection 10.4.1.

In this case, the optimal expected reward $\Phi_0 = \Phi_0(\mathcal{M}^{(0)}_{max,N})$ for the log-price process $\vec{Y}_{0,n}$ is defined by the following formula:

$$\Phi_0 = \mathsf{E}\phi_{0,0}(\vec{Y}_{0,0}) = \int_{\mathbb{R}_k} \phi_{0,0}(\vec{y}) P_0(d\vec{y}) . \tag{10.182}$$

It is natural to choose, for every $\varepsilon \in (0, \varepsilon_0]$, the initial distribution $P_\varepsilon(A)$ for the approximating log-price process $\vec{Y}_{\varepsilon,n}$ as the corresponding skeleton approximation for the distribution $P_0(A)$,

$$P_\varepsilon(A) = \sum_{\vec{y}_{\varepsilon,0,0,\bar{l}} \in A} P_0(\mathbb{I}_{\varepsilon,\bar{l}}) , \quad A \in \mathcal{B}_k , \tag{10.183}$$

where

$$\vec{y}_{\varepsilon,0,0,\bar{l}} = \left(l_1 \frac{\delta_1}{\sqrt{r_{\varepsilon,0}}}, \dots, l_k \frac{\delta_k}{\sqrt{r_{\varepsilon,0}}} \right) \bar{l} = (l_1, \dots, l_k) \in \mathcal{L}_{k,r_{\varepsilon,0}}, \tag{10.184}$$

and

$$\mathbb{I}_{\varepsilon,\bar{l}} = \mathbb{I}_{\varepsilon,1,l_1} \times \cdots \times \mathbb{I}_{\varepsilon,k,l_k} \bar{l} = (l_1, \dots, l_k) \in \mathcal{L}_{k,r_{\varepsilon,0}}, \tag{10.185}$$

where, for $i = 1, \dots, k$,

$$\mathbb{I}_{\varepsilon,i,l} = \begin{cases} (-\infty, \frac{\delta_i}{\sqrt{r_{\varepsilon,0}}}(-r_{\varepsilon,0} + \frac{1}{2})] & \text{if } l = -r_{\varepsilon,0}, \\ (\frac{\delta_i}{\sqrt{r_{\varepsilon,0}}}(l - \frac{1}{2}), \frac{\delta_i}{\sqrt{r_{\varepsilon,0}}}(l + \frac{1}{2})] & \text{if } -r_{\varepsilon,0} < l < r_{\varepsilon,0}, \\ \frac{\delta_i}{\sqrt{r_{\varepsilon,0}}}(r_{\varepsilon,0} - \frac{1}{2}), \infty) & \text{if } l = r_{\varepsilon,0}. \end{cases} \tag{10.186}$$

and $r_{\varepsilon,0}$ is a positive integer number.

In this case, the optimal expected reward $\Phi_\varepsilon = \Phi_\varepsilon(\mathcal{M}^{(\varepsilon)}_{max,N})$ for the log-price process $Y_{0,n}$ is defined, for every $\varepsilon \in (0, \varepsilon_0]$, by the following formula:

$$\Phi_\varepsilon = \sum_{\bar{l} \in \mathcal{L}_{k,r_{\varepsilon,0}}} \phi_{\varepsilon,0}(\vec{y}_{\varepsilon,0,0,\bar{l}}) P_0(\mathbb{I}_{\varepsilon,\bar{l}}). \tag{10.187}$$

Let us assume that the following condition holds for some vector parameter $\bar{\beta} = (\beta_1, \dots, \beta_k)$ with nonnegative components:
$\mathbf{D_{23}}[\bar{\beta}]$: $\mathsf{E}e^{\beta_i |Y_{0,0,i}|} < K_{97}$, for some $1 < K_{97} < \infty$,

and also that the following condition holds:
$\mathbf{K_{21}}$: $P_0(\overline{\mathbb{Y}}_0) = 0$, where $\mathbb{Y}_0$ is the set defined in condition $\mathbf{I_{10}}$.

Theorem 10.4.2. *Let the log-price process $\vec{Y}_{0,n}$ be represented by the multivariate inhomogeneous in space and time Gaussian random walk with transition probabilities given by relation* (10.157) *and the approximating log-price processes are represented, for every $\varepsilon \in (0, \varepsilon_0]$, by the multivariate inhomogeneous in space and time binomial random walks $Y_{\varepsilon,n}$ with transition probabilities given by relation* (10.158) *and parameters satisfying the conditions* $\mathbf{G_{16}}$, $\mathbf{G_{17}}$, $\mathbf{R_3}$ *and given by relations* (10.160) *and* (10.161). *Let also the conditions* $\mathbf{B_2}[\bar{\gamma}]$ *and* $\mathbf{D_{23}}[\bar{\beta}]$ *hold for the parameters $\bar{\gamma}$ and $\bar{\beta}$ such that, for every $i = 1, \dots, k$, either $\beta_i > \gamma_i > 0$ or $\beta_i = \gamma_i = 0$, and also the conditions* $\mathbf{I_{10}}$, $\mathbf{J_{17}}$ *and* $\mathbf{K_{21}}$ *hold, and $r_{\varepsilon,0}, r_\varepsilon \to \infty$ as $\varepsilon \to 0$. Then the following relation takes place:*

$$\Phi_\varepsilon \to \Phi_0 \quad \text{as } \varepsilon \to 0. \tag{10.188}$$

The proof of this theorem is analogous to the proof of Theorem 10.1.3.

Remark 10.4.1. If the model assumption (f) $\det(\Sigma_n(\vec{y})) \neq 0, \vec{y} \in \mathbb{R}_k, n = 1, 2, \dots$ may not hold, the condition $\mathbf{J_{17}}$ (b) in Theorems 10.4.1 and 10.4.2, can be replaced by the assumption that $\mathsf{P}\{\vec{y}_0 + \vec{W}_{0,n}(\vec{y}_0) \in \overline{\mathbb{Y}}'_n \cup \overline{\mathbb{Y}}_n\} = 0$, for $\vec{y}_0 \in \mathbb{Y}'_{n-1} \cap \mathbb{Y}_{n-1}$, $n = 1, \dots, N$.

Bibliographical Remarks

This book is devoted to systematical presentation of stochastic approximation methods for American type options for discrete time Markov log-price processes.

However, the bibliography contains references to works related to a wider circle of option pricing problems. It also includes references related to continuous time models, other types of options, and alternative approximation methods.

We would like to note that the bibliography for the second volume of this book will extend the present bibliography and will also include references to works related to optimal stopping problems for stochastic processes beyond the framework of option pricing problems.

Chapters 1–2 (multivariate modulated Markov price processes, American type options, option prices, reward functions, optimal expected rewards, optimal stopping times).

In Chapter 1, we present basic classes of discrete time multivariate modulated Markov log-price processes and consider main examples of such processes, including log-price processes represented by multivariate modulated random walks with Markov and semi-Markov modulation, and various autoregressive and autoregressive moving average type log-price processes and autoregressive and autoregressive and moving average stochastic volatility models, including modulated AR, ARMA, ARCH, AR/ARCH, CIR, and GARCH models. In Chapter 2, we define American type options, present examples of such options and introduce basic objects connected with American type options, namely, reward functions, optimal expected rewards, and optimal stopping times. We also formulate results concerning backward recurrence relations for reward functions and hitting structure of optimal stopping times for American type options, which create a basis for stochastic approximation methods studied in the book.

Below, we present references related to the above problems.

The overview of the Brownian motion and the geometrical Brownian motion can be found in Cox and Rubinstein (1985), Karatzas and Shreve (1988, 2005), Hull (1993, 2011), Stoll and Whaley (1993), Jarrow and Turnbull (2000), Borodin and Salminen (2002). For more detailed overviews of stochastic processes relevant for financial applications, we refer to books by Pliska (1997), Mikosch (1998), Rolski, Schmidli, Schmidt, and Teugels (1999), Shiryaev (1999), Jarrow and Turnbull (2000), Kijima (2003), Øksendal (2003), Schoutens (2003), Cont and Tankov (2004), Janssen and Manca (2007), Janssen, Manca, and Volpe (2007, 2008), Mishura (2008), Applebaum (2009), and Di Nunno, Øksendal, and Proske (2009).

Some well-known books on mathematical finance are Hull (1993, 2011), Wilmott, Howison, and Dewynne (1995), Embrechts, Klüppelberg, and Mikosch (1997), Pliska (1997), Karatzas and Shreve (1998), Mikosch (1998), Wilmott (1998), Elliott and Kopp (1999), Shiryaev (1999), Karatzas and Shreve (2001), Glasserman (2003), Prigent

(2003), Björk (2004), Cont and Tankov (2004), Shreve (2004a, 2004b), Zhu, Wu, and Chen (2004), Detemple (2005), Musiela and Rutkowski (2005), Hull (2006, 2011), Peskir and Shiryaev (2006), Janssen, Manca, and Volpe (2007, 2008), Föllmer and Schied (2011), and Melnikov (2011).

We would also like to refer to books specially devoted to option pricing and related problems that are Jarrow (1996), Wong (1996), Salopek (1997), Kholodnyi and Price (1998), Wilmott (1998), Dzhaparidze (2000), Kallianpur and Karandikar (2000), Zhu (2000), Korn, R. and Korn, E, (2001), Vollert (2003), Benth (2004), Higham (2004), Ziegler (2004), Achdou and Pironneau (2005), Herwig (2005), Jiang (2005), Kyprianou, Schoutens, and Wilmott (2005), Schulmerich (2005), Haug (2006), Hellermann (2006), Baecker (2007), Brosch (2008), Repplinger (2008), Maruhn (2009), Rostek (2009), Profeta, Roynette, and Yor (2010), Pascucci (2011), Buchen (2012), Junghenn (2012), and Swishchuk (2013).

The papers by McKean (1965) and Samuelson (1965) were the first papers connected to optimal stopping problems with valuation of American type options. Related originating results on optimal stopping for discrete and continuous time stochastic processes can also be found in Dynkin (1965), Chow, Robbins, and Siegmund (1971), Darling, Liggett, and Taylor (1972), Shiryaev (1976), and Van Moerbeke (1976).

Black and Scholes (1973) and Merton (1973) derived the celebrated valuation formula for European option in their seminal papers. Cox and Ross (1976), Harrison and Kreps (1979), and Harrison and Pliska (1981) invented the risk-neutral valuation approach, and gave its theoretical foundation. The justification of formula for optimal expected reward as supremum over all Markov stopping times for expected reward of American-type option can be found in Bensoussan (1984), Karatzas (1988), Pliska (1997), Kwock (1999), Shiryaev (1999), Hull (2006), and Föllmer and Schied (2011).

Structure of optimal stopping strategies represented by the first hitting times into the corresponding space–time domains and related free boundary problems have been studied in Merton (1973), Van Moerbeke (1976), Salminen (1985), Geske and Shastri (1985b), Omberg (1987), Dochviri (1988), Kim (1990), Jacka (1991), Myneni (1992), Boyle (1993), Dewynne, Howison, Rupf, and Wilmott (1993), Kim and Byun (1994), Shiryaev (1994), Barles, Burdeau, Romano, and Samsoen (1995), Broadie and Detemple (1997, 1999), Kuske and Keller (1998), Kukush and Silvestrov (1999, 2000, 2004), Kwok (1999), Touzi (1999), Villeneuve (1999), Jönsson (2001), Peskir (2001, 2005), Basso, Nordon, and Pianca (2002a, 2002b, 2004), Evans, Kuske, and Keller (2002), Jönsson, Kukush, and Silvestrov (2002, 2004, 2005), Kijima (2002), Meyer (2002), Dayanik and Karatzas (2003), Janson and Tysk (2003), Wu and Fu (2003), Battauz and Pratelli (2004), Ekström (2004a, 2004b), Shi and Jin (2004), Chan (2005), Rapuch (2005), Blanchet (2006), Chengrong, Lishang, and Baojun (2006), Vasin and Morozov (2006), Ekström and Tysk (2007), Ivanov (2007), Nyström (2007, 2008), Alfonsi and Jourdain (2008), Henderson and Hobson (2008a), Leduc (2008), Muthuraman (2008), Novikov (2008), Pascucci (2008), Wong and Zhao (2008), Bayraktar and Xing (2009), Iftimie and Marinescu (2009), Juneja and Kalra (2009a), Laurence and Salsa

(2009), Roux and Zastawniak (2009), Bayraktar, Karatzas, and Yao (2010), Borovkov, Downes, and Novikov (2010), Chevalier (2010), Chiarella, Ziogas, and Ziveyi (2010), Kimura (2010), Wong and Zhao (2010), Belopol'skaya and Romadanova (2011), Salminen (2011), Trabelsi (2011), Zhu and Chen (2011a), and Detemple, Tian, and Xiong (2012).

Analytical solutions for American-type options and related results can be found in McKean (1965), Roll (1977), Geske (1979), Whaley (1981), Johnson (1983), Geske and Jonsson (1984), Geske and Shastri (1985), Salminen (1985, 1999), Blomeyer (1986), Kim (1990), McDonald and Schroder (1990), El Karoui and Karatzas (1991), Jacka (1991), Carr, Jarrow, and Myneni (1992), Myneni (1992), Rutkowski (1994), Shiryaev (1994), Chesney and Gibson (1995), Iwaki, Kijima, and Yoshida (1995), Carr (1998), Karatzas and Kou (1998), Broadie and Detemple (1999), Karatzas and Sudderth (1999), McDonald and Schroder (2000), Björk and Landén (2002), Lamberton and Villeneuve (2003), Elliott and Chan (2004), Gukhal (2004), Cruz-Báez and González-Rodríguez (2005), Ekström and Tysk (2006), Novikov and Shiryaev (2007), Eberlein, Papapantoleon, and Shiryaev (2008), Karlsen and Wallin (2009), Chiarella, Ziogas, and Ziveyi (2010), Shevchenko (2010), Zhang and Li (2010), Christensen, Irle, and Novikov (2011), and Zhao and Wong (2012).

Option valuation for multivariate price processes has been studied in Stulz (1982), Johnson (1987), Boyle (1988, 1993), Boyle, Evnine, and Gibbs (1989), Madan, Milne, and Shefrin (1989), Cheyette (1990), He (1990), Kamrad and Ritchken (1991), Rubinstein (1994), Tan and Vetzal (1995), Celtner, Riddiough, and Stojanovic (1996), Gerber and Shin (1996), Broadie and Detemple (1997), Villeneuve (1999), Detemple, Feng, and Tian (2003), Novikov and Shiryaev (2004), Bally, Pagès, and Printems (2005), Arnarson and Eriksson (2007), Berridge and Schumacher (2008a, 2008b), Lundgren, Silvestrov, and Kukush (2008), De Vallière, Denis, and Kabanov (2009), Lundgren and Silvestrov (2009, 2011), Lundgren (2010b), Silvetsrov and Lundgren (2011), Khizhnyak (2011), and Kohler and Krzyżak (2012).

Kijima and Yoshida (1993) have been the first to considered the problem of option pricing with volatility modulated by a Markov chain. Further works on option pricing for modulated price processes are Di Masi, Kabanov, and Runggaldier (1995), Swishchuk (1995), Bollen (1998), Griego and Swishchuk (2000), Guo (2001a, 2001b), Buffingtion and Elliot (2002), Duan, Popova, and Ritchken (2002), Guo and Zhang (2004), Silvestrov and Stenberg (2004), Elliot, Chan, and Siu (2005), Aingworth, Das, and Motwani (2006), Di Graziano and Rogers (2006), Jobert and Rogers (2006), Silvestrov, Jönsson, and Stenberg (2006, 2008a, 2008b, 2009), Stenberg (2006a, 2006b), Yao, Zhang, and Zhou (2006), Boyarchenko and Levendorskiǐ (2009a), İlhan, Jonsson, and Sircar (2009), Liu and Li (2009), Yang (2010), Yip, Stephens, and Olhede (2010), Basak, Ghosh, and Goswami (2011), Elliott and Siu (2011), Elliott, Siu, and Badescu (2011), Huang, Forsyth, and Labahn (2011), Van der Hoek and Elliott (2012), and Holmes, Yang, and Zhang (2012).

Surveys and results of comparison studies for American-type contingent claims can be found in Detemple (2005), Pressacco, Gaudenzi, Zanette, and Ziani (2008), Ahn, Bae, Koo, and Lee (2011), and Zhu (2011).

Chapters 3, 9–10 (atomic Markov chains, multivariate binomial and trinomial models, convergence of rewards for inhomogeneous in time and space binomial and trinomial approximation models).

In Chapter 3, we investigate the structure of backward recurrence relations for log-price processes represented by atomic Markov chains, whose transition probabilities are concentrated at finite sets. Examples of such models are binomial and trinomial tree models, general discrete random walk models, and atomic Markov chain models. The so-called recombining conditions, which guarantee that the number of time–space nodes for trajectories of such atomic Markov chains have not more than polynomial rate of growth as functions of the number of time steps are thoroughly investigated for all the above mentioned models. In Chapter 9, we present results concerning exact and asymptotical fitting of parameters of binomial and trinomial approximating log-price processes to parameters log-price processes represented by univariate and multivariate homogeneous and inhomogeneous in time and space Gaussian random walks. In Chapter 10, we present results about convergence of rewards for binomial and trinomial approximating log-price processes to the corresponding rewards for log-price processes represented of univariate and multivariate homogeneous and inhomogeneous in time and space Gaussian random walks.

Below, we present references related to these problems.

The binomial and trinomial methods for valuation of American options were introduced by Parkinson (1977), Sharpe (1978), and Cox, Ross, and Rubinstein (1979), and Rendleman and Bartter (1979). Further development of these methods has been achieved in Jarrow and Rudd (1983), Boyle (1986, 1988), Hull and White (1988), Omberg (1988), Nelson and Ramaswamy (1990), Amin (1991), Breen (1991), Trigeorgis (1991), Tian (1993), Kim and Byun (1994), Pelsser and Vorst (1994), Curran (1995), Broadie and Detemple (1996b), Leisen and Reimer (1996), Lamberton (1998), Leisen (1998), Jiang and Dai (1999, 2004), Qian, Xu, Jiang, and Bian (2005), Ahn and Song (2007), Liang, Hu, Jiang, and Bian (2007), Lundgren, Silvestrov, and Kukush (2008), Joshi (2009), Lundgren and Silvestrov (2009, 2011), Liang, Hu, and Jiang (2010), Lundgren (2010a, 2010b), Silvetsrov and Lundgren (2011), and Prékopa and Szántai (2012).

Binomial, trinomial, and multinomial option approximations for multivariate and modulated geometric diffusion type models have been studied in Boyle (1986, 1988, 1993), Boyle, Evnine, and Gibbs (1989), Madan, Milne, and Shefrin (1989), Cheyette (1990), He (1990), Kamrad and Ritchken (1991), Boyle (1993, 1998), Kijima and Yoshida (1993), Rubinstein (1994), Broadie and Detemple (1997), Jönsson (2005a, 2005b), Silvestrov, Jönsson, and Stenberg (2006, 2008a, 2008b, 2009), Stenberg (2006a, 2006b), Lundgren, Silvestrov, and Kukush (2008), Khaliq and Liu (2009), Lundgren and Silvestrov (2009, 2011), Vellekoop and Nieuwenhuis (2009), AitSahlia, Goswami, and Guha (2010), Dolinsky (2010, 2011), Lundgren (2010a), and Silvetsrov and Lundgren (2011).

Surveys of numerical methods related to the tree approximations can be found in Geske and Shastri (1985a), Broadie and Detemple (1996a, 1996b), Rogers and Talay (1997), Kwok (1999), Glasserman (2003), and Hull (2006).

Chapters 4–8 (upper bound, convergence, and space-skeleton approximations of option rewards for multivariate modulated Markov log-price processes including log-price processes represented by multivariate, modulated, inhomogeneous in time and space Markov Gaussian random walks).

In Chapter 4, we present upper bounds for option rewards, which are used for getting the corresponding convergence results. In Chapter 5, we present general results about convergence of American type option rewards for multivariate modulated Markov log-price processes with bounded moment characteristics. In Chapter 6, we present analogous convergence results for multivariate modulated Markov log-price processes with unbounded moment characteristics. In Chapter 7, we give results about space-skeleton reward approximations for multivariate modulated Markov log-price processes. In Chapter 8, we present results about convergence of rewards for Markov Gaussian log-price processes with estimated drift and volatility coefficients, and space-skeleton reward approximations for log-price processes represented by multivariate, modulated, inhomogeneous in time and space Markov Gaussian random walks.

Below, we present references related to these problems.

Convergence of expected values for reward-type functionals for various time and space-skeleton stochastic approximations for different classes of price processes and general Markov processes have been studied in Mackjavičjus (1973, 1974), Aldous (1978, 1989), Fährmann (1978, 1979, 1982), Cox, Ross, and Rubinstein (1979), Dochviri and Shashiashvili (1982, 1983), Barone-Adesi and Whaley (1987), Omberg (1987), Dochviri (1988), Lamberton and Pagès (1990), Bjerksund and Stensland (1993), Lamberton (1993, 1998), Amin and Khanna (1994), Broadie and Detemple (1995), Iwaki, Kijima, and Yoshida (1995), Cutland, Kopp, Willinger, and Wyman (1997), Leisen (1998), Mulinacci and Pratelli (1998), Jiang and Dai (1999), Prignet (2003), Jiang and Dai (2004), Nieuwenhuis and Vellekoop (2004), Silvestrov and Stenberg (2004), Bermin, Kohatsu-Higa, and Perelló (2005), Dupuis and Wang (2005), Jönsson (2005a, 2005b), Qian, Xu, Jiang, and Bian (2005), Deng, Gu, Wang, and Zhang (2006), Ekström (2006a), Maller, Solomon, and Szimayer (2006), Silvestrov, Jönsson, and Stenberg (2006, 2008a, 2008b, 2009), Stenberg (2006a, 2006b), Chaudhury (2007), Chung and Chang (2007), Coquet and Toldo (2007), Lundgren, Silvestrov, and Kukush (2008), Alfonsi and Jourdain (2009), Lundgren and Silvestrov (2009, 2011), Vellekoop and Nieuwenhuis (2009), Lundgren (2010b), Ekström, Lötstedt, Von Sydow, and Tysk (2011), Li and Xing (2011), Silvetsrov and Lundgren (2011), Trabelsi (2011), and Zhang and Wang (2011).

Some results on the rate of convergence for reward approximations for European and American options are given in Broadie and Detemple (1996b), Leisen and Reimer (1996), Lamberton (1998), Leisen (1998), Jiang and Dai (1999), Heston and

Zhou (2000), Kukush and Silvestrov (2001), Liang, Hu, Jiang, and Bian (2007), Liang (2008), Chen and Zhu (2009), Hussain and Shashiashvili (2010), Liang, Hu, and Jiang (2010), Wang, G. and Wang, S. (2010), Dolinsky (2011), Kwon and Lee (2011), and Silvestrov and Li (2013).

Stochastic approximation methods including time and space-skeleton-type approximations for options and other securities for different classes of scalar and vector stochastic processes have been also developed and used in Nelson and Ramaswamy (1990), Tian (1992, 1994), Bjerksund and Stensland (1993), Naik (1993), Hull and White (1994a, 1994b), Amin (1995), Amin and Bodurtha (1995), Li, Ritchken, and Sankarasubramanian (1995), Broadie and Detemple (1996b), Whalley and Wilmott (1999), Achdou, Indragoby, and Pironneau (2004), Lai and Wong (2004), Pironneau (2004), Achdou (2005), Almendral (2005), Bally, Pagès, and Printems (2005), Bouchard and Temam (2005), Chevalier (2005), Maller, Solomon, and Szimayer (2006), Wang, G. and Wang, S. (2006), Chaudhary (2007), Chen and Glasserman (2007), Chevalier (2007), Bender, Sottinen, and Valkeila (2008), Berridge and Schumacher (2008b), Aguilar (2009), Bayraktar and Xing (2009), Boyarchenko and Levendorskiĭ (2009b), Carr and Madan (2009), D'Amico, Janssen, and Manca (2009), Egloff and Leippold (2009), Liu and Hong (2009), Prékopa and Szántai (2010), Badescu, Elliott, Kulperger, Miettinen, and Siu (2011), Dolinsky (2011), Elliott, Liew, and Siu (2011), Zhu and Chen (2011b), Kwok, Leung, and Wong (2012), Silvestrov and Li (2013).

For survey purposes related to valuation of options, we refer to works by Broadie and Detemple (1996a, 1996b, 2004), and Detemple (2005).

Supplementary references (other approximation methods for American and other type of options).

In order to balance the bibliography, we also give references to works on other types of approximation methods for American and other types options.

Monte Carlo-based approximation methods have been introduced for financial applications by Boyle (1977) and then have been applied to option pricing problems including American options by Tilley (1993), Barraquand and Martineau (1995), Boyle, Broadie, and Glasserman (1997, 2001), Broadie and Glasserman (1997), Silvestrov, Calochkin, and Sibirtsev (1999), Broadie, Glasserman, and Ha (2000), Jönsson (2001), Longstaff and Schwartz (2001), Silvestrov, Calochkin, and Malyarenko (2001), Basso, Nardon, and Pianca (2002a, 2004), Rogers (2002), Glasserman (2003), Schoutens and Symens (2003), Andersen and Broadie (2004), Caflisch and Chaudhary (2004), Chan, Chen, and Yeung (2004), Glasserman and Yu (2004a, 2004b), Jönsson, Kukush, and Silvestrov (2004, 2005), Kukush and Silvestrov (2004), Silvestrov and Stenberg (2004), Bally, Caramellino, and Zanette (2005), Belomestny and Milstein (2006), Muni Toke, and Girard (2006), Fujiwara and Kijima (2007), Bender, Kolodko, and Schoenmakers (2008), Berridge and Schumacher (2008a, 2008b), Broadie and Cao (2008), Kohler (2008, 2010), Juneja and Kalra (2009b), AitSahlia, Goswami, and Guha (2010), Doan, Gaikwad, Bossy, Baude, and Stokes-Rees (2010), Badescu, Elliott, Kulperger, Miettinen, and Siu (2011), Chang, Lu, and Chi (2012), Kan and Reesor (2012), and Kohler and Krzyżak (2012).

A survey of Monte Carlo-based methods for option pricing can be found in Glasserman (2003) and Bender, Kolodko, and Schoenmakers (2006).

Difference methods, numerical solutions for partial differential equations associated with prices for American options, and related techniques have been developed by Schwartz (1977) and Brennan and Schwartz (1977, 1978). Then studies of these methods were continued in Courtadon (1982), Marchuk and Shaidurov (1983), Bunch and Johnson (1984), Geske and Jonsson (1984), MacMillan (1986), Barone-Adesi and Whaley (1987), Hull and White (1990), Jaillet, Lamberton, and Lapeyre (1990), Breen (1991), Press, Teukolsky, Vetterling, and Flannery (1992), Ho, Stapleton, and Subrahmanyam (1994), AitSahlia (1995), Zhang (1997), Raible (2000), Boyle, Kolkiewicz, and Tan (2002), Benth, Karlsen, and Reikvam (2004), D'Halluin, Forsyth, and Labahn (2004), Boyle, Lai, and Tan (2005), Allegretto, Lin, and Yan (2006), Almendral and Oosterlee (2007a), Angermann and Wang (2007), Arnarson and Eriksson (2007), Detemple and Rindisbacher (2007), Kovalov, Linetsky, and Marcozzi (2007), Zhao, Davison, and Corless (2007), Holmes and Yang (2008), Muthuraman (2008), Chiarella, Kang, Meyer, and Ziogas, (2009), Hu, Liang, and Jiang (2009), Ikonen and Toivanen (2009), Ito and Toivanen (2009), Yousuf (2009), Zhang, Yang, Wang, and Teo (2009), Ballestra and Sgarra (2010), Chockalingam and Muthuraman (2010), Persson and Von Sydow (2010), Toivanen (2010a), Wang, G. and Wang, S. (2010), Yang (2010), Fujii, Matsumoto, and Tsubota (2011), Huang, Forsyth, and Labahn (2011), Kwon and Lee (2011), Saib, Tangman, and Bhuruth (2012), and Lee, J. and Lee, Y. (2013).

Other approximation methods, based on PDE for American option prices, integral representations of option prices, duality martingale relations, variational inequalities, linear programming, and related optimization techniques, have been developed by Kim (1990, 1994), Jaillet, Lamberton and Lapeyre (1990), Jacka (1991), Carr, Jarrow, and Myneni (1992), Cottle, Pang and Stone (1992), Wilmott, Dewynne, and Howison (1993), Dempster and Hutton (1995), Ju (1998), and then by Haugh and Kogan (2003), Denault and Pigeon (2004), Ikonen and Toivanen (2004, 2007a, 2007b, 2008), Reisinger and Wittum (2004), Berton and Eymard (2006), Wang, Yang, and Teo (2006), Zhang, Yang, and Teo (2006, 2008), Ikonen and Toivanen (2007a) Berridge and Schumacher (2008b), Ehrhardt and Mickens (2008), Khaliq, Voss, and Kazmi (2008), Tangman, Gopaul, and Bhuruth (2008), Toivanen (2008, 2010b), Wang and Yang (2008), Wong and Zhao (2008), Chawla and Kalla (2009), Lin, Liu, and Zhang (2009), Hintermüller and Tber (2010), Kohler, Krzyżak, and Todorovic (2010), Chockalingam and Muthuraman (2011), Feng, Linetsky, Morales, and Nocedal (2011), Hamatani and Fukushima (2011), Ishii and Omata (2011), Salmi and Toivanen (2011), Chiarella, Salmi, and Toivanen (2012), and Kunoth, Schneider, and Wiechers (2012).

Survey of related results can be found in Zhu, Wu, and Chen (2004), Detemple (2005), Zhu (2011), and Salmi and Toivanen (2012).

Finally, we give also selected references related to analytical and approximation methods for other type of options.

Knockout and related barrier-type options have been treated in Cox and Rubinstein (1985), Hull and White (1987), Rubinstein and Reiner (1991), Heston (1993), Boyle and Lau (1994), Rich (1994), Broadie and Detemple (1995), Derman, Kani, Ergener, and Bardhan (1995), Levy and Mantion (1995), Nelken (1995), Ritchken (1995), Jarrow (1996), Broadie, Glasserman, and Kou (1997), Roberts and Shortland (1997), Gau, Huang, and Subrahmanyam (2000), Karatzas and Wang (2000), Shepp, Shiryaev, and Sulem (2002), Dayanik and Karatzas (2003), Novikov, Frishling, and Kordzakhia (2003), Kyprianou, Schoutens, and Wilmott (2005), Carr and Hirsa (2007), Kijima and Suzuki (2007), Lundberg (2007a, 2007b, 2010a, 2010b), Henderson and Hobson (2008b), Maruhn (2009), Ekström, Lindberg, Tysk, and Wanntorp (2010), Ekström and Tysk (2010), Escobar, Krämer, Scheibl, Seco, and Zagst (2010), Ballestra and Pacelli (2011), Boyarchenko, M. and Boyarchenko, S. (2011), Carr (2011), Cox and Obłój (2011), Dia and Lamberton (2011), Dolinsky and Kifer (2011), Gotoh, Jin, and Sumita (2011), Griebsch and Wystup (2011), Henriksen (2011), Itkin and Carr (2011), Maruhn, Nalholm, and Fengler (2011), Mishura and Yukhnovs'kiĭ (2011), Ndogmo and Ntwiga (2011), Sagna (2011), Sun, Shi, and Gu (2011), Boyarchenko and Levendorskiĭ (2012), Chiarella, Kang, and Meyer (2012), Fourati (2012), Jun and Ku (2012), and Jun (2013).

Other type of pass-dependent options were studied in Goldnman, Sosin, and Gatto (1979), Conze and Viswanathan (1991), Hull and White (1993), Shepp and Shiryaev (1993, 1994), Chance (1994), Kramkov and Shiryaev (1994), Shiryaev, Kabanov, Kramkov, and Mel'nikov (1994a, 1994b), Akahori (1995), Dassios (1995), Kat (1995), Kat and Verdonk (1995), Levy and Mantion (1995), Broadie, Glasserman, and Kou (1997, 1999), Chesney, Jeanblanc-Picqué, and Yor (1997), Cheuk and Vorst (1997), Hugonnier (1999), Linetsky (1999), Babbs (2000), Kifer (2000), Borovkov and Novikov (2002), Denault and Pigeon (2004), Reisinger and Wittum (2004), Borovkov and Novikov (2005), Dai and Kwok (2005), Matache, Nitsche, and Schwab (2005), Szimayer (2005), Ekström (2006a, 2006b), Kukush, Mishura, and Shevchenko (2006), Almendral and Oosterlee (2007b), Boyle and Draviam (2007), Dolinsky and Kifer (2007), Lundgren, Silvestrov, and Kukush (2008), Pascucci (2008), Detemple and Emmerling (2009), Egloff and Leippold (2009), Lundgren and Silvestrov (2009, 2011), Basak, Ghosh, and Goswami (2011), Dolinsky, Iron, and Kifer (2011), Eberlein, Glau, and Papapantoleon (2011), Elliott, Liew, and Siu (2011), Schmitz Abe (2011), Silvestrov and Lundgren (2011), Agliardi (2012), Buchen (2012), Fourati (2012), Gerber, Shiu, and Yang (2012), Kwok, Leung, and Wong (2012), and Yousuf, Khaliq, and Kleefeld (2012).

The survey of related results can be found in Boyle and Potapchik (2008).

Some useful supplementary references to works, which contain results on backward recurrence reward algorithms for Markov-type processes, which play an important role in our studies, are Chung (1960), Kemeny and Snell (1961), Lamperty (1963), McLean and Neuts (1967), Pitman (1977), Silvestrov (1980a, 1980b, 1996), Stenberg, Manca and Silvestrov (2006, 2007), Stenberg (2006b), Janssen and Manca (2006,

2007), Janssen, Manca, and Volpe (2007, 2008), Gyllenberg and Silvestrov (2008), and Silvestrov, Manca, and Silvestrova (2013).

Finally, we would like to give some additional useful references to books on stochastic processes, related to models considered in this book. These are, Billingsley (1968, 1999), Gikhman and Skorokhod (1971, 1973, 1975, 1977, 1982), Rogers and Williams (1979), Stroock and Varadhan (1979), Silvestrov (1980b), Kijima (1997), Anulova, Veretennikov, Krylov, Liptser, and Shiryaev (1998), Janssen and Limnios (1999), Sato (1999), Balakrishnan, Charalambides, and Koutras (2000), Serfozo (2009), Limnios and Oprişan (2001), Kallenberg (2002), Janssen and Manca (2006), Anisimov (2008), Gyllenberg and Silvestrov (2008), and Durrett (2012).

In particular, we would like to refer to books related to limit theorems for stochastic processes, randomly stopped stochastic processes, and optimal stopping problems for stochastic processes, namely, books by Chow, Robbins, and Siegmund (1971), Silvestrov (1974, 2004), Shiryaev (1976), Gikhman and Skorokhod (1977), Gut (1988), Kruglov and Korolev (1990), Gnedenko and Korolev (1996), Kartashov (1996), Wong (1996), Bening and Korolev (2002), Whitt (2002), Jacod and Shiryaev (2003), and Peskir and Shiryaev (2006).

In conclusion, we would like to mention the results that will be presented in the second volume of this book. It will include results on skeleton reward approximations for autoregressive type log-price processes, structure of optimal stopping domains for American type options with general pay-off functions for multivariate modulated Markov price processes, Monte Carlo-based approximation algorithms for American option rewards based on structure information about optimal stopping domains, convergence and skeleton approximations for American-type knockout options, reselling of options, time skeleton approximations and convergence of time and space skeleton stochastic approximations for continuous time multivariate modulated Markov type price processes.

The present bibliography partly covers the above areas. It will be extended in the bibliography for the second volume of the book.

Bibliography

[1] Achdou, Y., An inverse problem for a parabolic variational inequality arising in volatility calibration with American options. *SIAM J. Control Optim.* **43** no. 5 (2005), 1583–1615.

[2] Achdou, Y., Pironneau, O., *Computational Methods for Option Pricing*. Frontiers in Applied Mathematics, **30**, Society for Industrial and Applied Mathematics (SIAM), Philadelphia, PA, 2005.

[3] Achdou, Y., Indragoby, G., Pironneau, O., Volatility calibration with American options. *Methods Appl. Anal.*, **11** no. 4 (2004), 533–556.

[4] Agliardi, R., A comprehensive mathematical approach to exotic option pricing. *Math. Methods Appl. Sci.*, **35** no. 11 (2012), 1256–1268.

[5] Aguilar, E. T., Robust efficient hedging for American options: the existence of worst case probability measures. *Statist. Decisions*, **27** no. 1 (2009), 1–23.

[6] Ahn, S., Bae, H., Koo, H., Lee, K., A survey on American options: old approaches and new trends. *Bull. Korean Math. Soc.*, **48** no. 4 (2011), 791–812.

[7] Ahn, J., Song, M., Convergence of the trinomial tree method for pricing European/American options. *Appl. Math. Comput.*, **189** no. 1 (2007), 575–582.

[8] Aingworth, D. D., Das, S. R., Motwani R., A simple approach for pricing equity options with Markov switching state variables. *Quant. Finance*, **6** no. 2 (2006), 95–105.

[9] AitSahlia, F., *Optimal Stopping and Weak Convergence Methods for Some Problems in Financial Economics*. Ph.D. Dissertation, Stanford University (1995).

[10] AitSahlia, F., Goswami, M., Guha, S., American option pricing under stochastic volatility: an efficient numerical approach. *Comput. Manag. Sci.*, **7** no. 2 (2010), 171–187.

[11] Akahori, J., Some formulae for a new type of path-dependent option. *Ann. Appl. Probab.* **5** no. 2 (1995), 383–388.

[12] Aldous, D., Stopping times and tightness. *Ann. Probab.*, **6** no. 2 (1978), 335–340.

[13] Aldous, D., Stopping times and tightness II. *Ann. Probab.*, **17** no. 2 (1989), 586–595.

[14] Alfonsi, A., Jourdain, B., General duality for perpetual American options. *Inter. J. Theor. Appl. Finance*, **11** no. 6 (2008), 545–566.

[15] Alfonsi, A., Jourdain, B., Exact volatility calibration based on a Dupire-type call-put duality for perpetual American options. *Nonlinear Diff. Eq. Appl.*, **16** no. 4 (2009), 523–554.

[16] Allegretto, W., Lin, Y., Yan, N., A posteriori error analysis for FEM of American options. *Discr. Cont. Dyn. Syst.* Ser. B, **6** no. 5 (2006), 957–978.

[17] Almendral, A., Numerical valuation of American options under the CGMY process. In: Kyprianou, A. E., Schoutens, W., Willmott, P. (eds) *Exotic Option Pricing and Advanced Lévy Models*. Wiley, Chichester, 2005, 259–276.

[18] Almendral, A., Oosterlee, C. W. (2007a), On American options under the variance gamma process. *Appl. Math. Finance*, **14** no. 2 (2007), 131–152.

[19] Almendral, A., Oosterlee, C. W. (2007b), Accurate evaluation of European and American options under the CGMY process. *SIAM J. Sci. Comput.*, **29** no. 1 (2007), 93–117.

[20] Amin, K., On the computation of continuous time option prices using discrete approximations. *J. Financ. Quant. Anal.*, **26** (1991), 477–496.

[21] Amin, K., Option pricing trees. *J. Derivat.*, **2** no. 4 (1995), 34–46.

[22] Amin, K., Bodurtha, J., Discrete-time American option valuation with stochastic interest rates. *Rev. Financ. Stud.*, **8** (1995), 193–234.

[23] Amin, K., Khanna, A., Convergence of American option values from discrete- to continuous-time financial models. *Math. Finance*, **4** no. 4 (1994), 289–304.

[24] Andersen, L., Broadie, M., Primal-dual simulation algorithm for pricing multidimensional American options. *Manag. Sci.*, **50** no. 9 (2004), 1222–1234.

[25] Angermann, L., Wang, S., Convergence of a fitted finite volume method for the penalized Black-Scholes equation governing European and American option pricing. *Numer. Math.*, **106** no. 1 (2007), 1–40.

[26] Anisimov, V. V., *Switching Processes in Queuing Models.* Applied Stochastic Models Series, Wiley-ISTE, Washington, 2008.

[27] Anulova, S. V., Veretennikov, A. Yu., Krylov, N. V., Liptser, R. Sh., Shiryaev, A. N., *Stochastic Calculus. Probability theory, III.* Encyclopaedia Math. Sci., **45**, Springer, Berlin (1998).

[28] Arnarson, T., Eriksson, J., On the size of the non-coincidence set of parabolic obstacle problems with applications to American option pricing. *Math. Scand.*, **101** no. 1 (2007), 148–160.

[29] Applebaum, D., *Lévy Processes and Stochastic Calculus.* Cambridge Studies in Advanced Mathematics. Cambridge University Press, Cambridge, 2009.

[30] Babbs, S., Binomial valuation of lookback options. *J. Econom. Dynam. Cont.*, **24** (2000), 1499–1525.

[31] Badescu, A., Elliott, R. J., Kulperger, R., Miettinen, J., Siu, T. K., A comparison of pricing kernels for GARCH option pricing with generalized hyperbolic distributions. *Inter. J. Theor. Appl. Finance*, **14** no. 5 (2011), 669–708.

[32] Baecker, P. N., *Real Options and Intellectual Property. Capital Budgeting under Imperfect Patent Protection.* Lecture Notes in Economics and Mathematical Systems, **587**, Springer, Berlin, 2007.

[33] Balakrishnan, N., Charalambides, C. H. A., Koutras, M. V., *Probability and Statistical Models with Applications.* Chapman & Hall, Boca Raton, 2000.

[34] Ballestra, L. V., Sgarra, C., The evaluation of American options in a stochastic volatility model with jumps: an efficient finite element approach. *Comput. Math. Appl.*, **60** no. 6 (2010), 1571–1590.

[35] Ballestra, L. V., Pacelli, G., A boundary element method to price time-dependent double barrier options. *Appl. Math. Comput.*, **218** no. 8 (2011), 4192–4210.

[36] Bally, V., Caramellino, L., Zanette, A., Pricing and hedging American options by Monte Carlo methods using a Malliavin calculus approach. *Monte Carlo Meth. Appl.*, **11** no. 2 (2005), 97–133.

[37] Bally, V., Pagés, G., Printems, J., A quantization tree method for pricing and hedging multidimensional American options. *Math. Finance*, **15** no. 1 (2005), 119–168.

[38] Barles, G., Burdeau, J., Romano, M., Samsoen, N., Critical stock price near expiration. *Math. Finance* **5** no. 2 (1995), 77–95.

[39] Barone-Adesi, G., Whaley, R., Efficient analytical approximation of American option values. *J. Finance*, **42** (1987), 301–320.

[40] Barraquand, J., Martineau, D., Numerical valuation of high dimensional multivariate American securities. *J. Financ. Quant. Anal.* **30** (1995), 383–405.

[41] Basak, G. K., Ghosh, M. K., Goswami, A., Risk minimizing option pricing for a class of exotic options in a Markov-modulated market. *Stoch. Anal. Appl.*, **29** no. 2 (2011), 259–281.

[42] Basso, A., Nardon, M., Pianca, P. (2002a), *Optimal exercise of American options.* Research Report 106/2002, University of Venice, Italy, 2002.

[43] Basso, A., Nardon, M., Pianca, P. (2002b), *An analysis of the effects of continuous dividends on the exercise of American options.* Research Report 109/2002, University of Venice, Italy, 2002.

[44] Basso, A., Nardon, M., Pianca, P., A two-step simulation procedure to analyze the exercise features of American options. *Decis. Econom. Finance.* **27** no. 1 (2004), 35–56.

[45] Battauz, A., Pratelli, M., Optimal stopping and American options with discrete dividends and exogenous risk. *Insur. Math. Econom.*, **35** no. 2 (2004), 255–265.

[46] Bayraktar, E., Karatzas, I., Yao, S., Optimal stopping for dynamic convex risk measures. *Illinois J. Math.*, **54** no. 3 (2010), 1025–1067.

[47] Bayraktar, E., Xing, H., Pricing American options for jump diffusions by iterating optimal stopping problems for diffusions. *Math. Methods Oper. Res.*, **70** no. 3 (2009), 505–525.

[48] Belomestny, D., Milstein, G. N., Monte Carlo evaluation of American options using consumption processes. *Inter. J. Theor. Appl. Finance*, **9** no. 4 (2006), 455–481.

[49] Belopol'skaya, Ya. I., Romadanova, M. M., A probabilistic approach to a problem with a free boundary and American option pricing. *Zap. Nauchn. Sem. S.-Peterburg. Otdel. Mat. Inst. Steklov.*, **384**, *Veroyatn. Stat.*, **16**, 40–77, 311 (English translation in *J. Math. Sci.*, **176** (2011) no. 2, 124–145) (2010).

[50] Bender, C., Kolodko, A., Schoenmakers, J., Policy iteration for American options: overview. *Monte Carlo Methods Appl.*, **12** no. 5–6 (2006), 347–362.

[51] Bender, C., Kolodko, A., Schoenmakers, J., Enhanced policy iteration for American options via scenario selection. *Quant. Finance*, **8** no. 2 (2008), 135–146.

[52] Bender, C., Sottinen, T., Valkeila, E., Pricing by hedging and no-arbitrage beyond semimartingales. *Finance Stoch.*, **12** no. 4 (2008), 441–468.

[53] Bening, V. E., Korolev, V. Yu., *Generalized Poisson Models and their Applications in Insurance and Finance*. Modern Probability and Statistics, VSP, Utrecht, 2002.

[54] Bensoussan, A., On the theory of option pricing. *Acta Appl. Math.*, **2** (1984), 139–158.

[55] Benth, F. E., *Option Theory with Stochastic Analysis. An Introduction to Mathematical Finance*. Springer, Berlin, 2004.

[56] Benth, F. E., Karlsen, K. H., Reikvam, K., A semilinear Black and Scholes partial differential equation for valuing American options: approximate solutions and convergence. *Interfaces Free Bound.*, **6** no. 4 (2004), 379–404.

[57] Berton, J., Eymard, R., Finite volume methods for the valuation of American options. *ESAIM: Math. Model. Numer. Anal.*, **40** no. 2 (2006), 311–330.

[58] Bermin, H., Kohatsu-Higa, A., Perelló, J., Hints for an extension of the early exercise premium formula for American options. *Physica A: Statist. Mech. Appl.*, **355** no. 1 (2005), 152–157.

[59] Berridge, S. J., Schumacher, J. M. (2008a), Pricing high-dimensional American options using local consistency conditions. In: Appleby, J. A. D., Edelman, D. C., Miller, J. J. H. (eds) *Numerical Methods for Finance*. Chapman & Hall/CRC Financ. Math. Ser., Chapman & Hall/CRC, Boca Raton, FL, 2008, 13–52.

[60] Berridge, S. J., Schumacher, J. M. (2008b), An irregular grid approach for pricing high-dimensional American options. *J. Comput. Appl. Math.*, **222** no. 1 (2008), 94–111.

[61] Billingsley, P. (1968, 1999), *Convergence of Probability Measures*. Wiley Series in Probability and Statistics, Wiley, New York.

[62] Bjerksund, P., Stensland, G., Closed form approximation of American option prices. *Scand. J. Manag.* **9**, Supp. 1 (1993), 87–99.

[63] Björk, T., *Arbitrage Theory in Continuous Time*. 2nd edition, Oxford University Press. Oxford, 2004.

[64] Björk, T., Landén, C., On the term structure of futures and forward prices. In: German, H., Madan, D., Pliska, S. R., Vorst, T. (eds) *Mathematical Finance – Bachelier Congress 2000. Selected Papers from the 1st World Congress of the Bachelier Finance Society, Paris, 2000*. Springer, Berlin (2002), 111–149.

[65] Black, F., Scholes, M., The pricing of options and corporate liabilities. *J. Polit. Economy*, **81** no. 3 (1973), 637–654.

[66] Blanchet, A., On the regularity of the free boundary in the parabolic obstacle problem. Application to American options. *Nonlinear Anal.*, **65** no. 7 (2006), 1362–1378.

[67] Blomeyer, E., An analytic approximation for the American put price for options on stocks with dividends. *J. Financ. Quant. Anal.*, **21**, 229 (1986), –233.

[68] Bollen, N., Valuing options in regime-switching models. *J. Derivat.* **6** no. 1 (1998), 38–49.

[69] Borodin, A. N., Salminen, P., *Handbook of Brownian Motion - Facts and Formulae.* 2nd edition, Birkhäuser, Basel, 2002.

[70] Borovkov, K. A., Downes, A. N., Novikov, A. A., Continuity theorems in boundary crossing problems for diffusion processes. In: Chiarella, C., Novikov, A. (eds) *Contemporary Quantitative Finance.* Springer, Berlin (2010), 335–351.

[71] Borovkov, K., Novikov, A., On a new approach to calculating expectations for option pricing. *J. Appl. Probab.*, **39** no. 4 (2002), 889–895.

[72] Borovkov, K., Novikov, A., Explicit bounds for approximation rates of boundary crossing probabilities for the Wiener process. *J. Appl. Probab.*, **42** no. 1 (2005), 82–92.

[73] Bouchard, B., Temam, E., On the hedging of American options in discrete time markets with proportional transaction costs. *Electron. J. Probab.*, **10** no. 22 (2005), 746–760.

[74] Boyarchenko, M., Boyarchenko, S., Double barrier options in regime-switching hyper-exponential jump-diffusion models. *Inter. J. Theor. Appl. Finance*, **14** no. 7 (2011), 1005–1043.

[75] Boyarchenko, M., Levendorskiĭ, S., Valuation of continuously monitored double barrier options and related securities. *Math. Finance*, **22** no. 3 (2012), 419–444.

[76] Boyarchenko, S., Levendorskiĭ, S. (2009a), American options in regime-switching models. *SIAM J. Cont. Optim.*, **48** no. 3 (2009), 1353–1376.

[77] Boyarchenko, S., Levendorskiĭ, S. (2009b), American options in Lévy models with stochastic interest rates. *J. Comput. Finance*, **12** no. 4 (2009), 51–89.

[78] Boyle, P., Options: a Monte Carlo approach. *J. Financ. Econom.*, **4** (1977), 323–338.

[79] Boyle, P., Option valuation using a three-jump process. *Inter. Options J.*, **3** (1986), 7–12.

[80] Boyle, P., A lattice framework for option pricing with two state variables. *J. Financ. Quant. Anal.*, **23** (1988), 1–12.

[81] Boyle, P., New life forms on the options landscape. *J. Financ. Engin.*, **2** (1993), 217–252.

[82] Boyle, P., Broadie, M., Glasserman, P., Monte Carlo methods for security pricing. *J. Econom. Dynam. Cont.*, **21** (1997), 1267–1321.

[83] Boyle, P., Broadie, M., Glasserman, P., Monte Carlo methods in security pricing. In: Jouini, E., Cvitanić, J., Musiela, M. (eds) *Option Pricing, Interest Rates and Risk Management.* Cambridge University Press, Cambridge (2001), 185–238.

[84] Boyle, P., Draviam, T., Pricing exotic options under regime switching. *Insur. Math. Econom.*, **40** no. 2 (2007), 267–282.

[85] Boyle, P., Evnine, J., Gibbs, S., Numerical evaluation of multivariate contingent claims. *Rev. Financ. Stud.*, **2** (1989), 241–250.

[86] Boyle, P. P., Kolkiewicz, A. W., Tan, K. S., Pricing American derivatives using simulation: a biased low approach. In: Fang, K., Hickernell, F. J., Niederreiter, H. (eds) *Monte Carlo and Quasi-Monte Carlo Methods.* Springer, Berlin, 2002 (2002), 181–200.

[87] Boyle, P., Lai, Y., Tan, K. S., Pricing options using lattice rules. *N. Amer. Actuar. J.*, **9** no. 3 (2005), 50–76.

[88] Boyle, P., Lau, S., Bumping up against the barrier with the binomial method. *J. Derivat.*, **1** no. 4 (1994), 6–14.

[89] Boyle, P., Potapchik, A., Prices and sensitivities of Asian options: a survey. *Insur. Math. Econom.*, **42** no. 1 (2008), 189–211.

[90] Breen, R., The accelerated binomial option pricing model. *J. Financ. Quant. Anal.*, **26** no. 2 (1991), 153–164.

[91] Brennan, M., Schwartz, E., The valuation of American put options. *J. Finance*, **32** no. 2 (1977), 449–462.

[92] Brennan, M., Schwartz, E., Finite difference methods and jump processes arising in the pricing of contingent claims: a synthesis. *J. Financ. Quant. Anal.*, **13** (1978), 461–474.

[93] Broadie, M., Cao, M., Improved lower and upper bound algorithms for pricing American options by simulation. *Quant. Finance*, **8** no. 8 (2008), 845–861.

[94] Broadie, M., Detemple, J., American capped call options on dividend paying assets. *Rev. Financ. Stud.*, **8** (1995), 161–191.

[95] Broadie, M., Detemple, J. (1996a), *Recent advances in numerical methods for pricing derivative securities.* CIRANO Scientific Series, 96s-17, 1996.

[96] Broadie, M., Detemple, J. (1996b), American option valuation: new bounds, approximations, and a comparison of existing methods. *Rev. Financ. Stud.*, **9** (1996), 1211–1250.

[97] Broadie, M., Detemple, J. (1997a), The valuation of American options on multiple assets. *Math. Finance*, **7** no. 3 (1997), 241–286.

[98] Broadie, M., Detemple, J., American options on dividend-paying assets. *Fields Inst. Comm.*, **22** (1999), 69–97.

[99] Broadie, M., Detemple, J. B., Option pricing: valuation models and applications. *Manag. Sci.*, **50** no. 9 (2004), 1145–1177.

[100] Broadie, M., Glasserman, P., Pricing American-style securities by simulation. *J. Econom. Dynam. Cont.*, **21** (1997), 1323–1352.

[101] Broadie, M., Glasserman, P., Ha, Z., Pricing American options by simulation using a stochastic mesh with optimized weights. In: Uryasev, P. (ed) *Probabilistic Constrained Optimization.* Nonconvex Optim. Appl., **49**, Kluwer, Dordrecht (2000), 26–44.

[102] Broadie, M., Glasserman, P., Kou, S., A continuity correction for discrete barrier options. *Math. Finance*, **7** no. 4 (1997), 325–348.

[103] Broadie, M., Glasserman, P., Kou, S., Connecting discrete and continuous path-dependent options. *Finance Stoch.*, **3** (1999), 55–82.

[104] Brosch, R., *Portfolios of Real Options*. Lecture Notes in Economics and Mathematical Systems, **611**, Springer, Berlin, 2008.

[105] Buchen, P., *An Introduction to Exotic Option Pricing*. Chapman & Hall/CRC Financial Mathematics Series. CRC Press, Boca Raton, FL, 2012.

[106] Buffington, J., Elliot, R. J., American options with regime switching. *Inter. J. Theor. Appl. Finance*, **5** no. 5 (2002), 497–514.

[107] Bunch, D., Johnson, H., A Simple and numerically efficient valuation method for American puts using a modified Geske-Johnson approach. *J. Finance*, **47** (1992), 809–816.

[108] Caflisch, R. E., Chaudhary, S., Monte Carlo simulation for American options. In: Givoli, D., Grote, M. J., Papanicolaou, G. C. (eds) *A Celebration of Mathematical Modeling*. Kluwer, Dordrecht (2004), 1–16.

[109] Carr, P., Randomization and the American put. *Rev. Financ. Stud.*, **11** no. 3 (1998), 597–626.

[110] Carr, P., Semi-static hedging of barrier options under Poisson jumps. *Inter. J. Theor. Appl. Finance*, **14** no. 7 (2011), 1091–1111.

[111] Carr, P., Hirsa, A., Forward evolution equations for knock-out options. In: Fu, M. C., Jarrow, R. A., Yen, J. J., Elliott, R. J. (eds) *Advances in Mathematical Finance.* Appl. Numer. Harmon. Anal., Birkhäuser, Boston (2007), 195–217.

[112] Carr, P., Jarrow, R., Myneni, R., Alternative characterizations of American put options. *Math. Finance*, **2** no. 2 (1992), 87–106.

[113] Carr, P., Madan, D., Saddlepoint methods for option pricing. *J. Comput. Finance.* **13** no. 1 (2009), 49–61.

[114] Celtner, D., Riddiough, T., Stojanovic, S., Insights on the effect on use choice: the perturbed option on the best of two underlying assets. *J. Urban Econom.*, **39** (1996), 20–50.

[115] Chan, T., Pricing perpetual American options driven by spectrally one-sided Lévy processes. In: Kyprianou, A. E., Schoutens, W., Willmott, P. (eds) *Exotic Option Pricing and Advanced Lévy Models.* Wiley, Chichester (2005), 195–216.

[116] Chan, R. H., Chen, Y., Yeung, K. M., A memory reduction method in pricing American options. *J. Stat. Comput. Simul.*, **74** no. 7 (2004), 501–511.

[117] Chance, D., The pricing and hedging of limited exercise caps and spreads. *J. Financ. Res.*, **17** (1994), 561–584.

[118] Chang, H., Lu, Z., Chi. X., Large-scale parallel simulation of high-dimensional American option pricing. *J. Algor. Comput. Technol.*, **6** no. 1 (2012), 1–16.

[119] Chaudhary, S. K., A simple American option pricing method using the fast Fourier transform. *Int. J. Theor. Appl. Finance*, **10** no. 7 (2007), 1191–1202.

[120] Chaudhury, M., Upper bounds for American options. *Res. Finance*, **23** (2007), 161–191.

[121] Chawla, M. M., Kalla, S. L., A Simpson-type scheme for the valuation of European and American options. *Bull. Pure Appl. Math.*, **3** no. 2 (2009), 212–218.

[122] Chen, N., Glasserman, P., Additive and multiplicative duals for American option pricing. *Finance Stoch.*, **11** no. 2 (2007), 153–179.

[123] Chen, W., Zhu, S., Optimal exercise price of American options near expiry. *ANZIAM J.*, **51** no. 2 (2009), 145–161.

[124] Chengrong, Y., Lishang, J., Baojun, B., Free boundary and American options in a jump-diffusion model. *Europ. J. Appl. Math.*, **17** no. 1 (2006), 95–127.

[125] Chesney, M., Gibson, R., State space symmetry and two-factor option pricing models. *Adv. Futures Options Res.*, **8** (1995), 85–112.

[126] Chesney, M., Jeanblanc-Picqué, M., Yor, M., Brownian excursions and Parisian barrier options. *Adv. Appl. Probab.*, **29** (1997), 165–184.

[127] Chevalier, E., Critical price near maturity for an American option on a dividend-paying stock in a local volatility model. *Math. Finance*, **15** no. 3 (2005), 439–463.

[128] Chevalier, E., Bermudean approximation of the free boundary associated with an American option. In: Figueiredo, I. N., Rodrigues, J. F., Santos, L. (eds) *Free Boundary Problems*. Inter. Ser. Numer. Math., **154**, Birkhäuser, Basel (2007), 137–147.

[129] Chevalier, E., Exercise boundary near maturity for an American option on several assets. *Stoch. Anal. Appl.*, **28** no. 4 (2010), 623–647.

[130] Cheuk, T., Vorst, T., Currency lookback options and the observation frequency: a binomial approach. *J. Inter. Money Finance*, **16** no. 2 (1997), 173–187.

[131] Cheyette, O., Pricing options on multiple assets. Adv.. *Futures Options Res.*, **4** (1990), 69–81.

[132] Chiarella, C., Kang, B., Meyer, G. H., The evaluation of barrier option prices under stochastic volatility. *Comput. Math. Appl.*, **64** no. 6 (2012), 2034–2048.

[133] Chiarella, C., Kang, B., Meyer, G. H., Ziogas, A., The evaluation of American option prices under stochastic volatility and jump-diffusion dynamics using the method of lines. *Inter. J. Theor. Appl. Finance*, **12** no. 3 (2009), 393–425.

[134] Chiarella, C., Ziogas, A., Ziveyi, J., Representation of American option prices under Heston stochastic volatility dynamics using integral transforms. In: Chiarella, C., Novikov, A. (eds) *Contemporary Quantitative Finance*. Springer, Berlin (2010), 281–315.

[135] Chockalingam, A., Muthuraman, K., Pricing American options when asset prices jump. *Oper. Res. Lett.*, **38** no. 2 (2010), 82–86.

[136] Chockalingam, A., Muthuraman, K., American options under stochastic volatility. *Oper. Res.*, **59** no. 4 (2011), 793–809.

[137] Chow, Y. S., Robbins, H., Siegmund, D., *Great Expectations: The Theory of Optimal Stopping*. Houghton Mifflin Company, Boston, 1971.

[138] Christensen, S., Irle, A., Novikov, A., An elementary approach to optimal stopping problems for AR(1) sequences. *Sequent. Anal.*, **30** no. 1 (2011), 79–93.

[139] Chung, K. L., *Markov Chains with Stationary Transition Probabilities*. Fundamental Principles of Mathematical Sciences, **104**, Springer, Berlin, 1960.

[140] Chung, S-L., Chang, H-C., Generalized analytical upper bounds for American option prices. *J. Financ. Quant. Anal.*, **42** no. 1 (2007), 209–228.

[141] Cont, R., Tankov, P., *Financial Modelling with Jump Processes*. Chapman & Hall/CRC, London, 2004.

[142] Conze, A., Viswanathan, R., Path dependent options: the case of lookback options. *J. Finance*, **46** (1991), 1893–1907.

[143] Coquet, F., Toldo, S., Convergence of values in optimal stopping and convergence in optimal stopping times. *Electron. J. Probab.*, **12** no. 8 (2007), 207–228.

[144] Cottle, R., Pang, J.-S., Stone, R., *The Linear Complementarity Problem*. Academic Press, Boston, 1992.

[145] Courtadon, G., A more accurate finite difference approximation for the valuation of options. *J. Financ. Quant. Anal.*, **17** (1982), 697–705.

[146] Cox, A. M. G., Obłój, J., Robust hedging of double touch barrier options. *SIAM J. Financ. Math.*, **2** (2011), 141–182.

[147] Cox, J., Ross, S., The valuation of options for alternative stochastic processes. *J. Financ. Econom.*, **3** (1976), 145–166.

[148] Cox, J., Ross, S., Rubinstein, M., Option pricing: a simplified approach. *J. Financ. Econom.*, **7** (1979), 229–263.

[149] Cox, J., Rubinstein, M., *Options Markets*. Prentice-Hall, Engelwood Clifs, NJ, 1985.

[150] Cruz-Báez, D. I., González-Rodríguez, J. M., A semigroup approach to American options. *J. Math. Anal. Appl.*, **302** no. 1 (2005), 157–165.

[151] Curran, M.,, Accelerating American option pricing in lattices. *J. Derivat.*, **3** no. 2 (1995), 8–17.

[152] Cutland, N. J., Kopp, P. E., Willinger, W., Wyman, M. C., Convergence of Snell envelopes and critical prices in the American put. In: Dempster, A., Pliska, S. (eds) *Mathematics of Derivative Securities*. Cambridge University Press, Cambridge (1997), 61–87.

[153] Dai, M., Kwok, Y. K., American options with lookback payoff. *SIAM J. Appl. Math.*, **66** no. 1 (2005), 206–227.

[154] D'Amico, G., Janssen, J., Manca, R., European and American options: the semi-Markov case. *Physica A: Statist. Mech. Appl.*, **388** no. 15–16 (2009), 3181–3194.

[155] Darling, D. A., Liggett, T., Taylor, H. M., Optimal stopping for partial sums. *Ann. Math. Statist.*, **43** no. 4 (1972), 1363–1368.

[156] Dassios, A., The distribution of the quantile of a Brownian motion with drift and the pricing of related path-dependent options. *Ann. Appl. Probab.*, **5** no. 2 (1995), 389–398.

[157] Dayanik, S., Karatzas, I., On the optimal stopping problem for one-dimensional diffusions. *Stoch. Process. Appl.*, **107** no. 2 (2003), 173–212.

[158] Dempster, M., Hutton, J. P., Fast numerical valuation of American, exotic and complex options. *Appl. Math. Finance*, **4** (1995), 1–20.

[159] Denault, M., Pigeon, B., Numerical solution of linear complementarity problems and American option pricing. *INFOR Inf. Syst. Oper. Res.*, **42** no. 1 (2004), 1–21.

[160] Deng, X., Gu, Y., Wang, S., Zhang, S., On convergence of a semi-analytical method for American option pricing. *J. Math. Anal. Appl.*, **313** no. 1 (2006), 353–365.

[161] Derman, E., Kani, I., Ergener, D., Bardhan, I., Enhanced numerical methods for options with barriers. *Financ. Analysts J.*, **51** no. 6 (1995), 65–74.

[162] Detemple, J., *American-Style Derivatives: Valuation and Computation*. Chapman & Hall/CRC Financial Mathematics Series, Chapman & Hall/CRC, Boca Raton, FL, 2005.

[163] Detemple, J., Emmerling, T., American chooser options. *J. Econom. Dynam. Contr.*, **33** no. 1 (2009), 128–153.

[164] Detemple, J., Feng, S., Tian, W., The valuation of American call options on the minimum of two dividend-paying assets. *Ann. Appl. Probab.*, **13** no. 3 (2003), 953–983.

[165] Detemple, J., Rindisbacher, M., Monte Carlo methods for derivatives of options with discontinuous payoffs. *Comput. Statist. Data Anal.*, **51** no. 7 (2007), 3393–3417.

[166] Detemple, J., Tian, W., Xiong, J., An optimal stopping problem with a reward constraint. *Finance Stoch.*, **16** no. 3 (2012), 423–448.

[167] De Vallière, D., Denis, E., Kabanov, Y., Hedging of American options under transaction costs. *Finance Stoch.*, **13** no. 1 (2009), 105–119.

[168] Dewynne, J. N., Howison, S. D., Rupf, I., Wilmott, P., Some mathematical results in the pricing of American options. *Europ. J. Appl. Math.*, **4** no. 4 (1993), 381–398.

[169] D'Halluin, Y., Forsyth, P. A., Labahn, G., A penalty method for American options with jump diffusion processes. *Numer. Math.*, **97** no. 2 (2004), 321–352.

[170] Dia, E. H. A., Lamberton, D., Continuity correction for barrier options in jump-diffusion models. *SIAM J. Financ. Math.*, **2** (2011), 866–900.

[171] Di Graziano, G., Rogers, L. C. G., Barrier option pricing for assets with Markov-modulated dividends. *J. Comput. Finance*, **9** no. 4 (2006), 75–88.

[172] Di Masi, G. B., Kabanov, Y. M., Runggaldier, W. J., Mean-variance hedging of options on stocks with Markov volatilities. *Teor. Veroyatn. Primenen.*, **39** no. 1, (1994), 211–222 (English translation in *Theory Probab. Appl.*, **39** no. 1, 172–182), 1994.

[173] Di Nunno, G., Øksendal, B., Proske, F., *Malliavin Calculus for Lévy Processes with Applications to Finance*. Universitext. Springer, Berlin, 2009.

[174] Doan, V. D., Gaikwad, A., Bossy, M., Baude, F., Stokes-Rees, I., Parallel pricing algorithms for multi-dimensional Bermudan/American options using Monte Carlo methods. *Math. Comput. Simul.*, **81** no. 3 (2010), 568–577.

[175] Dochviri, V. M., On optimal stopping with incomplete data. In: Watanabe, S., Prokhorov, Yu. V. (eds) *Probability Theory and Mathematical Statistics*. Lecture Notes in *Mathematics*, **1299**, Springer, New York (1988), 64–68.

[176] Dochviri, V. M., Shashiashvili, M. A., On the convergence of costs in the problem of optimal stopping of stochastic processes in the scheme of Kalman-Bucy. *Comm. Statist., C: Sequent. Anal.*, **1** no. 3 (1982), 163–176.

[177] Dochviri, V. M., Shashiashvili, M. A., Convergence of value functions in a discrete problem of optimal stopping with incomplete data. *Soobsh. Akad. Nauk Gruzin. SSR*, **109** no. 1 (1983), 33–35.

[178] Dolinsky, Y., Shortfall risk approximations for American options in the multidimensional Black-Scholes model. *J. Appl. Probab.*, **47** no. 4 (2010), 997–1012.

[179] Dolinsky, Y., Error estimates for multinomial approximations of American options in a class of jump diffusion models. *Stochastics*, **83** no. 4–6 (2011), 415–429.

[180] Dolinsky, Y., Iron, Y., Kifer, Yu., Perfect and partial hedging for swing game options in discrete time. *Math. Finance*, **21** no. 3 (2011), 447–474.

[181] Dolinsky, Y., Kifer, Yu., Hedging with risk for game options in discrete time. *Stochastics*, **79** no. 1–2 (2007), 169–195.

[182] Dolinsky, Y., Kifer, Yu., Binomial approximations for barrier options of Israeli style. In: Breton, M., Szajowski, K. (eds) *Advances in Dynamic Games*. Annals of the International Society of Dynamic Games, **11**, Birkhäuser and Springer, New York (2011), 447–467.

[183] Duan, J. C., Popova, I., Ritchken, P., Option pricing under regime switching. *Quant. Finance*, **2** (2002), 116–132.

[184] Dupuis, P., Wang, H., On the convergence from discrete time to continuous time in an optimal stopping problem. *Ann. Appl. Probab.*, **15** no. 2 (2005), 1339–1366.

[185] Durrett, R., *Essentials of Stochastic Processes*. 2nd edition, Springer Texts in Statistics. Springer, New York, 2012.

[186] Dynkin, E. B., Optimal choice of the stopping moment of a Markov process. *Dokl. Akad. Nauk SSSR*, **150** (1963), 238–240 (English translation in *Soviet Math. Doklady*, **4**, 627–629).

[187] Dzhaparidze, K. O., *Introduction to Option Pricing in a Securities Market*. CWI Syllabi, **47**, Stichting Mathematisch Centrum, Centrum voor Wiskunde en Informatica, Amsterdam, 2000.

[188] Eberlein, E., Glau, K., Papapantoleon, A., Analyticity of the Wiener-Hopf factors and valuation of exotic options in Lévy models. In: Di Nunno, G., Øksendal, B. (eds) *Advanced Mathematical Methods for Finance,* Springer, Heidelberg (2011), 223–245.

[189] Eberlein, E. Papapantoleon, A., Shiryaev, A. N., On the duality principle in option pricing: semimartingale setting. *Finance Stoch.*, **12** no. 2 (2008), 265–292.

[190] Egloff, D., Leippold, M., The valuation of American options with stochastic stopping time constraints. *Appl. Math. Finance*, **16** no. 3 (2009), 287–305.

[191] Ehrhardt, M., Mickens, R. E., A fast, stable and accurate numerical method for the Black-Scholes equation of American options. *Inter. J. Theor. Appl. Finance*, **11** no. 5 (2008), 471–501.

[192] Ekström, E. (2004a), Convexity of the optimal stopping boundary for the American put option. *J. Appl. Probab.*, **299** no. 1 (2004), 147–156.

[193] Ekström, E. (2004b), Properties of American option prices. *Stoch. Process. Appl.*, **114** no. 2 (2004), 265–278.

[194] Ekström, E. (2006a), Bounds for perpetual American option prices in a jump diffusion model. *J. Appl. Probab.*, **43** no. 3 (2006), 867–873.

[195] Ekström, E. (2006b), Properties of game options. *Math. Meth. Oper. Res.*, **63** (2006), 221–238.

[196] Ekström, E., Lindberg, C., Tysk, J., Wanntorp, H., Optimal liquidation of a call spread. *J. Appl. Probab.*, **47** no. 2 (2010), 586–593.

[197] Ekström, E., Lötstedt, P., Von Sydow, L., Tysk, J., Numerical option pricing in the presence of bubbles. *Quant. Finance*, **11** no. 8 (2011), 1125–1128.

[198] Ekström, E., Tysk, J., The American put is log-concave in the log-price. *J. Math. Anal. Appl.*, **314** no. 2 (2006), 710–723.

[199] Ekström, E., Tysk, J., Properties of option prices in models with jumps. *Math. Finance*, **17** no. 3 (2007), 381–397.

[200] Ekström, E., Tysk, J., The Black-Scholes equation in stochastic volatility models. *J. Math. Anal. Appl.*, **368** (2010), 498–507.

[201] El Karoui, N., Karatzas, I., A New approach to the Skorohod problem and its applications. *Stochastics Stoch. Rep.*, **34** no. 1–2 (1991), 57–82.

[202] Elliott, R. J., Chan, L., Perpetual American options with fractional Brownian motion. *Quant. Finance*, **4** no. 2 (2004), 123–128.

[203] Elliott, R. J., Chan, L., Siu, T. K., Option pricing and Esscher transform under regime switching. *Ann. Finance*, **1** no. 4 (2005), 423–432.

[204] Elliott, R. J., Kopp, P. E., *Mathematics of Financial Markets*. 2nd edition, Springer, New York, 1999.

[205] Elliott, R. J., Liew, C. C., Siu, T. K., Characteristic functions and option valuation in a Markov chain market. *Comput. Math. Appl.*, **62** no. 1 (2011), 65–74.

[206] Elliott, R. J., Siu, T., A risk-based approach for pricing American options under a generalized Markov regime-switching model. *Quant. Finance*, **11** no. 11 (2011), 1633–1646.

[207] Elliott, R. J., Siu, T. K., Badescu, A., On pricing and hedging options in regime-switching models with feedback effect. *J. Econom. Dynam. Contr.*, **35** no. 5 (2011), 694–713.

[208] Embrechts, P., Klüppelberg, C., Mikosch, T., *Modelling Extremal Events for Insurance and Finance*. Applications of *Mathematics*, **33**, Springer, Berlin, 1997.

[209] Escobar, M., Krämer, S., Scheibl, F., Seco, L., Zagst, R., Hedge funds as knock-out options. In: Menéndez, S. C., Pérez, J. L. F. (eds) *Mathematics in Finance*. Contemp. Math., **515**, Amer. Math. Soc., Providence, RI (2010), 1–15.

[210] Evans, J. D., Kuske, R., Keller, J. B., American options on assets with dividends near expiry. *Math. Finance*, **12** no. 3 (2002), 219–237.

[211] Fährmann, H., Zur konvergenz der optimalen werte der gewinnfunktion beim abbruch von zufallsprozessen im fallen von unvollständiger information. *Math. Oper. Statist.*, Series Statist., **9** no. 2 (1978), 241–253.

[212] Fährmann, H., On the convergence of the value in optimal stopping of random sequences with incomplete data. *Zastosow. Mat.*, **16** no. 3 (1979), 415–428.

[213] Fährmann, H., Convergence of values in optimal stopping of partially observable random sequences with quadratic rewards. *Theory Probab. Appl.*, **27** no. 2 (1982), 386–391.

[214] Feng, L., Linetsky, V. Morales, J. L., Nocedal, J., On the solution of complementarity problems arising in American options pricing. *Optim. Methods Soft.*, **26** no. 4–5 (2011), 813–825.

[215] Föllmer, H., Schied, A., *Stochastic Finance. An Introduction in Discrete Time.* 3rd revised and extended edition, Walter de Gruyter, Berlin, 2011.

[216] Fourati, S., Explicit solutions of the exit problem for a class of Lévy processes; applications to the pricing of double-barrier options. *Stoch. Process. Appl.*, **122** no. 3 (2012), 1034–1067.

[217] Fujii, M., Matsumoto, K., Tsubota, K., Simple improvement method for upper bound of American option. *Stochastics*, **83** no. 4–6 (2011), 449–466.

[218] Fujiwara, H., Kijima, M., Pricing of path-dependent American options by Monte Carlo simulation. *J. Econom. Dynam. Contr.*, **31** no. 11 (2007), 3478–3502.

[219] Gau, B., Huang, J. Z., Subrahmanyam, M., The valuation of American barrier options using the decomposition technique. J. Econom., *Dynam. Contr.*, **24** (2000), 1783–1827.

[220] Gerber, H. U., Shin, E. S. W., Martingale approach t.o pricing perpetual American options on two stocks. *Math. Finance*, **6** (1996), 303–322.

[221] Gerber, H. U., Shiu, E. S. W., Yang, H., Valuing equity-linked death benefits and other contingent options: a discounted density approach. *Insur. Math. Econom.*, **51** no. 1 (2012), 73–92.

[222] Geske, R., A Note on an analytical valuation formula for unprotected American options on stocks with known dividends. *J. Financ. Econom.*, **7** (1979), 375–380.

[223] Geske, R., Johnson, H., The American put options valued analytically. *J. Finance*, **39** no. 5 (1984), 1511–1524.

[224] Geske, R., Shastri, K. (1985a), Valuation by approximation: a comparison of alternative option valuation techniques. *J. Financ. Quant. Anal.*, **20** (1985), 45–71.

[225] Geske, R., Shastri, K (1985b), The early exercise of American puts. *J. Bank. Finance*, **9** (1985), 207–219.

[226] Gikhman, I. I., Skorokhod, A. V., *Theory of Random Processes. 1.* Probability Theory and Mathematical Statistics, Nauka, Moscow (1971) (English edition: *The Theory of Stochastic Processes. 1.* Fundamental Principles of Mathematical Sciences, **210**, and Classics in Mathematics, Springer, New York (1974, 2004)).

[227] Gikhman, I. I., Skorokhod, A. V., *Theory of Random Processes. 2.* Probability Theory and Mathematical Statistics, Nauka, Moscow (1973) (English edition: *The Theory of Stochastic Processes. 2.* Fundamental Principles of Mathematical Sciences, **218**, and Classics in Mathematics, Springer, New York (1975, 2004)).

[228] Gikhman, I. I., Skorokhod, A. V., *Theory of Random Processes. 3.* Probability Theory and Mathematical Statistics, Nauka, Moscow (1975) (English edition: *The Theory of Stochastic Processes. III.* Fundamental Principles of Mathematical Sciences, **232**, and Classics in Mathematics, Springer, Berlin (1979, 2007)).

[229] Gikhman, I. I., Skorokhod, A. V., *Controlled Stochastic Processes.* Naukova Dumka, Kiev (1977) (English edition: *Controlled Stochastic Processes.* Springer, New York (1979)).

[230] Gikhman, I. I., Skorokhod, A. V., *Stochastic Differential Equations and their Applications.* Naukova Dumka, Kiev, 1982.

[231] Glasserman, P., *Monte Carlo Methods in Financial Engineering.* Applications of Mathematics, **53**, Stochastic Modelling and Applied Probability, Springer, Berlin, 2003.

[232] Glasserman, P., Yu, B. (2004a), Number of paths versus number of basis functions in American option pricing. *Ann. Appl. Probab.*, **14** no. 4 (2004), 2090–2119.

[233] Glasserman, P., Yu, B. (2004b), Simulation for American options: regression now or regression later? In: Niederreiter, H. (ed) *Monte Carlo and Quasi-Monte Carlo Methods.* Springer, Berlin, 2004, 213–226.

[234] Goldman, M., Sosin, H., Gatto, M., Path dependent options: buy at the low, sell at the high. *J. Finance*, **34** (1979), 1111–1127.

[235] Gotoh, J., Jin, H., Sumita, U., Numerical evaluation of dynamic behavior of Ornstein-Uhlenbeck processes modified by various boundaries and its application to pricing barrier options. *Methodol. Comput. Appl. Probab.*, **13** no. 1 (2011), 193–219.

[236] Gnedenko, B. V., Korolev, V. Yu., *Random Summation. Limit Theorems and Applications.* CRC Press, Boca Raton, FL, 1996.

[237] Griebsch, S. A., Wystup, U., On the valuation of fader and discrete barrier options in Heston's stochastic volatility model. *Quant. Finance*, **11** no. 5 (2011), 693–709.

[238] Griego, R. J., Swishchuk, A. V., The Black-Scholes formula for a securities markets in a random environment. *Theor. Ĭmovir. Mat. Stat.*, **62**, (2000), 163–167 (English translation in *Theory Probab. Math. Statist.*, **62**, 171–175).

[239] Gukhal, C. R., The compound option approach to American options on jump-diffusions. *J. Econom. Dynam. Contr.*, **28** no. 10 (2004), 2055–2074.

[240] Guo, X. (2001a), An explicit solution to an optimal stopping problem with regime switching. *J. Appl. Probab.*, **38** no. 2 (2001), 464–481.

[241] Guo, X. (2001b), Information and option pricing. *Quant. Finance*, **1** no. 1 (2001), 38–44.

[242] Guo, X., Zhang, Q., Closed-form solutions for perpetual American put options with regime switching. *SIAM J. Appl. Math.*, **64** no. 6 (2004), 2034–2049.

[243] Gut, A., *Stopped Random Walks. Limit Theorems and Applications.* Applied Probability, A Series of the Applied Probability Trust, **5**, Springer, New York, 1988.

[244] Gyllenberg, M., Silvestrov, D., *Quasi-Stationary Phenomena in Nonlinearly Perturbed Stochastic Systems.* De Gruyter Expositions in Mathematics, **44**, Walter de Gruyter, Berlin, 2008.

[245] Hamatani, K., Fukushima, M., Pricing American options with uncertain volatility through stochastic linear complementarity models. *Comput. Optim. Appl.*, **50** no. 2 (2011), 263–286.

[246] Harrison, M., Kreps, D., Martingales and arbitrage in multiperiod security markets. *J. Econom. Theory*, **20** (1979), 381–408.

[247] Harrison, M., Pliska, S., Martingales and stochastic integrals in the theory of continuous trading. *Stoch. Process. Appl.*, **11** (1981), 215–260.

[248] Haug, E. G., *The Complete Guide to Option Pricing Formulas.* 2nd edition. McGraw-Hill, 2006.

[249] Haugh, M., Kogan, L., Pricing American options: a duality approach. *Oper. Res.*, **52** (2004), 258–270.

[250] He, H., Convergence from discrete- to continuous-time contingent claim prices. *Rev. Financ. Stud.*, **3** (1990), 523–546.

[251] Hellermann, R., *Capacity Options for Revenue Management. Theory and Applications in the Air Cargo Industry.* Lecture Notes in Economics and Mathematical Systems, **575**, Springer, Berlin, 2006.

[252] Henderson, V., Hobson, D. (2008a), An explicit solution for an optimal stopping/optimal control problem which models an asset sale. *Ann. Appl. Probab.*, **18** (2008), 1681–1705.

[253] Henderson, V., Hobson, D. (2008b), Perpetual American options in incomplete markets: the infinitely divisible case. *Quant. Finance*, **8** no. 5 (2008), 461–469.

[254] Henriksen, P. N., Pricing barrier options by a regime switching model. *Quant. Finance*, **11** no. 8 (2011), 1221–1231.

[255] Herwig, T., *Market-Conform Valuation of Options*. Lecture Notes in Economics and Mathematical Systems, **571**, Springer, Berlin, 2005.

[256] Heston, S. L., A closed-form solution for options with stochastic volatility with applications to bond and currency options. *Rev. Financ. Stud.*, **6** (1993), 327–343.

[257] Heston, S., Zhou, G., On the rate of convergence of discrete-time contingent claims. *Math. Finance*, **10** no. 1 (2000), 53–75.

[258] Higham, D. J., *An Introduction to Financial Option Valuation*. Mathematics, Stochastics and Computation, Cambridge University Press, Cambridge, 2004.

[259] Hintermüller, M., Tber, M. H., An inverse problem in American options as a mathematical program with equilibrium constraints: C-stationarity and an active-set-Newton solver. *SIAM J. Control Optim.*, **48** no. 7 (2010), 4419–4452.

[260] Ho, T. S., Stapleton, R., Subrahmanyam, M., A simple technique for the valuation and hedging of American options. *J. Derivat.*, **2** no. 1 (1994), 52–66.

[261] Holmes, A. D., Yang, H., A front-fixing finite element method for the valuation of American options. *SIAM J. Sci. Comput.*, **30** no. 4 (2008), 2158–2180.

[262] Holmes, A., Yang, H., Zhang, S., A front-fixing finite element method for the valuation of American options with regime switching. *Inter. J. Comput. Math.*, **89** no. 9 (2012), 1094–1111.

[263] Hu, B., Liang, J., Jiang, L., Optimal convergence rate of the explicit finite difference scheme for American option valuation. *J. Comput. Appl. Math.*, **230** no. 2 (2009), 583–599.

[264] Huang, Y., Forsyth, P. A., Labahn, G., Methods for pricing American options under regime switching. *SIAM J. Sci. Comput.*, **33** no. 5 (2011), 2144–2168.

[265] Hugonnier, J. N., The Feynman-Kac formula and pricing occupation time derivatives. *Inter. J. Theor. Appl. Finance*, **2** (1999), 153–178.

[266] Hull, J. C., *Options, Futures and Other Derivatives*. 6th edition, Pearson Prentice Hall, Upper Saddle River, NJ, (1993, 2006).

[267] Hull, J. C., *Options, Futures, and Other Derivatives and DerivaGem CD Package*. 8th Edition, Pearson Prentice Hall, Upper Saddle River, NJ, 2011.

[268] Hull, J., White, A., The pricing of options on assets with stochastic volatilities. *J. Finance*, **42**, 2 (1987), 281–300.

[269] Hull, J., White, A., The use of the control variate technique in option pricing. *J. Financ. Quant. Anal.*, **23** (1988), 237–251.

[270] Hull, J., White, A., Valuing derivative securities using the explicit finite difference method. *J. Financ. Quant. Anal.*, **25** (1990), 87–100.

[271] Hull, J., White, A., Efficient procedures for valuing European and American path-dependent options. *J. Derivat.*, **1** no. 1 (1993), 21–32.

[272] Hull, J., White, A. (1994a), Numerical procedures for implementing term structure models I: single-factor models. *J. Derivat.*, **2** no. 1 (1994), 7–16.

[273] Hull, J., White, A. (1994b), Numerical procedures for implementing term structure models II: two-factor models. *J. Derivat.*, **2** no. 2 (1994), 37–48.

[274] Hussain, S., Shashiashvili, M., Discrete time hedging of the American option. *Math. Finance*, **20** no. 4 (2010), 647–670.

[275] Iftimie, B., Marinescu, M., A construction of admissible strategies for American options associated with piecewise continuous processes. Stud. Univ. Babes-*Bolyai Math.*, **54** no. 2 (2009), 53–64.

[276] Ikonen, S., Toivanen, J., Operator splitting methods for American option pricing. *Appl. Math. Lett.*, **17** no. 7 (2004), 809–814.

[277] Ikonen, S., Toivanen, J. (2007a), Pricing American options using LU decomposition. *Appl. Math. Sci.*, **1** no. 51 (2007), 2529–2551.

[278] Ikonen, S., Toivanen, J. (2007b), Componentwise splitting methods for pricing American options under stochastic volatility. *Inter. J. Theor. Appl. Finance*, **10** no. 2 (2007), 331–361.

[279] Ikonen, S., Toivanen, J., Efficient numerical methods for pricing American options under stochastic volatility. *Numer. Meth. Partial Diff. Eq.*, **24** no. 1 (2008), 104–126.

[280] Ikonen, S., Toivanen, J., Operator splitting methods for pricing American options under stochastic volatility. *Numer. Math.*, **113** no. 2 (2009), 299–324.

[281] İlhan, A., Jonsson, M., Sircar, R., Optimal static-dynamic hedges for exotic options under convex risk measures. *Stoch. Process. Appl.*, **119** no. 10 (2009), 3608–3632.

[282] Ishii, K., Omata, S., Convergence of the approximation scheme to American option pricing via the discrete Morse semiflow. *Appl. Math. Optim.*, **64** no. 3 (2011), 363–415.

[283] Itkin, A., Carr, P., Jumps without tears: a new splitting technology for barrier options. *Inter. J. Numer. Anal. Model.*, **8** no. 4 (2011), 667–704.

[284] Ito, K., Toivanen, J., Lagrange multiplier approach with optimized finite difference stencils for pricing American options under stochastic volatility. *SIAM J. Sci. Comput.*, **31** no. 4 (2009), 2646–2664.

[285] Ivanov, R. V., On the pricing of American options in exponential Lévy markets. *J. Appl. Probab.*, **44** no. 2 (2007), 409–419.

[286] Iwaki, H., Kijima, M., Yoshida, T., American put options with a finite set of exercisable time epochs. *Math. Comput. Model.*, **22**, 10–12 (1995), 89–97.

[287] Jacka, S., Optimal stopping and the American put. *Math. Finance*, **1** (1991), 1–14.

[288] Jacod, J., Shiryaev, A. N., *Limit Theorems for Stochastic Processes.* 2nd edition, Fundamental Principles of Mathematical Sciences, **288**, Springer, Berlin, 2003.

[289] Jaillet, P., Lamberton, D., Lapeyre, B., Variational inequalities and the pricing of American options. *Acta Appl. Math.*, **21** (1990), 263–289.

[290] Janson, S., Tysk, J., Volatility time and properties of option prices. *Ann. Appl. Probab.*, **13** no. 3 (2003), 890–913.

[291] Janssen, J., Limnios, N., *Semi-Markov Models and Applications.* Springer, New York, 1999.

[292] Janssen, J., Manca, R., *Applied Semi-Markov Processes.* Springer, New York, 2006.

[293] Janssen, J., Manca, R., Semi-Markov *Risk Models for Finance, Insurance and Reliability.* Springer, New York, 2007.

[294] Janssen, J., Manca, R., Volpe, E., *Mathematical Finance: Deterministic and Stochastic Models.* Wiley, New York, 2007.

[295] Janssen, J., Manca, R., Volpe, E., *Mathematical Finance 2: Stochastic Models.* Wiley, New York, 2008.

[296] Jarrow, R., *Over the Rainbow: Developments in Exotic Options and Complex Swaps.* Risk Books, London, 1996.

[297] Jarrow, R., Rudd, A., *Option Pricing.* Dow Jones-Irwin, Homewood, Illinois, 1983.

[298] Jarrow, R. A., Turnbull, S. M., *Derivative Securities.* 2nd edition, South-Western College Publishing, Cincinnati, Ohio, 2000.

[299] Jiang, L., *Mathematical Modeling and Methods of Option Pricing.* World Scientific, River Edge, NJ, 2005.

[300] Jiang, L., Dai, M., Convergence of binomial tree method for American options. In: C. Hua, L. Rodino (eds) *Partial Differential Equations and their Applications.* World Scientific (1999), 106–118.

[301] Jiang, L., Dai, M., Convergence of the explicit difference scheme and the binomial tree method for American options. *J. Comput. Math.*, **22** no. 3 (2004), 371–380.

[302] Jobert, A., Rogers, L. C. G., Option pricing with Markov-modulated dynamics. *SIAM J. Contr. Optim.*, **44**, 6 (2006), 2063–2078.

[303] Johnson, H., An analytic approximation for the American put price. *J. Financ. Quant. Anal.*, **18** (1983), 141–148.

[304] Johnson, H., Options on the maximum or the minimum of several assets. *J. Financ. Quant. Anal.*, **22** (1987), 227–283.

[305] Jönsson, H., Monte Carlo studies of American type options with discrete time. *Theory Stoch. Process.*, **7(23)**, 1–2 (2001), 163–188.

[306] Jönsson, H. (2005a), *Convergence of reward functions in a discrete time optimal stopping problem: application to American type options.* Research Report 2005–3, Department of Mathematics and Physics, Mälardalen University, 2005, 31 pages.

[307] Jönsson, H. (2005b), *Optimal Stopping Domains and Reward Functions for Discrete Time American Type Options.* Doctoral Dissertation, No. **22**, Mälardalen University, 2005.

[308] Jönsson, H., Kukush, A. G., Silvestrov, D. S., Threshold structure of optimal stopping domains for American type options. *Theory Stoch. Process.*, **8(24)** no. 1–2 (2002), 169–177.

[309] Jönsson, H., Kukush, A. G., Silvestrov, D. S., Threshold structure of optimal stopping strategies for American type option I. *Theor. Ĭmovirn. Mat. Stat.*, **71**, 82–92 (English translation in *Theory Probab. Math. Statist.*, **71** (2004), 93–103).

[310] Jönsson, H., Kukush, A. G., Silvestrov, D. S., Threshold structure of optimal stopping strategies for American type option II. *Theor. Ĭmovirn. Mat. Stat.*, **72**, 42–53 (English translation in *Theory Probab. Math. Statist.*, **72** (2005), 47–58).

[311] Joshi, M. S., The convergence of binomial trees for pricing the American put. *J. Risk*, **11** no. 4 (2009), 87–108.

[312] Ju, N., Pricing an American option by approximating it early exercise boundary as a multiple exponential function. *Rev. Financ. Stud.*, **11** no. 2 (1998), 627–646.

[313] Jun, D., Continuity correction for discrete barrier options with two barriers. *J. Comput. Appl. Math.*, **237** no. 1 (2013), 520–528.

[314] Jun, D., Ku, H., Cross a barrier to reach barrier options. *J. Math. Anal. Appl.*, **389** no. 2 (2012), 968–978.

[315] Juneja, S., Kalra, H. (2009a), Analysis of the optimal exercise boundary of American options for jump diffusions. *SIAM J. Math. Anal.*, **41** no. 2 (2009), 825–860.

[316] Juneja, S., Kalra, H. (2009b), Variance reduction techniques for pricing American options using function approximations. *J. Comput. Finance*, **12** no. 3 (2009), 79–102.

[317] Junghenn, H. D., Option Valuation. *A First Course in Financial Mathematics.* Chapman & Hall/CRC Financial Mathematics Series. CRC Press, Boca Raton, FL, 2012.

[318] Kallenberg, O., *Foundations of Modern Probability.* 2nd edition, Probability and its Applications, Springer, New York, 2002.

[319] Kallianpur, G., Karandikar, R. L., *Introduction to Option Pricing Theory.* Birkhäuser, Boston, MA, 2000.

[320] Kamrad, B., Ritchken, P., Multinomial approximating models for options with k state variables. *Manag. Sci.*, **37** (1991), 1640–1652.

[321] Kan, K., Reesor, R., Bias reduction for pricing American options by least-squares *Monte Carlo. Appl. Math. Finance*, **19** no. 3 (2012), 195–217.

[322] Karatzas, I., On the pricing of American options. *Appl. Math. Optim.*, **17** (1988), 37–60.

[323] Karatzas, I., Kou, S. G., Hedging American contingent claims with constrained portfolios. *Finance Stoch.*, **2** no. 3 (1998), 215–258.

[324] Karatzas, I., Shreve, S. (1988, 2005), *Brownian Motion and Stochastic Calculus.* 2nd edition, Springer, New York.

[325] Karatzas, I. Shreve, S. E., *Methods of Mathematical Finance.* Applications of Mathematics, Stochastic Modelling and Applied Probability, **39**, Springer, New York, 2001.

[326] Karatzas, I., Sudderth, W. D., Control and stopping of a diffusion process on an interval. *Ann. Appl. Probab.*, **9** no. 1 (1999), 188–196.

[327] Karatzas, I., Wang, H., A barrier option of American type. *Appl. Math. Optim.*, **42** no. 3 (2000), 259–279.

[328] Karlsen, K. H., Wallin, O., A semilinear equation for the American option in a general jump market. *Interfaces Free Bound.*, **11** no. 4 (2009), 475–501.

[329] Kartashov, M. V., *Strong Stable Markov Chains.* VSP, Utrecht and TBiMC, Kiev, 1996.

[330] Kat, H., Pricing lookback options using binomial trees: an evaluation. *J. Financ. Engin.*, **4** (1995), 375–398.

[331] Kat, H., Verdonk, L., Tree surgery. *Risk*, **8** no. 2 (1995), 53–56.

[332] Kemeny, J. G., Snell, J. L. (1961) Finite continuous time Markov chains. *Theor. Veroyatn. Primen.*, **6**, 110–115 (Also in *Theory Probab. Appl.*, **6**, 101–105).

[333] Khaliq, A. Q. M., Liu, R. H., New numerical scheme for pricing American option with regime-switching. *Inter. J. Theor. Appl. Finance*, **12** no. 3 (2009), 319–340.

[334] Khaliq, A. Q. M., Voss, D. A., Kazmi, K., Adaptive θ-methods for pricing American options. *J. Comput. Appl. Math.*, **222** no. 1 (2008), 210–227.

[335] Khizhnyak, K. V., Pricing of a perpetual American option on the maximum of a risky and a riskless asset. *Vestnik Moskov. Univ.* Ser. XV Vychisl. Mat. Kibernet., **35** no. 3, 23–30 (English translation in *Moscow Univ. Comput. Math. Cybernet.*, **35** no. 3 (2011), 124–132).

[336] Kholodnyi, V. A., Price, J. F., *Foreign Exchange Option Symmetry.* World Scientific, River Edge, NJ, 1998.

[337] Kifer, Y., Game options. *Finance Stoch.*, **4** no. 4 (2000), 443–463.

[338] Kijima, M., *Markov Processes for Stochastic Modelling.* Stochastic Modeling Series. Chapman & Hall, London, 1997.

[339] Kijima, M., Monotonicity and convexity of option prices revisited. *Math. Finance*, **12** no. 4 (2002), 411–425.

[340] Kijima, M., *Stochastic Processes with Applications to Finance.* Chapman & Hall/CRC, Boca Raton, FL, 2003.

[341] Kijima, M., Suzuki, T., The pricing of options with stochastic boundaries in a Gaussian economy. *J. Oper. Res. Soc. Japan*, **50** no. 2 (2007), 137–150.

[342] Kijima, M., Yoshida, T., A simple option pricing model with Markovian volatilities. *J. Oper. Res. Soc. Japan*, **36** no. 3 (1993), 149–166.

[343] Kim, I. J., The analytic valuation of American options. *Rev. Financ. Stud.*, **3** (1990), 547–572.

[344] Kim, I., Analytical approximation of the optimal exercise boundaries for American futures option. *J. Futures Markets*, **14** (1994), 1–24.

[345] Kim, I., Byun, S., Optimal exercise boundary in a binomial option pricing model. *J. Financ. Engin.*, **3** (1994), 137–158.

[346] Kimura, T., Alternative randomization for valuing American options. *Asia-Pac. J. Oper. Res.*, **27** no. 2 (2010), 167–187.

[347] Kohler, M., A regression-based smoothing spline Monte Carlo algorithm for pricing American options in discrete time. *AStA Adv. Stat. Anal.*, **92** no. 2 (2008), 153–178.

[348] Kohler, M., A review on regression-based Monte Carlo methods for pricing American options. In: Devroye, L., Karasözen, B., Kohler, M., Korn, R. (eds) *Recent Developments in Applied Probability and Statistics.* Physica, Heidelberg (2010), 37–58.

[349] Kohler, M., Krzyżak, A., Pricing of American options in discrete time using least squares estimates with complexity penalties. *J. Statist. Plann. Infer.*, **142** no. 8 (2012), 2289–2307.

[350] Kohler, M., Krzyżak, A., Todorovic, N., Pricing of high-dimensional American options by neural networks. *Math. Finance*, **20** no. 3 (2010), 383–410.

[351] Korn, R., Korn, E., *Option Pricing and Portfolio Optimization*. Modern Methods of Financial Mathematics, Graduate Studies in Mathematics, **31**, American Mathematical Society, Providence, RI, 2001.

[352] Kovalov, P., Linetsky, V., Marcozzi, M., Pricing multi-asset American options: a finite element method-of-lines with smooth penalty. *J. Sci. Comput.*, **33** no. 3 (2007), 209–237.

[353] Kramkov, D. O., Shiryaev, A. N., Rational pricing of a "Russian option" in a symmetric binomial model of a (B,S)-market. *Teor. Veroyatn. Primenen.*, **39** no. 1 (1994), 191–200 (English translation in *Theory Probab. Appl.*, **39** no. 1, 209–237).

[354] Kruglov, V. M., Korolev, V. Yu., *Limit Theorems for Random Sums*. Izdatel'stvo Moskovskogo Universiteta, Moscow, 1990.

[355] Kukush, A. G., Mishura, Y. S., Shevchenko, G. M., On reselling of European option. *Theory Stoch. Process.*, **12(28)** no. 3–4 (2006), 75–87.

[356] Kukush, A. G., Silvestrov, D. S., Optimal stopping strategies for American type options with discrete and continuous time. *Theory Stoch. Process.*, **5(21)** no. 1–2 (1999), 71–79.

[357] Kukush, A. G., Silvestrov, D. S., Structure of optimal stopping strategies for American type options. In: Uryasev, P. (ed) *Probabilistic Constrained Optimization: Methodology and Applications*. Nonconvex Optim. Appl., **49**, Kluwer, Dordrecht (2000), 173–185.

[358] Kukush, A. G., Silvestrov, D. S., Skeleton approximation of optimal stopping strategies for American type options with continuous time. *Theory Stoch. Process.*, **7(23)** no. 1–2 (2001), 215–230.

[359] Kukush, A. G., Silvestrov, D. S., Optimal pricing of American type options with discrete time. *Theory Stoch. Process.*, **10(26)** no. 1–2 (2004), 72–96.

[360] Kunoth, A., Schneider, C., Wiechers, K., Multiscale methods for the valuation of American options with stochastic volatility. *Inter. J. Comput. Math.*, **89** no. 9 (2012), 1145–1163.

[361] Kuske, R., Keller, J. B., Optimal exercise boundary for an American put option. *Appl. Math. Finance*, **5** (1998), 107–116.

[362] Kwok, Y-K., *Mathematical Models of Financial Derivatives*. Springer, Heidelberg, 1999.

[363] Kwon, Y., Lee, Y., A second-order tridiagonal method for American options under jump-diffusion models. *SIAM J. Sci. Comput.*, **33** no. 4 (2011), 1860–1872.

[364] Kwok, Y. K., Leung, K. S., Wong, H. Y., Efficient options pricing using the fast Fourier transform. In: Duan, J., Härdle, W. K., Gentle, J. E. (eds) *Handbook of Computational Finance*. Springer Handbooks of Computational Statistics, Springer, Heidelberg (2012), 579–604.

[365] Kyprianou, A., Schoutens, W., Wilmott, P., *Exotic Option Pricing and Advanced Lévy Models*. Wilmott Collection, Wiley, 2005.

[366] Lai, T. L., Wong, S. P., Valuation of American options via basis functions. *IEEE Trans. Automat. Control*, **49** no. 3 (2004), 374–385.

[367] Lamberton, D., Convergence of the critical price in the approximation of American options. *Math. Finance*, **3** (1993), 179–190.

[368] Lamberton, D., Error estimates for the binomial approximation of American put option. *Ann. Appl. Probab.*, **8** no. 1 (1998), 206–233.

[369] Lamberton, D., Pagès, G., Sur l'approximation des réduites. *Ann. l'Inst. Henri Poincaré*, **26** no. 2 (1990), 331–355.

[370] Lamberton, D., Villeneuve, S., Critical price near maturity for an American option on a dividend-paying stock. *Ann. Appl. Probab.*, **13** no. 2 (2003), 800–815.

[371] Lamperty, J., Criteria for stochastic processes II: passage-time moments. *J. Math. Anal. Appl.*, **7** (1963), 127–145.

[372] Laurence, P., Salsa, S., Regularity of the free boundary of an American option on several assets. *Comm. Pure Appl. Math.*, **62** no. 7 (2009), 969–994.

[373] Leduc, G., Exercisability randomization of the American option. *Stoch. Anal. Appl.*, **26** no. 4 (2008), 832–855.

[374] Lee, J., Lee, Y., Tridiagonal implicit method to evaluate European and American options under infinite activity Lévy models. *J. Comput. Appl. Math.*, **237** no. 1 (2013), 234–243.

[375] Leisen, D. P., Pricing the American put option, a detailed convergence analysis for binomial models. *J. Econom. Dynam. Contr.*, **22** (1998), 1419–1444.

[376] Leisen, D., Reimer, M., Binomial models for option valuation – examining and improving convergence. *Appl. Math. Finance*, **3** no. 4 (1996), 319–346.

[377] Levy, E., Mantion, F., *Approximate Valuation of Discrete Lookback Options*. HSBC Midland, London, 1995.

[378] Li, A., Ritchken, P., Sankarasubramanian, L., Lattice models for pricing American interest rate claims. *J. Finance*, **50** (1995), 719–737.

[379] Li, W., Xing, M., Weak convergence for approximation of American option prices. *Commun. Stoch. Anal.*, **5** no. 3 (2011), 505–525.

[380] Liang, J., On the convergence rate of the binomial tree scheme for an American option with jump-diffusion. *Numer. Math. J. Chinese Univ.*, **30** no. 1 (2008), 76–96.

[381] Liang, J., Hu, B., Jiang, L., Optimal convergence rate of the binomial tree scheme for American options with jump diffusion and their free boundaries. *SIAM J. Financ. Math.*, **1** (2010), 30–65.

[382] Liang, J., Hu, B., Jiang, L., Bian, B., On the rate of convergence of the binomial tree scheme for American options. *Numer. Math.*, **107** no. 2 (2007), 333–352.

[383] Limnios, N., Oprişan, G., *Semi-Markov Processes and Reliability*. Statistics for Industry and Technology, Birkhäuser, Boston, MA, 2001.

[384] Lin, Q., Liu, T., Zhang, S., Superconvergence estimates of finite element methods for American options. *Appl. Math.*, **54** no. 3 (2009), 181–202.

[385] Linetsky, V., Step options. *Math. Finance*, **9** (1999), 55–96.

[386] Liu, G., Hong, L. J., Revisit of stochastic mesh method for pricing American options. *Oper. Res. Lett.*, **37** no. 6 (2009), 411–414.

[387] Liu, G., Li, V. W., Semiparametric bounds of mean and variance for exotic options. *Sci. China*, Ser. **A 52** no. 7 (2009), 1446–1458.

[388] Longstaff, F. A., Schwartz, E. S., Valuing American options by simulation: a simple least squares approach. *Rev. Financ. Stud.*, **14** (2001), 113–147.

[389] Lundgren, R. (2007a), Optimal stopping domains for discrete time knock out American options. In: Skiadas, C. H. (ed) *Recent Advances in Stochastic Modelling and Data Analysis*. World Scientific, 2007, 613–620.

[390] Lundgren, R. (2007b), Structure of optimal stopping domains for American options with knock out domains. *Theory Stoch. Process.*, **13(29)** no. 4 (2007), 98–129.

[391] Lundgren R. (2010a), *Convergence of American knock out options in discrete time*. Research Report 2010–1, School of Education, Culture, and Communication, Division of Applied Mathematics, Mälardalen University, 2010, 18 pages.

[392] Lundgren, R. (2010b), *Convergence of Option Rewards*. Doctoral Dissertation, No. **89**, Mälardalen University, 2010.

[393] Lundgren, R., Silvestrov, D., *Convergence of option rewards for multivariate price processes*. Research Report 2009:10, Department of Mathematics, Stockholm University, 53 pages, 2009.

[394] Lundgren, R., Silvestrov, D., Optimal stopping and reselling of European options. In: Rykov, V., Balakrishan, N., Nikulin, M. (eds) *Mathematical and Statistical Models and Methods in Reliability*. Birkhäuser, Boston (2011), 378–394.

[395] Lundgren, R., Silvestrov, D. S., Kukush, A. G., Reselling of options and convergence of option rewards. *J. Numer. Appl. Math.*, **96(1)** (2008), 149–172.

[396] Mackjavičjus, V., Passage to the limit in problems of optimal stopping of Markov Processes. *Lit. Mat. Sbornik*, **13** no. 1 (1973), 115–128.

[397] Mackjavičjus, V., Convergence of the rewards in problems of optimal stopping of Markov processes. *Lit. Mat. Sbornik*, **14** no. 1 (1974), 113–128.

[398] Madan, D., Milne, F., Shefrin, H., The multinomial option pricing model and its Brownian and Poisson limits. *Rev. Financ. Stud.*, **4** (1989), 251–266.

[399] Maller, R. A., Solomon, D. H., Szimayer, A., A multinomial approximation for American option prices in Lévy process models. *Math. Finance*, **16** no. 4 (2006), 613–633.

[400] Marchuk, G, Shaidurov, V., *Difference Methods and Their Extrapolations*. Springer, New York, 1983.

[401] Maruhn, J. H., *Robust Static Super-Replication of Barrier Options*. Radon Series on Computational and Applied Mathematics, **7**, Walter de Gruyter, Berlin, 2009.

[402] Maruhn, J. H., Nalholm, M., Fengler, M. R., Static hedges for reverse barrier options with robustness against skew risk: an empirical analysis. *Quant. Finance*, **11** no. 5 (2011), 711–727.

[403] Matache, A. M., Nitsche, P. A., Schwab, C., Wavelet Galerkin pricing of American options on Lévy driven assets. *Quant. Finance*, **5** no. 4 (2005), 403–424.

[404] McDonald, R., Schroder, M., A parity result for American options. *J. Comput. Finance*, **1** no. 3 (2000), 5–13.

[405] McKean, Jr., H., Appendix: a free boundary problem for the heat equation arising from a problem in mathematical economics. *Indust. Manag. Rev.*, **6** (1965), 32–39.

[406] McLean, R. A., Neuts, M. F., The integral of a step function defined on a semi-Markov process. *SIAM J. Appl. Math.*, **15** no. 3 (1967), 726–737.

[407] Melnikov, A., *Risk Analysis in Finance and Insurance*. 2nd edition. Chapman & Hall/CRC Financial Mathematics Series, CRC Press, Boca Raton, FL, 2011.

[408] Merton, R., Theory of rational option pricing. *Bell J. Econom. Manag. Sci.*, **4** (1973), 141–183.

[409] Meyer, G. H., Numerical investigation of early exercise in American puts with discrete dividends. *J. Comput. Finance*, **5** no. 2 (2002), 37–53.

[410] Mikosch, T., *Elementary Stochastic Calculus with Finance in View*. Advanced Series on Statistical Science & Applied Probability, **6**, World Scientific, River Edge, NJ, 1998.

[411] Mishura, Yu., *Stochastic Calculus for Fractional Brownian Motion and Related Processes*. Lecture Notes in Mathematics, **1929**, Springer, Berlin, 2008.

[412] Mishura, Yu. S., Yukhnovs'kiĭ, Yu. V. (2011) Limit behavior of prices of a barrier option in the Black-Scholes model with random drift and volatility. *Theor. Ĭmovirn. Mat. Stat.*, **84**, 94–101 (English translation in *Theory Probab. Math. Statist.*, **84**, 99–106).

[413] Mulinacci, S., Pratelli, M., Functional convergence of Snell envelopes: Applications to American options approximations. *Finance Stoch.*, **2** (1998), 311–327.

[414] Muni Toke, I., Girard, J., Monte Carlo valuation of multidimensional American options through grid computing. In: Lirkov, I., Margenov, S., Waśniewski, J. (eds) *Large-Scale Scientific Computing*. Lecture Notes in Comput. Sci., **3743**, Springer, Berlin (2006), 462–469.

[415] Musiela, M., Rutkowski, M., *Martingale Methods in Financial Modelling*. 2nd edition, Stochastic Modelling and Applied Probability, **36**, Springer, Berlin, 2005.

[416] Muthuraman, K., A moving boundary approach to American option pricing. *J. Econom. Dynam. Cont.*, **32** no. 11 (2008), 3520–3537.

[417] Myneni, R., The pricing of the American option. *Ann. Appl. Probab.*, **2** (1992), 1–23.

[418] Naik, V., Option valuation and hedging strategies with jumps in the volatility of asset returns. *J. Finance*, **48** no. 5 (1993), 1969–1984.

[419] Ndogmo, J. C., Ntwiga, D. B., High-order accurate implicit methods for barrier option pricing. *Appl. Math. Comput.*, **218** no. 5 (2011), 2210–2224.

[420] Nelken, I., (ed), *The Handbook of Exotic Options*. McGraw-Hill, 1995.

[421] Nelson, D., Ramaswamy, K., Simple binomial processes as diffusion approximations in financial models. *Rev. Financ. Stud.*, **3** (1990), 393–430.

[422] Nieuwenhuis, J. W., Vellekoop, M. H., Weak convergence of tree methods to price options on defaultable assets. *Decis. Econom. Finance*, **27** no. 2 (2004), 87–107.

[423] Novikov, A. A., Some remarks on the distribution of the first passage times and the optimal stopping of AR(1)-sequences. *Teor. Veroyatn. Primen.*, **53** no. 3 (2008), 458–471 (English translation in *Theory Probab. Appl.*, **53** no. 3, 419–429).

[424] Novikov, A., Frishling, V., Kordzakhia, N., Time-dependent barrier options and boundary crossing probabilities. *Georgian Math. J.*, **10** no. 2 (2003), 325–334.

[425] Novikov, A. A., Shiryaev, A. N., On an effective case of the solution of the optimal stopping problem for random walks. *Teor. Veroyatn. Primen.*, **49** no. 2 (2004), 373–382 (English translation in *Theory Probab. Appl.*, **49** no. 2, 344–354).

[426] Novikov, A., Shiryaev, A., On solution of the optimal stopping problem for processes with independent increments. *Stochastics*, **79** no. 3–4 (2007), 393–406.

[427] Nyström, K., Free boundary regularity for multi-dimensional American options through blow-ups and global solutions. *J. Comput. Math. Optim.*, **3** no. 1 (2007), 39–76.

[428] Nyström, K., On the behaviour near expiry for multi-dimensional American options. *J. Math. Anal. Appl.*, **339** no. 1 (2008), 644–654.

[429] Omberg, E., The valuation of American put options with exponential exercise policies. *Adv. Futures Options Res.*, **2** (1987), 117–142.

[430] Omberg, E., Efficient discrete time jump process models in option pricing. *J. Financ. Quant. Anal.*, **23** (1988), 161–174.

[431] Øksendal, B., *Stochastic Differential Equations. An Introduction with Applications.* 6th edition, Springer, Berlin, 2003.

[432] Parkinson, M., Option pricing: the American put. *J. Business*, **50** no. 1 (1977), 21–36.

[433] Pascucci, A., Free boundary and optimal stopping problems for American asian options. *Finance Stoch.*, **12** (2008), 21–41.

[434] Pascucci, A., *PDE and Martingale Methods in Option Pricing.* Bocconi & Springer Series, **2**, Springer, Milan and Bocconi University Press, Milan, 2011.

[435] Pelsser, A., Vorst, T., The binomial model and the *Greeks*. *J. Derivat.*, **1** no. 3 (1994), 45–49.

[436] Persson, J., Von Sydow, L., Pricing American options using a space-time adaptive finite difference method. *Math. Comput. Simul.*, **80** no. 9 (2010), 1922–1935.

[437] Peskir, G., *Principles of Optimal Stopping and Free-Boundary Problems.* Doctoral Dissertation, Lecture Notes Series, **68**, University of Aarhus, Aarhus, 2001.

[438] Peskir, G., On the American option problem. *Math. Finance*, **15** no. 1 (2005), 169–181.

[439] Peskir, G., Shiryaev, A. N., *Optimal Stopping and Free Boundary Problems.* Lectures in Mathematics, ETH Zürich, Birkhäuser, Basel, 2006.

[440] Pitman, J. W., Occupation measures for Markov chains. *Adv. Appl. Probab.*, **9** (1977), 69–86.

[441] Pliska, S. R., *Introduction to Mathematical Finance: Discrete Time Models.* Blackwell, Oxford, 1997.

[442] Prékopa, A., Szántai, T., On the analytical-numerical valuation of the Bermudan and American options. *Quant. Finance*, **10** no. 1 (2010), 59–74.

[443] Prékopa, A., Szántai, T., On the binomial tree method and other issues in connection with pricing Bermudan and American options. *Quant. Finance*, **12** no. 1 (2012), 21–26.

[444] Press, W., Teukolsky, S., Vetterling, W., Flannery, B., *Numerical Recipes in C: The Art of Scientific Computing.* 2nd edition, Cambridge University Press, Cambridge, 1992.

[445] Pressacco, F., Gaudenzi, M., Zanette, A., Ziani, L., New insights on testing the efficiency of methods of pricing and hedging American options. *Europ. J. Oper. Res.*, **185** no. 1 (2008), 235–254.

[446] Prigent, J. L., *Weak Convergence of Financial Markets.* Springer, New York, 2003.

[447] Profeta, C., Roynette, B., Yor, M., *Option Prices as Probabilities. A New Look at Generalized Black-Scholes Formulae.* Springer Finance, Springer, Berlin, 2010.

[448] Qian, X., Xu, C., Jiang, L., Bian, B., Convergence of the binomial tree method for American options in a jump-diffusion model. *SIAM J. Numer. Anal.*, **42** no. 5 (2005), 1899–1913.

[449] Raible, S., *Lévy Processes in Finance: Theory, Numerics, and Empirical Facts.* Dissertation zur Erlangung des Doktorgrades, Universität Freiburg, 2000.

[450] Rapuch, G., American options and the free boundary exercise region: a PDE approach. *Interfaces Free Bound.*, **7** no. 1 (2005), 79–98.

[451] Reisinger, C., Wittum, G., On multigrid for anisotropic equations and variational inequalities: pricing multi-dimensional European and American options. *Comput. Visual. Sci.*, **7** no. 3–4 (2004), 189–197.

[452] Rendleman, R., Bartter, B., Two-state option pricing. *J. Finance*, **34** (1979), 1093–1110.

[453] Repplinger, D., *Pricing of Bond Options. Unspanned Stochastic Volatility and Random Field Models.* Lecture Notes in Economics and Mathematical Systems, **615**, Springer, Berlin, 2008.

[454] Rich, D., The mathematical foundations of barrier option pricing theory. *Adv. Futures Options Res.*, **7** (1994), 267–312.

[455] Ritchken, P., On pricing barrier options. *J. Derivat.*, **3** no. 2 (1995), 19–28.

[456] Roberts, G. O., Shortland, C. F., Pricing barrier options with time-dependent coefficients. *Math. Finance*, **7** no. 1 (1997), 83–93.

[457] Rogers, L. G.G, Monte Carlo valuation of American options. *Math. Finance*, **12** (2002), 271–286.

[458] Rogers, L. C. G., Talay, D., *Numerical Methods in Finance.* Cambridge University Press, Cambridge, 1997.

[459] Rogers, L. C. G., Williams, D. (1979, 2000), *Diffusions, Markov Processes, and Martingales: Volume 1, Foundations.* Cambridge Mathematical Library Wiley, Chichester and Cambridge University Press, Cambridge.

[460] Rogers, L. C. G., Williams, D. (1979, 2000), *Diffusions, Markov Processes, and Martingales: Volume 2, Itô Calculus.* Cambridge Mathematical Library Wiley, Chichester and Cambridge University Press, Cambridge.

[461] Roll, R., An analytic valuation formula for unprotected American call options on stocks with known dividends. *J. Financ. Econom.*, **5** (1977), 251–258.

[462] Rolski, T., Schmidli, H., Schmidt, V., Teugels, J., *Stochastic Processes for Insurance and Finance.* Wiley Series in Probability and Statistics, Wiley, New York, 1999.

[463] Rostek, S., 2009. *Option Pricing in Fractional Brownian Markets.* Lecture Notes in Economics and Mathematical Systems, **622**, Springer, Berlin.

[464] Roux, A., Zastawniak, T., American options under proportional transaction costs: pricing, hedging and stopping algorithms for long and short positions. *Acta Appl. Math.*, **106** no. 2 (2009), 199–228.

[465] Rubinstein, M., Return to Oz. *Risk,* **7** no. 11 (1994), 67–71.

[466] Rubinstein, M., Reiner, E., Breaking down the barriers. *Risk*, **4** (1991), 28–35.

[467] Rutkowski, V. I., The early exercise premium representation of foreign market American options. *Math. Finance*, **4** (1994), 313–325.

[468] Sagna, A., Pricing of barrier options by marginal functional quantization. *Monte Carlo Meth. Appl.*, **17** no. 4 (2011), 371–398.

[469] Saib, A., Tangman, D. Y., Bhuruth, M., A new radial basis functions method for pricing American options under Merton's jump-diffusion model. *Inter. J. Comput. Math.*, **89** no. 9 (2012), 1164–1185.

[470] Salmi, S., Toivanen, J., An iterative method for pricing American options under jump-diffusion models. *Appl. Numer. Math.*, **61** no. 7 (2011), 821–831.

[471] Salmi, S., Toivanen, J., Comparison and survey of finite difference methods for pricing American options under finite activity jump-diffusion models. *Inter. J. Comput. Math.*, **89** no. 9 (2012), 1112–1134.

[472] Salminen, P., Optimal stopping of one-dimensional diffusions. *Math. Nachrich.*, **124** (1985), 85–101.

[473] Salminen, P., Optimal stopping and American put options. *Theory Stoch. Process.*, **5(21)** no. 1–2 (1999), 129–144.

[474] Salminen, P., Optimal stopping, Appell polynomials, and Wiener-Hopf factorization. *Stochastics*, **83** no. 4–6 (2011), 611–622.

[475] Salopek, D. M., *American Put Options.* Pitman Monographs and Surveys in Pure and Applied Mathematics, **84**, Longman, Harlow, 1997.

[476] Samuelson, P. A., Rational theory of warrant prices. *Indust. Manag. Rev.* **6** (1965), 13–31.

[477] Sato, K. *Lévy Processes and Infinitely Divisible Distributions.* Cambridge University Press, Cambridge, 1999.

[478] Schmitz Abe, K., Pricing exotic options using *MSL-MC. Quant. Finance*, **11** no. 9 (2011), 1379–1392.

[479] Schoutens, W., *Lévy Processes in Finance: Pricing Financial Derivatives.* Wiley Series in Probability and Statistics, Wiley, 2003.

[480] Schoutens, W., Symens, S., The pricing of exotic options by Monte-Carlo simulations in a Lévy market with stochastic volatility. *Inter. J. Theor. Appl. Finance*, **6** no. 8 (2003), 839–864.

[481] Schulmerich, M., *Real Options Valuation. The Importance of Interest Rate Modelling in Theory and Practice.* Lecture Notes in Economics and Mathematical Systems, **559**, Springer, Berlin, 2005.

[482] Schwartz, E., The valuation of warrants: implementing a new approach. *J. Financ. Econom.*, **4** (1977), 79–93.

[483] Serfozo, R., *Basics of Applied Stochastic Processes.* Probability and its Applications, Springer, Berlin, 2009.

[484] Sharpe, W. F., *Investments.* 2nd edition, Prentice-Hall, Englewood Cliffs, NJ, 1978.

[485] Shepp, L. A., Shiryaev, A. N., The Russian option: reduced regret. *Ann. Appl. Probab.*, **3** no. 3 (1993), 631–640.

[486] Shepp, L. A., Shiryaev, A. N., A new look at the "Russian option". *Teor. Veroyatn. Primenen.*, **39** no. 1, (1994), 130–149 (English translation in *Theory Probab. Appl.*, **39** no. 1, 103–119).

[487] Shepp, L. A., Shiryaev, A. N., Sulem, A., A barrier version of the Russian option. In: Sandmann, K., Schönbucher, P. J. (eds) *Advances in Finance and Stochastics.* Springer, Berlin (2002), 271–284.

[488] Shevchenko, G., On a constant associated with American options. *Theor. Ĭmovirn. Mat. Stat.*, **82**, (2010), 163–167 (English translation in *Theory Probab. Math. Statist.*, **82**, 171–175).

[489] Shi, Z. W., Jin, Z. M., Utility maximization problem on American options. *Math. Econom.*, **21** no. 1 (2004), 24–30.

[490] Shiryaev, A. N., *Sequential Analysis. Optimal Stopping Rules.* Nauka, Moscow (1976) (English edition: Optimal Stopping Rules, Springer, New York, 1978 and Stochastic Modelling and Applied Probability, **8**, Springer, Berlin, 2008).

[491] Shiryaev, A. N., On some concepts and stochastic models in financial mathematics. *Teor. Veroyatn. Primenen.*, **39** no. 1, (1994), 5–22 (English translation in *Theory Probab. Appl.* **39** no. 1, 1–13).

[492] Shiryaev, A. N., *Essentials of Stochastic Finance: Facts, Models, Theory.* Advanced Series on Statistical Science & Applied Probability, **3**, World Scientific, River Edge, NJ, 1999.

[493] Shiryaev, A. N., Kabanov, Yu. M., Kramkov, D. O., Mel'nikov, A. V. (1994a), Toward a theory of pricing options of European and American types. I. Discrete time. *Teor. Veroyatn. Primenen.*, **39** no. 1 (1994), 23–79 (English translation in *Theory Probab. Appl.*, **39** no. 1, 14–60).

[494] Shiryaev, A. N., Kabanov, Yu. M., Kramkov, D. O., Mel'nikov, A. V. (1994b), Toward a theory of pricing options of European and American types. II. Continuous time. *Teor. Veroyatn. Primenen.*, **39** no. 1 (1994), 80–129 (English translation in *Theory Probab. Appl.*, **39** no. 1, 61–102).

[495] Shreve, S. E. (2004a), *Stochastic Calculus for Finance I. The Binomial Asset Pricing Model.* Springer Finance, Springer, New York, 2004.

[496] Shreve, S. E. (2004b), *Stochastic Calculus for Finance II. Continuous-Time Models.* Springer Finance, Springer, New York, 2004.

[497] Silvestrov, D., *Limit Theorems for Composite Random Functions.* Vysshaya Shkola and Izdatelstvo Kievskogo Universiteta, Kiev, 1974.

[498] Silvestrov, D. S. (1980a), Mean hitting times for semi-Markov processes, and queueing networks. *Elektr. Info. Kybernetik*, **16** (1980), 399–415.

[499] Silvestrov, D. S. (1980b), *Semi-Markov Processes with a Discrete State Space.* Library for the Engineer in Reliability, Sovetskoe Radio, Moscow, 1980.

[500] Silvestrov, D. S., Recurrence relations for generalized hitting times for semi-Markov processes. *Ann. Appl. Probab.*, **6** no. 2 (1996), 617–649.

[501] Silvestrov, D., *Limit Theorems for Randomly Stopped Stochastic Processes.* Probability and Its Applications, Springer, London, 2004.

[502] Silvestrov, D. S., Galochkin, V. G., Malyarenko, A. A., OPTAN - a pilot program system for analysis of options. *Theory Stoch. Process.*, **7(23)** no. 1–2 (2001), 291–300.

[503] Silvestrov, D. S., Galochkin, V. G., Sibirtsev, V. G., Algorithms and programs for optimal Monte Carlo pricing of American type options. *Theory Stoch. Process.*, **5(21)** no. 1–2 (1999), 175–187.

[504] Silvestrov, D., Jönsson, H., Stenberg, F., *Convergence of option rewards for Markov type processes controlled by stochastic indices. 1.* Research Reports 2006–1, Department of Mathematics and Physics, Mälardalen University, 36 pages, 2006.

[505] Silvestrov, D., Jönsson, H., Stenberg, F. (2008a), Convergence of option rewards for Markov type price processes modulated by stochastic indices I. *Theor. Ĭmovirn. Mat. Stat.*, **79** (2008), 149–165 (English translation in *Theory Probab. Math. Statist.*, **79**, 153–170).

[506] Silvestrov, D. S., Jönsson, H., Stenberg, F. (2008b), *Convergence of option rewards for Markov type price processes modulated by stochastic indices.* Eurandom Report 2008–041 (2008), 34 pages.

[507] Silvestrov, D., Jönsson, H., Stenberg, F., Convergence of option rewards for Markov type price processes modulated by stochastic indices II. *Theor. Ĭmovirn. Mat. Stat.*, **80** (2009), 138–155 (English translation in *Theory Probab. Math. Statist.*, **80**, 153–172).

[508] Silvestrov, D., Li, Y., *Lattice approximation methods for American type options.* Research Report 2013:02, Department of Mathematics, Stockholm University, Sweden, 36 pages, 2013.

[509] Silvetsrov, D., Lundgren, R., Convergence of option rewards for multivariate price processes. *Theor. Ĭmovirn. Mat. Stat.*, **85** (2011), 102–116 (English translation in *Theory Probab. Math. Statist.*, **85**, 115–131).

[510] Silvestrov D., Manca, R. Silvestrova, E., Computational algorithms for moments of accumulated Markov and semi-Markov rewards (to appear in *Comm. Statist.*), 2013.

[511] Silvestrov, D., Stenberg, F., A pricing process with stochastic volatility controlled by a semi-Markov process. *Comm. Statist. Theory Meth.*, **33** no. 3 (2004), 591–610.

[512] Skorokhod, A. V., Limit theorems for stochastic processes. *Teor. Veroyatn. Primen.*, **1** no. 3, 289–319 (English translation in *Theory Probab. Appl.*, **1** no. 3, 261–290), 1956.

[513] Stenberg, F. (2006a), *Convergence of option rewards for exponential Lévy type price processes controlled by semi-Markov indices.* Research Reports 2006–2, Department of Mathematics and Physics, Mälardalen University, 2006, 42 pages.

[514] Stenberg, F. (2006b) *Semi-Markov Models for Insurance and Finance.* Doctoral Dissertation, No. 38, Mälardalen University, 2006.

[515] Stenberg, F., Manca, R., Silvestrov, D., Semi-Markov reward models for insurance. *Theory Stoch. Process.*, **12(28)** no. 3–4 (2006), 255–270.

[516] Stenberg, F., Manca, R., Silvestrov, D., An algorithmic approach to discrete time non-homogeneous backward semi-Markov reward process with an application to disability insurance. *Metodol. Comput. Appl. Probab.*, **9** (2007), 497–519.

[517] Stoll, H., Whaley, R., *Futures and Options: Theory and Applications.* South-Western Publishing, Cincinnati, Ohio, 1993.

[518] Stroock, D., Varadhan, S., *Multidimensional Diffusion Processes.* Springer, New York, 1979.

[519] Stulz, R., Options on the minimum or the maximum of two risky assets. *J. Financ. Econom.*, **10** (1982), 161–185.

[520] Sun, Y., Shi, Y., Gu, X., Barrier option pricing under the Vasicek model of the short rate. *J. Appl. Math. Inform.*, **29** no. 5–6 (2011), 1501–1509.

[521] Swishchuk, A. V., Hedging of options under mean-square criterion and semi-Markov volatility. *Ukr. Math. J.*, **47** no. 7 (1995), 1119–1127.

[522] Swishchuk, A. V., *Modeling and Pricing of Swaps for Financial and Energy Markets with Stochastic Volatilities.* World Scientific. New York, 2013.

[523] Szimayer, A., Valuation of American options in the presence of event risk. *Finance Stoch.*, **9** no. 1 (2005), 89–107.

[524] Tan, K. S., Vetzal, K. R., Early exercise regions for exotic options. *J. Derivat.*, **3** no. 1 (1995), 42–56.

[525] Tangman, D. Y., Gopaul, A., Bhuruth, M., A fast high-order finite difference algorithm for pricing American options. *J. Comput. Appl. Math.*, **222** no. 1 (2008), 17–29.

[526] Tian, Y., A Simplified binomial approach to the pricing of interest-rate contingent claims. *J. Financ. Engin.*, **1** (1992), 14–37.

[527] Tian, Y., A modified lattice approach to option pricing. *J. Futures Markets*, **13** (1993), 563–577.

[528] Tian, Y., A Reexamination of lattice procedures for interest rate-contingent claims. *Adv. Futures Options Res.*, **7** (1994), 87–111.

[529] Tilley, J., Valuing American options in a path simulation model. *Trans. Soc. Actuar.*, **45** (1993), 83–104.

[530] Toivanen, J., Numerical valuation of European and American options under Kou's jump-diffusion model. *SIAM J. Sci. Comput.*, **30** no. 4 (2008), 1949–1970.

[531] Toivanen, J. (2010a), A high-order front-tracking finite difference method for pricing American options under jump-diffusion models. *J. Comput. Finance*, **13** no. 3 (2010), 61–79.

[532] Toivanen, J. (2010b), A componentwise splitting method for pricing American options under the Bates model. In: Fitzgibbon, W., Kuznetsov, Yu., Neittaanmäki, P., Périaux, J., Pironneau, O. (eds) *Applied and Numerical Partial Differential Equations.* Comput. Methods Appl. Sci., **15**, Springer, New York, 2010, 213–227.

[533] Touzi, N., American options exercise boundary when the volatility changes randomly. *Appl. Math. Optim.*, **39** (1999), 411–422.

[534] Trabelsi, F., Asymptotic behavior of random maturity American options. *IAENG Inter. J. Appl. Math.*, **41** no. 2 (2011), 112–121.

[535] Trigeorgis, L., A log-transformed binomial numerical analysis method for valuing complex multi-option investments. *J. Financ. Quant. Anal.*, **26** (1991), 309–326.

[536] Van der Hoek, J., Elliott, R. J., American option prices in a Markov chain market model. *Appl. Stoch. Models Business. Industr.*, **28** no. 1 (2012), 35–59.

[537] Van Moerbeke, P., On optimal stopping and free boundary problems. *Arch. Ration. Mech. Anal.*, **60** (1976), 101–148.

[538] Van Moerbeke, P., On optimal stopping and free boundary problems. *Arch. Ration. Mech. Anal.*, **60** (1976), 101–148.

[539] Vasin, A. A., Morozov, V. V., Investment decisions under uncertainty and evaluation of American options. *Inter. J. Math. Game Theory Algebra*, **15** no. 3 (2006), 323–336.

[540] Vellekoop, M., Nieuwenhuis, H., A tree-based method to price American options in the Heston model. *J. Comput. Finance*, **13** no. 1 (2009), 1–21.

[541] Villeneuve, S., Exercise regions of American options on several assets. *Finance Stoch.*, **3** (1999), 295–322.

[542] Vollert, A., *A Stochastic Control Framework for Real Options in Strategic Evaluation*. Birkhäuser, Boston, MA, 2003.

[543] Wang, G., Wang, S., On stability and convergence of a finite difference approximation to a parabolic variational inequality arising from American option valuation. *Stoch. Anal. Appl.*, **24** no. 6 (2006), 1185–1204.

[544] Wang, G., Wang, S., Convergence of a finite element approximation to a degenerate parabolic variational inequality with non-smooth data arising from American option valuation. *Optim. Meth. Soft.*, **25** no. 4–6 (2010), 699–723.

[545] Wang, G., Yang, X. Q., The regularization method for a degenerate parabolic variational inequality arising from American option valuation. *Inter. J. Numer. Anal. Model.*, **5** no. 2 (2008), 222–238.

[546] Wang, S., Yang, X. Q., Teo, K. L., Power penalty method for a linear complementarity problem arising from American option valuation. *J. Optim. Theory Appl.*, **129** no. 2 (2006), 227–254.

[547] Whaley, R. E., On the valuation of American call options on stocks with known dividends. *J. Financ. Econom.*, **9** (1981), 207–211.

[548] Whalley, A. E., Wilmott, P., Optimal hedging of options with small but arbitrary transaction cost structure. *Europ. J. Appl. Math.*, **10** no. 2 (1999), 117–139.

[549] Whitt, W., *Stochastic-Processes Limits. An Introduction to Stochastic-Process Limits and their Applications to Queues*. Springer Series in Operations Research, Springer, New York, 2002.

[550] Wilmott, P., *Derivatives: The Theory and Practice of Financial Engineering*. Wiley Frontiers in Finance Series, Wiley, 1998.

[551] Wilmott, P., Dewynne, J., Howison, S., *Option Pricing: Mathematical Models and Computation*. Oxford Financial Press, Oxford, 1993.

[552] Wilmott, P., Howison, S., Dewynne, J., *The Mathematics of Financial Derivatives. A Student Introduction*. Cambridge University Press, Cambridge, 1995.

[553] Wong, D., *Generalised Optimal Stopping Problems and Financial Markets*. Pitman Research Notes in Mathematics Series, **358**, Longman, Harlow, 1996.

[554] Wong, H.Y, Zhao, J., An artificial boundary method for American option pricing under the CEV model. *SIAM J. Numer. Anal.*, **46** no. 4 (2008), 2183–2209.

[555] Wong, H., Zhao, J., Valuing American options under the CEV model by Laplace-Carson transforms. *Oper. Res. Lett.*, **38** no. 5 (2010), 474–481.

[556] Wu, R., Fu, M. C., Optimal exercise policies and simulation-based valuation for American-Asian options. *Oper. Res.*, **51**, 1 (2003), 52–66.

[557] Yang, H., A numerical analysis of American options with regime switching. *J. Sci. Comput.*, **44** no. 1 (2010), 69–91.

[558] Yao, D. D., Zhang, Q., Zhou, X. Y., A regime-switching model for European options. In: Yan, H., Yin, G., Zhang, Q. (eds) *Stochastic Processes, Optimization, and Control Theory Applications in Financial Engineering, Queueing Networks, and Manufacturing Systems*. Springer, London (2006), 281–300.

[559] Yip, W. Y., Stephens, D., Olhede, S., Hedging strategies and minimal variance portfolios for European and exotic options in a Lévy market. *Math. Finance*, **20** no. 4 (2010), 617–646.

[560] Yousuf, M., Efficient L-stable method for parabolic problems with application to pricing American options under stochastic volatility. *Appl. Math. Comput.*, **213** no. 1 (2009), 121–136.

[561] Yousuf, M., Khaliq, A., Kleefeld, B., The numerical approximation of nonlinear Black-Scholes model for exotic path-dependent American options with transaction cost. *Inter. J. Comput. Math.*, **89** no. 9 (2012), 1239–1254.

[562] Zhang, X., Numerical analysis of American option pricing in a jump-diffusion model. *Math. Oper. Res.*, **22** no. 3 (1997), 668–690.

[563] Zhang, J., Li, T., Pricing and hedging American options analytically: a perturbation method. *Math. Finance*, **20** no. 1 (2010), 59–87.

[564] Zhang, K., Wang, S., Convergence property of an interior penalty approach to pricing American option. *J. Ind. Manag. Optim.*, **7** no. 2 (2011), 435–447.

[565] Zhang, K., Wang, S., Yang, X. Q., Teo, K. L., A power penalty approach to numerical solutions of two-asset American options. *Numer. Math. Theory Meth. Appl.*, **2** no. 2 (2009), 202–223.

[566] Zhang, K., Yang, X. Q., Teo, K. L., Augmented Lagrangian method applied to American option pricing. *Automat. J. IFAC*, **42** no. 8 (2006), 1407–1416.

[567] Zhang, K., Yang, X., Teo, K. L., Convergence analysis of a monotonic penalty method for American option pricing. *J. Math. Anal. Appl.*, **348** no. 2 (2008), 915–926.

[568] Zhang, K., Yang, X. Q., Wang, S., Teo, K. L., Numerical performance of penalty method for American option pricing. *Optim. Methods Soft.*, **25** no. 4–6 (2010), 737–752.

[569] Zhao, J., Davison, M., Corless, R. M., Compact finite difference method for American option pricing. *J. Comput. Appl. Math.*, **206** no. 1 (2007), 306–321.

[570] Zhao, J., Wong, H., A closed-form solution to American options under general diffusion processes. *Quant. Finance*, **12** no. 5 (2012), 725–737.

[571] Zhu, J., *Modular Pricing of Options. An Application of Fourier Analysis.* Lecture Notes in Economics and Mathematical Systems, **493**, Springer, Berlin, 2000.

[572] Zhu, S., On various quantitative approaches for pricing American options. *New Math. Nat. Comput.*, **7** no. 2 (2011), 313–332.

[573] Zhu, S., Chen, W. (2011a), Should an American option be exercised earlier or later if volatility is not assumed to be a constant? *Inter. J. Theor. Appl. Finance*, **14** no. 8 (2011), 1279–1297.

[574] Zhu, S., Chen, W. (2011b), Pricing perpetual American options under a stochastic-volatility model with fast mean reversion. *Appl. Math. Lett.*, **24** no. 10 (2011), 1663–1669.

[575] Zhu, Y., Wu, X., Chen, I., *Derivative Securities and Difference Methods.* Springer Finance. Springer, New York, 2004.

[576] Ziegler, A., *A Game Theory Analysis of Options. Corporate Finance and Financial Intermediation in Continuous Time.* 2nd edition, Springer Finance, Springer, Berlin, 2004.

Index

R

S

De Gruyter Studies in Mathematics

Volume 55
Lucio Boccardo, Gisella Croce
Elliptic Partial Differential Equations, 2013
ISBN 978-3-11-031540-0, e-ISBN 978-3-11-031542-4, Set-ISBN 978-3-11-031543-1

Volume 54
Yasushi Ishikawa
Stochastic Calculus of Variations for Jump Processes, 2013
ISBN 978-3-11-028180-4, e-ISBN 978-3-11-028200-9, Set-ISBN 978-3-11-028201-6

Volume 53
Martin Schlichenmaier
KricheverNovikov Type Algebras: Theory and Applications, 2014
ISBN 978-3-11-026517-0, e-ISBN 978-3-11-027964-1, Set-ISBN 978-3-11-028025-8

Volume 52
Miroslav Pavlovic
Function Classes on the Unit Disc: An Introduction, 2014
ISBN 978-3-11-028123-1, e-ISBN 978-3-11-028190-3, Set-ISBN 978-3-11-028191-0

Volume 51
Daniel Groves
An Invitation to Hyperbolic Groups, 2014
ISBN 978-3-11-026277-3, e-ISBN 978-3-11-026280-3, Set-ISBN 978-3-11-916681-2

Volume 50
Zoltán Sasvári
Multivariate Characteristic and Correlation Functions, 2013
ISBN 978-3-11-022398-9, e-ISBN 978-3-11-022399-6, Set-ISBN 978-3-11-174044-7

Volume 49
Mikhail I. Ostrovskii
Metric Embeddings: Bilipschitz and Coarse Embeddings into Banach Spaces, 2013
ISBN 978-3-11-026340-4, e-ISBN 978-3-11-026401-2, Set-ISBN 978-3-11-916622-5

Bei Fragen zur Produktsicherheit wenden Sie sich bitte an:
If you have any questions regarding product safety,
please contact:

Walter de Gruyter GmbH
Genthiner Straße 13
10785 Berlin
productsafety@degruyterbrill.com